CALIXARENES 2001

Calixarenes 2001

Edited by

Zouhair Asfari
*ECPM-ULP-CNRS,
Strasbourg, France*

Volker Böhmer
*Johannes Gutenberg-Universität,
Mainz, Germany*

Jack Harrowfield
*UNA-ECPM,
Perth, Australia*

and

Jacques Vicens
*ECPM-ULP-CNRS,
Strasbourg, France*

Assistant Editor

Mohamed Saadioui
*Johannes Gutenberg-Universität,
Mainz, Germany*

KLUWER ACADEMIC PUBLISHERS
DORDRECHT / BOSTON / LONDON

A C.I.P. Catalogue record for this book is available from the Library of Congress.

ISBN 0-7923-6960-2

Published by Kluwer Academic Publishers,
P.O. Box 17, 3300 AA Dordrecht, The Netherlands.

Sold and distributed in North, Central and South America
by Kluwer Academic Publishers,
101 Philip Drive, Norwell, MA 02061, U.S.A.

In all other countries, sold and distributed
by Kluwer Academic Publishers,
P.O. Box 322, 3300 AH Dordrecht, The Netherlands.

Printed on acid-free paper

All Rights Reserved
© 2001 Kluwer Academic Publishers
No part of the material protected by this copyright notice may be reproduced or
utilized in any form or by any means, electronic or mechanical,
including photocopying, recording or by any information storage and
retrieval system, without written permission from the copyright owner.

Printed in the Netherlands.

CONTENTS

Preface   ix

1. **Synthesis of Calixarenes and Thiacalixarenes**   1
   C. David Gutsche

2. **Chemical Modification of Calix[4]arenes and Resorcarenes**   26
   Iris Thondorf, Alexander Shivanyuk, Volker Böhmer

3. **Selectively Modified Calix[5]arenes**   54
   Anna Notti, Melchiorre F. Parisi, Sebastiano Pappalardo

4. **Selective Modifications of Calix[6]arenes**   71
   Ulrich Lüning, Frank Löffler, Jan Eggert

5. **Chemistry of Larger Calix[n]arenes (n = 7, 8, 9)**   89
   Placido Neri, Grazia M. L. Consoli, Francesca Cunsolo, Corrada Geraci, Mario Piattelli

6. **Thia-, Mercapto-, and Thiamercapto-Calix[4]arenes**   110
   M. Wais Hosseini

7. **Double- and Multi-Calixarenes**   130
   Mohamed Saadioui, Volker Böhmer

8. **Self-Assembly in Solution**   155
   Dmitry M. Rudkevich

9. **Cavitands**   181
   Willem Verboom

10. **Carcerands**   199
    Christoph Naumann, John C. Sherman

11. **Homocalixarenes**   219
    Yosuke Nakamura, Takahiro Fujii, Seiichi Inokuma, Jun Nishimura

12. **Homooxa- and Homoaza-Calixarenes**   235
    Bernardo Masci

13. **Heterocalixarenes**   250
    Myroslav Vysotsky, Mohamed Saadioui, Volker Böhmer

14. **Oxidation and Reduction of Aromatic Rings**   266
    Silvio E. Biali

15. **Conformations and Stereodynamics**   280
    Iris Thondorf

16. **Dynamic Structures of Host-Guest Systems**   296
    Eric B. Brouwer, Gary D. Enright, Christopher I. Ratcliffe,
    John A. Ripmeester, Konstantin A. Udachin

17. **Molecular Dynamics of Cation Complexation and Extraction**   312
    Georges Wipff

18. **Quantum Chemical Calculations on Alkali Metal Complexes**   334
    Jérôme Golebiowski, Véronique Lamare,
    Manuel F. Ruiz-López

19. **Thermodynamics of Calixarene-Ion Interactions**   346
    Angela F. Danil de Namor

20. **Crown Ethers Derived from Calix[4]arenes**   365
    Alessandro Casnati, Rocco Ungaro, Zouhair Asfari,
    Jacques Vicens

21. **Metal-Ion Complexation by Narrow Rim Carbonyl Derivatives**   385
    Françoise Arnaud-Neu, M. Anthony McKervey,
    Marie-José Schwing-Weill

22. **Phase Transfer Extraction of Heavy Metals**   407
    D. Max Roundhill, Jin Yu Shen

23. **Calixarene-Based Anion Receptors**   421
    Susan E. Matthews, Paul D. Beer

24. **Water-Soluble Calixarenes**   440
    Alessandro Casnati, Domenico Sciotto, Giuseppe Arena

25. **Recognition of Neutral Molecules**   457
    Arturo Arduini, Andrea Pochini, Andrea Secchi,
    Franco Ugozzoli

26. **Complexation of Fullerenes**   476
    Zhen-Lin Zhong, Atsushi Ikeda, Seiji Shinkai

27.	**Calixarenes in Bioorganic and Biomimetic Chemistry** Francesco Sansone, Margarita Segura, Rocco Ungaro	**496**
28.	**Coordination Chemistry and Catalysis** Stéphane Steyer, Catherine Jeunesse, Dominique Armspach, Dominique Matt, Jack Harrowfield	**513**
29.	**Metal Reactivity on Oxo Surfaces Modeled by Calix[4]arenes** Carlo Floriani, Rita Floriani-Moro	**536**
30.	**Phenoxide Complexes of f-Elements** Pierre Thuéry, Martine Nierlich, Jack Harrowfield, Mark Ogden	**561**
31.	**Luminescent Probes** Nanda Sabbatini, Massimo Guardigli, Ilse Manet, Raymond Ziessel	**583**
32.	**Turning Ionophores into Chromo- and Fluoro-Ionophores** Rainer Ludwig	**598**
33.	**Mono- and Multi-Layers** Andrew Lucke, Charles J. M. Stirling, Volker Böhmer	**612**
34.	**Sensor Applications** Francis Cadogan, Kieran Nolan, Dermot Diamond	**627**
35.	**Calixarenes for Nuclear Waste Treatment** Françoise Arnaud-Neu, Marie-José Schwing-Weill, Jean-François Dozol	**642**
36.	**Calixarenes as Stationary Phases** Robert Milbradt, Volker Böhmer	**663**
	Subject Index	**677**

PREFACE

There is a parallel between the discovery, by Adolph von Bayer, of phenol/formaldehyde reactions some thirty years before the end of the nineteenth century and their exploitation at the beginning of the twentieth, and the rationalisation, by David Gutsche, of calixarene syntheses in the 1970s leading to the many applications being presently realised at the beginning of the twenty-first century. The intervening century has, of course, seen great changes in science and in the way it is conducted, and a striking feature of contemporary calixarene chemistry, reflected in the authors of the chapters in this book, is its truly international nature. A second obvious feature is its diversity, calixarenes finding applications from chelates to catalysts, from sensors to scaffolds. In this regard, it is fascinating to reflect upon the importance of the coining of the name "calixarene" as a means of unifying this diversity and bringing home to all of us who employ these remarkable molecules how broad their potential is.

Whilst calixarene chemistry may now well be regarded as a field approaching maturity, its development over the past three decades has proceeded at a pace which has not always left the language of the field clear, logical and universally understood. The very word "calixarene" has changed in meaning from one designating a macrocyclic polyphenol to one defining a $[1_n]$cyclophane (as discussed by David Gutsche in Chapter 1), though no doubt there are few of us who do not assume that the casual use of the word refers, in fact, to the polyphenol. More contentious issues concern just how far to extend the embrace of the term - "calixpyrrole", for example, is perhaps an unnecessary neologism to those familiar with the term "porphyrinogen" - and how, logically, to deal with the many known and possible variations on the calixarene structure.

The root "calyx" has been used to name other cup-shaped molecules, such as the proteins termed "calins" and here it is probably necessary to remark to any newcomer to calixarene chemistry that, despite the importance of so many cup-shaped, *cone* calix[4]arene derivatives, most calixarenes are not cup-shaped ! Nonetheless, while perhaps little confusion has resulted from the metamorphosis of the term calixarene to a name simply designating a particular group of cyclophanes, it is difficult to explain to any novice why, for example, calixarenes derived from resorcinol are given a name, "resorcarene" or "resorcinarene", in which the indication of their shape has been lost and the indication of the aromatic nature of their precursor has been doubled up ! Surely, calixresorcinol would be more logical ? The common use of "upper" and "lower", rather than, say, "wide" and "narrow" to designate the "rims" of a calixarene also reveals less of logic than of aesthetics in our view of the molecule. The plurality of terms in the field can also be confusing, since it is sometimes a little difficult to know exactly when a "calixarene" becomes a "cavitand", for example. Nonetheless, in chemistry and many other areas, common usage and whether or not a name rolls easily from the tongue have always been good bases for nomenclature. Hence, as editors, we have not tried to enforce a common or evidently logical nomenclature throughout *Calixarenes 2001*, although we have tried to ensure that every chapter is internally consistent in its usage.

In the time taken to write, edit and publish any book, science advances and it is impossible to ensure that the very latest facts are included. Given this qualification, however, we hope that what is to be found in *Calixarenes 2001* gives a good sense of the impact and vibrancy of calixarene chemistry. We believe that the many authors involved in devising the 36 chapters have provided a richness of perspectives that is the real value of the book and that will be of enduring utility even should "the facts" be largely modified. The book is not rigidly structured in the sense that one chapter will be incomprehensible if those preceding it have not been read but there is a rough sequence of synthesis-properties-applications, with a little theory mixed in. Calixarene synthesis is now at a high level of sophistication, with extremely elaborate functionalisation a common pursuit, though the relatively recent discovery of thiacalixarenes illustrates the continuing utility of synthesis at all levels. Properties of calixarene systems are defined with the full range of modern instrumental techniques, and provide the basis for applications in many areas of analytical, processing and materials chemistry.

Several books and reviews concerning calixarene chemistry have preceded this work and others may follow, though it is clear that to encompass the full range of the field in a single volume of moderate length is increasingly difficult and we are grateful as editors to the many authors who graciously reined in their enthusiasm for the sake of brevity. We have also been assisted by our colleagues in many ways but must make especial mention of Dr. Mohamed Saadioui for his tireless efforts in formatting and ensuring optimal graphical quality throughout the entire book. Issues ignored or inadequately treated are our responsibility, though we are confident that the strength of calixarene chemistry is such that these deficiencies will be remedied elsewhere !

Zouhair Asfari, Volker Böhmer, Jack Harrowfield, Jacques Vicens

Strasbourg, Mainz and Perth, January 2001

Chapter 1

SYNTHESIS OF CALIXARENES AND THIACALIXARENES

C. DAVID GUTSCHE

Department of Chemistry
Texas Christian University
Fort Worth, TX, 76129, USA.
e-mail: d.gutsche@tcu.edu

This chapter deals only with the synthesis of calixarene ring structures that can be generated from non-macrocyclic precursors. The many alterations that can be effected on the calixarene scaffolds so obtained are discussed in other chapters of the book.

1. Single-Step Condensations of Phenols with Formaldehyde

1.1. INTRODUCTION

Chemistry is often called the "Central Science", in recognition of its pivotal position among the panoply of sciences. Synthesis, in turn, might be called chemistry's "Central Topic", in recognition of the pivotal position it plays among chemistry's many facets. Calixarene chemistry is a good case in point, its burgeoning development being the result of the ease with which its starting materials can be prepared. This ease of synthesis, however, is a relatively recent development. Although it was discovered by Zinke in the 1940s that the base-induced reaction of *p*-alkylphenols with formaldehyde yields cyclic oligomers [1], this synthesis languished almost unnoticed [2] for 30 years until the 1970s when Gutsche and coworkers [3] reinterpreted the Zinke results and developed methods for synthesizing each of the three major cyclic oligomers comprising the original Zinke mixture in good and reproducible yields. It is the easy single-step accessibility of these materials, for which the name "calixarenes" was coined [4], that helped accelerate their subsequent exploration. A number of areas now comprise calixarene chemistry including, *inter alia,* cation, anion, and molecule complexation, ion transport phenomena, enzyme model building, and the construction of sensors for a wide variety of applications.

The term "calixarene" carries a degree of ambiguity. As originally conceived it applied to the phenol-derived cyclic oligomers and included the *endo* hydroxyl groups. To accommodate a systematic nomenclature, however, the calixarene descriptor is now taken to designate *only* the basic macrocyclic framework, and the hydroxyls and other substituents (*e.g. tert*-butyl) are designated as appendages. Thus, in colloquial usage the cyclic tetramer derived from *p-tert*-butylphenol can be called "*p-tert*-butylcalix[4]-arene", but its systematic name is 5,11,17,23-tetra-*tert*-butylcalix[4]arene-25,26,27,28-tetrol. Similarly, in colloquial usage the cyclic tetramer derived from resorcinol and acetaldehyde (see Sec 2) can be called a C-methylcalix[4]resorcarene, but its systematic name is 2,8,14,20-tetramethylcalix[4]arene-4,6,10,12,16,18,22,24-octol. In this context any [1$_n$]metacyclophane, substituted or not and regardless of the nature of the substitution, would fall in the umbra of the "calixarene" nomenclature. It is recommended, however, that the "calixarene" designation not be used in too wholesale a way but be restricted to those [1$_n$]metacyclophanes that are substituted in some fashion, generally with one or more heteroatom functional groups. Nevertheless, the nomenclature remains a bit unsettled, and we note with amusement admixed with concern that the designation (calixarene) is being applied with some abandon to almost any macrocyclic system in which a set of aromatic or heterocyclic moieties are joined together in a ring and are connected at their 1- and 3-positions by one or more bridging atoms.

Another aspect of calixarene nomenclature relates to the conformations that these macrocycles can adopt. It was recognized by Cornforth [2] that a calix[4]arene can exist in four main conformations, one with the aryl groups all *syn* to one another, one with three aryl groups *syn* and one *anti*, one with adjacent pairs of aryl groups *syn* and *anti*, and one with non-adjacent pairs of aryl groups *syn* and *anti*. These were later named by Gutsche as *cone, partial cone, 1,2-alternate,* and *1,3-alternate,* as depicted in Figure 1 with idealized structures having C_{4v}, C_s, C_{2h} and D_{2d} symmetry, respectively.

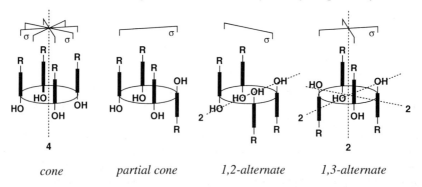

Figure 1. *Schematic representation of the four main conformations of calix[4]arenes; symmetry planes (σ) and axes (---) are indicated.*

Frequently, however, the actual molecules (in solution as well as in the solid state) possess conformations and symmetries different from those depicted in Figure 1 as the result of torsional changes in the aryl group orientations. For example, the *cone* conformer often assumes a *"pinched cone"* or *"flattened cone"* structure in which one pair of aryl residues becomes almost parallel while the other pair splays outward. As the number of aryl moieties in the macrocyclic array increases, the conformational specification and representation become increasingly difficult and less precise. Suggestions for addressing this problem have been published for the calix[5]arenes [5] and the calix[6]-arenes [6].

1.2. BASE-INDUCED SYNTHESES

1.2.1. tert-Butylphenol
The original Zinke synthesis employed NaOH as the base to induce the condensation of the *p*-alkylphenols with formaldehyde. Base-induction remains the method of choice for the single-step synthesis of all three of the "major" calixarenes (**2a, 4a, 6a**) as well as the two "minor" calixarenes (**3a, 5a**). Surprisingly, however, this synthesis, which

works with such controllable flexibility in the production of pure products from *p-tert*-butylphenol, generally works much less well with most other phenols which typically yield difficultly separable mixtures. The careful control of reaction conditions, which is the key to the success with *p-tert*-butylphenol, has failed so far to find comparable application with most of the other *p*-alkylphenols that have been studied.

$$1 \xrightarrow{\text{base, heat}} 2\text{-}6$$

n	compound
2	n = 4
3	n = 5
4	n = 6
5	n = 7
6	n = 8

R groups:
- a *t*-Bu
- b Me
- c Et
- d *i*-Pr
- e allyl
- f -(CH$_2$)$_{10}$Me
- g phenyl
- h OCH$_2$-phenyl (benzyloxy)
- i CH$_2$-phenyl
- j isopropenyl
- k *t*-amyl
- l C$_{10}$H$_{21}$
- m adamantyl
- n cyclohexyl
- o cyclohexyloxy
- p -C$_6$H$_4$-OCH$_3$

The essential features of the synthesis of the three major calixarenes from *p-tert*-butylphenol **1a** are
- (a) the condensation of the phenol with HCHO using NaOH (0.045 molar amount) as the base followed by dissolution in diphenyl ether and heating to reflux for 1.5 - 2 hours to produce *p-tert*-butylcalix[4]arene **2a** [7];
- (b) the heating of a solution of the phenol and formalin with KOH (0.34 molar amount) followed by dissolution in xylene and heating 3-4 hours [8] to produce *p-tert*-butylcalix[6]arene **4a** [9];
- (c) the heating of a solution of the phenol and paraformaldehyde in xylene with NaOH (0.030 molar amount) to produce *p-tert*-butylcalix[8]arene **6a** [10].

p-tert-Butylcalix[5]arene **3a**, originally isolated in only 3-5% yield [11], is now obtainable in 15-20% yield [12]; *p-tert*-butylcalix[7]arene **5a**, originally isolated in only 6% yield [13], is now obtainable in 11-17% yield, using LiOH as the base [14]. An alternative approach to the synthesis of **3a** involves the reaction of *p-tert*-butyldihomo-oxacalix[4]arene (see Chapter 12) with *p-tert*-butylphenol in the presence of KOH to give **3a** in yields up to 32%, isolated by chromatographic separation of a mixture also containing **2a**, **4a**, and **6a** [15].

1.2.2. Other p-substituted phenols
Calixarenes are not obtained through base-catalysed condensations from 4-nitro-, 4-cyano-, and 4-phenoxyphenol, 4-hydroxybenzoic acid, 4-hydroxyacetophenone, 4-hydroxybenzyl alcohol, and 1,4-dihydroxybenzene (hydroquinone) [16]. Some successes, however, have been obtained with phenols carrying *p*-substituents that are not electronically deactivating. For example, *p*-cresol **1b** has been reported to give **4b** in 74% yield [17] or **5b** in 22% yield [18]. Similarly, *p*-ethylphenol **1c** and *p*-isopropylphenol **1d** produce **5c** (27%) [18] and **6d** (3%) [19]. *p*-Isopropenylphenol **1e** yields a mixture of the calix[6]- and -[8]arenes **4e** and **6e**, using ethylene glycol as solvent and Na$_2$B$_4$O$_7$ as the base [20]. Phenols **1f** with long chain alkyl groups in the *p*-position are reported to give calix[6]- and -[8]arenes **4f** and **6f** [21]. *p*-Phenylphenol **1g** gives a mixture containing **4g**, **5g**, and **6g** [22]. *p*-Benzyloxyphenol **1h** gives **6h** in 48% yield [23] along with smaller amounts of **5h** and **2h**, whereas *p*-methoxyphenol produces a

complex mixture of oligomeric compounds. Under various conditions p-benzylphenol **1i** has been found to give a mixture of **5i** (33%), **4i** (16%) and **6i** (12%) [24,25] or **3i** and **5i** in 15-20% yield [26]. Subjecting a mixture containing **4i**, **6i**, paraformaldehyde, KOH and molecular sieves to ultrahigh intensity grinding for 4-16 hours results in a 10-15% yield of the cyclic pentamer **3i** and 5-10% of the cyclic heptamer **5i** [26,27].

Long chain p-alkyl groups [28] often give poor results, but those that are highly branched near the point of attachment to the phenol sometimes behave in a fashion somewhat comparable to p-tert-butylphenol. For example, p-(1,1,3,3-tetramethylbutyl)phenol **1k**, first investigated in 1955 [2], affords **2k** (10%) [29] and **6k** (30-40%), originally thought to be a conformational isomer of the cyclic tetramer. Similarly, p-tert-pentylphenol **1j** gives the major calixarenes **2j**, **4j**, and **6j**, in modest yields, while 21% of **6l** was obtained from **1l** [30]. The highly branched p-adamantylphenol **1m**, the chiral p-(2-isopropyl-5-methylcyclohexyl)-phenol **1n**, and p-[1,1-dimethyl-1-(p'-methoxyphenyl)-methyl]phenol **1o** give **6m** (72%) [31], chiral **6n** (30%) [32] and **6o** (51%) [33], respectively. The highly branched chiral 8-(p-hydroxyphenyl)-menthone **1p** reacts sluggishly with HCHO to give the cyclic pentamer **3p** (8.3%), hexamer **4p** (8.3%) and octamer **6p** (12.4%) [34].

1.2.3. Larger Calixarenes

For many years it was thought that the only cyclooligomers comprising the reaction mixtures obtained from the base-induced reaction of alkylphenols and HCHO were those containing four to eight aryl moieties. However, experiments carried out in the late 1980's [35] resulted in the isolation and characterization of p-tert-butylcalix[9]arene and p-tert-butylcalix[10]arene, with the obtention of mass spectral evidence for p-tert-butylcalix[11]arene and p-tert-butylcalix[12]arene. Subsequent work [36,37] has led to the isolation and characterization of all of the members of the series of "large" calixarenes from the cyclic nonamer (**7**, n = 9) to the cyclic eicosomer (**7**, n = 20). Other studies have also described syntheses of p-tert-butylcalix[n]arenes with n = 9-12 [38] as well as a calix[10]arene carrying a p-benzyl group [26]. The acid-catalyzed preparation of the p-tert-butylcalixarenes has proved to be a better route to the larger members than the base-induced condensation (see Sec 3).

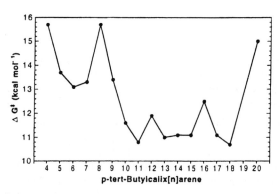

Figure 2. Plot of the $\Delta G^{\ddagger}$ values versus n for p-tert-butylcalix[n]arenes in CDCl$_3$ solution.

The calixarenes probably represent the most populous family of homologous (*i.e.*, benzylogous), well-defined cyclooligomers that is presently known. The conformational behavior of these compounds presents an interesting example of periodicity. As shown in Figure 2, the calixarenes that are more conformationally stable than their immediate neighbors are those containing 4, 8, 12, 16, and 20 aryl units. This "rule of 4" is ascribed to the more favorable intramolecular hydrogen bonding and intramolecular packing available to these members [35].

1.2.4. Naphthols
The base-induced reactions of naphthols with HCHO have been also explored. α-Naphthol **8** [39], for example, yields 9.6% of a calix[4]naphthalene (**9a**) containing *exo*-hydroxyl functions, in contrast to the calixarenes obtained from *p*-substituted phenols which yield calixarenes with all of the hydroxyl groups *endo*. Accompanying the symmetrical compound **9a** are two less symmetrical cyclic tetramers **9b** (5%), and **9c** (16%), while the fourth possible regioisomer could not be isolated.

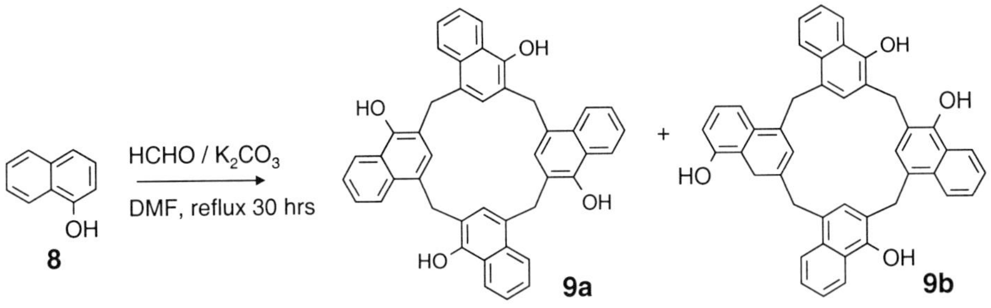

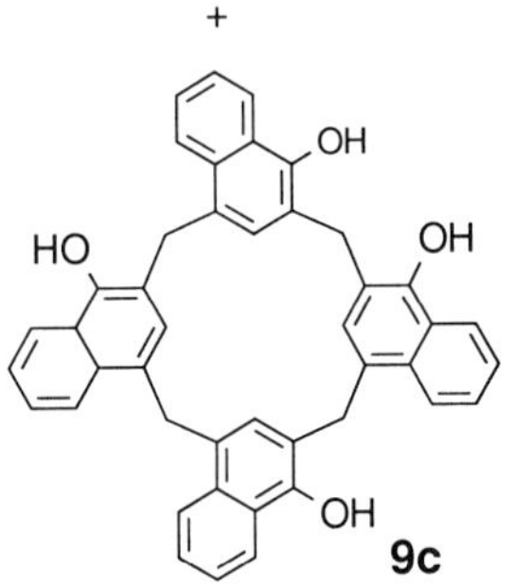

The tetrasulfonated analog **11** has been obtained in 15% yield by refluxing 1,8-naphthalene-sultone **10** with formalin and Cs_2CO_3 in DMF [40]. β-Naphthol reacts under similar conditions to yield a simple bis-naphthol.

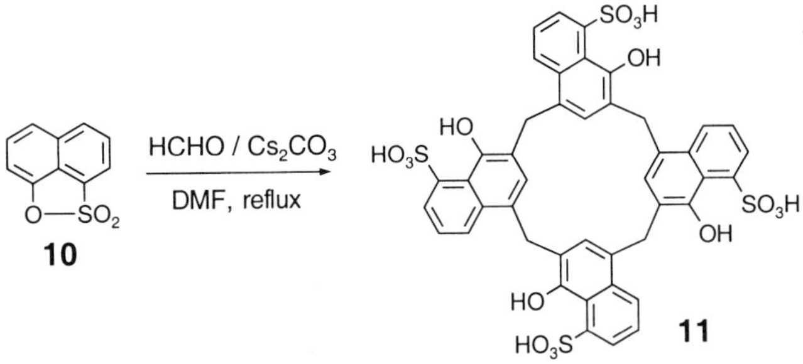

The disodium salt of chromotropic acid is reported to yield a tetra-sulfonated calix[4]arene **12** in which the eight OH groups are all *endo* [41], the deactivating effect of the sulfonic acid groups seemingly outweighed by the activating effect of the hydroxyl groups [42,43].

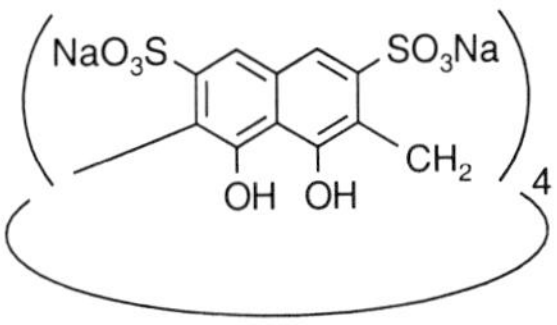

1.2.5. Bis-phenols
Calixarene-like compounds have been obtained from various bis-phenols in which the aryl groups are either directly connected or are separated by two or more methylene groups. For example [44], the α,ω-diarylalkyl compounds **13** and **15** yield **14** and **16** (with n = 2,3) with paraformaldehyde and NaOH, KOH, or CsOH. In the case of **13** the

product is a mixture of **14a** and **14b** in a ratio that varies from 29:38 for NaOH to 64:22 for CsOH.

13 m = 2
15 m = 3

14 m = 2
16 m = 3

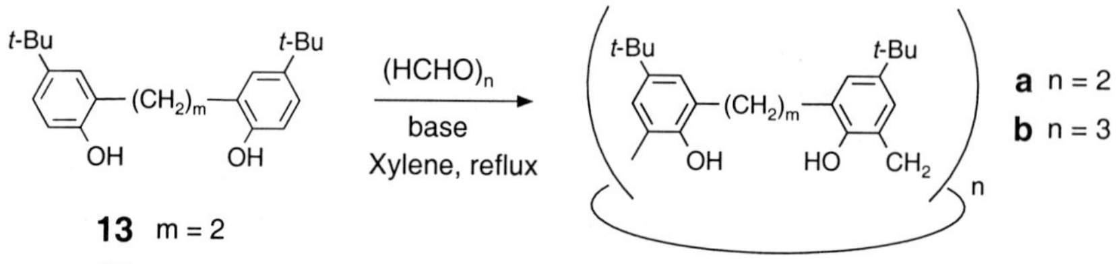

The cyclobutane-containing bisphenol **17b** gives **18b** (89%) with paraformaldehyde in the presence of LiOH [45], but in contrast to the previous example the reaction gives increasingly lower yields with the larger cations Na^+, K^+, and Rb^+ and fails completely with Cs^+, purportedly the result of a "template" effect. However, **17c** (n = 6) with CsOH gives a 78% yield of **18c**. The reaction fails with **17a** regardless of the base that is used [45].

17 **a** n = 4, **b** n = 5, **c** n = 6 **18**

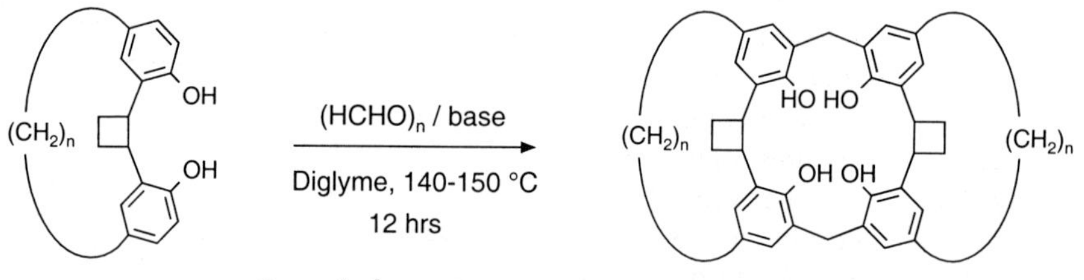

The effect of the base is again observed in the reaction of the bisphenol **19** which yields 49-52% of **20a** [46,47] with only a trace of **20b** with NaOH, but 66% of **20b** accompanied by only 4% of **20a** with CsOH. With KOH as the base, equal amounts of **20a** and **20b** are formed, but in no case was the highly strained cyclic dimer (n = 2) observed.

19

20 **a** n = 3
b n = 4

21

21/22 a X = Me_2C, R = *t*-Bu
b X = Me_2C, R = *n*-Pr
c X = CH_2, R = *t*-Bu
d X = CH_2, R = Me

22

23
a R = *t*-Bu
b R = Me

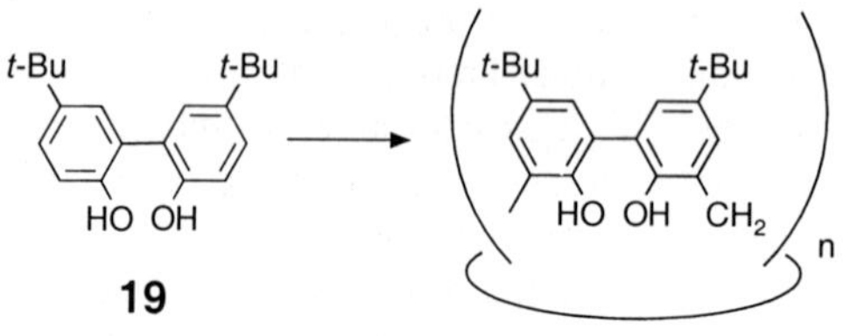

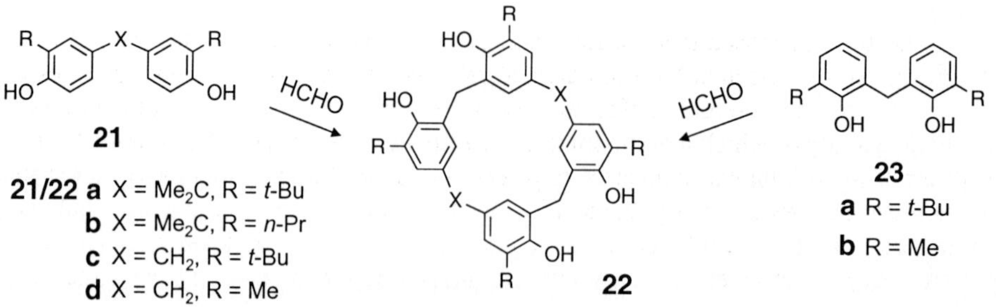

Calix[4]arenes carrying *exo*-OH groups can be prepared from bis-phenols **21** in which the OH groups are *para* to the bridge between the aryl rings. An early example is the conversion of **21a** to **22a** in 20% yield by simply heating a xylene solution of the bis-phenol with paraformaldehyde for 12 hours in an autoclave at 175°C [48]. The *n*-propyl analogue **22b** has been obtained in 30% yield from **21b** using $BF_3.Et_2O$ as the catalyst [49]. Using the thermally-induced reaction, the *p-tert*-butyl compound **21c** yields **22c**. However, the methyl-substituted compound **21d** gives much lower yields of **22d**, providing another example of the critical influence of the *tert*-butyl group [50].

1.2.6. Mechanistic studies

Although the base induced condensation of **1a** has been studied in some detail, the pathways along which the cyclooligomerizations proceed are far from being well understood. One proposal [51a] postulates that when oligomerization [52] has produced the linear tetramer in significant amount the tetramer forms an intramolecularly hydrogen-bonded dimer (a "hemicalixarene") in which the termini are proximate to one another and thus oriented for bond formation, producing the cyclic octamer **6a** (product of kinetic control). Under more strenuous conditions the cyclic octamer reverts to the cyclic tetramer **2a** (product of thermodynamic control). Template effects may explain the formation of cyclohexamer **4a** in the presence of KOH or RbOH, either through the interception and cyclodimerisation of the linear trimer assumed to be an intermediate in the formation of the linear tetramer or through the favouring of extended linear oligomerisation up to the hexamer, which then cyclises [51a].

Reaction mixtures obtained following the conventional procedure for synthesis of *p-tert*-butylcalix[6]arene **4a** [8] have been analyzed by GC and TLC [53]. At the end of the first reaction phase, small amounts of linear trimer and tetramer are present along with appreciable amounts of linear hexamer. This was interpreted to indicate that the precursor to the cyclic hexamer is the linear hexamer. Similar experiments, measuring the product composition as a function of time [54], were carried out for the standard preparation of the cyclic tetramer **2a** [7] and the cyclic octamer **6a** [10]. Since no linear octamer was detected the results were viewed as favoring the linear tetramer (i.e. a "pseudocalixarene") as the precursor to both the calix[4]arene **2a** and the calix[8]arene **6a**. The dihomooxacalix[4]arene (*cf.* Chapter 12) is postulated to play a role in the formation of the cyclic octamer **6a**, perhaps as a storage site for the linear tetramer.

It has been proposed that the reversion of the calix[8]arene **6a** to the calix[4]arene **2a** takes place by so-called "molecular mitosis" wherein the cyclic octamer pinches together to form a figure-eight conformer which then splits into a pair of cyclic tetramers. However, a study in which a 50:50 mixture of fully deuterated and fully protiated *p-tert*-butylcalix[8]arenes **6a** was converted to *p-tert*-butylcalix[4]arene **2a** has shown that this can, at most, be only a minor pathway [55]. If molecular mitosis were the only pathway, the reaction mixture would comprise equal amounts of fully deuterated and fully protiated cyclic tetramer. If complete fragmentation/recombination were the only pathway the reaction mixture would comprise fully and partially deuterated/protiated cyclic tetramers in a 1:4:6:4:1 ratio. The observed ratio of 1.1:1.2:1.7:1.2:1.0 falls between these extremes and, while not completely ruling out molecular mitosis, indicates that fragmentation/recombination must be a major pathway.

The reality of fragmentation/recombination under base-induced conditions is well illustrated:
- (a) by the KOH-induced reaction of *p-tert*-butylcalix[4]arene **2a** with benzylamine [56] to yield 8% of **24** and
- (b) by the NaOH-induced reaction of **25** with HCHO [57] to yield no **2a** or **6a** (a cyclic tetramer with one ethylene and three methylene bridges) but 35% of **26** and 25% of **27** (in which in both cases the ethylene bridge is conserved).

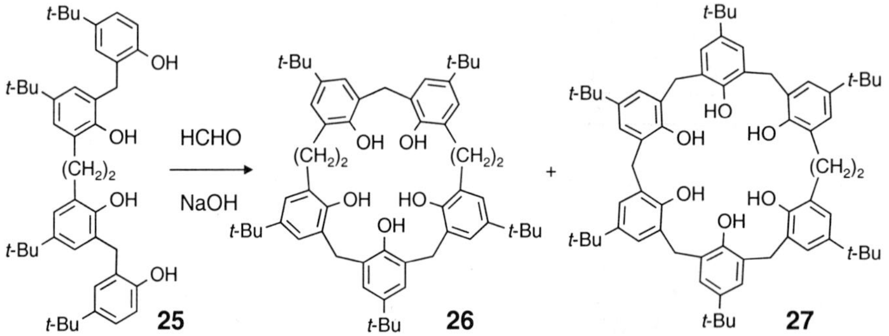

1.3. ACID-CATALYZED SYNTHESES

It was assumed for many years that the reaction of *p*-alkylphenols with HCHO under acidic conditions produces high yields of linear oligomers but only very small amounts of cyclic oligomers [58]. However, treatment of *p-tert*-butylphenol **1a** with s-trioxane and *p*-toluenesulfonic acid in CHCl$_3$ solution can produce an almost quantitative yield of calixarenes [35]. In contrast to the base-induced reaction (Sec 1.1), conditions have not been found that lead to high yields of individual cyclic oligomers. Instead, the reaction product is a mixture of all of the calixarenes ranging from the cyclic tetramer to the cyclic eicosomer, the large calixarenes (**7**, n > 8) being present in greater amounts than in the base-induced reaction. The acid-catalyzed preparation, therefore, is the procedure of choice for the isolation of the large calixarenes. Of pivotal importance in the separation and characterization of the large calixarenes was the development of an HPLC assay that allows close to baseline resolution of all of the cyclic oligomers (**7**, n = 4-20), as illustrated in Figure 3.

The conversion of **23a** to **22c**, which failed under the heat induced conditions (see Sec 1.1) [50], can be effected in 30% yield with BF$_3$.Et$_2$O as the catalyst [59]. In a study of this acid-catalyzed reaction it was found that with SnCl$_4$ in CH$_2$Cl$_2$ **23a** yields **22c** (46%) along with the larger ring compounds **28** (16%) and **29** (9%) [59].

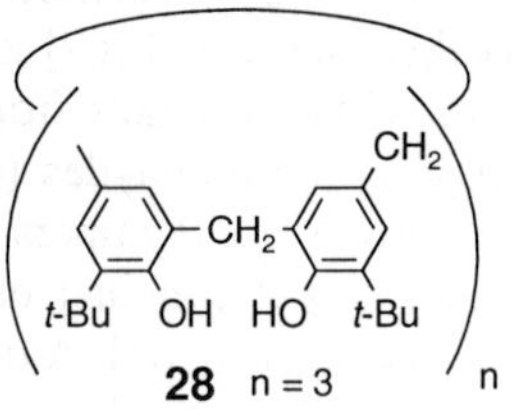

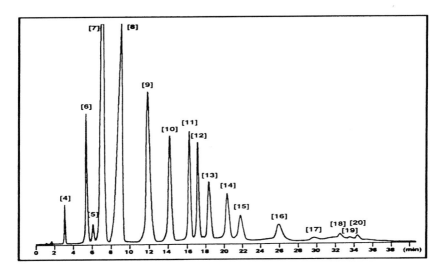

Figure 3. Chromatogram of an HPLC analysis of a crude product from an acid-catalyzed reaction of p-tert-butylphenol and s-trioxane.

Methylene bridge-substituted calix[4]arenes **31** have been obtained in 18-30% yields (in some cases as a mixture of the *cis* and *trans* isomers) from acid-catalyzed condensations starting with **30** [60]. Base-induction of the same reaction provides a 20% yield of product, along with the corresponding dihomo-oxacalix[4]arene [61].

2. Single-Step Condensation of Resorcinols with Aldehydes

2.1. INTRODUCTION

Almost simultaneously with Zinke's investigations of the base-induced condensation of p-alkylphenols and HCHO, Niederl and Vogel [62] in 1940 were reinterpreting the structures of the products from the acid-catalyzed reaction of resorcinol and aldehydes (except HCHO) and concluding that they are cyclic tetramers of the general structure **32**. This assignment was conclusively established in 1968 through X-ray crystallography [63]. A detailed investigation [64] established the synthesis procedures that have provided the starting materials for a large number of subsequent studies including the construction of the cavitands and carcerands [65] (See

Chapters 9, 10.). For example, heating a solution of resorcinol and acetaldehyde in aqueous ethanolic HCl at 80 °C for 16 h gives a 70% yield of C-methyl-calix[4]resorcarene (**32**, R = Me). For each particular aldehyde, however, the optimization of the conditions must be sought [66]. Lewis acid catalysts ($AlCl_3$, $SnCl_4$, BF_3, $Yb(OSO_2CF_3)_3$) in Et_2O have been shown also to be effective condensing agents [67].

Like the phenol-derived calixarenes, the ease of synthesis of the resorcinol-derived calixarenes has spurred a great deal of research activity [68]. One of the results of this plethora of investigations has been the assignment of a variety of names for these compounds, including Högberg compounds, octols, resorcarenes, resorcinarenes, calix[n]resorcarenes, calix[n]resorcinarenes, and calix[n]resorcinolarenes [69]. Since the compounds clearly are members the calixarene family as defined in Sec 1.1 this author prefers the name "calix[n]resorcarene" to reflect this fact.

2.2. STEREOCHEMICAL PROPERTIES

Calix[4]resorcarenes possess four prochiral centers at the bridging carbon atoms and, as a consequence, can exist in four different diastereomeric forms, as shown in Fig. 4. For the purpose of perceiving the stereochemical relationships between the prochiral centers the macrocyclic ring can be considered to be planar with the residues R of the CHR-bridges pointing to one or the other side. Assigning one of these residues on a prochiral center as the reference group (*r*) and then proceeding around the ring in sequential clockwise progression, the residues R of the other groups can be designated as *cis* (c) or *trans* (t) relative to the reference group (*r*). The reference group (*r*) is chosen so as to maximize the number of cis (c) designations.

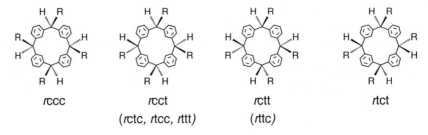

Figure 4. Stereochemical relationships among the four R groups at the methylene carbon atoms of calix[4]resorcarenes.

Of course, the calix[4]resorcarenes are not actually planar and, like the phenol-derived calix[4]arenes, have the potential for existing in a variety of conformations, as illustrated in Figure 5. In practice, the *r*ccc isomers have always been found to have either the C_4 symmetrical "*crown*" or "*bowl*" conformation (equivalent to a *cone* conformation) or the C_{2v} symmetrical "*boat*" conformation [70] (equivalent to a *flattened* or *pinched cone* conformation). The *r*ctt isomers have been found only in the C_{2h} symmetry "*chair*" conformation (equivalent to a transition state between a partial *cone* and a *1,3-alternate* conformation). The *r*cct isomer assumes a C_s symmetrical "*diamond*" conformation (equivalent to a *1,2-alternate* conformation), while the *r*tct isomer is predicted to assume a S_4 symmetrical *saddle* conformation (equivalent to a *1,3-alternate* conformation)."

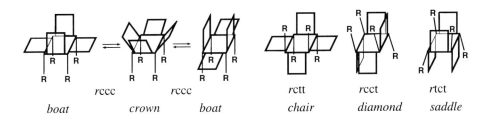

Figure 5. Conformations of calix[4]resorcarenes with different configurations.

In some of the conformers of the calix[4]resorcarenes (which can be designated as "*exo*-calixarenes"[71]) the orientation of the R group at the bridging carbon can be clearly designated as axial or equatorial. For example, in the *crown rccc* conformer all of the R groups are axial in the lower energy form and equatorial in the inverted higher energy form. This is in contrast to the phenol-derived calixarenes (characterized as "*endo*-calixarenes") where an R group on a bridging carbon prefers the equatorial orientation. The reason in both cases is the same, however, namely the avoidance of the neighborhood of the hydroxyl groups. For example, transformation of the *crown rccc* conformer to the *chair rccc* conformer places two of the R groups in equatorial orientations.

The conformational preferences among the calix[4]resorcarenes can be rationalized in terms of the distance between the OH groups on the aromatic rings and the R groups on the bridging carbon atoms. The greater their proximity, the higher the energy of the system. For example, the failure to obtain the *rtct* isomer in preparatively useful quantities can be ascribed to the fact that in all of its conformations the OH and R groups are in close proximity. Even though the R groups tend to confer a degree of rigidity on the calix[4]resorcarenes, conformational change remains possible, as demonstrated by the transformation of *chair rctt* isomers into *cone-shaped* cavitands carrying two axial and two equatorial R groups. (*cf.* Chapter 9)

The presence of both conformational and *cis/trans* isomerism in the calix[4]resorcarenes has led to some confusion in stereochemical discussions of these compounds. It should be remembered that conformational isomers are interconvertible (in theory, if not always in practice) simply by rotation around bonds, while the stereoisomers arising from substitution at the bridging methylene carbon atoms are interconvertible only by the breaking and remaking of bonds (if the tetrahedral geometry of the prochiral center is retained throughout the interchange).

As a consequence of the many stereoisomeric possibilties for the calix[4]resorcarenes it might have been supposed that the reaction products would be intractable mixtures. In practice, however, the product is frequently a single material or, at most, a mixture of two or three isomers with one predominating [72]. Investigations [64,73] have shown that the acid-catalyzed formation of the calixresorcarenes is reversible [74,75] and that the major product is:
- (a) the one that is most rapidly formed (the kinetic product - often the *rccc* (*boat*, *crown*) isomer, but sometimes the *rctt* (*chair*) isomer [76]), or
- (b) the one that is most stable (the thermodynamic product - generally the (*rccc*) isomer), or
- (c) the one that is the least soluble.

A detailed analysis of the reaction of resorcinol with acetaldehyde in MeOH/HCl showed [73] that the acetaldehyde reacts first with MeOH to form the dimethyl acetal, which then condenses with resorcinol to form linear dimers, trimers, tetramers, and even higher linear oligomers. The linear dimer and trimer are isolable, but the linear tetramer reacts very rapidly, though still reversibly, to form the cyclic tetramer seemingly because:
- (a) ring closure is as fast as chain propagation;
- (b) the macrocyclic products constitute the thermodynamic sink;
- (c) linear oligomers longer than four units degrade more rapidly than ring opening occurs, and
- (d) homogeneous reaction conditions are a prerequisite.

2.3. SCOPE AND LIMITS

A wide variety of aldehydes have been employed for the single step synthesis [77], including (a) simple aliphatic aldehydes from ethanal to undecanal [78,79], (b) functionalized aliphatic aldehydes such as 5-chloropentanal [77], 4-hydroxybutanal [80], and 2-sulfonatoethanal [81], (c) aralkyl aldehydes such as phenylethanal [77], (d) unsaturated aliphatic aldehydes such as 9-decenal [66], (e) benzaldehydes, including those with substituents such as alkyl or arylalkyl [82], OH [76,83], NO_2 [73], halogen [77], CN [77], NH_2 [84], MeCO [77], $B(OR)_2$ [85], *crown* ether [86], and glycosyl [87], (f) α-naphthaldehyde [88], (g) heterocyclic aldehydes [89], and (h) ferrocenylaldehyde [90]. The few types of aldehydes that fail to give calix[4]resorcarenes are those that are highly hindered (*e.g.* 2,4,6-trimethylbenzaldehyde) or that contain functionality proximate to the aldehyde function (*e.g.* $ClCH_2CHO$ [77]).

The versatility of the reaction is further enhanced by the use of resorcinols substituted at the 2-position which restricts reaction with the aldehyde to only the 4- and 6-positions. For example, 2-methylresorcinol, 2-bromoresorcinol, and 2-hydroxyresorcinol (pyrogallol) react with acetaldehyde to yield **33** (R^1 = Me; R^2 = Me, Br, OH) [77]. It even makes possible the reaction with formaldehyde or its equivalent. Thus, pyrogallol reacts with paraformaldehyde to give **33** (R^1 = OH; R^2 = H), isolated as the ester in 53% yield and also 2-methylresorcinol affords a calix[4]resorcarene [91,92]. 2-Propylresorcinol reacts with $H_2C(OEt)_2$ to give not only the cyclic tetramer **33** (R^1 = n-Pr; R^2 = H) but also the corresponding cyclic pentamer and hexamer. The latter two are present in larger amount after short reaction times, but the tetramer becomes the sole product after a longer time [93]. An electron-withdrawing group (*e.g.* NO_2) at the 2-position may prevent reaction, although 2-butyrylresorcinol and paraformaldehyde react in the presence of KO*t*-Bu to give a 58% yield of **33** (R^1 = H; R^2 = butyryl) [94].

2.4. OTHER SYNTHESES

Calix[4]resorcarenes with non-identical C-substituents can be prepared by condensing resorcinol with a mixture of aldehydes [95] or, in a more controlled fashion, by reacting dimers **34** with aldehydes to give **35** (X = $PhCH_2CH_2$, *p*-$PhNO_2$; Y = Et, *p*-PhOH, *p*-

PhOAc) [96]. Calix[4]resorcarenes containing a mixture of aryl units have also been prepared [97].

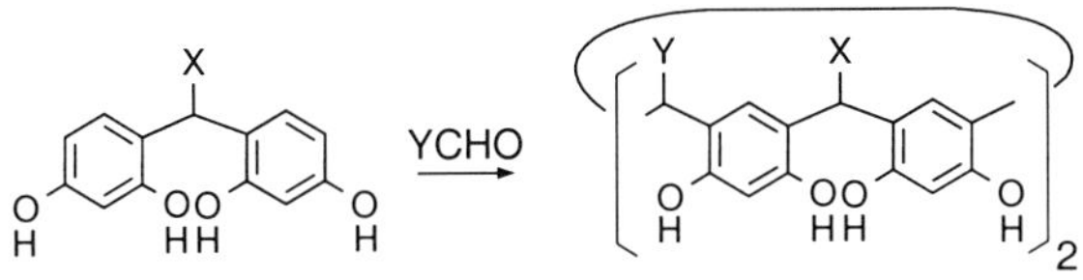

34 → YCHO → **35**

The bis-(carbethoxymethyl) ether of resorcinol yields 49% of the C-p-tolyl-calix[4]resorcarene tetraether by treatment with p-tolualdehyde and $AlCl_3$ in Et_2O [98]. The octamethyl ethers of calix[4]resorcarenes have been prepared in 40-85% yields by the $SnCl_4$-catalyzed reaction of 1,3-dimethoxybenzene with aliphatic aldehydes [99].

A one-step synthesis of the octamethyl ethers of calix[4]resorcarenes [100] involves treatment with $BF_3·Et_2O$ of (E)-2,4-dimethoxycinnamic acid esters or amides (**36a**) or (E)-2,6-dimethoxycinnamic acid esters or amides (**36b**) to give **37**, the 2,6-isomer rearranging during the course of the reaction. The R moieties include OMe, OEt, O-i-Pr, $NHCH_2Ph$, and $NHCH(CO_2Et)CH(CH_3)_2$. The products are obtained in ca 65-80% yields and comprise a mixture of stereoisomers, the proportion of which depends on the R group and the reaction conditions.

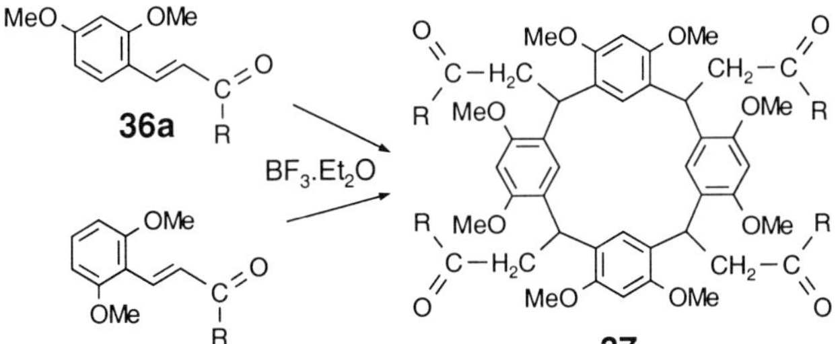

36a, **36b** —$BF_3.Et_2O$→ **37**

The octamethyl ether of the methylene-bridged calix[4]resorcarene can be also obtained in a one-step procedure [101] involving the trifluoroacetic acid-catalyzed reaction of 2,4-dimethoxy-benzyl alcohol **38** to give a 95% yield of **39** which was treated with BBr_3 to yield what is presumed to be **32** (R = H), unobtainable by direct condensation of resorcinol with HCHO.

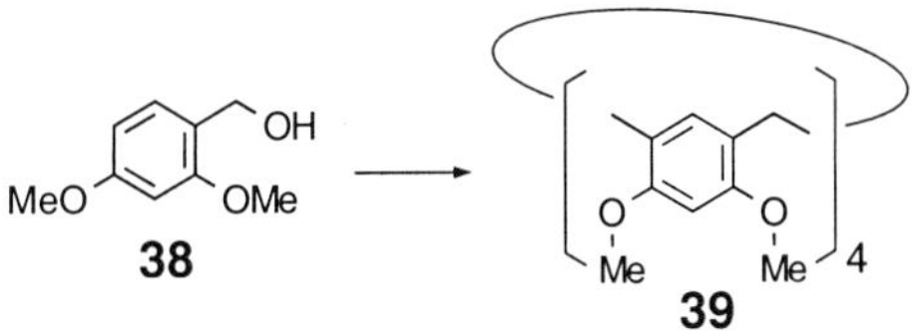

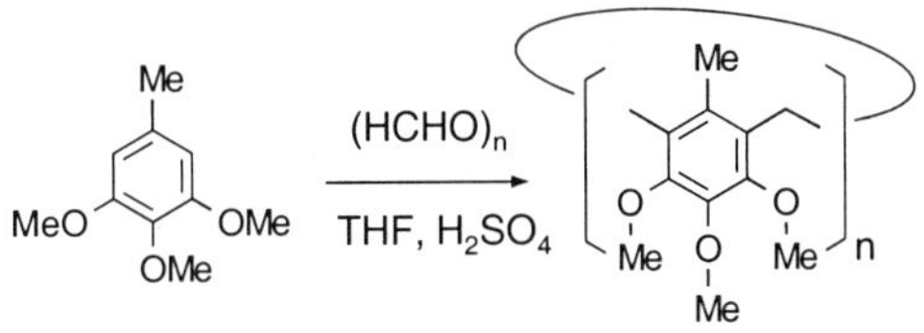

38 → **39**

Most of the acid-catalyzed cyclooligomerizations of resorcinols give the cyclic tetramer as the major or sole product. The acid-catalyzed reaction of **40**, however, yields a mixture of cyclo-oligomers **41** in which the cyclic tetramer (n = 4) is isolated in 9.4% yield and the larger oligomers (n = 5-13) in 1-2% yields [102], thus resembling the acid-catalyzed reaction of p-tert-butylphenol **1a** with HCHO (cf. Sec 1.2).

40 —$(HCHO)_n$, THF, H_2SO_4→ **41**

3. Non-Convergent Multi-Step Syntheses

Although multi-step syntheses [103,104] of calixarenes are often long and the yields sometimes modest (as, for example, in the steps leading to **42**), the method has the advantage of allowing the introduction of a variety of groups onto the *p*-positions. The cyclization step to form **2-5**, often carried out under high-dilution conditions, usually proceeds in high yield to give a single product, although side products sometimes are also formed [105].

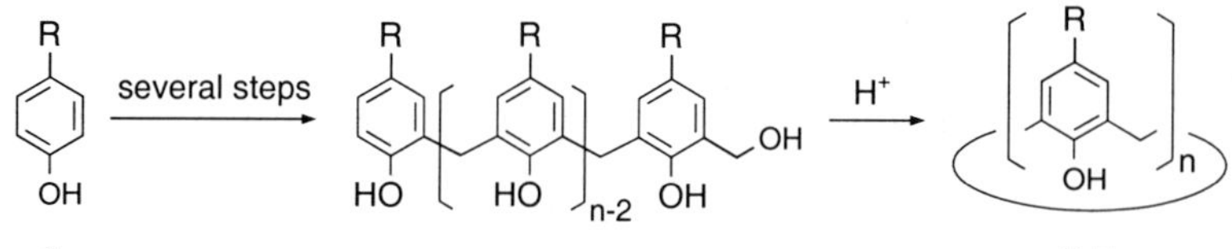

A recent example is the synthesis of carbonyl-containing calix[4,5,6] arenes **44** containing one carbonyl bridge in 37-61% yields by the acid-catalyzed cyclization of the corresponding carbonyl-containing hydroxymethylated linear oligomers **43** [106].

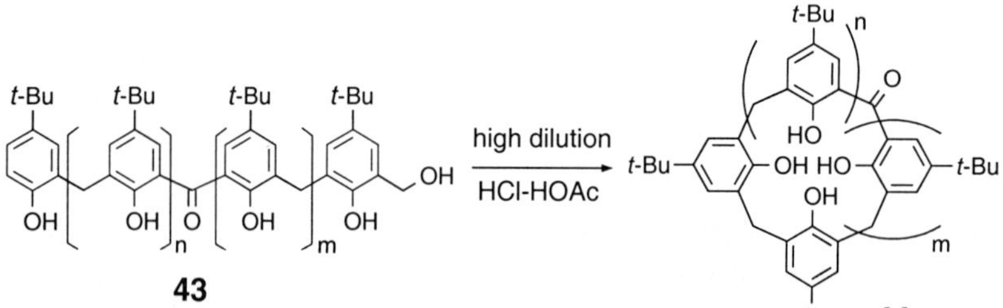

A recent modification of the stepwise approach uses the linear tetramer **45** lacking a hydroxymethyl group and employs acid catalysis in the presence of HCHO to give **46** (R^1 = Me or *t*-Bu; R^2 = H or CHO) in 8-25% yield [107].

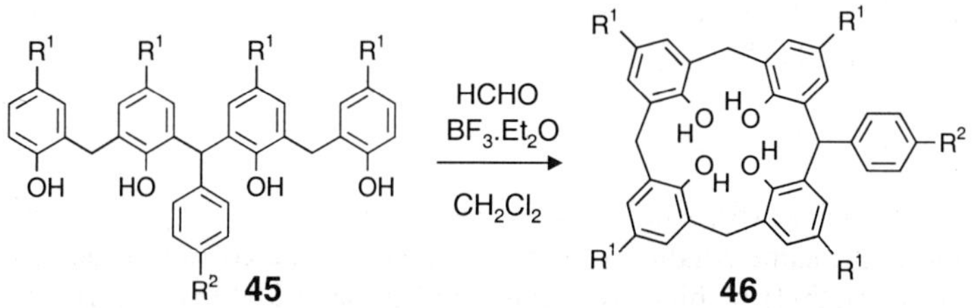

4. Convergent Syntheses (Fragment Condensation)

The non-convergent multi-step synthesis of calixarenes [51c] has now been largely abandoned in favor of more convergent processes for which the terms "fragment condensation" and "directed synthesis" have been coined [108]. In many instances it is perhaps a moot point as to whether these should be classified as single-step or multi-step syntheses, so the term "fragment condensation" provides an appropriate way to differentiate them from the single step syntheses discussed in Secs 1 and 2. Fragment condensation, in common with the non-convergent syntheses of calixarenes, offers the important potential of incorporating a variety of ring-substituents within the same calixarene molecule.

4.1. "3 + 1" FRAGMENT CONDENSATION SYNTHESES

Early examples [109,110] involve the condensation of a linear trimer **47** (Y^1 = H) with a bis-(halomethyl)-phenol **48** (Y^2 = CH_2Br or CH_2Cl) to form a cyclic tetramer **49** (Route A). Alternatively, the linear trimer (**47**, Y^1 = CH_2Br) rather than the monomeric unit (**48**, Y^2 = H) can carry the bromomethyl moieties (Route B). The corresponding hydroxymethylated compounds can also be used [111a], producing the cyclic tetramers in comparable yields of 15-38%. The reactions are often carried out in hot or refluxing dioxane [112] for extended periods of time using $TiCl_4$ as the catalyst. The yields vary considerably, and never exceed 40%.

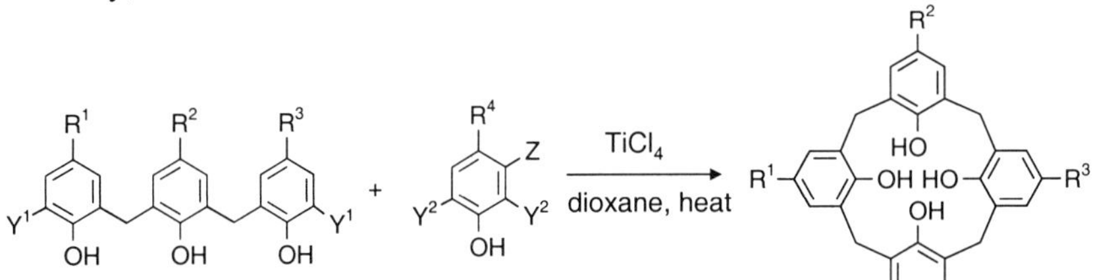

Route A: Y^1 = H, Y^2 = CH_2Br or CH_2OH
Route B: Y^1 = CH_2Br or CH_2OH, Y^2 = H

49 Z = H
50 Z = Me

A variety of p-substituents have been employed, including Me, t-Bu, Ph [111b], PhCO [113], Cl [109], CO_2Et [109], and NO_2 [109]. An interesting application is seen in the synthesis of bis-calixarenes in which a pair of calixarenes are joined at the p-position by a spanner (X = Me_2C, $(CH_2)_5$, $(CH_2)_8$, $(CH_2)_{12}$) [114]. An early example of a dissymmetric calix[4]arene **50** containing a single m-substituted phenolic unit was also prepared by this route [115].

4.2. "2 + 2" FRAGMENT CONDENSATION SYNTHESES

The "2 + 2" approach [116], similar in concept to the "3 + 1" approach and generally carried out under the same conditions, typically involves the condensation of a dimer **51** with a bis(bromomethyl) or bis-(hydroxymethyl) dimer **52** [116] to form **49** in which the p-substituents R can include those listed above.

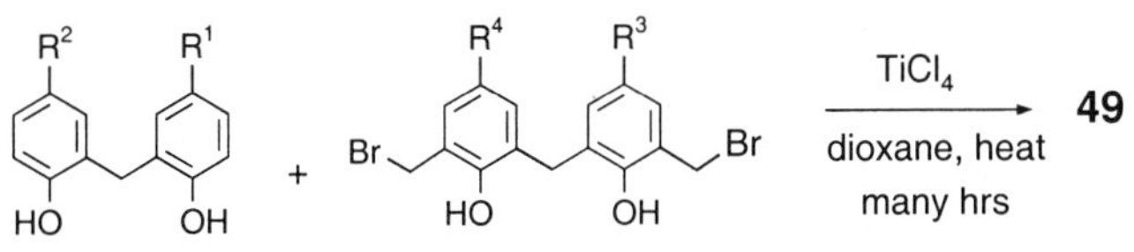

Alternatively, the process can involve the self-condensation of two molecules of a dimer ("2x2") containing a single hydroxymethyl moiety, as exemplified by the conversion of **53** to **54** [117,118] to produce calix[4]arenes with C_2 symmetry.

53 **54**

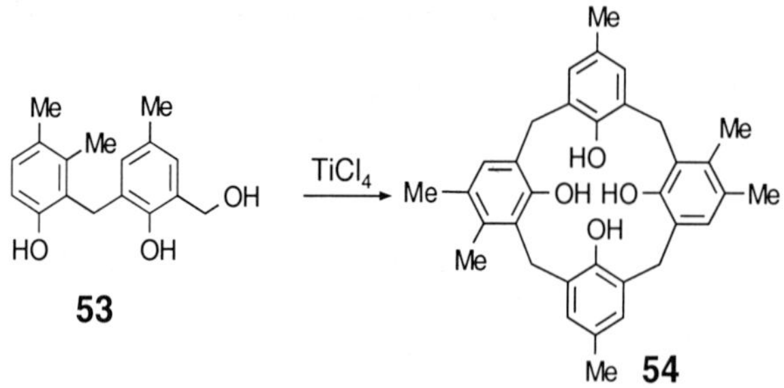

The "2 + 2" approach has been used in the synthesis of calix[4]arenes carrying substituents on one or more of the bridging methylene carbon atoms. Starting with **55** and **56** carrying R groups on the methylene bridges, compounds **57** (R^2 = Me, Et, *i*-Pr, n-Pr, *p*-MePh, *p*-NO_2Ph, 2-Py; R^4 = H or R^2) are formed in 20-35% yields [71 119]. In the same fashion the calixarenes **58** carrying both *exo*- and *endo*-OH groups were obtained [113. The azulocalix[4]arene **59** has also been prepared by this strategy [120].

55 **56** **57**

58

R = Cl, H, NO_2
R^3 = Me, Ph

59

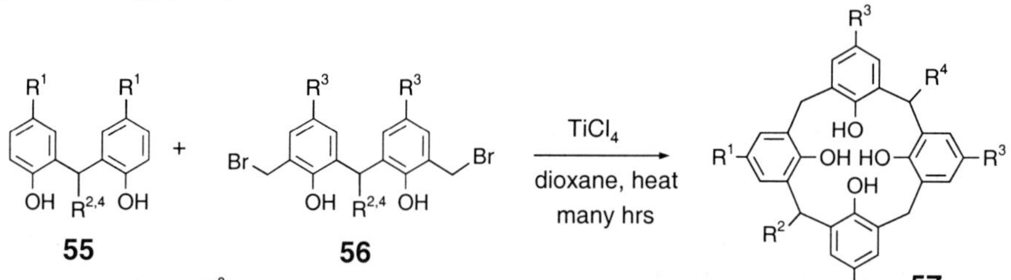

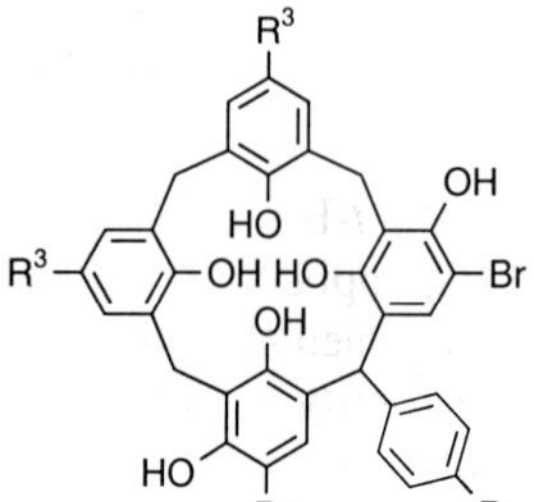

A quite different type of "2 + 2" process is illustrated by the treatment of the tetrabromo compound **60** with *t*-BuLi followed by diketone **61** to give **62** in 30-50% yields [121].

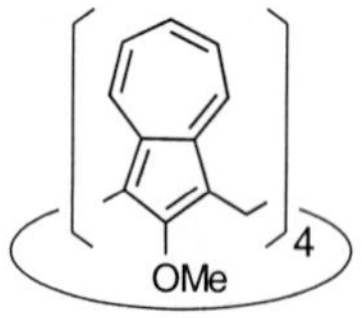

The "2 + 2" procedures described above for the syntheses of **49** and **62** have been used also to prepare extended calix[4]arenes with up to three annelated calix[4]arene rings [122,123] (see Chapter 7).

60 **61** **62**

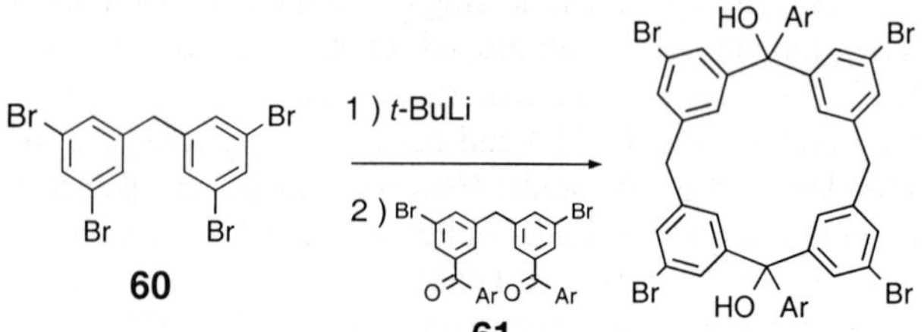

4.3. "2 + 1 + 1" FRAGMENT CONDENSATION SYNTHESES

A classic example of a "2 + 1 + 1" (or "2 + 2x1") process employs bis(*p*-hydroxyphenyl)alkanes **63** (X = $(CH_2)_n$) with **64** to produce the upper rim-bridged calixarene **65** (X = $(CH_2)_n$) in which n = 5-16 [124] in yields 2-20%. An interesting use of this approach is seen in the synthesis of compound **65** (X = $CH_2CH_2COCH_2CH_2$) obtained by the reaction of **63** (X = $CH_2CH_2COCH_2CH_2$) with **64** in which the $CH_2CH_2COCH_2CH_2$ bridge can be transformed into an additional aromatic ring by condensation with nitromalonaldehyde [125] (see Chapter 7). Another application of this approach to the construction of "head-to-tail" linked double calixarenes [126] is also described in Chapter 7.

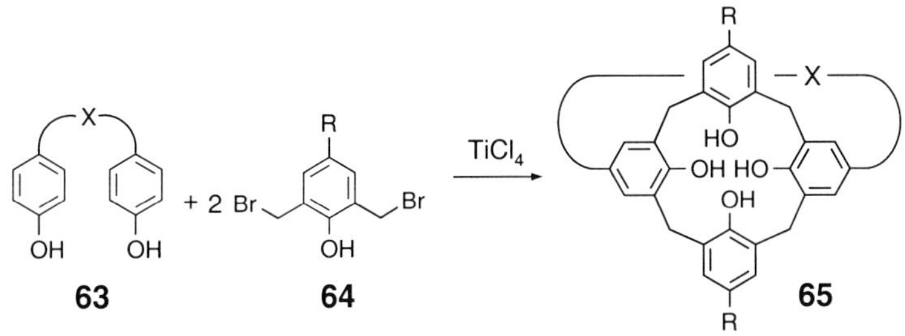

4.4. "1 + 1 + 1 + 1" FRAGMENT CONDENSATION SYNTHESES

"1 + 1 + 1 + 1" (or "4x1") Fragment condensations are formally analogous to the single step condensations discussed in Secs 1 and 2 except that the methylene unit is incorporated into the starting material instead of being introduced via HCHO during the reaction. For unsymmetrically substituted phenols this has the important implication that all phenol rings are incorporated in the same orientation.

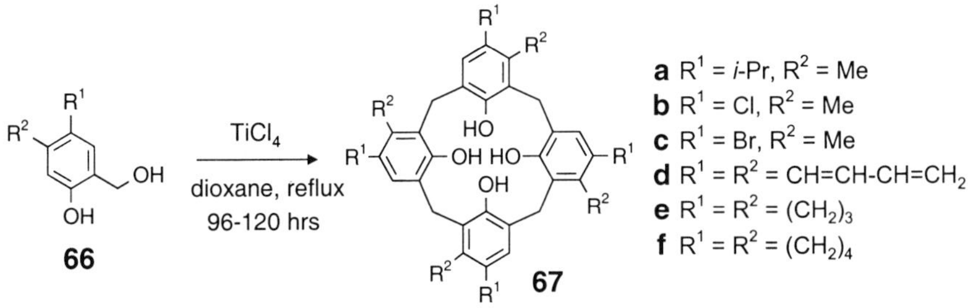

a R^1 = *i*-Pr, R^2 = Me
b R^1 = Cl, R^2 = Me
c R^1 = Br, R^2 = Me
d R^1 = R^2 = CH=CH-CH=CH$_2$
e R^1 = R^2 = $(CH_2)_3$
f R^1 = R^2 = $(CH_2)_4$

In one approach, a single component such as a hydroxymethylphenol **66** is employed to produce **67** in 18-30% yields [117,118], the starting materials being prepared (a) by hydroxymethylation with HCHO and base or (b) by reduction of the corresponding carboxylic acids. However, **66** (R^1 = *t*-Bu; R^2 = Me) fails to yield a calixarene under either acidic or basic conditions [118b].

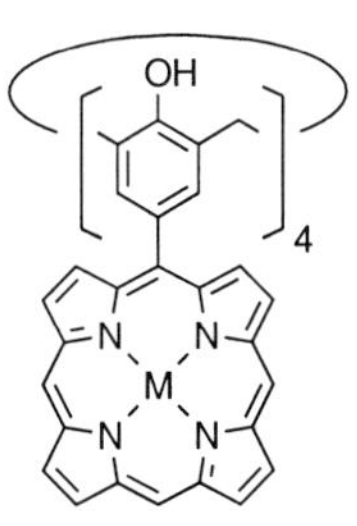

A 2-hydroxymethylphenol substituted at the 4-position with a porphyrin moiety has been converted to a calix[4]arene **68** in 60% yield by treatment with NaOH in diphenyl ether [127].

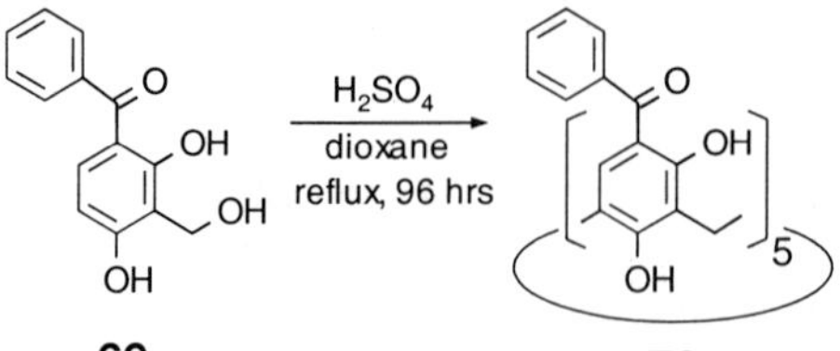

The use of hydroxymethylphenols has the potential for producing calixarenes with more than four aryl moieties in the cyclic array and, indeed, a small yield (4-6%) of the calix[5]arene **70** has been isolated from the acid-catalyzed dehydration of **69** in a "5x1" condensation [128]. Also noteworthy is the incorporation of the *m*-substituted aryl rings in a stereoregular fashion to give calixarenes with C_4 and C_5 symmetry [129].

Both hydroxymethyl compounds **71a** and **71b** yield the same octamethylcalix[4]arene **72**, the hydroxymethyl compound **71a** isomerising to **71b** before undergoing cyclization [130].

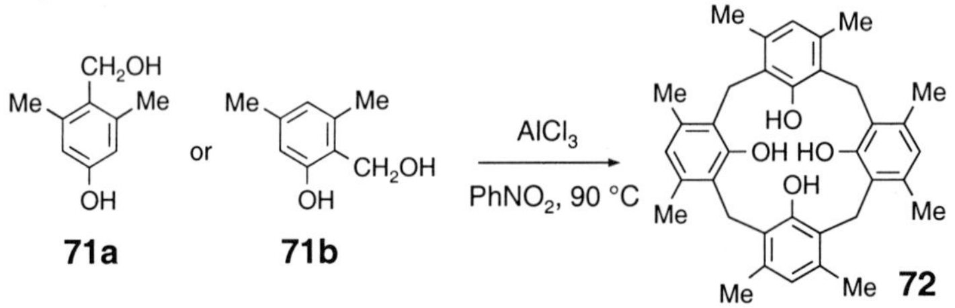

In a slightly different approach [131], ("2x1 + 2x1"), a pair of compounds is used, one a bis(halomethyl)phenol and one a simple phenol as, for example, in the reactions of **73** with **1** to provide 9-11% yields of **74** (R^1 = H; $R^{2,3}$ = *t*-Bu, Ph [132]) and 8-10% yields of **74** (R^1 = Me; R^2 = Cl; R^3 = Me, *t*-Bu).

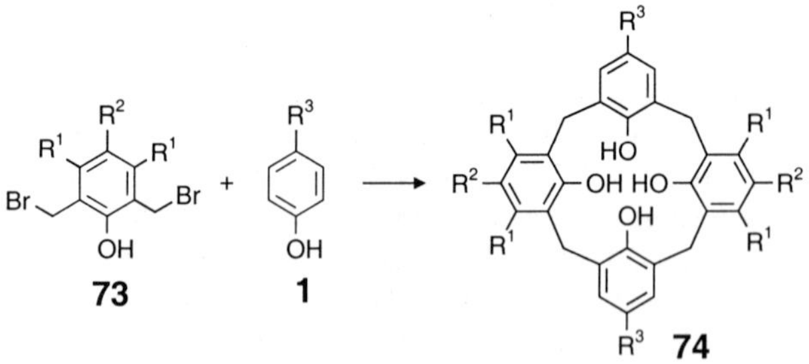

An interesting example leading to a multiply-connected bis-calixarene [114] is described in Chapter 7. A calixarene containing four *exo*-hydroxyl groups **77** is the result of the SnCl$_4$-catalyzed condensation of **75** with **76** (R = OH) to give a 66% yield of **77** [133]. When **76** (R = H) is used as the partner with **75** a mixture of calix[4]arenes is obtained that carry none (1%), one (9%), two (1,2-8%; 1,3-28%), or three (3%) *exo*-OH groups, illustrating the reversibility of the Friedel-Crafts condensation and the consequent fragmentation/recombination processes.

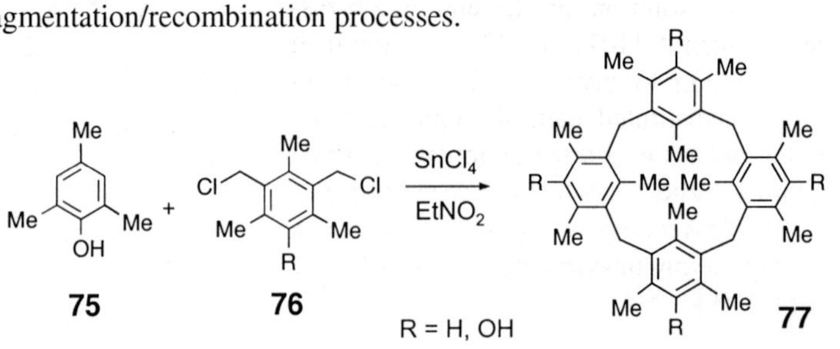

4.5. Fragment Condensation Syntheses of Calix[n]arenes in which [n] is other than 4

A "2 + 1" fragment condensation has been reported to yield quite highly strained calix[3]arenes [134], but attempts to reproduce this work have failed [22]. However, using **78** with *p-tert*-butylphenol in the presence of Nafion-H (perfluorinated sulfonic acid resin) the calix[3]arene **79** can be isolated in 17% yield [135].

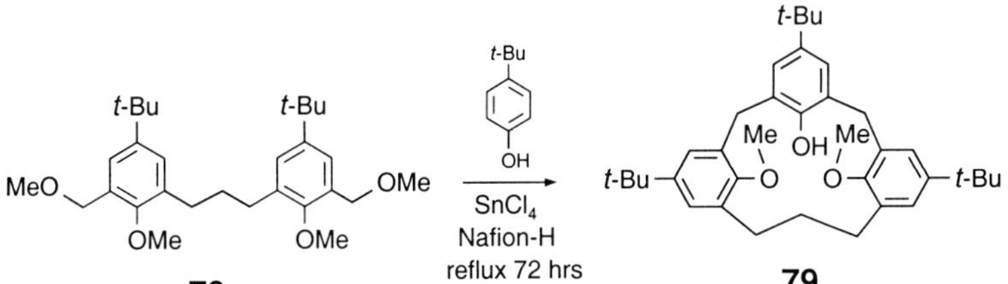

Reaction of **47** (Y^1 = H; $R^{1,3}$ = Me; R^2 = *t*-Bu) with **56** (R^3 = Me, Et, *i*-Pr, *t*-Bu, *p*-MePh, *p*-NO$_2$Ph; R^4 = H) [136], with the goal of producing a calix[5]arene by a "3 + 2" condensation, gave the desired product (11-13%) along with several others, including a calix[8]arene (3%). The tentative mechanistic conclusion was drawn that a single *ipso* attack of a benzyl cation is responsible for the formation of the calix[8]arene and that no complete scrambling of the phenolic units is involved.

Better yields (27-32%) are achieved by using **47** with the bis(hydroxymethyl) linear dimers and simply refluxing the reaction mixture in xylene without any added catalyst [137,138], giving calix[5]arenes with various substituents on the upper rim (Me, *t*-Bu, Br, C$_6$H$_5$). When using TiCl$_4$ or HCl as a catalyst much more complicated reaction mixtures are obtained. A monodeoxycalix[5]arene has been synthesized in a reaction that follows both a "3 + 1" and a "3 + 1 + 1" pathway to give a mixture containing 15-29% of the calix[4]arene and 0-7% of the calix[5]arene [139]. However, using a "3 + 2" process the monodeoxycalix[5]arene was obtained in 25% yield.

The TiCl$_4$-catalyzed "3 + 3" condensation of a linear trimer with a bis(bromo-methyl) trimer has been used to make **80** in 9% yield [110]. When using the analogous bis(hydroxy-methyl) trimer, some of the calix[4]arene was also obtained, but none of the calix[8]arene. In a "4 + 1" fragment condensation the calix[5]arenes **81** (R^1 = Me or *t*-Bu; R^2 = H, CHO, NO$_2$) are obtained in 18-28% yield by simply refluxing a xylene solution of **45** and a 2,6-bis-(hydroxymethyl)phenol [107]. A "7 + 1" fragment condensation reaction to give a 9.1% yield of an unsymmetrically substituted calix[8]arene has been achieved [140] using a *p-tert*-butylphenol-derived linear heptamer and 2,6-bis(hydroxymethyl)-4-carbethoxyphenol. Accompanying the calix[8]arene was the corresponding unsymmetrically-substituted calix[6]arene (7.3%).

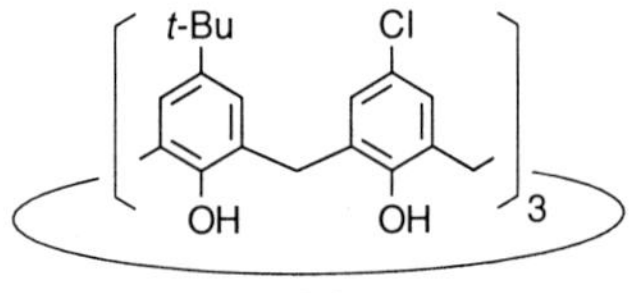

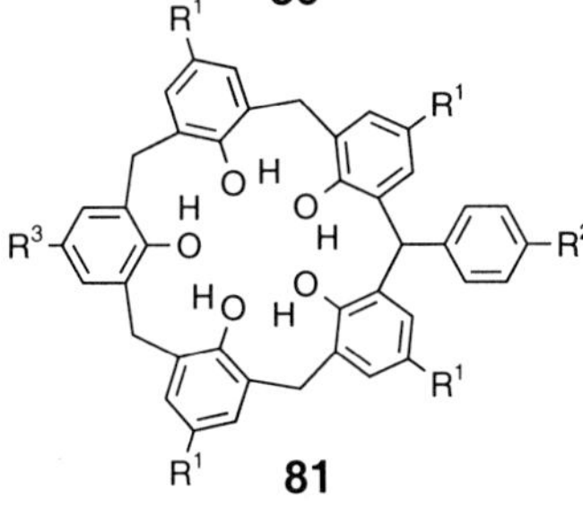

5. Thiacalixarenes

The thiacalixarenes **83** have attracted considerable recent interest as alternatives to the "classic" calixarenes, providing functionalisation potential not only on the rings but also on the bridging sulfur atoms. They have been made both by stepwise and single step procedures. Thus, linear tetramers containing one S and two CH$_2$, two S and one CH$_2$, or three S bridges yield, upon cyclization, the corresponding mono-, di-, and trithia-calix[4]arenes. The tetrathiacalix[4]arene **83** (n = 4) was prepared in 4.1% yield by treating the trithia linear tetramer **82** with SCl$_2$ [141].

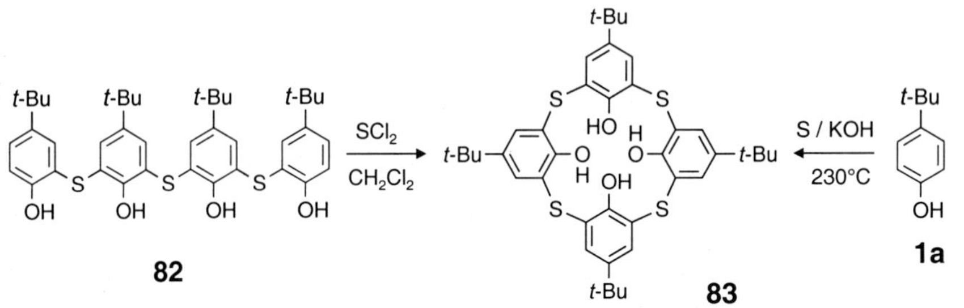

Much more efficient, however, is the single step synthesis [142] in which a mixture of *p-tert*-butylphenol, elemental sulfur, and NaOH is heated to 230 °C in a tetraethyleneglycol dimethyl ether solution to give **83** (n = 4) in yields up to 54%. Vanishingly small amounts (*ca* 0.03%) of the corresponding cyclic pentamer and hexamer were detected, but with CsOH as the base a ponderable amount (0.8%) of the hexamer was isolated and characterized [143]. Lower yields of the cyclic tetramer, however, are obtained with LiOH, KOH, or CsOH as the base.

Further examples of sulfur bridged metacyclophanes incorporating non phenolic aromatic units are described in Chapter 13.

6. Conclusion

All journeys start with the first few steps. Those in the journey of calixarene chemistry have involved learning how to make the basic calixarene frameworks, their scaffolds. This has been the topic of discussion in this first chapter which has focused on single-step syntheses, non-convergent stepwise syntheses, and convergent fragment condensation syntheses. Collectively, these procedures now provide a firm basis for moving on to the next steps of the journey which include learning how to add functional groups to the rims of the calixarenes, learning how to control and contour the shapes of these cyclo-oligomers, and learning how to employ these intriguing compounds in the great variety of applications for which they are being explored. Where the journey, whose margins change continually as we travel, are taking us provides the subject matter for the remaining chapters of this book.

7. References and Notes

[1] A. Zinke, E. Ziegler, *Ber.* **1941**, B74, 1729-1805; idem. *ibid.* **1944**, 77, 264-272.
[2] A notable exception is the work of Cornforth and coworkers: J. W. Cornforth, P. D'Arcy Hart, G. A. Nicholls, R. J. W. Rees, J. A. Stock, *Br. J. Pharmacol.* **1955**, *10*, 73-86; J. W. Cornforth, E. D. Morgan, K. T. Potts, R. J. W. Rees, *Tetrahedron* **1973**, *29*, 1659-1667.
[3] C. D. Gutsche, B. Dhawan, K. H. No, R. Muthukrishnan, *J. Am. Chem. Soc.* **1981**, *103*, 3782-3792.
[4] The resemblance of *p-tert*-butylcalix[4]arene to a Greek vase called a "calix crater" (variously spelled; *e.g.* kalyx krator) inspired the use of "calix" as a prefix attached to "arene" as a suffix to yield the word "calixarene" (C. D. Gutsche, R. Muthukrishnan, *J. Org. Chem.* **1978**, *43*, 4905-4906).
[5] D. R. Stewart, M. Krawiec, R. P. Kashyap, W. H. Watson, C. D. Gutsche, *J. Am. Chem. Soc.* **1995**, *117*, 586-601.
[6] S. Kanamathareddy, C. D. Gutsche, *J. Org. Chem.* **1994**, *59*, 3871-3879.
[7] C. D. Gutsche, M. Iqbal, *Org. Synth.* **1990**, *68*, 234-237.
[8] Microwave heating has been advocated for large scale production: V. Hedeboe, O. Jørgensen, H. D. Rasmussen, *Microwave and High Frequency Heating* **1997**, 155-156; O. Jørgensen, H. D. Rasmussen, M. Jørgensen, Abstracts from 7th *International Confernece on Microwave and High Frequency Heating, Valencia, Spain* **1999**, p. 323..
[9] C. D. Gutsche, B. Dhawan, M. Leonis, D. Stewart, *Org. Synth.* **1990**, *68*, 238-242.
[10] J. H. Munch, C. D. Gutsche, *ibid.* **1990**, *68*, 243-246.
[11] A. Ninagawa, H. Matsuda, *Makromol. Chem. Rapid Commun.* **1982**, *3*, 65-67; M. A. Markowitz, V. Janout, D. G. Castner, S. I. Regen, *J. Am. Chem. Soc.* **1989**, *111*, 8192-8200.
[12] D. R. Stewart, C. D. Gutsche, *Org. Prep. Proced. Int.* **1993**, *25*, 137-139; K. Iwamoto, K. Araki, S. Shinkai, *Bull. Chem. Soc. Jpn.* **1994**, *67*, 1499-1502
[13] Y. Nakamoto, S. Ishida, *Makromol. Chem. Rapid Commun.* **1982**, *3*, 705-707.
[14] F. Vocanson, R. Lamartine, P. Lanteri, R. Longeray, J. Y. Gauvrit, *New J. Chem.* **1995**, *19*, 825-829.
[15] I. Dumazet, N. Ehlinger, F. Vocanson, S. Lecocq, R. Lamartine, M. J. Perrin, *J. Incl. Phenom.* **1997**, *29*, 175-185. Phenols other than *p-tert*-butylphenol can also be used in this reaction to give calix[n]arenes carrying other *p*-alkyl groups in addition to the *p-tert*-butyl groups. However, the yields are vanishingly small.
[16] M. Yilmaz, U. S.Vural, *Synth. React. Inorg. Met. Org. Chem.* **1991**, *21*, 1231-1241.
[17] Y. Seki, Y. Morishige, N. Wamme, Y. Ohnishi, S. Kishida, *Appl. Phys. Lett.* **1993**,*62*, 3375-3376.
[18] Z. Asfari, J. Vicens, *Makromol. Chem. Rapid Commun.* **1989**, *10*, 181-183.
[19] J. Vicens, T. Pilot, D. Gamet, R. Lamartine, R. Perrin, *C. R. Acad. Sci. Paris* **1986**, *302*, 15-20.
[20] P. Novakov, S. Miloshev, P. Tuleshkov, I. Gitsov, M. Georgieva, *Angew. Makromol. Chem.* **1998**, *255*, 23-28.
[21] Z. Asfari, J. Vicens, *Tetrahedron Lett.* **1988**, *29*, 2659-2660. Recently reported is the calix[8]arene from *p*-(1'-methyltridecyl)-2'phenol (B. Gross, J. Jauch, V. Schurig, JU. *Microcolumn Sep.* **1999**, *11*, 313-317)
[22] D. R. Stewart, C. D. Gutsche, unpublished observations.
[23] A. Casnati, R. Ferdani, A. Pochini, R. Ungaro, *J. Org. Chem.* **1997**, *62*, 6236-6239.
[24] B. Souley, Z. Asfari, J. Vicens, *Polish J. Chem.* **1992**, *66*, 959-961.
[25] For a slightly modified procedure *cf.* J. L. Atwood, L. J. Barbour, P. J. Nichols, C. L. Raston, C. A. Sandoval, *Chem. Eur. J.* **1999**, *5*, 990-996.
[26] J. L. Atwood, M. J. Hardie, C. L. Raston, C. A. Sandoval, *Org. Lett.* **1999**, *1*, 1523-1526.
[27] Surprisingly, *p-tert*-butylcalix[6]- and -[8]arenes do not behave in comparable fashion.
[28] S. Shinkai, T. Nagasaki, K. Iwamoto, A. Ikeda, G.-X. He, T. Matsuda, M. Iwamoto, *Bull. Chem. Soc. Jpn.* **1991**, *64*, 381-386; Y. Nakamoto, G. Kallinowski, V. Böhmer, W. Vogt, *Langmuir* **1989**, *5*, 1116-1117.
[29] D. Foina, A. Pochini, R. Ungaro, G. D. Andreetti, *Makromol. Chem. Rapid Commun.* **1983**, *4*, 71-73; V. Bocchi, D. Foina, A. Pochini, R. Ungaro, G. D. Andreetti, *Tetrahedron* **1982**, *38*, 373-378.
[30] B. Gross, J. Jauch, V. Schurig, *J. Microcolumn Sep.* **1999**, *11*, 313-317.
[31] E. Lubitov, E. A. Shokova, V. V. Kovalev, *Synlett* **1993**, 647-648.
[32] J. Jauch, V. Schurig, *Tetrahedron: Asymmetry* **1997**, *8*, 169-172.
[33] C.-H. Tung, H.-F. Ji, *J. Chem. Soc., Perkin Trans. 2* **1997**, 185-188. The use of *p*-(hydroxyphenyl)phenol (bisphenol-A) failed to yield any trace of calixarenes.
[34] A. Soi, J. Pfeiffer, J. Jauch, V. Schurig, *Tetrahedron: Asymmetry* **1999**, *10*, 177-182.
[35] C. D. Gutsche, J. S. Rogers, D. R. Stewart, K.-A. See, *Pure Appl. Chem.* **1990**, *62*, 485-491.

[36] D. R. Stewart, C. D. Gutsche. Third International Calixarene Conference, Fort Worth, TX 1995, Abstract P-42; C. D. Gutsche, C. G. Gibbs, S. K. Sharma, D. R. Stewart, J. Wang, D. Xie. Fourth international Conference on Calixarenes, Parma, 1997; Abstract IL1;.
[37] D. R. Stewart, C. D. Gutsche, *J. Am. Chem. Soc.* **1999**, *121*, 4136-4146
[38] I. Dumazet, J.-B. Regnouf-de-Vains, R. Lamartine, *Synth. Commun.* **1997**, *27*, 2547-2555.
[39] a) P. E. Georghiou, Z. Li, *Tetrahedron Lett.* **1993**, *34*, 2887-2890; P. E. Georghiou, Z. Li, *J. Incl. Phenom.* **1994**, *19*, 55-66; P. E. Georghiou, M. Ashram, Z. Li, S. G. Chaulk, *J. Org. Chem.* **1995**, *60*, 7284-7289; b) P. E. Georghiou, Z. Li, M. Ashram, D. O. Miller, *ibid.* **1996**, *61*, 3865-3869.
[40] P. E. Georghiou, Z. Li, M. Ashram, *J. Org. Chem.* **1998**, *63*, 3748-3752.
[41] a) B.-L. Poh, C. S. Lim, K. S. Khoo, *Tetrahedron Lett.* **1989**, *30*, 1005-1008; B.-L. Poh, C. S. Lim, *Tetrahedron* **1990**, *46*, 3651-3658; b) B.-L. Poh, C. M. Tan, C. L. Loh, *ibid* **1993**, *49*, 3849-3856.
[42] 1-Amino-8-hydroxynaphthalenedisulfonic acid also condenses with HCHO to form a cyclic tetramer, but only one of the o-positions of each of the naphthalene rings is involved, the other bridges engaging the amino function at C-1 (B.-L. Poh, L. Y. Chin, C. W. Lee, *Tetrahedron Lett.* **1995**, *36*, 3877-3880.
[43] Some doubt concerning this outcome, however, has been expressed by other workers attempting to repeat the synthesis of **12**. Similar uncertainty has also been expressed concerning the single-step preparation of a tetra-sulfonated cyclic tetramer by refluxing an aqueous solution of ammonium 5-sulfonatotropolone, HCHO and NaOH: B.-L. Poh, Y. Y. Ng, *Tetrahedron* **1997**, *53*, 8635-8642; B.-L. Poh, C. S. Yue, *ibid.* **1999**, *55*, 5515-5518.
[44] a) T. Yamato, Y. Saruwatari, S. Nagayama, K. Meeda, M. Tashiro, *J. Chem. Soc., Chem. Commun.* **1992**, 861-862; b) T. Yamato, Y. Saruwatari, L. K. Doamekpor, K.-I. Hasegawa, M. Koike, *Chem. Ber.* **1993**, *126*, 2501-2504.
[45] a) Y. Okada, F. Ishii, Y. Kasai, J. Nishimura, *Chem. Lett.* **1992**, 755-758; Y. Okada, Y. Kasai, F. Ishii, J. Nishimura, *J. Chem. Soc., Commun.* **1993**, 976-978; b) Y. Okada, Y. Kasai, J. Nishimura, *Synlett* **1995**, 85-89.
[46] (a) T. Yamato, K.-i. Hasegawa, Y. Saruwatari, L. K. Doamekpor, *Chem. Ber.* **1993**, *126*, 1435-1439; T. Yamato, M. Yasumatsu, Y. Saruwatari, L. K. Doamekpor, *J. Incl. Phenom.* **1994**, *19*, 315-331; (b) T. Yamato, F. Zhang, M. Yasumatsu, *J. Chem. Res. Synop.* **1997**, 466.
[47] P. O'Sullivan, V. Böhmer, W. Vogt, E. F. Paulus, R. A. Jakobi, *Chem. Ber.* **1994**, *127*, 427-432.
[48] D. W. Chasar, *J. Org. Chem.* **1985**, *59*, 545-546.
[49] T. N. Sorrell, H. Yuan, *J. Org. Chem.* **1997**, *62*, 1899-1902.
[50] V. Böhmer, R. Dörrenbächer, M. Frings, M. Heydenreich. D. de Paoli, W. Vogt, G. Ferguson, I. Thondorf, *J. Org. Chem.* **1996**, *61*, 549-559.
[51] C. D. Gutsche, "Calixarenes", *Monographs in Supramolecular Chemistry*, J. F. Stoddart, ed. Royal Society of Chemistry, 1989, (a) p. 50-58, (b) p. 36-38. (c) p. 38-47.
[52] The oligomers are formed by the condensation of a phenolic unit (*i.e.* monomer, dimer, trimer, etc.) with HCHO to form an o-hydroxymethyl compound which then reacts with the parent phenol to form, probably via an *o*-quinonemethide intermediate, the next higher oligomer.
[53] F. Vocanson, R. Lamartine, R. Perrin, *Supramol. Chem.* **1994**, *4*, 153-157.
[54] F. Vocanson, R. Lamartine, *Supramol. Chem.* **1996**, *7*, 19-25.
[55] C. D. Gutsche, D. E. Johnston, Jr., D. R. Stewart, *J. Org. Chem.* **1999**, *64*, 3747-3750.
[56] H. Takemura, T. Shinmyozu, H. Miura, I. U. Khan, T. Inazu, *J. Incl. Phenom.* **1994**, *19*, 193-206.
[57] T. Yamato, M. Yasumatsu, L. K. Doamekpor, S. Nagayama, *Liebigs Ann. Chem.* **1995**, 285-289.
[58] J. F. Ludwig, A. G. Bailie, Jr., *Anal. Chem.* **1986**, *58*, 2069-2072.
[59] G. Sartori, C. Porta, F. Bigi, R. Maggi, F. Peri, E. Marzi, M. Lanfranchi, M. A. Pellinghelli, *Tetrahedron* **1997**, *53*, 3287-3300.
[60] a) G. Sartori, R. Maggi, F. Bigi, A. Arduini, A. Pastorio, C. Porta, *J. Chem. Soc., Perkin Trans. 1* **1994**, 1657-1658; b) G. Sartori, F. Bigi, C. Porta, R. Maggi, R. Mora, *Tetrahedron Lett.* **1995**, *36*, 2311-2314.
[61] G. Sartori, F. Bigi, C. Porta, R. Maggi, F. Peri, *Tetrahedron Lett.* **1995**, *36*, 8323-8326.
[62] J. B. Niederl, H. J. Vogel, *J. Am. Chem. Soc.* **1940**, *62*, 2512-2514; J. B. Niederl, J. S. McCoy, *ibid.* **1943**, *65*, 629-631.
[63] H. Erdtman, S. Högberg, S. Abramson, B. Nilsson, *Tetrahedron Lett.* **1968**, 1679-1682.
[64] a) A. G. S. Högberg, *J. Org. Chem.* **1980**, *45*, 4498; b) idem, *J. Am. Chem. Soc.* **1980**, *102*, 6046-6050.
[65] D. J. Cram, J. M. Cram, "Container Molecules and Their Guests" in *Monographs in Supramolecular Chemistry*, J. F. Stoddart, ed. Royal Society of Chemistry, 1994.

[66] E. U. Thoden van Velzen, J. F. J. Engbersen, D. N. Reinhoudt, *J. Am. Chem. Soc.* **1994**, *116*, 3597-3598.
[67] a) O. I. Pieroni, N. M. Rodriguez, B. M. Vuano, M. C. Cabaleiro, *J. Chem. Res.* **1994**, 188-189; b) A. G. M. Barrett, D. C. Braddock, J. P. Henschke, E. R. Walker, *J. Chem. Soc., Perkin Trans. 1* **1999**, 873-878.
[68] For a review of calixresorcarenes cf. P. Timmerman, W. Verboom, D. N. Reinhoudt, *Tetrahedron* **1996**, *52*, 2663-2704.
[69] Since the compounds clearly are members of the calixarene family, as defined in Sec 1.1, it is logical that this fact should be reflected in their name. Therefore, this author urges that "calix[n]resorcarene" be accepted as the colloquial designation for these compounds.
[70] Boat forms are interconvertible via a pseudorotation involving the crown/bowl conformation as an intermediate or transition state.
[71] S. E. Biali, V. Böhmer, J. Brenn, M. Frings, I. Thondorf, W. Vogt, J. Wöhnert, *J. Org. Chem.* **1997**, *62*, 8350-8360.
[72] In contrast fo the phenol-derived calixarenes, the calix[4]resorcarenes carrying a substituent on the bridging carbon may not undergo facile conformational interconversion, because the process requires the substituent(s) to pass by the *exo* hydroxyl groups. With bulky substituents and/or with O-alkyl or O-acyl derivatives this can be difficult, and there is at least one report of the isolation of a pair of conformational isomers (cf. ref. [100b] in which the configuration of compound **6c** is incorrectly assigned in the experimental section and should, instead, be designated as the *rccc* isomer, as is shown in the discussion section (B. Botta, private communication). Generally, however, conformational interconversion is rather rapid (L. Abis, E. Dalcanale, A. Du vosel, S. Spera, *J. Chem. Soc., Perkin Trans 2* **1990**, 2075-20890); see also the easy formation of cavitands from *rctt* and *rcct* isomers (Chapter 9).
[73] F. Weinelt, H.-J. Schneider, *J. Org. Chem.* **1991**, *56*, 5527-5535.
[74] As far back as 1968-1976 the structures of the boat (all-axial; *rccc*) and chair (all-axial; *rctt*) forms of the octabutyl esters of **28** (R^1 = *p*-bromophenyl; R^2 = H) were established by X-ray crystallography (H. Erdtman, S. Högberg, S. Abrahamsson, B. Nilsson, *Tetrahedron Lett.* **1968**, 1679-1682; B. Nilsson, *Acta. Chem. Scand.* **1968**, *22*, 732-747; K. J. Palmer, R. Y. Wong, L. Jurd, K. Stevens, *Acta Crystallogr. Sect. B* **1976**, *32*, 847-852). More recently, diamond isomers (*rctc*) have been isolated (L. Abis, E. Dalcanale, A. Du vosel, S. Spera, *J. Org. Chem.* **1988**, *53*, 5475-5479; Also, cf. ref [82].
[75] With Lewis acid catalysts in Et_2O, however, the reaction appears not to be reversible, the proportion of chair and boat isomers remaining constant regardless of reaction time.
[76] A. Shivanyuk, E. F. Paulus, V. Böhmer, W. Vogt, *Angew. Chem., Int. Ed. Eng.* **1997**, *36*, 1301-1303.
[77] L. M. Tunstad, J. A. Tucker, E. Dalcanale, J. Weiser, J. A. Bryant, J. C. Sherman, R. C. Helgeson, C. B. Knobler, D. J. Cram. *J. Org. Chem.* **1989**, *54*, 1305-1312.
[78] D. J. Cram, S. Karbach, H.-E. Kim, C. B. Knobler, E. F. Maverick, J. L. Ericson, R. C. Helgeson, *J. Am. Chem. Soc.* **1988**, *110*, 2229-2237.
[79] Y. Aoyama, Y. Tanaka, H. Toi, H. Ogoshi, *J. Am. Chem. Soc.* **1988**, *110*, 634-635; Y. Aoyama, Y. Tanaka, S. Sugahara, *ibid* **1989**, *111*, 5397-5404.
[80] B. C. Gibb, R. G. Chapman, J. C. Sherman, *J. Org. Chem.* **1996**, *61*, 1505-1509.
[81] K. Kobayashi, Y. Asakawa, Y. Kato, Y. Aoyama, *J. Am. Chem. Soc.* **1992**, *114*, 10307-10313.
[82] U. Schneider, H.-J. Schneider, *Chem. Ber.* **1994**, *127*, 2455-2469.
[83] Y. Yamakawa, M. Ueda, R. Nagahata, K. Takeuchi, M. Asai, *J. Chem. Soc., Perkin Trans. 1* **1998**, 4135-4139.
[84] T. Kijima, Y. Kato, K. Ohe, M. Machida, Y. Matsushita, T. Matsui, *Bull. Chem. Soc. Jpn.* **1994**, *67*, 2125-2129.
[85] P. T. Lewis, C. J. Davis, M. C. Saraiva, W. D. Treleaven, T. D. McCarley, R. M. Strongin, *J. Org. Chem.* **1997**, *62*, 6110-6111.
[86] P. D. Beer, E. L. Tite, A. Ibbotson, *J. Chem. Soc., Chem. Commun.* **1989**, 1874-1876.
[87] A. D. M. Curtis, *Tetrahedron Lett.* **1997**, *38*, 4295-4296.
[88] P. Sakhail, I. Neda, M. Freytag, H. Thönnessen, P. G. Jones, R. Schmutzler, *Z. Anorg. Allg. Chem.* **2000**, *626*, 1246-1254.
[89] R. J. M. Egberink, P. L. H. M. Cobben, W. Verboom, S. Harkema, D. N. Reinhoudt, *J. Incl. Phenom.* **1992**, *12*, 151-158.
[90] P. D. Beer, E. L. Tite, M. G. B. Drew, A. Ibbotson, *J. Chem. Soc., Dalton Trans.* **1990**, 2543-2550.
[91] H. Konishi, Y. Iwasaki, O. Morikawa, T. Okano, J. Kiji. *Chem. Express* **1990**, *5*, 869-872.
[92] G. Cometti, E. Dalcanale, A. Du Vosel, A.-M. Levelut, *Liquid Crystals* **1992**, *11*, 93-100.

[93] H. Konishi, K. Ohata, O. Morikawa, K. Kobayashi, *J. Chem. Soc., Chem. Commun.* **1995**, 309-310; H. Konishi, T. Nakamura, O.Kazunobu, K. Kobayashi, O. Morikawa, *Tetrahedron Lett.* **1996**, *37*, 7383-7386.
[94] H. Konishi, Y. Iwasaki, *Synlett* **1995**, 612.
[95] Y. Hayashi, T. Maruyama, T. Yachi, K. Kudo, K. Ichimura, *J. Chem. Soc., Perkin Trans. 2* **1998**, 981-987.
[96] G. Rumboldt, V. Böhmer, B. Botta, E. F. Paulus, *J. Org. Chem.* **1998**, *63*, 9618-9619.
[97] a) G. Cortes-Lopez, L. M. Gutierrez Tunstad, *Synlett* **1998**, 139-140; B. M. Vuano, O. I. Pieroni, *An. Asoc. Quim. Argent.* **1998**, *86*, 69-76 (*C. A.* **1998**, 129:260214b).
[98] O. I. Pieroni, M. Gonzalez Sierra, M. C. Cabaleiro, *J. Chem. Res.* **1994**, 455.
[99] W. Iwanek, *Tetrahedron* **1998**, *54*, 14089-14094.
[100] (a) B. Botta, P. Iacomacci, M. C. Di Giovanni, G. Delle Monache, E. Gacs-Baitz, M. Botta, A. Tafi, F. Corelli, D. Misiti, *J. Org. Chem.* **1992**, *57*, 3259-3261; (b) B. Botta, M. C. Di Giovanni, G. Delle Monache, M. C. De Rosa, E. Gacs-Batiz, M. Botta, F. Corelli, A. Tafi, A. Santini, E. Benedetti, C. Pedone, D. Misiti, *ibid.* **1994**, *59*, 1532-1541; (c) B. Botta, G. Delle Monache, M. C. De Rosa, A. Carbonetti, E. Gacs-Baitz, M. Botta, F. Corelli, D. Misiti, *ibid.* **1995**, *60*, 3657-3662; (d) B. Botta, G. Delle Monache, P. Salvatore, F. Gasparrini, C. Villani, M. Botta, F. Corelli, A. Tafi, E. Gacs-Baitz, A. Santini, C. F. Carvalho, D. Misiti, *ibid.* **1997**, *62*, 932-938; (e) B. Botta, G. Delle Monache, M. C. De Rosa, C. Seri, E. Benedetti, R. Iacovino, M. Botta, F. Corelli, V. Masignani, A. Tafi, E. Gacs-Baitz, A. Santini, D. Misiti, *ibid.* **1997**, *62*, 1788-1794.
[101] O. M. Falana, E. Al-Farhan, P. M. Keehn, R. Stevenson, *Tetrahedron Lett.* **1994**, *35*, 65-68.
[102] R. Schätz, C. Weber, G. Schilling, T. Oeser, U. Huber-Patz, H. Irngartinger, C.-W. von der Lieth, R. Pipkorn, *Liebigs Ann. Chem.* **1995**, 1401-1408.
[103] B. T. Hayes, R. F. Hunter, *Chem. Ind.* **1956**, 193-194; idem. *J. Appl. Chem.* **1958**, *8*, 743-748.
[104] H. Kämmerer, G. Happel, F. Caesar, *Makromol. Chem.* **1972**, *162*, 179-197; G. Happel, B. Mathiasch, H. Kämmerer, *ibid.* **1975**, *176*, 3317-3334; H. Kämmerer, G. Happel, *ibid.* **1978**, *179*, 1199-1207; H. Kämmerer, G. Happel, V. Böhmer, D. Rathay, *Monatsh. Chem.* **1978**, *109*, 767-773; H. Kämmerer, G Happel, *Makromol. Chem.* **1980**, *181*, 2049-2062; idem. *Monatsh. Chem.* **1981**, *112*, 759-768; H. Kämmerer, G. Happel, B. Mathiasch, *Makromol. Chem.* **1981**, *182*, 1685-1694; H. Kämmerer, G. Hapel in "*Weyerhauser Science Symposium on Phenolic Resins, 2*", Tacoma, Washington 1979, Weyerhauser Publishing Co, Tacoma **1981**, p. 143.
[105] K. H. No, C. D. Gutsche, *J. Org. Chem.* **1982**, *47*, 2713-2719.
[106] a) Y. Ohba, K. Irie, F. Zhang, T. Sone, *Bull. Chem. Soc. Jpn.* **1993**, *66*, 828-835 (describes the preparation of the linear oligomers); b) K. Ito, S. Izawa, T. Ohba, Y. Ohba, T. Sone, *Tetrahedron Lett.* **1996**, *37*, 5959-5962 (describes the cyclization but with sparse experimental details).
[107] M. Bergamaschi, F. Bigi, M. Lanfranchi, R. Maggi, A. Pastorio, M. A. Pellinghelli, F. Peri, C. Porta, G. Sartori, *Tetrahedron* **1997**, *53*, 13037-13052.
[108] For review articles *cf.* V. Böhmer, *Liebigs Ann./Recueil* **1997**, 2019-2030; V. Böhmer, "New Separation Chemistry Techniques for Radioactive Waste and Other Specific Applications", Elsivier Applied Science, **1991**, p. 133-141; V. Böhmer, W. Vogt. *Pure & Appl. Chem.* **1993**, *65*, 403-408.
[109] a) V. Böhmer, P. Chhim. H. Kämmerer, *Makromol. Chem.* **1979**, *180*, 2503-2506; V. Böhmer, F. Marschollek, L. Zetta, *J. Org. Chem.* **1987**, *52*, 3200-3205; b) L. Zetta, A. Wolff, W. Vogt, K.-L. Platt, V. Böhmer, *Tetrahedron* **1991**, *47*, 1911-1924.
[110] J. de Mendoza, P. M. Nieto, P. Prados, C. Sánchez, *Tetrahedron* **1990**, *46*, 671-682.
[111] (a) K. No, K. M. Kwon, J. E. Kim, *Bull. Korean Chem. Soc.* **1996**, *17*, 525-528; K. No, J. E. Kim, K. M.Kwon, *Tetrahedron Lett.* **1995**, *36*, 8453-8456; (b) K. No, K. L. Hwang, *Bull. Korean Chem. Soc.* **1993**, *14*, 753-755; K. No, J. E. Kim, S. J. Kim. *ibid.* **1996**, *17*, 869-872.
[112] The use of diethylene glycol dimethyl ether as the solvent is reported to afford yields up to 40% (Z.-L. Zhong, J.-S. Li, X.-R. Lu, Y-Y. Chen, *Gaodeng Xuexiao Huaxue Xuebao* **1998**, *19*, 83-85 (*C.A.* **1998**, 128:140471d)).
[113] M. Tabatabai, W. Vogt, V. Böhmer, *Tetrahedron Lett.* **1990**, *31*, 3295-3298.
[114] V. Böhmer, H. Goldmann, W. Vogt, J. Vicens, Z. Asfari, *Tetrahedron Lett.* **1989**, *30*, 1391-1394.
[115] H. Casabianca, J. Royer, A. Satrallah, A. Tatty-C, J. Vicens, *Tetrahedron Lett.* **1987**, *28*, 6595-6596.
[116] V. Böhmer, L. Merkel, U. Kunz, *J. Chem. Soc., Chem. Commun.* **1987**, 896-897.
[117] A. Wolff, V. Böhmer, W. Vogt, F. Ugozzoli, G. D. Andreetti, *J. Org. Chem.* **1990**, *55*, 5665-5667.
[118] (a) G. D. Andreetti, V. Böhmer, J. G. Jordon, M. Tabatabai, F. Ugozzoli, W. Vogt, A. Wolff, *J. Org. Chem..* **1993**, *58*, 4023-4032; (b) D.K. Fu, B. Xu, T. M. Swager, *J. Org. Chem.* **1996**, *61*, 802-804.
[119] C. Grüttner, V. Böhmer, W. Vogt, I. Thondorf, S. E. Biali, F. Grynszpan, *Tetrahedron Lett.* **1994**, *35*, 6267-6270.

[120] T. Asao, S. Ito, N. Morita, *Tetrahedron Lett.* **1988**, *29*, 2839-2842.
[121] a) A. Rajca, R. Padmakumar, D. J. Smithhisler, S. R. Desai, C. R. Ross II, J. J. Stezowski, *J. Org. Chem.* **1994**, *59*, 7701-7703; b) A. Rajca, S. Rajca, S. R. Desai, *J. Am. Chem. Soc.* **1995**, *117*, 806-816.
[122] a) V. Böhmer, R. Dörrenbächer, W. Vogt, L. Zetta, *Tetrahedron Lett.* **1992**, *33*, 769-772; b) E. F. Paulus, M. Frings, A. Shivanyuk, C. Schmidt, V. Böhmer, W. Vogt, *J. Chem. Soc., Perkin Trans. 2* **1998**, 2777-2782.
[123] A. Rajca, K. Lu, S. Rajca, *J. Am. Chem. Soc.* **1997**, 119, 10335-10345.
[124] V. Böhmer, H. Goldmann, W. Vogt, *J. Chem. Soc., Chem. Commun.* **1985**, 667-668; V. Böhmer, H. Goldmann, R. Kaptein, L. Zetta, *J. Chem. Soc., Chem. Commun.* **1987**, 1358-1360; E. Paulus, V. Böhmer, H. Goldmann, W. Vogt, *J. Chem. Soc., Perkin Trans 2* **1987**, 1609-1615; H. Goldmann, W. Vogt, E. Paulus, V. Böhmer, *J. Am. Chem. Soc.* **1988**, *110*, 6811-6817;H. Goldmann, W. Vogt, E. F. Paulus, F. L. Tobiason, M. J. Thielman, *J. Chem. Soc., Perkin Trans 1* **1990**, 1769-1775.
[125] B. Berger, V. Böhmer, E. Paulus, A. Rodriguez, W. Vogt, *Angew. Chem., Int. Ed. Engl.* **1992**, 31, 96-99.
[126] W. Wasikiewicz, G. Rokicki, J. Kielkiewicz, V. Böhmer, *Angew. Chem., Int. Ed. Engl.* **1994**, *33*, 214-216; W. Wasikiewicz, G. Rokicki, J. Kielkiewicz, E. F. Paulus, V. Böhmer, *Monatsch. Chemie* **1997**, *128*, 863-879.
[127] Khoury, R. G.; Jaquinod, L. Aoyagi, K.; Olmstead, M. M.; Fisher, A. J.; Smith, K. M. *Angew. Chem., Int. Ed. Engl.* **1997**, *36*, 2497-2500.
[128] M. Tabatabai, W. Vogt, V. Böhmer, G. Ferguson, E. Paulus, *Supramol. Chem.* **1994**, *4*, 147-152.
[129] An early example of a C$_4$-symmetrical calix[4]arene is most likely a mixture of regioisomers T.-T. Wu, J. R. Speas, *J. Org. Chem.* **1987**, *52*, 2330-2332.
[130] a) D. Dahan, S. E. Biali, *J. Org. Chem.* **1989**, *54*, 6003-6004; b) idem., *ibid.* **1991**, *56*, 7269-7274.
[131] V. Böhmer, K. Jung, M. Schön, A. Wolff. *J. Org. Chem.* **1992**, *57*, 790-792.
[132] The 1,3-dimethyl ether of **74** (R^1 = H; R^2 = *t*-Bu; R^3 = Ph) has been prepared in a 5-step sequence in 27% overall yield starting with *p-tert*-butylcalix[4]arene: J. D. van Loon, A. Arduini, L. Coppi, W. Verboom, A. Pochino, R. Ungaro, S. Harkema, D. N. Reinhoudt, *J. Org. Chem.* **1990**, *55*, 5639-5646.
[133] S. Pappalardo, G. Ferguson, J. F. Gallagher, *J. Org. Chem.* **1992**, *57*, 7102-7109.
[134] The fragment condensation of a variety of *p*-halocalixarenes has been reported by A. A. Moshfegh *et al* (*Helv. Chim. Acta.* **1982**, *65*, 1221-1228, 1229-1232, 1264-1270), including a "2 + 1" condensation leading to a calix[3]arene. However, CPK models of this compound indicate it to be quite severely strained, and it is probable that it is an oxacalixarene of undetermined structure.
[135] a) T. Yamoto, L. K. Doamekpor, H. Tsuzaki, M. Tashiro, *Chem. Lett.* **1995**, 89-90; b) T. Yamato, L. K. Doamekpor, H. Tsuzuki, *Liebigs Ann./Recu*eil **1997**, 1537-1544.
[136] S. E. Biali, V. Böhmer, I. Columbus, G. Ferguson, C. Grüttner, F. Grynszpan, E. F. Paulus, I. Thondorf, *J. Chem. Soc., Perkin Trans. 2* **1998**, 2261-2269.
[137] K. No, K. M. Kwon, *Synthesis* **1996**, 1293-1295.
[138] a) T. Haino, T. Harano, K. Matsumura, Y. Fukazawa, *Tetrahedron Letter.* **1995**, *36*, 5793-5796; b) T. Haino, K. Matsumura, T. Harano, K. Yamada, Y. Saijyo, Y. Fukazawa, *Tetrahedron* **1998**, *54*, 12185-12196.
[139] S. Usui, K. Deyema, R. Kinoshita, Y. Odegaki Y. Fukazawa, *Tetrahedron Lett.* **1993**, *34*, 8127-8130.
[140] H. Tsue, M. Ohmori, K.-I. Hirao, *J. Org. Chem.* **1998**, *63*, 4866-4867.
[141] T. Sone, Y. Ohba, K. Moriya, H. Kumada, K. Ito, *Tetrahedron* **1997**, *53*, 10689-10698.
[142] a) H. Kumagai, M. Hasegawa, S. Miyanari, Y. Sugawa, Y. Sato, T. Hori, S. Ueda, H. Kamiyama, S. Miyano, *Tetrahedron Lett.* **1997**, *38*, 3971-3972; b) Iki, H.; Kabuto, C. Fukushima, T.; Kumagai, H.; Takeya, H.; Miyanari, S.; Miyashi, T.; Miyano, S. *Tetrahedron* **2000**, *56*, 1437-1443.
[143] N. Iki, N. Morohashi, T. Suzuki, S. Ogawa, M. Aono, C. Kabuto, H. Kumagai, H. Takeya, S. Miyanari, S. Miyano, *Tetrahedron Lett.* **2000**, *41*, 2587-2590.

Chapter 2

CHEMICAL MODIFICATION OF CALIX[4]ARENES AND RESORCARENES

IRIS THONDORF[a], ALEXANDER SHIVANYUK[b], VOLKER BÖHMER[c]

[a] *Martin-Luther-Universität Halle-Wittenberg, Fachbereich Biochemie/ Biotechnologie, Kurt-Mothes-Str. 3, 06099 Halle, Germany.*
e-mail: thondorf@biochemtech.uni-halle.de
[b] *Department of Chemistry, University of Jyväskylä, P.O. Box 35, FIN-40351, Jyväskylä, Finland.*
[c] *Johannes Gutenberg-Universität, Fachbereich Chemie und Pharmazie, Abteilung Lehramt Chemie, Duesbergweg 10-14, D-55099 Mainz, Germany.*
e-mail: vboehmer@mail.uni-mainz.de

1. Introduction

There are two obvious places to modify a calixarene, namely the phenolic hydroxy groups (formation of ethers and esters) and the *p*-positions (electrophilic substitution, *ipso*-substitution). They may be addressed independently, which is the main advantage of calixarenes in comparison to *e.g.* cyclodextrins and one reason of the huge diversity of calixarene derivatives. Additionally, reactions may be carried out at the methylene bridges, may comprise the aromatic system of the phenolic units as a whole (oxidation, hydrogenation, see Chapter 14) or may lead to the replacement of the OH-function by other groups (see Chapter 6). Functional groups introduced in a first step may be further modified by subsequent reactions, including migrations.

Chemical reactions of resorcarenes [1] involve again the hydroxy groups and the 2-position in-between, both situated at the wide rim [2]. In addition, residues introduced in this way or present from the synthesis (*e.g.* introduced by the aldehyde) may be further modified subsequently.

A complete survey of all derivatives that have been prepared from calix[4]arenes **1** and resorcarenes **2** (although desirable as a kind of "dictionary") is beyond the frame of this chapter. We will concentrate instead on principles of the chemical modification of calix[4]arenes illustrated by selected examples of the recent literature [3]. Special emphasis will be given to selective (partial) conversions, since a complete substitution of all *p*-positions or all hydroxy groups is usually no longer a problem. The synthesis of multicalixarenes is covered in Chapter 7 while resorcarene-derived cavitands and (hemi)carcerands are treated in Chapters 9 and 10.

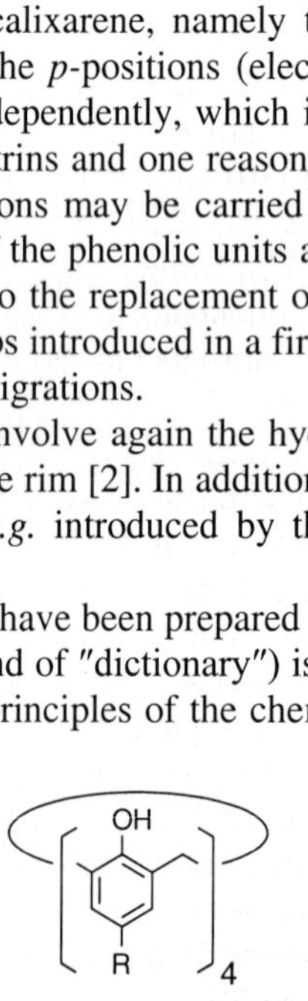

1a R = *t*-Bu
1b R = H

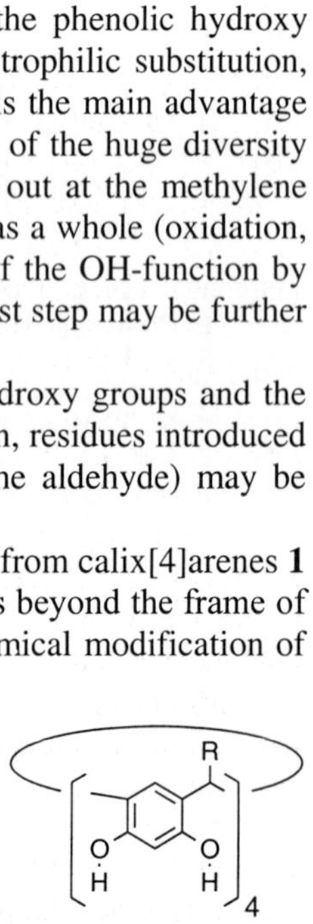

2

Z. Asfari et al. (eds.), Calixarenes 2001, 26–53.

2. *O*-Alkylation and *O*-Acylation of Calix[4]arenes

Two cases may be distinguished, when residues are attached to the phenolic oxygens of a calix[4]arene. Methoxy groups, like hydroxy groups can pass the annulus [4], while groups larger than ethoxy [5,6] cannot. In addition to regioisomers (1,2-di-, 1,3-di-substituted) stereoisomers (atropisomers) have to be considered, whenever two or more groups larger than ethoxy are present. Only if four of such larger groups are present the conformational description (*cone*, *partial cone*, *1,2-alternate*, *1,3-alternate*) may be solely used to characterise the isomer. In other cases one has to distinguish between the mutual arrangement of the residues (*syn*, *anti*) and the actual conformation of the molecule [7].

Most reactions discussed in the following have been done with either **1a** or **1b**, but should be possible also with other *p*-substituents. Where necessary we describe in the following compounds the residues Y^1-Y^4 and R^1-R^4 as indicated in the general formula **3**; substituents not mentioned are equal to H, the use of Y and R indicates four identical substituents.

2.1. MONO-ETHERS AND -ESTERS

Monoethers of calix[4]arenes have been prepared with an excess of the alkylating agent controlling the reaction by the amount of a weak base such as K_2CO_3 (0.6 mol in CH_3CN) or CsF (1-1.2 mol in DMF) [8]. However, direct *O*-alkylations with NaH in toluene [9] or CH_2Cl_2 [10] and $Ba(OH)_2$ in DMF [9,11] have been also described. Recently several examples have been prepared in methanol with sodium methylate as base in yields of 70-80 % [12], while using bis(butyltin)oxide with various alkylating agents in boiling toluene gave monoethers in 34-80 % yield [13]. A selective mono-*O*-benzylation (K_2CO_3) of one nitrophenol unit (82 %) has been reported for 5,17-dinitrocalix[4]arene [14].

The controlled cleavage of 1,3-diethers or tetraethers (see below) by trimethylsilyl iodide (1 or 3 mole) has been also described [15]. *O*-Alkylation of mono- or triesters (see below) and subsequent hydrolysis of the ester group(s) offers another rational access [16-18]. Reaction of **1a** with tris(dimethylamino)methylsilane or trichloromethylsilane has been used to protect three OH-functions. Methylation of the remaining fourth OH group (BuLi/TFA-OMe) and cleavage of the silyltriether gave the monomethylether in 83 % overall yield [19].

Monoesters have been obtained by direct acylation [20,21], from 1,3-diesters by reaction with imidazole [22] or from mono- and 1,3-dibenzyl ethers by acylation and subsequent hydrogenolysis of the benzyl protective groups [23].

A mono-glycoside adduct prepared by Mitsunobu reaction (DEAD/PPh$_3$/toluene/70°C) from **1b** with mannofuranose diacetonide in 71 % yield should be mentioned as a quite different example [24].

2.2. 1,3-DIFUNCTIONALISATION

Numerous 1,3-diethers with a *syn*-arrangement of the ether groups have been synthesised, and this derivatisation may be considered as one of the standard reactions in calix[4]arene chemistry [25]. Usual conditions are sodium or potassium carbonate as base in refluxing acetone or acetonitrile, but the reaction seems to tolerate many variations of these conditions. Various cyclisation products have been prepared using "divalent" alkylation agents (see Chapters 7 and 20). A second ether group different from the first one may be introduced also in this way [12,26,27].

The strong tendency for the 1,3-di-*O*-alkylation can be rationalised by the stabilisation of the corresponding monoanion of the monoether by two intramolecular hydrogen bonds and by sterical reasons. The tendency is not changed by moderately different substituents in *p*-position. Thus, calix[4]arenes of the AABB-type ($R^1 = R^2$, $R^3 = R^4$) may be desymmetrised to chiral 1,3-diethers in this way [28,29].

syn-1,3-Diesters are easily available for similar reasons [22,30,31], including examples with different ester groups [32]. And more or less expectedly derivatives with an ether and an ester group in 1,3-position have been prepared either by acylation of a monoether [33] or by *O*-alkylation of a monoester [18]. 1,3-*syn*-Sulfonates have been obtained in pyridine [34], by using the "proton sponge" 1,8-bis(dimethylamino)naphthalene in CH_2Cl_2 [35] or in THF with NaH as base [20,36]. The 1,3-dicarbonate of **1b** was obtained with benzyl chloroformate in acetonitrile/K_2CO_3 [36].

Again there may be a dependence on the *p*-substituents since 5,17-dinitro- and 5,17-dibutylcalix[4]arene ($R^1 = R^3 = NO_2$, *t*-butyl) have been *O*-benzylated at the unsubstituted phenolic units to give the 25,27-dibenzylether ($Y^2 = Y^4 =$ benzyl) (using Me_3SiOK and K_2CO_3, respectively, as base), while the 26,28-dibenzoate ($Y^1 = Y^3 =$ benzoyl) was obtained in the presence of $AlCl_3$ [14].

To the best of our knowledge, the direct formation of a 1,3-diether with an *anti*-orientation of the ether residues has not been reported. All known examples were obtained via protection/deprotection strategies [6]. Remarkably, however, the acylation with excess benzoyl chloride using NaH as base leads to the *anti*-isomer in boiling toluene, while the *syn*-isomer is formed in THF at 0°C [31].

2.3. 1,2-DIFUNCTIONALISATION

The *O*-alkylation of two adjacent hydroxy groups is less favoured. Usually it requires a very strong base in excess (*e.g.* NaH in DMF/THF) to form the dianion of the intermediate monoether, in which (most probably) for electrostatic reasons two opposite hydroxyl groups are deprotonated. (Even in the trianion the 1,2-diether would be statistically favoured over the 1,3-diether, which in turn might be sterically favoured.) The extent of the *O*-alkylation is controlled then by the alkylating agent, usually applied in stoichiometric (in practice often 2.2 moles) quantity. Various *syn*-1,2-diethers have been prepared by such direct dialkylation in yields of up to 90 % in special cases [27,37-39] including 1,2-crown ethers [40,41]. A monoether may be used also as starting material. Again the direct formation of an *anti*-1,2-diether has not been reported [42].

An alternative route to 1,2-diethers is again the selective cleavage of neighbouring (*syn* !) ether groups in tetraethers by $TiBr_4$ [43] or Me_3SiI [15], where a tetraether in the *partial cone* conformation yields the 1,2-*anti*-diether. Ether cleavage can be extended to a rational synthesis of 1,2-diethers as *syn* and also as *anti*-isomers by the use of protective groups such as benzylether groups [44].

A 1,2-bridged cyclic ester with phthalic acid was obtained from **1a** without dilution conditions in 36% yield, and could be used as starting material for further alkylation with ethylbromoacetate in 3,4-position [41a]. Thus, after hydrolysis an *anti*-1,2-diacid ($Y^1 = Y^2 = CH_2COOH$) was obtained.

While the preferred formation of 1,3- over 1,2-derivatives is easily explained by the most stable monoanion of the intermediate mono-*O*-alkyl/acyl derivative (two intramolecular hydrogen bonds), the monoanion of a 1,2-derivative should be more stable (one intramolecular hydrogen bond) than the monoanion of the analogous 1,3-derivative. Therefore, 1,3-derivatives should be rearranged into 1,2-derivatives under basic conditions, provided a reaction pathway is open. In fact this rearrangement has been observed for 1,3-diphosphates [45] and 1,3-ether/phosphates [46,47] as well as for the 1,3-benzyl/(3,5-dinitrobenzoate) [48] during the synthesis of further *O*-alkylation products (see below) (Scheme 1). It was mentioned already earlier by Gutsche, that the 1,2-bis(3,5-dinitrobenzoate) of **1a** can be obtained from the 1,3-isomer in the presence of imidazole [22]. An *intramolecular* migration of the acyl residue seems reasonable, at least for the phosphates, although it has not been strictly proved.

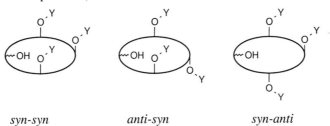

2.4. TRIFUNCTIONALISATION

Three stereoisomers may be distinguished for tri-*O*-alkyl/acyl derivatives having the same residue Y: *syn-syn, anti-syn, syn-anti*. (For different residues in 1,3-positions two *syn-anti* isomers are possible.)

        syn-syn   *anti-syn*   *syn-anti*

syn-syn-Triethers can be easily prepared by *O*-alkylation with alkyl iodides using BaO/Ba(OH)$_2$ as base in DMF [49,50]. Tri-*O*-alkylation of **1a** with *t*-butyl-bromoacetate was achieved (71 %) with CaH$_2$ in DMF [51] while the tri-*O*-alkylation of **1b** with ethyl bromoacetate using BaO or CaH$_2$ as base directly gave the free acid [52]. Even simpler is the tri-*O*-alkylation with an excess of alkyl bromides or iodides in refluxing acetone using 1,5 mole K$_2$CO$_3$ [53]. This is not necessarily surprising considering the easy access to the *syn*-1,3-diethers, which probably are the key intermediates.

Starting with a *syn*-1,2- or 1,3-diether it must be possible to obtain asymmetric or symmetric triethers AABH ($Y^1 = Y^2 \neq Y^3$) or ABAH ($Y^1 = Y^3 \neq Y^2$) under analogous conditions and to convert a mixed 1,3-diether (AHBH) into a triether with three different ether residues ACBH [33]. The former reaction was realised for a series of *syn*-triethers starting from the *syn*-1,2-di-picolyl ether of **1a** (1 equiv. RX/Cs$_2$CO$_3$) [27].

A chiral triether of the type ABBH (Y^1 = CH$_2$CO$_2$Et; Y^2 = Y^3 = CH$_2$CO$_2$CH$_2$-Pyrene) has been obtained (55 %), by alkylation of the corresponding monoether (K$_2$CO$_3$/THF) [54] and AHBH is the most likely intermediate [55,56].

Cleavage of an ether or ester group (eventually introduced as a protective group) must be mentioned as another strategy to obtain triethers, *e.g.* the triallyl ether of **1b** [57]. Various chiral triethers [58] with different ether residues were prepared via the triether monobenzoate of **1b** [12]. Trietheramides (Y^1-Y^3 = CH$_2$-C(O)NEt$_2$) have been synthesised, using the allylether groups (Y^4 = CH$_2$-CH=CH$_2$) as protective group (cleavage by Pd(PPh$_3$)$_4$ and Et$_3$N/HCOOH) in 77 % yield [59]. Reduction of the ester groups in tetraethers (Y = O-CH$_2$-CO$_2$Et) of **1a,b** with LiAlH$_4$ led to the *syn-syn*-triethers (Y^1- Y^3 = OCH$_2$CH$_2$OH) [60].

anti-syn- and *syn-anti-*Triether derivatives have been prepared mainly via tetraethers (section 2.5.), using the benzylether group as protective group [49]. The *anti-syn-*tripropylether of **1a** is available (48 %), by *O*-alkylation with propyl bromide using Cs$_2$CO$_3$ as base [50].

The tribenzoate of **1b** [16], one of the first examples of selectively derivatised calix[4]arenes, is obtained as the *anti-syn* isomer. There seems to be some generality in this stereochemical outcome (see also [52]) since the monobenzoylation of 1,3-diethers under similar conditions also leads to the *anti-syn* isomer [12]. The tribenzoate of **1a**, however, was obtained (*N*-methylimidazole, toluene, 70 %) as *syn-syn* isomer, assuming a *partial cone* conformation with an inverted phenol ring [61]. The dependence on the reaction conditions was shown for the *syn-*1,3-diallylether of **1b** which in CH$_3$CN gives *syn-syn*-diether/ester with PhCOCl/NaH while reaction with PhCOCl in the presence of pyridine led to the *anti-syn*-di-ether/ester [57]. An *anti-syn* orientation is also claimed for several tri-sulfonates [34]. The formation of an *anti-syn* 1,3-diether/ester derivative was also achieved by barium(II)-ion assisted monodeacylation of a 1,3-crown-5 diacetate in the *partial cone* conformation [62].

Several 1,2-diester-3-ether derivatives (*syn-syn*) have been prepared by alkylation of *syn-*1,3-diesters making use of the already mentioned rearrangement of 1,3- into 1,2-diesters prior to the *O*-alkylation step [48].

2.5. TETRAFUNCTIONALISATION

Direct tetra-*O*-alkylation of calix[4]arenes may lead to four stable stereoisomers (atropisomers), if residues larger than ethyl are introduced. Among them derivatives in the *cone* and in the *1,3-alternate* conformation are most attractive, as basic skeleton or as building blocks for more sophisticated structures (covalently linked [see Chapter 7] or self-assembled [see Chapter 8]), but all four possible isomers have been prepared in many cases [63-66]. Often, however, the isolation of a certain conformer requires chromatographic separation techniques.

The stereochemical result of a per-*O*-alkylation depends on *(i)* the residue Y to be attached, *(ii)* the calix[4]arene respectively its *p*-substituents R, *(iii)* the alkylation conditions (solvent, base, temperature). Template effects by metal cations (due to the base applied) have been used to direct the reaction towards a certain conformer.

Usually the *cone* isomer is formed in the presence of Na$^+$ cations. Na$_2$CO$_3$ in acetone or acetonitrile is often used for reactive alkylating agents such as bromo- or chloroacetates, -acetamides or -ketones (X-CH$_2$-CO-R), while NaH in DMF or THF/DMF (sometimes MeCN) is the standard for alkylbromides, -iodides or tosylates [37,49,67,68]. This is true also for the exhaustive *O*-alkylation of mono- [69], di- [68,70] and triethers [12,27,69c,71]. Starting from 1,2- and 1,3-*syn*-diethers AAHH (Y^1 = Y^2) and AHBH (Y^1 ≠ Y^3) chiral tetraethers (AABB, ACBC) in the *partial cone* conformation were obtained with Cs$_2$CO$_3$ as base [27, 72].

Larger alkali cations (K$^+$, Cs$^+$) favour the formation of *partial cone* and *1,3-alternate* isomers, although a general set of conditions is not available for the *O*-alkylation. For example, replacement of Na$_2$CO$_3$ by Cs$_2$CO$_3$ leads to the quantitative formation of the *partial cone* isomer instead of the *cone* isomer in the alkylation of **1a** with ethylbromoacetate (acetone/reflux) [64] while the effect is less pronounced for **1b**. Alkylation of **1a** with propylbromide (KO*t*-Bu/benzene/reflux) leads to a 1:1 mixture of *partial cone* and *1,3-alternate* [67]. Use of KOSiMe$_3$ or K$_2$CO$_3$ gives predominantly the tetra benzylether in the *partial cone* conformation [73]. Usually the *1,2-alternate* isomer is the most difficult to obtain [74,75]. In the case of propylethers of **1a** it was synthesised from the *syn*-1,3-dibenzylether by alkylation with propyliodide (NaH/THF/reflux; 67 % of the *partial cone* isomer), cleavage of the benzylether groups (Me$_3$SiCl/NaI/CHCl$_3$) and alkylation of the thus obtained *anti*-1,3-dipropylether with propylbromide (Cs$_2$CO$_3$/DMF/70°C) [63].

Reasonable to high yields of tetraethers in the *1,3-alternate* conformation have been recently obtained from the 1,3-di-ethoxyethyl ether of **1b** with chloroethoxyethyltosylate (Cs$_2$CO$_3$/acetone/reflux, 57 %) [76], from its 1,3-dibromo analogue (R^1 = R^3 = Br) with ethoxyethyltosylate (Cs$_2$CO$_3$/DMF/80°C, 82 %) [77] and from the 1,3-dibutylether with methyl bromoacetate (KH/DMF/RT, 88 %) [78]. The sequence of *O*-alkylation steps can also be important for the stereochemical result. Reaction of **1b** with alkenyl bromides (K$_2$CO$_3$/acetone/reflux) led to the *syn*-1,3-diethers which were subsequently *O*-alkylated with methyl bromoacetate (NaH/THF) to yield a mixture of *cone* and *1,3-alternate* isomers. The alternative *O*-alkylation sequence (ethylbromoacetate/K$_2$CO$_3$/acetone, reflux followed by alkenylbromide/NaH/ THF) gave the tetraethers in the *partial cone* conformation with *anti*-oriented alkenyl ether groups [79].

Derivatives in the *cone* conformation have been also obtained with four rather bulky substituents attached to the oxygen, such as Y = P(C$_6$H$_5$)$_2$ [80] or Y = CH(CO$_2$Et)$_2$ [81]. Calixarene derived dendrimers of the Frechet-type are probably the most impressive examples for the introduction of four bulky *O*-alkyl groups in a *cone* isomer [82].

Metal ion control has also been used to direct the stereochemical outcome of the diacetylation of 1,3-diethers toward the *cone* (Na$^+$) and *partial cone* (Tl$^+$) conformations [83]. Less is known about tetraesters of calix[4]arenes [84,85]. Tetrabutanoates in *partial cone, 1,2-* and *1,3-alternate* conformation have been obtained from **1a** [86], while all four isomers are known for the tetraacetates [66]. The tetratosylate (or *p*-bromophenylsulfonate) was used as protective group while modifying substituents in *p*-position, and derivatives in the *cone* [87,88] and *1,3-alternate* conformation [89] have been obtained in excellent yield [34], while the tetratriflate was recently prepared in low yield (11 %) as *cone* isomer [21].

3. Substitution on the Wide Rim

Although partial substitution of **1b** or *ipso*-substitutions of **1a** (or of their derivatives) have been described, partly even with a preparatively useful yield, each carefully directed, selective functionalisation [90] usually starts with the hydroxyl groups at the narrow rim [91]. The selectivity that can be achieved there can be transferred to the *p*-positions, using the fact that phenol units are more reactive than phenol ether or ester units. Fig. 1 gives a schematic overlook.

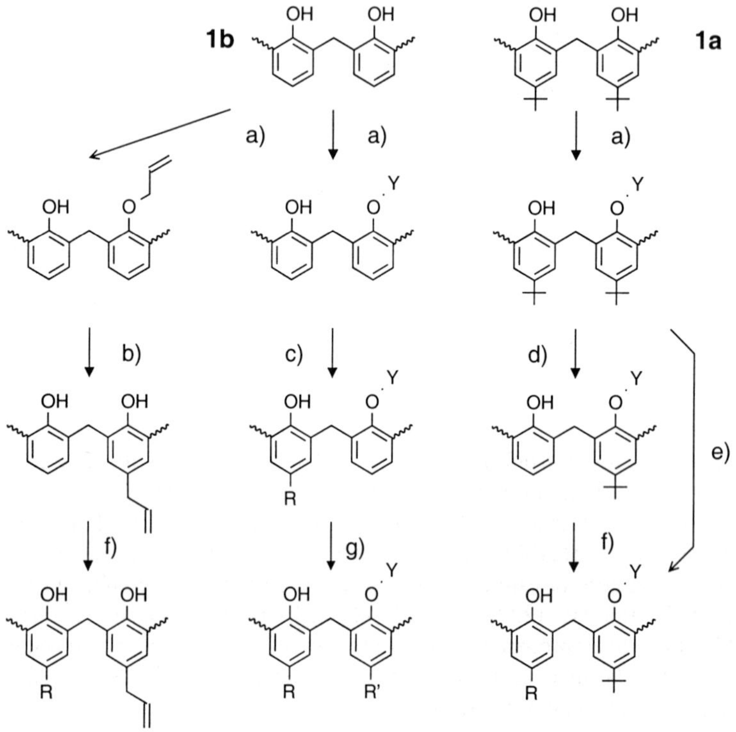

Figure 1 Selectivity transfer from the narrow to the wide rim, schematically presented for two units: **a**) selective introduction of ether residues; **b**) Claisen rearrangement; **c**) selective electrophilic substitution, **d**) selective debutylation, and **e**) selective ipso-substitution of phenol units; **f,g**) complete substitution of phenol or phenylether units, respectively. Eventually Y may be cleaved and Y as well as R can be further modified.

3.1. *Ipso*-Substitution of *p-t*-Butyl-Calix[4]arene

The direct sulfonation of **1a** with concentrated sulphuric acid leads to the tetrasulfonic acid (R = SO_3H) usually isolated as the pentasodium salt (80 %) [92]. No selective or partial *ipso*-sulfonation has been reported to our knowledge. The tetra *ipso*-acetylation of **1a** in CH_2Cl_2 with phenol as acceptor gave the tetraacetyl calix[4]arene (R = C(O)CH$_3$) in 42 % yield [93] and recently also the exhaustive nitration of **1a** (conc. HNO_3 or $KNO_3/AlCl_3$) in excellent yield was described [94].

The reaction of various tetraethers (*cone, partial cone, 1,3-alternate*) with 100 % HNO_3 in methylene chloride/acetic acid at room temperature (or below, *e.g.* 0°C) led to the corresponding tetranitro compounds in yields up to 85 % [67,95]. Quantities of 10 g or more are easily available on a laboratory scale. In special cases more drastic conditions are necessary, *e.g.* with the rigid 1,2;3,4-bis(crown-3-ether) which was *ipso*-nitrated in boiling acetic acid [96]. *Ipso*-nitration was also used to prepare various mononitro tri-*t*-butyl derivatives ($R^1 = NO_2$, R^2-$R^4 = t$-Bu) of tetraethers in preparative useful yields (up to 75 %) [97,98] while the two dinitro- and the trinitro derivatives have to be isolated by chromatography from the mixtures obtained by a partial *ipso*-nitration. A selective *ipso*-nitration has been reported for 1,3-diethers [99]. The dinitrophenol derivatives **4** were obtained in yields up to 75 %. *Ipso*-attack at the methylene bridges was found as a side reaction, proved by the X-ray structure of **5** with two 6-nitro-cyclohexa-2,4-dienone units [99a].

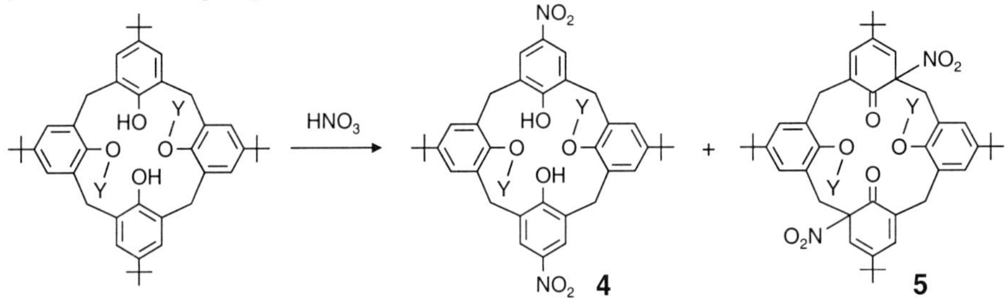

The elimination of *t*-butyl groups, a retro-Friedel-Crafts-alkylation, should be mentioned here also. Usually it is done with $AlCl_3$ as catalyst in toluene as solvent and acceptor for the *t*-butyl group [87], but phenol has been recommended as acceptor to facilitate the reaction [16,100]. The trans-butylation has been successfully applied to mono-, di- and triethers or esters of **1a**, where selectively the free phenolic units were debutylated [25,101]. Tri- [22,61], di- [22,25,102] and mono-*t*-butyl calix[4]arene derivatives [22] have been prepared in this way, from which the *O*-alkyl and especially the *O*-acyl groups can be cleaved again, if necessary or desired. A stepwise dealkylation was possible with Nafion in toluene [103].

3.2. ELECTROPHILIC SUBSTITUTIONS OF FREE *P*-POSITIONS

We concentrate in the following mainly on the (selective) introduction of halogen, nitro- and formyl groups, which are easily used or modified to attach various further functional groups. For complete conversions using the "Claisen rearrangement route", the "*p*-quinone methide route" and the "*p*-chloromethylation route" we refer to well known reviews [104].

3.2.1. Halogenation
The phenol units in partially *O*-alkylated calix[4]arenes can be selectively brominated. As recent examples we mention the selective bromination of the 1,3-dipropylether (NBS, dichloromethane, 0.5 h, 83 %) [105], 1,3-dibenzylether (NBS, methylethylketone, 40 h at RT, 99 % yield) [106] and the 1,3-ethoxyethyl ether (76 % after chromatographic separation of 1-4 % monobrominated compound) [77]. However, also the

selective monobromination of the tribenzoate of **1b** [105] or its tetrapropylether [107] was reported.

The bromination (as well as the nitration) of *p*-mono- and *p*-1,3-bis(acetamido) calix[4]arene tetraethers (NBS, methylethylketone, RT) occurs in *m*-position (*ortho* to the acetamido groups), leading to inherently chiral derivatives [108]. Surprisingly the *m*-substitution is preferred even over the *p*-substitution of free phenolether units. The exhaustive bromination (introducing 8 bromine atoms) was reported for the hydroquinone-octamethylether **6** yielding **7** in 32 % yield [109].

The complete iodination of all free *p*-positions in tetraethoxyethyl calix[4]arene (Y = (CH$_2$)$_2$OEt) or its 1,3-di-*t*-butyl analogue (R^1 = R^3 = *t*-Bu) with I$_2$/AgTFA with 90-100 % yield was reported [110,111]. Recently BTMA·ICl$_2$ was recommended for the iodination of **1b** (63 %), which allowed also the selective di-iodination of 1,3-diethers (70-80 %) [112]. ICl or NIS have been also used for selective iodinations [113].

3.2.2. Nitration

The selective mononitration was reported for the trimethyl ether of **1b** (NaNO$_3$/HCl, 80 %) [114] as well as for other triethers [52] and also for various triesters [33,115,116]. Diethers [14,91] as well as the calix[4]arene dicarbonate and ditosylate [36] have been selectively dinitrated in the *p*-position of their phenolic units and a selective trinitration of mono ethers or esters should be also possible. Ester and carbonate groups can be subsequently cleaved and the deprotected phenolic units can undergo further electrophilic substitution reactions.

Various publications deal with the partial nitration of tetraethers of **1b**, *e.g.* Y = CH$_2$CONEt$_2$ [117], Y = (CH$_2$)$_2$OEt [118], Y = C$_3$H$_7$ [67,119] (60-65 % HNO$_3$/acetic acid in CH$_2$Cl$_2$ at RT), where mono- to tri-nitro derivatives were isolated in yields between 10 and 90 % (a yield which could be called "selective" again). The tetra-nitration of 1,2;3,4-bis(crownethers) (NaNO$_3$/TFA, 60 %) may be mentioned as an example for the exhaustive nitration [120].

3.2.3. Formylation

The Gross formylation (TiCl$_4$/Cl$_2$CHOCH$_3$) of tetraethers of **1b** led at 40°C to the tetrasubstituted derivatives in 40 % while at RT the triformylated product was obtained in 55 % [121]. Under similar conditions (SnCl$_4$/Cl$_2$CHOCH$_3$/-10°C) also the mono- (50 %) and the 1,3-diformyl derivative (65 % + traces of the 1,2-diformylated product) were obtained for Y = (CH$_2$)$_2$OEt [122], which could be increased in the latter case to even 87 % using TiCl$_4$/Cl$_2$CHOCH$_3$ in large excess.

Recently also the Duff-reaction (hexamethylenetetraamine/TFA) was used for the triformylation (Y = (CH$_2$)$_2$OEt, 55 %) [123] and the tetraformylation (Y = C$_3$H$_7$, 75 %) [124]. The selective diformylation of 1,3-diethers (SnCl$_4$/Cl$_2$CHOCH$_3$/CHCl$_3$/-15°C) was also reported [125], as well as the monoformylation of tris-*t*-butylcalix[4]arene [126].

3.2.4. Further electrophilic substitutions

The alkylation of **1b** with propylene (NiSO$_4$/γ-Al$_2$O$_3$ as catalyst) led to the *p*-isopropyl calix[4]arene in 51 % [127,128]. Mixtures of adamantyl calix[4]arenes were obtained with 1-hydroxyadamantane, which were chromatographically separated [129] and (in the case of the triadamantyl calix[4]arene) further reacted with various electrophiles [130]. Hydroxyadamantanes bearing further functional groups may be also used to introduce functionalities [131]. Alkylation with 7-methoxy-1,3,5-cycloheptatriene gives cycloheptatriene substituents which can be converted to cationic tropylium groups [132].

Coupling of **1b** with *p*-carboxybenzene diazonium chloride [133] led to a tetraarylazo calix[4]arene [134] which subsequently was reduced (Na$_2$S$_2$O$_4$/NaOH) to the *p*-tetraamino calix[4]arene. Considering the selectivity by which diazonium cations can distinguish phenolate from phenolethers [135], it is surprising that this sequence has not yet been used to prepare selectively mono- to triamino calix[4]arenes.

Mono-, 1,2-di-, 1,3-di- and tri-*p*-arylazo-substituted calix[4]arenes were obtained by coupling **1b** with the diazonium salt of 6-amino-1,3-benzodioxin under controlled reaction conditions [136,137]. The coupling of 1,2- and 1,3-diallylcalix[4]arenes and *p*-substituted benzenediazonium salts yielded the corresponding bis(arylazo)calix[4]-arenes which bear two different substituents on the *p*-positions [138]. However, the formation of doubly bridged calix[4]arenes in the 1,3-alternate conformation by coupling of **1b** with bis-diazonium salts derived from 4,4'-diaminobiphenyls seems highly unlikely for steric reasons [139].

While **1b** undergoes aminomethylation at all four *p*-positions when treated with Me$_2$NH and HCHO in the presence of HOAc, aminomethylation occurs only at one *p*-position in the absence of HOAc. The resulting *p*-mono[(dimethylamino)methyl]-calix[4]arene can be further converted by the quinone methide route into various mono-*p*-substituted calix[4]arenes [140]. Partially or fully *O*-alkylated (propyl, ethoxyethyl) calix[4]arenes could be selectively or completely amidomethylated by condensation with various *N*-methylol-amides and -imides (TFA/CHCl$_3$) in yields of 30-97 % (Tscherniac-Einhorn reaction) [141].

Conditions for the selective chloromethylation of the phenolic units of the 1,3-di-methylether of **1b** have been also found [142].

4. Further Modifications

4.1. REARRANGEMENTS

The Claisen rearrangement and subsequent modifications of the *p*-allyl derivatives belong already to the "classical" strategies to functionalise calixarenes at their wide rim [87]. The reaction was used also for the selective introduction of one [16,17], two [143,144] or three [57] *p*-allyl functions. Recently an improved procedure in the presence of bis-(trimethylsilyl)urea (for an intermediate protection of the phenolic hydroxyl groups liberated by the rearrangement) increased the yield of *p*-allyl calix[4]arene to 99 % [10] making also available the larger *p*-allyl calixarenes and double calixarenes (see Chapter 7). The Fries rearrangement of calix[4]arene esters was also described for various examples [145] including the synthesis of inherently chiral derivatives [29].

4.2. MODIFICATION OF THE METHYLENE BRIDGES

Early attempts to modify the methylene bridges [146,147] by oxidation to ketones (in the tetraacetate of **1a**) and subsequent reduction to hydroxyl groups [148] did not find much resonance. More recent examples comprise the bromination to CHBr of the tetramethylether of **1a** [149] (yielding a single, *rccc* configured stereoisomer) and the homologous anionic *o*-Fries rearrangement of 1,3-biscarbamates (LDA/THF) in the *cone*, *partial cone*, and *1,3-alternate* conformation [150]. For the latter, reaction conditions have been found under which certain products are formed regio- and stereoselectively.

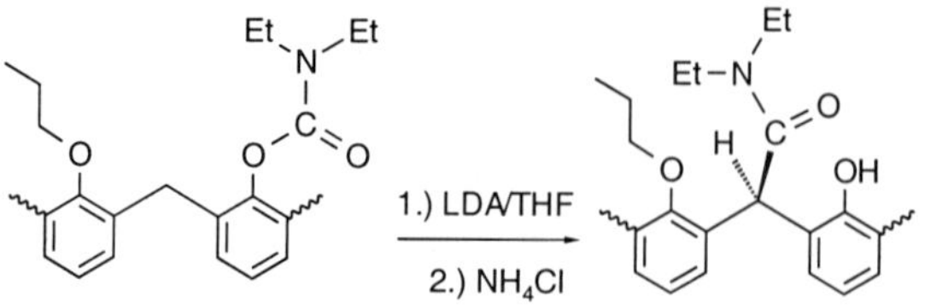

4.3. SELECTED (AND SELECTIVE) MODIFICATIONS OF SUBSTITUENTS AT THE NARROW RIM

Functional groups attached via ether (or less frequently ester) groups to the narrow rim can be further modified. Tetra- and 1,3-diesters [151] have been often used to attach further residues via ester or amide links [152], using the sequence

$$Y = CH_2\text{-}CO\text{-}OEt \rightarrow Y = CH_2\text{-}COOH \rightarrow Y = CH_2\text{-}CO\text{-}Cl \rightarrow Y = CH_2\text{-}CO\text{-}NR_2$$

In this connection the selective monohydrolysis of the tetraethylester ($Y = CH_2CO_2Et$) [153] in nonpolar solvents ($CHCl_3$/TFA, CH_2Cl_2/HNO_3) allows the modification of a single ester group [154], while the hydrolysis of two opposite ester groups in the more reactive tetra-*t*-butyl ester [155] should allow the same for two ester groups. The selective monohydrolysis of the 1,3-diethylester ($Y^1 = Y^2 = CH_2CO_2Et$) (1 mol KOH in EtOH, 80 %) [156] and subsequent reactions were also described [157].

Di- and tetraesters of **1a,b** have been reduced to the corresponding di- and tetraalcohols with LiAlH$_4$ in diethylether [158-160] or DIBAL in toluene [161], while LiAlH$_4$ in THF leads to the trialcohol by simultaneous cleavage of one ether link [60]. Tosylation of the O-CH$_2$CH$_2$-OH groups and further nucleophilic substitutions have been also described [158,159,162,163]. 3-Hydroxypropyl groups (O-CH$_2$CH$_2$CH$_2$-OH) were formed by hydroboration/hydrolysis of allylethers [157,159] and further converted via their tosylates analogously.

Two aminoethoxy groups in 1,3-position have been obtained by *O*-alkylation with bromoacetonitrile and subsequent reduction with BH$_3$/THF [164,165], while the corresponding tetra-(aminoethyl) ether was obtained from the tetratosylate (Y = (CH$_2$)$_2$OTs), by nucleophilic substitution with NaN$_3$ and subsequent catalytic hydrogenation of the azide [166]. *O*-Alkylation with *N*-(ω-bromoalkyl)phthalimide and subsequent hydrazinolysis seems to be the method of choice to attach longer aminoalkylether groups [166,167].

Aminoethoxy groups have been converted to amidine functions by reaction with MeAl(Cl)NH$_2$ (prepared in situ from Me$_3$Al and NH$_4$Cl) in toluene at 80° [168]. Glycidylether derivatives have been recommended for the synthesis of chiral calixarene based ligands by nucleophilic ring opening of the oxirane rings [169], but, difficulties have been met already with their stereoregular synthesis.

4.4. SELECTED (AND SELECTIVE) MODIFICATIONS OF SUBSTITUENTS AT THE WIDE RIM

4.4.1. Reactions of halogenated calix[4]arenes
p-Aryl substituted calix[4]arenes are available in yields between 15 and 73 % from palladium-catalysed Suzuki type cross-coupling reactions between tetrabromo calixarene tetraethers and arylboronic acids [170-173], by Negishi-type cross coupling reactions of *p*-iodocalix[4]arenes and phenylzinc chloride in the presence of Ni(PPh$_3$)$_4$ [110,174,175], and by reaction of tetrazinc tetrapropoxycalix[4]arene with aryl iodides (Pd(PPh$_3$)$_4$/THF) in yields up to 71 % [176]. Both the Suzuki and Negishi type cross-coupling reactions have been extended to selective functionalisations of the wider rim starting from the corresponding partially halogenated calix[4]arenes (see above) [176-179]. Negishi and Stille cross-coupling reactions of 1,3-dihalogenated calixarenes (R^1 = R^3 = Hal, Y^2 = Y^4 = C$_3$H$_7$) with *p*-phenylene-bis(boronic acid) and (*E*)-1,2-bis-(tributylstannyl)ethylene, respectively, have also been used to obtain calix[4]arene oligomers [178]. Further C-C coupling reactions starting from mono-, 1,3-di- or tetra-iodocalix[4]arene tetraethers comprise the Heck reaction leading to carboxylic esters [111,180], the Rosenmund-van Braun reaction furnishing *p*-cyanocalix[4]arenes [180] and the introduction of acetylene moieties by reaction with tri-methylsilylacetylene/TEA/CuI/Pd(PPh$_3$)$_2$Cl$_2$ [110]. 1,3-Dihalogenated calixarene tetraethers were converted into the corresponding 1,3-carbazole substituted compounds (76 % yield) by means of an Ullman type coupling reaction (Cu$_2$O/collidine/reflux) [179]. Copper-mediated coupling reactions of *p*-iodocalix[4]arenes with phthalimide followed by hydrazinolysis should be mentioned as an alternative strategy to obtain *p*-aminocalix[4]arenes [119,180-182].

The complete halogen-lithium exchange in tetrabromo tetraalkoxycalix[4]arenes (R = Br, Y = alkyl) can be achieved with an excess of *n*-BuLi [183] or *t*-BuLi in THF at −78 °C [184,185] while the selective mono- and 1,3-dilithiation has been accomplished with a controlled amount of *n*-BuLi in THF at −78 °C [185,186]. The subsequent quenching of the lithiated compounds with an electrophile has been used to introduce various *p*-substituents such as formyl, carboxy, methyl, methylthio, hydroxymethyl and deuterium, for which yields up to 90 % have been reported [183-185,187-189]. The reaction of the 1,3-di- and *p*-tetraboronic acids, which are available from the corresponding lithiated compounds, with H$_2$O$_2$/OH$^-$ has been used to introduce hydroxyl groups at the wider rim [184,186,190]. A direct oxidation of a monolithiated calix[4]arene by oxygen to the monohydroxy derivative has been recently reported [191]. The selective access to 1,3-diboronic derivatives of calixarene tetraethers has also been exploited for Suzuki-type cross-coupling reactions with iodoarenes [106] and with *p*-bromophenyl glycosides, the latter leading to carbohydrate functionalised calixarenes in moderate yields [177].

4.4.2. Conversion of nitro groups
Nitro groups, introduced by nitration or *ipso*-nitration are easily reduced to amino groups, which are starting materials for the attachment of various residues via acylation or Schiff-base formation. Three main procedures have been used for this reduction,

namely catalytic hydrogenation [95], reaction with hydrazine [114,118,192] or reduction with Sn(II) [193], but unfortunately none of these procedures could be used for a selective reduction.

By reaction with various amounts of Boc-anhydride it was possible, however, to obtain the mono- (36 %), 1,2-di- (48 %) and tri-protected (54 %) tetraamino calix[4]-arene tetraethers in preparatively useful amounts after chromatographic separations [194]. The fact that no 1,3-derivative is formed can be explained by a transcavity C(O)-NH···NH$_2$ hydrogen bond [195] which on the other hand allows an easy and efficient preparation of mono-Boc-protected derivatives of 1,3-diamino calix[4]arenes [196]. The Boc-protected compounds may be used as starting materials for various N-acylated compounds with mixed functionalities. The direct acylation of two opposite amino groups followed by complete acylation with a second acid has been reported for a special case [197].

Reaction of mono-, di- or tetra-aminocalix[4]arenes with various iso(thio)cyanates easily gives mono- [105,117], di- [198] or tetra(thio)urea derivatives [181,199,200], but two adjacent or three aminogroups of a tetraamine could also be selectively converted [201]. Reaction of aminogroups with triphosgene or thiophosgene led to iso(thio)cyanates which can be further reacted to (thio)urea derivatives (as an alternative way) or urethanes by treatment with amines or alcohols [198,202].

The transformation of a single amino group into the monoimide with Kemp's acid leads to an interesting compound **8** with an acid function pointing into the cavity [203].

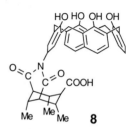

4.4.3. Conversion of formyl groups
Mono-, di- and tetraformyl derivatives of tetrapropoxy calix[4]arene have been used under Wittig-Horner or Horner-Wadsworth-Emmons conditions to synthesise various stilbene derivatives [134b,204,205]. The use of galactose-6- or ribose-5-phosphoranes and subsequent hydrogenation led to C-linked calixsugars [206].

While the Knoevenagel condensation of tetraformylcalix[4]arene tetrapropylether with malonic acid (pyridine, piperidine, reflux) led quantitatively to the tetra cinnamoic acid derivative, a product with two vinyl groups ($R^1 = R^3 = CH=CH_2$, $Y^2 = Y^4 = C_3H_7$) was surprisingly obtained from the 1,3-diformyl calix[4]arene dipropylether in 40 % [207]. Model reactions suggest that this decarboxylation is due to the free hydroxy groups.

Reductive amination (eventually via the isolated Schiff-bases) leads to secondary amines [123,208]. In this connection the monoprotection of a 1,3-diformyl calix[4]arene (56 % by refluxing with 1 equiv. of 2,2-dimethyl-1,3-propanediol in toluene) [209] allowed the amination of the second aldehyde function with a different amine.

Reduction of the mono- and 1,3-diformyl calix[4]arene tetraether led to the corresponding hydroxymethyl derivatives [105,210], while reaction with CF_3SiMe_3/TBAF furnished $CF_3C(OH)H$ groups (85 %) as strong H-bond donors [211]. McMurry-reaction on the other hand resulted in the intramolecular coupling of the formyl groups between opposite p-positions [212,213].

Mono-, di- and tetraformyl derivatives of tetraalkoxy calix[4]arenes have been also oxidised to the corresponding benzoic acid structures by H_2NSO_3H/$NaClO_2$

[125,214,215], which can be used for instance to attach aminoacids or peptides via the N-terminal end [216]. Baeyer-Villiger oxidation (m-CPBA) of mono- to tetraformyl derivatives gave the p-hydroxysubstituted calixarene derivatives [120], which are more easily accessible via the p-bromo compounds (see above).

4.4.4. Further modifications

Calix[4]arenes bearing at the wide rim two (opposite) or four -CH_2COOH functions (by hydrolysis of the cyanomethyl derivatives available via the "quinone methide route") can be converted to the mono- or bisanhydride in the *cone* and *1,3-alternate* conformation [87b,89]. The ring opening of these anhydride structures by various nucleophiles represents an especially elegant way to introduce selectively substituents at the wide rim, which in the case of the *1,3-alternate* derivatives **9** leads to chiral (C_2-symmetrical) derivatives **10**.

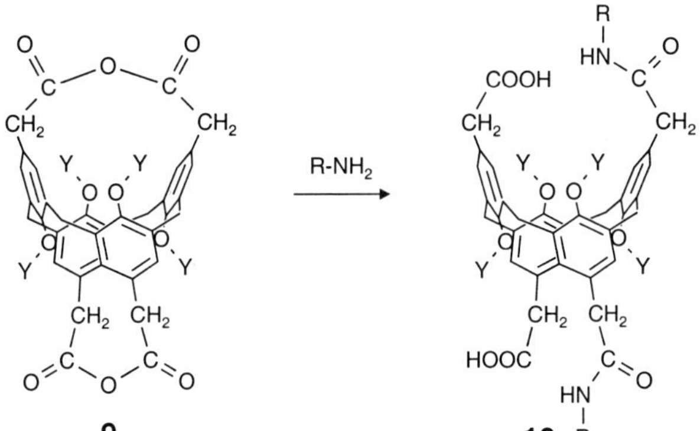

p-Cyanogroups have been converted to adamantyl amide functions by reaction with 1-adamantol in boiling CF_3COOH (Ritter reaction) which can be combined with the simultaneous alkylation of free p-positions [217].

5. Resorcarenes

5.1. REACTIONS OF THE HYDROXY GROUPS

The phenolic hydroxy groups of resorcarenes can be completely acylated [218]. Many examples of octaesters **11** [219] of *rccc*, *rctt* and *rtct* regioisomers have been reported. The reaction of **2** with an excess of diarylchlorophosphate, diphenylchlorophosphine, arylsulfonylchlorides, and trimethylchlorosilane or hexamethyldisilazane furnished octaphosphates [220], octaphosphinites [221], octasulfonates [222] and octatrimethylsilyl derivatives [8,223], respectively.

Complete *O*-alkylations of *rccc*-resorcarenes **2** result in octaethers **12** (Y = Me to Bu) [224]. The attachment of eight 3,5-dihydroxy benzyl ether groups to the *rccc*-resorcarene led to a first generation dendrimer **12a** [225]. Second generation dendrimers of the same type were prepared, starting with the mixture of *rccc*- and *rctt*-isomers obtained with *p*-hydroxy- and 3,5-dihydroxy-benzaldehyde, respectively [226].

Alkylation of **2** with excess of ethylbromoacetate led to octaesters **12b** which were transformed into corresponding octaacids **12c** [227]. Aminolysis of **12b** with chiral amines and aminoalcohols resulted in chiral octaamide derivatives **12d** [228,229]. Reduction of octaester **12b** with LiAlH$_4$ gave the octol **12e** which underwent a Mitsunobu reaction with phthalimide, DEAD and PPh$_3$ to give an octaphthalimide [230]. The hydrazinolysis of the phthalimido groups resulted in the corresponding octaamine **12f** which was reacted with lactono-lactone to give a water soluble resorcarene-sugar cluster (see Chapter 33). These examples show that virtually the same reactions are possible as have been used at the narrow rim of calix[4]arenes **1**.

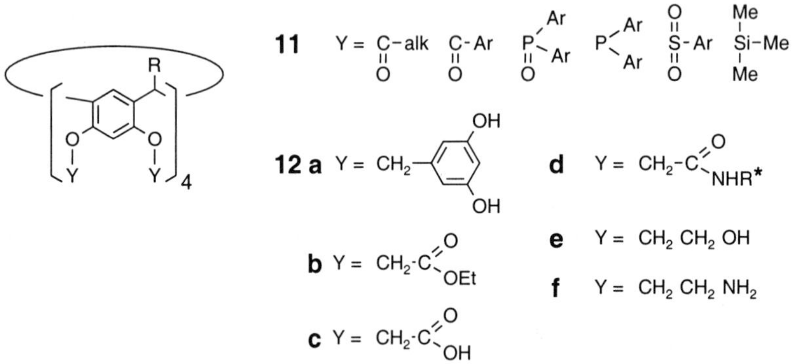

The reaction of **2** with (AlkO)$_2$P(O)Cl/Et$_3$N or the three-component phosphorylating agent ((AlkO)$_2$P(O)H/CCl$_4$/Et$_3$N) gave resorcarene tetraphosphates **13a** in yields of 25-40 % on a preparative scale. Initially, a chiral C_4-symmetrical arrangement of the phosphoryl groups was postulated [231,232], but later the C_{2v}-symmetrical structure was unambiguously proved by NMR spectroscopy and single crystal X-ray analysis [233]. The tetraphosphates **13a** can be converted into the water soluble tetraphosphoric acids **13b** by reaction with Me$_3$SiBr, followed by treatment with methanol [234].

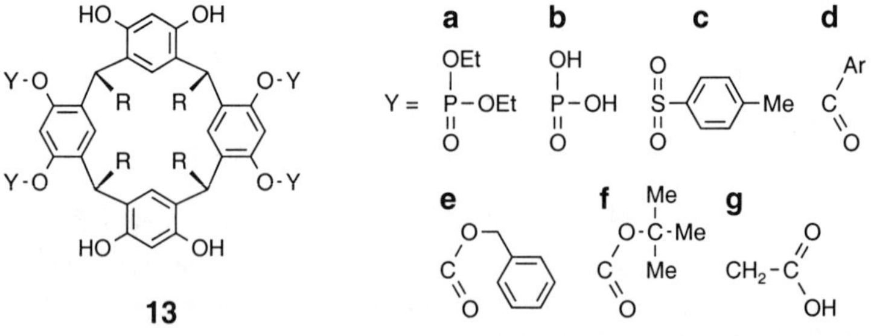

If **2** is reacted with four equivalents of an arylsulfonylchloride or an aroylchloride the C_{2v}-symmetrical tetrasulfonates **13c** [235] and tetraesters **13d** are obtained [236,237] while the regioselective acylation fails with aliphatic acid chlorides. The subsequent acylation or alkylation of the remaining hydroxy groups in **13c,d** resulted in C_{2v}-symmetrical derivatives containing two types of functional groups at the wide rim of the resorcarene, among which the tetra crown ethers obtained with benzo-15-crown-5-sulfonylchloride should be mentioned [238,235].

The regioselective acylation was successful also with Z-chloride (Et$_3$N, MeCN, RT) which allowed the partial protection of four hydroxy groups in **13e**. Exhaustive *O*-acylation of **13e** followed by mild removal of the Z-groups (H$_2$, Pd/C, dioxane) gave tetraacylated derivatives including the Boc-protected compound **13f** which are not available by direct acylation of the parent resorcarene **2**. The tetraacid **13g** was obtained in a similar way by *O*-alkylation of **13d** with ethylbromoacetate and subsequent hydrolysis [236].

One example of a monoalkylation was described so far. The reaction of **2** with *p*-methylbenzyl bromide in a 1:1 molar ratio resulted in a chiral resorcarene monoether which was further acylated to give a resorcarene containing one alkoxy and seven acetoxy groups [239].

5.2. ELECTROPHILIC SUBSTITUTIONS AT THE AROMATIC RINGS

The 2-position of the resorcinol rings may be substituted by mild electrophiles, *e.g.* by bromination, by coupling with diazonium salts and by Mannich type reactions. More drastic reactions such as nitration or sulfonation failed most probably due to disruption of resorcarene skeleton.

Aminomethylation of resorcarenes **2** with secondary amines and formaldehyde readily gives the corresponding tetraamines **14a** [240] which exist in apolar solvents in a chiral C_4-symmetrical conformation with left or right handed orientation of the pendant hydrogen bonded amino groups [241]. In this way various functional groups including chiral and cation binding functions [242] could be easily attached to the resorcarene platform. Thiomethylation of **2** with CH$_2$O and thiols in AcOH [243] was also possible while the attempted aminomethylation with TRIS furnished tetraalkoxymethylated products incorporating the alcohol present as solvent [244].

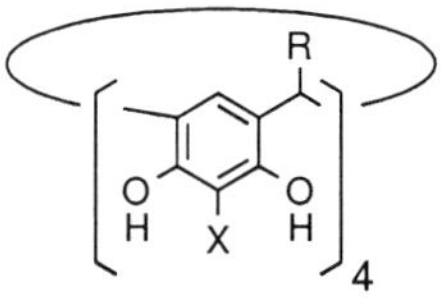

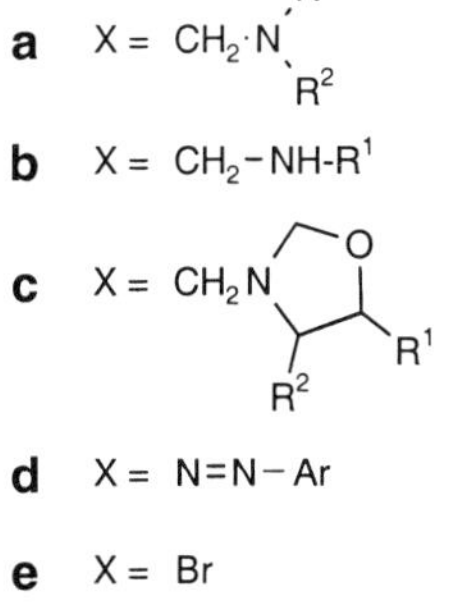

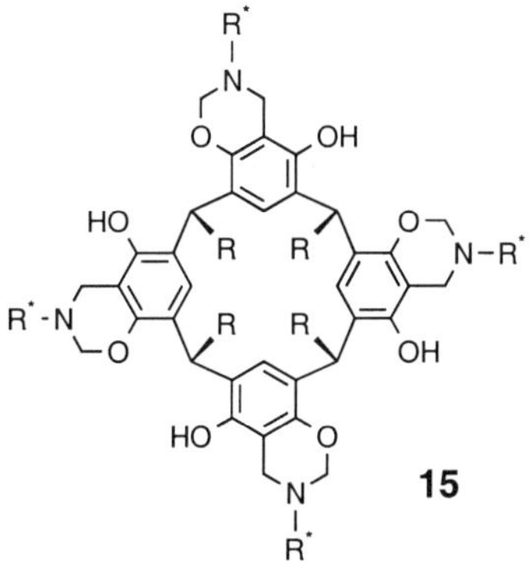

The aminomethylation of **2** with primary amines leads in an entirely regioselective reaction to chiral C_4-symmetrical tetrabenzoxazine derivatives **15** [245,246] which was proved by several crystal structures. The subsequent cleavage of the benzoxazine rings (HCl, BuOH, 100 °C) readily gives the corresponding secondary amines **14b** as hydrochlorides [247,248]. If the aminomethylation of **2** is carried out with chiral amines (*e.g.*

α–phenylethylamine or its *p*-substituted analogues) only one of the two possible diastereomeric tetrabenzoxazines **15** is formed in high yield [249], which was proved in two cases by X-ray analysis. In solution they undergo an acid-catalysed epimerisation as shown by ^{1}H NMR spectroscopy. Recent studies show that the high diastereoselectivity is due to the preferred crystallisation rather than to the preferred formation of a single epimer [250].

The diastereomerically pure tetrabenzoxazine derivatives **15** were used to synthesise other inherently chiral resorcarenes. Methylation [251] of the hydroxy groups of the chiral tetrabenzoxazines with dimethylsulfate or methyltriflate at –78 °C using BuLi as base led to the tetramethylated derivative **16** as a single diastereomer, for which an epimerisation is no longer possible. Further chemical modifications furnished various secondary (*e.g.* **17a,c**) or primary (**17b**) amines or benzoxazines directly as single enantiomers which remain chiral (**17b,c**) also after cleavage of the chiral auxiliary group [252].

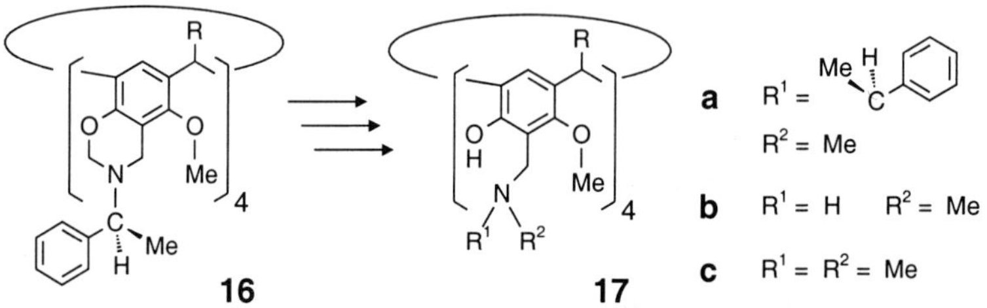

a $R^1 =$ (Me, H, C, phenyl) $R^2 =$ Me
b $R^1 =$ H $R^2 =$ Me
c $R^1 = R^2 =$ Me

The reaction of resorcarenes **2** with suitable diamines and CH$_2$O under high dilution conditions leads to 1,2-3,4-bis-bridged tetrabenzoxazines [253], or in the case of ethylenediamine to a carcerand-like head-to-head connected bis-resorcarene [254]. If 2-aminoalcohols are used in the Mannich reaction with resorcarenes either benzoxazine, oxazine or oxazolidine rings can be formed. In the case of aminoethanol, predominantly the benzoxazine **15** (R* = CH$_2$CH$_2$OH) was detected in solution, while with 2-alkyl aminoethanols, predominantly oxazolidines **14c** (R^1 = H) were obtained [255].

The exhaustive coupling of **2** with diazonium salts to **14d** [256] should make also tetraamino resorcarenes available by reduction of the azogroups, while the tetrabromo resorcarenes **14e** [257] are important starting materials for the synthesis of carcerands. The reaction of **2** with NBS in molar ratios from 1:1 to 1:3 resulted in a mixture of all possible partially brominated resorcarenes [258], in which the yield of the distal-dibromo derivative **18a** was much higher than statistically predicted, indicating some regioselectivity. Thiomethylation of **18a** with CH$_2$O and RSH in AcOH resulted in C_{2v}-symmetric derivatives **18b** containing two different functional groups at the wide rim of the resorcarene while the reaction with dithiols and CH$_2$O gave the distally bridged dibromo derivative **18c** [259].

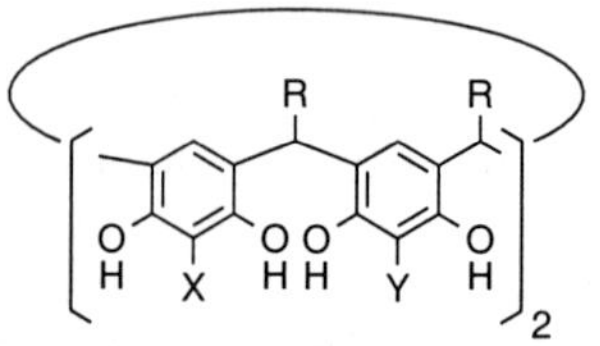

18 a X = Br   Y = H
   b X = Br   Y = CH$_2$-S-R^1
   c X = Br
      Y-Y = CH$_2$-S-R^2-S-CH$_2$

The different reactivity of *O*-acylated and unsubstituted resorcinol rings in the C_{2v}-symmetrical derivatives **13** allows their selective di-bromination and bis-aminomethylation to distally disubstituted derivatives [235,236]. The aminomethylation of resorcarene tetrasulfonates **13c** and tetrabenzoates **13d** with primary amines led to C_2-symmetrical bis-benzoxazine derivatives [260]. Various *trans*-cavity bridged compounds were obtained with diamines.

Aminomethylation of **13f** with secondary amines and subsequent removal of the Boc-protection gave 1,3-diaminomethylated derivatives which cannot be prepared directly by partial aminomethylation [236]. In the case of **13a-e** this synthesis was not possible due to problems with the complete removal of the protecting groups.

5.3. CHEMICAL MODIFICATIONS OF FUNCTIONAL GROUPS INTRODUCED BY THE ALDEHYDE

The acid catalysed condensation of resorcinol, 2-methylresorcinol and pyrogallol with aldehydes or their synthons containing hydroxy-, alkoxy-, aryldiazo- , sulfonyl- and $B(OH)_2$ groups, halogens and double bonds and crown ether fragments allows the introduction of additional functional groups into resorcarenes [261]. The functional groups introduced in this way can be further modified.

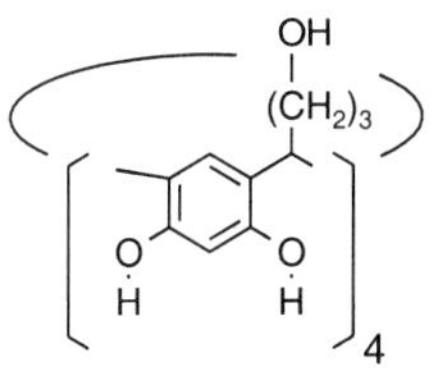

The tetra-boronic acid **2a** was used, for example, to extend the residues R by a phenyl or biphenyl unit [262]. Acylation of resorcarene **2b** containing four pendant double bonds followed by *anti*-Markovnikoff addition of $C_{10}H_{21}SH$ [263] resulted in octaacylated resorcarene derivatives containing four thioether fragments at the narrow rim. The smooth cleavage of acyl groups gave the free octol **2c** [263]. Photochemical addition of AcSH to the double bonds of **2b** analogously led to resorcarene **2d** footed with four SH groups [264]. These compounds and their derivatives were shown to form self-assembled mono-layers on gold surfaces (see Chapter 33).

a R = *p*-C_6H_4-$B(OH)_2$
b R = $(CH_2)_8$-CH=CH_2
c R = $(CH_2)_{10}$-S-$(CH_2)_9$-CH_3
d R = $(CH_2)_{10}$-SH
e R = $(CH_2)_3$-OH
f R = $(CH_2)_3$-NH_2
g R = $(CH_2)_3$-Cl

Partial epoxidation of octapivaloate **11** (R = $(CH_2)_8$CH=CH_2, Y = C(O)-*t*Bu) gave the monoepoxide. The hydrogenation of remaining double bonds (H_2, Pd/C), followed by acid-catalysed hydrolysis of the epoxide ring and removal of the protective ester groups led to a resorcarene containing one residue R ending in a hydroxy group [265].

The selective benzylation (K_2CO_3, NaI) of the phenolic hydroxy groups in **2e** led to the tetrahydroxy derivative **19a**. The subsequent acylation of the aliphatic hydroxy

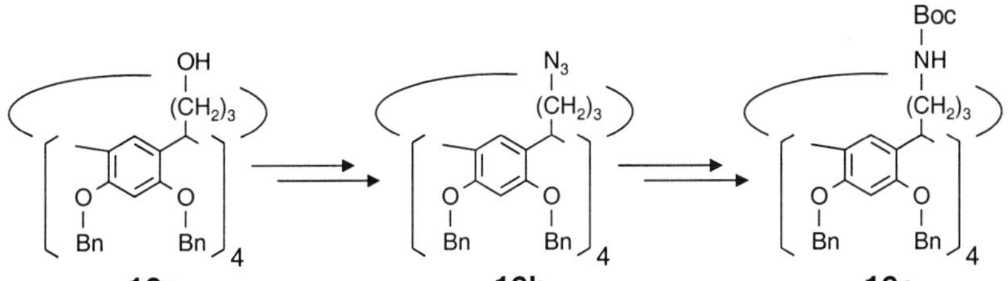

19a → **19b** → **19c**

groups with methane sulfonylchloride and reaction with NaN_3 resulted in the tetraazide **19b**. Catalytic hydrogenation (Raney-Ni) and reaction with $(Boc)_2O$ led to the N-protected amine **19c** from which the benzyl groups could be cleaved to give **2f** [266]. Alternatively compound **2f** could be prepared by Mitsunobu reaction of **19** with EtOC(O)C(O)NHBoc, DEAD and PPh_3 in CH_2Cl_2 followed by saponification/decarboxylation ($LiOH/THF-H_2O$) [266]. The analogue of **2e** containing four methyl groups at the 2-positions of the resorcinol rings was used to synthesise various cavitands footed with hydroxy, acetoxy and dihydroxyphosphoryl groups [267] (see Chapter 9). A series of amphiphilic resorcarenes with azobenzene residues was prepared starting with **2f** by etherification of the corresponding tetraiodide with *p*-hydroxyazobenzenes [268]

Reduction of *rccc*- and *rctc*-tetraesters **20a** by $LiAH_4$ in THF resulted in tetraols **20b** which by reaction with PPh_3 and CBr_4 could be transformed into corresponding tetrabromo derivatives [269]. The *rccc*-resorcarene **20b** was reacted under normal or high dilution conditions with two equivalents of glutaroyl-, adipoyl- or pimeloyl-dichloride (Et_3N, CH_2Cl_2) to give the doubly spanned resorcarenes **21** in moderate yields [269,270].

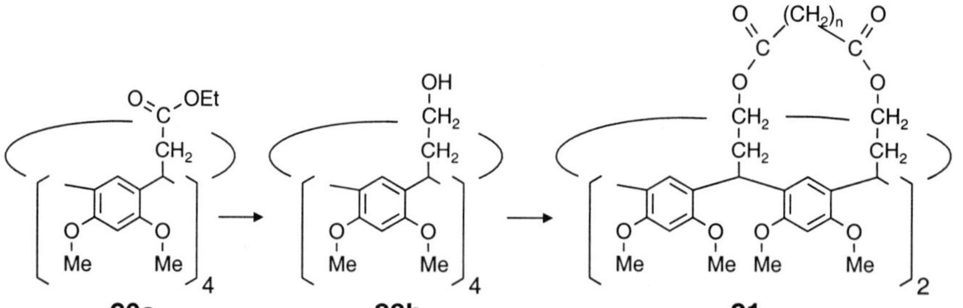

6. References and Notes

[1] For a general review on resorcarenes see: P. Timmerman, W. Verboom, D. N. Reinhoudt, *Tetrahedron* **1996**, *52*, 2663-2704.
[2] Substitutions at the aromatic *endo*-position are not known.
[3] Most of the reactions described for calix[4]arenes can be applied also to higher calixarenes; compare Chapters 3-5.
[4] The same applies for methyl, and mercapto groups: a) J. M. Van Gelder, J. Brenn, I. Thondorf, S. E. Biali, *J. Org. Chem.* **1997**, *62*, 3511-3519; b) C. G. Gibbs, P. K. Sujeeth, J. S. Rogers, G. G. Stanley, M. Krawiec, W. H. Watson, C. D. Gutsche, *J. Org. Chem.* **1995**, *60*, 8394-8402; c) P. Parzuchowski, V. Böhmer, S. E. Biali, I. Thondorf, *Tetrahedron: Asymmetry* **2000**, *11*, 2393-2402.
[5] Ethoxy groups are a borderline case. Stereoisomers may be isolated which, however, isomerize at higher temperatures or after longer times, see for instance: Z. B. Brzozka, B. Lammerink, D. N. Reinhoudt, E. Ghidini, R. Ungaro, *J. Chem. Soc., Perkin Trans. 2* **1993**, 1037-1040, and ref. [6].
[6] L. C. Groenen, J. D. Van Loon, W. Verboom, S. Harkema, A. Casnati, R. Ungaro, A. Pochini, F. Ugozzoli, D. N. Reinhoudt, *J. Am. Chem. Soc.* **1991**, *113*, 2385-2392.
[7] V. Böhmer, *Angew. Chem.* **1995**, *107*, 785-818; *Angew. Chem., Int. Ed. Engl.* **1995**, *34*, 713-745.
[8] L. C. Groenen, B. H. M. Ruel, A. Casnati, W. Verboom, A. Pochini, R. Ungaro, D. N. Reinhoudt, *Tetrahedron* **1991**, *47*, 8379-8384.
[9] K. Iwamoto, K. Araki, S. Shinkai, *Tetrahedron* **1991**, *47*, 4325-4342 and *ibid.* 7197 (corrigendum).

[10] Monoether linked double calixarenes were obtained with alkenyldichlorides in 51-84 % in CH$_2$Cl$_2$ under stoichiometric conditions in the presence of tetrabutylammonium iodide using NaH as base. J. Wang, C. D. Gutsche, *J. Am. Chem. Soc.* **1998**, *120*, 12226-12231.
[11] R. Milbradt, J. Weiss, *Tetrahedron Lett.* **1995**, *36*, 2999-3002.
[12] C.-M. Shu, W.-S. Chung, S.-H. Wu, Z.-C. Ho, L.-G. Lin, *J. Org. Chem.* **1999**, *64*, 2673-2679.
[13] F. Santoyo-González, A. Torres-Pinedo, A. Sanchéz-Ortega, *J. Org. Chem.* **2000**, *65*, 4409-4414.
[14] S. K. Sharma, C. D. Gutsche, *J. Org. Chem.* **1996**, *61*, 2564-2568.
[15] A. Casnati, A. Arduini, E. Ghidini, A. Pochini, R. Ungaro, *Tetrahedron* **1991**, *47*, 2221-2228.
[16] C. D. Gutsche, L. G. Lin, *Tetrahedron* **1986**, *42*, 1633-1640.
[17] E. M. Georgiev, J. T. Mague, D. M. Roundhill, *Supramol. Chem.* **1993**, *2*, 53-60.
[18] K. C. Nam, J. M. Kim, D. S. Kim, *Bull. Korean Chem. Soc.* **1995**, *16*, 186-189.
[19] S. Shang, D. V. Khasnis, J. M. Burton, C. J. Santini, M. Fan, A. C. Small, M. Lattman, *Organometallics* **1994**, *13*, 5157-5159. The analogous compound in which two adjacent oxygens are bridged by Si(CH$_3$)$_2$ could not be *O*-alkylated: M. Fan, H. Zhang, M. Lattman, *Organometallics* **1996**, *15*, 5216-5219.
[20] K. B. Ray, R. H. Weatherhead, N. Pirinccioglu, A. Williams, *J. Chem. Soc., Perkin Trans. 2* **1994**, 83-88.
[21] S. Chowdhury, J. N. Bridson, P. E. Georghiou, *J. Org. Chem.* **2000**, *65*, 3299-3302.
[22] K. A. See, F. R. Fronczek, W. H. Watson, R. P. Kashyap, C. D. Gutsche, *J. Org. Chem.* **1991**, *56*, 7256-7268.
[23] K. C. Nam, S. W. Ko, J. M. Kim, *Bull. Korean Chem. Soc.* **1998**, *19*, 345-348.
[24] A. Marra, M.-C. Scherrmann, A. Dondoni, A. Casnati, P. Minari, R. Ungaro, *Angew. Chem.* **1994**, *106*, 2533-2535; *Angew. Chem., Int. Ed. Engl.* **1994**, *33*, 2479-2481.
[25] J. D. Van Loon, A. Arduini, L. Coppi, W. Verboom, A. Pochini, R. Ungaro, S. Harkema, D. N. Reinhoudt, *J. Org. Chem.* **1990**, *55*, 5639-5646.
[26] G. Ferguson, J. F. Gallagher, A. J. Lough, A. Notti, S. Pappalardo, M. F. Parisi, *J. Org. Chem.* **1999**, *64*, 5876-5885.
[27] G. Ferguson, J. F. Gallagher, L. Giunta, P. Neri, S. Pappalardo, M. Parisi, *J. Org. Chem.* **1994**, *59*, 42-53.
[28] V. Böhmer, A. Wolff, W. Vogt, *J. Chem. Soc., Chem. Commun.* **1990**, 968-970.
[29] K. No, J. E. Kim, *Bull. Korean Chem. Soc.* **1995**, *16*, 1122-1125.
[30] C. D. Gutsche, K. A. See, *J. Org. Chem.* **1992**, *57*, 4527-4539.
[31] C. Shu, W. Liu, M. Ku, F. Tang, M. Yeh, L. Lin, *J. Org. Chem.* **1994**, *59*, 3730-3733.
[32] K. C. Nam, J. M. Kim, S. K. Kook, S. J. Lee, *Bull. Korean Chem. Soc.* **1996**, *17*, 499-502.
[33] K. C. Nam, J. M. Kim, Y. J. Park, *Bull. Korean Chem. Soc.* **1998**, *19*, 770-776.
[34] Z. Csók, G. Szalontai, G. Czira, L. Kollár, *Supramol. Chem.* **1998**, *10*, 69-77; surprisingly, the authors mention that the synthesis of tetrasulfonates was unsuccessful.
[35] J. J. González, P. M. Nieto, P. Prados, A. M. Echavarren, J. de Mendoza, *J. Org. Chem.* **1995**, *60*, 7419-7423.
[36] P. D. Beer, M. G. B. Drew, D. Hesek, M. Shade, F. Szemes, *Chem. Commun.* **1996**, 2161-2162.
[37] L. C. Groenen, B. H. M. Ruel, A. Casnati, P. Timmerman, W. Verboom, S. Harkema, A. Pochini, R. Ungaro, D. N. Reinhoudt, *Tetrahedron Lett.* **1991**, *32*, 2675-2678.
[38] F. Bottino, L. Giunta, S. Pappalardo, *J. Org. Chem.* **1989**, *54*, 5407-5409.
[39] For a chiral example derived from a calix[4]arene of the ABAB type ($R^1 = R^3$, $R^2 = R^4$) see: V. Böhmer, D. Kraft, M. Tabatabai, *J. Incl. Phenom.* **1994**, *19*, 17-39.
[40] A. Arduini, A. Casnati, M. Fabbi, P. Minari, A. Pochini, A. R. Sicuri, R. Ungaro, *Supramol. Chem.* **1993**, *1*, 235-246.
[41] For a single 1,2-OCH$_2$O-bridge see: a) D. Kraft, V. Böhmer, W. Vogt, G. Ferguson, J. F. Gallagher, *J. Chem. Soc., Perkin Trans. 1* **1994**, 1221-1230; b) J. Wöhnert, J. Brenn, M. Stoldt, O. Aleksiuk, F. Grynszpan, I. Thondorf, S. E. Biali, *J. Org. Chem.* **1998**, *63*, 3866-3874; For the analogous compound with two bridges see: c) P. Neri, G. Ferguson, J. F. Gallagher, S. Pappalardo, *Tetrahedron Lett.* **1992**, *33*, 7403-7406.
[42] Recently the simultaneous formation of the *syn* and *anti*-1,2-diether (9-10% each) was reported for the *O*-alkylation of **1a** with *N,N*-dibutyl-2-bromoacetamide (NaH, DMF, 60°C), however, the NMR-spectroscopic evidence is not quite convincing: O. M. Falana, H. F. Koch, D. M. Roundhill, G. J. Lumetta, B. P. Hay, *Chem. Commun.* **1998**, 503-504.

[43] A. Arduini, A. Casnati, L. Dodi, A. Pochini, R. Ungaro, *J. Chem. Soc., Chem. Commun.* **1990**, 1597-1598.
[44] See for instance: K. Iwamoto, H. Shimizu, K. Araki, S. Shinkai, *J. Am. Chem. Soc.* **1993**, *115*, 3997-4006.
[45] L. N. Markovsky, M. O. Vysotsky, V. V. Pirozhenko, V. I. Kalchenko, J. Lipkowski, Y. A. Simonov, *Chem. Commun.* **1996**, 69-71.
[46] M. O. Vysotsky, M. O. Tairov, V. V. Pirozhenko, V. I. Kalchenko, *Tetrahedron Lett.* **1998**, *39*, 6057-6060.
[47] See also the formation of the 1,2-derivative by *O*-methylation of the mono-dinitrobenzoate of **1b**: Y. J. Park, J. M. Shin, K. C. Nam, J. M. Kim, S.-K. Kook, *Bull. Korean Chem. Soc.* **1996**, *17*, 643-647.
[48] J. M. Kim, K. C. Nam, *Bull. Korean Chem. Soc.* **1997**, *18*, 1327-1330.
[49] K. Iwamoto, S. Shinkai, *J. Org. Chem.* **1992**, *57*, 7066-7073.
[50] K. Iwamoto, K. Araki, S. Shinkai, *J. Org. Chem.* **1991**, *56*, 4955-4962.
[51] W. S. Oh, T. D. Chung, J. Kim, H.-S. Kim, H. Kim, D. Hwang, K. Kim, S. G. Rha, J.-I. Choe, S.-K. Chang, *Supramol. Chem.* **1998**, *9*, 221-231.
[52] X. Chen, M. Ji, D. R. Fisher, C. M. Wai, *Synlett* **1999**, 1784-1786.
[53] Z. Asfari, unpublished.
[54] T. Jin, K. Monde, *Chem. Commun.* **1998**, 1357-1358.
[55] For a chiral triether of the ABBH type (Y^1 = α-picolyl, $Y^2 = Y^3$ = n-Pr) see: K. Iwamoto, A. Yanagi, T. Arimura, T. Matsuda, S. Shinkai, *Chem. Lett.* **1990**, 1901-1904.
[56] For chiral 1,2-crownethers see: F. Arnaud-Neu, S. Caccamese, S. Fuangswasdi, S. Pappalardo, M. F. Parisi, A. Petringa, G. Principato, *J. Org. Chem.* **1997**, *62*, 8041-8048.
[57] Z.-C. Ho, M.-C. Ku, C.-M. Shu, L.-G. Lin, *Tetrahedron* **1996**, *52*, 13189-13200.
[58] For various inherently chiral triethers see also: S. Caccamese, S. Pappalardo, *Chirality* **1993**, *5*, 159-163.
[59] M. P. Oude Wolbers, F. C. J. M. van Veggel, R. H. M. Heeringa, J. W. Hofstraat, F. A. J. Geurts, G. J. van Hummel, S. Harkema, D. N. Reinhoudt, *Liebigs Ann./Recueil* **1997**, 2587-2600.
[60] J. K. Browne, M. A. McKervey, M. Pitarch, J. A. Russell, J. S. Millership, *Tetrahedron Lett.* **1998**, *39*, 1787-1790. These triethers were converted into inherently chiral derivatives by lipase-catalyzed transesterification.
[61] S. Berthalon, J.-B. Regnouf de Vains, R. Lamartine, *Synth. Commun.* **1996**, *26*, 3103-3108.
[62] R. Cacciapaglia, A. Casnati, L. Mandolini, S. Schiavone, R. Ungaro, *J. Chem. Soc., Perkin Trans. 2* **1993**, 369-371.
[63] For all possible isomeric tetrapropylethers of **1a** or *p*-nitrocalix[4]arene see: P. J. A. Kenis, O. F. J. Noordman, S. Houbrechts, G. J. van Hummel, S. Harkema, F. C. J. M. van Veggel, K. Clays, J. F. J. Engbersen, A. Peersoons, N. F. van Hulst, D. N. Reinhoudt, *J. Am. Chem. Soc.* **1998**, *120*, 7875-7883 and references cited there.
[64] For the stereochemical outcome of the reaction of **1a** with ethyl bromoacetate under different reaction conditions see: a) K. Iwamoto, K. Fujimoto, T. Matsuda, S. Shinkai, *Tetrahedron* Lett. **1990**, *31*, 7169-7172 and ibidem **1991**, *32*, 830 (corrigendum); b) Ref. [49].
[65] For all possible isomeric α-picolylethers see: a) S. Pappalardo, L. Giunta, M. Foti, G. Ferguson, J. F. Gallagher, B. Kaitner, *J. Org. Chem.* **1992**, *57*, 2611-2624; b) S. Pappalardo, G. Ferguson, P. Neri, C. Rocco, *J. Org. Chem.* **1995**, *60*, 4576-4584.
[66] For all possible isomeric tetra-acetates see: S. Akabori, H. Sannohe, Y. Habata, Y. Mukoyama, T. Ishi, *Chem. Commun.* **1996**, 1467-1468.
[67] See *e.g.*: E. Kelderman, L. Derhaeg, G. J. T. Heesink, W. Verboom, J. F. J. Engbersen, N. F. Van Hulst, A. Persoons, D. N. Reinhoudt, *Angew. Chem.* **1992**, *104*, 1075-1077; *Angew. Chem., Int. Ed. Engl.* **1992**, *31*, 1107-1109.
[68] Phase transfer catalytic conditions (aq. 50% NaOH/toluene) have been recommended for the synthesis of *cone*-tetraethers of **1a**: I. Bitter, A. Grun, B. Agai, L. Toke, *Tetrahedron* **1995**, *51*, 7835-7840.
[69] For recent examples see: a) C. B. Dieleman, D. Matt, P. G. Jones, *J. Organomet. Chem.* **1997**, *545-546*, 461-473; b) G. Ulrich, R. Ziessel, I. Manet, M. Guardigli, N. Sabbatini, F. Fraternali, G. Wipff, *Chem. Eur. J.* **1997**, *3*, 1815-1822; c) S. E. Matthews, M. Saadioui, V. Böhmer, S. Barboso, F. Arnaud-Neu, M.-J. Schwing-Weill, A. G. Carrera, J.-F. Dozol, *J. Prakt. Chem.* **1999**, *341*, 264-273.
[70] See for instance: a) A. Casnati, A. Pochini, R. Ungaro, F. Ugozzoli, F. Arnaud, S. Fanni, M.-J. Schwing, R. J. M. Egberink, F. de Jong, D. N. Reinhoudt, *J. Am. Chem. Soc.* **1995**, *117*, 2767-2777;

b) G. Arena, A. Casnati, A. Contino, L. Mirone, D. Sciotto, R. Ungaro, *Chem. Commun.* **1996**, 2277-2278; c) A. Casnati, C. Fischer, M. Guardigli, A. Isernia, I. Manet, N. Sabbatini, R. Ungaro, *J. Chem. Soc., Perkin Trans. 2* **1996**, 395-399; d) A. Casnati, A. Pochini, R. Ungaro, C. Bocchi, F. Ugozzoli, R. J. M. Egberink, H. Struijk, R. Lugtenberg, F. de Jong, D. N. Reinhoudt, *Chem. Eur. J.* **1996**, *2*, 436-445.

[71]   a) P. Lhotak, S. Shinkai, *Tetrahedron* **1995**, *51*, 7681-7696; b) R. K. Castellano, J. J. Rebek, *J. Am. Chem. Soc.* **1998**, *120*, 3657-3663.
[72]   S. Pappalardo, S. Caccamese, L. Giunta, *Tetrahedron Lett.* **1991**, *32*, 7747-7750.
[73]   C. D. Gutsche, P. A. Reddy, *J. Org. Chem.* **1991**, *32*, 4783-4791.
[74]   K. Iwamoto, K. Araki, S. Shinkai, *J. Chem. Soc., Chem. Comun.* **1991**, 1611-1613.
[75]   For bis-crownethers in the *1,2-alternate* conformation see: a) A. Arduini, L. Domiano, A. Pochini, A. Secchi, R. Ungaro, F. Ugozzoli, O. Struck, W. Verboom, D. N. Reinhoudt, *Tetrahedron* **1997**, *53*, 3767-3776; b) G. Ferguson, A. J. Lough, A. Notti, S. Pappalardo, M. F. Parisi, A. Petringa, *J. Org. Chem.* **1998**, *63*, 9703-9710.
[76]   A. Ikeda, T. Tsudera, S. Shinkai, *J. Org. Chem.* **1997**, *62*, 3568-3574.
[77]   J.-A. Pérez-Adelmar, H. Abraham, C. Sánchez, K. Rissanen, P. Prados, J. de Mendoza, *Angew. Chem.* **1996**, *108*, 1088-1090; *Angew. Chem., Int. Ed. Engl.* **1996**, *35*, 1009-1011; see already: W. Verboom, S. Datta, Z. Asfari, S. Harkema, D. N. Reinhoudt, *J. Org. Chem.* **1992**, *57*, 5394-5398.
[78]   V. S. Talanov, R. A. Bartsch, *J. Chem. Soc., Perkin Trans. 1* **1999**, 1957-1961.
[79]   M. Pitarch, J. K. Browne, M. A. McKervey, *Tetrahedron* **1997**, *53*, 10503-10512 and 16195-16204.
[80]   C. Floriani, D. Jacoby, A. Chiesi-Villa, C. Guastini, *Angew. Chem.* **1989**, *101*, 1430-1431; *Angew. Chem., Int. Ed. Engl.* **1989**, *28*, 1376-1377.
[81]   L. Baklouti, R. Abidi, Z. Asfari, J. Harrowfield, R. Rokbani, J. Vicens, *Tetrahedron Lett.* **1998**, *39*, 5363-5366.
[82]   G. Ferguson, J. F. Gallagher, M. A. McKervey, E. Madigan, *J. Chem. Soc., Perkin Trans. 1* **1996**, 599-602.
[83]   A. Casnati, A. Pochini, R. Ungaro, R. Cacciapaglia, L. Mandolini, *J. Chem. Soc., Perkin Trans. 1* **1991**, 2052-2054.
[84]   For an early conformational study see: M. Iqbal, T. Mangiafico, C. D. Gutsche, *Tetrahedron* **1987**, *43*, 4917-4930.
[85]   See for instance: a) S. K. Sharma, C. D. Gutsche, *Synthesis* **1994**, 813-822; b) K. No, H. J. Koo, *Bull. Korean Chem. Soc.* **1994**, *15*, 483-488.
[86]   K. No, Y. J. Park, K. H. Kim, J. M. Shin, *Bull. Korean Chem. Soc.* **1996**, *17*, 447-452.
[87]   C. D. Gutsche, J. A. Levine, P. K. Sujeeth, *J. Org. Chem.* **1985**, *50*, 5802-5806.
[88]   D. Xie, C. D. Gutsche, *J. Org. Chem.* **1997**, *62*, 2280-2284.
[89]   S. K. Sharma, C. D. Gutsche, *J. Org. Chem.* **1999**, *64*, 3507-3512.
[90]   The expression "selective" is often abused for reactions in which not all of the potential functional groups are converted. It should be used only, if the product is obtained in (distinctly) higher yield than statistically expected.
[91]   For a first report see: J.-D. van Loon, A. Arduini, W. Verboom, R. Ungaro, G. J. van Hummel, S. Harkema, D. N. Reinhoudt, *Tetrahedron Lett.* **1989**, *30*, 2681-2684.
[92]   J. L. Atwood, S. G. Bott, in: *Calixarenes – A Versatile Class of Macrocyclic Compounds*, J. Vicens, V. Böhmer (Eds.), Kluwer, Dordrecht, 1990, pp. 199-210.
[93]   B. Yao, J. Bassus, R. Lamartine, *An. Quim. Int. Ed.* **1998**, *94*, 65-66.
[94]   P.-S. Wang, R.-S. Lin, H.-X. Zong, *Synth. Commun.* **1999**, *29*, 2225-2227.
[95]   R. A. Jakobi, V. Böhmer, C. Grüttner, D. Kraft, W. Vogt, *New. J. Chem.* **1996**, *20*, 493-501.
[96]   A. Arduini, V. Böhmer, L. Delmau, J.-F. Desreux, J.-F. Dozol, M. A. Garcia Carrera, B. Lambert, C. Musigmann, A. Pochini, A. Shivanyuk, F. Ugozzoli, *Chem. Eur. J.* **2000**, *6*, 2135-2144.
[97]   W. Verboom, A. Durie, R. J. M. Egberink, Z. Asfari, D. N. Reinhoudt, *J. Org. Chem.* **1992**, *57*, 1313-1316.
[98]   O. Mogck, P. Parzuchowski, M. Nissinen, V. Böhmer, G. Rokicki, K. Rissanen, *Tetrahedron* **1998**, *54*, 10053-10068 and references cited there.
[99]   a) O. Mogck, V. Böhmer, G. Ferguson, W. Vogt, *J. Chem. Soc., Perkin Trans. 1* **1996**, 1711-1715; see also: b) Q.-Y. Zheng, C.-F. Chen, Z.-T. Huang, *Tetrahedron* **1997**, *53*, 10345-10356; c) P.-S. Wang, R.-S. Lin, *Synth. Commun.* **1999**, *29*, 1661-1664; d) Q. Y. Zheng, C. F. Chen, Z. T. Huang, *Chin. J. Chem.* **2000**, *18*, 104-111.

[100] G. Mislin, E. Graf, M. W. Hosseini, A. D. Cian, N. Kyritsakas, J. Fischer, *Chem. Commun.* **1998**, 2545-2546.
[101] S. Kanamathareddy, C. D. Gutsche, *J. Org. Chem.* **1995**, *60*, 6070-6075.
[102] a) H. Kämmerer, G. Happel, V. Böhmer, D. Rathay, *Monatsh. Chem.* **1978**, *109*, 767-773; b) A. Casnati, E. Comelli, M. Fabbi, V. Bocchi, G. Mori, F. Ugozzoli, A. M. Manotti Lanfredi, A. Pochini, R. Ungaro, *Recl. Trav. Chim. Pays-Bas* **1993**, *112*, 384-392; c) W. Wasikiewicz, G. Rokicki, E. Rozniecka, J. Kielkiewicz, Z. Brzozka, V. Böhmer, *Monatsh. Chem.* **1998**, *129*, 1169-1181.
[103] S. G. Rha, S.-K. Chang, *J. Org. Chem.* **1998**, *63*, 2357-2359.
[104] C. D. Gutsche, in: *Calixarenes – A Versatile Class of Macrocyclic Compounds*, J. Vicens, V. Böhmer (Eds.), Kluwer, Dordrecht, 1990, pp. 3-37.
[105] A. Casnati, M. Fochi, P. Minari, A. Pochini, M. Reggiani, R. Ungaro, *Gazz. Chim. Ital.* **1996**, *126*, 99-106.
[106] S. Shimizu, S. Shirakawa, Y. Sasaki, C. Hirai, *Angew. Chem.* **2000**, *112*, 1313-1315; *Angew. Chem., Int. Ed.* **2000**, *39*, 1256-1259.
[107] A. Ikeda, M. Yoshimura, P. Lhotak, S. Shinkai, *J. Chem. Soc., Perkin Trans. 1* **1996**, 1945-1950.
[108] W. Verboom, P. J. Bodewes, G. van Essen, P. Timmerman, G. J. van Hummel, S. Harkema, D. N. Reinhoudt, *Tetrahedron* **1995**, *51*, 499-512.
[109] a) M. Mascal, R. T. Naven, R. Warmuth, *Tetrahedron Lett.* **1995**, *36*, 9361-9364; b) M. Mascal, R. Warmuth, R. T. Naven, R. A. Edwards, M. B. Hursthouse, D. E. Hibbs, *J. Chem. Soc., Perkin Trans. 1* **1999**, 3435-3441.
[110] A. Arduini, A. Pochini, A. R. Sicuri, A. Secchi, R. Ungaro, *Gazz. Chim. Ital.* **1994**, *124*, 129-132.
[111] P. Timmerman, W. Verboom, D. N. Reinhoudt, A. Arduini, S. Grandi, A. R. Sicuri, A. Pochini, R. Ungaro, *Synthesis* **1994**, 185-189.
[112] B. Klenke, W. Friedrichsen, *J. Chem. Soc., Perkin Trans. 1* **1998**, 3377-3379, and references cited there.
[113] A. Gunji, K. Takahashi, *Synth. Commun.* **1998**, *28*, 3933-3941.
[114] P. D. Beer, M. Shade, *Chem. Commun.* **1997**, 2377-2378.
[115] K. C. Nam, D. S. Kim, *Bull. Korean Chem. Soc.* **1994**, *15*, 284-286.
[116] P. D. Beer, J. B. Cooper, *Chem. Commun.* **1998**, 129-130.
[117] N. Pelizzi, A. Casnati, A. Friggeri, R. Ungaro, *J. Chem. Soc., Perkin Trans. 2* **1998**, 1307-1311.
[118] J. D. Van Loon, J. F. Heida, W. Verboom, D. N. Reinhoudt, *Recl. Trav. Chim. Pays-Bas* **1992**, *111*, 353-359.
[119] M. S. Brody, C. A. Schalley, D. M. Rudkevich, J. Rebek, Jr., *Angew. Chem.* **1999**, *111*, 1738-1742; *Angew. Chem., Int. Ed. Engl.* **1999**, *38*, 1640-1644.
[120] A. Arduini, L. Mirone, D. Paganuzzi, A. Pinalli, A. Pochini, A. Secchi, R. Ungaro, *Tetrahedron* **1996**, *52*, 6011-6018.
[121] a) A. Arduini, G. Manfredi, A. Pochini, A. R. Sicuri, R. Ungaro, *J. Chem. Soc., Chem. Commun.* **1991**, 936-937; b) A. Arduini, S. Fanni, G. Manfredi, A. Pochini, R. Ungaro, A. R. Sicuri, F. Ugozzoli, *J. Org. Chem.* **1995**, *60*, 1448-1453.
[122] P. Molenveld, J. F. J. Engbersen, H. Kooijman, A. L. Spek, D. N. Reinhoudt, *J. Am. Chem. Soc.* **1998**, *120*, 6726-6737.
[123] P. Molenveld, W. M. G. Stikvoort, H. Kooijman, A. L. Spek, J. F. J. Engbersen, D. N. Reinhoudt, *J. Org. Chem.* **1999**, *64*, 3896-3906.
[124] a) A. Dondoni, A. Marra, M.-C. Scherrmann, A. Casnati, F. Sansone, R. Ungaro, *Chem. Eur. J.* **1997**, *3*, 1774-1782; see already: b) T. Komori, S. Shinkai, *Chem. Lett.* **1992**, 901-904.
[125] A. Arduini, M. Fabbi, M. Mantovani, L. Mirone, A. Pochini, A. Secchi, R. Ungaro, *J. Org. Chem.* **1995**, *60*, 1454-1457.
[126] J.-B. Regnouf-de-Vains, R. Lamartine, *Tetrahedron Lett.* **1996**, *37*, 6311-6314. For further monofunctionalisations of tris-*t*-butylcalix[4]arene see: S. Berthalon, J.-B. Regnouf-de-Vains, R. Lamartine, *Tetrahedron Lett.* **1997**, *38*, 8527-8528. S. Berthalon, L. Motta-Viola, J.-B. Regnouf-de-Vains, R. Lamartine, S. Lecocq, M. Perrin, *Eur. J. Org. Chem.* **1999**, 2269-2274.
[127] B. Yao, J. Bassus, R. Lamartine, *New J. Chem.* **1996**, *20*, 913-915.
[128] While the $AlCl_3$-catalysed isopropylation with isopropylchloride in 1,2-dichloroethane was possible with calix[8]arene, hydroxyisopropylation occured with **1b** and both reactions were observed with calix[6]arene: B. Yao, J. Bassus, R. Lamartine, *Bull. Soc. Chim. Fr.* **1997**, *134*, 555-559.
[129] E. A. Shokova, A. N. Khomich, V. V. Kovalev, *Tetrahedron Lett.* **1996**, *37*, 543-546.

[130] V. Kovalev, E. Shokova, Y. Luzikov, *Synthesis* **1998**, 1003-1008.
[131] V. Kovalev, E. Shokova, A. Khomich, Y. Luzikov, *New J. Chem.* **1996**, *20*, 483-492.
[132] V. Wendel, W. Abraham, *Tetrahedron Lett.* **1997**, *38*, 1177-1180.
[133] Y. Morita, T. Agawa, E. Nomura, H. Taniguchi, *J. Org. Chem.* **1992**, *57*, 3658-3662.
[134] For further examples of tetraazo calix[4]arenes see: a) S. Shinkai, K. Araki, J. Shibata, D. Tsugawa, O. Manabe, *J. Chem. Soc., Perkin Trans. 1* **1990**, 3333-3337; b) E. Kelderman, L. Derhaeg, W. Verboom, J. F. J. Engbersen, S. Harkema, A. Persoons, D. N. Reinhoudt, *Supramol. Chem.* **1993**, *2*, 183-190.
[135] See for instance: N. J. van der Veen, R. J. M. Egberink, J. F. J. Engbersen, F. J. C. M. van Veggel, D. N. Reinhoudt, *Chem. Commun.* **1999**, 681-688.
[136] M.-L. Yeh, F.-S. Tang, S.-L. Chen, W.-C. Liu, L.-G. Lin, *J. Org. Chem.* **1994**, *59*, 754-757.
[137] For an example of a double calix[4]arene linked at the upper rim by a biphenyl-4,4-diazo group see: S. Bouoit-Montesinos, J. Bassus, M. Perrin, R. Lamartine, *Tetrahedron Lett.* **2000**, *41*, 2563-2567.
[138] C.-M. Shu, T.-S. Yuan, M.-C. Ku, Z.-C. Ho, W.-C. Liu, F.-S. Tang, L.-G. Lin, *Tetrahedron* **1996**, *52*, 9805-9818.
[139] H. M. Chawla, K. Srinivas, *J. Org. Chem.* **1996**, *61*, 8464-8467.
[140] I. Alam, S. K. Sharma, C. D. Gutsche, *J. Org. Chem.* **1994**, *59*, 3716-3720.
[141] K. J. C. van Bommel, F. Westerhof, W. Verboom, D. N. Reinhoudt, R. Hulst, *J. Prakt. Chem.* **1999**, *341*, 284-290.
[142] Z.-T. Huang, G.-Q. Wang, L.-M. Yang, Y.-X. Lou, *Synth. Commun.* **1995**, *25*, 1109-1118.
[143] K. C. Nam, T. H. Yoon, *Bull. Korean Chem. Soc.* **1993**, *14*, 169-171.
[144] C.-M. Shu, W.-L- Lin, G.-H. Lee, S.-M. Peng, W.-S. Chung, *J. Chin. Chem. Soc.* **2000**, *47*, 173-182. The *p*-allyl groups were converted into isoxazolinomethyl groups by 1,3-dipolar addition of nitrile oxides.
[145] For a recent study see: H. M. Chawla, Meena, *Indian J. Chem., Sect. B: Org. Chem. Incl. Med. Chem.* **1998**, *37B*, 28-33.
[146] For the synthesis of calixarenes with a single C=O group see: K. Ito, S. Izawa, T. Ohba, Y. Ohba, T. Sone, *Tetrahedron Lett.* **1996**, *37*, 5959-5962.
[147] For the preparation of calix[4]arenes with one or two CHR-groups by (2+2) fragment condensation see: C. Grüttner, V. Böhmer, W. Vogt, I. Thondorf, S. E. Biali, F. Grynszpan, *Tetrahedron Lett.* **1994**, *35*, 6267-6270. S. E. Biali, V. Böhmer, S. Cohen, G. Ferguson, C. Grüttner, F. Grynszpan, E. F. Paulus, I. Thondorf, W. Vogt, *J. Am. Chem. Soc.* **1996**, *118*, 12938-12949. M. Bergamaschi, F. Bigi, M. Lanfranchi, R. Maggi, M. A. Pastorio, M. A. Pellinghelli, F. Peri, C. Porta, G. Sartori, *Tetrahedron* **1997**, *53*, 13037-13052.
[148] a) A. Ninagawa, K. Cho, H. Matsuda *Makromol. Chem.* **1985**, *186*, 1379-1385; b) G. Görmar, K. Seiffarth, M. Schultz, J. Zimmermann, G. Fläming, *Makromol. Chem.* **1990**, *191*, 81-87.
[149] B. Klenke, C. Näther, W. Friedrichsen, *Tetrahedron Lett.* **1998**, *39*, 8967-8969.
[150] O. Middel, Z. Greff, N. J. Taylor, W. Verboom, D. N. Reinhoudt, V. Snieckus *J. Org. Chem.* **2000**, *65*, 667-675.
[151] Ether derivatives with Y = CH_2-CO-OR are colloquially often named "esters".
[152] For recent examples see: a) F. Arnaud-Neu, S. Fanni, L. Guerra, W. McGregor, K. Ziat, M.-J. Schwing-Weill, G. Barrett, M. A. McKervey, D. Marss, E. Seward, *J. Chem. Soc., Perkin Trans. 2* **1995**, 113-118; b) F. Arnaud-Neu, G. Barrett, S. Fanni, D. Marss, W. McGregor, M. A. McKervey, M.-J. Schwing-Weill, V. Vetrogon, S. Wechsler, *J. Chem. Soc., Perkin Trans. 2* **1995**, 453-461; c) L. Frkanec, A. Višnjevac, B. Kojic-Prodic, M. Zinic, *Chem. Eur. J.* **2000**, *6*, 442-453. d) R. Roy, J. M. Kim, *Angew. Chem.* **1999**, *111*, 380-384; *Angew. Chem., Int. Ed.* **1999**, *38*, 369-372.
[153] G. Barrett, V. Böhmer, G. Ferguson, J. F. Gallagher, S. J. Harris, R. G. Leonard, M. A. McKervey, M. Owens, M. Tabatabai, A. Vierengel, W. Vogt, *J. Chem. Soc., Perkin Trans. 2* **1992**, 1595-1601.
[154] For recent examples see: M. P. Oude Wolbers, F. C. J. M. Van Veggel, F. G. A. Peters, E. S. E. Van Beelen, J. W. Hofstraat, F. A. J. Geurts, D. N. Reinhoudt, *Chem. Eur. J.* **1998**, *4*, 772-780.
[155] F. Arnaud-Neu, G. Barrett, S. J. Harris, M. Owens, M. A. McKervey, M.-J. Schwing-Weill, P. Schwinté, *Inorg. Chem.* **1993**, *110*, 2644-2650.
[156] F. Arnaud-Neu, G. Barrett, G. Ferguson, J. F. Gallagher, M. A. McKervey, M. Moran, M.-J. Schwing-Weill, P. Schwinté, *Supramol. Chem.* **1996**, *7*, 215-222.
[157] P. D. Beer, M. G. B. Drew, A. Grieve, M. Kan, P. B. Leeson, G. Nicholson, M. I. Ogden, G. Williams, *Chem. Commun.* **1996**, 1117-1118.

[158] P. L. H. M. Cobben, R. J. M. Egberink, J. G. Bomer, P. Berfeld, W. Verboom, D. N. Reinhoudt, *J. Am. Chem. Soc.* **1992**, *114*, 10573-10582.
[159] J. K. Moran, E. M. Georgiev, A. T. Yordanov, J. T. Mague, D. M. Roundhill, *J. Org. Chem.* **1994**, *59*, 5990-5998.
[160] S. Knoblauch, O. M. Falana, J. Nam, D. M. Roundhill, H. Hennig, K. Zeckert, *Inorg. Chim. Acta* **2000**, *300-302*, 328-332.
[161] Mentioned in: F. F. Malone, D. J. Marrs, M. A. McKervey, P. O'Hagan, N. Thompson, A. Walker, F. Arnaud-Neu, O. Mauprivez, M.-J. Schwing-Weill, J.-F. Dozol, H. Rouquette, N. Simon, *J. Chem. Soc., Chem. Commun.* **1995**, 2151-2153.
[162] A. T. Yordanov, J. T. Mague, D. M. Roundhill, *Inorg. Chim. Acta* **1995**, *240*, 441-446.
[163] See *e.g.*: P. Schmitt, P. D. Beer, M. G. B. Drew, P. D. Sheen, *Angew. Chem.* **1997**, *109*, 1926-1928; *Angew. Chem., Int. Ed. Engl.* **1997**, *36*, 1840-1842.
[164] a) D. M. Roundhill, E. Georgiev, A. Yordanov, *J. Incl. Phenom.* **1994**, *19*, 101-109; b) E. M. Georgiev, N. Wolf, D. M. Roundhill, *Polyhedron* **1997**, *16*, 1581-1584; c) S. Smirnov, V. Sidorov, E. Pinkhassik, J. Havlicek, I. Stibor, *Supramol. Chem.* **1997**, *8*, 187-196; d) N. J. Wolf, E. M. Georgiev, A. T. Yordanov, B. R. Whittlesey, H. F. Koch, D. M. Roundhill, *Polyhedron* **1999**, *18*, 885-896.
[165] See also: K. C. Nam, S. O. Kang, S. W. Ko, *Bull. Korean Chem. Soc.* **1999**, *20*, 953-956.
[166] S. Barboso, A. Garcia Carrera, S. E. Matthews, F. Arnaud-Neu, V. Böhmer, J.-F. Dozol, H. Rouquette, M.-J. Schwing-Weill, *J. Chem. Soc., Perkin Trans. 2* **1999**, 719-723.
[167] L. A. J. Chrisstoffels, F. de Jong, D. N. Reinhoudt, S. Sivelli, S. Gazzola, A. Casnati, R. Ungaro, *J. Am. Chem. Soc.* **1999**, *121*, 10142-10151.
[168] P. A. Gale, *Tetrahedron Lett.* **1998**, *19*, 3873-3876.
[169] P. Neri, A. Bottino, C. Geraci, M. Piattelli, *Tetrahedron: Asymmetry* **1996**, *7*, 17-20.
[170] R. K. Juneja, K. D. Robinson, C. P. Johnson, J. L. Atwood, *J. Am. Chem. Soc.* **1993**, *115*, 3818-3819.
[171] M. S. Wong, J. F. Nicoud, *Tetrahedron Lett.* **1993**, *34*, 8237-8240.
[172] C. A. Gleave, I. O. Sutherland, *J. Chem. Soc., Chem. Commun.* **1994**, 1873-1874.
[173] K.-S. Paek, H. Ihm, K. No, *Bull. Korean Chem. Soc.* **1994**, *15*, 422-423.
[174] A. Arduini, A. Pochini, A. Rizzi, A. R. Sicuri, R. Ungaro, *Tetrahedron Lett.* **1990**, *31*, 4653-4656.
[175] A. Arduini, W. L. McGregor, D. Paganuzzi, A. Pochini, A. Secchi, F. Ugozzoli, R. Ungaro, *J. Chem. Soc., Perkin Trans 2* **1996**, 839-846.
[176] M. Larsen, M. Jorgensen, *J. Org. Chem.* **1997**, *62*, 4171-4173.
[177] C. Felix, H. Parrot-Lopez, V. Kalchenko, A. W. Coleman, *Tetrahedron Lett.* **1998**, *39*, 9171-9174.
[178] A. Dondoni, C. Ghiglione, A. Marra, M. Scoponi, *J. Org. Chem.* **1998**, *63*, 9535-9539.
[179] M. Larsen, F. C. Krebs, N. Harrit, M. Jorgensen, *J. Chem. Soc., Perkin Trans. 2* **1999**, 1749-1757.
[180] P. Timmerman, H. Boerrigter, W. Verboom, D. N. Reinhoudt, *Recl. Trav. Chim. Pays-Bas* **1995**, *114*, 103-111.
[181] K. D. Shimizu, J. J. Rebek, *Proc. Natl. Acad. Sci. U. S. A.* **1995**, *92*, 12403-12407.
[182] A. M. A. van Wageningen, P. Timmerman, J. P. M. van Duynhoven, W. Verboom, F. C. J. M. van Veggel, D. N. Reinhoudt, *Chem. Eur. J.* **1997**, *3*, 639-654.
[183] C. D. Gutsche, P. F. Pagoria, *J. Org. Chem.* **1985**, *50*, 5795-5802.
[184] H. Ihm, K. Paek, *Bull. Korean Chem. Soc.* **1995**, *16*, 71-73.
[185] M. Larsen, M. Jorgensen, *J. Org. Chem.* **1996**, *61*, 6651-6655.
[186] K.-S. Paek, H.-J. Kim, S.-K. Chang, *Supramol. Chem.* **1995**, *5*, 83-85.
[187] K. L. Hwang, S. H. Ham, K. H. No, *Bull. Korean Chem. Soc.* **1993**, *14*, 79-81.
[188] For a recent example of the introduction of two distal fluorenyl residues at the wider rim of a calix[4]arene via halogen-metal exchange see: A. Faldt, F. C. Krebs, M. Jorgensen, *Tetrahedron Lett.* **2000**, *41*, 1241-1244.
[189] J. J. Gonzalez, P. Prados, J. De Mendoza, *Angew. Chem.* **1999**, *111*, 546-549; *Angew. Chem., Int. Ed.* **1999**, *38*, 525-528.
[190] H. Ihm, H. Kim, K. Paek, *J. Chem. Soc., Perkin Trans. 1* **1997**, 1997-2003.
[191] J. Budka, M. Dudic, P. Lhotak, I. Stibor, *Tetrahedron* **1999**, *55*, 12647-12654.
[192] S. K. Sharma, C. D. Gutsche, *J. Org. Chem.* **1999**, *64*, 998-1003.
[193] D. M. Rudkevich, W. Verboom, D. N. Reinhoudt, *J. Org. Chem.* **1994**, *59*, 3683-3686.
[194] M. Saadioui, A. Shivanyuk, V. Böhmer, W. Vogt, *J. Org. Chem.* **1999**, *64*, 3774-3777.
[195] This may be the reason also for the difficulty to acylate two opposite amino groups under mild conditions.

[196] L. J. Prins, K. A. Jolliffe, R. Hulst, P. Timmerman, D. N. Reinhoudt, *J. Am. Chem. Soc.* **2000**, *122*, 3617-3627
[197] A. Soi, A. Hirsch, *New J. Chem.* **1998**, *22*, 1337-1339.
[198] J. Scheerder, R. H. Vreekamp, J. F. J. Engbersen, W. Verboom, J. P. M. van Duynhoven, D. N. Reinhoudt, *J. Org. Chem.* **1996**, *61*, 3476-3481.
[199] O. Mogck, V. Böhmer, W. Vogt, *Tetrahedron* **1996**, *52*, 8489-8496.
[200] J. Scheerder, J. P. M. van Duynhoven, J. F. J. Engbersen, D. N. Reinhoudt, *Angew. Chem.* **1996**, *108*, 1172-1175; *Angew. Chem., Int. Ed. Engl.* **1996**, *35*, 1090-1093.
[201] A. Pop, M. Saadioui, M. Vysotsky, V. Böhmer, *unpublished*
[202] A. M. A. Van Wageningen, E. Snip, W. Verboom, D. N. Reinhoudt, H. Boerrigter, *Liebigs Ann./Recl.* **1997**, 2235-2245.
[203] H. J. Cho, J. Y. Kim, S. K. Chang, *Chem. Lett.* **1999**, 493-494.
[204] M. Larsen, F. C. Krebs, M. Jorgensen, N. Harrit, *J. Org. Chem.* **1998**, *63*, 4420-4424.
[205] For Wittig reactions of the monoformyl-tris-*t*-butyl calix[4]arene see ref. [126].
[206] A. Dondoni, M. Kleban, A. Marra, *Tetrahedron Lett.* **1997**, *38*, 7801-7804.
[207] P. Lhotak, R. Nakamura, S. Shinkai, *Supramol. Chem.* **1997**, *8*, 333-344.
[208] P. Molenveld, J. F. J. Engbersen, D. N. Reinhoudt, *Eur. J. Org. Chem.* **1999**, 3269-3275.
[209] J. Bügler, J. F. J. Engbersen, D. N. Reinhoudt, *J. Org. Chem.* **1998**, *63*, 5339-5344.
[210] O. Struck, J. P. M. van Duynhoven, W. Verboom, S. Harkema, D. N. Reinhoudt, *Chem. Commun.* **1996**, 1517-1518.
[211] N. Pelizzi, A. Casnati, R. Ungaro, *Chem. Commun.* **1998**, 2607-2608.
[212] A. Arduini, S. Fanni, A. Pochini, A. R. Sicuri, R. Ungaro, *Tetrahedron* **1995**, *51*, 7951-7958.
[213] P. Lhotak, S. Shinkai, *Tetrahedron Lett.* **1996**, *37*, 645-648.
[214] R. H. Vreekamp, W. Verboom, D. N. Reinhoudt, *J. Org. Chem.* **1996**, *61*, 4282-4288.
[215] O. Struck, W. Verboom, W. J. J. Smeets, A. L. Spek, D. N. Reinhoudt, *J. Chem. Soc., Perkin Trans. 2* **1997**, 223-227.
[216] a) F. Sansone, S. Barboso, A. Casnati, M. Fabbi, A. Pochini, F. Ugozzoli, R. Ungaro, *Eur. J. Org. Chem.* **1998**, 897-905; see also: b) F. C. Krebs, M. Larsen, M. Jorgensen, P. R. Jensen, M. Bielecki, K. Schaumburg, *J. Org. Chem.* **1998**, *63*, 9872-9879.
[217] E. Pinkhassik, V. Sidorov, I. Stibor, *J. Org. Chem.* **1998**, *63*, 9644-9651.
[218] All twelve OH groups of pyrogallol derived compounds have been also esterified or etherified. See for example: G. Cometti., E. Dalcanale, A. Du Vosel, A.-M. Levelut, *J. Chem. Soc., Chem. Commun.* **1990**, 163-165.
[219] a) A. G. S. Högberg, *J. Am. Chem. Soc.* **1980**, *102*, 6046-6050; b) A. G. S. Högberg, *J. Org. Chem.* **1980**, *45*, 4498-4500; c) L. Abis, E. Dalcanale, A. Du Vosel, S. Spera, *J. Chem. Soc., Perkin Trans. 2* **1990**, 2075-2080; d) L. Abis, E. Dalcanale, A. Du Vosel, S. Spera, *J. Org. Chem.* **1988**, *53*, 5475-5479; e) R. J. M. Egberink, P. L. H. M. Cobben, W. Verboom, S. Harkema, D. N. Reinhoudt *J. Incl. Phenom.* **1992**, *12*, 151-158; f) P. D. Beer, E. L. Tite, *Tetrahedron Lett.* **1988**, *29*, 2349-2352.
[220] V. I. Kalchenko, D. M. Rudkevich, A. N. Shivanyuk, V. V. Pirozhenko, I. F. Tsymbal, L. N. Markovsky, *Zhurn. Obsch. Khim.* **1994**, *64*, 731-742; *Russ. J. Gen. Chem.* **1994**, *64*, 663-672.
[221] W. Hu, J. P. Rourke, J. J. Vital, R. J. Puddephatt, *Inorg. Chem.* **1995**, *34*, 323-329.
[222] V. I. Kalchenko, A. V. Solov'yov, N. R. Gladun, A. N. Shivanyuk, L. I. Atamas', V. V. Pirozhenko, L. N. Markowsky, J. Lipkowski, Yu. A. Simonov, *Supramol. Chem.* **1997**, *8*, 269-297.
[223] a) I. Neda, T. Siedentop, A. Vollbrecht, H. Thoennessen, P. G. Jones, R. Schmutzler, *Z. Naturforsch., B: Chem. Sci.* **1998**, *53*, 841-848; b) A. Vollbrecht, I. Neda, R. Schmutzler, *Phosphorus, Sulfur Silicon Relat. Elem.* **1995**, *107*, 173-179.
[224] a) M. Urbaniak, W. Iwanek, *Tetrahedron* **1999**, *55*, 14459-14466; b) G. Mann, L. Hennig, F. Weinelt, K. Müller, R. Meusinger, G. Zahn, T. Lippmann, *Supramol. Chem.* **1994**, *3*, 101-113; c) S. Pellet-Rostaing, J.-B. Regnouf de Vains, R. Lamartine, *Tetrahedron Lett.* **1995**, *36*, 5745-5748.
[225] O. Haba, K. Haga, M. Ueda, O. Morikawa, H. Konishi, *Chem. Mat.* **1999**, *11*, 427-432.
[226] Y. Yamakawa, M. Ueda, R. Nagahata, T. Takeuchi, M. Asai, *J. Chem. Soc., Perkin Trans. 1* **1998**, 4135-4139.
[227] J. R. Fransen, P. J. Dutton, *Can. J. Chem.* **1995**, *73*, 2217-2223.
[228] W. Iwanek, *Tetrahedron Asymm.* **1998**, *9*, 3171-3174.
[229] For their use in electrokinetic chromatography, see Chapter 34.

[230] T. Fujimoto, C. Shimizu. O. Hayashida, T. Fujimoto, C. Shimizu. O. Hayashida, Y. Aoyama, *J. Am. Chem. Soc.* **1997**, *119*, 6676-6677.
[231] a) L. N. Markovsky, D. M. Rudkevich, V. I. Kalchenko, *Zhurn. Obsch. Khim.* **1990**, *60*, 2813-2814; J. Gen. Chem. USSR (Engl. Transl.) **1990**, *60*, 2520-2521; b) L. N. Markovsky, V. I. Kalchenko, D. M. Rudkevich, A. N. Shivanyuk, *Mendeleev Commun.* **1992**, 106-108; c) L. N. Markovsky, V. I. Kalchenko, C. M. Rudkevich, A. N. Shivanyuk, *Phosph. Silicon. Sulf.* **1993**, *75*, 59-62.
[232] The reaction of the octamethylsilyl ether of resorcarene **2** with four equivalents of PF_2Cl afforded chiral C_4-symmetric tetrakis-difluorophosphites, comp. [223b].
[233] J. Lipkowski, O. I. Kalchenko, J. Slowikowska, V. I. Kalchenko, O. V. Lukin, L. N. Markovsky, R. Nowakowsky, *J. Phys. Org. Chem.* **1998**, *11*, 426-435.
[234] V. I . Kalchenko, A. N. Shivanyuk, V. V. Pirozhenko, L. N. Markovsky, *Zhurn. Obsch. Khim.* **1994**, *64*, 1562-1563; *Russ. J. Gen. Chem.* **1994**, *64*, 1397-1398.
[235] O. Lukin, A. Shivanyuk, V. V. Pirozhenko, I. F. Tsymbal, V. I Kalchenko, *J. Org. Chem.* **1998**, *63*, 9510-9516.
[236] A. Shivanyuk, E. F. Paulus, V. Böhmer, W. Vogt, *J. Org. Chem.* **1998**, *63*, 6448-6449.
[237] For the encapsulation of tropylium cations by hydrogen bonded dimers of **13d** see: A. Shivanyuk, E. F. Paulus, V. Böhmer, *Angew. Chem.* **1999**, *111*, 3091-3094; *Angew. Chem., Int. Ed.* **1999**, *38*, 2906-2909.
[238] A. N. Shivanyuk, V. I. Kalchenko, V. V. Pirozhenko, L. N. Markovsky *Zh. Obsh. Khim. (Russ)* **1994**, *64*, 1558-1559; *Russ. J. Gen. Chem.* **1994**, *64*, 1401-1402.
[239] H. Konishi, T. Tamura, H. Ohkubo, K. Kobayashi, O. Morikawa, *Chem. Lett.* **1996**, 685-686.
[240] U. Schneider, H.-J. Schneider, *Chem. Ber.* **1994**, *127*, 2455-2459.
[241] D. A. Leigh, P. Linnane, R. G. Pitchard, G. Jackson, *J. Chem. Soc., Chem. Commun.* **1994**, 389-390.
[242] P. Linnane, S. Shinkai, *Tetrahedron Lett.* **1995**, 36, 3865-3866.
[243] H. Konishi, H. Yamaguchi, M. Miyashiro, K. Kobayashi, O. Morikawa, *Tetrahedron Lett.* **1996**, *37*, 8547-8548.
[244] A. Shivanyuk, S. Numellin, K. Rissanen, *J. Org. Chem.*, submitted.
[245] a) Y. Matsushita, T. Matsui, *Tetrahedron Lett.* **1993**, *34*, 7433-7437; b) R. Arnecke, V. Böhmer, E. F. Paulus, W. Vogt, *J. Am. Chem. Soc.* **1995**, *117*, 3286-3288.
[246] A C_4-symmetrical derivative was obtained also during the attempt to synthesise a cavitand-derived tetraacetic acid: H.-J. Choi, D. Buhring, M. L. C. Quan, C. B. Knobler, D. J. Cram, *J. Chem. Soc., Chem. Commun.* **1992**, 1733-1735.
[247] K. Airola, V. Böhmer, E. F. Paulus, K. Rissanen, C. Schmidt, I. Thondorf, W. Vogt, *Tetrahedron* **1997**, *53*, 10709-10724.
[248] Cleavage of the benzoxazine unit occurs also in the reaction with 2-naphthol, which is alkylated to yield the respective tertiary amine: P. D. Woodgate, G. M. Horner, N. P. Maynard, *Tetrahedron Lett.* **1999**, *40*, 6507-6510.
[249] a) W. Iwanek, J. Mathay, *Liebigs Anal.* **1995**, 1463-1469; b) M. T. El Gihani, H. Heaney, A. M. Z. Slawin, *Tetrahedron Lett.* **1995**, *36*, 4905-4908; c) R. Arnecke, V. Böhmer, S. Friebe, S. Gebauer, G. J. Kraus, I. Thondorf, *Tetrahedron Lett.* **1995**, 36, 6221-6224.
[250] C. Schmidt, E. F. Paulus, V. Böhmer, W. Vogt, *New. J. Chem.*, in press.
[251] The *O*-acetylation was also reported, see references [249a,c] but could not be reproduced. Cleavage of the benzoxazines and *N*-acetylation occurred instead: C. Schmidt, E. F. Paulus, V. Böhmer, W. Vogt, *New J. Chem.* **2000**, *24*, 123-125.
[252] P. C. Bulman Page, H. Heaney, E. P. Sampler, *J. Am. Chem. Soc.* **1999**, *121*, 6751-6752.
[253] C. Schmidt, K. Airola, V. Böhmer, W. Vogt, K. Rissanen, Tetrahedron **1997**, *53*, 17691-17698.
[254] C. Schmidt, I. Thondorf, E. Kolehmainen, V. Böhmer, W. Vogt, K. Rissanen, *Tetrahedron Lett.* **1998**, *39*, 8833-8836.
[255] a) W. Iwanek, C. Wolf, J. Mattay, *Tetrahedron Lett.* **1995**, *36*, 8969-8972; b) C. Schmidt, T. Straub, D. Falabu, E. F. Paulus, E. Wegelius, E. Kolehmainen, V. Böhmer, K. Rissanen, W. Vogt, *Eur. J. Org. Chem.* **2000**, 3937-3944.
[256] O. Manabe, K. Asakura, T. Nishi, S. Shinkai, *Chem. Lett.* **1990**, 1219-1222.
[257] D. J. Cram, S. Karbach, H. Kim, C. B. Knobler, E. F. Maverick, J. L. Ericson, R. C. Helgeson, *J. Am. Chem. Soc.* **1988**, *110*, 2229-2237.
[258] H. Konishi, H. Nakamuro, H. Nakatani, T. Ueyama, K. Kobayashi, O. Morikawa, *Chem. Lett.* **1997**, 185-186.

[259] O. Morikawa, K. Nakanishi, M. Miyashiro, K. Kobayashi, H. Konishi, *Synthesis* **2000**, 233-236.
[260] A. Shivanyuk, C. Schmidt, V. Böhmer, E. F. Paulus, O. V. Lukin, W. Vogt, *J. Am. Chem. Soc.* **1998**, *120*, 4319-4326.
[261] a) L. M. Tunstad, J. A. Tucker, E. Dalcanale, J. Weiser, J. A. Bryant, J. C. Sherman, R. C. Helgeson, C. B. Knobler, D. J. Cram, *J. Org. Chem.* **1989**, *54*, 1305-1312; b) P. T. Lewis, C. J. Davis, M. C. Saraiva, W. D. Treleaven, T. D. McCarley, R. M. Strongin, *J. Org. Chem.* **1997**, *62*, 6110-6111; c) K. Kobayashi, M. Tominaga, Y. Asakawa, Y. Aoyama., *Tetrahedron Lett.* **1993**, *34*, 5121-5124; d) A. Shivanyuk, V. Böhmer, E. F. Paulus, *Gazz. Chim. Ital.* **1997**, *127*, 741-747; e) K. Ichimura, N. Fukushima, M. Fujimaki, S. Kawahara, Y. Matsuzawa, Y. Hayashi, K. Kudo, *Langmuir* **1997**, *13*, 6780-6786; f) P. D. Beer, E. L. Tite, A. Ibbotson, *J. Chem. Soc., Chem. Commun.* **1989**, 1874-1876.
[262] P. T. Lewis, R. M. Strongin, *J. Org. Chem.* **1998**, *63*, 6065-6067.
[263] E. U. Thoden van Velzen, J. F. J. Engbersen, P. J. de Lange, J. W. G Mahy, D. N. Reinhoudt, *J. Am. Chem. Soc.* **1995**, *117*, 6853-6862.
[264] F. Davis, C. Stirling, *Langmuir* **1996**, *12*, 5365-5374.
[265] S. Shoichi, D. M. Rudkevich, J. Rebek, *Org. Lett.* **1999**, *1*, 1241-1244.
[266] T. Haino, D. M. Rudkevich, A. Shivanyuk, K. Rissanen, J. Rebek, Jr., *Chem. Eur. J.* **2000**, in press
[267] A. R. Mezo, J. C. Sherman, *J. Org. Chem.* **1998**, *63*, 6824-6829.
[268] M. Fujimaki, S. Kawahara, Y. Matsuzawa, E. Kurita, Y. Hayashi, K. Ichimura, *Langmuir* **1998**, *14*, 4495-4502.
[269] B. Botta, G. Delle Monache, P. Ricciardi, G. Zappia, C. Seri, E. Gacs-Baitz, P. Csokasi, D. Misiti, *Eur. J. Org. Chem.* **2000**, *65*, 841-847.
[270] B. Botta, G. Delle Monache, M. C. De Rosa, C. Seri, E. Benedetti, R. Iacoviano, M. Botta, F. Correli, V. Masignani, A. Tafi, E. Gacs-Baitz, A. Santini, D. Misiti, *J. Org. Chem.* **1997**, *62*, 1788-1794.

Chapter 3

SELECTIVELY MODIFIED CALIX[5]ARENES

ANNA NOTTI[a], MELCHIORRE F. PARISI[a], SEBASTIANO PAPPALARDO[b]

[a]*Dipartimento di Chimica Organica e Biologica, Università di Messina, Salita Sperone 31, I-98166 Messina, Italy*
[b]*Dipartimento di Scienze Chimiche, Università di Catania, Viale A. Doria 6, I-95125 Catania, Italy. e-mail: spappalardo@dipchi.unict.it*

1. Introduction

Calix[5]arenes are potentially useful platforms for the design of molecular containers with peripheries of intermediate size between those of the smaller calix[4]arenes and the larger calix[n]arenes ($n \geq 6$). They resemble calix[4]arenes in having four basic up/down conformations (*cone, partial cone, 1,2-alternate, 1,3-alternate*). However, because of the wider dimension of their annulus, calix[5]arenes are more prone to conformational interconversions [1]. These generally occur *via* the lower-rim-through-the-annulus motion (phenolic oxygens swinging through the annulus) [2], even though the alternate upper-rim-through-the-annulus pathway (aryl groups swinging through the annulus) also seems to be possible when there is no *p*-substituent [3]. Since conformational mobility is detrimental to guest binding/inclusion, a mandatory step for the design of effective calix[5]arene-based receptors is to determine the size and nature of the substituents necessary to prevent any major conformational motion. This brief survey is intended to provide a background on the chemical manipulation of the upper and lower rims of calix[5]arenes in order to construct preorganized and permanent cavities for the inclusion of neutral molecules and positively charged species.

2. Synthesis of Calix[5]arenes and Chemical Modifications on Their Upper and Lower Rims

2.1. ONE-STEP SYNTHESIS

Until the last decade, calixarenes with an odd number of phenol units ('minor calixarenes' [4]) were obtained from the base-catalyzed one-step synthesis of *p*-alkylphenols and formaldehyde [5], in very low yields in comparison with their more accessible counterparts with an even number of phenol residues ('major calixarenes' [4]), for which specific procedures were established in the early 1980s [4]. In response to the escalating interest in the cyclic pentamers, improved synthetic procedures now provide moderately satisfactory yields of *p-tert*-butylcalix[5]arene **1** (up to 15–16% [6]) and *p*-benzylcalix[5]arene **2** (33% [7]).

2.2. (3 + 2) AND (4 + 1) FRAGMENT CONDENSATION

While the linear stepwise synthesis of calixarenes appears at present to be generally of only historical interest [8], a convergent approach aptly called 'fragment condensation'

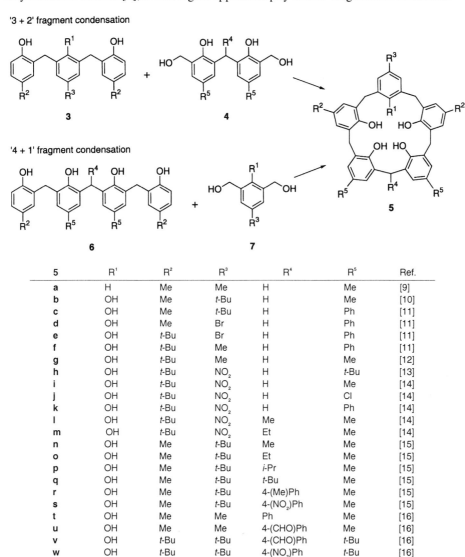

Figure 1. The (3 + 2) and (4 + 1) fragment condensation of calix[5]arenes.

5	R^1	R^2	R^3	R^4	R^5	Ref.
a	H	Me	Me	H	Me	[9]
b	OH	Me	t-Bu	H	Me	[10]
c	OH	Me	t-Bu	H	Ph	[11]
d	OH	Me	Br	H	Ph	[11]
e	OH	t-Bu	Br	H	Ph	[11]
f	OH	t-Bu	Me	H	Ph	[11]
g	OH	t-Bu	Me	H	Me	[12]
h	OH	t-Bu	NO_2	H	t-Bu	[13]
i	OH	t-Bu	NO_2	H	Me	[14]
j	OH	t-Bu	NO_2	H	Cl	[14]
k	OH	t-Bu	NO_2	H	Ph	[14]
l	OH	t-Bu	NO_2	Me	Me	[14]
m	OH	t-Bu	NO_2	Et	Me	[14]
n	OH	Me	t-Bu	Me	Me	[15]
o	OH	Me	t-Bu	Et	Me	[15]
p	OH	Me	t-Bu	i-Pr	Me	[15]
q	OH	Me	t-Bu	t-Bu	Me	[15]
r	OH	Me	t-Bu	4-(Me)Ph	Me	[15]
s	OH	Me	t-Bu	4-(NO_2)Ph	Me	[15]
t	OH	Me	Me	Ph	Me	[16]
u	OH	Me	Me	4-(CHO)Ph	Me	[16]
v	OH	t-Bu	t-Bu	4-(CHO)Ph	t-Bu	[16]
w	OH	t-Bu	t-Bu	4-(NO_2)Ph	t-Bu	[16]

is today skillfully exploited for the construction of calix[5]arenes carrying two to three different functionalities at the upper rim. It consists of the 'heat induced' (3 + 2) fragment condensation of an equimolar mixture of an appropriately *p*-substituted linear trimer **3** and a bis-hydroxymethylated dimer **4** in refluxing xylene to produce calix[5]-arenes **5a–k** in 19–32% yield [9–14]. The substituents introduced on the upper rim by this strategy include Me, *t*-Bu, Ph, halogen, and NO_2 group(s), as shown in Figure 1.

Calix[5]arenes **5l–w**, having an additional alkyl or aryl substituent (R^4) on a methylene bridge, have been prepared either by the (3 + 2) [14, 15] or (4 + 1) fragment condensation [16] of a suitably functionalized linear tetramer **6** with a 2,6-bis(hydroxymethyl)-4-alkylphenol **7** under similar conditions.

With the exception of mono-deoxycalix[5]arene **5a** (R^1 = H), which adopts a *partial cone* conformation both in the solid state and solution [9], calix[5]arene pentols **5b–w** exist in the *cone* conformation at ambient temperature, which is stabilized over the other conformations by an intramolecular cyclic array of hydrogen bonds between phenolic OH groups. The free energy barriers for the conformational interconversions are estimated to be very close to that measured for **1** ($\Delta G^{\neq}$ = 13.2 kcal mol^{-1} [1]). The ^{1}H NMR spectra of compounds **5l–w**, bearing the R^4 substituent on a methylene bridge, show the presence of two diastereomeric *cone* conformations in equilibrium (ratios up to 32:1). In the major isomer the R^4 substituent holds the equatorial position, as confirmed by X-ray structural analyses on **5q** and **5w** [15, 16].

2.3. Lower Rim Manipulation

The chemical modification at the lower rim of calix[5]arenes – *via* ether or ester formation – is a means to synthesize new host molecules by the introduction of additional functional groups, and to control the conformation and hindrance to ring inversion.

The exhaustive *O*-alkylation of calix[5]arenes to produce derivatives **8a–e** is accomplished in the presence of NaH (in refluxing 9:1 THF/DMF or DMF at room temperature) or K_2CO_3 (in refluxing acetonitrile or acetone), the latter base generally giving better yields [2, 3, 6b, 17–21].

Penta-amide **8f**, convertible to the corresponding penta-thioamide **8g** by treatment with Lawesson's reagent [22], and a calix[5]arene penta-ferrocene amide [23] have been prepared from the corresponding acid chloride and the appropriate amine. It is worth mentioning that pentamethylketone **8e** has been used as a template for the synthesis of a calix[5]arene-capped calix[5]pyrrole [24]. Penta-esters **8h** are prepared in excellent yield by heating **1** in the presence of the corresponding anhydride and a trace of acid, or by reacting **1** with an acid chloride in NaH/THF [2].

8	X
a	R^1
b	$(CH_2)_2OR^1$
c	CH_2Py
d	$CH_2CO_2R^1$
e	$CH_2C(O)R^1$
f	$CH_2C(O)NR^1_2$
g	$CH_2C(S)NR^1_2$
h	$C(O)R^1$

8a–h

R = H, *t*-Bu

Direct mono-*O*-alkylation of calix[5]arenes is realized in good yield using either a deficiency of alkylating agent and/or a weak base ($KHCO_3$ or CsF) in refluxing acetonitrile or acetone [2], or DMF at 60 °C [20]. In contrast to the smaller calix[4]arenes, for which regioselective di-*O*-substitution and stereoselective tri-*O*-substitution are well-established [4], no efficient procedures are so far available for the regioselective *O*-alkylation of calix[5]arenes.

The selective functionalization at the lower rim of calix[5]arenes works much better with (poly)functional reagents, mainly used to introduce a specific function (Me_2NP) bridging the 1,2-positions [25], to reduce the conformational freedom of the parent pentols *via* scaffolding elements such as *o*-phthaloyl [26] or polyether moieties [27], to shape their hydrophobic cavities [28, 29], as well as to confer inherent chirality to the systems [30–32].

1,3-Bridged calix[5]arene crown ethers **9** are obtained in reasonable to excellent yield by reacting calix[5]arene pentols with the appropriate oligoethylene glycol ditosylate and CsF in refluxing acetonitrile [27, 28]. The tosylates may incorporate into the polyether chain a phenylene [27b] or 1,1'-binaphthalene-2,2'-diyl [28] subunit. The reactions of **1** with hexaethylene glycol ditosylate and 2,2'-bis(5-tosyloxy-3-oxa-1-pentyloxy)-1,1'-binaphthalene also produce small amounts of the corresponding 1,2-bridged calix[5]crown regioisomers **10**. The distinction between the two regioisomers can be made by ^{1}H NMR spectroscopy, on the basis of the chemical shift of the residual phenolic OH protons. The conformation of *tert*-butylated calix[5]crowns can be fixed by attaching large substituents at the lower rim [27–29].

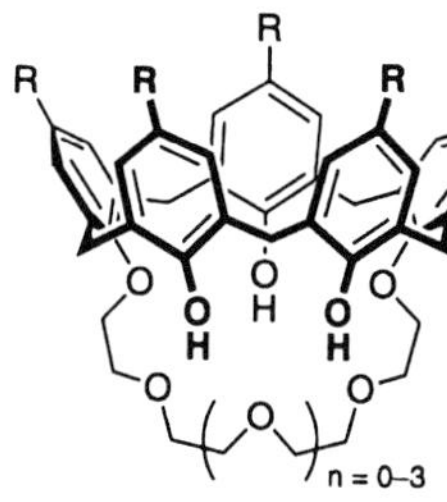

9

R = H, *t*-Bu

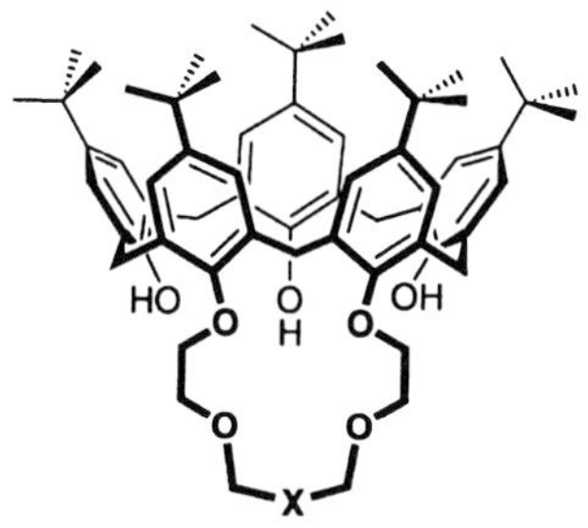

10

X = (CH$_2$OCH$_2$)$_3$, [naphthalene]$_2$OCH$_2$

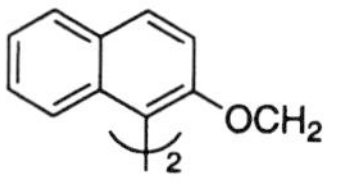

Both mono-*O*-alkylated calix[5]arenes and calix[5]crowns **9**, featuring a mirror plane as the only symmetry element, proved to be useful achiral sources for the production of inherently chiral calixarenes. Selective desymmetrization of these molecules is accomplished either by regioselective intrabridging (two contiguous OH groups adjacent to the *O*-substituent) of α-picolyl-*p*-*tert*-butylcalix[5]arene with tri- to pentaethylene glycol ditosylates and K$_2$CO$_3$ in DMF to produce unsymmetrical derivatives **11** (n = 0–2), or by selective *O*-alkylation or *O*-acylation of one of the two adjacent OH groups (4/5 positions) of **9** (n = 1) in the presence of weak bases to afford derivatives **12** [30, 31]. The two strategies are complementary, and a pair of inherently chiral regioisomers **11** (n = 2) and **12** (R^1 = α-picolyl; n = 2) have recently been synthesized and resolved into their enantiomers by a direct enantioselective HPLC method [31]. Compounds **11** and **12** adopt a *cone* conformation by virtue of hydrogen bond formation of the phenolic OH group with the adjacent hydroxyl or ethereal oxygen.

A reverse regioselectivity (*i.e.*, substitution on the 2-positioned OH group) is noted when calix[5]crowns **9** are acylated with *p*-nitrophenyl esters in the presence of Ba(II) ions. The reaction also shows a high stereoselectivity and depends on the size of the crown ether ring, producing exclusively the *partial cone* acetate to pentanoate **13** with crown-5 and the *cone* acetate **14** with crown-6 [33].

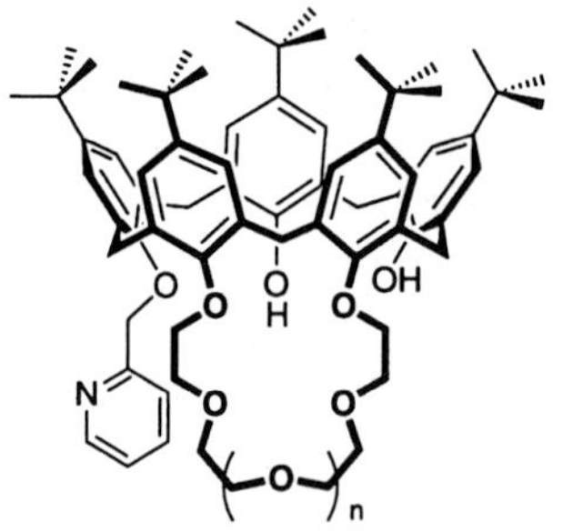

11

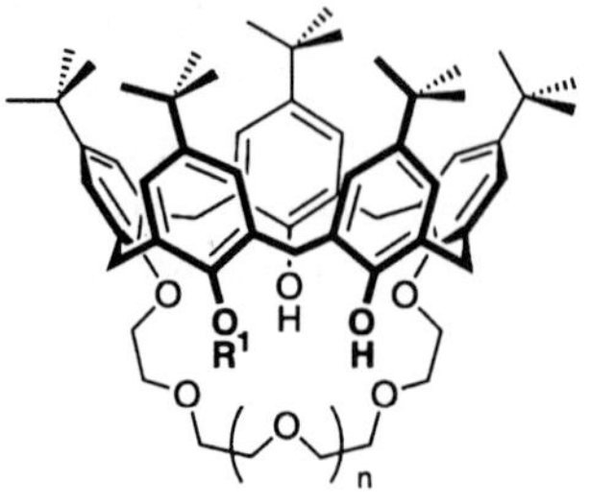

12

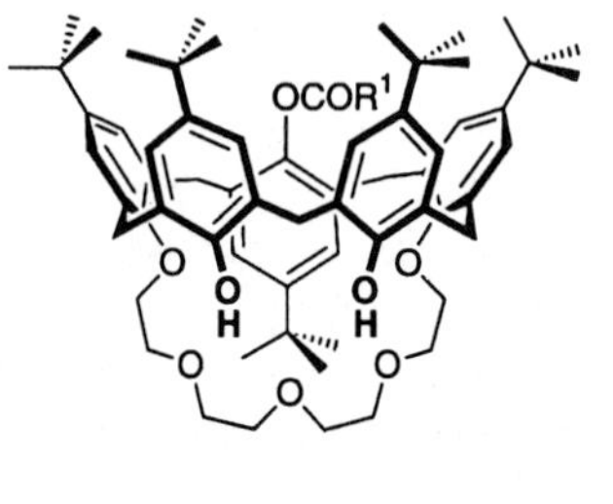

13

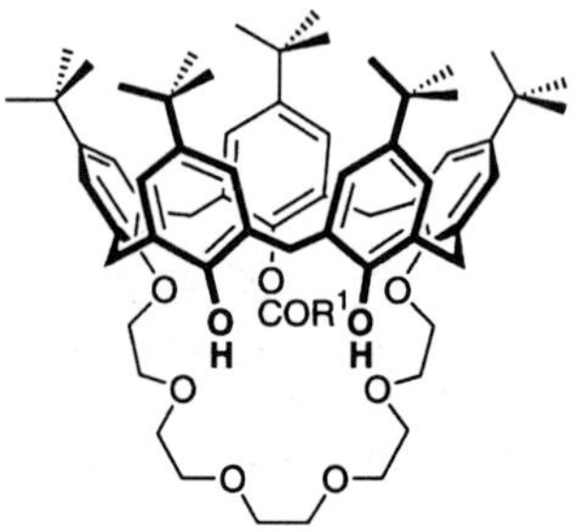

14

2.4. UPPER RIM MANIPULATION

The chemical modification at the upper rim of calix[5]arenes is triggered by the easy AlCl$_3$-catalyzed trans-alkylation of **1** in the presence of a suitable acceptor (phenol and/or toluene) [3b, 5b, 34]. This reaction is known to play a key role in calixarene chemistry, since a variety of calixarenes with different substituents in the *p*-positions can be subsequently obtained *via* the common electrophilic substitution on the phenol or phenol ether residues. Concerning the calix[5]arene system, sulfonation [35], mercuration [5b], aminomethylation [36], Gross formylation [37], coupling with diazonium and tetrazonium salts [38], *ipso*-sulfonation [39] and *ipso*-nitration [17] have been described. Nitrocalix[5]arenes are useful derivatives for the introduction of other functionalities onto the *p*-positions [40], generally *via* the aminocalix[5]arene intermediates obtained by reduction with H$_2$ and Raney Ni [17]. Other general procedures make use of the *p*-Claisen rearrangement of penta-*O*-allylcalix[5]arene [41] and the Mannich base/*p*-quinonemethide route [36].

No systematic studies have been made so far on the selective functionalization of the upper rim of calix[5]arenes having the same *p*-substituent. However, the chemistry of the more extensively studied calix[*n*]arenes (*n* = 4,6) has taught us that selective *p*-substitution (including trans-alkylation) can be realized by transferring to the upper rim the selective functionalization achieved at the lower rim, the phenol unit(s) generally being more reactive than phenol ether or ester units. By taking advantage of this strategy, penta-*O*-alkylated nitrocalix[5]arene **19** in a fixed *cone* conformation has been prepared in satisfactory overall yield through a sequence involving the exhaustive *O*-alkylation of mono-benzyl ether **15** [2] to produce **16** (80%), followed by debenzylation *via* hydrogenolysis to mono-hydroxy intermediate **17** (93%), selective *ipso*-nitration of the phenol unit to give **18** (57%), and final conversion to **19** (72%) by further alkylation of the residual OH group (Figure 2). This route is very promising, but limited by the lack of efficient procedures for the partial *O*-alkylation or *O*-acylation of calix[5]arenes with monofunctional electrophilic reagents.

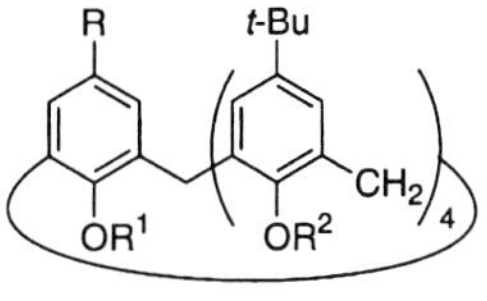

15–19

	R	R^1	R^2
15	*t*-Bu	Bn	H
16	*t*-Bu	Bn	$(CH_2)_3CH(CH_3)_2$
17	*t*-Bu	H	$(CH_2)_3CH(CH_3)_2$
18	NO_2	H	$(CH_2)_3CH(CH_3)_2$
19	NO_2	$(CH_2)_3CH(CH_3)_2$	$(CH_2)_3CH(CH_3)_2$

Figure 2. *Transfer of selectivity from the lower to the upper rim of calix[5]arenes [13].*

Alternate strategies employ readily available calix[5]arenes **5b** and **5g** (Figure 1), which can be selectively detertbutylated to derivatives **20** and **22** having one or two free *p*-positions, respectively. Compounds **20** and **22** proved to be particularly valuable intermediates for the synthesis of new host molecules *via* selective introduction of one or two specific functionalities at the upper rim (Figure 3).

20 R = H
21 R = $NHC(O)NH(CH_2)_5CH_3$

22–24

	X	R
22	H	H
23a	Me	Br
23b	Ac	I
23c	H	I
24a	H	
24b	Me	
24c	H	

25

	X
25a	
25b	
25c	

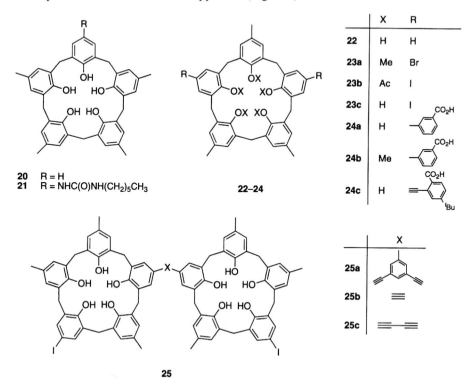

Figure 3. *Selected host molecules derived from **20** and **22**.*

Diazocoupling of **20** with *p*-aminobenzoic acid and subsequent reductive cleavage of the N=N double bond ($Na_2S_2O_4$) gives the aminocalix[5]arene, which has been converted into the hexylurea derivative **21** [10].

Compound **22** has been converted into molecular receptors **24a–c**, endowed with two converging benzoic acid residues connected to the upper rim by a direct Ar–Ar bond or through a –C≡C– bridge, by chemical manipulation on the intermediary *O*-protected dibromo or diiodo derivatives **23a,b**. The crucial step is the palladium-mediated Suzuki coupling of **23a,b** with the appropriate carboxy-protected arylboronic acid or arylacetylene fragment, followed by standard deprotections of phenolic OH and -CO_2H functions [12, 42]. Calix[5]arene **23b** has also served as the starting material for the synthesis of head-to-head bridged calix[5]arene receptors **25a–c**, in which the two cavities act cooperatively in the binding of C_{60} and C_{70} [43].

3. Preorganization and Shaping of the Calix[5]arene Cavity

3.1. IMMOBILE C_{5v} SYMMETRY STRUCTURES

Conformational interconversions of calix[5]arenes may occur *via* upper-rim and/or lower-rim-through-the-annulus pathways. In order to design preorganized permanent cavities, it is essential to assess the minimum size of substituents on both wide and narrow rims to prevent conformational interconversions. It is known that (i) the upper-rim-through-the-annulus motion in *p-tert*-butylcalix[5]arene penta-*O*-ethers is inoperative at ordinary temperatures, (ii) the $\Delta G^{\neq}$ values for conformational interconversions increase with increasing the bulkiness of R^1 groups at the lower rim, (iii) the penta-*n*-butyl ether derivative still shows a limited conformational mobility ($\Delta G^{\neq}$ = 15.3 kcal mol^{-1}), and (iv) sufficiently large groups (*i.e.*, benzyl) afford conformationally immobile penta-*O*-ethers [2]. Whereas penta-ether **26a** (Figure 4) at ambient temperature exists in $CDCl_3$ as a mixture of slowly interconverting *partial cone/cone* (19:1) conformers [19] {$\Delta G^{\neq}$ = 17.8 kcal mol^{-1} in tetrachloroethane-d_4 (TCE) [44]}, the threshold size for conformational immobilization of calix[5]arenes is likely reached with *cone* penta-ester **26g**. The added steric interaction arising from the carbonyl oxygen makes this group more effective than **26a** (of corresponding carbon content) in reducing the conformational mobility. Nevertheless, VT NMR studies have shown that, upon heating a TCE solution of **26g**, the pair of doublets for $ArCH_2Ar$ protons changes to broad singlets, but no hint of coalescence is observed at temperatures as high as 130 °C (estimated $\Delta G^{\neq}$ > 20 kcal mol^{-1} [44]). Thus, *p-tert*-butylcalix[5]arenes can be conformationally locked by endowing the lower rim with substituents slightly larger than the ethoxycarbonylmethyl group (*i.e.*, derivatives **26b,c** and **26h,i**).

The exhaustive alkylation of *p-tert*-butylcalix[5]arene **1** with large planar (benzyl, picolyl, 2-quinolylmethyl) or long-chain electrophiles is stereoselective, affording derivatives **26** in a fixed and quite regular C_{5v} *cone* conformation [45]. In sharp contrast, detertbutylated counterparts (*e.g.*, **26d,e** and **26j,k**) at ambient temperature exist in solution as a mixture of slowly interconverting *cone/non-cone* conformers [3]. Dynamic NMR studies at high temperatures have revealed the broadening of every region in their spectra, and the onset of coalescence at T_c > 110 °C. Since such bulky substituents surely inhibit the lower rim motion through the annulus, it follows that the slow interconversions necessarily occur by rotation of the upper rim substituent through the annulus, as earlier predicted on the basis of molecular modeling [2]. For this dynamic process, $\Delta G^{\neq}$ values in the range 17.9–18.8 kcal mol^{-1} have been measured [3a]. Therefore, overcrowding at the lower rim by bulky substituents, if not counterbalanced by the presence of *t*-butyl groups at the upper rim of calix[5]arenes, results in the loss of preorganized *cone* conformations. These findings emphasize the central role played by the upper rim *t*-butyl substituents both in the shaping of the cavity and in the control of the conformation upon lower rim derivatization.

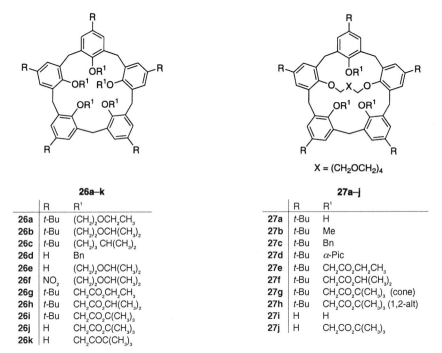

Figure 4. Selected calix[5]arene derivatives mentioned in sections 3.1, 3.2, and 5.3.

3.2. IMMOBILE C_s SYMMETRY STRUCTURES

The conformational mobility of calix[5]arenes can also be reduced by introducing scaffolding elements at the lower rim. Calix[5]crown triols **27a** and **27i** at ambient temperature adopt a distorted *cone*-like conformation (C_s symmetry) in solution [27], which can be rigidified (for R ≠ H) by attaching sufficiently large groups at the residual OH functionalities [28, 29]. Preorganized *cone* conformations are generally detected, except for the exhaustive alkylation of **27a** with BrCH$_2$CO$_2$C(CH$_3$)$_3$, which affords two inconvertible *cone* and *1,2-alternate* conformational isomers **27g** (62%) and **27h** (9%), respectively. In contrast, the reaction of detertbutylated **27i** with the same alkylating agent produces the corresponding *partial cone* (C_1 symmetry) **27j** as the sole product, which shows fluxional properties. The residual conformational mobility displayed by **27e** (slowly interconverting 9:1 *cone*/*partial cone* (C_1 symmetry) conformers in CDCl$_3$) further proves the necessity of introducing groups larger than $CH_2CO_2CH_2CH_3$ at the lower rim to obtain isolable conformational isomers [29].

The shape of the cavity of tri-*O*-alkylated **27b–h** is influenced by the size and nature of the substituents. The small-size methyl groups at the lower rim of **27b** favor a time-averaged *cone-out* conformation (the 'isolated' *p-tert*-butylaryl moiety tilted out of the calix cavity and the methoxy substituent linked to it pointing into the calix annulus), while large-size alkyl substituents induce *cone-in* (isolated *p-tert*-butylaryl moiety nestling into the cavity) (*i.e.*, **27c,d**) or regular *cone* conformations (*i.e.*, **27f,g**) (Figure 5).

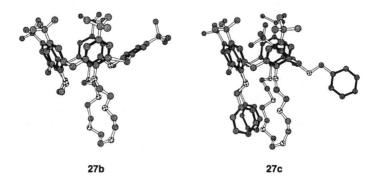

27b   **27c**

Figure 5. CS Chem3D™ *molecular models of calix[5]crown-6 derivatives* **27b** *and* **27c**, *showing their preferred* cone-out *and* cone-in *conformations*.

4. Host-Guest Complexes of Upper-Rim Functionalized Calix[5]arenes with Neutral Molecules

4.1. AROMATIC HYDROCARBONS

With the exception of the smaller calix[4]-arene congeners, water-soluble *p*-(2-diallylaminomethyl)calix[*n*]arenes **28a** and *p*-(2-carboxyethyl)calix[*n*]arenes **28b** (*n* = 5-8) form host-guest complexes with a series of aromatic hydrocarbons (durene, naphthalene, anthracene, phenanthrene, fluoranthene, pyrene and perylene), but not with the larger coronene and decacyclene [36]. Some size complementarity is noted between the guest molecule and the lower rim annulus of calixarenes. The association constants (K_a) of host-guest complexes of calix[5]arenes **28a,b** with naphthalene, anthracene and fluoranthene, determined by the solid-liquid extraction method [46], are in the range 2.0×10^3 to 9.1×10^3 M^{-1}. For a given guest the K_a values are almost the same irrespective of the upper rim substituent: this may suggest that the site of binding is not proximate to the water-solubilizing groups but is at the hydroxyl array at the lower rim. Complexation likely occurs by insertion of a portion of the aromatic hydrocarbon into the elliptical opening of the lower rim of the calixarene, but the nature of host-guest interactions is still unclear.

28a: R = $N(CH_2CH=CH_2)_2$
28b: R = CH_2CO_2H

4.2. FULLERENES

Calix[5]arene pentols are particularly suited for the inclusion of fullerenes, because they possess a preorganized *cone* cavity (determined by the cyclic array of hydrogen bonds), large enough to accommodate C_{60} by an induced-fit type adjustment of the host to the globular shape of the guest. The main driving force is π–π interaction (including charge-transfer-type interaction), and the perfect matching of the symmetry elements and curvatures of the interacting species maximizes the number of intermolecular contacts and the efficiency of the π–π interactions. Details on the use of calixarenes in the purification and separation of fullerenes can be found in Chapter 26.

Compounds **2** [7b, 47], **20**, **23c**, and *p*-methylcalix[5]arene [48], all form 1:1 complexes with C_{60} in organic solvents. The K_a values depend on the solvent and on the *p*-substituents. The K_a of **23c** with C_{60}, $(2.1\pm0.1) \times 10^3$ M^{-1} in toluene, is among the largest yet reported in organic solvents. However, in the solid state, the stoichiometry of host-guest complexes with **2** [7b, 47] and **23c** [48, 49] is 2:1. This apparent contradiction with the solution state has been explained in terms of stabilization of the 2:1 complex by the attractive van der Waals interactions between the two upper rims of the hosts [49].

A higher stability of the 2:1 complexes was achieved by connecting the two hosts with suitable linkages, as in head-to-head bridged calix[5]arenes **25a–c**. As a result, the K_a value of host **25a** with C_{60}, $(76\pm5) \times 10^3$ M^{-1} in toluene, is the largest yet reported in organic solvents, indicating an effective cooperation between the two cavities [43].

Interestingly, double calix[5]arenes **25a–c** bind C_{70} preferentially to C_{60}. The most efficient receptor for C_{70} is **25a**, $K_a = (163\pm16) \times 10^3$ M^{-1} in toluene, but the most selective is the ethynyl-bridged derivative **25b**, with a C_{70}/C_{60} selectivity value of 10.2 in toluene [43]. Since the guest is strongly encapsulated within the bis-capped hosts, the liberation of the guest from the complex is rather difficult. For separation purposes, better results were obtained with urea derivative **21**, which in non polar solvents provides a self-assembling molecular capsule (hydrogen bonded dimer), capable of binding fullerenes through the formation of a ternary complex. The free guest is then released from the complex by protonation of the urea moiety with CF_3CO_2H [10].

4.3. *N*-HETEROAROMATIC AMINES

Calix[5]arenes **24**, having two converging benzoic acid residues facing the cup-shaped cavity, behave as effective molecular receptors for 2-aminopyrimidine **29**, 9-ethyladenine **30**, and imidazole, by exploiting hydrogen bonding interactions as the main driving force for association. The stoichiometry and association constants of host-guest complexes in $CDCl_3$ have been assessed by standard ^1H NMR titration experiments. The binding mode of these guests to **24** and the structure of the resulting complexes were inferred from a combination of non-linear curve fitting analysis of the complexation induced shift (CIS) data of guest protons [50] and molecular modeling studies. The large downfield CIS values observed for the NH_2 protons of **29** and **30** upon addition of **24** are suggestive of a strong four-fold hydrogen bonding formation between the NH_2 protons of the guest and the carboxyl groups of **24**. Molecular modeling studies have shown two possible geometries for the host-guest complexes, according to whether the heteroaromatic portion of the guest resides outside (*outward* orientation) or inside the cavity (*inward* orientation) (Figure 6). The two geometries can be distinguished by ^1H NMR spectroscopy on the basis of the CIS values of the heteroaromatic protons of the guest, which are almost negligible in the former binding mode and substantial (upfield shifts up to about 2.4 ppm) in the latter [12, 42].

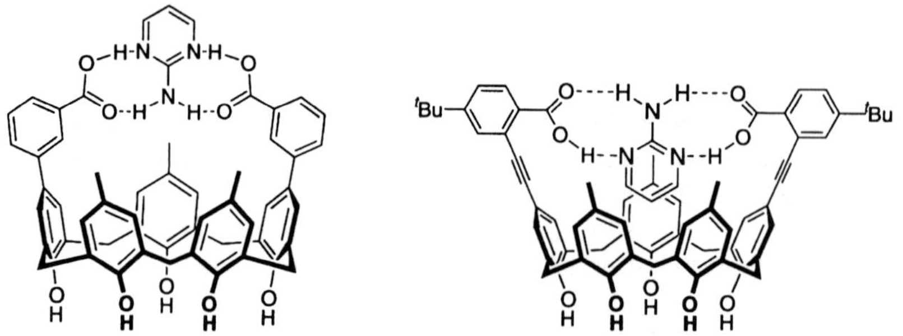

Figure 6. Outward *or* inward *binding mode of* **29** *to calix[5]arene receptors* **24a** *and* **24c**.

2-Aminopyrimidine **29** binds to hosts **24a** and **24b** with an *outward* orientation and to **24c** with an *inward* orientation. The affinity for **29** follows the order **24c** >> **24a** > **24b** (K_a values $(8.9\pm1.7)\times10^3$, $(8.7\pm1)\times10^2$, and $(3.1\pm0.3)\times10^2$ M^{-1}, respectively), emphasizing the importance of a preorganized cavity prior to complexation (cyclic array of intramolecular hydrogen bonding of phenolic hydroxyl groups, absent in **24b**), and the additional stabilization of the host-guest complex arising from cooperative van der Waals interactions (**24c** about 10 times more efficient than **24a**). Host **24a** binds 9-ethyl-adenine **30** ($K_a = (4.3\pm0.8)\times10^4$ M^{-1}) much more strongly than 2-amino pyrimidine **29** [12].

The large **30**/**29** selectivity (~ 50) shown by **24a** has been explained in terms of extra-stabilization of the complex, due to a strengthening of the cyclic array of intramolecular hydrogen bonding, through an induced-fit type movement of the two carboxyl groups in the guest binding process. In the **24a**·**29** complex the distance between the two -CO_2H groups is shorter than the corresponding distance in the **24a**·**30** complex, causing a weakening (breaking) of the hydrogen bonding pattern at the lower rim. Therefore, the shape selectivity is closely related to the extent of structural change in the host-guest complexes.

Imidazole binds to **24a** in a 2:1 ratio, the second guest molecule being bound much more strongly than the first one ($K_{a1} = 10^2$ and $K_{a2} = 4\times10^4$ M^{-1}) [12].

5. Host-Guest Complexes of Lower-Rim Functionalized Calix[5]arenes with Charged Guests

5.1. ALKALI METAL IONS

The alkali metal ion affinities of a number of penta-esters **8d** [21], penta-ketones **8e** [3b] and the trimethyl ethers of calix[5]crowns **9** [27b] have been assessed through phase-transfer experiments from water into CH_2Cl_2, and stability constant measurements in

MeOH. In general, ionophores **8d,e** are much more effective in extraction than their tetramer or hexamer counterparts, extraction favoring the larger cations without a particular preference. The efficiency of these ionophores mainly depends on the nature of the R^1 residue adjacent to the carbonyl group at the lower rim and, to a less extent, on the nature of the R substituent at the upper rim. The very high selectivities displayed by calix[4]arene crown ethers – determined by the size of the crown ether ring, the conformation and substitution patterns at their upper and lower rim [51] (See Chapter 20) – are lost with calix[5]crowns **9**, the highest Cs^+/Na^+ selectivity value being 630. The increased size of the annulus seems to be the main factor responsible for the low discriminating properties of **8d,e** and **9** toward alkali metal ions.

5.2. QUATERNARY AMMONIUM, PHOSPHONIUM AND IMINIUM IONS

The π-basic *cone* cavity of calix[5]crown triols **9** has proved to be a fairly efficient, but rather unselective, receptor site for a variety of quaternary ammonium (including acetylcholine), phosphonium, and iminium ions by exploiting cation–π interactions as the sole driving force for association [52]. Standard ^{1}H NMR titrations in $CDCl_3$ have evidenced fast complexation equilibria on the NMR time scale, with K_a values up to 210 M^{-1}. The calculated limiting upfield shifts ($\Delta\delta_\infty$) of the CH protons directly bound to the atom bearing the positive charge indicate that complexation involves a close contact between the polar head groups and the aromatic faces of the hosts. Whereas the length of the polyether bridge has little influence on the efficiency of the hosts, an adverse effect of the bulky *t*-butyl substituents on the upper rim is noticed. Host-guest associations are affected by the nature of the counterion and the polarity of the solvent, complex stability bearing as expected an inverse relationship to solvent polarity. In addition, up to five fold enhancements of K_a values are observed in TCE relative to $CDCl_3$. The large TCE molecule hardly fits into the calix cup, giving a poorly solvated binding site, which is more easily accessed by the guest.

5.3. PRIMARY ALKYLAMMONIUM IONS AND ω-AMINO ACID DERIVATIVES

1,3-Calix[5]crown-6 ethers are potentially heteroditopic receptors for alkylammonium ions, because they combine both a hydrophilic crown ether pocket at the lower rim and a preorganized electron-rich *cone* cavity on the opposite side. ^{1}H NMR titration experiments of C_s symmetric tri-ethers **27b–d** with 1 equiv. of the four isomeric butylammonium picrate salts, aimed at investigating the preferred binding site of these receptors, produced drastic spectral changes *only* in the case of n-$BuNH_3^+$ ions, indicating a selective recognition for the linear guest. The doubling of the peaks of the host and guest suggested the presence in solution of a host-guest complex in slow exchanging regime with the uncomplexed species on the NMR time scale. The *endo*-cavity nature of the complex is unambiguously supported by the dramatic upfield CIS ($\Delta\delta$ up to 3.87 ppm) of the protons of the included guest, which experience the diamagnetic shielding effect of the five aryl rings in a *cone* arrangement. The 1:1 stoichiometry of the complex and K_a have been determined by direct ^{1}H NMR analysis of equimolar host-guest solutions [28a, 53].

The unique ability of **27b–d** to discriminate the linear from branched primary alkylammonium ions, *via* formation of stable *endo*-cavity complexes, can be ascribed to a remarkable steric and electronic complementarity between the preorganized π-rich cavity of the calix[5]arene skeleton and the shape of the guest. In addition to CH–π and cation–π interactions [54], other non-covalent interactions such as hydrogen bonding formation between the cavity-included $-NH_3^+$ group and the ethereal oxygens of the host, may contribute to the stabilization of these inclusion complexes.

The selectivity observed likely depends on the presence of the *t*-butyl substituents at the upper rim, which sterically interfere with the branched alkyammonium guests. The site selectivity displayed by tri-ethers **27b–d** for the linear alkylammonium ions is somehow lost in the case of conformationally fixed *i*-propyl and *t*-butyl tri-esters **27f** and **27g**, which show with *n*-BuNH$_3^+$ ions two competitive modes of binding (Figure 7a), which can be unambiguously distinguished and monitored by ^{1}H NMR spectroscopy [29].

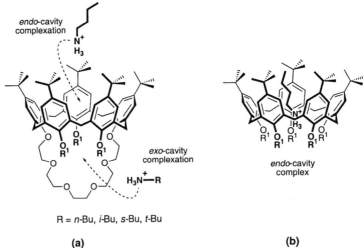

Figure 7. a) Dual binding mode of alkylammonium ions to p-tert-*butylcalix[5]crown-6* derivatives; b) geometry of the 1:1 endo-cavity complex between p-tert-*butylcalix[5]arene* derivatives and the n-BuNH$_3^+$ ion.

^{1}H NMR titration experiments of these receptors with *n*-BuNH$_3^+$ (up to 20 equiv.) feature two distinct sets of signals, none of which are compatible with the resonances of the free host. The one whose chemical shifts vary according to the amount of guest added identifies the host–*n*-BuNH$_3^+$ *exo*-cavity complex (fast exchanging regime).

In contrast, the chemical shifts of the second set of signals are guest concentration independent, and are consistent with the formation of an *endo*-cavity complex (slow exchanging regime [55]). Although both binding sites (crown ether moiety and calix[5]-arene cavity) actively recognize the *n*-BuNH$_3^+$ ion, the *endo*-complexation is favored, and its percentage increases with increasing amounts of salt. For instance, in the case of **27g** it levels off to about 80%, after 15 equiv. of *n*-BuNH$_3^+$ have been added. The ^{1}H NMR spectral profile of the titration experiment of tri-ester **27g** with *n*-BuNH$_3^+$ picrate is shown in Figure 8.

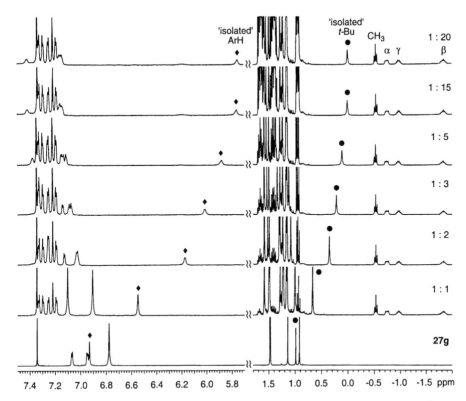

Figure 8. endo- *vs* exo-*Cavity complexation: selected regions (plotted on different scales) of the* ^{1}H *NMR (300 MHz; CDCl$_3$–CD$_3$OD, 9:1; 293 K) spectra of* **27g** *and increasing amounts of* n-BuNH$_3^+$ *picrate (1:1 to 1:20 host-guest ratios). The guest concentration dependent upfield drift of ArH (♦) and t-butyl (●) resonances of the 'isolated' aryl ring of the host identifies the* **27g**–n-BuNH$_3^+$ *exo-cavity complex (in fast exchange regime), while the invariant high field resonances (CH$_3$, α-, β-, and γ-CH$_2$) of the guest* n-*butyl chain show the presence of the* **27g**–n-BuNH$_3^+$ *endo-cavity complex (in slow exchange regime).*

For steric reasons, branched butylammonium ions can approach the crown ether binding site more easily and consequently form only *exo*-cavity complexes. In this case the host-guest recognition process presumably occurs *via* cooperative ion–dipole interactions and hydrogen bonding between the –NH$_3^+$ group of the guest and the oxygen atoms of the polyether chain [56], with a likely assistance of the carbonyl group(s) [57] of the lower rim alkoxycarbonylmethyl substituent(s).

Unlike the smaller calix[5]crown analogues, the inherently chiral 1,2-bridged calix[5]crown-6 derivative **11** (n = 2) [31] and the axially chiral 1,3-bridged *p-tert*-butylcalix[5]crown-6 triethers of **9** (incorporating a 1,1'-binaphthalene-2,2'-diyl subunit in the polyether chain [28]) also recognize and bind branched primary alkylammonium ions via their hydrophilic pocket. These results confirm that the presence of a six-oxygen pattern in the crown ether moiety is a prerequisite [56] for the *exo*-complexation of primary alkylammonium ions by calix[5]crowns, irrespective of the spanning (1,2- or 1,3-) and stiffness of the polyether chain. These chiral hosts do not seem to show enantioselective discrimination toward the enantiomerically pure (*R*)- or (*S*)-α-methylbenzyl-ammonium picrates, producing in both cases equal ratios of the corresponding 1:1 diastereomeric *exo*-cavity complexes.

The low association constants (K_a = 48–222 M^{-1} [28, 53]) observed for the selective inclusion of *n*-BuNH$_3^+$ ions by calix[5]crown-6 **27b–d** are probably related to the shape of the cavity of these molecules (C_s symmetry), to their tendency to fill the distorted *cone* cavity by canting the isolated *tert*-butylphenyl residue inward, and to the presence of a competitive recognition site represented by the the polyether chain at the lower rim.

More powerful receptors for the linear alkylammonium ions were obtained by switching to calix[5]arenes **26b,c** and **26h,i** in a fixed and quite regular C_{5v} *cone* conformation. As a result, these receptors bind linear RNH$_3^+$ ions much more efficiently and selectively than the neutral synthetic host systems reported to date [19]. The percentage of *endo*-cavity complex (Figure 7b) for each of the four butylammonium salts was determined by standard ^{1}H NMR analysis of equimolar solutions of host and guest. The hosts **26b,c** and **26h,i** show a remarkable affinity for the linear *n*-BuNH$_3^+$ (68-95%), and a much lower affinity for the branched *i*-BuNH$_3^+$ (6–38%), and *s*-BuNH$_3^+$ (5-28%) ions. No interaction could be detected with the bulky *t*-BuNH$_3^+$, even when a large excess of the salt was added. A direct NMR competition experiment has shown that when equimolar mixtures of the four butylammonium picrates are added to **26i**, only *n*-BuNH$_3^+$ (95%) and *i*-BuNH$_3^+$ (5%) undergo *endo*-complexation. Owing to the high degree of complexation for the *n*-BuNH$_3^+$ ion, relevant K_a values for **26b,c** and **26h,i** were determined in CDCl$_3$ saturated with water by using Cram's picrate extraction method [58]. High lgK_a values (4.63 < logK_a < 6.47) are always found with the *n*-BuNH$_3^+$ ion, while with the other branched ions and specifically with *t*-BuNH$_3^+$ they are significantly lower (lgK_a = 3.5). The most efficient calix[5]arenes **26c** and **26i** bind the *n*-BuNH$_3^+$ ion more selectively than the other isomeric ions. For instance, the *n*-BuNH$_3^+$/*t*-BuNH$_3^+$ selectivity of **26i** (~ 10^3) is two to three orders of magnitude higher than those reported for 18-crown-6 [59] and calix[6]arene hexaesters [57].

Receptor **26c** has found a useful application as a sensing agent in *n*-BuNH$_3^+$ liquid membrane ion selective electrodes (ISEs), in terms of both detection limit (3×10^{-6} M) and selectivity (*n*-BuNH$_3^+$ >> *i*-BuNH$_3^+$ > *s*-BuNH$_3^+$ > *t*-BuNH$_3^+$), due to the high degree of preorganization of its hydrophobic cavity and to the weak interaction of phenolic oxygens at the lower rim with interfering inorganic ions [60].

In addition to the size and shape complementarity between calix[5]arenes and linear alkylammonium ions, the electronic features of the host cavity also play a crucial role in the recognition process. Although *p*-NO$_2$-calix[5]arene **26f** possesses a regular *cone* conformation, no *endo*-cavity complexation occurs with alkylammonium ions, because of the electron withdrawing effect of the NO$_2$ groups at the upper rim, which strongly reduces the electron density of the cavity, and/or the tendency of phenolic oxygens to form hydrogen bonds with the polar head of the guest.

Since the structural motif of *n*-BuNH$_3^+$ is present in biologically important amines, the very powerful receptors **26c** and **26i** can also act as biomimetic host systems, forming 1:1 inclusion complexes with protected ω-amino acids (N^α-Ac-Lys-OMe·HCl), small peptides (Lys-Gly-OMe·2HCl), the methyl ester hydrochlorides of the neurotransmitter γ-aminobutyric acid, and the plasmin inhibitor ε-aminocaproic acid (**26c**: K_a = (8.3±0.5) ×10^3 M^{-1} in CDCl$_3$-CD$_3$OD 9:1) [13, 19] (See Chapter 27).

6. References and Notes

[1] C. D. Gutsche, L. J. Bauer, *J. Am. Chem. Soc.* **1985**, *107*, 6052–6059.
[2] D. R. Stewart, M. Krawiec, R. P. Kashyap, W. H. Watson, C. D. Gutsche, *J. Am. Chem. Soc.* **1995**, *117*, 586–601.
[3] a) G. Ferguson, A. Notti, S. Pappalardo, M. F. Parisi, A. L. Spek, *Tetrahedron Lett.* **1998**, *39*, 1965–1968; b) S. E. J. Bell, J. K. Browne, V. McKee, M. A. McKervey, J. F. Malone, M. O'Leary, A. Walker, *J. Org. Chem.* **1998**, *63*, 489–501.
[4] a) C. D. Gutsche, *Calixarenes Revisited in Monographs in Supramolecular Chemistry, Vol. 6* (Ed. J. F. Stoddart), The Royal Society of Chemistry: Cambridge, **1998**; b) V. Böhmer, *Angew. Chem., Int. Ed. Engl.* **1995**, *34*, 713–745.
[5] a) A. Nigakawa, H. Matsuda, *Makromol. Chem., Rapid Commun.* **1982**, *3*, 65–67; b) M. A. Markowitz, V. Janout, D. G. Castner, S. L. Regen, *J. Am. Chem. Soc.* **1989**, *111*, 8192–8200.
[6] a) D. R. Stewart, C. D. Gutsche, *Org. Prep. Proc. Int.* **1993**, *25*, 137–139; b) K. Iwamoto, K. Araki, S. Shinkai, *Bull. Chem. Soc. Jpn.* **1994**, *67*, 1499–1502.
[7] a) B. Souley, Z. Asfari, J. Vicens, *Polish J. Chem.* **1992**, *66*, 959–961; b) J. L. Atwood, L. J. Barbour, P. J. Nichols, C. L. Raston, C. A. Sandoval, *Chem. Eur. J.* **1999**, *5*, 990–996.
[8] H. Kammerer, G. Happel, B. Mathiasch, *Makromol. Chem.* **1981**, *182*, 1685–1694.
[9] S. Usui, K. Deyama, R. Kinoshita, Y. Odagaki, Y. Fukazawa, *Tetrahedron Lett.* **1993**, *34*, 8127–8130.
[10] M. Yanase, T. Haino, Y. Fukazawa, *Tetrahedron Lett.* **1999**, *40*, 2781–2784.
[11] K. No, K. M. Kwon, *Synthesis* **1996**, 1293–1295.
[12] a) T. Haino, T. Harano, K. Matsumura, Y. Fukazawa, *Tetrahedron Lett.* **1995**, *36*, 5793–5796; b) T. Haino, K. Matsumura, T. Harano, K. Yamada, Y. Saijyo, Y. Fukazawa, *Tetrahedron* **1998**, *54*, 12185–12196.
[13] A. Notti, S. Pappalardo, M. F. Parisi, *The 5th International Conference on Calixarene Chemistry*, Perth 19–23 Sept 1999, Book of Abstracts, L-28.
[14] C. Schmidt, M. Kumar, W. Vogt, V. Böhmer, *Tetrahedron* **1999**, *55*, 7819–7828.
[15] S. E. Biali, V. Böhmer, I. Columbus, G. Ferguson, C. Grüttner, F. Grynszpan, E. F. Paulus, I. Thondorf *J. Chem. Soc., Perkin Trans. 2*, **1998**, 2261–2269.
[16] M. Bergamaschi, F. Bigi, M. Lanfranchi, R. Maggi, A. Pastorio, M. A. Pellinghelli, F. Peri, C. Porta, G. Sartori, *Tetrahedron* **1997**, *38*, 13037–13052.
[17] R. A. Jakobi, V. Böhmer, C. Grüttner, D. Kraft, W. Vogt, *New J. Chem.* **1996**, *20*, 493–501.
[18] B. Souley, Z. Asfari, J. Vicens, *Polish J. Chem.* **1993**, *67*, 763–767.
[19] F. Arnaud-Neu, S. Fuangswasdi, A. Notti, S. Pappalardo, M. F. Parisi, *Angew. Chem., Int. Ed. Engl.* **1998**, *37*, 112–114.
[20] S. Pappalardo, G. Ferguson, *J. Org. Chem.* **1996**, *61*, 2407–2412.
[21] G. Barrett, M. A. McKervey, J. F. Malone, A. Walker, F. Arnaud-Neu, L. Guerra, M.-J. Schwing-Weill, *J. Chem. Soc., Perkin Trans. 2* **1993**, 1475–1479.
[22] F. Arnaud-Neu, G. Barrett, S. Corry, S. Cremin, G. Ferguson, J. F. Gallagher, S. J. Harris, M. A. McKervey, M.-J. Schwing-Weill, *J. Chem. Soc., Perkin Trans. 2* **1997**, 575–579.
[23] P. A. Gale, Z. Chen, M. G. B. Drew, J. A. Hearth, P. D. Beer, *Polyhedron* **1998**, *17*, 405–412.
[24] P. A. Gale, J. W. Genge, V. Král, M. A. McKervey, J. L. Sessler, A. Walker, *Tetrahedron Lett.* **1997**, *38*, 8443–8444.
[25] M. Fan, H. Zhang, M. Lattman, *Chem. Commun.* **1998**, 99–100.
[26] D. Kraft, V. Böhmer, W. Vogt, G. Ferguson, J. F. Gallagher, *J. Chem. Soc., Perkin Trans. 1* **1994**, 1221–1229.
[27] a) D. Kraft, R. Arnecke, V. Böhmer, W. Vogt, *Tetrahedron* **1993**, *49*, 6019–6024; b) F. Arnaud-Neu, R. Arnecke, V. Böhmer, S. Fanni, J. L. M. Gordon, M.-J. Schwing-Weill, W. Vogt, *J. Chem. Soc., Perkin Trans. 2* **1996**, 1855–1860.
[28] a) S. Pappalardo, M. F. Parisi, *J. Org. Chem.* **1996**, *61*, 8724–8725; b) S. Caccamese, A. Notti, S. Pappalardo, M. F. Parisi, G. Principato, *J. Incl. Phenom.* **2000**, *36*, 67–78.
[29] M. Gattuso, A. Notti, S. Pappalardo, M. F. Parisi, *Tetrahedron Lett.* **1998**, *39*, 1969–1972.
[30] R. Arnecke, V. Böhmer, G. Ferguson, S. Pappalardo, *Tetrahedron Lett.* **1996**, *37*, 1497–1500.
[31] S. Caccamese, A. Notti, S. Pappalardo, M. F. Parisi, G. Principato, *Tetrahedron* **1999**, *55*, 5505–5514.
[32] For a review on inherently chiral calixarenes see: V. Böhmer, D. Kraft, M. Tabatabai, *J. Incl. Phenom.* **1994**, *19*, 17–39.

[33] R. Cacciapaglia, L. Mandolini, R. Arnecke, V. Böhmer, W. Vogt, *J. Chem. Soc., Perkin Trans. 2* **1998**, 419–423.
[34] C. D. Gutsche, *Tetrahedron* **1986**, *42*, 1633–1640.
[35] S. Shinkai, H. Koreishi, K. Ueda, T. Arimura, O. Manabe, *J. Am. Chem. Soc.* **1987**, *109*, 6371–6376.
[36] C. D. Gutsche, I. Alam, *Tetrahedron* **1988**, *44*, 4689–4694.
[37] P. Dedek, V. Janout, S. L. Regen, *J. Org. Chem.* **1993**, *58*, 6553–6555.
[38] J. L. M. Gordon, V. Böhmer, W. Vogt, *Tetrahedron Lett.* **1995**, *36*, 2445–2448. S. Bouoit-Montesinos, J. Bassus, M. Perrin, R. Lamartine, *Tetrahedron Lett.* **2000**, *41*, 2563–2567.
[39] J. W. Steed, C. P. Johnson, C. L. Barnes, R. K. Juneja, J. L. Atwood, S. Reilly, R. L. Hollis, P. H. Smith, D. L. Clark, *J. Am. Chem. Soc.* **1995**, *117*, 11426–11433.
[40] F. Arnaud-Neu, V. Böhmer, J.-F. Dozol, C. Grüttner, R. A. Jakobi, D. Kraft, O. Mauprivez, H. Rouquette, M.-J. Schwing-Weill, N. Simon, W. Vogt, *J. Chem. Soc., Perkin Trans. 2* **1996**, 1175–1182.
[41] C. D. Gutsche, C. Gibbs, J. Wang, D. Xie, *The 5th International Conference on Calixarene Chemistry*, Perth 19–23 Sept 1999, Book of Abstracts, L-1; for details on the calix[4]arene system, see C. D. Gutsche, J. A. Levine, *J. Am. Chem. Soc.* **1982**, *104*, 2652–2653.
[42] T. Haino, K. Nitta, Y. Saijo, K. Matzumura, M. Hirakata, Y. Fukazawa, *Tetrahedron Lett.* **1999**, *40*, 6301–6304.
[43] T. Haino, M. Yanase, Y. Fukazawa, *Angew. Chem., Int. Ed. Engl.* **1998**, *37*, 997–998.
[44] A. Notti, S. Pappalardo, M. F. Parisi, submitted.
[45] Small amounts of immobile non-cone conformers of penta-esters **26** can be formed by esterification of penta-acetic acid **26** (R = *t*-Bu, R¹ = CH₂CO₂H). with a suitable alcohol and a trace of acid: A. Notti, S. Pappalardo, M. F. Parisi, unpublished observations.
[46] F. Diederich, K. Dick, *J. Am. Chem. Soc.* **1984**, *106*, 8024–8036.
[47] N. S. Isaacs, P. J. Nichols, C. L. Raston, C. A. Sandova, D. J. Young, *Chem. Commun.* **1997**, 1839–1840.
[48] T. Haino, M. Yanase, Y. Fukazawa, *Angew. Chem., Int. Ed. Engl.* **1997**, *36*, 259–260.
[49] T. Haino, M. Yanase, Y. Fukazawa, *Tetrahedron Lett.* **1997**, *38*, 3739–3742.
[50] a) K. A. Connors, *Binding Constants*, John Wiley & Son, New York, **1987**; b) D. J. Leggett, *Modern Inorganic Chemistry Series, Computational Methods for the Determination of Formation Constants*, Plenum Press, New York and London, **1985**.
[51] For leading references, see: a) H. Yamamoto, S. Shinkai, *Chem. Lett.* **1994**, 1115–1118; b) R. Ungaro, A. Casnati, F. Ugozzoli, A. Pochini, J.-F. Dozol, C. Hill, H. Rouquette, *Angew. Chem., Int. Ed. Engl.* **1994**, *33*, 1506–1509; c) A. Casnati, A. Pochini, R. Ungaro, C. Bocchi, F. Ugozzoli, R. J. M. Egberink, H. Struijk, R. Lugtenberg, F. de Jong, D. N. Reinhoudt, *Chem. Eur. J.* **1996**, *2*, 436–445; d) F. Arnaud-Neu, G. Ferguson, S. Fuangswasdi, A. Notti, S. Pappalardo, M. F. Parisi, A. Petringa, *J. Org. Chem.* **1998**, *63*, 7770–7779.
[52] R. Arnecke, V. Böhmer, R. Cacciapaglia, A. Dalla Cort, L. Mandolini, *Tetrahedron* **1997**, *53*, 4901–4908.
[53] G. Ferguson, A. Notti, S, Pappalardo, M. F. Parisi, A. L. Spek, *Proceedings of The 4th Internatinal Conference on Calixarenes*, Parma Aug 31–Sept 4, 1997, P35.
[54] J. C. Ma, D. A. Dougherty, *Chem. Rev.* **1997**, *97*, 1303–1324.
[55] Inclusion complexes between organic ammonium ions and calixarene systems in a slow exchanging regime have been reported for a calix[6]arene hexaester: K. Odashima, K. Yagi, K. Tohda, Y. Umezawa, *Anal. Chem.* **1993**, *65*, 1074–1083.
[56] a) J.-M. Lehn, *Angew. Chem., Int. Ed. Engl.* **1988**, *27*, 89–112; b) D. J. Cram, *Angew. Chem., Int. Ed. Engl.* **1988**, *27*, 1009–1020.
[57] S.-Y. Han, M.-H. Kang, Y.-E. Jung, S.-K. Chang, *J. Chem. Soc., Perkin Trans. 2* **1994**, 835–839.
[58] G. M. Lein, D. J. Cram, *J. Am. Chem. Soc.* **1985**, *107*, 448–455.
[59] S.-K. Chang, M. J. Jang, S.-Y. Han, *Chem. Lett.* **1992**, 1937–1940.
[60] M. Giannetto, G. Mori, A. Notti, S. Pappalardo, M. F. Parisi, *Anal. Chem.* **1998**, *70*, 4631–4635.

Chapter 4

SELECTIVE MODIFICATIONS OF CALIX[6]ARENES

ULRICH LÜNING, FRANK LÖFFLER, JAN EGGERT

Institut für Organische Chemie, Christian-Albrechts-Universität zu Kiel, Olshausenstr. 40, D-24098 Kiel, Germany.
e-mail: luening@oc.uni-kiel.de

Although efficient syntheses of calix[4]arene [1], calix[6]arene [2], and calix[8]-arene [3] have all long been known, studies of the functionalisation of calix[6]arene have until now been relatively limited [4]. This chapter surveys methods of functionalisation by the selective modification of the preformed macrocycle, the control of substitution by stepwise synthesis involving fragment condensation being far less practicable than with smaller calixarenes.

1. Regioselectively Functionalised Calix[6]arenes

Many hexa-ethers and -esters of calix[6]arene **1** have been obtained by exhaustive alkylation and acylation of the deprotonated calixarene [4]. If fewer than 5 phenol (and more than 1) groups react, regioisomers are possible and, distinguishing the phenolic rings by the letters A-F [5], the twelve substitution products may be designated:

monosubstitution:	A;
disubstitution:	A,B; A,C; A,D;
trisubstitution:	A,B,C; A,B,D; A,C,E;
tetrasubstitution:	A,B,C,D; A,B,C,E; A,B,D,E;
pentasubstitution:	A,B,C,D,E (= A-E);
and hexasubstitution:	A,B,C,D,E,F (= A-F).

While most partially alkylated species **2a** and **3a** can be obtained by direct synthesis [6-10], obtaining some regioisomers in acceptable yields requires two-step methods involving dealkylation or protection/deprotection. Exhaustive functionalisation of their residual phenolic groups can be used to convert them to regioisomers of mixed peralkyl calix[6]arenes. All 12 methyl ethers **3a** have been prepared [8] and converted [9] to the corresponding mixed methyl/ethoxycarbonylmethyl ethers **4a**. Ten of the possible 2'-pyridylmethyl ether isomers **2a** are known [10] and many particular isomers in the series **2a-6a** can be isolated in good yields (Table 1), though improved yields are desirable for species such as A,C-**3a** and A-E-**3a**.

Different alkylation conditions allow some of the species listed in Table 1 to be converted to new regioisomers. A,B,C-**2a**, for example, can be alkylated once more in DMF with NaH as the base, giving A,B,C,D-**2a** in 35 % yield [10]. Analogously, A,C,E-**3a** reacts in 39 % yield with MeI and K_2CO_3 to give A,B,C,E-**3a** [8].

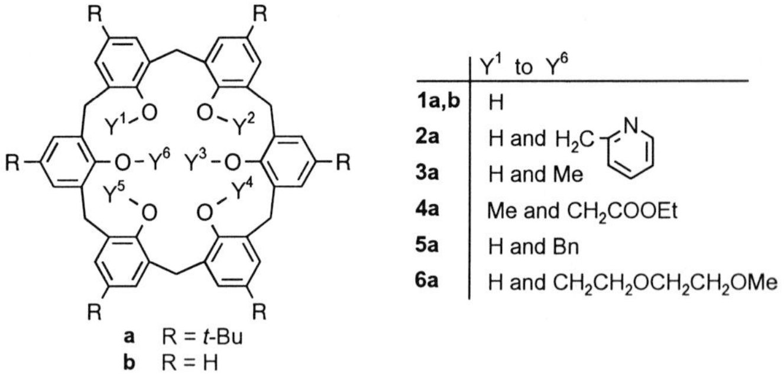

	Y^1 to Y^6
1a,b	H
2a	H and H_2C-pyridyl
3a	H and Me
4a	Me and CH_2COOEt
5a	H and Bn
6a	H and $CH_2CH_2OCH_2CH_2OMe$

a R = t-Bu
b R = H

TABLE 1. Some regioselectively alkylated calix[6]arenes obtained by direct alkylation of **1a**.

Regioisomer	Yield [%]	Reaction conditions[a]	equiv. base	equiv. YX[b]	Ref.
A-**3a**	82	acetone, 70 °C, 2 bar	1.1 K_2CO_3	1.1 MeI	[7]
A-**3a**	85	THF,))))	1.9 KH	15.5 MeI	[7]
A-**5a**	75-81	acetone, reflux	1.1 K_2CO_3	1.1 BnCl	[7]
A,B-**3a**	81	THF,))))	3.1 KH	20 Me_2SO_4	[7]
A,C-**3a**	26	DMF, 40 °C	3 CsF	10 MeI	[7]
A,D-**2a**	80[c]	DMF, 60 °C	4 BaO 2 $Ba(OH)_2$	2 PicCl • HCl	[10]
A,B,C-**2a**	51	DMF, 60 °C	8 K_2CO_3	4 PicCl • HCl	[10]
A,C,E-**3a**	72	acetone, 70 °C, 2 bar	3 K_2CO_3	4 MeI	[7]
A,B,D,E-**6a**	80	THF, reflux	7.7 NaH	7 MeO~O~OTs	[7]
A,B,D,E-**2a**	61	DMF, 60 °C	75 NaH	30 PicCl • HCl	[11]
A-E-**3a**[d]	15	acetone, 70 °C, 2 bar	4 K_2CO_3	5 MeI	[7]
A-F-**2a**	81	DMF, 70 °C	56 K_2CO_3	28 PicCl • HCl	[10]
A-F-**3a**	99	THF,))))	6 NaH	6 Me_2SO_4	[7]

[a] Room temp. if not stated otherwise. [b] PicCl = 2-(chloromethyl)-pyridine. [c] Barium complex. [d] A-E-**3a** can be obtained in 76 % yield from **1a** by mono-benzylation, subsequent exhaustive methylation and reductive cleavage of the benzyl ether [12].

The A,B,C,D-pattern also results using a protection/deprotection technique. Permethylation of A,B o-xylylene-bridged **13c** ([8]; Section 2.1.1.), followed by bridge cleavage with Me_3SiBr provides A,B,C,D-**3a** in 17 % yield. Use of a m-xylylene unit to achieve A,D-bridging (Section 2.1.3.) initially led by chance to A,B,D-**3a** (**7**) in that reaction with Me_3SiBr resulted to a reproducible extent in the loss of one methyl group as well as the bridge, though A,B,D,E-**3a** was still the major product [8].

Various systems such as the alkylation products from **1b** [7,13] and the phosphorylation products of **1a** [14] have been largely characterised, while some particular isomers, such as the B,D,F-N-methylimidazolylmethyl derivative of A,C,E-**3a** have been prepared for special uses, here as a biomimetic Zn(II)-binding ligand [15].

In addition to such narrow rim substitution, wide rim reactions, typically involving debutylation of **1a** followed by electrophilic attack [4], are well known [16]. Selective methylation of the narrow rim can be transferred to the wide through controlled debutylation [12,17,18]. A,C,E-**8a**, for instance, gives B,D,F-debutylated **8b** in 32 % yield and, after O-methylation to **9** (60 %), this can be chloromethylated to the wide rim

A,C,E-derivative **10** (87 %). (For further reactions of **10**, see Section 2.2.) Similarly, A,B,D,E-**3a** can be selectively debutylated then reacted to give wide rim disubstituted derivatives [12,18].

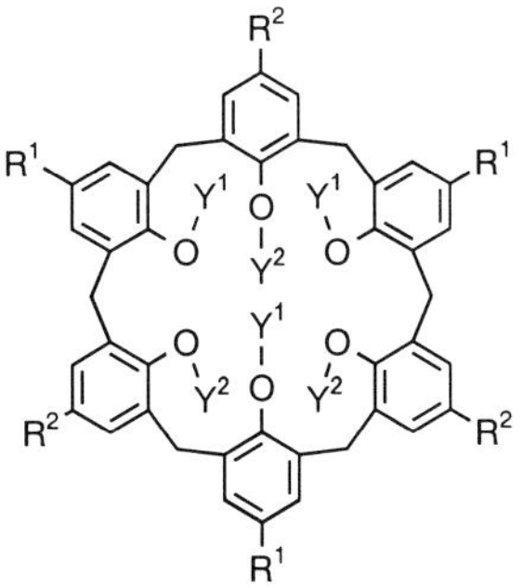

	R^1	Y^1	R^2	Y^2
7	t-Bu	Me	t-Bu	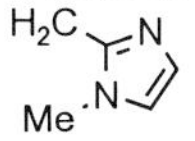 H_2C-N=N-Me (methylimidazole)
8a	t-Bu	Me	t-Bu	H
8b	t-Bu	Me	H	H
9	t-Bu	Me	H	Me
10	t-Bu	Me	CH_2Cl	Me
11a	t-Bu	Me	$NH-CO-NH_2$	$n-C_8H_{17}$
11b	t-Bu	Me	$NH-CO-NH_2$	CH_2CH_2OEt
11c	t-Bu	Me	$NH-CO-NH_2$	CH_2CONEt_2

Selective A,C,E-trisubstitution with urea groups at the wide rim of a calix[6]arene gives calixarenes **11** which form dimers in solution through multiple hydrogen bonds [19]. These are examples of calixarene aggregates which are discussed in detail in Chapter 8. More sophisticated examples of the control of calix[6]arene functionalisation are provided in the syntheses [16] of the calixquinones **12** by Tl(III) oxidation of partly debutylated species obtained from partially methylated or benzoylated forms of **1a**, and in the final wide rim aminomethylation of an A,D-wide rim allylated calix[6]arene formed via Claisen rearrangement of the corresponding narrow rim diether [13].

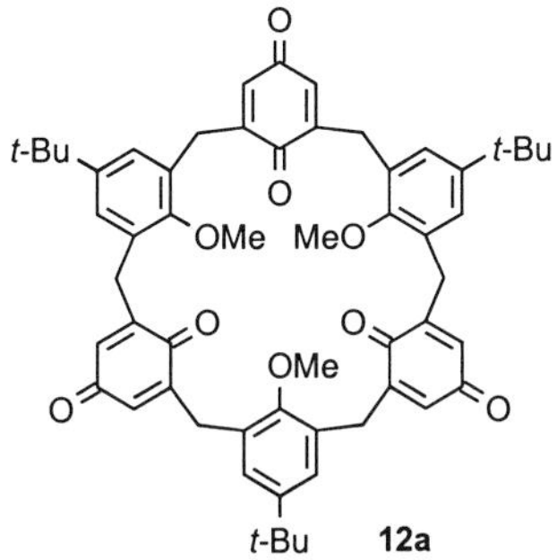

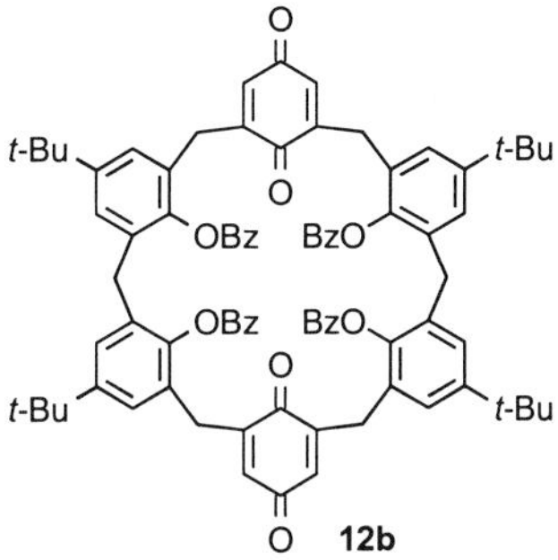

2. Bridged Calix[6]arenes

Reactions of calix[6]arenes with multi-electrophiles can involve both intra- and intermolecular reactions and only the processes leading to bridged, single calixarenes are considered here, the formation of multiple calixarenes in such ways being discussed in Chapter 7. As with simple partial alkylation, intramolecular bridging can occur in several ways.

2.1. SINGLY BRIDGED CALIX[6]ARENES

2.1.1. A,B-Bridged calix[6]arenes

A,B-Bridging of calix[6]arenes has been observed in at best moderate yields with short spanning groups such as methylene, ethylene, o-disubstituted benzene or phosphate. Thus, reaction of **1a** with BrCH$_2$Cl and Cs$_2$CO$_3$ affords 15 % of A,B-methylene-bridged **13a** [7], while reaction with ethylene glycol ditosylate and varying excesses of K$_2$CO$_3$ gives 13-17 % of **13b** [20], and the o-xylylene- [8] and o-phthaloyl-bridged [21] species **13c** and **13d** can be obtained in 47 and 45 % yields, respectively, through appropriate alkylation and acylation reactions.

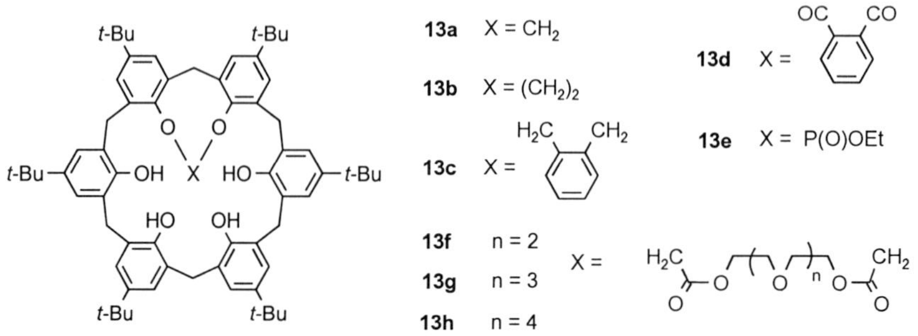

Reaction of **1a** with ethyl dichlorophosphate gives **13e** in low yield [22] but very recently the use of K$^+$ as a template ion, perhaps as in **14**, has been claimed to give satisfactory (up to 30 %) yields of A,B-bridged species involving long chains [23].

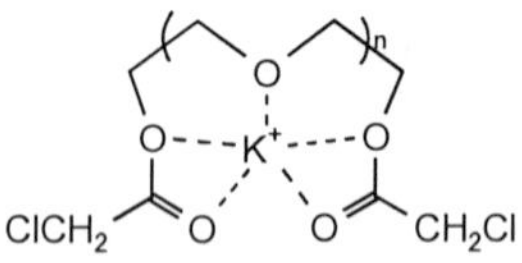

2.1.2. A,C-Bridged calix[6]arenes

Separation of electrophilic sites by more than one or two atoms should favour A,C- and A,D-bridging but known occurrences of the former are rare. The first reported case [24] was that of the formation of **15** in 24 % yield by reaction of **1a** with dipicolinoyl dichloride in the presence of KOt-Bu [24]. Better yields are obtained in the reactions with diethylene glycol ditosylate/Na$_2$CO$_3$ [25] to give **16** (41 %) and with 1,2-bis(chloroacetyl-amino)ethane/ K$_2$CO$_3$/KI [26] to give **17** (35 %).

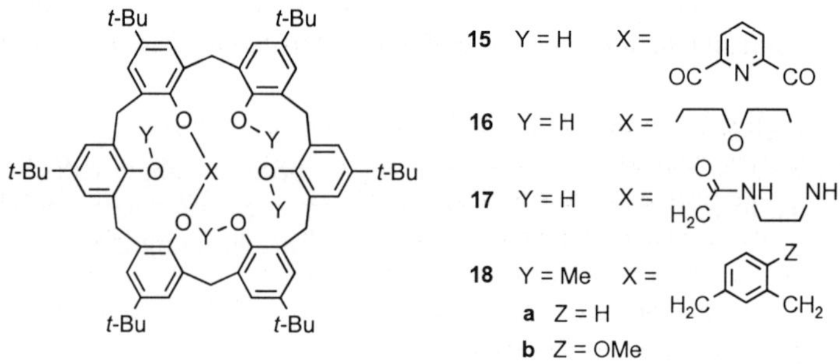

Partial methylation has been used to control sites available for bridging and D,F-bridged **18a** forms in 93 % yield when A,B,C,E-**3a** is reacted with 1,3-bis(bromomethyl)-benzene/Cs_2CO_3 [27]. The analogous chiral bicycle **18b**, stable to inversion at temperatures up to 100 °C, can be obtained with similar efficiency [28].

2.1.3. A,D-Bridged calix[6]arenes

A,D-Bridging of a calix[6]arene is easiest and has long been known [29,30]. Calix[6]-arenes **19** bridged by aliphatic diacids form in 28-55 % yield when A,D-bis(p-tolyl-methyl)ethers of calix[6]arene react with suberoyl, pimeloyl, adipoyl or succinoyl chlorides in the presence of triethylamine.

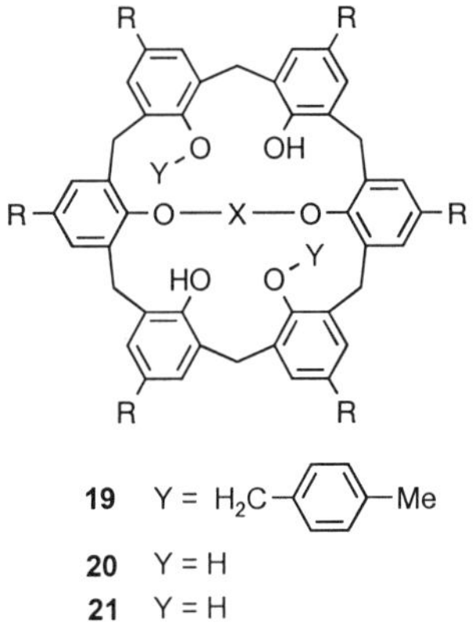

	X	R	yield [%]
19a	$OC-(CH_2)_2-CO$	t-Bu	55
19b	$OC-(CH_2)_4-CO$	t-Bu	37
19c	$OC-(CH_2)_5-CO$	t-Bu	33
19d	$OC-(CH_2)_6-CO$	t-Bu	28
20	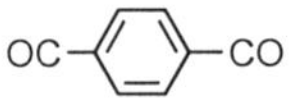 OC-⟨C6H4⟩-CO	t-Bu	42
21a	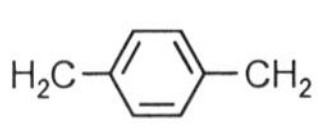 H_2C-⟨C6H4⟩-CH_2	t-Bu	75 (crude)
21b	H_2C-⟨C6H4⟩-CH_2	H	74 (crude)
21c	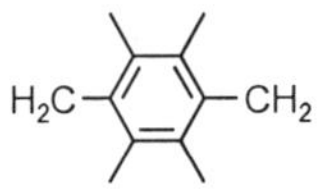 H_2C-⟨durene⟩-CH_2	t-Bu	73 (crude)
21d	H_2C-⟨durene⟩-CH_2	H	96 (crude)
21e	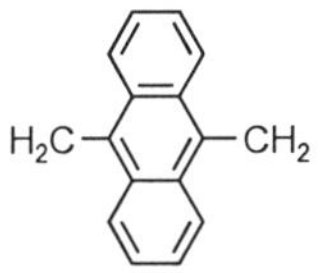 H_2C-⟨anthracene⟩-CH_2	t-Bu	40

19 Y = H_2C-⟨C6H4⟩-Me
20 Y = H
21 Y = H

The reason for the easy A,D-bridging of calix[6]arenes is probably their high flexibility. Thus, even rigid terephthaloyl chloride forms the A,D-bridged **20** in 42 % yield when triethylamine is used as a base [29] while its reactions with calix[4]arene are exclusively intermolecular. Diether bridges form as readily as diester and both **1a** and **1b** react with bis(halomethyl)arenes to give p-xylylene or bismethylenedurylene bridged calix[6]arenes **21a-d** in crude yields above 73 %. With very long reaction times, even the related 9,10-bismethylene-anthracenyl bridged calix[6]arene **21e** is accessible in 40 % yield.

Easy A,D-bridging by m-bis(bromomethyl)benzenes/pyridines is well established also [8,24]. Compounds such as **22** and **23** have been closely studied since the 2'-position of the m-xylylene bridge is protected by the calixarene. This can be exploited for the construction of new concave reagents [31] with centres, as in the case of pyridines **24**, of tunable basicity, or to increase the stability of a normally unstable functional group [32]. Table 2 summarises data concerning syntheses of **22** and **23**, **24** were obtained in 51-55 % yield using NaH as base [24].

TABLE 2. 2'-substituted A,D-m-xylylene bridged calix[6]arenes **22** and **23**, reaction conditions and yields.

	Z	Y	Base	Yield [%]	Ref.
22a	S-t-Bu	H	KOH	92	[32]
22b	S-t-Bu	Me			
22c	S(O)-t-Bu	Me	a)	81[a)]	[32]
22d	SOH	Me	b)	76-97[b)]	[32]
22e	S(O)-CH=CH-COOMe	Me	c)	86[c)]	[32]
23a	H	H	NaH	71	[8]
23b	Cl	H	NaH	77	[33]
23c	Br	H	KOH	73	[34]
23d	I	H	NaH	73	[33]
23e	CN	H	NaH	80	[33]
23f	COOMe	H	NaH	57	[33]
23g	NO_2	H	NaH	64	[35]
23h	NH_2	H	d)	67[d)]	[35]
23i	N_3	H	NaH	86	[45]
23j[e)]	OMe	H	KOH	73	[39]
23k	Se-n-Bu (all-up)	H	KOH	80	[46]
23l	Se-n-Bu (all-up)	Me			
23m	Se-n-Bu (all-up)	Bn			
23n	SeOH	Me	f)	55[f)]	[46]
23o	SeOH (all-up)	Bn	g)	74[g)]	[43]
23p	SeOH (uudddu)	Bn	g)	44[g)]	[43]
23q	SeO_2H (all-up)	Bn	h)	88[h)]	[43]
23r	SeO_2H (uudddu)	Bn	h)	94[h)]	[43]
23s	Se-S-n-Bu (all-up)	Bn	i)	78[i)]	[43]

22 R = H
23 R = t-Bu

[a)] From **22b**. [b)] From **22c**. [c)] From **22d**. [d)] By reduction of **23g** with Fe(CO)$_5$/NaOH. [e)] The bridge is p-bromo-substituted. [f)] From **23l**. [g)] From **23m** by oxidation with MCPBA. [h)] From **23o** and **23p**, respectively, by further oxidation with MCPBA. [i)] From **23o** by reaction with n-BuSH.

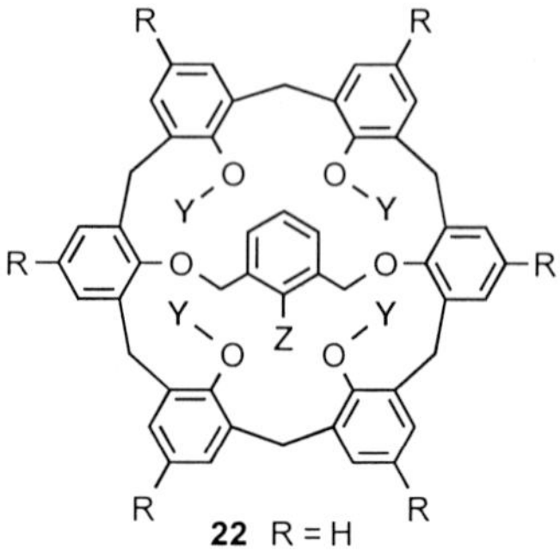

A,D-polyether bridged calix[6]arenes **25** form in 36-42 % yield from **1** and tetra-ethylene glycol ditosylate in the presence of KOt-Bu [36] and **26** is a by-product in the synthesis of the A,B,D-bridged calix[6]arene **42** [37], again indicating the ease with which A,D-bridges are formed first.

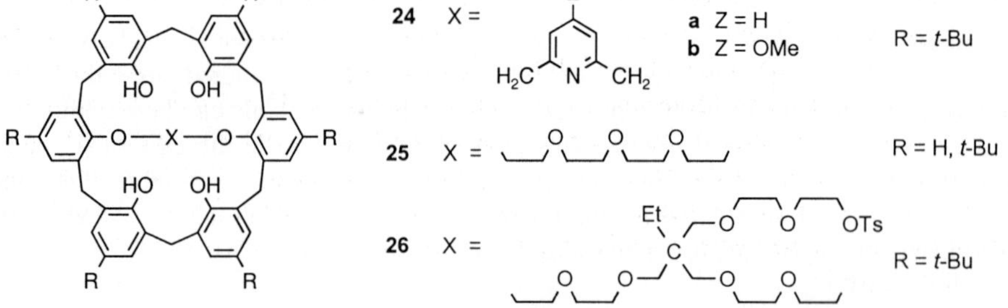

It is important to note, however, that unexpected bridging patterns have been observed in reactions of **1** with flexible oligoethylene ditosylates (Table 3) [20]. The product distribution depends on type and excess of the base as well as the solvent. Remarkable is the formation of the A,D-diethylene glycol bridged **29c**, which is strained and adopts a very flattened conformation in which the A- and D-rings are forced into the plane of the macrocycle by the short connecting chain. Attempts to further functionalise **29c** led to decomposition.

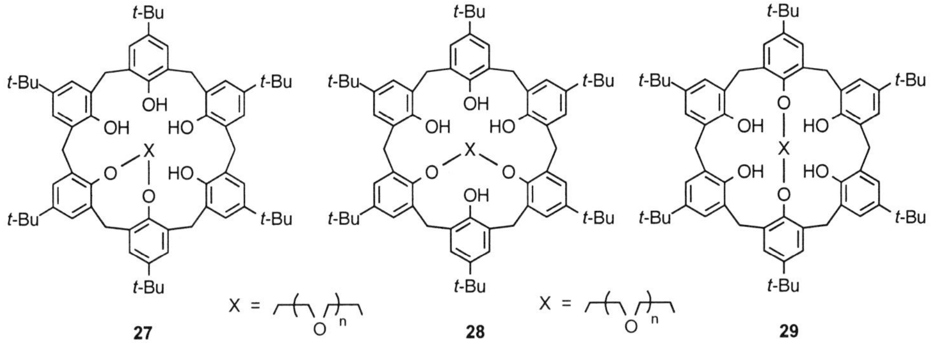

TABLE 3. Bridging modes of oligoethylene glycol ditosylates (mono to tetra, n = 0 to 3) and reaction conditions.

	n	Base	equiv. of base	AB (**27**)	AC (**28**)	AD (**29**)	Ref.
a	0	K_2CO_3	8-10	35 %	-	-	[20]
b	1	K_2CO_3	8-10	-	13%	-	[20]
c	1	K_2CO_3	40-50	-	17 % (+ 4%)[a]	9 %	[20]
d	1	Na_2CO_3	27-150	-	≤ 41 %	-	[25]
e	2	K_2CO_3	8-10	-	11 %	20 %[b]	[20]
f	(2)[c]	K_2CO_3	8-10	-	-	20 %[c,d]	[20]
g	3	K_2CO_3	8-10	3 %	19 %	-	[20]
h	3	KOt-Bu		-	-	36 %	[36]
i	3[e]	KOt-Bu		-	-	42 %	[36]

[a] As a partially alkylated derivative. [b] In toluene, **29e** is obtained in 40 % yield [57]. [c] 1,2-Bis(ethoxy)-benzene bridged derivative. [d] In toluene, **29f** is obtained in 55 % yield [57]. [e] Results for the bridging of calix[6]arene **1b**.

Hydrogen-bonding between residual phenolic groups and the ether oxygen atoms in A,D-bridged calix[6]arenes favours an "all-up" conformation though, as shown by variable-temperature NMR studies [33,34,41], not necessarily one in which the four phenolic groups are equivalent. Molecular mechanics calculations [33,34] indicate that a bent conformation (see **38** ahead) should be adopted by the bridge, and activation parameters have been determined for the process of its flipping between equivalent bent forms. Calculations [33] also indicate that in addition to symmetric *winged* and *pinched* conformations of the calixarene cycle, a *half-winged, half-pinched* form as shown below in structure **30**, is possible.

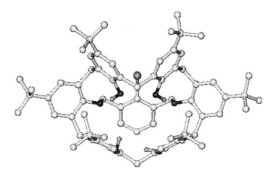

Figure 1. The 2'-bromoxylylene bridged calix[6]arene 23c adopts a half-winged/half-pinched conformation 30, with the pinched methylene group being next to the bromo substituent.

When the remaining phenols are alkylated the products become more flexible and a variety of conformations are found. Table 4 summarises a large number of per-alkylated calix[6]arenes and synthetic details for their syntheses.

Methylation of the phenolic groups lessens the stability of the all-*up* conformer but with increasing size of the substituent Y, conformer interconversion is slowed [41]. Benzylation allows all-*up* and *uudddu* conformers to be separated [42,43]. In the *uudddu* conformer, the bridge fills the cavity of the calix[6]arene, forming a " self-anchored rotaxane" [29]. X-ray-structures of *uudddu* conformers are known for the tetramethylated 2'-bromoxylylene calix[6]arene 31e [34] and for the analogous 2'-sulfenic acid 31c [32].

All-*up* 31i, bearing four positive charges, forms micelles at room temperature [44] and the tetraesters of the crown calix[6]arenes 33 have been investigated as metal ion complexants. 33a is sodium selective ($Li^+/Na^+/K^+$ = 1.4/36/<1), and 33d is lithium selective ($Li^+/Na^+/K^+$ = 35/1.3/<1) [40].

TABLE 4. Reagents and yields for the functionalisation of A,D-bridged calix[6]arenes at the free phenolic OH groups with different alkylating agents. For comparison related tetraethers are also listed.

Product (starting material in parentheses)		Base	Leaving group	Y	Yield [%]	Ref.
31a	(22a)	KH	I	Me	82	[32]
31b	(31a)	a)	a)	Me	81[a)]	[32]
31c	(31b)	b)	b)	Me	97[b)]	[32]
31c	(31b)	c)	c)	Me	76[c)]	[32]
31d	(31c)	d)	d)	Me	86[d)]	[32]
31e	(23c)	KH	I	Me	55	[34]
31f	(23c)	NaH	Br	CH_2Ph	58[e)] all-*up* 16[e)] *uudddu*	[42]
31g	(23c)	Cs_2CO_3	Cl	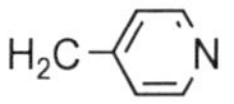 H_2C–CH$_2$–N (pyrrolidinyl)	55[e)] all-*up* 11[e)] *uudddu*	[38] [38]
31h	(23c)	NaH	Br	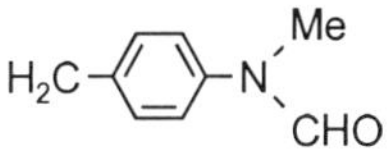 H_2C–C$_6H_4$–N(Me)(CHO)	83[e)] all-*up* 9[e)] *uudddu*	[44] [44]
31i	(31h)	f)	f)	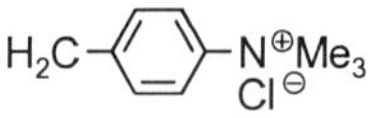 H_2C–C$_6H_4$–$N^⊕Me_3$ $Cl^⊖$	94[f)] all-up 68[f)] *uudddu*	[44] [44]
31j	(23i)	NaH	I	Me	77	[45]
31k	(23j)	Cs_2CO_3	Br	CH_2Ph	93	[39]
31l	(23k)	KOH	OSO_3Me	Me	90	[43]
31m	(23k)	NaH	Br	CH_2Ph	59 all-*up*	[46]
31m	(23k)	NaH	Br	CH_2Ph	+ 4 *uudddu*	[43]
32a	(24a)	NaH	OSO_3Me	Me	66	[41]
32b	(24a)	NaH	I	Et	73[g)]	[41]
32c	(24a)	K_2CO_3	Br	CH_2COOEt	72	[41]
32d	(24a)	NaH	Br	CH_2Ph	45[h)]	[41]
32d	(24a)	K_2CO_3	Br	CH_2Ph	43[i)]	[41]
32e	(24a)	pyridine	Cl	$SiMe_2iPr$	47[j)]	[47]
32f	(24b)	NaH	OSO_3Me	Me	73	[41]
32g	(24b)	NaH	I	Et	71	[41]
32h	(24b)	K_2CO_3	Br	CH_2COOEt	69	[41]
33a	(29e)	NaH	Br	CH_2COOEt	85	[40]
33b	(33a)	k)	k)	CH_2COOMe	95	[40]
33c	(29f)	NaH	Br	CH_2COOEt	83	[40]
33d	(33c)	k)	k)	CH_2COOMe	93	[40]
33e	(29h)	NaH	I	Me	> 95	[36]
33f	(29h)	NaH	Cl	CH_2-(CO)NEt_2	36	[36]

[a)] By oxidation of **31a**. [b)] 80°C in toluene from **31b**. [c)] 150°C in the solid state from **31b**. [d)] Reaction of **31c** with methyl propiolate at 50°C. [e)] Separated by chromatography. [f)] From the related calix[6]arene **31h** by reduction with borane and methylation. [g)] Conformer mixture at room temperature. [h)] Conformer mixture, one isomer could be isolated in 7 % yield. [i)] Conformer mixture. [j)] The A,C-bridged analogue **15** yielded 31%. [k)] Transesterification with methanol.

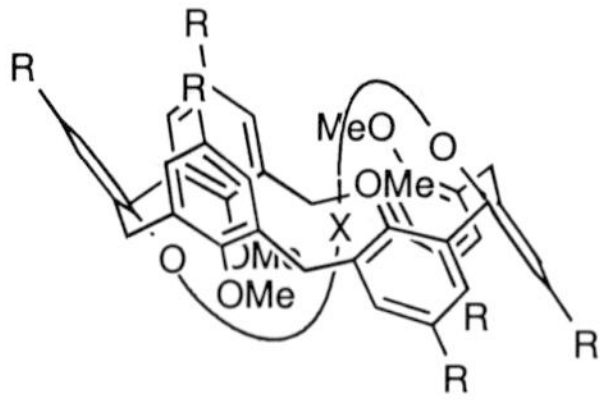

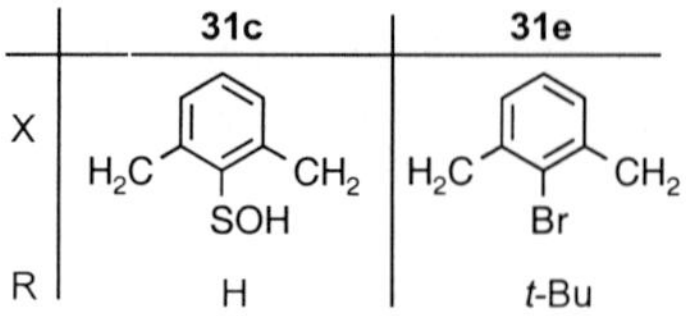

Extended functionalisation sometimes requires protection of residual phenol units by simple ether formation, as in nitrene insertion reactions resulting from irradiation of the azide **31j** [45] and in the sequential conversion of **31k** to the quinone-bridged **34** [39].

Sulfenic, selenenic and seleninic acid substituents on the bridge can be stabilised by being held within the calixarene cavity [32,43,46], and the tetrabenzylated selenenic acid analogue of **31c** is sufficiently stable to be structurally characterised by X-ray diffraction [46]. Remarkably, the *uudddu* conformation selenenic acid showed no reaction with *n*-butylthiol, while the all-*up* conformer reacted to give the coupling product **23s** [43] (see Table 2).

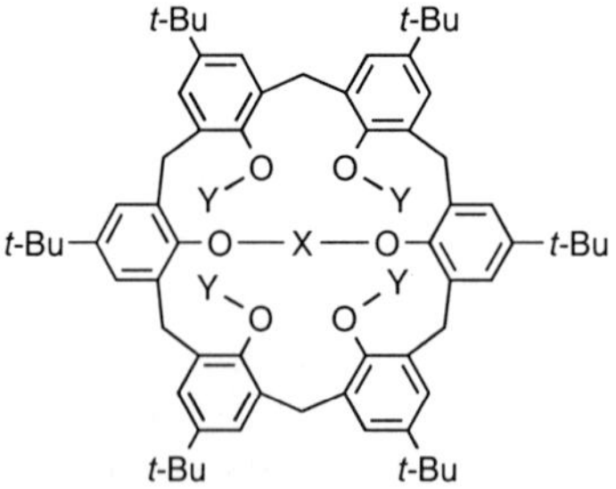

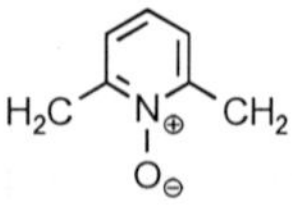

The reactivity of pyridine bridges, as in **24**, is also controlled by the calixarene environment, so that the interaction of the basic nitrogen with phenols depends not only on the phenol acidity but also upon any substitution adjacent to the OH group [24]. Nonetheless, an *N*-oxide **35** can be synthesised in 75 % yield by reacting the bimacrocyclic pyridine **24a** with *m*-chloroperbenzoic acid [33]. In comparison to their open-chain analogues, pyridines **24**, and **32** exhibit basicities higher by more than three orders of magnitude [41]. Only the acetals **57** (see section 2.4.) are an exception. Some pyridine bridged calix[6]arenes like **32e** have already been checked as stationary phases in capillary GC [47] (See Chapter 36.).

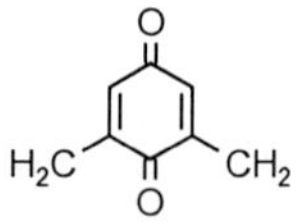

Another important basic unit for inclusion in concave reagents is 1,10-phenanthroline [48]. Although the distance between methylene groups in 2- and 9-position of a 1,10-phenanthroline is considerably larger than the distance in a 2,6-dimethylene pyridine the "bite angle" is smaller by 60° and this may explain why the yields are comparable in the reactions of both **36** (giving **38**) and **37** with **1** [24,49].

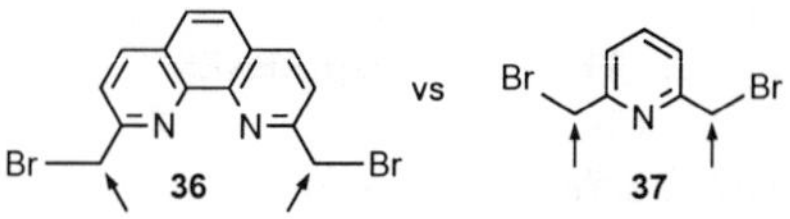

1,10-phenanthroline is not only a base but also a strong metal chelate and reagents **38** and **39** bind strongly to transition metal ions [48,49]. The resulting complexes exhibit catalytic properties, and the shielding in a concave reagent then alters selectivities [50].

With Cu(I) complexes of **38**, surprising selectivities have been found in the catalysed cyclopropanation of alkenes [51], for example, indene. While other concave 1,10-phenanthrolines **39** yielded *exo*-products with selectivities of up to 140 : 1, **38a** was *endo*-selective [51]. The use of methyl diazoacetate led to a 86 : 14 ratio of the respective *endo*- and *exo*-cyclopropanes. Thus, both isomers are accessible by using either **38a** or **39**. Mechanistic details are lacking but the orientation of the bridge in respect to the calix[6]arene ring seems to be important [33], as is the size of R in the diazo compound.

38a R = *t*-Bu

38b R = H

39

R^1	catalyst	*exo* : *endo*
t-Bu	**39**	140 : 1
Me	**38a**	14 : 86

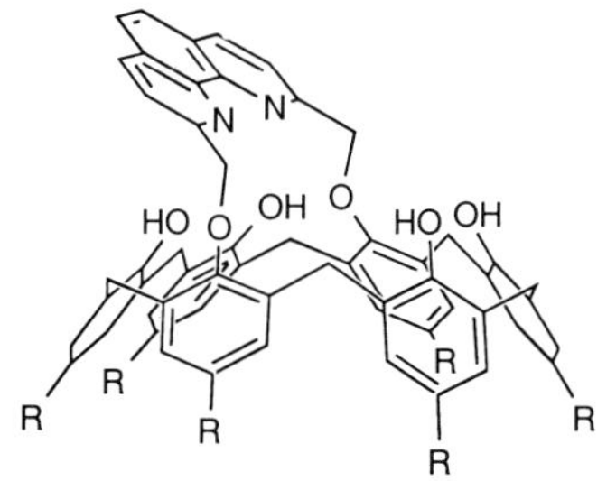

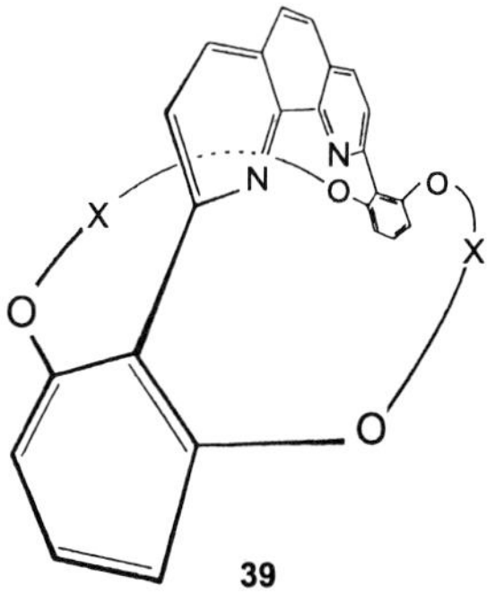

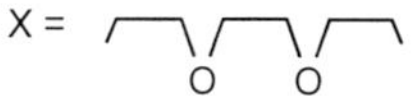

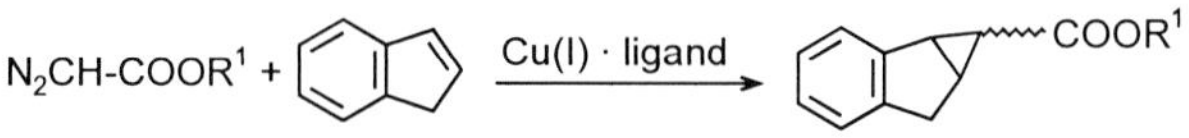

Coupling of pendent units is an alternative means of bridging and A,D-**40** is formed in moderate yield by Cu(II) oxidation of pendent acetylenic units in an A,D-diether precursor [18].

40

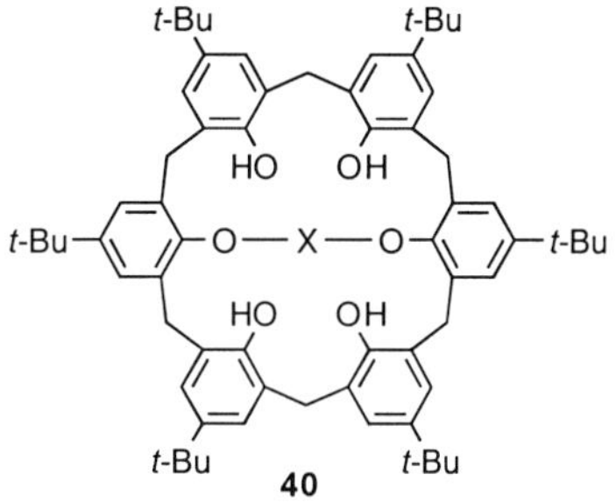

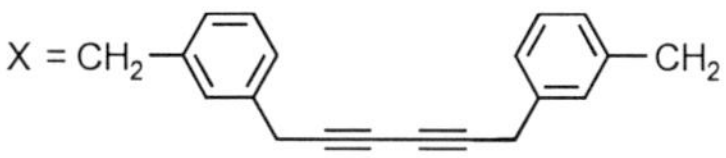

2.2. CAPPED CALIX[6]ARENES

2.2.1. Three legged caps

Given earlier results, it is not surprising that the capping of **1** with a flexible tripod like the tritosylate **41** using K_2CO_3 as a base results in an A,B,D-bridging pattern [37]. The capped product **42**, however, is only isolated in 5 % yield, with **26** being the major product (see section 2.1.3.). Using a tetratosylate, A,B,D-bridging was found in 15 % yield, with the fourth ethylene glycol chain binding a second calix[6]arene (see Chapter 7) [37].

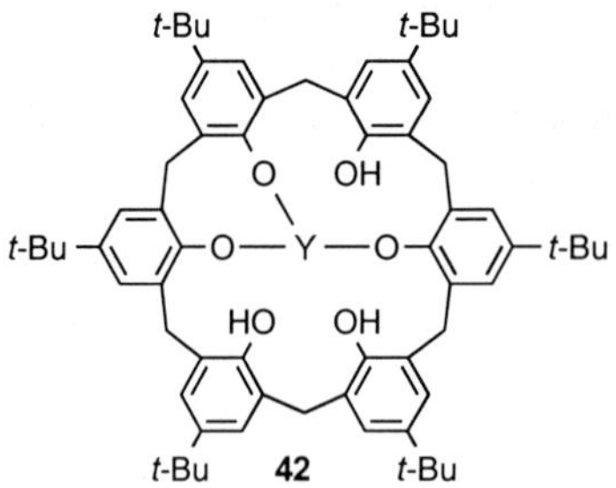

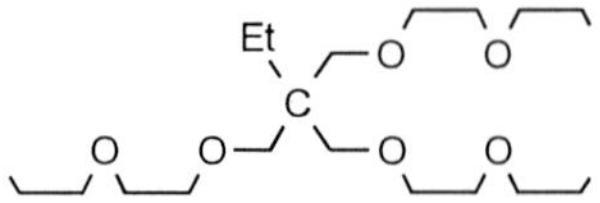

To achieve an A,C,E-capping pattern, as in **43**, B,D,F-trialkylated calixarenes were used as starting materials. Using 1,3,5-tris(bromomethylene)benzene and Cs_2CO_3 the capping of four B,D,F-trialkylated calix[6]arenes was possible in surprisingly high yields (up to 91 %) [27,52], probably due to the good geometric match of the calix[6]-arene and the capping unit.

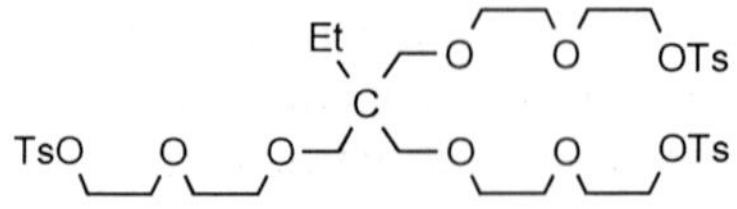

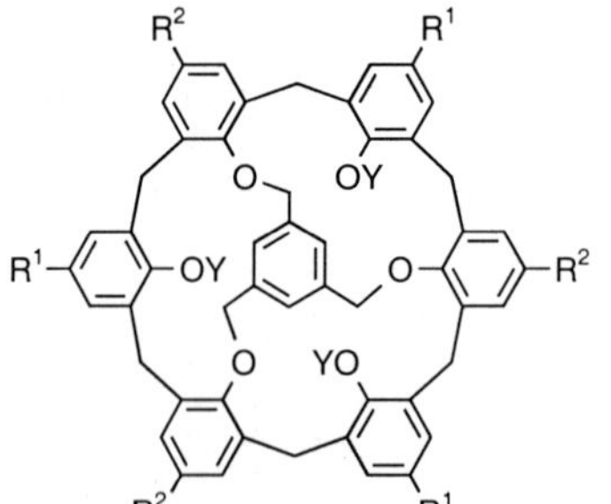

	R^1	R^2	Y	Yield [%]
a	t-Bu	t-Bu	Me	91
b	t-Bu	H	Me	80
c	t-Bu	t-Bu	n-Pr	60
d	H	H	Me	90

The A,C,E-trimethyl ether of **1a** can also be capped by a more flexible tritosylate in more than 50 % yield using NaH as a base [53] to give stereoisomers *in-* and *out-***44** (32 % *in*, 22 % *out*). Both isomers were used as GC stationary phases on an OV1701 column, and the selectivities for the separation of positional isomers of aromatic compounds proved to be different, which can only be explained by the differences between *in-* and *out-*stereochemistry [53].

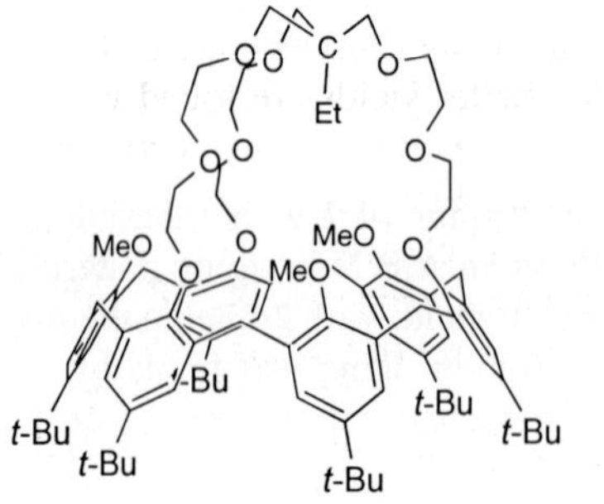

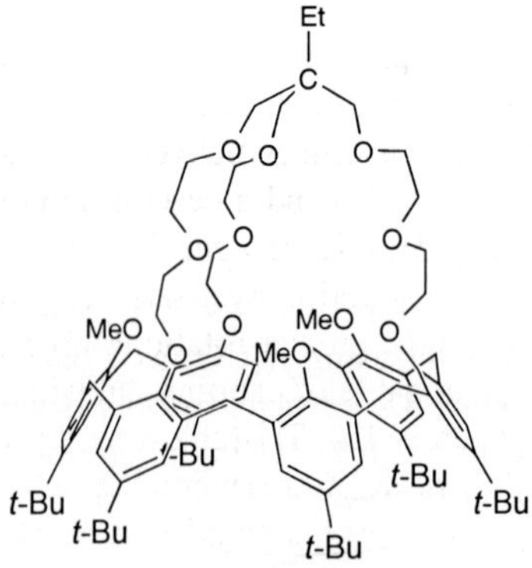

The linking of extended pendent groups can again be used to triply cap calix[6]-arenes in ways defined by the initial substitution pattern. The tri-acid chloride **45** derived from A,C,E-tris(alkoxycarbonylmethyl)-B,D,F-trimethyl ether of **1a** reacts under high dilution conditions with triazinetrione **47** to give the capped calix[6]arene **48**, though in only 2.5 % yield [54].

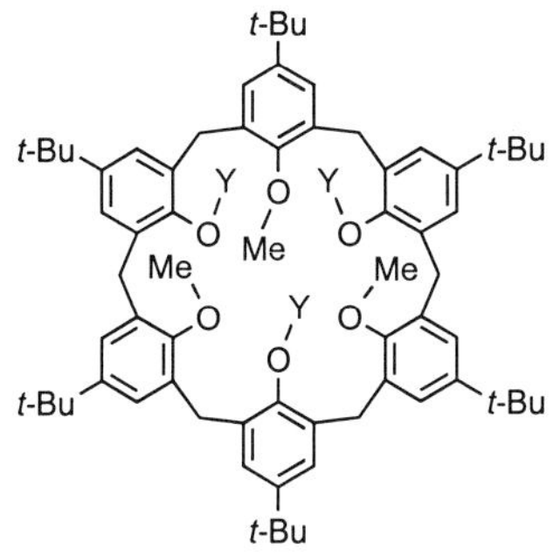

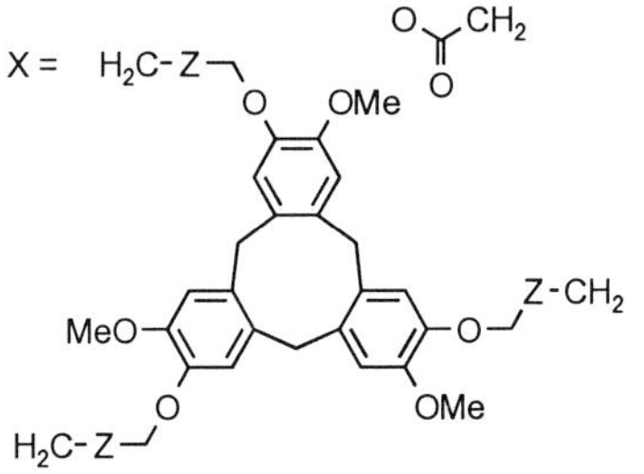

	49a	49b	49c	49d	49e	49f	49g
Z	—	CH$_2$	(CH$_2$)$_2$	(CH$_2$)$_3$	(CH$_2$)$_4$	CH$_2$OCH$_2$	(CH$_2$OCH$_2$)$_2$
Yield [%]	30	30	49	46	41	73	72

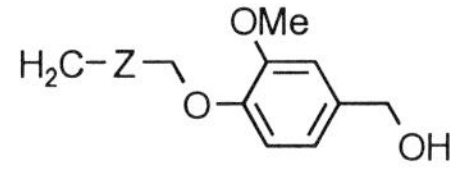

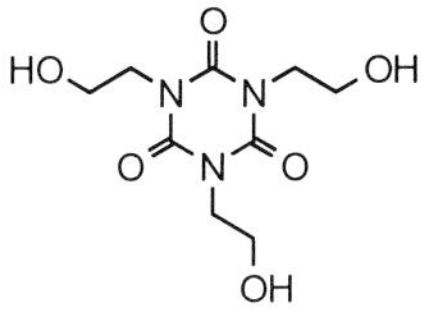

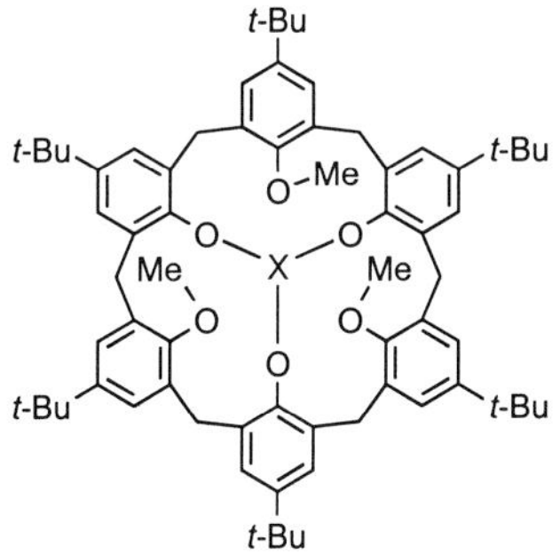

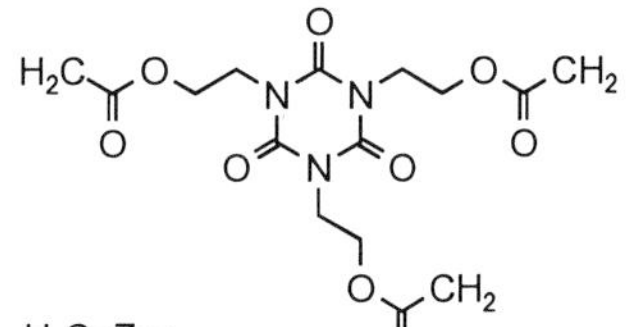

Similarly, proton-catalysed trimerisation of **46** gives a cyclotriveratrylene moiety and the formation of A,C,E-capped calix[6]arenes **49** (cryptocalix[6]arenes) in 30-73 % yield [55]. A tight linker between the calix[6]arene and the veratryl unit leads to only moderate yields. Beyond a certain length (**49c**), better yields are found which probably decrease for entropic reasons if the chains become even longer. Oxygen containing chains, as often found in cyclisations, give the best yields, supposedly due to the greater rotational freedom of these chains.

A rare example of capping the wide rim is based upon related procedures [17]. Tris(chloromethyl)-calix[6]arene **10** can be capped with 1,3,5-tris(mercaptomethyl)benzene and KOH in 28 % yield to give a calix[6]arene **50** A,C,E-capped at the wide rim.

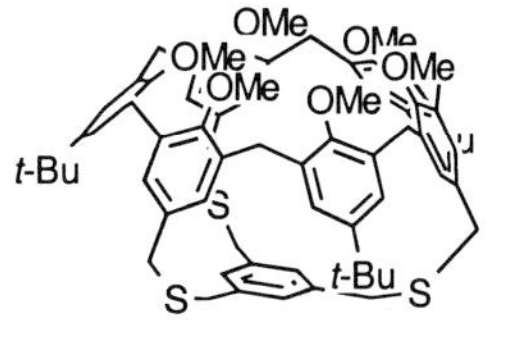

2.2.2. Four legged caps

Only a few examples of polymacrocycles exist in which more than three OH groups of a calix[6]arene are capped. Four-legged capped calix[6]arenes **51** are obtained in 40-72 % yield by reacting A,D-diprotected calix[6]arenes with 1,2,4,5-tetrakis(bromomethyl)benzene in the presence of Cs_2CO_3 [59]. Again the substitution pattern is determined by the starting material. No results of an attempt to cap the non-protected calix[6]arene with the same bridging unit have been reported.

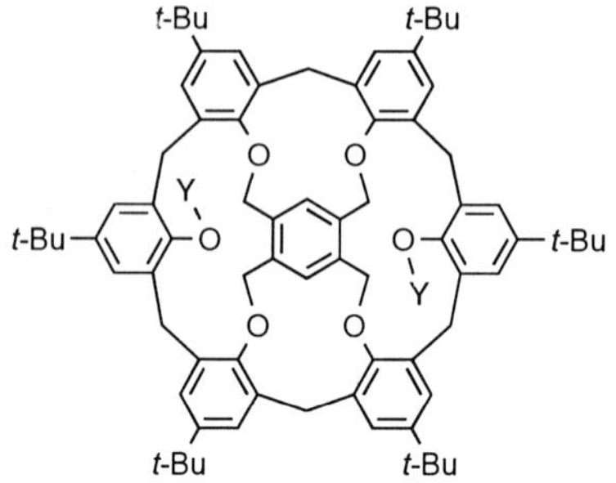

	Y	Yield [%]
51a	Me	66
51b	All	72
51c	n-Pr	65
51d	Bn	40

2.3. CALIXARENES WITH TWO BRIDGES

Double bridging follows patterns predictable from results for single bridging. A,D-Bridging is preferred, for instance, when triethylene glycol derivatives are used (see section 2.1.3.). Thus, macrocycles A,D;B,E-**52** form in good yield (66–70 %) on reacting calix[6]monocrowns with a mole of triethylene glycol ditosylate in the presence of NaH [57]. Trimacrocycles **52**, chiral though yet to be resolved, exhibit complexing properties for ammonium ions and **52b** is selective for *n*-propylammonium ions.

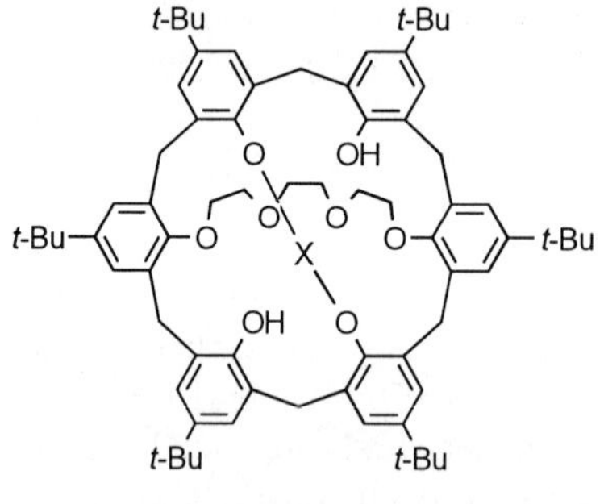

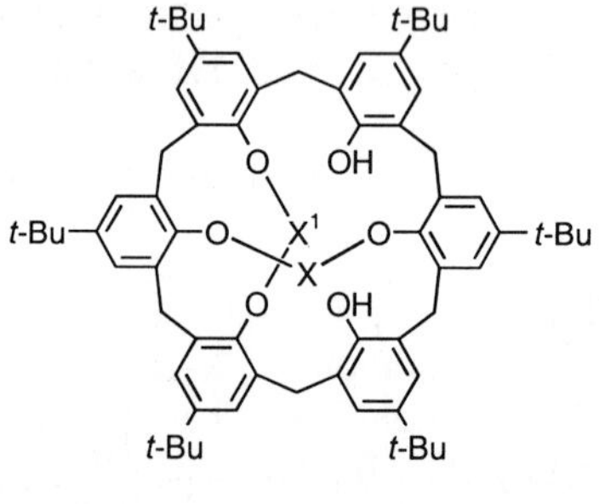

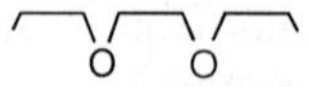

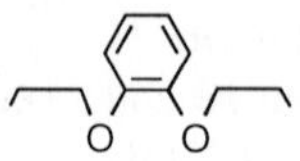

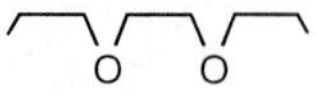

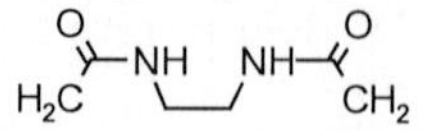

Starting from A,D-bridged calix[6]arenes **29e** and **29f**, bridging with tetraethylene glycol ditosylate resulted in A,C;B,E-bridged calixarene **53a** in 32 % yield and in the trimacrocycle **54** in 46 % yield, respectively [58]. While **53a** carries bridges on both sides of the calixarene macrocycle, **54** is well preorganised to complex and extract cations. An A,C;B,E-bridging pattern was also observed when **29e** was reacted with *N,N'*-bis(chloroacetyl)-ethylenediamine to give 32 % of **53b** [56].

The diethylene glycol unit is one of the few bridges leading to A,C-bridging and introduction of two bridges of this type converts A,D-diallyl calix[6]arene to A,C;D,F-**55** [60]. Two stable isomers form; an all-*up* conformer in 11 % yield, and a *uuuddd*-conformer in 15 % yield. As a metal ion complexant, all-*up*-**55** is Cs^+-selective ($Cs^+/Rb^+/K^+$ = 8000/810/21).

54

all-*up* - **55**

uuuddd - **55**

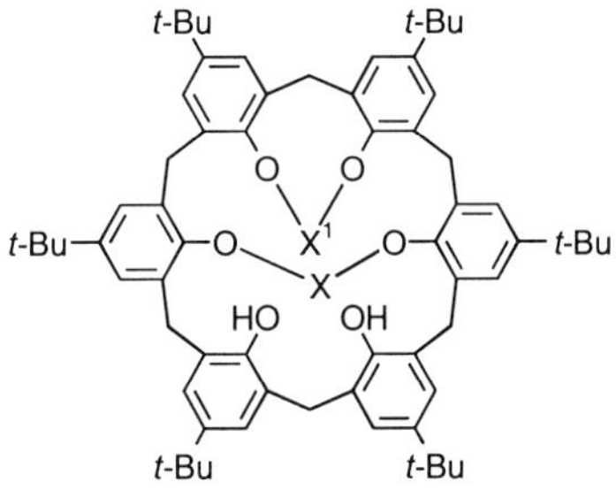

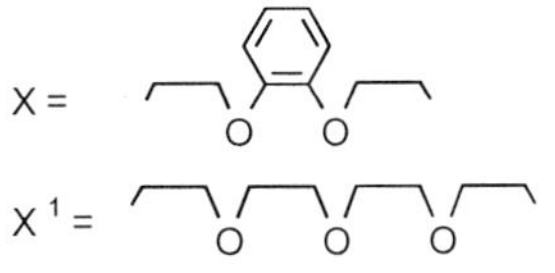

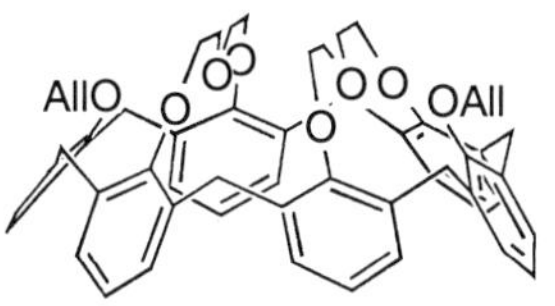

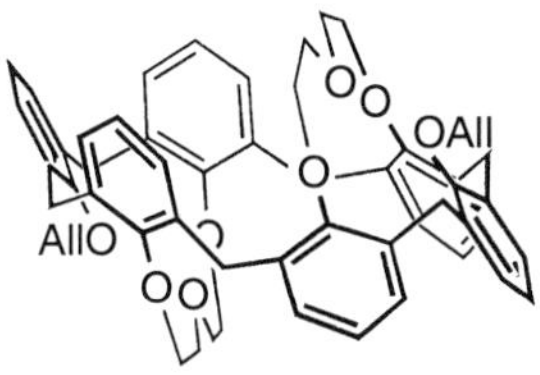

Double-bridging is possible with units favouring A,C- and A,B-spanning, though the production of A,C;D,E-**56** from A,C-**17** by reaction with a second bridging unit shows that the second step involves A,B-bridging [26]. This can be rationalised if the attachment of the second bridge starts at the E-position. Steric effects and the higher stability of the E-phenolate due to hydrogen bonding to the neighbouring D- and F-OH-groups explain the good yields of 67 % for the second bridging, and 40 % for the direct double bridging (using a tenfold excess of K_2CO_3 and 1.5 equivalents of the bridging reagents).

In contrast, double phosphate bridging gives a 9 % yield of A,B;D,E-**57** [22].

56 X = CH_2-C(=O)-NH-CH_2CH_2-NH-C(=O)-CH_2

57 X = P(O)OEt

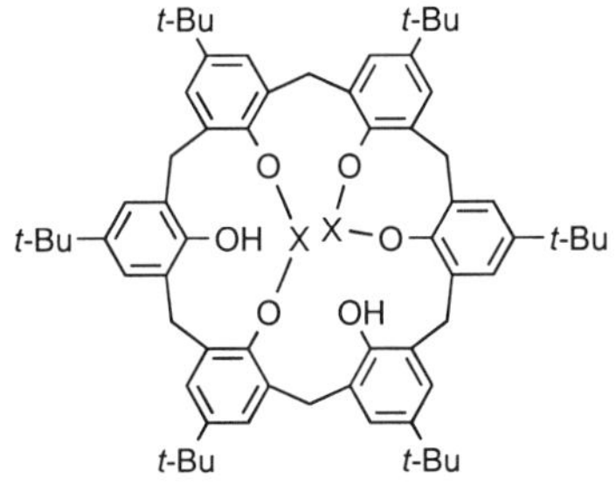

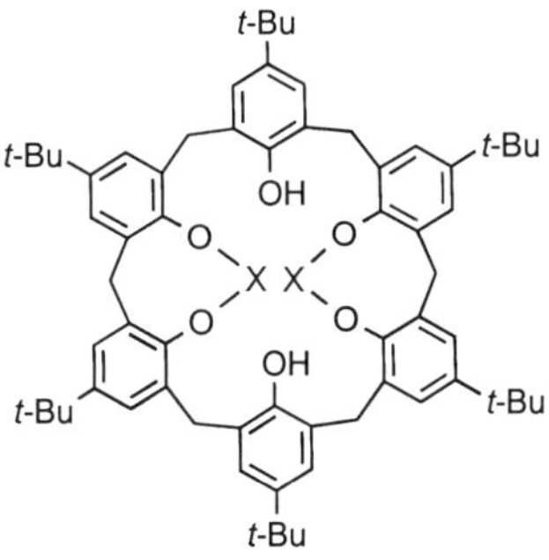

2.4. CALIXARENES WITH THREE BRIDGES

Reaction of calix[6]arene **1a** with excess of bromochloromethane in the presence of Cs_2CO_3 gives A,B;C,D;E,F-tris(methylene)bridged **58** in ~30 % yield [61]. Due to the lack of hydrogen bonds, **58** is very flexible at room temperature. A,B-bridging by bromochloromethane/Cs_2CO_3 of the A,D-pyridine bridged calix[6]arenes **24** gives A,D;B,C;E,F-triply bridged **59a** and **59b** in 92-93 % yield [41].

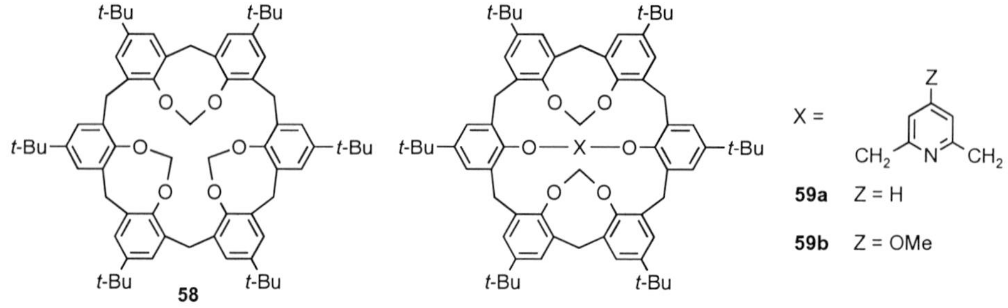

The hexakis(chloromethyl) compound **60** reacts with a *m*-phenylene diamine to give A,B;C,D;E,F-wide-rim-bridged calix[6]arene **61** in 9 % yield [62]. This acts as a host for C_{60} in toluene (see Chapter 26).

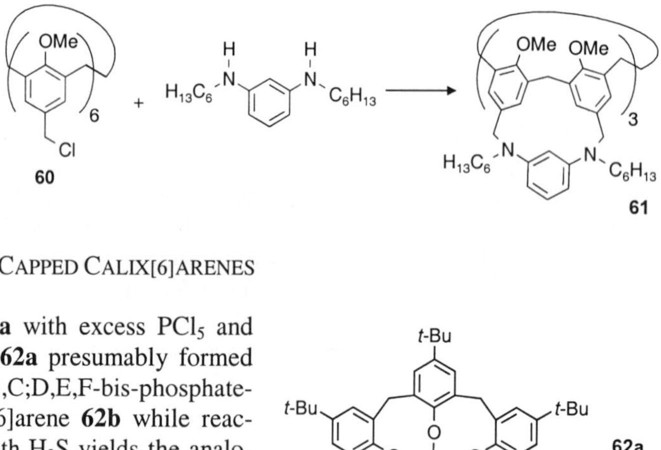

2.5. DOUBLY CAPPED CALIX[6]ARENES

Reaction of **1a** with excess PCl_5 and hydrolysis of **62a** presumably formed gives *syn*-A,B,C;D,E,F-bis-phosphate-capped calix[6]arene **62b** while reaction of **62a** with H_2S yields the analogous thiophosphate compound **62c** [63]. Thermolysis of an acyclic calix[6]arene diphosphate is a less efficient path to **62b** [64]. PCl_3 has also been used to cap **1a**, giving A,B,C;D,E,F-capped calix[6]arenes **63** as the *syn*- and *anti*-isomers. The *syn*-isomer has been isolated in 20-40 % yield by chromatography [65]. The phosphorus atoms act as donors for transition metals like Pt and Pd, and a PdMe(OTf) complex is catalytically active in copolymerisation of carbon monoxide and ethylene, giving polymers with molecular weights of > 30000 g/mol. (See Chapter 29.)

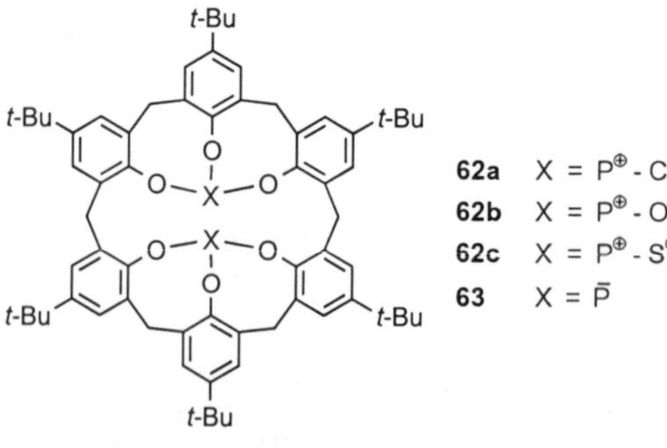

3. Conclusions and Outlook

Selective functionalisation of calix[6]arenes is possible at both the wide and narrow rims, although only bridging of the narrow rim has been widely investigated. The dominant bridging pattern is A,D but recently the number of examples of A,C- or A,B-bridging has increased. The selective bridging and capping is generally associated with greatly reduced conformational flexibility and hence with the creation of molecular cavities of new forms and sizes for which a range of applications may be possible.

4. References and Notes

[1] C. D Gutsche, M. Iqbal, *Org. Synth.* **1989**, *68*, 234-237.
[2] C. D. Gutsche, B. Dhawan, M. Leonis, D. Stewart, *Org. Synth.* **1989**, *68*, 238-242.
[3] J. H. Munch, C. D. Gutsche, *Org. Synth.* **1989**, *68*, 243-246.
[4] J. Vicens, V. Böhmer, *Calixarenes*, A Versatile Class of Macrocyclic Compounds, Kluwer Academic Press, Dordrecht **1991**. C. D. Gutsche, *Calixarenes*, The Royal Society of Chemistry, Cambridge, **1989** and **1992**. C. D. Gutsche, *Calixarenes Revisited*, The Royal Society of Chemistry, Cambridge, **1998**.
[5] Numbering, e.g. 1,2-disubstituted calix[6]arene, is misleading because positions 37 and 38, not 1 and 2, are substituted in such a compound.
[6] A. Casnati, P. Minari, A. Pochini, R. Ungaro, *J. Chem. Soc., Chem. Commun.* **1991**, 1413-1414.
[7] R. G. Janssen, W. Verboom, D. N. Reinhoudt, A. Casnati, M. Freriks, A. Pochini, F. Ugozzoli, R. Ungaro, P. M. Nieto, M. Carramolino, F. Cuevas, P. Prados, J. de Mendoza, *Synthesis* **1993**, 380-386.
[8] H. Otsuka, K. Araki, S. Shinkai, *J. Org. Chem.* **1994**, *59*, 1542-1547.
[9] H. Otsuka, K. Araki, S. Shinkai, *Tetrahedron* **1995**, *51*, 8757-8770.
[10] P. Neri, S. Pappalardo, *J. Org. Chem.* **1993**, *58*, 1048-1053.
[11] P. Neri, M. Foti, G. Ferguson, J. F. Gallagher, B. Kaitner, M. Pons, M. A. Molins, L. Giunta, S. Pappalardo, *J. Am. Chem. Soc.* **1992**, *114*, 7814-7821.
[12] J. de Mendoza, M. Carramolino, F. Cuevas, P. M. Nieto, P. Prados, D. N. Reinhoudt, W. Verboom, R. Ungaro, A. Casnati, *Synthesis* **1994**, 47-50.
[13] K. C. Nam, K. S. Park, *Bull. Korean Chem. Soc.* **1995**, *16*, 153-157.
[14] R. G. Janssen, W. Verboom, S. Harkema, G. J. van Hummel, D. N. Reinhoudt, A. Pochini, R. Ungaro, P. Prados, J. de Mendoza, *J. Chem. Soc., Chem. Commun.* **1993**, 506-508.
[15] O. Sénèque, M.-N. Rager, M. Giorgi, O. Reinaud, *J. Am. Chem. Soc.* **2000**, *122*, 6183-6189.
[16] A. Casnati, L. Domiano, A. Pochini, R. Ungaro, M. Carramolino, J. O. Magrans, P. M. Nieto, J. López-Prados, P. Prados, J. de Mendoza, R. G. Janssen, W. Verboom, D. N. Reinhoudt, *Tetrahedron* **1995**, *51*, 12699-12720.
[17] M. Takeshita, S. Nishio, S. Shinkai, *J. Org. Chem.* **1994**, *59*, 4032-4034.
[18] S. Kanamathareddy, C. D. Gutsche, *J. Org. Chem.* **1996**, *61*, 2511-2516.
[19] J. J. Gonzalez, R. Ferdani, E. Albertini, J. M. Blasco, A. Arduini, A. Pochini, P. Prados, J. de Mendoza, *Chem. Eur. J.* **2000**, *6*, 73-80.
[20] J. Li, Y. Chen, X. Lu, *Tetrahedron* **1999**, *55*, 10365-10374.
[21] D. Kraft, V. Böhmer, W. Vogt, G. Ferguson, J. F. Gallagher, *J. Chem. Soc., Perkin Trans. 1* **1994**, 1221-1230.
[22] J. K. Moran, D. M. Roundhill, *Phosphorus, Sulfur, and Silicon* **1992**, *71*, 7-12.
[23] Y. Chen, F.Yang, *Chem. Lett.* **2000**, 484-485.
[24] H. Ross, U. Lüning, *Angew. Chem.* **1995**, *107*, 2723-2725; *Angew. Chem., Int. Ed. Engl.* **1995**, *34*, 2555-2557.
[25] For A,C-Bridging with other oligoethylene glycol ditosylates see section 2.1.3. and ref. 20
[26] Y.-K. Chen, Y.-Y. Chen, *Org. Lett.* **2000**, *2*, 743-745.
[27] H. Otsuka, K. Araki, H. Matsumoto, T. Harada, S. Shinkai, *J. Org. Chem.* **1995**, *60*, 4862-4867.
[28] H. Otsuka, S. Shinkai, *J. Am. Chem. Soc.* **1996**, *118*, 4271-4275.
[29] S. Kanamathareddy, C. D. Gutsche, *J. Am. Chem. Soc.* **1993**, *115*, 6572-6579.
[30] S. Kanamathareddy, C. D. Gutsche, *J. Org. Chem.* **1992**, *57*, 3160-3166.
[31] a) U. Lüning, *Liebigs Ann. Chem.* **1987**, 949-955; b) U. Lüning, *Top. Curr. Chem.* **1995**, *175*, 57-99; c) U. Lüning, *J. Mater. Chem.* **1997**, *7*, 175-182.
[32] T. Saiki, K. Goto, N. Tokitoh, R. Okazaki, *J. Org. Chem.* **1996**, *61*, 2924-2925.

[33] U. Lüning, H. Ross, I. Thondorf, *J. Chem. Soc., Perkin Trans. 2* **1998**, 1313-1317.
[34] T. Saiki, K. Goto, N. Tokitoh, M. Goto, R. Okazaki, *Tetrahedron Lett.* **1996**, *37*, 4039-4042.
[35] H. Ross, U. Lüning, *Tetrahedron* **1996**, *52*, 10879-10882.
[36] A. Casnati, P. Jacopozzi, A. Pochini, F. Ugozzoli, R. Cacciapaglia, L. Mandolini, R. Ungaro, *Tetrahedron* **1995**, *51*, 591-598.
[37] a) J.-S. Li, Y.-Y. Chen, X.-R. Lu, *Eur. J. Org. Chem.* **2000**, 485-490; b) J.-S. Li, Y.-Y. Chen, Z.-L. Zhong, X.-R. Lu, T. Zhang, J.-L. Yan, *Chem. Lett.* **1999**, 881-882.
[38] S. Akine, K. Goto, T. Kawashima, *J. Incl. Phenom.* **2000**, *36*, 119-124.
[39] S. Akine, K. Goto, T. Kawashima, *Tetrahedron Lett.* **2000**, *41*, 897-901.
[40] Y. Chen, F. Yang, S. Gong, *Tetrahedron Lett.* **2000**, *41*, 4815-4818.
[41] H. Ross, U. Lüning, *Liebigs Ann.* **1996**, 1367-1373.
[42] T. Saiki, K. Goto, R. Okazaki, *Chem. Lett.* **1996**, 993-994.
[43] K. Goto, R. Okazaki, *Liebigs Ann./Recueil* **1997**, 2393-2407.
[44] S. Akine, K. Goto, T. Kawashima, R. Okazaki, *Bull. Chem. Soc. Jpn.* **1999**, *72*, 2781-2783.
[45] N. Tokitoh, T. Saiki, R. Okazaki, *J. Chem. Soc., Chem. Commun.* **1995**, 1899-1900.
[46] T. Saiki, K. Goto, R. Okazaki, *Angew. Chem.* **1997**, *109*, 2320-2322; *Angew. Chem., Int. Ed. Engl.* **1997**, *36*, 2223-2224.
[47] J. H. Park, H. J. Lim, Y. K. Lee, J. K. Park, B. E. Kim, J. J. Ryoo, K.-P. Lee, *J. High Resol. Chromatogr.* **1999**, *22*, 679-682.
[48] U. Lüning, M. Müller, M. Gelbert, K. Peters, H. G. von Schnering, M. Keller, *Chem. Ber.* **1994**, *127*, 2297-2306.
[49] H. Ross, U. Lüning, *Tetrahedron Lett.* **1997**, *38*, 4539-4542. F. Löffler, U. Lüning, G. Gohar, *New J. Chem.* **2000**, *24*, 935-938.
[50] M. Hagen, U. Lüning, *Chem. Ber./Recueil* **1997**, *130*, 231-234.
[51] F. Löffler, M. Hagen, U. Lüning, *Synlett* **1999**, 1826-1828.
[52] H. Otsuka, Y. Suzuki, A. Ikeda, K. Araki, S. Shinkai, *Tetrahedron* **1998**, *54*, 423-446.
[53] J. Xing, J. Li, C. Wu, Y. Chen, X. Lu, H. Han, *Anal. Lett.* **1999**, 3071-3081.
[54] K. Araki, K. Akao, H. Otsuka, K. Nakashima, F. Inokuchi, S. Shinkai, *Chem. Lett.* **1994**, 1251-1254.
[55] R. G. Janssen, W. Verboom, J. P. M. van Duynhoven, E. J. J. van Velzen, D. N. Reinhoudt, *Tetrahedron Lett.* **1994**, *35*, 6555-6558.
[56] K. C. Nam, Y. J. Choi, D. S. Kim, J. M. Kim, J. C. Chun, *J. Org. Chem.* **1997**, *62*, 6441-6443.
[57] Y. Chen, F. Yang, X. Lu, *Tetrahedron Lett.* **2000**, *41*, 1571-1574.
[58] Y.-Y. Chen, H.-B. Li, *Chem. Lett.* **2000**, 1208-1209.
[59] Y. Chen, Y. Chen, *Tetrahedron Lett.* **2000**, *41*, 9079-9082.
[60] M. T. Blanda, D. B. Farmer, J. S. Brodbelt, B. J. Goolsby, *J. Am. Chem. Soc.* **2000**, *122*, 1486-1491.
[61] P. Neri, G. Ferguson, J. F. Gallagher, S. Pappalardo, *Tetrahedron Lett.* **1992**, *33*, 7403-7406.
[62] K. Araki, K. Akao, A. Ikeda, T. Suzuki, S. Shinkai, *Tetrahedron Lett.* **1996**, *37*, 73-76.[63] J. Gloede, I. Keitel, *Phosphorus, Sulfur, and Silicon* **1995**, *104*, 103-112.
[64] F. Grynszpan, O. Aleksiuk, S. E. Biali, *J. Chem. Soc., Chem. Commun.* **1993**, 13-16.
[65] F. J. Parveliet, M. A. Zuideveld, C. Kiener, H. Kooijman, A. L. Spek, P. C. J. Kamer, P. W. N. M. Leeuwen, *Organometallics* **1999**, *18*, 3394-3405.

Chapter 5

CHEMISTRY OF LARGER CALIX[n]ARENES (n = 7, 8, 9)

PLACIDO NERI,[a] GRAZIA M. L. CONSOLI,[b] FRANCESCA CUNSOLO,[b]
CORRADA GERACI,[b] MARIO PIATTELLI[b]

[a]*Dipartimento di Chimica, Università di Salerno, Via S. Allende,
I-84081 Baronissi (SA), Italy, e-mail: neri@unisa.it*
[b]*Istituto per lo Studio delle Sostanze Naturali di Interesse Alimentare e
Chimico-Farmaceutico, C. N. R., Via del Santuario 110, I-95028 Valverde
(CT), Italy.*

1. Introduction

There is an increasing need for large receptors possessing two or more complexing sites that may act cooperatively or allosterically and which are able to host medium-sized organic molecules in addition to simple metal cations. This need can be met either by assembling two or more preformed concave modules (see Chapters 7-10) or by exploiting existing larger macrocycles.

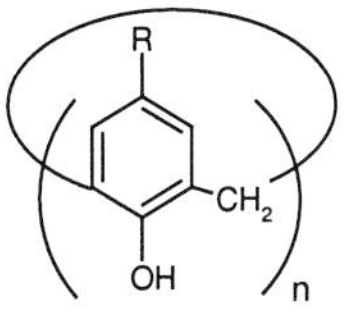

	R	n	Ref		R	n	Ref
a	*t*-butyl	7-9	[1,2,3]	j	ethyl	7	[11]
b	*t*-pentyl	8	[4]	k	phenyl	7,8	[12,13]
c	*t*-octyl	8	[5]	l	*p*-X-phenyl (X=C$_{1-10}$)	7,8	[12]
d	*i*-propyl	8	[6]	m	4'-X-4-biphenyl	7,8	[12]
e	*i*-propenyl	8	[7]	n	benzyl	7,8	[14]
f	*n*-alkyl (C$_{8,9,12}$)	7,8	[8]	o	*p*-CMe$_2$C$_6$H$_4$OR	8	[15]
g	*n*-alkyl (C$_{10,14,16,18}$)	8	[9]	p	benzyloxy	7,8	[16]
h	CMe$_2$(CH$_2$)$_{10}$Me	8	[10]	q	methylmenthyl	8	[17]
i	methyl	7	[11]	r	1-adamantyl	8	[18]

For the latter approach, calix[7]-, calix[8]-, and calix[9]arenes are very promising substrates but two basic problems have to be solved in order to use them: functionalization and preorganization of the macroring. Settling of the first problem requires the ability to introduce specific substituents at particular sites on the macrocycle. The second one concerns the high mobility of the large macrocycle, particularly detrimental to the establishment of host-guest interactions. Not long ago the solution of both problems was considered rather difficult but the work reviewed in this chapter shows that it has become an amenable task.

As for other members of the calixarene family (see Chapters 2-4), the usual starting compound for subsequent chemical manipulations is the readily-synthesized (see Chapter 1) *p-tert*-butyl-derivative, *e. g.*, *p-tert*-butylcalix[8]arene, **1a**. However, a different starting calixarene can sometimes be conveniently used and Chapter 1 provides a broad survey of the range of *p*-substituted larger-ring calixarenes, such as **1b-r**, which are available.

Large calixarenes with more complicated patterns of functionalization can be obtained under various conditions. Thus, calix[n]arenes bearing *meta*-OMe groups were obtained in low yields by acid-catalyzed direct condensation of 3,4,5-trimethoxytoluene [19]. Several calix[8]arenes selectively bearing either OH groups at the *meta* positions (**2**) [20] or other residues at the ArCH$_2$Ar bridges [21] have been obtained in low yields by fragment condensation procedures. An acid-promoted convergent "7 + 1" fragment condensation approach was used in the preparation in 9% yield of calix[8]arene **3** bearing two *p*-carboxyethyl ester functions at 1,5-positions of the macrocycle [22]. The first example of calix[7]arene bearing different substituents at the *para* positions was obtained using fragment condensation [23].

2. Exhaustive Functionalization of Calix[7-9]arenes

The easiest way to functionalize the parent calixarenes is at the phenolic OH groups (the *lower rim*) or at the *para* position of the aromatic rings (the *upper rim*).

In both instances, nearly all publications have concerned the more easily available calix[8]arenes, whereas only a few examples have been reported for calix[7]arenes and only one is presently available for calix[9]arenes (see Section 4). Less useful methods exist involving the *meta* positions or the ArCH$_2$Ar groups.

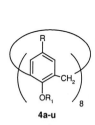

4a-u

	R$_1$	R
a	Me, Et	H [24], *t*-Bu [25], Ph [13], OBn [16], 1-adamantyl [18]
b	CH$_2$CHMeEt	SO$_3$Na [26]
c	*n*-C$_5$H$_{11}$, *n*-C$_3$H$_7$	OBn [16], H [59]
d	*n*-C$_4$H$_9$, *n*-C$_6$H$_{13}$, *n*-C$_{12}$H$_{25}$	SO$_3$Na [27]
e	CH$_2$CH=CH$_2$	H [24], *t*-Bu [25]
f	(CH$_2$CH$_2$O)$_n$X (n = 1,2; X = H, Me, CH=CH$_2$)	H [28,29], *t*-Bu [28,29,30], *t*-octyl [29], 1-adamantyl [18]
g	CH$_2$CO$_2$X (X = H, Me, Et, *i*-Pr, CH$_2$CH$_2$OMe)	H [31,32], *t*-Bu [32,33], *t*-octyl [31,34], OBn [16], 1-adamantyl [18]
h	CH$_2$COX (X = Me, *t*-Bu)	*t*-Bu [32]
i	(CH$_2$)$_3$X (X = NH$_2$, phthalimido)	H [31], *t*-Bu [31], *t*-octyl [34]
j	CH$_2$CONEt$_2$, CH$_2$CONH-*n*-Bu	*t*-Bu [35], OBn [16], 1-adamantyl [18]
l	α-CH$_2$Py	*t*-Bu [36]
m	*p*-CH$_2$C$_6$H$_4$X (X = H, Me, *t*-Bu, CN, NO$_2$)	*t*-Bu [25,37]
n	*p*-COC$_6$H$_4$X (X = H, Me, OMe, *t*-Bu, Br, CN, NO$_2$, C$_6$H$_5$)	*t*-Bu [25,38]
o	COX (X = Me, Et, CH=CH$_2$, C$_6$H$_5$, CHBrCH$_3$, CHCl$_2$, CF$_3$)	H [24,39], *t*-Bu [39,40,41], Ph [13], OBn [16], 1-adamantyl [18]
p	COCH=CH-C$_6$H$_3$-2,4-OMe	*t*-Bu [38]
q	SO$_2$Me	*t*-Bu [42]
r	(CH$_2$)$_3$SO$_3$Na	H [43]
s	PO(OEt)$_2$	*t*-Bu [44], *t*-octyl [44]
t	CH$_2$CH$_2$PO(C$_6$H$_5$)$_2$	*t*-Bu [45]
u	SiMe$_3$	*t*-Bu [2,25], Ph [13]

Introduction of chemical functions at the lower rim of calixarenes is usually obtained by exploiting the chemical reactivity of OH groups, easily converted to the corresponding ethers or esters. The exhaustive transformation of calix[n]arenes into the pertinent ethers or esters is conveniently obtained by using a large excess of strong reagents. In this way, calixarenes **4a-u** have been obtained in good to high yields. The introduced group has often been subjected to further transformations.

Variously substituted calixarenes (**5a-q**) have been obtained by successive chemical modifications of the upper rim of directly-synthesized compounds. The standard way starts with complete removal of the *para*-substituents followed by aromatic electrophilic substitution. Usually the *tert*-butyl group is removed by $AlCl_3$-mediated trans-*tert*-butylation to give the corresponding *p*-H-calixarenes [24, 46].

		R	n	R_1	Ref
a	H		7,8,9	H	[5,24,46]
b	Alkyl (Me, Et, *n*-Pr, *n*-Bu, *s*-Bu, *n*-Amyl)		8	H	[47]
c	CH_2CH_2COOH		7,8	H	[48]
d	$CH_2CH(COOEt)_2$		7,8	H	[48]
e	CMe_2OX (X = H, Me)		8	Me	[49]
f	$CH_2C(CO_2Et)_3$		8	H, Me	[50]
g	CH_2NX_2 (X = H, Me, $CH_2CH=CH_2$, CO_2Et)		7,8	H	[48,51]
h	$CH_2N^+Me_3Cl^-$		8	Me	[52]
i	CH_2Cl		8	H, Me	[53]
j	CH_2POXX' (X = X' = OH, Et; X = OH, X' = Et)		8	Me	[53]
k	$CH_2C[CONHC(CH_2OH)_3]_3$		8	Me	[50]
l	COX (X = Me, Et, C_6H_5)		7,8	H	[46,54]
m	COX (X = OH, Me, OMe)		8	Me, C_5H_{11}	[24,59]
n	SO_2X (X = OH, Cl)		8	H, Me	[55,56]
o	$SO_2N(CH_2CH_2OH)_2$		8	Me	[56]
p	N=N-Ar		7,9	H	[46,57]
q	NO_2		8	H	[55,58]

These are variously manipulated to introduce groups such as halomethyl, methylamino, nitro, diazo, etc. Often a protection of the OH groups at the lower rim is required before carrying out the required reaction at the upper rim. In several instances the introduced group has also been subjected to further transformations.

Removal of benzyl groups in *p*-benzyloxycalix[8]arene **1p** allows access to calix[8]arene-hydroquinone **6a** and to a subset of derivatives **6b-c**, in which the OH groups of the upper or lower rim are variously functionalized [16]. To the same subset belong compounds **6d-f**, obtained by means of Baeyer-Villiger oxidation of *p*-acetyl-calix[8]arenes **5m** to *p*-acetoxycalix[8]arenes **6d** [59]. These last were hydrolyzed to the corresponding *p*-OH-calix[8]arenes **6b**, which were subsequently alkylated to **6e-f**. Similarly, the p-$CMe_2C_6H_4OMe$ groups of **1o** can be demethylated by treatment with BBr_3 to give the corresponding p-$CMe_2C_6H_4OH$ phenols [15a].

	R	R_1
a	H	H
b	H	Me, C_5H_{11}, $COCH_3$, CH_2CONEt_2, C_3H_7
c	CH_2CONEt_2	CH_2CONEt_2
d	$COCH_3$	Me, C_5H_{11}
e	Me	Me
f	C_5H_{11}	C_5H_{11}

6a-f

7a R = Me
7b R = C_5H_{11}

A few calix[8]arenes bearing *meta* bromine atoms (**7a-b**) have been obtained by direct bromination of **6e-f** [59]. Because of the steric encumbrance by the Br atoms these compounds show a reduced conformational mobility.

3. Selective Functionalization of Calix[8]arenes

Selective functionalization of the larger calixarenes is of prime interest for the preparation of new hosts based on their skeleton. However, achievement of this goal has been considered discouraging because of the large number of possible partially, homogeneously substituted derivatives, namely 16, 28, and 46 for calix[7]-, calix[8]-, and calix[9]arenes, respectively (Figure 1). In addition, with these larger macrocycles assignment of the substitution pattern becomes more difficult. In fact, this often requires discrimination among several regioisomers with an identical molecular symmetry, thus precluding their identification from simple 1D NMR spectra. In these instances, it is mandatory to resort to techniques such as 2D NMR, X-ray crystallography or chemical correlation with known compounds.

These difficulties, coupled with scarcity of the parent macrocycles, explain the absence in the literature, up to now, of partial derivatives of seven and nine-terminus calixarenes. The ready availability of calix[8]arenes, however, has rendered more amenable their challenging selective functionalization.

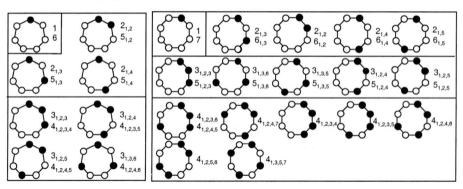

Figure 1. Schematic representation of the 16 and 28 partially substituted derivatives of a calix[7]arene (left) and a calix[8]arene (right). Regioisomers are grouped in boxes. The substitution pattern is indicated by a number corresponding to the number of substituents, which is followed by a subscript referring to their location. For simplicity, in the case of penta- and hexasubstituted compounds the subscripts indicate the unsubstituted aromatic rings. Filled or empty circles are related to differently substituted rings, thus each drawing may refer to two complementary degrees of substitution.

3.1. SELECTIVE FUNCTIONALIZATION AT THE LOWER RIM

3.1.1. "Alternate alkylation" pathway in the presence of weak bases
Apart from a few compounds either monosubstituted or with unknown substitution pattern [60], the first examples of partially *O*-substituted calix[8]arenes (**8a-i**) were reported in 1993 [37a, 61]. They were obtained in yields up to 50% by alkylation of *p*-*tert*-butylcalix[8]arene in the presence of weak bases such as K_2CO_3 or CsF. Their tetra-*O*-alkylation at alternate aromatic rings, namely 1,3,5,7, was firmly established on the basis of the unequivocal correspondence between observed NMR signal pattern and

molecular symmetry. Considering the variety of effective alkylating agents used, the mechanism of 1,3,5,7-tetrasubstitution of calix[8]arenes can be considered rather general and it was named "alternate alkylation" [37b].

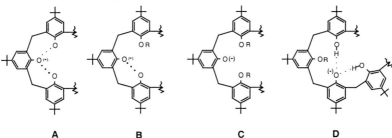

	R		
a	p-CH$_2$C$_6$H$_4$Bu^t	f	p-CH$_2$C$_6$H$_4$Br
b	p-CH$_2$C$_6$H$_4$Me	g	CH$_2$CO$_2$Bu^t
c	CH$_2$C$_6$H$_5$	h	CH$_2$CONMe$_2$
d	p-CH$_2$C$_6$H$_4$NO$_2$	i	n-Bu
e	p-CH$_2$C$_6$H$_4$CN		

8a-i

The preferential course of the reaction was explained in terms of consecutive steps of monodeprotonation and monoalkylation through preferential formation of monoanions stabilized by flanking hydrogen bonds. In fact, because of the weakness of the base (K$_2$CO$_3$ or CsF) only monodeprotonations are expected. Thus, among the three types of monoanions attainable at various stages of the alkylation, stability should follow the order **A > B > C** (Figure 2).

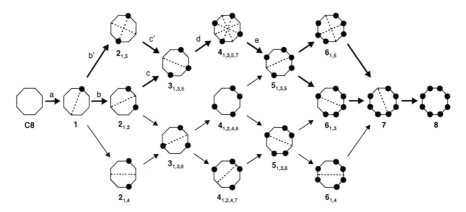

Figure 2. Plausible hydrogen bond stabilization for calix[n]arenes monoanions.

Therefore, a hypothetical pathway can be drawn simply by assuming that in each step the more stabilized anions are preferentially alkylated (Figure 3).

Figure 3. Possible routes for "alternate alkylation" of a calix[8]arene in the presence of weak bases. Thick arrows refer to the main pathway leading to the products isolated in the alkylation of p-tert-butylcalix[8]-arene with p-methylbenzyl bromide in the presence of CsF. Symmetry planes are indicated by dotted lines. The symbols used refer to Figure 1.

In order to test this hypothesis the alkylation course was investigated using a typical electrophile, *p*-methylbenzyl bromide, in the presence of CsF as the base.

The composition of the reaction mixture as a function of time was followed by chromatography [37b]. As seen in Figure 4, after the formation of the monobenzyl derivative the reaction goes through the successive formation of 1,5-, 1,3,5- and finally 1,3,5,7-tetrasubstituted compounds. This last is accumulated to some extent probably because further reaction requires a slow passage through less stable anions of type **C** (Figure 2), making its isolation an easy task. This holds true for different alkylating agents and in general 1,3,5,7-tetraethers are obtained in reasonable yields without difficulties.

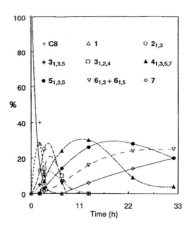

Figure 4. Time-course of product distribution in the alkylation of p-tert-butylcalix[8]arene in THF/DMF with p-methylbenzyl bromide and CsF (1:8:16 molar ratio).

During the above studies 1,2,4-tri-*O*-alkylated calix[8]arenes were also isolated in variable amounts. Their formation does not fit the proposed pathway. Apparently, less stable anions of type **B** must be formed concurrently with those of type **A** (Figure 2). This may be due to the flexibility of calix[8]arene macrocycle allowing their stabilization through seemingly remote interactions like that in anion **D** [62].

An interesting case of *alternate alkylation* is the methylation of *p-tert*-butylcalix[8]arene which again proceeds through formation of the 1,3,5,7-tetramethoxy derivative [63]. Identification of this insoluble product was based on its full benzylation that gave a sufficiently soluble derivative [62]. The low solubility of many methoxycalix[8]arenes can turned to advantage for their removal from the reaction mixture simply by filtration. In this way the remaining soluble compounds can be more easily isolated [62].

3.1.2. Alkylation products obtained in the presence of stronger bases

It was anticipated that products with different substitution patterns could be formed in the presence of stronger bases able to carry out multiple deprotonation. An investigation of the methylation of *p-tert*-butylcalix[8]arene (**1a**) in the presence of NaH, BaO/Ba(OH)$_2$ or Cs$_2$CO$_3$ showed that, as expected, other methoxycalix[8]arenes with different substitution pattern and greater solubility were formed (Figure 5) [64]. In this way, a total of 19 out of the 28 possible partial methoxycalix[8]arenes have been identified.

In a different approach, the effect of a strong base was investigated by using a well-defined preformed calix[8]arene trianion. In principle, it is conceivable that such a polyanion should behave differently with respect to a monoanion, thus giving rise to products with different substitution patterns. The calix[8]arene trianion used in the study was prepared by reaction of *p-tert*-butylcalix[8]arene with Et$_4$NOH to give the easily crystallized tris(tetraethylammonium) salt [65]. This was subjected to alkylation with typical electrophiles (*p*-methylbenzyl bromide or *p-tert*-butylbenzyl bromide) under standard conditions to give, in order of decreasing yields, 1,3,5,7-tetra-, 1,3-di-,

1,3,5-tri-, 1,2,4-tri-, and monobenzyl-calix[8]arene, besides trace amounts of 1,4- and 1,5-disubstituted compounds [66].

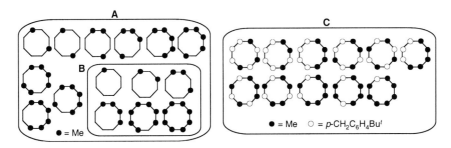

*Figure 5. Isolated methoxycalix[8]arenes under various conditions. The compounds obtained in the presence of strong bases are grouped in **A**. Among them those sub-grouped in **B** were also isolated as CH_2Cl_2-soluble derivatives from weak-bases-promoted methylations. Benzylation products of weak-bases-promoted methylation mixtures are grouped in **C**.*

This result can be explained in terms of the alkylation pathway in Figure 6, in which it is assumed that after each alkylation step the left negative charge is redistributed, by exchange equilibria, over the remaining phenolic groups. Subsequent alkylation occurs at those positions where the charge is preferentially localized, because contiguous H-bond stabilization is possible and the electrostatic repulsion is minimized. Thus, the occurrence of protonation/deprotonation equilibria allows the formation of 1,3,5,7-tetrasubstituted calix[8]arene.

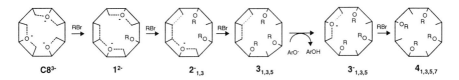

Figure 6. Proposed alkylation pathway for calix[8]arene trianion $C8^{3-}$. The symbols used refer to Figure 3.

3.1.3. Comparative properties of partially substituted calix[8]arenes: NMR chemical shifts of OH groups and quaternary aromatic carbons

The availability of a large number of methoxycalix[8]arenes allows a comparison of some of their properties. In this regard of particular interest is the dependence of the chemical shift of OH groups from the degree and pattern of substitution. In fact, it is well documented that its value increases with increased strength of the H-bond with proximal groups which, in turn, increases with the number of consecutive hydroxyls [67]. The maximum value of *ca.* 10 ppm is reached in the parent calixarenes wherein a "circular-H-bond" can be formed. In partially substituted derivatives an OH group can be classified as "isolated", "singly-H-bonded" or "doubly-H-bonded", if it is flanked by none, one or two hydroxyls, respectively, and accordingly its chemical shift increases in the same order.

The observations on methoxycalix[8]arenes lead to the following generalizations: "isolated" OH groups usually resonate at $\delta < 7.7$; "singly-bonded" hydroxyls give signals at $7.7 < \delta < 8.7$; "doubly-bonded" OH groups resonate at $\delta > 8.7$ [64]. However, these limits move to lower field if stronger cooperativity occurs, due to the presence of an extended sequence of contiguous hydroxyls. This effect appears to involve isolated OH groups as well, if they can give a *quasi*-circular" hydrogen bond through some kind of interaction with non-adjacent OH groups.

In a similar way, the ^{13}C NMR chemical shift changes exhibited by quaternary aromatic carbons upon methylation lead to the following observations: oxygen-bearing carbons undergo an average downfield $\Delta\delta$ of 4.0 ppm, *t*-butyl-bearing carbons show an average $\Delta\delta$ of 3.8 ppm, bridgehead carbons (C-CH_2) exhibit an average $\Delta\delta$ of 5.5 ppm [64]. Therefore, these last move from the crowded region at 126-128 ppm to an empty zone at 132-134 ppm. This downfield shift rule is valid in the entire range of partially methylated calix[8]arenes and, on the basis of literature data, for several other alkylated calix[n]arenes. Consequently, this rule appears to be confidently applicable in ^{13}C NMR signal assignment and structural elucidation of any alkylated calixarene.

3.1.4. Strategies for structure assignments
Discrimination between regioisomers having the same kind of symmetry can be difficult if sophisticated techniques such as 2D NMR or X-ray crystallography are inapplicable. In these instances, a relatively simple alternative is offered by chemical correlation.

A pentaalkylated calix[8]arene possessing a single binary symmetry element bisecting aromatic rings (Ar—Ar symmetry) could have three different substitution pattern (1,3,5, 1,3,6, 1,2,3). The 1,3,5-substitution may be proved simply by monoalkylation of the corresponding 1,3,5,7-tetraalkylated compound, which can only give rise to 1,3,5-pentaalkylated derivative because of its high symmetry (Figure 7A) [37b, 62].

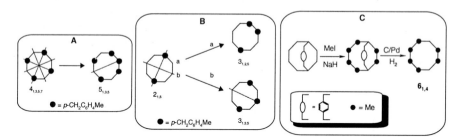

Figure 7. Examples of structure assignment through chemical correlations: *A*) 1,3,5-pentalkylation of $5_{1,3,5}$ proved by monoalkylation of $4_{1,3,5,7}$; *B*) monoalkylation of $2_{1,5}$ allowed unequivocal structure assignment of $3_{1,3,5}$; *C*) use of a protecting group for unequivocal synthesis of hexamethoxy derivative $6_{1,4}$.

Similarly, discrimination of the 1,3,5-trisubstitution from those with the same Ar—Ar symmetry, namely 1,2,3 or 1,3,6, can be reached by monoalkylation of the corresponding 1,5-dialkylated calix[8]arene, which can only give two trialkylated compounds having different symmetry (Figure 7B) [37b].

A different approach makes use of specifically placed protecting groups removed after the desired alkylation. Thus, in order to prepare 1,4-hexamethoxycalix[8]arene a known 1,4-(*p*-xylylene)-bridged calix[8]arene [68] was methylated and subjected to hydrogenolysis to give the desired compound (Figure 7C) [64].

3.1.5. Other partially alkylated calix[8]arenes

It is evident that monoalkylation of calix[8]arenes can be attained using limiting amounts of weak bases or strictly controlled reaction times. Thus an interesting monoalkylated calix[8]arene, calixfullerene bearing a C_{60}-fullerene moiety was prepared from **1a** [69]. Because of the complementarity between calix[8]arene and buckminsterfullerene this compound exists in equilibrium between self-included and non-self-included species (see Chapter 22).

A group of partially alkylated calix[8]arenes (**9-10**) with diverse substituents at the lower rim has been prepared [62-64] by stepwise reactions.

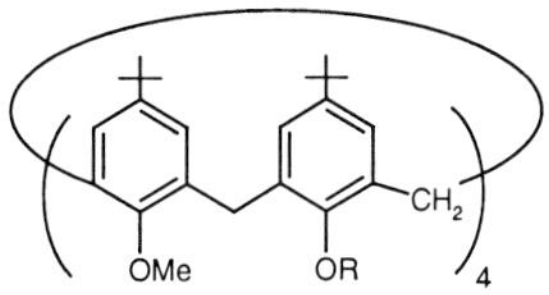

	R
a	H
b	$p\text{-}CH_2C_6H_4Br$
c	$p\text{-}CH_2CO_2Et$
d	$p\text{-}COC_6H_4Me$

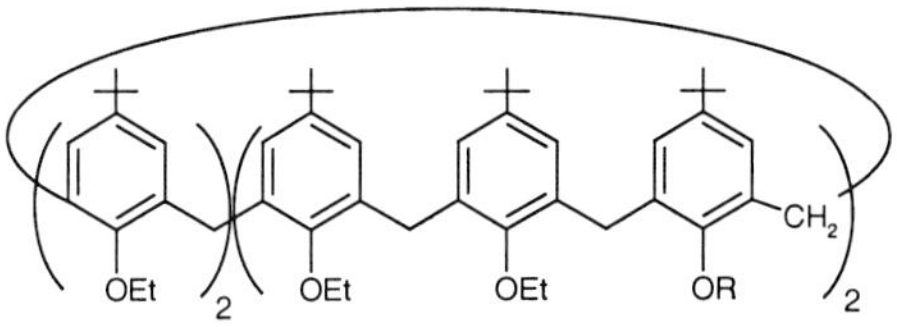

9a-d

10a R = H
10b R = Me

3.1.6. Partial esterification

Given the success of partial alkylation, partial acylation might also be expected to be possible but study of benzoylation has shown that reduction of either reaction time or relative amount of acyl halide usually gives only complex mixtures of various derivatives, unresolvable by chromatography. However, the preparation of heptaesters (40-80% yields) [38] could be optimized. Interestingly, these compounds (**11a-g**), in analogy to calix[4]arene tribenzoate [24], can be regarded as "protected" *p-tert*-butylcalix[8]arene in the synthesis of mono-functionalized derivatives (see next section).

Partial phosphorylation of *p-tert*-butylcalix[8]arene by chloro-diethylphosphate gives mono-, 1,3-bis- and 1,5-bis(diethoxyphosphoryl)calix[8]arenes **12a-c** in 12-31% yield [44]. It is expected that further progress with regard to partial esterification could be made in the near future.

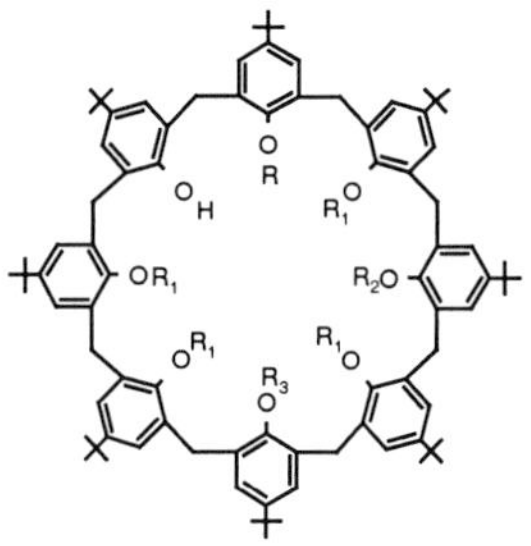

11a-g

$R = R_1 = R_2 = R_3$

a	$p\text{-}COC_6H_4Bu^t$
b	$p\text{-}COC_6H_4Me$
c	$p\text{-}COC_6H_4OMe$
d	COC_6H_5
e	$p\text{-}COC_6H_4Br$
f	$p\text{-}COC_6H_4CN$
g	$p\text{-}COC_6H_4NO_2$

12a-c

	R	R_1	R_2	R_3
a	$PO(OEt)_2$	H	H	H
b	$PO(OEt)_2$	H	$PO(OEt)_2$	H
c	$PO(OEt)_2$	H	H	$PO(OEt)_2$

3.2. SELECTIVE FUNCTIONALIZATION AT THE UPPER RIM

Several methods are currently available for partial derivatization at the upper rim of calix[4]- and calix[6]arenes (see Chapters 2 and 4), usually relying on lower reactivity in the electrophilic aromatic substitution of substituted *versus* unsubstituted phenolic rings [70]. Therefore, under favorable conditions the substitution pattern of partial *O*-

derivatized calixarenes can be transferred to the upper rim. For example, from the easily available heptabenzoyl derivative **11e**, it is possible to selectively remove the *tert*-butyl group of the OH-bearing ring by AlCl$_3$-mediated *trans*-butylation to toluene [71]. The resulting mono-de-*tert*-butylated calix[8]arene **13** on oxidative coupling at the freed *para* position gave the first example of a double calix[8]arene, named *p-tert*-butyl-5,5'-bicalix[8]arene **14** [71] (see Chapter 7).

Beyond this example, selective functionalization at the upper rim of calix[8]arenes still remains largely unexplored.

A few examples of oxidative reactions of *p-tert*-butylcalix[8]arene (**1a**) have been reported (see Chapter 14). For instance, treatment with CrO$_3$ results in the oxidation of three ArCH$_2$Ar groups to ArCOAr, the exact sequence along the macroring being uncertain [72]. Milder oxidation of **1a** with PhN$^+$Me$_3$ Br$_3^-$ affords monospirodienone calix[8]arene, while oxidation with K$_3$Fe(CN)$_6$ gives two isomeric tetrakis(spirodienone) derivatives (1-2% yield), again of uncertain structure [73].

4. Shaping of the Calix[n]arene Skeleton

The larger calix[n]arenes undergo conformational interconversion by means of the "oxygen-through-the-annulus" or "*p*-substituent-through-the-annulus" passages (Figure 8). To obtain conformationally defined derivatives both mechanisms have to be blocked.
This can be achieved by introducing bulky-enough groups at both rims (the usual *p-tert*-butyl group is too small). However, this approach does not appear to be particularly rewarding because of the inherent flexibility of the calix[n]arene skeleton which could still give rise to a multitude of conformations depending on the relative orientation of

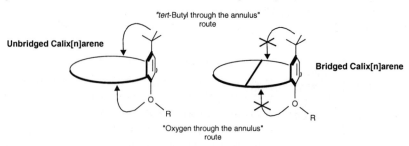

Figure 8. Possible pathways for conformational interconversion in p-tert-butylcalix[n]arenes and their blockage following intramolecular bridging.

each aryl ring. This is shown by the number of different conformations observed in the solid state for native calix[n]arenes and their simple derivatives or metallic complexes.

Thus, *p-tert*-butylcalix[8]arene crystallizes from pyridine either in the *pleated-loop* [74a] or in the *chair-like* [74b] conformation (Figure 9), whereas it adopts the *pinched* conformation in some of its alkali or alkali-earth metal salts [75]. The *chair-like* conformation was also recently found for octapropoxy-*p*-hydroxycalix[8]arene [16b, 74c]. Other geometries for the calix[8]arene skeleton are found in various transition and lanthanide metal ion complexes [76, see also Chapters 28 and 30].

p-tert-Butylcalix[7]arene [77a,b] crystallizes in a conformation which can be described as a combination of three aryl rings in a *cone* arrangement and four ones in a *pleated-loop* geometry, whereas *p*-benzylcalix[7]arene and *p*-ethylcalix[7]arene both adopt a pinched conformation having two hydroxy groups inverted with respect to the others and appearing relatively flat in projection [14b]. The X-ray structure of the *p-tert*-butylcalix[7]arene is consistent with the MM3 most stable conformation [77c]. An X-ray crystal structure of nonaethylacetate-*p-tert*-butylcalix[9]arene has been recently reported [77d].

Figure 9. *Side views of pleated-loop (left) [74a] and chair-like (right) [74b] X-ray crystal structures of p-tert-butylcalix[8]arene (tert-butyl groups and solvent molecules are omitted).*

Considering this large conformational variability, the introduction of suitable scaffolding elements by intramolecular bridging is considered essential in order to obtain well-defined shapes suitable for molecular recognition (Figure 8). To date, this prospect has really only been explored for calix[8]arenes.

4.1. SINGLY-BRIDGED CALIX[8]ARENES

The introduction of a single bridge in the calix[8]arene skeleton can give rise to four regioisomers, namely the 1,2-, 1,3-, 1,4- or 1,5-bridged compounds.

4.1.1. Aromatic bridges
The first bridged calix[8]arenes (**15a-c**), reported in 1994, were obtained by alkylation of 1,3,5,7-tetrasubstituted calix[8]arenes with 1,4-bis(bromomethyl)benzene [78]. On the basis of VT NMR studies they were found to be conformationally locked in a

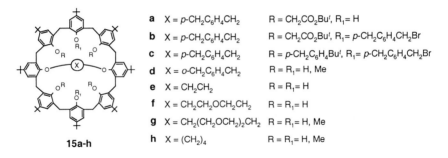

15a-h

a X = *p*-CH$_2$C$_6$H$_4$CH$_2$ R = CH$_2$CO$_2$Bu^t, R$_1$= H
b X = *p*-CH$_2$C$_6$H$_4$CH$_2$ R = CH$_2$CO$_2$Bu^t, R$_1$= *p*-CH$_2$C$_6$H$_4$CH$_2$Br
c X = *p*-CH$_2$C$_6$H$_4$CH$_2$ R = *p*-CH$_2$C$_6$H$_4$Bu^t, R$_1$= *p*-CH$_2$C$_6$H$_4$CH$_2$Br
d X = *o*-CH$_2$C$_6$H$_4$CH$_2$ R = R$_1$= H, Me
e X = CH$_2$CH$_2$ R = R$_1$= H
f X = CH$_2$CH$_2$OCH$_2$CH$_2$ R = R$_1$= H
g X = CH$_2$(CH$_2$OCH$_2$)$_2$CH$_2$ R = R$_1$= H, Me
h X = (CH$_2$)$_4$ R = R$_1$= H, Me

structure composed by two *syn*-oriented subunits, each viewable as a *cone* calix[4]arene in which an aryl ring is replaced by the bridging elements.

Other arene-bridged calix[8]arenes (**15d** and **16a-c**) were obtained by NaH-promoted direct alkylation of *p-tert*-butylcalix[8]arene with various bis(bromomethyl) arenes [68]. In this way, the 1,5- and/or 1,4-isomers were isolated in yields ranging from 6 to 47%.

In a further example (**16d**), *p-tert*-butylcalix[8]arene was 1,4-bridged in 5-10% yield by direct alkylation with an acridone-based linker prepared trough a multistep synthesis [79]. Molecular modeling indicated that **16d** adopts a *pinched* conformation with two non-symmetrical cavities.

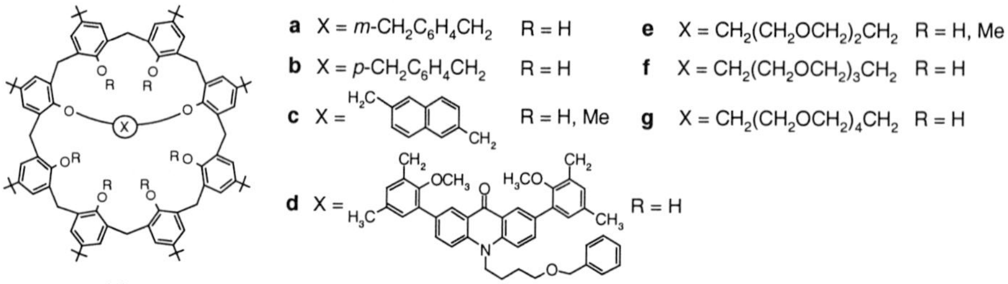

a $X = m\text{-}CH_2C_6H_4CH_2$ $R = H$
b $X = p\text{-}CH_2C_6H_4CH_2$ $R = H$
c $X = $ (naphthalene-CH$_2$) $R = H, Me$
d $X = $ (acridone linker) $R = H$
e $X = CH_2(CH_2OCH_2)_2CH_2$ $R = H, Me$
f $X = CH_2(CH_2OCH_2)_3CH_2$ $R = H$
g $X = CH_2(CH_2OCH_2)_4CH_2$ $R = H$

16a-g

4.1.2. Crown-ether and related aliphatic bridges

Introduction of crown ether bridges gives rise to compounds collectively named calix[8]crown-*n*, where *n* represents the number of oxygen atoms in the bridge (see Chapter 20). The four possible singly-bridged calix[8]crowns (**15-18**) have been obtained by direct regioselective alkylation of *p-tert*-butylcalix[8]arene under various conditions [80]. The most useful preparative procedures give access to 1,5-crown-n (n = 2, **15e**, 88%; n = 3, **15f**, 78% yield) or 1,4-crown-n (n = 4-6) **16e-g** in 25-46% yields, in the presence of Cs_2CO_3 or NaH as base, respectively. 1,3- and 1,2-crowned calix[8]-arenes **17** and **18** are obtained in the presence of K_2CO_3, but in lower yields.

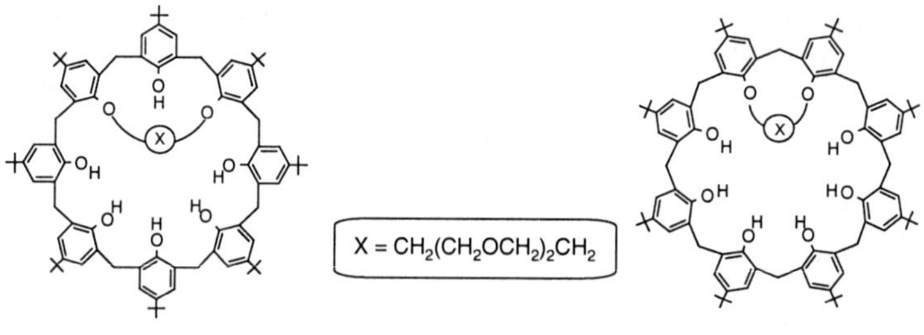

$X = CH_2(CH_2OCH_2)_2CH_2$

17   **18**

The regiochemical outcome in these reactions has been explained in terms of strength of the base, which gives rise to anions with the charge preferentially localized at specific positions (Figure 10). Thus, the preference for 1,4-bridging should be dictated by the formation of a 2,4,6,8-tetraanion (**20**), while a 3,5,7-trianion (**21**) should favor 1,5-bridging. The 1,3- and 1,2-bridging are possibly promoted by a 3-monoanion (**22**) or, respectively, by a cation template effect. It is worth noting that the yields in 1,5-

bridging are higher with shorter bridges, indicating that in the closure step the macrocycle adopts a conformation having those positions very close to each other (*e. g. pinched* or *chair-like* conformation). This is corroborated by the high yield of 1,5-tetramethylene-bridged calix[8]arene **15h** upon alkylation with I(CH$_2$)$_4$I [80c].

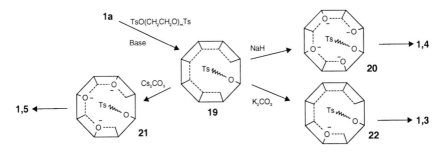

Figure 10. Proposed explanation for the regiochemical outcome in the synthesis of calix[8]crowns.

4.2. DOUBLY-BRIDGED CALIX[8]ARENES

Introduction of two identical bridging elements in the calix[8]arene skeleton gives rise to twenty-two possible regioisomers, seven of them completely asymmetric, ten with a single symmetry element and the remaining ones with higher symmetry. They will be discussed on the basis of the currently available bridging patterns.

4.2.1. 1,3:5,7-Double-bridging

The first doubly-bridged calix[8]arenes (**23a-c**) were obtained by alkylation, in the presence of Cs$_2$CO$_3$ as the base, of 1,3,5,7-tetrasubstituted calix[8]arenes with tetraethylenglycol ditosylate [82], 1,2- or 1,3-bis(bromomethyl)benzene in yields up to 96% [83]. The surprisingly high yield indicates that the aromatic rings at position 1 and 3 are spatially very close to each other so even a short *ortho*-xylene spacer can bridge them. VT NMR studies indicated that, in the 230-390 K range, intramolecular bridging inhibits the flipping motion of the aromatic rings. The conformation adopted, deduced from diagnostic signals in the ^{1}H and ^{13}C-NMR, involves a *syn* orientation of the eight calix[8]arene aromatic rings. Debenzylation of calix[8]bis-crown derivative **23a** afforded the conformationally mobile tetrahydroxy compound **23d**.

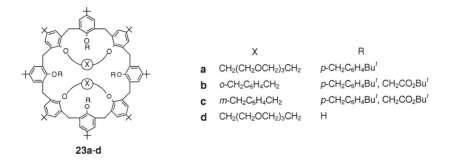

4.2.2. 1,5:3,7-Double-bridging
Compound **24a** with 1,5:3,7-bridging pattern was also obtained in the above cited synthesis of calix[8]bis-crowns [82]. It was shown to adopt a basket-shaped conformation in which one of the polyether chains fills the cavity. At high temperature the shielded chain comes out from the cavity, thus allowing free inversion of benzylated rings. Debenzylation of **24a** afforded the highly symmetrical compound **24b** having freely rotating phenol rings at room temperature.

	X	R
a	$CH_2(CH_2OCH_2)_3CH_2$	$p\text{-}CH_2C_6H_4Bu^t$
b	$CH_2(CH_2OCH_2)_3CH_2$	H
c	$o\text{-}CH_2C_6H_4CH_2$	H
d	$o\text{-}CH_2C_6H_4CH_2$	Me
e	$CH_2CH_2OCH_2CH_2$	H
f	$CH_2CH_2OCH_2CH_2$	Me
g	$CH_2CH_2OCH_2CH_2$	Et
h	$(CH_2)_4$	H

24a-h

Further examples of 1,5:3,7-doubly-bridged derivatives were obtained by direct alkylation of *p-tert*-butylcalix[8]arene with 1,2-bis(bromomethyl)benzene (**24c**) [68], diethylene glycol ditosylate (**24e**) [84], or 1,4-diiodobutane (**24h**) [85]. In the first instance NaH was used as base in a single step (19% yield). In the other two cases a two step reaction (Cs_2CO_3 followed by NaH) allowed increasing the yields up to 68%.

Molecular modeling predicted for the obtained compounds an intriguing D_{2d} structure composed of four *¾-cone* clefts in which the eight phenolic oxygen atoms converge toward the central polyhedral cavity (Figure 11) [68, 84]. The bridged oxygens at 1,3,5,7 positions adopt a tetrahedral arrangement, while the residual OH groups are placed in a planar square bisecting the tetrahedron. This geometry was confirmed by single crystal X-ray structure of **24h**, which differed from the model in that one aryl ring is "inverted", probably due to favorable interactions in the crystal packing [85].

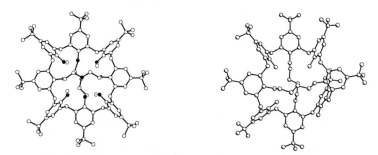

Figure 11. Idealized D_{2d} conformation of **24e** (left) and X-ray crystal structure of **24h** (right). H atoms and crystallization solvent molecules are omitted.

4.2.3. Other double-bridging patterns
Various bis-crowns derivatives have been obtain in workable amounts by direct alkylation of *p-tert*-butylcalix[8]arene, using more vigorous conditions than those adopted for

the preparation of mono-crowns. Thus, reaction with triethylene glycol ditosylate in the presence of Cs_2CO_3 afforded 1,4:2,5-bis-crown-4 **25** [86], 1,4:2,3-bis-crown-4 **26** and 1,4:3,5-bis-crown-4 **27** [81]. In addition, alkylation of the easily available 1,4-crown-4 in presence of KH gave the highly symmetrical 1,4:5,8-bis-crown-4 **28** [86]. Analogously, 1,2-crown-4 afforded 1,2:3,4-bis-crown-4 **29** in the presence of K_2CO_3 [81].

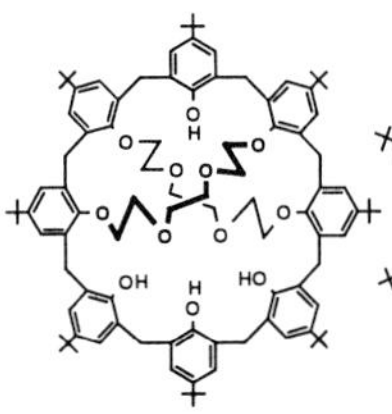

25 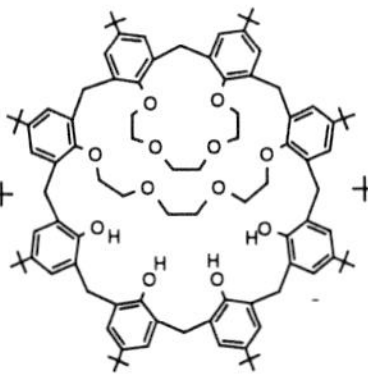   **26** 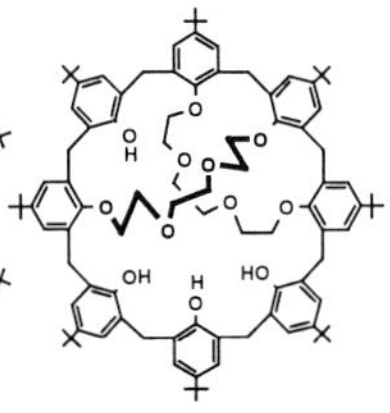   **27** 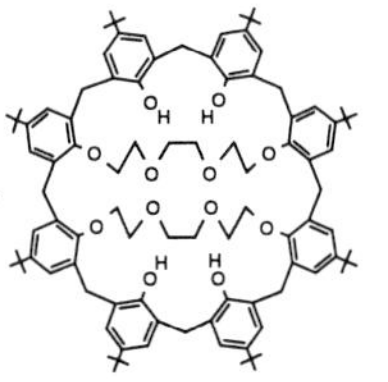   **28** 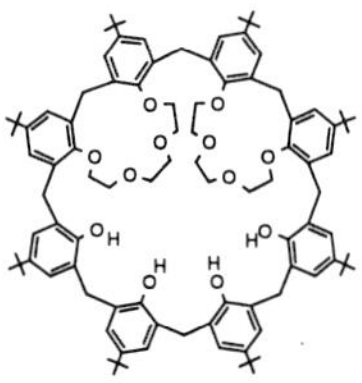   **29**

4.3. TRIPLY- AND QUADRUPLY-BRIDGED CALIX[8]ARENES

A few examples of triply- and quadruply-phosphorus-bridged calix[8]arenes (**30-33**) have been recently obtained in 18-80% yield [87]. In the case of **30** and **31** two stereoisomers (*exo* or *endo* forms) were separated by HPLC.

Finally, multiple intrabridging at the *para* positions of *p*-H-calix[8]arene with diazotized 4,4'-diamminodiphenyls has been reported. However, improbable structures were tentatively proposed for the obtained products which, therefore, have to be considered as uncharacterized [88].

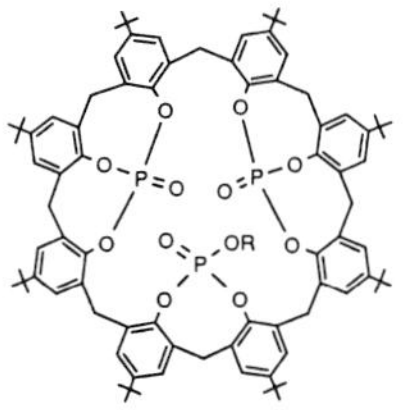

30
R = H
R = Et

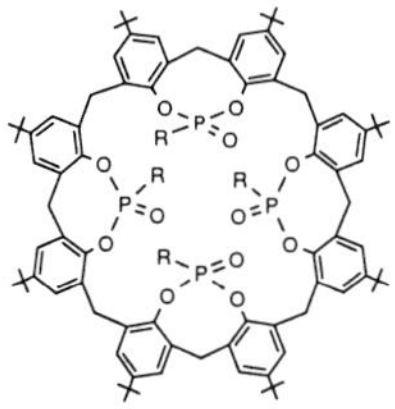

31
R = Cl
R = OH
R = OEt

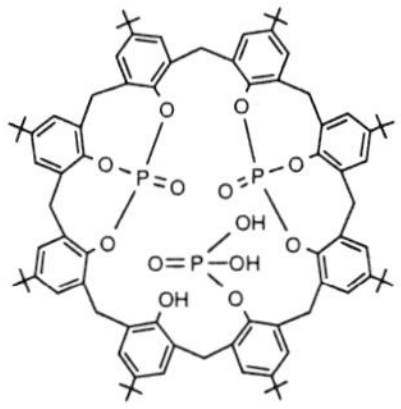

32

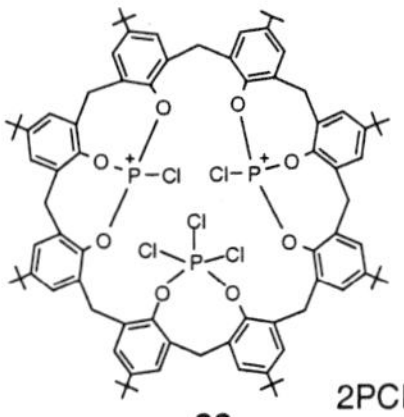

33
$2PCl_6^-$

4.4. CAPPED CALIX[8]ARENES

The first example of a lower rim capped calix[8]arene, **34**, was obtained by "head-to-tail" coupling of *p-tert*-butylcalix[8]arene with tetramethoxy-*p*-tetrachloromethyl-calix[4]arene in the presence of CsF as the base [89]. Evidently, the dimension of the calix[4]arene unit was suitable to cap *p-tert*-butylcalix[8]arene at the 1,3,5,7-positions. VT NMR spectra showed only a partial rigidification of the structure, which moves from C_{2v} to C_{4v} symmetry upon heating from 330 to 370 K. This behavior was ascribed to the passage from slow to fast exchange between two equivalent elliptical structures assumed by the calix[8]arene macrocycle due to the presence of the capping calix[4]-arene unit [89].

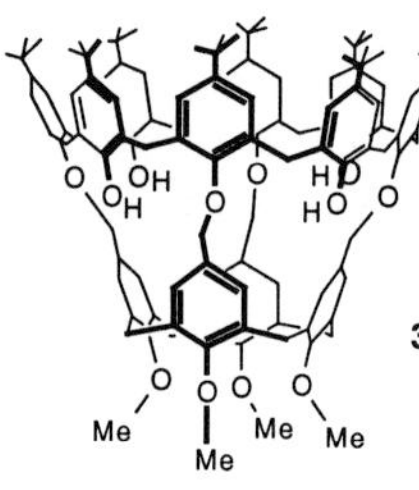

A different example (**35**) was obtained from 1,3,5,7-calix[8]arene derivatives using a durene unit as capping element [83]. Dynamic NMR in $CDCl_3$ or $C_6D_5NO_2$ evidenced the absence of conformational interconversion in the 230-390 K range. This result indicates that the flipping motion of the calixarene aromatic rings is effectively inhibited. On the basis of a computational study, the benzylated aromatic rings are *syn*-oriented to give a fixed "*pseudo*-pleated-loop" conformation.

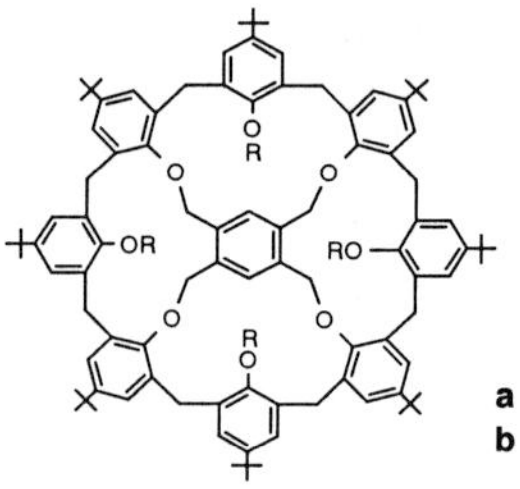

5. Chiral Calix[8]arenes

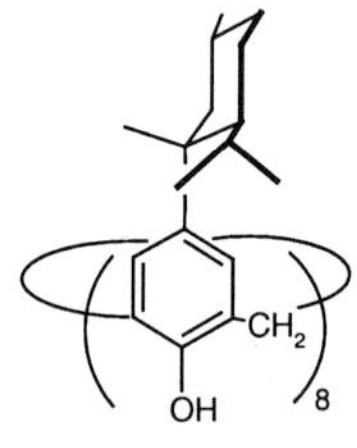

As for the smaller homologues, chiral calix[8]-arenes can be obtained by attaching chiral groups at the macrocyclic framework or eliminating any symmetry plane or inversion center in their three-dimensional structure, thus producing inherent chirality. Four derivatives of the first approach are known, in which one or two camphorsulphonyl [60a] or eight (*S*)-2-methyl-butyl groups [26], **4b**, have been bonded to the phenolic hydroxyls, and calix[8]arene **1s** bearing a methylmenthyl moiety, which was attached at the *para* position prior of the macrocyclization step [17].

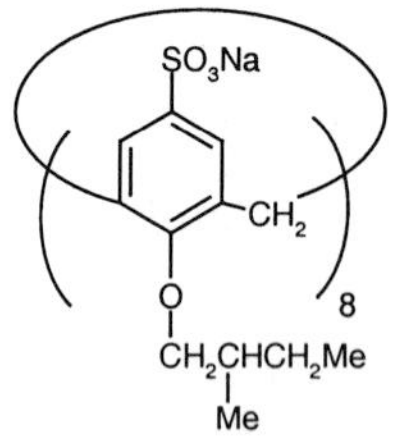

For inherently chiral calixarenes, the problem of conformational interconversion has to be overcome and this implies intramolecular bridging as mandatory step. The bridges can be used as source of asymmetry or as simple scaffolding elements, in conjunction with other monofunctional substituents arranged asymmetrically at the remaining positions. The former possibility was exploited with the synthesis of 1,4:2,5- and 1,4:3,5-calix[8]bis-crown-4 (**25** and **27**, respectively) whose chirality is due to steric constraint of the two intercrossing identical polyether chains [86, 81]. The inherent chirality of **25** was first evidenced by signal splitting in the ^1H NMR spectrum after addition of Pirkle reagent and definitely proved by the specular CD spectra of the enantiomers, resolved through enantioselective HPLC [90]. This result demonstrates that the topological constraint of intercrossing chains can be sufficient for chirality generation notwithstanding the residual mobility of unsubstituted phenolic rings.

The second possibility found realization through proper alkylation of cage 1,5:3,7-doubly-bridged calix[8]arenes. In fact, because of the tridimensionality of their structure, introduction of one, two- or three substituents can give rise to chiral derivatives. Among the five possible methoxy derivatives of **24c**, the inherent chirality of **36a**, **38a** and **39a** was demonstrated by NMR spectroscopy (Pirkle reagent), though HPLC enantioresolution was ineffective [91].

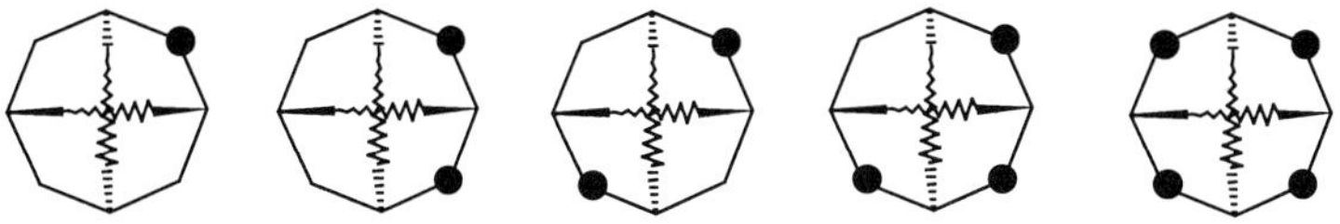

36a	37a	38a-b	39a	24d
Chiral	Achiral	Chiral	Chiral	Achiral

a
$\sim\!\!\sim\!\!\sim$ = o-CH$_2$C$_6$H$_4$CH$_2$
● = Me

b
$\sim\!\!\sim\!\!\sim$ = CH$_2$CH$_2$OCH$_2$CH$_2$
● = p-CH$_2$C$_6$H$_4$Bu^t

Note that the above picture would became more complicated if one takes into account the possibility that groups larger than methyl may give rise to *syn/anti* isomerism like that observed for calix[4]arenes. Indeed, the introduction of two *p-tert*-butylbenzyl groups at the 1,5-positions of **24e** afforded the C_2-symmetric derivative **38b** which represents the first example of a calix[8]arene with two aryl rings locked in *anti* orientation [85]. This was evidenced by a typical ^{13}C NMR resonance at 39.3 ppm for the pertinent ArCH$_2$Ar carbons, in accordance with the de Mendoza single rule [92]. The chirality of **38b** was indicated by addition of Pirkle reagent [85].

In a different approach, chirality was found in transition and lanthanide metal ion complexes as a consequence of the coordination geometry of the metal [76, see also Chapters 28 and 30].

6. Complexing Properties of Calix[8]arenes

The selective inclusion of buckminsterfullerene-C$_{60}$ by *p-tert*-butylcalix[8]arene is discussed in detail in Chapter 26. Inclusion properties of water-soluble derivatives are discussed in Chapter 24, and metal ion coordination chemistry is discussed in Chapters 28 and 30.

Solvent extraction (see Chapter 21, 22) experiments have demonstrated the affinity of (2-pyridylmethoxy)-calix[8]arene **41** for Ag$^+$, Cu^{2+} and UO$_2^{2+}$ [36]. The complexing abilities toward alkali, alkaline-earth and rare-earth metal cations of other octafunc-tionalized calix[8]arenes **4h** [32], **4j** [35, 93], and **40a-e** [32, 93] have been examined as a completion of the studies conducted on the smaller homologues (see

	R	R$_1$
a	t-Bu, H	CH$_2$CO$_2$Et
b	t-Bu, H	CH$_2$CO$_2$Me
c	t-Bu	CH$_2$CH$_2$OMe
d	t-octyl	CH$_2$CH$_2$OMe
e	t-octyl	(CH$_2$CH$_2$O)$_2$Me

40a-e

Chapters 21 and 35). In comparison with their tetrameric or hexameric counterparts, they are generally less efficient and/or less discriminating. This has been ascribed to minimal preorganization of the complexing site due to greater conformational mobility. A similar result was also obtained with calix[8]mono-crown derivatives **15-18** which, in contrast to calix[4]crowns (see Chapter 20), have no complexing ability worth of note because they are still conformationally mobile notwithstanding the presence of the bridge [80b,c].

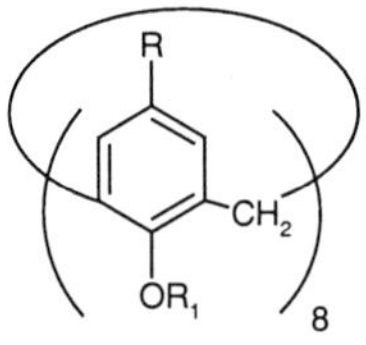

On moving to doubly-bridged derivatives an increased rigidity is observed, which is particularly significant in the case of the 1,5:3,7-doubly-bridged-calix[8]arenes **24d-f** [68, 84]. In their D_{2d} structure a polyhedral, almost spheroidal cavity delimitated by eight (or ten) oxygens is present, of a size matching that of cesium cation. Complexation tests either using ^{1}H NMR, standard two-phase picrate extraction or UV titration have demonstrated high association constants for Cs$^+$ [81, 84, 94]. The induced complexation shifts in ^{1}H NMR clearly demonstrated that the cesium cation is encapsulated within the polar spheroidal cavity (Figure 12). These compounds are also able to complex other alkali metal cations but they show high selectivity for cesium. In particular, the Cs$^+$/Na$^+$ selectivity is close to that of 1,3-alternate calix[4]crown-6 derivatives currently known as the most selective cesium ionophores (see Chapters 20 and 35).

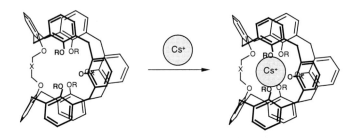

Figure 12. Cesium cation encapsulation by 1,5:3,7-bis-bridged calix[8]arenes 24d-f.

In contrast with other ionophores the rate of complexation of Cs$^+$ by **24d** is slow enough to be monitored by spectrophotometry. In fact, the cation uptake is significantly longer than the mixing deadtime of *ca.* 5 s [95]. This behavior, comparable to that observed for spherands [96a], calixspherands [96b] and cavitands [96c], is attributable to hindered access to the closed ionophoric cavity.

7. Conclusions

Clearly, the chemistry of larger calix[n]arenes is much less explored than that of the smaller homologues. In the case of heptamer and nonamer this is partly due to lack of efficient large-scale preparation procedures for the basic macrocycles. This is not true for calix[8]arenes which are easily obtained in high yields with standard procedures (up to the point that sometimes they are considered unwanted, unavoidable cyclization products). In fact, for all the three macrocycles the main reason has to be the discouraging prospect of an intricate chemistry affording conformationally mobile, unshaped products.

On the other hand, the large dimensions that make their chemistry difficult, are somewhat attractive for the preparation of large receptors able to host molecules bigger than those of simple organic solvents. However, this requires the investment of research efforts to disclose the more amenable paths toward properly functionalized, tridimensionally preorganized host molecules. The work reviewed in this chapter demonstrates that this approach is worthwhile since very interesting molecules can be easily obtained by adapting to the particular chemical and conformational propensity of these macrocycles.

Of course, much work is still to be done requiring the imagination and ingenuity of researchers to further advance the field of supramolecular chemistry.

8. References and Notes

[1] F. Vocanson, R. Lamartine, P. Lanteri, R. Longeray, J. Y. Gauvrit, *New J. Chem.* **1995**, *19*, 825-829.
[2] a) C. D. Gutsche, B. Dhawan, K. H. No, R. Muthukrishnan, *J. Am. Chem. Soc.* **1981**, *103*, 3782-3792 (*Corrigendum* **1984**, *106*, 1891); b) J. H. Munch, C. D. Gutsche, *Org. Synth.* **1990**, *68*, 243-246.
[3] a) D. R. Stewart, C. D. Gutsche, *J. Am. Chem. Soc.* **1999**, *121*, 4136-4146; b) I. Dumazet, J.-B. Regnouf de Vains, R. Lamartine, *Synth. Commun.* **1997**, *27*, 2547-2555.
[4] S. R. Izatt, R. T. Hawkins, J. J. Christensen, R. M. Izatt, *J. Am. Chem. Soc.* **1985**, *107*, 63-66.
[5] a) J. W. Cornforth, P. D'Arcy Hart, G. A. Nicholls, R. J. W. Rees, J. A. Stock, *Brit. J. Pharmacol.* **1955**, *10*, 73-86; b) J. W. Cornforth, E. D. Morgan, K. T. Potts, R. J. W. Rees, *Tetrahedron* **1973**, *29*, 1659-1667.
[6] J. Vicens, T. Pilot, D. Gamet, R. Lamartine, R. Perrin, *C. R. Acad. Sci. Paris* **1986**, *392*, 15.
[7] P. Novakov, S. Miloshev, P. Tuleshkov, I. Gitsov, M. Georgieva, *Angew. Makromol. Chem.* **1998**, *255*, 23-28.
[8] Y. Nakamoto, T. Kozu, S. Oya, S. Ishida, *Netsu. Kokasei Jushi* **1985**, *6*, 78.
[9] Z. Asfari, J. Vicens, *Tetrahedron Lett.* **1988**, *29*, 2659-2660.
[10] B. Gross, J. Jauch, V. Schurig, *J. Microcolumn Sep.* **1999**, *11*, 313-317.
[11] Z. Asfari, J. Vicens, *Makromol. Chem. Rapid Commun.* **1989**, *10*, 181-183.
[12] a) Hitachi Chemical Co., Ltd., Jpn. Kokai Tokkyo Koho, JP 59 104 332, **1984** [*Chem. Abstr.* **1984**, *101*, 191409b]; b) *ibid.* JP 59 104 333, **1984** [*Chem. Abstr.* **1984**, *101*, 191411w]; c) *ibid.* JP 59 12 913, **1984** [*Chem. Abstr.* **1984**, *101*, 24477]; d) *ibid.* JP 59 12 915, **1984** [*Chem. Abstr.* **1984**, *101*, 8163]; e) *ibid.* JP 60 202 113, **1984** [*Chem. Abstr.* **1986**, *104*, 207904]; f) *ibid.* JP 61 106 775, **1986** [*Chem. Abstr.* **1986**, *105*, 231158k].
[13] C. D. Gutsche, P. F. Pagoria, *J. Org. Chem.* **1985**, *50*, 5795-5802.
[14] a) B. Souley, Z. Asfari, J. Vicens, *Pol. J. Chem.* **1992**, *66*, 959-961; b) J. L. Atwood, M. J. Hardie, C. L. Raston, C. A. Sandoval, *Org. Lett.* **1999**, *1*, 1523-1526.
[15] a) C.-H. Tung, H.-F. Ji, *J. Chem. Soc., Perkin Trans. 2* **1997**, 185-188; b) K. H. Ahn, S. G. Kim, J. S. U, *Bull. Korean Chem. Soc.* **2000**, *21*, 813-816.
[16] a) A. Casnati, R. Ferdani, A. Pochini, R. Ungaro, *J. Org. Chem.* **1997**, *62*, 6236-6239; b) P. C. Leverd, V. Huc, S. Palacin, M. Nierlich, *J. Incl. Phenom.* **2000**, *36*, 259-266.
[17] J. Jauch, V. Schurig, *Tetrahedron: Asymmetry* **1997**, *8*, 169-172.
[18] I. E. Lubitov, E. A. Shokova, V. V. Kovalev, *Synlett* **1993**, 647-648.
[19] R. Schätz, C. Weber, G. Schilling, T. Oeser, U. Huber-Patz, H. Irngartinger, C.-W. von der Lieth, R. Pipkorn, *Liebigs Ann.* **1995**, 1401-1408.
[20] G. Sartori, C. Porta, F. Bigi, R. Maggi, F. Peri, E. Marzi, M. Lanfranchi, M. A. Pellinghelli, *Tetrahedron* **1997**, *53*, 3287-3300.
[21] S. E. Biali, V. Böhmer, I. Columbus, G. Ferguson, C. Grüttner, F. Grynszpan, E. F. Paulus, I. Thondorf, *J. Chem. Soc., Perkin Trans. 2* **1998**, 2261-2269.
[22] M. Tsue, M. Ohmori, K.-i. Hirao, *J. Org. Chem.* **1998**, *63*, 4866-4867.
[23] H. Kämmerer, G. Happel, *Makromol. Chem.* **1980**, *181*, 2049-2062.
[24] C. D. Gutsche, L-G. Lin, *Tetrahedron* **1986**, *42*, 1633-1640.
[25] a) C. D. Gutsche, L. J. Bauer, *J. Am. Chem. Soc.* **1985**, *107*, 6059-6063; b) b) W. G. Wang, Q. Y. Zheng, Z. T. Huang, *Synth. Commun.* **1999**, *29*, 3711-3718.
[26] I. Arimura, H. Kawabata, T. Matsuda, T. Muramatsu, H. Satoh, K. Fujio, O. Manabe, S. Shinkai, *J. Org. Chem.*, **1991**, *56*, 301-306.
[27] a) S. Shinkai, K. Araki, O. Manabe, *J. Chem. Soc., Chem. Commun.* **1988**, 187-189; b) J. He, X. W. An, Z. J. Wang, *Acta Chim. Sin.* **1999**, *57*, 1332-1337.
[28] J. K. Moran, E. M. Georgiev, A. T. Yordanov, J. T. Mague, D. M. Roundhill, *J. Org. Chem.* **1994**, *59*, 5990-5998.
[29] V. Bocchi, O. Foina, A. Pochini, R. Ungaro, G. D. Andreetti, *Tetrahedron* **1982**, *38*, 373-378.
[30] T. Nishkubo, A. Kameyama, K. Tsutsui, M. Iyo, *J. Polym. Sci. A: Polym. Chem.* **1999**, *37*, 1805-1814.
[31] C. M. McCartney, T. Richardson, M. B. Greenwood, N. Cowlam, F. Davis, C. J. M. Stirling, *Supramol. Science* **1997**, *4*, 385-390.

[32] a) M. A. McKervey, E. M. Seward, G. Ferguson, B. L. Ruhl, S. J. Harris, *J. Chem. Soc., Chem. Commun.* **1985**, 388-390; b) F. Arnaud-Neu, E. M. Collins, M. Deasy, O. Ferguson, J. J. Harris, B. Kaitner, A. J. Lough, M. A. McKervey, E. Marques, B. L. Ruhl, M. J. Schwing-Weill, E. M. Seward, *J. Am. Chem. Soc.* **1989**, *111*, 8681-8691.
[33] a) S.-K. Chang, I. Cho, *Chem. Lett.* **1984**, 477-478; b) S.-K. Chang, I. Cho, *J. Chem. Soc., Perkin Trans. 1* **1986**, 211-214.
[34] a) S. Wolfe, S. K. Hasan, *Can. J. Chem.* **1970**, *48*, 3572-3579; b) A. Arduini, A. Pochini, S. Reverberi, R. Ungaro, *J. Chem. Soc., Chem. Commun.* **1984**, 981-982.
[35] S.-K.Chang, S.-K. Kwon, I. Cho, *Chem. Lett.* **1987**, 947-948.
[36] S. Shinkai, T. Otsuka, K. Araki, T. Matsuda, *Bull. Chem. Soc. Jpn.* **1989**, *62*, 4055-4057.
[37] a) P. Neri, E. Battocolo, F. Cunsolo, C. Geraci, M. Piattelli, *J. Org. Chem.* **1994**, *59*, 3880-3889; b) P. Neri, C. Geraci, M. Piattelli, *J. Org. Chem.* **1995**, *60*, 4126-4135.
[38] G. M. L. Consoli, F. Cunsolo, M. Piattelli, P. Neri, *J. Org. Chem.* **1996**, *61*, 2195-2198.
[39] Z.-T. Huang, G.-Q. Wang, *Synth. Commun.* **1994**, *24*, 11-22.
[40] S. Angot, K. S. Murthy, D. Taton, Y. Gnanou, *Macromolecules* **1998**, *31*, 7218-7225.
[41] J. Ueda, M. Kamigaito, M. Sawamoto, *Macromolecules* **1998**, *31*, 6762-6768.
[42] Z. Csok, G. Szalontai, G. Czira, L. Kollar, *Supramol. Chem.* **1998**, *10*, 69-77.
[43] S. Shinkai, T. Arimura, K. Araki, H. Kawabata, *J. Chem. Soc., Perkin Trans. 1* **1989**, 2039-2045.
[44] J. M. Harrowfield, M. Mocerino, B. J. Peachey, B. W. Schelton, A. H. White, *J. Chem. Soc., Dalton Trans.* **1996**, 1687-1699.
[45] J. F. Malone, D. J. Marrs, M. A. McKervey, P. O'Hagan, N. Thompson, A. Walker, F. Arnaud-Neu, O. Mauprivez, M.-J. Schwing-Weill, J. F. Dozol, H. Rouquette, N. Simon, *J. Chem. Soc., Chem. Commun.* **1995**, 2151-2153.
[46] a) B. Yao, J. Bassus, R. Lamartine, *An. Quim., Int. Ed.* **1998**, *94*, 65-66; b) S. Bouoit-Montésinos, F. Vocanson, J. Bassus, R. Lamartine, *Synth. Commun.* **2000**, *30*, 911-915.
[47] T. Suzuki, K. Nakashima, S. Shinkai, *Tetrahedron Lett.* **1995**, *36*, 249-252.
[48] C. D. Gutsche, I. Alam, *Tetrahedron* **1988**, *44*, 4689-4694.
[49] S. Jacob, I. Majoros, J. P. Kennedy, *Macromolecules* **1996**, *29*, 8631-8641.
[50] G. R. Newkome, Y. Hu, M. J. Saunders, F. R. Fronczek, *Tetrahedron Lett.* **1991**, *32*, 1133-1136.
[51] C. D. Gutsche, K. C. Nam, *J. Am. Chem. Soc.* **1988**, *110*, 6153-6162.
[52] Y. Shi, H.-J. Schneider, *J. Chem. Soc., Perkin Trans. 2* **1999**, 1797-1803.
[53] M. Almi, A. Arduini, A. Casnati, A. Pochini, R. Ungaro, *Tetrahedron* **1989**, *45*, 2177-2182.
[54] Z.-T. Huang, G.-Q. Wang, *Chem. Ber.* **1994**, *127*, 519-523.
[55] S. Shinkai, K. Araki, T. Tsubaki, T. Arimura, O. Manabe, *J. Chem. Soc., Perkin Trans. 1* **1987**, 2297-2299.
[56] S. Shinkai, H. Kawabata, T. Matsuda, H. Kawaguchi, *Bull. Chem. Soc. Jpn.* **1990**, *63*, 1272-1274.
[57] S. Bouoit, J. Bassus, R. Lamartine, *An. Quim., Int. Ed* **1998**, *94*, 342-344.
[58] J.-C. G. Bünzli, F. Ihringer, *Inorg. Chem. Acta* **1996**, *246*, 195-205.
[59] a) M. Mascal, R. T. Naven, R. Warmuth, *Tetrahedron Lett.* **1995**, *36*, 9361-9364; b) M. Mascal R. Warmuth, R. T. Naven, R. A. Edwards, M. B. Hursthouse, D. E. Hibbs, *J. Chem. Soc., Perkin Trans. 1* **1999**, 3435-3441.
[60] a) R. Muthukrishnan, C. D. Gutsche, *J. Org. Chem.* **1979**, *44*, 3962-3964; b) C. D. Gutsche, *Acc. Chem. Res.* **1983**, *16*, 161-170; c) H. Taniguchi, E. Nomura, R. Maeda, Jpn. Kokai Tokkyo Koho, JP 62 23 156, **1987** [*Chem. Abstr.* **1988**, *109*, 55418s].
[61] P. Neri, C. Geraci, M. Piattelli, *Tetrahedron Lett.* **1993**, *34*, 3319-3322.
[62] G. M. L. Consoli, F. Cunsolo, P. Neri, *Gazz. Chim. Ital.* **1996**, *126*, 791-798.
[63] F. Cunsolo, G. M. L. Consoli, M. Piattelli, P. Neri, *Tetrahedron Lett.* **1995**, *36*, 3751-3754.
[64] F. Cunsolo, G. M. L. Consoli, M. Piattelli, P. Neri, *J. Org. Chem.* **1998**, *63*, 6852-6858.
[65] J. M. Harrowfield, M. I. Ogden, W. R. Richmond, B. W. Skelton, A. H. White, *J. Chem. Soc., Perkin Trans. 2* **1993**, 2183-2190.
[66] P. Neri, G. M. L. Consoli, F. Cunsolo, C. Rocco, M. Piattelli, *J. Org. Chem.* **1997**, *62*, 4236-4239.
[67] a) K. Araki, K. Iwamoto, S. Shinkai, T. Matsuda, *Bull. Chem. Soc. Jpn.* **1990**, *63*, 3480-3485; b) R. G. Janssen, W. Verboom, B. T. G. Lutz, J. H. van der Maas, M. Maczka, J. P. M. van Duynhoven, D. N. Reinhoudt, *J. Chem. Soc., Perkin Trans. 2* **1996**, 1869-1876.
[68] A. Ikeda, K. Akao, T. Harada, S. Shinkai, *Tetrahedron Lett.* **1996**, *37*, 1621-1624.
[69] M. Takeshita, T. Suzuki, S. Shinkai, *J. Chem. Soc., Chem. Commun.* **1994**, 2587-2588.
[70] a) J. D. van Loon, A. Arduini, L. Coppi, W. Verboom, A. Pochini, R. Ungaro, S. Harkema, D. N. Reinhoudt, *J. Org. Chem.* **1990**, *55*, 5639-5646; b) J. de Mendoza, M. Carramolino, F. Cuevas, P. M. Nieto, P. Prados, D. N. Reinhoudt, W. Verboom, R. Ungaro, A. Casnati, *Synthesis* **1994**, 47-50.

[71] A. Bottino, F. Cunsolo, M. Piattelli, D. Garozzo, P. Neri, *J. Org. Chem.* **1999**, *64*, 8018-8020.
[72] A. Ninagawa, K. Cho, H. Matsuda, *Makromol. Chem.* **1985**, *186*, 1379-1385.
[73] F. Grynszpan, O. Aleksiuk, S. E. Biali, *Pure Appl. Chem.* **1996**, *68*, 1249-1254.
[74] a) C. D. Gutsche, A. E. Gutsche, A. I. Karaulov, *J. Incl. Phenom.* **1985**, *3*, 447-451; b) M. Czugler, S. Tisza, G. Speier, *J. Incl. Phenom.* **1991**, *11*, 323-331; c) P. C. Leverd, V. Huc, S. Palacin, M. Nierlich, *Z. Krist.-New Cryst. Struct.* **2000**, *215*, 549-552.
[75] a) N. P. Clague, W. Clegg, S. J. Coles, J. D. Crane, D. J. Moreton, E. Sinn, S. J. Teat, N. A. Young, *Chem. Commun.* **1999**, 379-380; b) N. P. Clague, J. D. Crane, D. J. Moreton, E. Sinn, S. J. Teat, N. A. Young, *J. Chem. Soc., Dalton Trans.* **1999**, 3535-3536.
[76] a) G. E. Hofmeister, E. Alvarado, J. A. Leary, D. I. Yoon, S. F. Pedersen, *J. Am. Chem. Soc.* **1990**, *112*, 8843-8851; b) V. C. Gibson, C. Redshaw, W. Clegg, M. R. Elsegood, *J. Chem. Soc., Chem. Commun.* **1995**, 2371-2372; c) J.-C. Bünzli, F. Ihringer, P. Dumy, C. Sager, R. D. Rogers, *J. Chem. Soc., Dalton Trans.* **1998**, 497-503.
[77] a) M. Perrin, S. Lecocq, *C. R. Acad. Sci. Paris* **1990**, *310*, 515-; b) G. D. Andreetti, F. Ugozzoli, Y. Nakamoto, S.-I. Ishida, *J. Incl. Phenom.* **1991**, *10*, 241-253; c) T. Harada, S. Shinkai, *J. Chem. Soc., Perkin Trans. 2* **1995**, 2231-2242; d) I. Dumazet, S. Bouoit-Montesinos, M. Perrin, R. Lamartine, *Abstracts of The Fifth International Conference on Calixarene Chemistry*, University of Western Australia (Perth, Australia) **1999**, P65.
[78] F. Cunsolo, M. Piattelli, P. Neri, *J. Chem. Soc., Chem. Commun.* **1994**, 1917-1918.
[79] Y. S. Tsantrizos, W. Chew, L. D. Colebrook, F. Sauriol, *Tetrahedron Lett.* **1997**, *38*, 5411-5414.
[80] a) C. Geraci, M. Piattelli, P. Neri, *Tetrahedron Lett.* **1996**, *37*, 3899-3902; b) C. Geraci, M. Piattelli, G. Chessari, P. Neri, *J. Org. Chem.* **2000**, *65*, 5143-5151; c) C. Geraci, M. Piattelli, P. Neri, unpublished results.
[81] C. Geraci, G. L. M. Consoli, M. Piattelli, P. Neri, submitted.
[82] C. Geraci, M. Piattelli, P. Neri, *Tetrahedron Lett.* **1995**, *36*, 5429-5432.
[83] F. Cunsolo, G. M. L. Consoli, M. Piattelli, P. Neri, *Tetrahedron Lett.* **1996**, *37*, 715-718.
[84] C. Geraci, G. Chessari, M. Piattelli, P. Neri, *Chem. Commun.* **1997**, 921-922.
[85] C. Geraci, A. Bottino, M. Piattelli, E. Gavuzzo, P. Neri, *J. Chem. Soc., Perkin Trans. 2* **2000**, 185-187.
[86] C. Geraci, M. Piattelli, P. Neri, *Tetrahedron Lett.* **1996**, *37*, 7627-7630.
[87] a) J. Gloede, *Abstracts of The Fifth International Conference on Calixarene Chemistry*, University of Western Australia (Perth, Australia) **1999**, P29; b) J. Gloede, B. Costisella, *ibid.* **1999**, P30.
[88] a) H. M. Chawla, K. Srinivas, *J. Chem. Soc., Chem. Commun.* **1994**, 2593-2594; b) H. M. Chawla, K. Srinivas *Tetrahedron Lett.* **1994**, *35*, 2925-2928; c) H. M. Chawla, personal communication.
[89] A. Arduini, A. Pochini, A. Secchi, R. Ungaro, *J. Chem. Soc., Chem. Commun.* **1995**, 879-880.
[90] S. Caccamese, G. Principato, C. Geraci, P. Neri, *Tetrahedron: Asymmetry* **1997**, *8*, 1169-1173.
[91] A. Ikeda, Y. Suzuki, S. Shinkai, *Tetrahedron: Asymmetry* **1998**, *9*, 97-105.
[92] C. Jaime, J. de Mendoza, P. Prados, P. M. Nieto, C. Sanchez, *J. Org. Chem.* **1991**, *56*, 3372-3376.
[93] R. Ungaro, A. Pochini in *Calixarenes, a Versatile Class of Macrocyclic Compounds* (Eds. J. Vicens, V. Böhmer), Kluwer, Dordrecht, **1991**, pp. 127-147.
[94] A. Ikeda, Y Suzuki, K. Akao, S. Shinkai, *Chem. Lett.* **1996**, 963-964.
[95] Y. Suzuki, H. Otsuka, A. Ikeda, S. Shinkai, *Tetrahedron Lett.* **1997**, *38*, 421-424.
[96] a) D. J. Cram, *Science* **1983**, *219*, 1177-1183; b) W. I. Iwema-Bakker, M. Haas, C. Khoo-Beattie, R. Ostaszewski, S. M. Franken, H. J. Jr. den Hertog, W. Verboom, D. de Zeeuw, S. Harkema, D. N. Reinhoudt, *J. Am. Chem. Soc.* **1994**, *116*, 123-133; c) D. J. Cram, J. M. Cram, *Container Molecules and Their Guests*, Royal Society of Chemistry, Cambridge, **1994**.

Chapter 6

THIA-, MERCAPTO-, AND THIAMERCAPTO-CALIX[4]ARENES

MIR WAIS HOSSEINI

Université Louis Pasteur, Institut Le Bel, 4, rue Blaise Pascal, F-67000 Strasbourg, France. e-mail: hosseini@chimie.u-strasbg.fr

1. Introduction

The increasing interest in calixarenes over the last two decades results from the ease and reproducibility of basic synthetic procedures [1-4], and the ease of structural and functional modifications. Also significant is the conformational isomerism of calixarenes, a property currently positively explored in the design of receptors, catalysts, and building blocks. Thiacalixarenes (*e.g.*,**1**-**3**) represent a new class of calixarene derivatives for which the replacement of CH_2 junctions by S atoms introduces additional coordination sites and modifies the dimensions of the cavity [5,6].

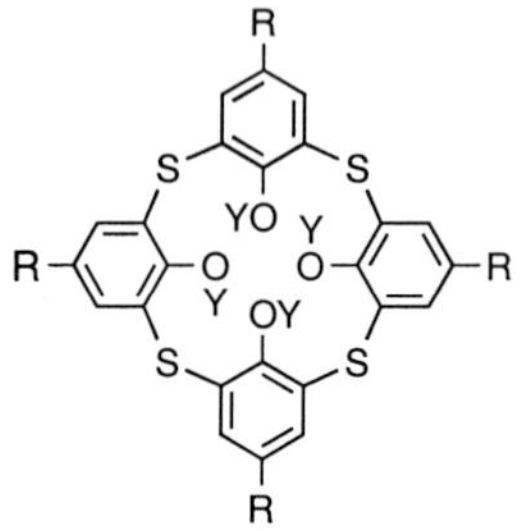

1 Y = H, R = *t*-Bu
2 Y = H, R = *t*-Oct
3 Y = H, R = H

2. Synthesis of Thiacalix[4]arenes

The *p*-*t*-butyl derivative **1** of tetrathiacalix[4]arene, as for the methylene calixarene series, is of especial importance as a readily synthesised starting material. The *p*-*t*-octyl derivative **2** can be obtained similarly [6] and relatively tedious multi-step syntheses [5-7] offer considerable possibilities for other variations in the structures (See Chapter 1). The following summary concerns methods known for the synthesis of thiacalix[4]arene **3** itself and an already large range of derivatives.

2.1. Synthesis of Tetrathiacalix[4]arene

Although first attempted unsuccessfully by reaction of phenol with elemental sulfur, synthesis of **3** was readily achieved [8] by de-*tert*-butylation of **1** using the procedure developed for the synthesis of calix[4]arene [1].

2.2. Synthesis of Tetrasulfinylcalix[4]arene

Oxidation at sulfur in **1** and **3** can generate either sulfoxide or sulfone entities. Conversion of all four sulfur atoms in **1** and **3** to sulfoxide leads to new sulfinyl-type ligands **4** and **5** respectively. The first synthesis of **4** by partial oxidation of **1** using sodium perborate in $CHCl_3/H_2O$ was reported in 1998 [7]. H_2O_2 in acetic acid was later found to be effective also [9]. **5** was obtained by partial oxidation of **3** by *m*-chloroperbenzoic acid (*m*-CPBA) in CH_2Cl_2 [9].

Z. Asfari et al. (eds.), Calixarenes 2001, 110–129.

In principle, the tetrasulfoxides **4** and **5** may adopt many isomeric forms [9]. Four labile conformations, idealised as *cone*, *partial cone*, *1,2-alternate* and *1,3-alternate*, are expected, and each of these can be associated with various configurations defined in terms of relative orientations of the substituents on the configurationally stable S atoms. **4** and **5**, seemingly single isomers in the solid state, have been structurally characterised by X-ray crystallography [9]. Both adopt a *1,3-alternate* conformation with the oxygen atoms of the SO groups alternately above and below the plane defined by four S atoms (Sections 3.4, 3.5).

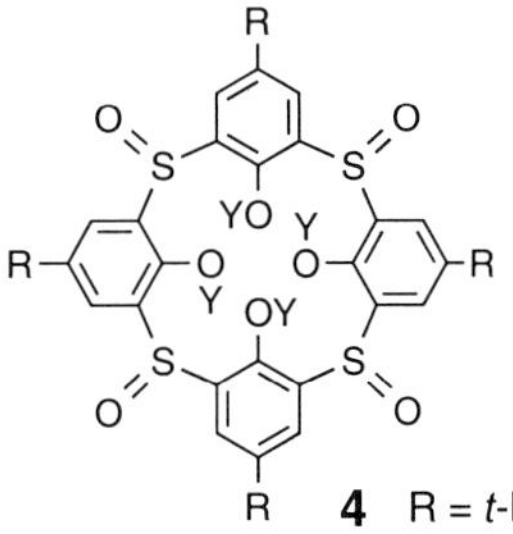

The tetrasulfinyl derivative **7** in the *cone* conformation has been prepared [10] by a strategy based on benzylation of the hydroxyl moieties by treatment with benzyl bromide in a THF-DMF mixture in the presence of NaH. Oxidation using sodium perborate in $CHCl_3$-acetic acid gave the *cone* tetrasulfinyl derivative **6**, shown by X-ray crystallography to have all four S=O groups oriented towards the same face of the molecule. Removal of the benzyl groups provided the tetrasulfinyl stereoisomer **7** for which the S=O groups retain the *syn-syn-syn–syn* configuration.

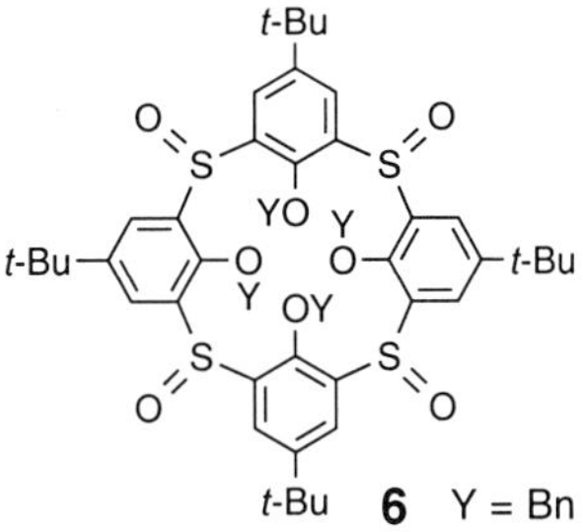

4 R = *t*-Bu
5 R = H

6 Y = Bn
7 Y = H

2.3. SYNTHESIS OF TETRASULFONYLCALIX[4]ARENES

The first synthetic strategy [7] for the complete oxidation of **2** to **8** was based on the protection of OH moieties as methoxy groups (in **9**) prior to the oxidation of the sulfide linkages to sulfones. Initial efforts to perform the oxidation with 30 % H_2O_2 in acetic acid by heating at reflux for 8 days gave a very poor yield (10 %), though this could be increased to 88 % using *m*-CPBA. Demethylation of the product **10** with BBr_3 in CH_2Cl_2, however, again gave a poor yield, so that direct reaction of **5** with 30 % H_2O_2 in acetic acid (59 % yield) proved a simpler route to **8** [11].

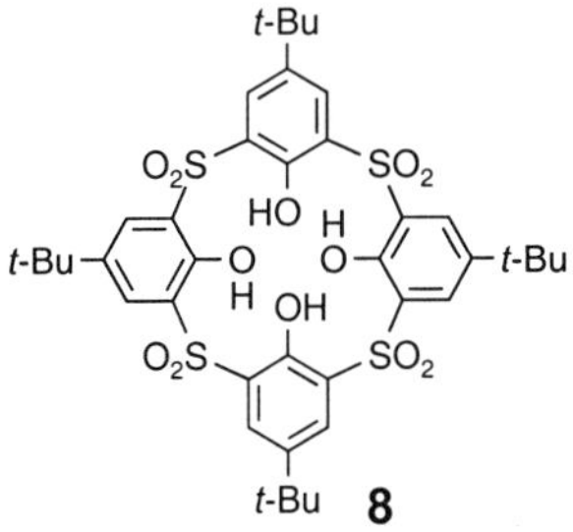

8 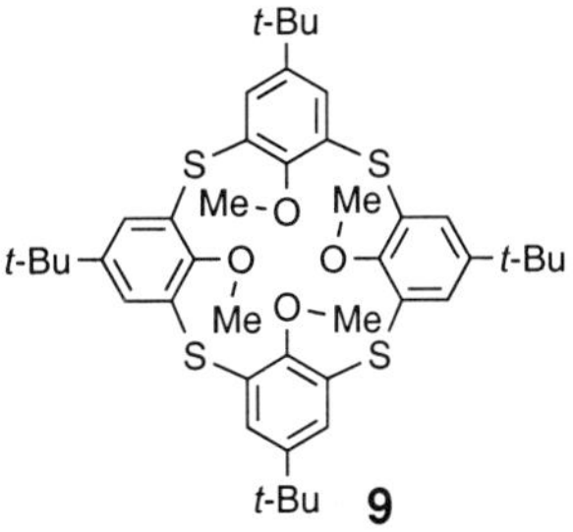   **9** 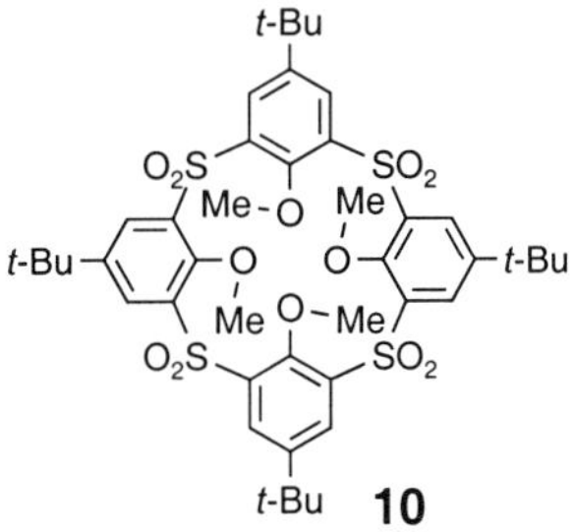   **10**

2.4. FUNCTIONALISATION OF THIACALIX[4]ARENE DERIVATIVES AT THE LOWER RIM

The methylation of **1**, giving **9**, was first achieved during synthesis of the tetrasulfonyl derivative **8** [11]. In systematic studies of reactions of both **1** and **3** [12, 13], alkylations were performed using RI or RBr in acetone or acetonitrile at reflux in the presence of M_2CO_3 (M = Na, K, Cs) [13]. The better solvent appeared to be acetone, where both K_2CO_3 and Cs_2CO_3 gave high yields. **12-14** as well as **17-18** were all found to adopt the *1,3-alternate* conformation, also established in the solid state for **12** by an X-ray diffraction study [13]. For **16**, the conformational behaviour in $CDCl_3$ solution has been studied by NMR spectroscopy [12].

11 Y = Me, R = *t*-Bu	**17** Y = *n*-Pr, R = H	
12 Y = Et, R = *t*-Bu	**18** Y = *n*-Bu, R = H	
13 Y = *n*-Pr, R = *t*-Bu	**19** Y = $(CH_2)_2$OMe, R = H	
14 Y = *n*-Bu, R = *t*-Bu	**20** Y = CH_2CO_2Et, R = *t*-Bu	
15 Y = Me, R = H	**21** Y = CH_2CO_2Et, R = H	
16 Y = Et, R = H	**22** Y = CH_2CO_2H, R = H	**23** Y = $(CH_2)_2$OMe

The synthesis of **19** was achieved in 58 % yield upon treatment of **3** by 1-methoxy-2-(toluene-4-sulfonyloxy)ethane in DMF in the presence of Cs_2CO_3 [14]. Again, the *1,3-alternate* conformation of **19** was established in the solid state by X-ray crystallography [14]. *m*-CPBA in $CHCl_3$ oxidises **19** to the tetrasulfone **23** (59 % yield) [14].

Several syntheses are known for the tetra(ethyl ester) derivatives **20-22** [14-17]. The role played by alkali metal cations in the synthesis as well as in the distribution of conformers of **20** was studied by treatment of **1** in refluxing acetone by excess of ethylbromoacetate in the presence of excess of M_2CO_3. The highest overall yield of *ca* 76-80 % (all yields are based on isolated compounds, though, when the yields were less than *ca* 3 %, they were considered negligible) was obtained after 7 days [17]. Except for the $20_{1,2-A}$ (1,2-A and 1,3-A refer to *1,2-* and *1,3-alternate* conformers, whereas C and PC refer to *cone* and *partial cone* conformers) isomer which was not isolated, the other three isomers were purified and structurally studied both in solution and in the solid state (see ahead). Under similar conditions for *p-t*-butyl-calix[4]arene, no trace of the *1,2-* or *1,3-alternate* conformers could be detected [18]. Although no $20_{1,2-A}$ isomer could be isolated, up to 85 % of the $20_{1,3-A}$ isomer was obtained. Using Li_2CO_3, no trace of the tetrasubstituted **20** was detected. Na_2CO_3 gave exclusively the 20_C conformer, with an overall yield of 80 %. In the case of K_2CO_3, 70 % of 20_{PC} and 30 % of $20_{1,3-A}$ were formed, while in the presence of Cs_2CO_3, $20_{1,3-A}$ (85 %) and 20_{PC} conformer (15 %) were obtained.

The tetra-ester **21** was obtained in 85 % yield upon refluxing a solution of **3** and ethylbromoacetate in dry acetone in the presence of Cs_2CO_3 and the tetra-acid derivative **22** was obtained quantitatively by hydrolysis of **21** using LiOH in THF/H_2O [14]. Again, the *1,3-alternate* conformation for both compounds **21** and **22** was assigned in the solid state by X-ray structure determinations.

Introduction of dansyl groups at the lower rim of **1** provides water soluble **24-27**, useful for detection of metal ions in aqueous solution [19]. The dansylation process using dansylsulfonyl chloride in the presence of NaH afforded mainly the di- and tri-dansyl derivatives **25** and **26** respectively, with the mono- and tetra-substituted compound **24** and **27** present in trace amounts.

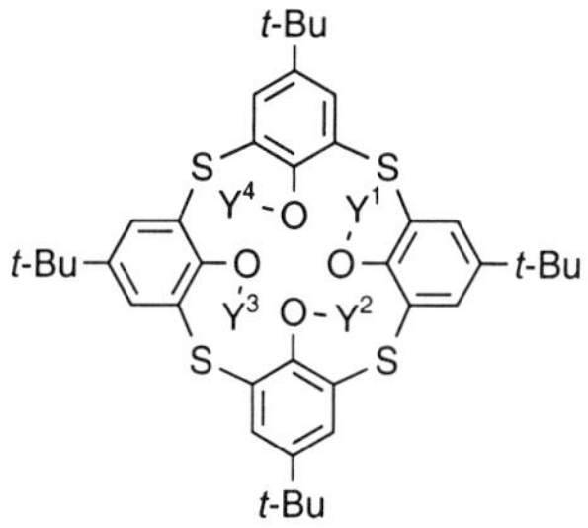

24 Y¹ = Y² = Y³ = H, Y⁴ = dansyl
25 Y¹ = Y³ = H, Y² = Y⁴ = dansyl
26 Y¹ = H, Y² = Y³ = Y⁴ = dansyl
27 Y¹-Y⁴ = dansyl

dansyl = (naphthalene with NMe$_2$ and SO$_2$)

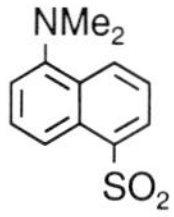

The amidation of *cone* **22** by (*S*)-1-phenylethylamine was used to provide the tetra-substituted derivative **28** as a chiral stationary phase for capillary gas chromatography. Using (*S*)-1-phenylethylamine, the reaction (CH$_2$Cl$_2$, Et$_3$N) was performed in 76 % yield on the acid chloride derivative **30** of **29** [20] (see Chapter 36).

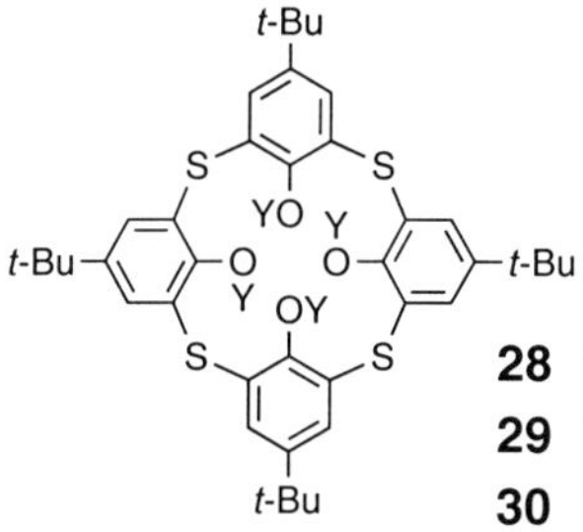

28 Y = H$_2$C-C(O)-NH-CH(Ph)
29 Y = CH$_2$CO$_2$H
30 Y = CH$_2$COCl

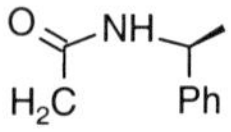

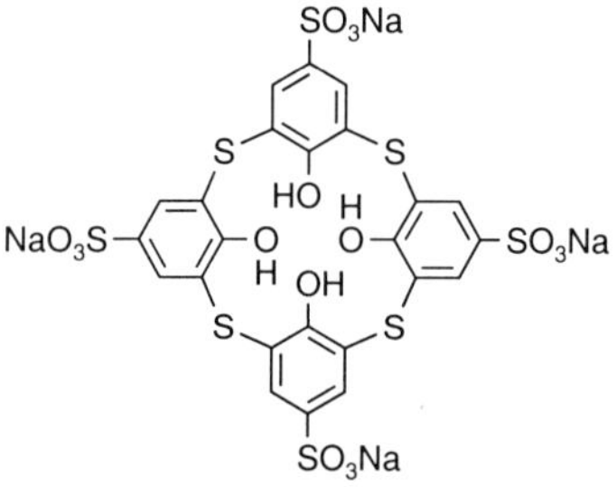

31

Recently, the activation of an aromatic nucleus towards nucleophilic substitution produced by sulfinyl or sulfonyl groups adjacent to the reaction site has been exploited in the efficient synthesis of amino calixarenes from the tetramethyl ethers of both the sulfinyl and sulfonyl derivatives of *p-t*-butyl-tetrathiacalix[4]arene [21].

2.5. FUNCTIONALISATION OF THIACALIX[4]ARENE DERIVATIVES AT THE UPPER RIM

Sulfonation of **1** by concentrated H$_2$SO$_4$ at 80 °C gave the tetrasulfonate **31** in 63 % yield [20]. The conformational behaviour of **31** has not been studied but its ability to extract halogenated organic substrates from water has been reported [22, 23].

3. Solid State Structural Investigations

The lack of CH$_2$ group signals in thiacalixarenes renders NMR spectroscopy much less useful in assignment of conformations than it is for ordinary calixarenes, so that structure determinations by crystallography are correspondingly more important.

3.1. STRUCTURE OF *p-t*-BUTYL-TETRATHIACALIX[4]ARENE **1** IN THE SOLID STATE

In the solid state, **1** adopts a cone conformation and forms various solvates involving cavity inclusion [6c, 8] with some selectivity, as seen in the crystallisation of the CH$_2$Cl$_2$ and CHCl$_3$ inclusates on addition of MeOH to solutions of **1** in these two solvents but a MeOH inclusate on diffusion of MeOH into a toluene solution. Structure determinations [8a] (Figure 1) on these 3 inclusates showed as common features that: i) in all three cases, the substrate penetrates deeply into the cavity of the calix; ii) both CH_2Cl_2 and $CHCl_3$ substrates as well as the *p-t*-butyl groups were found to be disordered in the crystalline phase; iii) the average distance between two adjacent oxygen atoms was 2.85 Å. Although the stabilising role of the included substrate cannot be excluded, the rather short distance between adjacent oxygen atoms for all three structures indicates the existence of an intramolecular H-bonds array stabilising the cone conformation; iv) as expected, the dimensions of the thiacalix **1** (*ca* 5.5 Å x 7.8 Å) were found to be larger than for *p-t*-butyl-calix[4]arene (*ca* 5.1 Å x 7.2 Å).

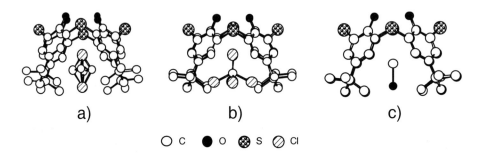

Figure 1. Lateral views of the crystal structures of the inclusion complexes formed by compound **1** with CH_2Cl_2 (a), $CHCl_3$ (b), and MeOH (c). For clarity, H atoms are not represented. Both CH_2Cl_2 and $CHCl_3$ substrate as well as the p-t-butyl groups of the receptor molecule were found to be disordered.

In the case of CH_2Cl_2 inclusion complex, a centrosymmetric array is formed by alternate columns composed of inclusion complexes (Figure 2a). Interestingly, within each infinite column, the inclusion complexes are packed one on the top of the other, leading thus to a distance of 3.31 Å between one chlorine atom of the substrate pointing out of the cavity towards the four oxygen atoms belonging to the next thiacalix unit (Figure 2b). For the other two cases, the same type of packing was observed.

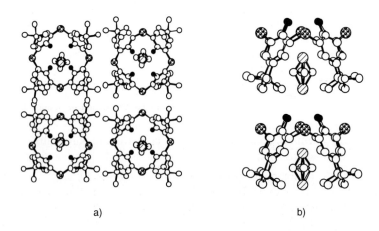

Figure 2. Portions of the structure of the inclusion complex formed by **1** with CH_2Cl_2.

3.2. STRUCTURE AND PACKING OF TETRATHIACALIX[4]ARENE 3 CRYSTALS

For **3**, an X-ray structure determination revealed that: (i) **3** crystallised in the hexagonal crystal system with $P6_3/m$ as the space group and the unit cell contained both **3** and 2 water molecules not localised within the cavity of the host molecule; (ii) **3** adopts a *cone* conformation; (iii) the average C-S and C-O distances are *ca* 1.78 Å and 1.37 Å respectively; (iv) the average distance between two adjacent sulfur atoms is 5.51 Å; v) the average distance between two adjacent oxygen atoms is 2.64 Å, which may again be indicative of an intramolecular H-bonded array; (vi) in marked contrast with **1**, **3** forms, by self-inclusion, trimeric units (Figure 3). In the case of calix[4]arene and p-*t*-butyl-calix[4]-arene, similar structural parallels exist [24]. For **1**, with bulky groups at the upper rim, self-inclusion seems to be unfavourable. Consequently, **1** forms inclusion complexes with appropriate substrates. For **3**, self-inclusion precludes the inclusion of small guest molecules.

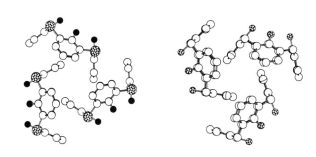

Figure 3. Crystal structures of the compounds 3 (left) and calix[4]-arene (right): lateral views of the trimeric inclusion complex.

3.3. *p-t-* BUTYL-TETRASULFINYL TETRATHIACALIX[4]ARENE 4 IN THE SOLID STATE

Unlike *cone* **1** and **3**, compound **4** adopts the *1,3-alternate* conformation (Figure 4) [9]. The oxygen atoms of the SO groups (d_{SO} = 2.62-2.72 Å) alternate above and below the plane defined by four S atoms. The *1,3-alternate* conformation seems to be stabilised by four strong intramolecular hydrogen bonds between OH and SO groups (<OH···OS> 2.67 Å).

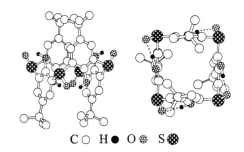

C ○ H ● O ⊛ S ●

Figure 4. Lateral and top views of the solid state structure of 4 showing the 1,3-alternate conformer of the calix unit and the alternate orientation of the SO groups.

3.4. TETRASULFINYL TETRATHIACALIX[4]ARENE 5 IN THE SOLID STATE

Crystals of **5** were tetragonal, space group $P4_2/n$, with **5** adopting a conformation very similar to that of **3**, with <OH···OS> 2.64 Å [9]. Interestingly, the packing pattern showed a 3-D network obtained by strong stacking interactions between all four aromatic groups of the molecular units (the distance between centroids of the aromatic rings is 3.49 Å).

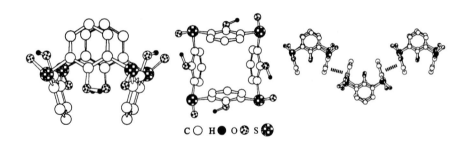

Figure 5. Lateral (left) and top (middle) views of the solid state structure of **5** showing the *1,3-alternate* conformer of the calix unit and the alternate orientation of the SO groups, as well as a portion of the structure (right) showing the formation of a 3-D network through stacking of the aromatic groups.

The *1,3-alternate* conformation has also recently been found in a chiral, partially-oxidised disulfinyl derivative of *p-t*-butyl-tetrathiacalix[4]arene where the two SO groups are adjacent [25].

3.5. *p-t*- BUTYL-TETRASULFONYLCALIX[4]ARENE 8 IN THE SOLID STATE

For the sulfone **8** (Figure 6), again the conformation was found to be *1,3-alternate* with oxygen atoms of the sulfones pointing outward [11]. The average SO and CS bond distances are *ca* 1.43 Å *ca* 1.77 Å respectively. The average distance between two adjacent sulfur atoms is *ca* 5.52 Å.

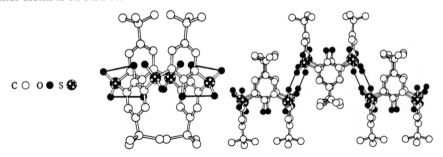

Figure 6. A portion of the X-ray structure of the 3-D network formed between consecutive compounds **8**. The 3-D network is obtained by intermolecular H-bonding between OH and SO groups. H atoms are not represented.

The *1,3-alternate* conformation for compound **8** was rationalised in terms of H-bonding between the OH and SO_2 groups (O···OS distances varying from 2.81 to 3.02 Å). Compound **8** was also found to form a 3-D network through intermolecular H-bonding between the OH and SO_2 groups belonging to adjacent units, with an average O···OS distance of *ca* 2.87 Å (Figure 6).

3.6. *p-t*- BUTYL-TETRASULFONYLCALIX[4]ARENE ETHER 10 IN THE SOLID STATE

For **10** (Figure 7), the *1,3-alternate* conformation was again observed in the solid state [11], showing that S···OH bonding is not necessary to stabilise this form, as is also indicated by the finding of the *1,3-alternate* conformation for the thiacalixarene ether **11** [11].

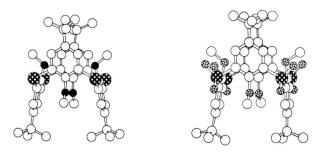

*Figure 7. X-ray structures of **10** (left) and **11** (right) demonstrating the 1,3-alternate conformation adopted in the crystalline phase.*

3.7. STRUCTURES OF TETRATHIACALIX[4]ARENE TETRA ETHER **19** AND TETRA-SULFONYLCALIX[4]ARENE ETHER **23** IN THE SOLID STATE

Both **19** and **23** adopt the *1,3-alternate* conformation in the solid state (Figure 8) [14]. Suitable single crystals of both were obtained from a CH_2Cl_2/MeOH mixture. Unsolvated **19** crystallises in a rhombohedral form, space group $P3$. The average CS distance and CSC angle are 1.782 Å and 106.4°, respectively. The average C-OPh and C-OMe distances are 1.377 Å and 1.320 Å, respectively. Unsolvated **23** crystallises in a tetragonal form, space group $P4_2/n$. The average CS distance, CSC and OSO angles are 1.785 Å, 104.5° and 117.7°, respectively. The average C-OPh and C-OMe distances are 1.377 Å and 1.402 Å, respectively.

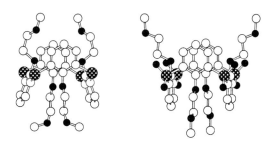

*Figure 8. X-ray structures of **19** (left) and **23** (right) showing the 1,3-alternate conformation.*

3.8. *p-t-* BUTYL-TETTRATHIACALIX[4]ARENE ESTER **20** IN THE SOLID STATE

The ^1H and ^{13}C nmr spectra of all three isolated conformers of compound **20** (**20**$_C$, **20**$_{PC}$ and **20**$_{1,3-A}$) show sharp signals indicating the presence of conformationally rigid isomers. As expected for their symmetry, the same number of signals and the same pattern are observed for both **20**$_C$ and **20**$_{1,3-A}$ conformers. The lower symmetry **20**$_{PC}$ isomer, however, gave disctinctive spectra. The solid state conformations of all were established by X-ray diffraction (Figure 9) [17]. Unsolvated **20**$_C$ crystallised in a monoclinic form, space group $P2_1/n$. The average CS distance and CSC angle are 1.785 Å and 100.1°, respectively, the average C=O distance is 1.190 Å, and the average C-O (arom) and C-O (ethyl) distances are 1.376 Å and 1.326 Å, respectively.

20$_{PC}$ crystallised in the monoclinic system, space group $P2_1/n$. In the unit cell two crystallographically slightly different molecules of **20**$_{PC}$ as well as a CH_2Cl_2 molecule (not included in the cavity of the calix) are present. Only one of the two molecules is shown in Figure 9. The average CS distance and CSC angle are 1.784 Å and 100.5°, respectively. The average C=O distance is 1.191 Å whereas the average C-O (arom) and C-O (ethyl) distances are found to be 1.377 Å and 1.333 Å, respectively. Unsolvated $20_{1,3-A}$ crystallised in monoclinic form, space group $I2/a$. The average CS distance and CSC angle are 1.777 Å and 102.25°, respectively. The average C=O distance is 1.169 Å whereas the average C-O (arom) and C-O (ethyl) distances are 1.377 Å and 1.268 Å, respectively.

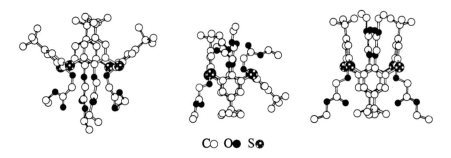

Figure 9. X-ray structures of 20_C (left), 20_{PC} (middle) and $20_{1,3-A}$.

3.9. STRUCTURE OF TETRATHIACALIX[4]ARENE TETRA ESTER 21 IN THE SOLID STATE.

Unsolvated **21** crystallises from CH_2Cl_2/MeOH in a monoclinic form, space group $C2/c$ (Figure 10) [14]. It adopts the *1,3-alternate* conformation. The average CS distance and CSC angle are 1.780 Å and 101.7°, respectively. The average C=O distance is 1.186 Å whereas the average C-O (arom) and C-O (ethyl) distances are 1.420 Å and 1.330 Å, respectively.

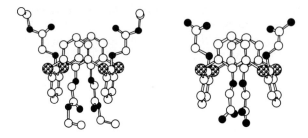

Figure 10. X-ray structures of **21** (left) and **22** (right) showing the 1,3-alternate conformation.

3.10. TETRATHIACALIX[4]ARENE TETRA ACID 22 IN THE SOLID STATE

22, in *1,3-alternate* conformation, crystallises from MeOH/H_2O in a triclinic form, space group P-1 (Figure 10). In the lattice, in addition to two crystallographically different calix units, two H_2O and one CH_3OH molecules are present. The average CS distance and CSC angle are 1.782 Å and 105.7°, respectively. The average C=O and C-OH distances are 1.207 Å and 1.310 Å, respectively, whereas the average C-O (arom) and C-O (ethyl) distances are 1.433 Å and 1.381 Å, respectively.

4. Synthesis of Thiamercaptocalix[4]arene

The first attempt to prepare *p-t*-butyl-tetramercaptotetrathiacalix[4]arene **32**, containing the thiacalixarene framework bearing four SH groups, was based on treatment of *p-t*-butyl-thiophenol by elemental sulfur under the same conditions as those used for the preparation of **1** [5] but failed. Finally, compound **32** was obtained by a stepwise strategy starting from compound **1** and subsequent transformation of all four OH groups into SH functionalities [26].

32	$Y^1 - Y^4 = SH$
33 or **34**	$Y^1 = Y^2 = OH, Y^3 = Y^4 = OCSNMe_2$
35, 36 or **37**	$Y^1 = OH, Y^2 - Y^4 = OCSNMe_2$
38 or **39**	$Y^1 - Y^4 = OCSNMe_2$
40 or **41**	$Y^1 - Y^4 = SCONMe_2$

The key step in the strategy used was the thermal rearrangement of the $OCSNMe_2$ groups into $SCONMe_2$ [27]. This method has been also employed for the preparation of mercaptocalix[4]arene derivatives [28-33]. When compound **1** was treated with N,N-dimethylthiacarbamoyl chloride ($ClCSNMe_2$) in diglyme at 180 °C and in the presence of NaH, the condensation produced mainly a variety of di- and tri- substituted derivatives and no trace of the tetrasubstituted conformer. However, all bis(thiocarbamoyl) compounds **33** (10 %) and **34** (10 %) as well as tris(thiocarbamoyl) compounds **35** (10 %), **36** (10 %) and **37** (8 %) were isolated and structurally characterised either by 2-D NMR or by X-ray diffraction on single-crystals in the case of **34**, **35** and **36** (unpublished). The tetra substituted derivatives **38** (12 %) and **39** (80 %) were obtained upon treatment of **1** by $ClCSNMe_2$ in refluxing acetone and in the presence of K_2CO_3. Whereas the structure of **38** was elucidated in the solid state by X-ray diffraction (unpublished), the conformation of compound **39** was established by 2-D NMR methods [26].

The effect of the cation in M_2CO_3 (M = Li, Na, K, and Cs) on the yield of formation of the tetrasubstituted derivatives was also studied. Whereas for Li^+ cation, 30 % of the di-substituted compound **34** and 5 % of the tri-substituted conformer **36** were obtained, in the presence of Na^+, 30 % of **4** and 3 % of **7** could be isolated. In the case of Cs^+ cation, as for K^+, 12 % of **38** and 80 % of **39** were obtained. In marked contrast with *p-t*-butyl-calix[4]arene for which under same conditions no tetra- substituted derivative was reported, in the case of **1**, a mixture of tetrasubstituted conformers was obtained in 92 % yield.

Since for the final compound **32**, the SH groups may freely pass though the annulus of the calix, the separation of the *1,2-alternate* **38** and the *1,3-alternate* **39** conformers was not attempted and the thermal rearrangement was performed in quantitative yield on the mixture of **38** and **39** at 300 °C in vacuum leading to both transposed conformers **40** and **41**. Whereas the deprotection of the sulfur groups leading to the desired compound **32** failed using $LiAlH_4$ in refluxing THF, the treatment of the mixture of **40** and **41** by hydrazine hydrate at 100 °C for 18 h afforded the desired compound **32** in 90 % yield. Starting from **1**, the overall yield for the synthesis of **32** was 80 %.

5. Structural Analysis of *p-t*-Butyl-Tetrathiatetramercaptocalix[4]arene 32

As expected, in CDCl$_3$ solution at 25 °C, the ^{1}H-NMR spectrum of **32** was extremely simple and comprised a singlet corresponding to the aromatic protons and two other singlets corresponding to the *t*-butyl and SH groups. This does not allow differentiation of the *cone* and *1,3-alternate* conformers, so the solid state conformation of **32** was studied by X-ray diffraction on a single-crystal (Figure 11) [26]. In contrast with the parent compound **1** which adopts the *cone* conformation in the crystalline phase [8], compound **32** adopts the *1,3-alternate* conformation. The aromatic cycles located on the same face of the ligand are almost parallel. The average C-S and C-SH distances are 1.78 Å and 1.76 Å, respectively.

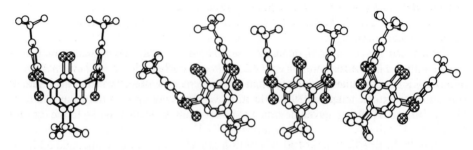

Figure 11. X-ray structure of 32 and disposition of different units in the lattice.

6. Synthesis of Mercaptocalix[4]arenes

Many binuclear complexes, based on ditopic receptors possessing convergent sets of binding sites (endoditopic ligands), have been reported. In these complexes, intrinsic molecular proprieties such as magnetic coupling, redox activity and optical features may be tuned with remarkable precision. In principle, these molecular properties may be extended to polynuclear species. Although functionalised polymers (complexing polymers), bearing binding sites such as amino groups, carboxylate moieties and many other functional groups capable of binding metal cations are well known, in recent years much attention has been focused on polymeric species (coordination polymers) in which the metal plays a structural role.

By imposing the *1,3-alternate* conformation on a calix[4]arene, one may design a variety of exo-ligands based on different coordination sites, and one dimensional coordination networks, for example, may be designed using exo-ditopic species. In particular, with simple coordination units such as unidentate thiols and nitriles, tetra coordinating ligands with four coordination sites located at the summit of a pseudo-tetrahedron may be obtained [34-35].

6.1. SYNTHESIS OF *p-t*-BUTYL-DIMERCAPTOCALIX[4]ARENE 42

The rationale behind the design of **42** [28,29,33] was that it would behave as a heteroquadridentate exo-ditopic ligand with differentiation of the two faces when in the *1,3-alternate* conformation. The synthesis of **42** was based on partial carbamoylation of *p-t*-butyl-calix[4]arene **43**, achieved by its treatment with (Me)$_2$NCSCl in the presence of NaH in DMF at 25 °C, leading thus to *O*-phenyl-dimethylthiocarbamate **44**. To avoid decomposition accompanying thermally-induced rearrangement, the two hydroxy groups on **44** were protected either by methyl (**45**) or by benzyl groups (**46**). The rearrangement of **45** to **47** was achieved by heating the solid at 360 °C under argon for 20-30 minutes. The reduction of **47** by LiAlH$_4$ in dry THF afforded compound **48**. Finally, the deprotection of **48** by BBr$_3$ in CH$_2$Cl$_2$ gave the desired compound **42**.

42 Y^1 = OH, Y^2 = SH
43 Y^1 = Y^2 = OH
44 Y^1 = OH, Y^2 = OCSNMe$_2$
45 Y^1 = OMe, Y^2 = OCSNMe$_2$
46 Y^1 = OBn, Y^2 = OCSNMe$_2$
47 Y^1 = OMe, Y^2 = SCONMe$_2$
48 Y^1 = OMe, Y^2 = SH

For compound **42**, the ^{1}H-NMR spectrum in both CDCl$_3$ and in C$_2$D$_2$Cl$_4$ was composed of rather sharp signals at 25 °C demonstrating probably the presence of a single conformer and a slow rate of interconversion. NOESY-NMR experiments were consistent with a *cone* conformation. As in **43**, the *cone* conformation of **42** is probably stabilised by intramolecular H-bonds. Supporting this was the observation of a mixture of conformers at room temperature for **45** in which the two hydroxy moieties were replaced by two methoxy groups, thus preventing the formation of OH⋯S hydrogen bonds.

Unlike **43**, for **42** five conformations may be envisaged (Figure 12).

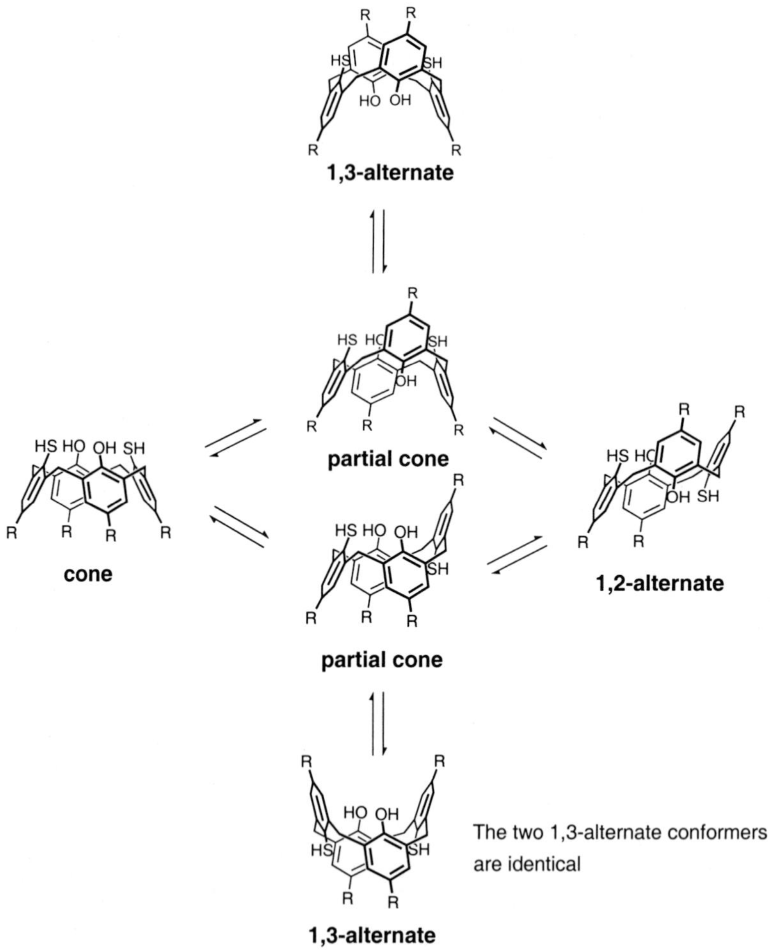

*Figure 12. The five possible conformers for compound **42**.*

Variable-temperature ^{1}H-NMR measurements on **42** in $C_2D_2Cl_4$ showed coalescence of signals corresponding to the H_A and H_B protons of the CH_2 groups to a singlet occurred at *ca* 67 °C (at 200 MHz) with a rate constant of interconversion k_c of *ca* 350 s^{-1}, and a free energy of activation $\Delta G^{\#}$ of *ca* 16 kcal/mol. In comparison, for **43** in $CDCl_3$: T_c = 52 °C, $\Delta G^{\#}$ = 15.7 Kcal/mole, and k_c=150 s^{-1}. Although for **43** a unique process was postulated for the cone inversion process leading to equivalence of the two protons, one cannot assume an exchange process involving only two states for **42**. Indeed, above the coalescence temperature, the δ value observed for the singlet did not correspond to the average value for H_A and H_B protons.

6.2. MERCAPTO CALIXARENES WITH LONG ALKYL CHAINS

The lipophilic dimercaptocalix derivatives **49** and **50** have been prepared recently [30]. The synthetic strategy used to prepare **49** and **50** was again based on dithiacarbamoylation of *p-t*-butyl-calix[4]arene **43** at the 1 and 3 positions, followed by transformation of the remaining two OH groups into ether fragments by treatment with bromododecane, the thermal rearrangement of the OCSNMe$_2$ fragments into the SCONMe$_2$ groups and the reductive deprotection of the sulphur atoms using LiAlH$_4$.

Thiacarbamoylation of **43** with (Me)$_2$NCSCl in the presence of K$_2$CO$_3$ in refluxing acetone for 24 h. yielded 5-10 % of the tri substituted derivative, as well as the di substituted conformers **51** (13 %) and **55** (40 %), non inter-convertible at room temperature (Figure 13). Whereas the treatment of **51** with bromododecane in the presence of NaH in refluxing THF afforded the tetrasubstituted compound **52** in *1,3-alternate* conformation in quantitative yield after 24 h, surprisingly, under the same conditions, the reaction of **55** afforded the compound **56** in the *1,2-alternate* conformation. Both

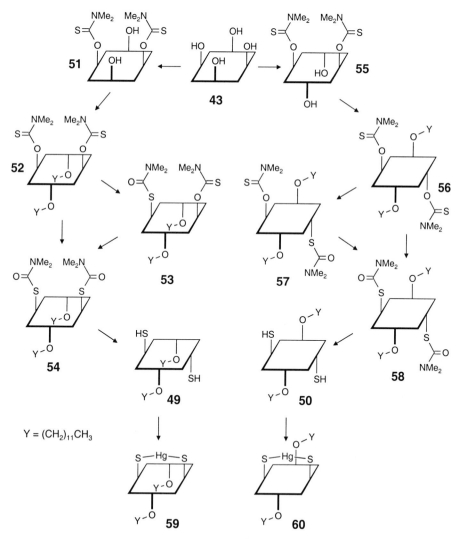

Figure 13. Dodecyl ethers of p-t-butyl-dimercaptocalix[4]arene (schematic representation).

compounds **52** and **56** were characterized in the solid state by single-crystal X-ray diffraction methods (Figure 14).

The thermal rearrangement of **52** in vacuum on the melt at 340 °C yielded the mono-rearranged compound **53** (50 %), the bis-rearranged compound **54** (25 %) and unreacted starting material **52** (25 %). Subsequently, compound **53** could be rearranged in 20 % yield to **54** under similar conditions. Attempts to increase the yield by performing the reaction in refluxing ditolyl ether failed. Reductive deprotection of **54** with LiAlH$_4$ in refluxing THF yielded the dimercapto compound **49** (80 %) in the *partial cone* conformation. The same sequence of reactions on **56** afforded **57** (40 %), the bis-rearranged derivative **58** (20 %) and the starting compound **56** (40 %). As before, **57** could be rearranged into **58** under the same conditions in 15 % yield. The latter, in the *1,2-alternate* conformation, gave **50** in 74 % yield by treatment with LiAlH$_4$ in refluxing THF. The final structural assignments for both **49** and **50** were based on 2D-ROESY NMR experiments.

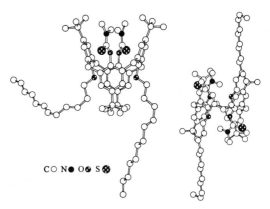

*Figure 14. Molecular structures of **52** (left) and **56** (right). For the sake of clarity, H atoms and solvent molecules are not represented.*

Both compounds **49** and **50** were able to efficiently extract the Hg^{2+} cation from acidic aqueous solution into chloroform with a remarkable selectivity over Cd^{2+} and Pb^{2+} cations [30]. Hg(II) complexes **59** and **60** have been isolated and studied by 2D-ROESY as well as by ^{199}Hg-NMR experiments.

6.3. SYNTHESIS OF *p-t*-BUTYL-TETRAMERCAPTOCALIX[4]ARENE

The inefficient synthesis [1] first descibed for **61** has subsequently been improved to make this a readily available compound [31,33]. The syntheses of a variety of mercaptocalixarenes have now been investigated in some detail [33]. Thermal rearrangement of thiacarbamoyl compounds, as described several times above is the basis of all.

Thus, **43** can be transformed to **61** in an overall high yield. Although in solution compound **43** may adopt *cone, partial cone, 1,2-alternate* and *1,3-alternate* conformations, both ^{1}H and ^{13}C NMR data indicate that **61** adopts preferentially the *1,3-alternate* conformation.

7. Structural Analysis of Mercaptocalix[4]arenes

7.1. STRUCTURE OF *p-t*-BUTYL-DIMERCAPTOCALIX[4]ARENE **42**

In the solid state, compound **42** adopts a *cone* conformation (Figure 15) [29] with: i) O1-O2 and S1-S2 distances of 3.469 Å and 5.317 Å, respectively, ii) an average O···S separation of *ca* 3.29 Å, iii) in respect to the mean plane defined by all four CH$_2$ groups, average angles between planes containing the aromatic moieties bearing SH or OH groups of *ca* 93° and 137°, respectively, showing that, while the thiophenol rings are almost perpendicular to the mean plane described above, the two phenol rings are considerably tilted towards the exterior of the calix, iv) average O-H

Figure 15. Crystal structure of 42. For the sake of clarity, H atoms and solvent molecules are not represented.

and S-H distances, using isotropically refined hydrogen atoms, of *ca* 1.03 Å and 1.29 Å respectively, v) $d_{SH\cdots O}$ and $d_{OH\cdots S}$ distances and angles of O1H44···S1 = 2.374 Å, 159.73°; S1H45···O2 = 2.203 Å, 132.04°; O2H46···S2 = 2.249 Å, 158.49°; S2H47···O1 = 2.200 Å, 145.87°. For **42**, although the OH and SH hydrogen atoms are properly oriented, the average distance between O and S atoms of *ca* 3.29 Å and the observed angles indicate that if H-bonds exist, they are rather weak.

7.2. STRUCTURE OF *p-t*-BUTYL-TETRAMERCAPTOCALIX[4]ARENE 61

The crystal structure of **61** defined its *1,3-alternate* conformation (Figure 16) [31] and showed that the two phenyl moieties disposed on the same side were almost parallel to each other, whereas the two sets of two aromatics were almost perpendicular. The average distance separating the two thiophenol groups located on the same side was *ca* 4.96 Å.

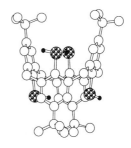

Figure 16. Molecular structure of 61 showing the 1,3-alternate conformation. For the sake of clarity, H atoms and solvent molecules are not represented.

7.3. STRUCTURE OF MONONUCLEAR MERCURY *p-t*-BUTYL-DIMERCAPTOCALIX[4]ARENE 42

The preferential binding of mercury to **42** at the sites bearing sulfur atoms is expected due to the thiophilicity of Hg(II). The crystal structure of **42**-Hg complex [29] showed that: i) **42** adopts a *1,3-alternate* conformation in which the mercury ion is coordinated to two sulfur atoms, ii) the average Hg-S distance is *ca* 2.35 Å and S-Hg-S angle is *ca* 176.0°, iii) the distance between two S atoms in the S-Hg-S unit (4.710 Å) is shorter than that in the free ligand **42**

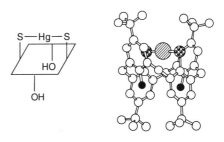

Figure 17. Structure of Hg-42. H atoms and solvent molecules are not represented.

(5.317 Å), showing that the ligand must bow by *ca* 0.60 Å in order to accommodate the metal ion. Consequently, with respect to the mean plane defined by all four CH_2 groups, the average angles between planes containing the aromatic moieties bearing SH or OH groups are *ca* 99.3° and 101.2° respectively. The conformational change from the *cone* form of the free ligand means that the distance between Hg ions and the centroids of the aromatic rings located on the same side is *ca* 3.05 Å, indicative of the "2+4" coordination mode for Hg^{2+} cation (Figure 17).

7.4. STRUCTURE OF DINUCLEAR MERCURY *p-t*-BUTYL-TETRAMERCAPTOCALIX[4]ARENE

A crystal structure determination for the complex formed between Hg(II) and **61** (Figure 18) [31] showed it to be a binuclear complex **61**-Hg_2 in which each Hg ion is coordinated to two thiolate moieties with an average Hg-S distance of *ca* 2.34 Å and an average S-Hg-S angle of *ca* 178.5°. One linear S-Hg-S unit projects upon the other at an angle of 89.3°, with the Hg-Hg distance being 3.66 Å. The average distance between two S atoms in the S-Hg-S units (4.59 Å) is shorter than that in the free ligand **61** (4.96 Å), showing that the ligand must bow by 0.38Å in order to accommodate the metal ions. In the binuclear complex the average distance between Hg ions and the centroids of the aromatic rings located on the same side is *ca* 3.1Å. Again, due to the *1,3-alternate* conformation of the ligand, it appears possible that a mercury(II) species may have its secondary coordination requirements satisfied by polyhapto (π) interactions with two of the phenyl rings.

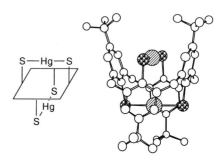

Figure 18. Structure of Hg_2-61 complex. H atoms and solvent molecules are not represented.

8. Coordination Chemistry of Thiacalix[4]arene Derivatives

The coordination ability of calix[4]arenes such as **43** towards metallic elements of all types has been established. Although in few cases the contribution of the aromatic π cloud to the binding process was demonstrated, for the majority of reported examples the binding of metal cations takes place at the OH groups [1]. The introduction of S into thiacalixarenes and the conversion of this into sulfoxide and sulfone entities provides opportunities for new modes of coordination, possibly suited, for example, to the design of new polynuclear complexes.

Inclusion-based molecular networks (koilates) may be formed in the solid state using hollow molecular units (koilands) possessing at least two divergent cavities and connector molecules capable of undergoing inclusion processes [36]. Calix[4]arene derivatives offer particular promise in this regard, as the fusion of calixarene units by bridging through entities such as Si(IV) readily enables the divergent orientation of cavities of a size and nature sufficient to accommodate a variety of guests [37-43]. The fact that a number of different metals (including Ti, Nb, Al, Zn, Mg and Eu) [44-48] may be used similarly as bridges adds a further dimension to the possible utility of these building blocks.

Thiacalix[4]arenes have been exploited for the design of new hollow molecular building blocks of interest for the formation of inclusion molecular networks. For example, the structurally characterised (Figure 19) Cu(II) complex of **1** provides a magnetically active tecton [49]. Measurements of magnetic susceptibility (4-298 K) could be interpreted, in the light of the structural information described below, in terms of antiferromagnetic coupling (J = -103±1 cm^{-1}) of four equivalent Cu(II) ions arranged in a square.

The crystal structure of the Cu$_4$(**1**)$_2$ complex shows two inequivalent units are found within

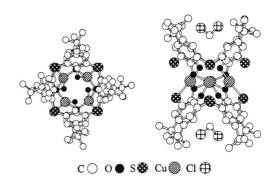

C ○ O ● S ✱ Cu ▨ Cl ⊕

Figure 19. Top (left) and lateral (right) views of the crystal structure of the tetranuclear copper complex showing the square arrangement of the metal centres and their bridging by phenoxide groups (left) and the inclusion of solvent molecules (right). For clarity, H atoms and solvent molecules are not represented.

the unit cell. In both, the Cu$_4$ array, sandwiched between two thiacalix entities in a *cone* conformation similar to that of the free ligand **1** [8], is close to exactly square, with each copper in a six-coordinate O$_4$S$_2$ donor-atom environment (Figure 20). The coordination sphere is far from regular, with one Cu-O bond *ca* 0.5 Å longer than the other three and one Cu-S bond *ca* 0.3 Å longer than the other. Interestingly, this copper koiland, as in the case of silicon koilands obtained with **43** [36-37], forms an inclusion complex with two molecule of CH$_2$Cl$_2$. Each one of the solvent molecule penetrates deeply into the cavity of the koiland.

Other structures obtained have shown that this mode of binding is certainly not restricted to Cu(II). For example, solid state structural analysis of the Zn(II) and Co(II) complexes of **1** revealed the formation of koilands in which the two calix units are linked by three metal centres (Figure 20). In both cases, again, both cavities were occupied by solvent molecules, demonstrating the inclusion ability of the new koilands [50].

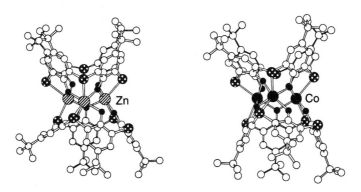

Figure 20. Lateral views of the solid state structures of the trinuclear zinc (left) and cobalt (right) complexes possessing two cavities occupied each by one CH$_2$Cl$_2$ molecule. For clarity, H atoms and solvent molecules are not represented.

Using other conditions, a tetranuclear Zn complex engaging three calix units **1** has been obtained and structurally characterised in the solid state [51]. The extraction ability of thiacalixarene derivatives [7,15,16,52], including the dansyl-modified compounds **24-27** [19] towards alkali, alkaline-earth and transition metals has been shown to be unusual, though structural results such as those discussed above indicate that establishment of the exact nature of extracted species may be extremely complicated.

9. References and Notes

[1]   C. D. Gutsche, *"Calixarenes"*, Monographs in Supramolecular Chemistry, Ed. J. F. Stoddart, R.S.C., London, 1989; C. D. Gutsche, *"Calixarenes revisited"*, Monographs in Supramolecular Chemistry, Ed. J. F. Stoddart, R.S.C., London, 1998
[2]   *Calixarenes. A Versatile Class of Macrocyclic Compounds,* Ed. J. Vicens and V. Böhmer, Kluwer Academic Publishers, Dordrecht, 1991.
[3]   V. Böhmer, *Angew. Chem., Int. Ed. Engl.* **1995**, *34*, 713-745.
[4]   C. D. Gutsche, M. Iqbal, *Organic Syntheses* **1989**, *68*, 234-237.
[5]   a) T. Sone, Y. Ohba, K. Moriya, H. Kumada, K. Ito, *Tetrahedron* **1997**, *53*, 10689-10698; b) H. Kumagai, M. Hasegawa, S. Miyanari, Y. Sugawa, Y. Sato, T. Hori, S. Ueda, H. Kamiyama, S. Miyano, *Tetrahedron Lett.* **1997**, *38*, 3971-3972.
[6]   a) N. Iki, C. Kabuto, T. Fukushima, H. Kumagai, H. Takeya, S. Miyanari, T. Miyashi, S. Myano, *Tetrahedron* **2000**, *56*, 1437-1443; b) N. Iki, N. Morohashi, T. Suzuki, S; Ogawa, M. Aono, C. Kabuto, H. Kumagai, H. Takeya, S. Miyanari, S. Miyano, *Tetrahedron Lett.* **2000**, *41*, 2587-2590; c) A. Bilyk, A. K. Hall, J.M. Harrowfield, M. W. Hosseini, B. W. Skelton, A. H. White, *Inorg. Chem.* **2001**, *40*, 672-686.
[7]   N. Iki, H. Kumagai, N. Morohashi, K. Ajima, M. Hasegawa, S. Miyano, *Tetrahedron Lett.* **1998**, *39*, 7559-7562.
[8]   H. Akdas, L. Bringel, E. Graf, M. W. Hosseini, G. Mislin, J. Pansanel, A. De Cian, J. Fischer, *Tetrahedron Lett.* **1998**, *39*, 2311-2314.
[9]   G. Mislin, E. Graf, M. W. Hosseini, A. De Cian, J. Fischer, *Tetrahedron Lett.* **1999**, *40*, 1129-1132.
[10] N. Morohashi, N. Iki, C. Kabuto, S. Miyano, *Tetrahedron Lett.* **2000**, *41*, 2933-2937.
[11] G. Mislin, E. Graf, M. W. Hosseini, A. De Cian, J. Fischer, *Chem. Commun.* **1998**, 1345-1346.
[12] P. Lhotak, M. Himl, S. Pakhomova, I. Stibor, *Tetrahedron Lett.* **1998**, *39*, 8915-8918.
[13] J. Lang, H. Dvorakova, I. Bartosova, P. Lhotak, I. Stibor, *Tetrahedron Lett.* **1999**, *40*, 373-376.
[14] H. Akdas, W. Jaunky, E. Graf, M. W. Hosseini, J.-M. Planeix, A. De Cian, J. Fischer, *Tetrahedron Lett.* **2000**, *41*, 3601-3606.
[15] N. Iki, F. Narumi, T. Fujimoto, N. Morohashi, S. Miyano, *J. Chem. Soc., Perkin Trans. 2* **1998**, 2745-2750.
[16] Iki, N. Morohashi, F. Narumi, T. Fujimoto, T. Suzuki, S. Miyano, *Tetrahedron Lett.* **1999**, *40*, 7337-7341.
[17] H. Akdas, G. Mislin, E. Graf, M. W. Hosseini, A. De Cian, J. Fischer, *Tetrahedron Lett.* **1999**, *40*, 2113-2116.
[18] K. Iwamoto, S. Shinkai, *J. Org. Chem.* **1992**, *57*, 7066-7073.
[19] M. Narita, Y. Higuchi, F. Hamada, H. Kumagai, *Tetrahedron Lett.* **1998**, *39*, 8687-8690.
[20] N. Iki, F. Narumi, T. Suzuki, A. Sugawara, S. Miyano, *Chem. Lett.* **1998**, 1065-1066.
[21] H. Katagiri, N. Iki, T. Hatori, C. Kabuto, S. Miyano, *J. Am. Chem. Soc.* **2001**, *123*, 779-780.
[22] N. Iki, T. Fujimoto, S. Miyano, *Chem. Lett.* **1998**, 625-626.
[23] N. Iki, T. Fujimoto, T. Shindo, K. Koyama, S. Miyano, *Chem. Lett.* **1999**, 777-778.
[24] R. Ungaro, A. Pochini, G. D. Andreetti, V. Sangermano, *J. Chem. Soc., Perkin Trans. 2* **1984**, 1979-1985.
[25] N. Morohashi, N. Iki, T. Onodera, C. Kabuto, S. Miyano, *Tetrahedron Lett.* **2000**, *41*, 5093-5097.
[26] P. Rao, M. W. Hosseini, A. De Cian, J. Fischer, *Chem. Commun.* **1999**, 2169-2170.
[27] M. S. Newman, H. A. Karnes, *J. Org. Chem.* **1966**, *31*, 3980-3984.
[28] Y. Ting, W. Verboom, L. G. Groenen, J.-D. van Loon, D. N. Reinhoudt, *J. Chem. Soc., Chem. Commun.* **1990**, 1432-1433.
[29] X. Delaigue, M. W. Hosseini, A. De Cian, N. Kyritsakas, J. Fischer, *J. Chem. Soc., Chem. Commun.*

1995, 609-610.
[30] P. Rao, O. Enger, E. Graf, M. W. Hosseini, A. De Cian, J. Fischer, *Eur. J. Inorg. Chem.* **2000**, 1503-1508.
[31] X. Delaigue, J. McB. Harrowfield, M. W. Hosseini, A. De Cian, J. Fischer, N. Kyritsakas, *J. Chem. Soc., Chem. Commun.* **1994**, 1579-1580.
[32] X. Delaigue, M. W. Hosseini, *Tetrahedron Lett.* **1993**, *34*, 8111-8112.
[33] a) C. G. Gibbs, C. D. Gutsche, *J. Am. Chem. Soc.* **1993**, *115*, 5338-5339; b) C. G. Gibbs, P. K. Sujeeth, J. S. Rogers, G. G. Stanley, M. Krawiec, W. H. Watson, C. D. Gutsche, *J. Org. Chem.* **1995**, *60*, 8394-8402.
[34] G. Mislin, E. Graf, M. W. Hosseini, A. De Cian, N. Kyritsakas, J. Fischer, *Chem. Commun.* **1998**, 2545-2546.
[35] C. Klein, E. Graf, M. W. Hosseini, A. De Cian, J. Fischer, *Chem. Commun.* **2000**, 239-240.
[36] M. W. Hosseini, A. De Cian, *Chem. Commun.* **1998**, 727-733 and references therein.
[37] X. Delaigue, M. W. Hosseini, A. De Cian, J. Fischer, E. Leize, S. Kieffer, A. Van Dorsselaer, *Tetrahedron Lett.* **1993**, *34*, 3285-3288.
[38] F. Hajek, E. Graf, M. W. Hosseini, *Tetrahedron Lett.* **1996**, *37*, 1409-1412.
[39] F. Hajek, M. W. Hosseini, E. Graf, A. De Cian, J. Fischer, *Tetrahedron Lett.* **1997**, *38*, 4555-4558.
[40] F. Hajek, E. Graf, M. W. Hosseini, X. Delaigue, A. De Cian, J. Fischer, *Tetrahedron Lett.* **1996**, *37*, 1401-1404.
[41] F. Hajek, M. W. Hosseini, E. Graf, A. De Cian, J. Fischer, *Angew. Chem., Int. Ed. Engl.* **1997**, *36*, 1760-1762.
[42] F. Hajek, E. Graf, M. W. Hosseini, A. De Cian, J. Fischer, *Cryst. Engin.* **1998**, *1*, 79-85.
[43] J. Martz, E. Graf, M. W. Hosseini, A. De Cian, N. Kyritsakas, J. Fischer, *J. Mat. Chem.* **1998**, *8*, 2331-2333.
[44] M. M. Olmstead, G. Sigel, H. Hope, X. Xu, P. Power, *J. Am. Chem. Soc.* **1985**, *107*, 8087-8091.
[45] F. Corazza, C. Floriani, A. Chiesi-Villa, C. Guastini, *J. Chem. Soc., Chem. Commun.* **1990**, 1083-1084.
[46] J. L. Atwood, S. G. Bott, C. Jones, C. L. Raston, *J. Chem. Soc., Chem. Commun.* **1992**, 1349-1351.
[47] J. L. Atwood, P. C. Junk, S. M. Lawrence, C. L. Raston, *Supramol. Chem.* **1996**, *7*, 15-17.
[48] A. Bilyk, J.M. Harrowfield, B.W. Skelton, A.H. White, *J. Chem. Soc., Dalton Trans.* **1997**, 4251-4256.
[49] G. Mislin, E. Graf, M. W. Hosseini, A. Bilyk, A. K. Hall, J. McB. Harrowfield, B. W. Skelton, A. H. White, *Chem. Commun.* **1999**, 373-374.
[50] A. Bilyk, A. K. Hall, J. McB. Harrowfield, M. W. Hosseini, G. Mislin, B. W. Skelton, C. Taylor, A. White, *Eur. J. Chem.* **2000**, 823-826.
[51] N. Iki, N. Morohashi, C. Kabuto, S. Miyano, *Chem. Lett.* **1999**, 219-220.
[52] N. Iki, N. Morohashi, F. Narumi, S. Miyano, *Bull. Chem. Soc. Jpn.* **1998**, *71*, 1597-1603.

Chapter 7

DOUBLE- AND MULTI-CALIXARENES

MOHAMED SAADIOUI, VOLKER BÖHMER

Johannes Gutenberg-Universität, Fachbereich Chemie und Pharmazie, Abteilung Lehramt Chemie, Duesbergweg 10-14, D-55099 Mainz Germany; e-mail: vboehmer@mail.uni-mainz.de

1. Introduction

Calixarenes are easily (and often selectively) functionalized and therefore they can be combined in various ways to larger molecules containing more than one calixarene substructure. In the following chapter we will try to give an overview on such multicalixarenes held together by "conventional" covalent links. Self-assembled structures are treated separately in Chapter 8.

The earliest examples for double calixarenes of importance are the carcerands and hemicarcerands of D. Cram in which two resorcarene derived cavitands are combined. Due to their unique properties to include (more or less permanently) smaller molecules, they are treated separately in Chapter 10. Further combinations of cavitands and calixarenes are described in Chapter 9.

The following survey is ordered mainly under structural aspects, which is not possible without some arbitrary classifications. Some frequently used building blocks for the construction of multicalixarenes are shown in Figure 1. Their synthesis is described in Chapter 2.

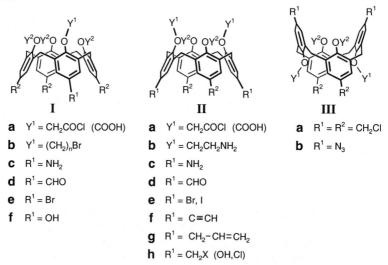

Figure 1. Frequently used building blocks for the synthesis of multicalix[4]arenes. Only the functional groups used for the covalent linkage are indicated. Residues not explicitly mentioned are equal to H, alkyl (Y^2) or t-Bu (R^2) in many cases.

2. Double Calixarenes

2.1. CONNECTION VIA THE NARROW RIM

2.1.1. Single bridged compounds
The first bis-calix[4]arenes connected via a single bridge between the narrow rims were prepared by reacting the triester monoacid chloride **Ia** ($Y^2 = CH_2COOEt$, $R^1 = R^2 = t$-Bu) with 1,2-diaminoethane or ethylene glycol in benzene [1]. The molecules **1/2** possess two metal binding sites each of which consists of four ester (amide) groups. ^{1}H NMR studies showed that they form a stable 1:2 complex with Na^+ in solution. Intramolecular vibration of the Na^+ cation between the two metal binding sites has been observed later [2] in the case of the 1:1 complex with **1** ($R^1 = R^2 = H$) by ^{1}H NMR.

Various amide-bridged double calixarenes of type **2** (with spacers differing in length and flexibility) were later synthesized using the same strategy [3] and confirmed for one example by an X-ray structure which shows the usual pinched *cone* conformation of the two calix[4]arene units. The synthesis was extended to symmetrical octa-amide biscalix[4]arenes **3** (prepared in four steps from *p-t*-butylcalix[4]arene via the triamide **I** ($Y^1 = OH$, $Y^2 = CH_2CONEt_2$) which was etherified with the corresponding bis-chlorocetamide). The non symmetric biscalix[4]arene **4** was obtained by *O*-alkylation with a BOC-protected chloro-acetamide and acylation of the mono amine obtained by deprotection **I** ($Y^1 = CH_2CO-NH(CH_2)_nNH_2$) with **Ia**.

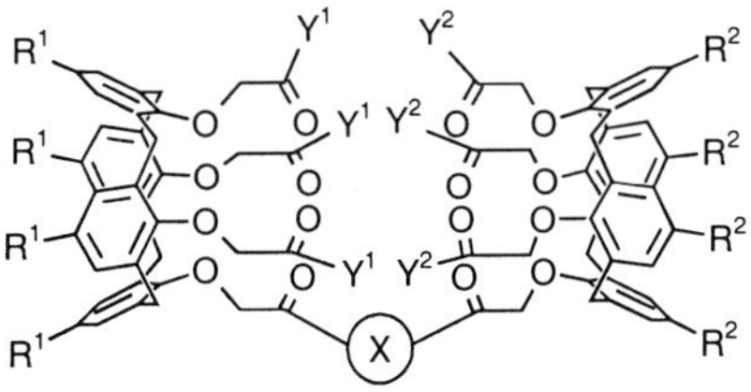

	Y^1	Y^2	X
1	OEt	OEt	OCH_2CH_2O
2	OEt	OEt	$HN(CH_2)_nNH$
3	NEt_2	NEt_2	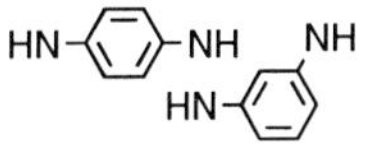 $HN-\langle\rangle-NH$, $HN-\langle\rangle-NH$
4	NEt_2	OEt	$HN(CH_2)_nNH$

$R^1, R^2 = H$ or *t*-Bu

Dinuclear Eu^{3+}/Eu^{3+} and Eu^{3+}/Nd^{3+} complexes were formed by **3** connected by *p*-phenylene in contrast to the hexaester **2**. Photophysical studies made an energy transfer $Eu^{3+} \rightarrow Nd^{3+}$ with > 50% efficiency likely and showed a strong solvent dependence of the luminescence properties.

Various singly bridged double calix[4]arenes **5a,b** have been synthesized from di(propyl)ethers directly by *O*-alkylation with α,ω-dibromoalkanes [4] or from tri(propyl)-ethers with monobromoalkyl tripropoxy calix[4]arenes [5].

Singly-bridged double-calix[6]arenes have been obtained similarly by reaction of *p-t*-butylcalix[6]arene pentamethylether with *p*-xylylene dibromide or by oxidative coupling of the monopropargyl-pentamethylether [6]. Octa-urea derivatives **6**, in which the two calixarenes are linked via amide functions, were prepared from the corresponding monoacid **Ia** ($Y^2 = $ alkyl, $R^1 = R^2 = $ NH-CO-NH-R'), obtained from *t*-butyl-calix[4]arene triethers in five steps: *O*-alkylation with ethylbromoacetate, ipso-nitration, reduction, *N*-acylation with an isocyanate, and ester hydrolysis) [7]. For their self-assembly to polymers see Chapter 8.

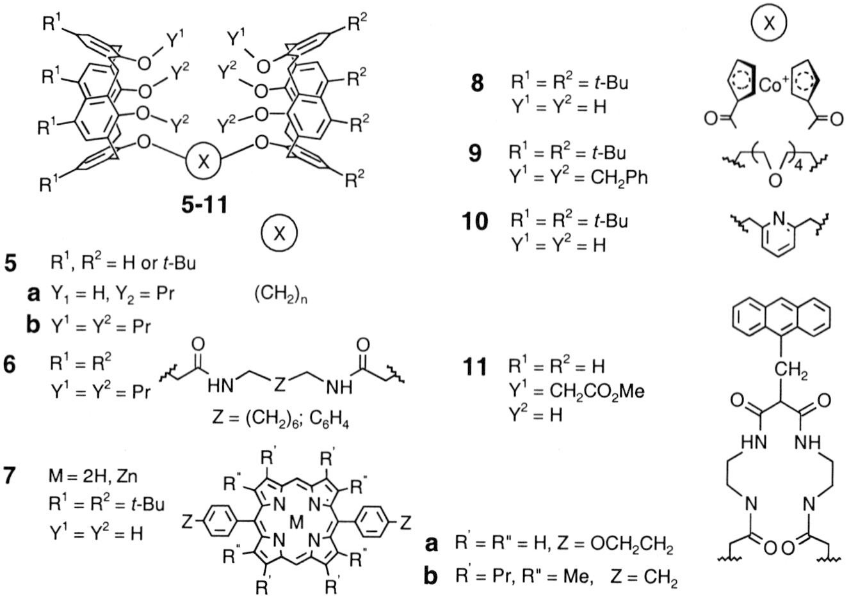

5-11

5 $R^1, R^2 = H$ or t-Bu
  a $Y_1 = H, Y_2 = Pr$
  b $Y^1 = Y^2 = Pr$

6 $R^1 = R^2$
   $Y^1 = Y^2 = Pr$
   $Z = (CH_2)_6; C_6H_4$

7 $M = 2H, Zn$
   $R^1 = R^2 = t$-Bu
   $Y^1 = Y^2 = H$

8 $R^1 = R^2 = t$-Bu
   $Y^1 = Y^2 = H$

9 $R^1 = R^2 = t$-Bu
   $Y^1 = Y^2 = CH_2Ph$

10 $R^1 = R^2 = t$-Bu
   $Y^1 = Y^2 = H$

11 $R^1 = R^2 = H$
   $Y^1 = CH_2CO_2Me$
   $Y^2 = H$

a $R' = R'' = H, Z = OCH_2CH_2$
b $R' = Pr, R'' = Me, Z = CH_2$

Two double calixarenes **7** in which two calix[4]arenes are linked by a single porphyrin were prepared [8], along with their metal complexes, (*e.g.* with Zn^{2+}). Their fluorescence quenching was studied in the presence of *p*-quinone in water/THF [8a] and in $CDCl_3$ [8b]. The quinone acceptor is complexed in the latter case by the phenolic OH groups via hydrogen bonding and the photoinduced electron transfer occurs with a rate constant of $(8.0 \pm 0.2) \cdot 10^8$ s^{-1}.

Further monobridged double calix[4]arenes (**8-11**), partly isolated only as side products, can be only mentioned [9-13].

2.1.2. Double bridged compounds
The preferred formation of *syn*-1,3-diethers or 1,3-diesters of calix[4]arenes has been used for the rational synthesis of doubly bridged double calix[4]arenes (type **A**) in various ways. In the simplest way, the calix[4]arene may be directly reacted with (rigid) difunctional reagents. The remaining hydroxyl groups can be subsequently *O*-alkylated (*O*-acylated) leading to compounds which can be obtained also (in principle) from the analogous 1,3-diether (1,3-diester). Larger cyclic structures (see below) or 1,3-bridged compounds (see the calixcrowns, Chapter 20) are necessarily competing reaction products, as well as singly bridged linear oligomers in the case of incomplete conversion. Alternatively, 1,3-dietherification may be used to attach functional groups to the calix[4]arene (*e.g.* **IIa,b**) which subsequently are connected using a bifunctional reagent.

Compounds **12a-c** [11,14,15] are early examples of type **A** (confirmed in one case [11] by a crystal structure), while compounds **13-15** were prepared more recently (by condensation of *p-t*-butylcalix[4]arene with the respective bisbromomethylated anisole, benzoic acid or pyridine derivatives and further methylation or hydrolysis [12,16,17]). No pronounced selectivity towards alkali metal picrates has been found in the case of **13** and **15** [16], while receptor **14b** displayed binding and extraction abilities towards uranyl cation although a crystal structure of the free ligand shows the two benzoic acid groups directed away from the cavity delimited by the two calix[4]arene moieties.

The doubly-bridged biscalixarene **16c** has two metal complexing sites [18]. Temperature-dependent ^{1}H NMR studies provided clear evidence for intermolecular metal oscillation of alkali metal cations such Na$^+$ or K$^+$ between these two binding sites. In contrast to bis-calixarene **1**, the measurements allowed distinction between an intramolecular and an intermolecular exchange process. Similar double calix[4]arenes **16a,b** were isolated as side products in the synthesis of calix[4]-crowns [19] (see Chapter 20) by *O*-alkylation of *p-t*-butylcalix[4]arene or its 1,3-dimethyl ether with triethylene glycol ditosylate using an excess of K$_2$CO$_3$ in refluxing acetonitrile [19].

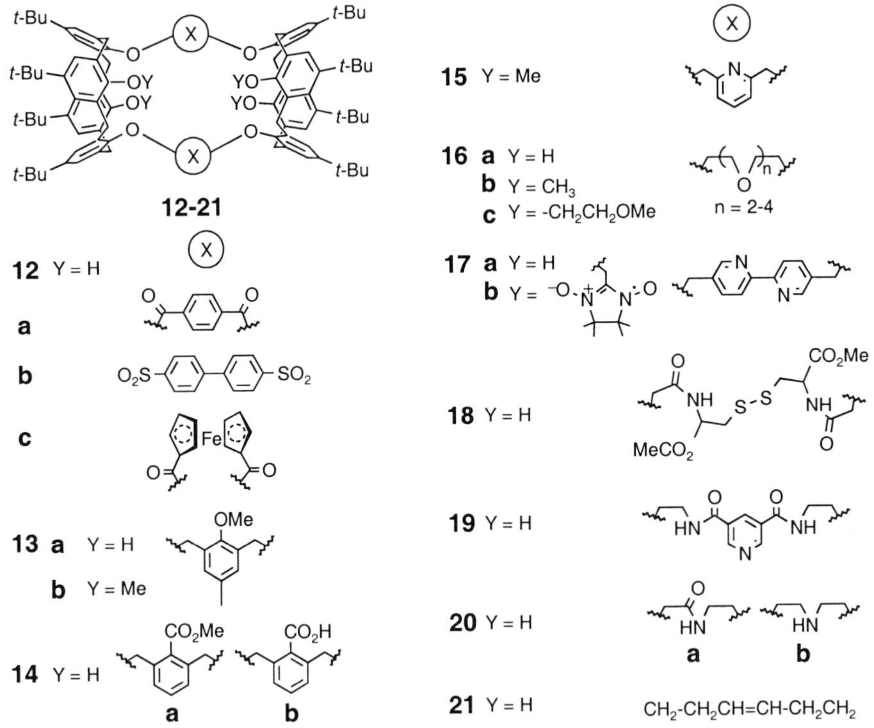

Reaction of calix[4]arenes with bis-bromomethylbipyridine leads to double calixarene **17a** in 23% yield [20] which was further converted to **17b**, bearing stable nitrosylnitroxide radicals for which a through-space exchange interaction is exhibited [21].

The "cyclopeptide" **18** was obtained by reaction of the 1,3-diacidchloride **IIa** with the respective cystine derivative [22], while an analogously prepared cyclic amide with ethylenediamine [14] was later shown (X-ray) to be the corresponding 1,3-bridged mono-calixarene [23].

Reaction of the 1,3-bis(aminoethyl)ether **IIb** with 3,5-bis(chlorocarbonyl)pyridine gave the cyclic tetraamide **19** (double calixarene, 17%) along with the cyclic diamide (1,3-bridged calixarene, 19%) [24] and further cyclic compounds should be available with other diacid dichlorides analogously. Acylation of **IIb** with diacidchloride **IIa** gave the biscalixarene diamide **20a** in 21% which was subsequently reduced to the diamine **20b** in 80% [25].

Recently the metathesis reaction of various alkenyl ethers of calix[4]arenes (and of *p*-allylcalix[4]arenes) was studied [26]. From the bis-(4-butenyl ether) the doubly-bridged macrocycle **21** was obtained in 53% yield containing most probably *E* and *Z* isomeric structures of the alkenyl bridges. With the longer bis-alkenyl ethers only the products of an intramolecular metathesis reaction were found.

The porphyrin bridged biscalix[4]arene **22** (and subsequently its Zn-complex) in which two calixarene molecules are covalently linked via two bridges to a porphyrin spacer were prepared in 3-5% yield by condensation of the respective calix[4]arene based dialdehydes with pyrrole in refluxing propionic acid as one of three possible isomers [27].

The spirobiscalix[4]crown-6 **23** obtained by treatment of *p-t*-butylcalix[4]arene with tetrakis(tosyloxyethoxyethoxymethyl)methane [28,29] may be classified as "doubly bridged", while **24** (obtained with tris(tosyloxyethoxyethoxymethyl)propane in 35% yield) is a hybrid between "singly" and "doubly bridged" [30]. Bi-macrocycles **23** and **24** were found to extract alkali metal cations such as Li^+, Na^+ and NH_4^+ very effectively but with low selectivity.

Acylation of *p-t*-butylcalix[4]arene by benzene-1,2,4,5-tetracarboxylicacid tetrachloride led to a double calix[4]arene **25** (dilution conditions, 29%) where two adjacent hydroxyl functions are linked to the connecting unit [31]. The NMR-spectrum suggested a single compound, but it was not possible to distinguish between the C_{2v}-symmetrical U-shaped isomer (as suggested by the formula) and the C_{2h}-symmetrical S-shaped isomer.

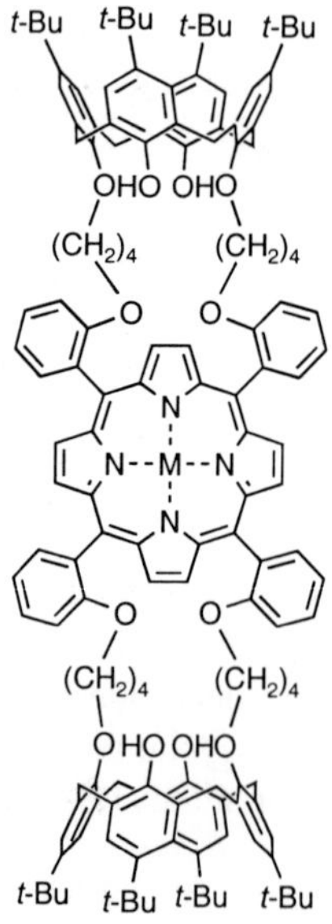

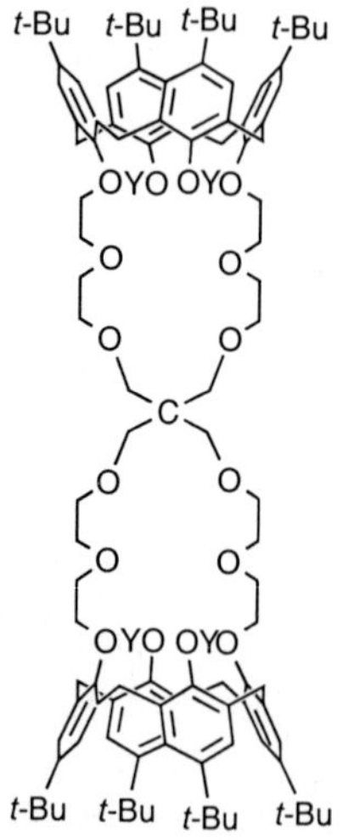

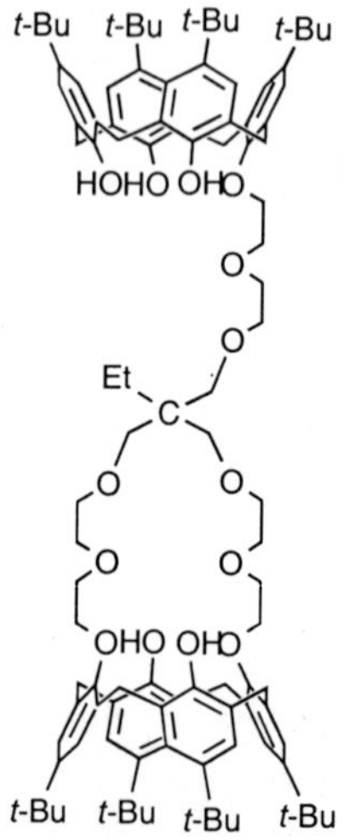

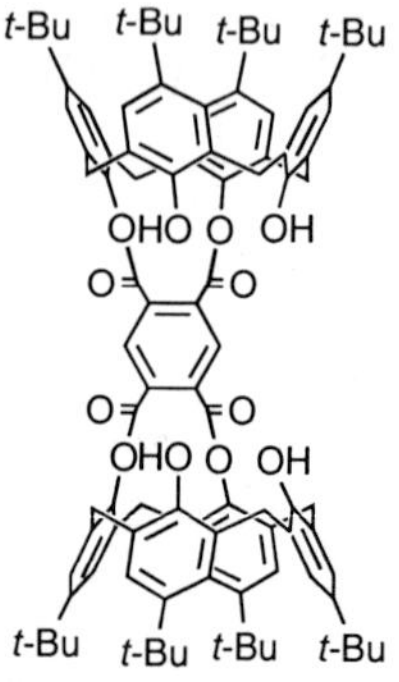

Biscalixarenes bridged via spacers containing thiourea or amide/thioether links were recently reported [32].

2.1.3. More than two bridges

A double calix[4]arene linked by three glycolic chains was obtained in 38% yield by treatment of *p-t*-butylcalix[4]arene with diethylene glycol ditosylate in refluxing acetonitrile. The ^{1}H- and ^{13}C NMR spectrum is not in agreement with a C_{2v}-symmetrical molecule due to the presence of several conformations of the two calix[4]arene parts in addition to the possible *anti* or *syn* linkage [19].

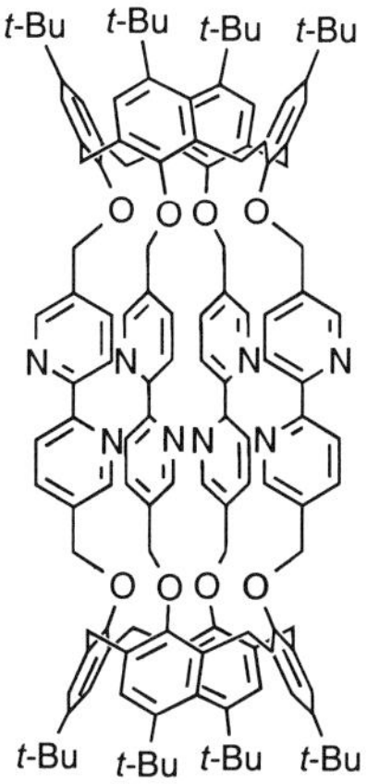

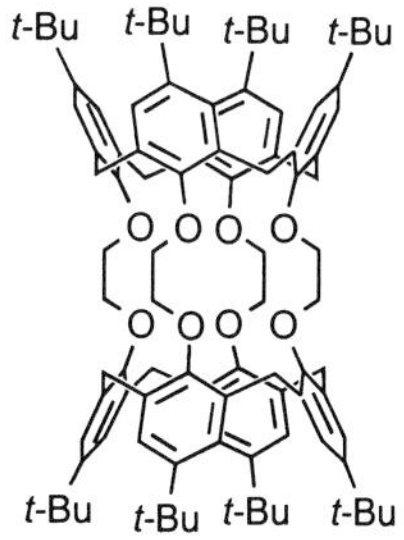

Four bipyridine bridges between the narrow rims are present in the *calix[4]barreland* **26**, prepared in 4% yield from **17a**. These compounds (**17a**, **26**) were designed as ligands for lanthanides, but due to their high rigidity the diffusion of the lanthanide ion into the cavity is inhibited [33] and open chain di- and tetraethers with pipyridine groups are more effective (see Chapter 31).

A rather rigid double calixarene **27** (*calix[4]tube*) connected via four ethylene bridges was also obtained in 51% yield by the reaction of the tetrakis(tosyloxyethoxy) *p-t*-butylcalix[4]arene with the parent *p-t*-butylcalix[4]arene in the presence of K$_2$CO$_3$. [34] Obviously the K$^+$-ion acts as a template, since **27** is a selective ligand for K$^+$, forming kinetically and thermodynamically stable complexes, in which the K$^+$ is coordinated by the eight oxygens in a cubic fashion (confirmed by a single crystal X-ray structure of the complex and free **27**). It was suggested that the K$^+$-ion enters the ligand via the calix[4]arene cavity.

Reaction of various calix[4]arenes with SiCl$_4$ (NaH, dry THF, 52% yield for R = *t*-Bu) leads to double calix[4]arenes **28a**. Compound **29** could be isolated as a (more or less desired) side-product [35] which can be converted with TiCl$_4$ into the "mixed" derivative(s) **28b**. Similar structures where three neighboring phenolic units of a single calix[4]arene are covalently bound to the metal are known with aluminium [36], titanium [37] and various structures with other connecting metals are reported in Chapters 29 and 30. The molecules of compounds **28** (*koilands*) posses two cavities arranged in divergent fashion which cannot collapse due to the rather rigid bridging. Using a suitable connector (*e.g.* hexadiyne) one-dimensional networks (*koilates*) have been constructed in the solid state by inclusion in their cavities [38].

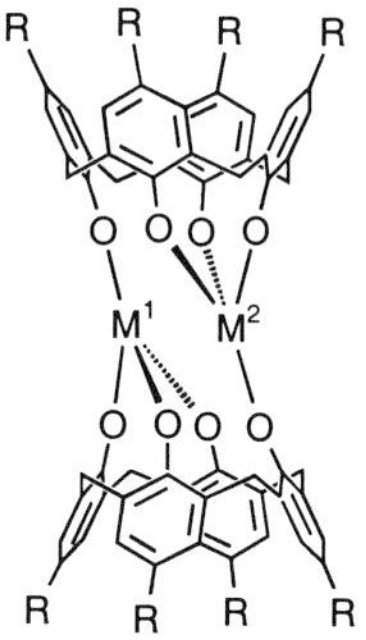

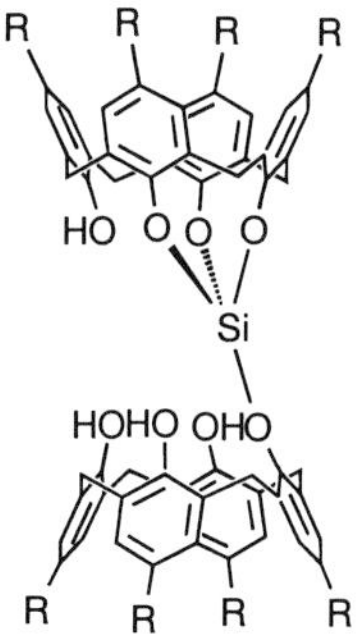

2.1.4. Double calix[4]arenes in the 1,3-alternate conformation
If the condensation of *p-t*-butylcalix[4]arene is carried out with an excess of tetraethylene glycol ditosylate (10 equiv.) and K_2CO_3 a double calix[4]-biscrown-5 **30** is formed directly (52% yield) instead of **16a**, in which the two calixarenic units are additionally crowned and thus fixed in the 1,3-alternate conformation [39]. Selectivity for potassium and rubidium was shown for **30** in the solid state and in solution, the metal cation being complexed in the central cavity of the tritopic receptor. Similar structures (doubly-crowned double calix[4]arenes) were obtained from a mesitol-derived calix[4]arene known to exist in the *1,3-alternate* conformation [40], along with the expected double-calix[4]-crowns [41]. The formation of the dimer is often even favoured over the intramolecular bridging as evidenced by the formation of 34% **31** from 1,3-dipropyl calix[4]arene [42].

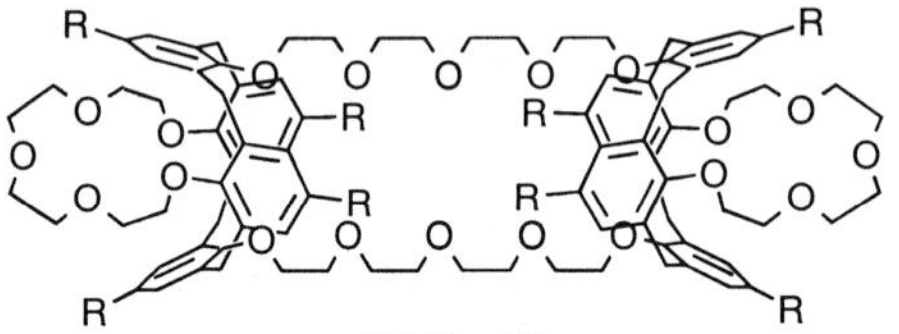

30 R = *t*-Bu 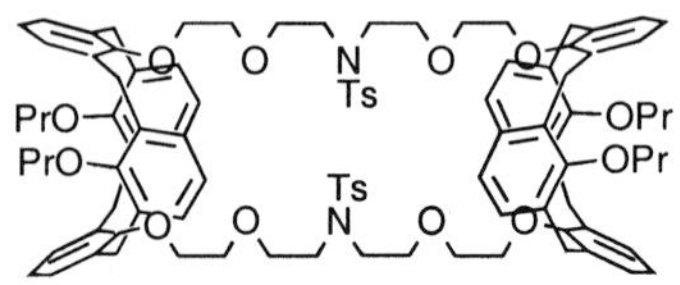   **31**

2.2. CONNECTION VIA THE WIDE RIM

The first double calix[4]arenes of type **32** were prepared by (3+1) fragment condensation, *e.g.* of bromomethylated, *para*-linked bisphenols with linear trimers, in yields up to 13% [43,44]. Subsequently all further compounds were synthezised by the connection of calixarenes **I-III** suitably functionalised at the wide rim. Among those functional groups formyl (CHO, and its reduction and oxidation products CH_2OH, COOH), amino (available by reduction of the nitro group), hydroxyl and halogen (Br, I) are most frequently used.

2.2.1 Single bridged compounds
Reductive coupling of monoformyl tetrapropylether **Ic** ($Y^1 = Y^2 = C_3H_7$) gave the dimeric structure **33a** in 35% yield [45], while the similar compound **33b** was obtained in even lower yield (16%) by a Stille cross-coupling reaction of the monobromo tripropyl-ether **Id** ($Y^2 = C_3H_7$) with (*E*)-1,2-bis(tributylstannyl)ethylene [46]. The *trans* configuration of the double bond was assumed in both cases. **33a** was found to be a good host molecule for quaternary ammonium salts, with higher complexation constants in comparison to the similar double-calixarene with a $X = CH_2$ group [47]. Biscalix[4]arenes **34** with aromatic spacers were prepared in moderate yield [48] from the tetrapropylether **Id** ($Y^1 = Y^2 = C_3H_7$).

A single amino function on the wide rim like in **Ic**, easily introduced by controlled ipso-nitration of tetraethers ($Y^1 = Y^2 =$ alkyl) or selective nitration of triethers ($Y^2 =$ alkyl) and subsequent reduction, can be used to connect two calix[4]arene substructures via one bridge by reaction with various diacid chlorides. Compounds **35** [49], **36a** [50], **36b,c** [51], **36d,e** [52] can be seen as examples. (See Chapter 23 for **36d** as anion receptor.)

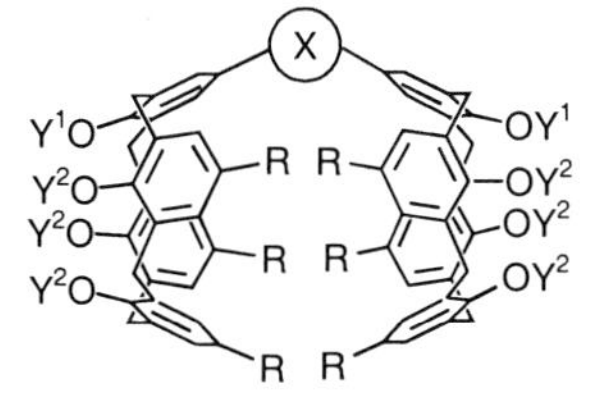

DOUBLE- AND MULTI-CALIXARENES

32-40

32 $Y^1 = Y^2 = H$, $R = Me$ (X)   **a** $-(CH_2)_n-$
                                                      b $Me-C-Me$

33 a $Y^1 = Y^2 = Pr$, $R = H$ (X) $CH=CH$
    b $Y^1 = R = H$, $Y^2 = Pr$

34 $Y^1 = Y^2 = Pr$, $R = H$ (X)   **a** $o-C_6H_4$
                                                      b $m-C_6H_4$
                                                      c $p-C_6H_4$

35 $Y^1 = Y^2 = C_5H_{11}$ **a** $-COHNCH_2OCH_2NHCO-$ **c** $-COHN-\text{(m-C}_6H_4\text{)}-NHCO-$ **d** $-COHN-\text{(pyridyl)}-NHCO-$
   $R = t-Bu$ **b** $-COHN-\text{(p-C}_6H_4\text{)}-NHCO-$

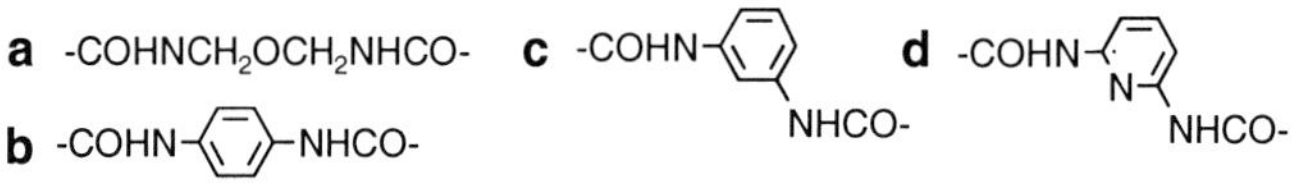

36 $R = H$ **a** $Y^1 = H$, $Y^2 = CH_2CO_2Et$ **b** $Y^1 = Y^2 = CH_2CO_2Et$ **c-e** $Y^1 = H$, $Y^2 = Me$

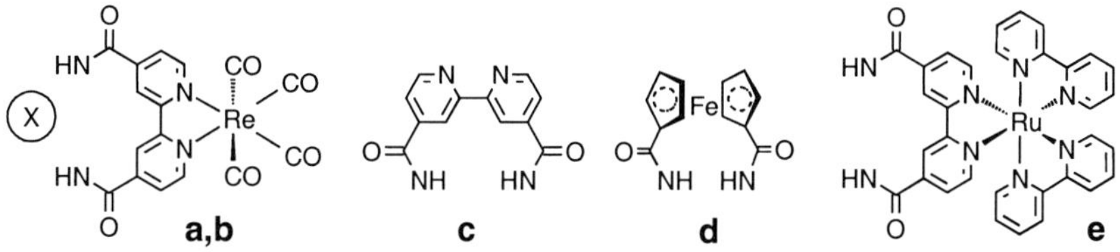

37 a,c $Y^1 = Y^2 = C_5H_{11}$, $R = t-Bu$, $n = 0$
    b,d $Y^1 = Y^2 = C_8H_{17}$, $R = H$, $n = 1$

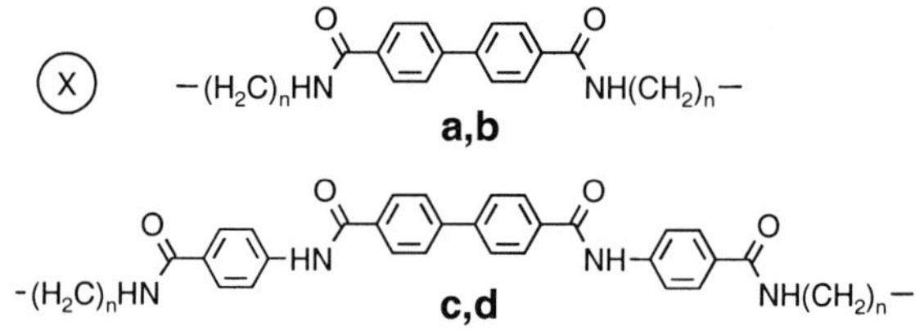

38 $Y^1 = Y^2 = H$, $R = $ adamantyl-Ts   **a** adamantyl   **b** azobiphenyl-azo   **c** $-H_2C-N\text{(piperazine)}N-CH_2-$

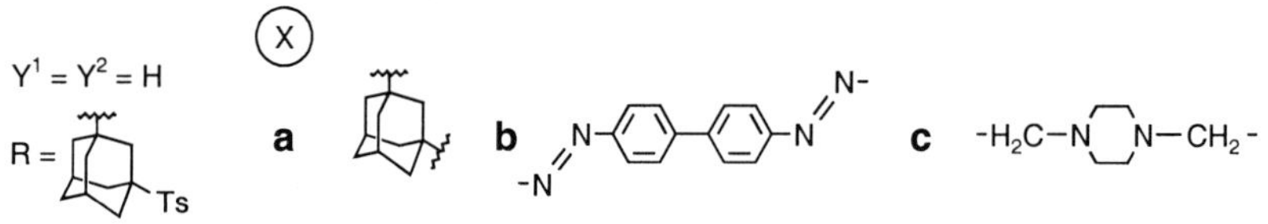

39 $Y^1 = Y^2 = H$
   a $R^1 = OMe$
   b $R^1 = H$

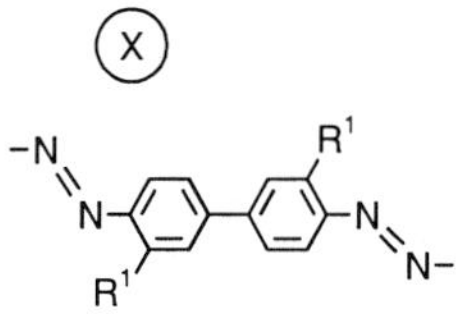

40 $Y^1 = Y^2 = Pr$, $R = H$ (X = porphyrin with R' substituents)

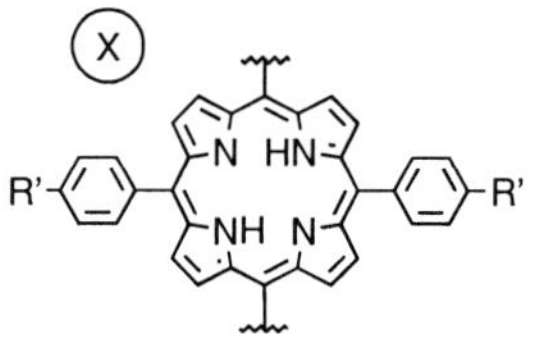

This synthetic strategy could be also extended to the preparation of rotaxanes with calix[4]arenes as stoppers [53]. To this purpose the monoamino calix[4]arene tetraether is extended by reaction with *p*-nitrobenzoylchloride followed by reduction with H_2/PtO_2. Acylation with 4,4'-biphenyl-1,1'-dicarbonyl dichloride in the presence of the cyclic amide **41** led to the rotaxane **42** in 15% yield [54]. In all other cases only the dumbbell **37** could be isolated.

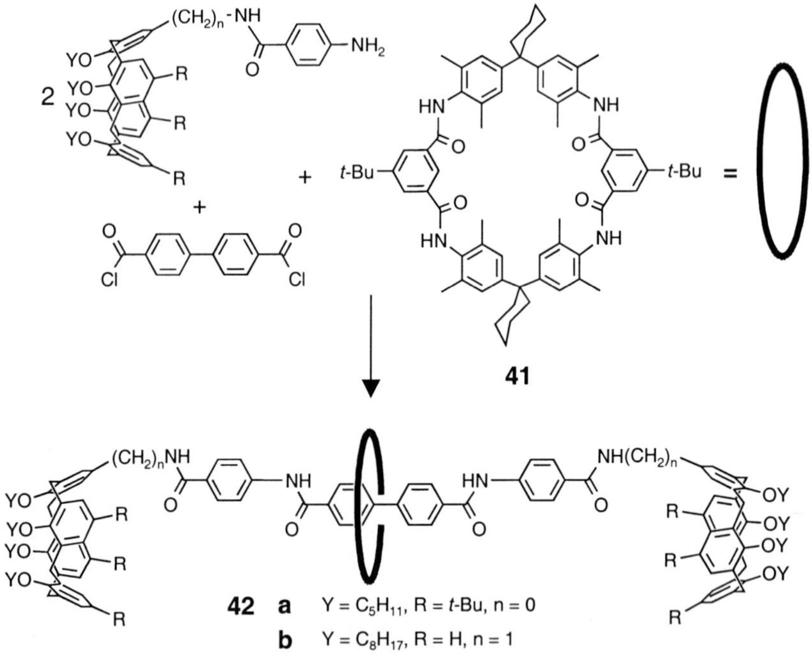

Double calixarenes **38** in which two trisadamantylcalix[4]arenes are linked via various spacers were also reported [55,56]. Calix[4]arenes **39** (as well as the analogous calix[5]arenes) connected by a bisazobiphenyl linker [57] were synthesised in 8-15% yield by reacting 1 equiv. of calix[4/5]arene with 0.5 equiv. of the tetrazonium salts in THF/pyridine. The *cone* conformation and *anti*-position of the two calix[4]arene units was proved for one case by X-ray analysis.

The porphyrin bridged biscalix[4]arenes **40** were prepared by condensation of **Id** with various aryldipyrrolylmethanes in yields up to 54% [58], while the inverse reaction sequence (condensation of **Ic** with pyrrole followed by condensation with a benzaldehyde) was less succesful.

The tandem Claisen rearrangement of double calix[4-6]arenes singly bridged at the narrow rim by a 2-butenyl or 2-methylenepropyl linker is an especially elegant methhod to convert them in one step and in yields of 61-93% into the corresponding double calixarenes singly bridged at the wide rim [59]. As shown for calix[5]arene **43** in Eq. 2 eight allylether groups in the nonconnected phenolic units may be simultaneously rearranged to give **44** with a yield of 61%. (For the interaction of these compounds with fullerenes see Chapter 26).

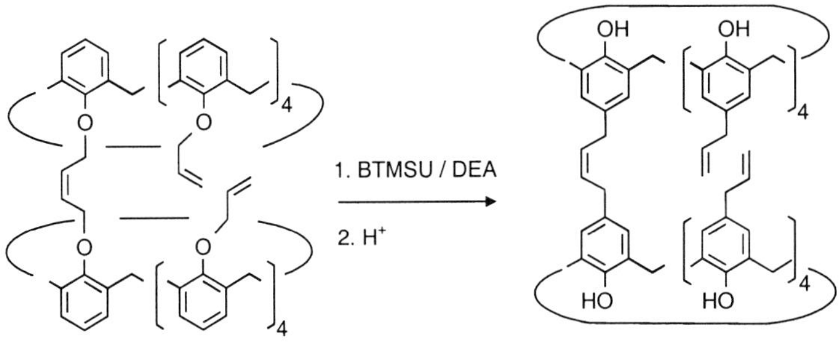

43   **44**

(Eq. 2)

1. BTMSU / DEA
2. H⁺

Double calix[5]arenes with various spacers were obtained by Pd(0) catalysed coupling of the diiodo-calix[5]arene (in form of its pentaacetate) [60]. They are excellent hosts for fullerenes with a rare preference for C_{70} in one case (see Chapters 3 and 26).

To improve the host-properties the double-calix[4]arenes **45** were recently prepared in which two rigid bis-crown-3 units (compare Chapter 20) are connected via various spacers [61]. The mono-functionalisation of the C_{2v}-symmetrical starting compound makes it chiral (C_1), and hence **45** was formed as a mixture of two diastereomers (a meso-form and a pair of enantiomers) which could not be separated. Depending on the connecting spacer, **45** was in fact a better host for quaternary ammonium cations than the single bis-crown-3 calixarene.

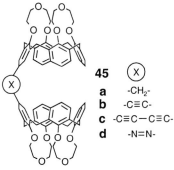

45   X
a   -CH₂-
b   -C≡C-
c   -C≡C-C≡C-
d   -N=N-

A direct *para-para* linkage of two calix[4]arene substructures by oxidative coupling with FeCl₃·6H₂O leads to the double calixarene **46** containing a biphenol unit [62]. The yield is much better when the coupling is carried out with the tribenzoate [63]. The X-ray structure shows the expected *cone* conformation for each calixarene part, but, the two cavities point in different directions. *O*-Alkylation of the free hydroxy groups leads to various derivatives in which the two moieties may exist in different conformations [64]. The structural features of **46** suggest a potential cooperativity of the two host-cavities which was confirmed by preliminary complexation studies [62]. This prompted the authors to extend the synthesis also to larger biscalix[6/8]arenes [65] (see Chapter 5). Very recently, an analogous biscalix[5]arene and its complexes with C_{60} (confirmed by X-ray analysis) and C_{70} were described [66].

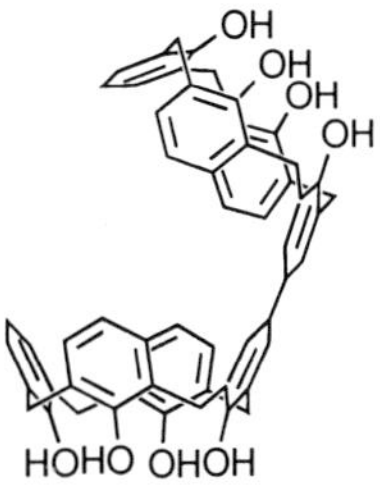

46

2.2.2. Double bridged compounds
In most of the examples described, the connection occurs between opposite phenolic units of a calix[4]arene, leading to molecules of type **B**.

The double calix[4]arenes **47** with two oxa-propylene bridges were prepared by reaction of the corresponding di-hydroxymethyl and di-chloromethyl calix[4]arenes **IIh** (R^1 = CH$_2$OH or CH$_2$Cl) in DMF in the presence of CsOH in 38-42% yield [67]. An X-ray structure shows that the two calixarene subunits are present in a flattened *cone* conformation not very favourable for the inclusion of guests.

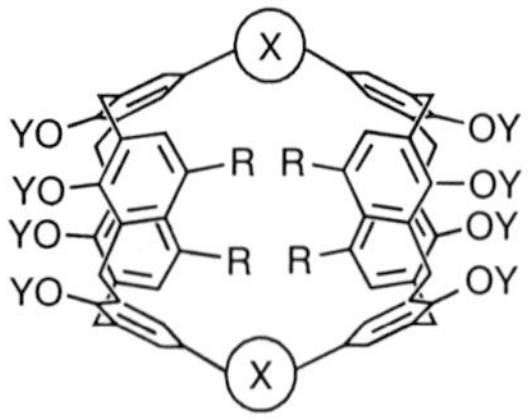

	R	Y	X
47 a	H	CH$_2$CH$_2$OEt	
b	t-Bu	CH$_2$CH$_2$OEt	-CH$_2$OCH$_2$-
48 a	H	Pr	-OCH$_2$-
b	H	Pr	-OCH$_2$CH$_2$O-
49	t-Bu	CH$_2$CH$_2$OEt	-C≡C-C≡C-
50	H	Pr	-OC-HN-⟨⟩-NH-CO-

A high inclusion ability for various quaternary ammonium ions such as *N*-methylpyridinium and its analogues was reported for the biscalixarenes **48** connected by two shorter (-OCH$_2$-) [68] or longer [69] bridges. The very slow dynamic process of complexation-decomplexation (= separation of cation and counter anion) in the case of **48a** is supposed to be due to the encapsulation of the cation in the biscalixarene.

The oxidative coupling reaction of the respective di-ethynyl calix[4]arene **IIf** using copper(II) acetate gave the doublecalix[4]arene **49** in 20% yield [70]. No binding abilities were observed for this compound, or for the amide bridged **50** [71].

The tandem Claisen rearrangement mentioned already for singly bridged calixarenes [59] works also in the case of doubly bridged calixarenes, with yields of 15-22% for calix[4]arenes **51** and 33-70% for calix[6]arenes. Considering the intramolecular mechanism (indicated in Eq. 3) with six-membered cyclic intermediates/transition states, this is a quite astonishing result for a macro-macrocyclic double calixarene.

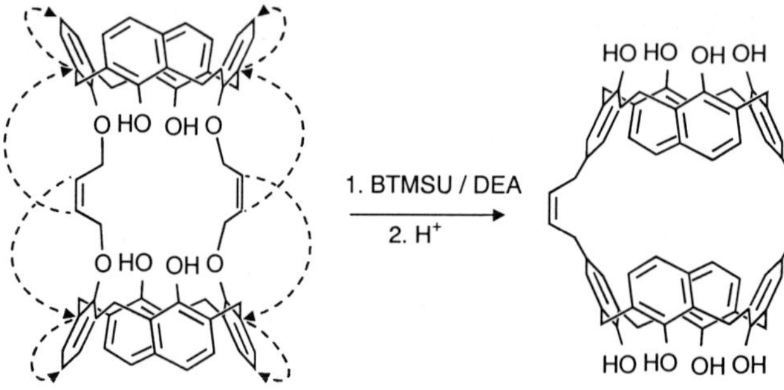

(Eq. 3)

51

Non-symmetric bridges are found in the Schiff-base derivatives **52** which were prepared in 74-92% yield by condensation of readily available calix[4]arenes **IIc** and **IId** bearing distal amino and formyl groups in CH$_2$Cl$_2$ without high dilution conditions [72]. Reduction of the imine functions by an excess of NaBH$_4$ at room temperature in EtOH/THF led to more flexible amines **53** in 87 and 95% yield. Their structures were confirmed by ^{1}H, ^{13}C-NMR and Mass (FAB) spectroscopy and for **53a** by X-ray analysis. The amines **53** show a high affinity for silver cations, which is confirmed by membrane transport and CHEMFET studies. ^{1}H NMR titration experiments in CDCl$_3$ as well as FAB-mass spectra suggest a 1:1 complex.

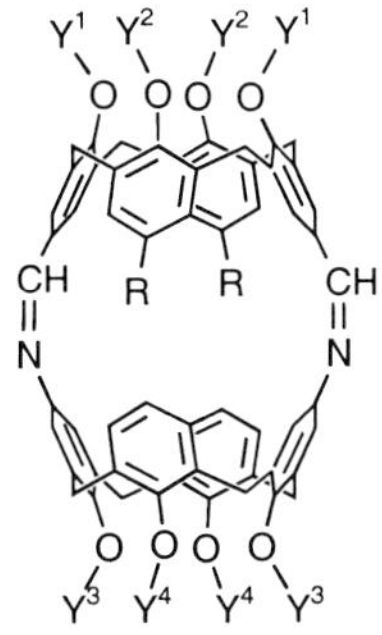

52a-c

a R = NO$_2$ Y^1 = Y^2 = Y^3 = Y^4 = Pr
b R = H   Y^1 = Y^2 = Pr
          Y^3 = Y^4 = CH$_2$CH$_2$OEt
c R = H   Y^1 = Y^3 = H, Y^2 = Y^4 = Pr

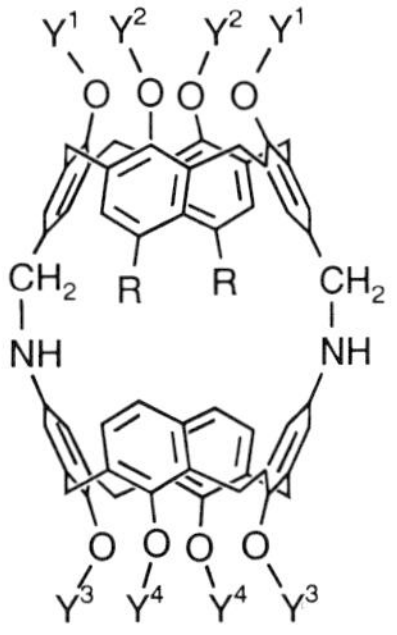

53a,b

Condensation of diamine **IIc** with various aromatic dialdehydes in CH$_2$Cl$_2$-MeOH led to biscalix[4]arenes in 19-98% yield. Molecular recognition of viologens and analogues was also reported [73].

Wide rim dialdehydes **II** (R^1 = -NH-CO-CH$_2$-O-C$_6$H$_4$-CHO), derived from **IIc** were used as starting material for the synthesis of compounds where two calix[4]arenes are linked via two bridges to a porphyrin [74]. Direct condensation with pyrrole gave double-calix[4]arene **54** (where the connection involves adjacent methine bridges of the porphyrin) [27] while a condensation of **II** with excess of pyrrole led to a bis-aryldipyrrolylmethane which upon further condensation with **II** finally gave **55** (in which opposite methine bridges are connected to a calix[4]arene) [75]. Consequently the calixarenes are situated above and below the porphyrin plane in **55** providing rather rigid hydrophobic cavities for guest inclusion. The Zn-complex of **55** strongly binds guests such as pyridine or 4-methylpyridine with association constants higher by factors of 10-1000 as compared to the simple Zn-tetra phenylporphyrin, while 4-*tert*-butylpyridine is not bound, obviously since the bulky *t*-butyl group prevents its inclusion to the cavity.

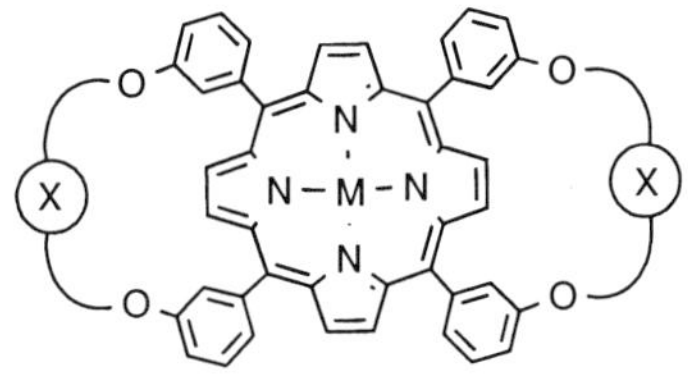

54 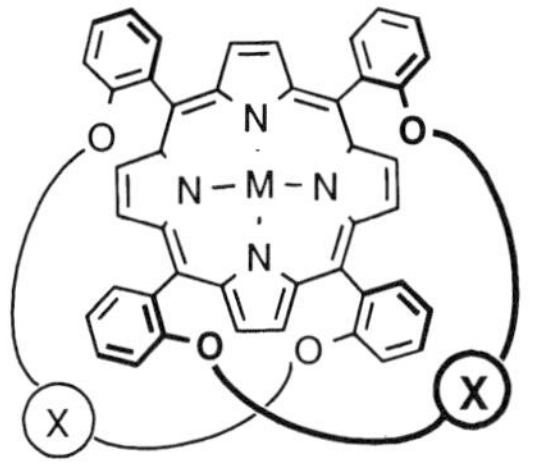   **55**

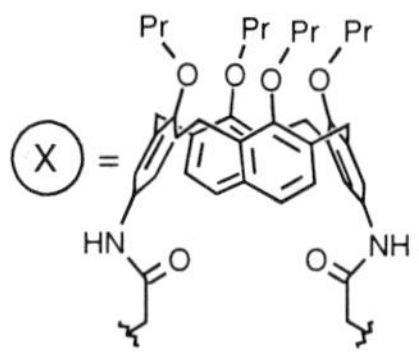

Recently a fourfold 1,3-dipolar cycloaddition was used to prepare biscalix[4]-arenes from diamides (R^1 = NH-CO-CH=CH$_2$ or NH-CO-C≡CH) derived from **IIc** [76].

2.2.3. More than two bridges
Compared to the diversity of carcerands in which two cavitands are connected via four bridges (see Chapter 10), double calixarenes connected at their wide rims by more than

two bridges are rare. This may be caused by difficulties faced during their synthesis due to the flexibility of the calixarenes in comparison to cavitands.

Surprisingly the first example of a calixarene-based "*carcerand*" was a capsule in which two calix[6]arene hexamethylether units are linked by six -CH$_2$SCH$_2$- bridges, which was obtained by high dilution condensation of the respective chloro- and thiomethyl derivatives. Some evidence was reported that this capsule is capable of *constrictive binding* of *N*-methylformanilide [77]. More recently triply bridged calix[6]arenes **56** were synthesized [78] in analogy to doubly bridged calix[4]arenes **52**, while calix[4]arene analogues to **56b** obviously have not been described [78].

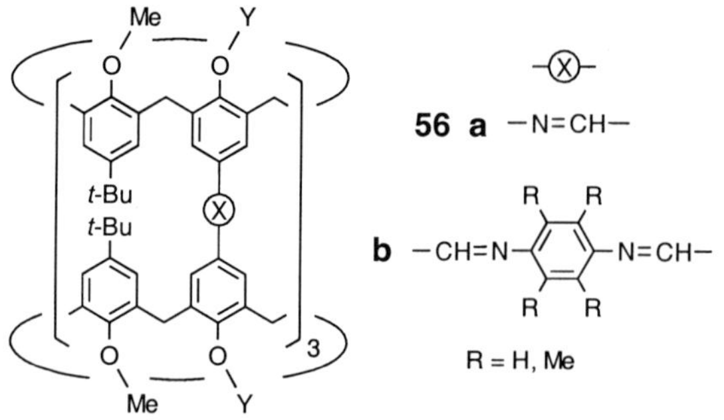

56 a —N=CH—

b —CH=N-⟨⟩-N=CH—

R = H, Me

Double calix[4]arene **57a** was prepared by reacting the tetrachloromethylated tetrapropylether (fixed in the *cone*-conformation) with ethylene glycol in excess (to avoid intramolecular bridging) and subsequent condensation of the product with *p*-chloromethyl calix[4]arene tetramethylether (its conformational flexibility avoids some of the potential wrong connections) under high dilution in THF in the presence of NaH (12% yield) [79]. Organic guests such as RNMe$_3^+$ were found to be included in the cavity and the *exo*-cage-*endo*-cage exchange rate is slow on the NMR time scale.

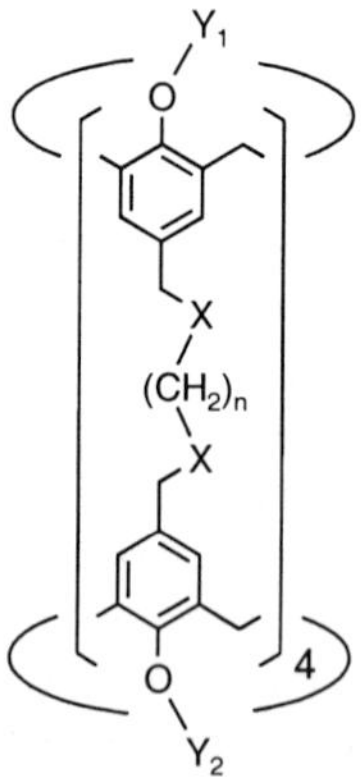

Symmetric octathio bis-calix[4]arenes **57b** and **57c** have been obtained similarily from the *p*-chloromethylated calix[4]arene with thioglycol or the *p*-thiomethylated precursor with diiodomethane [80].

57a X = O Y$_1$ = Pr, Y$_2$ = Me, n = 2
b X = S Y$_1$ = Y$_2$ = CH$_2$CH$_2$OEt, n = 2
c X = S Y$_1$ = Y$_2$ = CH$_2$CH$_2$OEt, n = 1

A surprising result was reported for the reaction of a *p*-chloromethylated calix[4]-arene tetraether in the *cone* conformation with 4,4'-biphenol. Instead of the intended double-calix[4]arene with four bridges between the two calixarenes a compound was isolated in 0.7% yield, for which the spectroscopic data (^{1}H NMR) suggest a structure with a transannular bridge between the distal positions of each calix[4]arene and two bridges between the two calix[4]arene subunits [81].

2.2.4. Double calix[4]arenes in the 1,3-alternate conformation

For all the examples described up to now the connection of the two calixarenes via the wide rim occurs, while ether groups at the narrow rim are used to keep the building blocks in the *cone* conformation.

"Nano-tubes" **58** consisting of two calix[4]arene units in the *1,3-alternate* conformation were prepared in 10% yield by condensation of the *p*-chloromethylated tetrapropylether **IIIa** ($Y^1 = Y^2 = Pr$), fixed already in the *1,3-alternate* conformation, with 1.3 equiv. of catechol in refluxing acetone in the presence of potassium carbonate and NaI and further treatment of the mixture with 1 equiv. of a corresponding bisphenol [82]. The complexation ability towards silver cations Ag^+ was studied by ^{1}H NMR spectroscopy and complexes with one and two Ag^+-cations could be found. Although from other studies it is likely that a Ag^+ can pass the annulus of an *1,3-alternate* calix[4]arene [83], no evidence for an intramolecular exchange process between the two π-basic cavities has been found in the 1:1 complex.

The preparation of a similar doubly bridged *bis*-calix[4]arene in *1,3-alternate* conformation [84] was reported by direct oxidative coupling of the corresponding tetra propargyl derivative **III** ($R^1 = R^2 = CH_2$-O-CH_2-C≡CH, $Y^1 = Y^2 = Bz$). Two intra- and two intermolecular condensation steps led to **59** in 7% yield.

Aza-Wittig condensation of the calix[4]arene di-azide **IIIb** ($Y^1 = Y^2 = CH_2$-CH_2-O-Et, prepared in four steps from *p*-H-calix[4]arene) with tetraphthaldehyde led in 30% yield to the double calix[4]arene **60** in which the calixarene subunits are connected via imino functions [85]. The structure was confirmed by both, ^{1}H-NMR-spectroscopy and X-ray analysis. Complexation studies with silver triflate show that the first two silver cations are complexed near the imine groups. Excess of silver triflate leads to the uptake of two further silver cations between the glycolic chains at the outer end of the calix[4]arene parts. The tetranuclear complex was also detected by FAB-MS.

2.3. CONNECTION BETWEEN THE WIDE AND NARROW RIM ("HEAD- TO-TAIL")

Compounds in which two calix[4]arenes are connected by two bridges between the wide and the narrow rim represent the third possibility for "doubly bridged" double calix[4]arenes (type **C**). The first examples of this type **61** were prepared as shown in Eq. 4 using a 2+1+1 condensation in the last step [86], similar to the synthesis of *p*-bridged calix[4]arenes [87] (see Chapter 1). However, the yield is lower (up to 8%), since $TiCl_4$ cannot be used as catalyst. This makes it inappropriate to extend the synthesis to triple-calixarenes by repeating the reaction sequence, although a regioselective *O*-alkylation of the *p*-bridged phenolic units in **61** seems likely. For one example, the structure could be confirmed by X-ray analysis [86].

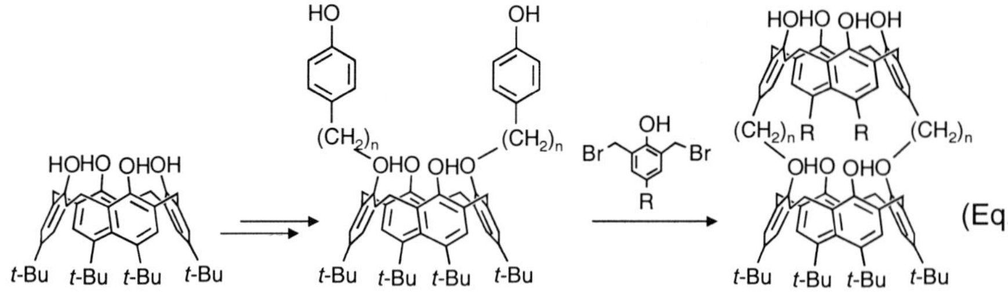

Bis-calix[4]arenes **62**, on the other hand were prepared by reacting the respective diamino derivative **IIc** with the 1,3-di_acidchloride **IIa** in 60 and 20% yield respectively. [88] **62b** forms a 1:1 complex with various anions with a selectivity of one order of magnitude for fluoride over chloride, while much weaker complexes were formed with HSO_4^- and $H_2PO_4^-$. (For details see Chapter 23)

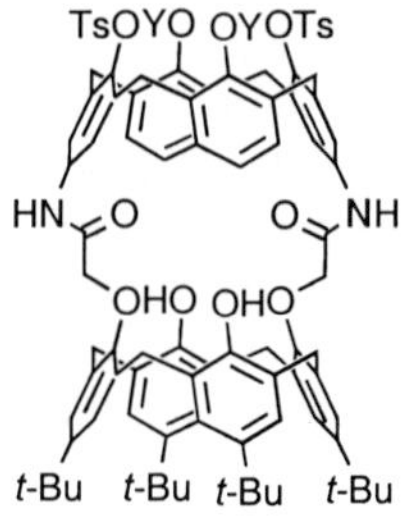

The tendency of p-t-butylcalix[8]arene to form partial O-alkylation products with an "alternating" order (1,3,5,7-tetraethers, see Chapter 5) was used in the synthesis of the double calixarene **63**, in which a calix[4]arene and a calix[8]-arene are linked in head-to-tail fashion [89]. This macrocavitand with a deep cavity was prepared in 30% yield by reacting p-t-butylcalix[8]arene with p-chloromethyl-calix[4]-arene tetramethylether in the presence of CsF and NaI under high dilution conditions. ^{1}H NMR spectroscopy shows in the polar solvent DMSO at 370 K two AX systems (pairs of doublets) for the Ar-CH$_2$-Ar groups in the calix[4]- and calix[8]arene part, reflecting a C_{4v}-symmetry. In the apolar solvent $C_2D_2Cl_4$ at 330 K, the presence of two signals for the methoxy groups and the aromatic protons belonging to the calix[4]arene indicate a flattened *cone* conformation with C_{2v}-symmetry. This causes an *elliptical shape* of the whole biscalixarene and thus, no complexation ability towards aromatic guest molecules was found.

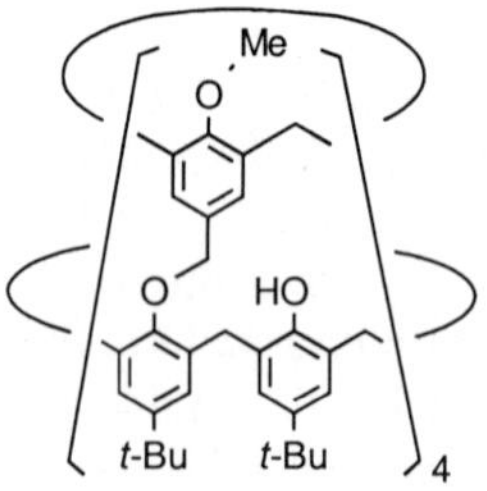

3. Molecules Consisting of more than Two Calixarene Subunits

3.1. Linear Oligomers

Rather than the anticipated bis(chloromethylated)calix[4]arene, linear oligomers **64** in which up to 5 calix[4]arenes are connected via methylene bridges have been obtained by reaction of the di-t-butylcalix[4]arene di-methyl ether **II** (Y^1 = Me, R^1 = t-Bu, $Y^2 = R^2$ = H) with chloromethyl methyl ether (ZnCl$_2$ as catalyst, rt.) [90]. Obviously, under these reaction conditions the chloromethylated intermediates react with further molecules to give a mixture of oligomers, which could be separated chromatographically (3%, 3.5% and 4% yield for the linear trimer to the pentamer). The major product is the cyclic trimer **72a** formed in 12% yield. The single oligomers are well characterized by ^{1}H NMR spectra which indicate that each calix[4]arene unit assumes the *cone* conformation (as usual for 1,3-dimethylethers).

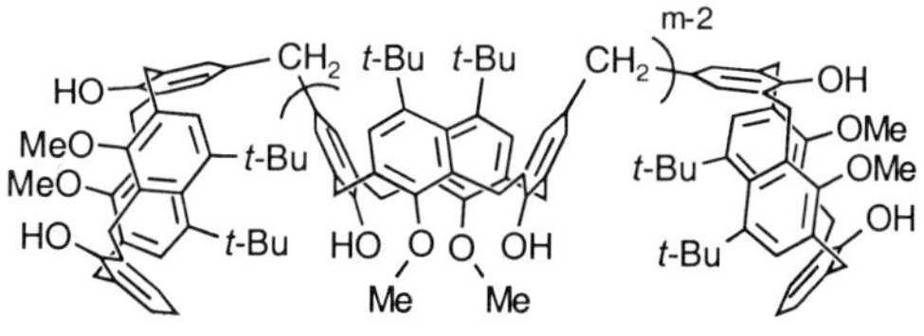

64
m = 3-5

Mixtures of linear oligomers (and polymers) having the calix[4]arenes linked through vinylene or phenylene moieties were prepared by Stille (using (*E*)-1,2-bis-(tributylstannyl)ethylene) or Suzuki (using boronic acid) cross-coupling reactions with the di-(*p*-bromo or *p*-iodo) calix[4]arene di-propylether **IIe** (Y^1 = Pr) [46].

Intermolecular metathesis of the 1,3-*p*-allyl calix[4]arene (Grubbs' catalyst, benzene) led to the linear trimer **65** isolated in 20% yield (from the mixture containing also a linear and a cyclic dimer) [26]. This structure was proved by mass and ^{1}H NMR spectroscopy, showing that both *E* and *Z* isomers for the alkene groups are present.

Triple calixarenes **66** connected via the narrow rim by aliphatic chains were prepared in a more rational way. *O*-Alkylation of a tripropyl ether with excess of an α,ω-dibromoalkane led to a monobromoalkyl derivative **Ib** (Y^2 = Pr), which was used to alkylate a 1,3-di-propyl ether (R = H, *t*-Bu; 27-53%) [5].

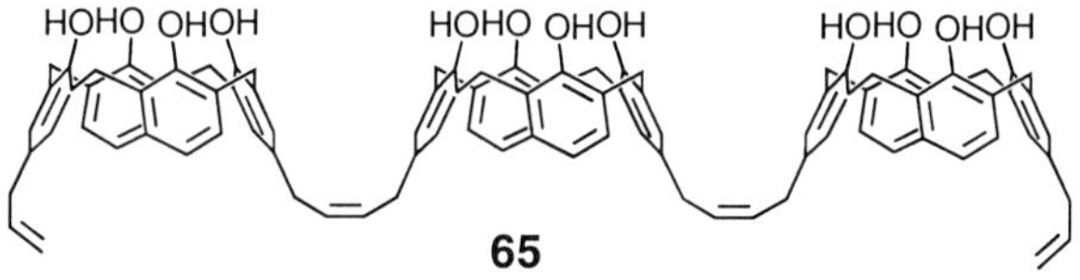

65

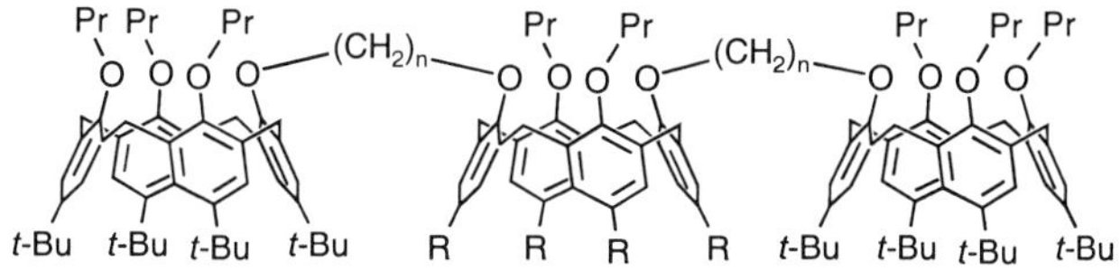

66 R = H or *t*-Bu n = 3, 6

Three calix[4]arene units are directly fused by two silicon atoms in **67**, which was prepared in 60-69% yield by dropwise addition of a dilute SiCl$_4$ solution (1.2 mole) to a suspension of *p*-*t*-butylcalix[4]arene and NaH (4 mole) in dry THF [35b]. NMR spectra (^{29}Si, ^{1}H, ^{13}C) revealed that the three calixarene subunits exist in the *cone* conformation and upon cooling (from 40 to –80 °C) the triply bound calixarenes become non equivalent due a hindered rotation around the Si-O bond [35b].

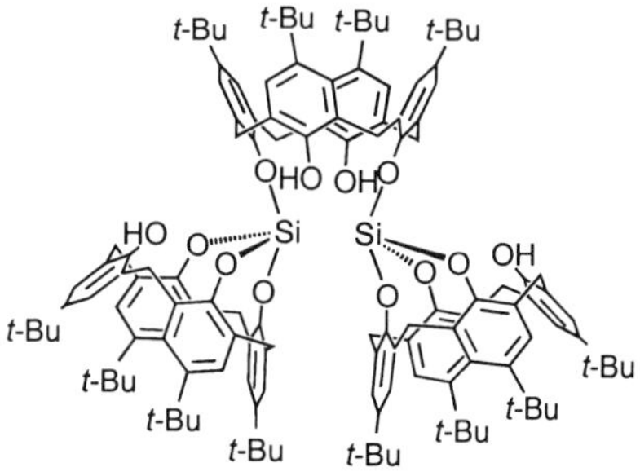

67

A trimeric structure **68** in which the calix[4]arenes are 1,3-capped and connected in the *1,3-alternate* conformation was prepared in 17% yield similar to the preparation of double calixarene **58**, using instead of bisphenol, 0.4 mole of a calix[4]arene with four terminal hydroxy groups (obtained in 39% yield by reacting the tetrachloromethylated derivative **IIIa** ($Y^1 = Y^2 = Pr$) with an excess of bisphenol (80 equiv.) in the last condensation step [82].

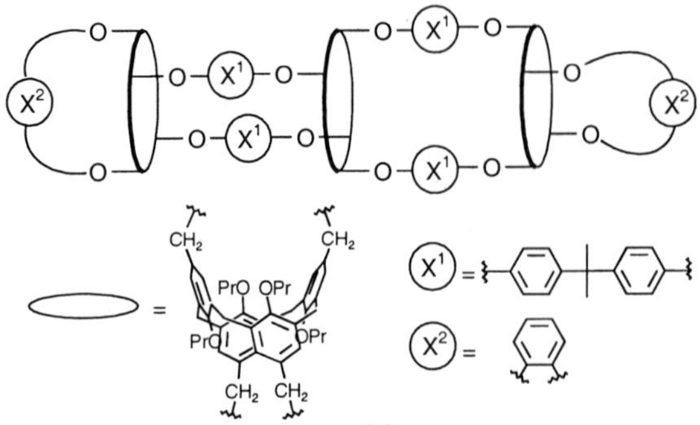

3.2. CYCLIC OLIGOMERS

Compounds **69a** [14] and **69b** [11] are early examples of cyclic trimers obtained in 18 and 34% yield in analogy to **12b,c**, while **70** was obtained more recently from *t*-butylcalix[4]arene by condensation with the bis-bromomethylated benzoic ester in 17% yield in analogy to **14a** [17].

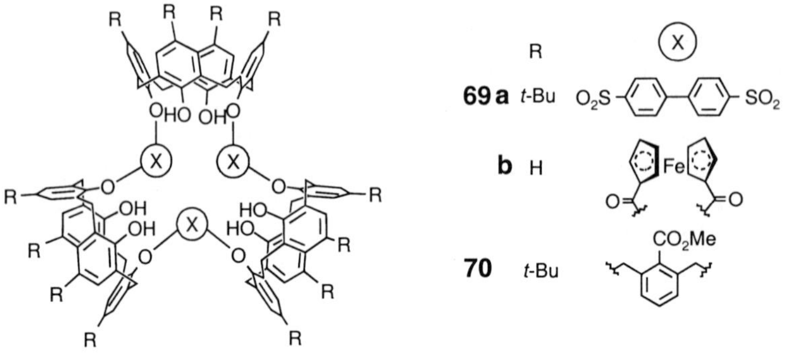

A series of cyclic oligomers ("macrocycles of macrocycles") was prepared using the reactions (*O*-alkylation in DMF/NaH) and intermediates mentioned for the synthesis of **5** and **66**. The cyclic trimer **71a** was obtained in 20% yield by 1:1 condensation of the dihydroxy dipropyl ether with a biscalix[4]arene having two bromohexyl ether residues on the narrow rim, while the cyclic tetramer **71b** was formed in 32% yield when a biscalix[4]arene having two hydroxy groups was used. The hexamer **71c** and octamer **71d** were also isolated in 5 and 4% yield. Mass spectoscopy (SIMS mode) shows for the cyclic trimer **71a** a peak for [M+Li$^+$] and [M+K$^+$] for the cyclic tetramer **71b**. In the case of the hexamer and octamer the molecular weights have been estimated by gel permeation chromatography GPC (vide supra). Simple ^{1}H NMR spectra prove the highly symmetrical structures of these compound.

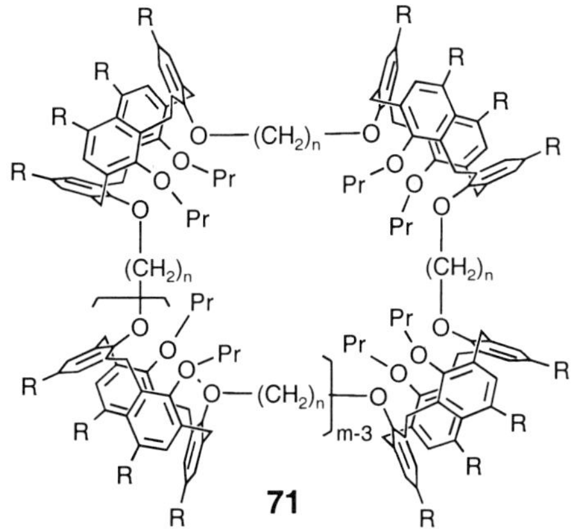

71
- a m = 3
- b m = 4
- c m = 6
- d m = 8

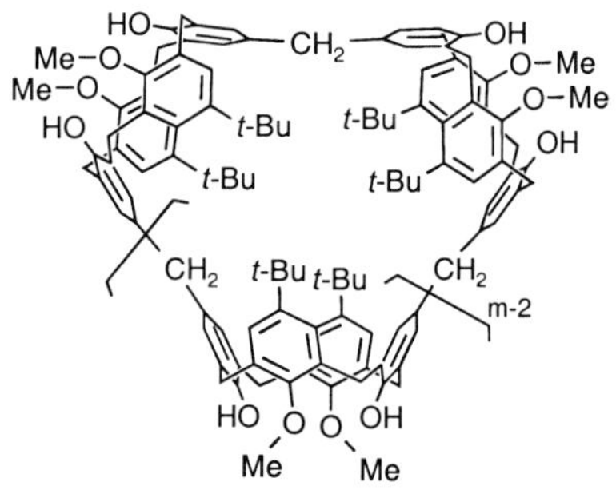

72
- a m = 3
- b m = 4
- c m = 5

The cyclic trimer **72a** in which the calix units are connected via the wide rim by methylene linkages was isolated as the major product (12%, see above [90]) from the mixture of linear (**64**) and cyclic oligomers, while the tetramer **72b** and pentamer **72c** were obtained in about 4% yield respectively [90].

When the metathesis reaction of diallyl calix[4]arene **IIg** is carried out in dichloromethane, a cyclic trimer **73** is formed quantitatively arising from the macrocylization of the linear trimer (formed in benzene see section 3.1) [26].

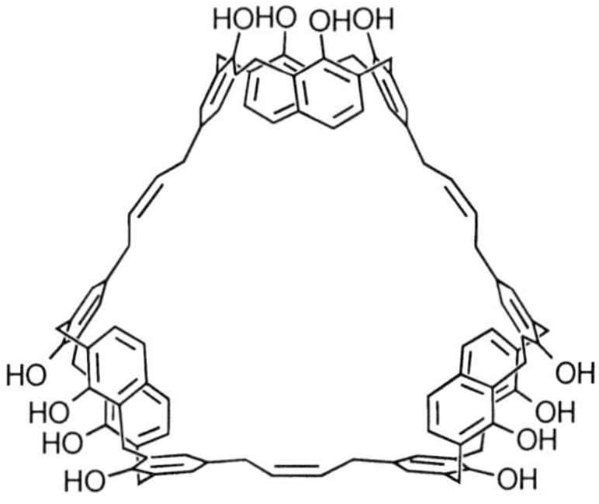

3.3. BRANCHED MOLECULES (CALIX-DENDRIMERS)

Alkylation of the monohydroxy tetrapropyl ether **If** ($Y^1 = Y^2 = Pr$, available in three steps from tetra propoxycalix[4]arene) with 1,3,5-tris(bromomethyl)mesitylene or hexakis(bromomethyl)benzene in DMF in the presence of NaH led to the triple calix[4]arene **74a** and the hexacalix[4]arene **74b** in 54% and 51% yield, respectively [48]. The structure of **74b** is proved by mass spectroscopy and by a ^{1}H NMR spectrum reflecting a high symmetry together with the typical splitting of the monosubstituted calix[4]arene. An alternating up-down arrangement of the calixarene units around the hexa-substituted benzene is supposed.

A triple tetraureacalix[4]arene (compare **6**) in which the calixarene units are attached via their narrow rim to a branching benzene was studied in selfassembly processes [7] (see Chapter 8).

74 a R = Me
b R = cal

75

76

	X	Y
a	cal	C_5H_{11}
b	cal	CH_2CO_2Et
c	mes	CH_2CO_2Et

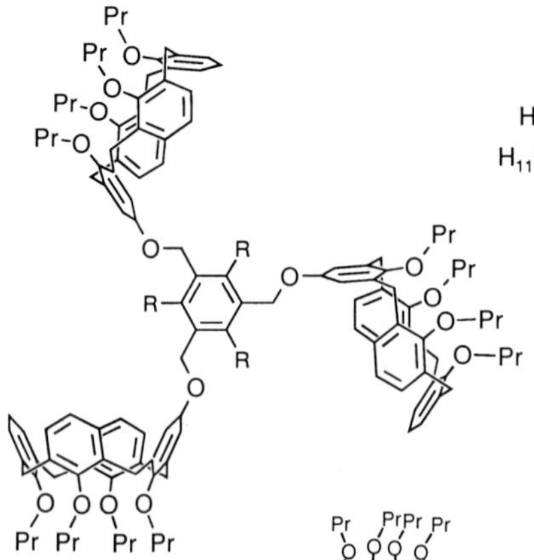

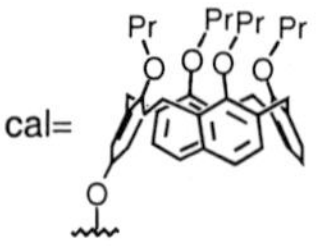

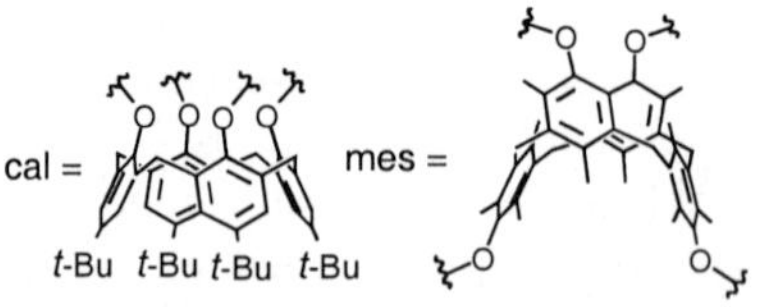

In analogy to **35/36** acylation of a monoamino calix[4]arene **Ic** ($R^2 = t$-Bu) with trimesoyl chloride led to the triple calix[4]arenes **75** in 74% yield [49]. A calix[4]arene itself may be taken as branching point as in the penta-calixarenes **76** which were obtained from the respective tetraacidchlorides in the *cone* or *1,3-alternate* conformation. Molecules **76a** may be considered as a first generation of dendrimers built up by calixarenes as core and as branching points [49].

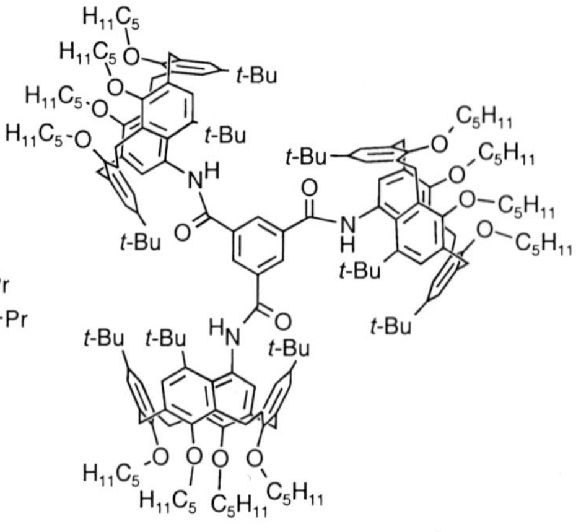

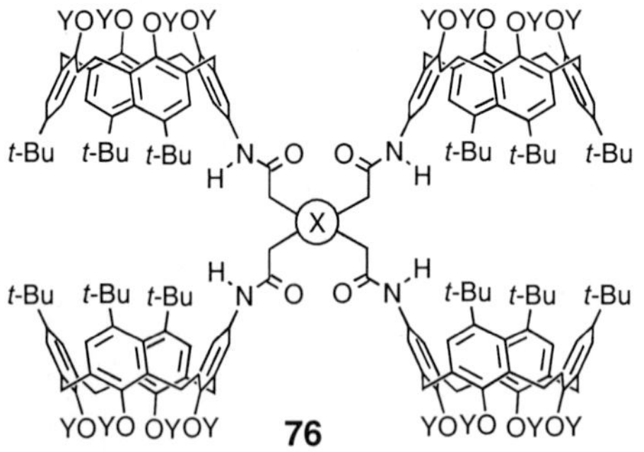

This has been claimed also for the penta-calix[4]arene **77**, which was synthesized in analogy to similar double and triple calixarenes [5]. However, due to the head-to-head (or tail-to-tail) connection, the same principle of the connection cannot be used for the next generation.

77

Y = Pr	R	n
a	H	3
b	t-Bu	3
c	H	6
d	t-Bu	6

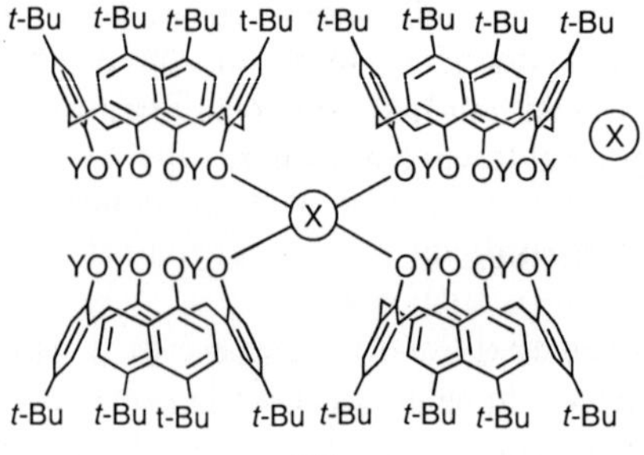

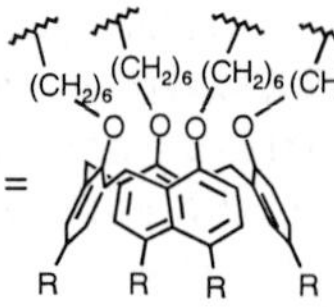

4. Special Structures

The compounds discussed up to now consist of molecules in which substructures consisting of complete calix[n]arenes are connected covalently via bridges of various length and constitution. There is, however, also the possibility to connect calixarenes via their methylene bridges and to construct macrobi- and -tricyclic molecules in which a given aromatic unit belongs to more than one calix[n]arene substructure.

4.1. BICYCLO-CALIXARENES

The first molecules of this type were prepared as illustrated in Eq. 5. The condensation of 1,5-bis-(4-hydroxyphenyl)pentanone-3 with bisbromomethylated phenols leads, as in similar cases [91] to bridged calix[4]arenes **78** in yields up to 30%. After *O*-alkylation of two opposite hydroxyl functions (from two possible 1,3-diethers exclusively the isomer **79** is formed in which the *p*-bridged phenolic units are *O*-alkylated) the alkaline condensation with nitromalonedialdehyde leads to the formation of a *p*-nitrophenol unit (E), linked via methylene bridges to the units A/C [92]. Thus, in addition to the original calix[4]arene system ABCD, two further [1$_4$]metacyclophane (= calix[4]arene) systems ABCE and AECD exist in **80a**, a structure which explains the name chosen for these compounds.

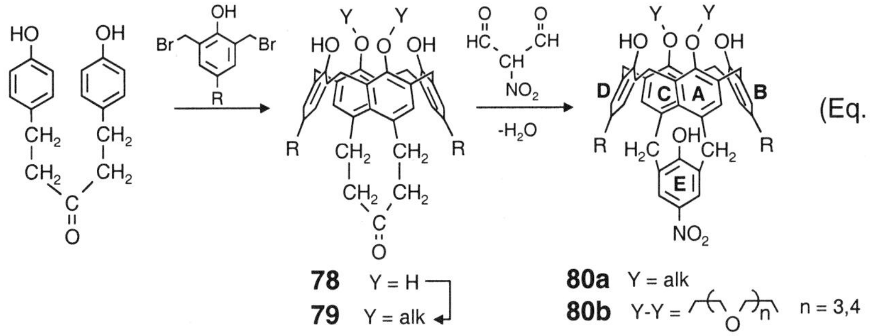

(Eq. 5)

78 Y = H
79 Y = alk
80a Y = alk
80b Y-Y = (crown ether), n = 3,4

Various compounds of this type with or without the O-alkyl groups (they may be easily cleaved in the case of benzylethers) have been prepared, including also the crown ether derivatives **80b** as potassium or cesium selective chromoionophores [93]. Some of them have been confirmed by single crystal X-ray analysis [94]. In all cases the original (*endo*) calix[4]arene assumes the *cone* conformation, while the newly formed (*exo-endo*) calix[4]arenes are found in the *partial cone* and *1,3-alternate* conformation. In solution a conformational equilibrium partial *cone*/*1,3-alternate* exists which is frozen on the NMR-time scale at lower temperature.

4.2. ANNELATED CALIXARENES

Annelated double **82** and triple calix[4]arenes **83** in which two adjacent phenolic units belong to two [1$_4$]metacyclophane systems have been prepared by fragment condensation (2+2) (see Chapter 1) of *exo*-calix[4]arenes **81** with bisbromomethylated dimers as shown in Eq. 6 [95]. Yields of pure products were in the range of 15–25% for **82** and 6–

10% for **83**. Remarkably, the four hydroxyl groups of the newly formed calix[4]arene structures are exclusively in *endo* position, and structures with *exo/endo* OH groups, which would be also possible, have never been observed [96].

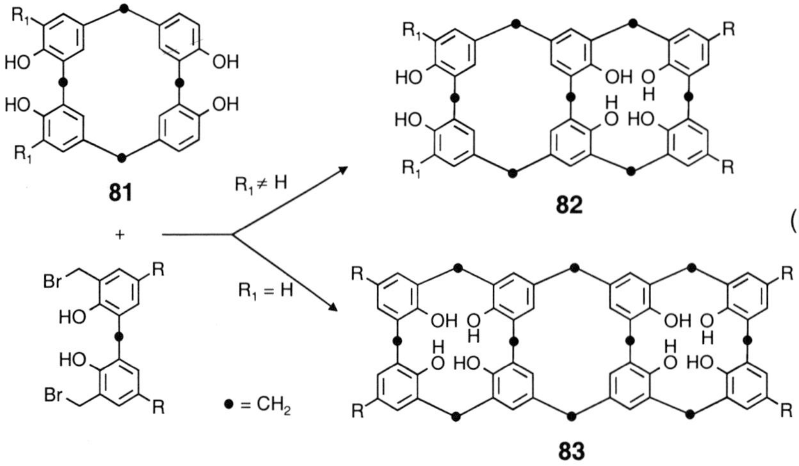

(Eq. 6)

MD-simulations suggest that both calix[4]arene parts in **82** can interconvert only between *cone*, *partial cone* and *1,2-alternate* conformations, among which the *cone* conformation is preferred for the *endo* part. This is in agreement with NMR studies, which show diastereotopic methylene protons (pairs of doublets) up to the highest temperatures available, indicating the impossiblity of a complete ring inversion (topomerisation of the methylene protons). For **83** this limited mobility is restricted to the outer calix[4]arene parts, while the (*exo*) calix[4]arene section in the middle must be fixed either in the *1,2-alternate* or the *cone* conformation. From these two possible conformational isomers only the former was obtained. An X-ray structure (Ca-salt of **83**, R = NO_2) shows the both *endo*-calix[4]arene parts in the expected *cone* conformation [97].

An annelated triple-calix[4]arene **85a** was also obtained in a convergent synthesis using the addition of dilithiated fragments to diketones as shown for the last steps in Eq. 7 [98].

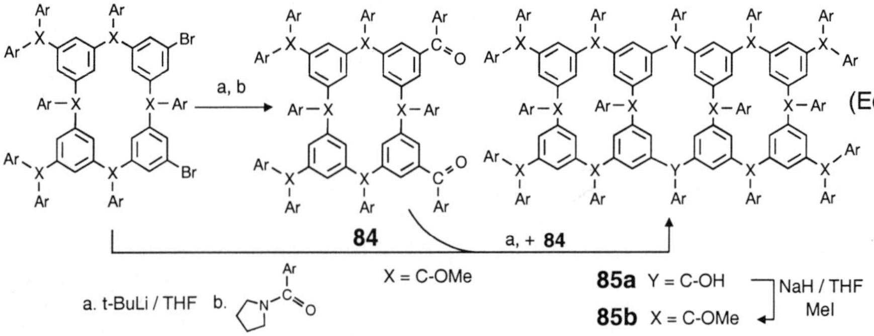

(Eq. 7)

The situation here is complicated by that fact, that various cis/trans isomers are possible due to the ArCOH bridges formed during the cyclisation. **85b** was reduced to a polyanion (Na/K in THF) from which by quenching with methanol the corresponding hydrocarbon and by oxidation with I_2 (<168K) a tetradecaradical with 14 "unpaired" electrons were obtained.

4.3. CONNECTION VIA THE BRIDGES

It should be noted that calix[4]arene units cannot only be connected via the wide or the narrow rim but also via the methylene bridges. Compounds **86** where obtained by condensation of the respective (iso or tere)phthalaldehyde with an excess of p-cresol (R^2 = Me) or p-t-butylphenol (R^2 = t-Bu) and subsequent double (2+2) fragment condensation with a bisbromomethylated dimer [99].

Two calix[6]arene units connected via dixanthylene bridges **87** were also reported [100].

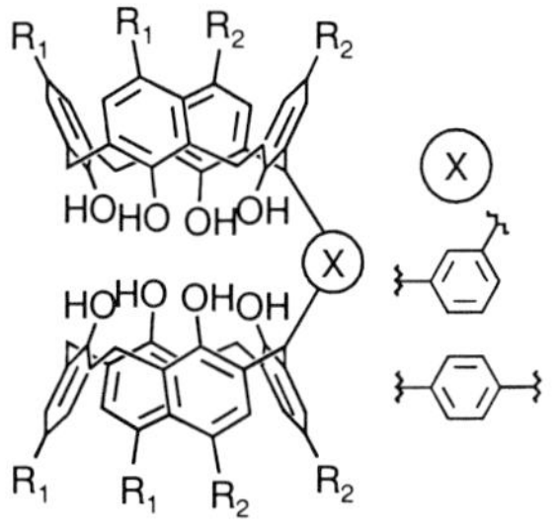

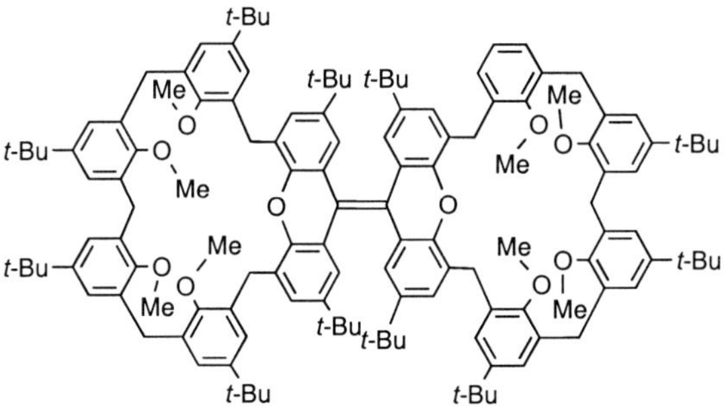

5. Outlook

Oligomers built up by calix[4]arene subunits are clearly predominant among the covalently linked multicalixarenes described so far, since the selective derivatisation of calix[4]arenes is most developed (see Chapter 2). However, the recent progress in the selective functionalisation of calix[5,6,8]arenes (see Chapters 3-5) will make these larger members more and more available as building blocks for the construction of oligomers and polymers held together via covalent bonds or via reversible bonds used in self-assembly processes (see Chapter 8). Mechanical interlocking may be considered as a further possibility, and the first example for a double calix[4]arene of the catenane-type **88** [101] is a suitable structure to close this chapter with an outlook to future possibilities.

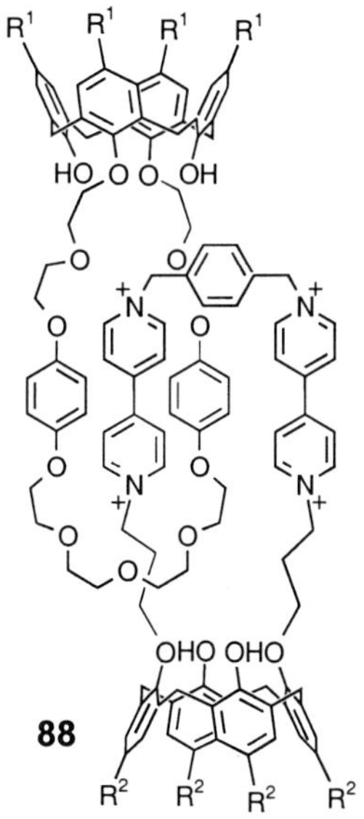

6. References and Notes

[1]   M. A. McKervey, M. Owens, H.-R. Schulten, W. Vogt, V. Böhmer, *Angew. Chem.* **1990**, *102*, 326-328; *Angew. Chem., Int. Ed. Engl.* **1990**, *29*, 280-282.
[2]   F. Ohseto, T. Sakaki, K. Araki, S. Shinkai, *Tetrahedron Lett.* **1993**, 34, 2149-2152.
[3]   M. P. Oude Wolbers, F. C. J. M. van Veggel, R. H. J. W. Hofstraat, F. A. J. Geurts, J. van Hummel, S. Harkema, D. N. Reinhoudt, *Liebigs Ann./Recueil* **1997**, 2587-2600
[4]   P. Lhoták. M. Kawaguchi, A. Ikeda, S. Shinkai, *Tetrahedron* **1996**, *38*, 12399-12408.
[5]   P. Lhoták S. Shinkai, *Tetrahedron* **1995**, *51*, 7681-7696.
[6]   S. Kanamathareddy, C. D. Gutsche, *J. Org. Chem.* **1994**, *59*, 3871-3879.
[7]   R. K. Castellano, J. Rebek, Jr., *J. Am. Chem. Soc.* **1998**, *120*, 3657-3663.
[8]   a) R. Milbradt, J. Weiss, *Tetrahedron Lett.* **1995**, *36*, 2999-3002; b) T. Arimura, C. T. Brown, S. L. Springs, J. L. Sessler, *Chem. Commun.* **1996**, 2293-2294.
[9]   F. Arnaud-Neu, S. Caccamese, S. Fuangswasdi, S. Pappalardo, M. F. Parisi, A. Petringa, G. Principato, *J. Org. Chem.* **1997**, *62*, 8041-8048.
[10] F. Unob, Z. Asfari, J. Vicens, *Tetrahedron Lett.* **1998**, *39*, 2951-2954.
[11] P. D. Beer, A. D. Keefe, A. M. Z. Slawin, D. J. Williams, *J. Chem. Soc., Dalton Trans.* **1990**, 3675-3682.
[12] A. P. Marchand, H.-S. Chong, M. Takhi, T. D. Power, *Tetrahedron* **2000**, *56*, 3121-3126.
[13] For further examples of monobridged calix[4]- and -[6]arenes see: F. Santoyo-González, A. Torres-Pinedo, A. Sanchéz-Ortega, *J. Org. Chem.* **2000**, *65*, 4409-4414.
[14] D. Kraft, J. D. Van Loon, M. Owens, W. Verboom, W. Vogt, M. A. McKervey, V. Böhmer, D. N. Reinhoudt, *Tetrahedron Lett.* **1990**, *31*, 4941-4944.
[15] J.-D. van Loon, D. Kraft, M. J. K. Ankone, W. Verboom, S. Harkema, W. Vogt, V. Böhmer, D. N. Reinhoudt, *J. Org. Chem.* **1990**, *55*, 5176-5179.
[16] Z. L. Zhong, Y. Y. Chen. X. R. Lu, *Tetrahedron Lett.* **1995**, 36, 6735-6738.
[17] P. Schmitt, P. D. Beer, M. G. B. Drew, P. D. Sheen, *Tetrahedron Lett.* **1998**, *39*, 6383-6386.
[18] a) F. Ohseto, S. Shinkai, *Chem. Lett.* **1993**, 2045-2048; b) F. Ohseto, S. Shinkai, *J. Chem. Soc., Perkin Trans. 2* **1995**, 1103-1109.
[19] Z. Asfari, J. Weiss, S. Pappalardo, J. Vicens, *Pure. Appl. Chem.* **1993**, *65*, 585-590.
[20] G. Ulrich, R. Ziessel, *Tetrahedron Lett.* **1994**, *35*, 6299-6302.
[21] G. Ulrich, P. Turek, R. Ziessel, *Tetrahedron Lett.* **1996**, 37, 8755-8758.
[22] X. Hu, A. S. Chan, X. Han, J. He, J.-P. Cheng, *Tetrahedron Lett.* **1990**, *40*, 7115-7118.
[23] V. Böhmer, G. Ferguson, J. F. Gallagher, A. J. Lough, M. A. McKervey, E. Madigan, M. B. Moran, J. Phillips, G. Williams, *J. Chem. Soc., Perkin Trans. 1* **1993**, 1521-1527.
[24] P. D. Beer, M. G. B. Drew, K. Gradwell, *J. Chem. Soc., Perkin Trans. 1* **2000**, 511-519.
[25] E. Pinkhassik, I. Stibor, V. Havlícek, *Collect. Czech. Chem. Commun*, **1996**, *61*, 1182-1189.
[26] a) M. A. McKervey, M. Pitarch, *Chem. Commun.* **1996**, 1689-1690; b) M. Pitarch,V. McKee, M. Nieuwenhuyzen, M. A. McKervey, *J. Org. Chem.* **1998**, *63*, 946-951.
[27] D. Rudkevich, W. Verboom, D. N. Reinhoudt, *Tetrahedron Lett.* **1994**, *35*, 7131-7134.
[28] J.-S. Li, Y.-Y. Chen, X.-R. Lu, *Eur. J. Org. Chem.* **2000**, 485-490.
[29] J.-S. Li, Z. Zhong, Y.-Y. Chen, X.-R. Lu,. *Tetrahedron Lett.* **1998**, *39*, 6507-6510.
[30] Reaction of tetrakis(tosyloxyethoxyethoxymethyl)methane with *p-t*-butylcalix[6]arene led to a "singly-triple" bridged double-calix[6]arene. This shows that such a classification should not be "overdone".
[31] D. Kraft, V. Böhmer, W. Vogt, G. Ferguson, J. F. Gallagher, *J. Chem. Soc., Perkin Trans. 1* **1994**, 1221-1230.
[32] F. González, A. T. Pinedo, C. S. Barria, *Eur. J. Org. Chem.* **2000**, 3587-3593.
[33] G. Ulrich, R. Ziessel, I. Manet, M. Guardigli, N. Sabbatini, F. Fraternali, G. Wipff, *Chem. Eur. J.* **1997**, *3*, 1815-1822.
[34] P. Schmitt, P. D. Beer, M. G. B. Drew, P. D. Sheen, *Angew. Chem.* **1997**, *109*, 1926-1928; *Angew. Chem., Int. Ed. Engl.* **1997**, *36*, 1840-1842.
[35] a) X. Delaigue, M. W. Hosseini, *Tetrahedron Lett.* **1993**, *47*, 7561-7564; b) X. Delaigue, M. W. Hosseini, *Tetrahedron Lett.* **1994**, *35*, 1711-1714; c) J. Martz, E. Graf, M. W. Hosseini, A. De Cian, J. Fisher, *J. Mater Chem.* **1998**, 8 (11), 2331-2333; d) M. W. Hosseini, A. De Cian, *Chem. Commun.* **1998**, 727-733.
[36] J. L. Atwood, S. G. Bott, C. Jones, C. L. Raston, *J. Chem. Soc., Chem. Commun.* **1992**, 1349-1351.
[37] M. M. Olmstead, G. Sigel, H. Hope, X. Xu, P. P. Power, *J. Am. Chem. Soc.* **1985**, *107*, 8087-8091.

[38] F. Hajek, E. Graf, M. W. Hosseini, *Tetrahedron Lett.* **1996**, *37*, 1401-1404; F. Hajek, E. Graf, M. W. Hosseini, A. De Cian, J. Fischer, *Angew. Chem.* **1997**, *109*, 1830-1832; *Angew. Chem., Int. Ed. Engl.* **1997**, *36*, 1760-1762.
[39] Z. Asfari, R. Abidi, F. Arnaud, J. Vicens, *J. Incl. Phenom.* **1992**, *13*, 163-169.
[40] S. Pappalardo, G. Ferguson, J. F. Gallagher, *J. Org. Chem.* **1992**, 57, 7102-7109.
[41] Z. Asfari, S. Pappalardo, J. Vicens, *J. Incl. Phenom.* **1992**, *14*, 189-192.
[42] J. S. Kim, O. J. Shon, J. W. Ko, M. H. Cho, I. Y. Yu, J. Vicens, *J. Org. Chem.* **2000**, *65*, 2386-2392.
[43] V. Böhmer, H. Goldmann, W. Vogt, J. Vicens, Z. Asfari, *Tetrahedron Lett.* **1989**, *30*, 1391-1394.
[44] Examples for double calix[4]arenes connected via two [-(CH$_2$)$_4$-] and four [-(CH$_2$)$_{10}$-] aliphatic chains were also obtained in very low yields. They could be increased later to about 10% for a doubly bridged [-(CH$_2$)$_{10}$-] compound: M. Tabatabai, V. Böhmer, unpublished results
[45] P. Lhoták S. Shinkai, *Tetrahedron Lett.* **1996**, *37*, 645-648.
[46] A. Dondoni, C. Ghiglione, A. Marra, M. Scoponi, *J. Org. Chem.* **1998**, *63*, 9535-9539.
[47] K. Araki, K. Hisaichi, T. Kanai, S. Shinkai, *Chem. Lett.* **1995**, 569-570.
[48] J. Budka, M. Dudic, P. Lhotak, I. Stibor, *Tetrahedron* **1999**, *55*, 12647-12654.
[49] O. Mogck, P. Parzuchowski, M. Nissinen, V. Böhmer, G. Rokicki, K. Rissanen, *Tetrahedron* **1998**, *54*, 10053-10068.
[50] P. Behr, M. Shade, *Chem. Commun.* **1997**, 2377-2378.
[51] P. Behr, M. Shade, *Gazz. Chim. Ital.* **1997**, *127*, 651-652.
[52] P. Behr, J. Cooper, *Chem. Commun.* **1998**, 129-130.
[53] C. Fischer, M. Nieger, O. Mogck, V. Böhmer, R. Ungaro, F. Vögtle, *Eur. J. Org. Chem.* **1998**, 155-161
[54] For a rotaxane with a calix[6]arene as a wheel see: A. Arduini, R. Ferdani, A. Pochini, A. Secchi, F. Ugozzoli, *Angew. Chem.* **2000**, *112*, 3595-3598; *Angew. Chem., Int. Ed. Engl.* **2000**, *39*, 3453-3456.
[55] V. Kovalev, E. Shokova, A. Khomich, Y. Luzikov, *New J. Chem.* **1996**, *20*, 483-492.
[56] V. Kovalev, E. Shokova, E. Shokova, Y. Luzikov, *Synthesis* **1998**, 1003-1008.
[57] S. Bouoit-Montésinos, J. Bassus, M. Perrin, R. Lamartine, *Tetrahedron Lett.* **2000**, *41*, 2563-2567.
[58] M. Dudic, P. Lhoták, V. Král, K. Lang, I. Stibor, *Tetrahedron Lett.* **1999**, *40*, 5949-5952.
[59] J. Wang, C. D. Gutsche, *J. Am. Chem. Soc.* **1998**, *120*, 12226-12231.
[60] T. Haino, M. Yanase, Y. Fukazawa, *Angew. Chem.* **1998**, *110*, 1044-1046; *Angew. Chem., Int. Ed. Engl.* **1998**, *37*, 997-998.
[61] A. Arduini, A. Pochini, A. Secchi, *Eur. J. Org. Chem.* **2000**, 2325-2334.
[62] P. Neri, A. Bottino, F. Cunsolo, M. Piattelli, E. Gavuzzo, *Angew. Chem.* **1998**, *110*, 175-178; *Angew. Chem., Int. Ed. Engl.* **1998**, *37*, 166-169; For a koilate of 46 see: A. Bottino, F. Cunsolo, M. Piattelli, E. Gavuzzo, P. Neri, *Tetrahedron Lett.* **2000**, *41*, 10065-10069.
[63] C. D. Gutsche, L. G. Lin, *Tetrahedron* **1986**, *42*, 1633-1640.
[64] A. Bottino, F. Cunsolo, M. Piattelli, P. Neri, *Tetrahedron Lett.* **1998**, *39*, 9549-9552.
[65] A. Bottino, F. Cunsolo, M. Piattelli, D. Garozzo, P. Neri, *J. Org. Chem.* **1999**, *64*, 8018-8020.
[66] J. Wang, S. G. Bodige, W. H. Watson, C. D. Gutsche, *J. Org. Chem.* **2000**, *65*, 8260-8263.
[67] A. Arduini, S. Fanni, G. Manfredi, A. Pochini, R. Ungaro, A. R. Sicuri, F. Ugozzoli, *J. Org. Chem.* **1995**, *60*, 1448-1453.
[68] K. Araki, H. Hayashida, *Tetrahedron Lett.* **2000**, *41*, 1209-1213.
[69] K. Araki, H. Hayashida, *Chem. Lett.* **2000**, 20-21.
[70] A. Arduini, A. Pochini, A. R. Sicuri, A. Secchi, R. Ungaro, *Gazz. Chim. Ital.* **1994**, *124*, 129-132.
[71] M. Jorgensen, M. Larsen, P. Sommer-Larsen, W. B. Petersen, H. Eggert, *J. Chem. Soc., Perkin Trans. 1* **1997**, 2851-2855.
[72] O. Struck, L. A. J. Christoffels, R. J. W. Lugtenberg, W. Verboom, G. J. van Hummel, S. Harkema, D. N. Reinhoudt, *J. Org. Chem.* **1997**, *62*, 2487-2493.
[73] G. T. Hwang, B. H. Kim, *Tetrahedron Lett.* **2000**, *41*, 5917-5921.
[74] A double calix[4]arene connected via two porphyrin bridges was prepared in 0.4 % yield, see: Z. Asfari, J. Vicens, J. Weiss, *Tetrahedron Lett.* **1993**, *34*, 627-628.
[75] D. Rudkevich, W. Verboom, D. N. Reinhoudt, *J. Org. Chem.* **1995**, *60*, 6585-6587.
[76] G. T. Hwang, B. H. Kim, *Tetrahedron Lett.* **2000**, *41*, 10055-10060.
[77] T. Arimura, S. Matsumoto, O. Teshima, T. Nagasaki, S. Shinkai, *Tetrahedron Lett.* **1991**, *32*, 5111-5114.
[78] A. Arduini, R. Ferdani, A. Pochini, A. Secchi, *Tetrahedron* **2000**, *56*, 8573-8577.
[79] K. A. Araki, K. Sisido, K. Hisaichi, S. Shinkai, *Tetrahedron Lett.* **1993**, *34*, 8297-8300.
[80] M. T. Blanda, K. E. Griswold, *J. Org. Chem.* **1994**, *59*, 4313-4315.
[81] A. Siepen, A. Zett, F. Vögtle, *Liebigs Ann.* **1996**, 757-760.

[82] A. Ikeda S. Shinkai, *J. Chem. Soc., Chem. Commun.* **1994**, 2375-2376.
[83] A. Ikeda, S. Shinkai, *J. Am. Chem. Soc.* **1994**, *116*, 3102-3310.
[84] S. Kanamathareddy, C. D. Gutsche, *J. Org. Chem.* **1995**, *60*, 6070-6075.
[85] J.-A. Pérez-Adelmar, H. Abraham, C. Sánchez, K. Rissanen, P. Prados, J. de Mendoza, *Angew. Chem.* **1996**, *108*, 1088-1090; *Angew. Chem., Int. Ed. Engl.* **1996**, *35*, 1009-1011.
[86] a) W. Wasikiewicz, G. Rokicki, J. Kielkiewicz, V. Böhmer, *Angew. Chem.* **1994**, *106*, 230-232; *Angew. Chem., Int. Ed. Engl.* **1994**, *33*, 214-216; b) W. Wasikiewicz, G. Rokicki, J. Kielkiewicz, E. F. Paulus, V. Böhmer, *Monatsh. Chem.* **1997**, *128*, 863-879.
[87] H. Goldmann, W. Vogt, E. Paulus, V. Böhmer, *J. Am. Chem. Soc.* **1988**, *110*, 6811-6817.
[88] P. D. Beer, P. A. Gale, D. Hesek, *Tetrahedron. Lett.* **1995**, *36*, 767-770.
[89] A. Arduini, A. Pochini, A. Secchi, R. Ungaro, *J. Chem. Soc., Chem. Commun.* **1995**, 879-880.
[90] Y. S. Zheng, Z. T. Huang, *Tetrahedron Lett.* **1998**, *39*, 5811-5814.
[91] V. Böhmer, *Liebigs Ann./Recueil* **1997**, 2019-2030.
[92] B. Berger, V. Böhmer, E. Paulus, A. Rodriguez, W. Vogt, *Angew. Chem.* **1992**, *104*, 89-92; *Angew. Chem., Int. Ed. Engl.* **1992**, *31*, 96-99.
[93] W. Wasikiewicz, M. Slaski, G. Rokicki, V. Böhmer, C. Schmidt, E. F. Paulus, *New J. Chem.* **2001**, in press.
[94] P. O'Sullivan, E. F. Paulus, V. Böhmer, unpublished results.
[95] a) V. Böhmer, R. Dörrenbächer, W. Vogt, L. Zetta, *Tetrahedron Lett.* **1992**, *33*, 769-772; b) V. Böhmer, R. Dörrenbächer, M. Frings, M. Heydenreich, D. de Paoli, W. Vogt, G. Ferguson, I. Thondorf, *J. Org. Chem.* **1996**, *61*, 549-559.
[96] Most probably this is due to the template effect of the $TiCl_4$ applied in this synthesis.
[97] E. F. Paulus, M. Frings, A. Shivanyuk, C. Schmidt, V. Böhmer, *J. Chem. Soc., Perkin Trans. 2* **1998**, *39*, 2777-2782.
[98] A. Rajca, K. Lu, S. Rajca, *J. Am. Chem. Soc.* **1997**, *119*, 10335-10345.
[99] V. Böhmer, C. Grüttner, R. Assmus, unpublished results.
[100] O. Aleksiuk, S. Biali, *J. Org. Chem,* **1993**, *61*, 5670-5673.
[101] Z.-T. Li, G.-Z. Ji, C.-X. Zhao, S.-D. Yuan, H. Ding, C. Huang, A.-L. Du, M. Wei, *J. Org. Chem.* **1999**, *64*, 3572-3584.

Chapter 8

SELF-ASSEMBLY IN SOLUTION

DMITRY M. RUDKEVICH

The Skaggs Institute for Chemical Biology at The Scripps Research Institute, 10550 North Torrey Pines Road, La Jolla, California 92037, USA.
e-mail: dmitry@scripps.edu

1. Introduction

Self-assembly is the spontaneous noncovalent association of two or more molecules under equilibrium conditions into stable, well-defined aggregates [1]. Intermolecular forces are capable of effectively organizing multicomponent supramolecular assemblies in a reversible and accurate fashion using error-correcting mechanisms. Within the last decade, many excellent reviews and original publications describing the steady progress in this field have appeared [2]. In Nature, self-assembly is a ubiquitous strategy involved in the formation of, for example, cell membranes, double-stranded nucleic acids and viruses. In chemistry, self-assembly provides a novel, rapid way to construct complex nanostructures, receptor systems, catalysts and new materials. Calixarenes, in particular, have had a great impact in the history of self-assembly [3]. Their three dimensional, concave surface, commercial availability, relatively rigid structure and well-developed synthetic chemistry have made them extremely convenient platforms for elaboration [4].

The discussion in this chapter is developed on the basis of the increasing complexity of the assembled systems. Many different forces can be recognised as controlling molecular assembly and a crucial issue is how they may be used in concert to give both desired structures and functions to the product supermolecules. Of particular interest are reversibly formed capsules, or "molecule-within-molecule" complexes, which represent an extreme form of molecular recognition.

2. Assembly through Hydrogen Bonding

2.1. DIMERS WITHOUT GUESTS

Hydrogen bonding is near-ubiquitous as a force in self-assembly. It is highly directional, specific and, of course, biologically relevant. That a calixarene platform could be functionalized with multiple hydrogen bonding sites and used in the design of well-defined assemblies was recognized in the early 1990s (Figure 1) [5]. Self-complementary 2-pyridone groups attached at the upper rim of calixarene interact to form a labile dimer **1** ($K_D = 100 \pm 20$ M^{-1} in $CDCl_3$, ^1H NMR); two diametrically attached 2-pyridones caused the formation of much larger aggregates. When two uracil moieties were attached to the lower rim of calix[4]arene as in **2**, dimer formation was even more favourable ($K_D = 3.4 \times 10^3$ M^{-1} in $CDCl_3$) [6]. This is, probably, due to the contribution of intramolecular hydrogen bonding.

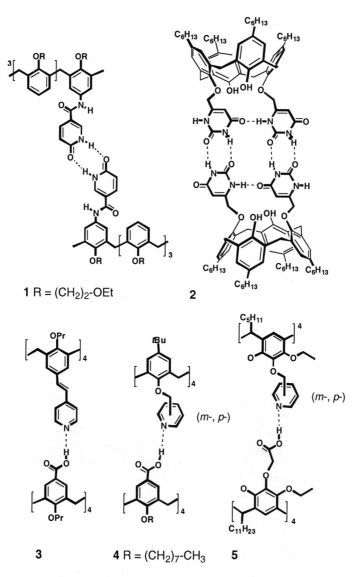

Figure 1. Hydrogen-bonded dimers 1-5 of calixarenes.

Multiple pyridine-carboxylic acid interactions in CDCl$_3$ were employed in the design of heterotopic calix[4]arene **3** [7], **4** [8a] and resorcinarene **5** dimers [8b]. To establish the 1:1 stochiometry, ^{1}H NMR spectroscopy and VPO measurements were employed. For dimers **4**, the association constant values of K_D = 7.6 x 10^3 and 1.3 x 10^3 M^{-1} for the *p*- and *m*-pyridyl derivatives, respectively, were determined. For assemblies **5**, association constants greater than 10^7 M^{-1} were found.

Self-complementary calix[4]arene dicarboxylic acids **6a,b** [9,10] were shown to form dimeric assemblies in apolar solution and in the solid state (X-ray crystallography) (Figure 2). In all cases mentioned above however, the calixarene skeleton adopts a C_{2v}

symmetrical *pinched cone* conformation, with the two opposite aromatic rings parallel in a face-to-face arrangement and the other two flipped outward. This arrangement minimizes the internal cavity's dimensions and no solvent/guest inclusion was ever detected inside such dimers. Moreover, due to the dimerisation, the pinched structures appear kinetically stable on the NMR time scale.

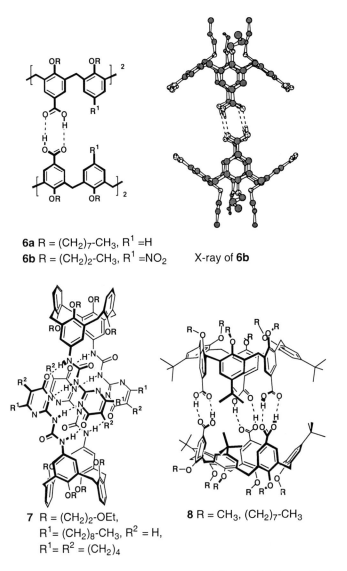

6a R = $(CH_2)_7$-CH_3, R^1 =H
6b R = $(CH_2)_2$-CH_3, R^1 =NO_2 X-ray of **6b**

7 R = $(CH_2)_2$-OEt,
R^1= $(CH_2)_8$-CH_3, R^2 = H,
R^1= R^2 = $(CH_2)_4$

8 R = CH_3, $(CH_2)_7$-CH_3

Figure 2. Self-assembled dimers **6-8**. In the X-ray structure of dinitrocalix[4]arene dicarboxylic acid **6b**, aromatic units bearing the carboxylic acid groups are oriented face-to-face and the other two aromatic rings are flattened [10]. In calix[6]arene dimer **8**, two connected aromatics belonging to different monomers adopt an almost coplanar orientation. Together, the six aromatic rings locked by hydrogen bonding define a trigonal prism perpendicular to the plane of the calixarene methylene units. This also forces the remaining aromatic rings to diverge.

Much more stable dimers **7** were obtained from *1,3-alternate* calix[4]arenes functionalised with two 2-ureido-4-pyrimidinone moieties [11] (Figure 2). Eight hydrogen bonds participate in the assembly process, and the association constant K_D is $>10^6$ M^{-1} in CDCl$_3$. Dimer **7** is stable even in CDCl$_3$-(CD$_3$)$_2$SO mixtures; K_D = 570 ± 110 M^{-1} at χ_{DMSO} = 0.73.

2.2. CAPSULES

With appropriate curvature and carefully engineered positioning of hydrogen bonding sites, calixarene self-assembling systems generate capsules. Their design in the early 1990s was inspired by work of Cram and others on covalently sealed container-molecules, the *carcerands* [12]. By definition, self-assembling capsules are receptors with enclosed cavities which are formed through the reversible noncovalent interactions between two or more subunits [2a]. They form reversibly on time-scales of 10^{-3} to 10^3 s, somewhere in between diffusion complexes (~10^{-10} s) and carceplexes (~10^{10} s). Molecular recognition within the assembled capsules - the encapsulation phenomenon - is largely determined by the host and guest volumes. A recent compilation of occupancy factors, or packing coefficients, of molecule-within-molecule complexes in solution produced the optimal value of ~55% occupancy, which is believed to be also a feature of natural biological cavities such of enzymes [13].

Upon dimerisation of the calix[6]arene tricarboxylic acid, the usually flexible skeleton adopts a rigid flattened C_{3v} *cone*-shaped conformation and dimer **8** results [14] (Figure 2). In this, inclusion of small heteroaromatic molecules was detected, with K_{ass} values of ~350 and ~230 M^{-1} for the 1:1 complexes with *N*-methylpyridinium iodide and *N*-methyl-4-picolinium iodide, respectively (in CDCl$_3$). Fast exchange on the NMR time scale between the complexed and free guest species was observed at ambient temperatures. As expected for the shielded environment, upfield shifts ($\Delta\delta \approx 0.5 - 0.8$) for all groups of the guest protons were recorded.

Recently, several groups published solid-state X-ray structures of capsule-like dimers **9** in which two resorcinarene molecules are linked together through hydrogen bonding with multiple H$_2$O [15,16] or 2-propanol molecules [17], respectively (Figure 3). Tetraethylammonium Et$_4$N$^+$ cation [15] and even the hydrogen bonded complex Et$_3$NH$^+$---OH$_2$ [16] were found encapsulated within **9**. In the ^{1}H NMR spectrum of the former complex in CD$_3$OD, the Et$_4$N$^+$ protons were seen shifted upfield ($\Delta\delta \approx 0.2$), which may indicate that the complex is stable in this solution as well. In the ^{1}H NMR spectrum of the latter complex in wet CDCl$_3$, the CH protons of the Et$_3$NH$^+$ cation shifted upfield ($\Delta\delta \approx 1.3$), and the exchange between the encapsulated and free Et$_3$NH$^+$ cation was slow on the NMR time scale (303 K) [16].

Upon partial deprotonation of phenolic hydroxyls in the cavitand tetraols, charged O$^-$---H-O hydrogen bonds form and bridge the two host molecules in a kinetically stable capsular assembly around neutral guests species (pyrazine, etc.) in apolar solvents (see Chapter 10).

A self-complementary tetraacylated C_{2v} symmetrical resorcinarene was shown to assemble in a well-defined capsule **10** around tropylium cation in CH$_2$Cl$_2$ and CHCl$_3$. Both host-guest charge-transfer interactions and eight C=O---H-O hydrogen bonds between two resorcinarene halves are responsible for this feature [18]. The tropylium

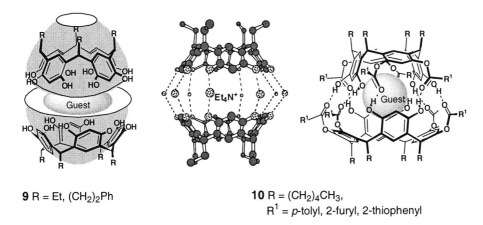

9 R = Et, (CH$_2$)$_2$Ph

10 R = (CH$_2$)$_4$CH$_3$,
R^1 = *p*-tolyl, 2-furyl, 2-thiophenyl

*Figure 3. Hydrogen-bonded dimeric structures of resorcinarenes **9** (cartoon representation used elsewhere). The assembly is mediated by H$_2$O or iso-propanol molecules [15-17]. In the middle: adopted from the X-ray molecular structure of Aoki's capsule with Et$_4$N$^+$ cation inside. Capsule **10** assembles only in the presence of tropylium salts (BF$_4^-$, PF$_6^-$). Both homo- and heterodimeric capsules were detected in CDCl$_3$ solution. The X-ray structure of **10** (Guest = tropylium cation) was also studied [18].*

complex **10** is kinetically stable and its exchange with the corresponding resorcinarene monomer is slow on the NMR time scale (293 K).

Dimeric calix[4]arene tetraurea capsules **11** are those perhaps most widely studied to date [19]. In these, a seam of sixteen intermolecular hydrogen bonds at the upper rim is formed. Eight urea moieties, four from each hemisphere, assemble head-to-tail as shown in Figure 4. This gives a rigid cavity of ~200 Å^3 [20] which can accommodate one benzene-sized guest molecule. The encapsulated guest exchange is slow on the NMR time scale [21]. Capsules **11** include benzene, toluene, some halobenzenes, especially fluorobenzenes, and polycyclic aliphatics (*e.g.*, camphor derivatives, 3-methylcyclopentanone) [22]. The encapsulated benzene molecule is observed at $\delta \approx 4$ in [D$_{10}$]*p*-xylene; for the encapsulated fluorobenzene the CH protons are seen between $\delta \approx$ 5.5 and 3 and the *para*- and *ortho*-hydrogens are separated by more than $\Delta\delta = 2$; encapsulated aliphatic guests are seen upfield of $\delta = 0$.

From NOESY experiments with a lower symmetry capsule of type **11**, a rate constant for benzene molecule exchange of 0.47 ± 0.1 s^{-1} was calculated [23]. The guest exchange time can however be significantly prolonged by introducing sterically bulky substituents (e.g. trityl, p-tritylphenyl) to the urea moieties [24]. Electrospray ionization mass spectrometry (ESI MS) was used to characterize the capsules in solution and in the gas phase [25]. Ion labeling in this case was achieved through the encapsulation of quaternary ammonium ions in CHCl$_3$.

When two different tetraureas are mixed together, heterodimers usually form along with the corresponding homodimers [19b]. From a mixture of tetratolylurea and tetratolylsulfonyl urea, heterodimers form exclusively [26]. The increased acidity of the -SO$_2$NH urea proton may be the reason for such selection. There is also some indication (NOESY) of close contacts between the aryl groups of both ureas in the heterodimer.

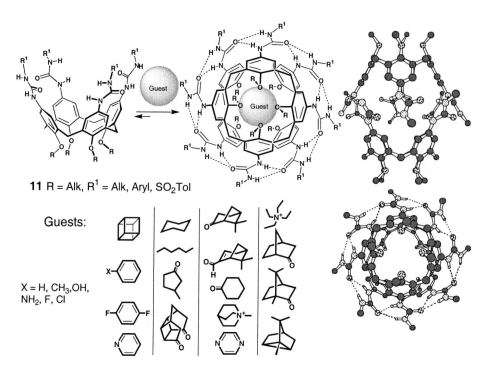

Figure 4. Hydrogen-bonded dimeric capsules **11** of calix[4]arene tetraureas and their guests. Right: the X-ray structure of **11** ($R = CH_2C(O)OEt$, R^1 = p-tolyl) (side and top view) [20].

The preference of the aryl urea/sulfonyl urea could possibly be the basis of a binary molecular code.

Homodimeric capsules **11** are achiral and S_8-symmetric, heterodimeric capsules are chiral due to the orientation of the urea functions and possess C_4-symmetry [27]. Calixarene tetraureas derived from optically active amino acids also show a very high preference for forming heterodimers with aryl urea derivatives [27]. Within these, the head-to-tail arrangement of the eight urea moieties is oriented entirely in one direction, with the amino-acid configuration determining the clockwise or counter-clockwise arrangement of the ureas. This heterodimerisation was shown to occur exclusively with amino acids that have β-branched side chains, like isoleucine and valine, and CD spectroscopy was used to assign the absolute configuration of these capsules. Chiral guests entrapped within the heterodimeric capsules sense the hydrogen bond's directionality; a racemic mixture of enantiomeric guests produces two diastereomeric complexes in a ratio different from 1. For example, with racemic norcamphor as a guest, the heterodimer shows two sets of assemblies in a ratio 1.3:1.

Similar capsular structures can be obtained from other concave molecules. For example, charged hydrogen bond interactions between a cyclotriveratrylene (CTV) based tricarboxylate and a CTV tris(ammonium) salt give rise to heterodimeric capsule formation even in pure $(CD_3)_2SO$ [28]. Kinetically stable complexes with tetramethylsilane, t-butyl chloride, 1,1,1-trichloroethane, $CHCl_3$ and $CHBr_3$ were detected in this solvent.

2.3. STRAPPED DIMERS

In unimolecular capsule **12** (Figure 5), two calix[4]arene tetraureas are covalently connected via their upper rims [29]. When the linker is short enough to minimize the loss of entropy due to restriction of freely rotating single bonds, stronger dimerisation is expected. With the hexamethylene -$(CH_2)_6$- spacer, only the intramolecularly folded capsule **12** was detected by ESI MS. For detection, the *N*-methyl quinuclidinium cation (as its BF_4^- salt) was encapsulated and provided a charge for the whole complex in the gas phase. In $CDCl_3$ solution, signals for the encapsulated quinuclidinium CH protons appeared upfield at δ = -0.2 to -0.4 in the ^{1}H NMR spectrum.

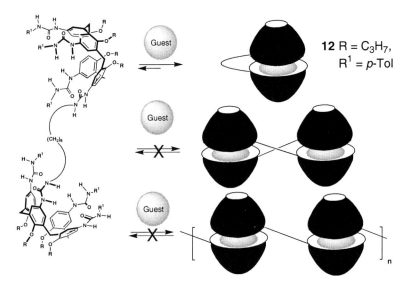

Figure 5. Covalent connection of the two calix[4]arene ureas at the upper rim results in unimolecular capsule **12**. Neither intermolecular dimers (middle) nor polymeric aggregates (bottom) are formed in chloroform solution. At the same time, sulfonyl urea **11** ($R = (CH_2)_9CH_3$, $R^1 = SO_2$-p-Tol) breaks capsule **12** with the formation of a dumbbell-like aggregate [29].

2.4. LARGE DIMERIC CAPSULES

1,3,5-Triureidocalix[6]arenes possessing *N*-unsubstituted urea moieties were shown to dimerise through a cyclic array of hydrogen bonds into a cylindrical capsule [30]. There, benzene, fluorobenzene and dichloromethane were successfully entrapped. Encapsulation of fullerene C_{60} within a self-assembling dimer based on a calix[5]arene mono-urea was described [31]. Subsequently, assemblies which are capable of accommodating larger or even two guest species, were introduced. For example, self-complementary cyclic tetraimides built on the resorcinarene platform dimerise through hydrogen bonding to afford a cylindrical capsule **13** with an estimated internal volume of 460 Å^3 and internal dimensions of ~ 6 x 15 Å^2 [32,33] (Figure 6). The shape of the C_{4v}-symmetrical monomer is vase-like and the dimerisation takes place in the rim-to-rim manner. The imides are stabilized by a seam of eight bifurcated C=O---H-N hydrogen bonds. Two aromatic molecules (benzene, toluene, etc.) were easily encapsulated inside [33]. The complexes formed are tetramolecular species, and the entropic costs of assembly are obviously overcome by intermolecular forces and subtle volume-filling effects.

The complexation processes are slow on the NMR time scale. When both benzene and p-xylene were added in a 1:1 ratio to a [D_{12}]mesitylene solution of **13**, an unsymmetrically filled capsule was formed predominantly [32]. Apparently, two guests can not squeeze past each other to exchange positions in the capsule. Benzene also paired with p-trifluoromethyltoluene, p-chlorotoluene, 2,5-lutidine, and p-methylbenzyl alcohol to give new species with one of each guest inside. In a competition experiment with three different solvent guests, in which toluene, benzene and p-xylene were added in a 2:1:1 ratio to the [D_{12}]mesitylene solution of **13**, again, the capsule was filled preferentially (~ 90%) with benzene and p-xylene, and only ~ 10% of the capsule with two toluenes inside was observed. Moreover, benzene and p-xylene in [D_{12}]mesitylene replaced encapsulated toluenes within a few minutes at room temperature when 1:1:1 ratio of toluene, benzene and p-xylene was employed [32]. Such selectivity perfectly matches the overall length of two guests with the dimensions of the cavity.

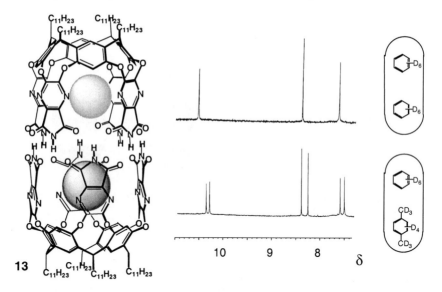

Figure 6. Cylindrical capsule **13**. Middle: portions of the ^{1}H NMR spectra of **13** in [D_6]benzene (top) and in [D_6]benzene-[D_{10}]p-xylene, 1:1 mixture. The imide NH signal is situated downfield of δ = 10 and the aromatic CH signals of the resorcinarene skeleton are situated between δ = 8.5 and 7. For the asymmetrically filled capsule, doubled sets of the signals are observed [32].

Capsule **13** exhibits complexation of smaller hydrogen bonded aggregates, e.g. 2-pyridone/2-hydroxypyridine dimer, benzamide dimer and benzoic acid dimer [34]. Diastereomeric "complexes within complexes", using chiral guests, were also observed. Two different species were detected in the presence of the racemic *trans*-1,2-cyclohexanediol while one appears when only the single enantiomer is available [34] (Figure 7). Integration indicates the presence of two guest species inside each capsule, but the intensity of the signals for the enantiopure vs the racemic guests indicates more of the

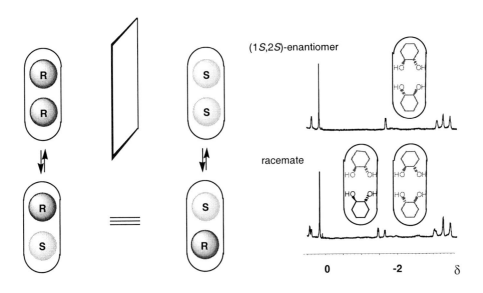

Figure 7. Encapsulation of two chiral molecules (cartoon) results in diastereomeric complexes. Right: upfield portions of the ^{1}H NMR spectra of **13** in [D$_{12}$]mesitylene with (1S,2S)-trans-1,2-cyclohexanediol (top) and with racemic trans-1,2-cyclohexanediol. For the encapsulated cyclohexane, the CH signals are situated between δ = 0.5 and –3 ppm [34].

latter - the capsule prefers to be filled with a guest and its mirror image rather than two identical molecules.

The chemical shift differences between guests in the bulk solution and those in capsule **13** are related to their positions in the dimeric assembly. In [D$_{12}$]mesitylene solution, α-, β- and γ-picolines gave homo-complexes with two identical guests per capsule [35]. When γ-picoline is encapsulated, the chemical shift of the methyl groups in the NMR spectrum places them near the ends of the cavity: at δ = –2.79. The dipole involving the nitrogen, therefore, prefers the middle of the cavity, and at the methyl ends the change in shift is maximal: $\Delta\delta$ = 4.55. For encapsulated β-picoline, the methyl chemical shift is δ = –1.65 ($\Delta\delta$ = 3.43), whereas with α-picoline, the methyl group is seen at δ = 0.58 ($\Delta\delta$ = 1.33). These guests spin along the long axis of the capsule but apparently do not tumble about other axes. When equal amounts of two different picolines were added to a [D$_{12}$]mesitylene solution of **13**, nonsymmetric hetero-complexes with capsules filled by two different guests were formed in addition to the corresponding homo-complexes. Moreover, when all three picolines were added to **13**, two homo- γ-picoline and β-picoline filled) and three different hetero-complexes were clearly detected [35].

Upon encapsulation, the tertiary anilide, *p*-[*N*-methyl-*N*-(*p*-tolyl)]toluamide, is fixed in its *Z* configuration [34] (Figure 8). In bulk solution, these anilides exist as mixtures of *E* and *Z* isomers, with the *E* configuration favoured. The *E* isomer, however, cannot fit inside capsule **13** and its inherent stability is overcome by the CH-π, van der Waals and dipolar interactions offered by the interior surface of the capsule.

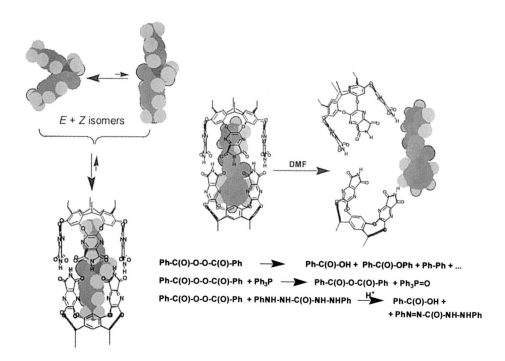

Figure 8. Encapsulation experiments with **13** in [D_{12}]mesitylene (NMR analysis). Left: capsule **13** accepts exclusively Z isomer of tertiary p-[N-methyl-N-(p-tolyl)]toluamide [34]. Right: benzoyl peroxide is stable inside capsule **13** but can be rapidly released with DMF. Once released, it can be either thermally decomposed or can undergo chemical reactions with Ph$_3$P or 1,5-diphenylcarbazide [33].

Benzoyl peroxide is of ~ 4.3 x 13.2 Å^2 dimensions and it is readily encapsulated by **13** in [D_{12}]mesitylene solution [33]. The encapsulated benzoyl peroxide possesses chemical properties *different* from those in a bulk solution (Figure 8). For example, inside capsule **13** it is stable for at least 3 days at 70 °C in [D_{12}]mesitylene solution, while in the absence of **13** it decomposes within ~3 hrs under those conditions. The encapsulated benzoyl peroxide does not oxidize Ph$_3$P, which is present in bulk [D_{12}]mesitylene solution, while free benzoyl peroxide easily does (50% conversion within ~2 hrs, ^{1}H and ^{31}P NMR controls). The encapsulated benzoyl peroxide can be slowly replaced by p-[N-(p-tolyl)]toluamide within several days, by p-[N-methyl-N-(p-tolyl)]toluamide within weeks or by the highly competitive hydrogen bond acceptor DMF – within seconds. The DMF competes for the internal hydrogen-bond donors in the assembly and only the monomeric cavitand can be detected; no encapsulated benzoyl peroxide remains. When ~1-2% (v/v) of DMF was added to the [D_{12}]mesitylene solution of [**13**·benzoyl peroxide], containing 10 eq of Ph$_3$P, a redox reaction occurred within minutes and the formation of Ph$_3$P=O was readily detected by ^{31}P NMR [33]. Larger quantities of acetic acid (≥ 5% vol) can also open the capsule within seconds, and the released benzoyl peroxide can oxidise 1,5-diphenylcarbazide, producing an intense bright-pink color λ_{max} = 555 nm) [33]. This is a widely used colorimetric test for peroxide determination. Most probably, dissociation to two benzoyloxy radicals is prevented inside the capsule. Alternatively, the recombination is much faster than decarboxylation and hydrogen abstraction processes.

More of these useful encapsulation phenomena are likely to emerge as the capsules become larger and able to accommodate more guest molecules. Capsule **14** (Figure 9) with an internal volume of 950 Å^3 (!) was recently constructed [36]. This is one of the largest dimeric self-assembled cavities synthesised to date. Four preformed glycoluril modules were coupled to the well-known resorcinarene-based cavitand tetrol; the resulting molecules self-assemble through 16 hydrogen bonds into a rigid dimeric structure. The internal cavity in **14** is large enough to accommodate K$^+$, Sr^{2+} and Ba^{2+} cryptates (^{1}H NMR, ESI MS). From the desymmetrisation of the ^{1}H NMR spectrum upon encapsulation of K$^+$-cryptand-SCN$^-$, it was concluded that the anion is also encapsulated. The exchange of positions between the cryptated cation and the anion is slow on the NMR time scale [36].

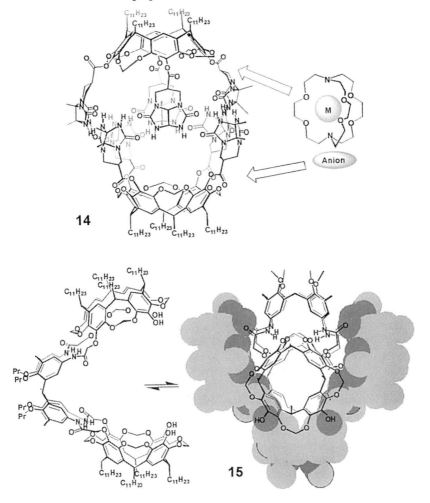

*Figure 9. Extra-large capsular assemblies **14** and **15**.*

A rigid and preorganised extended molecule, a combination of one calix[4]arene and two resorcinarene cavitand units, shows a tendency to dimerize into a spherical capsule **15** [37]. Calculations indicate that the cavitand hydroxyls of one monomer should be in close proximity to the carbonyl oxygens of the amide bridges (Figure 9), allowing the formation of eight hydrogen bonds and the generation of a capsule large enough to accommodate up to nine $CHCl_3$ molecules. However, no kinetically stable capsules were detected, and the dimerisation constant was rather low : $K_D = 11$ M^{-1} in $CDCl_3$.

2.5. CAPSULES FORMED BY HIGHER DEGREES OF OLIGOMERISATION

Hydrogen bonding assemblies of more than just two components have also been engineered. For example, two cavitand tetracarboxylic acids and four 2-aminopyrimidines form a capsular structure **16** of 9 x 15 $Å^2$ dimensions (Figure 10) both in chloroform solution and in the solid state [38]. Two nitrobenzene molecules fit inside.

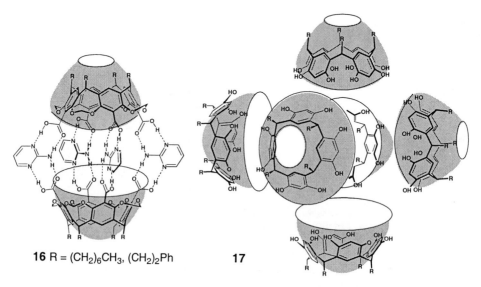

16 R = $(CH_2)_6CH_3$, $(CH_2)_2Ph$ **17**

*Figure 10. Self-assembled cavities **16** and **17** constructed by multiple hydrogen bonds. Capsule **16** is stable not only in the solid state (X-ray analysis) but also in solution ($K_{ass} = 3.7 \times 10^{19}$ M^{-5}, $CDCl_3$, 298 K) [38]. Nanostructure **17** is assembled from six resorcinarene and eight H_2O molecules.*

Six molecules of resorcinarene assemble into a spectacular spherical cavity **17** with a diameter of 17.7 Å and an internal volume of ~1375 $Å^3$ [39]. This is the biggest cavity synthesized to date from organic materials [40]. Sixty hydrogen bonds hold the cavity together and eight water molecules also participate (Figure 10). It was deduced that the assembly involves both C_{4v}- and C_{2v}-symmetrical resorcinarene conformers in a ratio 4:2 (2:1). ^{1}H NMR spectroscopy gives some indication that the assembly persists in benzene solution as well. Although the presence of guest species was inferred from electron density maxima, they were not identified from the X-ray experiment. Molecular modeling suggests that the assembly is large enough to accommodate very big guests such as fullerenes or porphyrins.

When two calix[4]arene tetraureas are covalently linked at their lower rims (see **18**, Figure 11), hydrogen bonding results in a polymer chain of capsules, or *polycaps* [41]. The polycaps form only when a guest of proper size and shape is present; in competitive solvents such DMSO or MeOH, the polycaps dissociate. By direct analogy with parent capsules **11**, the formation of heterodimeric systems was explored. Specifically, the polycap rapidly broke down to a dumbbell-shaped assembly when treated with an excess of the simple dimeric capsule (e.g. **11**). The dumbbell featured a well resolved NMR spectrum that showed all of the expected resonances. Combination of aryl- and sulfonylurea biscalixarenes effectively afforded heteromeric polycaps [42].

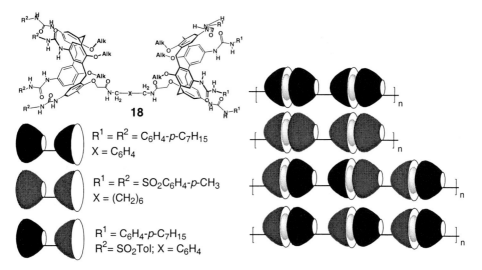

*Figure 11. Covalently linked calix[4]arene ureas **18** form polycaps in apolar solution. Both urea and sulfonylurea derivatives were prepared and showed homo- and heteromeric assemblies [41,42].*

To help fill the space between polymer chains, the monomers were functionalized with long alkyl groups capable of forming a liquid-like sheath around them. The resulting polycaps (at high concentrations) self-organize into polymeric liquid crystals [43]. Lyotropic, nematic liquid crystalline phases were generally observed. Molecules like difluorobenzene and nopinone were readily encapsulated in these liquid crystalline phases. Further characterization by X-ray diffraction showed peaks at 2.4 and 1.6 nm that match well the repeat distances and the dimensions of the polycaps.

2.6. ROSETTES

Calix[4]arenes have been used as a platform to build series of so-called *double-rosettes* composed from complementary melamines and barbiturates/(iso)cyanurates [44]. While the calixarene skeleton adopts a *pinched cone* conformation, without a cavity, it provides a necessary structural rigidity and also chemical diversity to study some fundamental properties and behavior of very complex self-assembled structures. In particular, the control over informational content of aggregates (*e.g.*, chirality, combinatorial selection, increasing structural complexity) has been achieved.

Thus, calix[4]arenes diametrically substituted at their upper rim with two melamine moieties assemble into well-defined box-like, double-rosette structures **19** upon addition of two equivalents of barbiturates/isocyanurates (Figure 12). These assemblies consist of nine components that are held together by 36 hydrogen bonds; they are stable in apolar solvents at ≥ 0.1 mM concentrations. In addition to high resolution ^{1}H NMR spectroscopy and, in one spectacular case, even X-ray crystallography, matrix assisted laser desorption ionization time of flight (MALDI-TOF) mass spectrometry was extensively used to characterise the double-rosettes [45]. In this case, ion labeling was achieved through the complexation of soft Ag^+ cations with aromatic units of the rosettes as well as with nitrile substituents (in $CHCl_3$) [46].

In principle, three isomeric double-rosettes can be formed by combination of melamines and barbiturates/cyanurates : the staggered D_3-symmetrical isomer, the symmetrically eclipsed C_{3h}-isomer, and the unsymmetrically eclipsed C_s-isomer [47]. When 5,5-disubstituted barbituric acid derivatives (*e.g.*, 5,5-diethylbarbiturate, DEB) are employed, the D_3-symmetrical structure forms exclusively, with antiparallel orientation of the two rosette motifs. It is chiral and exists as a racemic mixture in the absence of other chiral sources. When chiral melamines were attached to the calixarene upper rim, complete induction of (supramolecular) chirality was achieved [47].

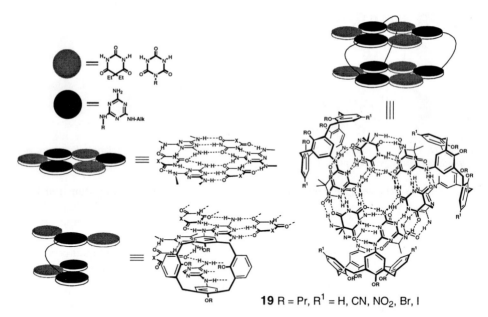

*Figure 12. Calix[4]arene bismelamine based double-rosette **19** and its components – barbiturates, isocyanurates and melamines (cartoon representation used elsewhere in this chapter).*

By mixing different calixarene bismelamines with DEB, libraries of self-assembled double-rosettes were generated under thermodynamic control [48]. Under conditions that allow the reversible exchange of the components, a mixture of homomeric and heteromeric assemblies is formed (Figure 13). For two different calixarene bismelamines **20** and **21** and DEB, two homomeric and two heteromeric assemblies are formed in a statistically expected ratio of 1:1:3:3 (±10%) in [D$_8$]toluene. In principle, by mixing

10 different bismelamines and DEB, a dynamic library of 220 different double-rosettes is expected. When the Zn-porphyrin containing bismelamine **22** was employed in combination with bismelamine **21**, it was possible to achieve the guest-templated selection of the most strongly binding double-rosette [49]. In this case, 1,3,5-tris(4'-pyridyl)benzene **23** interacted with the three Zn-porphyrin fragments of the rosette $(22)_3$-$(DEB)_6$, thus serving as a thermodynamic template for its preferable formation. The thermodynamic equilibrium in the four-component library was thus shifted from an almost statistical one to a 1:1 mixture of the homomeric assemblies $(21)_3$-$(DEB)_6$ and $(22)_3$-$(DEB)_6$.

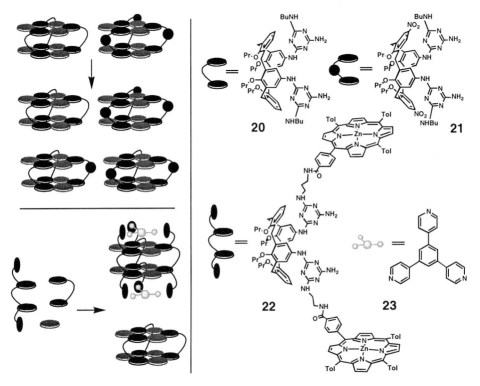

Figure 13. Dynamic combinatorial libraries of calixarene double-rosettes $(20)_3$-$(DEB)_6$, $(21)_3$-$(DEB)_6$ and $(22)_3$-$(DEB)_6$ and their heteromers. Guest-templated selection of the strongest binding assembly $(22)_3$-$(DEB)_6$ from the mixtures $(21)_3$-$(DEB)_6$, $(22)_3$-$(DEB)_6$ and their heteromers was effectively achieved by trispyridyl **23**.

Once the dynamic combinatorial libraries were formed, their covalent capture could be effectively achieved by a ring-closing metathesis reaction [50]. Reaction of assembly $(24)_3$-$(DEB)_6$ functionalised with oct-7-enyl side chains at the melamines with Grubbs' catalyst in CD_2Cl_2 resulted in a covalent linking of three calixarene fragments into the 123-membered (!) macrocycle **25** as the double-rosette assembly in 96% yield (Figure 14). ^{1}H NMR spectroscopy shows that the reaction goes without dissociation of the rosettes. The same reaction on the calixarene bismelamines in the absence of the assembly, without DEB added, gave no trace of macrocycle **25**. The dynamic library of homomeric and heteromeric assemblies $(24)_3$-$(DEB)_6$ (R = H, Br; n = 6,8) was also successfuly formed and analysed by ^{1}H NMR and HPLC.

24 n = 6,8; R = H, Br

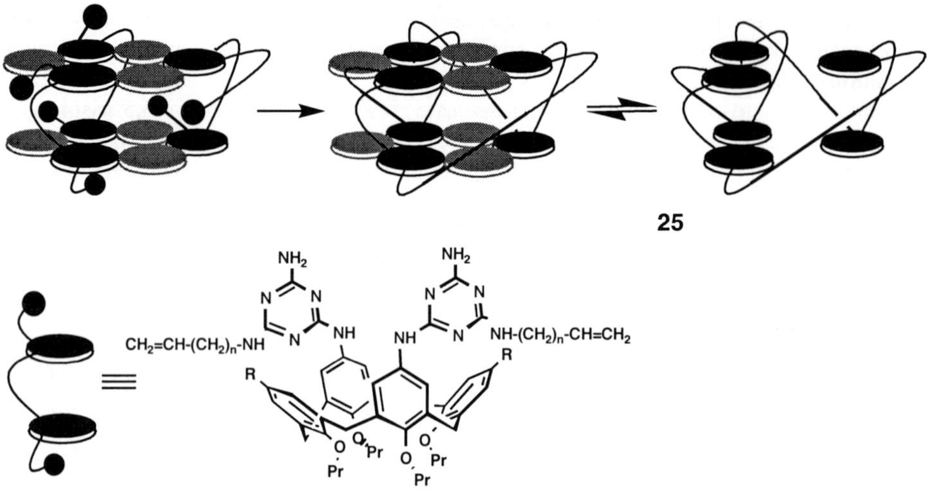

Figure14. *Covalent capture of noncovalent combinatorial libraries.*

More complex assemblies $(\mathbf{26})_3$-$(DEB)_{12}$ (Figure 15), consisting of 15 components held together by 72 hydrogen bonds, were constructed from calixarene-supported tetramelamines and four equivalents of DEB or 5-ethyl-5-phenylbarbituric acid and were characterised by ^1H NMR spectroscopy and MALDI-TOF mass spectrometry [51]. Whereas calixarene bismelamines **20** in double-rosettes $(\mathbf{20})_3$-$(DEB)_6$ exhibit positive cooperativity, the tetramelamine **26** displays negative cooperativity. At a ratio of **26**:DEB < 1:3, the ^1H NMR spectrum reflects no rosette formation but rather an ill-defined aggregation. Only at a ratio **26**:DEB > 1:3, the rosettes start to form. In contrast,

26 X = $CH_2(m\text{-}C_6H_4)CH_2$, $(CH_2)_6$

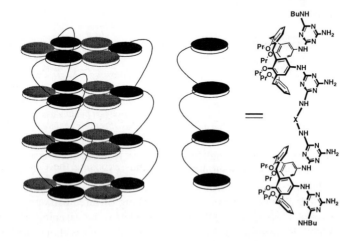

Figure 15. *Fifteen-component hydrogen-bonded nanostructure* $(\mathbf{26})_3$-$(DEB)_{12}$.

titrations of bismelamines with DEB showed rosette formation already at a rather low concentration of DEB. In addition, mixing bismelamine-DEB and tetramelamine-DEB assemblies gave no heteromers.

The knowledge obtained from these studies may eventually be used in the design of self-assembled nanostructures with binding properties that mimic those of antibodies [52]. First, full control over the structure and stereochemistry of highly functional molecules, capable of recognition and organizing themselves spontaneously into well-defined nanostructure, should be achieved. Then, the dynamic noncovalent libraries of such structures can be prepared. Upon screening of these combinatorial libraries, template effects of the guest can select the strongest binding receptor under thermodynamic control. After that, covalent post-modification can be performed (Figure 16).

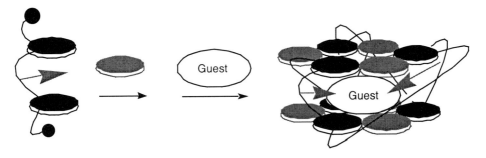

Figure 16. Template-directed assembly and covalent post-assembly modification of the best-binding receptor structure from a supramolecular library.

3. Metal-Induced Self-Assembly

3.1. ASSEMBLY OF RECEPTORS

The high affinity of some calixarene receptors towards cations has been effectively used to control the hydrogen bonding assembly processes and switch the equilibrium between the monomer and polymeric self-assembled clusters. Thus, sodium cation (as perchlorate salt) triggers the assembly of a calix[4]arene **27** (Figure 17) functionalized with two 2,6-diamidopyridine fragments with complementary dialkylbarbiturates in $CDCl_3$ and 9:1 $CDCl_3$-CD_3CN mixture [53]. In the absence of Na^+, diamidopyridine moieties are involved in intramolecular hydrogen bonding with the calixarene ether oxygens. Addition of Na^+ disrupts the intramolecular hydrogen bonding ; the carbonyl oxygen atoms turn around to coordinate the cation, thus exposing the diamidopyridine for intermolecular binding.

The melamine-barbiturate motif has been used to assemble together both anion and cation binding functions [54]. The Na^+ complexes (as SCN^-, N_3^- and I^- salts) of calix[4]-arene triester monoamide, possessing a melamine moiety at the lower rim, assembled with the Zn-porphyrin, attached to a barbiturate moiety. Metalloporphyrins are known to bind anions. In complex **28**, the anion from the Na-calixarene complex coordinates to the Zn-porphyrin held in close proximity (Figure 17). The whole aggregation process can be disrupted by addition of competitive MeOH.

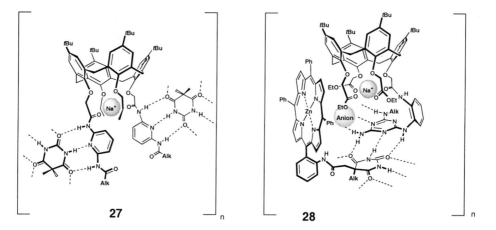

Figure 17. Hydrogen bonding assemblies of calix[4]arene based receptors.

3.2. ASSOCIATION VIA COORDINATION

Metal-ligand binding is much stronger than hydrogen bonding and very well suited for self-assembly. Metal-induced assemblies are usually robust, although their strength can be controlled by solvent polarity and pH. They can be charged and redox-active. In addition, they may operate in much more polar solution or even in water. Early examples of this type includes bis(acac) calix[4]arene dimer **29** held together through the chelation of Cu^{2+} cations [55]. Tungsten-oxo calix[4]arenes **30** aggregate in a head-to-tail columnar fashion with a W=O group protruding into the cavity of another calixarene [56] (Figure 18).

Tetranuclear molecular square **31** (Figure 18), assembled in almost quantitative yield from the corresponding calix[4]arene platinum(II) bistriflate and bipyridine in CH_2Cl_2, was shown to transport some arylsulfonate salts through the liquid chloroform membrane in a U-type cell [57].

Further progress in this field was, probably, stimulated by the successful synthesis of hydrogen-bonded capsules. Metal-induced capsular self-assembly **32** (Figure 19) of deep cavitands was first described by Jacopozzi and Dalcanale [58]. Two tetracyanocavitands were connected through four square-planar Pd(II) or Pt(II) entities in CH_2Cl_2, $CHCl_3$ and acetone. Evidence of encapsulation of one triflate anion upon dimerisation was obtained (^{19}F NMR). For the Pt(II) case, the assembly process was shown to be reversible: Et_3N dissociated the capsule **32**, while the addition of trifluoroacetic acid restored it.

Water-soluble Co(II) and Fe(II) resorcinarene-based cages **33** (Figure 19) were recently described [59]. These are formed by combining cobalt(II) or iron(II) chlorides in aqueous solution at pH > 5 with the resorcinarene cavitand functionalised with four iminodiacetic acid moieties. Upon synthesis, up to six water molecules occupy the inner cavity. When the appropriate guest molecule is present in the reaction mixture, it usually goes inside, replacing the water molecules. Benzene, halobenzenes, chlorinated hydrocarbons, (cyclo)alkanes and alcohols were successfully trapped – in aqueous solution! Due to their paramagnetic nature, these complexes may act as NMR shift reagents as they cause very substantial upfield isotropic hydrogen shifts ($\Delta\delta \sim 25\text{-}40$) in the guest molecules with a signal separation of δ ~12. At lower pH, the ligand becomes protonated, the metal ion is no longer coordinated to the cavitand, and the assembly falls apart [59].

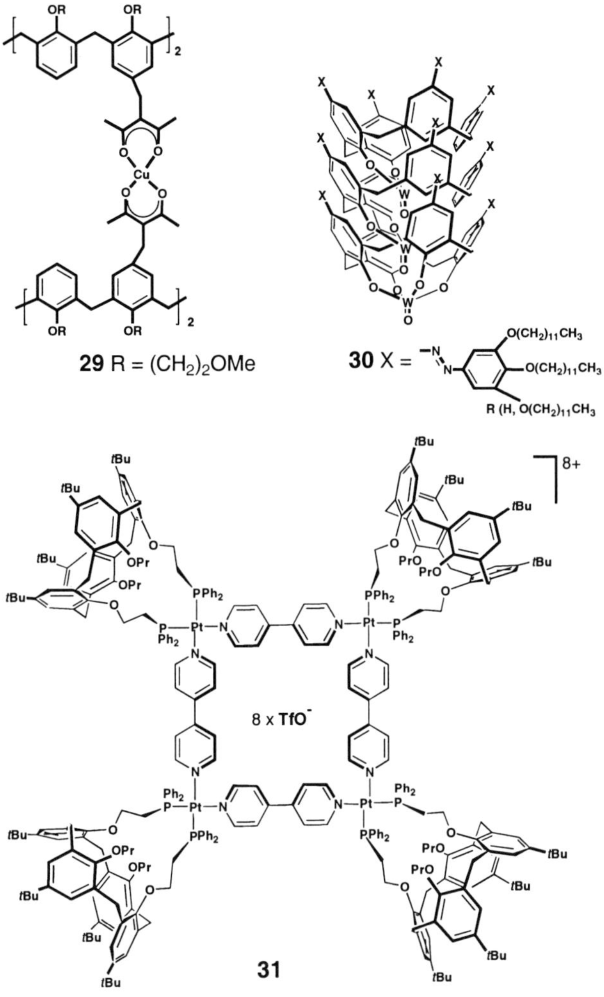

Figure 18. Early examples of metal-assisted self-assembly of calixarenes [55-57].

Homooxacalix[3]arene **34** bearing 4-pyridyl groups at the upper rim forms dimeric capsules with Pd(II)(Ph$_2$PCH$_2$CH$_2$PPh$_2$)(OTf)$_2$ in a 2:3 ratio in dichloromethane [60,61] (Figure 20). A kinetically stable complex between capsule **34** and fullerene C$_{60}$ was obtained and characterised by ^{1}H and ^{13}C NMR spectroscopy in Cl$_2$CDCDCl$_2$; association constants K_{ass} = 39 M^{-1} at 30 °C and 54 M^{-1} at 60 °C were obtained. Complexation of Li$^+$ cation at the calixarene lower rims induces their more flattened conformation. Apparently, this is more suitable for the C$_{60}$ inclusion and changes K_{ass} to 2100 M^{-1} (30 °C) [61].

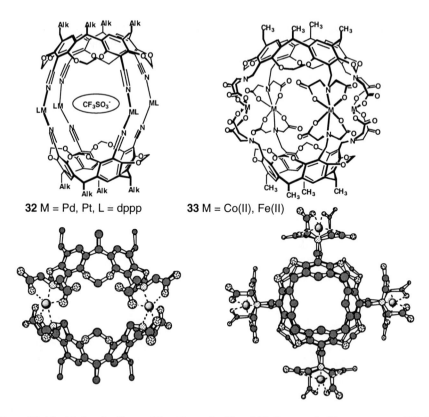

Figure 19. Metal-induced self-assemblies of capsules **32** and **33**. Bottom: the X-ray structure of **33** (M = Co(II)) [59] (side and top view).

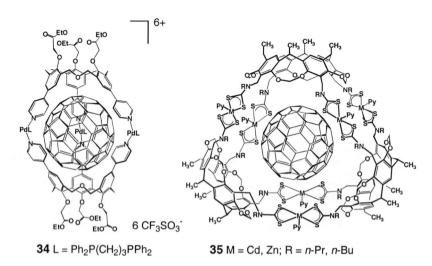

Figure 20. Self-assembled capsular structures **34** and **35** and their complexes with C_{60} fullerene.

Cavitands functionalized with four dithiocarbamate units assemble around Zn(II), Cd(II) and Cu(III) cations in EtOH/H_2O with the formation of spectacular nanosized structures: trimeric (for Zn(II) and Cd(II)) **35** and even tetrameric (for Cu(III)) [62]. Although the ^{1}H NMR spectra of the assemblies are broad in apolar solvents, the complexes were readily crystallised from pyridine/H_2O and characterised by X-ray crystallography. The trimers (M = Zn(II), Cd(II)) form very stable 1:1 complexes with fullerene C_{60} in benzene and toluene solutions (Figure 20); the association constants were determined by spectrophotometry at 295 K and were within the range of $\log K_{ass}$ = 3.5 – 6 [63]. Characteristically, purple solutions of fullerene turn red-brown upon addition of the colourless host compounds.

4. Assembly through Solvophobic Forces

Even weaker forces - van der Waals, CH/π interactions, solvophobic forces, etc. - can cause quite strong self-assembly. Solvophobic dimerisation of Cram's velcrands was, perhaps, the earliest example of calixarene self-assembly [64].

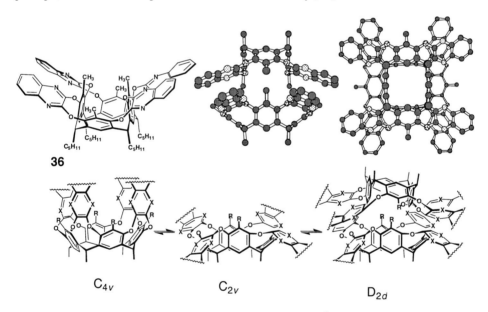

Figure 21. Top: velcrand **36** possesses an almost planar (15 x 20 Å) rectangular structure with two protruding up-methyl groups and two methyl-sized holes. In the X-ray structure of the velcraplex (**36**)$_2$, there are 70 intermolecular atom-to-atom van der Waals contacts; 44 more are within contact distance plus 0.1-0.2 Å [64a]. Bottom view: representation of the conformational equilibria in cavitands and velcrands leading to the formation of velcraplexes.

When the oxygen atoms in 2'-methyl-resorcinarenes are bridged with heteroarylene units (*e.g.*, pyrazines or quinoxalines), extended structures result (see for example **36**, Figure 21). Only the C_{2v}-symmetrical conformer is observed: the presence of the 2'-methyl substituent sterically destabilizes the C_{4v}-symmetrical vase shape. The kite-shaped C_{2v}-symmetrical conformer **36** – a *velcrand* - tends to dimerize in organic solution through solvophobic interactions; the D_{2d}-symmetrical dimers are known as *velcraplexes*. The two molecules of **36** are rotated 90° with respect to each other so extensive surface contacts are made and the 2'-methyl groups fill the holes. Thus, two molecules in the dimer share a large preorganized surface composed of four methyls inserted into four complementary cavities, and four heteroarylenes in close contact with one another. As a result, the values for $-\Delta G$ of the dimerisation in some cases exceed 9 kcal mol^{-1} in solution. Some dimerisation processes are entropy driven and enthalpy opposed. The values for $-\Delta H$ range from -5 to 8 kcal mol^{-1} and those for ΔS from -6 to $+40$ cal mol^{-1} K^{-1}, respectively. In the ^{1}H NMR spectra, separate signals can be seen for both the dimer and the monomer; their ratio is dependent on temperature and concentration. When two different velcrands are mixed together, their heterodimers can be observed. Velcraplex formation was confirmed by FAB-MS and VPO measurements, and X-ray crystallography.

More recently, D_{2d} velcraplexes were observed in aqueous solution [65]. Velcrands **37** functionalized with eight secondary C(O)-NHAlk carboxamide groups on their upper rim tend to dimerise through both solvophobic interactions and intermolecular hydrogen bonding [66]. In the dimers, four NH donors of one molecule appear in close proximity to the diaryl ether oxygen acceptors of the other (Figure 22). The N-H---O angle of ~170° is nearly ideal for the formation of eight intermolecular hydrogen bonds.

The "deep cavity" tetraester **38** prepared by Cram and co-workers formed a dimeric species in the solid state and in solution at high concentrations (50 mM) [67]. "Deep cavity" tetraamides **39** ($R^1 = (CH_2)_6CH_3$ and $(CH_2)_7CH_3$) dimerize via intermolecular hydrogen bonding in toluene [68]. At the same time, one of the four amide alkyl

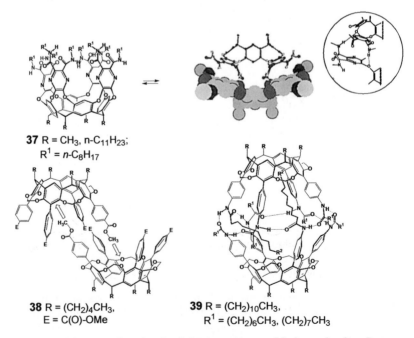

37 R = CH$_3$, n-C$_{11}$H$_{23}$;
R^1 = n-C$_8$H$_{17}$

38 R = (CH$_2$)$_4$CH$_3$,
E = C(O)-OMe

39 R = (CH$_2$)$_{10}$CH$_3$,
R^1 = (CH$_2$)$_6$CH$_3$, (CH$_2$)$_7$CH$_3$

Figure 22. Top: velcrand **37** dimerises through solvophobic interactions and hydrogen bonding. Bottom: dimerisation with self-inclusion: early examples of self-complementary assembled "deep cavity" dimers **38** and **39**.

chains from each cavitand is encapsulated in the cavity of the opposing dimer (Figure 22). The terminal CH$_3$ groups are situated in the deepest part of the cavity and their resonances are shifted upfield (up to $\delta = -1.8$) in the ^{1}H NMR spectra in [D$_8$]toluene. Due to this self-inclusion, the solvent/guest molecules inside the cavity are replaced. Calculations indicate that ~57% of the cavity is filled by self-inclusion, and there is not enough room to accommodate any other molecules. Neither the short chain *n*-propionyl-amide derivative (R^1 = CH$_2$CH$_3$) nor the long chain palmitoylamide derivative (R^1 = (CH$_2$)$_{14}$CH$_3$) show dimerisation to **39**; both exclusively exist in the monomeric state in [D$_8$]toluene. While the same pattern of intermolecular hydrogen bonds are possible in their respective dimers, neither side chain is appropriate for self-inclusion [68].

Analogously, cavitands **40** (n = 0, 1) [69] with covalently attached complementary guest molecule (*e.g.* adamantane) at the upper rim dimerise in apolar solution (Figure 23). These cavitands possess a cooperative seam of intramolecular C=O---H-N hydrogen bonds that stabilises the folded conformation and influences the kinetics of guest exchange in apolar solvents. Their self-complementary shapes result in the formation of noncovalent dimers of considerable kinetic and thermodynamic stability : K_{ass} values up to 3000 M^{-1} ($-\Delta G^{295}$ = 4.7 kcal mol^{-1}) in [D$_{10}$]*p*-xylene and up to 325 M^{-1} ($-\Delta G^{295}$ = 3.3 kcal mol^{-1}) in [D$_6$]benzene were observed. At the same time, no dimerisation was detected in CDCl$_3$ [69].

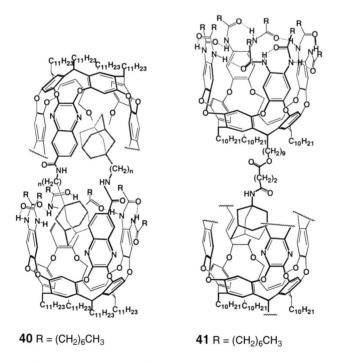

40 R = (CH$_2$)$_6$CH$_3$ **41** R = (CH$_2$)$_6$CH$_3$

*Figure 23. Self-complementary cavitands **40** and **41** dimerizes through solvophobic interactions.*

Upon assembly, the adamantyl signals were clearly seen upfield at $\delta = 0$, characteristically shielded by the cavitand's aromatic walls. The dimerisation is reversible and subject to control by solvent and temperature; the process is enthalpically favored and entropy-opposed [69]. The complementary adamantyl guest was also attached to the lower rim of "self-folding" cavitand **41** [70]. The ^{1}H NMR spectrum of **41** in [D_{10}]*p*-xylene solution clearly showed adamantane encapsulation, the CH signals for the adamantane fragment being observed upfield of $\delta = 0$. Two sets of signals -for the cavitands and the corresponding oligomeric caviplexes- were observed. Because the association constant was not high (~100 M^{-1}), only ~15-20% of the dimeric form was seen under these conditions and higher oligomers are expected to be present only at high concentrations. Even so, this may lead to new noncovalently assembled polymeric aggregates brought together by the weakest of intermolecular interactions [70].

5. Concluding Remarks

The behavior and functions of calixarene-based self-assembled structures can now be predicted and, to a certain extent, controlled. Using different driving forces -hydrogen bonding, metal-ligand interactions, solvophobic effects, *etc*.- a sizable number of noncovalently organized receptors and sensors, potential reaction/catalytic chambers, polymers and materials has been prepared. Combined with the extremely rich and well-developed chemistry of calixarenes and its broad applications, the field of self-assembly undoubtedly has opened new horizons to explore.

6. Acknowledgement: The Skaggs Research Foundation is greatly acknowledged for financial support. I am grateful to Professor J. Rebek, Jr. and Dr. V. Böhmer for encouraging discussions and to Dr. S. D. Starnes for proofreading the manuscript.

7. References and Notes

[1] General reviews: a) G. M. Whitesides, J. P. Mathias, C. T. Seto, *Science* **1991**, *254*, 1312-1319; b) D. S. Lawrence, T. Jiang, M. Levett, *Chem. Rev.* **1995**, *95*, 2229-2260; (c) D. Philp, J. F. Stoddart, *Angew. Chem., Int. Ed. Engl.* **1996**, *35*, 1154-1196.

[2] a) M. M. Conn, J. Rebek, Jr., *Chem. Rev.* **1997**, *97*, 1647-1668; b) B. Linton, A. D. Hamilton, *Chem. Rev.* **1997**, *97*, 1669-1680; c) J. de Mendoza, *Chem. Eur. J.* **1998**, *4*, 1373-1377; d) J. Rebek, Jr., *Acc. Chem. Res.* **1999**, *32*, 278-286.

[3] Monographs: a) C. D. Gutsche, *Calixarenes*, Royal Society of Chemistry, Cambridge, 1989; b) *Calixarenes. A Versatile Class of Macrocyclic Compounds*, Vicens, J., Böhmer, V., Eds., Kluwer, Dordrecht, 1991; c) C. D. Gutsche, *Calixarenes Revisited*, Royal Society of Chemistry, Cambridge, 1998. Review articles: a) S. Shinkai, *Tetrahedron* **1993**, *49*, 8933-8968; b) V. Böhmer, *Angew. Chem., Int. Ed. Engl.* **1995**, *34*, 713-745; c) P. Timmerman, W. Verboom, D. N. Reinhoudt, *Tetrahedron* **1996**, *52*, 2663-2704; d) V. Böhmer, D. Kraft, M. Tabatabai, *J. Incl. Phenom.* **1994**, *19*, 17-39. Homocalixarenes: S. Ibach, V. Prautzsch, F. Vögtle, C. Chartroux, K. Gloe, *Acc. Chem. Res.* **1999**, *32*, 729-740. Calixpyrroles: P. Gale, J. L. Sessler, V. Kral, *Chem. Commun.* **1998**, 1-8.

[4] a) I. Higler, P. Timmerman, W. Verboom, D. N. Reinhoudt, *Eur. J. Org. Chem.* **1998**, 2689-2702; b) D. M. Rudkevich, J. Rebek, Jr., *Eur. J. Org. Chem.* **1999**, 1991-2005.

[5] J.-D. van Loon, R. G. Janssen, W. Verboom, D. N. Reinhoudt, *Tetrahedron Lett.* **1992**, *33*, 5125-5128. For the early prediction of the calix[4]arene based capsule see: C. Andreu, R. Beerli, N. Branda, M.

Conn, J. de Mendoza, A. Galan, I. Huc, Y. Kato, M. Tymoschenko, C. Valdez, E. Wintner, R. Wyler, J. Rebek, Jr., *Pure Appl. Chem.* **1993**, *65*, 2313-2318.
[6]   R. H. Vreekamp, W. Verboom, D. N. Reinhoudt, *Recl. Trav. Chim. Pays-Bas* **1996**, *115*, 363-370.
[7]   K. Koh, K. Araki, S. Shinkai, *Tetrahedron Lett.* **1994**, *35*, 8255-8258.
[8]   a) R. H. Vreekamp, W. Verboom, D. N. Reinhoudt, *J. Org. Chem.* **1996**, *61*, 4282-4288; b) I. Higler, L. Grave, E. Breuning, W. Verboom, F. de Jong, T. M. Fyles, D. N. Reinhoudt, *Eur. J. Org. Chem.* **2000**, 1727-1734.
[9]   A. Arduini, M. Fabbi, M. Mantovani, L. Mirone, A. Pochini, A. Secchi, R. Ungaro, *J. Org. Chem.* **1995**, *60*, 1454-1457.
[10] O. Struck, W. Verboom, W. J. J. Smeets, A. L. Spek, D. N. Reinhoudt, *J. Chem. Soc., Perkin Trans. 2* **1997**, 223-227.
[11] J. J. Gonzalez, P. Prados, J. de Mendoza, *Angew. Chem., Int. Ed. Engl.* **1999**, *38*, 525-528.
[12] a) D. J. Cram, J. M. Cram, *Container Molecules and their Guests*, Royal Society of Chemistry, Cambridge, 1994, pp. 131-148; b) A. Collet, J.-P. Dutasta, B. Lozach, J. Canceill, *Top. Curr. Chem.* **1993**, *165*, 103-129; c) A. Jasat, J. C. Sherman, *Chem. Rev.* **1999**, *99*, 931-967.
[13] S. Mecozzi, J. Rebek, Jr., *Chem. Eur. J.* **1998**, *4*, 1016-1022.
[14] A. Arduini, L. Domiano, L. Ogliosi, A. Pochini, A. Secchi, R. Ungaro, *J. Org. Chem.* **1997**, *62*, 7866-7868.
[15] K. Murayama, K. Aoki, *Chem. Commun.* **1998**, 607-608.
[16] A. Shivanyuk, K. Rissanen, E. Kolehmainen, *Chem. Commun.* **2000**, 1107-1108.
[17] K. N. Rose, L. J. Barbour, W. Orr, J. L. Atwood, *Chem. Commun.* **1998**, 407-408.
[18] A. Shivanyuk, E. F. Paulus, V. Böhmer, *Angew. Chem., Int. Ed. Engl.* **1999**, *38*, 2906-2909.
[19] a) K. D. Shimizu, J. Rebek, Jr., *Proc. Natl. Acad. Sci. USA* **1995**, *92*, 12403-12407; b) O. Mogck, V. Böhmer, W. Vogt, *Tetrahedron* **1996**, *52*, 8489-8496; c) R. K. Castellano, D. M. Rudkevich, J. Rebek, Jr., *J. Am. Chem. Soc.* **1996**, *118*, 10002-10003. For the *pinched cone* calix[4]arene diurea dimers, see: J. Scheerder, R. H. Vreekamp, J. F. J. Engbersen, W. Verboom, D. N. Reinhoudt, *J. Org. Chem.* **1996**, *61*, 3476-3481.
[20] X-ray structure: O. Mogck, E. F. Paulus, V. Böhmer, I. Thondorf, W. Vogt, *Chem. Commun.* **1996**, 2533-2534.
[21] B. C. Hamann, K. D. Shimizu, J. Rebek, Jr., *Angew. Chem., Int. Ed. Engl.* **1996**, *35*, 1326-1329.
[22] Review : J. Rebek, Jr., *Chem. Commun.* **2000**, 637-643.
[23] O. Mogck, M. Pons, V. Böhmer, W. Vogt, *J. Am. Chem. Soc.* **1997**, *119*, 5706-5712.
[24] M. O. Vysotsky, I. Thondorf, V. Böhmer, *Angew. Chem., Int. Ed. Engl.* **2000**, *39*, 1264-1267.
[25] C. A. Schalley, R. K. Castellano, M. S. Brody, D. M. Rudkevich, G. Siuzdak, J. Rebek, Jr., *J. Am. Chem. Soc.* **1999**, *121*, 4568-4579.
[26] R. K. Castellano, B. H. Kim, J. Rebek, Jr., *J. Am. Chem. Soc.* **1997**, *119*, 12671-12672.
[27] R. K. Castellano, C. Nuckolls, J. Rebek, Jr., *J. Am. Chem. Soc.* **1999**, *121*, 11156-11163.
[28] S. B. Lee, J.-I. Hong, *Tetrahedron Lett.* **1996**, *37*, 8501-8504.
[29] M. S. Brody, C. A. Schalley, D. M. Rudkevich, J. Rebek, Jr., *Angew. Chem., Int. Ed. Engl.* **1999**, *38*, 1640-1644.
[30] J. J. Gonzalez, R. Ferdani, E. Albertini, J. M. Blasco, A. Arduini, A. Pochini, P. Prados, J. de Mendoza, *Chem. Eur. J.* **2000**, *6*, 73-80.
[31] M. Yanase, T. Haino, Y. Fukazawa, *Tetrahedron Lett.* **1999**, *40*, 2781-2784.
[32] T. Heinz, D. M. Rudkevich, J. Rebek, Jr., *Nature* **1998**, *394*, 764-766. A self-assembled capsule from the resorcinarene-based cyclic tetraurea has also been reported: M. H. K. Ebbing, M. J. Villa, J. M. Valpuesta, P. Prados, J. de Mendoza, In : Supramolecular Chemistry, A European Research Conference, Rolduc, The Netherlands, 1998.
[33] S. K. Körner, F. C. Tucci, D. M. Rudkevich, T. Heinz, J. Rebek, Jr., *Chem. Eur. J.* **2000**, *6*, 187-195.
[34] T. Heinz, D. M. Rudkevich, J. Rebek, Jr., *Angew. Chem., Int. Ed. Engl.* **1999**, *38*, 1136-1139.
[35] F. C. Tucci, D. M. Rudkevich, J. Rebek, Jr., *J. Am. Chem. Soc.* **1999**, *121*, 4928-4929.
[36] A. Lützen, A. R. Renslo, C. A. Schalley, B. M. O'Leary, J. Rebek, Jr., *J. Am. Chem. Soc.* **1999**, *121*, 7455-7456.
[37] I. Higler, W. Verboom, F. C. J. M. van Veggel, F. de Jong, D. N. Reinhoudt, *Liebigs Ann./Recueil* **1997**, 1577-1586.
[38] K. Kobayashi, T. Shirasaka, K. Yamaguchi, S. Sakamoto, E. Horn, N. Furukawa, *Chem. Commun.* **2000**, 41-42.
[39] L. R. MacGillivray, J. L. Atwood, *Nature* **1997**, *389*, 469-472.
[40] An even larger (1520 Å^3 volume) hexameric assembly was obtained in the solid state (X-ray crystallography) from tetrahydroxyresorcinarene. However, there is no evidence of this assembly in solution:

[41] T. Gerkensmeier, W. Iwanek, C. Agena, R. Fröhlich, S. Kotila, C. Näther, J. Mattay, *Eur. J. Org. Chem.* **1999**, 2257-2262.
[42] R. K. Castellano, D. M. Rudkevich, J. Rebek, Jr., *Proc. Natl. Acad. Sci. USA* **1997**, *94*, 7132-7137.
[43] R. K. Castellano, J. Rebek, Jr., *J. Am. Chem. Soc.* **1998**, *120*, 3657-3663.
[44] R. K. Castellano, C. Nuckolls, S. H. Eichhorn, M. R. Wood, A. J. Lovinger, J. Rebek, Jr., *Angew. Chem., Int. Ed. Engl.* **1999**, *38*, 2603-2606.
[45] R. H. Vreekamp, J. P. M. van Duynhoven, M. Hubert, W. Verboom, D. N. Reinhoudt, *Angew. Chem., Int. Ed. Engl.* **1996**, *35*, 1215-1218. The rosette motif was first introduced by Whitesides and Lehn: a) G. M. Whitesides, E. E. Simanek, J. P. Mathias, C. T. Seto, D. N. Chin, M. Mammen, D. M. Gordon, *Acc. Chem. Res.* **1995**, *28*, 37-44; b) K. C. Russel, E. Leize, A. van Dorsselaer, J.-M. Lehn, *Angew. Chem., Int. Ed. Engl.* **1995**, *34*, 209-213.
[46] P. Timmerman, R. H. Vreekamp, R. Hulst, W. Verboom, D. N. Reinhoudt, K. Rissanen, K. A. Udachin, J. Ripmeester, *Chem. Eur. J.* **1997**, *3*, 1823-1832.
[47] K. A. Jolliffe, M. Crego Calama, R. Fokkens, N. M. M. Nibbering, P. Timmerman, D. N. Reinhoudt, *Angew. Chem., Int. Ed. Engl.* **1998**, *37*, 1247-1250.
[48] a) L. J. Prins, J. Huskens, F. de Jong, P. Timmerman, D. N. Reinhoudt, D. N. *Nature* **1999**, *398*, 498-502. For the structural isomerism in double rosettes, see: L. J. Prins, K. A. Jolliffe, R. Hulst, P. Timmerman, D. N. Reinhoudt, *J. Am. Chem. Soc.* **2000**, *122*, 3617-3627.
[49] M. Crego Calama, R. Hulst, R. Fokkens, N. M. M. Nibbering, P. Timmerman, D. N. Reinhoudt, *Chem.Commun.* **1998**, 1021-1022.
[50] M. Crego Calama, P. Timmerman, D. N. Reinhoudt, *Angew. Chem., Int. Ed. Engl.* **2000**, *39*, 755-758.
[51] F. Cardullo, M. Crego Calama, B. H. M. Snellink-Ruël, J.-L. Weidmann, A. Bielejewska, R. Fokkens, N. M. M. Nibbering, P. Timmerman, D. N. Reinhoudt, *Chem. Commun.* **2000**, 367-368.
[52] K. A. Jolliffe, P. Timmerman, D. N. Reinhoudt, *Angew. Chem., Int. Ed. Engl.* **1999**, *38*, 933-937. Polymeric rod-like nanostructures were obtained from calix[4]arene bismelamines ans biscyanurates, see: H.-A. Klok, K. A. Jolliffe, C. L. Schauer, L. J. Prins, J. P. Spatz, M. Möller, P. Timmerman, D. N. Reinhoudt, *J. Am. Chem. Soc.* **1999**, *121*, 7154-7155.
[52] P. Timmerman, D. N. Reinhoudt, *Adv. Mater.* **1999**, *11*, 71-74.
[53] P. Lhotak, S. Shinkai, *Tetrahedron Lett.* **1995**, *36*, 4829-4832.
[54] A. N. Shivanyuk, D. M. Rudkevich, D. N. Reinhoudt, *Tetrahedron Lett.* **1996**, *37*, 9341-9344.
[55] K. Fujimoto, S. Shinkai, *Tetrahedron Lett.* **1994**, *35*, 2915-2918.
[56] B. Xu, T. M. Swager, *J. Am. Chem. Soc.* **1993**, *115*, 1159-1160.
[57] P. J. Stang, D. H. Cao, K. Chen, G. M. Gray, D. C. Muddiman, R. D. Smith, *J. Am. Chem. Soc.* **1997**, *119*, 5163-5168.
[58] P. Jacopozzi, E. Dalcanale, *Angew. Chem., Int. Ed. Engl.* **1997**, *36*, 613-615.
[59] a) O. D. Fox, N. K. Dalley, R. G. Harrison, *J. Am. Chem. Soc.* **1998**, *120*, 7111-7112; b) O. D. Fox, J. F.-Y. Leung, J. M. Hunter, N. K. Dalley, R. G. Harrison, *Inorg. Chem.* **2000**, *39*, 783-790; c) O. D. Fox, N. K. Dalley, R. G. Harrison, *Inorg. Chem.* **1999**, *38*, 5860-5863.
[60] a) A. Ikeda, M. Yoshimura, F. Tani, Y. Naruta, S. Shinkai, *Chem. Lett.* **1998**, 587-588; b) A. Ikeda, M. Yoshimura, H. Udzu, C. Fukuhara, S. Shinkai, *J. Am. Chem. Soc.* **1999**, *121*, 4296-4297.
[61] A. Ikeda, H. Udzu, M. Yoshimura, S. Shinkai, *Tetrahedron* **2000**, *56*, 1825-1832.
[62] O. D. Fox, M. G. B. Drew, P. D. Beer, *Angew. Chem., Int. Ed. Engl.* **2000**, *39*, 136-140.
[63] O. D. Fox, M. G. B. Drew, E. J. S. Wilkinson, P. D. Beer, *Chem. Commun.* **2000**, 391-392.
[64] a) J. A. Bryant, C. B. Knobler, D. J. Cram, *J. Am. Chem. Soc.* **1990**, *112*, 1254-1255 and 1255-1256; b) D. J. Cram, H.-J. Choi, J. A. Bryant, C. B. Knobler, *J. Am. Chem. Soc.* **1992**, *114*, 7748-7765.
[65] T. Haino, D. M. Rudkevich, J. Rebek, Jr., *J. Am. Chem. Soc.* **1999**, *121*, 11253-11254.
[66] F. C. Tucci, D. M. Rudkevich, J. Rebek, Jr., *Chem. Eur. J.* **2000**, *6*, 1007-1016.
[67] C. von dem Bussche-Hünnefeld, R. C. Helgeson, D. Bühring, C. B. Knobler, D. J. Cram, *Croat. Chem. Acta* **1996**, *69*, 447-458.
[68] S. Ma, D. M. Rudkevich, J. Rebek, Jr., *J. Am. Chem. Soc.* **1998**, *120*, 4977-4981.
[69] a) A. R. Renslo, D. M. Rudkevich, J. Rebek, Jr., *J. Am. Chem. Soc.* **1999**, *121*, 7459-7460; b) A. R. Renslo, F. C. Tucci, D. M. Rudkevich, J. Rebek, Jr., *J. Am. Chem. Soc.* **2000**, *122*, 4573-4582.
[70] S. Saito, D. M. Rudkevich, J. Rebek, Jr., *Org. Lett.* **1999**, *1*, 1241-1244.

Chapter 9

CAVITANDS

WILLEM VERBOOM

Laboratory of Supramolecular Chemistry and Technology
MESA$^+$ Research Institute, University of Twente, P.O. Box 217
7500 AE Enschede, The Netherlands.
e-mail: w.verboom@ct.utwente.nl

1. Introduction

Cavitands have been defined as "synthetic organic compounds that contain an enforced cavity large enough to accommodate simple organic compounds or ions" [1]. This chapter only deals with resorcarene-based cavitands **I**, henceforth simply termed cavitands.

Compared to their close analogues, the calix[4]arenes, the cavitands are much more rigid, since the aromatic rings are interconnected at two positions. Importantly, the size of the cavity can be varied using different types of bridges Y. Functional groups can be introduced at different positions, both at the rim of the bowl (A) and in the substituents on the methylene bridges R.

Initially, the cavitands were used mainly for the preparation of (hemi)carcerands and (hemi)carceplexes by (non-)covalent linking of two cavitands to give molecules with a closed surface. Several reviews have appeared on this topic [2-4]. Recent developments are given in Chapter 10. Increasingly, cavitands have been recognized as suitable molecular building blocks for different applications in supramolecular chemistry. This Chapter elaborates on earlier reviews [5] of such applications.

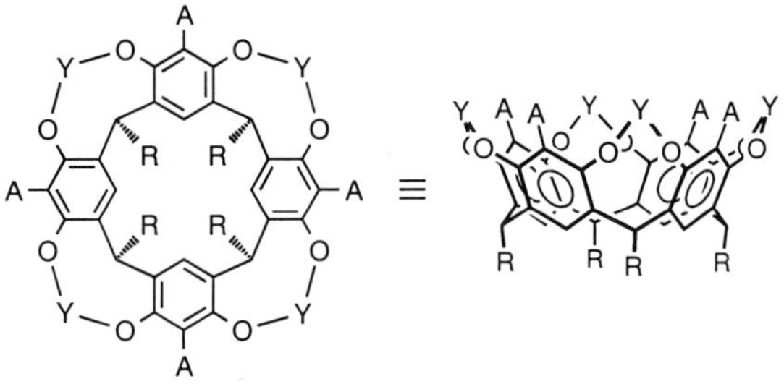

general formula **I**

2. Synthesis

The classical procedure for the preparation of cavitands **2** involves the bridging of resorc[4]arenes **1**, the condensation products of the reaction of a resorcinol and an aldehyde [6] with a large excess CH_2BrCl at 60–70 °C for 24 h. When A = Br or Me, higher yields are obtained than when A = H. In nearly all cases chromatography is necessary to separate (undesired) partially bridged cavitands and other byproducts, although it was recently found [7] that performing the reaction (**1**: R = Me, C_5H_{11}, CH_2CH_2Ph, A = Br, Me) at a higher temperature (88 °C) in a sealed tube leads to a much shorter reaction time (3 h) and almost quantitative yields.

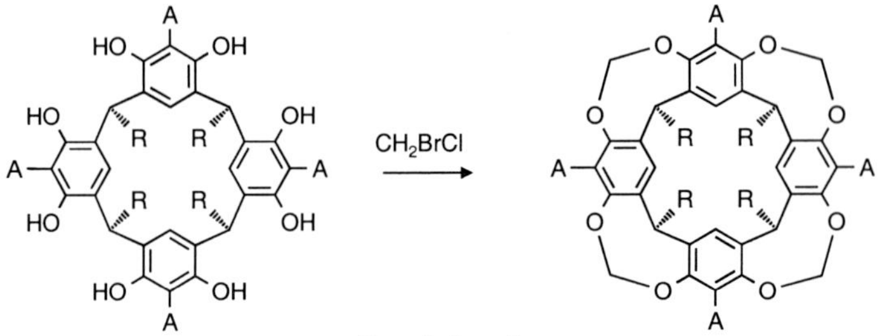

R = alkyl, aryl
A = H, Br, Me

1 → **2** (CH$_2$BrCl)

The stereoisomerism of resorcarenes has been discussed in Chapter 1. So far, mainly *rccc*-resorc[4]arenes in the crown conformation have been bridged to give the corresponding cavitands with all substituents in the axial position [8]. Condensation of 2-methylresorcinol with aryl aldehydes gives exclusively the *rctt* methylresorc[4]arenes **3**. Despite of their unfavourable chair conformation, treatment of **3** with 6-8 equivalents of CH$_2$BrCl in the presence of K$_2$CO$_3$ in DMF at 65 °C allowed the rapid (10 h) formation of the cavitands **4**, having two adjacent aryl substituents in the axial position and the two others in the equatorial position [9]. The formation of the cavitands **4** may be the result of an equilibrium between resorc[4]arenes **3** in the chair conformation and the corresponding crown conformers (not detected by NMR), the latter being trapped by reaction with CH$_2$BrCl. Using only 2.5 equivalents of CH$_2$BrCl, no partly bridged intermediates could be detected; only fully bridged cavitand and starting material were isolated. This suggests that the conformational inversion from chair to crown is connected with the formation of the first bridge, which is also the rate-determining step for cavitand formation. In contrast, for unsubstituted *rccc*-resorc[4]arenes (lacking a methyl group between the hydroxyl groups), introduction of the final bridge is the most difficult, allowing the isolation of partially bridged cavitands [10].

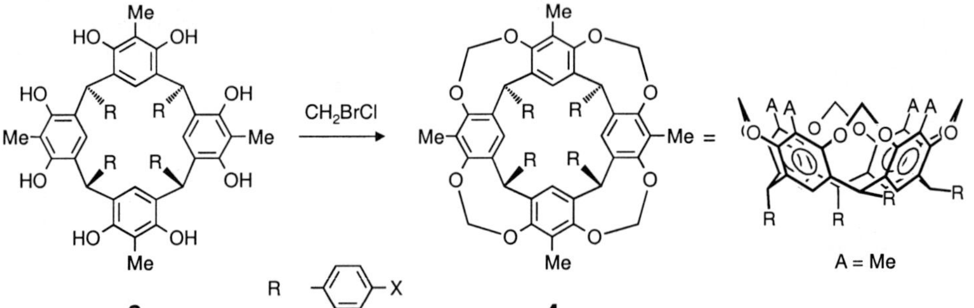

3 → **4** (CH$_2$BrCl)   A = Me

R = –C$_6$H$_4$–X

X = H, Br, F, CN, OMe, NO$_2$, p-C$_6$H$_4$-Me

3. Variation in the Bridges

Cavitands have been prepared with various bridges (I: Y = (CH$_2$)$_n$ and SiAlk$_2$). Much recent work has been focussed on the introduction of phosphorus-based bridges. Reaction of resorc[4]arene **1** (R = CH$_2$CH$_2$Ph, A = H) with phenyldichlorophosphine in the presence of pyridine as a base gave the phosphonitocavitand **5**. Since the lone pair of electrons on the phosphorus atoms of **5** can be directed either outward (o) or inward (i) with respect to the middle of the bowl, there are six possible isomeric forms for the cavitand **5**: iiii, oiii, ooii, oioi, oooi, oooo. Molecular mechanics calculations indicated that the isomer with all phenyl groups directed outward and lone pairs directed inward would be preferred. This stereochemistry is consistent with the observation of a single signal in the ^{31}P NMR spectrum.

With the quadridentate ligand **5**, Cu(I), Ag(I), Au(I) and Pt(II) complexes have been prepared [11]. The X-ray crystal structure of the chlorogold(I) complex of **5** shows the presence of three AuCl units around the upper rim of the bowl with the fourth AuCl unit folded inside the cavity. The gold(I)-rimmed bowl forms inclusion complexes with amines. In liquid-liquid extraction studies both the Au(I)- and Pt(II)-cavitand complexes showed a high affinity for alkali metal cations with a strong preference for K$^+$ [12]. The Cu(I) and Ag(I) complexes show size-selective inclusion of halides, stabilized by a unique µ$_4$-face-bridged binding mode. The chloride-occluded silver(I) cavitand complex acts as a nucleophile to convert alkyl iodides to alkyl chlorides. The high yielding reaction follows the reactivity sequence *t*-BuI > *i*-PrI > MeI. The course of these nucleophilic substitution reactions is strongly influenced by the anion inclusion, since iodide is more strongly bound than chloride. (See also Chapter 23)

	R	X
5	CH$_2$CH$_2$Ph	Ph
6	Me	OEt
7	Me, C$_6$H$_{13}$, C$_7$H$_{15}$, C$_9$H$_{19}$	NMe$_2$, NEt$_2$
8	C$_6$H$_{13}$, C$_7$H$_{15}$	OC$_6$H$_{13}$, OC$_8$H$_{17}$
9	Me	Cl
10	Me	Me

Bridging of the eight phenolic hydroxyl groups in resorc[4]arene **1** (R = Me, A = H, Br) with dichloroarylphosphonate gave the phosphonate cavitands **11**, which have four dioxaphosphocin rings with four stereogenic centres on the phosphorus(V) atoms. The reaction leads to all the six possible diastereoisomers having different orientations of the P=O groups either outward or inward with respect to the cavity. The product ratio depends on the aryl group used. The different diastereoisomers were separated by chromatographic methods and the configuration of each was elucidated using ^{1}H, ^{13}C, and ^{31}P NMR spectroscopy. For a firm assignment it was necessary to study cavitands having both methylene and P(O)Ph bridges [13,14].

The phosphonate cavitand **11** (Ar = Ph) [15] with the four P=O moieties pointing inward is an efficient extractant and a powerful ligand for alkali metals and ammonium ions. Stability constants were found ranging from $K = 2.7 \times 10^8$ M^{-1} for *t*-BuNH$_3^+$ to $K = 8.1 \times 10^9$ M^{-1} for MeNH$_3^+$ in chloroform saturated with water at 291 K. It was suggested that the large K-values reflect the combined effect of ion-dipole, hydrogen bonding, van der Waals, and π-cation interactions both at the phosphorylated upper rim and within the aromatic cavity of the receptor.

Reaction of resorc[4]arene **1** with phosphorus di- and triamides gave the corresponding cavitands **6** and **7** in yields of 70-80% [16,17]. Cavitands **7** form 1:2 complexes with diethyl- and triethylamine, H$_2$NCH$_2$CH$_2$OH, Me$_2$NCH$_2$CH$_2$OH, and HOOC-CH(NH$_2$)(CH$_2$)$_4$NH$_2$. Heating of the amidophosphites **7** with sulfur afforded the corresponding amidothiophosphates **12** with an axial orientation of the thiophosphoryl groups. Bridging with phenyl-, methyl, and ethyldiamidophosphites resulted in the formation of the phosphatocavitands **13**, while phosphitocavitands **8** were obtained using hexyl- and octyldiamidophosphite.

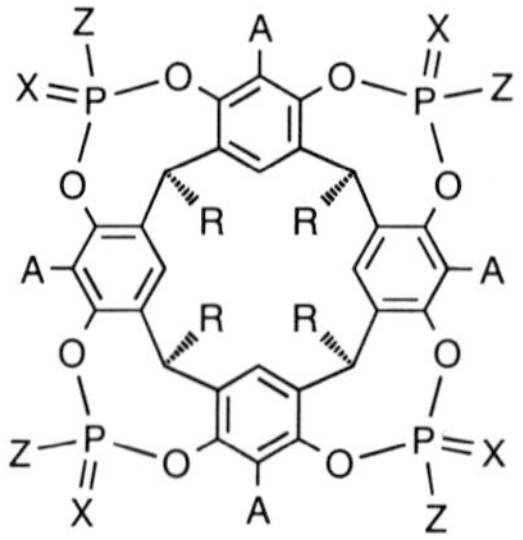

	R	A	X	Z
11	Ph	H, Br	O	Ar
12	Me	H	S	NMe$_2$, NEt$_2$
13	C$_6$H$_{13}$, C$_7$H$_{15}$	H	O	OPh, OMe, OEt
14	C$_9$H$_{19}$	H	O	CH$_2$Cl

Bridging with chloromethyldichlorophosphonate afforded cavitand **14** in 75% yield, having four chloro atoms for further functionalization [17]. The ClP-bridged cavitand **9**, prepared in quantitative yield from resorc[4]arene **1** (R = Me, A = H) and PCl$_3$, provides various derivatives [18]. Reaction of **9** with methylmagnesium iodide and Me$_2$SiNMe$_2$ furnished the cavitands **10** and **7** (R = Me, X = NMe$_2$) in 88% and 77% yield, respectively. Oxidative addition of tetrachloro-*o*-benzoquinone and hexafluoroacetone to the latter cavitand gave the cavitands **15** and **16** as a mixture of stereoisomers.

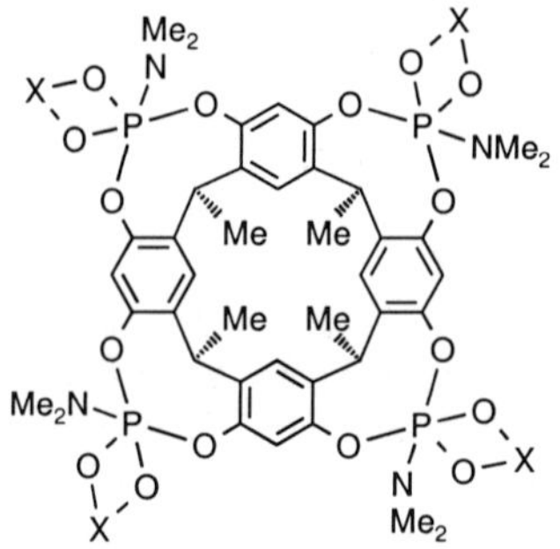

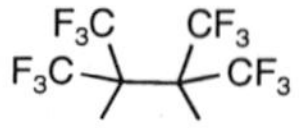

4. Functionalization

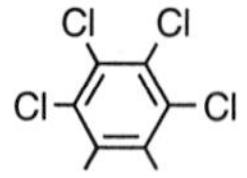

The synthesis of many functionalized cavitands starts from the tetrabromocavitand **2** (A = Br). The bromo atom can be substituted by a cyano group [19], and with bromo-lithium exchange and subsequent quenching with the appropriate reagents, hydroxyl, thiol, formyl, and carboxylic ester groups can be introduced. Reduction of the ester functionalities gives hydroxymethyl groups which can be converted to chloromethyl and subsequently to thiomethyl groups [2]. On the other hand, tetramethylcavitands (**2**: A = Me) can be brominated to afford bromomethyl groups. The yield can be improved by the use of AIBN [20] rather than benzoyl peroxide [21] as a catalyst for the bromination with NBS. This provides a faster route to the corresponding tetrakis(mercaptomethyl)cavitand on treatment of the tetrakis(bromomethyl)cavitand with thiourea [22,23] and is also a superior route [24] to the well-known tetrakis(hydroxymethyl)cavitand [25]. The so-called *caviteins*, introduced as a family of *de novo* proteins, can be obtained by attachment of four peptide chains to the tetrakis(mercaptomethyl)cavitand [23,26]. (See Chapters 10, 27)

Reaction of tetramethylcavitand (**2**: $R = C_5H_{11}$, A = Me) with one equivalent of NBS gave the mono(bromomethyl)cavitand as the main reaction product [27]. The bis(chloromethylated) cavitand **17** [28], prepared by chloromethylation of cavitand **2** (R = Me, A = H) in 35% yield, has been used for the synthesis of the bis-aza-15-crown-5-functionalized cavitands **18**. (The corresponding monosubstituted cavitands have been also obtained).

Partially functionalized cavitands have been prepared from tetrakis(bromomethyl)cavitand **2** ($R = C_5H_{11}$, A = CH_2Br) by partial substitution of bromo atoms by phthalimido groups [29,30]. All four partially functionalized products were formed by reaction with less than four equivalents of potassium phthalimide, using tributylhexadecylphosphonium bromide as phase transfer catalyst. Depending on the reaction conditions, varying ratios of the different compounds were formed (Table 1). In acetonitrile (entries 4 and 5) the relative distribution of both disubstituted compounds is the inverse of the 1:2 ratio observed in DMF (entry 3), involving a 2-3 fold excess of the distal over the proximal diphthalimido cavitand. This was explained as due to the lower solubility of the distal compound resulting in partial precipitation. The remaining CH_2Br moieties could be converted into methyl groups by treatment with $NaBH_4$ in DMF at -10 ^{0}C, while the amino groups formed upon deprotection (NH_2NH_2) are a suitable handle for further functionalization [31].

Mono, 1,2-di- and 1,3-di- and trihydroxycavitands, can be isolated as side-products in the tetrahydroxycavitand synthesis but these compounds do not contain functionalities at the other aromatic units [32]. On the 1,2-positions of the tetrahydroxycavitand **2** ($R = C_5H_{11}$, A = OH) a single crown ether bridge was introduced to give crowncavitand **19**, leaving the 3,4-dihydroxy positions open for further functionalisation [31]. Reaction of tetrakis(bromomethyl)cavitand with resorcinol under stoichiometric conditions in a Cs_2CO_3-DMA or DMF mixture gave the bridged cavitand **20** in 27% yield [33]. On the other hand, reaction with an excess of catechol or resorcinol in a mixture of K_2CO_3/DMF gave the corresponding tetraaryl-fenced cavitands, which were subsequently capped with 1,2,4,5-tetrakis(bromomethyl)benzene [34].

TABLE 1. Partial functionalization of *tetrakis*(bromomethyl)cavitand **2** ($R = C_5H_{11}$, A = CH_2Br) by reaction with potassium phthalimide (K-Phth).

entry	solvent	K-Phth [equiv]	time [hours]	mono	1,3-bis	1,2-bis	tris	tetrakis	total
1	toluene	2.0	5	28	12	25	8	2	75
2	toluene	2.4	17	7	8	20	37	--	72
3	DMF	2.2	4.5	21	2	5	14	10	52
4	CH_3CN	2.2	4	12	22	7	17	11	69
5	CH_3CN	2.2	5	13	26	11	20	5	75

The isolated yield columns are: mono, 1,3-bis, 1,2-bis, tris, tetrakis, total.

Cavitands with four arylboronic acid substituents [**2**: R = 4-B(OR)$_2$-Ph, A = H] [35], may possibly afford access to an array of heterofunctional architectures. So far only one example is known of the selective functionalization of the residues R on the cavitand bridges. A cavitand with four (CH$_2$)$_8$CH=CH$_2$ chains on the methylene bridges, upon treatment with *m*-chloroperbenzoic acid, gave the *mono*-epoxide in 30% yield along with bis-epoxides. Subsequently, the three remaining double bonds were hydrogenated, the epoxide was hydrated under acidic conditions, the resulting diol oxidatively cleaved and the acid produced reduced to the alcohol [36]. The analogous reaction sequence was carried out for the O-protected resorcarene (compare Chapter 2) which, after deprotection, was converted to the cavitand.

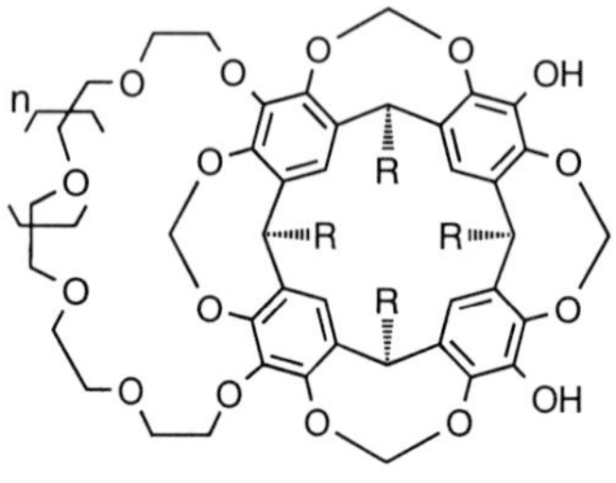

19 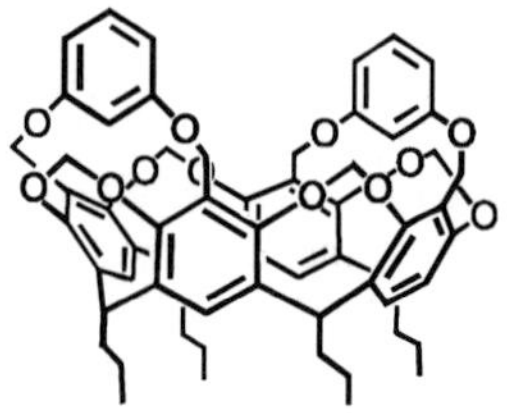   **20**

5. Cavitands with Deepened Cavities

The synthesis of deep open-ended cavities, based on calix[4]arene and resorc[4]arene modules, has been reviewed elsewhere [37] and only a few recent examples of such cavitands will be considered here.

One approach for the formation of larger cavities involves the enlargement of the aromatic walls by aromatic substituents A in the general formula **I**. Two strategies can be envisaged: The cavitand **21a** was prepared in two steps from 2-(4-bromophenyl)resorcinol [38], whereas the cavitand **21b** was obtained in 71% yield by a Suzuki reaction of the corresponding tetrabromocavitand with (4-nitrophenyl)boronic acid [39]. The latter method provides easy access to a variety of cavitands with functionalized deep cavities from a common precursor. A corresponding Suzuki coupling with **2** (R = *p*-Br-Ph, A = H) gave the cavitands **22** [40].

Another approach involves the use of specific bridging reagents. The resorc[4]-arenes **1** can be bridged with a range of substituted benzal bromides to give the cavitands **23** [41,42]. In all cases, although a total of six diastereoisomers could potentially be formed, only one isomer was separated from the varying amounts of oligomeric material. The stereoselective bridging is highly dependent on both the solvent for the reaction and the R groups of the starting resorc[4]arene **1**. The best yields (up to 56%) were obtained with a polar aprotic solvent, while cavitand formation improved, and the yields became less solvent-dependent as the size of the pendent R groups (methyl, butyl, phenylethyl) increased.

The cavitands **24** can be prepared by bridging the resorcin[4]arene hydroxyl groups with 2,3-dichloroquinoxaline or its 6,7-disubstituted analogs [43]. These molecules flutter between C_{4v} and C_{2v} symmetries dependent on the temperature. The former, having a vase-like shape with all walls up, is preferred at higher temperatures. The latter, with a kite-like shape, has the walls flipped outward and is the dominant conformation below room temperature.

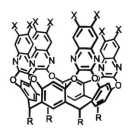

21a R = CH$_2$CH$_2$Ph, X = Br
b R = CH$_{11}$H$_{23}$, X = NO$_2$

22a X^1 = H, X^2 = H
b X^1 = NO$_2$, X^2 = OMe

23 R = Me, C$_4$H$_9$, (CH$_2$)$_{10}$Me, CH$_2$CH$_2$Ph

24 R = alkyl, X = H, Br, Me

An alternative strategy [44] involves the use of a simple building block in the bridging reaction followed by an extension of the rim through heterocyclic synthesis. The octanitrocavitand **25** is formed in 80% yield by reaction of resorc[4]arene **1** (R = C$_{11}$H$_{23}$, A = H) with 1,2-difluoro-4,5-dinitrobenzene. On hydrogenation of the nitro groups, the resulting phenylenediamine units were condensed with 1,2-diketones. The fused pyrazine products provided the deepened cavities. In this way, condensation with diethyl 2,3-dioxosuccinate and acenaphthenequinone gave the cavitands **26** (X = OEt) and **27** in 13 and 49% yield, respectively. The cavities feature wide openings with guest/solvent exchange fast on the NMR time scale. The uptake and release of guests involves the folding and unfolding of the host walls, motions that are influenced by solvent size and polarity. For the 1:1 complexation of cavitand **27** with the fullerene C$_{60}$ in toluene at 293 K a complexation constant K = 900 ± 250 M^{-1} was found.

Starting from octanitrocavitand **25**, a series of cavitands **28** that fold into a deep open-ended cavity stabilized by intramolecular hydrogen bonds was synthesised. In apolar solvents, a seam of eight intramolecular C=O···H-N hydrogen bonds can be stitched along the upper rim, four of which bridge adjacent rings and are held in place by four intraannular hydrogen bonds in seven-membered cyclic structures. The self-folding is reversible and controlled by solvent and temperature [45]. Complexation of molecules such as adamantanes, lactams, and cyclohexane derivatives by the self-folding cavitands **28** has been demonstrated. The exchange between complexed and free guest species is slow on the NMR time scale. Probably the ring of hydrogen bonds is responsible for this, since it impedes the opening to the so-called kite conformation [46].

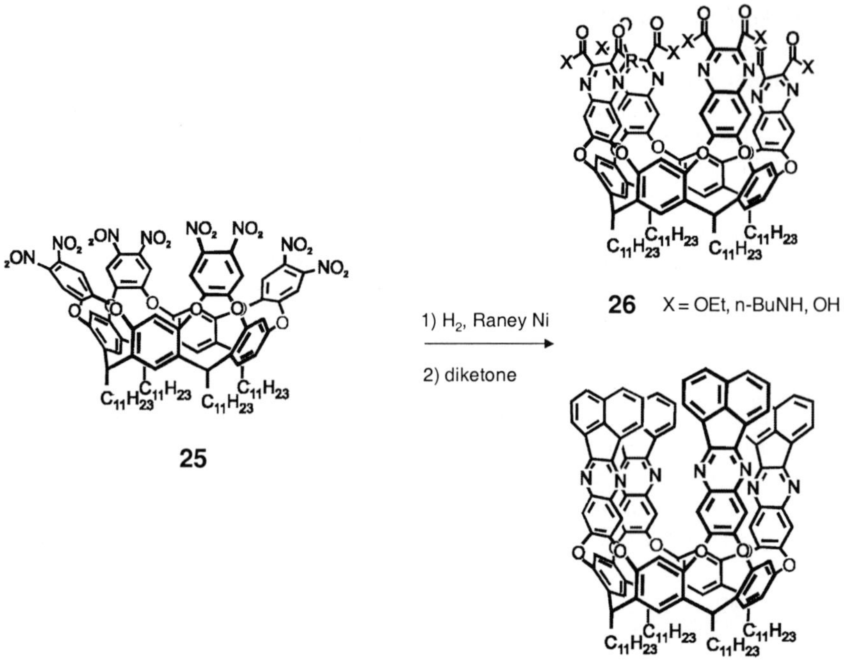

25

1) H₂, Raney Ni
2) diketone

26 X = OEt, n-BuNH, OH

27

Complexation of bis-adamantyl derivatives [Ad-NH-C(O)-Ad] gives rise to the formation of two isomeric complexes in which one or the other adamantyl residue is included in the cavity of **28**. As is the case with, *e.g.,* calix[4]arene-based carceplexes [47], these complexes differ by orientation of the guest molecule inside the cavity, as the guests are too large to tumble freely within it [48]. The complexation of adamantane has been used for the design of self-complementary cavitands by attachment of an adamantane residue either at the upper or at the lower rim of the cavitand; for an example see cavitand **29** [49,36]. Recently, the synthesis of a deep cavitand by bridging of resorc[4]arene **1** (R = $C_{11}H_{23}$, A = H) with 5,6-dichloropyrazine-2,3-dicarboxylic acid imide was reported. Dimerisation of this compound results in a large cylindrical capsule of nanometer dimensions [50]. (See Chapter 8)

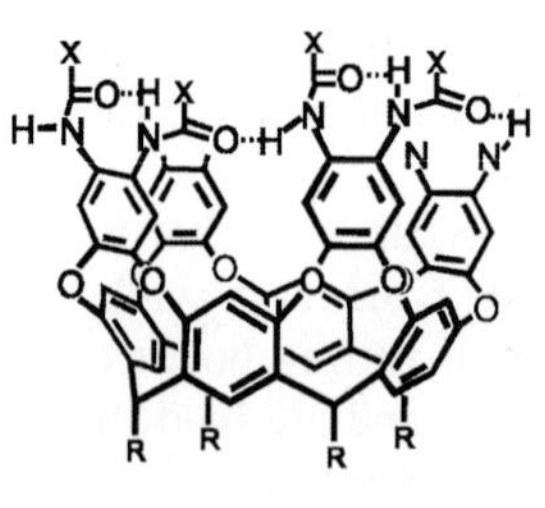

28

R = C_9H_{19}, $C_{11}H_{23}$, C_4H_8-Z

R = *n*-C_7H_{15}, cyclo-C_6H_{11}, CH_2Cl

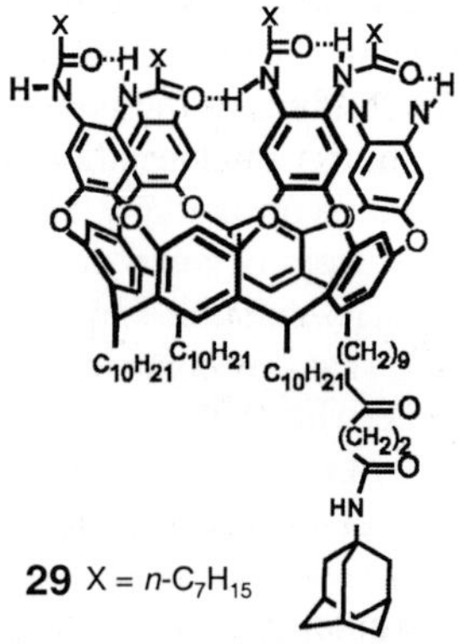

29 X = *n*-C_7H_{15}

6. Water-Soluble Cavitands (See also Chapter 24, 34, 35)

The sodium salts of the tetrahydroxycavitands **30** are water-soluble with solubilities of 0.3 and 0.4 M at ambient temperatures [51]. The ^{1}H NMR spectrum of cavitand **30** (R = Me) is sharp, well-resolved, and concentration independent, whereas the spectrum of cavitand **30** (R = CH$_2$CH$_2$Ph) is broad, suggesting that the pendent phenylethyl groups induce aggregation in water.

Reaction of tetrakis(bromomethyl)cavitands with pyridine and hexamethylenetetraamine gave the water-soluble cavitands **31** and **32**, respectively. In the case of **31** the water-solubility depends on the length of the alkyl chain (R = Me, C$_5$H$_{11}$, and C$_{11}$H$_{23}$) 16, 53, and >140 mM, respectively). Undecylcavitand **31** (R = C$_{11}$H$_{23}$) forms distinct, rod-shaped aggregates, as was observed by transmission electron microscopy (TEM) [52]. The cavitands **32** show a good 1:1 binding affinity to dianionic aromatic guests with binding constants of $K = 10^2 - 10^4$ M^{-1} [53]. Using the synthetic methodology for hydroxyl-footed cavitands (R = (CH$_2$)$_3$OH), the water-soluble phosphate-footed cavitands **33** were prepared and the functional groups A used for the synthesis of *caviteins* (see above) [54,55]. The cavitands **30** show a high affinity for Cs$^+$ and the analogue tetrakis-(diethanolaminomethyl)cavitand **34** is soluble both in neutral and basic aqueous media [56].

	R	A
30	Me, CH$_2$CH$_2$Ph	O$^-$Na$^+$
31	Me, C$_5$H$_{11}$, C$_{11}$H$_{23}$	pyridinium
32	Me	hexamethylenetetraammonium
33	(CH$_2$)$_3$OPO$_3$H$_2$	CH$_2$Br, CH$_2$SH
34	Me	CH$_2$N(CH$_2$CH$_2$OH)$_2$

In analogy to calix[4]arenes [57], the cavitands **28** have been made water-soluble by the attachment of four tris(hydroxymethyl)aminomethane moieties. The persilylated cavitand **28** [X = Et, Z = OC(O)NHCH$_2$C(O)NHC(CH$_2$OSiMe$_3$)$_3$] binds guests such as N-(1-adamantyl)acetamide in *p*-xylene-*d*$_{10}$ as a solvent. The guest-free species exists as two cycloenantiomers with clockwise and counterclockwise orientation of the C(O)NH···O=CNH groups and overall C$_4$ symmetry. Desilylation of this compound gave a dodecahydroxycavitand **28** [Z = OC(O)NHCH$_2$C(O)NHC(CH$_2$OH)$_3$] that rearranges to the corresponding octahydroxy tetraammonium salt exhibiting excellent water solubility. Addition of appropriate guests, such as aminomethyladamantane and aminomethylcyclohexane hydrochlorides and *N*-methyl quinuclidinium trifluoroacetate, resulted in significant changes in the NMR spectra indicative of complexation [58].

7. Complexation of Neutral Molecules (See also Chapter 25)

It has been known for a long time that the cavity of cavitands **2** can complex small solvent molecules [2]. Extensive studies have been made of complexation in the gas phase of cavitands **24** having four quinoxaline walls. The cavitand was evaporated from a desorption chemical ionization (DCI) probe within an ion source containing a mixture of a buffer gas (methane) and small amounts of specific aromatic or functionalized aliphatic molecules. A mechanism involving the preliminary ionization of the host molecule and its subsequent interaction with a neutral guest is predominant, but simultaneous interaction between neutral species might also occur to some extent. In the case of aromatic guests (benzene, toluene, anisole), CH-π interactions are dominant [59,60].

This type of cavitand has also been used as a chemically sensitive layer in quartz-crystal-microbalance (QCM) sensors when specific interactions, such as CH-π interactions, are present between the preorganized cavity of the receptor and the analyte. These interactions perturb the selectivity pattern expected on the basis of purely dispersive interactions between the analyte and the cavitand layer by adding their contribution to the sensor response. CH-π interactions alone provide significant specificity only in the presence of analytes with rather acidic methyl groups such as acetonitrile, nitromethane, and ethyl acetate [61]. However, the mixed-bridged cavitands **35** [62] are capable of two synergistic interactions with the analyte, viz. hydrogen bonding with the P=O group and CH-π interactions with the π-basic cavity. Only when the P=O group is oriented inward are simultaneous interactions possible and this is reflected in a remarkable increase in the response towards linear alcohols in mass sensors. The interactions can even be fine-tuned by chemical modification. Replacing the four bromo substituents (A) with methyl groups enhances the π-basic character of the cavity and thus intensifies the CH-π interaction. This strategy may lead to highly selective supramolecular mass sensors [63]. Several groups have also applied cavitands in monolayers on acoustic wave sensors for the detection of organic vapors [64,65].

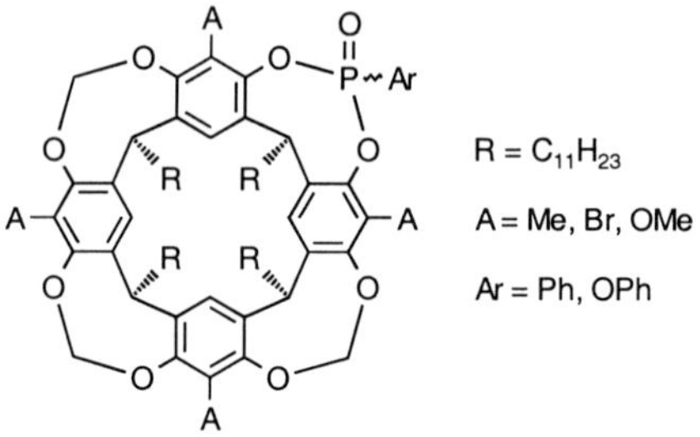

Cavitands have also been used as a molecular platform to attach ligating sites. Attachment of four carbohydrate branches composed of glucose, maltose, or maltotriose derivatives gives sugar clusters **36** [66]. The fluorescent guest 8-anilinonaphthalene-1-sulphonate (ANS) was bound with $K = 900 - 2600$ M^{-1} in aqueous HEPES buffer at pH 7.2. The cavitand sugar clusters having maltose or maltotriose residues also strongly bind concanavalin A. (See Chapter 27)

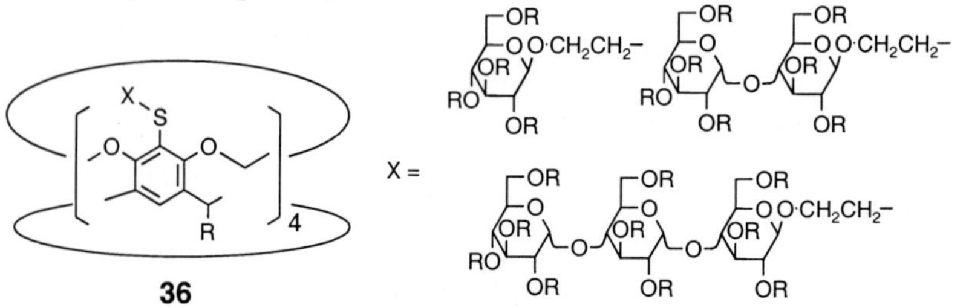

8. Complexation of Cations (See also Chapters 28, 34, 35)

In contrast to the calixarenes, there are few cavitand-based receptors for cations. Attachment of ligating sites to the rigid cavitand frame allows, in principle, a tight preorganization of the coordinating sites. Furthermore, the limited flexibility of the ligating sites is favorable in cases where the coordinating atoms of the ligand have to compete with solvent molecules or anions.

In analogy with the corresponding calix[4]arenes, the cavitand (thio)amides **37** and **38** were prepared starting from the corresponding tetrahydroxycavitand **2** (R = C_5H_{11}, A = OH) and tetrakis(hydroxymethyl)cavitand **2** (R = C_5H_{11}, A = CH_2OH) and *N,N*-dimethyl-2-chloroacetamide followed by treatment with Lawesson's reagent. Chemically modified field effect transistors (CHEMFETs) based on thioamide **38** exhibit a Pb^{2+} response under acidic conditions (pH = 4), with a detection limit of log $a(Pb^{2+})$ of –4.6. Unfortunately, no selectivity was found in the presence of different interfering ions. The distance between the donating thiocarbonyl groups is probably too large for a strong interaction with the Pb^{2+} ion. However, in comparison with **37**, cavitand **38** has an extra methylene spacer between the building block and the donating thiocarbonyl groups which may result in a smaller cavity for the formation of a Pb^{2+} complex. CHEMFETs with receptor **38** show Pb^{2+} responses in the presence of K^+, Cu^{2+}, Cd^{2+}, and Ca^{2+} ions (log $K_{Pb,j}$ = -1.9, -2.1, -2.4, and –4.3, respectively) [24].

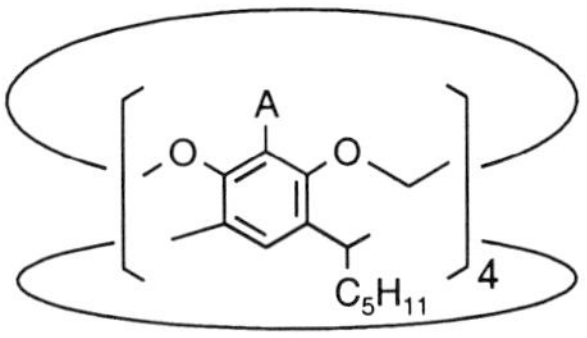

37 A = $OCH_2C(S)NMe_2$

38 A = $CH_2OCH_2C(S)NMe_2$

In the treatment of nuclear waste (carbamoylmethyl)-phosphoryl or oxophosphinyl derivatives are used as extractants for the radiotoxic, trivalent trans-plutonium actinides. In order to improve the extraction properties, the ligating sites have been grouped together on a cavitand platform, in analogy to calixarenes substituted by CMPO functions at the wide rim [67]. The (carbamoylmethyl)phosphonate (CMP) and –phosphine oxide (CMPO) cavitands **41a,b** and **42a,b**, respectively, were prepared via two different routes as summarised in Scheme 1. In an extraction study using radiotracers, europium(III) was used as a general representative for the trivalent actinides. The CMP(O) cavitands **41a,b** and **42a,b** are very effective europium extractants, cavitand **41b** having the highest extraction constant for 1:1 complexation with Eu(picrate)$_3$ (K_{ex} = 2.7 x 10^{12} M^{-4}). Although uranium is extracted to the same extent as Eu^{3+} and the selectivity over Fe^{3+} is small, the selectivities are better than with simple CMP(O) derivatives. Model cavitand

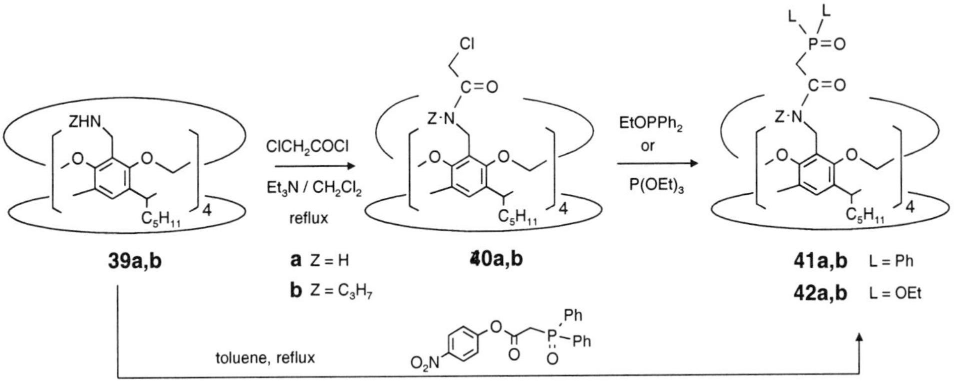

39a,b

a Z = H
b Z = C_3H_7

CICH$_2$COCl
Et$_3$N / CH$_2$Cl$_2$
reflux

40a,b

EtOPPh$_2$
or
P(OEt)$_3$

41a,b L = Ph
42a,b L = OEt

toluene, reflux

43, easily prepared from the tetrakis(bromomethyl)cavitand, extracts Sr^{2+} very efficiently and is a moderate extractant for the other elements studied (Na^+, Cs^+, UO_2^{2+}, Fe^{3+}, Eu^{3+}) [20]. (See Chapter 35)

The properties of the compounds **41a,b** and **42b** as ligands for trivalent lanthanide cations in acidic solutions with high nitrate concentrations, have been studied in detail. It was found that at $LiNO_3$ concentrations > 1-2 M, and with excess ligand compared to the metal concentration, $Eu(NO_3)_3$ is extracted by these ligands as a 1:1 complex. At higher nitrate concentrations (> 2.5 M), one $LiNO_3$ associates with the Eu complex, giving rise to increasing distribution ratios. At low acidity, the ligands are insensitive to variations in the H^+ concentration, but at higher acidity protonation of the ligands results in a decrease of the distribution coefficients. Variation of the anions (Cl^-, ClO_4^-, NO_3^-) strongly influences the extractions due to the different anion lipophilicities [68]. The same compounds have also been used for the removal of Eu^{3+} from moderately acidic nitrate solutions by transport through supported liquid membranes (SLMs). The experimental time-dependent Eu^{3+} concentrations were monitored on-line with a radiotracer and the concentration profiles could be described with a non-steady-state model. Evaluation of the rate-limiting step showed that aqueous diffusion plays a significant role and that the SLM and the solvent extraction models are in good agreement [69].

Starting from differently substituted aminomethylcavitands the ligands **43**, with one to three CMPO moieties, were prepared to study the influence of the number of ligating sites on the extraction properties and complexation strength. Compared to the tetrakis-CMPO cavitand **41a**, the partially CMPO-substituted ligands **43** are less effective extractants for Eu^{3+}. The difference in extraction behavior of both disubstituted ligands is striking, the value of the extraction constant for **43c** being 5.5 times larger than that for **43b**, illustrating the effect of preorganization. Although the individual contributions of the CMPO moieties are not equally essential for the complex formation and strength, the complexation constant increases going from mono- (**43a**) to tetrakis-CMPO ligand **41a**, illustrating a cooperative effect of the ligating sites [30].

Cavitands **45** and **46** with phosphine sulfide moieties were prepared as potential (more stable) alternatives for the well-known CYANEX 301 [bis(2,4,4-trimethylpentyl)-dithiophosphinic acid]. However, these compounds did not extract Am^{3+} or Eu^{3+}, not even in the presence of synergists, indicating that ionic interactions, as in CYANEX 301, are essential for the appli-cation of sulfur-containing extrac-

44 A = $CH_2P(O)(OEt)_2$
45 A = $CH_2P(S)Ph_2$
46 A = $(CH_2)_2P(S)Ph_2$
47 A = $CH_2P(O)OHPh$

tants. The tetraphosphinic acid cavitand **47** extracts Eu^{3+} as a 2:1 complex when excess ligand is present and as a 1:1 complex when excess metal is present. With cavitand **47**, compared to its "simple" dialkyl analogue CYANEX 272, similar extraction efficiencies are established with ca. 10^3 times lower extractant concentration (although in dichloromethane *vs.* in the less polar dodecane). The selectivity for Eu^{3+} over Am^{3+} remains of the same order ($S_{Eu/Am}$ = 5) [70,71].

9. Complexation of Anions (See also Chapter 23)

The rigid cavitand platform has also been used for the development of anion receptors. It is known that (thio)urea-functionalized calix[4]arenes complex halide anions in chloroform. The complexation constants are modest due to strong intramolecular hydrogen bonding and to the competitive formation of dimers (see *e.g.* refs [72,73]) via an array of eight (thio)urea moieties [74]. In the case of the corresponding cavitand-based receptors, such intramolecular hydrogen bonds cannot be formed and dimer formation is prevented by the methyleneoxy bridges of the cavitand.

A series of tetrakis[(thio)ureamethyl]cavitands **48** was prepared by reaction of tetrakis(aminomethyl)cavitand with the appropriate iso(thio)cyanates. The solubility of the (thio)ureacavitands was enhanced for use in supported liquid membrane transport experiments by introducing long aliphatic chains that resemble the membrane solvent (*o*-nitrophenyl *n*-octyl ether; NPOE) to give cavitands **49**. In the membrane transport experiments, a selectivity for chloride over bromide was observed. These experiments represent the first example of carrier-mediated transport through supported liquid membranes with neutral carriers that complex halide anions exclusively by hydrogen bonds. From NMR titration experiments, it was seen that the cavitands **48** and **49** show a small preference for chloride over the other halides with the highest binding constant (K = 4.7 x 10^5 M^{-1} in $CDCl_3$) for the *p*-fluorophenylthiourea cavitand. In general, the association constants for the complexation of halides are about 10^2 times higher than for the corresponding tetrakis(thio)ureacalix[4]arenes [75].

A novel method for the determination of the association constants of halide complexation was developed using infrared spectroscopy. The non-associated NH moieties give a complex band profile with a maximum at 3408 cm^{-1}. Upon complexation of (thio)ureacavitands **48** (Z = Ph, *p*-F-Ph, X = S) and **49** (X = O) with a halide, a second broad band emerges at a lower frequency. Both the position and bandwidth of this band indicate complexation of the halide by hydrogen bonding [76]. The association constants determined with infrared and ^{1}H NMR spectroscopy are in excellent agreement. Near-infrared surface-enhanced Raman spectroscopy and infrared reflection-absorption spectroscopy have been used to investigate adsorption and ordering processes of cavitands **48** (Z = *n*-Pr, X = O and Z = Ph, X = S) at interfaces of colloidal gold and planar gold substrates, respectively [77]. Making use of the affinity of the (thio)ureacavitands **48** (Z = *n*-Pr, X = O and Z = Ph , *p*-F-Ph, X = S) for chloride anions, their association when dissolved in acetone with a chloride-containing monolayer on gold has been investigated. The layers formed are stable in ethyl acetate and acetone, and can be desorbed by hydrochloric acid in ethanol (pH = 1). The adsorption of the thioureacavitands is partially irreversible due to adsorption of the sulfur-containing compounds onto the gold through defects in the self-assembled monolayer. The process of adsorption was readily followed using surface plasmon resonance [78].

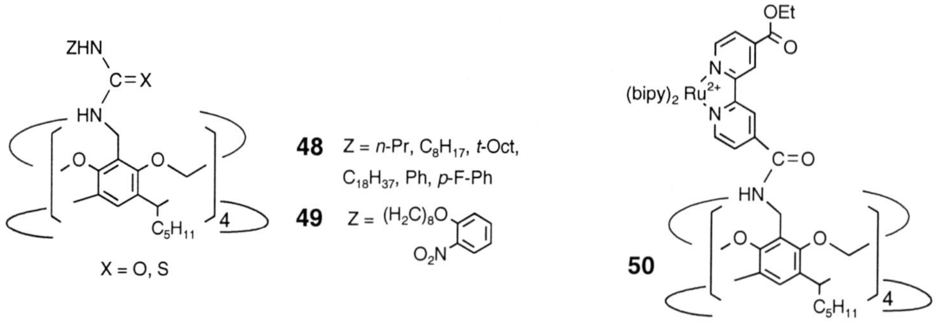

48 Z = n-Pr, C$_8$H$_{17}$, t-Oct, C$_{18}$H$_{37}$, Ph, p-F-Ph

49 Z = (H$_2$C)$_8$O—(o-NO$_2$-C$_6$H$_4$)

X = O, S

The tetrakis(ruthenium(II) bipyridyl)cavitand **50** was prepared from tetrakis(aminomethyl)cavitand. From UV-visible absorption and fluorescence emission spectroscopy it was found that this compound binds chloride, acetate, and benzoate anions with a preference for carboxylate anions (for benzoate log K is 4.4 in acetonitrile at 296 K) [79].

10. Cavitands as Molecular Building Blocks for the Construction of Larger Assemblies

Cavitands have also been used as molecular building blocks for the modular construction of larger receptor molecules. This modular approach [80] is based on the appropriate combination of (different) molecular building blocks to which functional groups can be attached. Coupling of these building blocks via covalent and non-covalent bonds gives rise to molecules with large well-defined cavities and hydrophobic surfaces. The coupling of two cavitands to each other is described in Chapter 10. This section will only deal with a few examples of covalent assembly.

Cavitands have been combined with calix[4]arenes in different ways, in a 1:1 fashion giving calix[4]arene-based carceplexes **51**, in a 1:2 (**53**) and a 2:1 fashion (**52**), and finally in a 2:2 combination to give **54**. The most straightforward method for the synthesis of the carceplexes **51** comprises the closure of the final two bridges in the presence of a suitable guest as a template. Guests such as e.g. DMA and NMP can adopt two different orientations in the cavity of the calix[4]arene-based carceplexes **51** [47]. The 2:2 compound **54**, the so-called holand, has a shielded cavity of nanosize dimensions (1000 Å^3). Although a computer simulation program revealed a good fit for different steroids, aromatic compounds, and sugar derivatives amongst others, no complexation of these molecules in CDCl$_3$ could be detected. Molecular modeling showed the presence of four solvent molecules (chloroform or tetrahydrofuran) in the cavity. The interaction energy of the four solvent molecules has to be overcome by the binding energy of a guest molecule. Furthermore, the rigidity of holand **54** might prevent the structural deformations required for complexation [81].

U-shaped receptor molecules **53** and **52** have a cavity similar to that of holand **54**, but are more flexible. This allows them to accommodate the structural deformations probably necessary for complexation. They can be considered as a holand minus a cavitand or a calix[4]arene module, respectively. Receptor molecules **53** were obtained by reaction of 1,2-bis(chloroacetamido)calix[4]arenes with tetrahydroxycavitand **2** (R = C$_{11}$H$_{23}$, A = OH). Polar substituents R on the remaining p-positions of the calix[4]arene such as nitro, acetamido, and phthalimido groups, favor the exclusive *endo* orientation (with the two cavities oriented to each other) of the first coupled calix[4]arene. The 2:1 receptor molecules **53** selectively complex certain corticosteroids (having an acetate group at C^{21} and hydroxy groups at C^{11} and C^{17}) with association constants of $K = (0.9-9.5) \times 10^2$ M^{-1} in CDCl$_3$ [82]. By combining 1:1 calix-resorc[4]arenes, connected with two spacers, with a second cavitand, 1:2 coupled products **52** were obtained. The stereochemistry of compounds **52** is perhaps influenced by intramolecular interactions leading exclusively to *endo-endo* isomers. The 1:2 calix[4]arene-cavitands **52** selectively complex certain corticosteroids, acylated five-membered sugars and quinine derivatives with association constants of $(1.0-6.0) \times 10^2$ M^{-1} in CDCl$_3$ [83]. In the compounds **51**- **54** the modules are connected via NHC(O)CH$_2$O spacers. Another 2:2 compound was prepared having NHC(S)NHCH$_2$ spacers. However, also in this case no complexation was observed [84]. For more details on calixarene-cavitand combinations, see Chapter 8.

Replacement of the C$_{11}$H$_{23}$ chains by di-decyl sulfide groups in the compounds **51** and **53** makes them very suitable for self-assembly in monolayers on gold [85,86]. For details, see Chapter 33.

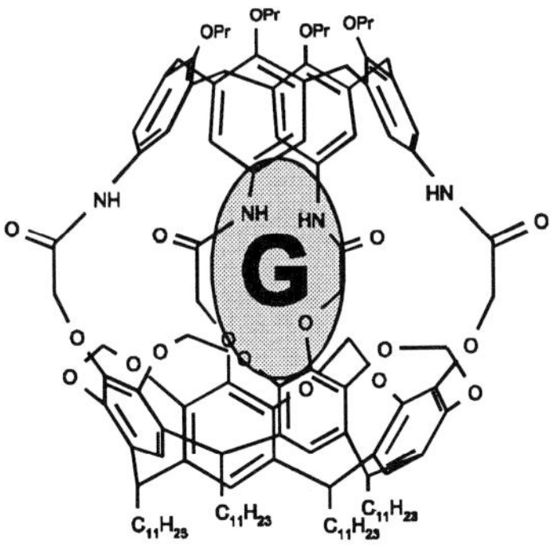

51

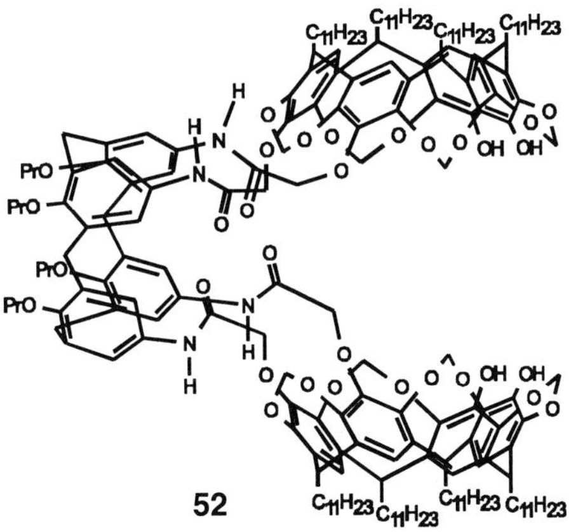

52

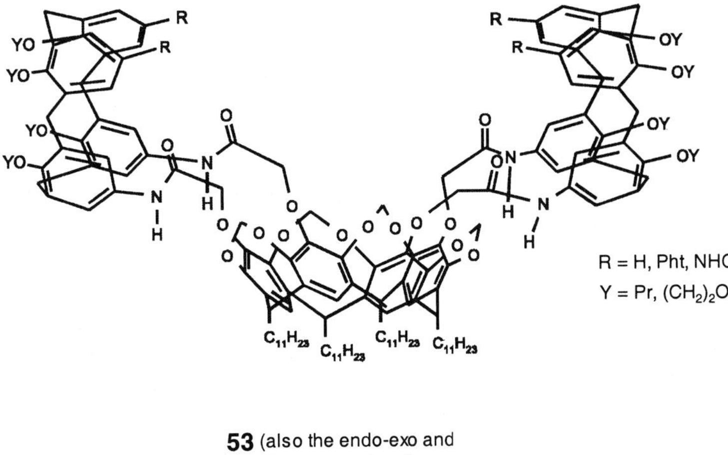

R = H, Pht, NHC(O)Me
Y = Pr, (CH$_2$)$_2$OEt

53 (also the endo-exo and the exo-exo form)

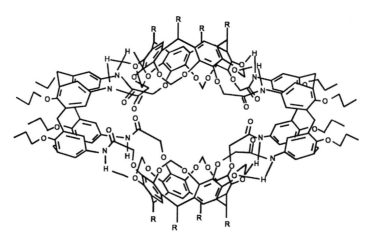

54 R = C$_{11}$H$_{23}$

11. References and Notes

[1]   J. R. Moran, S. Karbach, D. J. Cram, *J. Am. Chem. Soc.* **1982**, *104*, 5826-5828.
[2]   D. J. Cram, J. M. Cram, *Container Molecules and their Guests*, Monographs in Supramolecular Chemistry, vol. 4 (Ed.: J. F. Stoddart), The Royal Society of Chemistry, Cambridge, **1994**.
[3]   J. C. Sherman, *Tetrahedron* **1995**, *51*, 3395-3422.
[4]   A. Jasat, J. C. Sherman, *Chem. Rev.* **1999**, *99*, 931-967.
[5]   P. Timmerman, W. Verboom, D. N. Reinhoudt, *Tetrahedron* **1996**, *52*, 2663-2704.
[6]   For a few recent syntheses see: G. Rumboldt, V. Böhmer, B. Botta, E. F. Paulus, *J. Org. Chem.* **1998**, *63*, 9618-9619; B. Vuano, O. I. Pieroni, *Synthesis* **1999**, 72-73; A. G. M. Barrett, D. C. Braddock, J. P. Henschke, E. R. Walker, *J. Chem. Soc., Perkin Trans 1* **1999**, 873-878; H. Konishi, H. Sakakibara, K. Kobayashi, O. Morikawa, *J. Chem. Soc., Perkin Trans 1* **1999**, 2583-2584;
[7]   E. Román, C. Peinador, S. Mendoza, A. E. Kaifer, *J. Org. Chem.* **1999**, *64*, 2577-2578.
[8]   For early examples of cavitands derived from *rcct*- and *rctt*-resorcarenes see: L. Abis, E. Dalcanale, A. Du vosel, S. Spera, *J. Chem. Soc., Perkin Trans. 2* **1990**, 2075-2080.
[9]   O. Middel, W. Verboom, R. Hulst, H. Kooijman, A. L. Spek, D. N. Reinhoudt, *J. Org. Chem.* **1998**, *63*, 8259-8265.
[10] P. Timmerman, H. Boerrigter, W. Verboom, G. J. van Hummel, S. Harkema, D. N. Reinhoudt, *J. Incl. Phenom.* **1994**, *19*, 167-191.
[11] W. Xu, J. J. Vittal, R. J. Puddephatt, *J. Am. Chem. Soc.* **1995**, *117*, 8362-8371.
[12] W. Xu, J. P. Rourke, J. J. Vittal, R. J. Pudddephatt, *Inorg. Chem.* **1995**, *34*, 323-329.
[13] T. Lippmann, H. Wilde, E. Dalcanale, L. Mavilla, G. Mann, U. Heyer, S. Spera, *J. Org. Chem.* **1995**, *60*, 235-242.
[14] P. Jacopozzi, E. Dalcanale, S. Spera, L. A. J. Chrisstoffels, D. N. Reinhoudt, T. Lippmann, G. Mann, *J. Chem. Soc., Perkin Trans. 2* **1998**, 671-677.
[15] P. Delangle, J.-P. Dutasta, *Tetrahedron Lett.* **1995**, *36*, 9325-9328.
[16] V. I. Maslennikova, E. V. Panina, A. R. Bekker, L. K. Vasyanina, E. E. Nifantyev, *Phosphorus Sulfur Silicon* **1996**, *113*, 219-223.
[17] A. I. Konovalov, V. S. Reznik, M. A. Pudovik, E. K. Kazakova, A. R. Burilov, I. L. Nikolaeva, N. A. Makarova, G. R. Davletschina, L. V. Ermolaeva, R. D. Galimov, A. R. Mustafina, *Phosphorus Sulfur Silicon* **1997**, *123*, 277-292.
[18] A. Vollbrecht, I. Neda, H. Thönnessen, P. G. Jones, R. K. Harris, L. A. Crowe, R. Schmutzler, *Chem. Ber./Recueil* **1997**, *130*, 1715-1720.
[19] P. Jacopozzi, E. Dalcanale, *Angew. Chem., Int. Ed. Engl.* **1997**, *36*, 613-615.
[20] H. Boerrigter, W. Verboom, D. N. Reinhoudt, *J. Org. Chem.* **1997**, *62*, 7148-7155.
[21] T. N. Sorrell, F. C. Pigge, *J. Org. Chem.* **1993**, *58*, 784-785.

[22]   J. Lee, K. Choi, K. Paek, *Tetrahedron Lett.* **1997**, *38*, 8203-8206.
[23]   A. R. Mezo, J. C. Sherman, *J. Am. Chem. Soc.* **1999**, *121*, 8983-8994.
[24]   H. Boerrigter, R. J. W. Lugtenburg, R. J. M. Egberink, W. Verboom, D. N. Reinhoudt, A. L. Spek, *Gazz. Chim. Ital.* **1997**, *127*, 709-716.
[25]   J. A. Bryant, M. T. Blanda, M. Vincenti, D. J. Cram, *J. Am. Chem. Soc.* **1991**, *113*, 2167-2172.
[26]   B. C. Gibb, A. R. Mezo, J. C. Sherman, *Tetrahedron Lett.* **1995**, *36*, 7587-7590.
[27]   C. Peinador, E. Román, K. Abboud, A. E. Kaifer, *Chem. Commun.* **1999**, 1887-1888.
[28]   F. Hamada, S. Ito, M. Narita, N. Nashirozawa, *Tetrahedron Lett.* **1999**, *40*, 1527-1530.
[29]   H. Boerrigter, W. Verboom, G. J. van Hummel, S. Harkema, D. N. Reinhoudt, *Tetrahedron Lett.* **1996**, *37*, 5167-5170.
[30]   H. Boerrigter, W. Verboom, D. N. Reinhoudt, *Liebigs Ann./Recueil* **1997**, 2247-2254.
[31]   I. Higler, H. Boerrigter, W. Verboom, H. Kooijman, A. L. Spek, D. N. Reinhoudt, *Eur. J. Org. Chem.* **1998**, 1597-1607.
[32]   T. A. Robbins, D. J. Cram, *J. Chem. Soc., Chem. Commun.* **1995**, 1515-1516.
[33]   C. Ihm, M. Kim, H. Ihm, K. Paek, *J. Chem. Soc., Perkin Trans. 2* **1999**, 1569-1575.
[34]   K. Paek, C. Ihm, H. Ihm, *Tetrahedron Lett.* **1999**, *40*, 4697-4700.
[35]   P. T. Lewis, C. J. Davis, M. C. Saraiva, W. D. Treleaven, T. D. McCarley, R. M. Strongin, *J. Org. Chem.* **1997**, *62*, 6110-6111.
[36]   S. Saito, D. M. Rudkevich, J. Rebek, Jr., *Org. Lett.* **1999**, *1*, 1241-1244.
[37]   D. M. Rudkevich, J. Rebek, Jr., *Eur. J. Org. Chem.* **1999**, 1991-2005.
[38]   C. von dem Bussche-Hünnefeld, D. Bühring, C. B. Knobler, D. J. Cram, *J. Chem. Soc., Chem. Commun.* **1995**, 1085-1087.
[39]   S. Ma, D. M. Rudkevich, J. Rebek, Jr., *J. Am. Chem. Soc.* **1998**, *120*, 4977-4981.
[40]   P. T. Lewis, R. M. Strongin, *J. Org. Chem.* **1998**, *63*, 6065-6067.
[41]   H. Xi, C. L. D. Gibb, E. D. Stevens, B. C. Gibb, *Chem. Commun.* **1998**, 1743-1744.
[42]   H. Xi, C. L. D. Gibb, B. C. Gibb, *J. Org. Chem.* **1999**, *64*, 9286-9288.
[43]   P. Soncini, S. Bonsignore, E. Dalcanale, F. Ugozzoli, *J. Org. Chem.* **1992**, *57*, 4608-4612.
[44]   F. C. Tucci, D. M. Rudkevich, J. Rebek, Jr., *J. Org. Chem.* **1999**, *64*, 4555-4559.
[45]   F. C. Tucci, D. M. Rudkevich, J. Rebek, Jr. *Chem. Eur. J.* accepted for publication.
[46]   D. M. Rudkevich, G. Hilmersson, J. Rebek, Jr., *J. Am. Chem. Soc.* **1998**, *120*, 12216-12225.
[47]   A. M. A. van Wageningen, P. Timmerman, J. P. M. van Duynhoven, W. Verboom, F. C. J. M. van Veggel, D. N. Reinhoudt, *Chem. Eur. J.* **1997**, *3*, 639-654.
[48]   S. Ma, D. M. Rudkevich, J. Rebek, Jr., *Angew. Chem., Int. Ed. Engl.* **1999**, *38*, 2600-2602.
[49]   A. R. Renslo, D. M. Rudkevich, J. Rebek, Jr., *J. Am. Chem. Soc.* **1999**, *121*, 7459-7460.
[50]   S. K. Körner, F. C. Tucci, D. M. Rudkevich, T. Heinz, J. Rebek, Jr. *Chem. Eur. J.* **2000**, *6*, 187-195.
[51]   J. R. Fraser, B. Borecka, J. Trotter, J. C. Sherman, *J. Org. Chem.* **1995**, *60*, 1207-1213.
[52]   M. H. B. Grote Gansey, F. K. G. Bakker, M. C. Feiters, H. P. M. Geurts, W. Verboom, D. N. Reinhoudt, *Tetrahedron Lett.* **1998**, *39*, 5447-5450.
[53]   D.-R. Ahn, T. W. Kim, J.-I. Hong, *Tetrahedron Lett.* **1999**, *40*, 6045-6048.
[54]   B. C. Gibb, R. G. Chapman, J. C. Sherman, *J. Org. Chem.* **1996**, *61*, 1505-1509.
[55]   A. R. Mezo, J. C. Sherman, *J. Org. Chem.* **1998**, *63*, 6824-6829.
[56]   S. Pellet-Rostaing, L. Nicod, F. Chitry, M. Lemaire, *Tetrahedron Lett.* **1999**, *40*, 8793-8796.
[57]   M. H. B. Grote Gansey, F. J. Steemers, W. Verboom, D. N. Reinhoudt, *Synthesis* **1997**, 643-648.
[58]   T. Haino, D. M. Rudkevich, J. Rebek, Jr., *J. Am. Chem. Soc.* **1999**, *121*, 11253-11254.
[59]   M. Vincenti, E. Dalcanale, *J. Chem. Soc., Perkin Trans. 2* **1995**, 1069-1076.
[60]   M. Vincenti, C. Minero, E. Pelizzetti, A. Secchi, E. Dalcanale, *Pure Appl. Chem.* **1995**, *67*, 1075-1084.
[61]   J. Hartmann, P. Hauptmann, S. Levi, E. Dalcanale, *Sens. Actuators B* **1996**, *35-36*, 154-157. J. Hartmann, J. Auge, R. Lucklum, S. Rösler, P. Hauptmann, B. Adler, E. Dalcanale, *Sens. Actuators B* **1996**, *34*, 305-311.
[62]   E. Dalcanale, P. Jacopozzi, F. Ugozzoli, G. Mann, *Supramol. Chem.* **1998**, *9*, 305-316.
[63]   R. Pinalli, F. F. Nachtigall, F. Ugozzoli, E. Dalcanale, *Angew. Chem., Int. Ed. Engl.* **1999**, *38*, 2377-2380.
[64]   J. W. Grate, S. J. Patrash, M. H. Abraham, C. M. Du, *Anal. Chem.* **1996**, *68*, 913-917.
[65]   F. L. Dickert, U. P. A. Bäumler, H. Stathopulos, *Anal. Chem.* **1997**, *69*, 1000-1005.
[66]   O. Hayashida, K. Nishiyama, Y. Matsuda, Y. Aoyama, *Tetrahedron Lett.* **1999**, *40*, 3407-3410.
[67]   a) F. Arnaud-Neu, V. Böhmer, J.-F. Dozol, C. Grüttner, R. A. Jakobi, D. Kraft, O. Mauprivez, H. Rouquette, M.-J. Schwing-Weill, N. Simon, W. Vogt, *J. Chem. Soc., Perkin Trans. 2* **1996**, 1175-1182; b) for calix[4]arenes bearing CMPO-functions on the narrow rim see: S. Barboso, A. Garcia Carrera, S. E. Matthews, F. Arnaud-Neu, V. Böhmer, J.-F. Dozol, H. Rouqette, M.-J. Schwing-Weill, *J. Chem. Soc., Perkin Trans. 2* **1999**, 719-723.

[68] H. Boerrigter, W. Verboom, F. de Jong, D. N. Reinhoudt, *Radiochim. Acta* **1998**, *81*, 39-45.
[69] H. Boerrigter, T. Tomasberger, A. S. Booij, W. Verboom, D. N. Reinhoudt, F. de Jong, *J. Membr. Sci.* **2000**, *165*, 273-291.
[70] H. Boerrigter, T. Tomasberger, W. Verboom, D. N. Reinhoudt, *Eur. J. Org. Chem.* **1999**, 665-674.
[71] G. Modolo, R. Odoj, *J. Radioan. Nucl. Chem.* **1998**, *228*, 83-88.
[72] R. K. Castellano, B. H. Kim, J. Rebek, Jr., *J. Am. Chem. Soc.* **1997**, *119*, 12671-12672.
[73] O. Mogck, M. Pons, V. Böhmer, W. Vogt, *J. Am. Chem. Soc.* **1997**, *119*, 5706-5712.
[74] J. Scheerder, M. Fochi, J. F. J. Engbersen, D. N. Reinhoudt, *J. Org. Chem.* **1994**, *59*, 7815-7820.
[75] H. Boerrigter, L. Grave, J. W. M. Nissink, L. A. J. Chrisstoffels, J. H. van der Maas, W. Verboom, F. de Jong, D. N. Reinhoudt, *J. Org. Chem.* **1998**, *63*, 4174-4180.
[76] J. W. M. Nissink, H. Boerrigter, W. Verboom, D. N. Reinhoudt, J. H. van der Maas, *J. Chem. Soc., Perkin Trans. 2* **1998**, 1671-1675, 2541-2546, and 2623-2630.
[77] J. W. M. Nissink, J. H. van der Maas, *Appl. Spectrosc.* **1999**, 53, 528-539.
[78] J. W. M. Nissink, Th. Wink, J. H. van der Maas, *J. Mol. Struct.* **1999**, *479*, 65-73.
[79] I. Dumazet, P. D. Beer, *Tetrahedron Lett.* **1999**, *40*, 785-788.
[80] I. Higler, P. Timmerman, W. Verboom, D. N. Reinhoudt, *Eur. J. Org. Chem.* **1998**, 2689-2702.
[81] P. Timmerman, K. G. A. Nierop, E. A. Brinks, W. Verboom, F. C. J. M. van Veggel, W. P. van Hoorn, D. N. Reinhoudt, *Chem. Eur. J.* **1995**, *1*, 132-143.
[82] I. Higler, P. Timmerman, W. Verboom, D. N. Reinhoudt, *J. Org. Chem.* **1996**, *61*, 5920-5931.
[83] I. Higler, W. Verboom, F. C. J. M. van Veggel, F. de Jong, D. N. Reinhoudt, *Liebigs Ann./Recueil* **1997**, 1577-1586.
[84] A. M. A. van Wageningen, E. Snip, W. Verboom, D. N. Reinhoudt, H. Boerrigter, *Liebigs Ann./Recueil* **1997**, 2235-2245.
[85] B.-H. Huisman, D. M. Rudkevich, A. Farrán, W. Verboom, F. C. J. M. van Veggel, D. N. Reinhoudt, *Eur. J. Org. Chem.* **2000**, 269-274.
[86] A. Friggeri, F. C. J. M. van Veggel, D. N. Reinhoudt, *Chem. Eur. J.* **1999**, *5*, 3595-3602.

Chapter 10

CARCERANDS

CHRISTOPH NAUMANN, JOHN C. SHERMAN
Department of Chemistry, 2036 Main Mall
University of British Columbia
Vancouver, V6T 1Z1, B.C. Canada. e-mail: sherman@chem.ubc.ca

1. Introduction

Carcerands are globular, closed-surface molecules with an empty internal cavity, the first possible structure (Figure 1) being proposed in 1983 [1,2]. Carceplexes are carcerands that contain entrapped molecules (Figure 2). The guest molecules can only escape by breaking covalent bonds of their host. Hemicarcerands are similar to carcerands, differing only by the presence of portals in the shell that are large enough to allow guest exchange. Hemicarcerands with an entrapped guest are called hemicarceplexes and hemicarceplexes are typically kinetically stable in solution at room temperature without loss of guest; thus, they can be isolated and characterized spectroscopically [1]. The distinction between a carceplex and a hemicarceplex depends on portal size, guest size and shape, solvent, temperature, and even the number of guests. Since the same non-covalent interactions that drive the formation of carceplexes are also at play in natural recognition and self-assembly processes, analysis of host-guest interactions involved in less complicated systems such as carceplexes can enhance our understanding of more complex natural phenomena.

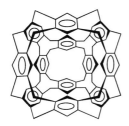

Figure 1. First proposed carcerand.

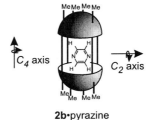

2b•pyrazine

Figure 2. Schematic representation of carceplex 2b•guest.

2. Carceplexes

The first fully characterized soluble carceplex was synthesized over ten years ago. It consisted of two bowl-shaped cavitands that were covalently linked (Scheme 1) [3]. By linking two rigid molecules containing enforced cavities, a large closed-off space was created inside the new host molecule that was occupied by guest molecule(s). Due to the shielding of the aromatic rings of the two bowl molecules, the ^{1}H NMR spectra of carceplexes **2•**guests show chemical shifts for the signals of the entrapped guest molecules of up to 4.5 ppm upfield from their normal position free in solution.

2.1. ACETAL-BRIDGED CARCEPLEXES

Tetrol **1**, a cavitand functionalized with four phenolic groups, proved to be one of the most useful precursors for the synthesis of carceplexes. Acetal-bridged carceplexes **2•**guests were prepared by reacting two tetrols and four molecules of bromochloromethane under basic conditions at high dilution in dipolar, aprotic solvents (Scheme 1) [3].

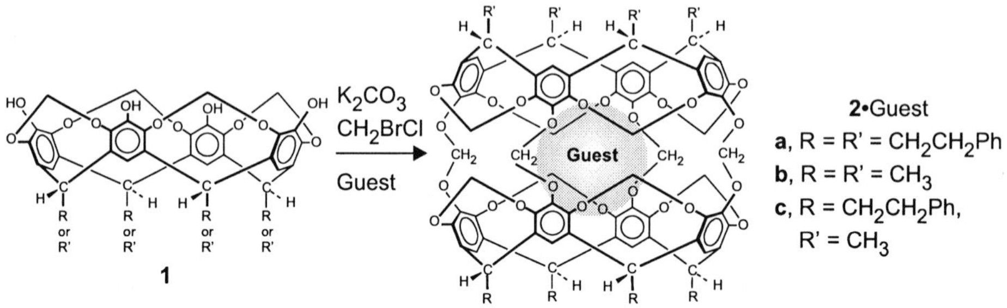

Scheme 1

2.1.1. Synthesis, template ratios, and mechanism of formation

Carceplexes **2a•**guest have been synthesized in yields of 87%, which is very high for the formation of eight new bonds, and joining of seven molecules (including the guest) [4,5]. Initial guests were solvents such as dimethylacetamide (DMA), dimethylformamide (DMF), and dimethylsulfoxide (DMSO). For N-formylpipyridine (NFP) as solvent, no carceplex or carcerand was isolated. Molecular models indicate that NFP is too bulky a molecule to fit in the interior of carcerand **2**. However carceplex **2a•**DMA was formed in NFP containing 5 % DMA [3b]. It was concluded that the synthesis of carceplex **2a•**guest needs a template such as DMA that can fit inside the formed cavity. Lack of a template prevents the formation of carceplex **2a•**guest in neat NFP. When two suitable templates compete against each other, guest selectivity by the forming host is observed, such that one guest is preferentially incarcerated over the other [4]. The range of this selectivity was determined under standardized conditions for **2•**guest formation [4,5]. The reactions were run in a solvent that is a poor guest itself in the presence of a known amount of two guests (templates 1 and 2). Carceplex formation is an irreversible reaction. The product ratio of the two formed carceplexes was determined by integration of the guest signals in the ^{1}H NMR spectrum. The product ratio represents the relative rates of the guest determining step (GDS) in the formation of the two carceplexes. The GDS was found to be the formation of the second acetal bridge after which guest exchange is no longer possible [4,5]. The product ratio of carceplex **2•**template 1/carceplex **2•**template 2 yields the relative templating ability of template 1 over template 2, and has been named the *template ratio*.

Note that the GDS is not the rate determining step (RDS). The RDS is the formation of the fourth bridge. The template that yields the most energetically favored transition state during formation of the second bridge is entrapped preferentially by the forming host and subsequently yields a higher product ratio. The best guest (pyrazine) was found to be a million fold better guest than the poorest guest, N-methyl-pyrrolidinone (NMP) (see Table 1) [4,5]. Small differences in size and shape lead to huge differences in templating ability. Pyrazine was found to be 30 times better than pyridine and 420 times better than benzene; the template ratio of 1,4-dioxane is 1,400 times bigger than that of 1,3-dioxane [4].

Why is pyrazine such a good template? The crystal structure of pyrazine carceplex **2a** shows that the two cavitand bowls are parallel and twisted (21°) with respect to one another [6]. The guest pyrazine orients itself inside the cavity with hydrogen atoms pointing into the two cavitand bowls and the nitrogen atoms aligned to the equator of the carceplex (Figure 2). This orientation of the guest in addition to the observed parallel and twisted geometries of the two bowls maximizes the noncovalent interactions between host and guest as well as between the parts of the host itself.

2.1.2. Reversible charged hydrogen bonded (CHB) capsules

In the GDS for forming **2**•guest, the templating guest organizes the monolinked tetrols around itself. In fact, two tetrols **1** self-assemble when a base such as 1,8-diazabicyclo[5.4.0]undecene-7 (DBU) is added in the presence of a suitable template. Addition of four equivalents of DBU to two equivalents of tetrol **1** in CDCl$_3$ yields a significantly changed ^{1}H NMR spectrum in CDCl$_3$ that resembles the one found for carceplex **2**•CHCl$_3$ but contains no guest signal [7]. Deuterium NMR showed the emergence of a peak corresponding to incarcerated CDCl$_3$ within the complex. Chloroform is encapsulated between two tetrol cavitands held together by four charged hydrogen bonds (CHB) to yield reversible capsule **3**•CDCl$_3$ (Scheme 2) [7, 8, 9]. Addition of pyrazine caused only small shifts of bowl-proton signals but the singlet for pyrazine protons appeared at δ 4.3 rather than at the δ 8.6 of the "free" molecule.

Guest exchange had taken place, such that no **3**•CDCl$_3$ remained. Thus capsule **3**•pyrazine has a higher relative stability than capsule **3**•CDCl$_3$. The relative thermodynamic stabilities (K_{rel}) of capsule **3b**•guest were measured for a number of templates of the carceplex **2**•guest reaction, and it was found that the guest selectivities for the complexes **3**•guest match the template ratios for carceplex **2**•guest (Table 1) [7].

A series of capsules where cavitands were linked by 0-3 CHBs and/or 0-2 OCH$_2$O linkers gave similar guest selectivity as capsule **3**•guest. A crystal structure of complex **3b**•pyrazine was solved and closely resembles the one for the analogous carceplex **2a**•pyrazine [9].

TABLE 1. Selected template ratios for carceplex **2a**•guest and K$_{rel}$ values for complex **3**•guest.

Guest	Template Ratio for 2a•guest	K_{rel} of complex 3•guest in nitrobenzene-d$_5$
pyrazine	1,000,000	980,000
methyl acetate	470,000	420,000
1,4-dioxane	290,000	240,000
DMSO	70,000	58,000
pyridine	34,000	7,100
acetone	6,700	1,300
benzene	2,400	540
1,3-dioxane	200	140
DMA	20	9
NMP	1	1

The same non-covalent interactions that drive formation of reversible complex **3**•guest also dictate the kinetic template effect in forming carceplex **2**•guest. Therefore it is possible to consider the former as a transition state model for the GDS in formation of the latter.

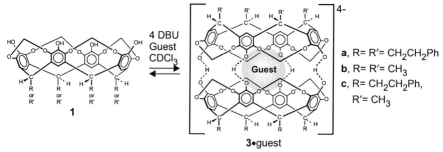

Scheme 2

2.2. CARCEPLEXES CONTAINING LARGER OR MULTIPLE CAVITIES

CPK modeling indicates that bigger cavities are possible by linking several tetrol molecules in a cyclic array in 1,3 (AC) positions. When AC-bis-benzylated diol **4** (obtained in about 15% yield from tetrol **1**) was reacted in DMF with bromochloromethane under dilute, basic conditions, several cyclic arrays were obtained after hydrogenolysis (Scheme 3) [10].

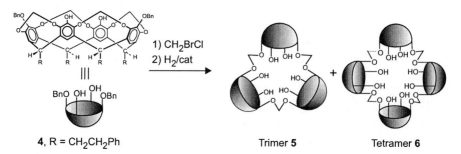

Scheme 3

2.2.1. Trimer carceplex

Molecules larger than benzene are too large to be incarcerated by **2** and carceplexes **2**•guest show only 1:1 stoichiometry. Carceplexes that can encapsulate larger guests bearing several functional groups are very interesting due to the increased potential for doing chemistry within the confined area. Larger cavities can also be filled with several guest molecules. An interesting outcome with multiple molecules as templates is whether one can distinguish a solvent effect from a template effect. Trimer **5** possesses a sizable cavity but one that contains large portals on the top and bottom. When **5** is reacted with a suitable cap, a large carceplex can be synthesized. Thus, capping of trimer **5** in DMF as solvent yields carceplex **7**•(DMF)$_3$ (Scheme 4) [11]. The ^{1}H NMR spectrum at 298 K shows only one set of signals for the encapsulated DMF molecules indicating fast movement on the NMR time scale. CPK models show that the three

DMF molecules should enjoy a great deal of freedom, more so than one molecule of DMF enjoys inside carcerand **2** [11], even though it is considered that in **2**•guest, the guests have a gas-phase-like environment [3]. High temperature ^{1}H NMR experiments in nitrobenzene-d_5 showed that the DMF molecules inside trimer carceplex **7**•(DMF)$_3$ have a lower barrier of rotation about the C-N bond than free DMF but a higher one than in **2**•DMF [11]. This indicated that the DMFs are either less gas-phase-like in **7**•(DMF)$_3$ than in **2**•DMF, or are experiencing a polar micro-solvent environment. Since CPK models indicate a looser fit in **7**•(DMF)$_3$ versus **2**•DMF, the latter explanation must hold.

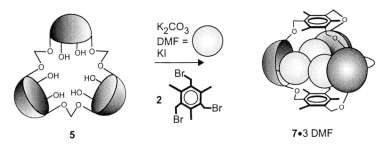

Scheme 4

2.2.2. Tetramer bis-carceplex and bis-capsules

Tetramer **6** is a much more flexible molecule than trimer **5**. Its potential cavity is large enough to accommodate a C$_{60}$ molecule (according to CPK models) but it can also form two much smaller cavities comparable in size to the one of carceplex **2**•guest. When tetramer **6** was subjected to the optimized carceplex **2**•pyrazine reaction conditions, bis-carceplex **8**•(pyrazine)$_2$ was isolated in 74% yield [10]. The ^{1}H NMR spectrum in CDCl$_3$ shows a signal for incarcerated pyrazine at δ 4.30 as compared to δ 4.07 in carceplex **2**•pyrazine. (Scheme 5).

When tetramer **6** was treated with DBU in CDCl$_3$ in presence of pyrazine, a complex similar to the tetrol CHB capsule **3**•pyrazine was observed by ^{1}H NMR spectroscopy [12]. However, in the case of the tetramer **6**, bis-capsule **9**•(pyrazine)$_2$ was formed (Scheme 6). The host had lost the C_4 axis of symmetry evident from the NMR spectrum of **6**. In addition, a signal for encapsulated pyrazine appeared at δ 4.42 in CDCl$_3$ and integration of the signals confirmed a 1:2 ratio of tetramer:pyrazine.

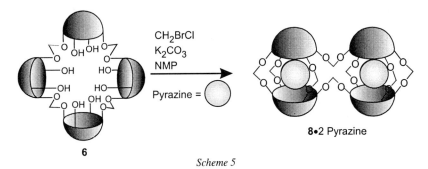

Scheme 5

Nitrobenzene is too big to be entrapped and tetramer **6** was not observed to form bis-capsule **9**•(guest)$_2$ in nitrobenzene unless a suitable guest was added; once again, an empty cavity is energetically disfavored [12]. CPK modeling reveals that the formation of one capsule by a template also forces the other two remaining bowls to clamp down on each other due to the short tether between the four cavitand bowls. No complex formed composed of one capsule filled by a guest, while the other is left empty. Mixed bis-capsules **9**•guest 1•guest 2 were explored [12]. For instance, in CDCl$_3$ at 298 K, in the presence of 12.5 M:32.2 mM:1.6 mM of CDCl$_3$:methyl acetate (MeOAc):tetramer **6**, a mixture of bis-capsules **9**•(CDCl$_3$)$_2$, **9**•CDCl$_3$•MeOAc, and **9**•(MeOAc)$_2$ were observed in a ratio of 1:4.7:8.6 at equilibrium. Addition of 100 equivalents of MeOAc to that mixture formed bis-capsule **9**•(MeOAc)$_2$ exclusively. This is a reversible process guided by the templating abilities of different guests, methyl acetate being a better template than CDCl$_3$.

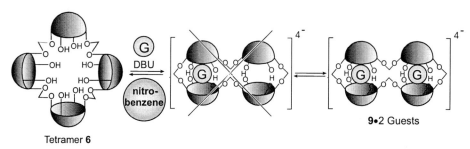

Scheme 6

Communication between the two capsules leads to chemical shift differences for encapsulated methyl acetate (also found for CDCl$_3$) when its neighboring encapsulated guest is changed. Poorer templating neighboring guests (CDCl$_3$) lead to upfield shifts for the encapsulated MeOAc, whereas better guests (pyrazine) cause a downfield shift for MeOAc. The overall measured range is δ 0.31 for MeOAc depending on the neighboring guest. The neighbor-dependent chemical shifts for the encapsulated guests are most likely a result of differential clamping down of opposing bowls as transmitted through the tether, where better guests clamp their bowls down tighter, tug harder on the other capsule, and cause it to unravel more [12].

2.3. BENZYLTHIA-BRIDGED CARCEPLEXES

The first benzylthia-bridged carceplex was found to be insoluble and could not be fully characterized [13]. Later work provided ways of imparting better solubility to these carceplexes. Reaction of tetrabenzylthiol cavitand **11** with tetrabenzylchloride cavitand **10** in solvents such as DMF, 2-butanone, 3-pentanone, ethanol/benzene, and methanol/benzene yielded carceplexes **12**•guest having one solvent molecule incarcerated, and two methanol molecules incarcerated when a 2:1 mixture of methanol/benzene was used as solvent (Scheme 7) [14]. When the reaction was run in neat benzene, no carceplex or carcerand was isolated. This reaction was found to be templated since a suitable guest had to be present at the formation of the carceplex **12**•guest for it to form [14]. In preliminary experiments, the template ratio of this reaction was found to be bigger than the one observed for the formation of the tetrol carceplex **2**, with about a two-million fold range between the best and worst template [15].

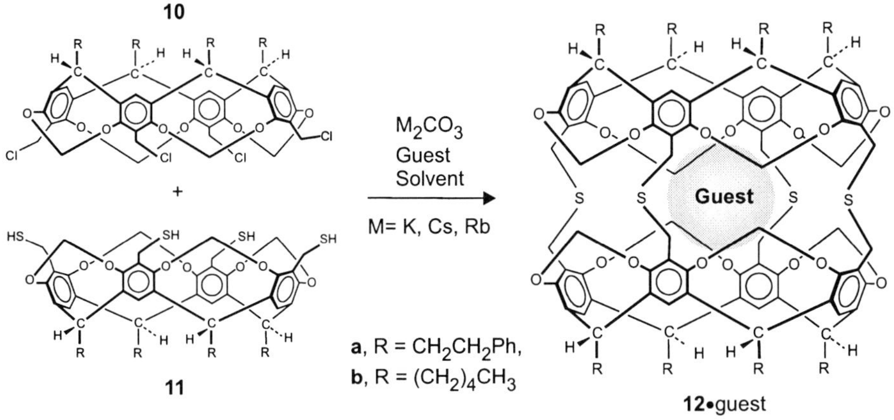

Scheme 7

2.4. LOWER SYMMETRY CARCEPLEXES

Lower symmetry carceplexes can be formed by linking two different bowl molecules or by introducing a guest that orients itself preferentially within the cavity. Thus new types of supramolecular stereoisomerism can be created.

2.4.1. Calix[4]arene-calixresorcinarene carceplexes

The route to lower symmetry carceplexes was opened by the intramolecular cyclization of an *endo*-1,2-coupled calix[4]arene-calix[4]resorcinarene **13** in DMA, DMF, and NMP to form carceplexes **14**•guest in high yield (Scheme 8) [16,17].

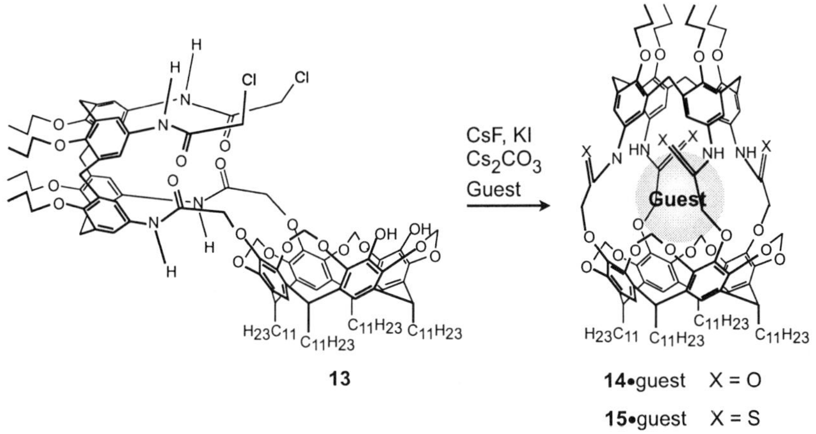

Scheme 8

The templating abilities of different guests were subsequently investigated using 1,5-dimethyl-2-pyrrolidinone as a solvent (itself a guest, but a poor template) and doping it with 5 vol. % of each competing guest. [17] The template effects found were 3.7: 1 for DMA: 2-butanone. Due to the lower symmetry of the hybrid carceplexes **14**•guest, and the restricted rotation of the incarcerated guest molecule around the pseudo C_2 axes of the carceplex, diastereomeric carceplexes **14**•guest were observed for the larger guests such as DMA and NMP where the isomers differ in how the guest orients itself inside the cavity. This stereoisomerism was named carceroisomerism [18]. The activation energy for the interconversion of the different stereoisomers was found to be between 12.7 and 17.5 kcal/mol, for DMA and NMP, respectively. This barrier was increased when the carbonyl oxygens were exchanged by sulfur atoms, presumably because the cavity size of the new carceplex **15**•guest decreases [16,17].

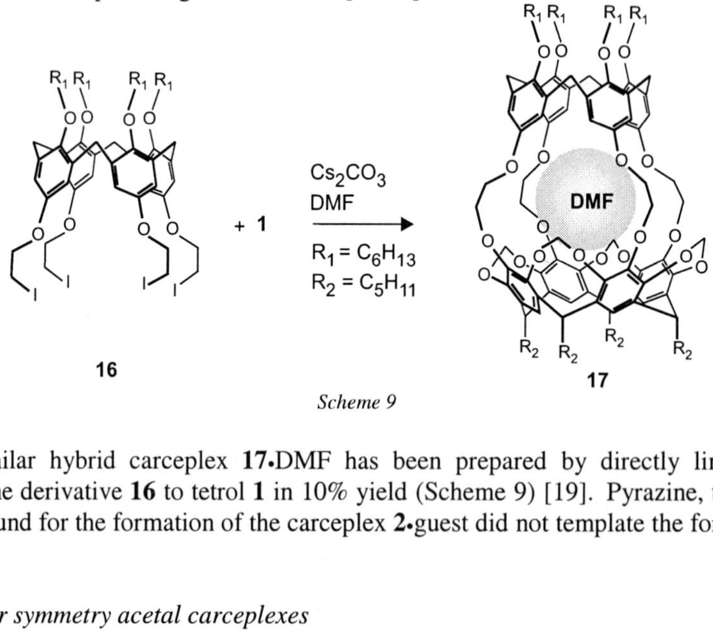

Scheme 9

A similar hybrid carceplex **17**•DMF has been prepared by directly linking a calix[4]arene derivative **16** to tetrol **1** in 10% yield (Scheme 9) [19]. Pyrazine, the best template found for the formation of the carceplex **2**•guest did not template the formation of **17**•guest.

2.4.2. Lower symmetry acetal carceplexes

Guest rotation in carceplexes **2**•guest is restricted at least in certain directions, and carceplex diastereoisomerism can also arise as a result of restricted internal rotations of the guest and of twisting of one half of the host with respect to the other. Crystal structure determinations of a carceplex **2**•guest and a complex **3**•guest show they are chiral due to interbowl twists of 13-21°, respectively [3a, 6, 9]. In solution, the helical conformational stereoisomers (twistomers) of carceplex **2**•guest containing a chiral guest or a guest with enantiotopic hydrogens can be observed on the NMR time scale at low temperature [20]. At 223 K two sets of host and guest signals are observed for carceplex **2b**•(R)-(-)-2-butanol, one for each diastereomer, left-handed twistomer•(R)-(-)-2-butanol and right-handed twistomer•(R)-(-)-2-butanol (Figure 3). The energy barrier for interconversion between these twistomers was determined to be 12.6 kcal/mol by variable temperature ^{1}H NMR spectroscopy [20].

In the case of carceplex **2b**•DMSO the energy barrier for the interconversion of the twistomers is 13.6 kcal/mol, and 12.7 kcal/mol for the interconversion of the carceroisomers due to the slowed rotation of DMSO about the C_2 axes of carceplex **2b**•guest (Figure 4) [3a, 20]. The energy barriers for the two processes are only slightly different (13.4 kcal/mol and 12.6 kcal/mol) for **3a**•DMSO, indicating that differences in the interbowl linkages (acetal methylenes vs. CHBs) have no significant effects [20]. With carceplex **2c**•DMSO, at 293 K two sets of signals could be observed for the bowl H_{in} protons and one guest signal for DMSO in the ^{1}H NMR spectrum. At 223 K four signals each for the DMSO and for H_{in} were observed, showing both the twistomer and the carceroisomer interconversion are frozen out on the NMR time scale at 223 K.

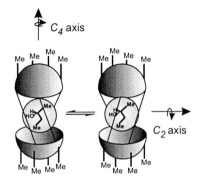

Figure 3. Schematic representation of the diastereomeric twistomers of **2b**•guest.

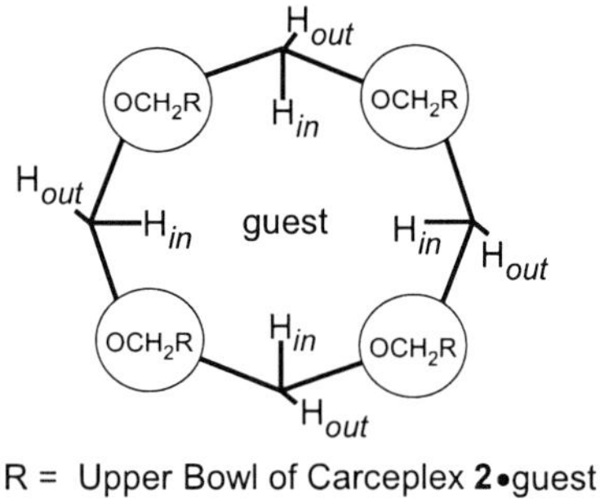

Figure 4. Schematic representation of the lower bowl of **2**•guest.

NMR measurements gave values for the free energy change for interconversion of twistomers in **2b**•1,4-thioxane. It is 16.3 and 16.5 kcal/mol based on the coalescence of the OCH$_2$ (T_c = 333 K) and SCH$_2$ (T_c = 323 K) protons, respectively, of the guest. 1,4-Thioxane raises the barrier for interconversion of the twistomers as compared to DMSO and 2-butanol. In addition, the value for ring-flipping of 1,4-thioxane in carceplex **2b**•guest was found to be 1.8 kcal/mol higher than the one measured in solution. Clearly, restrictions are put on the conformational mobility of guests due to the rigidity of the host molecule [21]. Similar restraints on conformational mobility of the incarcerated guest were measured for carceplex **2b**•1,4-dioxane and capsule **3**•1,4-dioxane (about 1.6 kcal/mol higher barrier for the interconversion of the encapsulated chair-isomers than in solution) [21]. Interestingly, the determination of the ring-flipping barrier of 1,4-dioxane free in solution requires a more complicated experiment than the one necessary for encapsulated 1,4-dioxane [21].

3. Hemicarceplexes

3.1. HEMICARCEPLEXES CONTAINING DISTINCT PORTALS

The reaction of two triol bowls **18** [3a] with bromochloromethane under standard carceplex conditions afforded hemicarceplexes **19**•guest containing a single hole in their outer shell. Incarcerated guests include DMSO (51% yield, much higher than expected considering the possible ways of misalignment), DMA, DMF, and pyrazine [22-24]. Two triols **18** form a CHB capsule **20**•guest after addition of DBU and in presence of a template [5, 25]. Unlike with tetrol **1,** there are a number of different ways that two triol molecules could align themselves. The most stable complex (**20**•guest) predominates since the process is under thermodynamic control [25]. The formation of capsule **20**•guest facilitates the synthesis of hemicarceplex **19**•DMSO allowing it to be obtained in higher than statistical yield. Because the only difference between the observed CHB capsule of the triol and the tetrol molecules is the number of CHB bonds formed (three versus four), guest selectivities are similar for capsules **20**•guest and **3**•guest as well as for hemicarceplex **19**•guest and carceplex **2**•guest [5, 24, 25].

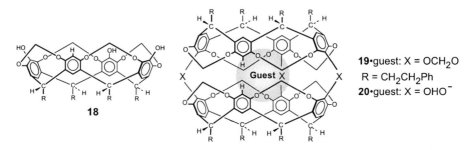

19•guest: X = OCH$_2$O
R = CH$_2$CH$_2$Ph
20•guest: X = OHO$^-$

3.2. HEMICARCEPLEXES CONTAINING FOUR SLOTTED PORTALS

Many different hemicarceplexes have been prepared by varying the size and shape of the four intrahemisphere linkages [26,1e]. Complexation properties of many hemicarcerands have been studied extensively and reviewed in detail recently [1e]. For instance, hemicarcerand **21** was shown to be able to incarcerate diphenyl ether, one of the largest guests observed [27]. On the ^{1}H NMR time scale at ambient temperature, diphenylether does not rotate within the host molecule [27]. Hemicarcerand **21** was synthesized in 50% yield by reacting four linker molecules (*m*-ClCH$_2$C$_6$H$_4$CH$_2$Cl) with two molecules of tetrol **1** (a 4:2 approach, eight bonds formed) in NMP-Cs$_2$CO$_3$. When the reaction was attempted by reacting one equivalent of tetrachloride cavitand **22** (made in 60% yield from **1**) with one equivalent of tetrol **1** (a 1:1 approach, four bonds formed), only 2.2% yield of hemicarcerand **21** was isolated [26]. In the first case, capsule **3**•NMP is preorganized by the template NMP before reacting with the bridging molecules to yield the desired product. After the first bridge is made, the resulting intermediate can still form a capsule, to some extent via CHBs. In the second case, NMP forms the same intermediate capsule **3**•NMP. Once one linkage is made between **3**•NMP and **22**, preorganization is lost (no CHBs are possible) and polymerization becomes favorable. In addition, NMP acts as a negative template because it keeps the concentration of free tetrol **1** low.

Tetrol **23** differs from **1** by its longer intra-bowl OCH$_2$CH$_2$O bridges and in the eight-bond-4:2-approach gave **24** in only 8% yield. However, synthesis of the tetrachloride **25** in 69% yield and reaction of one equivalent of **25** with one equivalent of tetrol **23** yielded 43% of hemicarcerand **24**. Features of the crystal structures of **1** and **23** explain the yield differences for hemicarcerands **21** and **24** by the two different approaches. The lengthening of the intrahemispheric links from OCH$_2$O in **1** to

OCH$_2$CH$_2$O in **23** yields a roughly rectangular cavity deviating slightly from C_4 symmetry in **23**. There is also steric hindrance in bringing two molecules of **23** together and **23** is poorly preorganized to form a CHB capsule analogues to **3**•guest (possessing effective C_4 symmetry). The concentration of free **23** thus remains high. On a statistical basis, a four bond making process is expected to give a higher yield than an eight bond making process, unless there is a template effect involved, which is not the case for **23** [26].

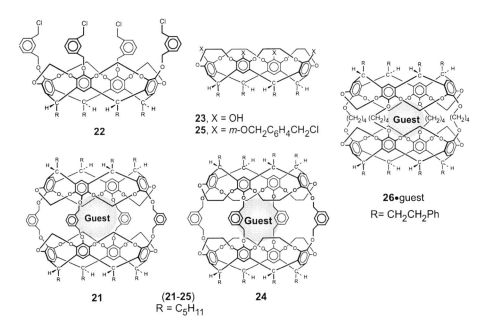

The tetramethylene bridged hemicarcerand **26** is one of the most interesting hemicarcerands because it forms the widest variety of hemicarceplexes of all the hemicarcerands studied by the Cram group [28]. A study of the template-formation of **26**•guest was published recently [29]. Two molecules of tetrol **1** were reacted with four molecules of 1,4-dibromobutane. Clearly, after addition of base and in presence of a suitable template, CHB capsule **3**•guest can form, its relative thermodynamic stability mirroring the kinetic template ratios for formation of carceplex **2**•guest and hemicarceplex **19**•guest. Capsule **3**•guest represents a good transition state model for the GDS in the formation of carceplex **2**•guest. In the case of the tetramethylene bridged hemicarceplex **26**•guest such a correlation was not found since many of the best templates for **26**•guest are too large for the cavity of capsule **3**•guest. The range of template ratios for 30 suitable guest molecules was found to be 3,600 between the best guest, *p*-xylene, and the poorest measurable one, NFP (Table 2) [29]. Most of the best templates are *p*-disubstituted benzenes. The para substituents were found to be situated deep inside the bottom and top cavity whereas the guest aromatic hydrogens are close to the equator. There is restricted rotation about the C_2 axis of symmetry of hemicarceplex **26**•guest on the ^{1}H NMR timescale leading in the case of asymmetric guests to two sets of signals for the host molecules [29]. It is likely that capsule 3•guest gets disrupted after formation of the first tetramethylene bridge. Free guest exchange is presumably possible until the formation of the third or fourth inter bowl bridge [29].

TABLE 2. Selected template ratios for hemicarceplex 26•guest and carceplex 2•guest.

Guest	Template Ratios for 26•guest	Template ratios for carceplex 2•guest
p-xylene	3,600	
4-bromotoluene	2,800	
p-dibromobenzene	2,100	
4-chlorotoluene	2,000	
anisole	1,900	
1-bromo-4-chlorobenzene	1,700	
p-dichlorobenzene	840	
1-bromo-4-iodobenzene	730	
1-chloro-4-iodobenzene	620	
2-butanol	200	2,800
benzene	110	2,400
DMA	26	20
DMSO	17	180,000

3.3. LOWER SYMMETRY HEMICARCEPLEXES

Hemicarceplexes having different spheres have been made by reacting different cavitands with each other. Once again, the importance of the ability of cavitand **1** to form complex **3**•guest when treated with base and a suitable template must be underscored. As mentioned above, when the intra-bowl link was extended from methylene in **1** to ethylene and propylene units, complex formation was not found to happen [26]. Thus, the reaction of tetrachloride **22** with tetrol **23** (1:1 approach) in NMP yielded 21% of hemicarcerand **27** while the corresponding switched reaction of tetrachloride **25** with tetrol **1** (1:1 approach) yielded only 2% of **27** [26]. Two tetrols **23** are sterically precluded from capsular self-assembly by NMP under basic conditions prior to the reaction with the tetrachloride, and thus **27** is formed in statistical yield. The low yield for the reaction of **1** with **25** can be explained again by the formation of capsule **3**•NMP leaving low concentrations of free tetrol **1** to react with tetrachloride **25**. These observations together with those reported for hemicarcerand **26** emphasize the importance and power of preorganization due to weak non-covalent interactions.

Hemicarcerand **27** was found to bind a number of guests and the hemicarceplex **27**•1,2,3-(MeO)$_3$C$_6$H$_3$ shows separate signals for the methoxyl groups in 1,3 positions due to the slow rate of rotation of the guest around the C_2 axis of the host at ambient temperature in CDCl$_3$. One signal is at δ –0.44 for the terminal methoxyl pointing into the hemisphere based on **23**, the other at δ 0.32 for the methoxyl group that points towards the **1**-based hemisphere [27].

Synthesis of other lower symmetry hemicarceplexes [30-32,1e] are based on the obser-

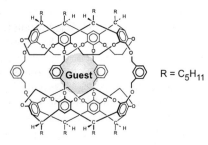

27•guest

vation that the introduction of the fourth bridge connecting the two cavitands is the slowest step [22] making the preparation of diol **28** (Scheme 10) fairly easy [30]. A different fourth bridge may be introduced, as in the reaction of **28** with (S)-(-)-2,2'-bis(bromomethyl)-1,1'-binaphthyl in Cs_2CO_3-Me_2NCOMe at 40 °C to give the chiral (S) hemicarceplex **29**•$CHCl_3$ in 79% yield (after chromatographic purification with $CHCl_3$) [32]. Chiral hemicarcerand **30** was isolated by reacting **28** with (S,S)-1,4-di-O-tosyl-2,3-O-isopropylidene-L-threitol following a similar protocol [32]. In both cases it is assumed that a hemicarceplex was formed during the reaction but that during the work-up period guest exchange or decomplexation occurred. Thus, strictly speaking, **30** is not a hemicarcerand when DMA is the guest.

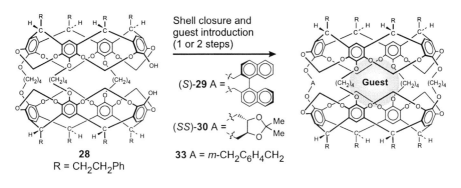

Scheme 10

Racemic guests were presented to **29**•$CHCl_3$ and **30** by heating **29**•$CHCl_3$ and **30** to 115-160 °C dissolved in neat racemic guests or in guests-Ph_2O mixtures. Another method was to carry out the shell closure procedure, the introduction of the fourth bridge, in the presence of a 100-fold excess of the racemic guest. This was done at a much lower temperature, 40 °C. Diastereomeric hemicarceplexes **29**•guest were obtained in the following ratios: 20(R):1(S) for the guest methyl p-tolyl sulfoxide, 1.6(R):1(S) for methyl phenyl sulfoxide, unity for 1,2-dichloropropane, and 2.5(S):1(R) for the guest 1-phenyl ethanol [32]. Lower chiral recognition factors were observed for hemicarceplexes **30**•guest. The highest diastereomeric ratio, 1.4(R):1(S), was observed for 2-butanol as a guest. The pronounced difference in chiral recognition between **29**•guest and **30**•guest, two compounds differing only in their fourth linker bridge, is due to the difference in rigidity of the chiral linkers. Only the bismethylenebinaphthyl bridge can adapt its shape in response to the configuration of the guest by changing its naphthyl to naphthyl dihedral angle. This capability leads to a lower energy for diastereomers that have a higher number of interactions to the host. Only the (R,S)-hemicarceplex **29**•methyl p-tolyl sulfoxide (R is from the guest; S is from the host) was isolated from a racemic mixture of the guest at 398 K and only starting material when pure (S)-methyl p-tolyl sulfoxide was used. The free energy difference between the two diastereomeric hemicarceplexes **29**•methyl p-tolyl sulfoxide was calculated to be >2.4 kcal/mol at 398 K. When the lower temperature 'sealing-in' method was used, lower chiral recognition factors were observed. Changing the dihedral angle of the chiral bridge of **29**•guest can change the overall shape of the outer shell of the diastereomeric hemicarceplexes **29**•guest. This was observed for the diastereomers (*R,S*) and (*S,S*)-**29**•methyl phenyl sulfoxide, which have different R_f values on thin layer silica gel plates, probably reflecting differences in the shape of their outer shell caused by the host's ability to accommodate its guest's spacial needs. Guest chemical shift differences varied for the two diastereomeric hemicarceplexes **29**•guest from δ 0.01 to δ 0.36 depending on the guest [32].

Four C_{4v} tetraoxatetrathiahemicarceplexes **32**•guest (Scheme 11) have been obtained by reacting benzyl thiol cavitand **11** with a phenethyl footed tetraiodo tetrol derivative **31** under high dilution and basic conditions in various solvents [33]. Hemicarceplex **32**•DMA with oxa- and thiahemispheres was obtained in 11% yield. Its ^{1}H NMR spectrum in CDCl$_3$ at room temperature showed three singlets, at δ 1.69 (cis N-Me to the acetyl group), δ -0.31 (anti N-Me), and δ -1.33 (acetyl). The spectrum was the same in CD$_2$Cl$_2$ at room temperature but at -90 °C the singlets for anti N-Me and acetyl had split into major and minor signals in a 3:1 ratio. Similar results were obtained for hemicarceplex **32**•NMP except that the major:minor ratio was 2:1 at -21 °C. The N-Me resonances remained two unequal singlets even at 141 °C [33].

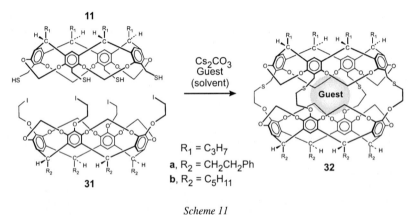

Scheme 11

3.4. DECOMPLEXATION OF HEMICARCEPLEXES

The size and rigidity of a guest especially in correlation to the size and shape of the portal(s) primarily determines the rate of decomplexation. The term *constrictive binding* has been used to describe the inhibition of decomplexation of guests that have larger cross-sectional areas than that of the host's portals [22]. Small guests such as H$_2$O exchange rapidly at room temperature on the NMR time scale. In general, at higher temperatures, as conformational mobility of the host molecule increases it becomes more feasible for guests to exit the cavity. *Gating* is a concept used to describe the loss of one acetonitrile molecule from hemicarceplex **12**•(CH$_3$CN)$_2$ by conformational changes of the host's portals at high temperatures (110°C) to yield carceplex **12**•CH$_3$CN [16a]. Interestingly, the designation of carceplex versus hemicarceplex depends not only on portal size and guest size/shape, but also on the number of encapsulated guests. It was calculated that intrabowl acetal groups of **12** can undergo a conformational flip from an inward to an outward position in effect enlarging the adjacent portal size to make the escape of one acetonitrile molecule possible. The calculated activation energy for *gating* is 22 kcal/mol [34]. For hemicarceplex **19**•guest, at 140 °C, the half-lives for decomplexation is 14 h for DMF, and 34 h for DMA as guest. DMSO needs even more energy to overcome the decomplexation barrier (at 195°C, $t_{1/2}$ is 24 h) [22]. For chiral hemicarceplexes **29**•guest, chiral decomplexation rate differences (due to the diastereomeric transition state differences for decomplexation) have been measured. 1,2-Dibromo propane had a diastereomeric rate factor of 2.8 ($t_{1/2}$ slow/ $t_{1/2}$ fast), whereas 1,2-dichloro propane had a diastereomeric rate difference of 1.6 in $CDCl_3$ at 25 °C [32]. In absolute terms, the decomplexation rate was found to be between 26 ($t_{1/2}$ fast) and 49 ($t_{1/2}$ slow) times faster for the smaller 1,2-dichloro propane than for 1,2-dibromo propane. The solvent also influences the rate of decomplexation. For example, the decomplexation of **33**•p-(MeO)$_2$C$_6$H$_4$ takes 800 times longer in 1,1,2,2-tetrachloroethane-d$_2$ than in $CDCl_3$ at 35 °C, perhaps because the larger solvent cannot solvate the large guest very well [35].

3.5. REACTIONS WITHIN THE ENCLOSED CAVITY

One of the most exciting aspects of carcerand chemistry is that the sheltered space inside the cavity can stabilize reactive intermediates. A compound such as the Lawesson reagent that readily replaces carbonyl oxygens by sulfur atoms on the shell of hybrid carceplex **14**•guest was not able to react with carbonyl compounds that were incarcerated inside the cavity [17]. Only a few reactions inside carceplexes and hemicarceplexes are known since needed reagents often cannot be brought inside the cavity [1g, 30, 36].

A dramatic example of a cavity reaction is the formation of cyclobutadiene by irradiation of hemicarceplex **19a**•α-pyrone. The singlet ground state cyclobutadiene subsequently dissociates to two molecules of acetylene and hemicarcerand **19a** is regenerated [37].

Hydroquinone guests inside hemicarceplex **26**•guest have been oxidized to incarcerated benzoquinones, the latter ones could be reduced back to hydroquinones. In the same hemicarceplex **26**•guest, nitrobenzene was reduced to hydroxylaminobenzene (PhNHOH) [36]. Hydroxyl groups of incarcerated disubstituted benzene derivatives have been methylated in *ortho* and *meta* positions which reside near the host's portals; presumably the *para* positions are too sheltered by the bottom or top cavitands to be accessible to reagents [30].

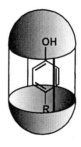
Sheltered

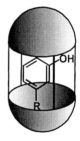
Accessible

Benzyne has been reported to be stabilized inside hemicarcerand **26** at low temperatures [38]. Hemicarceplex **26**•benzocyclobutenedione was irradiated above 400 nm yielding hemicarceplex **26**•benzocyclopropenone (Scheme 12). Benzocyclopropenone is highly strained and easily hydrolyzes to form benzoic acid; in fact, in solution it is not stable above –78 °C. However, as a guest within **26** it was stable at room temperature, its carbonyl group buried by one of the cavitand bowls, protected from water even though water is small enough to enter the hydrophobic cavity [1g].

Photolysis of **26**•benzocyclopropenone at −196 °C yielded the benzyne hemicarceplex, and its ^{1}H NMR spectrum was recorded at −75 °C. Two guest signals at δ 4.99 and δ 4.31 were found significantly shifted upfield compared to the calculated chemical shifts of "free" benzyne (δ 7.6 and δ 7.0) [39, 1g]. After warming the benzyne hemicarceplex to room temperature, benzyne undergoes a Diels-Alder reaction with one of the aromatic units of the host [40].

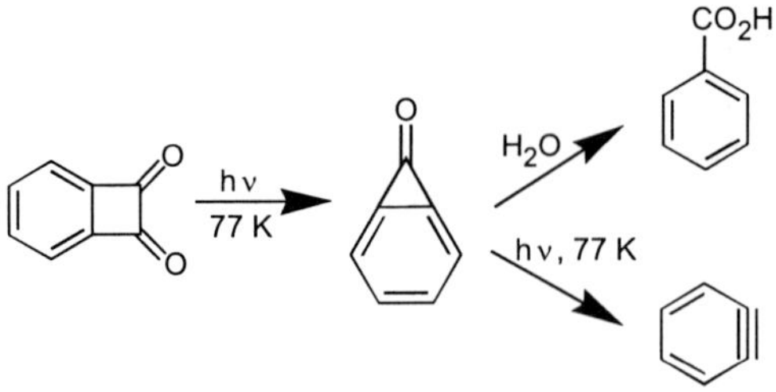

Scheme 12

Recently the room temperature stabilization of 1,2,4,6-cycloheptatetraene **34** inside **26** has been reported [41]. It was generated by a photochemical phenylcarbene rearrangement from **26**•phenyldiazirine and was characterized by ^{1}H NMR and FTIR spectroscopy as well as by the hemicarceplexes obtained after reaction with O_2 and HCl gas/methanol (Scheme 13). Interestingly, methanol itself did not react with encapsulated **34** after 3 days at 60 °C even though it is small enough to enter the cavity as shown by the conversion of **26**•7-chlorocyclohepta-1,3,5-triene to **26**•7-methoxycyclohepta-1,3,5-triene. The non-polar cavity of **26** seems to stabilize the allene **34** and prevent it from equilibrating to cycloheptatrienylidene **35** that would likely react quickly with methanol. A side product is an intermolecular insertion product into one of the host's acetal bridges by the reactive phenylcarbene intermediate [41].

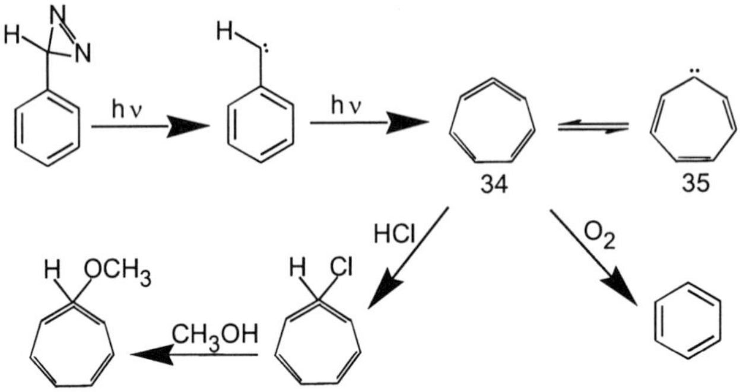

Scheme 13

The stabilization of 3-sulfolene inside carcerand **14** has been reported [17]. "Free" 3-sulfolene decomposes into SO_2 and butadiene by means of a thermal retro Diels-Alder reaction at 100-130 °C. When carceplex **14**•3-sulfolene was heated inside a mass spectrometer, only intact carceplex was detected at up to 180 °C. At higher temperatures, carcerand **14** was detected as well as SO_2 and butadiene. It was suggested that SO_2 and butadiene formed inside carcerand **14** by breakup of 3-sulfolene react with each other to form 3-sulfolene faster than they escape through the shell of **14**. Only at temperatures above 180 °C does the rate of escape from the inner phase become enough fast to allow the detection of the fragmentation products [17].

4. Related Species

The lantern shaped molecules **36** and **37**, based on a single cavitand molecule that is capped by conformationally flexible lid molecules bearing a group **X** are capsules that can undergo drastic conformational changes depending on the nature of **X** (Scheme 14) [42-44].

The two major isomers are described as concave (*C*) having a larger cavity and convex (*V*), which barely has a cavity. The *V* isomer is thermodynamically more favored than the *C* isomer [43] except when the lid contains a very bulky group such as a triphenylsilyl group. Desilylation of **36** gave predominantly (10:1) the thermodynamically disfavored *C* isomer (**37**-*C*) over (**37**-*V*). Lantern shaped **38** exists as the *V* isomer [44], and photolysis generated the β-unsubstituted enol **39**. Such enols are usually very unstable in solution and tautomerize rapidly, but **39** proved to be very stable, even in the presence of TFA, persisting for three days at room temperature in CDCl$_3$ before yielding the ketone **40** [44].

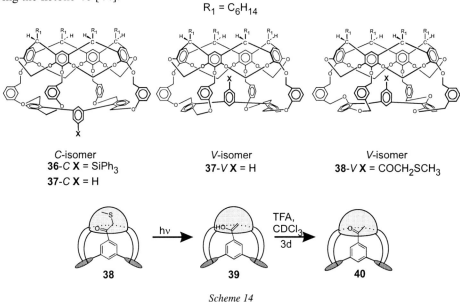

Scheme 14

Large hemicarcerand-looking hosts **42-44** based on extended cavitands have recently been synthesized (Scheme 15) but no guests have been complexed within those hosts. This could be due to both the large portals that make for easy decomplexation of smaller guests and the high activation energy barrier for complexation of larger guests such as C$_{60}$ [45].

Another approach to nano-scale hosts has been described using deep-cavity cavitands (DCC) **45** [46, 47] and linking benzyl alcohol groups, instead of the predominately used phenol groups, with CH$_2$BrCl. Two DCC bowls **45** were reacted with 8.4 equivalents of CH$_2$BrCl, 10 equivalents of *t*-BuOK and 15 equivalents of H$_2$O in DMSO at room temperature for 24 h in the absence of a single large templating molecule to obtain host **46** in 80% yield. No encapsulated guests were observed in solution.

It was suggested that the remarkable yield reflects the templating effort of bulk DMSO solvent molecules. DMA, DMF, NMP, and tetramethylurea as solvent(s) and the use of bases weaker than *t*-BuOK gave poor yields of **46** [47]. The importance of self-assembled capsules based on the ability of the phenol group to form a CHB with its conjugate base has been emphasized earlier in this chapter. It is not yet clear if the much less acidic benzyl alcohol group(s) forms CHBs as well.

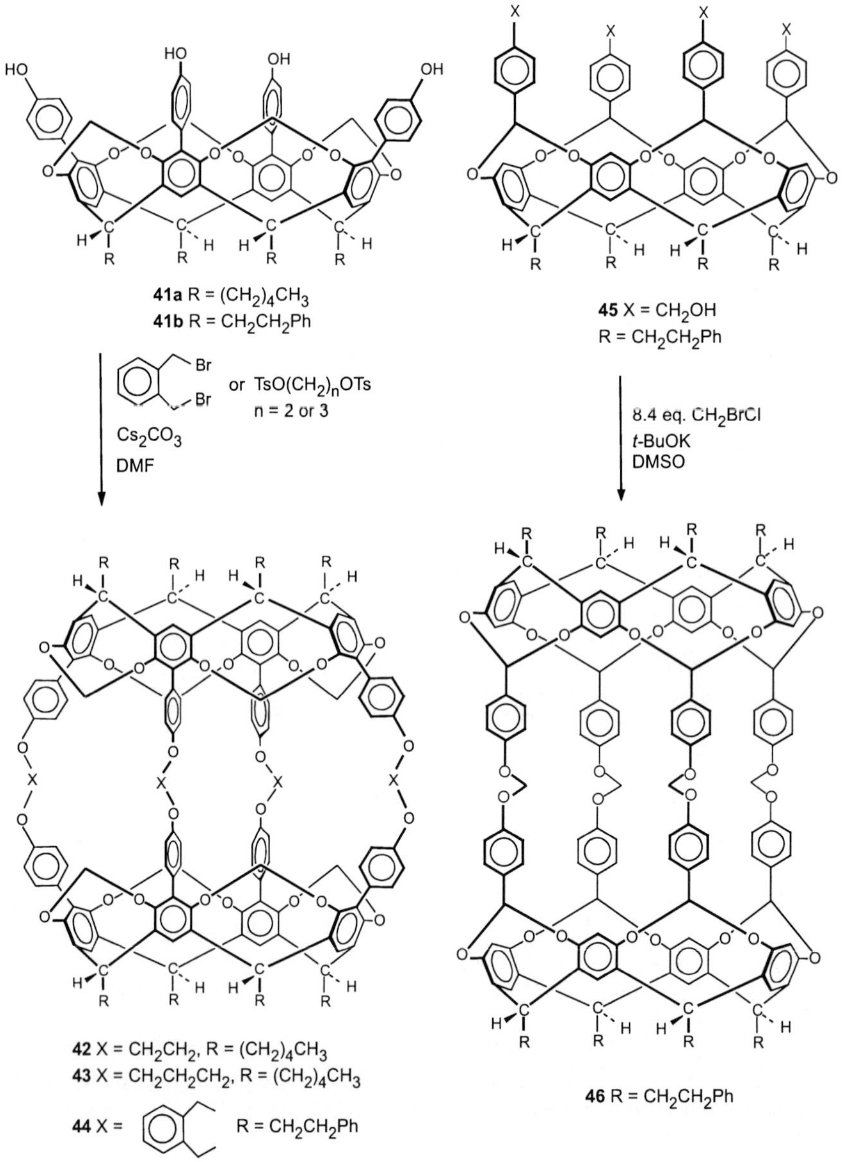

41a R = $(CH_2)_4CH_3$
41b R = CH_2CH_2Ph

45 X = CH_2OH
R = CH_2CH_2Ph

or TsO$(CH_2)_n$OTs
n = 2 or 3

Cs_2CO_3
DMF

8.4 eq. CH_2BrCl
t-BuOK
DMSO

42 X = CH_2CH_2, R = $(CH_2)_4CH_3$
43 X = $CH_2CH_2CH_2$, R = $(CH_2)_4CH_3$
44 X = 　　　　R = CH_2CH_2Ph

46 R = CH_2CH_2Ph

Scheme 15

5. Outlook

The future of carceplexes and hemicarceplexes lies in the creation of vessels on a more grand scale [48]. Creation of larger vessels will enable the stabilization and study of yet larger and more esoteric reactive species. As well, larger vessels will facilitate the study of template effects where very large molecules or multiple molecules can be used as templates. The latter remains largely unexplored and begs the question of when a template effect becomes a solvent effect. Encapsulation of multiple molecules also provides a unique perspective on the more general question of when a group of molecules constitutes a phase. Can there be a phase transition inside a carceplex? Can one, two, or even more solvent spheres about an analyte be studied inside a carceplex? Such studies will benefit from an interdisciplinary approach as they will present significant synthetic challenge and will require rigorous physical and spectroscopic analysis.

6. References and Notes

[1] a) E. Maverick, D. J. Cram, In *Comprehensive Supramolecular Chemistry*; J.-M. Lehn, J. L. Atwood, J. E. D. Davies, D. D. MacNicol, F. Vögtle, Series Ed.; Pergamon: New York, **1996**; Vol. 2, 367-418; b) D. J. Cram, J. M. Cram, *Container Molecules and Their Guests*; Royal Society of Chemistry: Cambridge, **1994**, Vol. 4; c) J. C. Sherman, In *Large Ring Molecules*; J. A. Semlyen, Ed.; J. Wiley & sons: West Sussex, England, **1996**; 507-524; d) J. C. Sherman, *Tetrahedron* **1995**, *51*, 3395-3422; e) A. Jasat, J. C. Sherman, *Chem. Rev.* **1999**, *99*, 931; f) D. A. Makeiff, J. C. Sherman, In *Templated Organic Synthesis*; F. Diederich, P. J. Stang, Ed.; Wiley-VCH: Weilheim, Germany, **2000**; 105-131; g) R. Warmuth, *J. Incl. Phenom.* **2000**, *37*, 1.
[2] D. J. Cram, *Science* **1983**, *219*, 1177-1183.
[3] a) J. C. Sherman, C. B. Knobler, D. J. Cram, *J. Am. Chem. Soc.* **1991**, *113*, 2194-2204; b) J. C. Sherman, D. J. Cram, *J. Am. Chem. Soc.* **1989**, *111*, 4527-4528.
[4] R. G. Chapman, N. Chopra, E. D. Cochien, J. C. Sherman, *J. Am. Chem. Soc.* **1994**, *116*, 369-370.
[5] R. G. Chapman, J. C. Sherman, *J. Org. Chem.* **1998**, *63*, 4103-4110.
[6] J. R. Fraser, B. Borecka, J. Trotter, J. C. Sherman, *J. Org. Chem.* **1995**, *60*, 1207-1213.
[7] R. G. Chapman, J. C. Sherman, *J. Am. Chem. Soc.* **1995**, *117*, 9081-9082.
[8] For reviews of other reversible capsules, see M. M. Conn, J. Rebek Jr., *Chem Rev.* **1997**, *97*, 1647-1668; and Chapters 8 and 9 of this book.
[9] R. G. Chapman, G. Olovsson, J. Trotter, J. C. Sherman, *J. Am. Chem. Soc.* **1998**, *120*, 6252-6260.
[10] N. Chopra, J. C. Sherman, *Angew. Chem., Int. Ed. Engl.* **1997**, *36*, 1727-1729.
[11] N. Chopra, J. C. Sherman, *Angew. Chem. Int. Ed. Engl.* **1999**, *38*, 1995-1997.
[12] N. Chopra, C. Naumann, J. C. Sherman, *Angew. Chem. Int., Ed. Engl.* **2000**, *39*, 194-196.
[13] a) D. J. Cram, S. Karbach, Y. H. Kim, L. Baczynskyj, K. Marti, R. M. Sampson, G. W. Kalleymeyn, *J. Am. Chem. Soc.* **1988**, *110*, 2554-2560; (b) D. J. Cram, S. Karbach, Y. H. Kim, L. Baczynskyj, G. W. Kalleymeyn, *J. Am. Chem. Soc.* **1985**, *107*, 2575-2576.
[14] a) J. A. Bryant, M. T. Blanda, M. Vincenti, D. J. Cram, *J. Am. Chem. Soc.* **1991**, *113*, 2167-2172; b) J. A. Bryant, M. T. Blanda, M. Vincenti, D. J. Cram, *J. Chem. Soc., Chem. Commun.* **1990**, 1403-1405.
[15] A. Jasat, J. C. Sherman, unpublished results.
[16] A. M. A. van Wageningen, J. P. M. van Duynhoven, W. Verboom, D. N. Reinhoudt, *J. Chem. Soc., Chem. Commun.* **1995**, 1941-1942.
[17] A. M. A. van Wageningen, P. Timmerman, J. P. M. van Duynhoven, W. Verboom, F. C. J. M. van Veggel, D. N. Reinhoudt, *Chem. Eur. J.* **1997**, *3*, 639-654.
[18] P. Timmerman, W. Verboom, F. C. J. M. van Veggel, J. P. M. van Duynhoven, D. N. Reinhoudt, *Angew. Chem., Int. Ed. Engl.* **1994**, *33*, 2345-2348.
[19] H. Ihm, K. Paek, *Bull. Korean Chem. Soc.* **1999**, *20*, 757-760.
[20] R. G. Chapman, J. C. Sherman, *J. Am. Chem. Soc.* **1999**, *121*, 1962-1963.
[21] R. G. Chapman, J. C. Sherman, *J. Org. Chem.* **2000**, *65*. 513-516.
[22] D. J. Cram, M. E. Tanner, C. B. Knobler, *J. Am. Chem. Soc.* **1991**, *113*, 7717-7727.

[23] M. E. Tanner, C. B. Knobler, D. J. Cram, *J. Am. Chem. Soc.* **1990**, *112*, 1659-1660.
[24] N. Chopra, J. C. Sherman, *Supramol. Chem.* **1995**, *5*, 31-37.
[25] R. G. Chapman, J. C. Sherman, *J. Am. Chem. Soc.* **1998**, *120*, 9818-9826.
[26] R. C. Helgeson, K. Paek, C. B. Knobler, E. F. Maverick, D. J. Cram, *J. Am. Chem. Soc.* **1996**, *118*, 5590-5604.
[27] R. C. Helgeson, C. B. Knobler, D. J. Cram, *J. Am. Chem. Soc.* **1997**, *119*, 3229-3244.
[28] T. A. Robbins, C. B. Knobler, D. R. Bellew, D. J. Cram, *J. Am. Chem. Soc.* **1994**, *116*, 111-122.
[29] D. A. Makeiff, D. J. Pope, J. C. Sherman, *J. Am. Chem. Soc.* **2000**, *122*, 1337-1342.
[30] S. K. Kurdistani, R. C. Helgeson, D. J. Cram, *J. Am. Chem. Soc.* **1995**, *117*, 1659-1660.
[31] J. Yoon, C. Sheu, K. N. Houk, C. B. Knobler, D. J. Cram, *J. Org. Chem.* **1996**, *61*, 9323-9339.
[32] J. Yoon, D. J. Cram, *J. Am. Chem. Soc.* **1997**, *119*, 11796-11806.
[33] K. Paek, H. Ihm, S. Yun, H. C. Lee, *Tetrahedron Lett.* **1999**, *40*, 8905-8909.
[34] a) K. Nakamura, K. N. Houk, *J. Am. Chem. Soc.* **1995**, *117*, 1853-1854; b) K. N. Houk, K. Nakamura, C. Sheu, A. E. Keating, *Science* **1996**, *273*, 627-629.
[35] J. Yoon, D. J. Cram, *Chem. Commun.* **1997**, 1505-1506.
[36] T. A. Robbins, D. J. Cram, *J. Am. Chem. Soc.* **1993**, *115*, 12199.
[37] D. J. Cram, M. E. Tanner, R.Thomas, *Angew. Chem., Int. Ed. Engl.* **1991**, *30*, 1024-1027.
[38] R. Warmuth, *Angew. Chem., Int. Ed. Engl.* **1997**, *36*, 1347-1350.
[39] H. Jiao, P. von Rague Schleyer, B. R. Beno, K. N. Houk, R. Warmuth, *Angew. Chem., Int. Ed. Engl.* **1997**, *36*, 2761-2764.
[40] R. Warmuth,. *Chem. Commun.* **1998**, 59-60.
[41] R. Warmuth, M. A. Marvel, *Angew. Chem., Int. Ed. Engl.* **2000**, *39*, 1117-1119.
[42] S. Watanabe, K. Goto, T. Kawashima, R. Okazaki, *Tetrahedron Lett.* **1995**, *36*, 7677-7680.
[43] S. Watanabe, K. Goto, T. Kawashima, R. Okazaki, *J. Am. Chem. Soc.* **1997**, *119*, 3195-3196.
[44] S. Watanabe, K. Goto, T. Kawashima, R. Okazaki, *Tetrahedron Lett.* **1999**, *40*, 3569-3572.
[45] C. von dem Busche-Hünnefeld, D. Bühring, C. B. Knobler, D. J. Cram, *J. Chem. Soc., Chem. Commun.* **1995**, 1085-1087.
[46] H. Xi, C. L. D. Gibb, E. D. Stevens, B. C. Gibb, *Chem. Commun.* **1998**, 1743-1744.
[47] C. L. D. Gibb, E. D. Stevens, B. C. Gibb, *Chem. Commun.* **2000**, 363-364.
[48] For reviews of larger cavitands see. D. M. Rudkevich, J. Rebek Jr., *Eur. J. Org. Chem.* **1999**, 1991-2005; and Chapter 9 of this book.

Chapter 11

HOMOCALIXARENES

YOSUKE NAKAMURA, TAKAHIRO FUJII, SEIICHI INOKUMA, JUN NISHIMURA

Department of Chemistry, Gunma University, Kiryu, Gunma 376-8515, Japan. e-mail: nisimura@chem.gunma-u.ac.jp

1. Introduction

In the strictest sense, "homocalixarenes" [1] should contain a $[2_n]$metacyclophane skeleton in which all the benzene rings are bridged by ethano linkages. Here, the name is used in a broader sense to refer to the compounds **I** - **III** shown below (*m, n, o, p* = 1–3, but mainly 2; X = OH, OR; Y = H, alkyl), where at least one, but not necessarily all, of the bridges is larger than methylene. Their conformational and ionophoric properties often differ significantly from those of their calixarene analogues.

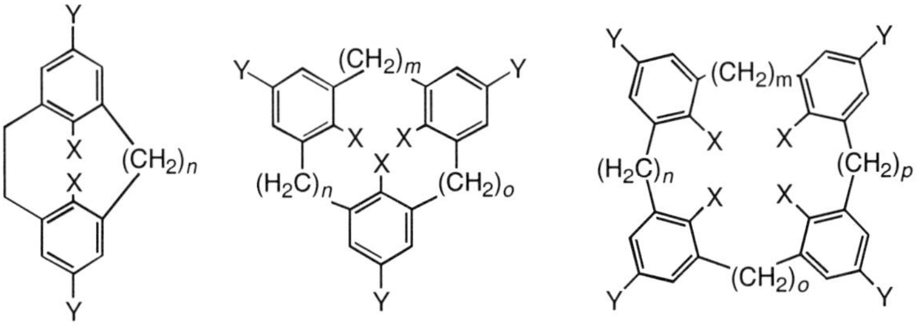

I II III

(X = OH, OR; Y = H, alkyl)

2. Synthesis

Typical methods of synthesis of homocalixarenes fall into two categories, "one pot" and "convergent", analogous to those well-known for calixarenes (See Chapter 1).

2.1. One-Pot Methods

2.1.1. Müller-Röscheisen cyclization

$[2_n]$Metacyclophanes were obtained by reaction of 1,3-bis(bromomethyl)benzene and sodium tetraphenylethene in THF at –80 °C (Müller-Röscheisen Cyclization) [2]. As extension of this procedure, homocalix[*n*]arenes **3–7** with *endo*-directed methoxy groups were obtained from **1** [3]. These products were isolated by the chromatographic technique. Each of **4–7** was readily demethylated with BBr_3 to afford **13–16** with hydroxy

Z. Asfari et al. (eds.), Calixarenes 2001, 219–234.

groups. Similarly, **2** afforded **8–12** bearing *exo*-directed methoxy groups. In general, homocalix[*n*]arenes are obtained in smaller yields than the corresponding calix[*n*]arenes. The distribution of ring size in the homocalix[*n*]arenes obtained apparently depended on the position of methoxy substituent in the starting materials **1** and **2**; *endo*-directed substituents favour the formation of relatively large macrocyclic systems.

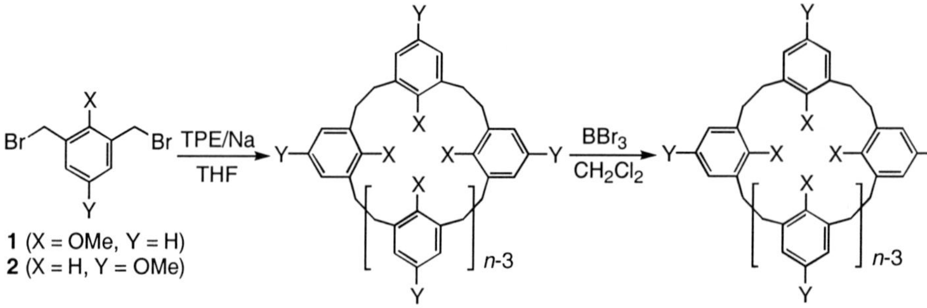

n	X = OMe Y = H	X = H Y = OMe	n	X = OH Y = H	X = H Y = OH
2	—	8 (21%)	2	—	17 (93%)
3	—	9 (10%)	3	—	18 (97%)
4	3 (trace)	10 (11%)	4	—	19 (99%)
5	4 (6%)	11 (3%)	5	13 (67%)	20 (87%)
6	5 (8%)	12 (2%)	6	14 (93%)	21 (93%)
7	6 (4%)	—	7	15 (84%)	—
8	7 (2%)	—	8	16 (74%)	—

Homocalix[*n*]pyridines **22–25** were also obtained from 2,6-bis(bromomethyl)-4-methoxypyridine in a manner similar to that mentioned above [4].

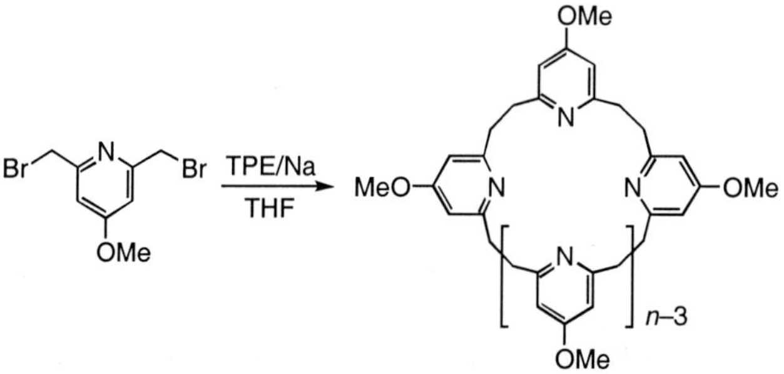

22 (*n* = 3) 12%
23 (*n* = 4) 0.6%   (2)
24 (*n* = 5) 2%
25 (*n* = 6) 0.6%

2.1.2. Cyclization with TosMIC ((p-toluenesulfonyl)methyl isocyanide)

The TosMIC cyclization has been frequently employed as a method for preparation of trimethylene-bridged cyclophanes, such as [3.3]para- and -metacyclophanes [5]. This method has been applied to the synthesis of triketone **27** [6]. Wolff-Kishner reduction of **27** followed by demethylation with BBr$_3$ afforded homocalix[3]arene **29**. The introduction of various substituents onto the phenolic OH groups of **29** gave **28b–g**, which are of interest form the viewpoints of conformational behavior and ionophoric properties, as described below.

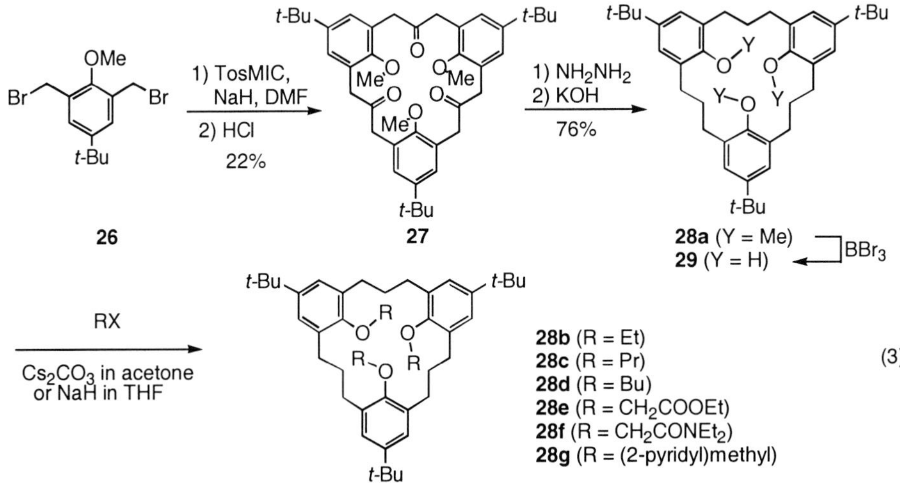

2.1.3. Malonate cyclization

Homocalix[*n*]arenes **31** and **32** have been obtained by the simple condensation of **30** with **1** [7].

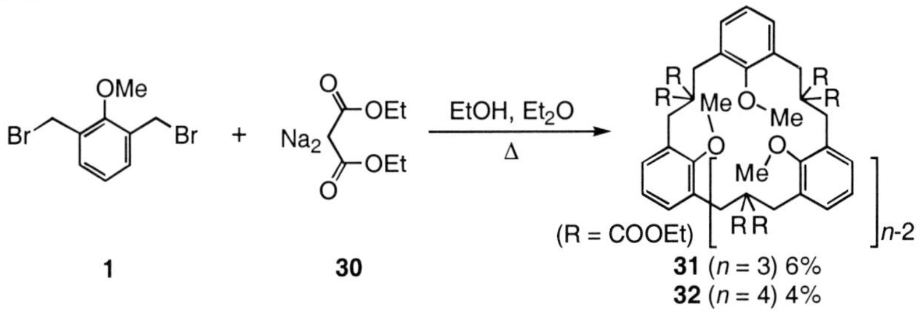

2.2. CONVERGENT METHODS

2.2.1. Sulfur extrusion

Dihydroxy-substituted [2.*n*]metacyclophanes **37** may be regarded as a kind of homocalixarene containing two benzene rings. Their methods of synthesis can be widely applied to larger macrocyclic systems. Thus, thia[3.*n*]metacyclophanes **34**, obtained by the reactions of bischloromethyl compounds **33** with Na_2S, were oxidized with *m*-CPBA into sulfone **35**. The pyrolysis of **35** readily afforded *syn*- and *anti*-**36** [8]. All *anti*-**36a–e** were demethylated to give the *anti*-**37a–e**, respectively. Under similar conditions, *syn*-**36b** and **c** gave the corresponding *syn*-isomers, whereas *syn*-**36d** and **e** gave only *anti*-**37d** and **e**, respectively.

Unsymmetrical homocalix[3]arene **41** possessing a [2.2.1]metacyclophane skeleton was obtained similarly [9]. One of the two conformers of **41** was further transformed into the trihydroxy derivative **42**.

Unsymmetrical homocalix[4]arene **46** with a [2.*n*.2.*n*]metacyclophane (*n* = 1–3) skeleton has also been synthesized by a procedure similar to **42** [10].

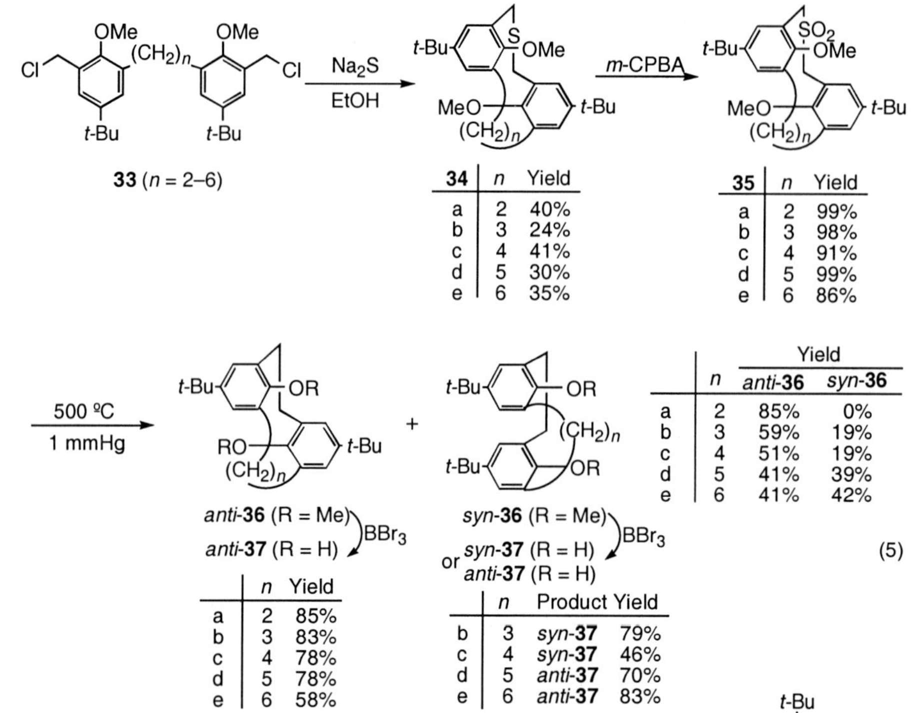

34	n	Yield
a	2	40%
b	3	24%
c	4	41%
d	5	30%
e	6	35%

35	n	Yield
a	2	99%
b	3	98%
c	4	91%
d	5	99%
e	6	86%

	n	Yield anti-36	syn-36
a	2	85%	0%
b	3	59%	19%
c	4	51%	19%
d	5	41%	39%
e	6	41%	42%

(5)

	n	Yield
a	2	85%
b	3	83%
c	4	78%
d	5	78%
e	6	58%

	n	Product	Yield
b	3	syn-37	79%
c	4	syn-37	46%
d	5	anti-37	70%
e	6	anti-37	83%

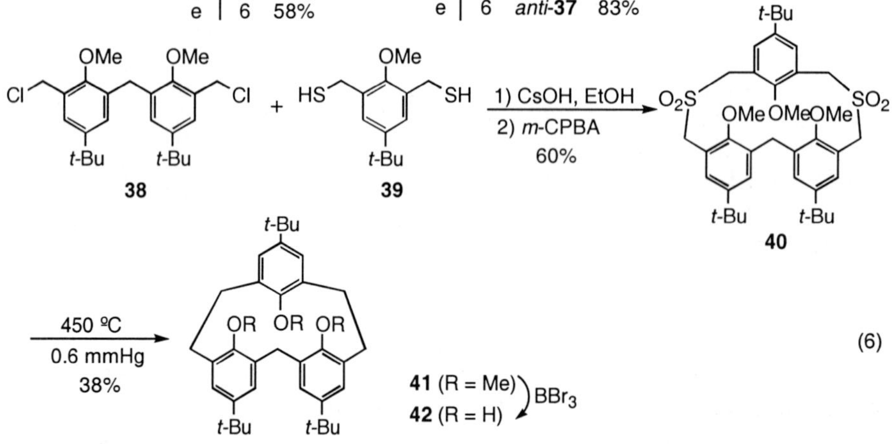

(6)

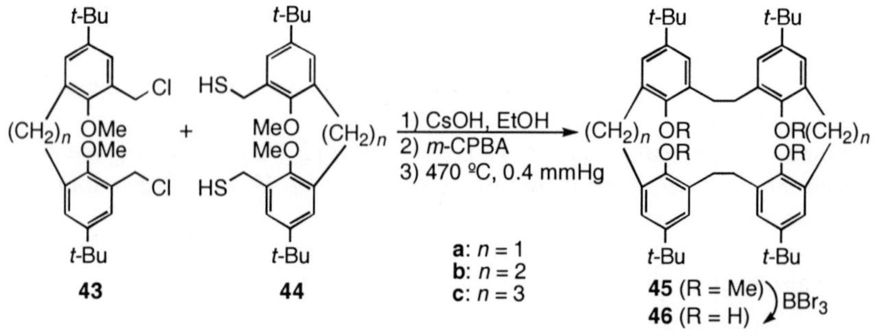

(7)

a: $n = 1$
b: $n = 2$
c: $n = 3$

2.2.2. Nafion-H catalyzed cyclobenzylation

Cyclobenzylation of **47** with **48** in the presence of Nafion-H (solid perfluorinated resin–sulfonic acid) followed by conventional demethylation provided the [3.1.1]metacyclophane **50** [11].

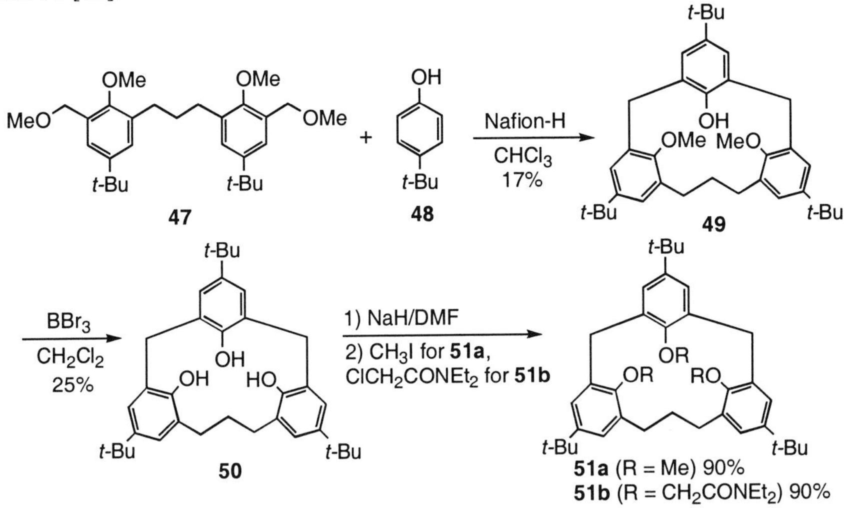

(8)

2.2.3. Müller-Röscheisen cyclization

In a convergent stepwise procedure, the intermolecularly-cyclised product **53**, with a [2.2.2.2]metacyclophane skeleton, was obtained along with a substituted [2.2]metacyclophane via intramolecular ring closure [12]. The demethylation of **53** afforded homocalix[4]arene **54**.

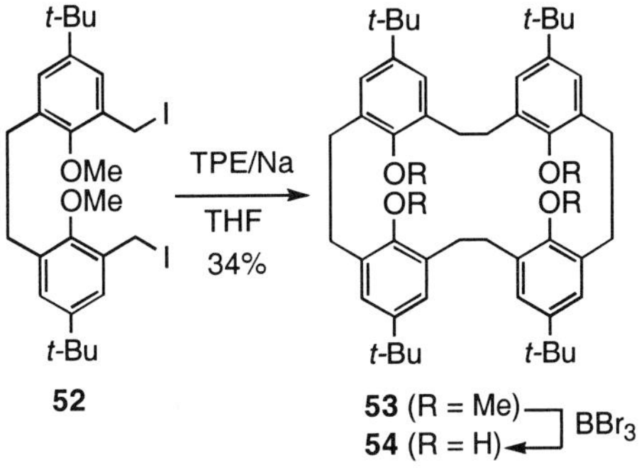

(9)

2.2.4. Cross-coupling reactions using organometallic reagents

The cross-coupling reaction of the dianion of **55**, generated by *t*-BuLi, with **56** yielded homocalix[4]arene **57** along with several linear by-products [13].

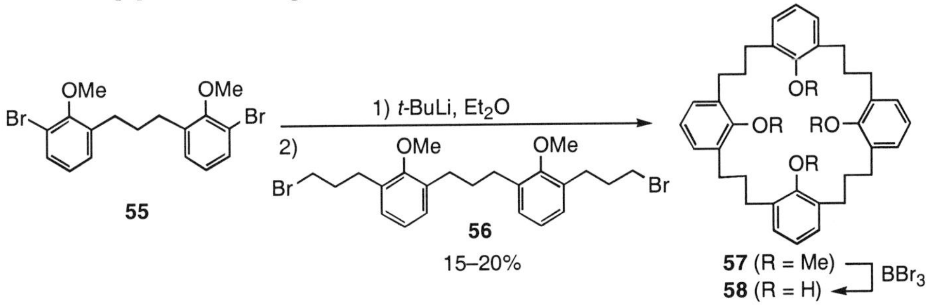

(10)

2.2.5. Condensation with aldehydes

The base-catalysed condensation of phenols with formaldehyde, commonly utilized for the preparation of normal calixarenes, has been successfully applied to the syntheses of [3.1.3.1]metacyclophane **60** from **59** [14] and [2.1.2.1.2.1]- **63** and [2.1.2.1.2.1.2.1]-metacyclophane **64** from **62** [15]. The yields and ratios of **63** and **64** depended on the nature of alkali hydroxide (MOH) employed as a catalyst. [3.1.3.1]Metacyclophane **60** was further transformed into **61** bearing various substituents [14,16–18].

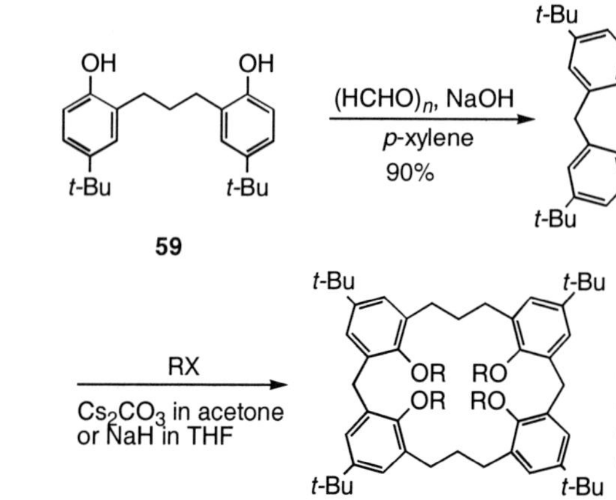

(11)

61a (R = Bu)
61b (R = Bzl)
61c (R = (2-pyridyl)methyl)
61d (R = CH_2COOMe)
61e (R = CH_2COOEt)
61f (R = CH_2COOt-Bu)

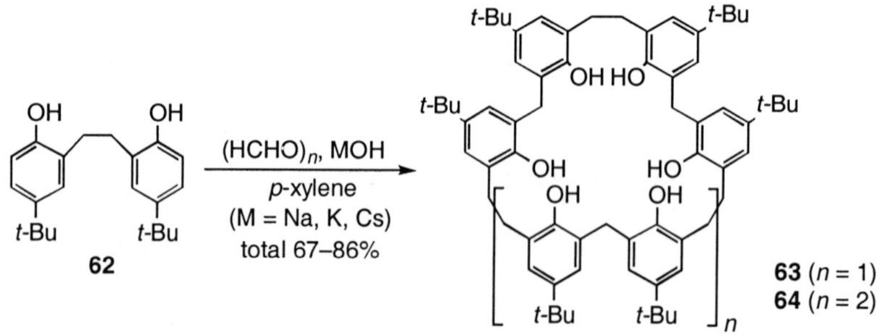

(12)

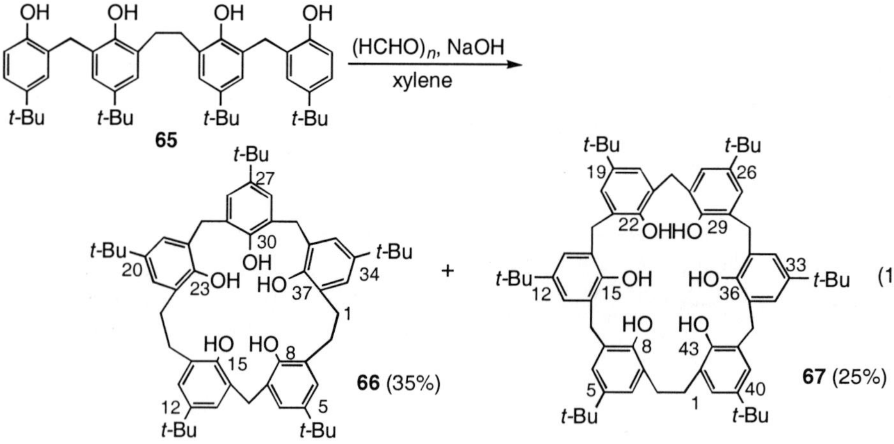

63 (n = 1)
64 (n = 2)

(13)

66 (35%)

67 (25%)

Surprisingly, base-catalysed condensation of **65** with formaldehyde gave the [2.1.2.1.1]- and [2.1.1.1.1.1]metacyclophanes (**66** and **67**) instead of the desired octahydroxy[2.1.1.1.2.1.1.1]metacyclophane, although the mechanism of their formation was not fully elucidated [19].

Homocalix[4]arenes **70** and **71** with a [2.1.2.1]metacyclophane skeleton were synthesized using *syn*-[2.*n*]metacyclophanes **68** and **69**, respectively, as building blocks [20]. Both **70** and **71** were further transformed into [4.1.4.1]metacyclophanes **72** and **73**, respectively, by Birch reduction.

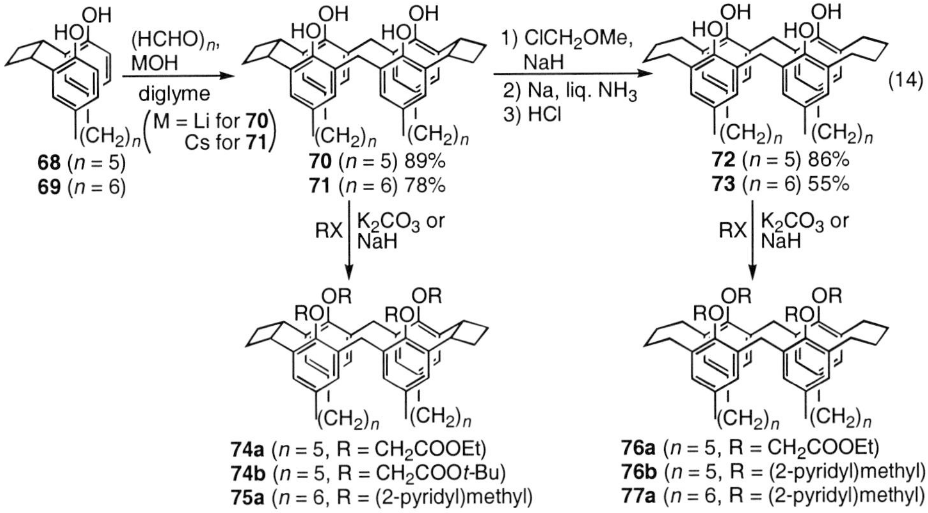

(14)

3. Structure and Conformational Behavior

Homocalixarenes may be designated in terms of the number of benzene rings as homocalix[2]arenes, homocalix[3]arenes, homocalix[4]arenes, and higher homocalix[*n*]-arenes. For homocalix[2]arenes, only *syn*- and *anti*-conformers are possible, as illustrated in Eq. 5. In homocalix[3]arenes, if all the three benzene rings are symmetrically bridged, there are only two conformers, *cone* and *partial-cone*. However, if one of the bridges differs from the other two, for example, as in [3.1.1]metacyclophane **50**, two *partial-cone* conformations are possible, *2-partial-cone* and *3-partial-cone*, depending on the position of substitutent directed to the opposite side. Scheme 1 illustrates the total three possible conformers, where a bold line represents the different bridge.

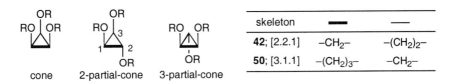

skeleton		
42; [2.2.1]	–CH$_2$–	–(CH$_2$)$_2$–
50; [3.1.1]	–(CH$_2$)$_3$–	–CH$_2$–

Scheme 1

For homocalix[4]arenes with identical bridges, four conformers (*cone, partial-cone, 1,2-alternate*, and *1,3-alternate*) are possible, as for normal calix[4]arene derivatives. In contrast, unsymmetrical species such as [2.1.2.1]metacyclophane derivatives allow two inequivalent *1,2-alternate* conformers, differing in the location of the symmetry plane. In order to distinguish them, conformers with a symmetrical plane parallel to the longer bridge and the shorter bridge can be defined as *1,2-alternate* and *1,4-alternate*, respectively. The total five conformers are illustrated in Scheme 2, where bold lines represent the longer bridges.

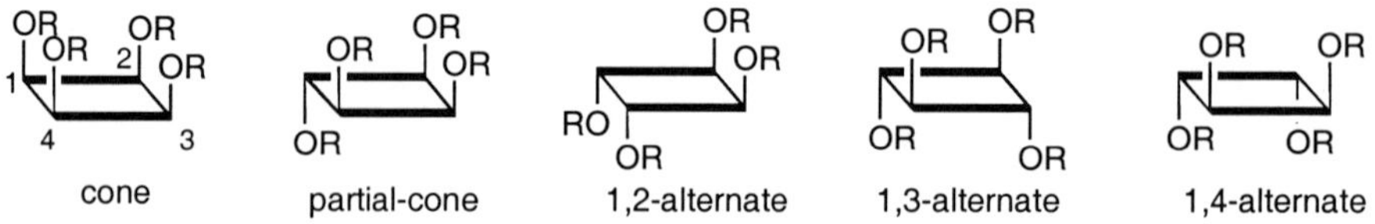

Scheme 2

skeleton		
46a; [2.1.2.1]	$-(CH_2)_2-$	$-CH_2-$
60; [3.1.3.1]	$-(CH_2)_3-$	$-CH_2-$

As with normal calixarenes the conformational interconversion of homocalixarenes has been investigated by dynamic ^{1}H NMR, and the activation free energy ($\Delta G^{\ddagger}$) has been derived mainly from coalescence temperatures (T_c) of suitable signals by conventional methods.

3.1. [2.*n*]METACYCLOPHANES

The ring inversion barriers of dihydroxy compounds **37** have been found to decrease with increasing length of the bridges [8]. The conformations of **37a–c** are rigid, whereas those of **37d,e** are flexible. In **37e**, the ΔG for the conformational ring flipping is estimated to be 20.6 kcal/mol (in CDCl$_3$). On the other hand, dimethoxy compounds **36a–e** are rigid and the $\Delta G^{\ddagger}$ for flipping is estimated to be > 25 kcal/mol (T_c > 150 °C).

It has also been found that the ratio of *anti* to *syn* conformers in **37e** is strongly solvent dependent. The portion of the *syn* conformer increases with increasing dielectric constant of the solvent. Greater solvent polarity favours the much more polar *syn* conformer with its strong intramolecular hydrogen bond.

3.2. HOMOCALIX[3]ARENES

3.2.1. [2.2.1]Metacyclophanes
In the trihydroxy derivative **42**, both '*2-partial-cone*' and '*3-partial-cone*' conformers were detected by ^{1}H NMR spectroscopy [9]. At higher temperatures (60–100 °C) broadening of *tert*-butyl and aromatic protons was observed, supposedly because of the conformational interconversion. Trimethoxy derivative **41** also adopts both '*2-partial-cone*' and '*3-partial-cone*' conformations, but no interconversion occurred between them.

3.2.2. [3.1.1]Metacyclophanes
Studies of the conformational properties of [3.1.1]metacyclophanes **49–51** [11,21] have shown that triol **50** adopts a symmetric 'cone' conformation at room temperature. The $\Delta G^{\ddagger}$ for ring inversion is 19.5 kcal/mol (T_c = 140 °C in hexachloro-1,3-butadiene-CDCl$_3$ (3:1)). Trimethoxy derivative **51a** adopts an unsymmetrical '2-partial-cone' conformation. It was also found, based on dynamic ^{1}H NMR spectroscopy, that the methyl groups are large enough to inhibit the oxygen-through-the annulus rotation ($\Delta G^{\ddagger}$ > 25 kcal/mol, T_c > 140 °C in hexachloro-1,3-butadiene-CDCl$_3$ (3:1)). Dimethoxy derivative **49** is also fixed in a '2-partial-cone' conformation at room temperature. The $\Delta G^{\ddagger}$ for ring inversion is 16.7 kcal/mol (T_c = 80 °C in CDBr$_3$). For this conformation, the OH group may act like a "molecular pendulum" in forming hydrogen bonds to each of the methoxy group oxygen atoms. On the other hand, triamide **51b** adopts a symmetrical '3-partial-cone' conformation at room temperature.

3.2.3. [3.3.3]Metacyclophanes
Studies of the conformational properties of [3.3.3]metacyclophanes **28** and **29** [6,21] have shown that triol **29** is flexible; the protons of the ArCH$_2$CH$_2$CH$_2$Ar methylene group appear each as a singlet even below –60 °C (CDCl$_3$/CS$_2$ (1:3)).

The conformer distribution in *O*-alkylation products of **29** depends on the reaction conditions (Table 1). When **29** was *O*-alkylated with alkyl halides (RX: R = Et, Pr, and Bu) in the presence of Cs$_2$CO$_3$, the *partial-cone* conformer was obtained in quantitative yield. On the other hand, the *cone* conformer was predominant in the *O*-substitution with ethyl bromoacetate and *N,N*-diethylchloroacetamide when the stronger base NaH was employed.

TABLE 1. Conformer distribution for the reaction of **29** with alkyl halides[a]

Run	RX	Base	Product (Yield/%[b])	Distribution[c]	
				cone	partial-cone
1	EtBr	Cs$_2$CO$_3$	**28b** (98)	0	100
2	PrBr	Cs$_2$CO$_3$	**28c** (91)	0	100
3	BuBr	Cs$_2$CO$_3$	**28d** (82)	0	100
4	EtOOCCH$_2$Br	Cs$_2$CO$_3$	**28e** (90)	5	95
5	Et$_2$NOCCH$_2$Cl	Cs$_2$CO$_3$	**28f** (98)	33	67
6	PrBr	NaH[d]	**28c** (10)[e]	0	100
7	EtOOCCH$_2$Br	NaH[d]	**28e** (97)	100	0
8	Et$_2$NOCCH$_2$Cl	NaH[d]	**28f** (98)	100	0

[a] Reaction conditions: In acetone at reflux for 3 h unless indicated otherwise. [b] Isolated yields. [c] Relative yields determined by ^{1}H NMR spectroscopy. [d] THF/reflux for 3 h. [e] Starting compound was recovered in 90% yield.

The influence of *O*-substituents on the oxygen-through-the-annulus rotation of *O*-alkylated homocalix[3]arenes **28** is compared with that in the corresponding homooxacalix[3]arene in Table 2. It is clear that in the homooxacalix[3]arenes the rotation is inhibited only by groups bulkier than butyl group, while in homocalix[3]arenes **28** the propyl group is bulky enough to inhibit the rotation. These results suggest that it is more difficult to inhibit the rotation in *O*-alkylated homooxacalix[3]arene than in **28** in spite of the difference in the bond distances between $C(sp^3)$-$C(sp^3)$, 1.53 Å and $C(sp^3)$-O, 1.43 Å. This is possibly explained by both the staggered conformation of the diarylpropane in **28** and the flexibility of the ether linkages in homooxacalix[3]arene.

TABLE 2. Influence of *O*-substituents on the oxygen-through-the annulus rotation in homocalix[3]arene **28** and homooxacalix[3]arene

O-Substituent (R)	Homocalix[3]arene **28**	Homooxacalix[3]arene
Me	Flexible (T_c < –50 °C)[a]	Flexible (T_c < –50 °C)
Et	Flexible (T_c = 90 °C)[b]	Flexible (T_c = 50 °C)[a]
Pr	Rigid[b]	Flexible[c]
Bu	Rigid[b]	Rigid

[a] Solvent: $CDCl_3/CS_2$ = 1/3. [b] Solvent: $CDBr_3/CDCl_3$ (3/1). [c] The oxygen-through-the-annulus rotation is slower than the NMR time scale.

3.3. HOMOCALIX[4]ARENES

3.3.1. [2.n.2.n]Metacyclophanes

In the series of [2.*n*.2.*n*]metacyclophanes **46a–c** [10], only **46b** shows intramolecular hydrogen-bonding as in calix[4]arene. The ^{1}H NMR spectrum in $CDCl_3$ shows the signals for the hydroxy groups around δ 10.40 as a broad singlet and the IR spectrum in KBr shows an absorption for OH stretching around 3220 cm^{-1} (calix[4]arene; δ_{OH} 10.19 in ^{1}H NMR, ν_{OH} 3160 cm^{-1} in IR). However, the ^{1}H NMR spectrum is essentially unchanged in the temperature range of –40 to +60 °C. These observations indicate that **46b** is conformationally flexible (T_c < –40 °C), even though the hydroxy groups are close to each other. It was concluded that the calix[4]arene-like intramolecular hydrogen bonds could not fix the conformation of **46b**. Both **46a** and **46c**, having weaker intramolecular hydrogen bonds, are also flexible. In the (ethoxycarbonyl)methyl derivative prepared from **46a** and ethyl bromoacetate, only the *1,3-alternate* was not obtained among the possible five conformers, presumably because of its lesser stability [22].

Homocalix[4]arenes **70–73** maintain a *cone* conformation between r.t. and 140 °C in DMSO and r.t. and 100 °C in pyridine according to dynamic NMR measurements [20].

3.3.2. [3.1.3.1]Metacyclophanes

The ^{1}H NMR spectra of tetrahydroxy[3.1.3.1]metacyclophane **60** show single peaks for *tert*-butyl, methylene, aromatic, and phenolic OH protons at room temperature due to rapid conformational flipping [14]. The $\Delta G^\ddagger$ for ring inversion is 12.5 kcal/mol (T_c = 0 °C in $CDCl_3$). This value is smaller than that of calix[4]arene (15.7 kcal/mol). This is attributed to the increase of ring size by introduction of the two trimethylene-bridges. However, as far as judged from the coalescence temperature, **60** is expected to be more rigid than the corresponding [2.1.2.1]metacyclophane **46a** (T_c < –40 °C) in spite of its larger ring size. This is explained by the staggered conformation of the diarylpropane-like [3.3]metacyclophanes, which adopt a *syn*-conformation. Thus, **60** can form a stronger intramolecular hydrogen bond (δ_{OH} 9.35 in the ^{1}H NMR spectrum, ν_{OH} = 3254 cm^{-1} in the IR spectrum) than **46a** (δ_{OH} 8.8, ν_{OH} = 3418 cm^{-1}).

Below –40 °C, the signal for phenolic-OH at δ 9.35 in **60** splits into two singlets at δ 9.15 and 10.12. This phenomenon can be attributed to the formation of two sets of nonequivalent phenolic-OHs because the conformational fluctuation of the cyclophane ring is frozen below this temperature by intramolecular hydrogen bonding between the OH groups on facing benzene rings (Scheme 3). The estimated $\Delta G^{\ddagger}$ for fluctuation is 10.5 kcal/mol (T_c = –40 °C). Dynamic ^{1}H NMR studies and consideration of a CPK model indicate that below 0 °C the conformation of **60** is in a 'stepped-flattened *1,3-alternate* form' as a result of the calix[4]arene-like intramolecular hydrogen bonding.

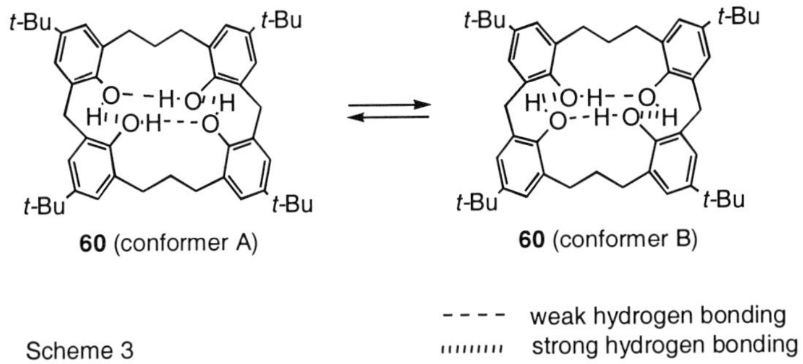

60 (conformer A)   **60** (conformer B)

Scheme 3

- - - - weak hydrogen bonding
∣∣∣∣∣∣∣ strong hydrogen bonding

The *O*-alkylation of **60** with butyl bromide in the presence of Cs$_2$CO$_3$ yields exclusively *1,2-alternate*-**61a** in quantitative yield (Table 3). In contrast, the reaction with benzyl bromide affords *cone*-**61b** exclusively in the presence of NaH and preferentially in the presence of Cs$_2$CO$_3$.

TABLE 3. *O*-Substitution of tetraol **60** with butyl bromide and benzyl bromide in the presence of NaH and Cs$_2$CO$_3$

				Yield (%)[a,b]	
Run	RX	Product	Base	*cone*	*1,2-alternate*
1	BuBr	61a	Cs$_2$CO$_3$	0	100 (90)
2	BzlBr	61b	NaH	100 (90)	0
3	BzlBr	61b	Cs$_2$CO$_3$	80 (70)	20 (13)

[a] Relative yields determined by ^{1}H NMR spectroscopy. [b] Isolated yields are shown in parentheses.

The influence of *O*-substitution on the oxygen-through-the-annulus rotation of tetra-*O*-alkylated [3.1.3.1]metacyclophanes **61** is summarized in Table 4. From Table 4, it is clear that the rotation is completely inhibited by the butyl group in [3.1.3.1]metacyclophane **61**. The propyl group is also bulky enough to inhibit the rotation, but the conformational ring inversion can still occur above 90 °C. In calix[4]arene, on the other hand, the propyl group completely inhibits the rotation. According to these data, it is more difficult to inhibit the rotation in [3.1.3.1]metacyclophane **61** than in *O*-alkylated calix[4]arene because of the longer bridges. The ^{1}H NMR results and consideration of CPK models justify the assertion that both tetrapropoxy- and butoxy[3.1.3.1]metacyclophanes adopt an '*anti*-stepped' *1,2-alternate* conformation.

TABLE 4. Influence of O-substituents on the oxygen-through-the-annulus rotation in [3.1.3.1]metacyclophanes **61**[a]

O-Substituent (R)	
Me	Flexible ($T_c < -60$ °C)
Et	Flexible ($T_c < -60$ °C)
Pr	Flexible ($T_c = 90$ °C, $\Delta G_c^{\ddagger} = 14.7$ kcal/mol)[b]
Bu	Rigid[b]
Bzl	Rigid[b]

[a] T_c and $\Delta G_c^{\ddagger}$ were determined in CDCl$_3$-CS$_2$ (1:3) by using SiMe$_4$ as reference unless otherwise indicated. [b] Solvent: CDBr$_3$-CDCl$_3$ (6:1).

3.4. HIGHER HOMOCALIX[n]ARENES ($n = 5, 6, 8$)

The OH proton chemical shifts and OH stretching vibration frequencies of several macrocyclic metacyclophanes are given in Table 5 [15,19]. From these data it is obvious that these compounds can form intramolecular hydrogen bonds.

[2.1.2.1.1]Metacyclophane **66** is assumed to adopt a rigid conformation having a plane of symmetry (C_s symmetry) at room temperature; a conformation with all OH groups in the same side of the ring 'cone' or a conformation with the OH groups in positions 8 and 15 'inverted' would be reasonable [19]. $\Delta G^{\ddagger}$ for ring inversion of **66** is 16.7 kcal/mol ($T_c = 85$ °C in CDBr$_3$). This value is larger than that of calix[n]arene ($n = 4$: 15.7 kcal/mol, $n = 5$: 13.2 kcal/mol in CDCl$_3$). This difference may be due to a stronger intramolecular hydrogen bond in **66** than in calix[n]arenes.

TABLE 5. Selected ^{1}H NMR and IR spectral data of calixarenes

Compounds	IR [KBr]	^{1}H NMR [CDCl$_3$]	T_c ($\Delta G_c^{\ddagger}$)
	ν_{OH} (cm^{-1})	δ_{OH}	[°C (kcal/mol)]
Calix[4]arene	3160	10.19	52 (15.7)[a]
66; [2.1.2.1.1]	3250	9.43, 9.83	85 (16.7)[b]
67; [2.1.1.1.1.1]	3175, 3250, 3450	8.15, 9.76, 10.52	35 (14.5)[b]
63; [2.1.2.1.2.1]	3298	8.90	−60 (ca. 10)[c]
64; [2.1.2.1.2.1.2.1]	3355	9.80	40 (14.4)[a]

[a] Solvent: CDCl$_3$. [b] Solvent: CDBr$_3$. [c] Solvent: CDCl$_3$/CS$_2$=1/3.

For [2.1.1.1.1.1]metacyclophane **67**, $\Delta G^{\ddagger}$ for ring inversion is 14.5 kcal/mol ($T_c = 35$ °C in CDBr$_3$) [19]. This value is slightly larger than that of calix[6]arene (13.3 kcal/mol). It is also observed that below −20 °C all phenolic hydroxyl groups are inequivalent. The conformational fluctuation of the cyclophane ring is frozen below this temperature by the intramolecular hydrogen bond. On the basis of the dynamic ^{1}H NMR studies and consideration of a CPK model, it is concluded that below −20 °C **67** adopts an asymmetric 'winged' conformation with respect to the diarylethane unit due to the intramolecular hydrogen bond.

In [2.1.2.1.2.1]metacyclophane **63** at −60 °C in CDCl$_3$ the signal of ArCH$_2$Ar methylene protons splits into two broad singlets ($T_c = -60$ °C, $\Delta G^{\ddagger} = $ ca. 10 kcal/mol), but that of the ArCH$_2$CH$_2$Ar ethylene protons remains unsplit [15]. This result suggests that the ethylene chains are still flexible even at this temperature. In comparison, in [2.1.2.1.2.1.2.1]metacyclophane **64** at 0 °C the singlet signal of the ArCH$_2$Ar methylene protons splits into two sets of doublets and the ArCH$_2$CH$_2$Ar ethylene protons are also observed to be split [15]. $\Delta G^{\ddagger}$ for inversion of methylene protons is estimated to be 14.4 kcal/mol, which is smaller than that of calix[8]arene (15.7 kcal/mol). The higher $\Delta G^{\ddagger}$ value for **64** than for **63** indicates that the former is more rigid due to its much stronger intramolecular hydrogen bond.

In the case of **64** it has also been found that below −38 °C the signal for the phenolic hydroxyl groups splits into two singlets. This phenomenon may be attributed to the formation of two sets of inequivalent phenolic hydroxyl groups because the conformational fluctuation of the cyclophane ring is frozen below this temperature by intramolecular hydrogen bonds in the two sets of four hydroxyl groups. The estimated $\Delta G^{\ddagger}$ for fluctuation of **64** is 11.0 kcal/mol (T_c = −38 °C). On the basis of the dynamic ^{1}H NMR studies and model construction, it is concluded that below −38 °C, **64** should adopt the 'pleated-loop' conformation due to intramolecular hydrogen bonds.

4. Ionophoric Properties

Excellent ionophores derived from homocalixarenes have been obtained by the introduction of functional groups similar to those found to be very effective on calixarenes [23–29]. (See Chapters 21, 22.)

4.1. [2.n]METACYCLOPHANES

The efficiency and selectivity of calixarenes as ionophores depend upon both their stereochemistry and the nature of their substituents [27]. The series (**78–79** and **80–82**) of simple and rigid homocalixarenes possess a variety of ligating sites. The binding ability of triply-bridged cyclophanes such as **78** and **79** toward alkali metal and heavy metal cations was examined by the liquid-liquid extraction method [30]. Most of these compounds acted as efficient cation-binding clefts. In particular, **79** showed both selectivity and high extraction efficiency for the Ag$^+$ cation, suggesting that the proper arrangement of only two ligating groups on a rigid framework can be sufficient.

Other triply-bridged compounds ("crownopaddlanes" **80–82**) with two cyclobutane rings have been successfully prepared by the intramolecular [2 + 2] photocycloaddition, and their complexing abilities toward Li$^+$, Na$^+$, and K$^+$ were evaluated by solid-liquid extraction [31]. **80–82** all quantitatively extracted Li$^+$ in single solid-liquid extraction.

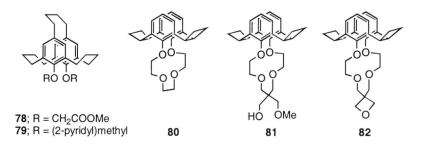

78; R = CH$_2$COOMe
79; R = (2-pyridyl)methyl   **80**   **81**   **82**

Upon competitive extraction, **80** showed a higher selectivity toward Li$^+$ than towards Na$^+$ or K$^+$ (Li$^+$/Na$^+$ = 610, Li$^+$/K$^+$ = 976). The solid-state structure of **80** was determined by X-ray crystallography and considered to exhibit preorganization for complexation with the smallest group I metal, Li$^+$, *viz.*, the polyether ring was curved toward one of the cyclobutane rings and the cavity diameter of **80** was ca. 1.22 Å, appropriate for binding to Li$^+$. Furthermore, the cyclobutane blades of **80** may act as a steric barrier to 2:1 sandwich complexation. These results suggest that these factors may act cooperatively in solution and result in the quantitative and highly selective extraction of Li$^+$.

4.2. HOMOCALIX[3]ARENES

In liquid-liquid extraction, the extractabilities of homocalixarenes **28e** and **28f** [6,21] were lower than those of the corresponding homooxacalix[3]arenes, suggesting the contribution of the three ethereal oxygens of the latter to the complexation. Although both *cone*- and *partial cone*-**28g** barely extracted either alkali metal cations or *n*-butylammonium cation, they exhibited higher extractabilities for Ag$^+$ than dibenzopyridino-18-crown-6 [32]. Both **28e** and **28f** in their *cone* conformation also showed high affinity for ammonium cations due to their threefold symmetry. The ^{1}H NMR titration of triamide **51b** with NaSCN clearly demonstrates that a 1:1 complex is formed which is stable on the NMR time scale [11].

4.3. HOMOCALIX[4]ARENES

The rigid *cone*-homocalixarenes **74–77** with various functional groups have been prepared [20]. Compounds **74a** and **74b** efficiently extracted K$^+$, Rb$^+$, and Cs$^+$. Their affinity was similar to that of the corresponding calix[6]arene derivatives [25]. Compound **74b** selectively extracted Hg^{2+} and Ag$^+$ above some transition metal cations examined. Both **74a** and **74b** showed higher affinity toward La^{3+} than Sm^{3+} and Yb^{3+}. The extractability of **76a** was not so great, though **76b** possessing pyridylmethyl moieties again showed high affinity to Ag$^+$ and Hg^{2+}.

In liquid-liquid extraction, **75** showed selectivity for K$^+$ and Rb$^+$ in the alkali metals group, for Sr^{2+} in the alkaline earth group and for Ag$^+$ above various transition metals. Of the ligands **75**, **75a** exhibited both high extraction efficiency and significant selectivity within the alkali, alkaline earth and transition metal groups. The extractant strength of **77** was dramatically diminished by the increase in the length of linkages between the aromatic nuclei of **75**, though **77a** again efficiently extracted Ag$^+$ and Hg^{2+}.

Of the four conformers of the (ethoxycarbonyl)methyl derivative prepared from **46a** only the *cone*-conformer appeared to form complexes with alkali metal cations [22]. This compound showed moderate selectivity toward Na$^+$ and K$^+$ over Cs$^+$.

Of the [3.1.3.1]metacyclophane-based ionophores **61c** [16] and **61d–f** [17], *cone*-**61c** showed selectivity toward Rb$^+$ and Cs$^+$ with moderate extractability for all the alkali metal cations, indicating that the binding sites of the host compound exist apart from each other. Both *cone*- and *partial cone*-**61c** exhibited high selectivity and extractability for Ag$^+$ which were superior to those of dibenzopyridino-18-crown-6. Tetraalkyl esters *1,2-alternate*-**61e** and **61f** showed strong Rb$^+$ affinities comparable with that for 18-crown-6. Depending on the bulkiness of alkyl groups, a high Rb$^+$ selectivity was observed for **61e**, though no significant ion selectivity was observed with **61d** and **61f**.

4.4. Higher Homocalix[n]arenes ($n = 5, 6, 8$)

The homocalixarenes **13** and **16** strongly discriminate in favor of the alkaline earth metal cations over the alkali metal cations in liquid-liquid extraction under basic conditions, and the efficiency of **16** was higher than that of **13** due to its greater number of phenol units [3,33]. The stability constants for complexes of alkali and alkaline earth metal cations of both homocalixarenes in water were higher than those of conventional crown ethers. The amide derivative prepared from **14** and N,N-diethylchloroacetamide showed high selectivity toward Ba^{2+} in the H_2O-toluene extraction system.

5. Concluding Remarks

An eclectic but illustrative selection of homocalixarene chemistry has been described above. Homocalixarene syntheses are generally somewhat more difficult than those of calixarenes, usually requiring the multiple-step, "convergent" approach. Nonetheless, these molecules commonly show less conformational mobility than their dihomo-oxacalixarene analogues and several have provided efficient and selective ionophores.

6. References and Notes

[1] S. Ibach, V. Prautzsch, F. Vögtle, C. Charteoux, K. Gloe, *Acc. Chem. Res.* **1999**, *32*, 729–740.
[2] K. Burri, W. Jenny, *Helv. Chim. Acta* **1967**, *50*, 1978–1993.
[3] a) F. Vögtle, J. Schmitz, M. Nieger, *Chem. Ber.* **1992**, *125*, 2523–2531; b) J. Schmitz, F. Vögtle, M. Nieger, K. Gloe, H. Stephan, O. Heitzsch, H.-J. Buschmann, W. Hasse, K. Cammann, *Chem. Ber.* **1993**, *126*, 2483–2491.
[4] a) F. Vögtle, G. Brodesser, M. Nieger, K. Rissanen, *Recl. Trav. Chim. Pays-Bas* **1993**, *112*, 325–329; b) H. Stephan, T. Krüger-Rambusch, K. Gloe, W. Hasse, B. Ahlers, K. Cammamm, K. Rissanen, G. Brodesser, F. Vögtle, *Chem. Eur. J.* **1998**, *4*, 434–440.
[5] a) K. Kurosawa, M. Suenaga, T. Inazu, T. Yoshino, *Tetrahedron Lett.* **1982**, *23*, 5335–5338; b) T. Shinmyozu, Y. Hirai, T. Inazu, *J. Org. Chem.* **1986**, *51*, 1551–1555.
[6] T. Yamato, L. K. Doamekpor, K. Koizumi, K. Kishi, M. Haraguchi, M. Tashiro, *Liebigs Ann.* **1995**, 1259–1267.
[7] F. Vögtle, M. Zuber, *Synthesis* **1972**, 543; C. Meiners, M. Nieger, F. Vögtle, *Liebigs Ann.* **1996**, 297–302.
[8] T. Yamato, J. Matsumoto, K. Tokuhisa, M. Kajihara, K. Suehiro, M. Tashiro, *Chem. Ber.* **1992**, *125*, 2443–2454.
[9] a) A. Tsuge, T. Sawada, S. Mataka, N. Nishiyama, H. Sakashita, M. Tashiro, *J. Chem. Soc., Chem. Commun.* **1990**, 1066–1068; b) A. Tsuge, T. Sawada, S. Mataka, N. Nishiyama, H. Sakashita, M. Tashiro, *J. Chem. Soc., Perkin Trans. 1* **1992**, 1489–1494.
[10] M. Tashiro, A. Tsuge, T. Sawada, T. Makishima, S. Horie, T. Arimura, S. Mataka, T. Yamato, *J. Org. Chem.* **1990**, *55*, 2404–2409.
[11] a) T. Yamato, L. K. Doamekpor, H. Tsuzuki, M. Tashiro, *Chem. Lett.* **1995**, 89–90; b) T. Yamato, L. K. Doamekpor, H. Tsuzuki, *Liebigs Ann.* **1997**, 1537–1544.
[12] M. Tashiro, T. Yamato, *J. Org. Chem.* **1981**, *46*, 1543–1552.
[13] D. H. Burns, J. D. Miller, J. Santana, *J. Org. Chem.* **1993**, *58*, 6526–6528.
[14] a) T. Yamato, Y. Saruwatari, S. Nagayama, K. Maeda, M. Tashiro, *J. Chem. Soc., Chem. Commun.* **1992**, 861–862; T. Yamato, Y. Saruwatari, M. Yasumatsu, *J. Chem. Soc., Perkin Trans. 1* **1997**, 1725–1730; b) T. Yamato, Y. Saruwatari, M. Yasumatsu, *J. Chem. Soc., Perkin Trans. 1* **1997**, 1731–1737.
[15] T. Yamato, Y. Saruwatari, L. K. Doamekpor, K. Hasegawa, M. Koike, *Chem. Ber.* **1993**, *126*, 2501–2504.
[16] T. Yamato, M. Haraguchi, T. Isawa, H. Tsuzuki, S. Ide, *An. Quím., Int. Ed.* **1997**, *93*, 301–309.

[17] T. Yamato, Y. Saruwatari, M. Yasumatsu, H. Tsuzuki, *New J. Chem.* **1998**, 1351–1358.
[18] T. Yamato, Y. Saruwatari, M. Ysaumatsu, S. Ide, *Eur. J. Org. Chem.* **1998**, 309–316.
[19] T. Yamato, M. Yasumatsu, L. K, Doamekpor, S. Nagayama, *Liebigs Ann.* **1995**, 285–289.
[20] a) Y. Okada, J. Nishimura, *J. Incl. Phenom.* **1994**, *19*, 41–53; b) Y. Okada, Y. Kasai, J. Nishimura, *Synlett* **1995**, 85–89.
[21] T. Yamato, *J. Incl. Phenom.* **1998**, *32*, 195–207.
[22] T. Sawada, A. Tsuge, T. Thiemann, S. Mataka, M. Tashiro, *J. Incl. Phenom.* **1994**, *19*, 301–313.
[23] S. K. Chang, I. Cho, *J. Chem. Soc., Perkin Trans. 1* **1986**, 211–214.
[24] M. A. McKervey, E. M. Seward, G. Ferguson, B. Ruhl, S. Harris, *J. Chem. Soc., Chem. Commun.* **1985**, 388–390.
[25] F. Arnaud-Neu, E. M. Collins, M. Deasy, G. Ferguson, S. Harris, J. B. Kaitner, A. J. Lough, M. A. McKervey, E. Marques, B. Ruhl, M. J. Schwing-Weill, E. M. Seward, *J. Am. Chem. Soc.* **1989**, *111*, 8681–8691.
[26] T. Arimura, M. Kuborta, T. Matsuda, O. Manabe, S. Shinkai, *Bull. Chem. Soc. Jpn.* **1989**, *62*, 1674–1676.
[27] K. Iwamoto, S. Shinkai, *J. Org. Chem.* **1992**, *57*, 7066–7073.
[28] K. Araki, N. Hashimoto, H. Otsuka, S. Shinkai, *J. Org. Chem.* **1993**, *58*, 5958–5963.
[29] N. Sato, S. Shinkai, *J. Chem. Soc., Perkin Trans. 1* **1993**, 2671–2673.
[30] S. Gao, S. Inokuma, J. Nishimura, *J. Incl. Phenom.* **1996**, *23*, 329–341.
[31] S. Inokuma, M. Takezawa, H. Satoh, Y. Nakamura, T. Sasaki, J. Nishimura, *J. Org. Chem.* **1998**, *63*, 5791–5796.
[32] T. Yamato, M. Haraguchi, J. Nishikawa, S. Ide, *J. Chem. Soc., Perkin Trans. 1* **1998**, 609–614.
[33] G. Brodesser, F. Vögtle, *J. Incl. Phenom.* **1994**, *19*, 111–135.

Chapter 12

HOMOOXA- AND HOMOAZA-CALIXARENES

BERNARDO MASCI

Dipartimento di Chimica and Centro CNR di Studio sui Meccanismi di Reazione, Università La Sapienza, Box 34-Roma 62 - P.le Aldo Moro 5, 00185 Roma, Italy. e-mail: bernardo.masci@uniroma1.it

1. Introduction

The names homooxacalixarene and homoazacalixarene (sometimes also oxacalixarene and azacalixarene) are currently used to indicate in a specific manner the calixarene analogues in which CH_2 groups are partly or completely replaced by CH_2OCH_2 or CH_2NRCH_2 groups, respectively [1]. The abridged names are not devoid of ambiguity (see Section 2.1.1.), but work well in practice provided direct reference to the basic structures **1**-**8**, is also made. Chapters 11 and 13 describe other classes of calixarene analogues and homologues.

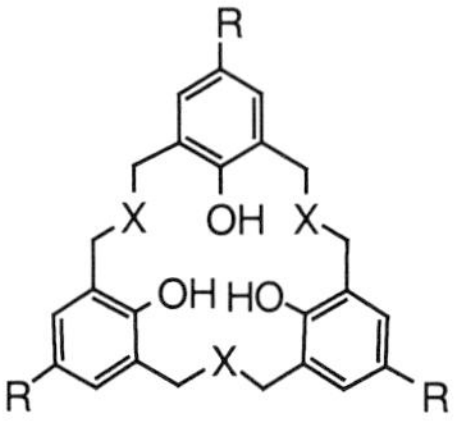

1, X=O hexahomotrioxacalix[3]arene
2, X=NR' hexahomotriazacalix[3]arene

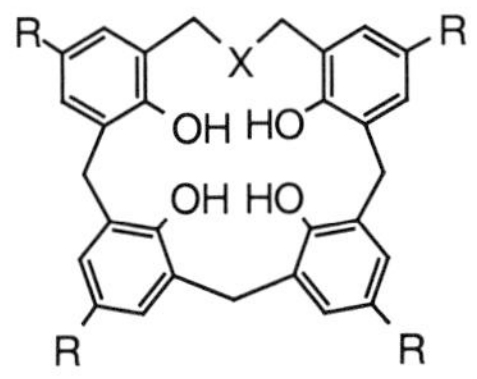

3, X=O dihomooxacalix[4]arene
4, X=NR' dihomoazacalix[4]arene

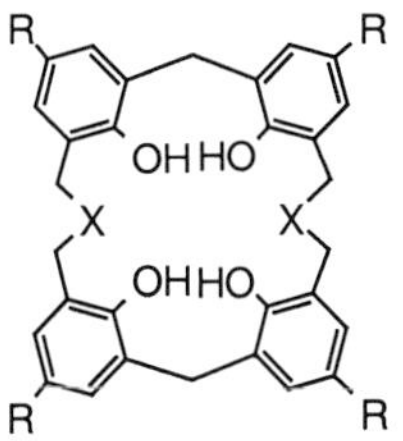

5, X=O tetrahomodioxacalix[4]arene
6, X=NR' tetrahomodiazacalix[4]arene

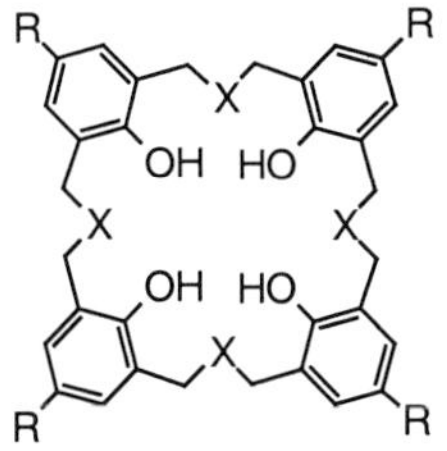

7, X=O octahomotetraoxacalix[4]arene
8, X=NR' octahomotetraazacalix[4]arene

Letters added to compound numbers **1**-**8** identify *p*-substituents as follows:

R	*t*-Bu	Me	Et	*i*-Pr	Cl	OMe	H	Br	Ph	CO_2Et
Letter	**a**	**b**	**c**	**d**	**e**	**f**	**g**	**h**	**i**	**l**

Z. Asfari et al. (eds.), Calixarenes 2001, 235–249.

2. Homooxacalixarenes

2.1. FORMATION OF PARENT COMPOUNDS

A few reports on compounds with the ring system of **1**, **3**, **5**, and **7** can be found in the early literature [2-5], but the chemistry of homooxacalixarenes as a class essentially originated with a paper [6] on the formation of **1a**, **5a**, and **3a** through thermal dehydration of concentrated solutions (0.8-0.2 mol L^{-1}) of bishydroxymethylathed phenols in boiling xylene. Yields were satisfactory from **9** and **10** and almost quantitative from **12**, whilst no characterizable material could be obtained from **11**. Thermal dehydration has been followed to a limited extent in subsequent research on homooxacalixarenes. Compound **3a**, in particular, is much better prepared under conditions like those used for common calixarenes involving a reaction between p-tert-butylphenol, NaOH and formaldehyde [7].

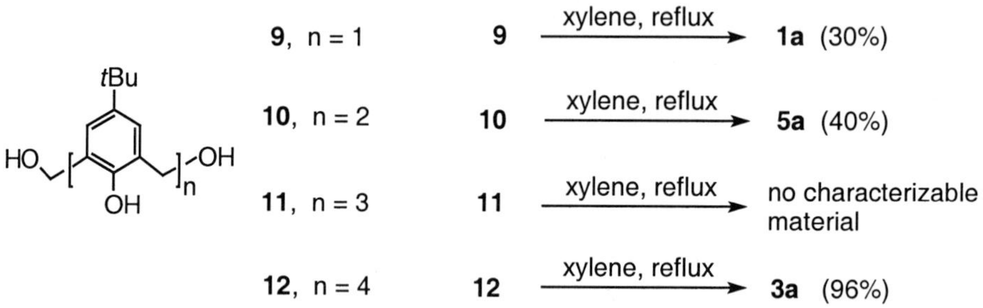

Variable results, including the isolation of the interesting compound **7a** (1%) along with **1a** (6%) [8] have been reported on thermally dehydrating **9** and an alternative acid catalysed procedure, requiring high dilution and a dehydrating agent, has been developed [9] to prepare **1a** and some analogues. Purification was carried out through crystallization of sodium salts of the monoanions of the ligands and the yields were up to 32%. Small quantities of **7a** and analogues were also apparently formed but not isolated [9]. According to another procedure [10], compounds **13** undergo deprotection and acid catalysed high dilution cyclization to **14** in wet chloroform with yields up to 50%. The stepwise construction of **13** from the ArOMOM derivative of **9** or analogues and bromomethylated acetonides also derived from compound **9** or analogues allowed compounds **14** with up to three different R groups to be obtained [10]. Nonetheless, in several instances thermal dehydration appears to have been successfully employed in the preparation of **1a** [11,12] as well as of some interesting analogues [13,14].

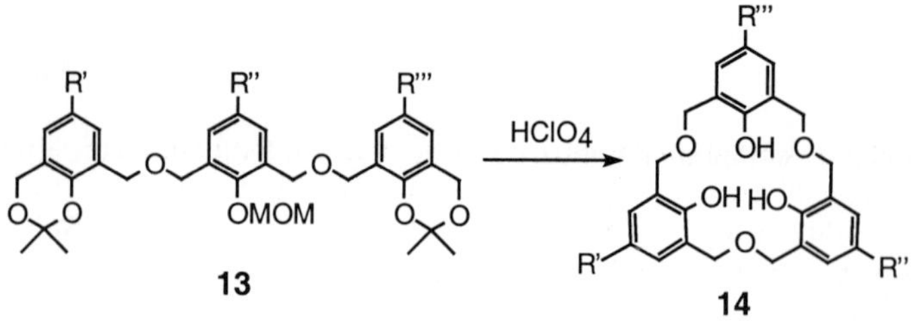

For **5a**, thermal dehydration is the only reported method of synthesis [6] but only very limited study has been made of this compound.

2.1.1. Comparison of synthetic methods
The nature of the *p*-substituent is expected to be important in synthesis in general but only in the case of **1** has the point been investigated in some detail. Acid-catalysed methods [9,10] seem superior if small quantities are needed of a compound with a special *p*-substitution pattern, but thermal dehydration appears to be far more convenient for the preparation of relatively large amounts of the popular **1a**. Acid-catalysed methods are reported to require high dilution (about 200 mL solvent for 1 g precursor) while thermal dehydration works on the scale of 5 mL solvent for 1 g precursor. A rather flat cyclooligomer distribution would be expected for such high concentrations of bifunctional compounds in the absence of special effects. It is evident that intramolecular hydrogen bonding strongly favours the first feasible ring closure in the thermal dehydration, actually acting as a template, not only in the case of **1a**, but also in the case of **3a** and of **5a**. Such effects do not apparently operate in the acid catalysed reactions.

An extensive investigation aimed at controlling thermal dehydration and making available homooxacalixarenes of various structure gave important results in the formation of large homooxacalixarenes [15]. In a modification of an earlier method [6], **10** provided the dimer **5a** along with the trimer **15** and the tetramer **16**. "Irregular" compounds with one CH$_2$OCH$_2$ bridge less than expected [16], namely the well known **3a** and compound **17**, for which a directed synthesis should be very difficult, were also obtained. These large ligands were easily available in quantities of 0.25-1.0 g from 10 g **10** in 27 mL solvent. ^{1}H NMR spectroscopy provided insight into the dynamics of the dehydration process and of the subsequent degradation and transformation of the large oligomers formed [15]. Control of the dehydration reaction has made available many new parent structures, in particular the conditions were found to successfully dehydrate **11** to its dimer **18** and other large compounds [17].

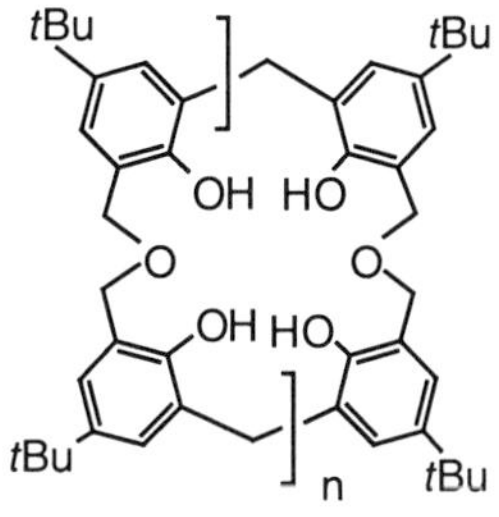

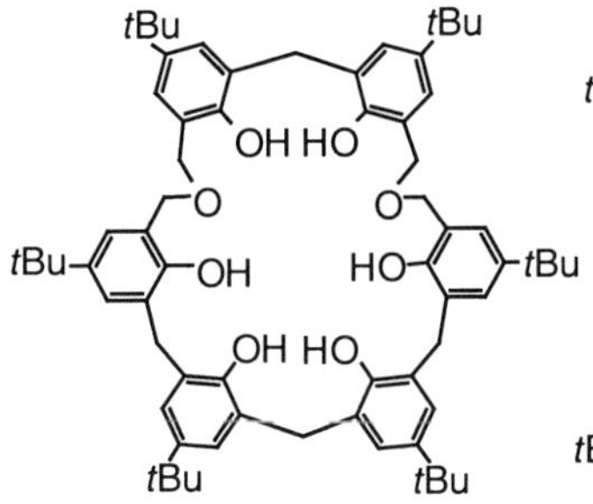

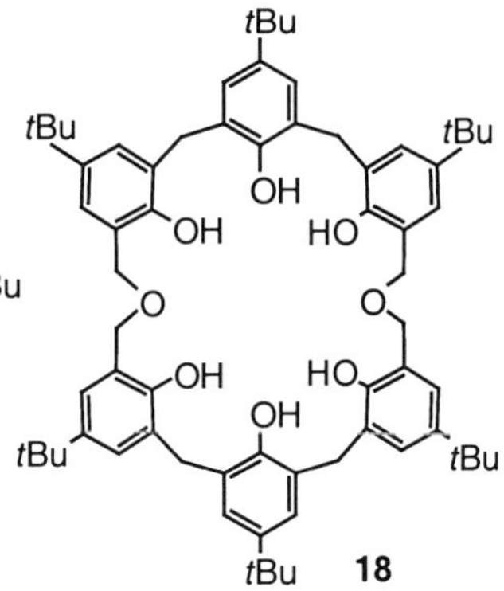

The naming system used for **1**, **3**, **5**, and **7** is not suitable to deal with many more recently characterised structures. Compounds **17** and **18**, for instance, would both be termed tetrahomodioxacalix[6]arenes. A proposed quick and precise identification of the ring system [18] is based on the length of the CH$_2$OCH$_2$ and CH$_2$ bridges, following cyclophane nomenclature. So **17**, which is a [3.1.3.1.1.1]metacyclophane, could be indicated in short as a [3.1.3.1.1.1]homooxacalixarene and the isomeric **18** as a [3.1.1.3.1.1]homooxacalixarene.

2.1.2. Parent homooxacalixarenes and calixarenes
The relationship between homooxacalixarenes and calixarenes is more than purely formal, since they can be both present as the products of a single reaction. In the reactions of phenols with formaldehyde, a perfectly alternating sequence of hydroxymethylation and alkylation steps gives calixarene compounds, while intervening dehydration from two hydroxymethylated functions gives rise to compounds with CH_2OCH_2 bridges. The reversibility of the several steps makes interconversions possible. Compound **3a** is well known [5] to be formed along with typical calixarenes but is not significantly transformed into calixarenes when heated in alkaline solution [6], so that its possible role as an intermediate is unlikely, but **5a** gives *p-tert*-butylcalix[4]arene, *p-tert*-butylcalix[6]arene, *p-tert*-butylcalix[8]arene, and **3a** under the same conditions [6]. Large homooxacalixarenes appear to be easily transformed into lower homologues and give products corresponding to loss of formaldehyde even upon heating in neutral solution [15].

2.2. SUBSTITUENTS ON THE UPPER RIM

As with the simple calixarenes, the *tert*-butyl substituent is commonly encountered. The sensitivity of the benzyl ether link in homooxacalixarenes to electrophiles means, however, that it is much more difficult to replace this group with other substituents by conventional debutylation/realkylation procedures. It is therefore generally necessary to vary the *p*-substituent in the starting phenol to obtain different homooxacalixarenes.

The synthesis has been reported of **1a-1e** [9], **1f-1g** [10], **1h** [10,13], **1i** [8,19], **1l** [14], and also of compound **1** with R = $CMe_2C_6H_4$-*p*-OMOM [20], whilst for the other ring systems only **3d** [21], **3i**, and **5i** [22] have been reported. Several compounds of the type **14** with two or three different *p*-substituents have also been prepared [10].

A few upper rim transformations have been conducted on *O*-substituted **1** as part of the synthesis of encapsulating ligands, namely conversion of triesters to triamides [14] and of a tribromoderivative to **19** [23]. The sequence of transformations from **20** to **23** has also been carried out to obtain cage compounds, namely **24** from **22** and 1,3,5-tris(bromomethyl)benzene, and **25** from **23** and 1,3,5-tris(mercaptomethyl)benzene [24].

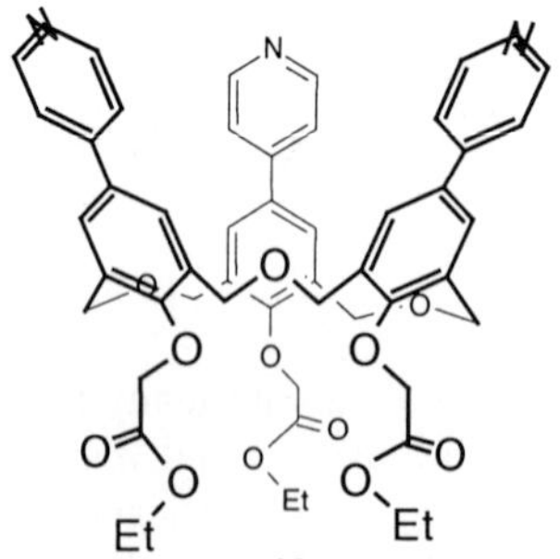

19

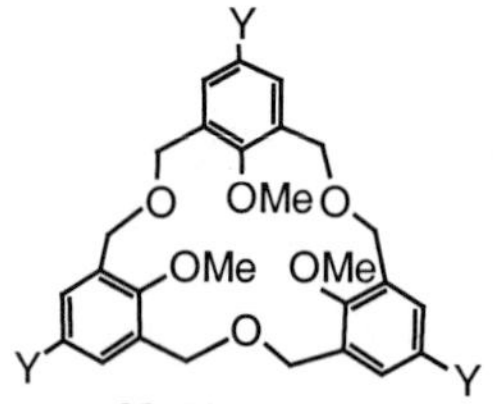

20, Y = Br
21, Y = CHO
22, Y = CH_2OH
23, Y = CH_2Br

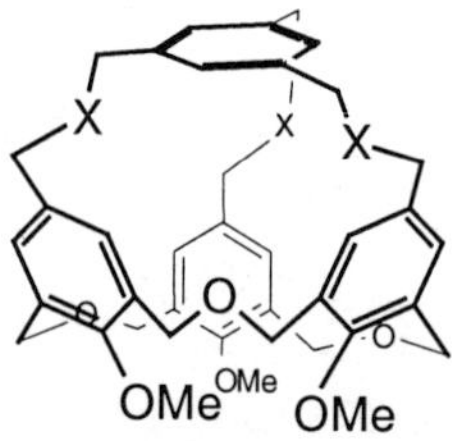

24, X = O
25, X = S

2.3. Substituents on the Lower Rim

In reported tri-*O*-substituted **1a**, groups range from the simple alkyls in **26-29** [25] to groups with additional binding sites for metal ions **30** [11], **31** [26], and **32** [27], to groups containing silicon or phosphorus **33** [28], **34** and **35** [29], **36**, **37**, **42**, and **43** [30], to the peculiar ones of structures **38** [31], **39** [32], **40** [33], and **41** [34]. Compound **44** was also obtained from the intermediate dibutylether compound [35].

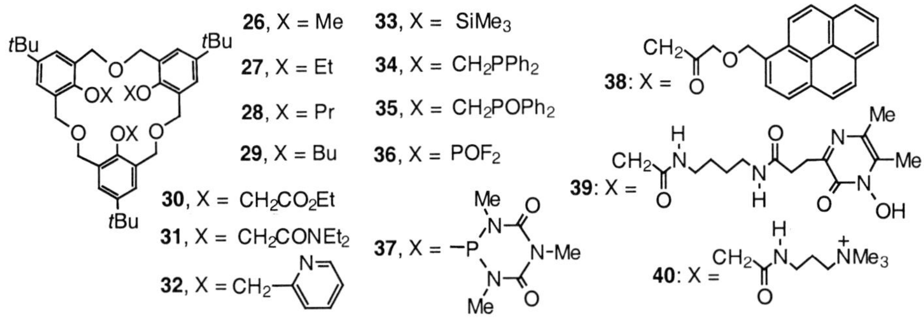

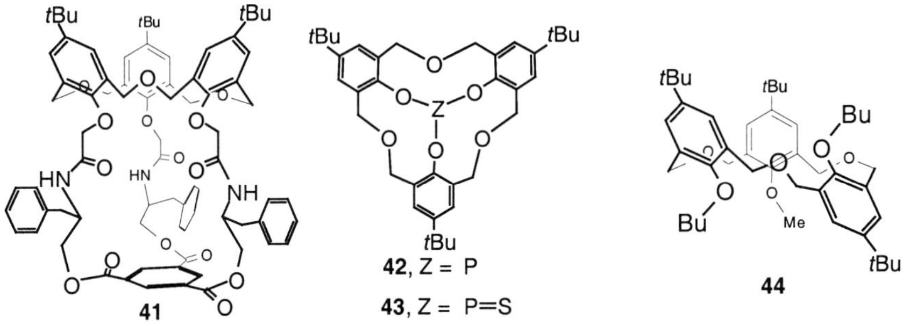

Phenolic-*O* acylation and alkylation reactions of the readily available **3a** have been used to prepare **45** [5], **46-49** [36], **50-53** [37], the calixcrowns **54** and **55** [38] and monoalkyl ethers [39]. Compounds **56** [18] and **57** [40] have been obtained from **5a**. The *p*-phenyl analogue of **57** is also known [41].

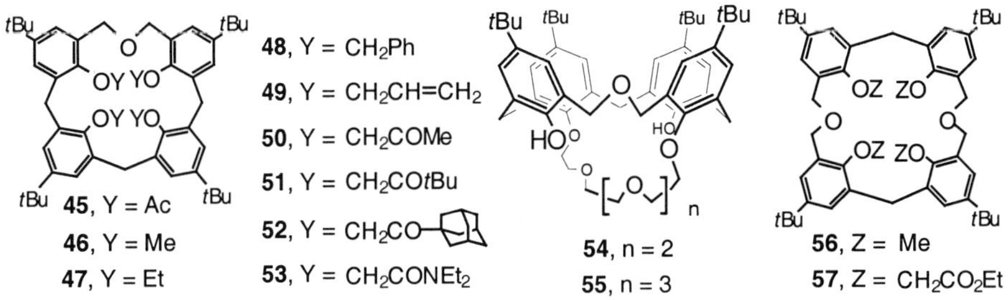

2.3.1. Ether derivatives by-passing the parent macrocycle

O-alkylated homooxacalixarenes can be obtained directly through reactions of precursor ethers and monocyclic, bridged, and doubly bridged derivatives of **7** have been easily prepared [42] by the procedure illustrated for compounds **58-60**.

The method has been extended to obtain **66** and other tricyclic derivatives of **5a** with controlled regiochemistry [43] and also to obtain derivatives **67** and **68** in the unprecedented [3.3.1.1] and [3.3.3.1] systems, respectively [18]. Thus, tetramethylethers are available for all the possible *p-tert*-butylhomooxacalix[4]arenes [18].

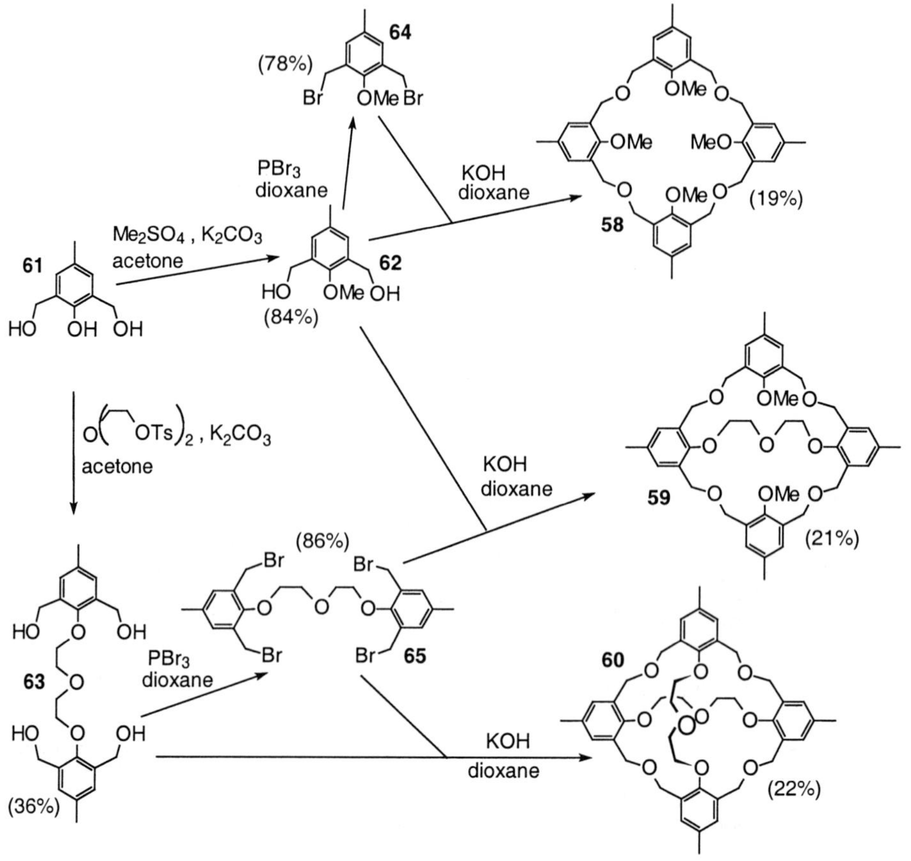

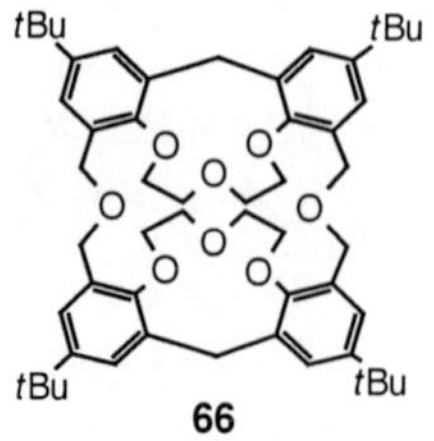

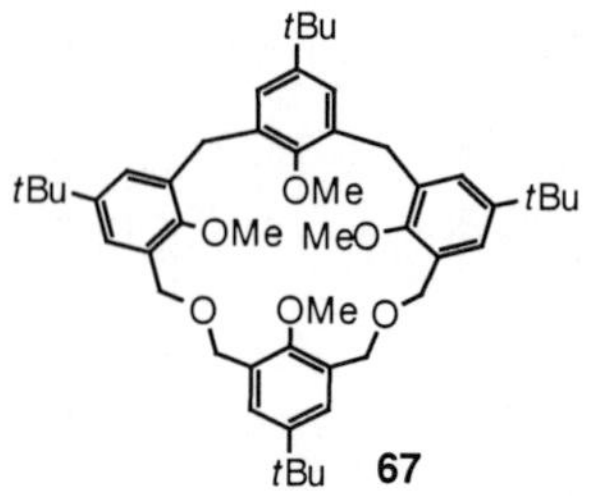

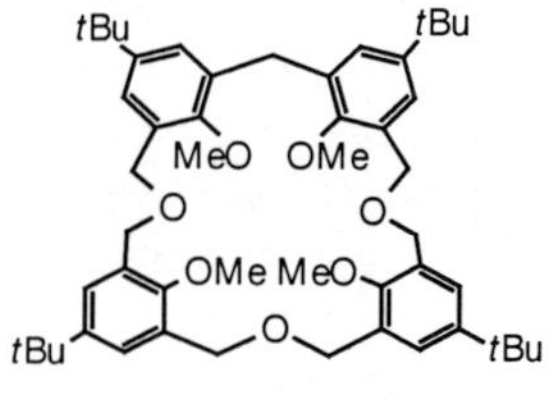

2.4. CONFORMATIONAL STUDIES IN SOLUTION

The substitution of methylene with dimethyleneoxa bridges increases the size of the annulus of the molecule and its conformational mobility. In CDCl$_3$, reported [44] $\Delta G^{\neq}$ barriers for conformational inversion are 12.9, 11.9 and < 9 kcal mol^{-1} for **3a**, **5a**, and **1a**, respectively, *cf.* 15.7, 13.2, 13.3, 12.3, and 15.7 kcal mol^{-1} for *p-t*-Bu-calix[n]arenes, n = 4-8, respectively. Compounds **1a** and **3a** were deduced to be in a somewhat "flattened" *cone* conformation [44]. As with calixarenes, intramolecular H-bonding, whose strength is indicated by the values of δ_{OH} in the ^{1}H NMR spectra, appears to determine conformation in parent compounds [6,8,15,17]. The nature of conformational isomerism for homooxacalixarenes is more complicated than for calixarenes but the main conformations in the whole homooxacalix[4]arenes family are shown in Figure 1 [18].

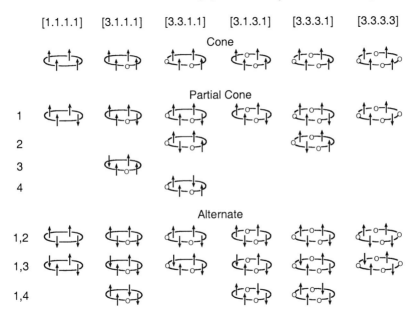

Figure 1. Schematic representation of the main conformations of the various homooxacalix[4]arene systems [18]. Aromatic units were numbered 1-4 starting from the ring containing carbon 1 in calixarene nomenclature and the direction was chosen giving the lowest locants for the oxygens and then for the antiparallel unit in partial cone or parallel units in alternate conformations.

For the tetramethylethers of the *p-tert*-butyl-substituted compounds, the ease of inversion has been found to increase regularly on increasing the number of the dimethyleneoxa bridges from the [1.1.1.1] (calixarene) system to the [3.3.3.3] system of octahomotetraoxacalix[4]arene. In this series, although the ArCH$_2$Ar subunits appear to be *syn* in all cases but one, as deduced from ^{13}C NMR chemical shifts [45], the *cone* conformation is not favoured, *anti* arrangement occurring in ArCH$_2$OCH$_2$Ar moieties [18].

Tetra-alkyl derivatives of **3a** have been isolated in the fixed *cone* conformation with large groups, namely in **48** [36] and **51-53** [37], while with smaller groups that can swing through the annulus the preferred conformation is the *1,4-alternate* (Figure 1), elsewhere designated *1,2-alternate B* [36,37]. The two *partial cone* conformers have been isolated in the case of **50** [37], while inherently chiral monoalkyl [39] and calix-crown species **54** and **55** [40] appear to adopt the *cone* conformation.

In **28** the *partial cone/cone* conformation ratio has been determined by NMR to vary from 6 to 8 in the temperature range 30-100 °C [25], but rotation through the annulus is suppressed in derivatives of **1a** with substituents on the oxygen atoms bulkier than propyl [25,35]. In tri-*O*-substituted derivatives **29** [25], **30** [11], and **32** [27] the *partial cone/cone* yield ratio is higher than one and changes on changing the base counterion from Na$^+$ to K$^+$ to Cs$^+$. The template effects on the stereoselectivity of the alkylation reaction are particularly strong in the case of **31** for which the above ratio has been reported to be 0:100 with NaH in THF and 99:0 with K$_2$CO$_3$ or Cs$_2$CO$_3$ in acetone [26].

Theoretical studies have been carried out on the conformation of **16** and derivatives [46]. Experimental study of a conformational equilibrium has been made in the case of the interconversion of the "capsule" **69** and the "self-threaded rotaxane" **70** [14].

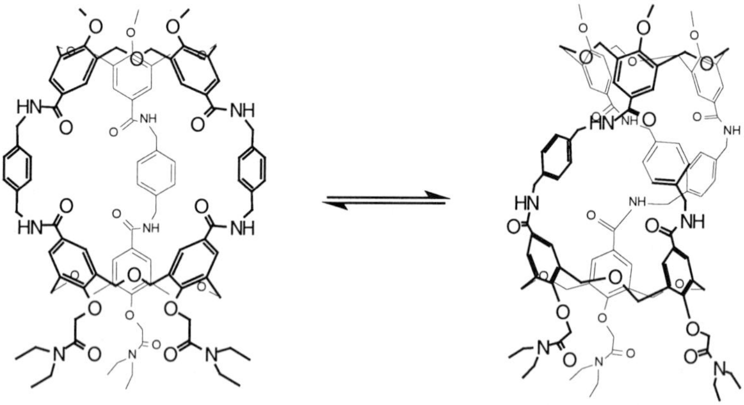

69 ⇌ **70**

2.5. SOLID STATE INVESTIGATIONS

Intramolecular H-bonding is considered to determine the *cone* conformation of **1a**, **5a**, and **3a** found for the solids by X-ray crystallography. The presence of guest solvent molecules changes the depth of the bowl and the symmetry in **3a** and **3d** [47,48,21]. Compound **1a** was found to exist as a shallow nest due to its three intramolecular hydrogen bonds [49], while in two analogues an extended network of six intramolecular hydrogen bonds has been observed, each phenolic hydrogen atom being shared with two oxygen atoms (a phenolic oxygen and a dibenzyl ether oxygen) [10]. Such networks of bifurcated intramolecular hydrogen bonds have been recently observed also in the distorted *cone* of **5a** [50] and may be of common occurrence with parent homooxacalixarenes. Compounds **46**, **47**, and **49** are reported [51] to exhibit the *1,4-alternate* conformation (Figure 1) in the crystal. The crystal structure has been determined of the *partial cone* isomer of **33** [28] and of a triester derivative of **1i** [19]. The *1,4-alternate* conformation was found in crystals of **57** [40] and of its *p*-phenyl analogue [41]. The same conformation was found in the crystals of **66** [52].

The nature of the cation-anion interaction in the solid state complex of tetraethylammonium salt of the monoanion of **16** as well as of the corresponding typical calix[4]arene compound has been explored [53]. Complexes with Rh, Au and Mo have been obtained from **34** [29]. Crystal structures are known for complexes of Ti(IV) [54], Sc(III) [55], V(V) [56], lanthanides [57] and uranyl ion [58] with the trianion of **10** or *p*-substituted analogues. (See Chapters 28, 29, 30).

2.6. COMPLEXATION IN SOLUTION

The oxygen atoms in the ring are potential binding sites and this prompted study of complexation and liquid-liquid extraction of metal ions. The parent compounds **1** do not bind alkali metal ions [9], while simple derivatives **26-29** are effective but do not show marked selectivity in the extraction of Na^+, K^+, and Cs^+ ions [26]. Among derivatives **30-32** [11,26,27] with additional binding sites on the lower rim, **31** in the fixed *cone* conformation appears to be a particularly strong and selective ligand with association constants in $THF/CHCl_3$ 1:1 at 25 °C larger than 10^7 M^{-1} for Na^+, Ca^{2+}, and Ba^{2+} ions [26]. Inclusion in the cage, with four cation-π and several cation-oxygen dipole interactions, probably occurs in the complex between **24** and Cs^+ ion, for which a high extraction from water into dichloromethane and a high selectivity are observed [24].

The shallow nest of hexahomotrioxacalix[3]arenes, both as the simple ligand and as subunits in capsular hosts [20,23,33], has proved to be particularly effective in complexation of fullerenes (see also Chapter 26).

The C_3 symmetry of the oxygen atom array in derivatives of **1** is suitable for complexation of primary ammonium ions [25]. The *cone* form tributylether of **1a** showed a high affinity for $BuNH_3^+$ ion [25], while chiral recognition of optically active primary alkylammonium ions was possible with **44** having fixed, opposed butyl groups [35]. A fluorimetric sensing system (see Chapter 34) for primary ammonium ions was developed on **38** [31] and the C_3 symmetrically capped derivative **41** appears to be a well preorganized host molecule for these ions at low temperatures [34]. Cooperative binding of Ga(III) and alkylammonium ions occurred for **39** [32].

2.6.1. Complexation of quaternary ammonium ions
Homooxacalixarenes strongly bind quaternary ammonium ions in lipophilic solvents. In the absence of such relatively strong interactions as hydrophobic forces, hydrogen bonding and ion pairing with the ligand, this binding essentially relies on the cation-π interaction [59]. Published results indicate [12] that the simple parent compounds **1a**, **5a**, and **3a** bind acetylcholine and several other quaternary ammonium ions in $CDCl_3$, with marked upfield shift of the ^1H NMR signals of the guest protons. Compound **1a** proves to be the most efficient ($-\Delta G°$ up to 2.7 kcal mol^{-1}) and the most interesting ligand in the set. Complexation apparently occurs thank to intramolecular hydrogen bonding that organises the cavities in *cone*-like arrangements [12]. Capsule molecules analogous to **69** have been developed for complexation of large organic cations. Methyl viologen ditosylate and similar salts were complexed with $-\Delta G°$ up to 3.6 kcal mol^{-1} [14]. A $-\Delta G°$ value as large as 4.4 kcal mol^{-1} was observed for complexation of tetramethylammonium picrate in $CDCl_3$ at 303 K with ligand **66** [43]. Contrary to initial

supposition, this ligand is not in a *cone* conformation [52], but the latter can be obtained with different bridging groups [43,60].

Even in the absence of hydrogen bonding, bridging and other specific elements of preorganization, homooxacalix[4]arenes are effective ligands for quaternary ammonium ions. All the tetramethylethers of *p-tert*-butylhomooxacalix[4]arenes, which do not adopt a *cone* conformation when free, effectively bind tetramethylammonium picrate and *N*-methylpyridinium iodide in halogenated solvents [18]. Changes in the strength of the complexation as a function of the cavity size are shown in Figure 2. There is a marked preference for tetramethylammonium picrate in the case of the [3.1.3.1] system [18]. Both values of the association constants and the location of the guest in the complex could be determined by ^{1}H NMR spectroscopy. The relative importance of the shielding in the various positions of *N*-methylpyridinium ion protons indicates a deeper and deeper inclusion on increasing the cavity size [18].

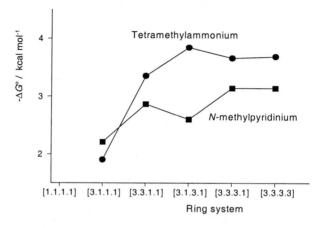

Figure 2. Free energy of complexation of tetramethylammonium picrate and N-methylpyridinium iodide in CDCl$_2$CDCl$_2$ at 303 K for the whole family of tetramethylethers of p-tert-butylhomooxacalix[4]-arenes, indicated according to the metacyclophane ring system [18].

2.7. CONCLUDING COMMENT

Interesting further developments of homooxacalixarene chemistry can be expected as a result of: *i*) making available several parent compounds and *ii*) directly obtaining simple or highly preorganized derivatives. A high potential in host-guest chemistry resides partly in the preorganization of the shallow cup of **1** and derivatives, and in the insertion into a calixarene-like structure of flexible dimethyleneoxa bridges. Systems already known have larger cavities and greater flexibility than their simple calixarene analogues. The relatively smooth change in the cavity size on increasing the number of dimethyleneoxa bridges should be compared with the coarse regulation available in the typical calix[n]arene series.

3. Homoazacalixarenes

3.1. FORMATION OF HOMOAZACALIXARENES

In homoazacalixarene compounds, substituents can be present in side arms as well as on the upper and lower rim. *N*-substituted compounds in the series of **2, 6**, and **4** have been obtained through the condensation of bishydroxymethylated phenols and polyphenols with alkylamines [61,62], as shown for **72-74**. Water was azeotropically removed while heating relatively concentrated (about 0.1 mol L^{-1}) solutions of compounds **71, 10** or **12** and macrocycles were obtained in fairly good yields, apparently due to intramolecular hydrogen-bonding promoting cyclization. The *p*-chloro analogue of **72** also has been reported [63], while other successfully employed amines are methylamine [64], 2-picolylamine and (S)-(-)-α-methylbenzylamine [65]. Compound **76** can be prepared by a high dilution procedure [66] from **75** and methylglycinate, though it is difficult to separate from some tetrameric product. The last method has been extended to obtain several compounds incorporating chiral amino acid residues in the series of **6a** and **6b**, as reported in the case of **78** [67].

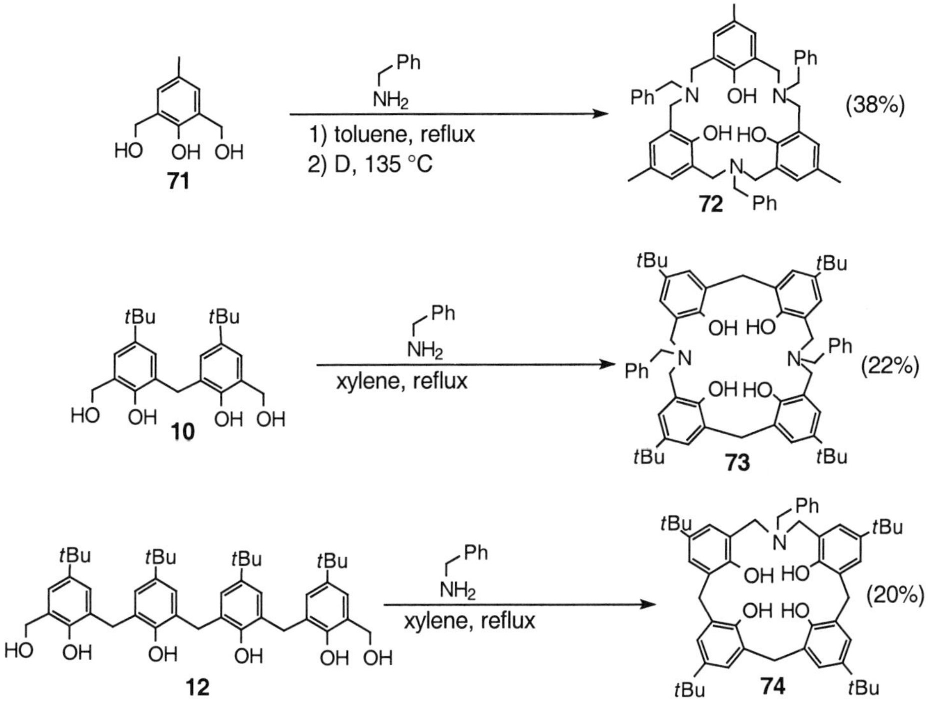

Interesting, from a mechanistic point of view, is the formation of **74** in 8% yield from *p-tert*-butylcalix[4]arene after boiling for 150 h in xylene with benzylamine, paraformaldehyde and KOH in a catalytic amount [65]. An *N*-unsubstituted homoazacalixarene, **82,** was obtained through reaction of **79** with **80** in the presence of Ni(II) or Zn(II) salts, and reduction with NaBH$_4$ of the Ni$_4$ or Zn$_4$ complexes of the intermediate macrocyclic Shiff base **81** [68-70]. *O*-silylation has been carried out on compound **76** [71], while *O*-methylated compounds can be formed by reacting 2,6-bis(bromomethyl)-4-*tert*-butylanisole with *p*-toluensulfonamide and NaH in DMF [65]. Besides the [3.3.3.3]metacyclophane, the apparently very strained [3.3]metacyclophane is reported to be formed. Hydrolysis of the methoxy and *N*-tosylamide functions could not be carried out satisfactorily [65].

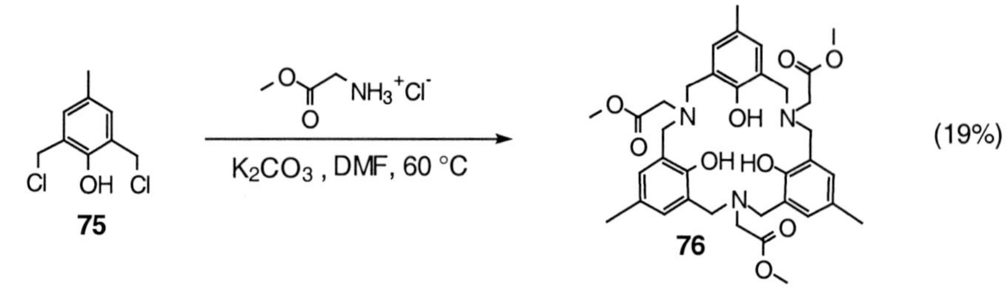

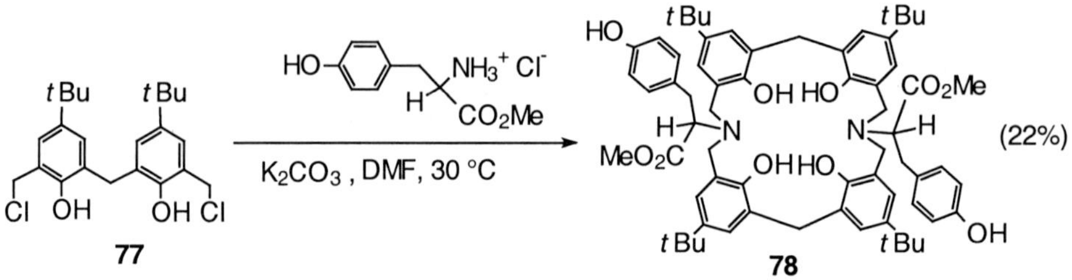

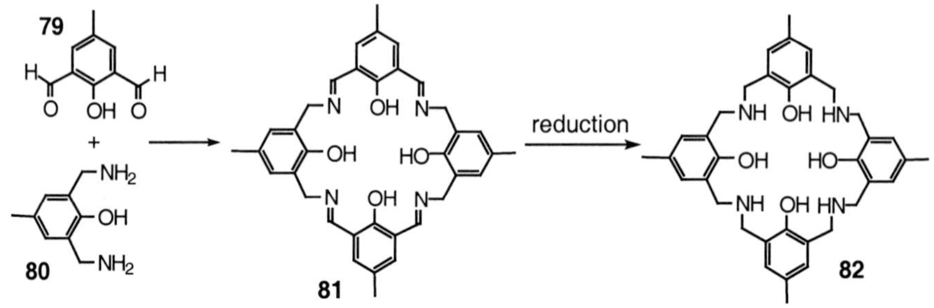

3.1.1. Special homoazacalixarenes

The literature on homoazacalixarenes is more sparse than that on homooxacalixarenes, but their chemistry is in principle even richer. Both the reactivity of the amino groups and the interaction of the substituents in the side arms with those on the upper and on the lower rim should be considered.

Compound **83**, an intermediate that can be isolated during the reduction of **81** with NaBH$_4$, and **84**, a product obtained from **82** and formaldehyde, have side arms covalently linked to the lower rim [70]. Modification of the known synthesis [62] of tetra-homodiazacalix[4]arenes provided compound **86** and also compound **87**, which is the first reported homoazaoxacalixarene [64]. Also derived was the salt **88**. Treatment of **88** with NaHCO$_3$ in methanol provided the azacalix betaine **89** [64]. Complex **90** (see Chapter 31 for the full structure) can also be considered as deriving from the triply zwitterionic form of a homoazacalixarene [63].

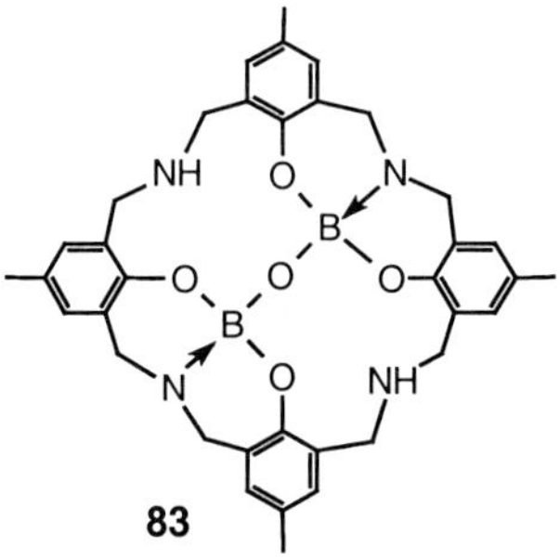

3.2. PROPERTIES OF HOMOAZACALIXARENES

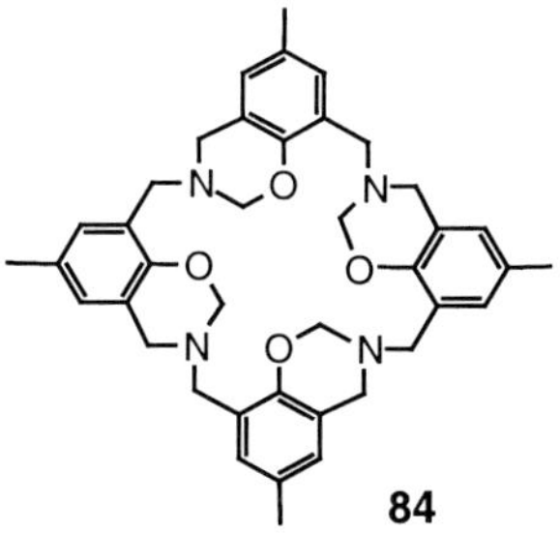

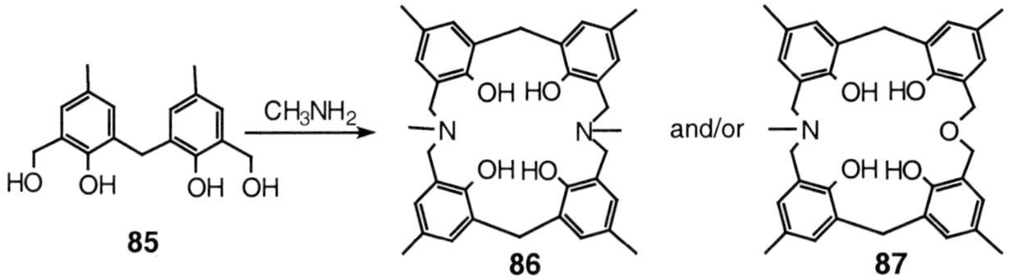

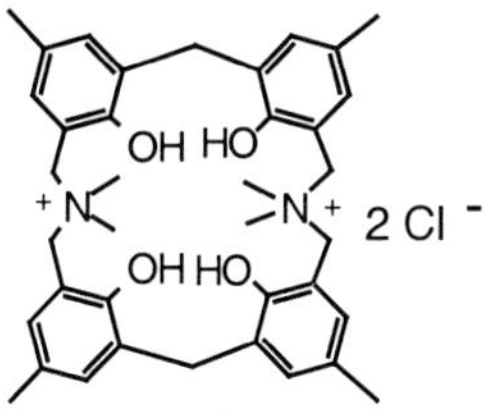

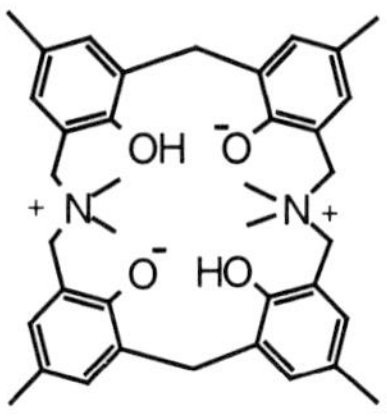

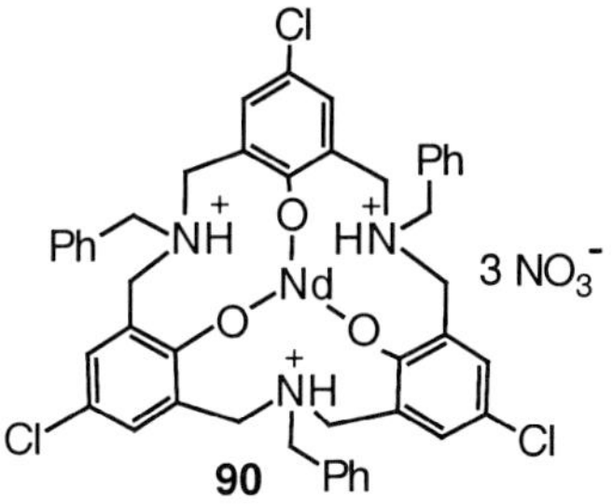

N-benzyl substituted homoazacalixarenes have very strong intramolecular OH⋯N hydrogen bonds, as indicated by IR absorptions in the range 2700-3000 cm^{-1} and by ^{1}H NMR spectra showing abnormally low-field-shifted OH signals in the range 10.7-11.6 ppm. The strength of the hydrogen bonds has also been considered the origin of the high barrier for conformational inversion of **74** (15.9 and 17.8 kcal mol^{-1} in Me$_2$SO-d_6 and toluene-d_{10}, respectively) [72]. At low temperatures, **74** showed six ^{1}H NMR signals with shifts down to 17.1 ppm, assigned to the predominant *alternate* and *cone* species [65]. In liquid-liquid alkali metal picrate extraction experiments, **72** proved to be a weak complexant with marked K$^+$ selectivity [61], while the analogue with 2-picolyl substituents in the side arms was much more effective although less selective [65]. Complexation of racemic α-methylbenzyl trimethylammonium iodide in CDCl$_3$ by **78** resulted in signal splitting in the ^{1}H NMR spectrum of the salt [67]. The four pK_a values have been determined in MeOH for compound **88**: pK_1 = 4.50, pK_2 = 6.17, pK_3 = 11.88, and pK_4 = 12.01, the doubly zwitterionic form **89** predominating in the pH range 8-10 [64].

Structure determinations by X-ray crystallography reveal a shallow cup-like structure for two parent hexahomotriazacalix[3]arenes [63,66]. In the reported structure of a parent tetrahomodiazacalix[4]arene the same overall shape as observed in a homo-oxa-analogue is found, with OH···N interactions prevailing over OH···O [50]. The crystal structure of an O-silylated derivative of **76** has been also reported [71].

A dome shaped cation is observed in the X-ray crystal structure of the solvated salt of **82** with the composition **82**·4HCl·H_2O·CH_3OH·C_2H_5OH, the water molecule and two chloride ions being H-bonded inside the cavity [70]. The structure has also been reported of complexes of **82** with four Zn(II) and with three Co(III) cations inside the cavity [70].

4. References and Notes

[1] a) C. D. Gutsche, *"Calixarenes"* Royal Society of Chemistry, Cambridge, England, **1989**, Ch. 2; b) C. D. Gutsche, *"Calixarenes Revisited"* Royal Society of Chemistry, Cambridge, England, **1997**, Ch. 2.
[2] H. von Euler, E. Adler, B. Bergstrom, *Ark. Chemi, Mineral. Geol.* **1941**, *14B, Nr. 30*, 1.
[3] H. Kämmerer, M. Dahm, *KunstPlast. (Solothurn, Switz.)* **1959**, *6*, 20-25.
[4] K. Hultzsch, *Kunststoffe* **1962**, *52*, 19-24.
[5] C. D. Gutsche, B. Dhawan, K. H. No, R. Muthukrishnan, *J. Am. Chem. Soc.* **1981**, *103*, 3782-3792.
[6] B. Dhawan, C. D. Gutsche, *J. Org. Chem.* **1983**, *48*, 1536-1539.
[7] C. Bavoux, F. Vocanson, M. Perrin, R. Lamartine, *J. Incl. Phenom.* **1995**, *22*, 119-130.
[8] P. Zerr, M. Mussrabi, J. Vicens, *Tetrahedron Lett.* **1991**, *32*, 1879-1880.
[9] P. D. Hampton, Z. Bencze, W. Tong, C. E. Daitch, *J. Org. Chem.* **1994**, *59*, 4838-4843.
[10] K. Tsubaki, T. Otsubo, K. Tanaka, K. Fuji, *J. Org. Chem.* **1998**, *63*, 3260-3265.
[11] K. Araki, N. Hashimoto, H. Otsuka, S. Shinkai, *J. Org. Chem.* **1993**, *58*, 5958-5963.
[12] B. Masci, *Tetrahedron* **1995**, *51*, 5459-5464.
[13] A. Ikeda, Y. Suzuki, M. Yoshimura, S. Shinkai, *Tetrahedron* **1998**, *54*, 2497-2508.
[14] Z. Zhong, A. Ikeda, S. Shinkai, *J. Am. Chem. Soc.* **1999**, *121*, 11906-11907.
[15] B. Masci, *Tetrahedron*, in press.
[16] Loss of formaldehyde is long known to take place along with dehydration in the hardening of Bakelite and related phenol-formaldehyde chemistry, see for instance: a) K. Hultzsch, *Ber.* **1941**, *74*, 898-904; b) A. Zinke, E. Ziegler, I. Hontschik, *Monats. Chem.* **1948**, *78*, 317-324.
[17] B. Masci, *J. Org. Chem.*, in press.
[18] B. Masci, M. Finelli, M. Varrone, *Chem. Eur. J.* **1998**, *4*, 2018-2030.
[19] S. Khrifi, A. Guelzim, F. Baert, M. Mussrabi, Z. Asfari, J. Vicens, *Acta Cryst.* **1995**, *51 C*, 153-157.
[20] K. Tsubaki, K. Tanaka, T. Kinoshita, K. Fuji, *Chem. Commun.* **1998**, 895-896.
[21] K. Suzuki, A. E. Armah, S. Fujii, K. Tomita, Z. Asfari, J. Vicens, *Chem. Lett.* **1991**, 1699-1702.
[22] K. No, *Bull. Korean Chem. Soc.* **1999**, *20*, 33-34.
[23] A. Ikeda, M. Yoshimura, F. Tani, Y. Naruta, S. Shinkai, *Chem. Lett.* **1998**, 587-588; b) A. Ikeda, M. Yoshimura, H. Udzu, K. Fukuhara, S. Shinkai, *J. Am. Chem. Soc.* **1999**, *121*, 4296-4297.
[24] K. Araki, H. Hayashida, *Tetrahedron Lett.* **2000**, *41*, 1807-1810.
[25] K. Araki, K. Inada, H. Otsuka, S. Shinkai, *Tetrahedron* **1993**, *49*, 9465-9478.
[26] H. Matsumoto, S. Nishio, M. Takeshita, S. Shinkai, *Tetrahedron* **1995**, *51*, 4647-4654.
[27] T. Yamato, M. Haraguchi, J.-I. Nishikawa, S. Ide, H. Tsuzuki, *Can. J. Chem.* **1998**, *76*, 989-996.
[28] P. D. Hampton, C. E. Daitch, E. N. Duesler, *New. J. Chem.* **1996**, *20*, 427-432.
[29] C. B. Dieleman, D. Matt, I. Neda, R. Schmutzler, A. Harriman, R. Yaftian, *Chem. Commun.* **1999**, 1911-1912.
[30] I. Neda, T. Kaukorat, R. Schmutzler, *Main Group Chem. News* **1998**, *63*, 4-29.
[31] M. Takeshita, S. Shinkai, *Chem. Lett.* **1994**, 125-128.
[32] J. Ohkanda, H. Shibui, A. Katoh, *Chem. Commun.* **1998**, 375-376.
[33] A. Ikeda, T. Hatano, M. Kawaguchi, H. Suenaga, S. Shinkai, *Chem. Commun.* **1999**, 1403-1404.
[34] M. Takeshita, F. Inokuchi, S. Shinkai, *Tetrahedron Lett.* **1995**, *36*, 3341-3344.
[35] K. Araki, K. Inada, S. Shinkai, *Angew. Chem., Int. Ed. Engl.* **1996**, *35*, 72-74.

[36] a) P. M. Marcos, J. R. Ascenso, R. Lamartine, J. L. C. Pereira, *Supramol. Chem.* **1996**, *6*, 303-306; b) P. M. Marcos, J. R. Ascenso, R. Lamartine, J. L. C. Pereira, *Tetrahedron* **1997**, *53*, 11791-11802.
[37] S. Félix, J. R. Ascenso, R. Lamartine, J. L.C. Pereira, *Tetrahedron* **1999**, *55*, 8539-8540.
[38] S. Félix, J. R. Ascenso, R. Lamartine, J. L.C. Pereira, *Synth. Commun.* **1998**, *28*, 1793-1799.
[39] P. M. Marcos, J. R. Ascenso, R. Lamartine, J. L.C. Pereira, *J. Org. Chem.* **1998**, *63*, 69-74.
[40] F. Arnaud-Neu, S. Cremin, D. Cunningham, S. J. Harris, P. McArdle, M. A. McKervey, M. McManus, M.-J. Schwing-Weill, K. Ziat, *J. Incl. Phenom.* **1991**, *10*, 329-339.
[41] K. No, Y. J. Park, E. J. Choi, *Bull. Korean Chem. Soc.* **1999**, *20*, 905-909.
[42] B. Masci, S. Saccheo, *Tetrahedron* **1993**, *49*, 10739-10748.
[43] G. De Iasi, B. Masci, *Tetrahedron Lett.* **1993**, *34*, 6635-6638.
[44] C. D. Gutsche, L. J. Bauer, *J. Am. Chem. Soc.* **1985**, *107*, 6052-6059.
[45] C. Jaime, J. De Mendoza, P. Prados, P. M. Nieto, C. Sanchez, *J. Org. Chem.* **1991**, *56*, 3372-3376.
[46] a) R. J. Bernardino, B. J. C. Cabral, J. L. C. Pereira, *Theochem.* **1998**, *455*, 23-32; b) M. A. Santos, P. M. Marcos, J. L. C. Pereira, *Theochem.* **1999**, *463*, 21-26.
[47] a) M. Perrin, C. Bavoux, S. Lecocq, *Supramol. Chem.* **1996**, *8*, 23-29; b) C. Bavoux, M. Perrin, *J. Incl. Phenom.* **1992**, *14*, 247-256; c) J. M. Harrowfield, M. I. Ogden, A. H. White, *J. Chem. Soc., Dalton Trans.* **1991**, 979-985; d) A. E. Armah, K. Suzuki, S. Fujii, K. Tomita, Z. Asfari, J. Vicens, *Acta Cryst.* **1992**, *48 C*, 1474-1476.
[48] K. Tomita, K. Suzuki, H. Ohishi, I. Nakanishi, *J. Incl. Phenom.* **2000**, *37*, 341-357.
[49] K. Suzuki, H. Minami, Y. Yamagata, S. Fujii, K. Tomita, Z. Asfari, J. Vicens, *Acta Cryst.* **1992**, *48 C*, 350-352.
[50] P. Thuéry, M. Nierlich, J. Vicens, B. Masci, H. Takemura, *Eur. J. Inorg. Chem.*, in press.
[51] M. Perrin, S. Lecocq, P. M. Marcos, J. L. C. Pereira, *Supramol. Chem.* **1998**, *9*, 137-141.
[52] B. Masci, G. Portalone: unpublished results.
[53] J. M. Harrowfield, M. I. Ogden, W. R. Richmond, B. W. Skelton, A. H. White, *J. Chem. Soc., Perkin Trans. 2* **1993**, 2183-2190.
[54] P. D. Hampton, C. E. Daitch, T. M. Alam, Z. Bencze, M. Rosay, *Inorg. Chem.* **1994**, *33*, 4750-4758.
[55] C. E. Daitch, P. D. Hampton, E. N. Duesler, *Inorg. Chem.* **1995**, *34*, 5641-5645.
[56] P. D. Hampton, C. E. Daitch, T. M. Alam, E. A. Pruss, *Inorg. Chem.* **1997**, *36*, 2879-2883.
[57] C. E. Daitch, P. D. Hampton, E. N. Duesler, T. D. Alam, *J. Am. Chem. Soc.* **1996**, *118*, 7769-7773.
[58] P. Thuéry, M. Nierlich, B. Masci, Z. Asfari, J. Vicens, *J. Chem. Soc., Dalton Trans.* **1999**, 3151-3152.
[59] J. C. Ma, D. A. Dougherty, *Chem. Rev.* **1997**, *97*, 1303-1324.
[60] B. Masci, G. De Iasi, M. Gabrielli, R. Cacciapaglia, unpublished results.
[61] H. Takemura, K. Yoshimura, I. U. Khan, T. Shinmyozu, T. Inazu, *Tetrahedron Lett.* **1992**, *33*, 5775-5778.
[62] I. U. Khan, H. Takemura, M. Suenaga, T. Shinmyozu, T. Inazu, *J. Org. Chem.* **1993**, *58*, 3158-3161.
[63] P. Thuéry, M. Nierlich, J. Vicens, H. Takemura, *J. Chem. Soc., Dalton Trans.* **2000**, 279-283.
[64] H. Takemura, K. Yoshimura, T. Inazu, *Tetrahedron Lett.* **1999**, *40*, 6431-6434.
[65] H. Takemura, T. Shinmyozu, H. Miura, I. U. Khan, *J. Incl. Phenom.* **1994**, *19*, 189-206.
[66] P. D. Hampton, W. Tong, S. Wu, E. N. Duesler, *J. Chem. Soc., Perkin Trans. 2* **1996**, 1127-1130.
[67] K. Ito, T. Ohta, Y. Ohba, T. Sone, *J. Heterocyclic Chem.* **2000**, *37*, 79-85.
[68] M. Bell, A. J. Edwards, B. F. Hoskins, E. H. Kachab, R. Robson, *J. Am Chem. Soc.* **1989**, *111*, 3603-3610.
[69] M. J. Grannas, B. F. Hoskins, R. Robson, *J. Chem. Soc., Chem. Commun.* **1990**, 1644-1646.
[70] M. J. Grannas, B. F. Hoskins, R. Robson, *Inorg. Chem.* **1994**, *33*, 1071-1079.
[71] P. Chirakul, P. D. Hampton, E. N. Duesler, *Tetrahedron Lett.* **1998**, *39*, 5473-5476.
[72] H. Takemura, T. Shinmyozu, T. Inazu, *Coord. Chem. Rev.* **1996**, *156*, 183-200.

Chapter 13

HETEROCALIXARENES

MYROSLAV VYSOTSKY, MOHAMED SAADIOUI, VOLKER BÖHMER
*Johannes Gutenberg-Universität, Fachbereich Chemie und Pharmazie,
Abteilung Lehramt Chemie, Duesbergweg 10-14,
D-55099 Mainz, Germany.
e-mail: vboehmer@mail.uni-mainz.de*

1. Introduction

The name calixarene, originally coined for the family of methylene bridged cyclic oligomers of phenols, has been gradually extended to compounds with similar aromatic units (e.g., resorcinols, pyrogallol, naphthols) or more or less similar bridges (CHR, S etc.). In recent years the prefix "calix" has been used frequently and sometimes inappropriately in reference to cyclic oligomers which are not $[1_n]$metacyclophanes. This chapter concerns compounds of these types not otherwise discussed in chapters 1, 6, 11 and 12.

2. Calixarenes from Heterocycles

2.1. CALIXPYRROLES

Calixpyrroles **1** (also known as porphyrinogens) are among the oldest of the synthetic cyclo-oligomers, having first been obtained by Baeyer in 1886 by condensation of acetone with pyrrole [1]. Later, the synthesis was refined by several other research groups [2] and the calixpyrroles have been recently resurrected in a number of laboratories [3-6] as putative anion binders and metal complexing agents. A variety of ketones, including alkanones, cycloalkanones, and acetophenones (which fail in the furan condensation, see section 2.2) have been used [7] in place of acetone. The products in most cases are the calix[4]pyrroles **1**, but when diaryl-di-(2-pyrrolyl)methanes are condensed with acetone in the presence of CF_3CO_2H calix[6]pyrroles **2b** are isolated in yields as high as 25% [4]. (For another access to calix[6]pyrroles see section 2.4.) Other acid catalysts such as $BF_3 \cdot Et_2O$ and $MeSO_3H$, however, give only the calix[4]pyrroles **2a** [4,8].

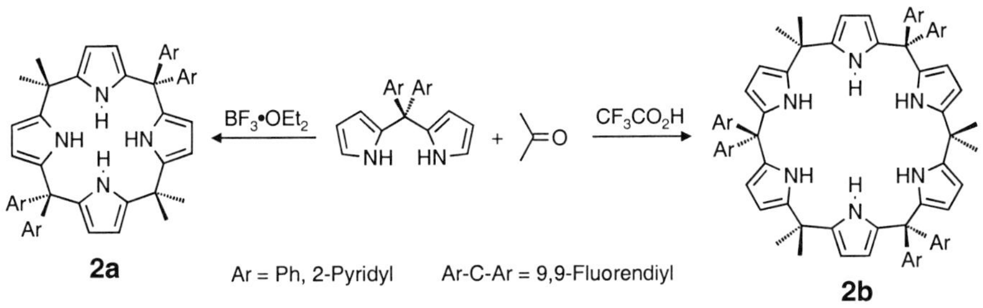

Although the condensation of pyrrole generally proceeds via the 2,5-positions leading to structure **1**, isomeric structures have also been found in which one or more of the pyrrole rings are incorporated *via* the 2,4-positions [9]. The distribution of isomers is strongly dependent on solvent, catalyst and temperature. With cyclohexanone, for example, MeSO$_3$H in EtOH (reflux, 4 h) gives up to 22% of calix[4]pyrrole **3** with one inverted pyrrole ring, while *p*-CH$_3$C$_6$H$_4$SO$_3$H in CHCl$_3$ (reflux, 60 h) gives 57% of calix[4]pyrrole **4** with two adjacent inverted pyrrole rings [9,10]. Condensations of pyrrole with hydroxy acetophenones in the presence of MeSO$_3$H led to mixtures of all possible configurational isomers [11], which were separated by column chromatography [3a,5d]. The hydroxy groups at the aryl moieties can be used to introduce additional functionalities [6]. Condensation of pyrrole with ketone mixtures occurs randomly, leading to products with two different bridges [12], for example, **5**.

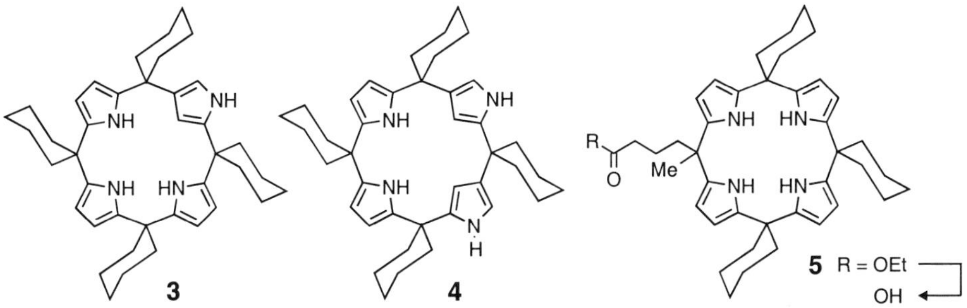

A novel series of macrocycles **6-8**, called *calixphyrins*, intermediate between porphyrins and calixpyrroles, have been prepared by acid-catalysed condensation of mesityldipyrrylmethane with acetone followed by DDQ oxidation. Calix[4]- and -[8]phyrin (**6, 8**) were characterised by X-ray crystallography [13,14].

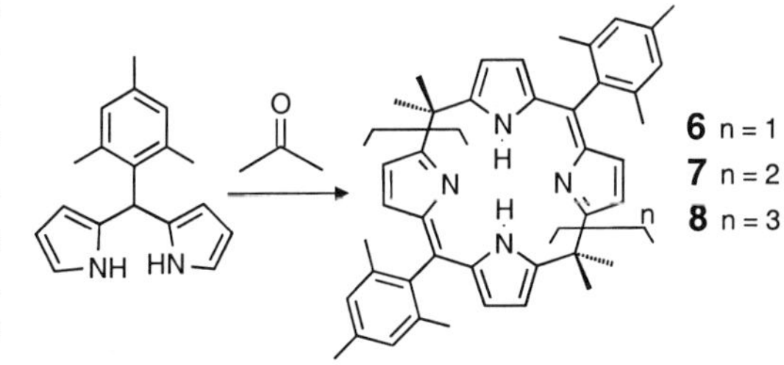

Particularly interesting examples are
- (a) the cylindrical compound **10a** obtained in 32% yield from the condensation of pyrrole with the *p*-*t*-butylcalix[4]arene 2-ketopropyl ether **9a** [15]. In like fashion, using

the calix[5]arene counterpart **9b**, a cylindrical compound **10b** containing calix[5]-arene and calix[5]pyrrole rings is obtained in 10% yield [16],
- (b) the “strapped calixpyrrole” **11** was obtained using a “pre-strapped” diketone [7]. Several partially but usefully functionalized calix[4]-pyrroles **12** have been synthesised *via* lithiation of one or two β-positions followed by quenching with electrophiles like ethylbromoacetate [3b] or CO_2 [17], or *via* iodination [18] of one β-position [19]. The later one was used in a synthesis of the mono-ethynyl calix[4]pyrrole **12b** (R^2 = C≡CH) which then was converted into bis-calix[4]pyrroles [20].

Various applications of calix-pyrroles have been based on their capacity to form NH hydrogen bonds to anions [21,22]. (See Chapters 23 and 36.)

9 **a** n = 4 **10**
b n = 5

11

12 a $R^1 = R^2 = CH_2CO_2Et$
b $R^1 = H$
$R^2 = CH_2CO_2Et$, CO_2H, CO_2Ar, I, C≡CH, C≡C-Ar

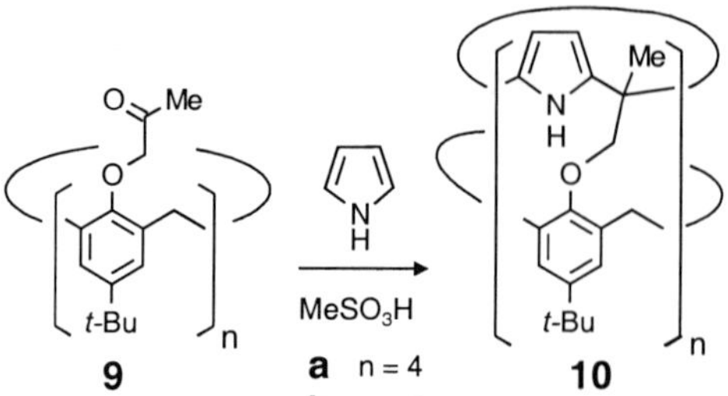

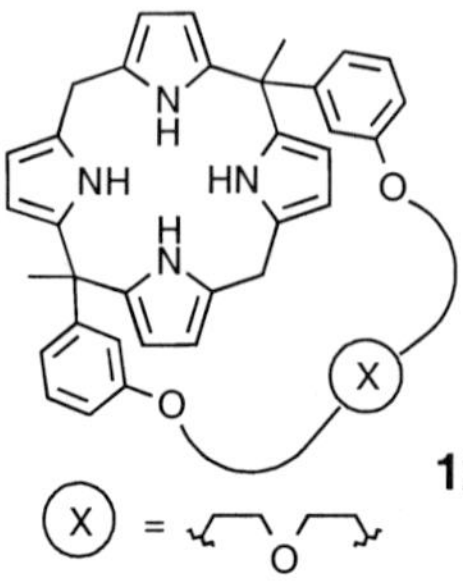

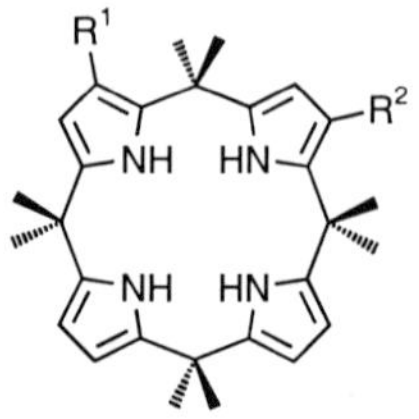

2.2. Calixfurans

Both macrocycles **14** and linear oligomers **13** have been known since the 1950s as products of the condensation of furan with aldehydes and ketones [23-26]. The principal directly formed macrocyclic product is the tetramer **14a** but the linear tetra-, penta- and hexamer all cyclize in workable yields (34-52%) under acidic conditions to **14a-c**, respectively [27,28]. A careful study [29] of the reaction between furan and a variety of ketones has shown that the yield enhancement from the addition of metal salts is due not to a template effect, as previously supposed [30], but to a decrease of pH [29a,c].

3,4-Disubstitution of furan or the use of acetophenone or cyclopentanone as the carbonyl reagent prevents cyclisation [29]. Although the acid-catalyzed condensation of furan with HCHO gives largely linear oligomers [27b], small amounts of **14a** with $R^1 = R^2 = H$ are obtained, as they are

13

14a n = 4
b n = 5
c n = 6

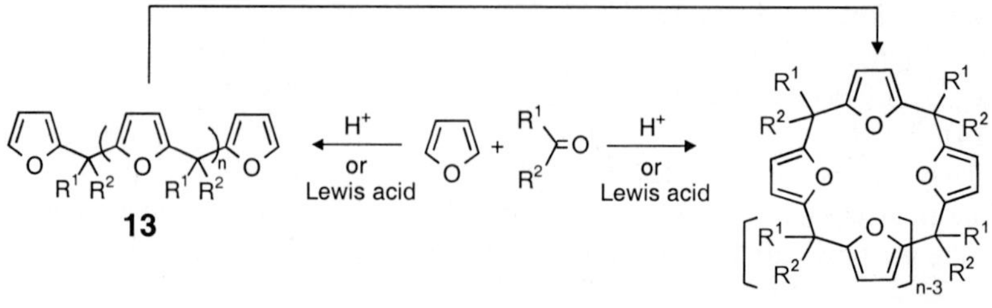

(a) by condensation of 5,5'-methylenedi-2-furaldehyde with 2,2'-methylenedifuran [31],
(b) by condensation of 2,2'-methylenedifuran with HCHO [32] and
(c) by the ZnCl$_2$/HCl-catalyzed condensation of furfuryl alcohol [33,34].

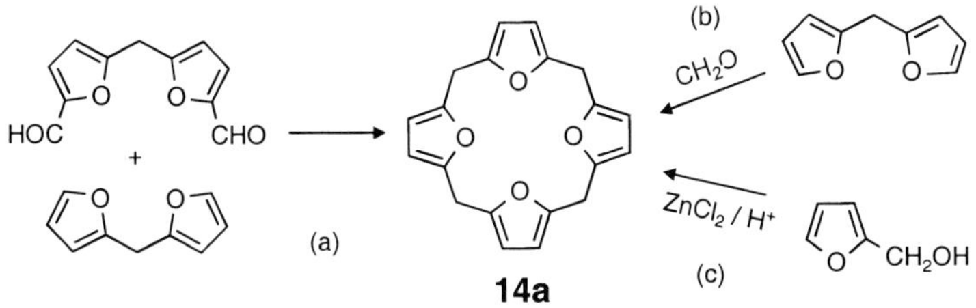

Full and partial functionalization of calix[n]furans on the basis of Diels-Alder reactions of the furan moieties with benzyne or dimethylacetylenedicarboxylate have been described. Subsequent transformations of the reacted furan groups led to naphthalene [35,36], o-phthalic ester or 3,4-furandicarboxylate units [36]. Hydrogenation of **14a-c** on Pd-C catalyst leads to corresponding macrocycles with completely reduced tetrahydrofuran moieties [23,25], while oxidation of **14a** ($R^1 = R^2 = H$) leads to a tetraoxaporphyrin dication [31,27].

2.3. CALIXTHIOPHENES

Calixthiophenes are not as easily prepared as their calixfuran counterparts. One approach employs 2-hydroxymethyl-3,4-diethylthiophene which reacts in the presence of p-CH$_3$C$_6$H$_4$SO$_3$H to give 30% of **15a** (n = 4) and 16% of **15b** (n = 5) [37].

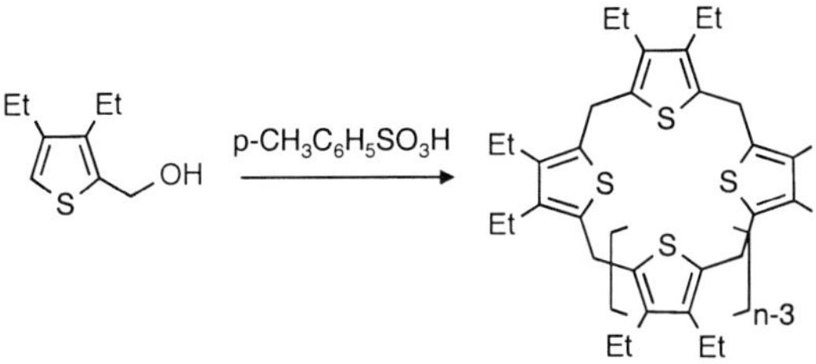

Homooxacalix[n]thiophenes (**16**, n = 3-7) result from the treatment of 2,5-bis-(hydroxymethyl)thiophene with CF$_3$SO$_3$H in THF [38], the product mixture being separable by GPC into individual components in yields ranging from 4-11%. A "2 + 2" approach [39] involves the condensation of **17a** with **17b** to give a 4% yield of **18**. Reaction of **14a** with H$_2$S under strong acid catalysis produces, via S for O exchange, the thia-analogue of **14a** in 66% yield [33,40].

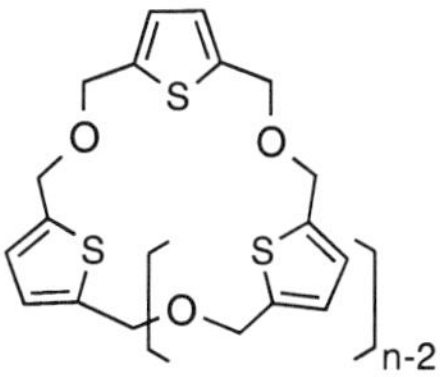

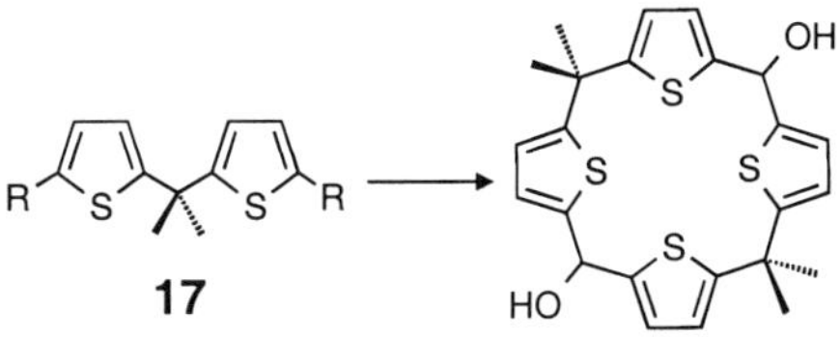

A calix[5]arene **20** consisting of four phenol and a single thiophene moiety has been synthesised by high-dilution acid-catalysed cyclisation of the linear precursor **19** in 40-50% yields [41], similar to those of the stepwise syntheses discussed in Chapter 1.

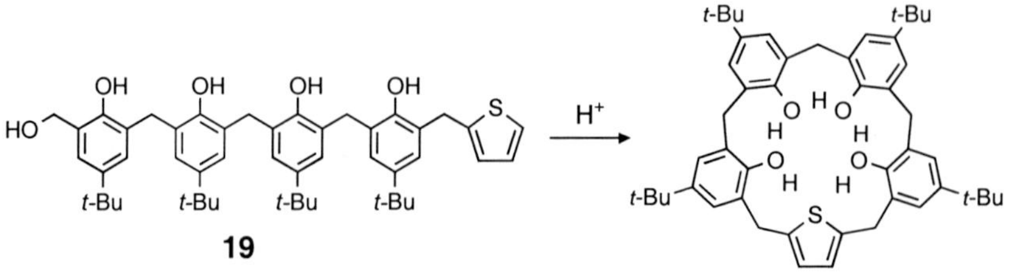

2.4. MIXED MACROCYCLES

Starting from calix[6]furan **14c** hexa- and octaketones **21** and **22** (as well as a deca-ketone) were obtained by oxidation with 4 mol MCPBA [42], while the application of 6.2 mol MCPBA led to the dodecaketone **23**. Subsequent reduction of the olefinic double bonds with Zn/AcOH and treatment with AcONH$_4$/EtOH resulted in the formation of calix[3]furan[3]pyrrole **24** (X-ray structure determination) and calix[2]-furan[4]pyrrole **25** respectively while the dodecaketone **23** gave the calix[6]pyrrole **26** [43].

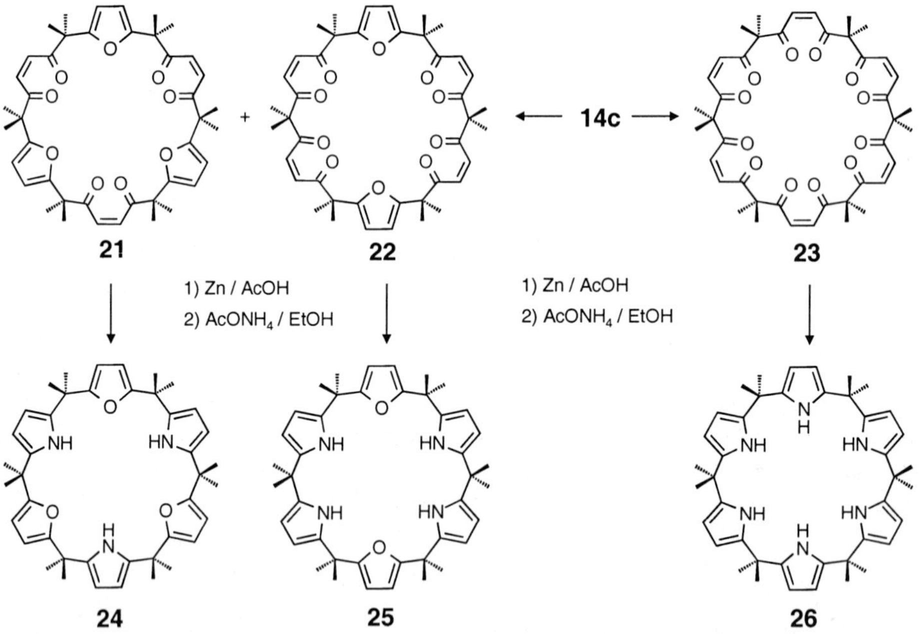

Another approach based on "3 + 1" condensation resulted in cylic alternating oligomers of furan (thiophene) and pyrrole. Tetramers **27a-c** were isolated in 40-60 % yield and the octamers **28a-c** in 10-20 % yield. Traces of the hexamers **29**, which can be formed only *via* bond breaking, were also isolated [44,45].

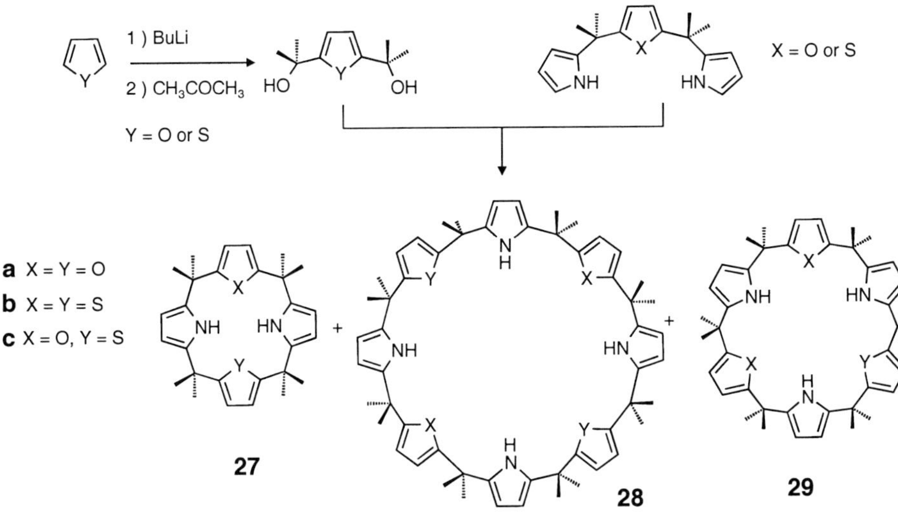

2.5. CALIXPYRIDINES

Less direct routes than through reactions of simple substituted pyridines are required for calixpyridine synthesis. Thus, reaction of dipyridinemethane **30** with MeCN and LiH produced the tetrapyridine macrocycle **31**, which was then converted into **32** [46]. Structural analogs of resorcarenes named octahydroxypyridine[4]arenes, have been synthesised by condensation of 2,6-dihydroxypyridine with various aldehydes [47].

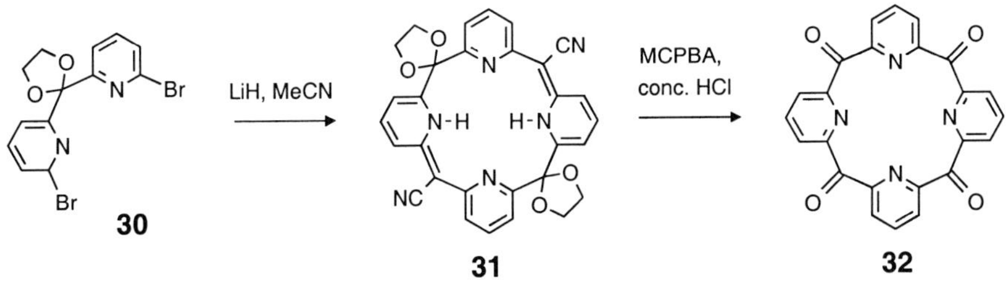

Ring enlargement reaction of pyrrole rings in **1** ($R^1 = R^2 =$ Me) by reaction with dichlorocarbene led to the mixture of macrocyclic compounds **33** with four chloropyridine units (two positions of chlorine *a* and *b* are possible) and macrocycles with three and two chloropyridine units (calix[m]pyridino[n]pyrroles, m + n = 4) [48]. Use of dioxane as the solvent instead of 1,2-dimethoxyethane resulted in the formation of macrocycles with only one or two pyridine moieties [48].

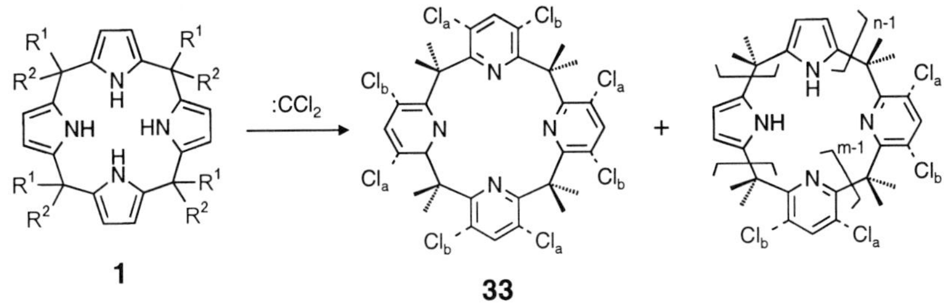

Starting from calix[4]pyrrole **1** ($R^1 = R^2 = Et$), compounds containing only one pyridine unit (e.g. **34**) as well two pyridine units (e.g. **35**), were successfully synthesized *via* a sequence of organometallic reactions [49].

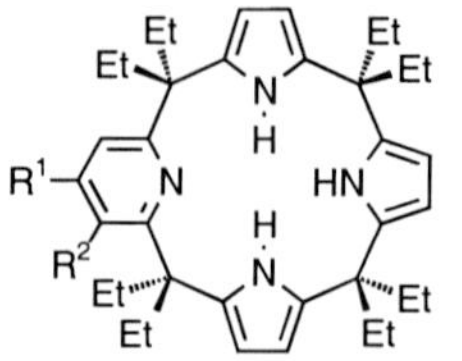

$R^1 = R^2 = H$
$R^1 = H, R^2 = Et, CH=CH_2$
$R^1 = CH_3, R^2 = H$

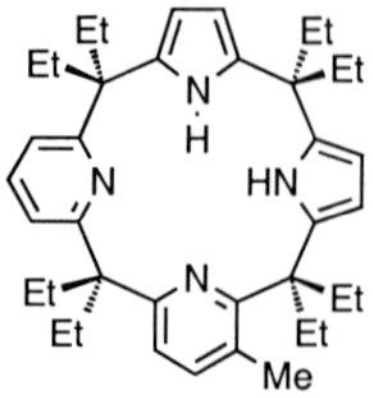

2.6. CATIONIC HETEROCALIXARENES

Positively charged macrocycles are interesting from the viewpoint of anion recognition. Calixarene analogues have been mainly obtained by N-benzylation of pyridines or pyrimidines. For instance the calix[4]arene analogue **36** with four pyridine moieties connected via methylene bridges, was produced by condensation of 3-bromomethylpyridine in 10% yield [50,51]; **36** assumes a *1,2-alternate* conformation in the crystalline state.

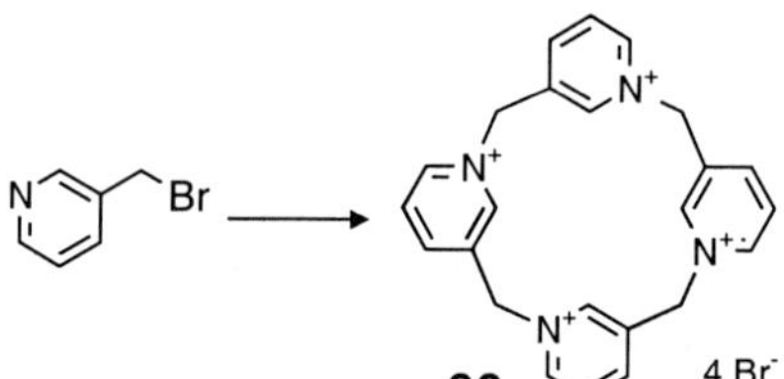

Thermal decomposition of **37** (vitamin B1) in 15-50 % water-methanol solutions led to the pyrimidine-based calix[4]arene analogue **38** [52]. If methanol is used as solvent, the calix[6]arene **39** analogue is formed in 34% [53].

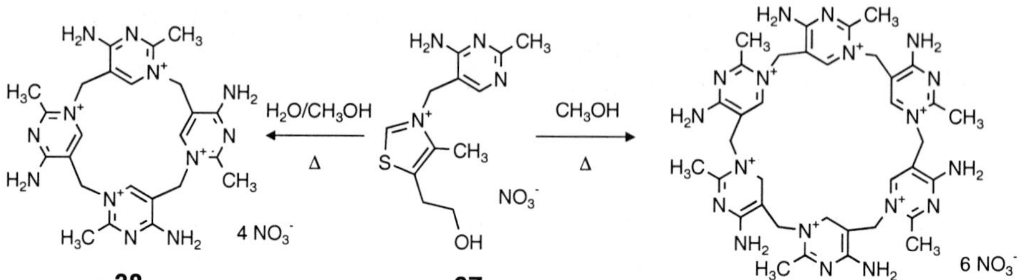

The constitution of these calix-pyrimidines were proved by X-ray structure determination, which revealed a *1,3-alternate* conformation for **38**. There were reported as well structures of **39** with anions like $[Hg_2I_7]^{3-}$ [54], $[Ba(NO_3)_6]^{4-}$ and $[Pb(NO_3)_6]^{4-}$ [55]. The ability to bind tricarboxylate anions was demonstrated for **36** [50].

2.7. CALIXINDOLES

Calixindoles [56] **42** have been prepared by
- (a) the POCl$_3$-catalyzed condensation of 4,6-dimethoxy-3-R^2-indoles **40** with aryl al-

dehydes which give **42** (R^1 = Ar, R^2 = Me, Ar) [57] (reactions with formaldehyde failed to give a defined product), and

- (b) the acid-catalyzed condensation of indoles **41a,b** carrying either CH_2OH [58] or R^1CHOH [59] groups at the 2- or 7-positions to give **42** (R^1 = H or RNHCO, R^2 = Ar). The major products are the cyclic trimers (**42**, n = 1) in *ca* 60% yield, but the cyclic tetramer (**42**, n = 2) has been isolated in 25% yield in one case [58], and a cyclic pentamer (**42**, n = 3) has been detected in very small amount [59].

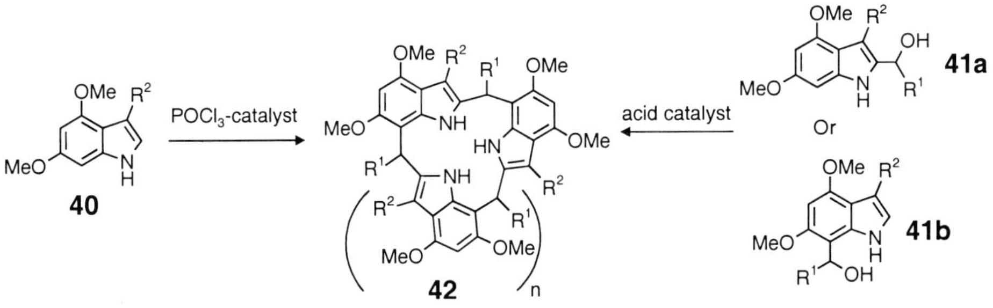

Since they are linked through inequivalent positions, different orientations of the indole units with respect to each other are possible within a macrocycle. Either all indole units are 2:7 linked or 2:2, 2:7 and 7:7 linkages exist simultaneously in a trimer. While strategy (b) implies a unique orientation (compare the directed synthesis of C_4-symmetrical calix[4]arenes, Chapter 1) of the indole systems, the first possibility is realized exclusively also in the one step condensations according to (a). Calixindoles with 2:2, 2:7 and 7:7 linkages were obtained only *via* a multistep approach [58].

In addition CHR^1 bridges with $R^1 \neq H$ lead to the existence of configurational isomers for which the nomenclature of resorcarenes would be appropriate (rcc(c), rct(t), etc.). Usually a mixture of all possible isomers is formed in these cases. The directionality of the indole moieties (or the existence of 2:7 linkages) makes the calixindoles chiral, though they are yet to be used for enantiomer separation and recognition.

2.8. CALIXBENZOFURANS

In contrast to the reaction of indoles, analogous condensations of 3,4,6-trisubstituted benzofurans with formaldehyde or aromatic aldehydes led to cyclic trimers **43a-g**, and **44a-g** with different directionality of the heterocyclic units within the macrocyclic structure. In one case a cyclic tetramer was identified as well [60].

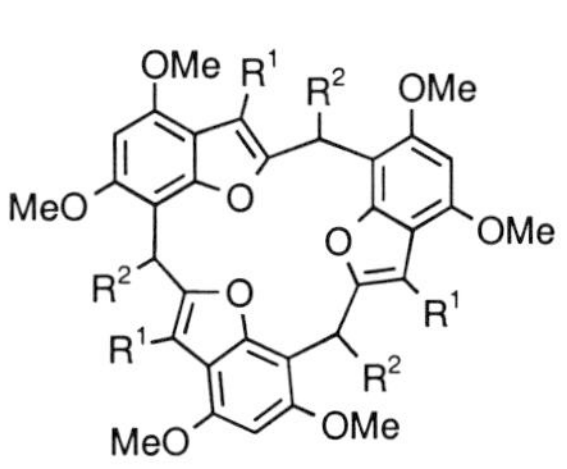

43/44a R^1 = Ph, R^2 = H
b R^1 = *p*-BrC_6H_4, R^2 = H
c R^1 = *t*-Bu, R^2 = H
d R^1 = R^2 = Ph
e R^1 = Ph, R^2 = Tol
f R^1 = *p*-BrC_6H_4, R^2 = Ph
g R^1 = *t*-Bu, R^2 = Ph

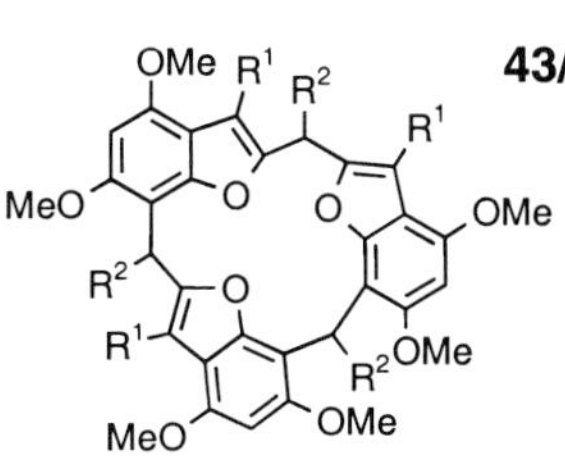

2.9. CALIXUREAS

Treatment of the disodium salt of the triazone derivative **45** with 1,3-bisbromomethylbenzenes yields the calix[4]arene compounds **46** [61,62].

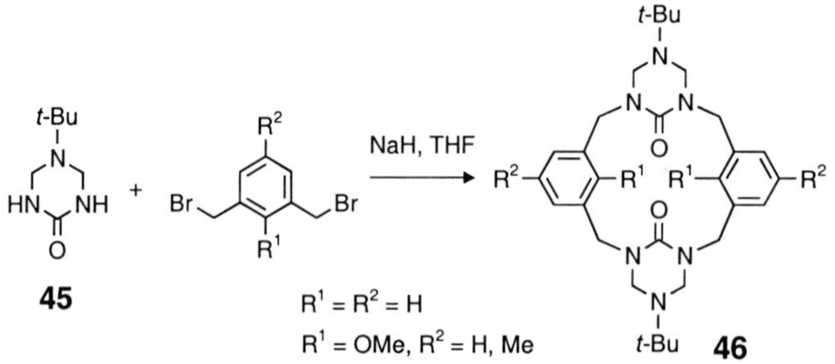

In a similar way, the bis-2-imidazolidone derivative **47** has been synthesised [61b]. A *partial cone* conformation has been demonstrated for the macrocycles **46** in the crystalline state, while ^{1}H NMR studies indicated a *partial cone – cone* conformational equilibrium in solution.

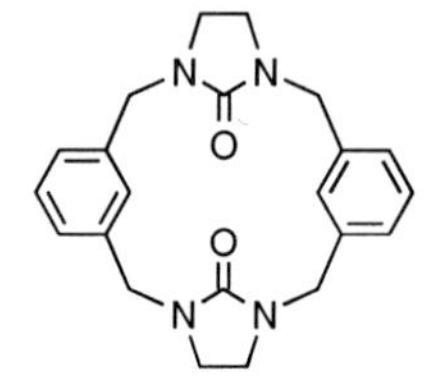

Calix[2]uracil[2]arenes **48** have been prepared in 20-30% yields in a two-step sequence starting from the simple pyrimidine and uracil derivatives [63]. As shown by X-ray structure determination and NMR studies, these macrocycles assume either the *cone* (when $R^1 = R^3 = OH$) or a *partial cone* conformation (with *anti*-orientation of the R^1 and R^3 residues) in the crystalline state and in solution. In general, the shorter C-N bonds make the dimensions of the cavity in **48** smaller than in conventional calix[4]-arenes. This is reflected in the lower conformational mobility.

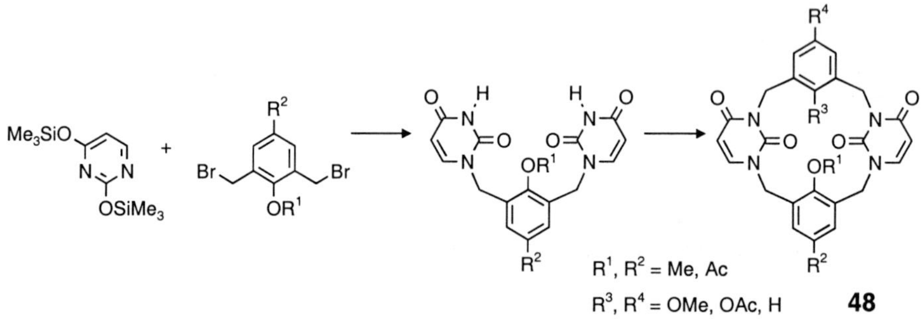

Bis-ketobridged calix[4]arene analogs **49**, containing only one uracil, benzimidazol-2(1*H*)-one or quinazoline-2,4(1*H*,3*H*)-dione unit, have also been synthesised [64,65]. In one case, a crystal structure shows the macrocycle possesses a *partial cone* like conformation. Various calix[1]heterocycle[2]arenes have been prepared analogously, but the "analogy" of such compounds with calixarenes becomes more and more contrived, which is true also for the names created for them [66].

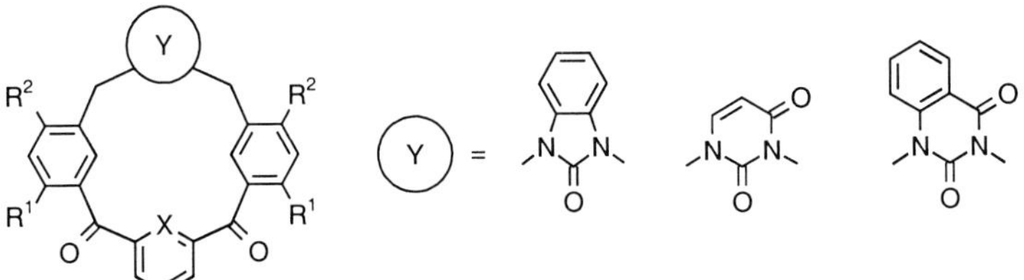

49

X = CH, N
R¹ = H, R² = OAlk; R² = H, R¹ = OMe

Condensation of bis(bromomethyl)-anisoles with the linear trimer **50** led to calix[8]benzimidazolones **52** with yields 7-35% [67] (up to 3% of the corresponding calix[4]arenes similar to **48** could be isolated) while, the calix[9]arene analogue **53** was prepared in 26% yield by a two-step fragment condensation with methylene-1,1'-bis-(benzimidazol-2-one) **51** using protection/deprotection and high dilution techniques [68].

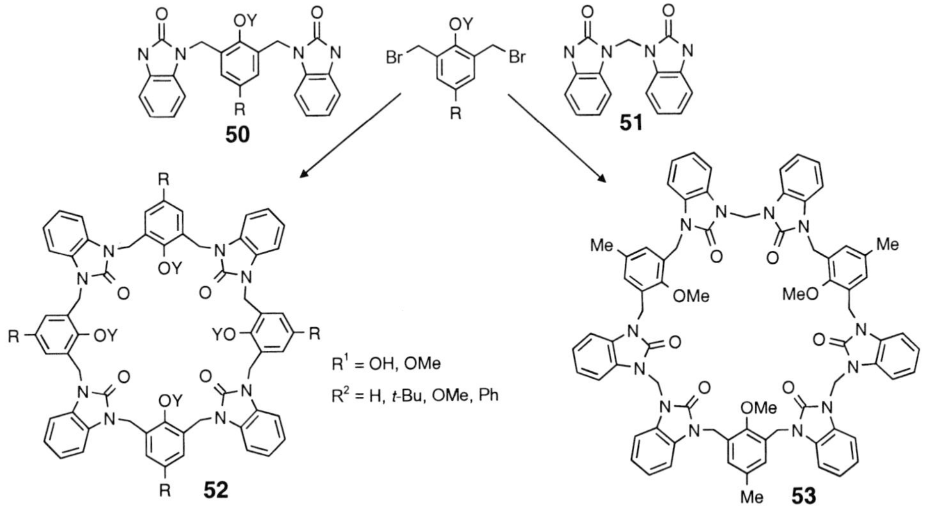

50 **51**

52
R¹ = OH, OMe
R² = H, t-Bu, OMe, Ph

53

3. Calixarenes with Bridges other than CRR'

Calixarenes with sulfur bridges are discussed in detail in Chapters 1 and 6, while homo-calixarenes, homo-oxa and homoaza-calixarenes are discussed in Chapters 11 and 12 [69,70]. Other examples of this growing family of compounds are the tetraoxacalix[4]-arenes **54a** (X=O) [71], the diazadioxacalix[4]arenes **54b** (X=NH) [72], the tetraaza-calix[4]arene **55** [73], pentaaza calix[5]- and hexaaza calix[6]arene analogues **56a** [74] and **56b** [75] as well as S-linked calix[3]triazine **57** [76,77].

X = O, NH; R^1 = H, NO_2
R^2 = H, CH_3, benzo[d]

54

55

R = $NAlk_2$, OCH_3

56a

56b

57

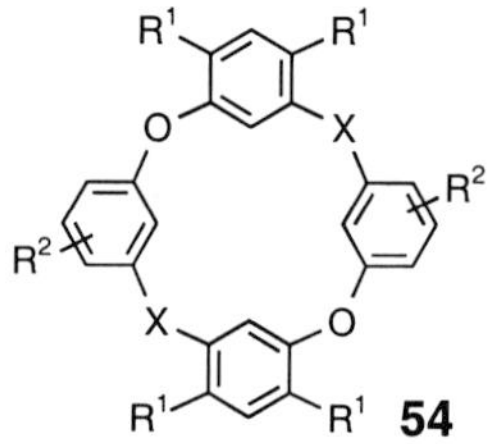

The base-induced condensation of **58** with **59** gave 26-90% of the cyclic tetramer **60** (R = H, Me) as the major product and the analogous cyclic hexamer as a minor product [78].

58 **59** **60**

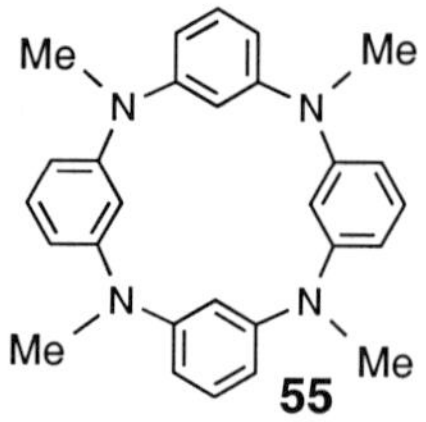

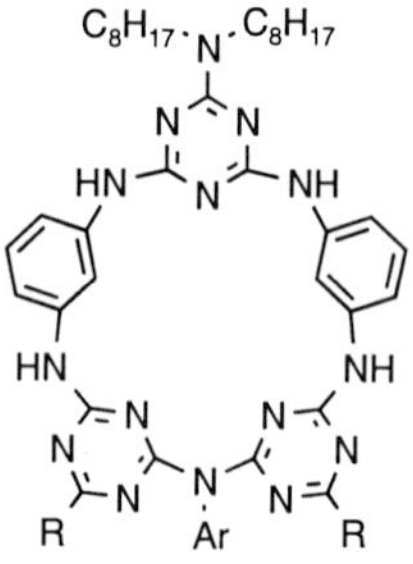

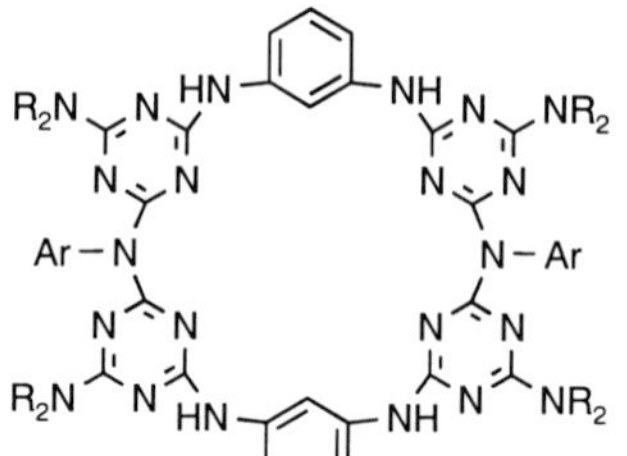

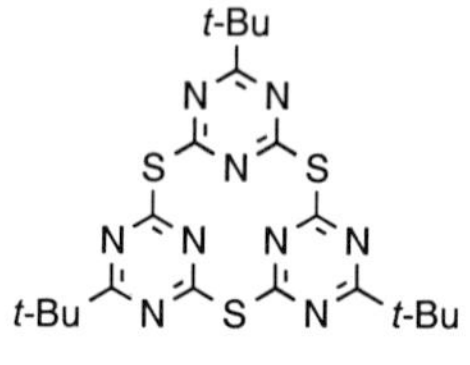

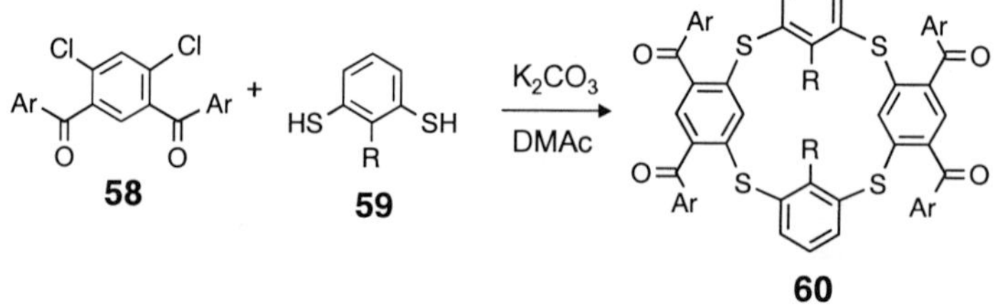

The generation of dianions from furan, thiophene and N-methylpyrrole by reaction of the corresponding heterocycles with n-BuLi/TMEDA and their subsequent treatment with electrophiles like Me_2SiCl_2, Me_2SnCl_2 or $PhPCl_2$ provided calix[4]- and -[6]arene analogues **61** [79].The similar derivatives with BNR_2 [80] and $GeMe_2$ [81] bridges were prepared from the corresponding heteroatom bridged dimers in 62% and 4% yields, respectively. Only mono-deprotonation takes place under similar conditions with 4-*t*-butylanisole and the silacalix[4]arene tetramethyl ether **62** was prepared in two consecutive deprotonation/reaction steps [79]. In comparison to the tetramethyl ether of *t*-butylcalix[4]arene, the macrocycle **62** is conformationally more mobile and defined conformers could not be observed in the ^1H NMR spectra at low temperatures. MM3 calculations predict the following stability of the possible conformers: *partial cone* > *1,3-alternate* > *1,2-alternate* >> *cone*.

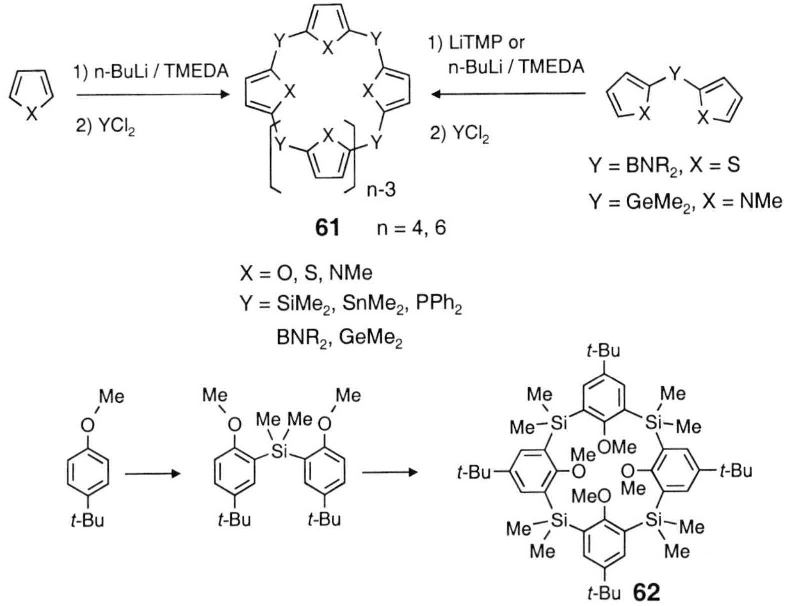

61 n = 4, 6

X = O, S, NMe
Y = SiMe₂, SnMe₂, PPh₂
   BNR₂, GeMe₂

Y = BNR₂, X = S
Y = GeMe₂, X = NMe

62

The one step condensation of 1,3,2-diazaphosphinine **63** with bis(phenylethynyl)-dimethylsilane **64** (toluene, 110 °C, 10 h) led to the silacalix[3]- and -[4]phosphinines **67**, **68** formed together with some undefined oligomers. X-ray structure determinations were performed for both [82]. The reaction can be used also for a stepwise synthesis of **68**, for instance via **65** and **66** and can be extended to the calix[4]arene analogue **69**. All compounds assume *partial cone* like conformations in the solid state.

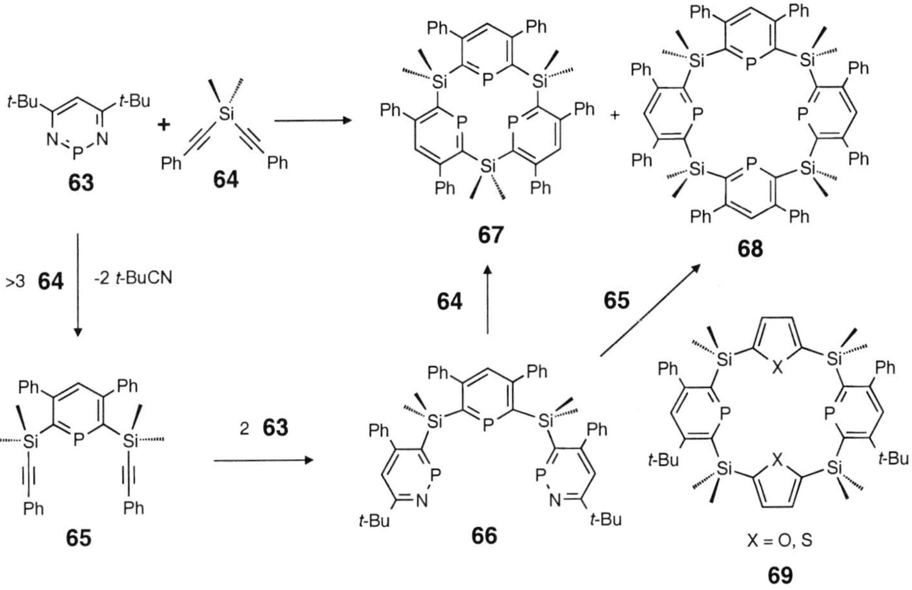

X = O, S

The thiacalix[4]thiophene **70** ($[1_4](2,5)$thiophenophane) with sulfur bridges was prepared by dilithiation of thiophene and subsequent reaction with SCl_2 in less than 1% yield [83]. A cyclisation starting from the linear sulfur bridged oligomer (n = 1) gives much better yields (24%) of the tetramer **70** [84]. Similar condensations of linear oligomers with five and six thiophene units led to corresponding cyclic compounds with yields 10-16%. All these calixarene analogues possess a high conformational mobility.

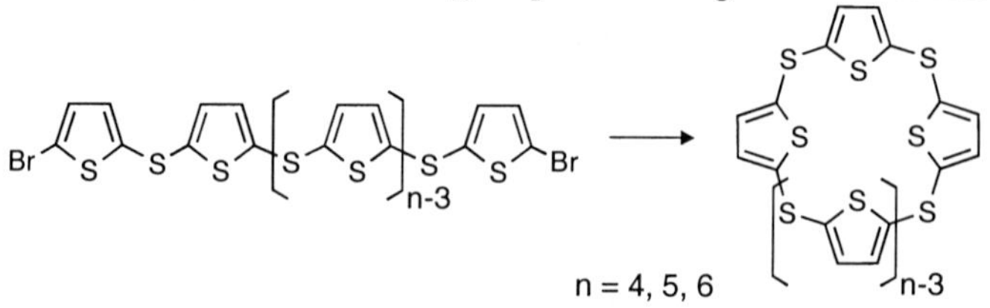

n = 4, 5, 6

70

Oligomerisation of the monomer [(en)Pt(UH-*N1*)(H_2O)]$NO_3 \cdot H_2O$ (UH = uracil monoanion) led to the metallacalix[4]arene **71** (27%) along with other oligomers [85,86]. An X-ray structure determination showed that **71** assumes a *1,3-alternate* conformation (S_4-symmetry) stabilized by four hydrogen bonds between the carbonyl oxygen of one uracil moiety and the hydroxyl group of the neighboring one. In contrast to the solid state structure, both *cone* (C_4-symmetry) and *1,3-alternate*, along with further minor species, seem to be present in solution. A *cone* conformation was also suggested for complexes with cations like Ag(I), Zn(II), Be(II) and La(III) [87]. The complexes with Zn(II) and Be(II) are able to include organic anions like aryl- or alkyl sulfonates in their cavities.

As with **71**, the reaction of [(en)M(H_2O)$_2$]$^{2+}$ (M = Pt, Pd) with 2-hydroxypyrimidine led to metallacalix[4]arenes **72** [88].

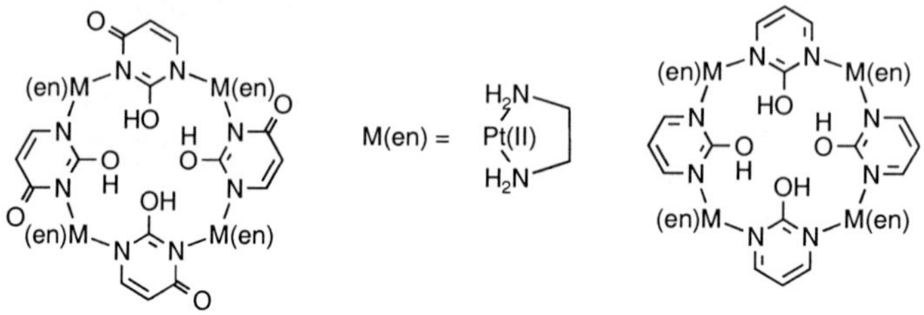

M(en) = Pt(II)

71

72

Acknowledgements: We are grateful to Prof. C. D. Gutsche for his encouragement and assistance.

4. References and Notes

[1] A. Baeyer, *Ber. Dtsch. Chem. Ges.* **1886**, *19*, 2184-2185.
[2] a) M. Dennstedt, J. Zimmerman, *Ber. Dtsch. Chem. Ges.* **1887**, *20*, 850-857; b) M. Dennstedt, *Ber Dtsch. Chem. Ges.* **1890**, *23*, 1370-1374; c) P. Rothemund, C. L. Gage, *J. Am. Chem. Soc.* **1955**, *77*, 3340-3342; d) W. Brown, B. J. Hutchinson, M. H. MacKinnon, *Can. J. Chem.* **1971**, *49*, 4017-4022.

[3] a) P. Anzenbacher, K. Jursíková, V. M. Lynch, P. A. Gale, J. L. Sessler, *J. Am. Chem. Soc.* **1999**, *121*, 11020-11021; b) P. A. Gale, J. L. Sessler, W. E. Allen, N. A. Tvermoes, V. Lynch, *Chem. Commun.* **1997**, 665-666; c) P. A. Gale, J. L. Sessler, V. Král, V. Lynch, *J. Am. Chem. Soc.* **1996**, *118*, 5140-5141; d) P. A. Gale, J. L. Sessler, V. Král, *Chem. Commun.* **1998**, 1-8.
[4] B. Turner, M. Botoshansky, Y. Eichen, *Angew. Chem., Int. Ed.* **1998**, *37*, 2475-2478.
[5] a) D. Jacoby, C. Floriani, A. Chiesi-Villa, C. Rizzoli, *J. Chem. Soc., Chem. Commun.* **1991**, 220-221; b) D. Jacoby, S. Isoz, C. Floriani, A. Chiesi-Villa, C. Rizzoli, *J. Chem. Soc., Chem. Commun.* **1991**, 790-792; c) D. Jacoby, C. Floriani, A. Chiesi-Villa, C. Rizzoli, *J. Am. Chem. Soc.* **1995**, *117*, 2805-2816; d) L. Bonomo, E. Solari, G. Toraman, R. Scopelliti, M. Latronico, C. Floriani, *Chem. Commun.* **1999**, 2413-2414.
[6] S. Camiolo, P. Gale, *Chem. Commun.* **2000**, 1129-1130.
[7] J. L. Sessler, P. Anzenbacher, Jr., K. Jurisková, H. Miyaji, J. W. Genge, N. A. Tyemoes, W. E. Allen, J. A. Shriver, *Pure & Appl. Chem.* **1998**, *70*, 2401-2408.
[8] For the synthesis of fluorinated calix[5]- and –[8]pyrroles see: J. L. Sessler, P. Anzenbacher, Jr., J. A. Shriver, K. Jursíková, V. M. Lynch, M. Marques *J. Am. Chem. Soc.* **2000**, *122*, 12061-12062.
[9] S. Depraetere, M. Smet, W. Dehaen, *Angew. Chem., Int. Ed.* **1999**, *38*, 3359-3361.
[10] Only the C_1-symmetrical derivative **4** is presented, but three further derivatives with C_s and one with C_2 symmetry are possible, see [9].
[11] Compare the rccc, rcct, rctt and rtct isomers of resorcarenes, Chapter 1.
[12] a) J. L. Sessler, A. Andrievsky, P. A. Gale, V. Lynch, *Angew. Chem., Int. Ed. Engl.* **1996**, *35*, 2782-2785; b) P. Anzenbacher, Jr., K. Jursíková, J. L. Sessler, *J. Am. Chem. Soc.* **2000**, *122*, 9350-9351.
[13] V. Král, J. L. Sessler, R. S. Zimmerman, D. Seidel, V. Lynch, B. Andrioletti, *Angew. Chem., Int. Ed.* **2000**, *39*, 1055-1058.
[14] See also: C. Bucher, D. Seidel, V. Lynch, V. Král, J. L. Sessler, *Org. Lett.*, **2000**; *2*, 3103-3106.
[15] P. A. Gale, J. L. Sessler, V. Lynch, P. I. Sansom, *Tetrahedron Lett.* **1996**, *37*, 7881-7884.
[16] P. A. Gale, J. W. Genge, V. Král, M. A. McKervey, J. L. Sessler, A. Walker, *Tetrahedron Lett.* **1997**, *38*, 8443-8444.
[17] H. Miyaji, P. Anzenbacher, Jr., J. L. Sessler, E. R. Bleasdale, P. A. Gale, *Chem. Commun.* **1999**, 1723-1724.
[18] H. Miyaji, W. Sato, J. L. Sessler, V. M. Lynch, *Tetrahedron Lett.* **2000**, *41*, 1369-1373.
[19] See also: P. Anzenbacher, Jr., K. Jursíková, J. Shriver, H. Miyaji, V. M. Lynch, J. L. Sessler, P. A. Gale, *J. Org. Chem.* **2000**, *65*, 7641-7645.
[20] W. Sato, H. Miyaji, J. L. Sessler, *Tetrahedron Lett.* **2000**, *41*, 6731-6736.
[21] a) J. L. Sessler, A. Gebauer, P. A. Gale, *Gazz. Chim. Ital.* **1997**, *127*, 723-726; b) P. A. Gale, L. J. Twyman, C. I. Handlin, J. L. Sessler, *Chem. Commun.* **1999**, 1851-1852; c) J. L. Sessler, P. A. Gale, J. W. Genge, *Chem. Eur. J.* **1998**, *4*, 1095-1099; d) G. Cafeo, F. H. Kohnke, G. L. La Torre, A. J. P. White, D. J. Williams, *Chem. Commun.* **2000**, 1207-1208; e) H. Miyaji, W. Sato, J. L. Sessler, *Angew. Chem., Int. Ed.* **2000**, *39*, 1777-1780; f) B. Turner, A. Shterenberg, M. Kapon, K. Suwinska, Y. Eichen, *Chem. Commun.* **2001**, 13-14.
[22] For anion receptors based on fluorinated calix[4]pyrroles, with augmented affinities and enhanced selectivities see: P. Anzenbacher, A. C. Try, H. Miyaji, K. Jursíková, V. M. Lynch, M. Marques, J. L. Sessler, *J. Am. Chem. Soc.* **2000**, *122*, 10268-10272.
[23] M. Chastrette, F. Chastrette, J. Sabadie, *Org. Synth.* **1977**, *57*, 74-78.
[24] a) R. G. Ackman, W. H. Brown, G. F. Wright, *J. Org. Chem.* **1955**, *20*, 1147-1158; b) R. E. Beals, W. H. Brown, *J. Org. Chem.* **1956**, *21*, 447; c) W. H. Brown, W. N. French, *Can. J. Chem.* **1958**, *36*, 537; d) W. H. Brown, B. Hutchinson, *Can. J. Chem.* **1978**, *56*, 617-621.
[25] Y. Kobuke, K. Hanji, K. Horiguchi, M. Asada, Y. Nakayama, J. Furukawa, *J. Am. Chem. Soc.* **1976**, *98*, 7414-7419.
[26] The name Tetraoxaquaterenes was proposed for cyclic tetramers, see [24a].
[27] a) R. M. Musau, W. Whiting, *J. Chem. Soc., Chem. Commun.* **1993**, 1029-1031; b) S. Tanaka, H. Tomokuni, *J. Heterocyclic Chem.* **1991**, *28*, 991-994.

[28] Calix[6]furans with C=O or CH₂-bridges (*via* reduction of C=O) beside of CMe₂ have been also prepared in 16-29% yield from linear dimers or trimers and ethyl N,N-dimethylcarbamate: A. Gast, E. Breitmaier, *Chem. Ber.* **1991**, *124*, 233-235.
[29] a) M. de Sousa Healy, A. J. Rest, *J. Chem. Soc., Perkin Trans. 1*, **1985**, 973-982; b) M. de Sousa Healy, A. J. Rest, *J. Chem. Soc., Chem. Commun.* **1981**, 149-150; c) A. J. Rest, S. A. Smith, I. D. Tyler, *Inorg. Chim. Acta.* **1976**, 16, L1-L2; d) A. J. Rest, *Adv. Inorg. Chem. Radiochem.* **1978**, *21*, 1-40.
[30] M. Chastrette, F. Chastrette, *J. Chem. Soc., Chem. Commun.* **1973**, 534-535.
[31] E. Vogel, W. Haas, B. Knipp, J. Lex, H. Schmickler, *Angew. Chem., Int. Ed. Engl.* **1988**, 27, 406-410.
[32] a) W. H. Brown, H. Sawatsky, *Can. J. Chem.* **1956**, *34*, 1147; b) W. Haas, B. Knipp, M. Sicken, J. Lex, E. Vogel, *Angew. Chem., Int. Ed. Engl.* **1988**, 27, 409-502.
[33] E. Vogel, P. Röhrig, M. Sicken, B. Knipp, A. Hermann, M. Poho, H. Schmickler, J. Lex, *Angew. Chem., Int. Ed. Engl.* **1989**, *28*, 1651-1655.
[34] In the similar reaction (BF₃ as a catalyst) traces of cyclic penta-, hexa-, hepta- and octamers have been also observed: R. M. Musau, A. Whiting, *J. Chem. Soc., Perkin Trans. 1* **1994**, 2881-2888.
[35] F. H. Kohnke, M. F. Parisi, F. M. Raymo, P. A. O'Neil, D. J. Williams, *Tetrahedron* **1994**, *50*, 9113-9124.
[36] G. Cafeo, M. Giannetto, F. H. Kohnke, G. L. La Torre, M. F. Parisi, S. Menzer, A. J. P. White, D. J. Williams, *Chem. Eur. J.* **1999**, *5*, 356-368.
[37] E. Vogel, M. Pohl, A. Herrmann, T. Wiss, C. König, J. Lex, M. Gross, J. P. Gisselbrecht, *Angew. Chem., Int. Ed. Engl.* **1996**, *35*, 1520-1524.
[38] N. Komatsu, A. Taniguchi, H. Suzuki, *Tetrahedron Lett.* **1999**, *40*, 3749-3752.
[39] M. Ahmed, O. Meth-Cohn, *J. Chem. Soc. (C)* **1971**, 2104-2111.
[40] In similar way, calixselenophene was prepared by the reaction of **14a** with H₂Se, see [31].
[41] K. Ito, Y. Ohba, T. Tamura, T. Ogata, H. Watanabe, Y. Suzuki, T. Hara, Y. Morisawa, T. Sone, *Heterocycles* **1999**, *51*, 2807-2813.
[42] P. D. Williams, E. Le Goff, *J. Org. Chem.* **1981**, *46*, 4143-4147.
[43] G. Cafeo, F. H. Kohnke, G. L. La Torre, A. J. P. White, D. J. Williams, *Angew. Chem., Int. Ed.* **2000**, *39*, 1496-1498.
[44] Y. S. Jang, H.J. Kim, P.H. Lee, C. H. Lee, *Tetrahedron Lett.* **2000**, *41*, 2919-2923.
[45] For the synthesis of super-expanded calix[n]furano[m]pyrroles (n + m = 5, 6, 10, 12) see: N. Arumugam, Y.-S. Jang, C.-H. Lee, *Org. Lett.*, **2000**, *2*, 3115-3117.
[46] G. R. Newkome, Y. J. Joo, F. R. Fronczek, *J. Chem. Soc., Chem.Commun.* **1987**, 854-856.
[47] T. Gerkensmeier, J. Mattay, C. Näther, *Chem. Eur. J.*, **2001**, *7*, 465-474.
[48] V. Král, P. A. Gale, P. Anzenbacher, P. Jr., K. Jursíková, V. Lynch, J. L. Sessler, *Chem. Commun.* **1998**, 9-10.
[49] D. Jacoby, S. Isoz, C. Floriani, A. Chiesi-Villa, C. Rizzoli, *J. Am. Chem. Soc.* **1995**, *117*, 2793-2804.
[50] S. Shinoda, M. Tadokoro, H. Tsukube, R. Arakawa, *Chem. Commun.* **1998**, 181-182.
[51] Imidazolium-linked cyclophanes were also synthesized from imidazol derivatives and (bromomethyl)-benzenes: M. V. Baker, M. J. Bosnich, C. C. Williams, B. W. Skelton, A. H. White, *Aust. J. Chem.* **1999**, *52*, 823-825.
[52] R. E. Cramer, V. Fermin, E. Kuwabara, R. Kirkup, M. Selman, K. Aoki, A. Adeyemo, H. Yamazaki, *J. Am. Chem. Soc.* **1991**, *113*, 7033-7034.
[53] R. E. Cramer, C. A. Waddling, C. H. Fujimoto, D. W. Smith, K. E. Kim, *J. Chem. Soc., Dalton Trans.* **1997**, 1675-1683.
[54] R. E. Cramer, M. J. Carrie, *Inorg. Chem.* **1994**, *29*, 3902-3904.
[55] R. E. Cramer, K. A. Mitchell, A. Y. Hirazumi, S. L. Smith, *J. Chem. Soc., Dalton Trans.* **1994**, 563-569.
[56] These macrocycles are not [1ₙ]metacyclophanes.
[57] a) D. St.C. Black, D. C. Craig, N. Kumar, *J. Chem. Soc., Chem. Commun.* **1989**, 425-426; b) D. St.C. Black, D. C. Craig, N. Kumar, *Aust. J. Chem.* **1996**, *49*, 311-318.
[58] a) D. St.C. Black, M. Bowyer, N. Kumar, P. S. R. Mitchell, *J. Chem. Soc., Chem.Commun.* **1993**, 819-823; b) D. St.C. Black, D. C. Craig, N. Kumar, *Tetrahedron Lett.* **1995**, *36*, 8075-8078.

[59] D. St.C. Black, D. C. Craig, N. Kumar, D. B. McConnell, *Tetrahedron Lett.* **1996**, *37*, 241-244.
[60] D. St. C. Black, D. C. Craig, N. Kumar, R. Rezaie, *Tetrahedron* **1999**, *55*, 4803-4814.
[61] a) P. R. Dave, G. Doyle, T. Axenrod, H. Yazdekhasti, H. L. Ammon, *Tetrahedron Lett.* **1992**, *33*, 1021-1024; b) P. R. Dave, G. Doyle, T. Axenrod, H. Yazdekhasti, H. L. Ammon,. *J. Org. Chem.* **1995**, *60*, 6946-6952.
[62] Similar derivatives starting from 1,4-bisbromomethylbenzenes have been also reported in [61].
[63] a) S. Kumar, G. Hundal, H. Singh, *Tetrahedron Lett.* **1997**, *38*, 3607-3608; b) S. Kumar, G. Hundal, D. Paul, M. S. Hundal, H. Singh, *J. Org. Chem.* **1999**, *64*, 7717-7726.
[64] S. Kumar, D. Paul, G. Hundal, M. S. Hundal, H. Singh, *J. Chem. Soc., Perkin Trans. 1* **2000**, 1037-1043.
[65] The name proposed for **49** (X = N), calix[1]cyclicurea[1]pyridine[2]arenes, is too long to make its use convenient, especially for structures which less and less resemble calixarenes.
[66] S. Kumar, G. Hundal, D. Paul, M. Singh Hundal, H. Singh, *J. Chem. Soc., Perkin Trans. 1* **2000**, 2295-2301.
[67] E. Weber, J. Trepte, K. Gloe, M. Piel, M. Czugler, V. C. Kravtsov, Y. A. Simonov, J. Lipkowski, E. V. Ganin, *J. Chem. Soc., Perkin Trans. 2*, **1996**, 2359-2366.
[68] J. Trepte, M. Czugler, K. Gloe, E. Weber, *Chem. Commun.* **1997**, 1461-1462.
[69] For a synthesis of capped thiocalix[6]arene analog see: A. P. West, Jr., D. V. Engen, R. A. Pascal, Jr., *J. Am. Chem. Soc.* **1989**, *111*, 6846-6847.
[70] Heteroatom-bridged calixarene analoges have been recently reviewed: B. König, M. H. Fonseca, *Eur. J. Inorg. Chem.* **2000**, 2303-2310.
[71] N. Sommer, H. A. Staab, *Tetrahedron Lett.* **1966**, 2837-2841.
[72] E. E. Gilbert, *J. Heterocyclic Chem.* **1974**, *11*, 899; P. A. Lehmann, *Tetrahedron* **1974**, *30*, 727-733.
[73] A. Ito, Y. Ono, K. Tanaka, *New J. Chem.* **1998**, 779-781.
[74] H. Graubaum, G. Lutze, B. Costisella, *J. Prakt. Chem.* **1997**, *339*, 266-271.
[75] H. Graubaum, G. Lutze, B. Costisella, B. Zur Linden, *J. Prakt. Chem.* **1997**, *339*, 672-674.
[76] M. Mascal, J. L. Richardson, A. J. Blake, W.-S. Li, *Tetrahedron Lett.* **1997**, *38*, 7639-7640.
[77] Calixsalens have been also reported: Z. Li, C. Jablonski, *Chem. Commun.* **1999**, 1531-1532.
[78] T. Freund, C. Kübel, M. Baumgarten, V. Enkelmann, L. Cherghei, R Reuter, K. Müllen, *Eur. J. Org. Chem.* **1998**, 555-564.
[79] a) B. König, M. Rödel, P. Budenitschek, P. G. Jones, *Angew. Chem., Int. Ed. Engl.* **1995**, *34*, 661-662; b) B. König, M. Rödel, P. Budenitschek, P. G. Jones, I. Thondorf, *J. Org. Chem.* **1995**, *60*, 7406-7410.
[80] F. H. Carré, R. J.-P. Corriu, T. Deforth, W. E. Douglas, W. S. Siebert, W. Weinmann, *Angew. Chem., Int. Ed.* **1998**, *37*, 652-654.
[81] B. König, M. Rödel, *Chem. Ber.* **1997**, *130*, 412-423.
[82] a) N. Avarvari, N. Maigrot, L. Ricard, F. Mathey, P. Le Floch, *Chem. Eur. J.* **1999**, *5*, 2109-2118; b) N. Avarvari, N. Mezailles, L. Ricard, P. Le Floch, *Science* **1998**, *280*, 1587-1589.
[83] B. König, M. Rödel, I. Dix, P. G. Jones, *J. Chem. Res. (S)* **1997**, 69-71.
[84] a) J. Nakayama, N. Katano, Y. Sugihara, A. Ishii, *Chemistry Lett.* **1997**, 897-898; b) N. Katano, Y. Sugihara, A. Ishii, J. Nakayama, *Bull. Chem. Soc. Jpn.* **1998**, *71*, 2695-2700.
[85] H. Rauter, E. C. Hillgeris, B. Lippert, *J. Chem. Soc., Chem. Commun.* **1992**, 1385-1386.
[86] H. Rauter, E. C. Hillgeris, A. Erxleben, B. Lippert, *J. Am. Chem. Soc.* **1994**, *116*, 616-624.
[87] J. A. R. Navarro, M. B. L. Janik, E. Freisinger, B. Lippert, *Inorg. Chem.* **1999**, *38*, 426-432.
[88] J. A. R. Navarro, E. Freisinger, B. Lippert, *Inorg. Chem.* **2000**, *39*, 2301-2305.

Chapter 14

OXIDATION AND REDUCTION OF AROMATIC RINGS

SILVIO E. BIALI

Department of Organic Chemistry,
The Hebrew University of Jerusalem
Jerusalem 91904, Israel, e-mail: silvio@vms.huji.ac.il

1. Introduction

Calixarenes have been subjected to a large number of chemical modifications over the last two decades [1]. These modifications, intended to alter their properties and/or general shape, usually have been based on the functionalization of the calix macrocycle at the *intra*annular ("narrow rim") and/or *extra*annular ("wide rim") positions [1]. In this chapter a less common chemical transformation involving the oxidation or reduction of the calixarene phenolic rings will be reviewed. The oxidation of the bridging methylene groups [2], the reductive cleavage of aryl substituents [3], and the oxidative coupling of calix[*n*]arenes to give bicalix[*n*]arenes [4] are treated in Chapters 5 and 7.

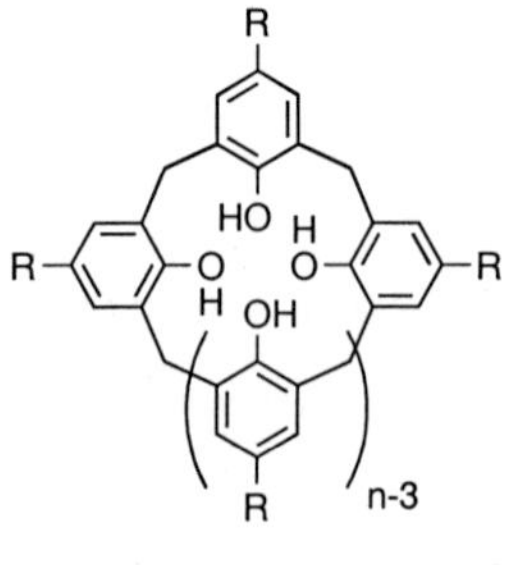

1 R = *t*-Bu
2 R = H
a n = 4
b n = 5
c n = 6
d n = 8

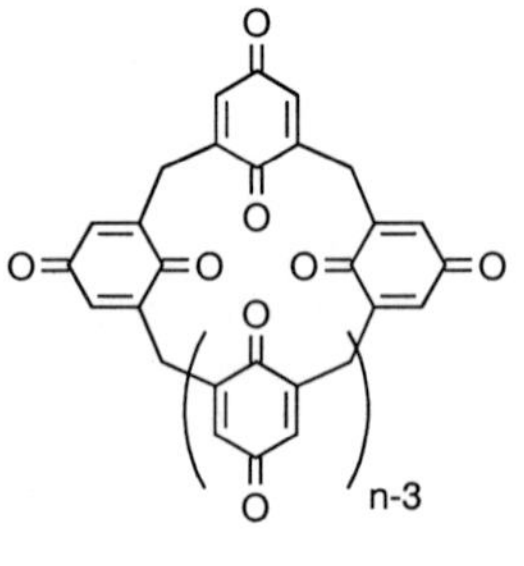

3a n = 4
3b n = 5
3c n = 6

2. Calixquinones

The reaction of calixarenes with strong oxidizing agents may afford derivatives in which one or several phenolic moieties have been oxidized to quinone groups ("calixquinones") [5,6]. Calixquinones are of interest *per se* as complexing agents with redox and/or charge transfer properties, and as synthetic intermediates for the preparation of functionalized calixarenes [7].

2.1. Synthesis of Calixquinones

Early work on calix[4]quinone (**3a**) was surveyed by Gutsche in 1992 [8]. The preparation of **3a** by three different multistep routes was reported by Taniguchi and coworkers in 1989 [9]. The most convenient route involved azo coupling of calixarene **2a** with *p*-

Z. Asfari et al. (eds.), Calixarenes 2001, 266–279.

carboxybenzenediazonium chloride, reduction of the azo derivative (which is soluble in aqueous alkaline solutions) with $Na_2S_2O_4$ and oxidation of the resulting *p*-amino-calixarene with $FeCl_3/K_2CrO_4$. The yield of **3a** by this route was 95% [9].

Calix[n]arenes **2a-c** can be oxidized in one step to the corresponding polyquinones **3a-c** by reaction with ClO_2 (generated by reaction of sodium chloride with aqueous potassium persulfite) [8]. The use of $Tl(OCOCF_3)_3$ [10] enables the direct oxidation of **1a** (containing *p-tert*-butyl groups which are lost during the oxidation) but the reaction fails for the larger-ring *p-tert*-butylcalix[*n*]arenes **1b-d** [8].

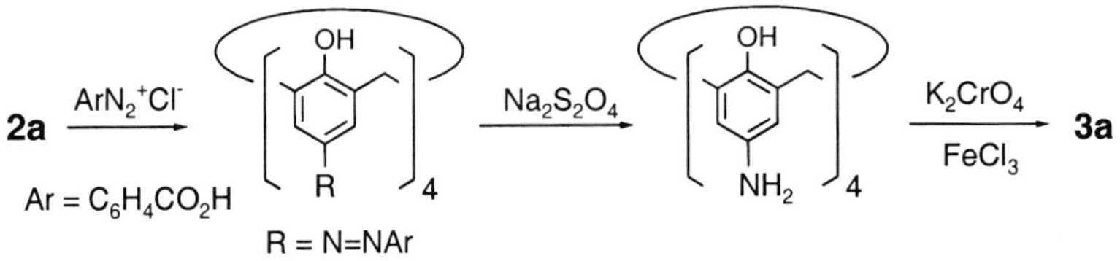

$$2a \xrightarrow{ArN_2^+Cl^-} \text{[structure, R = N=NAr]} \xrightarrow{Na_2S_2O_4} \text{[structure, NH}_2\text{]} \xrightarrow{K_2CrO_4, FeCl_3} 3a$$

$Ar = C_6H_4CO_2H$

The selective oxidation of some of the phenol rings of a calixarene can be achieved by protection of one or more of the phenolic OH groups by etherification or esterification prior to the oxidation step. Using this strategy calix[4]arenes with one, two or three quinone groups [8] and calix[6]arene diquinones [8] and triquinones [11] have been prepared. $Tl(OCOCF_3)_3$ is the reagent most commonly used for the oxidation [8, 11-13] but other reagents (ClO_2 [8,14], $Tl(NO_3)_3 \cdot 3H_2O$ [11,15,16] and $NaBO_3 \cdot 4H_2O$ ("sodium perborate") [17]) have also been used occasionally. Usually the number of quinone groups in the final product is identical to the number of unprotected phenol groups in the calixarene. However, oxidation of a calixcrown (with two "unprotected" phenol rings) with excess $NaBO_3 \cdot 4H_2O$ afforded a monoquinone derivative instead of the expected diquinone product [17].

In some cases attempted electrophilic substitution of a calixarene using an oxidizing electrophile has led to calixquinones instead of (or in addition to) the substitution product(s). Treatment of a didehydroxylated calixarene with NO_2BF_4 afforded a calixdiquinone instead of the dinitration product [18]. Nitration of a calix[4]arene tetraether afforded, in addition to the major products (the mononitro and dinitro derivatives) small amounts of a calixarene hemiketal (characterized by X-ray crystallography) derived from the oxidation of one aryl group [19]. Attempted iodination of a bis(benzoyloxy)-calixarene afforded the expected diiodo compound **4**, together with monoquinone (**5**) and diquinone (**6a**) calixarene derivatives [20].

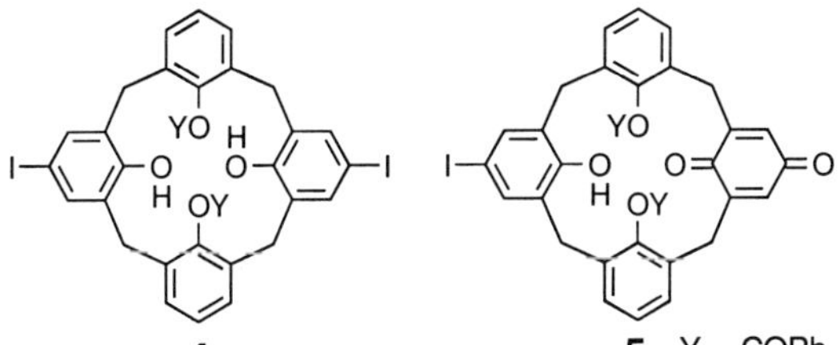

4

5 Y = COPh

2.1.1. Conformation in the solid state

X-ray crystallography of **3a** indicated that in the crystal the compound adopts a *partial cone* conformation [9a]. On the basis of dipole-dipole repulsions between the carbonyl groups an *1,3-alternate* conformation with an antiparallel disposition of the *intra*annular carbonyl groups could have been expected. The presence of the *partial cone* conformation in the crystal has been ascribed to van der Waals interactions between quinone rings of neighboring molecules related by the crystallographic center of symmetry [9a].

The molecular structure of calix[6]quinone **3c** has been determined by X-ray crystallography [8]. The macrocycle adopts an "up-down-out-down-up-out" conformation in which four rings (those denoted "up" and "down") are nearly perpendicular to the main macrocyclic plane and two rings (designated "out") are approximately parallel to the plane [8].

The structure of several distal calixdiquinones has been determined by X-ray crystallography. In the crystal calixdiquinones **6b** and **6e** adopt a *partial cone* conformation in which the two aryl groups point in opposite directions [21] whereas in **6c** the *partial cone* is also adopted, but the quinone rings are those oriented *anti*. The X-ray structures of several diquinones in free and complexed form were reported by Beer and coworkers [12b,12c,22]. Diquinones **6d**, **6g**, and **6h** adopt *partial cone, cone* and *1,3-alternate* conformations, respectively. In all the complexes examined (e.g., **6g**-Sr(ClO$_4$)$_2$, **6g**-KClO$_4$, **6h**-NaClO$_4$, **6i**-*n*-BuNH$_3$BF$_4$) the macrocycle adopted a *cone* conformation, which is the arrangement most suitable for complexation since the four *intra*annular oxygens are oriented in a convergent fashion. In some cases coordination of the cation with both the *intra*annular and *extra*annular oxygens was observed [12c]. In the crystal a calix[4]arene triester monoquinone adopts a *partial cone* conformation in which the quinone group is oriented *anti* [16c].

a R = H, Y = COPh
b R = H, Y = Me
c R = H, Y = *i*-Pr
d R = *t*-Bu, Y = Me
e R = *t*-Bu, Y = Et
f R = *t*-Bu, Y = *n*-Pr
g R = *t*-Bu, Y = CH$_2$CO$_2$Et
h R = *t*-Bu, Y = CH$_2$CONH$_2$
i R = *t*-Bu, Y = CH$_2$CONEt$_2$
j R = *t*-Bu, Y = (CH$_2$)$_2$NHCONHPh

6

7

2.1.2. Solution conformation of calix[4]diquinones

The solution conformation of distal calixdiquinones has been studied in detail. At room temperature the quinone groups undergo a fast (on the NMR timescale) rotation through the annulus process [12a,21]. One of the calix[4]diquinones which has been investigated most intensively is the diethoxy derivative **6e** [23]. In solution at room temperature the quinone rings rotate rapidly on the NMR timescale. The aryl rings possess a larger rotational barrier since the ethoxy group must pass through the ring annulus. Initial studies assigned a rigid *cone* conformation to **6e** [15] (necessarily requiring a *syn* orientation of all rings) but later studies have indicated that the two aryl rings are oriented *anti* [21,23]. Addition of Na$^+$ resulted in slow isomerization (over several hours at room temperature) of the *anti* into the *syn cone* form. A kinetic analysis showed that the ion does not affect the *anti/syn* interconversion rate but rather affects the position of the conformational equilibrium [23]. Addition of Ba^{2+} to **6g** [12b] or to a ruthenated calixdiquinone [13a] led to spectral changes consistent with an increased inhibition of the mobility of the quinone rings due to the ligation with the metal ion.

2.2. REACTIONS OF CALIXQUINONES

Calixquinones can react with nucleophiles at the carbonyl group or at the vinyl carbons [24]. The systems which have been studied most extensively are the diquinone and monoquinone calix[4]arene derivatives **6f** and **7**. Reaction of **6f** with malononitrile or pyrrolidine afforded products derived from an initial 1,2-nucleophilic addition to the *exo* carbonyl group [24]. The quinone group of a calixmonoquinone reacts with 2,4-dinitrophenylhydrazine at the *exo* carbonyl group to give the chromogenic *p*-2,4-dinitrophenylazo derivative [16a]. In a related reaction, oxidation of calixarenes in the presence of aniline affords indoanilinocalixarenes [25]. Monoquinone **7** reacts with mercaptoacetic acid, acetate, thiourea, *p*-thiocresol or the sodium salt of diethyl malonate, via a 1,4-conjugated addition to the quinone double bond yielding chiral *meta*-monosubstituted calix[4]arenes [24].

Reduction of the calix[*n*]quinones **3a-c** with Zn/HCl or $Na_2S_2O_4$ yields the calixhydroquinones **8a-c** [8]. An alternative method for the preparation of calix[*n*]hydroquinones involves Baeyer-Villiger oxidation of *p*-acetyl [9b,26] or *p*-formyl [27] calix[*n*]arenes followed by hydrolysis of the resulting acetate or formate groups. Calixhydroquinones derivatives can be obtained also by base catalyzed condensation of a suitable phenol with formaldehyde [28], in a reaction analogous to the well-established procedures for the preparation of **1a-d** [29]. Reaction of *p*-(benzyloxy)-phenol with paraformaldehyde and NaOH gave a mixture of *p*-benzyloxy calix[6]-, calix[7]-, and calix[8]arene (main reaction product). The use of potassium *tert*-butoxide as base gave a better yield of the octamer **9a** [30,31]. Debenzylation of **9a** by treatment with iodotrimethylsilane or $Pd(OH)_2/C$ and cyclohexene afforded the calix[8]hydroquinone **8d** [28]. Calix[4]hydroquinone **8a** can be reoxidized to calix[4]quinone **3a** by reaction with ferric chloride [9b]. The crystal structure of a calix[8]hydroquinone derivative (**9b**) has been recently determined [31]. The OH groups of **9b** do not form intramolecular hydrogen bonds, but they are involved in intermolecular hydrogen bonds with pyridine or water molecules [31]. The calixhydroquinone derivatives are strongly activated towards electrophilic aromatic substitution, and upon reaction with bromine yield the perbrominated derivatives [26b].

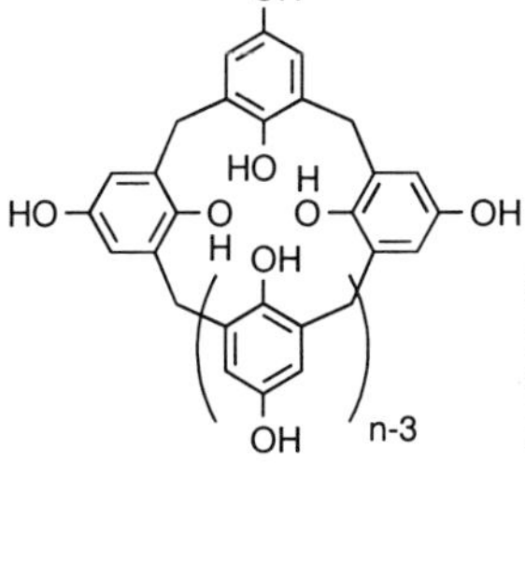

8a n = 4
8b n = 5
8c n = 6
8d n = 8

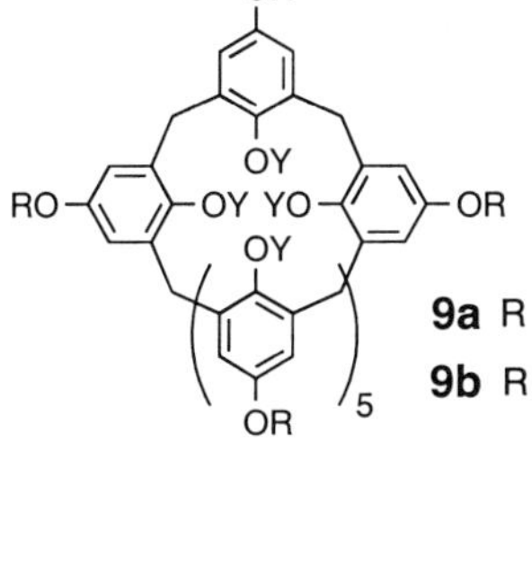

9a R = $CH_2C_6H_5$, Y = H
9b R = H, Y = Pr

2.3. REDOX AND BINDING PROPERTIES

The redox and binding properties of calixquinones have been the subject of several studies [11,14,32-40]. Electrochemical studies on a series of calix[4]quinones have indicate that increasing the number of quinone subunits in the calix skeleton causes a shift of the first half-wave potential to more positive values (*i.e.*, a greater ease of reduction) [34]. Each quinone group in diquinone and triquinone calix[4]arene derivatives is reduced at a different potential, but in contrast, the three quinone groups in a 1,3,5-triquinone calix[6]arene are reduced at the same potential [32b,34]. Since the quinone group can participate in binding processes, complexation can be detected by electrochemical means in the form of perturbations of the reduction waves [12,32]. Monoquinone **7** binds Na$^+$ in its neutral state as evidenced by its ^{1}H NMR in CD$_3$CN. From the shift in the $E_{1/2}$ value, it was estimated that the reduction of the calixquinone **7** to its monoanion results in a 10^6 binding enhancement [34]. Calixdiquinones **6g** and **6i** form very stable complexes with Na$^+$ and Ba^{2+}, and weaker adducts with ammonium ions [12a]. Semiempirical calculations indicated that the two *intra*annular carbonyl oxygens of the quinone groups and the two carbonyl oxygens of the ester groups of **6g** form a nearly tetrahedral arrangement, which bind NH$_4^+$ by virtue of four hydrogen bonds [36]. Calixdiquinones substituted by ruthenium (II) and rhenium (I) bipyridyl complexes have been shown to selectively bind acetate ions. The binding results in a remarkable increase (up to 500%) in the intensity of emission from the ligand [38]. A urea derivative of the calixdiquinone (**6j**) shows high selectivity for HSO$_4^-$ over other anions [39].

3. Oxidation to Calixarenes Spirodienone Derivatives

The calixarenes are polyphenolic compounds and the presence of several phenolic subunits in steric proximity promotes reactions which otherwise would be unlikely to occur in simple phenols. In contrast to strong oxidants, which react with calixarenes affording calixquinone derivatives, mild oxidation of calixarenes affords the corresponding spirodienone derivatives [41]. This reaction is analogous to the reported oxidation of bis(2-hydroxy-1-naphthyl)methane **10** which gives the spironaphthalenone **11** (eq. 1) [42]. The spirodienone calixarene derivatives are of interest as potential ligands due to the presence of a cyclic array of ether and carbonyl binding groups and of the spiro stereocenters which may render the systems chiral.

(Eq. 1)

3.1. PREPARATION OF SPIRODIENONE CALIXARENE DERIVATIVES

Oxidation of the parent *p-tert*-butylcalix[4]arene (**1a**) with phenyltrimethylammonium tribromide/base afforded a mixture of three isomeric bis(spirodienone) calixarene derivatives (**12a, 12b** and **12c**, eq. 2) differing in the arrangement of the carbonyl and ether groups (alternate or non alternate) and/or the configurations (*R* or *S*) of the spiro stereocenters [43]. The isomers (which were characterized by X-ray crystallography) interconvert in solution and in toluene at 80 °C, the equilibrium mixture consisted of 65% **12a**, 10% **12b** and 25% **12c** [43]. Huang and coworkers have shown that the alternate *meso* bis(spirodienone) derivative **12a** can be obtained in a regio- and stereoselective fashion in 95% yield by oxidation of **1a** with I_2/PEG 200/25% aq. KOH/CHCl$_3$ [44]. Other *p*-alkyl-calix[4]arenes yielded bis(spirodienone) derivatives in good yields, but in the case of *p-iso*-propyl-calix[4]arene small amounts of the isomer analogous to **12c** and of a mono(spirodienone) derivative were also obtained. $K_3Fe(CN)_6$/toluene/base has also been used in some cases for the preparation of spirodienone derivatives [48b]. Treatment of calixnaphthols having *exo* hydroxyl groups [45] with phenyltrimethylammonium tribromide resulted in an intractable mixture of products, but the calixnaphthols **13** with *endo* hydroxyls afforded their corresponding bis-(spirodienone)derivatives [46]. Notably, **13a** gave the chiral derivative **14a** (analogous to **12b**) which was characterized by X-ray crystallography. Oxidation of **13b** afforded the *meso* (**14b**) and racemic (**14c**) forms of the bis(spirodienone) derivative with an alternate disposition of the ether and carbonyl groups [46].

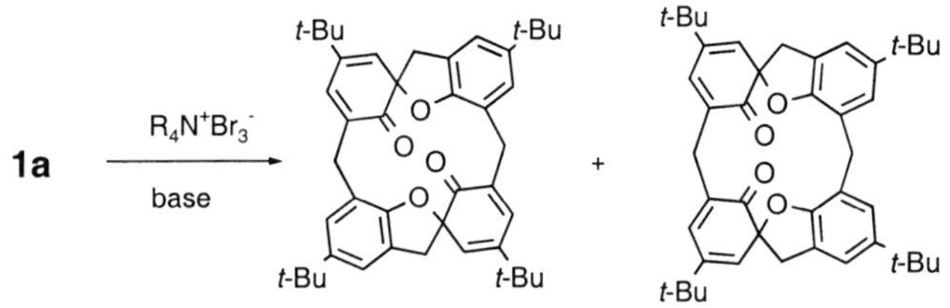

1a $\xrightarrow{R_4N^+Br_3^-, \text{ base}}$ **12a** (*RS*) **12b** (*RR/SS*) + **12c** (*RS*) (Eq. 2)

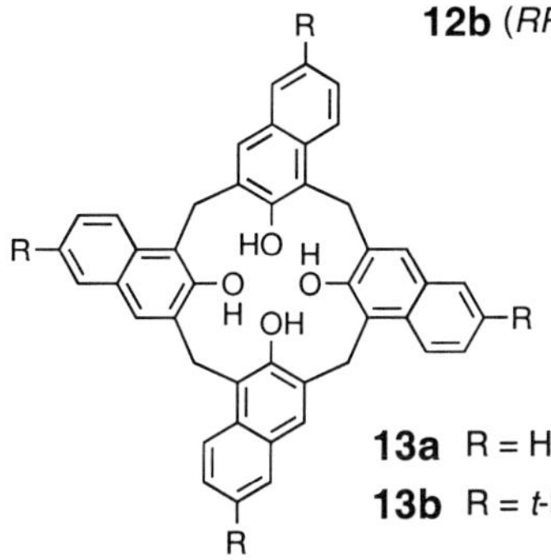

13a R = H
13b R = *t*-Bu

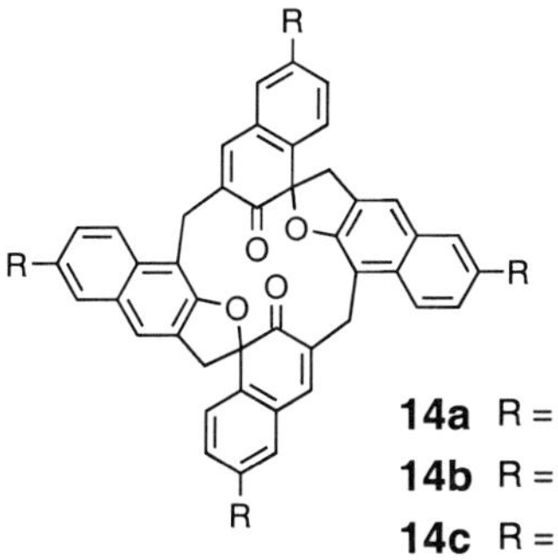

14a R = H, *RR/SS*
14b R = *t*-Bu, *RS*
14c R = *t*-Bu, *RR/SS*

The larger the calixarene, the larger the number of potential isomers of the poly(spirodienone) derivative but products with alternate arrangements of the ether and carbonyl groups are usually formed preferentially [47]. Oxidation of **1c** afforded the tris(spirodienone) derivative with alternate disposition of stereocenters and dissimilar configurations of the three stereocenters (*RRS/SSR*, **15**) and a tris(spirodienone) derivative (**16**) with non alternate functional groups [48]. X-ray crystallography indicated that in the crystal **15** adopts a conformation of nearly triangular shape in which the dihydrofuran oxygens point to the center of the cavity and the carbonyl groups are oriented nearly perpendicular to the mean macrocyclic plane. Equilibration studies in toluene at 85 °C indicated that **15** is the lowest energy form. From the oxidation of **1b** with

$K_3Fe(CN)_6$/base a bis(spirodienone)calixarene derivative (**17**) with alternate arrangement of the carbonyl and ether groups was isolated [48b]. Oxidation of **1a-c** with an equimolar amount of the oxidation reagent and a weaker base afforded the chiral mono-(spirodienone) calixarene derivatives **18a-c** [49-51].

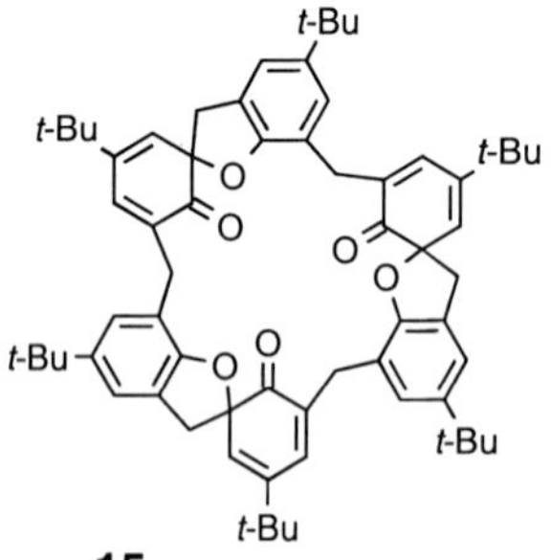

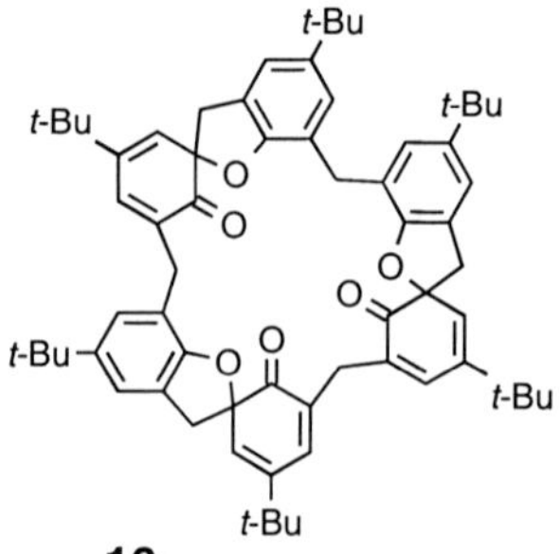

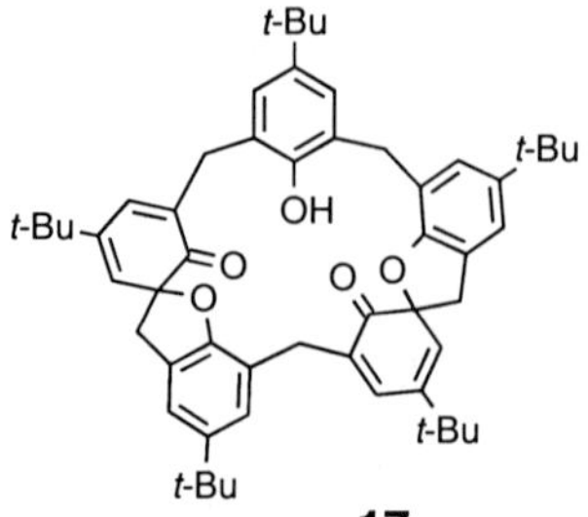

3.2. REACTIONS OF THE CALIXARENE SPIRODIENONE DERIVATIVES

The mono(spirodienone) calixarenes have proven useful as key intermediates for the preparation of aminocalixarenes [51,52] and monodehydroxylated calixarenes [51]. Treatment of the monospirodienone calixarene derivatives **18b** and **18c** with MeOH/H$^+$ resulted in ring expansion, and yielded the xanthenocalixarenes **19a** and **19b** [51]. The reaction failed for the smallest spirodienone **18a**, probably due to the steric strain of the resulting xanthenocalix[4]arene.

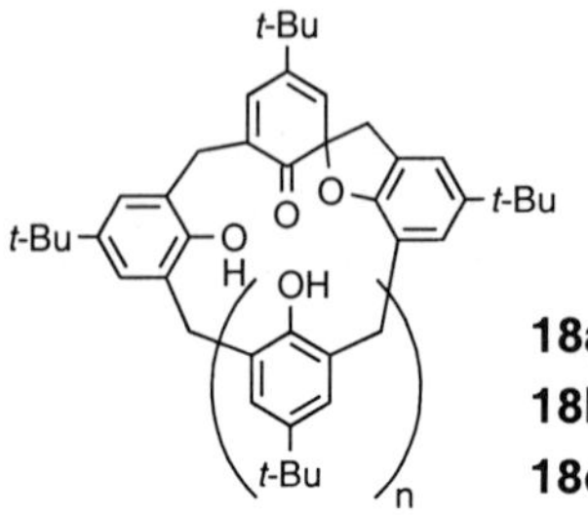

The reaction of mono(spirodienone) calixarene **18a** with bromine afforded a dispiro derivative [50]. Treatment of **18a** with HCl resulted in *ipso* substitution of a *p-tert*-butyl group by chlorine and reduction of the spiro bond [43b].

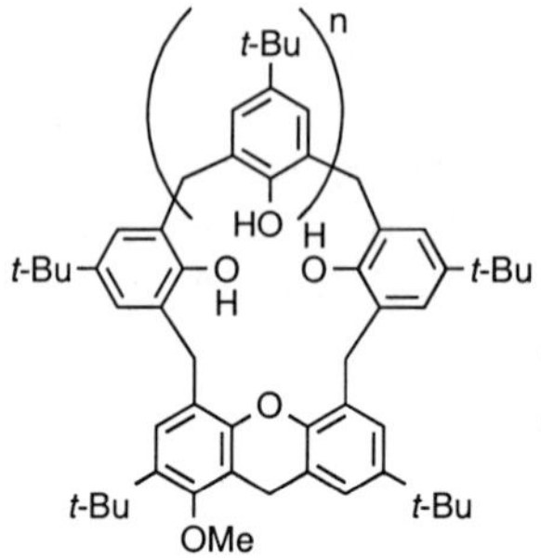

Since the spirodienone readily reverts to phenol, the monospirodienone derivative **18a** can be used as an intermediate for the preparation of proximally disubstituted calixarene derivatives [49]. Although CH$_2$Cl$_2$ is not a good alkylating agent, treatment of **18a** with CH$_2$Cl$_2$/aq NaOH (30%) in the presence of a phase transfer catalyst yielded the bridged monospirodienone derivative **20a** which was characterized by X-ray crystallography [53]. NaBH$_4$ reduction of the carbonyl group occurred stereoselectively by attack of the *exo* face of the carbonyl (the face *anti* to the C-O bond), the same selectivity observed in the addition of MeLi to **12c** [54]. Thermal rearrangement of the resulting spirodienol afforded the monodioxamethylene bridged calix[4]arene **20b** (eq. 3)[53].

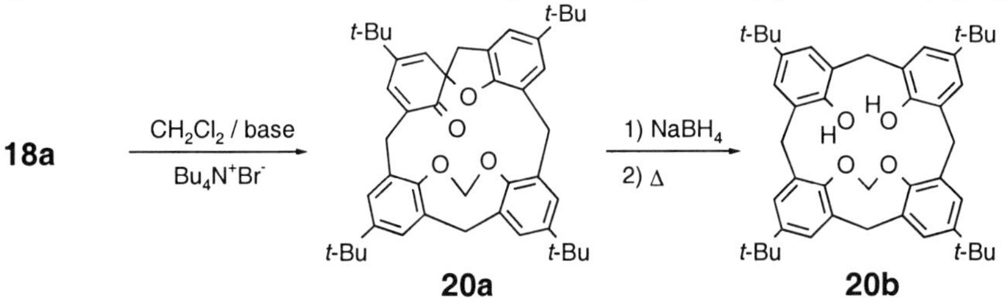

18a $\xrightarrow[\text{Bu}_4\text{N}^+\text{Br}^-]{\text{CH}_2\text{Cl}_2 / \text{base}}$ **20a** $\xrightarrow[\text{2) } \Delta]{\text{1) NaBH}_4}$ **20b** (Eq. 3)

The replacement of two distal OH groups of **1a** by methyls can be achieved using a bis(spirodienone) derivative as key intermediate. Reaction of **12a** with a large excess of MeLi afforded **21**. The organolithium reagent acts both as a nucleophile (adding to the carbonyl groups of **12a**) and as a base, abstracting a methylene proton in a β–elimination reaction. Ionic hydrogenation of **21** yielded a calixarene with two *intra*annular methyl groups (eq. 4) [54].

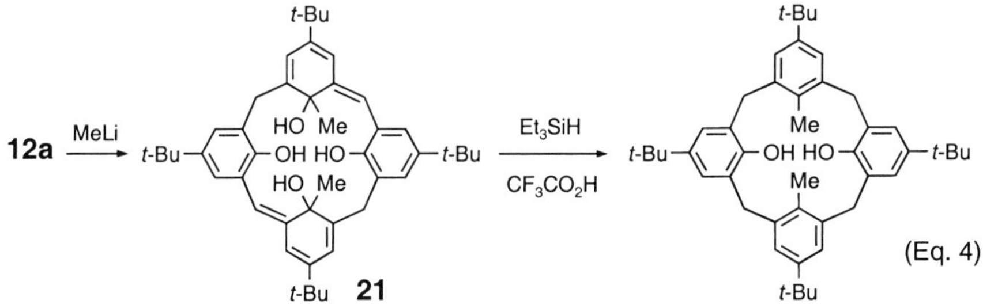

12a $\xrightarrow{\text{MeLi}}$ **21** $\xrightarrow[\text{CF}_3\text{CO}_2\text{H}]{\text{Et}_3\text{SiH}}$ (Eq. 4)

The Diels-Alder reaction of the diene subunits of the bis(spirodienones) and tris(spirodienone) calixarene derivatives has been studied [43,48]. In all cases, the reaction takes place by attack on the *exo* face of the diene. The Diels-Alder adducts are configurationally stable and do not mutually isomerize in solution.

4. Hydrogenation of the Aromatic Rings

Aromatic rings are usually more resistant to hydrogenation than isolated double bonds. Catalytic hydrogenation of phenol can afford, depending on the nature of the catalyst and the reaction conditions, cyclohexanone, cyclohexanol, or cyclohexane as products [55,56].

4.1. Isomerism in Saturated Calixarenes

The full hydrogenation of **1a** creates (assuming that the C-O bonds are not cleaved) 16 stereocenters (cf, **22**), and therefore a large of number of isomers exists. All of the experimental studies to date have been conducted on the de-*tert*-butylated-calix[4]arene **2a**. This compound presents the obvious advantage that the number of stereocenters in the fully hydrogenated calix[4]cyclohexanol **23** is reduced to twelve, and in addition, since the arene rings are less shielded, less drastic hydrogenation conditions are required than for **1a**. Four configurational isomers are possible for a calix[4]arene incorporating a single cyclohexanol ring (**24a-d**). The *cis-trans* isomers **24a** and **24b** (*RR/SS*) represent an enantiomeric pair, while the *cis-cis* (**24c**) and *trans-trans* (**24d**) forms (*RrS* and *RsS*, respectively) are achiral *meso* forms. The number of configurational isomers increases with the number of stereocenters and for the distal dodecahydrocalixarene and a fully hydrogenated derivative lacking OH groups it has been calculated (using the configurational matrix method [57]) that twenty and forty-three configurational isomers are possible [58].

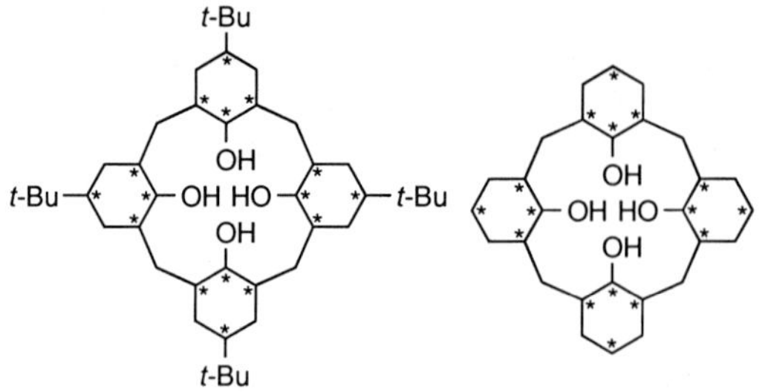

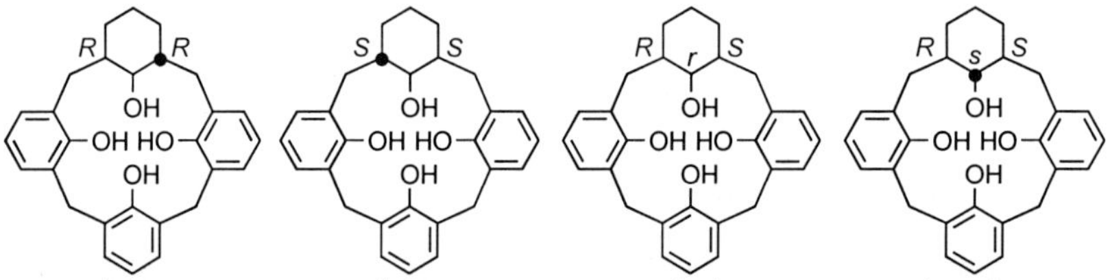

cis-trans **24a** *cis-trans* **24b** *cis-cis* **24c** *trans-trans* **24d**

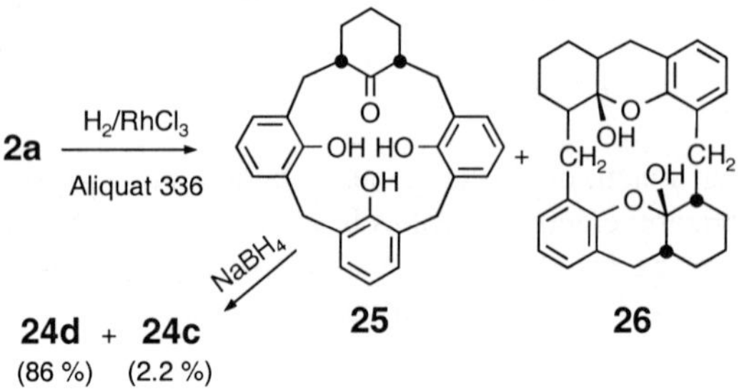

2a → (H₂/RhCl₃, Aliquat 336) → **25** + **26**

↓ NaBH₄

24d + **24c**
(86 %) (2.2 %)

4.2. PARTIALLY HYDROGENATED CALIXARENES

A single phenol ring in **2a** can be stereospecifically hydrogenated at room temperature in a biphasic system using $RhCl_3$/Aliquat 336 as catalyst [59,60]. The chemoselectivity of the transformation is remarkable, since with the same catalytic system, phenol is reduced into a mixture of 27% cyclohexanol, 67% cyclohexanone and 6% cyclohexane [61]. In addition to **25**, small amounts of the asymmetric bis(hemiketal) **26** were also obtained. X-ray crystallography indicated that the compound obtained (**25**) is the *meso* (*RS*) form which adopts a *cone* like conformation [59]. Reduction of the carbonyl group of **25** with $NaBH_4$ afforded the *RsS* calixcyclohexanol **24d** as the major (95%) isomer, with the minor product being the *RrS* form **24c**. The major and minor products were characterized by X-ray crystallography which indicated that in both compounds all the methylenes are connected to the equatorial positions of the cyclohexyl ring [62]. Both

24c [62] and **24d** [58,62] adopt *cone*-like conformations on which the cyclohexanol OH group (which is located in an equatorial position in **24d**, and in an axial position in **24c**) and the phenolic OH groups are involved in a circular array of hydrogen bonds. The calixcyclohexanol **24d** readily forms an europium(III) complex [62].

The preparation of partially hydrogenated calixarenes with intact binding groups using Pd/C as catalyst proved difficult. Calix[4]arenes derivatives incorporating one (**24c**) or two distal cyclohexanols rings (**27**) were isolated by conducting the hydrogenation of **2a** at relatively low temperatures (100 °C) and stopping the reaction before completion (eq. 5) [58]. X-ray crystallography indicated that **27** possess one axial and one equatorial hydroxyl group with all the OH groups involved in a circular array of hydrogen bonds.

4.3. FULLY HYDROGENATED DERIVATIVES

The product of the full hydrogenation of **2a** depends on the reaction conditions and/or the catalyst used. Hydrogenation with Raney Ni (1450 psi, *i*-PrOH, 240 °C) affords in 15% yield the saturated diether **28** in which the two perhydroxanthene subunits possess a *trans-syn-trans* configuration and with all the C-O bonds located at equatorial positions of the cyclohexyl rings [63]. MM2 calculations indicate that the product obtained corresponds to the lowest energy isomer. Hydrogenation of **2a** with Pd/C at 120 °C afforded the saturated diether **29** in which the perhydroxanthene subunits possess a *cis-syn-cis* fusion and all the C-O bonds are located at axial position of the rings [54]. Calculations indicate that this isomer is 13.5 kcal mol^{-1} less stable than **28**. The diethers **28** and **29** probably arise from the dehydration of a calixcyclohexanol.

Hydrogenation of **2a** with Pd/C using higher temperatures (250 °C, 600 psi H_2) afforded pentacyclo[19,3,1,1^{3,7},1^{9,13},1^{15,19}]octacosane (**30**). X-ray crystallography indicated that in the saturated metacyclophane **30** (of approximate D_{2d} symmetry) all the bridging methylene groups (the substituents on the cyclohexyl rings) are located at equatorial positions of the rings, and the configuration of the stereocenters is such that the pairs of methine protons on the four rings are alternately disposed above and below the mean macrocyclic plane [58].

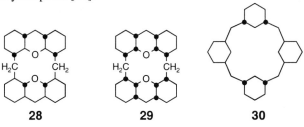

4.4. FACE SELECTIVITY OF THE HYDROGENATION

The analysis of the face selectivity of the hydrogenation of the parent **2a** is complicated by its flexibility. The hydrogenation of the four rings is not a simultaneous process, and since the structure and conformation of the intermediates is unknown, it is possible that they adopt different conformations displaying different face selectivities.

In order to determine the face selectivity of the hydrogenation using Pd/C as a catalyst, a calix[4]-arene propyl ether was chosen as substrate. Since the bulky propyl groups precludes the rotation through the annulus of the rings, the compound can be prepared in four atropisomeric forms (**31**-*cone*, **31**-*partial cone*, **31**-*1,3-alternate* and **31**-*1,2-alternate*) which do not interconvert even at high temperatures [64]. In each atropisomer the two faces of any ring are different since one is located inside the cavity ("*endo*" face) while the second is on the outside ("*exo*" face). Separate hydrogenations of **31**-*cone*, **31**-*partial cone* and **31**-*1,3-alternate* were conducted using Pd/C as a catalyst (600 psi H_2, 120 ^{0}C). In each case a different product (**32a**, **32b** and **32c**, respectively) was formed exclusively indicating that the reaction proceeded with high stereoselectivity and that the *up-down* arrangement of the rings determines the pattern of stereocenters in the hydrogenated products. NMR and X-ray data indicated that the hydrogenation proceeded in an all-*exo* fashion, and that in the final products the bridging methylenes are connected to the axial positions of the cyclohexyl rings ("axial" form). Since interconversion of the axial form to the lowest energy equatorial form requires both fourfold chair inversion and rotation through the annulus of the saturated rings, and the latter process is precluded by the bulky groups, the high energy axial forms obtained in the reaction are kinetically stable [65].

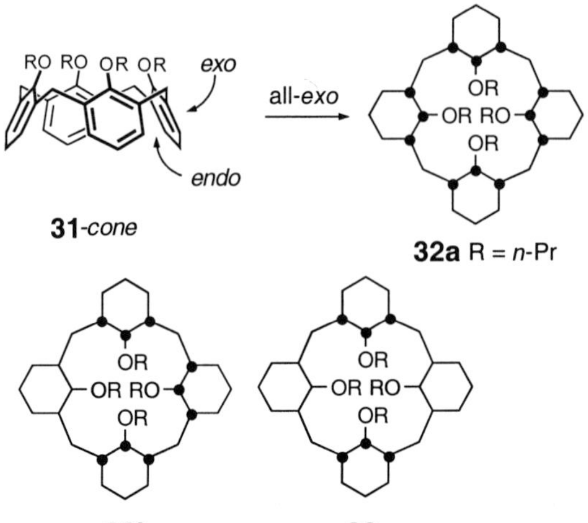

4.5. CALIX[4]CYCLOHEXANONE AND CALIX[4]CYCLOHEXANOL

Single stereoisomers of the fully hydrogenated calix[4]cyclohexanone and calix[4]cyclohexanol were prepared taking advantage of the acidity of the carbonyl α protons. Hydrogenation of **2a** with $RhCl_3 \cdot 3H_2O$/Aliquat 336 (200 psi H_2, 90 ^{0}C) in a biphasic system afforded a mixture a calixcyclohexanones. This mixture was epimerized into the lowest energy isomer by treatment with base. X-ray crystallography of calix[4]cyclohexanone (**33**) indicated that it possesses *RS SR RS SR* configurations of the stereocenters and that the molecule adopts a conformation in which the carbonyls are oriented in antiparallel fashion [66].

Reaction of the carbonyl groups of **33** with $NaBH_4$ proceeded with high stereospecificity and yielded calix[4]cyclohexanol **34** (with *RsS SsR RsS SsR* configurations).

X-ray crystallography indicated that the compound adopts a *cone*-like conformation in which, similarly to the parent **2a**, the four hydroxyl groups are engaged in a circular array of hydrogen bonds. However, in contrast to **2a** in the *cone* conformation of **34** are present two symmetry unequivalent rings, with their methine protons pointing "in" (towards the cavity) or "out". Both rings interconvert by a *cone*-to-*cone* inversion process.

Calix[4]cyclohexanol **34** is substantially more rigid than **2a**, and saturation transfer experiments indicate that the barrier of the *cone*-to-*cone* inversion is 22.1 kcal mol^{-1} [66].

5. Conclusions

Synthetic methodologies for the oxidation and reduction of the phenol rings of the calix[*n*]arenes are now well established. The oxidation and reduction of these groups affect the "walls" which delimit the calix cavity and therefore substantially alter the conformation, rigidity and binding properties of the parent systems. The oxidized and reduced derivatives provide new building blocks for the construction of host systems and further extend the wide range of synthetic modifications possible for the calix skeleton.

6. References and Notes

[1] For recent reviews on calixarenes see: a) V. Böhmer, *Angew. Chem., Int. Ed. Engl.* **1995**, *34*, 713-745. b) C. D. Gutsche, *Aldrichimica Acta* **1995**, *28*, 1-9 c) C. D. Gutsche, *Calixarenes Revisited* Royal Society of Chemistry, Cambridge, **1998**.

[2] a) G. Görmar, K. Seiffarth, M. Schultz, J. Zimmermann, G. Flamig, *Makromol. Chem.* **1990**, *191*, 81-87. b) O. Aleksiuk and S. E. Biali, *J. Org. Chem.* **1996**, *61*, 5670-5673.

[3] See for example: Z. Goren, S. E. Biali, *J. Chem. Soc., Perkin Trans. 1* **1990**, 1484-1487. For a review on the hydroxyl replacement in calixarenes see: S. E. Biali, *Isr. J. Chem.* **1997**, *37*, 131-139.

[4] a) Bicalix[n]arenes derivatives (*n*=4,6,8) connected by a biphenyl-like bond at the *para*-position of one ring are obtained by oxidation of calixarenes derivatives with FeCl$_3 \cdot$ 6H$_2$O in refluxing acetonitrile; b) P. Neri, A. Bottino, F. Cunsolo, M. Piatelli, E. Gavuzzo, *Angew. Chem., Int. Ed. Engl.* **1998**, *37*, 166-169; c) A. Bottino, F. Cunsolo, M. Piatelli, D. Garozzo, P. Neri, *J. Org. Chem.* **1999**, *64*, 8018-8020.

[5] For reviews on the quinone group see: *The chemistry of functional groups: The chemistry of quinonoid compounds*, Part 1, S. Patai Ed, Wiley, Chichester (1974), Part 2, S. Patai, Z. Rappoport, Eds, Wiley, Chichester (1988).

[6] Calixarenes covalently linked to a quinone have been described in the literature. See a) D. Bethell, G. Dougherty, D. C. Coupertino, *J. Chem. Soc., Chem. Commun.* **1995**, 675-676; b) idem, *Acta Chem. Scand.* **1998**, *52*, 407; c) S. Akine, K. Goto, T. Kawashima, *Tetrahedron Lett.* **2000**, *41*, 897-901.

[7] For a review on calixquinones see ref 1c, pp 57-58, 130-132, 149.

[8] P. A. Reddy, R. P. Kashyap, W. H. Watson, C. D. Gutsche, *Isr. J. Chem.* **1992**, *32*, 89-96.

[9] a) Y. Morita, T. Agawa, Y. Kai, N. Kanehisa, N. Kasai, E. Nomura, H. Taniguchi, *Chem. Lett.* **1989**, 1349-1352. b) Y. Morita, T. Agawa, Y. Kai, E. Nomura, H. Taniguchi, *J. Org. Chem.* **1992**, *57*, 3658-3662.
[10] A. McKillop, B. P. Swann, E. C. Taylor, *Tetrahedron* **1970**, *26*, 4031-4039.
[11] A. Casnati, L. Domiano, A. Pochini, R. Ungaro, M. Carramolino, J. O. Magrans, P. M. Nieto, J. Lopez-Prados, P. Prados, J. de Mendoza, R. G. Janssen, W. Verboom, D. N. Reinhoudt, *Tetrahedron* **1995**, *51*, 12699-12720.
[12] a) P. D. Beer, Z. Chen, P. A. Gale, *Tetrahedron*, **1994**, *50*, 931-940. b) P. D. Beer, Z. Chen, P. A. Gale, J. A. Heath, R. J. Knubley, M. I. Ogden, M. G. B. Drew, *J. Incl. Phenom.* **1994**, *19*, 343-359. c) P. D. Beer, P. A. Gale, Z. Chen, M. G. B. Drew, J. A. Heath, M. I. Ogden, H. R. Powell, *Inorg. Chem.* **1997**, *36*, 5880-5893.
[13] a) A. Harriman, M. Hissler, P. Jost, G. Wipff, R. Ziessel, *J. Am. Chem. Soc.* **1999**, *121*, 14-27. b) H. C.-Y. Bettega, J.-C. Moutet, G. Ulrich, R. Ziessel, *J. Electroanal. Chem.* **1996**, *406*, 247-250.
[14] K.-C. Nam, D.-S. Kim, S.-J. Yang, *Bull. Korean Chem. Soc.* **1992**, *13*, 105-107.
[15] J. -D. van Loon, A. Arduini, L. Coppi, W. Verboom, A. Pochini, R. Ungaro, S. Harkema, D. N. Reinhoudt, *J. Org. Chem.* **1990**, *55*, 5639-5646.
[16] a) H. Yamamoto, K. Ueda, K. R. A. S. Sandanayake, S. Shinkai, *Chem. Lett.* **1995**, 497-498. b) H. Yamamoto, K. Ueda, H. Suenaga, T. Sakaki, S. Shinkai, *Chem. Lett.* **1996**, 39-40. c) W. S. Oh, T. D. Chung, J. Kim, H.-S. Kim, H. Kim, D. Hwang, K. Kim, S. G. Rha, J.-I. Choe, S.-K. Chang, *Supramol. Chem.* **1998**, *9*, 221-229. c) T. D. Chung, S. K. Kang, H. Kim, J. R. Kim, W. S. Oh, S.-K. Chang, *Chem. Lett.* **1998**, 1225-1226.
[17] a) C.-F. Chen, Q.-Y. Zheng, Z.-T. Huang, *Synthesis* **1999**, 69-71; b) For a review on calixcrowns see: A. Casnati, R. Ungaro, Z. Asfari, J. Vicens, Chapter 20.
[18] F. Grynszpan, N. Dinoor, S. E. Biali, *Tetrahedron Lett.* **1991**, *32*, 1909-1912.
[19] P. Timmerman, S. Harkema, G. J. Van Hummel, W. Verboom, D. N. Reinhoudt, *J. Incl. Phenom.* **1993**, *16*, 189-197.
[20] B. Klenke, W. Friedrichsen, *J. Chem. Soc., Perkin Trans. 1* **1998**, 3377-3379.
[21] A. Casnati, E. Comelli, M. Fabbi, V. Bocchi, G. Mori, F. Uggozoli, A. M. Manotii Lanfredi, A. Pochini, R. Ungaro, *Recl. Trav. Chim. Pays-Bas* **1993**, *112*, 384-392.
[22] P. D. Beer, Z. Chen, M. G. B. Drew, P. A. Gale, *J. Chem. Soc., Chem. Commun.* **1994**, 2207-2208.
[23] M. Gomez-Kaifer, P. A. Reddy, C. D. Gutsche, L. Echegoyen, *J. Am. Chem. Soc.* **1997**, *119*, 5222-5229.
[24] P. A. Reddy, C. D. Gutsche, *J. Org. Chem.* **1993**, *58*, 3245-3251.
[25] a) Y. Kubo, S. Tokita, Y. Kojima, Y. T. Osano, T. Matsuzaki, *J. Org. Chem.* **1996**, *61*, 3758-3765. b) Y. Kubo, S. Maeda, M. Nakamura, S. Tokita, *J. Chem. Soc., Chem. Commun.* **1994**, 1725-1726. c) Y. Kubo, S. Maruyama, N. Ohhara, M. Nakamura, S. Tokita, *J. Chem. Soc., Chem. Commun.* **1995**, 1727-1728.
[26] a) M. Mascal, R. T. Naven, R. Warmuth, *Tetrahedron Lett.* **1995**, *36*, 9361-9364. b) M. Mascal, R. Warmuth, R. T. Naven, R. A. Edwards, M. B. Hursthouse, D. E. Hibbs, *J. Chem. Soc., Perkin Trans 1.* **1999**, 3435-3441.
[27] A. Arduini, L. Mirone, D. Paganuzzi, A. Pinalli, A. Pochini, A. Secchi, R. Ungaro, *Tetrahedron* **1996**, *52*, 6011-6018.
[28] A. Casnati, R. Ferdani, A. Pochini, R. Ungaro, *J. Org. Chem.* **1997**, *62*, 6236-6239.
[29] a) C. D. Gutsche, M. Iqbal, *Org. Synth.* **1990**, *68*, 234-237; b) C. D. Gutsche, B. Dhawan, M. Leonis, D. Stewart, *ibid.* **1990**, *68*, 238-242; c) J. H. Munch, C. D. Gutsche, *ibid.* **1990**, *68*, 243-246.
[30] T. Nakayama, M. Ueda, *J. Mater. Chem.* **1999**, *9*, 697-702.
[31] P. C. Leverd, V. Huc, S. Palacin, M. Nierlich, *J. Incl. Phenom.* **2000**, *36*, 259-266.
[32] For a review on the electrochemical recognition of guest molecules by redox-active receptors see: a) P. D. Beer, P. A. Gale, Z. Chen, *Adv. Phys. Org. Chem* **1998**, *31*, 1-90. b) P. L. Boulas, M. Gomez-Kaifer, L. Echegoyen, *Angew. Chem. Int. Ed.,* **1998**, *37*, 216-247. The intraannular and extraannular substituents on the aryl rings of triquinone **27** (p. 235) should be mutually exchanged. c) For a review on the electrochemistry of calixarenes see: T. D. Chung, H. Kim, *J. Incl. Phenom.* **1998**, *32*, 179-193.
[33] K. Suga, M. Fujihira, Y. Morita, T. Agawa, *J. Chem. Soc., Faraday Trans.* **1991**, *87*, 1575-1578.
[34] M. Gomez-Kaifer, P. A. Reddy, C. D. Gutsche, L. Echegoyen, *J. Am. Chem. Soc.* **1994**, *116*, 3580-3587.
[35] a) D. Choi, T. D. Chung, S. K. Kang, S. K. Lee, T. Kim, S.-K. Chang, H. Kim, *J. Electroanal. Chem.* **1995**, *387*, 133-134; b) T. D. Chung, D. Choi, S. K. Kang, S. K. Lee, S.-K. Chang, H. Kim, *J. Electroanal. Chem.* **1995**, *396*, 431-439.
[36] T. D. Chung, S. K. Kang, J. Kim, H.-S. Kim, H. Kim, *J. Electroanal. Chem.* **1997**, *438*, 71-78.

[37]   M. Hissler, A. Harriman, P. Jost, G. Wipff, R. Ziessel, *Angew. Chem., Int. Ed.* **1998**, *37*, 3249-3252.
[38]   P. D. Beer, V. Timoshenko, M. Maestri, P. Passaniti, V. Balzani, *Chem. Commun.* **1999**, 1755-1756.
[39]   K. C. Nam, S. O. Kang, H. S. Jeong, S. Jeon, *Tetrahedron Lett.* **1999**, *40*, 7343-7346.
[40]   Z. Chen, P. A. Gale, J. A. Heath, P. D. Beer, *J. Chem. Soc., Faraday Trans.* **1994**, *90,* 2931-2938.
[41]   For a review on spirodienone calixarene derivatives see: O. Aleksiuk, F. Grynszpan; A. M. Litwak, S. E. Biali, *New J. Chem.* **1996**, *20*, 473-482.
[42]   a) J. Abel, *Ber.* **1892**, *25*, 3477-3484. b) E. A. Shearing, S. Smiles, *J. Chem. Soc.* **1937**, 1931-1936. c) For an example of a spirodienone derivative of a dihydroxymetacyclophane see: T. Yamato, J. Matsumoto, M. Sato, K. Fujita, Y. Nagano, *J. Chem. Research (S)* **1997**, 74-75; *J. Chem. Research (M)* **1997**, 518-529.
[43]   a) A. M. Litwak, S. E. Biali, *J. Org. Chem.* **1992**, *57*, 1943-1945. b) A. M. Litwak, F. Grynszpan, O. Aleksiuk, S. Cohen, S. E. Biali, *J. Org. Chem.* **1993**, *58*, 393-402.
[44]   W.-G. Wang, W.-C. Zhang, Z.-T. Huang, *J. Chem. Res. (S)* **1998**, 462-463.
[45]   P.E. Georghiou, Z. Li, *Tetrahedron Lett.* **1993**, *34*, 2887-2890.
[46]   P. E. Georghiou, M. Ashram, H. J. Clase, J. N. Bridson, *J. Org. Chem.* **1998**, *63*, 1819-1826.
[47]   F. Grynszpan, O. Aleksiuk, S. E. Biali, *Pure Appl. Chem.* **1996**, *68*, 1249-1254.
[48]   a) F. Grynszpan, S. E. Biali, *J. Chem. Soc., Chem. Commun.* **1994**, 2545-2546. b) F. Grynszpan, S. E. Biali, *J. Org. Chem.* **1996**, *61*, 9512-9521.
[49]   O. Aleksiuk, F. Grynszpan, S. E. Biali, *J. Chem. Soc., Chem. Commun.* **1993**, 11-13.
[50]   F. Grynszpan, O. Aleksiuk, S. E. Biali, *J. Org. Chem.* **1994**, *59,* 2070-2074.
[51]   O. Aleksiuk, S. Cohen, S. E. Biali, *J. Am. Chem. Soc.* **1995**, *117*, 9645-9652.
[52]   O. Aleksiuk, F. Grynszpan, S. E. Biali, *J. Org. Chem.* **1993**, *58,* 1994-1996.
[53]   J. Wöhnert, J. Brenn, M. Stoldt, O. Aleksiuk, F. Grynszpan, I. Thondorf, S. E. Biali, *J. Org. Chem.* **1998**, *63*, 3866-3874.
[54]   J. M. Van Gelder, J. Brenn, I. Thondorf, S. E. Biali, *J. Org. Chem.* **1997**, *62*, 3511-3519.
[55]   P. Rylander, *Catalytic Hydrogenation in Organic Chemistry*, Academic Press, N.Y., **1979**, p. 192.
[56]   For studies on the hydrogenation of [2.2]paracyclophane see: H. Hopf, R. Savinsky, B. Diesselkamper, R. G. Daniels, A. de Meijere, *J. Org. Chem.* **1997**, *62*, 8941-8943.
[57]   R. Willem, H. Pepermans, C. Hoogzand, K. Hallenga, M. Gielen, *J. Am. Chem. Soc.* **1981**, *103*, 2297-2306.
[58]   I. Columbus, S. E. Biali, *J. Am. Chem. Soc.* **1998**, *120*, 3060-3067.
[59]   A. Bilyk, J. M. Harrowfield, B. W. Skelton, A. H. White, *An. Quim.* **1997**, *93*, 363-370.
[60]   Y. Sasson, A. Zoran, J. Blum, *J. Mol. Catal.* **1981**, *11*, 293-300.
[61]   J. Blum, I. Amer, A. Zoran, Y. Sasson, *Tetrahedron Lett.* **1983**, *24,* 4139-4142.
[62]   A. Bilyk, J. M. Harrowfield, B. W. Skelton, A. H. White, *J. Chem. Soc., Dalton Trans.* **1997**, 4251-4256 .
[63]   F. Grynszpan, S. E. Biali, *Chem. Commun.* **1996**, 195-196.
[64]   a) W. Verboom, S. Datta, Z. Asfari, S. Harkema, D. N. Reinhoudt, *J. Org. Chem.* **1992**, *57*, 5394-5398. b) A. Ikeda, T. Nagasaki, K. Araki, S. Shinkai, *Tetrahedron* **1992**, *48*, 1059-1070. c) K. Iwamoto, K. Araki, S. Shinkai, *J. Org. Chem.* **1991**, *56*, 4955-4962. d) H. Iki, T. Kikuchi, S. Shinkai, *J. Chem. Soc., Perkin Trans 1* **1993**, 205-210. [65] I. Columbus, M. Haj-Zaroubi, J. S. Siegel, S. E. Biali, *J. Org. Chem.* **1998**, *63*, 9148-9149.
[66]   I. Columbus, M. Haj-Zaroubi, S. E. Biali, *J. Am. Chem. Soc.* **1998**, *120*, 11806-11807.

Chapter 15

CONFORMATIONS AND STEREODYNAMICS

IRIS THONDORF

Martin-Luther-Universität Halle-Wittenberg, Fachbereich Biochemie/Biotechnologie, Kurt-Mothes-Str. 3, 06099 Halle, Germany. e-mail: thondorf@biochemtech.uni-halle.de

1. Introduction

Understanding of the conformational properties of calixarenes is of fundamental importance with respect to their utilisation as building blocks for the construction of supramolecular systems and artificial receptors. Molecular models can provide relevant information about molecular stereochemistry as well as intra- and intermolecular interactions. Over the past few years, virtual molecular models based on the combination of molecular graphics with quantum chemical and molecular mechanics (MM) calculations, have increasingly entered chemical laboratories. Although these models are simplified or idealised systems neglecting essential aspects of chemical reality, they enable a deeper understanding of the molecular properties on the microscopic level.

This brief overview of calixarene conformations and stereodynamics is based largely on the results of molecular mechanics calculations. Literature results are summarised in several tables to allow focus on the synergism of experiment and theory for solving problems related to calixarene stereochemistry.

2. Calix[4]arenes

2.1. [1$_4$]-METACYCLOPHANES

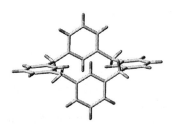

Figure 1. The crystal structure of 1a (CSD-Refcode: SIVMOZ [1]).

In the crystalline state, the unsubstituted [1$_4$]-metacyclophane **1a** adopts the *chair* conformation [1]. This differs from the basic *cone*, *partial cone* (*paco*), *1,2-alternate* (*1,2-alt*) and *1,3-alternate* (*1,3-alt*) conformations usually discussed for calix[4]arenes in the coplanar arrangement of two opposite phenyl rings. Although it resembles the *paco* and *1,2-alt* conformations, the sequence of the signs of the dihedral angles ϕ/χ around the methylene bridges [2] is completely different (*chair*: +−,+−,−+,−+; *paco*: +−,+−, ++,−−; *1,2-alt*: +−,++,−+,−−).

MM calculations indicate (Table 1) that this chair-like form is not the most stable conformation for isolated **1a**, nor for **1b** or **1c**. Although most calculations gave the *cone* as a global minimum, the small energy differences suggested **1a-c** would be conformational mixtures in solution.

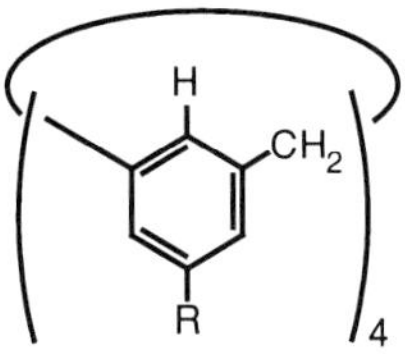

	1a	R = H
	1b	R = Me
	1c	R = *t*-Bu
	1d	R = Br

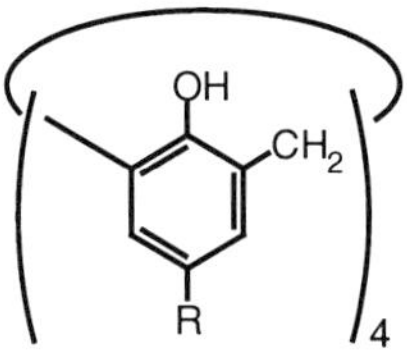

	2a	R = *t*-Bu
	2b	R = *i*-Pr
	2c	R = H
	2d	R = Me

TABLE 1. Calculated relative energies (kcal/mol) of the characteristic conformers of [1₄]-metacyclophanes

Entry	Compd.	Method	*cone*	*paco*	*1,2-alt*	*1,3-alt*	Ref.
1	**1a**	MMX	0.0	0.7	0.8	1.5	[1]
2		MM2	0.0	1.5	1.4	0.5	[3]
3		MM3(89)	0.0	1.2	1.8	1.7	[4]
4		MM3(92)	0.0	1.2	1.8	1.7	[5]
5		CHARMM	0.0	0.5	1.5	1.9	[6]
6	**1b**	MM3(89)	0.0	1.1	1.8	1.3	[4]
7	**1c**	MM2	0.6	1.1	3.1	0.0	[7]
8		MM3(89)	0.0	1.0	4.7	2.7	[4]
9		MM3(92)	0.0	1.0	2.5	1.1	[5]
10		MM3(92)	0.0	1.1	3.2	1.6	[8]

Low temperature ^{1}H NMR data for **1a**, **1c** and **1d** show they are flexible molecules [1,7,9]. For **1c**, an upper limit of 5.9 kcal/mol for the rotational barrier was estimated [10]. CHARMM [11] and MM3 calculations [12] for **1a** and **1c** indicated that this barrier could be lower by 1.5-2.8 kcal/mol [6,8].

2.2. CALIX[4]ARENE-25,26,27,28-TETROLS

The tetrols **2** exist both in solution and in the crystalline state exclusively in the *cone* conformation. The dominance of the *cone* is commonly ascribed to its stabilisation by the homodromic hydrogen bonding system in which each OH group acts simultaneously as donor and acceptor. MM3 calculations [5] indicated, however, that the OH groups actually induce transannular strain in the *cone* conformation.

MM calculations, summarised in Table 2, can usually reproduce the extraordinary stability of the *cone* conformation. However, the results obtained by different force fields do not provide a clear picture of the order and overall energy range of the basic conformations, which are also unavailable from experimental studies.

A problem for molecular mechanics calculations concerns the geometry of the most stable *cone* conformer. Some force fields [4,15,18] favour the ideal, C_4-symmetrical *cone* conformer[*] for **2a-c** but MM3(92) calculations [5,13] furnished this form as transi-

---

[*] When the direction of the OH groups is taken into account, the *cone* conformer is classified as being of C_4 symmetry while in solution an averaged C_{4v} symmetrical structure is observed.

tion state lying 0.6 kcal/mol (**2c**) above the C_2-symmetrical, *pinched cone* conformer [14]. The symmetry reduction was attributed to repulsive van der Waals contacts between the axial methylene protons and the adjacent OH groups being diminished in the C_2-symmetrical form [5]. A fast $C_2 \rightleftharpoons C_2$ pseudorotation would then explain the spectroscopically observed fourfold symmetry of the *cone* conformation [5].

TABLE 2. Calculated relative energies (kcal/mol) of the characteristic conformers of calix[4]arene tetrols

Entry	Compd.	Method	*cone*	*paco*	*1,2-alt*	*1,3-alt*	Ref.
1	**2a**	AMBER[a]	0.0	7.7	11.3	12.6	[15]
2		AMBER[b]	0.0	10.0	11.8	10.0	[16]
3		OPLS-AMBER[b]	0.0	19.6	23.1	19.4	[16]
4		MM2[b]	0.0	14.7	19.2	23.9	[16]
5		MM3(92)	0.0	6.1	7.5	12.8	[5]
6		MM3(92)	0.0	6.1	7.7	11.2	[8]
7		CHARMm[c]	73.8	0.0	16.4	28.7	[16]
8		CHARMm[d]	0.0	3.4	7.2	4.6	[17]
9	**2b**	MM2	0.0	0.6	3.3	1.0	[18]
10	**2c**	MM3(89)	0.0	9.9	11.7	18.7	[5]
11		MM3(92)	0.0	5.6	6.1	10.6	[5]
12		CHARMM	0.0	8.9	11.5	17.2	[6]

[a] as implemented in MacroModel V.1.5, [b] as implemented in MacroModel V.3.5, [c] as implemented in QUANTA, [d] as implemented in QUANTA 3.2/CHARMm 21.3.

The *cone* $\rightleftharpoons$ *cone* interconversion (topomerization) can be easily monitored by variable temperature ^{1}H NMR since it exchanges the methylene proton environments. Depending on the solvent, energy barriers in the range of 14.5-16 kcal/mol have been derived by the coalescence method. A "broken chain pathway" involving stepwise rotation of the phenolic rings and a "continuous chain pathway" proceeding via a single transition state resembling a screwed *1,2-alt* conformation have been suggested as limiting mechanisms [19].

Molecular dynamics (MD) simulations [15] using the AMBER force field [20-22] indicated that the pathways of conformational transitions of **2c** proceed via a stepwise mechanism involving the *paco* conformation as key intermediate. At high simulation temperatures, enough kinetic energy was available to surmount also the activation barriers for concerted interconversions, a presumption supported by CHARMM calculations [6] and by studies using the MM3 force field [8] which indicated high barriers for concerted ring inversion processes. For the stepwise rotation of the four phenolic rings resulting in *cone* topomerisation, the lowest energy pathway calculated involved the *paco* and *1,2-alt* conformations, while a higher barrier occurred for passage through the *1,3-alt* conformation [6,8,23-27]. The barriers calculated with the CHARMM and MM3 force fields [6,8] ($\Delta E^{\ddagger}$ = 13.8 and 14.5 kcal/mol for **2a** and **2c**, respectively) are in good agreement with experimental values of $\Delta H^{\ddagger}$ = 15.9 (**2a**) and 14.2 kcal/mol (**2c**) [28,29] while MM2 calculations [24,30] provided drastic underestimates.

The conformational properties can be influenced by substituents. Different substituents in the *p*-positions of **2** have only minor influence on the stability of the *cone* conformation (*cf.*, Table 2, entries 5 and 10) [5] and on the rotational barriers [31] but bridging of opposite *p*-positions by alkyl chains prevents the topomerisation of the *cone* conformation [32,33]. Moreover, short alkyl bridges cause deformation of the meta-cyclophane skeleton to a *pinched cone* conformation and weakening of the hydrogen bonding system. The same conformational changes result when eight methyl groups are introduced in the *m*-positions of a calixarene [34,35] while for a calixarene bearing four *m*-methyl substituents fourfold symmetry is apparent on the NMR timescale [36]. Consistent with MM calculations [37], the latter compound exists, however, in the crystalline state in a C_2-symmetrical *cone* conformation [36].

The alkanediyl calix[4]arenes **3** and **4** adopt a *cone* conformation both in solution and in the crystal [38-41]. Stereoisomers exist depending on the substituent orientations on the methine bridges. MM calculations [38,40] predict a preference for the equatorial position, markedly more so for alkyl groups than for aryl groups. Strong repulsions between the spherical alkyl substituents and the adjacent OH groups are calculated to destabilise the axial arrangement while the flat shape of aryl substituents results in similar steric requirements in their axial and equatorial disposition.

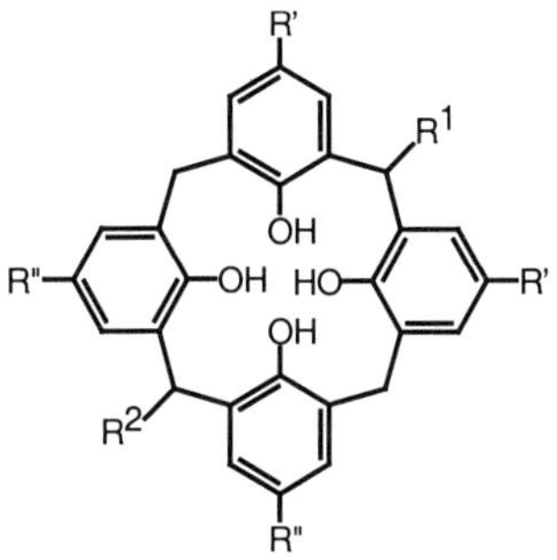

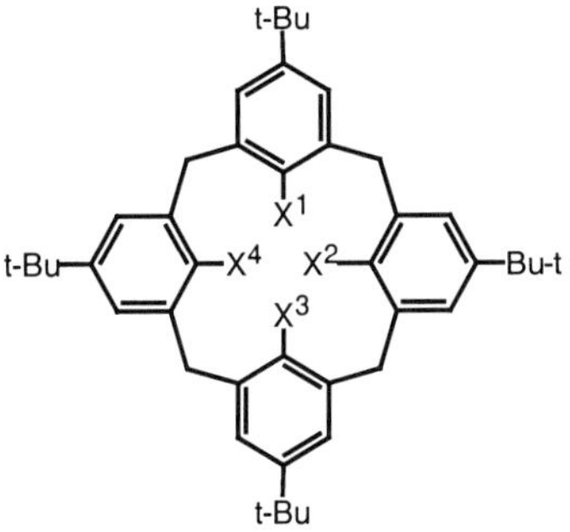

	X^1	X^2	X^3	X^4
5	OH	OH	OH	H
6	OH	OH	H	H
7	OH	H	OH	H
8	OH	H	H	H
9	NH$_2$	OH	OH	OH
10	NH$_2$	OH	NH$_2$	OH
11	NH$_2$	NH$_2$	NH$_2$	NH$_2$
12	SH	SH	SH	SH
18	Me	Me	Me	Me
19	Me	OH	Me	OH
20	SH	OH	SH	OH

3 R^1 = alkyl, aryl, R^2 = H
4 R^1 = R^2 = alkyl, aryl

3. Calix[4]arenes Modified at the *intra*-Annular Positions

3.1. PARTIALLY HYDROXYLATED CALIX[4]ARENES

The partial replacement of the OH groups in **2** by hydrogen interrupts the circular hydrogen bonding system while simultaneously the transannular strain is reduced. Both effects should result in destabilisation of the *cone* conformation and decrease of the rotational barriers.

Conformational interconversions of **5-8** are slow on the NMR timescale at low temperatures. From spectroscopic data it has been concluded that **5** exists in the *cone* conformation [5,42], **6** in the *cone* or *1,2-alt* conformation [43], **7** in the *1,3-alt* [44] or as a mixture of the *cone* and *1,3-alt* conformations [5] and **8** as a mixture of the *cone*, *paco* and *1,3-alt* conformations [5].

MM3 calculations indicate that inversion of the phenyl rings in **5-8** involves much lower barriers than the rotation of the phenol rings [45]. Thus, it has been conjectured that the compounds should not be completely rigid at low temperatures and the observed spectra should correspond to the weighted average of rapidly interconverting conformations [45].

The calculated relative stability of the *cone* conformation of **5-8** (Table 3) decreases with decreasing number of OH groups. The small energy differences between the conformations, particularly for **6-8**, suggest, as experimentally observed, that the compounds should exist in solution as a mixture of conformations [5,45].

Remarkably, the barriers for conformational interconversions of **5-8** estimated from variable temperature ^{1}H NMR ($\Delta G^\ddagger$ = 9.6 (**5**), 10.6 (**6**), 9.5 (**7**) and 9.7 kcal/mol (**8**)) seem to be nearly independent of the number of OH groups [5,43,46]. MM3 calculations indicate that similar steric restrictions in the rate-limiting steps of these interconversion processes should give similar activation barriers, while the breakage and formation of hydrogen bonds during the rotation of the phenol rings hardly affects the barriers [45].

TABLE 3. Calculated relative energies (kcal/mol) of the characteristic conformers of partially dehydroxylated calix[4]arenes [45]

Compd.	cone	paco^a	1,2-alt	1,3-alt
5	0.0	2.4	5.5	7.4
6	0.0	0.4	1.4	3.3
7	0.0	0.6	3.0	2.8
8	0.4	0.0	2.0	1.6

^{a} From the possible stereoisomers of the *paco* and *1,2-alt* conformations only the lowest energy conformers are listed.

3.2. AMINOCALIX[4]ARENES

An amino group confers basic properties to the molecule [44,46-48] but does not interrupt the circular hydrogen bonding array, so that the conformational properties should resemble those of the tetrols **2**. In fact, both the monoamino compound **9** and the 1,3-diaminocalix[4]arene **10** exist in solution at low temperatures in a *cone* conformation which is frozen on the NMR timescale [44,47]. The somewhat lower rotational barriers for the topomerisation of the *cone* together with the high field shift of the OH protons and the higher frequencies of the ν_{OH} vibration bands have been interpreted as due to weaker intramolecular hydrogen bonding interactions than in **2**.

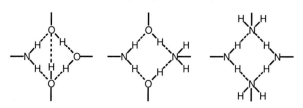

Figure 2. Hydrogen bonding patterns in the most stable cone conformers of **9-11**.

MM3 calculations indicate that the homodromic hydrogen pattern present in **2** is replaced by an antidromic pattern in **9** and **10** (Figure 2) [49]. A homodromic pattern in **9** and **10** would require the energetically unfavourable coplanar arrangement of the nitrogen lone electron pair and the aniline ring. Consequently, *intra*-annular amino groups should act either as twofold donor or as twofold acceptor of hydrogen bonds. Thus, in **9** and **10**, preference is given to weaker hydrogen bonds (NH···O) to the disadvantage of stronger hydrogen bonds (OH···N). Interestingly, MM3 calculations predict for the hypothetical calix[4]arene **11** that the C_4-symmetrical *cone* conformer with homodromic hydrogen bonds represents the transition state of the pseudorotation of the C_2-symmetrical *cone* form with heterodromic hydrogen bonds (Figure 2) [8,49].

3.3. CALIX[4]ARENES BEARING *ENDO*-METHYL AND *ENDO*-MERCAPTO GROUPS

Replacement of the OH groups in **2** by SH groups favours the *1,3-alt* conformation for **12** in solution and in the solid state [50-52]. Analogously, mesitylene derivatives **13** and **14** as well as the dodeca-methoxy compound **16** have been found to exist in the *1,3-alt* conformation [53-57]. There is some uncertainty in the assignment [58] of the *paco* conformation to **17** as the published ^{1}H NMR data are compatible with the presence in solution of a mixture of regioisomers [59].

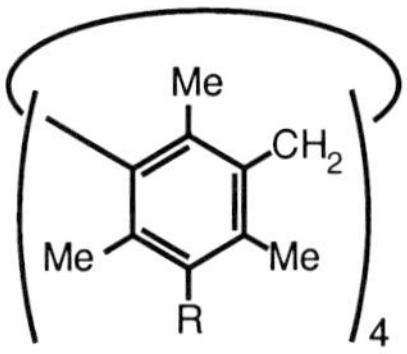

13 R = H, SH, OH
14 R = OCH$_2$CH(CH$_3$)C$_2$H$_5$, CH$_2$C(O)NHCH(CH$_3$)Ph
15 R = OMe

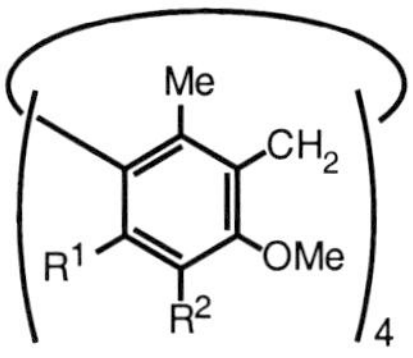

16 R^1 = R^2 = OMe
17 R^1 = Me, R^2 = H

MM calculations for **12** and **16** and the model calix[4]arenes **15** and **18** (Table 4), confirm the *1,3-alt* as most stable conformation. This has been ascribed to accommodation of the steric bulk of the *endo*-mercapto and *endo*-methyl groups in the *1,3-alt* conformation [8,57].

The ^{1}H NMR spectrum of **12** displays a singlet for the protons of the methylene bridge in a broad temperature range, consistent with a fixed *1,3-alt* conformation [50-52] or a flexible conformation in which rapid rotation of the aryl rings is taking place. As for **12**, it has been suggested that the substituent pattern in **13** and **16** renders the molecules conformationally rigid [55]. MM3 calculations suggest, however, that *endo*-methyl and -mercapto groups can pass through a calix[4]arene annulus [8,60]. This has been corroborated by NMR experiments for **19** [60].

TABLE 4. Calculated relative energies (kcal/mol) of the characteristic conformers of calix[4]arenes bearing *intra*-annular mercapto and methyl groups

Entry	Compd.	Method	cone	paco[a]	1,2-alt	1,3-alt	Ref.
1	12	CHARMm[b]	4.2	0.0	2.9	0.0	[52]
2		MM3(92)	5.8	4.6	5.1	0.0	[8]
3	15	MM3(96)	8.2	6.2	13.4	0.0	[57]
4	16	MM2	1.1	5.2	5.2	0.0	[56]
5	18	MM3(92)	8.5	3.8	5.8	0.0	[8]
6		MM3(89)	5.9	2.1	3.8	0.0	[4]
7	19	MM3(92)	0.2	0.2	3.9	0.0	[60]
8	20	CHARMm[b]	0.0	2.4	4.9	3.3	[52]

[a] from the possible stereoisomers of the *paco* and *1,2-alt* conformations of **19** and **20** only the lowest energy conformers are listed, [b] as implemented in the QUANTA3.2/CHARMm 21.3 program package.

In the chiral calix[4]arenes **14** the enantiotopic protons of the central scaffold become diastereotopic [57]. Hence, these compounds enable distinction by NMR spectroscopy between the slow and fast exchange regimes of symmetric calix[4]arene systems adopting the *1,3-alt* conformation. Although **14** cannot be viewed as completely rigid due to relatively low barriers calculated for the NMR silent *1,3-alt* → *paco* → *1,3-alt* process of **15**, probably repulsive steric interactions between the *exo*-methyl groups at vicinal rings prevent the *1,3-alt* → *1,3-alt** process and render the molecules rigid on the NMR timescale.

Interestingly, **19** and **20** differ in their conformational preferences [52,60]. **19** exists both in solution and in the crystal in the *1,3-alt* conformation, while the 1,3-dimercapto derivative **20** adopts the *cone* conformation, possibly due to weak hydrogen-bonding.

3.4. ALKOXYCALIX[4]ARENES

Tetramethoxycalix[4]arenes **21** in solution at room temperature exist as rapidly equilibrating mixtures of all possible isomers [61-63]. The key intermediate in the conformational interconversions is the *paco* conformation which can be converted to the *cone*, *1,2-alt* and *1,3-alt* by rotation of a single ring. These three one-step processes are under entropy control, as the nearly equal activation enthalpies point to similar steric hindrance associated with the aromatic ring flip [64]. The pathway of lowest energy for the topomerisation of the thermodynamically most stable *paco* conformation proceeds via the least stable conformation, *1,3-alt*, while the *1,2-alt* conformation is kinetically most stable. Simulations of the ring inversions of **21a** by means of the CHARMM force field overestimated the barriers but the results are in qualitative agreement with the experimental data [6].

21a R = *t*-Bu, R' = H
21b R = Me, R' = H
21c R = H, R' = H
21d R = Br, R' = H
21e R = Me, R' = Me

In the crystalline state, tetramethoxy calix[4]arenes are usually present in the *paco* conformation [15,67-72]. In solution, the *paco* conformation also predominates except for water soluble tetramethoxy calix[4]arenes found in the *1,3-alt* conformation [73]. The conformational equilibrium is strongly influenced by the solvent and by the *p*-substituents [61,63,73-76]. This may be why force field calculations (Table 5) often fail to reproduce the conformer distribution. MD simulations [77] show that in dichloromethane a solvent molecule is included in the cavity of **21a**.

	R	R'	X^1	X^2	X^3	X^4
22a	*t*-Bu	H	OMe	OH	OH	OH
22b	*t*-Bu	H	OMe	OH	OMe	OH
22c	*t*-Bu	H	OMe	OMe	OH	OH
22d	*t*-Bu	H	OMe	OMe	OMe	OH
23a	Me	Me	OMe	OH	OH	OH
23b	Me	Me	OMe	OH	OMe	OH
23c	Me	Me	OMe	OMe	OH	OH
23d	Me	Me	OMe	OMe	OMe	OH

Partially O-methylated calix[4]arenes **22a-d** and **23a,b,d** exist in solution preferentially in the *cone* conformation [36,62,73-79]. MM calculations (Table 6) reproduce the thermodynamic stability of the *cone* only in a limited sense. This applies particularly to the MM2 and AMBER force fields while the relative energies calculated with TRIPOS [80] are in qualitative agreement with the experimental results. The calculated distribution of conformers of the debutylated **22a-d** also is in general accordance with the NMR data [81].

TABLE 5. Calculated relative energies (kcal/mol) of the characteristic conformers of tetramethoxy calix[4]arenes

Entry	Compd.	Method	*cone*	*paco*	*1,2-alt*	*1,3-alt*	Ref.
1	**21a**	AMBER[a]	6.4	1.8	6.7	0.0	[15]
2		AMBER[b]	2.5	2.5	9.0	0.0	[16]
3		OPLS-AMBER[a]	5.9	6.0	15.1	0.0	[16]
4		MM2	2.8	1.7	6.0	0.0	[75]
5		MM2[a]	3.6	3.6	7.4	0.0	[16]
6		MM3(89)	1.5	0.0	6.1	1.5	[74]
7		MM3(92)	1.3	0.0	6.4	1.9	[65]
8		CHARMM[b]	4.5	0.0	0.7	0.8	[6]
9		CHARMm[c]	6.0	0.0	3.8	1.8	[61]
10		CHARMm[c]	75.9	0.0	16.3	33.9	[16]
11	**21b**	AMBER 3.0	9.3	4.2	4.2	0.0	[15]
12		AMBER[a]	5.4	0.9	5.9	0.0	[15]
13		MM2[a]	1.9	0.1	2.7	0.0	[15]
14		MM2	1.9	0.0	5.1	0.5	[15]
15		MM3(89)	0.7	0.0	4.5	1.6	[4]
16		CHARMm[c]	2.5	0.0	8.5	1.5	[15]
17	**21c**	MM2	6.2	0.0	5.1	2.6	[63]
18		MM3(89)	0.3	0.0	4.1	1.5	[74]
19		MM3(92)	0.3	0.0	4.1	1.1	[65]
20	**21d**	MM3(92)	2.4	0.4	6.0	0.0	[65]
21		PIMM[66]	4.7	0.0	5.6	1.0	[68]
22	**21e**	MM2	2.5	0.0	0.4	0.1	[37]
23		TRIPOS	1.7	0.0	3.3	2.0	[37]

[a] as implemented in MacroModel V.1.5, [b] as implemented in MacroModel V.3.5, [c] as implemented in QUANTA.

Remarkably, although tetramethoxy calix[4]arenes **21** and the 1,3-dimethyl ether of **19** [60] invert rapidly, the partially O-methylated derivatives are less flexible. Thus, the *cone* ⇌ *cone* interconversion of **22a** and **22b** is completely blocked on the NMR timescale even at high temperatures [76]. The conjecture that the combination of intramolecular hydrogen bonding and transannular strain is responsible for the low flexibility of partially O-alkylated calix[4]arenes [82] is supported by MM calculations on **22a-d** showing that the breakage of hydrogen bonds and the passage of the methoxy groups through the macrocyclic annulus exerts a cooperative effect on the activation barriers [81]. The effect decreases in the order monomethyl ether > 1,2-dimethyl ether > 1,3-dimethyl ether > trimethyl ether. Calculated activation barriers exceed those determined recently [83,84] for the enantiomerisation of the chiral calixarene derivatives **23a**, **23b** and **23d** but fall in the same order.

The monodioxamethylene calix[4]arene **24**, which may be regarded as simplest 1,2-diether of a calixarene, is flexible at room temperature in solution [85] and MM3 calculations indicate the rate-limiting steps for the conformational interconversions of **24** involve transitions, *e.g.* *boat-chair* ⇌ *distorted-boat*, of the dioxocine subunits rather than rotations of the aromatic rings.

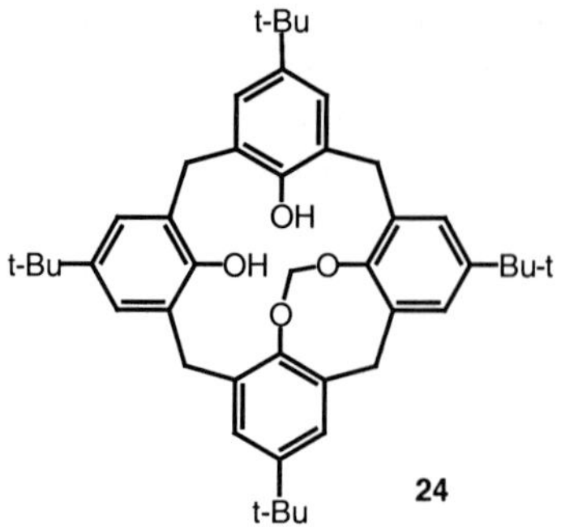

TABLE 6. Calculated relative energies (kcal/mol) of the characteristic conformers of partially O-methylated calix[4]arenes

Entry	Compd.	Method	*cone*	*paco*	*1,2-alt*	*1,3-alt*	Ref.
1	22a	AMBER	0.0	0.7	5.4	5.4	[15]
2	22b	AMBER	0.3	0.2	1.0	0.0	[15]
3	22c	AMBER	2.8	0.0	5.2	2.9	[15]
4	22d	AMBER	3.5	0.7	3.8	0.0	[15]
5	23a	TRIPOS	0.0	1.6	4.4	6.1	[37]
6		MM2	1.8	0.0	1.9	0.0	[37]
7	23b	TRIPOS	0.0	1.2	1.9	4.3	[37]
8		MM2	0.0	2.0	3.5	2.3	[37]
9	23c	TRIPOS	0.0	0.5	3.1	4.5	[37]
10		MM2	0.0	0.0	1.9	0.3	[37]
11		AMBER 3.0	0.0	2.4	3.3	4.6	[37]
12	23d	TRIPOS	0.0	0.6	5.6	2.8	[37]
13		MM2	1.4	0.1	0.0	0.0	[37]

a From the possible stereoisomers of the *paco* and *1,2-alt* conformations only the lowest energy conformers are listed.

3.5. CALIX[4]ARENES BEARING *EXTRA*-ANNULAR OH-GROUPS

Resorc[4]arenes **25**, bearing eight *exo*-OH groups, can exist as four different configurational isomers (diastereomers) depending on the relative orientation of the substituents R at the methine bridges [86]. The situation is complicated by the fact that each of the diastereomers can adopt different conformations. Unlike calix[4]arenes, five characteristic conformations are discussed for the resorc[4]arenes: *crown* (= *cone*), *boat* (= *pinched cone*), *chair*, *diamond* and *saddle* (= *1,3-alt*) [86].

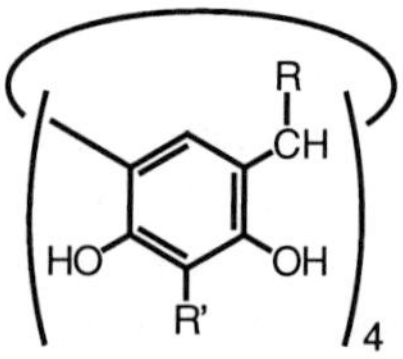

25 R = alkyl, aryl, R' = H
26a R = H, R' = n-hexyl
26b R = H, R' = Me

All-*cis* (rccc) resorc[4]arenes **25** exist both in the crystal and in solution in the *cone* conformation with axial (*endo*) disposition of the substituents R. This conformation is stabilised by four isolated intramolecular hydrogen bonds between the *exo*-OH groups. In contrast to the calix[4]arene tetrols **2** no indication of ring inversion processes has been found by ^{1}H NMR, suggesting that either the compound is conformationally locked or that the conformational equilibrium is strongly biased toward the *cone* conformation with axial arrangement of substituents [87,88].

The methylene-bridged resorc[4]arene **26a** adopts at low temperature in solution a *cone* conformation and undergoes the *cone* ⇌ *cone* topomerisation process at elevated temperatures [89]. The lower barrier for this process ($\Delta G^{\ddagger}_{298}$ = 12.0 kcal/mol) compared to **2** was attributed to the weaker intramolecular hydrogen bonds but MM3 calculations for **26b** suggest that the main contribution for this decrease in the barriers is the absence of the *intra*-annular substituents [90]. The calculated relative energies of the four characteristic conformations (*cone*: 0.0, *paco*: 5.8, *1,2-alt*: 6.6, *1,3-alt*: 11.7 kcal/mol) resemble those of the calix[4]arenes **2** (Table 2), indicating that the weaker hydrogen bonding in **26** is compensated by the reduction of the transannular strain.

The reduction of the number of *exo*-OH groups to four as in **27** causes a marked change in the conformational properties [91-93]. The *exo*-calix[4]arenes **27** are highly flexible compounds which are not frozen on the NMR timescale even at low temperatures. The low barrier calculated for the topomerisation of the *cone* conformation of **27c** ($\Delta E^{\ddagger}$ = 6.7 kcal/mol) reflects this enhanced flexibility [23].

27a	R = Me, R' = *t*-Bu, R" = H
27b	R = R' = *t*-Bu, R" = H
27c	R = R' = R" = H
27d	R = Me, R' = R" = H
28a	R = R" = Me, R' = *t*-Bu
28b	R = R' = *t*-Bu, R" = Me

In chloroform solution, **27a** exists as a mixture of the *cone* and *1,2-alt* conformations, while in the crystalline state it adopts the *cone* conformation and **27c** and **27d** a *diamond* structure (Figure 3) [91-95]. Both conformations contain the maximum number of possible intramolecular hydrogen bonds and it is assumed that they do not significantly differ in their steric demands. Force field calculations favour, however, the *cone* conformation over the *diamond* due to the more favourable bonding interactions [23,91,92]. In the *diamond* conformation the usual reference plane of the methylene carbons is folded and this leads to a change in the sequence of the signs of the ϕ/χ torsion angles which is (+−,−+,−+,−−) for **27d** and (+−,++,−+,−−) for a regular *1,2-alt* conformation.

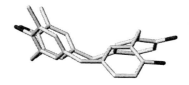

*Figure 3. The crystal structure of **27d** [92].*

The replacement of a methylene bridge in **27** by an ethane-1,1-diyl bridge (**28**) can result in diastereomeric *cone* or *1,2-alt* conformations with the bridge substituent either axial or equatorial. In agreement with MM3 calculations, the methyl group favours the axial position, avoiding repulsive van der Waals contacts to the *exo*-OH groups [92]. NOESY experiments have shown that **28a** exists in CDCl$_3$ solution in the *1,2-alt* conformation while MM3 calculations favour the *cone* over the *1,2-alt* form.

In principle, an inversion process in which all rings pass through the macrocyclic annulus should interconvert the preferred *1,2-alt* conformation of **28** with the methyl group located in an axial position with a diastereomeric *1,2-alt* conformation in which the methyl group is located in equatorial position. Although MM3 calculations indicated that the barrier for this process should be little higher than for the unsubstituted *exo*-calix[4]arenes **27**, no line broadening, indicating a dynamic exchange process with a second, less populated conformation, was found [92]. This suggests that the equilibrium is strongly biased toward one conformer and/or that the barrier is very low.

4. Calix[5]arene-31,32,33,34,35-pentols and their Derivatives

Figure 4. *Crystal structure of 5,11,17,23,29-Pentamethyl-35-deoxy-31,32,33,34-tetrahydroxycalix[5]-arene (CSD-Refcode: WEJYOZ [96]). All hydrogens, except the endo-H, are omitted for clarity. The dark lines specify the major and the minor plane used to define the conformation.*

The conformation of a calixarene is commonly defined by the orientation of the aromatic nuclei with regard relative to a reference plane passing through the methylene carbon atoms. With the exception of the *diamond* conformation, this is feasible for calix[4]arenes while an inspection of the crystal structures of calix[5]arenes available in the Cambridge Crystallographic Database [97,98] reveals that the five methylene carbon atoms deviate considerably (between 0.3 Å and 1.0 Å) from a common plane in nearly 50 % of these structures. This is illustrated for one example (rms 0.93 Å) in Figure 4. It has also been reported [17,99] that the four up/down designations used for calix[4]arenes are insufficient to accommodate the conformers of calix[5]arenes obtained from MM calculations. Hence, there is a need for a proper definition of calix[5]arene conformations.

TABLE 7. Calculated relative energies (kcal/mol) of the characteristic conformers of calix[5]arenes

Entry	Compd.	Method	*cone*	*paco*	*1,2-alt*	*1,3-alt*	Ref.
1	**29a**	CHARMm[a]	0.0	2.0	6.2	1.3	[17]
2	**29b**	MM3(92)	0.0	5.2	2.9	10.9	[99]
3		TRIPOS	0.7	2.1	0.0	2.6	[99]
4	**30a**	MM3(92)	3.2	1.6	1.8	0.0	[99]
5		TRIPOS	2.5	1.8	0.0	2.4	[99]

[a] as implemented in QUANTA3.2/CHARMm 21.3.

In principle, this can be accurately done by specifying the torsion angles around the Ar-CH$_2$-Ar bonds [2]. More pictorial but less unequivocal is the iconographic representation of calix[5]arene conformations [17]. It seems to be of more practical use to describe the conformations of calix[5]arenes by means of two reference planes [99]. Among the five possible planes defined by four of the methylene carbon atoms, the one with the lowest rms deviation is chosen as the 'major' plane and the remaining methylene carbon together with its two adjoining methylene carbons is used to define the 'minor' plane. The conformation of a calix[5]arene is finally specified by the arrangement of the *endo*-substituents with respect to the concave region defined by the two planes. The notations 'T' or 't' (depending whether the ring belongs to the 'major' or 'minor' plane, respectively) indicate that the substituent is pointing towards this region and 'A' and 'a' if it points away from it. The conformation of the molecule shown in Figure 4 should be therefore designated as AAAtt. The transformation of a folded arrangement of the two planes into a pseudoplanar arrangement then enables the assignment of the 32 individual conformations resulting from the 'A'/'T' notation to the four basic conformations *cone* (2 subclasses), *paco* (10), *1,2-alt* (10) and *1,3-alt* (10). MD simulations of **29b** in vacuum indicate that a fast pseudorotational process interconverts the different subclasses of a given basic conformation [99].

29a R = *t*-Bu, Y = H
29b R = Me, Y = H
29c R = Y = H
30a R = Me, Y = Me
30b R = *t*-Bu, Y = Me

MM3 calculations of **29b** and **29c** furnished a nearly C_5-symmetrical *cone* conformer as global energy minimum [13,99]. The energies of the conformers calculated with this force field decrease with decreasing number of hydrogen bonds [99]: *cone* (5) > *1,2-alt* (4) > *paco* (3) > *1,3-alt* (1), while calculations using the CHARMm and TRIPOS force fields predict that **29a** and **29b** should exist in solution as a mixture of rotamers (Table 7), thus conflicting with the experimental observation that calix[5]arene-31,32,33,34,35-pentols exist both in solution and in the crystal in the *cone* conformation [19,31,82].

As in **3** and **28**, an alkyl substituent at the methine bridge in **31** assumes a position remote from the hydroxy groups. In contrast to the alkanediyl calix[4]arenes **3**, aryl substituents R^1 in **31** also show a pronounced preference for the equatorial position [41,100]. This cannot be explained by repulsive interactions between the OH-groups and the substituent in the axial arrangement since the steric situation at the methine bridge is very similar for **3** and **31**. MM calculations suggest, however, that small differences in the geometries around the methine bridges of **3** and **31** with equatorial disposition of R^1 may account for a reduction of the steric strain in **31** and thus for the larger energy difference between the two diastereomers [100].

31 R = Me, R' = *t*-Bu, R^1 = alkyl, aryl
32 R = R' = *t*-Bu, R^1 = alkyl, aryl

Force field calculations of the pentamethylether **30a** (Table 7, entries 4 and 5), although differing in the estimation of the most stable conformer, have indicated that the molecule should exist in an equilibrium of several conformations which is commensurate with the ^{1}H NMR data published [17] for **30b**.

5. Calix[6]arene-37,38,39,40,41,42-hexols and Higher Calixarenes

In crystals, calix[6]arene-37,38,39,49,41,42-hexols adopt either a *pinched cone* conformation in which all of the oxygen atoms lie on the same side of the molecule and two opposite methylene groups point towards the center of the cavity or a conformation named "*1,2,3-alternate*" or "*double partial cone*", in which three adjacent oxygen atoms lie on the one side, and the other three on the other side of the molecule [101-108]. The solid state conformation appears to depend on the crystallisation solvent [106]. In solution at slow exchange conditions, the ^{1}H NMR spectrum is commensurate with a conformation possessing a plane of symmetry, a centre of symmetry, or a twofold axis [13,109,110]. It has been concluded [110] from low-temperature NMR experiments that the conformation of **33a** in solution corresponds to a C_2-symmetrical *winged cone* which is characterised by the outward orientation of all methylene groups and the bending out of two opposite phenol rings while the four remaining aryl groups are in an up alignment. CHARMm and MM3 calculations on **33b** [13,111] have, however, revealed that the *pinched cone* of ideal C_2-symmetry is most stable for calix[6]arene hexols and it has been proposed, based on the comparison between the experimentally observed and calculated proton-proton distances that the structure in solution is also the *pinched cone*.

33a $X^1 = X^2 = H$, R = *t*-Bu
33b $X^1 = X^2 = R = H$
34 X^1, X^2 = alkyl, acyl

In contrast to the calix[4]arenes, conformational interconversions of O-substituted calix[6]arenes **34** can proceed via two different pathways, in which either the *endo*-substituents or the *p*-substituents can pass through the macrocyclic annulus [112-116]. In agreement with the experimental observations, MM calculations [117] indicate that the former pathway is preferred for smaller substituents Y^1 and Y^2 and the latter for calix[6]arenes bearing *endo*-substituents that are too large to rotate through the annulus.

The increasing flexibility of the calix[n]arene scaffold with increasing n dramatically enlarges the number of possible structures. Thus, it has been reported [13] that 90 distinct conformers (corresponding to local energy minima) were obtained from a conformational search of **33b**, while 651 low-energy structures resulted for the corresponding heptamer and about 4800 conformers were anticipated for calix[8]arene. If one assumes that these calix[n]arenes may carry substituents both at the phenolic oxygens and in *p*-position, the number of possible conformers further increases and may surpass the limit of the present computational methods.

The multi-minima problem is not only valid for the higher or heavily substituted calix[n]arenes. As shown for simple calix[4]arenes [1,37,91] there is a necessity for a critical evaluation of the energy hypersurface for the molecule of interest by means of a suitable search method [118]. Caution is advised for the interpretation of the results obtained from MM optimisations on individual, manually generated structures as often found in the calixarene literature.

6. References and Notes

[1]   J. E. McMurry, J. C. Phelan, *Tetrahedron Lett.* **1991**, *32*, 5655-5658.
[2]   F. Ugozzoli, G. D. Andreetti, *J. Incl. Phenom.* **1992**, *13*, 337-348.
[3]   C. Jaime, J. De Mendoza, P. Prados, P. M. Nieto, C. Sanchez, *J. Org. Chem.* **1991**, *56*, 3372-3376.
[4]   T. Harada, J. M. Rudzinski, E. Osawa, S. Shinkai, *Tetrahedron* **1993**, *49*, 5941-5954.
[5]   T. Harada, F. Ohseto, S. Shinkai, *Tetrahedron* **1994**, *50*, 13377-13394.
[6]   S. Fischer, P. D. J. Grootenhuis, L. C. Groenen, W. P. van Hoorn, F. C. J. M. van Veggel, D. N. Reinhoudt, M. Karplus, *J. Am. Chem. Soc.* **1995**, *117*, 1611-1620.

[7]   Z. Goren, S. E. Biali, *J. Chem. Soc., Perkin Trans. 1* **1990**, 1484-1487.
[8]   I. Thondorf, J. Brenn, *J. Mol. Struct. (THEOCHEM)* **1997**, *398-399*, 307-314.
[9]   A. Rajca, R. Padmakuma, D. J. Smithhisler, S. R. Desai, C. R. Ross, J. J. Stezowski, *J. Org. Chem.* **1994**, *59*, 7701-7703.
[10] F. Grynszpan, S. E. Biali, *Tetrahedron Lett.* **1991**, *32*, 5155-5158.
[11] B. R. Brooks, R. E. Bruccoleri, B. D. Olafson, D. J. States, S. Swaminathan, M. Karplus, *J. Comput. Chem.* **1983**, *4*, 187-217.
[12] N. L. Allinger, Y. H. Yuh, J.-H. Lii, *J. Am. Chem. Soc.* **1989**, *111*, 8551-8566, 8566-8575, 8576-8582.
[13] T. Harada, S. Shinkai, *J. Chem. Soc., Perkin Trans. 2* **1995**, 2231-2242.
[14] See also: F. L. Dickert, O. Schuster, *Adv. Mater.* **1993**, *5*, 826-829.
[15] P. D. J. Grootenhuis, P. A. Kollman, L. C. Groenen, D. N. Reinhoudt, G. J. van Hummel, F. Ugozzoli, G. D. Andreetti, *J. Am. Chem. Soc.* **1990**, *112*, 4165-4176.
[16] K. B. Lipkowitz, G. Pearl, *J. Org. Chem.* **1993**, *58*, 6729-6736.
[17] D. R. Stewart, M. Krawiec, R. P. Kashyap, W. H. Watson, C. D. Gutsche, *J. Am. Chem. Soc.* **1995**, *117*, 586-601.
[18] F. Bayard, C. Decoret, D. Pattou, J. Royer, A. Satrallah, J. Vicens, *J. Chim. Phys. Phys.-Chim. Biol.* **1989**, *86*, 945-954.
[19] C. D. Gutsche, „Calixarenes", *Monographs in Supramolecular Chemistry*, J. F. Stoddart (ed.), Cambridge, The Royal Society of Chemistry, **1989**.
[20] P. K. Weiner, P. A. Kollman, *J. Comput. Chem.* **1981**, *2*, 287-303.
[21] S. J. Weiner, P. A. Kollman, D. A. Case, U. C. Singh, C. Ghio, G. Alagona, S. Profeta, P. Weiner, *J. Am. Chem. Soc.* **1984**, *106*, 765-784.
[22] S. J. Weiner, P. A. Kollman, D. T. Nguyen, D. A. Case, *J. Comput. Chem.* **1986**, *7*, 230-252.
[23] I. Thondorf, J. Brenn, W. Brandt, V. Böhmer, *Tetrahedron Lett.* **1995**, *36*, 6665-6668.
[24] J. Royer, F. Bayard, C. Decoret, *J. Chim. Phys. Phys.-Chim. Biol.* **1990**, *87*, 1695-1700.
[25] W. K. den Otter, W. J. Briels, *J. Chem. Phys.* **1997**, *106*, 5494-5508.
[26] W. K. den Otter, W. J. Briels, *J. Chem. Phys.*, **1997**, *107*, 4968-4978.
[27] W. K. den Otter, W. J. Briels, *J. Am. Chem. Soc.* **1998**, *120*, 13167-13175.
[28] K. Araki, S. Shinkai, T. Matsuda, *Chem. Lett.* **1989**, 581-584.
[29] The values published in ref. [28] are probably incorrect. Compare ref. [60].
[30] J. T. Sprague, J. C. Tai, Y. Yuh, N. L. Allinger, *J. Comput. Chem.* **1987**, *8*, 581-594.
[31] V. Böhmer, *Angew. Chem., Int. Ed. Engl.* **1995**, *34*, 713-745.
[32] H. Goldmann, W. Vogt, E. Paulus, V. Böhmer, *J. Am. Chem. Soc.* **1988**, *110*, 6811-6817.
[33] F. Arnaud-Neu, V. Böhmer, L. Guerra, M. A. McKervey, E. F. Paulus, A. Rodriguez, M. J. Schwing-Weill, M. Tabatabai, W. Vogt, *J. Phys. Org. Chem.* **1992**, *5*, 471-481.
[34] E. Dahan, S. E. Biali, *J. Org. Chem.* **1989**, *54*, 6003-6004.
[35] E. Dahan, S. E. Biali, *J. Org. Chem.* **1991**, *56*, 7269-7274.
[36] G. D. Andreetti, V. Böhmer, J. G. Jordon, M. Tabatabai, F. Ugozzoli, W. Vogt, A. Wolff, *J. Org. Chem.* **1993**, *58*, 4023-4032.
[37] I. Thondorf, G. Hillig, W. Brandt, J. Brenn, A. Barth, V. Böhmer, *J. Chem. Soc., Perkin Trans. 2* **1994**, 2259-2267.
[38] C. Grüttner, V. Böhmer, W. Vogt, I. Thondorf, S. E. Biali, F. Grynszpan, *Tetrahedron Lett.* **1994**, *35*, 6267-6270.
[39] G. Sartori, R. Maggi, F. Bigi, A. Arduini, A. Pastorio, C. Porta, *J. Chem. Soc., Perkin Trans. 1* **1994**, 1657-1658.
[40] S. E. Biali, V. Böhmer, S. Cohen, G. Ferguson, C. Grüttner, F. Grynszpan, E. F. Paulus, I. Thondorf, W. Vogt, *J. Am. Chem. Soc.* **1996**, *118*, 12938-12949.
[41] M. Bergamaschi, F. Bigi, M. Lanfranchi, R. Maggi, A. Pastorio, M. A. Pellinghelli, C. Peri, C. Porta, G. Sartori, *Tetrahedron* **1997**, *53*, 13037-13052.
[42] Y. Fukazawa, K. Deyama, S. Usui, *Tetrahedron Lett.* **1992**, *33*, 5803-5806.
[43] O. Aleksiuk, F. Grynszpan, S. E. Biali, *J. Chem. Soc., Chem. Commun.* **1993**, 11-13.
[44] K. Araki, H. Murakami, F. Ohseto, S. Shinkai, *Chem. Lett.* **1992**, 539-542.
[45] I. Thondorf, *J. Chem. Soc., Perkin Trans. 2* **1999**, 1791-1796.
[46] F. Grynszpan, O. Aleksiuk, S. E. Biali, *J. Org. Chem.* **1994**, *59*, 2070-2074.
[47] O. Aleksiuk, F. Grynszpan, S. E. Biali, *J. Org. Chem.* **1993**, *58*, 1994-1996.
[48] F. Ohseto, H. Murakami, K. Araki, S. Shinkai, *Tetrahedron Lett.* **1992**, *33*, 1217-1220.
[49] I. Thondorf, J. Brenn, manuscript in preparation.

[50] C. G. Gibbs, C. D. Gutsche, *J. Am. Chem. Soc.* **1993**, *115*, 5338-5339.
[51] X. Delaigue, J. M. Harrowfield, M. W. Hosseini, A. De Cian, J. Fischer, N. Kyritsakas, *J. Chem. Soc. Chem. Commun.* **1994**, 1579-1580.
[52] C. G. Gibbs, P. K. Sujeeth, J. S. Rogers, G. G. Stanley, M. Krawiec, W. H. Watson, C. D. Gutsche, *J. Org. Chem.* **1995**, *60*, 8394-8402.
[53] X. Delaigue, M. W. Hosseini, *Tetrahedron Lett.* **1993**, *34*, 8111-8112.
[54] S. Pappalardo, F. Bottino, G. Ronsisvalle, *Phosphorus Sulfur* **1984**, *19*, 327-333.
[55] S. Pappalardo, G. Ferguson, J. F. Gallagher, *J. Org. Chem.* **1992**, *57*, 7102-7109.
[56] R. Schätz, C. Weber, G. Schilling, T. Öser, U. Huber-Patz, H. Irngartinger, C.-W. von der Lieth, R. Pipkorn, *Liebigs Ann.* **1995**, 1401-1408.
[57] P. Parzuchowski, V. Böhmer, S. E. Biali, I. Thondorf, *Tetrahedron: Asymmetry* **2000**, *11*, 2393-2402.
[58] T. T. Wu, J. R. Speas, *J. Org. Chem.* **1987**, *52*, 2330-2332.
[59] V. Böhmer, personal communication.
[60] J. M. van Gelder, J. Brenn, I. Thondorf, S. E. Biali, *J. Org. Chem.* **1997**, *62*, 3511-3519.
[61] L. C. Groenen, J. D. van Loon, W. Verboom, S. Harkema, A. Casnati, R. Ungaro, A. Pochini, F. Ugozzoli, D. N. Reinhoudt, *J. Am. Chem. Soc.* **1991**, *113*, 2385-2392.
[62] C. D. Gutsche, B. Dhawan, J. A. Levine, H. N. Kwang, L. J. Bauer, *Tetrahedron* **1983**, *39*, 409-426.
[63] K. Iwamoto, K. Araki, S. Shinkai, *J. Org. Chem.* **1991**, *56*, 4955-4962.
[64] J. Blixt, C. Detellier, *J. Am. Chem. Soc.* **1994**, *116*, 11957-11960.
[65] K. Iwamoto, K. Araki, S. Shinkai, *J. Org. Chem.* **1991**, *56*, 4955-4962.
[66] A. E. Smith, H. J. Lindner, *J. Comput.-Aided Mol. Des.* **1991**, *5*, 235-262
[67] F. Hamada, S. G. Bott, G. W. Orr, A. W. Coleman, H. Zhang, J. L. Atwood, *J. Incl. Phenom.* **1990**, *9*, 195-206.
[68] A. Soi, W. Bauer, H. Mauser, C. Moll, F. Hampel, A. Hirsch, *J. Chem. Soc., Perkin Trans. 2* **1998**, 1471-1478.
[69] F. Hamada, G. W. Orr, H. Zhang, J. L. Atwood, *J. Crystallogr. Spectrosc. Res.* **1993**, *23*, 681-684.
[70] R. K. Juneja, K. D. Robinson, C. P. Johnson, J. L. Atwood, *J. Am. Chem. Soc.* **1993**, *115*, 3818-3819.
[71] For exceptions, see: S. G. Bott, A. W. Coleman, J. L. Atwood, *J. Incl. Phenom.* **1987**, *5*, 747-758 and ref. [63].
[72] For the solvent dependence of the conformational distribution and MM3 calculations of trimethyl ethers of monodeoxycalix[4]arenes see: Y. Fukazawa, K. Yoshimura, S. Sasaki, M. Yamazaki, T. Okajima, *Tetrahedron* **1996**, *52*, 2301-2318.
[73] T. Nagasaki, K. Sisido, T. Arimura, S. Shinkai, *Tetrahedron* **1992**, *48*, 797-804.
[74] T. Harada, J. M. Rudzinski, S. Shinkai, *J. Chem. Soc., Perkin Trans. 2* **1992**, 2109-2115.
[75] S. Shinkai, K. Iwamoto, K. Araki, T. Matsuda, *Chem. Lett.* **1990**, 1263-1266.
[76] L. C. Groenen, E. Steinwender, B. T. G. Lutz, J. H. van der Maas, D. N. Reinhoudt, *J. Chem. Soc., Perkin Trans. 2* **1992**, 1893-1898.
[77] W. P. van Hoorn, W. J. Briels, J. P. M. van Duynhoven, F. C. J. M. van Veggel, D. N. Reinhoudt, *J. Org. Chem.* **1998**, *63*, 1299-1308.
[78] C. Alfieri, E. Dradi, A. Pochini, R. Ungaro, *Gazz. Chim. Ital.* **1989**, *119*, 335-338.
[79] V. Böhmer, M. Tabatabai, unpublished results.
[80] M. Clark, R. D. Cramer, N. van Opdenbusch, *J. Comp. Chem.* **1989**, *10*, 982-1012.
[81] W. P. van Hoorn, M. G. H. Morshuis, F. C. J. M. van Veggel, D. N. Reinhoudt, *J. Phys. Chem. A* **1998**, *102*, 1130-1138.
[82] C. D. Gutsche, „Calixarenes Revisited", Cambridge, The Royal Society of Chemistry, **1998**.
[83] T. Kusano, M. Tabatabai, Y. Okamoto, V. Böhmer, *J. Am. Chem. Soc.* **1999**, *121*, 3789-3790.
[84] V. Böhmer, M. Tabatabai, T. Kusano, Y. Okamoto, "The Conformational Inversion of Partially O-Methylated Calix[4]arenes, a Long Lasting Problem in Calixarene Chemistry", Lecture-22, 5th International Conference on Calixarene Chemistry, Perth, Australia, 19-23. 09. 1999.
[85] J. Wöhnert, J. Brenn, M. Stoldt, O. Aleksiuk, F. Grynszpan, I. Thondorf, S. E. Biali, *J. Org. Chem.* **1998**, *63*, 3866-3874.
[86] For a review on resorcarenes, see: P. Timmerman, W. Verboom, D. N. Reinhoudt, *Tetrahedron* **1996**, *52*, 2663-2704.
[87] L. Abis, E. Dalcanale, A. Du vosel, S. Spera, *J. Chem. Soc., Perkin Trans. 2* **1990**, 2075-2080.
[88] *Diamond* ⇌ *cone* and *chair* ⇌ *cone* interconversions have been observed for the rcct and rctt isomers: see ref. [87] and O. Middel, W. Verboom, R. Hulst, H. Kooijman, A. L. Spek, D. N. Reinhoudt, *J. Org. Chem.* **1998**, *63*, 8259-8265.

[89] H. Konishi, O. Morikawa, *J. Chem. Soc., Chem. Commun.* **1993**, 34-35.
[90] I. Thondorf, J. Brenn, V. Böhmer, *Tetrahedron* **1998**, *54*, 12823-12828.
[91] V. Böhmer, R. Dörrenbächer, M. Frings, M. Heydenreich, D. de Paoli, W. Vogt, G. Ferguson, I. Thondorf, *J. Org. Chem.* **1996**, *61*, 549-559.
[92] S. E. Biali, V. Böhmer, J. Brenn, M. Frings, I. Thondorf, W. Vogt, J. Wöhnert, *J. Org. Chem.* **1997**, *62*, 8350-8360.
[93] G. Sartori, C. Porta, F. Bigi, R. Maggi, F. Peri, E. Marzi, M. Lanfranchi, M. A. Pellinghelli, *Tetrahedron* **1997**, *53*, 3287-3300.
[94] G. Sartori, F. Bigi, C. Porta, R. Maggi, R. Mora, *Tetrahedron Lett.* **1995**, *36*, 2311-2314.
[95] V. Böhmer, G. Ferguson, M. Frings, *Acta Crystallogr., Sect. C: Cryst. Struct. Commun.* **1997**, *C53*, 1293-1295.
[96] S. Usui, K. Deyama, R. Kinoshita, Y. Odagaki, Y. Fukazawa, *Tetrahedron Lett.* **1993**, *34*, 8127-8130.
[97] F. H. Allen, O. Kennard, *Chemical Design Automation News* **1993**, *8*, 31-37.
[98] Cambridge Crystallographic Database, release 5.18.
[99] I. Thondorf, J. Brenn, *J. Chem. Soc., Perkin Trans. 2* **1997**, 2293-2300.
[100] S. E. Biali, V. Böhmer, I. Columbus, G. Ferguson, C. Grüttner, F. Grynszpan, E. F. Paulus, I. Thondorf, *J. Chem. Soc., Perkin Trans. 2* **1998**, 2261-2270.
[101] G. D. Andreetti, G. Calestani, F. Ugozzoli, A. Arduini, E. Ghidini, A. Pochini, R. Ungaro, *J. Incl. Phenom.* **1987**, *5*, 123-126.
[102] M. Halit, D. Oehler, M. Perrin, A. Thozet, R. Perrin, J. Vicens, M. Bourakhouadar, *J. Incl. Phenom.* **1988**, *6*, 613-623.
[103] G. D. Andreetti, F. Ugozzoli, A. Casnati, E. Ghidini, A. Pochini, R. Ungaro, *Gazz. Chim. Ital.* **1989**, *119*, 47-50.
[104] J. L. Atwood, D. L. Clark, R. K. Juneja, G. W. Orr, K. D. Robinson, R. L. Vincent, *J. Am. Chem. Soc.* **1992**, *114*, 7558-7559.
[105] P. Thuery, N. Keller, M. Lance, J.-D. Vigner, M. Nierlich, *J. Incl. Phenom.* **1995**, *20*, 373-379.
[106] W. J. Wolfgong, L. K. Talafuse, J. M. Smith, M. J. Adams, F. Adeogba, M. Valenzuela, E. Rodriguez, K. Contreras, D. M. Carter, A. Bacchus, A. R. McGuffey, S. G. Bott, *Supramol. Chem.* **1996**, *7*, 67-78.
[107] J. L. Atwood, L. J. Barbour, C. L. Raston, I. B. N. Sudria, *Angew. Chem., Int. Ed.* **1998**, *37*, 981-983.
[108] M. Munakata, L. P. Wu, T. Kuroda-Sowa, M. Maekawa, Y. Suenaga, K. Sugimoto, I. Ino, *J. Chem. Soc., Dalton Trans.* **1999**, 373-378.
[109] C. D. Gutsche, L. J. Bauer, *J. Am. Chem. Soc.* **1985**, *107*, 6052-6059.
[110] M. A. Molins, P. M. Nieto, C. Sanchez, P. Prados, J. De Mendoza, M. Pons, *J. Org. Chem.* **1992**, *57*, 6924-6931.
[111] W. P. van Hoorn, F. C. J. M. van Veggel, D. N. Reinhoudt, *J. Org. Chem.* **1996**, *61*, 7180-7184.
[112] The passage of p-H through the macrocyclic annulus has been recently proved for calix[5]arenes (G. Ferguson, A. Notti, S. Pappalardo, M. F. Parisi, A. L. Spek, *Tetrahedron Lett.* **1998**, *39*, 1965-1968.) while larger p-substituents cannot pass the interior of a calix[5]arene.
[113] C. D. Gutsche, L. J. Bauer, *J. Am. Chem. Soc.* **1985**, *107*, 6059-6063.
[114] H. Otsuka, K. Araki, S. Shinkai, *Chem. Express* **1993**, *8*, 479-482.
[115] H. Otsuka, K. Araki, T. Sakaki, K. Nakashima, S. Shinkai, *Tetrahedron Lett.* **1993**, *34*, 7275-7278.
[116] J. P. M. van Duynhoven, R. G. Janssen, W. Verboom, S. M. Franken, A. Casnati, A. Pochini, R. Ungaro, J. de Mendoza, P. M. Nieto, P. Prados, D. N. Reinhoudt, *J. Am. Chem. Soc.* **1994**, *116*, 5814-5822.
[117] W. P. van Hoorn, F. C. J. M. van Veggel, D. N. Reinhoudt, *J. Phys. Chem. A* **1998**, *102*, 6676-6681.
[118] For an overview on conformational search methods, see *e.g.*, A. R. Leach, in K. B. Lipkowitz, D. B. Boyd (eds.), *Reviews in Computational Chemistry*, Vol. 2, VCH Publishers Inc., **1991**, pp. 1-55.

Chapter 16

DYNAMIC STRUCTURES OF HOST-GUEST SYSTEMS

ERIC B. BROUWER, GARY D. ENRIGHT, CHRISTOPHER
I. RATCLIFFE, JOHN A. RIPMEESTER, KONSTANTIN A. UDACHIN

Steacie Institute for Molecular Sciences, National Research Council of Canada, 100 Sussex Drive, Ottawa, Canada K1A 0R6.
e-mail: jar@ned1.sims.nrc.ca

1. Introduction

As illustrated elsewhere in this volume, calixarenes have proven to be a remarkably versatile class of materials. Numerous schemes have been devised that employ the simple calixarenes, modified molecules, surface-bound species or larger assemblies for a variety of functions. With the large amount of information now available, it is quite appropriate to assess our state of knowledge, as ultimately the rational design of the next generation of materials depends on our level of understanding of structural and dynamic details of the currently known species. In this work we summarise studies that have contributed to our understanding of calixarenes.

In previous reviews we have noted the difficulty that structural studies of even the simplest calixarenes has caused [1,2]: in general, the compounds are disordered, often with involvement of both guest and host sub-lattices and there is evidence for dynamic coupling between them. We have also advanced the use of synoptic approaches, where we use complementary techniques to develop improved structural models. Without a doubt, the combination of crystallography and solid-state nuclear magnetic resonance (NMR) spectroscopy has proven to be remarkably effective in solving a number of complex problems, and continuing progress in technique development will advance these capabilities further.

During the course of our work on calixarenes we have stayed with the simple *p-tert*-butylcalix[4]arenes (tBC), starting with the very first tBC structure reported, that with toluene as guest. This compound has been under study in our laboratory for nearly ten years, and it has shown some very unusual effects, such as superlattice diffraction peaks that fade with time and a wavelength-dependent structure. Also, a good structural model was found that refined very well but turned out to have the wrong symmetry as shown by NMR spectroscopy.

At the start of this work, only a few guest species were known for the simple tBCs, presumably because of the low solubility in many solvents. Many commonly held views on calixarenes derive from a very limited database and are easily shown to be vast oversimplifications, *e.g.*, the suitability of specific guest types for the tBCs, general structural features of the host-guest complexes and the contributing factors to compound stability. With some less conventional approaches we were able to prepare compounds with a wide range of different guest species, including unsubstituted and halogenated aliphatics and alicyclics, and to study these to extract trends in structural features.

A systematic examination shows that the often-observed high symmetry of the host lattice normally reflects the shape and dynamics of molecular guests. Lattice distortions and disorder can be introduced by steric interactions between guest and host or inter-guest polar interactions. The cavity in the simplest tBC structure is quite asymmetric and hence it is a good site for testing the importance of weak interactions in orienting guest molecules in the cavities, information of critical importance in identifying salient features leading to molecular recognition.

By the same token, from the range in size and functionality of guests studied it would appear that there is little compelling evidence that specific interactions are required for stable compound formation, and that perhaps many of these materials should be described as clathrates rather than stoichiometric host-guest compounds. This will become more clear once accurate stoichiometries and thermodynamic data become available. Of course, this will have considerable bearing on the future development of tBCs for applications such as sorbents and sensor films.

2. Structural Approaches

The tBC host molecule has a flexible upper rim defined by the t-butyl groups. In part, this flexibility arises from various orientational possibilities of the latter as influenced by the guest. In contrast, the lower rim is fixed by in-plane hydrogen bonding. By combining NMR spectroscopy with single crystal X-ray diffraction (XRD) we have been able to study the long- and short- range order and the dynamic disorder inherent in these compounds and to obtain greatly improved structural data.

The first structural determination of a calixarene inclusion compound was the single-crystal XRD study of Andreetti in 1979 [3]. In this study, the toluene guest molecule was determined to be inserted methyl group first into the conical tBC host molecule along a 4-fold crystallographic axis. In order to satisfy the 4-fold crystallographic symmetry, the guest molecules were disordered over two symmetry-related sites. In addition, the t-butyl groups of the host molecule were disordered over two inequivalent sites with a 77:23 occupancy ratio, something difficult to understand. In this system—as with most high symmetry calixarene compounds—the inherent disorder masks details of the host-guest interactions, and hampers the derivation of reliable structural models.

Solid-state ^{13}C NMR studies in our laboratory had shown that at room temperature the toluene guest rapidly reorients between equivalent guest sites but that below 248K, a splitting of many of the ^{13}C calixarene resonance lines indicated a transition to a lower symmetry host lattice [4]. It was surprising therefore when a 150K structural determination revealed the same high symmetry space group ($P4/n$) as observed at 293K. Upon refinement (incorporating a lower local symmetry as suggested by NMR data), new structural details are resolved in the same space group: The t-butyl disorder is explained as a guest-induced distortion of the host calixarene molecule, which appears as a 50:50 disorder in the host molecule. Two opposing aromatic rings of the calixarene pivot outward and the attached t-butyl group takes on a different orientation. This distortion from 4-fold symmetry is, in all likelihood, correlated with the orientation of the toluene guest (Figure 1) as steric effects induce the flexible upper rim to distort away from the toluene guest. The long axis of the toluene guest is tilted off the host 4-fold axis of symmetry by ~7°. A re-determination of the room temperature structure revealed that the structural model derived from the 150K data greatly improves the agreement factor. The apparent high (tetragonal) symmetry of the structure determined from XRD apparently is due to the spatial averaging over many sites occupied by the distorted (lower-symmetry) calixarene molecules.

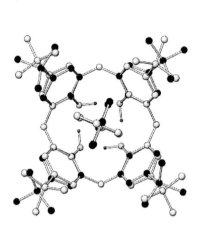

Figure 1. The tBC-toluene structure with framework lattice distortion. The superposition of two disordered host-guest units with C_{2v} symmetry is illustrated by the black and white colouration.

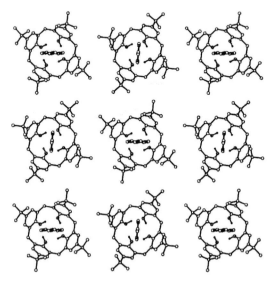

Figure 2. Packing of tBC-toluene at 150K illustrating the guest correlations in the ab plane.

At low temperatures the guest-induced distortion of individual host lattice molecules persists long enough to establish a correlation in the orientation of guests in adjacent host molecules (in a plane perpendicular to the 4-fold axis of symmetry). This additional order results in superlattice reflections in the XRD pattern. In our initial studies the data were collected with Cu K_α radiation (λ_{Cu} = 1.54056 Å). The superlattice reflections initially appeared as weak peaks, half-integral in h and k, which disappeared over a period of several hours. Upon recollecting the low temperature data with a CCD diffractometer using Mo K_α radiation (λ_{Mo} = 0.70926 Å), the weak superlattice reflections again are evident, but now are persistent [5]. The permanence of the superlattice reflections in the latter case arises from the correlations in guest orientation that persist over the much smaller crystal volume needed to give reflections for Mo radiation (V_{Cu}/V_{Mo} = 10). The low temperature structure of the host calixarene lattice appears to be metrically tetragonal with a unit cell volume double that observed with Cu radiation. However, the structure is best described as a highly twinned monoclinic lattice (space group $P2/c$) with tBC molecules centred on 2-fold axes [5]. There are two crystallographically-distinct tBC molecules, each with a toluene guest that is disordered over two equivalent sites (related by a 180° rotation about the calixarene's 2-fold axis of symmetry). The packing in the *ab* plane perpendicular to the unique axis is shown in Figure 2. The observed phase transition at 248K can then be associated with the onset of long-range order in the orientation of the guest toluene molecules.

Most 1:1 tBC-guest inclusion compounds appear to have tetragonal symmetry. The structures are isostructural with the room temperature tBC-toluene structure, space group *P*4/*n* ($a \sim 12.8$ Å, $c \sim 13.8$ Å) with the host molecules situated on the 4-fold axis of symmetry. Probably all of these tetragonal structures are better described as twinned monoclinic structures (space group *P*2/*n*, with the *c* axis unique). However, only those compounds with guests containing heavy atoms, such as chloro- and bromo-benzene, and 1-chloro- and 1-bromo-butane, show significantly improved agreement factors when this model is applied. The low temperature tBC-toluene structure can be described as a twinned monoclinic lattice, space group *P*2/*c* with $a' = a + b$, $b' = c$, $c' = a - b$ (where *a*, *b* and *c* are the room temperature lattice vectors). The new unit cell has $a' \cong c' \sim 18.1$ Å, $b' \sim 13.8$ Å with $\beta \cong 90.0°$, and the host calixarene molecules are centred on the 2-fold axes of symmetry. In addition, we have observed two other space groups in which the tBC molecules have no symmetry. In both cases the unit cell is similar to the tBC-toluene unit cells except that the length of the unique axis is doubled. For nitrobenzene and mesitylene guests, the unit cell is similar to that of the room temperature tBC-toluene structure (with a doubled *c* axis): the structure is orthorhombic, space group *Pc*2$_1$*n* (alt. #33), with all guests approximately aligned along the *a* axis ($a \leq b$). For the *o*-dichlorobenzene guest, the cell is similar to that of the low temperature tBC-toluene structure (with the *b'* axis doubled), space group twinned in *P*2$_1$/*c* with guests alternately aligned approximately $\pm 45°$ to the *a'* axis (similar to *Figure 2*).

3. ^{2}H NMR Dynamic Studies

3.1. DEUTERIUM NMR IN THE SOLID STATE

NMR studies of ^{2}H, perhaps more than any other NMR nucleus, provide a very powerful means of obtaining information about the molecular dynamics of calixarene compounds in the solid state, particularly of the guest but also of the host. This information in turn has played an important complementary role in improving structural refinement models, especially since many calixarene crystals show disorder.

^{2}H is a spin $I = 1$ quadrupolar nucleus with three spin states which give rise to two allowed NMR transitions. The interaction of the nuclear quadrupole moment with the electric field gradient (*efg*) around the ^{2}H nucleus, perturbs these transitions to first order, giving two lines equally displaced on either side of the NMR frequency, with an orientation-dependent splitting. This quadrupole coupling is the dominant interaction in ^{2}H NMR spectroscopy of solids. In a powder with random crystallite orientations, the sum of all such pairs of lines gives rise to a characteristic lineshape, with pairs of major features, *i.e.*, edges, shoulders and peaks whose widths are proportional to the three principal components of the *efg* tensor (Figure 3). The *efg* tensor can be visualised as the three axes of an ellipsoid surrounding the ^{2}H nucleus, with the largest axis along the chemical bond involving the ^{2}H atom (Figure 3).

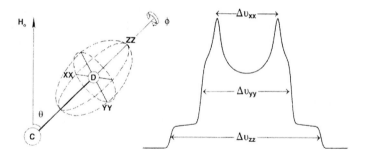

Figure 3. Left: efg ellipsoid surrounding a ^2H nucleus in a C-D bond. The orientation in the field H_o is defined by the angles (θ, φ). Right: a ^2H NMR powder lineshape indicating the width of the features Δv_{ii} which are proportional to the principal efg components XX,YY,ZZ.

If the molecule undergoes motion between sites such that the *efg* changes orientation at rates comparable to the linewidth, the powder lineshape distorts and narrows in a manner dependent on the rate and details of the motion. The rate, of course, increases with increasing temperature. The observed static linewidths are of the order of 2×10^5 Hz. In practice this means that for jump rates < 10^3 Hz, the static lineshape is barely affected. For intermediate rates between 10^3 and 10^7 Hz, the lineshapes undergo narrowing and distortion, and for rates > 10^7 Hz, narrowing is complete. The latter is referred to as the fast motion limit (*fml*), and represents an effective smaller *efg*, which is an average over all the orientations visited during the motion. By studying changes in the ^2H NMR lineshapes, it is possible to develop models for the molecular dynamics and, in some cases, obtain independent structural information in terms of molecular orientation. Intermediate rate lineshapes can be simulated numerically for a specific model and matched to the observed lineshapes at specific temperatures, allowing for the determination of the activation energy (E_a) for the motion from an Arrhenius plot of rate *vs.* inverse temperature: rate $\propto e^{-E_a/RT}$. ^2H NMR also detects phase transitions that are accompanied by drastic changes in dynamics. Detailed descriptions of ^2H NMR have been published [6].

3.2. Deuterium NMR Studies of Calixarene Adducts

The following examples illustrate the great utility of ^2H NMR in the study of 1:1 tBC-guest compounds:

3.2.1. p-tert-Butylcalix[4]arene-toluene

Toluene (both -d_5 and -d_8) in tBC provides a case study in how our understanding of structure and dynamics in this system has progressed over time, and how ^2H NMR has influenced the structural determinations in a complementary fashion. The early room temperature XRD determination showed the toluene C_2 molecular axis aligned with the calixarene 4-fold axis of symmetry [3]. Room temperature ^2H spectra, showing three superimposed axial lineshapes from CD_3, *ortho*- and *meta*-D and *para*-D, are compatible with 4-fold guest reorientation about the molecular axis, but with an additional off-axis motion [4]. However, the lineshapes are equally compatible with a 4-fold motion of the toluene C_2-symmetry axis at an angle of 12° with respect to the 4-fold calix symmetry axis, and the re-determined XRD structure indeed shows that the toluene is tilted off the C_4 symmetry axis by 7°. Furthermore, upon cooling the ^{2}H NMR data indicated a phase transition below 248K, which was confirmed later by ^{13}C NMR, calorimetry and the 150K XRD structure. In the low temperature phase, the ^{2}H NMR lineshapes [7] show that the toluene is static at 129K, and that it undergoes rapid 180° flips about the molecular C_2 symmetry axis by 180K (E_a = 34 ±3 kJ mol^{-1}). This motion is incompatible with the presence of a 4-fold calixarene axis of symmetry suggested in the preliminary interpretation of the 150K XRD data. However it is consistent with the final structural analysis where each of two crystallographically-different tBC-toluene units has only a 2-fold axis of symmetry. Detailed analysis of the low temperature spectra suggests that there is also a small angle 2-site jump associated with the 2 disordered positions of each toluene [8].

The motion of the host t-butyl-d_9 groups in tBC-toluene has also been studied [7,8]. The t-butyl groups are essentially static below 129K. A much narrowed ^{2}H NMR fml lineshape at 204K and above can be interpreted in terms of a combination of three motions: methyl C_3, t-butyl C_3' and a hopping motion of the C_3' axis between two sites, where one site is tilted 15° further away from the axis of the calix (XRD: 10°). The 2-site motion of the t-butyl groups and the 4-fold toluene motion appear to be coupled in the room temperature phase so that two t-butyl groups on opposite sides of the host are "in", the other two are "out", depending on the instantaneous orientation of the toluene.

3.2.2. p-tert-Butylcalix[4]arene-(benzene or pyridine)
Benzene-d_6 and pyridine-d_5 have similar shapes, and comparable positions and orientations in the tBC cavity, yet their dynamic behaviour shows a striking difference [9].

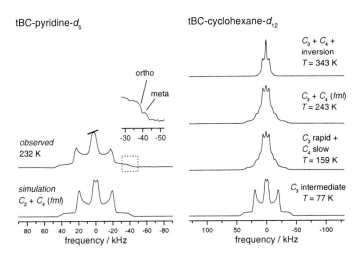

Figure 4. ^{2}H NMR spectra: tBC-pyridine-d_5 (left) and tBC-cyclohexane-d_{12} (right); the narrow isotropic signal in the 232 K spectrum of tBC-pyridine-d_5 is due to a small amount of uncomplexed pyridine-d_5 and is truncated for presentation purposes.

Benzene undergoes rapid in-plane reorientation about its molecular C_6 symmetry axis even at 77K. At higher temperatures the C_6 axis rotates about the C_4 symmetry axis of the calix at an angle of 79° at 195K, (XRD at 150K: 79°). Pyridine, on the other hand, shows no in-plane reorientation, since the *ortho*-, *meta*- and *para*-deuterons retain distinct ^{2}H lineshape features. At low temperature, pyridine undergoes 180° flips about the molecular C_2 axis. At higher temperature this C_2 axis, like the C_6 axis of benzene, also rotates about the calix C_4 symmetry axis at an angle of 53° at 210K (XRD at 150K: 47°) (Figure 4). This information about the C_2 axis orientation assisted with the determination of the guest nitrogen position in the XRD structural refinement. Note that for both guests the dynamic 4-fold disorder at 295K becomes a static disorder at low temperature. The guest disorder is also intertwined with the 1:7 disorder observed in the positions of the 4 *t*-butyl groups of the host. With the guest in one of its four symmetry-related positions, one host *t*-butyl group must have a second orientation of similar energy to the first, whereas the remaining three *t*-butyl groups have only a single low energy orientation. At low temperature (when the 4-fold motion is too slow) the observed symmetry is due to space averaging. As the guest reorientation becomes activated, the two-site disorder switches from one *t*-butyl group to another, *i.e.*, the *t*-butyl group positional disorder becomes dynamic.

3.2.3. p-tert-Butylcalix[4]arene-nitrobenzene

For nitrobenzene-d_5 in tBC, the guest molecular C_2 axis is 68° off the calix axis, but in a mixed inclusion complex with propane (at 36%) the C_2 axis aligns with the calix axis. This alignment has a profound effect on the motion of the nitrobenzene [10]. In the pure complex, ^{2}H NMR shows only 180° flips of the nitrobenzene about its C_2 symmetry axis. The motion is static below 200K and fast by 290K, with $E_a = 56.6 \pm 1.3$ kJ mol^{-1}. No 4-fold motion occurs in this complex, so it is not possible to determine an orientation from the ^{2}H NMR results. In sharp contrast, the same 2-fold motion is already rapid at 136K in the mixed complex, indicating a much lower barrier; as the temperature increases, the 2-fold motion gradually becomes a 4-fold motion that is still not in the *fml* at 319K. In this and other cases where a low temperature 2-fold reorientation gradually gives way to a high temperature 4-fold reorientation, it is likely that there are secondary potential minima between and at higher energy than the 180° positions. As the temperature increases, these secondary sites are increasingly populated. More importantly, the separation in energy between the two pairs of sites must gradually diminish in order to achieve 4-fold symmetry. It is likely that by increasing the populations of the intermediate sites, the host lattice undergoes subtle structural adjustments, which in turn reduce the energy separation between the two pairs of sites in a synergistic manner. We see here the makings of a phase transition.

3.2.4. p-tert-Butylcalix[4]arene-cyclohexane

Cyclohexane-d_{12} in tBC already shows reorientation about the 3-fold molecular symmetry axis of the chair-form molecule at intermediate rates at 77K (the lineshape of the axial deuterons, D_{ax}, is virtually unchanged from static, whereas for the equatorial deuterons, D_{eq}, the line is narrowed by a factor of ~3) [8,11]. This motion is fast by 160K. Before this temperature is reached, a second motion, reorientation of the C_3 molecular axis about the calix 4-fold symmetry axis, has already begun. This combined C_3+C_4 motion is rapid by 243K. At this point, D_{ax} and D_{eq} still give two distinct narrowed lineshapes. Inversion of the cyclohexane molecule occurs above 250K and is almost in the *fml* by 343K. The inversion motion interchanges D_{ax} and D_{eq} positions leading to a single lineshape for all deuterons (Figure 4). Again XRD shows disorder of cyclohexane about the host C_4 symmetry axis, and there is exact agreement between the angle of tilt of the C_3 molecular symmetry axis away from the C_4 host symmetry axis obtained by XRD and calculated from the ^{2}H NMR lineshapes (67°).

It is clear from these examples that motions are intimately linked with structure, symmetry, dynamic disorder, phase changes and temperature.

4. Solid-state NMR Spectroscopy and Other Characterisation Techniques

4.1. ^{13}C SOLID-STATE NMR STUDIES

Structural and dynamic data complementary to diffraction and ^{2}H NMR techniques are available from solid-state NMR spectroscopy of spin-½ nuclei such as ^{13}C. The first solid-state ^{13}C CP-MAS (cross-polarisation, magic angle spinning) NMR spectrum of a calixarene host-guest inclusion compound was that of tBC-toluene [12]. Subsequent reports have also included ^{13}C solid-state NMR spectra [4,7,13,14]. Due to the 4-fold symmetry of the structure at room temperature, the spectrum shows a single resonance for each chemically distinct carbon in the host *t*-butylphenol repeat unit (*Figure 5*). Lowering the temperature decreases the symmetry element and increases the multiplicity of each carbon resonance, as the four repeat units are no longer equivalent symmetrically. Of all the guest carbons, the 1' carbon is most diagnostic of the guest symmetry: a singlet at room temperature, it becomes a doublet at the phase transition, and increases further in multiplicity down to 115K (*Figure 5*). These features are consistent with the guest site symmetry and the nature of the guest molecular motion being strongly correlated.

The chemical shift of the toluene methyl carbon (5') is lowered by 6 ppm from the value observed for toluene in solution. Because of its location deep inside the cavity, the guest methyl carbon is sensitive to the ring current effects of the aromatic "walls" of the host [12]. In general, the difference between solution and inclusion chemical shift values of guest molecules is an indicator of the orientation of the guest inside the calixarene cavity [15,16] (as modified by guest dynamics).

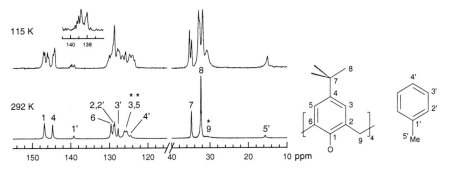

Figure 5. Partial ^{13}C CP-MAS NMR spectra of tBC-toluene at 292 and 115K; * *indicates signals that disappear with dipolar dephasing. Note the increased multiplicity of the carbon signals at 115K. Inset: expansion of the toluene C-1' carbon resonance at 115K.*

The ^{13}C CP-MAS NMR experiment is also sensitive to dynamic effects when a short delay (~40 μs) is inserted prior to proton decoupling in the pulse sequence. The dipolar dephasing modification [17] causes all immobile carbons with attached protons to disappear from the resulting spectrum due to the strong ^{1}H-^{13}C dipolar interaction. For tBC-toluene, the host bridging methylene (9) and proton-bearing ring (3,5) carbons disappear, indicating that these carbons are rigid whereas the methyl carbons (8) are dynamic. A special case arises when the C–H bond lies along an axis of motion such as the *para*-carbon of toluene (4'). While the toluene rotates about its axis of symmetry at room temperature, this does not average the C–H dipolar interaction tensor. Consequently, the *p*-carbon disappears from the dipolar-dephased spectrum. Such information is valuable in determining the geometry of motional averaging and, in special cases, the orientation of motionally-averaged guests inside the cavity.

4.2. SOLID-STATE NMR STUDIES OF OTHER SPIN-½ NUCLEI

In principle, ^{1}H solid-state NMR spectra will give similar structural and dynamic information on supramolecular compounds as ^{13}C NMR. However, the strong ^{1}H-^{1}H dipolar interaction gives broad and featureless lines. Line-narrowing methods such as magic angle sample spinning at high speeds (>30 kHz) and combined rotation and multiple pulse (CRAMP) techniques [18] give narrower ^{1}H NMR spectra, but these techniques are not widely accessible at present. The second technique has been used to study the change to higher field of the toluene methyl ^{1}H chemical shift upon inclusion in the tBC host: the relative change in the ^{1}H chemical shift was even greater than that in the ^{13}C chemical shift (Figure 6).

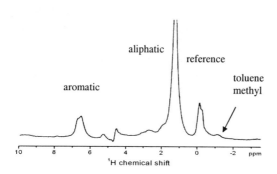

Figure 6. ^{1}H CRAMP NMR spectrum of tBC-toluene.

Non-^{13}C NMR studies have been most useful for the study of guest structure and dynamics. ^{15}N-labeled pyridine in tBC gives a single isotropic peak indicating one guest environment; the chemical shift tensor is diagnostic of the pyridine guest motion and orientation [9]. A single ^{19}F NMR peak is observed for 1-fluorobenzene at room temperature, and the chemical shift tensor is similarly indicative of guest motion and orientation [19]. The dipolar interaction between the guest ^{19}F and host t-butyl ^{13}C nuclei, which is proportional to the inverse cube of the distance, can be detected by the rotational echo double resonance (REDOR) NMR experiment in order to gain host-guest distances [19]. This technique has potential for giving distance information in the absence of diffraction structural information. The tBC-Xe clathrate allows one to use the Xe guest as a direct NMR probe of the compound structure and dynamics: ^{129}Xe NMR spectra indicate the number of different Xe environments, and the chemical shift anisotropy indicates cavity symmetry and dynamic information [20].

4.3. OTHER SOLID-STATE CHARACTERISATION TECHNIQUES

The various techniques employed in the solid-state characterisation of calixarenes have been reviewed [2], and include computer modeling, vibrational spectroscopy, neutron scattering and calorimetric studies. Molecular recognition, guest orientation and the role of various weak intramolecular forces have been probed by computer modelling. Modelling is vastly underdeveloped in the area of supramolecular chemistry, especially in the solid state, as more than the 1:1 guest-host interactions need to be taken into account. We have argued that the quality of information derived from modelling is directly related to the level of confidence in the experimental data, and that structural and dynamic models often have significant potential for improvement. π-Methyl interactions [21], while present, are likely not to be the primary interactions to account for observed data such as the on-axis toluene orientation within the host cavity when one considers other guests such as benzene and pyridine.

Infrared and Raman vibrational studies have been reported for solid-state inclusions involving the calixarene host and various guests, and appear to be not very informative for specific molecular interactions and structural features. In the case of p-*tert*-butylcalix[8]arene-C_{60}, the vibrational spectra indicate little interaction between the two components [22]. Low temperature inelastic neutron scattering has been used to probe the guest methyl reorientation in the 2:1 tBC-*p*-xylene, and the host-guest interactions [23]; not surprisingly, there is no barrier to methyl reorientation in addition to the very low intra-molecular barrier upon inclusion in the host cavity.

Macroscopic measurements such as thermogravimetric analysis (TGA) and calorimetry probe the structure, composition and stoichiometry of inclusion compounds and how these change under various conditions. In TGA, the mass of an inclusion is monitored as temperature increases, giving a weight-loss curve as the guest is released from the inclusion compound, as is illustrated in Figure 7 for tBC-pentane.

TGA is particularly useful for small, volatile guests [14], and gives the determination of the host-guest ratio and a measure of compound stability. Calorimetric techniques such as differential scanning calorimetry (DSC) measure the heat flow into a sample as a function of temperature, and give information about phase transitions and

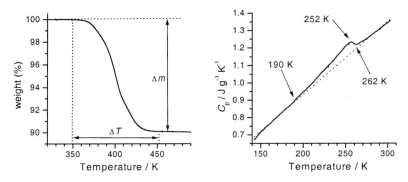

Figure 7. Left: Thermogravimetric analysis of tBC-pentane, heating rate of 1 °C min^{-1}. The weight loss corresponds to the loss of one pentane molecule per host molecule. Right: The DSC trace of tBC-toluene; the heat capacity derived from integration of the exotherm of the phase transition is $C_p = 1.93$ kJ mol^{-1}.

decomposition. In the case of tBC-toluene, a broad exotherm with a maximum at ~248 K is observed characteristic of a phase transition. We have ascribed this to the onset of long-range ordering in the orientation of the toluene guest molecules and a change in the average symmetry of the material (upon cooling).

5. *p-tert*-Butylcalix[4]arene Hosts with Long-chain Guests

Small molecules such as *n*-pentane and toluene form tBC clathrates in a 1:1 host-guest ratio [24]. Starting with *n*-hexane, the longer linear alkane guest molecules are too large to form 1:1 clathrates, and instead stabilise inclusion compounds with two tBC host molecules encapsulating a guest. The 2:1 host-guest clathrates form tetragonal crystals (*P4/nnc*, a = 12.9 Å, c = 25.3 Å). The clathrate cavity is defined by two host molecules connected head-to-head and rotated 45° with respect to each other about a 4-fold axis of symmetry. The calixarene dimer has 4_2 symmetry. We have studied several 2:1 host-guest clathrates with different alkane guest molecules in an attempt to: (a) find the size limit for paraffin guest molecules that form 2:1 clathrates, (b) probe the influence of guest molecule size and substituents on the guest's conformation, and (c) describe the influence of large guests on the host structure [25]. In all cases, the guest molecules are disordered about the 4-fold axis of symmetry and the structures are refined using standard bond lengths

For *n*-pentane and the alkane derivatives of similar lengths X–$(CH_2)_3$–Y (where X,Y = CH_3, OH, Cl, Br), the terminal halogen substituent is excluded from the cavity in the 1:1 clathrates [16]. This effect is most strikingly illustrated for the 1,3-dichloropropane guest molecule in which the central methylene group penetrates the cavity, while forcing both chlorine atoms outside. XRD studies of 2:1 tBC-*n*-hexane show that the guest molecule is inserted along the calixarene four-fold axis of symmetry in an all-*trans* conformation with both methyl groups deeply inserted in their respective calixarene cavities (Figure 8).

Replacement of one *n*-hexane methyl group by chlorine is accompanied by a change in the conformation of the guest molecule inside the host cavity. The methyl group of 1-chloropentane sits deep in one calixarene cavity, but the chlorine atom is less deeply inserted in the other, perhaps due to the chlorophobicity of the calixarene cavity [16]. As a result of the host aversion for chlorine, the 1-chloropentane guest is "compressed" along the *c*-axis of the crystal. The conformation of 1-chloropentane is *gauche-anticlinal-trans* (*g-ac-t*) with torsion angles 76, 140 and 167°. In contrast, a rigid, chlorine-containing guest such as 1,4-dichlorobenzene is not adjustable, and is thus oriented with both chlorine atoms deep inside the two tBC cavities (Figure 8).

The *n*-hexane and 1-pentanol molecules are not compressed inside the cavity, and possess the energetically-favoured all-*trans* conformation. Paraffin and monoalcohol molecules larger than *n*-hexane are unable to adopt the low-energy, all-*trans* conformation inside the cavity because of their length, and so acquire less energetically favourable conformations. 1-Heptanol has a conformation close to *s-g-ac-s-g* (where *s* = *syn-peri-planar*), octanol *g-s-g-ag-g* (where *ag* = *anti-gauche*), and dodecane *ac-g-s-g-ac-g-ac-g-ac*. Dodecane ($C_{12}H_{26}$) appears to be the largest paraffin molecule that fits inside the 2:1 clathrate cavity, since no clathrate has yet been found for the larger tetradecane ($C_{14}H_{30}$).

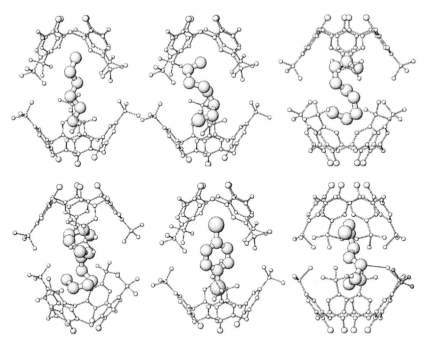

Figure 8. 2:1 tBC compounds with: (top row) nearly all-trans n-hexane, 1-heptanol, 1-octanol, (bottom row) dodecane, 1,4-dichlorobenzene, 1-octanol (rotated by 45° about the 4-fold symmetry axis with respect to the rendering above it; the dark line indicates the shortest distance from the guest to the disorderd host t-Bu group).

The symmetry of the guest molecules in these 2:1 clathrates is not the same as the symmetry of the cavity. If the cavity has the symmetry 4_2, the guest molecule has symmetry 2 only in the case of *n*-hexane in the all-*trans* conformation. For 1-heptanol, 1-octanol, and dodecane, the guest molecules have no symmetry elements and are disordered equally over eight orientations related by the symmetry of the cavity. However, well-defined positions for all guest atoms are found in each instance.

Positional disorder in the host molecule depends on the size of the guest molecules. For clathrates formed with *n*-hexane, 1-pentanol and 1-chloropentane, no disordering of the host molecule is observed. However, for the larger 1-heptanol, 1-octanol and dodecane molecules, *t*-butyl groups in the host tBC molecule are disordered between two positions with the site occupancy ratio 1:7. The disordering ratio can be attributed to the steric influence of the guest molecule: a methylene group of the guest molecule shows a close approach to one of eight *t*-butyl groups of the cavity. This *t*-butyl group, normally pointing into the cavity, pushes outside the cavity and is rotated by approximately 30° about the *t*-butyl local axis of symmetry to achieve a normal intermolecular distance (r_{C-C} ~3.5 Å) from the guest molecule.

6. Hydrogen-bonding Interactions between tBC Host and Amine Guests

The calixarene host can be modified synthetically at the lower phenol or the upper alkyl rims to tailor the host's inclusion properties, to enhance its solubility properties and to locate functional groups for specific intermolecular interactions [24]. So far, all these elaborations have been covalent in nature. The phenolic hydrogen bonds at the lower rim, in principle, can interact with hydrogen bond acceptors such as amines to form a supramolecular complex with a non-covalent modification of the calixarene. In solution, amines interact with calixarenes by being inserted into the cavity or, if too large, by interacting with the cavity's outside wall [26]. Both *endo* and *exo* amine sites involve a hydrogen bonded interaction between the guest and the host. This motif persists in the solid state, although with little change in the overall arrangement of the host and guest in the crystal [27]. In our examination of the relative roles of hydrogen-bond strength (as indicated by amine pK_a values) and steric factors, we discovered that some amines cause significant changes in the host-guest structure [28]. In particular, 1,4-butanediamine occupies sites both *endo* and *exo* to the cavity. There are two distinct host molecules, labelled **A** and **B** in Figure 9.

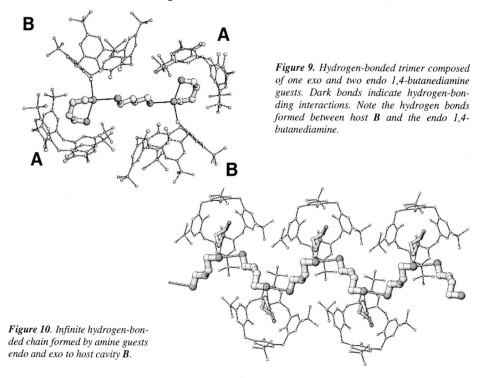

Figure 9. Hydrogen-bonded trimer composed of one exo and two endo 1,4-butanediamine guests. Dark bonds indicate hydrogen-bonding interactions. Note the hydrogen bonds formed between host **B** and the endo 1,4-butanediamine.

Figure 10. Infinite hydrogen-bonded chain formed by amine guests endo and exo to host cavity **B**.

The *endo* amine in cavity **A** forms hydrogen bonds with itself, the phenolic OH of a neighbouring host **B** cavity and with an *exo* amine. This arrangement gives rise to a discrete hydrogen-bonded trimer of guests. Guests occupying cavity **B** also hydrogen-bond to themselves, and interact with a second *exo* guest to form an infinite chain of hydrogen-bonded *exo* and *endo* guests (Figure 10). The net result is a novel structure with a 2:3.5 host-guest ratio in which the guests participate in two independent hydrogen-bonding networks: a discrete trimer, and an infinite chain.

An interesting aspect of this structure is the presence of an infinite, hydrogen-bonded chain of amine guests occupying a 1-dimensional channel in the crystal. Upon heating, the ^{13}C NMR shows dramatic changes in the guest resonances consistent with either a decrease in guest rigidity or even loss of guest from some sites. Close examination of the hydrogen bonding distances—and thus strengths—suggests that the channel guests may be loosely held, and easily removed.

The unusual structural arrangement in the host-guest inclusions with amines is reflected in the ^{13}C CP-MAS NMR spectra, which show an increase in the multiplicity (from 1 to 8) of the host lines, most notably in the C-1 and C-4 carbons of the host. However, not all amines form a low-symmetry structure, and the ones with low steric requirements appear to favour the 1:1 high symmetry case. In all cases, there does not appear to be a strong dependence of structure on the amine pK_a values. The ^{13}C NMR indicates that, in addition to 1,4-butanediamine, low-symmetry structures with host-guest hydrogen bonding are formed with ammonia, benzylamine, 2-aminoethanol and 1-cyclohexylethylamine [28].

7. Concluding Remarks

Based on the results discussed in this chapter we can come to a number of conclusions and suggestions. First, some general observations on solid tBC compounds.

It can be stated categorically that the availability of information on local order from NMR has been crucial in arriving at correct structural models. For many tBC compounds it is possible to obtain good structural models once the nature of the t-butyl disorder is recognised. At low temperatures, the symmetry of the tBC host lattice generally reflects the guest symmetry, that is, departure of the guest from axial symmetry gives rise to low symmetry calixarene units. At higher temperatures, this is modified by the introduction of molecular motion, leading to structures of time-averaged higher symmetry as observed from diffraction. Many guests have their orientations "locked in" during the crystallisation process, and for toluene and chlorobenzene guests it is possible to see both a favoured and a minor orientation in the host lattice. For these guests one must conclude that steric interactions are the primary reason for the guest position in the cavity. Other guests are free to reorient as shown by NMR, and in these cases the guest positions determined by diffraction can be used to give information on molecular recognition processes in the cavity.

Most, if not all, molecules of appropriate size form reasonably stable inclusions with tBC, and this observation suggests that considerable caution should be exercised in interpreting compound stability in terms of specific interactions. By examining series of structurally-related guest compounds it can be seen that the deep cavity is chlorophobic, but that space filling seems to be more important than the avoidance of unfavourable interactions in determining whether compound formation will take place. A similar conclusion can be drawn from the fact that long molecules will "curl up" in order to fill space effectively in an inclusion compound rather than exist in a favoured all-*trans* conformation without compound formation. The host-guest interactions must be more than enough to compensate for a number of unfavourable contacts. At this stage it is a moot point whether to describe tBC compounds as clathrate lattices, with mainly non-specific interactions contributing to the stability of essentially non-stoichiometric compounds, or as stoichiometric molecular host-guest compounds with compositions dependent on specific guest-host interactions.

In suggesting further work, we note that in order to understand the details of tBC host-guest interactions it is necessary to complete a comprehensive data set of structures that catalogue the various possible structural types and host-guest geometries. Beside the structural details, it also is necessary to measure compound stabilities in order to gauge the relative importance of specific *vs.* non-specific interactions. With the availability of this information it will become possible to develop, test and improve potential functions with the use of modeling calculations.

Also of general importance for structure determination are the observations on the relationship between domain size and radiation wavelength and the distinction between twinned ordered structures *vs.* disordered non-twinned structures. Again we emphasise that the use of complementary techniques is of prime importance in providing quality control and in giving additional insight for solving the increasingly difficult structural problems encountered in supramolecular structural chemistry.

8. References and Notes

[1] a) E. B. Brouwer, J. A. Ripmeester, G. D. Enright, *J. Incl. Phenom.* **1996**, *24*, 1-17; b) J. A. Ripmeester, E. B. Brouwer, G. D. Enright, C. I. Ratcliffe, K. A. Udachin in *Supramolecular Engineering of Synthetic Metallic Materials* (Ed. J. Veciana, C. Rovira, D. B. Amabilino) NATO ASI Series **1999**, *C518*, 83-104.

[2] E. B. Brouwer, J. A. Ripmeester, *Adv. Supramol. Chem.* **1999**, *5*, 121-155.

[3] G. D. Andreetti, R. Ungaro, A. Pochini, *J. Chem. Soc., Chem. Commun.* **1979**, 1005-1007.

[4] G. A. Facey, R. H. Dubois, M. Zakrzewski, C. I. Ratcliffe, J. L. Atwood, J. A. Ripmeester, *Supramol. Chem.* **1993**, *1*, 199-200.

[5] G. D. Enright, E. B. Brouwer, J. A. Ripmeester, manuscript in preparation.

[6] a) J. A. Ripmeester, C. I. Ratcliffe in *Comprehensive Supramolecular Chemistry, Vol. 8* (Eds.: J. L. Atwood, J. E. D. Davies, D. D. MacNicol, F. Vogtle, J.-M. Lehn), Pergamon, New York, **1996**, pp. 323-380; b) R. J. Wittebort, E. T. Olejniczak, R. G. Griffin, *J. Chem. Phys.* **1987**, *86*, 5411-5420; c) M. S. Greenfield, A. D. Ronemus, R. L. Vold, R. R. Vold, P. D. Ellis, T. E. Raidy, *J. Magn. Reson.* **1987**, *72*, 89-107; d) J. H. Davies, K. R. Jeffrey, M. Bloom, M. I. Valic, T. P. Higgs, *Chem. Phys. Lett.* **1976**, *42*, 390-394.

[7] E. B. Brouwer, G. D. Enright, J. A. Ripmeester, *Supramol. Chem.* **1996**, *7*, 79-83.

[8] E. B. Brouwer, PhD thesis, Carleton University (Ottawa, Canada), **1996**.

[9] E. B. Brouwer, G. D. Enright, C. I. Ratcliffe, G. A. Facey, J. A. Ripmeester, *J. Phys. Chem. B* **1999**, *103*, 10604-10616.

[10] a) E. B. Brouwer, G. D. Enright, J. A. Ripmeester, *Supramol. Chem.* **1996**, *7*, 7-9; b) *J. Am. Chem. Soc.* **1997**, *119*, 5404-5412.

[11] E. B. Brouwer, G. D. Enright, J. A. Ripmeester, *Supramol. Chem.* **1996**, *7*, 143-145.

[12] T. Komoto, I. Ando, Y. Nakamoto, S. I. Ishida, *J. Chem. Soc., Chem. Commun.* **1988**, 135-137.

[13] T. Liang, K. K. Laali, *Chem. Ber.* **1991**, *124*, 2637-2640.

[14] J. Schatz, F. Schildbach, A. Lentz, S. Rastätter, *J. Chem. Soc., Perkin Trans. 2* **1998**, 75-77.

[15] T. Yamanobe, I. Nakamura, K. Hibino, T. Komoto, H. Kurosu, I. Ando, Y. Nakamoto, S.-I. Ishida, *J. Mol. Struct.* **1995**, *355*, 15-20.

[16] E. B. Brouwer, K. A. Udachin, G. D. Enright, C. I. Ratcliffe, J. A. Ripmeester, *Chem. Commun.*, **1998**, 587-588.

[17] S. J. Opella, M. H. Frey, *J. Am. Chem. Soc.* **1979**, *101*, 5854-5856.

[18] R. K. Harris, P. Jackson, L. H. Merwin, B. J. Say, G. Hagele, *J. Chem. Soc., Faraday Trans. 1* **1988**, *84*, 3649-3672.

[19] E. B. Brouwer, R. D. M. Gougeon, J. Hirschinger, K. A. Udachin, R. K. Harris, J. A. Ripmeester, *Phys. Chem. Chem. Phys.*, **1999**, 1, 4043-4050.

[20] E. B. Brouwer, G. D. Enright, J. A. Ripmeester, *Chem. Commun.* **1997**, 939-940.

[21] M. Nishio, M. Hirota, Y. Umezawa, *The CH-π interaction: Evidence, Nature, and Consequences*, Wiley-VCH, New York, **1998**.

[22] B. Paci, G. Amoretti, G. Arduini, G. Ruani, S. Shinkai, T. Suzuki, F. Ugozzoli, R. Caciuffo, *Phys. Rev. B* **1997**, *55*, 5566-5569.
[23] P. Schiebel, G. Amoretti, C. Ferrero, B. Paci, M. Prager, R. Caciuffo, *J. Phys.: Condens. Matter.* **1998**, *10*, 2221-2231.
[24] C. D. Gutsche, *Calixarenes Revisited*, RSC, Cambridge, **1998**.
[25] K. A. Udachin, G. D. Enright, C. I. Ratcliffe, E. B. Brouwer, J. A. Ripmeester, manuscript submitted to *Supramol. Chem.*
[26] a) L. J. Bauer, C. D. Gutsche, *J. Am. Chem. Soc.* **1985**, *107*, 6063-6069; b) C. D. Gutsche, M. Iqbal, A. Alam, *ibid.* **1987**, *109*, 4314-4320; c) G. Görmar, K. Seiffarth, M. Schulz, C. L. Chachimbombo, *J. Prakt. Chem.* **1991**, *333*, 475-479; d) A. F. Danil de Namor, R. M. Cleverley, M. L. Zapata-Ormachea, *Chem. Rev.* **1998**, *98*, 2495-2525.
[27] a) G. D. Andreetti, F. Ugozzoli, Y. Nakamoto, S.-I. Ishida, *J. Incl. Phenom.* **1991**, *10*, 241-253; b) M. Czugler, S. Tisza, G. Speier, *ibid.* **1991**, *11*, 323-331; c) C. Bavoux, M. Perrin, *ibid.* **1992**, *14*, 247-256; d) J. M. Harrowfield, M. I. Ogden, W. R. Richmond, B. W. Skelton, A. H. White, *J. Chem. Soc., Perkin Trans. 2* **1993**, 2183-2190; e) J. M. Harrowfield, W. R. Richmond, A. N. Sobolev, A. H. White, *ibid.* **1994**, 5-9.
[28] E. B. Brouwer, K. A. Udachin, G. D. Enright, J. A. Ripmeester, *Chem. Commun.* **2000**, 1905-1906.

Chapter 17

MOLECULAR DYNAMICS OF CATION COMPLEXATION AND EXTRACTION

GEORGES WIPFF

Institut de Chimie, 4, rue B. Pascal, 67 000 Strasbourg, France.
e-mail: wipff@chimie.u-strasbg.fr

1. Introduction

Numerous modeling studies, facilitated by recent spectacular increases in computer power and the implementation of new theoretical tools, have been stimulated by exploration of the chemistry of cation binding by calixarenes. Compared to other important ion-binding agents like crown ethers and cryptands, calixarenes have a number of novel properties. Their phenolic building units are unsymmetrical and usually hydrophobic, while their pendent ion-binding sites are hydrophilic. They are thus expected to display solvent dependent conformational and binding properties. Another important feature concerns preorganization. In contrast to macrocyclic ligands with topologically connected binding sites, calixarenes provide a platform attached to which are mobile binding sites which do not form a topologically connected preorganized cavity. The latter is induced upon complexation. On the experimental side, there is a relative scarcity of X-ray structures, particularly those enabling comparison of how different cations bind to a given host. It is thus difficult to validate computer modeled structures of calixarene complexes. Whether structures in the solid state are representative of those in solution, a recurrent question, generally difficult to assess by experiment only, can, however, be tackled by simulations.

This chapter concerns systems which have been modelled by molecular dynamics (MD) simulations in solution in attempts to account for solvent and conformational effects on the binding and recognition properties. The most extensive studies deal with alkali metal cation complexes in aqueous solution, involving the questions of cation shielding from solvent and of structural fit and specific interactions between the host and the cation. For several systems, the choice of water as solvent may seem unrealistic, considering the very low solubility in water of the calixarenes involved. The simulations may, nonetheless, provide models of water saturated phases and insights into the effect of a solvent which strongly competes for cation coordination. Simulations of non-aqueous solutions are more recent and raise the question of status of accompanying counterions. The next step is to consider mixed liquid phases such as an organic phase containing traces of water, or mixed miscible solvents. What happens at the interface between water and an immiscible solvent is of particular importance in the context of assisted liquid-liquid extraction ("LLE"). Recent modeling studies of the interfacial activity of calixarenes provided new insights into this question and led to consideration of collective and concentration effects in LLE. Experimental aspects of complexation and extraction properties of the modeled systems are reported in several chapters of this book and in the cited references. For general reviews, see [1-5]. For brevity throughout this chapter, we will denote metallic cations as M^+ (for alkali metals), M^{2+} (alkaline-earth metals) and M^{3+} (lanthanide metals), and the calixarene ligands of interest as **L1** to **L11** (Figure 1).

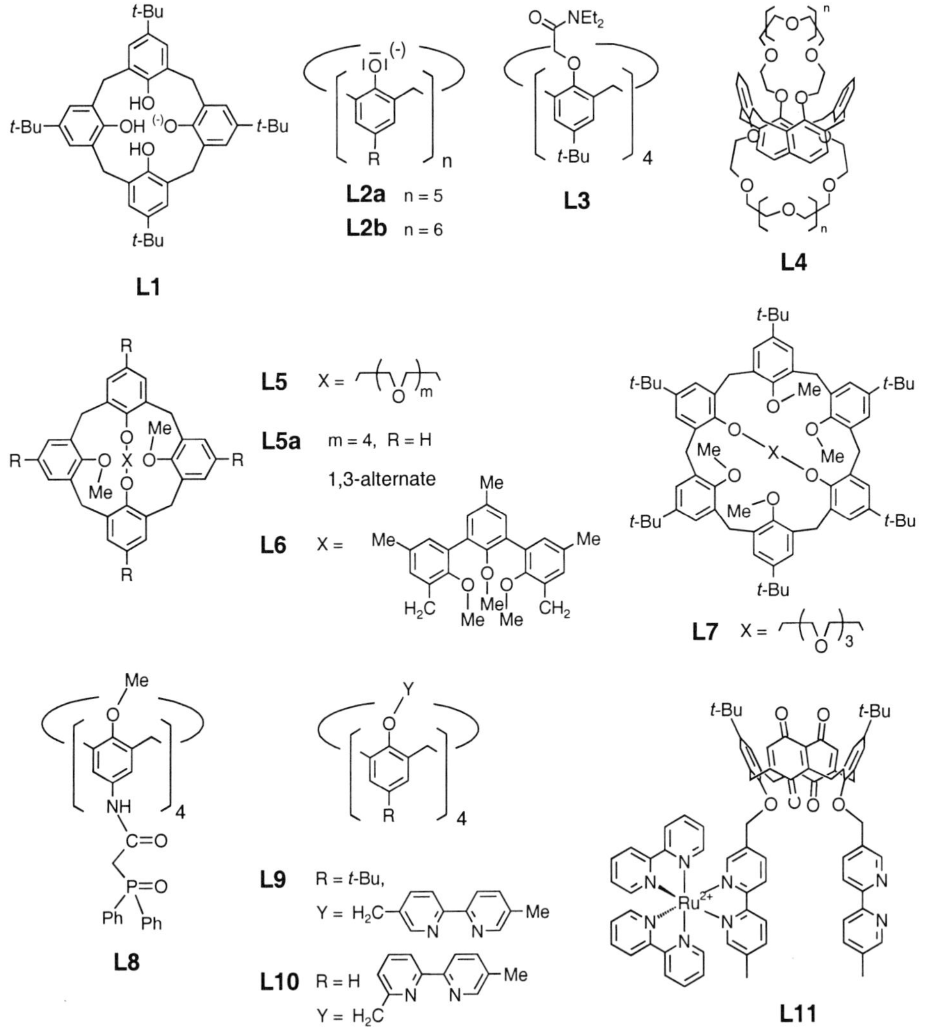

Figure 1. Typical simulated systems.

2. MD and FEP Simulation Methods

The two most important problems in molecular modeling are the adequate sampling of the configurations of the systems and the correct evaluation of the potential energy of the corresponding configurations. In practice, for a selected size of the system, a compromise has to be found between these requirements and the available computer resources. General presentations of the methods can be found in refs [6,7].

2.1. THE FORCE FIELD

Current standard representations of the potential energy U are of molecular mechanics type, based on a ball-and-stick representation of the system (Force Field models). A typical equation is given below, consistent with the AMBER [8], CHARMM [9] or BOSS [10] programs. Deformations of bonds and bond angles from their equilibrium positions r_{eq} and θ_{eq} are assumed to be harmonic, while torsional terms are expressed as a function of the dihedral angles ϕ. Most crucial is the representation of non-bonded interactions, and in particular those involving the complexed cation. MM2 derived models where the cation makes pseudo-bonds with the ligand may valuably depict static systems for a selected cation coordination type and number, but do not allow for dynamic exchange of ligands [11,12]. Thus, a non-covalent model of cation coordination is preferable, generally assuming pairwise additive coulombic and van der Waals interactions between the non-bonded atoms. Charge transfer and polarization effects are neglected, and the formal charge is +1 for alkali metal, +2 for alkaline-earth metal and +3 for lanthanide metal cations. As the cation hardness increases, the neglect of electronic reorganization effects becomes more problematic [13].

$$U = \Sigma_{bonds} K_r (r-r_{eq})^2 + \Sigma_{angles} K_\theta(\theta-\theta_{eq})^2 + \Sigma_{dihedrals}\Sigma_n V_n(1+\cos n\phi)$$
$$+ \Sigma_{i<j} [q_iq_j/R_{ij} - 2\varepsilon_{ij}(R_{ij}^*/R_{ij})^6 + \varepsilon_{ij}(R_{ij}^*/R_{ij})^{12}]$$

The force field is characterized by the energy formula, the selection of related parameters, and the protocol used for the calculations. Some common features of many calixarene simulations are the following. The parameters used for the solvents have been fitted from Monte Carlo simulations (of, *e.g.*,density, heat of vaporization) on the pure liquid phases: the models are TIP3P [14] for water and OPLS for acetonitrile, methanol and chloroform [15], where the H atoms of CH_n groups are represented implicitly (united atom approximation). The van der Waals ε and R* parameters of M^+ and M^{2+} cations have been fitted empirically to account for free energies of hydration in pure water [16]. Recently, a similar procedure has been used to fit lanthanide(III) ion parameters [17]. The atomic charges of the calixarenes and of molecular ions are generally obtained from the electrostatic potentials (ESP) calculated by quantum mechanical methods (*ab initio* at the HF level with a 6-31G* basis set, or semi-empirical methods) on small molecules which mimic constitutive fragments (*e.g.* anisole, amide, pyridine, ethers, CMPO). Atomic charges have no physical meaning, no unique definition, and their choice is often a matter of controversy. In practice, the sensitivity of calculated results to the choice of charges has to be assessed by comparative calculations. However, results for other macrocylic systems show there is a reasonable consistency between the water solvent model, the cation representation, and the atomic ESP charges [18]. Alternative models using calixarene charges derived from semi-empirical (*e.g.* AM1, PM3) charges yield optimized structures of the complexes very close to the experimental ones [19, 20] but have not been tested, to our knowledge, by FEP simulations in water.

2.2. MOLECULAR DYNAMICS (MD)

The host-guest complex is immersed in a box of explicitly represented solvent molecules, energy minimized, and its motions simulated by MD. Newton's equations are "solved" iteratively every 1 or 2 fs (10^{-15} s). Total simulation times were about 100 ps ten years ago, and are up to a few ns (10^{-9} s) presently. This is more than enough for solvent molecules to relax and properly solvate the solute or for the system to escape from metastable situations, but too short to observe slower processes like conformational interconversions of calix[4]arenes. MD is typically performed at 300K in (N,V,T) or (N, P, T) ensembles, where N is the number of particles in the simulation box (kept constant by using periodic boundary conditions), V is the volume and P is the pressure.

2.3. CALCULATION OF ION BINDING SELECTIVITIES BY FREE ENERGY PERTURBATION (FEP) SIMULATIONS

The generation of ensembles of configurations, coupled with statistical mechanics, enables calculation of the ion binding selectivity by a given ligand **L** [21]. Comparison of the binding of cations M_1^+ and M_2^+ by **L**, to form the **L**·M_1^+ and **L**·M_2^+ complexes in the "liquid" solvent is based on the thermodynamic cycle shown below. The selectivity, defined experimentally as $\Delta\Delta G = \Delta G_1 - \Delta G_2$, is equal to $\Delta G_{3\text{-liquid}} - \Delta G_{4\text{-liquid}}$, obtained by the "alchemical route". $\Delta G_{3\text{-liquid}}$ represents the difference in solvation free energies between the two ions, while $\Delta G_{4\text{-liquid}}$ is the difference in free energies between the corresponding complexes. The "alchemical route", where M_1^+ is mutated into M_2^+, leads to smaller statistical uncertainties than the "physical route" (direct calculation of ΔG_1 and ΔG_2).

$$\begin{array}{ccc}
 & \Delta G_1 & \\
M_1^+(X^-)_{\text{liquid}} + L_{\text{liquid}} & \xrightarrow{} & L \cdot M_1^+(X^-)_{\text{liquid}} \\
\Delta G_{3\text{-liquid}} \downarrow & & \downarrow \Delta G_{4\text{-liquid}} \\
M_2^+(X^-)_{\text{liquid}} + L_{\text{liquid}} & \xrightarrow{} & L \cdot M_2^+(X^-)_{\text{liquid}} \\
 & \Delta G_2 &
\end{array}$$

Intermediate steps ("windows") are used to allow for stepwise small perturbations between the initial and final states. They correspond to a hybrid potential energy $U_\lambda = \lambda U_1 + (1-\lambda)U_0$, where U_1 and U_0 are calculated with the corresponding R_1^*, ε_1 and R_0^*, ε_0 parameters of the ions and λ ranges from 0 to 1. At each window, the differences in free energy between the states λ and $\lambda + \Delta\lambda$ ("forward calculation") and the states λ and $\lambda - \Delta\lambda$ ("backward calculation") are obtained from:

$$\Delta G_{\lambda_i} = G_{\lambda_{i+1}} - G_{\lambda_i} = -RT \ln \left\langle \exp{-\frac{U_{\lambda_{i+1}} - U_{\lambda_i}}{RT}} \right\rangle_{\lambda_i}$$

where R is the molar gas constant and T is the absolute temperature. $\langle \ \rangle_{\lambda_i}$ stands for the ensemble average at the state λ_i where the potential energy is U_{λ_i}. At each window, a few picoseconds of equilibration are followed by data collection and averaging. Alternative integration schemes ("slow growth" and "thermodynamic integration") have been used in some cases [21].

3. Simulations of Cation Complexes of Calixarenes in Homogeneous Pure Liquid Solutions

The main questions adressed by modeling concern the nature of the complex (conformation of the ligand **L**, interactions of the cation with **L**, the solvent and counterions) and the binding selectivity. Typical examples, assuming $L \cdot M^{n+}$ species, are described in this section. Generally speaking, intrinsic cation-ligand interactions increase with the charge of the cation and, for a given charge, with the decreasing size of the ion: $Li^+ > Na^+ > K^+ > Rb^+ > Cs^+$ in the alkali metal series, $Mg^{2+} > Ca^{2+} > Sr^{2+} > Ba^{2+}$ in the alkaline earth metal series and $Yb^{2+} > Eu^{2+} > La^{2+}$ in the lanthanide series. Thus, binding selectivities mostly result from competitive solvent effects which determine the precise nature of the complexes, as well as from the "effective" ion-ligand interactions. In all cases, the structures are found to be dynamic in nature. For instance, the *cone* of calix[4]arene complexes, often of average C_{4v} symmetry, oscillates between C_{2v}-like forms, with frequencies depending on the guest and environment. Dynamic features, as revealed by computer graphics systems [22], are difficult to present in traditional publication formats. Representation via the Internet offers a promising alternative [23].

3.1. INFLUENCE OF SOLVENT ON THE CATION BINDING MODE BY CALIX[4]ARENE MONOANIONS

The *p-t*-butylcalix[4]arene oxyanion **L1**, locked in the *cone* conformation by hydrogen bonds at the narrow rim, extracts alkali metal cations from a basic aqueous phase into organic liquids. In analogy with the solid state structure of the $L1 \cdot Cs^+$ complex [24] it has been assumed that the M^+ cation sits inside the *cone* of **L1** (*endo* complex). However, *exo* complexation cannot be precluded, as shown by the solid state structure of the Na^+ complex [25]. Both types of complexes were simulated in three liquid environments [26,27] of increasing polarity (chloroform, acetonitrile and water) with Li^+, Na^+, K^+, and Cs^+ as potential guests. Only $L1 \cdot Li^+$ converged to a unique type of structure where Li^+ is slightly *exo* in the three solvents. In water, the systems differ markedly, as the Na^+, K^+ and Cs^+ *endo* and *exo* ions dissociate, leading to fully separated $L1 \cdot M^+$ ions pairs and filling of the *cone* of **L1** by 2 to 3 water molecules. In chloroform and acetonitrile, distinct *endo* and *exo* complexes of Na^+, K^+ and Cs^+ are observed (Figure 2). The interaction energy between **L1** and M^+ is somewhat more attractive for *endo* that for *exo* complexes (by 15-30 kJ/mol). On the other hand, *exo* complexes display better interactions with the solvent, due to the cation-solvent and **L1**-solvent interactions. Noteworthy is the stabilization brought about by solvent guest molecules (one chloroform, or two acetonitrile or two water molecules) in the *cone* of $L1 \cdot M^+$ *exo*. Acetonitrile stabilizes *exo* complexes (by -90 to -170 kJ/mol, compared to *endo*) more than chloroform and calculations predict that in weakly polar solvents like chloroform, *endo* complexation is preferred, while in more polar ones like acetonitrile or acetone, *exo* complexation is preferred. This is consistent with experimental trends. Polarization effects have also been investigated, and shown to further stabilize the *endo* complexes by about

50 kJ/mol. The calculated cation binding selectivities lead to a preference for the smallest ions in acetonitrile and dry chloroform ($Li^+ > Na^+ > K^+ > Rb^+ > Cs^+$), following the experimental trends.

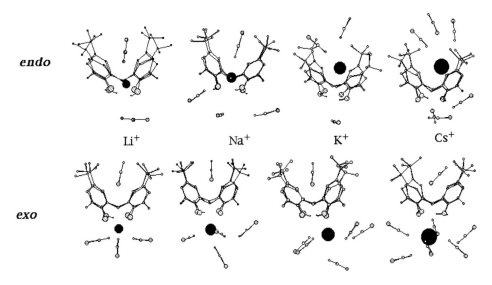

Figure 2. The $L1·M^+$ endo (top) and exo (bottom) complexes in acetonitrile.

There has been much interest in cation-π interactions as a possible source of stabilization of *endo* complexes. According to the force field results, there is no need to explicitly include these interactions to account for the *endo* nature of the Cs^+ complex. As cation-π interactions are larger for Na^+ than for Cs^+ [28], the observation of **L1**·Na^+ *exo* in the crystal indicates that the binding mode is not determined by these interactions. Noteworthy in this solid state structure is the additional coordination of polar solvent molecules to Na^+ *exo*.

The form of ammonium ion (NR_4^+) complexes of **L1** monoanion has also been investigated by simulation in the gas phase and in acetonitrile solution [29], where NH_4^+, NMe_4^+ and NEt_4^+ complexes have been compared. *Exo* complexation was predicted, mainly due to solvation effects. In contrast to alkali metal cations, *exo* NMe_4^+ and NEt_4^+ cations oscillate between positions involving binding to the oxygen atoms at narrow rim of **L1** and *exo* π-stacking with the phenyl rings.

3.2. COMPLEXES OF *t*-BUTYL-CALIX[4]TETRADIETHYL-AMIDE: INFLUENCE OF SOLVENT ENVIRONMENT AND CATION SIZE AND CHARGE

Na^+ is both efficiently bound and extracted by the *t*-butylcalix[4]arene-tetradiethyl-amide ligand **L3**, the complex adopting presumably the same structure as **L3**·K^+ in the solid state, where the cation is surrounded by four carbonyl and four ether oxygen atoms of the ligand in an encapsulating, fourfold-symmetrical arrangement. Similar binding modes are expected with other amide or ester derivatives.

Comparative MD simulations on **L3**·M$^+$ (M$^+$ = Li$^+$, Na$^+$, K$^+$, Cs$^+$) complexes in water [31] and in acetonitrile [30] demonstrated the effect of solvent on the ligand wrapping around the cation (Figure 3). For each amide arm, two typical orientations, respectively convergent and divergent, were considered. Convergent C=O dipoles (C=O$_{conv}$) delineate a polar pseudo-cavity suitable for cation complexation, while divergent carbonyls (C=O$_{div}$) cannot simultaneously bind a single cation.

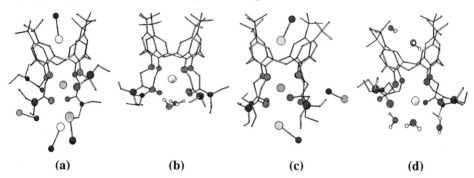

(a) (b) (c) (d)

Figure 3. *L3·K$^+$ complex simulated in acetonitrile (a),(c) and in water (b),(d) [30]. Coordination types are 4+0 (a) and (b), 3+1 (c) and 2+2 (d).*

In water, **L3**·K$^+$ rapidly exchanges between 4+0, 3+1 and 2+2 combinations of C=O$_{conv}$ and C=O$_{div}$. In acetonitrile, where solvent-amide interactions are weaker than in water, the most populated forms are of 3+1 type, instead of 4+0 as in the solid. Thus, weaker binding of K$^+$ in water, compared to methanol or acetonitrile solutions, does not solely result from the higher cation desolvation energy but also from a different structure where K$^+$ is more loosely bound by **L3**.

The effect of cation size has also been investigated. Li$^+$, too small to simultaneously bind to the eight oxygens of **L3**, rapidly exchanges in acetonitrile between carbonyl and phenolic oxygen sites. In water, the arms of **L3** are divergent, and it binds weakly to Li$^+$ along with water molecules. Conversely, Cs$^+$ is too large for the pseudo cavity of **L3**, which opens, allowing water to coordinate to the cation.

Simulation of the complexes **L3**·M^{2+} (Mg^{2+}, Ca^{2+}, Sr^{2+} and Ba^{2+}) in water and acetonitrile showed a greater preference for convergent forms than with **L3**·M$^+$, due to stronger cation-ligand attractions [32,33]. In the case of **L3**·Eu^{3+} [31], the ligand is not strong enough, however, to "solvate" the cation, which is captured by water. An interesting feature of **L3**·M^{n+} complexes is the solvent content of the *cone* of **L3**. With M$^+$ cations as guests, water or acetonitrile molecules fill the *cone*, adopting the same orientation as in **L3** uncomplexed, *i.e.* with the *cone* and solvent dipoles antiparallel. In complexes of M^{2+} or M^{3+}, the solvent molecules of the *cone* are "inverted" and oriented by the cationic charge.

The cation binding selectivity by **L3** in acetonitrile was calculated to be Li$^+$ > Na$^+$ > K$^+$ > Cs$^+$, as found by experiment [30,31]. Note that these FEP simulations, run for 55 to 105 ps were perhaps too short to correctly sample all possible conformers. In the alkaline-earth metal ion series, the calculated binding sequence in water was Ca^{2+} > Sr^{2+} > Ba^{2+} > Mg^{2+}, also in agreement with experiment [32]. As these complexes are more "convergent", the question of sampling is less crucial.

3.3. SOLVENT AND CONFORMATION EFFECTS ON THE Na^+/Cs^+ BINDING SELECTIVITY OF CALIXCROWN HOSTS

Simulations for the alkali metal cation complexes of calix[4]crown hosts **L4** and **L5** illustrate the effects of ligand conformation, solvent and counterion on the Na^+ vs Cs^+ binding selectivity. Early, purely theoretical studies [34,35] compared the *cone, partial cone* and *1,3-alternate* forms of dimethoxy substituted ligands **L5**, bearing four R = H vs *t*-butyl substituents, and crown-5 vs crown-6 bridging chains in the para positions. In all cases, gas phase interaction energies decreased in the order $Na^+ > K^+ > Rb^+ > Cs^+$. FEP simulations on *p-t*-butylcalix[4]crown-6 in the gas phase predicted a clear preference for Na^+ for all systems in the three conformations, decreasing in the order *cone > partial cone > 1,3-alternate*. This conclusion was assessed with four different sets of charges. As calix[4]crown-5 was somewhat too small for Cs^+, simulations in water focused on calix[4]crown-6 and revealed a marked solvent effect on the nature of the complexes. With R = *t*-butyl, all complexes remained of inclusive type during the dynamics (50 to 100 ps), but with R = H, the *1,3-alternate* Na^+ and *cone* Cs^+ systems dissociated, due to weaker cation-ligand and stronger cation-solvent interactions in these forms. Thus, in water, Cs^+ preferred the *1,3-alternate* host, while Na^+ preferred the *cone*. FEP simulations on the three forms of *t*-butycalix[4]crown-6 showed that the *cone* form prefers Na^+ (by about 25 kJ/mol), while the *1,3-alternate* form prefers Cs^+ (by about 40 kJ/mol) and the *partial cone* displays a weak preference for Na^+. Based on the fact that the OMe goups are not directly involved in the cation binding, it was predicted that other O-alkyl derivatives would behave similarly. For the *1,3-alternate* complexes, a preference for Cs^+ over Na^+ binding was also predicted in methanol and acetonitrile solutions, where these calixarenes are soluble. This contrasted with the gas phase or dry organic solutions, where Na^+ was predicted to be preferred by the three conformers of the ligand [34].

These predictions were confirmed by experiment and subsequent theoretical investigations. Comparison of different solvent models [34-36] confirmed the preference for Na^+ over Cs^+ in chloroform for all three conformers (decreasing in the order *cone > partial cone > 1,3-alternate*), with marked counterion effects at the quantitative level. Modelling of the effect of wetting the organic phase [36] revealed that a few added water molecules aggregate around the picrate (Pic^-) anion (Figure 4c and f) and play a crucial role on the binding selectivity of *1,3-alternate* calix[4]crown-6·M^+Pic^- complexes, which behave more as if they are in pure water than in dry chloroform solution, inverting therefore the preference for Cs^+.

The cation binding mode by this ligand was shown to be quite versatile (Figure 4). For Cs^+, which fits the host's cavity, the calculated structure is the same in water as in chloroform and in the solid state, with different relationships with the Pic^- counterions, however (Figure 4). In chloroform, as in the crystal, Pic^- forms a loose ion pair with complexed Cs^+. In water, there is a π-stacking interaction of Pic^- with one aromatic unit, while Cs^+ is weakly hydrated. This contrasts with Na^+ as guest. In dry conditions and without counterion, Na^+ sits deeply in the host, interacting with four phenoxide oxygens. Adding Pic^- to the system shifts Na^+ to the crown ether region, where it also binds to Pic^-. In aqueous solution and in wet chloroform, Na^+ co-complexes with one H_2O molecule, with two alternative modes: *endo* binding with H_2O in the crown or *exo* binding with H_2O fixed to the phenolic oxygens (Figure 4). Thus, as noted for cavitand systems [37] and cryptates [38], ions may adopt multiple binding modes involving co-complexation of solvent molecules. Similar features have been found more recently in the *1,3-alternate* calix[4]-*bis*crown **L4**·Na$^+$ system with Pic$^-$ [39,40] or NO$_3^-$ [41] as counterions.

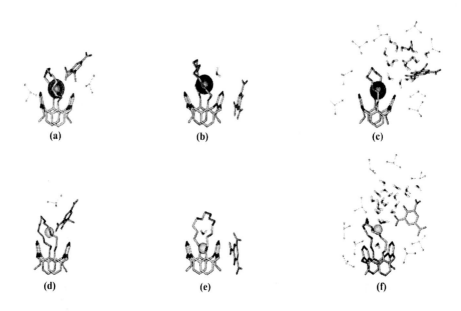

Figure 4. *L5a·Cs$^+$Pic$^-$ (top) and L5a·Na$^+$Pic$^-$ (bottom): in dry chloroform (a) and (d), in water (b) and (e), and in "wet" chloroform solution containing 20 H$_2$O molecules (c) and (f).*

MD studies on Na$^+$/Cs$^+$ complexes of calix[4]-*bis*crown-6 **L4** in water, methanol, chloroform and acetonitrile solutions, confirmed, based on FEP simulations, the preference for Cs$^+$ and the role of counterions [39,40]. Other simulations [42] involving the di-*iso*propoxy analogue of **L5a**, where two OMe are replaced by O-*iso*propoxy groups, and *bis*(1,3-crown ether) derivatives of calix[4]arene, and using different electrostatic representations of the ligand, also indicated a better fit for Cs$^+$ than for Na$^+$. Other forms of the crown ether moiety such as benzo- [42], naphtho- [43] and dibenzo- [44] crowns of various lengths have been considered in relation to "finetuning" of Cs$^+$/Na$^+$ selectivity [45, 46].

3.4. AMMONIUM ION COMPLEXES OF CALIX[6]CROWN-6 IN CHLOROFORM

Complexation of NH$_4^+$ and NMe$_4^+$ ammonium cations by *t*-butyl-calix[6]arene-crown-6 (**L7**) has been simulated in chloroform, assuming a *cone* conformation for the ligand [47]. Several binding modes, involving the *cone* moiety or the crown moiety of **L7** were compared without counterion, and with CH$_3$CO$_2^-$ as counterion. The binding mode turned out to be quite versatile, and markedly determined by the counterion, which remained in contact with the complexed NR$_4^+$ guest.

3.5. ALKALI METAL CATION BINDING TO CALIXSPHERANDS

Calixspherand **L6**, with a calix[4]arene platform bridged by anisole units, belongs to a class of highly preorganized ligands which may form kinetically stable complexes with alkali metal cations. Gas phase modelling indicated that the complexes adopt a "flattened" *cone* conformation [48]. Comparison of the Na^+, K^+ and Rb^+ complexes showed that larger cations force the bridge to bend away, making the cation more accessible to solvent molecules, and presumably making the Rb^+ complex kinetically more labile than the Na^+ and K^+ complexes.

In water, FEP calculations on the absolute and relative free energies of Na^+ *vs* K^+ binding correctly reproduced the tighter binding of K^+ [49]. An elegant analysis of energy components and structural features showed how the delicate balance between ion-water and ion-host interactions leads to the observed selectivity.

3.6. LANTHANIDE COMPLEXES OF CALIX[4]ARENE-CMPO IN SOLUTION

Calix[4]arenes substituted at the wide rim by CMPO arms display remarkable extraction properties toward trivalent lanthanide and actinide cations [50, 51]. These systems possess flexible arms, each able to achieve "converging" orientations and bidentate coordination of the cation. No structural information is available on these systems. Molecular modelling investigations were performed on the **L8**·M^{3+} complexes (M = La, Eu and Yb) in the presence of NO_3^- counterions, with the aim of determining whether the four arms could simultaneously bind to M^{3+}, and whether M^{3+} would be sufficiently shielded from the solvent to be extracted to an organic phase [52]. Several types of structures, where the NO_3^- anions were either absent, in close contact with **L8**, remote from **L8**, or partially coordinated to M^{3+} were analysed. In water, methanol or acetonitrile solutions, after relatively long simulations (about 1 ns each), no convergence to a single type of complex was achieved, probably because the energy barriers between different conformers, higher than those with M^+ or M^{2+} guests, were too high to be overcome at 300 K within one nanosecond.

These studies make clear that counterions and solvation effects markedly influence the extent of ligand wrapping around trivalent cations, as supported by NMR investigations [53]. This is, in addition to problems of energy representation, a specific feature of trivalent ion binding by neutral hosts. As suggested by X-ray structures [54], NMR [53] and SANS [55] experiments on related systems, peculiar binding modes and stoichiometries different from 1:1 with formation of aggregates or polymeric species are likely involved in the coordination of M^{3+} cations by phosphoryl derivatives of calixarenes.

3.7. URANYL ION BINDING BY CALIX[n]ARENE^{n-} ANIONS IN WATER (n = 5 AND 6)

Complexation of the linear UO_2^{2+} and spherical Cu^{2+} cations by the calix[5]arene pentaanion **L2a** and calix[6]arene hexaanion **L2b** (R = Me) has been simulated in water [56] in relation to understanding the very large differences in the stability constants of the complexes of the two metals with the corresponding *para*-sulfonato derivatives. It was assumed that the cation binds to the phenoxide, rather than the sulfonate oxygens, which were replaced by methyl groups in the simulations. A first MD study [56] used "hand made" parameters for UO_2^{2+}, while a more recent one [57] used parameters based on the UO_2^{2+} vs Sr^{2+} free energies of hydration. Both ligands were shown to coordinate in the equatorial plane of UO_2^{2+}, preventing coordination of water to the U atom (Figure 5). For Cu^{2+}, however, retention of some coordinated water was possible. It was proposed that the entropy gain resulting from the dehydration of the cations and of the calixarene anions brings an important contribution to the free energy of complexation in water. The FEP comparison of UO_2^{2+} and Sr^{2+} binding to **L2b** predicted the larger stability of the former complexes [57]. This contrasts with the UO_2^{2+} vs Sr^{2+} complexes of the calix[4]amide **L3**, where FEP predicted a preference for Sr^{2+} [32].

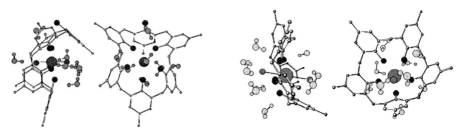

Figure 5. The UO_2^{2+} complexes of the calix[6]arene^{6-} (left) and of calix[5]arene^{5-} (right) polyanions. Orthogonal views.

UO_2^{2+} complexed by preorganized macrocycles may selectively bind additional ligands. This question has been assessed by Monte Carlo simulations which reproduced the binding of ammonia and *n*-propylamine quite well [58].

3.8. LUMINESCENT COMPLEXES OF CALIX[4]ARENE PODANDS CONTAINING 2,2'-BIPYRIDINE MOIETIES

To assess the consequences of introducing bipyridine units through different points of functionalisation, the Eu^{3+} complexes of calix[4]arene-bipyridine ligands **L9** and **L10** and analogues were simulated in water and acetonitrile [59,60]. The MD simulations account for the greater stability and stronger luminescence of **L9**, compared to **L10**, as due to a better complementarity between Eu^{3+} and the bipyridines, and cation shielding from solvent and NO_3^- counterions in **L9**. Only in the **L9** complex can Eu^{3+} bind to the eight nitrogen atoms of the ligand. Similar conclusions have been obtained by Monte Carlo simulations on these complexes in acetonitrile solution [61]. The importance of the calixarene platform for preorganizing the bipyridine sites has also been analyzed by MD simulations where this platform was removed [59]. In relation to the fluorescence quenching by water, it is worth mentioning the Monte Carlo investigation of a terphenyl-based Eu^{3+} complex in methanol, to which small amounts of water were added [62].

MD simulations on calix[4]diquinones offered an explanation of why they function as sensitive photoionic receptors [63, 64]. Light absorbed by the $[Ru(bpy)_3]^{2+}$ fragment(s) covalently linked to a calix[4]arene platform can be quenched via light induced electron transfer to the nearby quinones. Complexation of cations like Ba^{2+} at the lower rim of the calixarene restores the luminescence. MD simulations on **L11** uncomplexed and on its Ba^{2+} complex, as well as on the corresponding systems with two opposite

[Ru(bpy)$_3$]$^{2+}$ fragments revealed that, in the absence of Ba^{2+}, pyridines and quinones display conformational interconversions between *cone, partial cone* and *1,3-alternate* forms, involving transient structures with π–π orbital overlap suitable for energy transfer. In the Ba^{2+} complexes, such orbital overlap and electron transfer are prevented by restricted conformational mobility of the quinones and pendent bipyridine arms.

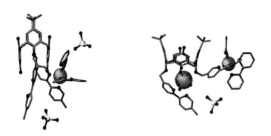

*Figure 6. The **L11** ligand uncomplexed (left) and complexed with Ba^{2+} in acetonitrile (right).*

3.9. ANION COMPLEXATION IN CHLOROFORM SOLUTION

Though most simulations have centered on cation complexes of calixarenes, anion coordination has also been considered. The selective halide anion, X$^-$, complexation by a bis(phenyl)urea *t*-butylcalix[4]arene in the gas phase and in chloroform solution, where X$^-$ forms strong hydrogen bonds with four N-H protons, has been studied by Monte Carlo simulations [65]. The observed affinity order (Cl$^-$ > Br$^-$ > I$^-$) was quantitatively reproduced by FEP simulations. Some apparent conflict between calculated and experimental results concerned F$^-$, which was predicted to bind with by far the greatest affinity to the calixarene, while no F$^-$ complexation was observed experimentally. The difference was attributed to the possible influence of water. According to the calculations, complexation of F$^-$ by two water molecules is sufficient to overcome complexation by the host. Thus, as simulated for alkali metal cation complexes [36], small amounts of water in an apolar liquid may be sufficient to inverse binding selectivities.

4. Computing Selectivities in Liquid-Liquid Extraction: Counterion and Solvent Effects

Combining FEP mutations of cation uncomplexed in the source phase (water) and complexed in the receiving (organic) phase allows calculation of the extraction selectivities in solvent extraction, based on the thermodynamic cycle below:

$$\Delta G_{extr-1} - \Delta G_{extr-2} = \Delta G_{3-aq} - \Delta G_{4-org}.$$

Applications deal with alkali metal cation extraction into chloroform by calix[4]arenes, and the preference for Cs$^+$ *vs* Na$^+$ by the negatively charged **L1** ligand, [27] the neutral calix-crown **L5** [36, 66] and calix-*bis*-crown **L4** ligands [40, 67].

$$M_1^+(X^-)_{aq} + L_{org} \xrightarrow{\Delta G_{extr-1}} L \cdot M_1^+(X^-)_{org}$$

$$\Delta G_{3\text{-aq}} \downarrow \qquad\qquad \downarrow \Delta G_{4\text{-org}}$$

$$M_2^+(X^-)_{aq} + L_{org} \xrightarrow{\Delta G_{extr-2}} L \cdot M_2^+(X^-)_{org}$$

For the three types of ligands, the calculations correctly reproduced the extraction sequence, but quantitative agreement with experiment was not achieved, due to approximations in the potential energy, to sampling limitations and simplification of the system. Traces of water and counterions in the organic phase also may have dramatic effects and simulations [36] have shown that *the binding selectivity in wet chloroform is similar to that in pure water, and opposite to the selectivity in dry chloroform*. Recently, binding selectivities have been calculated at the water - organic interface (*vide infra*), where the cationic moiety of the complex is also partially hydrated [68].

5. Adsorption of Calixarenes at the Liquid-Liquid Interfaces. From Monomeric Species to Supramolecular Assemblies

The precise mechanism of ion capture by extractants in liquid-liquid extraction (LLE) systems is unknown from experiment, but new insights emerged from MD simulations at a liquid-liquid interface, selecting calixarene systems known from experiment to extract or transport cations from water to an organic phase. Questions concern the size and nature of the interface, the extent of solvent mixing in this peculiar region, the role of counterions, the distribution of solute species and "what happens at the interface" in LLE [69].

5.1. "COMPUTER DISCOVERY" OF THE SURFACE ACTIVITY OF EXTRACTANT MOLECULES

We conducted MD experiments where different solutes were initially placed at the interface between a box of water molecules and a box of organic solvent (chloroform) molecules. Some solutes migrated to the solvent where they are more soluble (*e.g.* Cs^+ ions to the bulk water phase and *n*-butane to the organic phase). Calixarenes used in extraction systems are soluble in chloroform and quasi-insoluble in water. Surprisingly, instead of diffusing into bulk chloroform during the dynamics, they remained "adsorbed" at the interface, mostly on the chloroform side, like surfactants. Adsorption was also found computationally with extractant molecules like crown ethers [70], alkylphosphates [71] and calixarene derivatives such as **L1** and **L3** to **L5**, in their uncomplexed, as well in their complexed states [34,40,66,72]. At first sight, this seemed inconsistent with the solubility of these species in the organic solvents.

Methodological tests confirmed that adsorption does not artefactually result from the starting situation. For instance, **L5a**, initially positioned in the organic phase, within the cutoff distance from the interface, moved back to the interface during the simulation, even when substituted by lipophilic O-alkyl substituents [66]. Increasing the polarity of chloroform or using different simulation conditions did not lead to spontaneous extraction of the complex [66].

From the simulations, it became clear that what was being observed was that extractant molecules really do adsorb at the interface. This was rationalized on two main bases. First, all extractant molecules display specific interactions with water. This is observed for symmetrical ones like 18-crown-6 [73] or cryptands [74] and, *a fortiori*, for calixarenes of amphiphilic topology. At the interface, they enjoy attractive interactions with water, due to specific hydration patterns (*e.g.* hydrogen bonds to the crown ether moiety of calixcrowns, to carbonyl or phosphoryl groups of **L3** or **L8**, or phenoxy oxygen of **L1**). Another anchoring mode was observed involving water molecules which fill the *cone* of the calixarene [72]. In some cases, the hydrogen bonded water molecules were sitting right at the interface in contact with the bulk. In other cases, they built a relay ("water finger" [75,76]) between the extractant and aqueous phase. Polar interactions with water are more favorable than van der Waals interactions with the organic molecules. Voluminous solutes do not move to bulk water because of the high corresponding "cavitation energy" (energy price to create a cavity in the solvent, related to its surface tension [77]). As a result, extractant molecules and their complexes are "expelled" from water to the water surface. This is not proven by direct experiments, but is consistent with the formation of calixarene monolayers at the water-air interface, where selective uptake of cations from water has been observed [78-81]. Surface activity is required in phase-transfer catalysed reactions often used in calixarene synthesis [82].

Complexed calixarenes are more surface active than the uncomplexed ones, due to their increased amphiphilic character. Indeed, all complexes simulated at the interface remained there, instead of spontaneously diffusing to chloroform, as anticipated from the experimental results. When initially placed in the organic phase, they move back to the interface during the simulation [66, 83].

5.2. INSIGHTS INTO THE FREE ENERGY PROFILE FOR INTERFACE CROSSING

FEP simulations have been used to calculate the change in free energy (ΔG) resulting from a z-translation away from the interface for typical species involved in the Cs^+ extraction by **L5a** [84]. Moving the free or complexed calixarene 10 Å from the interface to water is indeed a high energy process, while moving to chloroform is somewhat more easy (Figure 7). An energy minimum is observed on the chloroform side, at a few Ångströms from the interface, for uncomplexed **L5a** and for its Cs^+ complex. This minimum was confirmed by methodological tests involving enhanced sampling of the system and increased cutoff distances. Because of computer time limitations, the displacement remained too short, however, to determine whether further migration to chloroform corresponds to an energy plateau, or if it involves an energy barrier followed by a downhill process.

Displacement to both directions also revealed a high affinity of the Pic^- anion, widely used in LLE experiments, for the interface. Thus, *adsorption of ions at the interface does not require amphiphilic topologies*, as confirmed by other computations on this ion, guanidinium$^+$, [70, 85-88] symmetrical ions like BPh_4^-, $AsPh_4^+$ or large spherical ions [88], dicarbollides [89] and ClO_4^- [86].

The free energy profile for the calixarene system contrasts with the profile for uncomplexed Cs^+, which rises steeply towards the chloroform side and slightly decreases towards the water side. Again, counterion effects were demonstrated by comparing Pic^- vs Cl^- as counterions [84].

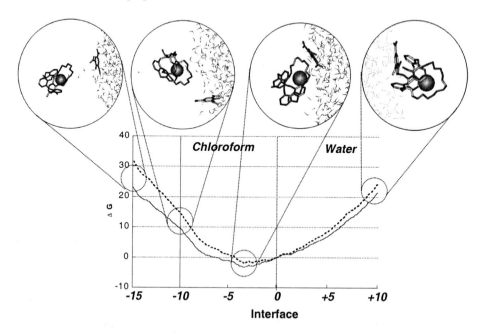

Figure 7. Change in free energy ΔG (kcal/mol) as the $L5a \cdot Cs^+ Pic^-$ complex is moved from the interface to the water (right) and chloroform (left) phases.

5.3. SIMULATIONS OF SOLVENT DEMIXING

MD simulation of demixing confirms the interfacial activity of calixarenes [68]. Completely mixed water/chloroform solutions of uncomplexed **L5a** and of its Cs^+Pic^- complex were prepared and their time evolution simulated by MD. For comparison, neat mixtures and binary mixtures containing salts alone were also considered. The separation of the two liquids was found to be very rapid (about 0.5 ns), leading to completely separated liquid phases, as in simulations which started with adjacent boxes of the liquids. The kinetics of demixing was of exponential type, similar for all studied systems, mostly driven by the solvent-solvent cohesive forces.

An example is given in Figure 8, for a solution containing species involved in the Cs^+ extraction by **L5a**: two uncomplexed ligands, one Cs^+ complex, four Cs^+Pic^- and four $Cs^+NO_3^-$ ion pairs. In less than 0.2 ns, water aggregates around the hydrophilic Cs^+ and Cl^- species, while the calixarenes and Pic^- anions sit at the border which gradually forms between the two liquids. After 0.5 ns the two liquids are almost separated, and the Cs^+ complex is adsorbed at the interface. Adsorption of the uncomplexed calixarenes, a slower process, is complete after 1 ns. One calixarene is anchored by its crown ether and the other by the calixarene scaffold, via water fingers.

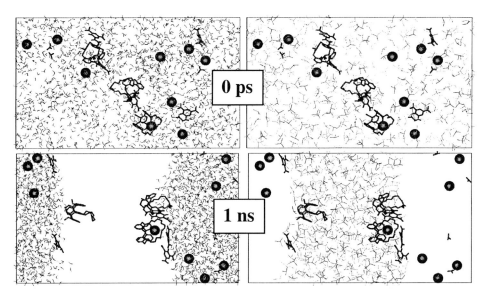

*Figure 8. Binary water / chloroform mixture containing 2 **L5a**, 1 **L5a**•Cs^+ Pic^-, 4 Cs^+ NO_3^- et 4 Cs^+ Pic^-. The water (left) and chloroform (right) liquids are shown side by side, instead of superposed, for clarity. Completely mixed liquids (0 ps) and separated phases (1 ns) [68].*

All Pic^- anions sit at the interface, some of them attracting Cs^+ cations close to it. We suggested that *the negative potential which results from the interfacial anions is a key feature which facilitates the approach of cation close enough to be interface to be captured* [70,85,86,88]. Similar features were obtained in independent demixing simulations on binary solutions of the uncomplexed salts and of the calixarenes. Thus, contrary to expectations from macroscopic data, uncomplexed calixarenes, their complexes and Pic^- anions sit at the interface, rather than in the organic phase. Similar conclusions have been drawn from demixing simulations with TBP (tri-*n*-butyl phosphate) uncomplexed, or complexed with $UO_2(NO_3)_2$ salts [87]. TBP, an amphiphilic molecule, acts simultaneously as (co)-solvent, surface active molecule and ligand for lanthanide or actinide cations.

Concerning the mechanism of ion extraction by calixarenes, the MD results demonstrate that *water and "oil" are never completely mixed at the microscopic level*. They are separated by an interface, where extractant molecules adsorb. Lipophilic anions used in LLE also concentrate at the interface and attract cations close enough to the interfacial ligands to be captured. Thus, *ion capture and "recognition" take place at the interface*. Interfacial recognition is supported by FEP simulations on the Na^+/Cs^+ binding selectivity by **L5a** at the interface, which showed that Cs^+ is preferred. The role of interfacial Pic^- anion was also demonstrated by *ad hoc* simulations [68].

5.4. SIMULATION OF CONCENTRATION EFFECTS : MOLECULAR ASSEMBLIES OF CALIXARENES AT THE LIQUID-LIQUID INTERFACE

Though simulations with dilute solutions (about 0.05 mol/l) show that calixarenes concentrate at the interface, the nature of the resulting assemblies is unknown. Do they form a more or less saturated monolayer ? What happens beyond this layer ? Is there some gradient of concentration ? Since uncomplexed and complexed ligands are both surface active, what is the nature of the interface with mixed species ? What are the driving forces for diffusing from the interface to the organic phase in LLE ?

Recent simulations on calixarenes and other extractant systems demonstrate the *importance of aggregation effects* at the interface. For instance, the monomeric 18-crown-6·K^+Pic^- complex was found to stick at the interface for 1 ns of MD simulation [70] while simulation of a more concentrated system led to an equilibrium involving decomplexation of some K^+ complexes and the diffusion of one K^+ complex to the organic phase, driven by a Pic^- counterion. Thus, *accumulation of extractants at the interface modifies (reduces) the surface tension, facilitating desorption of complexes from the interface*. Another driving force results from the accumulation and mutual repulsion of anions at the interface, pushing some of them to the organic phase, followed by one K^+ complex. According to MD simulations [83,85-87], calixarenes are more surface active than crown ethers.

Several situations related to the Cs^+ extraction by **L5a** are shown in Figure 9 a-f,

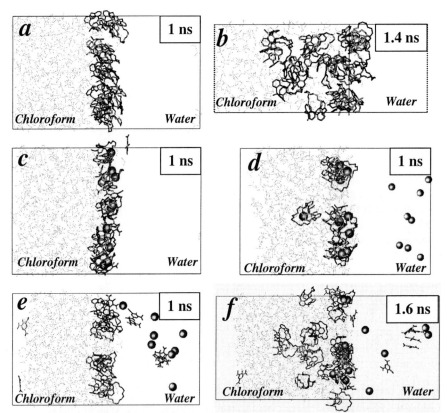

Figure 9. *The water- chloroform interface (water molecules not shown), at the end of MD simulations a-f. **a**: monolayer of 15 **L5a**; **b**: two layers of 9 **L5a** each; **c**: one layer of **L5a**·Cs^+Pic^- complexes; **d**: one layer of **L5a**·Cs^+Cl^- complexes; **e**: one layer of **L5a** ligands + 9 uncomplexed Cs^+Pic^- ions; **f**: one layer of 9 **L5a**·Cs^+Pic^- complexes at the interface + 9 **L5a** ligands in chloroform + 5 Cs^+ Pic^- ions in water [83].*

after 1 ns or more of dynamics. (a) and (b) correspond to uncomplexed **L5a**, forming initially one layer of 15 molecules (a) and a bilayer of 9+9 molecules (b). In no case is the interface saturated at the end of the dynamics, as shown by the water surface in Figure 10. The interface is flat on the average, of about 10 Å width, but locally rough and displays troughs and mounds onto which calixarenes are anchored. Like wavelets, these surface heterogeneities dynamically exchange in a few picoseconds, likely facilitating the ion uptake process. Simulation (b) shows that there is no strong driving force for molecules to diffuse from the second layer to the unsaturated first layer. They form irregular arrangements, close to the first layer. Some of them drag water molecules into the organic phase, which would otherwise be dry.

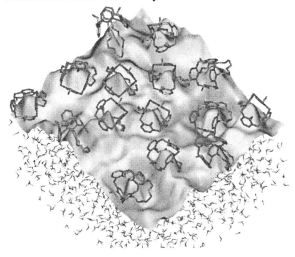

*Figure 10. Surface of water at the chloroform-water interface where 15 **L5a** molecules are adsorbed. Chloroform (top side) not shown.*

Simulations (c) and (d) demonstrate the *role of counterions* at the interface. Both started with a monolayer of 9 **L5a**·Cs^+ complexes, with Pic^- and Cl^- as counterions, respectively, but displayed contrasting behaviour. In (c) the Pic^- anions remained at the interface where electroneutrality was achieved, while in (d) hydrophilic Cl^- anions were captured by water, creating a marked interfacial potential. The accumulation of positively charged complexes at the interface pushes one of them into the organic phase. No such diffusion is observed with a single complex, again demonstrating the *importance of collective effects*.

Simulations (e) and (f) deal with alternative key situations in LLE. In (e) the interfacial layer of calixarene is contacted with the aqueous solution of Cs^+Pic^- ions to be extracted. Although complexation does not spontaneously take place during the dynamics, a remarkable situation is observed, where one Cs^+ cation has been attracted by the interfacial anions close to calixarene ligands. This would lead, via a least motion pathway, to complexation. Similar situations have been characterized for the 18-crown-6· K^+Pic^- system [70]. In system (f) an interfacial layer of 9 **L5a**·Cs^+Pic^- complexes is initially surrounded by 8 uncomplexed ligands in chloroform and by 5 Cs^+Pic^- ions in water. One ligand moved during the dynamics to the interface, which remained unsaturated, while one Cs^+ decomplexed and migrated to water. These simulations remain too short to converge to similar situations and to show spontaneous cation complexation. Taken together, they demonstrate the importance of interfacial phenomena in LLE assisted by calixarenes.

5.5. INTERFACIAL ASPECTS IN THE EXTRACTION OF DIVALENT AND TRIVALENT CATIONS

As the cation charge increases, it is more strongly "repulsed by the interface", which prevents its capture by interfacial ligands. Again, neutralizing or synergistic counterions are expected to play a major role. Simulation of the Sr^{2+} complex of the calix[4]arene tetraamide **L3**, and of the more hydrophilic **L3'** analogue where two opposite $CONEt_2$ arms are replaced by $CONH_2$ [33] showed the latter complex to be more surface active, due to the reduced shielding and better hydration of complexed Sr^{2+}, rather than to specific hydration of primary amides (Figure 11).

*Figure 11. The **L3**·$Sr(Pic)_2$ (left) and **L3'**·$Sr(Pic)_2$ (right) complexes at the water-choroform interface. The interface is displayed horizontally, with water on top and chloroform below (not shown).*

Europium complexes of calixarenes have also been calculated to be surface active, as in the study of the $Eu(NO_3)_3$ complex of the CMPO-calixarene **L8** [52]. Here, it seems that the interfacial activity is modulated by the size and nature of the O-alkyl substituents at the narrow rim. The preorganizing role of the calixarene platform at the interface was illustrated by comparison of 4 CMPOs (no calixarene) and the CMPO-calixarene. The former were found to dilute at the interface, which is unfavorable for ion capture, in contrast to the calixarene-anchored CMPOs. The high surface activity of trivalent cation complexes prevents their extraction in the absence of other features (salting out and acidity or synergistic effects). These effects are presently under study in related systems [90].

6. Conclusion

Cation binding by calixarenes represents a particularly rich and stimulating field for computer simulations, extending from selective cation complexation in pure liquid phases to heterogeneous liquids and liquid-liquid extraction systems. It illustrates the versatile character of cation recognition where solvent environment, counterions, and dynamics features markedly determine the complementarity between the cation receptor and its ionic guest.

Another facet has recently emerged from simulations at liquid-liquid interfaces: calixarenes display in their uncomplexed and complexed states remarkable surface activity, a feature which has important consequences concerning the mechanism of ion capture and extraction. Microscopic interfaces such as those we have simulated may correspond to the liquids at rest, as well as to the surface of droplets in metastable microemulsions. The simulated surface activity also likely relates to the formation of aggregates, micelles and supramolecular arrangements in heterogeneous liquid phases, including supercritical fluids [91], which will be studied by computations and experiments in the near future.

On the computational side, words of caution are necessary concerning all assumptions and limitations inherent in any implementation of theoretical approaches. Most of the reported studies aim at exploring the limitations and possibilities of current methodologies, and to simultaneously gain insights into structural and energy features of complex systems. As discussed in most cited references, key questions concern the choice and simplification of physical models, the energy representation of the system and the sampling of its representative states. Rather than answering difficult questions purely computationally, MD simulations provide (hopefully) fertilizing pictures whose status ranges from microscopic representations of "reality" to mere heuristic views which support the understanding and interpretation of experimental (mostly thermodynamic and structural) data. Like traditional models in chemistry, computer simulations stimulate reasoning, interpretation and prediction.

Acknowledgements: The author is grateful to his colleagues and students, cited as co-authors in the references, and to Dr. M. Baaden for technical assistance. Calculations used the computer resources from Université Louis Pasteur and CNRS IDRIS. Much of this work has been stimulated by EEC collaborative projects on nuclear waste management (coordinated by Drs M. Hugon and J.-F. Dozol) and by PRACTIS in France.

7. References and Notes

[1] M.A. McKervey, M.-J. Schwing, F. Arnaud-Neu in *Comprehensive Supramolecular Chemistry* , **1996**, J. L. Atwood, J. E. D. Davies, D. D. McNicol, F. Vögtle and J.-M. Lehn Eds., Pergamon, New York pp 537-603.
[2] V. Böhmer, *Angew. Chem., Int. Ed. Engl.* **1995**, *34,* 713-745.
[3] G. D. Andreetti, F. Ugozzoli, R. Ungaro, A. Pochini in *Inclusion Compounds* , **1991**, J. L. Atwood, J. E. D. Davies and D. D. McNicol Eds., Oxford University Press, Oxford pp 65.
[4] R. Ungaro, A. Arduini, A. Casnati, O. Ori, A. Pochini, F. Ugozzoli in *Computational Approaches in Supramolecular Chemistry* , **1994**, G. Wipff Ed., Kluwer, Dordrecht pp 277-300.
[5] J.L. Atwood in *Cation Binding by Macrocycles* , **1991**, Y. Inoue and G. Gokel Eds., M. Dekker, Inc., New York and Basel pp 581-597.
[6] M. P. Allen, D. J. Tildesley *Computer Simulation of Liquids.*, **1987**, W. F. van Gunsteren and P. K. Weiner Ed., Clarendon press: Oxford.
[7] W. F. van Gunsteren, P. K. Weiner, A. J. Wilkinson *Computer simulations of biomolecular systems. Theoretical and experimental appplications. Vol. 2,* **1993**, ESCOM: Leiden.
[8] D. A. Case, D. A. Pearlman, J. C. Caldwell, T. E. Cheatham III, W. S. Ross, C. L. Simmerling, T. A. Darden, K. M. Merz, R. V. Stanton, A. L. Cheng, J. J. Vincent, M. Crowley, D. M. Ferguson, R. J. Radmer, G. L. Seibel, U. C. Singh, P. K. Weiner, P. A. Kollman, *AMBER5, University of California, San Francisco* **1997**.

[9] B. R. Brooks, R. E. Bruccoleri, B. D. Olafson, D. J. States, S. Swaminathan, M. Karplus, *J. Comput. Chem.* **1983**, *4,* 187-217.
[10] W. L. Jorgensen, *BOSS, Version 3.8, Yale University: New Haven, CT.* **1997**.
[11] P. Comba, *Coordin. Chem. Rev.* **1999**, *185-186,* 81-98.
[12] B. P. Hay, *Coordin. Chem. Rev.* **1993**, *126,* 177-236 and references cited therein.
[13] F. Berny, N. Muzet, L. Troxler, A. Dedieu, G. Wipff, *Inorg. Chem.* **1999**, *38,* 1244-1252.
[14] W. L. Jorgensen, J. Chandrasekhar, J. D. Madura, *J. Chem. Phys.* **1983**, *79,* 926-936.
[15] W. L. Jorgensen, J. M. Briggs, M. L. Contreras, *J. Phys. Chem.* **1990**, *94,* 1683-1686.
[16] J. Åqvist, *J. Phys. Chem.* **1990**, *94,* 8021-8024.
[17] a) F. C. J. M. van Veggel, D. Reinhoudt, *Chem. Eur. J.* **1999**, *5,* 90-95; b) S. Durand, J.-P. Dognon, P. Guilbaud, C. Rabbe, G. Wipff, *J. Chem. Soc., Perkin Trans. 2* **2000**, 705-714.
[18] W. D. Cornell, P. Cieplak, C. I. Bayly, P. A. Kollman, *J. Am. Chem. Soc.* **1993**, *115,* 9620-9631.
[19] P. Kane, D. Fayne, D. Diamond, S. E. J. Bell, M. A. McKervey, *J. Mol. Model.* **1998**, *4,* 259-267.
[20] S. E. J. Bell, M. A. McKervey, D. Fayne, P. Kane, D. Diamond, *J. Mol. Model.* **1998**, *4,* 44-52.
[21] P. Kollman, *Chem. Rev.* **1993**, *93,* 2395-2417.
[22] E. Engler, G. Wipff in *Crystallography of Supramolecular Compounds,* **1996**, G. Tsoucaris Ed., Kluwer, Dordrecht pp 471-476.
[23] A. Varnek, B. Dietrich, G. Wipff, J.-M. Lehn, E. V. Boldyreva, *J. Chem. Educ.* **2000**, *77,* 222-226.
[24] J. M. Harrowfield, M. I. Ogden, W. R. Richmond, A. H. White, *J. Chem. Soc., Chem. Commun.* **1991**, 1159-1161.
[25] F. Hamada, K. D. Robinson, G. W. Orr, J. L. Atwood, *Supramol. Chem.* **1993**, *2,* 19-24.
[26] R. Abidi, M. V. Baker, J. M. Harrowfield, D. S.-C. Ho, W. R. Richmond, B. W. Skelton, A. H. White, A. Varnek, G. Wipff, *Inorg.Chim. Acta* **1996**, *246,* 275-286.
[27] A. Varnek, C. Sirlin, G. Wipff in *Crystallography of Supramolecular Compounds,* **1995**, G. Tsoucaris Ed., Kluwer, Dordrecht pp 67-99.
[28] J. C. Ma, D. A. Dougherty, *Chem. Rev.* **1997**, *97,* 1303-1324.
[29] F. Fraternali, G. Wipff, *J. Incl. Phenom.* **1997**, *28,* 63-78
[30] A. Varnek, G. Wipff, *J. Phys. Chem.* **1993**, *97,* 10840-10848.
[31] P. Guilbaud, A. Varnek, G. Wipff, *J. Am. Chem. Soc.* **1993**, *115,* 8298-8213.
[32] N. Muzet, G. Wipff, A. Casnati, L. Domiano, R. Ungaro, F. Ugozzoli *J. Chem. Soc., Perkin Trans. 2* **1996**, 1065-1075.
[33] F. Arnaud-Neu, S. Barboso, F. Berny, A. Casnati, N. Muzet, A. Pinalli, R. Ungaro, M. J. Schwing-Weil, G. Wipff, *J. Chem. Soc., Perkin Trans. 2* **1999**, 1727-1738.
[34] G. Wipff, M. Lauterbach, *Supramol. Chem.* **1995**, *6,* 187-207.
[35] M. Lauterbach, Thesis, Université Louis Pasteur Strasbourg, **1997**.
[36] M. Lauterbach, A. Mark, W. van Gunsteren, G. Wipff, *Gazz. Chim. Ital.* **1997**, *127,* 699-709.
[37] P. A. Kollman, C. I. Bayly, *J. Am. Chem. Soc.* **1994**, *116,* 697-703.
[38] G. Wipff, L. Troxler in *Computational Approaches in Supramolecular Chemistry* **1994**, G. Wipff Ed., Kluwer, Dordrecht pp 319-348.
[39] A. Varnek, G. Wipff, *J. Mol. Struct. (THEOCHEM)* **1996**, *363,* 67-85.
[40] A. Varnek, G. Wipff, *J. Comput. Chem.* **1996**, *17,* 1520-1531.
[41] P. Thuéry, M. Nierlich, V. Lamare, J.-F. Dozol, Z. Asfari, J. Vicens, *Supramol. Chem.* **1997**, *8,* 319-332.
[42] V. Lamare, C. Bressot, J.-F. Dozol, J. Vicens, Z. Asfari, R. Ungaro, A. Casnati, *Sep. Sci. Technol.* **1997**, *32,* 175-191.
[43] P. Thuéry, M. Nierlich, J. C. Bryan, V. Lamare, J.-F. Dozol, Z. Asfari, J. Vicens, *J. Chem. Soc., Dalton Trans.* **1997**, 4191-4202.
[44] V. Lamare, J.-F. Dozol, F. Ugozzoli, A. Casnati, R. Ungaro, *Eur. J. Org. Chem.* **1998**, 1559-1568.
[45] V. Lamare, J.-F. Dozol, S. Fuangswasdi, F. Arnaud-Neu, P. Thuéry, M. Nierlich, Z. Asfari, J. Vicens, *J. Chem. Soc., Perkin Trans. 2* **1999**, 271-284.
[46] P. Thuéry, M. Nierlich, V. Lamare, J.-F. Dozol, Z. Asfari, J. Vicens, *J. Incl. Phenom.* **2000**, *36,* 375-408.
[47] F. Fraternali, G. Wipff, *An. Quim., Int. Ed.* **1997**, *93,* 376-384.
[48] L. C. Groenen, J. A. J. Brunink, W. I. W. Bakker, S. Harkema, S. S. Wijmenga, D. N. Reinhoudt, *J. Chem. Soc., Perkin Trans. 2* **1992**, 1899-1906.
[49] S. Miyamoto, P. A. Kollman, *J. Am. Chem. Soc.* **1992**, *114,* 3668-3674.
[50] F. Arnaud-Neu, V. Böhmer, J.-F. Dozol, C. Grüttner, R. A. Jakobi, D. Kraft, O. Mauprivez, H. Rouquette, M.-J. Schwing-Weil, N. Simon, W. Vogt, *J. Chem. Soc., Perkin Trans. 2* **1996**, 1175-1182.

[51] L. H. Delmau, N. Simon, M.-J. Schwing-Weill, F. Arnaud-Neu, J.-F. Dozol, S. Eymard, B. Tournois, V. Böhmer, C. Grüttner, C. Musigmann, A. Tunayar, *Chem. Commun.* **1998**, 1627-1628.
[52] a) L. Troxler, M. Baaden, V. Böhmer, G. Wipff, *Supramol. Chem.* **2000**, *12*, 27-51; b) for cation binding by phosphoryl analogues see M. Baaden; G. Wipff; M. R. Yaftian; M. Burgard; D. Matt, *J. Chem. Soc., Perkin Trans. 2* **2000**, 1315-1321.
[53] B. Lambert, V. Jacques, A. Shivanyuk, S. E. Matthews, A. Tunayar, M. Baaden, G. Wipff, V. Böhmer, J.-F. Desreux, *Inorg. Chem.* **2000**, *39*, 2033-2041.
[54] M. Nierlich, J.-F. Dozol, private communication. S. Cherfa, Thesis, Université Paris-Sud, **1998**.
[55] R. Chiariza; V. Urban; P. Thiyagarajan; A. W. Herlinger, *Solv. Ext. Ion Exch.* **1999**, *17*, 113-132.
[56] P. Guilbaud, G. Wipff, *J. Incl. Phenom.* **1993**, *16*, 169-188.
[57] P. Guilbaud, G. Wipff, *New J. Chem.* **1996**, *366*, 55-63.
[58] F. C. J. M. van Veggel, H. G. Noorlander-Bunt, W. L. Jorgensen, D. N. Reinhoudt, *J. Org. Chem.* **1999**, *63*, 3554-3559.
[59] F. Fraternali, G. Wipff, *J. Phys. Org. Chem.* **1997**, *10*, 292-304.
[60] G. Uhlrich, R. Ziessel, I. Manet, M. Guardigli, N. Sabbatini, F. Fraternali, G. Wipff, *Chem. Eur. J.* **1997**, *3*, 1815-1822.
[61] F. C. J. M. van Veggel *J. Phys. Chem. A* **1997**, *101*, 2755-2765.
[62] F. C. J. M. van Veggel, M. P. O. Wolbers, D. N. Reinhoudt *J. Phys. Chem. A* **1998**, *102*, 3060-3066.
[63] M. Hissler, A. Harriman, P. Jost, G. Wipff, R. Ziessel, *Angew. Chem., Int. Ed. Engl.* **1998**, *37*, 3249
[64] A. Harriman, M. Hissler, P. Jost, G. Wipff, R. Ziessel, *J. Am. Chem. Soc.* **1998**, *121*, 14-27.
[65] a) N. A. McDonald, W. P. v. Hoorn, E. M. Duffy, W. L. Jorgensen in *Supramolecular Science : Where It Is and Where It Is Going*, **1999**, R. Ungaro and E. Dalcanale Ed., Kluwer Academic Publishers, Dordrecht pp 147-156; b) N. A. McDonald; E. M. Duffy; W. L. Jorgensen, *J. Am. Chem. Soc.* **1998**, *120*, 5104-5111.
[66] M. Lauterbach, G. Wipff in *Physical Supramolecular Chemistry*, **1996**, L. Echegoyen and A. Kaifer Ed., Kluwer Acad. Pub., Dordrecht pp 65-102.
[67] A. Varnek, G. Wipff, *Solv. Ext. Ion Exch.* **1999**, *17*, 1493-1505.
[68] N. Muzet, E. Engler, G. Wipff, *J. Phys. Chem. B* **1998**, *102*, 10772-10788.
[69] H. Watarai, *Trends Anal. Chem.* **1993**, *12*, 313-318.
[70] L. Troxler, G. Wipff, *Anal. Sci.* **1998**, *14*, 43-56.
[71] P. Beudaert, V. Lamare, J.-F. Dozol, L. Troxler, G. Wipff, *Solv. Extr. Ion Exch.* **1998**, *16*, 597-618.
[72] G. Wipff, E. Engler, P. Guilbaud, M. Lauterbach, L. Troxler, A. Varnek, *New J. Chem.* **1996**, *20*, 403-417.
[73] G. Ranghino, S. Romano, J. M. Lehn, G. Wipff, *J. Am. Chem. Soc.* **1985**, *107*, 7873-7877.
[74] P. Auffinger, G. Wipff, *J. Am. Chem. Soc.* **1991**, *113*, 5976-5988.
[75] K. J. Schweighofer, I. Benjamin, *J. Phys. Chem.* **1995**, *99*, 9974-9985.
[76] I. Benjamin, *Annu. Rev. Phys. Chem.* **1997**, *48*, 407-451 and references cited therein.
[77] A. W. Adamson, *Physical Chemistry of Surfaces*. *5th ed.*, **1990**, Ed., Wiley: New York.
[78] L. Dei, A. Casnati, P. L. Nostro, A. Pochini, R. Ungaro, P. Baglioni, *Langmuir* **1996**, *12*, 1589-1593.
[79] P. L. Nostro, A. Casnati, L. Bossoletti, L. Dei, P. Baglioni, *Colloids Surf. A* **1996**, *116*, 203-209.
[80] L. Dei, A. Casnati, P. L. Nostro, P. Baglioni, *Langmuir* **1995**, *11*, 1268-1272.
[81] Y. Ishikawa, T. Kunitake, T. Matsuda, T. Otsuka, S. Shinkai, *J. Chem. Soc., Chem. Commun.* **1989**, 736-737.
[82] L. T. Byrne, J. M. Harrowfield, D. C. R. Hockless, B. J. Peachey, B. W. Skelton, A. H. White, *Aust. J. Chem.* **1993**, *46*, 1673-1683.
[83] N. Muzet, Thesis, Université Louis Pasteur, Strasbourg, **1999**.
[84] M. Lauterbach, E. Engler, N. Muzet, L. Troxler, G. Wipff, *J. Phys. Chem. B* **1998**, *102*, 225-256.
[85] F. Berny, N. Muzet, R. Schurhammer, L. Troxler, G. Wipff in *Current Challenges in Supramolecular Assemblies*, *NATO ARW Athens*, **1998**, G. Tsoucaris Ed., Kluwer Acad. Pub., Dordrecht pp 221-248.
[86] F. Berny, N. Muzet, L. Troxler, G. Wipff in *Supramolecular Science: where it is and where it is going*, **1999**, R. Ungaro and E. Dalcanale Ed., Kluwer Acad. Pub., Dordrecht pp 95-125.
[87] M. Baaden, F. Berny, N. Muzet, L. Troxler, G. Wipff in *Calixarenes for Separation. ACS Symposium series 757*, **2000**, G. Lumetta, R. Rogers and A. Gopolan Ed., ACS, Washington DC pp 71-85.
[88] a) F. Berny, R. Schurhammer, G. Wipff, *Inorg. Chim. Acta* **2000**, *303-304*, 104-111; b) R. Schurhammer, G. Wipff *New J. Chem.* **1999**, 381-391.
[89] E. Stoyanov, I. Smirnov, A. Varnek, G. Wipff in *Radioactive Waste Management Strategies and Issues*, **2000**, C. Davies Ed., European Commission EUR 19143 EN Brussels, pp 519-522.
[90] M. Baaden, F. Berny, G. Wipff, *J. Mol. Liquids* **2001**, *90*, 3-12.
[91] R. Schurhammer; F. Berny; G. Wipff, *Phys. Chem. Chem. Phys.* **2001**, in press.

Chapter 18

QUANTUM CHEMICAL CALCULATIONS ON ALKALI METAL COMPLEXES

JÉRÔME GOLEBIOWSKI[a], VÉRONIQUE LAMARE[a], MANUEL F. RUIZ-LÓPEZ[b]

[a] *CEA Cadarache, DESD/SEP, 13108 Saint Paul Lez Durance Cedex, France. e-mail: veronique.lamare@cea.fr*
[b] *Laboratoire de Chimie Théorique, UMR CNRS-UHP No.7565, Université Henri Poincaré, BP 239, 54506 Vandœuvre-les-Nancy, France.*

1. Introduction

Currently, the most effective known ligands for the selective extraction of cesium from liquid nuclear wastes (see Chapter 35) are the *1,3-alternate* calix[4]arene crown ether derivatives described in detail in Chapter 20. The importance of this property has led to many studies of these materials and in particular to many attempts to provide a theoretical explanation of the selectivity in their binding to alkali metal cations. To date, most calculations have been done using molecular mechanics (MM) and molecular dynamics (MD) methods, which are based on the use of an empirical force field model to describe the system. These approaches are described in detail in Chapter 17. Important influences on selectivity thereby identified include the calixarene conformation [1], differences in cation hydration energies and cation hydration in the complex [2].

The main limitation of computations using classical force fields is the fact that electronic effects are not taken into account explicitly. For instance, the interaction between the cation and calixarene aromatic cavity cannot be accurately described with pairwise additive force field models. More generally, polarization effects are expected to be substantial, due to the strong cation-macrocycle electrostatic interaction. A rigorous description of all these features can only be achieved by means of quantum mechanical (QM) methods. Then, a comparative analysis of polarization, steric and pure electrostatic effects becomes possible and is useful in enabling improvement of available force field schemes by defining, for instance, appropriate polarizabilities in some centres.

QM calculations have been undertaken successfully for crown-ethers [3-7]. However, only a few quantum mechanical studies on calixarene compounds have been published. Initially, semiempirical methods were used to investigate the properties of calix[4]arenes in the *cone* conformation, focusing on the equilibrium geometry [8,9] and hydrogen-bonds and charge delocalization in polyanions [10]. Semiempirical calculations have also been used to investigate the influence of conformation on second-order nonlinear optical properties of calix[4]arene [11,12] and to assess the importance of π-π interactions in C_{60}-fullerene/*p-t*-butylcalix[8]arene 1:1 adducts [13]. Recently, AM1 semiempirical calculations were performed on *p-t*-butylcalix[4]arene-crown-6 and some of its alkyl ammonium cation complexes, to estimate the binding energy and enthalpy of formation of such compounds [14]. The first density functional study of the structure, conformational equilibria and proton affinity of tetrahydroxycalix[4]arene was reported in 1999 [15], and as-yet unpublished work [16] at the density functional and *ab initio* levels on tetramethoxycalix[4]arene has revealed the inadequacy of force-field and semiempirical calculations in predicting relative conformer energies. Improved MD simulations based on combined QM/MM potentials have been applied to the complexes of Na$^+$ and Cs$^+$ with 1,3-alternate-calix[4]arene-bis-crown-6 [17].

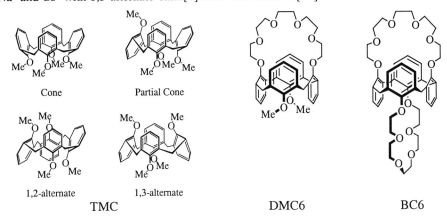

Figure 1. The four conformations of tetramethoxycalix[4]arene (TMC); 1,3-alternate-dimethoxycalix[4]-arene-crown-6 (DMC6); and 1,3-alternate-calix[4]arene-biscrown-6 (BC6).

In the following section, we outline the different levels of theory applicable to such large chemical systems. We then present a few applications that illustrate current research in the field. We first consider the simplest model, *1,3-alternate*-tetramethoxy-calix[4]arene, and its complexes with the alkali metal cations Na$^+$, K$^+$, Rb$^+$ and Cs$^+$. Complexes of the crown-ether derivative, *1,3-alternate*-dimethoxy-calix[4]arene-crown-6, with Na$^+$ - Cs$^+$ are then considered. Finally, the *1,3-alternate*-calix[4]arene-bis-crown-6 complex with Cs$^+$ is described. These ligands are shown in Figure 1. The role of the solvent is discussed in the latter case through the use of a dielectric continuum model. Finally, we conclude this chapter by making a brief presentation of several promising methodological developments which could contribute to make significant advances on the theoretical description of the structure, energetics and chemical properties of calixarene-cation complexes.

2. Calixarene Macrocycle Models and Computational Methods

A wide variety of QM approaches may be used to investigate calixarene-cation complexes. The choice of the method is usually dictated by the size of the system investigated, the computational capabilities and the required accuracy. For some purposes, the macrocycle may be modelled by small units, thus allowing high-level calculations. The basic building-blocks of calixarenes are phenolic groups and therefore analysis of the interaction of alkali metal cations with benzene [18-21] or anisole [22] has received some attention in earlier theoretical studies. More generally, cation-π interactions have been investigated by several authors [18-21,23,24]. In the case of calixarene-crown ethers, analysis of cation binding to dimethylether [3,25], 1,2-dimethoxyethane [4,26], or small crown-ethers [5-7,26] has been used to model the calixcrown interactions.

Standard QM techniques include *ab initio*, density functional and semiempirical levels of theory. *Ab initio* methods allow the most sophisticated calculations since all electron-electron interactions are evaluated rigorously [27]. One must note, however, that the quality of the results is extremely dependent on the size of the atomic orbital basis set and the inclusion or not of electron correlation energy. A reasonable choice appears to be the use of double-ζ basis sets with polarization orbitals for heavy atoms and perturbation theory at second order (Moller-Plesset, MP2 method) to account for correlation effects. Note that computational time in *ab initio* methods increases as N^4 or N^5, N being the number of atomic orbitals in the system. Hence, correlated *ab initio* calculations are only possible for model systems of moderate size.

Density Functional Theory (DFT) is an interesting alternative to *ab initio* calculations [28]. The exchange-correlation energy, which is the computationally expensive part in the *ab initio* formalism, is evaluated here through functionals of the electron density and electron density gradient. Compared to MP2, for instance, the method is much faster (especially for large systems), with comparable accuracy. The computational time increases as N^3. Among the currently available functionals, that known as B3LYP [29] performs nicely and has become quite popular. It is in fact a hybrid functional, since the exchange energy is obtained from a parametric combination of the exact Hartree-Fock value and that predicted by Becke's functional.

Semiempirical calculations [30] are much simpler and faster than either *ab initio* or DFT and the increase of computational time with system size is $\sim N^2$. Many integrals are neglected or parametrised and only valence electrons are treated explicitly. Nowadays, semiempirical calculations can be carried out for very large systems (several hundreds of atoms). However, they display some well-known artifacts, in particular for hydrogen bonds [31,32], and therefore must be used with care. Another limitation is that atomic parameters for some elements of the periodic table are not yet available. This is the case for most alkali metal atoms.

In semiempirical calculations for calixarene-alkali metal cation complexes, following previous work [25,33] we treated M^+ as a monovalent core with negligible electron affinity, *i. e.*, with no atomic orbitals assigned to it. One must then derive a set of semiempirical core parameters to reproduce as well as possible the structure of some model systems, for instance M^+–water, M^+–dimethylether and M^+–benzene. However, one may also need to correct the standard core-core expression in order to get good binding energies, as illustrated in the case of hydrogen-bonded systems [34]. Suitable results were achieved by using the core parameter $\alpha = 0.95$ for Cs (all other core parameters being zero [35]) and an additional core-core term having the form of a Lennard-Jones potential:

$$E = \sum_S \frac{A_{MS}}{R_{MS}^{12}} - \frac{B_{MS}}{R_{MS}^6}$$

where R_{MS} is the interatomic distance between the cation M and the S atom in the ligand. Values of A and B for Cs^+ are given in Table 1. Note that these parameters correspond

TABLE 1. Optimized parameters (a.u.) for Cs$^+$ using the AM1 method.

Core-Core function	Lennard-Jones parameters	
α	A × 10^{-6}	B
0.95	24.80899 (C)	424.6196 (C)
	10.87851 (O)	334.3767 (O)
	0.0 (H)	0.0 (H)

to calculations using the AM1 method [36] and they are not necessary equivalent to those used in classical force-fields or another semiempirical method.

A realistic study of calixarene-cation complexes needs to account for solvent effects. The simplest model consists in the treatment of the liquid as a polarisable dielectric continuum that is characterized by its dielectric constant ε°. The solute is placed in a cavity created in the continuum. It polarizes the dielectric that in turn creates an electrostatic potential inside the cavity, the so-called reaction field. In this work, the reaction field, the solute-solvent interaction energy and the free energy of solvation are obtained through a multipolar expansion of the solute charge distribution [37]. The solute Hamiltonian is modified to account for the corresponding interaction and the resulting self-consistent reaction field (SCRF) equations are solved iteratively. The model allows calculation of optimal geometry, the computational time being only slightly increased with respect to gas phase studies [38].

Finally, we must emphasize that quantum mechanical calculations can be combined with molecular mechanics (MM) methods in different ways [39,40]. Hybrid QM/MM techniques are promising because they allow investigation of very large systems and QM/MM methods can easily be coupled to Molecular Dynamics (MD). The idea is to describe quantum mechanically the part of the system with chemical interest, a classical force-field representation being assumed for the remaining part. The classical MM part may include, for example, the solvent molecules and/or some substituents of the calixarene. In the latter case, the QM/MM frontier chemical bonds must be described using the Localised-SCF method [41] or related algorithms. MD simulations of calixarene-alkali cation complexes in aqueous solution using QM/MM potentials have recently been reported, demonstrating the capabilities of this approach [17].

The *ab initio* and DFT calculations presented in this chapter have been done using the Gaussian 98 program [42] whereas the semiempirical computations have been carried out using the GEOMOS program [43] modified to allow calculations with Cs atom.

3. Study of Some Calixarene-Alkali Cation Complexes

3.1. TETRAMETHOXY-CALIX[4]ARENE

Tetramethoxycalix[4]arene (TMC) has received some attention because this compound is one of the simplest calixarene derivatives. Nicholas *et al.* performed theoretical studies, at the BLYP/6-31G** and MP2/aVDZ levels, on TMC-Na$^+$ and TMC-Cs$^+$ complexes in order to investigate the possible interaction sites in the calixarene cavity [44]. An earlier study on the anisole building block had proven that the interaction with the oxygen atom was of the same order of magnitude as the interaction with the arene system [22].

The starting geometries were built by assuming a quadridentate coordination of the cation in which each of the four methoxybenzene components is coordinated to the cation. By combining the oxygen atoms (O) and the aromatic rings (A), this led to eight distinct coordination modes for Na^+ and six for Cs^+. In several cases, conformations with predominant arene coordination formed stronger complexes than those with oxygen coordination. For the Na^+ cation, important differences appeared for relative stabilities of the complexes with respect to the level of theory used, and some BLYP results were discarded as they might not properly account for repulsive non bonded interactions. MP2 results, allowing for the cost of conformational reorganization, indicate that the Na^+ complex in the gas phase would be a mixture of 88 % of *partial cone* 2O+2A and 12 % of *cone* 2O+2A,. For TMC-Cs^+, the most stable gas phase conformation was predicted to be *partial cone* 1O+3A (99.99 %), followed by *1,3-alternate* 2O+2A (0.01 %). When locking the conformation by appropriate substituents, some binding modes appeared to be particularly suitable to optimize the selectivity amongst cations. Thus, as new Cs^+/Na^+ selective ligands, the authors suggested that rigid calixarene derivatives with *1,2-alternate* 2O+2A or *partial cone* 1O+3A binding sites could be interesting alternatives to the *1,3-alternate* 2O+2A binding mode in calixcrowns. This result is in agreement with solvent extraction data on tetra-paratertiobutyl n-propoxy calix[4]arene, whose *partial cone* conformer shows the best Cs^+/Na^+ selectivity [45].

We have performed further gas phase calculations for the *1,3-alternate* TMC complex with Na^+, K^+, Rb^+ or Cs^+ in order to evaluate the contribution of the calixarene cavity to the cation affinity of calix-crowns [46]. These calculations were made using the hybrid DFT B3LYP functional, the 6-31+G* basis set [27] for TMC atoms and Na^+ cation, and the double ζ quality basis set associated with a Hay and Wadt core pseudopotential in the case of K^+, Rb^+ and Cs^+ [47]. Geometries were optimized at the Hartree-Fock level using the same basis set.

The optimized M^+-O and M^+-G_π distances (G_π is the centroid of the aromatic ring) in *1,3-alternate*-TMC alkali metal ion complexes are close to those found respectively for anisole [22] and benzene [18] complexes. The largest deviation occurs for Na^+, for which the average M^+-O distance is 2.36 Å and M^+-G_π is equal to 2.848 Å. This confirms the remarkable flexibility of TMC, which adapts to the cation size. The predicted complex structures are plotted in Figure 2 in the case of Na^+ and Cs^+, the location of K^+ and Rb^+ being intermediate between these two extrema.

Figure 2. Na^+ and Cs^+ complexes of *1,3-alternate* TMC, obtained at the Hartree-Fock level.

TABLE 2. Strain energies, binding enthalpies and free energies (kcal/mol) for 1,3-alternate TMC alkali metal ion complexes. Comparative binding enthalpies are given for 12C4 complexes.

Cation	E_{Strain}	ΔH^{298}	ΔG^{298}	ΔH^{298} 12C4 [4]
Na^+	1.2	-73.5	-66.1	-61.7
K^+	0.4	-54.7	-46.6	-46.9
Rb^+	2.2	-44.0	-35.9	-39.1
Cs^+	3.7	-34.8	-27.0	-33.5

Some energy data are reported in Table 2. The calculated binding enthalpies are compared to those predicted for the 12C4 crown-ether [4] that has the same number of donor sites (either four oxygen atoms or combination of two oxygen atoms and 2 arene rings). This comparison shows that TMC has the strongest affinity for alkali metal cations. Hence, the calixarene cavity does not appear to be only a molecular platform on which donor groups can be grafted but it significantly contributes to the overall stability of the complexes.

The strain energy is a positive term which represents the energy required to modify the geometry of the ligand (*i.e.*, it is the difference between the ligand energy in the complex geometry and that for the free ligand in the *1,3-alternate* conformation). It is low in all cases, showing that the calixarene ligand is not greatly distorted by complexation. The role of the electrostatic energy may be evaluated roughly by a calculation in which the cation is substituted by a single point charge. In such a calculation, the ligand has the complex geometry but its wavefunction is not allowed to be polarised by the point charge. The resulting electrostatic term varies between -88.2 kcal/mol for Na^+ and -49.3 kcal/mol for Cs^+, being about 15 kcal/mol greater (in absolute value) than the corresponding binding energies. Since the strain energy is small, the role of other electronic effects appears to be substantial. The polarization energy may be evaluated using again the point-charge complex but allowing the electronic cloud of the ligand to relax. The corresponding electronic polarization contributions are quite important and represent about 30-40 % of the electrostatic energies given above.

An evaluation of the activation energy for Cs^+ and Na^+ cations to cross the π tube of the aromatic cavity has been carried out. This process requires an energy increase of about 45 kcal/mol for Cs^+ whereas for Na^+ the potential energy surface is very flat so that no significant energy barrier is predicted.

3.2. 1,3-ALTERNATE-DIMETHOXY-CALIX[4]ARENE-CROWN-6

Experimentally, the calix[4]arene-crown-6 ligand shows a very high Cs^+/Na^+ selectivity and some derivatives are being considered for use in Cs^+ recovery from acidic nuclear liquid waste usually containing high nitrate anion concentrations (Chapter 35). Calculations have been done at the B3LYP/6-31+G* level, using optimised structures at the Hartree-Fock level, for the alkali metal ion complexes (from Na^+ to Cs^+) with dimethoxy-calix[4]arene-crown-6, (DMC6) in the 1,3-alternate conformation, and the effect of nitrate counter ion has been assessed in the case of the Na^+ complex. The starting

TABLE 3. Geometrical data for the DMC6 alkali complexes, in Å. Comparison with X-Ray structures for the Cs⁺ complex.

Cation/Counter ion	M-G_O [a]	M-C_π [b]	M-O $_{average}$ [d]
Na⁺ / NO_3^-	1.61	4.16	3.60 ± 1.32
Na⁺	0.54	4.17	2.79 ± 0.80
K⁺	1.17	2.64	3.22 ± 0.59
Rb⁺	0.86	2.72	3.27 ± 0.43
Cs⁺	0.65	3.32	3.38 ± 0.04
Cs⁺ / Pic⁻ [c]	-	-	3.35 ± 0.16
Cs⁺ / Pic⁻ [c]	-	-	3.34 ± 0.10

[a] : G_O, centroid of the six oxygen atoms of the crown.
[b] : C_π, centroid of the four aromatic rings.
[c] : X-ray structures, two independent molecules, see ref. [51].
[d] : For the free DMC6, the average G_O-O distance is 3.54 ± 0.70 Å.

structures for Na⁺ to Cs⁺ complexes were built on the basis of X-ray data for the K⁺-calix[4]arene-bis-crown-6 complex after removing the second crown [48]. The DMC6-NaNO₃ complex was also built starting from the corresponding bis-crown complex.

Table 3 reports some geometrical data calculated for alkali metal cation complexes with DMC6. In Figure 3, selected optimized structures are plotted. Inspection of structural data shows that the crown of the free ligand is rather well adapted for Cs⁺ binding. Indeed, the distance between the oxygen atoms and their centroid is 3.54 Å, quite close to 3.19 Å, the sum of the cation ionic radius and van der Waals oxygen atom radius [49,50]. The Cs⁺ complex displays C_2 symmetry, while the Na⁺ complex is highly distorted. Na⁺, K⁺, Rb⁺ and Cs⁺ are in close contact with three, four, five and six oxygen atoms of the crown, respectively. The three larger cations also interact with the aromatic cavity, with a strength depending on the cation size. Independently of the presence or not of the nitrate counter ion, Na⁺ is in interaction with the crown-ether part rather than with aromatic carbon atoms. The counter ion pulls Na⁺ outside the mean plane of the crown oxygen atoms and tends to open the crown, due to repulsive interactions between the oxygen atoms of the crown and the nitrate ion.

Figure 3. Structures of the Cs⁺, Na⁺ and NaNO₃ - DMC6 complexes obtained at the Hartree-Fock level.

TABLE 4. Binding enthalpies, free energies and strain contributions for the DMC6 and TB21C7 complexes, in kcal/mol.

Cation	DMC6			TB21C7 [a]	
	ΔH^{298}	ΔG^{298}	E_{strain}	ΔG^{298}	E_{strain}
Na^+/NO_3^-	-60.0	-46.1	2.3	-	-
Na^+	-82.3	-73.1	15.7	-65.1	22.4
K^+	-72.8	-64.1	6.0	-	-
Rb^+	-65.7	-57.2	3.0	-49.4	7.0
Cs^+	-67.6	-63.4	1.3	-42.6	5.4

[a]: See ref. [54], all binding energies are calculated at the same level than that of DMC6. However, DFT optimised structures for Rb^+ and Cs^+ complexes are considered whereas the TB21C7-Na^+ was optimised at the Hartree-Fock level (TB21C7 = tribenzo-21-crown-7).

Binding energy data are summarised in Table 4. Free energies are substantial but they appear to be less dependent on cation size than in the case of TMC complexes. We observe a large decrease from Na^+ to Rb^+ complex but Cs^+ complex exhibits a binding free energy similar to the one of K^+ complex. Note that the presence of the nitrate counter ion lowers the binding enthalpy of Na^+ by ~20 kcal/mol, in absolute value.

Thus, in the gas phase, DMC6 has a larger affinity for Na^+ than for the other alkali metal cations. This agrees well with experimental ESI/MS results on mono and bis-crown derivatives [52]. However, the DMC6 Na^+/Cs^+ selectivity in the gas phase is predicted to be 9.7 kcal/mol ($\Delta\Delta G^{298}$) which is small compared to other systems like TMC. The reason for that could be the specific characteristics of the binding site. As opposed to TMC or crown-ethers, the number of interacting sites in DMC6 is not the same for all the cations. For instance, in *1,3-alternate*-TMC, there are two oxygen atoms and two aromatic rings interacting with each cation. In crown ethers of the 18C6 or 21C7 type, because of the high flexibility of the macrocycles, all the oxygen atoms interact with the charged species [53]. In contrast, in DMC6, the geometric constraints of the calixarene cavity substantially limit the deformation of the crown and this forces Na^+ to be in close contact and to have a strong electrostatic interaction with only three (or four) oxygen atoms. For Cs^+, the cation size is appropriate to interact with all the available sites, *i.e.*, the six oxygen atoms and the two aromatic rings.

Strain energies are quite different along the series. For Cs^+, this quantity is small and lower than that predicted above for either TMC or TB21C7 (tribenzo-21-crown-7), which is a large and quite rigid crown-ether. Conversely, for Na^+, the strain energy is significant in the absence of the counter ion. Indeed, the calix-crown DMC6 appears to combine advantages of both ligand moieties. It shows a strong affinity for the cations due to the crown-ether unit and has a cavity preorganised for Cs^+ binding, thanks to the presence of the calixarene ring.

The electronic polarisation has been estimated as explained above. As for TMC, it is quite important, representing 25-35 % of the total electrostatic contribution to the binding energy for Na^+ and Cs^+, respectively.

3.3. 1,3-ALTERNATE-CALIX[4]ARENE-BIS-CROWN-6

The *1,3-alternate*-calix[4]arene-bis-crown-6 (BC6) ligand is the simplest biscrown derivative having a high affinity for Cs^+, and its binding properties have been extensively studied in several solvents [55-58]. The size of this system precludes *ab initio* or DFT computations. Thus, we have used the semiempirical AM1 method to estimate the binding energy and related properties of mononuclear Cs^+-BC6 and binuclear $2Cs^+$-BC6 complexes that have been observed experimentally. In these calculations, Cs^+ was assumed to be a monovalent core described by the parameters given in Table 1. The solvent effect was estimated with the help of a continuum model by assuming an ellipsoidal shape for the molecular cavity and $\varepsilon^\circ = 80$ for the solvent (*i.e.*, the dielectric constant of water).

The structures of the mononuclear and binuclear complexes are shown in Figure 4. The computed Cs-O distances (O in the interacting loop) for the gas phase mononuclear complex are close to those obtained for DMC6 : the smallest value is 3.139 Å and the largest 3.524 Å. The average distance is 3.337 Å, which compares well with the experimental X-ray average of 3.37 Å [58]. Similar values are found for the binuclear complex in which the average computed value is 3.250 Å. The solvent effect on the geometrical parameters is small, the mean Cs-O distance for the solvated mononuclear complex being slightly smaller (by ~ 0.026 Å). The Cs-Cs distance in the binuclear complex is 6.830 and 6.832 Å in the gas phase and in solution, respectively.

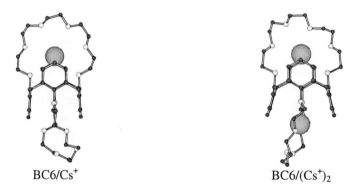

BC6/Cs^+ BC6/$(Cs^+)_2$

Figure 4. Mono and binuclear BC6/Cs^+ complexes, obtained at the AM1 level.

Calculated energies are shown in Table 5. For comparison, we have carried out further AM1 calculations for the complex formed by Cs^+ and DMC6. The corresponding binding energy is $\Delta E^\circ = -67.8$ kcal/mol, in good agreement with the DFT value given above. Significant aspects are that the binding of Cs^+ by BC6 appears to be more exothermic than binding by DMC6 by slightly more than 10 kcal/mol and that binuclear complexes are stable both in the gas phase and in solution. However, free energy favours the formation of two mononuclear complexes rather than that of a binuclear complex and a free BC6 molecule. This preference is very high in gas phase and decreases in aqueous solution but in the latter medium our results still predict a free energy difference of 12.4 kcal/mol. Therefore, binuclear complexes are expected to be detected only in

TABLE 5. Predicted solvation energy (ΔG^{solv}), gas phase binding energy ($\Delta E^{binding}$) and binding free energy ($\Delta G^{binding}$) for BC6-Cs$^+$ complexes, in kcal/mol.

	ΔG^{solv}	$\Delta E^{binding}$		ΔG
		gas	gas [b]	solution
Cs$^+$	-59.8[a]			
BC6	-2.4			
Cs$^+$-BC6	-29.8			
2Cs$^+$-BC6	-99.5			
Cs$^+$ + BC6 → Cs$^+$-BC6		-78.4	-75.2	-46.0
Cs$^+$ + Cs$^+$-BC6 → 2Cs$^+$-BC6		-23.6	-20.4	-33.6
2 (Cs$^+$-BC6) → 2Cs$^+$-BC6 + BC6		+54.8	+54.8	+12.4

[a] : Experimental value [59]
[b] : Assuming a correction of -3.2 kcal/mol to $\Delta E^{binding}$ for each Cs$^+$ binding, as obtained for DMC6 above.

presence of an excess of cesium ions. This is in good agreement with both solvent extraction experiments [55] and results of MD simulations [58].

Finally, we have estimated the energy change due to the displacement of the cations from their equilibrium position in 1,3-alternate TMC to their position in the calix-crowns DMC6 or BC6. If one defines the cation position as the d(M-Cπ) distance (M = cation, Cπ = centroid of the aromatic cavity), the following remarks may be made. First, the Cs$^+$ position is quite similar in all the calixarenes. Thus, the value of d(M-Cπ) changes by only 0.29 Å and 0.25 Å from TMC to DMC6 and BC6, respectively. Conversely, for Na$^+$, there is a large displacement, d(M-Cπ) increasing by 2.62 Å in both calixcrowns. Clearly, the cavity size is particularly well adapted to Cs$^+$: the position occupied by the cation in TMC is not far from the optimum position for the interaction with the crown-6. Hence, the cation-π interaction energy loss when moving from the optimal place in TMC to the position in calix-crown-6 derivative is about 1 kcal/mol for Cs$^+$ and 50 kcal/mol for Na$^+$.

4. Perspectives

The quantitative prediction of absolute selectivities for calixarene alkali metal complexes by quantum chemical computations is still difficult. The size of systems of experimental interest and the necessity to account for the solvation effects are the main problems for an accurate theoretical evaluation of the binding energies. Nevertheless, we have shown in this chapter that experimental trends may be adequately reproduced with DFT or semiempirical methodologies. The main interest in these computations is that they enable accounting for electronic effects, which are not explicitly considered in standard molecular mechanics approaches.

The role of the solvent has been evaluated in the QM calculations by using a continuum model, which is a crude approximation. More reliable calculations would require taking into account the discrete nature of the solvent and the estimation of statistical averages. Within this scope, the development of hybrid quantum/classical methods

[39] and techniques that scale linearly with the system size [60] are quite promising since they may be coupled to Molecular Dynamics simulations. Such studies are only possible at the semiempirical level since the quantum calculation must be made for a large number of structures (~10^6). Therefore, progress in the description of the cations and their interaction with the ligands is essential. This is now feasible and applications to the dynamics of the complexation of Cs^+ by *1,3-alternate*-calix[4]arene-bis-crown-6 in aqueous solution will be reported soon [17].

Acknowledgements: The authors thank B. Hay for the communication of the manuscript of his papers.

5. References and Notes

[1] G. Wipff, M. Lauterbach *Supramol. Chem.* **1995**, *6*, 187-207.
[2] P. Thuery, M. Nierlich, V. Lamare, J. F. Dozol, Z. Asfari, J. Vicens, *J. Incl. Phenom.* **2000**, *36*, 375-408, and references therein.
[3] S. E. Hill, E. D. Glendening, D. Feller, *J. Phys. Chem. A* **1997**, *101*, 6125-6131.
[4] S. E. Hill, D. Feller, E. D. Glendening, *J. Phys. Chem. A* **1998**, *102*, 3813-3819.
[5] M. S. Islam, R. A. Pethrick, D. Pugh, M. J. Wilson, *J. Chem. Soc. Faraday Trans.* **1997**, *93*, 387-392.
[6] M. S. Islam, R. A. Pethrick, D. Pugh, M. J. Wilson, *J. Chem. Soc. Faraday Trans.* **1998**, *94*, 39-46.
[7] M. J. Wilson, R. A. Pethrick, D. Pugh, M. S. Islam, *J. Chem. Soc. Faraday Trans.* **1997**, *93*, 2097-2104.
[8] K. B. Lipkowitz, G. Pearl, *J. Org. Chem.* **1993**, *58*, 6729-6736.
[9] Y. Li, Y. Z. Zheng, J. L. Guo, X. Q. Song *Chin, Chem. Lett.* **1994**, *5*, 781-784.
[10] D. J. Grootenhuis, P. A. Kollman, L. C. Groenen, D. N. Reinhoudt, G. Van Hummel, F. Ugozzoli, G. D.,Andreetti, *J. Am. Chem. Soc.* **1990**, *112*, 4165-4176.
[11] E. Brouyere, J. L. Bredas, *Synth. Metals* **1995**, *71*, 1699-1700.
[12] E. Brouyere, A. Persoons, J. L. Bredas, *J. Phys. Chem. A* **1997**, *101*, 4142-4148.
[13] F. Lara-Ochoa, J. A. Cogordan, R. Cruz, M. Martinez, I. Silaghi-Dumitrescu, *Fullerene Sci. Technol.* **1999**, *7*, 411-419.
[14] J. I. Choe, K. Kim, S. K. Chang, *Bull. Korean Chem. Soc.* **2000**, *21*, 465-470.
[15] R. J. Bernardino, B. J. C. Cabral, *J. Phys. Chem. A* **1999**, *103*, 9080-9085.
[16] J. B. Nicholas, D. E. Bernholdt, B. P. Hay, private communication.
[17] J. Golebiowski, V. Lamare, J. F. Dozol, M. T. C. Martins-Costa, C. Millot, M. F. Ruiz-López Poster presented at *5th World Congress of Theoretically Oriented Chemists (WATOC 99)*, London, U. K., 1-6 August **1999**.
[18] J. B. Nicholas, B. P. Hay, D. A. Dixon, *J. Phys. Chem. A* **1999**, *103*, 1394-1400.
[19] E. Cubero, F. J. Luque, M. Orozco, *Proc. Natl. Acad. Sci. USA* **1998**, *95*, 5976-5980.
[20] J. W. Caldwell, P. A. Kollman, *J. Am. Chem. Soc.* **1995**, *117*, 4177-4178.
[21] H. L. Jiang, W. L. Zhu, X. J. Tan, J. Z. Chen, Y. F. Zhai, D. X. Liu, K. X. Chen, R. Y. Ji, *Acta Chim. Sin.* **1999**, *57*, 860-868.
[22] J. B. Nicholas, B. P. Hay, *J. Phys. Chem. A* **1999**, *103*, 9815-9820.
[23] H. Minoux, C. Chipot, *J. Am. Chem. Soc.* **1999**, *121*, 10366-10372.
[24] J. C. Ma, D. A. Dougherty, *Chem. Rev.* **1997**, *97*, 1303-1324.
[25] M. A. Thompson, *J. Am. Chem. Soc.* **1995**, *117*, 11341-11364.
[26] P. Bultinck, A. Goeminne, D. Van de Vondel, *J. Mol. Struct. (THEOCHEM)*, **1999**, *467*, 211-222.
[27] A. Szabo, N. S. Ostlund in *Modern Quantum Chemistry : Introduction to Advanced Electronic Structure Theory*, Dover Publication, Inc. New York, **1996**, p. 111-230.
[28] L. J. Bartolotti, K. Flurchick in *Reviews in Computational Chemistry, Vol. 7*, (K. B. Lipkowitz, D. B. Boyd Eds.), VCH Publishers, Inc. New York, **1996**, p. 187-216.

[29] A. D. Becke, *J. Chem. Phys.* **1993**, *98*, 5648-5652.
[30] W. Thiel in *Advances in Chemical Physics, Vol. 93*, (I. Prigogine, S. A. Rice Eds.), Wiley, J., **1996**, .
[31] G. I. Csonka, *J. Comp. Chem.* **1993**, *14*, 895-898.
[32] G. I. Csonka, J. G. Angyan, *J. Mol. Struct. (THEOCHEM)*, **1997**, *393*, 31-38.
[33] M. A. Thompson, E. D. Glendening, D. Feller, *J. Phys. Chem.* **1994**, *98*, 10465-10476.
[34] M. Bernal, M. T. C. Martins-Costa, C. Millot, M. F. Ruiz-Lopez, *J. Comput. Chem.* **2000**, *21*, 572-581.
[35] M. J. Field, P. A. Bash, M. Karplus, *J. Comp. Chem.* **1990**, *11*, 700-733.
[36] M. J. S. Dewar, E. G. Zoebisch, E. F. Healy, J. J. P. Stewart, *J. Am. Chem. Soc.* **1985**, *107*, 3902-3909.
[37] D. Rinaldi, M. F. Ruiz-Lopez, J. L. Rivail, *J. Chem. Phys.* **1983**, *78*, 834-838.
[38] D. Rinaldi, J. L. Rivail, N. Rguini, *J. Comput. Chem.* **1992**, *13*, 675-680.
[39] J. Gao in *Reviews in Computational Chemistry, Vol. 7*, (K. B. Lipkowitz, D. B. Boyd Eds.), VCH Publishers Inc., N. Y., **1996**, p. 119-185.
[40] M. F. Ruiz-Lopez, J. L. Rivail in *Encyclopedia of Computational Chemistry, Vol. 1*, (P. von Rague-Schleyer Ed., Wiley, Sons, **1998**, p. 437-448.
[41] V. Théry, D. Rinaldi, J. L. Rivail, B. Maigret, G. Ferenczy, *J. Comp. Chem.* **1994**, *15*, 269-282.
[42] M. J. Frisch, G. W. Trucks, H. B. Schlegel, G. E. Scuseria, M. A. Robb, J. R. Cheeseman, V. G. Zakrzewski, J. A. Montgomery, R. E. Stratmann, J. C. Burant, S. Dapprich, J. M. Millam, A. D. Daniles, K. N. Kudin, M. C. Strain, O. Farkas, J. Tomasi, V. Barone, M. Cossi, R. Cammi, B. Mennucci, C. Pomelli, C. Adamo, S. Clifford, J. Ochterski, G. A. Petersson, P. Y. Ayala, Q. Cui, K. Morokuma, D. K. Malick, A. D. Rabuck, K. Raghavachari, J. B. Foresman, J. Cioslowski, J. V. Ortiz, B. B. Stefanov, G. Liu, A. Liashenko, P. Piskorz, I. Komaromi, R. Gomperts, R. L. Martin, D. J. Fox, T. Keith, M. A. Al-Laham, C. Y. Peng, A. Nanayakkara, C. Gonzalez, M. Challacombe, P. M. W. Gill, B. G. Johnson, W. Chen, M. W. Wong, J. L. Andres, M. Head-Gordon, E. S. Replogle, J. A. Pople Gaussian 98, Revision E.7, Gaussian, Inc., Pittsburg, PA **1998**.
[43] D. Rinaldi, P. E. Hoggan, A. Cartier, *QCPE Bull.*, **1989**, *(QCPE 584a)*, 128.
[44] B. P. Hay, J. B. Nicholas, *J. Am. Chem. Soc.* **2000**, *122*, 10083-10089.
[45] A. Ikeda, S. Shinkai, *J. Am. Chem. Soc.* **1994**, *116*, 3102-3110.
[46] J. Golebiowski, V. Lamare, M. F. Ruiz-Lopez, manuscript in preparation.
[47] P. J. Hay, W. R. Wadt, *J. Chem. Phys.* **1985**, *82*, 299-310.
[48] P. Thuery, M. Nierlich, V. Lamare, J. F. Dozol, Z. Asfari, J. Vicens, *Supramol. Chem.* **1997**, *8*, 319-332.
[49] A. M. James, M. P. Lord, *Macmillan's Chemical, Physical Data Book*, London, UK **1992**.
[50] R. D. Shannon, *Acta Crystallogr.* **1976**, *A32*, 751.
[51] A. Casnati, A. Pochini, R. Ungaro, F. Ugozzoli, F. Arnaud, S. Fanni, M. J. Schwing, R. J. M. Egberink, F. De Jong, D. N. Reinhoudt, *J. Am. Chem. Soc.* **1995**, *117*, 2767-2777.
[52] F. Allain, H. Virelizier, C. Moulin, V. Lamare, J. F. Dozol in *Rapport Scientifique 1998, CEA-R-5835*, Direction du Cycle du Combustible, **1999**, p. 94-99.
[53] V. Lamare, D. Haubertin, J. Golebiowski, J. F. Dozol, *J. Chem. Soc. Perkin Trans 2.* **2001**, 121-127.
[54] J. Golebiowski, V. Lamare, J. F. Dozol, M. F. Ruiz-Lopez manuscript in preparation.
[55] Z. Asfari, C. Bressot, J. Vicens, C. Hill, J. F. Dozol, H. Rouquette, S. Eymard, V. Lamare, B. Tournois, *Anal. Chem.* **1995**, *67*, 3133-3139.
[56] F. Arnaud-Neu, Z. Asfari, B. Souley, J. Vicens, *New J. Chem.* **1996**, *20*, 453-463.
[57] A. D'Aprano, J. Vicens, Z. Asfari, M. Salomon, M. Iammarino, *J. Sol. Chem.* **1996**, *25*, 955-970.
[58] Z. Asfari, C. Naumann, J. Vicens, M. Nierlich, P. Thuery, C. Bressot, V. Lamare, J. F. Dozol, *New J. Chem.* **1996**, *20*, 1183-1194.
[59] Y. Marcus, *Biophys. Chem.* **1994**, *51*, 111-127.
[60] W. Yang, *Phys. Rev. Lett.* **1991**, *66*, 1438-1441.

Chapter 19

THERMODYNAMICS OF CALIXARENE-ION INTERACTIONS

ANGELA F. DANIL DE NAMOR

Laboratory of Thermochemistry
Department of Chemistry, School of Physics and Chemistry
University of Surrey, Guildford, Surrey GU2 5XH, UK.
e-mail: a.danil-de-namor@surrey.ac.uk

1. Introduction

Synthetic developments in the field of macrocycle chemistry have been largely motivated by the need to find ligands capable of selective complexation of a given neutral or ionic species. In solution, the complexation of a macrocycle with a given guest is largely controlled by solvation [1]. Thermodynamic transfer functions, $\Delta_t P^\circ$ ($P^\circ = G^\circ, H^\circ, S^\circ$) from a reference solvent (s_1) to another (s_2) for the host, the guest and the complex are the parameters to consider in assessing the medium effect on the thermodynamics of complexation. This is illustrated in eq. 1 where the differences in $\Delta_c P^o$ for systems involving a metal cation, M^{n+}, and a ligand, L, to give a 1:1 metal-ion complex, ML^{n+}, in two solvents, s_1 and s_2, are dependent on the $\Delta_t P^\circ$ values for the transfer of M^{n+}, L and ML^{n+} from s_1, to s_2.

$$\Delta_c P^o(s_2) - \Delta_c P^o(s_1) = \Delta_t P^o(ML^{n+})(s_1 \rightarrow s_2) - \Delta_t P^o(M^{n+})(s_1 \rightarrow s_2) - \Delta_t P^o(L)(s_1 \rightarrow s_2) \quad (1)$$

This chapter contains a critical discussion of the various parameters for systems involving calixarene derivatives and ionic species (cations or anions) in different media. General information on calixarenes is readily obtained [2-5] and only structures of derivatives directly discussed are given here. (See Chart 1)

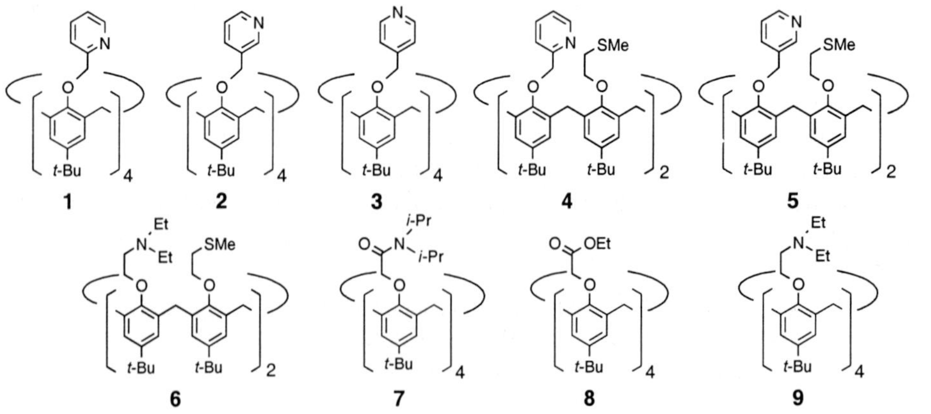

Chart 1. Narrow rim functionalised calix[4]arenes.

Z. Asfari et al. (eds.), Calixarenes 2001, 346–364.

2. Solution Thermodynamics of Calixarene Derivatives. Transfer Functions

2.1. SOLUBILITIES AND DERIVED STANDARD GIBBS ENERGIES. TRANSFER GIBBS ENERGIES

Provided that the composition of the solid in equilibrium with a saturated solution of the ligand is the same, solubility data referred to the process described in eq. 2 can be used to derive the standard solution Gibbs energy changes, $\Delta_{soln}G°$, of the ligand in a given solvent and at a given temperature.

$$L_{(sol.)} \rightarrow L_{(soln)} \quad (2)$$

For the solid in equilibrium with a saturated solution of the ligand, $\Delta_{soln}G = 0$, so that:

$$\Delta_{soln}G° = -RT \ln[L]\delta_L \quad (3)$$

When low solubilities are involved, the activity coefficient of the ligand, δ_L (molar scale; standard state for solubility 1 mol dm^{-3}) may be taken as unity. Since the solution process involves the contribution of the crystal lattice and the solvation Gibbs energies, the former is eliminated by the calculation of the standard transfer Gibbs energy, $\Delta_t G°$, of the solute from a reference solvent s_1 to another s_2. The data concern the process:

$$L_{(s_1)} \xrightarrow{K_t} L_{(s_2)} \quad (4)$$

The thermodynamic transfer constant, K_t, is given by

$$K_t = \frac{[L]\delta_L(s_2)}{[L]\delta_L(s_1)} = \frac{[L](s_2)}{[L](s_1)} \quad (5)$$

and K_t is obtained from the ratio of the solubilities of the ligand in the two solvents. Thus, the standard transfer Gibbs energy is calculated from

$$\Delta_t G° = -RT \ln K_t = \Delta_{soln}G°(s_2) - \Delta_{soln}G°(s_1) \quad (6)$$

Representative thermodynamic transfer functions for calixarene derivatives are listed in Table 1 [6-9].

2.1.1. Solvent effects

$\Delta_t G°$ values (Table 1) provide quantitative information about the extent of solvation of a given ligand in the solvent relative to CH_3CN at 298.15 K. Ligand-solvent interactions are selective, with more effective solvation leading to more negative $\Delta_t G^0$ values, and the capacity of the solvent to solvate each of the ligands listed in Table 1 follows the sequence:

1. 1-BuOH > 1-PrOH > PhCN > EtOH > DMF > CH_3OH > CH_3CN
2. CH_3OH > PhCN > EtOH > DMF > CH_3CN > 1-BuOH
3. CH_3OH > PhCN > EtOH > CH_3CN > DMF
4. THF > PhCN > DMF > 1-BuOH > 1-PrOH > EtOH > CH_3CN > CH_3OH
5. 1-PrOH > 1-BuOH > EtOH > CH_3OH > CH_3CN
6. 1-BuOH > EtOH > DMF > CH_3OH > CH_3CN
7. 1-BuOH ≅ 1-BuOH$_{(H_2O\ satd.)}$ > CH_3CN

Derivatives **1-7** are highly soluble in dichloromethane (CH_2Cl_2) and chloroform ($CHCl_3$), to the extent that solvate formation is observed when these are placed in a saturated atmosphere of these solvents. Consequently, $\Delta_{soln}G°$ cannot be derived. The different sequences found in the $\Delta_t G°$ values from acetonitrile to the various solvents for these ligands provide experimental evidence that these data cannot be correlated with any single solvent property [1].

TABLE 1. Representative thermodynamic transfer function values of narrow rim functionalised calix[4]-arenes from acetonitrile to another solvent (s_2) at 298.15 K.

Calixarene Derivative	s_2	K_t	$\Delta_t G°/kJ\ mol^{-1}$	$\Delta_t H°/kJ\ mol^{-1}$	$\Delta_t S°/J\ K^{-1}\ mol^{-1}$
			($CH_3CN \rightarrow s_2$)		
1	CH_3CN	1.00	0[b]	0	0
	CH_3OH	3.04	-2.76[b]		
	EtOH	14.0	-6.55[b]		
	1-PrOH	55.6	-9.96[b]		
	1-BuOH	91.6	-11.20[b]		
	DMF	4.60	-3.79[b]		
	PhCN	25.7	-8.05[b]		
2	CH_3CN	1.00	0[b]	0	0
	CH_3OH	11.90	-6.14[b]		
	EtOH	7.95	-5.14[b]		
	1-BuOH	0.66	1.03[b]		
	DMF	2.32	-2.09[b]		
	PhCN	8.55	-5.32[b]		
3	CH_3CN	1.00	0[b]	0	0
	CH_3OH	10.70	-5.88[b]	10.69	55.6
	EtOH	5.49	-4.22[b]		
	DMF	0.77	0.66[b]		
	PhCN	6.19	-4.52[b]		
4	CH_3CN	1.00	0[c]		
	CH_3OH	0.69	0.92[c]		
	EtOH	3.91	-3.38[c]		
	1-PrOH	8.83	-5.40[c]		
	1-BuOH	10.98	-5.94[c]		
	DMF	15.98	-6.87[c]		
	THF	327.8	-14.36[c]		
	PhCN	122.5	-11.92[c]		
5	CH_3CN	1.00	0[c]		
	CH_3OH	1.48	-0.97[c]		
	EtOH	3.39	-3.03[c]		
	1-PrOH	7.04	-4.84[c]		
	1-BuOH	5.83	-4.37[c]		
6	CH_3CN	1.00	0[d]	0	0
	CH_3OH	3.51	-3.11[d]	-11.43	-27.9
	EtOH	6.27	-4.55[d]	-12.68	-27.3
	1.BuOH	41.0	-9.21[d]	-15.23	-20.2
	DMF	5.08	-4.03[d]	-1.34	9.1
7	CH_3CN	1.00	0	0	0
	1-BuOH	4.67	-3.82[e]		
	1-BuOH[a]	3.50	-3.11[e]		

Abbreviations used: CH_3CN; acetonitrile; CH_3OH; methanol; EtOH; ethanol; 1-PrOH, 1-propanol; 1-BuOH, 1-butanol; DMF; N,N-dimethylformanide; PhCN; benzonitrile; THF; tetrahydrofuran; [a]H_2O satd., [b]Ref. [6], [c]Ref. [7], [d]Ref. [8], [e]Ref. [1]

2.1.2. Ligand effects

The most contrasting behaviour is shown in the solvation changes that **1** undergoes in the alcohols (protic solvents) relative to its geometrical isomers, **2** and **3**. For **1**, solvation increases from CH_3OH to 1-BuOH whereas the opposite behaviour is found for **2** and **3**.

Thus, replacement of **2** by **1** in the CH_3OH–1-BuOH solvent system results in an increase of K_t by a factor of 548. Partial replacement of pendent arms containing 2-pyridyl groups in **1** by sulphonyl-methyl substituents **4**, decreases the K_t value by a factor of 2 in the same solvent system. This may reflect the contribution of the oxymethylene pyridyl groups to the higher solvation observed for **1** relative to **4** but the opposite behaviour is found for **2** relative to **5**. In this case, partial replacement of pyridyl groups by sulphonylmethyl substituents leads to an increase in the K_t value by a factor of 72 in the CH_3OH–1-BuOH solvent system. Similar assessments can be carried out to evaluate the effect of replacing aromatic (**4** and **5**) by aliphatic (**6**) amines in the pendent arms of the ligand. In this case, the K_t value in the CH_3CN-1-BuOH system increases for **6** relative to **4** and **5** by factors of about 4 and 7, respectively. This may be attributed to an increase in the basicity of the amine, which enhances its capability to enter hydrogen bond formation with the protic solvent. Ligand effects involving two dipolar aprotic solvents (CH_3CN-PhCN) may be quite substantial. In order to achieve a better understanding on the solvent effect on the transfer process, enthalpy and entropy data are required.

2.2. STANDARD ENTHALPIES AND ENTROPIES OF SOLUTION. TRANSFER FUNCTIONS

Transfer enthalpies, $\Delta_t H^\circ$, are calculated from the standard enthalpies of solution, $\Delta_{soln} H^\circ$, of a solute in the appropriate solvents:

$$\Delta_t H^\circ (L)(s_1 \rightarrow s_2) = \Delta_{soln} H^\circ (s_2) - \Delta_{soln} H^\circ (s_1) \qquad (7)$$

For the derivation of $\Delta_{soln} H^\circ$ direct calorimetric studies are desirable [10]. For slightly soluble non-electrolytes (uncharged ligands), and provided that very accurate solubility data can be obtained over a moderate temperature range, enthalpy data can be obtained, with caution [1,11], from the variation of solubility with temperature by the use of the van't Hoff equation. $\Delta_{soln} H^\circ$ values for calixarene derivatives have been obtained [12] for cases in which the reliability of the data had been checked by independent means (see Section 4.2). Availability of $\Delta_{soln} H^\circ$ and $\Delta_{soln} G^\circ$ data allows the calculation of the standard entropy of solution, $\Delta_{soln} S^\circ$ through the relationship.

$$\Delta_{soln} S^\circ = \frac{\Delta_{soln} H^\circ - \Delta_{soln} G^\circ}{T} \qquad (8)$$

For analysis of solute-solvent interactions, the crystal lattice contribution is removed by the calculation of transfer enthalpies, $\Delta_t H^\circ$ and entropies, $\Delta_t S^\circ$ from acetonitrile (reference solvent) to another medium. $\Delta_t H^\circ$ and $\Delta_t S^\circ$ values for calixarene derivatives are relatively limited, with some representative figures given in Table 1.

For **3**, the transfer from CH_3CN to CH_3OH is entropically controlled since $\Delta_t H^\circ$ ($CH_3CN \rightarrow CH_3OH$) is positive, while $\Delta_t G^\circ$ is negative. Analyses of ^1H chemical shift values in the NMR spectra of **3** in a variety of solvents have been used to assess possible solvation modes and they suggest, for example, that acetonitrile interacts with the hydrophobic cavity of the ligand as in other calixarene derivatives [13]. A possible visualisation of the transfer process is one in which, although **3** is better solvated in CH_3OH than in CH_3CN, the inclusion of acetonitrile in the hydrophobic cavity of the ligand has pronounced consequences in terms of enthalpy ($\Delta_c H°$ more negative) and mainly in entropy (more negative, due to loss of conformational freedom). Consequently, the transfer process from CH_3CN to CH_3OH occurs with a gain in entropy and a loss of enthalpic stability.

Another instance of relevance to solvent extraction processes (organic phase saturated with water) is provided by the transfer thermodynamic data for ligand **7** from the dry to the water-saturated alcohol [1]. Thus, the $\Delta_t G°$ (1-BuOH-water satd. 1-BuOH) value of 0.71 kJ mol^{-1} shows little effect of the presence of water in the organic phase while $\Delta_t H°$ (-14.65 kJ mol^{-1}) and $\Delta_t S°$ (-51.5 J K^{-1} mol^{-1}) values [1] show major effects. Useful comparison may be made of the transfer of **3** from CH_3CN to CH_3OH and of **7** from CH_3CN to water-saturated 1-BuOH. Opposite effects in terms of enthalpy and entropy are observed and this is due to the fact that for ligand **3**, the reference (CH_3CN) rather than the receiving solvent provides the medium for specific ligand-solvent interactions, while for **7** it is the receiving medium. Thus, the gain in enthalpy (negative $\Delta_t H°$) and the loss of entropy (negative $\Delta_t S°$) must be attributed to interactions between water and the ligand. Explanations based on molecular mechanics calculations [14] for p-*tert*-butylcalix[4]arene tetradiethylamide and water have been offered but they await experimental tests. Unlike the transfer of ligand **3** from CH_3CN to CH_3OH (in which favourable entropy changes overcome unfavourable enthalpic factors, to result in a favourable $\Delta_t G°$), in the transfer of **7** from 1-BuOH to the water saturated solvent the $\Delta_t G°$ value is close to 0 kJ mol^{-1}. This is consistent with an analysis [15] suggesting that there is an entropy/enthalpy compensation in solvent reorganisation.

The essence of current understanding is therefore:

i) Transfer Gibbs energies provide a quantitative measure of the differences in the solvation of the ligand in two solvents while transfer enthalpy and entropy parameters are good reporters of specific solute-solvent interaction differences.

ii) Although enthalpy and entropy parameters are good reporters of solute-solvent interactions, these data are not expected to provide information regarding the site of interactions between the ligand and the solvent.

3. Solution Thermodynamics of Electrolytes (free and complex)

3.1. SOLUBILITIES AND STANDARD GIBBS ENERGIES. TRANSFER FUNCTIONS

Analysis of solution equilibria involving electrolytes is difficult, though the issues involved have been the subject of numerous authoritative publications [16-20]. Provided that the composition of the solid in equilibrium with its saturated solution is the same and ionic species are predominant in solution, $\Delta_{soln} G°$, for a dissociated 1:1 electrolyte can be calculated from solubility measurements (eq. 9) at a given temperature,

$$MX_{(sol.)} \rightarrow M^+_{(s)} + X^-_{(s)} \qquad (9)$$

through the relationship,

$$\Delta_{soln} G° = -RT \ln K°_{sp} \qquad (10)$$

In eq. 10, $K°_{sp}$ expressed in terms of ionic activities is the thermodynamic solubility product referred to the standard state of 1 mol dm^{-3}. In non-aqueous media, ion-association can be very important and if so, knowledge of the ion pair formation constant, K_a for the process represented in eq. 11 is required for the calculation of $K°_{sp}$ and therefore, $\Delta_{soln}G°$.

$$M^+_{(s)} + X^-_{(s)} \xrightarrow{K_a} MX_{(s)} \quad (11)$$

From solubility data, the K_a value and the ion size parameter, å, of the electrolyte in the appropriate solvent, the concentration, c_i and the mean ionic activity coefficient, $\delta_\pm$ of the ionic species in solution can be calculated by the method of successive approximations using the Debye-Hückel theory [21,22]. The activity coefficient of the ion pair MX is usually taken to be unity. Thus, using c_i, $\delta_\pm$ and the $K°_{sp}$, the $\Delta_{soln}G°$ values are calculated. Ion-pair formation constants in non-aqueous media for a large variety of salts in a wide range of solvents, mostly at 298.15 K have been reported [23,24]. Particular attention should be paid to this issue in the selection of the solvent to be used in the determination of thermodynamic parameters of complexation of ions (anions or cations) in non-aqueous media [25].

For salts involving multiply-charged ions, several species may be present in solution, and it is important to have information of the speciation in solution prior to proceeding with thermodynamic investigations. In studies, for example, on lanthanide salts in dipolar aprotic media an initial report [26] that in acetonitrile, trifluoromethane sulphonate (triflate) salts are fully dissociated was rejected in subsequent investigations [27-29]. Conductometric studies [30] on lanthanide triflates in acetonitrile revealed that concentrations of 10^{-5} mol dm^{-3} or even lower are required to safely consider that the tripositive cations are predominant in acetonitrile. Given such problems, it is hardly surprising to find that information on $\Delta_{soln}G°$ of electrolytes other than 1:1 in non-aqueous media is relatively scarce and speculation rather common.

TABLE 2. Transfer thermodynamic parameters for 1:1 (free and complex) electrolytes from acetonitrile to methanol and to N,N-dimethylformamide at 298.15 K. Single-ion values based on the Ph$_4$AsPh$_4$B convention.

	$\Delta_t G°$/kJ mol^{-1} [a]	$\Delta_t H°$/kJ mol^{-1} [b]	$\Delta_t S°$/J K^{-1} mol^{-1}
		CH$_3$CN → CH$_3$OH	
Electrolyte			
Ag$^+$ + ClO$_4^-$	30.78	33.28	8.4
Ag$^+$ **6** + ClO$_4^-$	13.60	0.33	-44.5
Single-Ion			
Ag$^+$	30.12	18.85	-37.8
Ag$^+$ **6**	12.94	-14.10	-90.7
		CH$_3$CN → DMF	
Electrolyte			
Ag$^+$ + ClO$_4^-$	-0.13	-3.69	-11.9
Ag$^+$ **6** ClO$_4^-$	-8.47	-21.90	-45.0
Single Ion			
Ag$^+$	4.65	2.18	-8.3
Ag$^+$ **6**	-3.69	-16.03	-41.4

[a]Ref. [31] [b]Ref. [8]

Combination of $\Delta_{soln}G°$ of a dissociated 1:1 calixarene based electrolyte in two solvents yields $\Delta_t G°$ values of $ML^+ + X^-$ from a reference solvent to another. These data can also be derived from Eq. 1 provided that $\Delta_c G°$ in both solvents and transfer data for the free cation salt and the ligand are available. Table 2 lists transfer thermodynamic data from CH_3CN to CH_3OH (protic) and to DMF (dipolar aprotic) for electrolytes and single ions involved in the free and the **6**-complexed species at 298.15 K. Since anions and cations cannot be separately studied, single ion values are often derived following the "extra-thermodynamic" Ph_4AsPh_4B (Parker) convention [16].

In Eq. 1, provided that both the free and complex cations have the same counter-ion, $\Delta(\Delta_c P°)$ is independent of any extra-thermodynamic convention since the anion contribution is cancelled out. These data reveal:

(i) Changes in transfer parameter values for both "free" and complexed cations can be large to the extent that in some cases they may produce a reverse pattern, indicating that the medium effect on complexation can not be merely attributed to the solvation changes of the free cation.

(ii) The availability of transfer functions for the free and complex cations (Table 2) and the ligand (Table 1) allows unambiguous calculation of the relative stability of complex formation for a given system (cation-ligand or anion-ligand) in one solvent with respect to another (since $\Delta(\Delta_c G°) = -RT \ln K_s(s_2)/K_s(s_1)$). Thus, combination of transfer data for the Ag^+-**6** system leads to a $\Delta(\Delta_c G°)$ (CH_3CN-CH_3OH) of -14.07 kJ mol^{-1} and $\Delta(\Delta_c G°)$ of -4.31 kJ mol^{-1} for Ag^+-**6** in CH_3CN-DMF.

3.2. ENTHALPIES AND ENTROPIES OF SOLUTION OF FREE AND COMPLEX ELECTROLYTES. TRANSFER FUNCTIONS

$\Delta_{soln}H°$ values of electrolytes are obtained by extrapolation from finite concentrations to infinite dilution. Methods, particularly for data obtained in non-aqueous solvents, have been extensively discussed in the literature [32,33]. Most $\Delta_{soln}H°$ have been obtained from a plot of heats of solution against the square root of the ionic strength, $I^{1/2}$. Provided that low electrolyte concentrations are used, even in relatively apolar media where ion-pair formation is likely, this extrapolation method seems suitable, based on the agreement currently found between $\Delta_{soln}H°$ values obtained by this method and those derived from a plot of measured $\Delta_{soln}H$ at various concentrations ($< 10^{-2}$ mol dm^{-3}) against the degree of ionisation, α- of the electrolyte. Here, $\Delta_{soln}H°$ is the value at $\alpha = 1$ (full dissociation of the electrolyte) and this extrapolation method requires knowledge of the ion-pair formation constant, K_a, of the electrolyte in the appropriate solvent. Although great care must be taken in the determination of $\Delta_{soln}H°$ of electrolytes in non-aqueous solvents by calorimetry [17], the derivation of these data does not require as much ancillary information as that for the calculation of $\Delta_{soln}G°$ of electrolytes from solubility data. For reliable values of $\Delta_{soln}H°$ of calixarene based electrolytes, high stability of the complex is necessary to avoid heat effects due to its dissociation.

Data known for $\Delta_{soln}H°$ of calixarene based electrolytes mainly concern 1:1 electrolytes derived from calixarene esters and alkali-metal cations and silver **6** salts [8,12,31,34]. In assessing the medium effect on the complexation process, $\Delta_t P°$ values are required. Some representative data for free and complex salts are shown in Table 2.

Presently relevant single-ion values based on the Ph$_4$AsPh$_4$B convention at 298.15 K were calculated and are listed in Table 2. For the Ag$^+$/6 system, for example, these data and those of Table 1 show that the stability of the complex in enthalpic terms is greater in CH$_3$OH ($\Delta\Delta_c H° = -21.52$ kJ mol^{-1}) and in DMF ($\Delta\Delta_c H° = -16.87$ kJ mol^{-1}) than in CH$_3$CN. In terms of entropy, there is a greater loss in entropy upon complexation in CH$_3$OH ($\Delta\Delta_c S° \cong -25$ J K^{-1} mol^{-1}) and in DMF ($\Delta\Delta_c S° = -42.2$ J K^{-1} mol^{-1}) than in CH$_3$CN. The scope of transfer functions in establishing the factors controlling the medium effect on the binding process is demonstrated in the following section.

4. Thermodynamics of Complexation of Calixarene Derivatives and Ionic Species

4.1. STABILITY CONSTANTS AND DERIVED STANDARD GIBBS ENERGIES OF COMPLEXATION

Complexation of a cation (M^{n+}) or an anion (X^{n-}) by a neutral calixarene (L) in a solvent (s) may be represented in it simplest forms as

$$M^{n+}(s) + L(s) \xrightarrow{K_s} ML^{n+}(s) \quad (12)$$

$$X^{n-}(s) + L(s) \xrightarrow{K_s} LX^{n-}(s) \quad (13)$$

The thermodynamic equilibrium constant, K_s, for cation complexation is

$$K_s = \frac{a_{ML^{n+}}}{a_{M^{n+}} a_L} = \frac{[ML^{n+}]\delta_{\pm ML^{n+}}}{[M^{n+}]\delta_{\pm M^{n+}}[L]\delta_L} \cong \frac{[ML^{n+}]}{[M^{n+}][L]} \quad (14)$$

and for an anion complexation

$$K_s = \frac{a_{LX^{n-}}}{a_{X^{n-}} a_L} = \frac{[XL^{n-}]\delta_{\pm XL^{n-}}}{[X^{n-}]\delta_{\pm X^{n-}}[L]\delta_L} \cong \frac{[XL^{n-}]}{[X^{n-}][L]} \quad (15)$$

a, $\delta_\pm$ and δ denote activity, mean activity (in the case of ionic species) and activity coefficient (neutral ligand), respectively. The approximations (molar scale) hold in dilute solutions where $\delta_L \cong 1$; $\delta_{\pm ML}^{n+} = \delta_{\pm M}^{n+}$ and $\delta_{\pm LX}^{n-} = \delta_{\pm X}^{n-}$. In very dilute solutions, all activity coefficients may be set equal to unity.

If changes in the K_s values are observed by altering the counter-ion, additional processes (ion pair formation between the free or the complex species or both with the counter-ion) are taking place besides complexation. Derivation of the stability constant then requires corrections to be applied. The preferred choice would be to select a solvent for which there is experimental evidence that the electrolyte is predominantly dissociated. It is often assumed that if a free salt in a given medium and at given concentration and temperature is ionised, the complex salt will also generate ionic species in solution. This assumption is based on the fact that the latter involves a large ion (complex cation or anion) expected to have a lower tendency to enter ion-pair formation with the counter-ion than the former. Although this may be often the case, it must be verified in order to ensure that the equilibrium constant is a true stability constant, representative of the complexation process.

In complexation reactions involving bi-valent and tri-valent ions, ion association may be expected to be relatively strong and the difficulties involved in allowing for its effects have been described in, for example, in the case of some Ba(II) complexes

[35,36]. As discussed elsewhere [1], the lack of such analysis for alkaline earth metal complexes of calixarenes in methanol means that the stability constants reported for these systems can only be considered as "conditional". For some cryptand complexes of tripositive cations, it has even been shown that ion pairing is enhanced by the complexation [37] but there is hardly any investigation in which the behaviour of calixarene based electrolytes in solution has been considered [38]. Complex formation by an ion pair may be written

$$M^+X^-_{(s)} + L_{(s)} \xrightarrow{K_{assn}} ML^+X^-_{(s)} \qquad (16)$$

where the equilibrium constant K_{assn} is clearly not the same as K_s [1].

Conductance measurements have proved very useful not only to assess the extent of ion pair formation but also to determine the composition of the complex (which may vary with the solvent), and the strength of complexation as well as the extent of solvation of the free relative to the complex ions. Illustrative examples of curves showing conductance as a function of ligand:metal ratio in which some interesting features appear are given in Figure 1 [30]. For the titration of Eu^{3+} with the tetraethylcalix[4]arene ester, **8**, in acetonitrile at 298.15 K, there is only a slight slope of the curve and the point of inflection is not well defined, suggesting that the complex formed is relatively weak, though location of the point of inflection indicates it is a 1:1 species. A significant feature of these data is that the conductance increases as the complex is (inefficiently) formed, contrary to expectation if simply the sizes of the free and complexed species are considered. This can be explained if the free lanthanide cation is much more solvated than the complex cation and, therefore, more mobile. A similar rationalisation of the effects of the addition of crown ethers to lithium salts in CH_3CN and PC [39,40] was later verified by thermodynamic studies of these electrolytes in these solvents. An increase in conductance by the addition of the ligand is also observed for the interaction of Gd^{3+} and **6** in CH_3CN at 298.15 K (Fig. 1b). However, in this case the curve shows well-defined break at the reaction stoichiometry of 1:1, consistent with the formation of a stronger complex. By far the most interesting conductimetric titration is that of Tb(III) and a

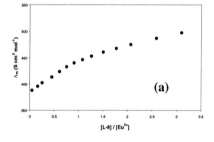
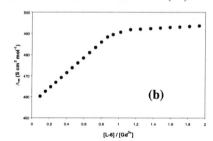
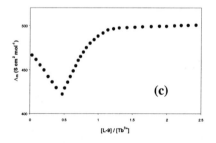

*Figure. 1. Conductometric titrations of a) Eu^{3+} by **8**, b) Gd^{3+} by **6**, c)Tb^{3+} by **9** in acetonitrile at 298.15 K.*

tetra(diethylaminoethyl)calix[4]arene, **9**, in CH_3CN [30] (Fig. 1c). Addition of the ligand to the electrolyte results in an initial decrease in conductance due to complex formation. The first end point is observed at a ligand/metal cation ratio of 0.5, indicating formation of a M_2L species. Further addition of the ligand causes an increase in conductance but for L:M >1, the variations in the conductance are negligible, indicating the second species to be a 1:1 complex. The concentrations specifically chosen for this work were such that the Ln^{3+} cation was the predominant species in CH_3CN. Knowing the solution speciation, selection of the method of determination of stability constants is the next critical step [1,35].

Whenever possible, it is desirable to use more than a single experimental technique to quantitatively characterise solution equilibria. Efforts to resolve ambiguities in earlier studies of complexation of unipositive cations by calixarene 1 were the basis of recent work (see Table 3). Study of pyridinocalixarenes **1** and **3**, using a variety of techniques (^1H NMR, conductometry, potentiometry, calorimetry, X-ray crystallography) has shown the importance of selecting a suitable solvent for complexation studies involving ionic species [6,25,41]. The results provided evidence rectifying conclusions [42] in error because of inadequate consideration of the behaviour of electrolytes in apolar media. Another important issue [1] concerns the use of the "extraction method" to characterise complexation. This has been described as a convenient though semi-quantitative tool for establishing comparisons between ion binding properties [43]. Thus, the degree of extraction of a metal cation from water to a non-aqueous phase has been used to assess the selectivity of calixarene esters and ketones for alkali-metal cations, and to assess the complexing abilities of **1** for alkali-metal cations. It should be explicitly stated that, in general, extraction data are not a suitable means to assess the affinity of a ligand for a particular ionic species in a given solvent (see Section 5). This assessment must be based on the stability constant data. log K_s values of **1** and alkali-metal cations (Table 3) follow the sequence $Li^+ > Na^+ > K^+ > Rb^+$ (>Cs^+, though values for this cation were immeasurably small). For complexes of various cations and **1** in the CH_3CN-PhCN solvent system, factors affecting their stability can be analysed in detail. Contributions to higher complex stabilities in s_2 relative to s_1 are those for which $\Delta_t G°$ functions are unfavourable for the reactants (better solvated in s_1 than in s_2) and favourable for the product (better solvated in s_2 than in s_1). This is best illustrated in the form of a thermodynamic cycle (expressed in terms of G) [44]:

$$
\begin{array}{ccccc}
M^{n+}(s_1) & + & L(s_1) & \xrightarrow{\Delta_c G°} & ML^{n+}(s_1) \\
\Big\downarrow \Delta_t G° & & \Big\downarrow \Delta_t G° & & \Big\downarrow \Delta_t G° \\
M^{n+}(s_2) & + & L(s_2) & \xrightarrow{\Delta_c G°} & ML^{n+}(s_2)
\end{array}
\quad (17)
$$

Using data from Tables 1 and 3, the $\Delta_t G°$ value (kJ mol^{-1}) for the complex can be calculated. Thus:

$$
\begin{array}{ccccc}
Li^+(CH_3CN) & + & \mathbf{1}(CH_3CN) & \xrightarrow{-33.96} & Li\mathbf{1}^+(CH_3CN) \\
\Big\downarrow 5.52 & & \Big\downarrow -8.05 & & \Big\downarrow 2.32 \\
Li^+(PhCN) & + & \mathbf{1}(PhCN) & \xrightarrow{-29.11} & Li\mathbf{1}^+(PhCN)
\end{array}
\quad (18)
$$

$$\text{Na}^+_{(\text{MeCN})} + \mathbf{1}_{(\text{MeCN})} \xrightarrow{-31.31} \text{Na}\mathbf{1}^+_{(\text{MeCN})}$$

$$\downarrow 6.77 \quad \downarrow -8.05 \quad \downarrow 2.97 \qquad (19)$$

$$\text{Na}^+_{(\text{PhCN})} + \mathbf{1}_{(\text{PhCN})} \xrightarrow{-27.06} \text{Na}\mathbf{1}^+_{(\text{PhCN})}$$

$$\text{K}^+_{(\text{MeCN})} + \mathbf{1}_{(\text{MeCN})} \xrightarrow{-17.90} \text{K}\mathbf{1}^+_{(\text{MeCN})}$$

$$\downarrow 11.17 \quad \downarrow -8.05 \quad \downarrow 0.41 \qquad (20)$$

$$\text{K}^+_{(\text{PhCN})} + \mathbf{1}_{(\text{PhCN})} \xrightarrow{-20.61} \text{K}\mathbf{1}^+_{(\text{PhCN})}$$

This shows that while the ligand and the metal-ion complex values favour complexation in CH_3CN, those for the free metal cation do not. For Li^+ and Na^+, the unfavourable contribution of the free cation for complexation in CH_3CN is overcome by that of the ligand and the complex cation and their complexes are slightly more stable in CH_3CN than in PhCN, while the opposite is true for K^+ and **1**

Concerning the medium effect on the thermodynamics of ion complexation by calixarenes, the Ag(I)/**6** system [8,31] is by far the best understood. $\Delta_t G°$ values for the reactants and the product can be used to calculate $\Delta(\Delta_c G°)$ in the CH_3CN-CH_3OH (-14.07 kJ mol^{-1}) and in the CH_3CN-DMF (-4.31 kJ mol^{-1}) systems, values corroborated by $\Delta_c G°$ values (see Table 3) derived from direct measurements of stability constants in the appropriate solvents. In the CH_3CN-DMF system, the strength of complexation is slightly favoured in DMF. This is attributed to both the free and the complex cation factors contributing favourably, these being only partially negated by ligand factors.

TABLE 3. Thermodynamics of complexation of calixarene derivatives and univalent cations in various solvents at 298.15 K.

Cation	Ligand	Solvent	Log K_s	$\Delta_c G°$/kJ mol^{-1}	$\Delta_c H°$/kJ mol^{-1}	$\Delta_c S°$/J K^{-1} mol^{-1}
Ag$^+$	6	CH$_3$CN[a]	3.41	-19.44	-23.16	-12.5
Ag$^+$	6	DMF[a]	4.16	-23.75	-39.39	-52.4
Ag$^+$	6	PhCN[a]	4.00	-22.83	-44.85	-73.8
Ag$^+$	6	CH$_3$OH[a]	5.87	-33.51	-44.39	-36.5
Ag$^+$	6	EtOH[a]	4.65	-26.54	-36.34	-32.9
Ag$^+$	6	1-PrOH[a]	4.58	-26.14	-36.92	-36.2
Li$^+$	1	CH$_3$CN[b]	5.95	-33.96	-23.91	33.7
Na$^+$	1	CH$_3$CN[b]	5.48	-31.31	-25.61	19.1
K$^+$	1	CH$_3$CN[b]	3.14	-17.90	-18.47	-1.9
Rb$^+$	1	CH$_3$CN[b]	2.58	-14.70	-15.50	-2.7
Ag$^+$	1	CH$_3$CN[b]	5.09	-29.06	-19.04	33.6
Li$^+$	1	PhCN[b]	5.10	-29.11	-33.77	-15.6
Na$^+$	1	PhCN[b]	4.74	-27.06	-24.43	8.8
K$^+$	1	PhCN[b]	3.61	-20.61	-8.78	39.7
Ag$^+$	3	CH$_3$CN[c]	2.63	-15.03	-25.63	-35.6

[a] Refs.[8,31], [b] Ref. [6], [c] Refs. [25,41]

$$\begin{array}{ccc}
\text{Ag}^+(\text{CH}_3\text{CN}) + \mathbf{6}\,(\text{CH}_3\text{CN}) & \xrightarrow{-19.44} & \text{Ag}\mathbf{6}^+(\text{CH}_3\text{CN}) \\
\downarrow 30.12 & \downarrow -3.11 & \downarrow 12.94 \\
\text{Ag}^+(\text{CH}_3\text{OH}) + \mathbf{6}\,(\text{CH}_3\text{OH}) & \xrightarrow{-33.51} & \text{Ag}\mathbf{6}^+(\text{CH}_3\text{OH})
\end{array} \qquad (21)$$

$$\begin{array}{ccc}
\text{Ag}^+(\text{CH}_3\text{CN}) + \mathbf{6}\,(\text{CH}_3\text{CN}) & \xrightarrow{-19.44} & \text{Ag}\mathbf{6}^+(\text{CH}_3\text{CN}) \\
\downarrow 4.65 & \downarrow -4.03 & \downarrow -3.69 \\
\text{Ag}^+(\text{DMF}) + \mathbf{6}\,(\text{DMF}) & \xrightarrow{-23.75} & \text{Ag}\mathbf{6}^+(\text{DMF})
\end{array} \qquad (22)$$

For anions, there is insufficient information on $\Delta_t G°$ values to proceed with useful discussion of complexation.

4.2. ENTHALPIES AND ENTROPIES OF COMPLEXATION

Titration calorimetry [45] is a relatively simple technique for measurement of thermodynamic quantities in solution. It can, however, involve large systematic errors if improperly conducted [10]. The successful application of titration calorimetry to a given system requires that i) the stability of the complex and the reaction conditions are such that a measurable amount of reaction takes place and ii) the enthalpy change for the reaction to be investigated is measurably different from zero. The common use of this technique is thus limited to log K_s values between 1 and 6.5 and enthalpy changes large enough so a measurable temperature change is generated in the reaction [46]. When log K_s values are higher than 6.5, log K_s can not be derived from direct calorimetry titrations but $\Delta_c H°$ can. Competitive calorimetry can be used to determine high stability constants, although errors are likely to be large. In titration macrocalorimetry, there are two problems often encountered in the field of calixarene chemistry. These are due firstly to the low solubility of the ligand in a given solvent, which leads to concentrations that are too low to generate measurable heats on complexation with the ion. In the complexation of **1** and monovalent cations (see Table 3) in CH_3CN at 298.15, for example, attempts to use classical titration calorimetry failed for this reason. Secondly, slow complexation kinetics cause problems with equipment designed on the assumption of rapid reaction. These problems are overcome in modern titration microcalorimetry, which is a much more sensitive technique and enables measurement of heats associated with slow processes [47]. In reactions of calixarene derivatives including **7** with lanthanide cations in CH_3CN, the complexation process is slow and very significant differences were found between the data obtained from microcalorimetry and those from macrocalorimetry. For some systems, log K_s values differed by 0.7 log units and $\Delta_c H°$ values differed by up to 40 kJ mol^{-1} [30]. To overcome this problem, most pronounced for highly-charged cations, it is advisable to find ancillary experimental evidence (*e.g.* conductance measurements) in order to determine the concentration range at which the effects of ion-pair formation are minimised. Although enthalpy data have been reported for dipositive ions and calixarene derivatives in polar solvents, neglect of ion association means they are likely to be in error [1].

Even for systems involving 1+ cations and calixarene derivatives in polar media, ion pair formation at high temperatures has rarely been studied and no information is available for 2+ and 3+ ions. Consequently, enthalpy data derived from equilibrium measurements for such species at different temperatures [48] are meaningless

Although studies [6,25,41] of complexation of alkali-metal and silver cations by pyridinocalix[4]arenes have not provided enthalpy and therefore no entropy data for transfer from CH_3CN to $PhCN$, the endothermic character of the $\Delta_t H°$ for these metal cations shows that the stability in enthalpic terms is greater in CH_3CN than in $PhCN$ and therefore, this factor contributes favourably to a higher enthalpic stability in $PhCN$ than in CH_3CN. The fact that $\Delta\Delta_c H^0$ values are negative for Li^+, approximately zero for Na^+ and positive for K^+ gives a clear indication that the transfer enthalpy of the ligand and of the metal-ion complex need to be taken into account in determining the factors which contribute to the enthalpic stability of complex formation. The latter statement is also valid in terms of entropy, where striking differences are observed in acetonitrile relative to benzonitrile. As far as the thermodynamic functions for Ag^+ and **6** in the various solvents are concerned (Table 3), enthalpy (kJ mol^{-1}) and entropy (J K^{-1} mol^{-1}) data can be inserted in the following thermodynamic cycle, using as representative examples the solvent systems discussed in terms of Gibbs energy.

$\Delta(\Delta_c H°)$ values of -21.23 kJ mol^{-1} and -16.23 kJ mol^{-1} in the CH_3CN-CH_3OH and in the CH_3CN-DMF solvent systems, respectively, agree well with those obtained from transfer data (section 3.2). This applies in terms of entropy also. Individual parameters contribute to medium effects in two distinctive patterns. Thus, in the CH_3CN-CH_3OH solvent system, the higher enthalpic stability in the alcohol is due to the contribution of both the free and complex cations.

MeCN-MeOH

$$Ag^+ (MeCN) + \mathbf{6}\,(MeCN) \xrightarrow{\Delta_c H° = -23.16,\ \Delta_c S° = -12.5} Ag\mathbf{6}^+ (MeCN)$$

$\Delta_t H° = 18.85$, $\Delta_t S° = -37.8$ $\quad$ $\Delta_t H° = -11.43$, $\Delta_t S° = -27.9$ $\quad$ $\Delta_t H° = -14.10$, $\Delta_t S° = -90.7$ $\qquad$ (23)

$$Ag^+ (MeOH) + \mathbf{6}\,(MeOH) \xrightarrow{\Delta_c H° = -44.39,\ \Delta_c S° = -36.5} Ag\mathbf{6}^+ (MeOH)$$

MeCN-DMF

$$Ag^+ (MeCN) + \mathbf{6}\,(MeCN) \xrightarrow{\Delta_c H° = -3.6,\ \Delta_c S° = -.5} Ag\mathbf{6}^+ (MeCN)$$

$\Delta_t H° = 2.18$, $\Delta_t S° = -8.3$ $\quad$ $\Delta_t H° = -1.34$, $\Delta_t S° = 9.1$ $\quad$ $\Delta_t H° = -16.03$, $\Delta_t S° = -41.4$ $\qquad$ (24)

$$Ag^+ (DMF) + \mathbf{6}\,(DMF) \xrightarrow{\Delta_c H° = -39.39,\ \Delta_c S° = -52.4} Ag\mathbf{6}^+ (DMF)$$

The ligand being enthalpically more stable in CH_3OH does not contribute favourably to the $\Delta_c H°$. For CH_3CN-DMF, the complexation process is enthalpically more favoured in DMF than in CH_3CN. This is mainly due to the higher enthalpic stability of the metal–ion complex in DMF. Indeed, the small transfer enthalpies of the free metal cation (favourable) and the ligand (unfavourable) barely contribute to the relatively large variations observed in the complexation enthalpies in these two solvents.

Similar analysis of entropy terms shows that complexation in CH_3OH and in DMF involves a greater loss of entropy than in CH_3CN, largely due to the unfavourable transfer entropies of the metal-ion complex that overcome those of the ligand and the free metal cation. Although moving from CH_3CN to a protic solvent (CH_3OH) and a dipolar aprotic solvent (DMF) leads to $\Delta(\Delta_c H°)$ values (CH_3CN-CH_3OH = -21.23 kJ mol^{-1}; CH_3CN-DMF = -16.23 kJ mol^{-1}) that are not dramatically different, the thermochemical origins of these differences (reflected in $\Delta_t H°$ values) are.

4.3. ENTHALPIES OF CO-ORDINATION

The standard enthalpy of co-ordination, $\Delta_{coord} H°$ involving 1:1 salts (MX and MLX) and L refers to the process

$$MX_{(sol.)} + L_{(sol.)} \xrightarrow{\Delta_{coord} H°} MLX_{(sol.)} \quad (25)$$

where the reactants and the product are in their pure (solid) physical state [1,35]. From the combination of solution and complexation thermodynamic functions, this parameter can be calculated by the use of eq. 26,

$$\Delta_{coord} H° = \Delta_{soln} H°(M^+ + X^-)_{(s)} + \Delta_{soln} H°(L)_{(s)} + \Delta_c H°_{(s)} - \Delta_{soln} H°(ML^+ + X^-)_{(s)} \quad (26)$$

Eq. 26 can be also formulated in terms of Gibbs energies and entropies but there is little information regarding $\Delta_{coord} G°$ and $\Delta_{coord} S°$ for systems involving calixarene derivatives. This is mainly due to the lack of ancillary information that is required for the derivation of $\Delta_{soln} G°$ of calixarene based salts. For a given system and provided that no side reactions occur, the value of $\Delta_{coord} H°$ should be the same, regardless of the solvent involved. This is illustrated in Table 4 (see footnote), where the standard enthalpy of co-ordination for the Ag6ClO$_4$ system has been derived from solution and complexation data in four different solvents. Therefore, $\Delta_{coord} H°$ provides an useful mean of checking the accuracy of the various parameters involved in eq, 26.

For the solid state, the counter-ion participates in the co-ordination process. This is illustrated in the data shown in Table 4 involving Li and Na salts and ligand **8**. Thus, for systems containing Li, co-ordination enthalpies decrease in the following order,

$$I^- > ClO_4^- > Br^- > CF_3SO_3^- > AsF_6^- > BF_4^- \quad (27)$$

The same sequence was found for the system investigated with sodium. It is possible that this parameter correlates with anion polarisability.

The ligand effect on $\Delta_{coord} H°$ may be compared to corresponding data involving crown ethers (12-crown-4 (12C4); 15-crown-5 (15C5)) and cryptand 222 (222) (Table 4) [34]. Thus, for systems containing Li$^+$ and the same anion, coordination enthalpies for cryptate salts are very close to those involving *p-tert*-butylcalix[4]arene tetraester, **8**. However, much higher enthalpic stabilities are observed for Li/crown ether complexes, particularly for systems involving fluorine-containing anions. This is likely to be due to a direct interaction between Li$^+$ and the counter-ion. The data indicate considerable weakening of the anion-cation interaction in Li cryptates and calixarene complexes which may be due to the more effective shielding of the cation by the hydrophilic cavities of these ligands relative to crown ethers, which cause the metal ion to be bound in a hole rather than in a cavity [34].

TABLE 4. Enthalpies of Co-ordination of Calixarene Salts at 298.15 K.

System	$\Delta_{coord}H°/kJ\ mol^{-1}$
$AgClO_{4(sol.)} + 6_{(sol.)} \rightarrow Ag6ClO_{4(sol.)}$ [a]	-71.78[b]
$AgClO_{4(sol.)} + 6_{(sol.)} \rightarrow Ag6ClO_{4(sol.)}$ [a]	-72.40[c]
$AgClO_{4(sol.)} + 6_{(sol.)} \rightarrow Ag6ClO_{4(sol.)}$ [a]	-69.47[d]
$AgClO_{4(sol.)} + 6_{(sol.)} \rightarrow Ag6ClO_{4(sol.)}$ [a]	-72.07[e]
$LiAsF_{6(sol.)} + 8_{(sol.)} \rightarrow Li8AsF_{6(sol.)}$	-22.1[f]
$LiBF_{4(sol.)} + 8_{(sol.)} \rightarrow Li8BF_{4(sol.)}$	-19.7[f]
$LiCF_3SO_{3(sol.)} + 8_{(sol.)} \rightarrow Li8CF_3SO_{3(sol.)}$	-27.7[f]
$LiClO_{4(sol.)} + 8_{(sol.)} \rightarrow Li8ClO_{4(sol.)}$	-60.2[f]
$LiI_{(sol.)} + 8_{(sol.)} \rightarrow Li8I_{(sol.)}$	-77.4[f]
$LiBr_{4(sol.)} + 8_{(sol.)} \rightarrow Li8Br_{(sol.)}$	-37.4[f]
$NaBF_{4(sol.)} + 8_{(sol.)} \rightarrow Na8BF_{4(sol.)}$	-5.6[f]
$NaCF_3SO_{3(sol.)} + 8_{(sol.)} \rightarrow Na8CF_3SO_{3(sol.)}$	-20.0[f]
$NaClO_{4(sol.)} + 8_{(sol.)} \rightarrow Na8ClO_{4(sol.)}$	-31.9[f]
$NaI_{(sol.)} + 8_{(sol.)} \rightarrow Na8I_{(sol.)}$	-49.5[f]
$LiBF_{4(sol.)} + 12C4_{(sol.)} \rightarrow Li12C4BF_{4(sol.)}$	-45.0[f]
$LiBF_{4(sol.)} + 15C5_{(sol.)} \rightarrow Li15C5BF_{4(sol.)}$	-47.1[f]
$LiBF_{4(sol.)} + 222_{(sol.)} \rightarrow Li222BF_{4(sol.)}$	-18.1[f]
$LiClO_{4(sol.)} + 222_{(sol.)} \rightarrow Li222ClO_{4(sol.)}$	-61.0
$NaClO_{4(sol.)} + 222_{(sol.)} \rightarrow Na222ClO_{4(sol.)}$	-58.1[f]
$NaBr_{(sol.)} + 222_{(sol.)} \rightarrow Na222Br_{(sol.)}$	-43.1[f]
$NaI_{(sol.)} + 222_{(sol.)} \rightarrow Na222I_{(sol.)}$	-61.4

[a] Ref. [31], [b] Eq. 27 using CH_3OH, [c] Eq. 27 using EtOH, [d] Eq. 27 using 1-PrOH, [e] Eq. 27 using CH_3CN, [f] Ref. [34]

The cation effect is reflected in the co-ordination enthalpies, and systems containing **8** and the same anion show a decrease in enthalpic stability in moving from lithium to sodium. These data are therefore good reporters of (i) the anion effect for systems containing the same ligand and metal cation and (ii) the cation effect for systems containing the same ligand and anion. However, the availability of crystal lattice enthalpies for these ligands and their ionic complex salts would substantially improve our present understanding of these systems.

5. Extraction Processes

Solvent extraction of metal ions by calixarenes has been widely studied and discussed [1,43,49,50] (See also Chapters 21, 22.). In general, other systems arise [51], but here the most common situation is one in which the calixarene is highly insoluble in water, so that the simplest case of extraction of a 1:1 electrolyte by formation of a 1:1 complex can be represented as

$$M^+(w) + X^-(w) + L(s) \xrightarrow{K_{ex}} MLX(s) \tag{28}$$

Ionic species are predominant in the aqueous phase saturated with the solvent and ion pairs are the species present in the water saturated organic phase. It has been stated that K_{ex} is related to the stability constant, K_s of the metal-ion-ligand complex in water and to the extracting ability of the system under the conditions described by eq. 29 [43]. To clarify this statement, it is useful to consider the more complete representation of the system shown in Scheme 1. In this scheme K_p, K'_p and K_{pL} denote the partition constants of the free and complex electrolyte and the ligand respectively in the mutually saturated solvents. K_a and K'_a are the ion pair formation constants between the free and complex cation with the anion in the water saturated organic phase. Therefore, the distribution constants of the free ($K_d = K_p \cdot K_a$) and the complex species ($K'_d = K'_p \cdot K'_a$) can be obtained.

$$
\begin{array}{ccccc}
M^+(w) + X^-(w) + L(w) & \xrightarrow{K_{s(w)}} & ML^+(w) + X^-(w) \\
\downarrow K_p \quad \downarrow K_{pL} & & \downarrow K'_p \\
M^+(s) + X^-(s) + L(s) & \xrightarrow{K_{s(s)}} & ML^+(s) + X^-(s) \\
\downarrow K_a \quad \downarrow K'_{pL}=1 & & \downarrow K'_a \\
MX(s) + L(s) & \xrightarrow{K_{assn}} & MLX(s)
\end{array}
\tag{29}
$$

The stability constant in water, $K_{s(w)}$ and organic phase, $K_{s(s)}$ (eq. 14) and the association constant, K_{assn} (eq. 16) are referred to water (saturated with the organic phase) and to the organic phase (saturated with water). Thus, the extraction constant, K_{ex}, can therefore be expressed in terms of the individual equilibrium constants, .

$$K_{ex} = K_{s(w)} \cdot K'_d / K_{pL} = K_d \cdot K_{ass} = K_p \cdot K_{s(s)} \cdot K'_a \tag{30}$$

It follows from eq. 30 that K_{ex} is related to $K_{s(w)}$ but also to K'_d and K_{pL}. For the same ligand and different cations the value of K'_d may be approximately constant *if the complexed cations are well shielded by the* ligand. Since the K_{pL} value is the same, K_{ex} will be proportional to K_s (in the water saturated solvent), on the assumption that eq. 28 is representative of the extraction process, if (i) the complex salt is equally extracted in the organic phase, (ii) the metal-ion complex salt, MLX is not strongly dissociated in the organic phase. If dissociation occurs, eq. 30 has to be reformulated on the basis of the above cycle (eq. 29). Again provided that the cation is well shielded by the ligand, this dissociation may not depend much on the cation involved. Fundamentally, *provided the complexed cation is sufficiently shielded from the solvent,* K_{ex} will be approximately proportional to the complexation in the aqueous phase saturated with the solvent. The problem is that eq. 28 has been formulated without checking that this is representative of the process in the solvent system considered. A typical example is the water-dichloromethane (CH_2Cl_2) solvent system [43]. In fact, ion-pair formation constants of lithium and sodium ethyl *p-tert*-butylcalix[6]arene hexanoate picrates in CH_2Cl_2 saturated with water show that the extent of dissociation of these complexes even at concentrations of 10^{-2} mol dm^{-3} is quite considerable in this solvent system saturated with water. On the other hand, there is no evidence that in CH_2Cl_2 saturated with water, the alkali-metal complexed cations are well shielded by the ligand. Furthermore, neither $K_{s(w)}$ nor K_{pL} or K'_d can be experimentally determined due to solubility limitations.

In the face of such deficiencies, correlations have been made between extraction data (%E) in the water-dichloromethane solvent system with stability constants in methanol [43]. In the context of calixarene chemistry, the use of methanol as representative of water is hardly justified. In this case, alternative routes would be to assess K_{ex} in terms of K_{assn} [1] and K_d or indeed in terms of K_p, $K_{s(s)}$ and K'_a (see eq. 30) depending whether the extraction takes place through the ion-pair or the free metal cation interacting with the ligand in the water saturated organic phase. Most of these parameters can be experimentally determined. In fact K_d values of alkali-metal picrates in the mutually saturated water-dichloromethane solvent system have been recently determined from distribution data in the presence and in the absence of **8** together with K'_a values in water saturated dichloromethane [52].

6. Conclusions

Calixarenes are indeed versatile macrocycles but most systems in which they act as complexing agents are exceedingly complicated. In a single medium, more than one process may occur and these demand delicate thermodynamic analysis. The extrapolation of findings from ion complexation processes involving other macrocycles such as crown ethers and cryptands to the field of calixarene coordination chemistry is far from simple and obvious.

In the general area of thermodynamics of calixarene chemistry involving ionic species in different media there is plenty of scope for further developments. Areas which require more research efforts are:

i) Anion complexation processes. No detailed investigations have been carried out on these systems. (See, however, Chapters 23 and 34.)

ii) Solution behaviour of calixarene derivatives and their ionic complexes.

iii) Heat capacity measurements. Particular attention should be focused on the factors governing selectivity as a function of the temperature across a wide variety of derivatives and ionic species in different media.

iv) Ionic complexation in mixed solvents.[53]

In addition, some of the systems investigated need to be revisited. The complexing abilities of currently used ligands such as narrow rim calixarenes esters, ketones and amides for ionic species continue to demand careful attention [53].

Acknowledgements: A. F. D. de N. thanks Dr. B. G. Cox (Zeneca) for useful discussions. The author is grateful to many of her co-workers whose names appear in the list of references. The financial support provided by the EU under contract ERBICI8CT970140 is gratefully acknowledged.

7. References and Notes

[1] A. F. Danil de Namor, R. M. Cleverly, M. L. Zapata-Ormachea, *Chem. Rev.* **1998**, *98*, 2495-2525.
[2] C. D. Gutsche, *Calixarenes*, Monographs in Supramolecular Chemistry, Cambridge, **1989**.
[3] J. Vicens, V. Böhmer, (Eds.) *Calixarenes. A Versatile Class of Macrocyclic Compounds*, Kluwer Academic Publishers, Dordrecht, **1991**.
[4] J. Vicens, Z. Asfari, J. M. Harrowfield, (Eds.) *Calixarenes 50th Anniversary:* Commemorative Volume, Kluwer Academic Publishers, Dordrecht, **1994**
[5] C. D. Gutsche, *Aldrichim. Acta* **1995** *28*, 3-9.
[6] A. F. Danil de Namor, O. E. Piro, L. E. Pulcha Salazar, A. Aguilar Cornejo, N. Al-Rawi, E. E. Castellano, F. J. Sueros Velarde, *J. Chem. Soc., Faraday Trans.* **1998**, *20*, 3097-3104.
[7] A. F. Danil de Namor, N. Al-Rawi, O. E. Piro, E. E. Castellano, to be submitted for publication, **2000**.
[8] A. F. Danil de Namor, M. L. Zapata-Ormachea, R. G. Hutcherson, *J. Phys. Chem.* **1998**, *102*, 7839-7844.
[9] A. F. Danil de Namor, R. G. Hutcherson, F. J. Sueros Velarde, A. Alvarez-Larena, J. L. Briansó, *J. Chem. Soc., Perkin Trans. 1* **1998**, 2933-2938.
[10] I. Wadsö, *J. Chem. Soc.; Chem. Soc. Rev.* **1997**, 79-86 and references therein.
[11] M. Stödeman, I. Wadsö, *Pure Appl. Chem.* **1995**, *67*, 1059-1068.
[12] A. F. Danil de Namor, M. L. Zapata-Ormachea, O. Jafou, N. Al-Rawi, *J. Phys. Chem. B.* **1997** *101*, 6772-6779.
[13] A. F. Danil de Namor, E. Gil, M. A. Llosa Tanco, D. A. Pacheco Tanaka, L. E. Pulcha Salazar, R. A. Schulz, J. Wang, *J. Phys. Chem.* **1995**, *99*, 16781-16785.
[14] A. Varnek, G. Wipff, *J. Phys. Chem.* **1993**, *97*, 10840-10848.
[15] E. Grunwald, C. Steel, *J. Am. Chem. Soc.* **1989**, *117*, 5687-5692.
[16] B. G. Cox, G. R. Hedwig, A. J. Parker, D. W. Watts, *Aust. J. Chem.* **1974**, *27*, 477-501.
[17] C. M. Criss in *Physical Chemistry of Organic Solvent Systems*, (Eds.: A.K. Covington, T. Dickinson) Plenum Press, London and New York, **1973**, pp 23-135.
[18] C. M. Criss, M. Salomon in *Physical Chemistry of Organic Solvent Systems*, (Eds.: A.K. Covington, T. Dickinson) Plenum Press, London and New York, **1973**, pp 253-329.
[19] O. Popovich, R. P. T. Tomkins, *Nonaqueous Solution Chemistry*, J. Wiley & Sons Inc., **1981**
[20] Y. Marcus, *Ion Solvation*, Wiley, Chichester, **1985**.
[21] M. H. Abraham, A. F. Danil de Namor, *J. Chem. Soc., Faraday Trans. 1* **1976**, *72*, 955-962.
[22] M. H. Abraham, A. F. Danil de Namor, *J. Chem. Soc., Faraday Trans. 1* **1978**, *74*, 2101-2110.
[23] G. Janz, R. P. T. Tomkins, *Nonaqueous Electrolytes Handbook*, Academic Press, New York, **1972**.
[24] B. Kratochvil, H. L. Yeager, *Top. Curr. Chem.* **1972**, *27*, 1.
[25] A. F. Danil de Namor, *Coord. Chem. Rev.* **1999**, *190-192*, 283-295.
[26] A. Seminara, E. Rizzarelli, *Inorg. Chim. Acta* **1980**, *40*, 249-256.
[27] J.-C. G. Bünzli, V. Kasparev, *Inorg. Chim. Acta* **1991**, *182*, 101-107.
[28] F. Pilloud, J-C. G. Bünzli, *Inorg. Chim. Acta* **1987**, *139*, 153-154.
[29] P. Di Bernardo, G. R. Choppin, R. Portanova, P. L. Zanonato, *Inorg. Chim. Acta* **1993**, *207*, 85-91.
[30] a) A. F. Danil de Namor, O. Jafou, O. Sueros Velarde, **2000** unpublished work; b) A. F. Danil de Namor, Inaugural Lecture, Natl. Academy of Sciences, Argentina, **1999**.
[31] A. F. Danil de Namor, M. L. Zapata-Ormachea, R. G. Hutcherson, *J. Phys. Chem. B.* **1999**, *103*, 366-371.
[32] Y.-C. Wu, H. L. Friedman, *J. Phys. Chem.* **1966**, *70*, 501-509.
[33] M. H. Abraham in *Thermochemistry and its Applications to Chemical and Biochemical Systems*, (Ed. M. A. V. Ribeiro da Silva), Reidel Publishing Company, Dordrecht, **1984**, pp. 393-409.
[34] A. F. Danil de Namor, L. E. Pulcha Salazar, M. A. Llosa Tanco, D. Kowalska, J. Villanueva-Salas, R. A. Schulz, *J. Chem. Soc., Faraday Trans.* **1998**, *94*, 3111-3115.
[35] B. G. Cox, H. Schneider, *Coordination and Transport Properties of Macrocyclic Compounds in Solution*, Elsevier Science Publishers, N. Y., **1992**.
[36] M. K. Chantooni, I. M. Kolthoff, *Proc. Natl. Acad. Sci. USA* **1981**, *78*, 7245-7247.

[37] O. H. Gansow, A. R. Kausar, K. M. Triplett, M. J. Weaver, E. L. Yee, *J. Am. Chem. Soc.* **1977**, *99*, 7087-7088.
[38] A. F. Danil de Namor, M. C. Cabaleiro, B. M. Vuano, M. Salomon, O. Pieroni, D. Pacheco Tanaka, C. Y. Ng, M. A. Llosa Tanco, N. M. Rodriguez, J. D. Cardenas-Garcia, A. R. Casal, *Pure & Appl. Chem.* **1994**, *66*, 436-440.
[39] A. F. Danil de Namor, M. A. Llosa Tanco, M. Salomon, J. C. Y. Ng, *J. Phys. Chem.* **1994**, *98*, 11796-11802.
[40] A. F. Danil de Namor, J. C. Y. Ng, M. A. Llosa Tanco, M. Salomon, *J. Phys. Chem.* **1996**, *100*, 14485-14491.
[41] A. F. Danil de Namor, E. E. Castellano, L. E. Pulcha Salazar, O. E. Piro, O. Jafou, *Phys. Chem. Chem. Phys.* **1999**, *1*, 285-293.
[42] S. Pappalardo, G. Ferguson, P. Neri, C. Rocco, *J. Org. Chem.* **1995**, *68*, 4576-4584.
[43] M. A. McKervey, M-J. Schwing-Weill, F. Arnaud in *Comprehensive Supramolecular Chemistry*, (Ed. G. W. Gokel), Elsevier Science Publisher, N. Y., **1996**.
[44] M. H. Abraham, A. F. Danil de Namor, W. H. Lee, *J. Chem. Soc., Chem. Commun.* **1977**, 893-894.
[45] D. J. Eatough, J. J. Christensen, R. M. Izatt, *Experiments in Thermometric Titrimetry and Titration Calorimetry,* Brigham University Press, Provo, Utah, USA, **1974**.
[46] J. J. Christensen, J. Ruckman, D. J. Eatough, R. M. Izatt, *Thermochim. Acta* **1972**, *3*, 203-218.
[47] J. Suurkuusk, I. Wadsö, *Chem. Scripta* **1982**, *20*, 155-163.
[48] S. Shinkai, S. Nisho, M. Takeshita, *J. Org. Chem.* **1994**, *59*, 4032-4034.
[49] A. F. Danil de Namor, J. F. Sueros Velarde, A. R. Casal, A. Pugliese, M. T. Goitia, M. Montero, F. Fraga Lopez, *J. Chem. Soc., Farady Trans.* **1997**, *93*, 3955-3959.
[50] A. F. Danil de Namor, M. T. Goitia, A. R. Casal, F. J. Sueros Velarde, M. I. Gonzalez Barja, J. A. Villanueva-Salas, M. L. Zapata-Ormachea, *Phys. Chem. Chem. Phys.* **1999**, *1*, 3633-3638.
[51] M. H. Abraham, A. F. Danil de Namor, R. A. Schulz, *J. Chem. Soc., Faraday Trans. 1* **1980**, *76*, 869-884.
[52] A. F. Danil de Namor, A. Pugliese, A. R. Casal, M. Barrios Llerena, J. F. Sueros Velarde, P. A. Aymonino, *Phys. Chem. Chem. Phys.* **2000**, *2*, 4355-4360.
[53] A. F. Danil de Namor, D. Kowalska, Y. Marcus, **2000**; A. F. Danil de Namor, S. Shahine, D. Kowalska, **2000**; work in progress.

Chapter 20

CROWN ETHERS DERIVED FROM CALIX[4]ARENES

ALESSANDRO CASNATI,[a] ROCCO UNGARO,[a] ZOUHAIR ASFARI,[b]
JACQUES VICENS[b]

[a] *Dipartimento di Chimica Organica e Industriale dell'Università, Parco Area delle Scienze 17/A, I-43100 Parma, Italy. e-mail : ungaro@unipr.it*
[b] *ECPM, Groupe de Chimie des Interactions Moléculaires Spécifiques, UMR 7512 du CNRS, 25 rue Becquerel F-67087 Strasbourg, France.
e-mail : vicens@chimie.u-strasbg.fr*

1. Introduction

Calix[4]arene-crown ethers or calix[4]crowns are one of the most widely investigated class of cation ligands based on calixarenes [1]. They show binding properties towards alkali metal, alkaline earth metal and ammonium cations which can be tuned by subtle conformational changes around the binding region. In some cases, binding appears to involve not just the ether-oxygen donors, but also the calixarene aromatic nuclei, providing further experimental evidence for the operation of cation/π interactions [2,3].

The first calix[4]arene-crown ethers synthesised were the 1,3-dihydroxycalix[4]-crowns **1a** and **2a** derived from *p-tert*-butylcalix[4]arene [4,5]. These ligands have two OH groups located on opposite sides of but in close proximity to the polyether ring. They were designed to act as *ionisable ligands* by loss of the protons in basic conditions and thus to form neutral complexes with monovalent or divalent cations. U-tube transport experiments proved that these two prototypes were able indeed to transport monovalent cations from a basic source phase to a receiving acidic phase, **1a** showing selectivity for potassium ion and **2a** for cesium ion. These early experiments showed that the introduction of the crown ether loop at the lower rim of calix[4]arenes not only increased the cation binding ability of the parent *p-tert*-butylcalix[4]arene, but allowed control of the selectivity through the modulation of the crown ether size. This chapter is focussed on advances in this area of calixarene chemistry concerning synthesis and fundamental aspects of the coordination chemistry of sophisticated calix[4]crowns. Applications in sensors and other devices, and in chromatography are discussed in Chapters 15, 34 and 35.

2. Calix[4]arene-mono-Crowns

2.1. SYNTHESIS AND CONFORMATIONAL PROPERTIES

The synthesis of 1,3-dihydroxy-*p-tert*-butylcalix[4]arene-crown ethers **1a** and **2a** was originally performed by reacting *p-tert*-butylcalix[4]arene with tetra- or pentaethylene glycol ditosylates in benzene in the presence of *t*-BuOK as base [6,7]. However, this reaction gives several by-products including calix[4]bis-crowns (*vide infra*) and a better and more general synthetic route was later developed (Scheme 1) [8]. It exploits the selective 1,3-dimethylation of the calix[4]arenes at the lower rim [6]. The bridging of the two remaining OH groups with the appropriate polyethylene glycol ditosylates using Cs_2CO_3 as base, followed by the selective removal of the methoxy groups with trimethylsilyl iodide in $CHCl_3$ gives all calix[4]crowns in good yields [8].

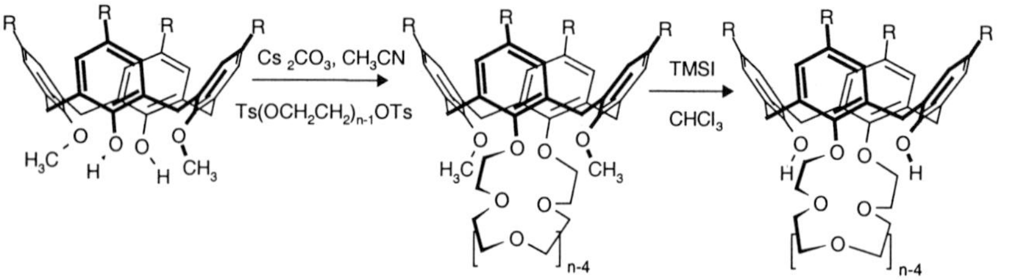

Scheme 1

1: R = *t*-Bu, n = 5
2: R = *t*-Bu, n = 6
3: R = H, n = 4
4: R = H, n = 5
5: R = H, n = 6
6: R = H, n = 7

1a: R = *t*-Bu, n = 5
2a: R = *t*-Bu, n = 6

A very important role in building an understanding of complexation features by calix[4]crowns has been played by the 1,3-dimethoxycalix[4]crowns **1-5**. The X-ray crystal structures of these ligands show that they adopt a *pinched cone* conformation in the solid state, which is also the preferred one in solution [9]. However, in the complexes the conformational outcome is dependent upon several factors such as the solvent, the nature of the cation salt and of the upper rim substituents. In chloroform solution, binding of potassium picrate [K(pic)] induces a change to the *partial cone* for ligand **1** [7] and to the *1,3-alternate* conformation for **4** [9]. The *tert*-butyl groups at the upper rim of ligand **1** inhibit the complexation of potassium by the *1,3-alternate* conformer which, however, is most suited for binding when the aromatic nuclei have no substituents. Complexation of Na(pic) leaves compound **5** in the *cone* conformation whereas Cs(pic) converts it to the *1,3-alternate* structure [8,10]. This different behaviour is clear from the form of the ^{1}H NMR spectra of ligand **5** (Figure 1a) and its cation complexes (Figure 1b,c). The spectrum of the sodium complex (1c) displays an AX system (J = 14Hz) for the diastereotopic *pseudo*-axial (●) and *pseudo*-equatorial (■) methylene bridge protons, typical of a calix[4]arene *cone* structure, whereas that of the cesium complex (Figure 1b) shows only a singlet for the two protons (◆), reflecting their much more similar environments in a *1,3-alternate* structure. Conclusive evidence for this conformational change comes from the X-ray crystal structure of the cesium picrate complex of receptor **5** [10]. Interestingly, this structure shows that the cesium ion is bound not only to the oxygen atoms of the polyether ring but also to the two opposite aromatic rings of the inverted phenolic units of the calix[4]arene. Evidence for this interaction was also obtained in solution by NMR studies.

The 1,3-dimethoxycalix[4]arene-crown-4 (**3**) was also synthesised using a procedure identical to that of Scheme 1. This compound, due to the reduced length of the polyether bridge, is conformationally locked on the NMR time-scale at room temperature as a mixture of the three possible conformations [11].

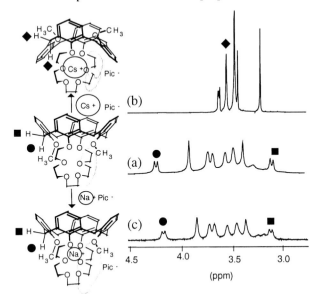

Figure 1. ^1H NMR spectrum in CD_3CN at room temperature of (a) ligand **5**, (b) its cesium complex and (c) its sodium complex.

An interesting example of a conformationally mobile compound is the calix[4]-arene-diquinone crown-5 (**7**) synthesised through the oxidation of calix[4]crown **1** with thallium (III) trifluoroacetate [12]. Compound **7**, as for most of the calix[4]arene-1,3-diquinone derivatives [12,13] is conformationally mobile in solution but assumes a *cone* structure in its 1:1 complexes with sodium and potassium cations. The dependence of calix[4]arene conformation upon metal ion binding has been recently exploited to develop ionophores that exhibit both chromogenic and electrochemical recognition properties [14]. The redox-switch *p-tert*-butylcalix[4]arene-crown ether **8**, which is in the *cone* conformation as free ligand, adopts the *partial cone* upon binding of potassium ion [15]. Similar behaviour is also shown by the *p-tert*-butylcalix[4]arene-crown-5 (**9**) bearing a 3-thienyl and a methyl group on the two remaining opposite phenolic oxygens [16].

The synthesis and complexing properties of bis-deoxycalix[4]arene-crown-6 derivatives (*e.g.* **10**) with two opposite aromatic nuclei devoid of OR groups, have been described [17]. These two nuclei can freely rotate inside the calix[4]arene annulus, showing a conformational behaviour very similar to that observed in 1,3-calix[4]arene diquinones [13] or for OH-depleted calix[4]arenes [18-20]. Compound **10** has been proposed as a new type of cesium selective ionophore for radioactive waste treatment [17].

The information obtained in the study of conformationally mobile 1,3-dimethoxycalix[4]arene-crowns **1-5** and their metal ion complexes has been used for the design and synthesis of preorganized receptors locked in the preferred conformation for binding. This was easily accomplished by introducing alkyl groups larger than CH_3 in the 1,3-positions. The 1,3-diethoxycalix[4]crowns give different stereoisomers which have been isolated and fully characterised [7]. The structural and binding properties of the different conformers of 1,3-diethoxy-*p-tert*-butylcalix[4]arene-crown-5 have been discussed in several review articles [21-23]. More recently, attention has been devoted to the *p*-H-calix[4]crowns. The general synthesis of the *1,3-alternate* conformers is outlined in Scheme 2.

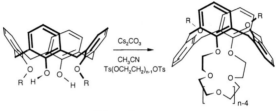

Scheme 2

11: R = *i*-Pr, n = 5
12: R = *i*-Pr, n = 6

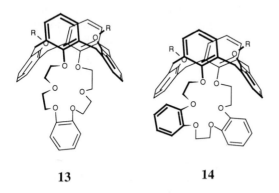

13   14

Yet more recently, based on assessment of factors influencing selectivity in metal ion binding by molecular modeling and analysis of X-ray structures, syntheses of *1,3-alternate* calix[4]-arene-benzo- (**13**) and –dibenzocrown-6 (**14**) incorporating one or two aromatic units in the crown ether loop, were devised [24-26].

In order to compare the binding properties of different stereoisomers, the *cone* **15-17** and the *partial cone* **18-19** isomers were also synthesised [8,9,27] using the strategies depicted in Scheme 3, which exploit general synthetic routes [28] for the selective obtainment of calix[4]arene conformational isomers. The synthesis of the three isomers of calix[4]arene-crown-6 bearing branched aliphatic chains has been also described [29].

15: R = Et, n = 5
16: R = *i*-Pr, n = 5
17: R = *i*-Pr, n = 6

18: R = *i*-Pr, n = 5
19: R = *i*-Pr, n = 6

Scheme 3

All three possible isomers of calix[4]crowns bearing two α-picolyl pendent groups have been synthesised [30]. The *cone* (30% yield) and the *partial cone* (44% yield) isomers are obtained in mixture by treating *p-tert*-butylcalix[4]arene-crown-5 **1a** with excess of 2-chloromethylpyridine hydrochloride in the presence of Cs_2CO_3 in DMF, whereas the *1,3-alternate* isomer is produced in 60% yield following the synthetic procedure described in Scheme 2. Fluorescent calix[4]crowns-4 **20** and **21** locked in the *cone* conformation have also been synthesised [31,32]. An interesting phenomenon has been disclosed using the calix[4]azacrown-5 **22** [33]. ^{1}H NMR spectroscopic studies show that Ag^+ is bound to the crown-capped side but when the nitrogen atom in the crown ring is protonated with trifluoroacetic acid, Ag^+ is drawn to the bis-ethoxyethoxy side through the π basic tube formed by the *1,3-alternate* calix[4]arene. When the NH^+ in the crown ring is deprotonated with Li_2CO_3 or diazabicycloundecene, Ag^+ is drawn back to the crown-capped side.

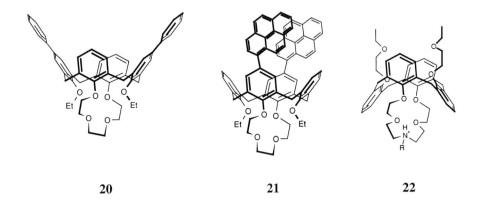

20   21   22

A novel family of calix[4]arene-azacrowns has also been synthesised [34,35]. In this case the polyaza chain is attached to the calix through amido functions, but their complexation properties have not been extensively studied. Some of them (*e.g.* **23**) bear additional binding groups on the N atom in the crown bridge, giving rise to *N*-pivoted azacrowns [35]. An interesting synthetic strategy aimed at obtaining calix[4]crowns to which one or two additional binding groups might be readily attached on the polyether ring has been recently reported [36]. Monomethylol- (*e.g.* **24**) or bismethylol- (*e.g.* **25**) *lariat*-calix[4]crown ethers have been obtained starting from racemic or enantiopure 1-

allyloxy glycerol, respectively. Introduction of 6-methyl-2,2'-bipyridine units on these *lariat*-crown ethers enabled formation of highly luminescent lanthanide complexes [36] (see Chapters 25, 31).

Several azocalix[4]crowns, designed to act as photoresponsive cation ligands, have been synthesised. In one series of compounds (*e.g.* **26**) the azobenzene unit on the bridge is linked in a 4,4'-fashion (N and O atoms in *para* positions), whereas in other compounds (*e.g.* **27**), the linkage occurs in a 2,2'-fashion [37-39]. In compounds **26**, the calix adopts a *cone* conformation in solution and the azo moiety is in the *trans* form when the glycol chain is long (n = 2) and in the *cis* form when this chain is short (n = 0) [37]. Compound **27** is fixed in the *1,3-alternate* conformation and the azo moiety is mainly in the *trans* form in solution [38]. The X-ray crystal structure of a *cone* azocalix[4]arene-crown-6 similar to **27** shows the azo group to be in the *trans* form [40]. Upon irradiation of ligand **27**, isomerisation from *trans* to *cis* occurs and the photostationary state (*cis/trans* = 55/45) is strongly affected by cation complexation, which increases the percentage of the *cis*-**27** in the following order $Na^+ < K^+ < Rb^+ < Cs^+$. The complexed cation inhibits the thermal isomerisation from *cis* to *trans* [38].

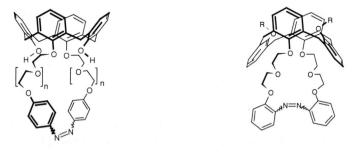

26   **27**: R = CH$_2$CH$_2$OCH$_2$CH$_2$OCH$_3$

The (proximal) calix[4]arene-1,2-mono-crowns are known, although they have been much less studied in comparison with the (distal) 1,3-counterparts. The first example was the 1,2-dimethoxycalix[4]arene-crown-5 **28** synthesised by exploiting the Ti(IV) assisted proximal 1,2-bisdemethylation of tetramethoxy-*p-tert*-butylcalix[4]arene [41].

Later it was shown that calix[4]arene-1,2-mono-crown ethers can be obtained in reasonably good yield by direct condensation of calix[4]arenes and oligoethylene glycol ditosylates with NaH/DMF [42-44].

The reaction of 1,2-dipicolyloxycalix[4]arene with several oligoethylene glycol ditosylates gives 1,2-bridged calix[4]arene-mono-crowns in *1,2-alternate* and *cone* conformations [45]. Mono-O-alkylated *p-tert*-butylcalix[4]arenes give mainly inherently chiral calix[4]arene-1,2-mono-crown ethers, which have been resolved by HPLC on chiral stationary phases [46] and used to synthesise calix[4]arene-1,2-monocrowns in the fixed *partial cone* conformation (*e.g.* **29**) [30].

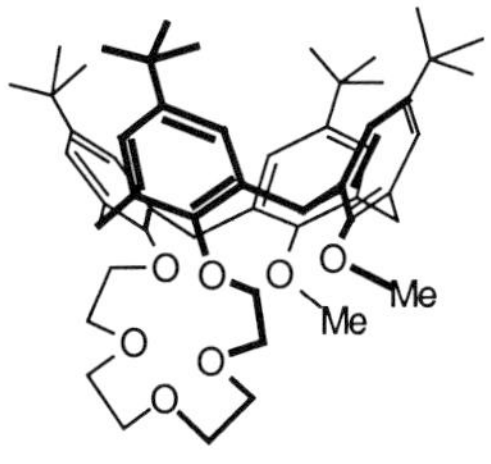

28 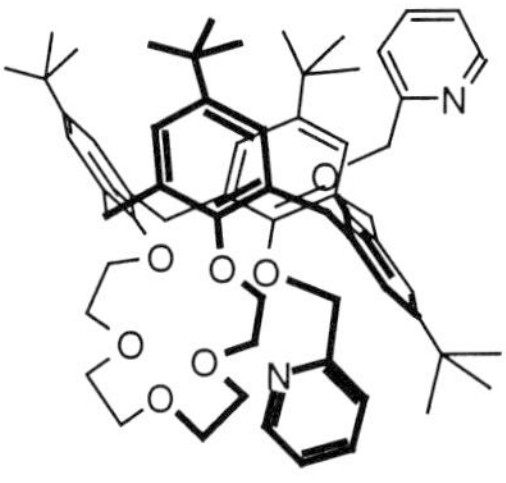   **29**

2.2. COMPLEXATION PROPERTIES

A critical evaluation of the different methods used to study the thermodynamics of calixarene chemistry including formation of metal ion complexes of calix[4]crowns, has been published [47]. Recently, fluorescence [48] and ESI-MS [49] spectroscopy has been added to these methods. Some selected complexation data of the calix[4]crowns described in the previous section have been collected in TABLE 1 with the aim of highlighting the more general features of the host properties of this class of ionophores.

The first general remark is that, as expected from the size complementarity, 1,3-calix[4]crowns-4 (*e.g.* **21**) are selective for sodium, the 1,3-calix[4]crowns-5 (*e.g.* **4, 7, 11, 18**) for potassium (*cf.* also ref. [50]) and 1,3-calix[4]crowns-6 **5** and **12** for cesium, among the alkali metal ions. The conformationally mobile compounds **4** and **5** are less effective and also less selective than the corresponding more preorganized receptors **11** and **12** which are fixed in the *1,3-alternate* structure.

It is worth noting that compound **11** is more selective than valinomycin in binding potassium ion and that the very high Cs^+/Na^+ selectivity shown by compound **12** led to interesting applications in the removal of cesium from radioactive waste (see Chapter 35). The cesium selectivity can be further improved by incorporating phenyl rings in the polyether bridge (*e.g.* **13** and **14**). The calix[4]crown-diquinone **7** shows binding efficiency comparable to that of compound **4** but the K^+/Na^+ selectivity is lower. This can be explained by the fact that, in the case of ligand **7**, both sodium and potassium are complexed by the ligand in the *cone* conformation, allowing the quinone carbonyls to enter the co-ordination sphere, whereas sodium is complexed by **4** in the *cone* conformation and potassium in the *1,3-alternate*. The possibility of interaction between the complexed ion and the quinone carbonyl groups strongly influences the electrochemistry of these nuclei. In all cation complexes the reduction potentials are significantly shifted anodically, allowing these derivatives to be used as possible potentiometric sensing devices. Discrimination between the enantiomers of various chiral amines and amino acids has been shown to be possible using 1,3-calixcrown-5 or -6 derivatives having a chiral binaphthyl moiety within the crown ether loop [52-55]. The 1,2-calix[4]crown (**29**) is a much less effective ligand than the 1,3-calix[4]crowns, but the K^+/Na^+ and the Rb^+/Na^+ selectivity is good.

In most cases, the selectivity data obtained in homogeneous solution have been confirmed by ion transport through supported liquid membranes [8,9,56-58] and polymer inclusion membranes [59], chromatography [60] and ion sensors (see Chapters 32, 34 and 36) [8,9,27,61-64].

TABLE 1. Association constants ($LogK_a$) of different calix[4]arene-crowns at 25°C.

Ligands	Solvent	Li^+	Na^+	K^+	Rb^+	Cs^+	Ref.
4	MeOH	-	2.10	6.6	5.7	2.0	[51]
7	MeOH	-	5.0	>6.0	-	-	[12]
11	MeOH	-	2.6	>9	6.8	5.07	[51]
16	MeOH	0.3	0.7	3.8	2.3	1.7	[51]
18	MeOH	-	2.3	7.49	6.2	2.67	[51]
5	MeOH	<1	<1	2.13	3.18	4.2	[8]
12	MeOH	<1	<1	4.5	5.93	6.1	[8]
29	MeOH	<1	<1	3.38	3.46	2.40	[30]
21	MeCN/THF[a)]	4.65	5.16	-	-	-	[32]
4	CHCl$_3$	4.71	4.31	7.20	7.09	5.13	[9]
11	CHCl$_3$	4.78	4.30	9.83	9.41	6.87	[9]
5	CHCl$_3$	-	4.3	4.2	5.2	6.0	[8]
12	CHCl$_3$	-	5.2	6.4	7.9	8.8	[8]
14	CHCl$_3$	5.0	4.7	7.8	8.9	9.0	[24]

[a)] MeCN/THF = 1000/1

3. Calix[4]arene-bis-Crowns

The generic term calix[4]arene-bis-crowns or simply calix[4]bis-crowns refers to the family of macrotricycles constructed from one calix[4]arene and two polyethylene glycol units. The polyether chains are usually attached *via* the phenolic oxygens of the calixarene platform [65,66]. As with calix[4]arene-mono-crowns, they are of two types depending on whether attachment is proximal or distal: calix[4]arene-1,2;3,4-bis-crowns, in which each crown ether is attached to vicinal (proximal) oxygens, and calix[4]arene-1,3;2,4-bis-crowns in which the crown ether loops are attached to opposite (distal) oxygen atoms of the calix[4]arene unit.

3.1. CALIX[4]ARENE-1,2;3,4-BIS-CROWNS

Due to the vicinal O-linkages, calix[4]arene-1,2;3,4-bis-crowns have been isolated in the *cone* and *1,2-alternate* conformations. In 1990, the stepwise synthesis of the first member of this series, the *p-tert*-butylcalix[4]arene-1,2;3,4-bis-crown-5 **30** in *cone* conformation, was reported [41]. Initially, the regioselective 1,2-di-demethylation of the tetramethoxy-*p-tert*-butylcalix[4]arene with TiBr$_4$ in ethanol was performed, followed by the 1,2-bridging with tetraethylene glycol ditosylate in the presence of t-BuOK in refluxing benzene. The repetition of both reactions leads to the desired molecule in the *cone* conformation. Preliminary extraction data showed that 30 is able to bind alkali metal cations with the selectivity order Rb$^+$ ≈ K$^+$ >> Na$^+$ > Cs$^+$ [41]. Subsequently, the reaction of the calix[4]arene with tetraethylene glycol ditosylate in the presence of cesium carbonate in acetonitrile leading to the formation of calix[4]arene-1,2;3,4-bis-crown-5 (31) in *cone* conformation in 9% yield was described [67]. The 1,2-substitution, which was proved by solving the X-ray crystal structure of 31 [68], was attributed to *a cesium effect* during ring closure [67]. Calix[4]arene-1,2;3,4-bis-crown-3 32 and calix[4]arene-1,2;3,4-bis-crown-4 33 *p*-substituted by H, *tert*-butyl, cyclohexyl and phenyl were synthesised in the *cone* conformation by using NaH in DMF [69].

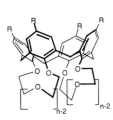

30: n = 5, R = *t*-Bu
31: n = 5, R = H
32: n = 3, R = H, *t*-Bu, C$_6$H$_5$, C$_6$H$_{11}$
33: n = 4, R = H, *t*-Bu

These receptors were shown to complex neutral organic molecules and ammonium cations in apolar organic media. The X-ray crystal structure of the 1:1 *endo* complex of *p*-cyclohexylcalix[4]arene-1,2;3,4-bis-crown-3 (32: R = C$_6$H$_{11}$) with nitromethane has been solved [69]. It has recently been reported that calix[4]arene-1,2;3,4-bis-crowns can be obtained directly from calix[4]arenes (with or without *tert*-butyl groups on the calix unit) in the *cone* or in the *1,2-alternate* conformations depending on the base/solvent systems (NaH/DMF, *t*-BuOK/toluene, *t*-BuORb/toluene, *t*-BuOCs/toluene) and on the length of the polyethylene glycol ditosylate. By using a low polarity solvent and a soft cation it is possible to direct the synthesis of calix[4]arene-1,2;3,4-bis-crowns towards the *1,2-alternate* conformation [44].

3.2. CALIX[4]ARENE-1,3;2,4-BIS-CROWNS

Because of 1,3- and 2,4-bridges, the calix[4]arene-1,3;2,4-bis-crowns are constrained to the *1,3-alternate* conformation. The first member of the series is the *p-tert*-butyl calix[4]arene-1,3;2,4-bis-crown-5 (34) and was obtained as a by-product during the preparation of calix[4]arene mono-crown 1a (*vide supra*) [7]. The *1,3-alternate* conformation of 34 was ascertained by the determination of its crystal structure [70].

34: R = *t*-Bu
35: R = H

While the crystal structure determination of the related calix[4]arene-1,3;2,4-bis-crown-5 (35) showed that this molecule has a closely similar solid state conformation to that of 34, the two ligands differ significantly in their interactions with alkali metal picrates in chloroform solution [71]. Most obviously, the *tert*-butyl substituents of 34 appear to inhibit metal binding [71]. Such a steric hindrance was also noted earlier for a double calix crown *p*-substituted with *tert*-butyl groups [72] and led to efforts to prepare calix[4]arene-1,3;2,4-bis-crowns without *p-tert*-butyl groups.

In general, calix[4]arene-1,3;2,4-bis-crowns **36-40** were obtained in a one-step procedure in ~ 50-80% yields by reacting calix[4]arene with 2-4 equivalents of the appropriate polyethylene glycol ditosylates in the presence of an excess of potassium carbonate in refluxing acetonitrile for 6-14 days [56,73]. The ditosylate and the potassium carbonate were added in two portions. The first step of the reaction is the selective 1,3-bridging of the calix[4]arene, resulting in a 1,3-calix[4]arene mono-crown which must adopt a *1,3-alternate* conformation to form a second bridge by reaction with the second portion of the reagent. This finding eventually led to the synthesis of *unsymmetrical* calix[4]arene-1,3;2,4-bis-crowns in which the two crown ether units are different (*vide infra*).

In separate work, the calix[4]arene-1,3;2,4-bis-crown (**39**) and a related calix bis-crown-7, in which the two benzene units are separated by three oxygen atoms, have been synthesised in over 70% yield *via* the reaction of calix[4]arene with the appropriate dibenzo *dimesylate* in refluxing acetonitrile in the presence of cesium carbonate [74].

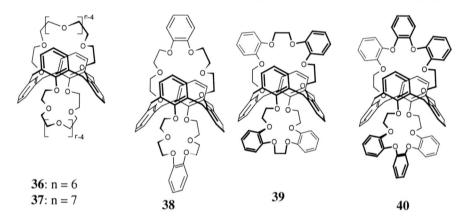

36: n = 6
37: n = 7
38
39
40

3.3. X-RAY STRUCTURES OF CALIX[4]ARENE-1,3;2,4-BIS-CROWNS AND THEIR ALKALI METAL COMPLEXES

The X-ray crystal structures of calix[4]arene-1,3;2,4-bis-crowns and their alkali metal complexes have been recently reviewed [75]. The ligands have a globular shape, with the polyether chains defining two possible sites for cation coordination. Both *mono-* and *binuclear* complexes can be obtained, and the following generalisations may be drawn from the structural observations:
1. Ligands **34** and **35** containing 5 oxygen atoms in the glycol chains have a loop size suitable for the complexation of potassium and rubidium. The ligands **36, 38-40** containing 6 oxygens are convenient for complexation of cesium. Ligand **37** with 7 oxygens has a crown ether loop too large for effective complexation of the alkali metal ions.
2. The complexation of cesium by **36** occurs without change of the conformation of the glycol chains, showing the ligand to be highly preorganized for capturing this metal.
3. There are cation-π interactions in the complexes of **36** with potassium and cesium.

3.4. COMPLEXATION STUDIES

Thermodynamic studies of complexation for calix[4]arene-bis-crowns are not presented in this chapter and can be found in Chapters 21 and 35. This section concerns some mechanistic features of complex formation and some properties of the complexes in solution. Only the bis-crown-6 compounds **36, 38-40** showed good complexation properties for cesium, with high Cs^+/Na^+ selectivity in the order **36** < **38** < **39** < **40** [76-82]. Ligand **40**, in which the crown ethers are composed of fused phenyl ethers, did not complex sodium at all [83]. The increasing Cs^+/Na^+ selectivity along with the increase of the number of phenyl groups in the crown ether could be due to an increasing difficulty in binding Na^+ and/or to an increasing ability to complex Cs^+ attributable to a rigidification of the crown-6 because of the replacement of sp^3 carbons by sp^2 ones. This problem has been investigated by molecular dynamics calculations (see Chapter 18) [80]. Interestingly, simultaneous removal of technetium and cesium has been achieved by using **36**·Cs^+ as an ionised coextractant of TcO_4^- followed by a separation from the cesium by a second calixarene in a double supported liquid membrane process [84].

As for calix[4]mono-crowns, the selectivity data obtained in homogeneous solution have been exploited to prepare potassium sensing agents [85], cesium-selective optical sensors [86,87] and cesium-selective electrodes [88] (see Chapter 32 and 34).

Generally, calix[4]arenes in the *1,3-alternate* conformation possess two independent binding sites on each side of the calix unit and several examples show that cations can exchange from one site to the other by both *intra-* and *inter*molecular processes. Intramolecular exchange implies a cation tunnelling through the π-basic tube of the calix in the *1,3-alternate* conformation [89,90]. NMR techniques were used to obtain information on the behaviour and structure of alkali metal complexes of **35** and **36** in solutions. Ligand **35** formed 1:1 and 1:2 complexes with alkali metal picrates in CD_2Cl_2-DMF-d_7 [91] and $CDCl_3$ [92]. 1:1 Complexes have been observed for K^+, Rb^+, Cs^+ and NH_4^+ with $LogK_a$ 2.9 - 4.6 [91]. 1:2 Complexes were observed for Na^+ and NH_4^+ with $LogK_a$ 2.0 and 2.7 respectively in CD_2Cl_2-DMF-d_7. Variable temperature ^1H NMR spectra of 1:1 complexes with K^+, Rb^+, Cs^+ and NH_4^+ showed the existence of two coalescence temperatures T_c corresponding to both *intra-* and *inter*molecular metal-ligand exchanges. The *intra*molecular exchange pathway was identified from the independence of the coalescence temperature on the concentration of the sample [91]. The replacement of the central O atoms in the crown-5 loops by NH groups stops the cation-ligand exchanges in the case of ammonium picrate in $CDCl_3$ [93].

^1H and ^{133}Cs NMR techniques were used to investigate the mode of binding of **36** with cesium cation [94]. In 1:1 $CDCl_3$-CD_3OD, **36** formed 1:1 and 1:2 complexes with CsSCN. A study in function of the concentration of the metal indicated a cation-ligand exchange between the mononuclear and binuclear species [92]. This metal-ligand exchange is probably due to cesium-cesium electrostatic repulsion because of the presence of the first cesium in the 1:1 complex preventing from the second to enter [92]. In a recent paper good evidence has been given of the 1:1 and 2:1 metal : ligand species by analysing the equilibria of extraction of cesium nitrate by calix[4]arene-bis(*tert*-octylbenzo)-crown-6 in 1,2-dichloroethane [93]. Binding of the second cation by the ligand was found to be approximatively two orders of magnitude weaker than binding of the first cation [93]. The synthesis and the crystal structure of the 2:1 Ag^+ complex of **36** showed that the two silver ions, located in both sites of the ditopic calix[4]crown, are disordered over two positions, one of which corresponds mainly to a bonding to ether oxygen atoms and the other to a polyhapto bonding with calixarene phenyl rings. Three metal ion positions are located near the crown ether extremity, whereas one polyhapto-bonded ions is displaced towards the calixarene cavity [94]. These results showed the amphiphilic character of silver ion.

It was recently reported that a calix[4]arene-biscrown-6 bearing one ionisable group on the aromatic ring shows a pH dependent extraction efficiency [95].

3.5. UNSYMMETRICAL CALIX[4]ARENE-1,3;2,4-BIS-CROWNS

In the preceding section 3.3. enhanced Cs^+/Na^+ selectivity of calix[4]arene-1,3;2,4-bis-crown-6 was obtained by introduction of 1,2-phenyl or 2,3-naphthyl groups in the middle of the crown-6 loops. To prove that the observed selectivity was due to a better binding of cesium ion by the crown ether loop containing the aromatic ring, *unsymmetrical* calix[4]arene-1,3;2,4-bis-crowns **41** and **42** in which the two crown loops are different were prepared [83]. The ^{1}H NMR spectra of **41**˙Cs^+ and of **42**˙Cs^+ and the X-ray crystal structure of **42**˙Cs^+ showed indeed that the cesium ion prefers to bind to the modified polyether loops [83]. An unsymmetrical bis-crown-6 bearing one crown-6 and one crown-6 modified by a 1,3-phenylene group was shown by ^{1}H NMR to complex only one cesium ion in the unmodified crown-6 loop [96]. Recently unsymmetrical calix[4]-crown-5-crown-6 derivatives were utilised to prepare the 1:1:1 heterobinuclear complex with one potassium in the crown-5 loop and one cesium in the crown-6 loop as evidenced by an X-ray structure determination [97].

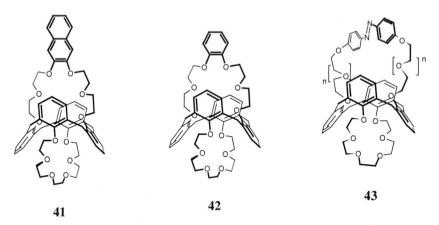

41   42   43

The synthesis of unsymmetrical azobenzene-modified calix[4]arene-1,3;2,4-bis-crowns **43** as potential allosteric systems, was recently reported [83]. They consist of calix[4]arene-1,3;2,4-bis-crowns combining crown-6 and azobenzene modified crown-4 and crown–6 (**43**, n = 1 and 2 respectively). In the case of n = 1, preliminary complexation studies of alkali metal and ammonium picrates showed the cations to be located in the unmodified crown ether of the 1:1 complexes. Complexation induces changes in the *trans/cis* ratio of the azobenzene unit, probably as a result of conformational changes of the calixarene [98].

3.6. MIXED RECEPTORS AND CALIXCRYPTANDS

Mixed receptors consist of ligands containing one polyether ring and one other loop different from a crown ether on the two sides of the calixarene unit in the *1,3-alternate* conformation. These mixed receptors were mainly developed with the aim of preparing *hard* and *soft* co-receptors. The first member of this family was constructed from a calix[4]arene in the *1,3-alternate* conformation with one crown-6 for complexing cations and one bridge having a binding site for anions [99]. Transport experiments showed that the crown-6 bridge complexed cesium while the other unit complexed anions such as Cl⁻ and NO_3^- [99].

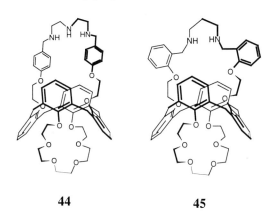

44   **45**

Receptor **44** containing both a crown-6 and a polyaza-oxa ether chain formed 1:1 complexes with Ni^{2+} and Zn^{2+} with the cations located in the polyaza- oxa ether loop and with Cs^+ with no attributable location [100]. Similarly, **45** formed a 1:1 complex with Zn^{2+} with a location of this metal between the nitrogen atoms and with Cs^+ included in the crown-6 chain [101]. Attempts to prepare 1:1:1 heterodinuclear complexes with these *hard* and *soft* cations failed.

Ligands **46** and **47** belong to the family of *calixcryptands*: they combine a calixarene unit in the *1,3-alternate* conformation with one crown-6 and one cryptand unit [102-105]. In CDCl₃, receptor **46** formed 1:1 complexes with K^+ and Cs^+ picrates which were also detected by FAB(+)MS [102,103]. The potassium ion was shown to be located in the cryptand unit whereas the cesium was shown to be located in the crown-6 loop. The 1:1:1 **46**·K^+·Cs^+ heterobinuclear complex was observed by ¹H NMR. This complex could not be detected by FAB(+)MS spectrometry but instead the **46**·Na^+·Cs^+ 1:1:1 complex was observed probably due to the presence of sodium in the matrix [102].

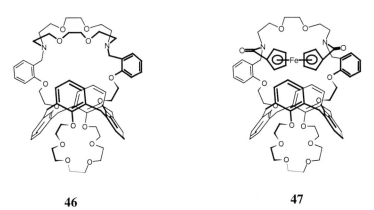

46   **47**

Ligand **47** possesses one crown-6 unit and one cryptand unit with a ferrocenyl redox active center. Cyclic voltammogramms of **47** in the presence of alkali and alkaline earth metal ions indicated that discernible effects are observable for only Ca^{2+}, Sr^{2+} and Ba^{2+}. It was suggested that the cations are bound loosely to the carbonyl groups rather than being encapsulated within the calixarene [104].

4. Calix[4]crowns in Action

In this section we will briefly cover other properties of calix[4]arene-crown ethers different but related to their better known ionophoric ability.

4.1. TRANSACYLASE MIMICS

The alkaline-earth complexes of calix[4]mono-crowns are efficient turnover catalysts for the cleavage of aryl esters, thus acting as artificial esterases. A 3.0 mM solution of *p*-nitrophenylacetate (pNPOAc) in MeCN/MeOH 9:1 (v/v) at 25°C containing 40mM di-*iso*-propylethyl amine/perchlorate salt buffer in a [B]/[BH$^+$] ratio of 3:1 was kinetically evaluated in the absence (background reaction) and in the presence of 0.39 mM of the barium salt of *p-tert*-butylcalix[4]crown-5 **1a** [106].

The catalytic species **48** is obtained, together with its monoprotonated form, by mixing an equimolar amount of calix[4]crown **1a** and Ba(ClO$_4$)$_2$ in the buffer solution. Figure 2 reports the spectrophotometrically determined concentration of *p*-nitrophenol (pNPOH) from pNPOAc in the absence (curve a) and presence (curve b) of metal catalyst.

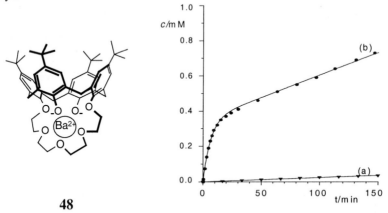

48

*Figure 2. Liberation of p-nitrophenol in the methanolysis of p-nitrophenyl acetate. (▼) background reaction: buffer alone plus 0.39 mM of **1a**; (•) buffer plus 0.39 mM of **1a** and 0.39 mM Ba(ClO$_4$)$_2$.*

The background methanolysis of pNPOAc in these conditions (curve a) is very slow ($k_{bg} = 8.8 \times 10^{-5}$ min^{-1}) whereas a dramatic increase of the initial *p*-nitrophenol release is observed in the presence of the catalyst ("burst kinetics"), followed by a slower linear release (curve b). The biphasic kinetics are typical for a double displacement mechanism [107] in which fast acylation of the catalyst (cat) ($k_1 = 0.15$ min^{-1}) is followed by slower deacylation of the acetylated form (catAc) ($k_2 = 7.0 \times 10^{-3}$ min^{-1}).

The solid line in Figure 2 (curve b) is the best fit of data points to the equation derived from this kinetic (Scheme 4). The catAc intermediate was identified by HPLC in acid quenched samples of the reaction mixture in a catalysis experiment corresponding to curve b of Figure 2. It undergoes decomposition to the products with a rate constant ($k_2 = 7.7 \times 10^{-3}$ min^{-1}) in good agreement with that evaluated from the overall reaction ($k_2 = 7.0 \times 10^{-3}$ min^{-1}). The behaviour of the barium complex of calix[4]crown-5 (**48**) is reminiscent of that of proteolytic enzymes (*e.g.*, α-chymotrypsin) which are able to hydrolyse substrate esters *via* a similar double displacement (ping-pong) mechanism [108]. Through coordination of the reactants, the complexed barium ion acts both to activate the carbonyl group as an electrophile and to enhance nucleophile attack by ensuring its proximity to the electrophile, in both the acylation and deacylation steps (nucleophilic-electrophilic catalysis). The scope of this catalyst is restricted to the acetate esters with reactivity in the range approximately defined by the phenyl acetate – *p*-nitrophenylacetate pair, with a maximum efficiency for *p*-chlorophenylacetate.

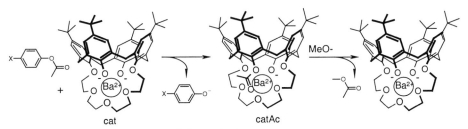

Scheme 4. Double displacement transacylation mechanism.

With the aim of improving the catalytic activity, the *lariat* calix[4]crown barium complex **49** was synthesised and tested in the catalytic transacilation of several aryl acetates. Although this potential trifunctional catalyst did not show improved activity when compared with **48**, evidence was obtained that the protonated side arm of **49**, was able to act as a built-in intramolecular acid catalyst (Scheme 5) assisting the departure of poor leaving groups such as those of aryl acetate (ArOAc). The overall catalytic efficiency however was negatively affected by unfavourable steric effects of the pendent group present in the binding region [106].

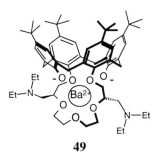

49

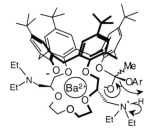

Scheme 5. Intramolecular general acid catalysis in aryl acetate transacylation by **49**.

4.2. MOLECULAR MACHINES

Molecular machines are in vogue and scattered examples are found in literature of thermally-, chemically-, electrochemically- and /or photochemically-labile molecular systems or devices, sensors, logic gates *etc.* [109]. These molecular machines use thermal, chemical or electrical energy or light to work. The globular shape of calix[4]arene-1,3;2,4-bis-crowns has been exploited to prepare *molecular mappemondes* (*globes*) and *gyroscopes*. They are constructed from calix[4]arene-1,3;2,4-bis-crowns-6 held in the arms of a polyether loop *via* a 1+1 condensation [110]. Depending on the length of the polyether arm, a 2+2 dimer was also isolated, leading to a *molecular mill* [111]. ^{1}H NMR was used to show that the spinning of the calixarene can be stopped in the presence of large amounts of ammonium picrate.

5. Conclusions

In this chapter we have summarised the tremendous development experienced by calix[4]arene-crown ethers in the last 10 years, confining ourselves to compounds in which the phenolic oxygen atoms of the calix[4]arene unit are part of the crown ether loop. These are the classical calix[4]crowns, whose most important feature is that a fine tuning of their complexation properties towards metal ions can be obtained by selecting the most appropriate conformation of the calix[4]arene and the crown ether size. Probably the main interest in these ligands derives from their applications in ion sensors (see Chapter 32) and as selective cesium extractants in radioactive waste treatment (see Chapter 35).

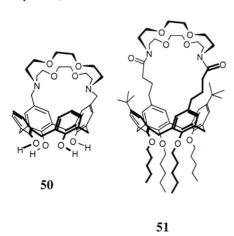

50

51

However, there are other ways of combining calix[4]arenes and crown ethers. For example, upper-rim calix[4]-crowns used for conformational studies [112] and to enhance the binding properties [113] of primary ammonium ions compared to alkali metal cations should be mentioned. Lariat-type upper-rim calix[4]crowns have been shown to extract alkyl ammonium ions with efficiency which depends on the nature of the side-chains [114]. The Mannich reaction has been used to attach diaza-18-crown-6 on calix[4]arenes at the upper rim in a 1,2-manner leading to a calixcryptand, (*e.g.* **50**) [115].

New calix[4]amidocrowns and calix[4]amidocryptands (*e.g.* **51**) have been obtained by reaction of a calix[4]arene containing 2 distal acid chloride groups at the upper rim with appropriate diamines. The calix[4]amidocryptands showed significant Li$^+$ and Na$^+$ selectivity [116].

A calix[4]dithiacrown-5 has been prepared by 1,3-selective O-alkylation of *p-tert*-butyl calix[4]arene with chlorethyl tosylate followed by capping with ß,ß'-dimercaptoethylether in basic conditions. The ligand formed a platinum complex able to catalyze the hydrosilylation of olefins with triethoxysilane [117]. A *p-tert*-butylcalix[4]arene-diaza-dithia-crown has been obtained, as its Ni complex, by S alkylation reactions. Demetallation resulted in a calix[4]thiacrown containing 2 N-, 2 O- and 2 S-donor atoms [118]. The synthesis of a calix[4]selenacrown, a novel type of calix[4]crown has been reported [119]. Crowncavitands were synthesised by alkylation of tetrahydroxycavitands with polyethylene glycol ditosylates. The bridging of two adjacent hydroxy groups is favoured when using pentaethylene glycol unit. The combination of the obtained crowncavitands with calix[4]arenes or resorcin[4]arenes resulted in potential receptor molecules with large hydrophobic surfaces [120]. The assembly of a calix[4]arene and a crown ether unit through a catenane architecture, which opens an interesting new research field, has been recently reported [121-123].

6. References and Notes

[1] A. Casnati, R. Ungaro, in *Calixarenes in Action*, Imperial College Press (Eds.: L. Mandolini, R. Ungaro), London, **2000**, p. 62-84.
[2] J. C. Ma, D. A. Dougherty, *Chem. Rev.* **1997**, *97*, 1303-1324.
[3] P. Lhotak, S. Shinkai, *J. Phys. Org. Chem.* **1997**, *10*, 273-285.
[4] C. Alfieri, E. Dradi, A. Pochini, R. Ungaro, G. D. Andreetti, *J. Chem. Soc., Chem. Commun.* **1983**, 1075-1077.
[5] R. Ungaro, A. Pochini, G. D. Andreetti, *J. Incl. Phenom.* **1984**, *2*, 199-206.
[6] P. J. Dijkstra, J. A. J. Brunink, K. E. Bugge, D. N. Reinhoudt, S. Harkema, R. Ungaro, F. Ugozzoli, E. Ghidini, *J. Am. Chem. Soc.* **1989**, *111*, 7567-7575.
[7] E. Ghidini, F. Ugozzoli, R. Ungaro, S. Harkema, A. A. El-Fadl, D. N. Reinhoudt, *J. Am. Chem. Soc.* **1990**, *112*, 6979-6985.
[8] A. Casnati, A. Pochini, R. Ungaro, F. Ugozzoli, F. Arnaud, S. Fanni, M. J. Schwing, R. J. M. Egberink, F. de Jong, D. N. Reinhoudt, *J. Am. Chem. Soc.* **1995**, *117*, 2767-2777.
[9] A. Casnati, A. Pochini, R. Ungaro, C. Bocchi, F. Ugozzoli, R. J. M. Egberink, H. Struijk, R. Lugtenberg, F. de Jong, D. N. Reinhoudt, *Chem. Eur. J.* **1996**, *2*, 436-445.
[10] R. Ungaro, A. Casnati, F. Ugozzoli, A. Pochini, J. F. Dozol, C. Hill, H. Rouquette, *Angew. Chem. Int. Ed. Engl.* **1994**, *33*, 1506-1509.
[11] R. Ungaro, A. Casnati, *unpublished results*.
[12] P. D. Beer, P. A. Gale, Z. Chen, M. G. B. Drew, J. A. Heath, M. I. Ogden, H. R. Powell, *Inorg. Chem.* **1997**, *36*, 5880-5893.
[13] A. Casnati, E. Comelli, M. Fabbi, V. Bocchi, G. Mori, F. Ugozzoli, A. M. Manotti Lanfredi, A. Pochini, R. Ungaro, *Recl. Trav. Chim. Pays-Bas* **1993**, *112*, 384-392.
[14] K. Takahashi, A. Gunji, D. Guillaumont, F. Pichierri, S. Nakamura, *Angew. Chem. Int. Ed. Engl.* **2000**, *39*, 2925-2928.
[15] D. Bethell, G. Dougherty, D. C. Cupertino, *J. Chem. Soc., Chem. Commun.* **1995**, 675-676.
[16] G. Ferguson, J. F. Gallagher, A. J. Lough, A. Notti, S. Pappalardo, M. F. Parisi, *J. Org. Chem.* **1999**, *64*, 5876-5885.
[17] R. A. Sachleben, A. Urvoas, J. C. Bryan, T. J. Haverlock, B. P. Hay, B. A. Moyer, *Chem. Commun.* **1999**, 1751-1752.
[18] S. E. Biali, V. Böhmer, S. Cohen, G. Ferguson, C. Gruttner, F. Grynszpan, E. F. Paulus, I. Thondorf, W. Vogt, *J. Am. Chem. Soc.* **1996**, *118*, 12938-12949.
[19] Y. Ting, W. Verboom, L. C. Groenen, J.-D. van Loon, D. N. Reinhoudt, *J. Chem. Soc., Chem. Commun.* **1990**, 1432-1433.
[20] F. Grynszpan, S. E. Biali, *Tetrahedron Lett.* **1991**, *32*, 5155-5158.
[21] R. Ungaro, A. Arduini, A. Casnati, O. Ori, A. Pochini, F. Ugozzoli, *NATO ASI Ser. C 426* **1994**, 277-300.
[22] F. Ugozzoli, O. Ori, A. Casnati, A. Pochini, R. Ungaro, D. N. Reinhoudt, *Supramol. Chem.* **1995**, *5*, 179-184.

[23] R. Ungaro, A. Arduini, A. Casnati, A. Pochini, F. Ugozzoli, *Pure Appl. Chem.* **1996**, *68*, 1213-1218.
[24] V. Lamare, J. F. Dozol, F. Ugozzoli, A. Casnati, R. Ungaro, *Eur. J. Org. Chem.* **1998**, 1559-1568.
[25] J. S. Kim, J. H. Pang, I. H. Suh, D. W. Kim, D. W. Kim, *Synth. Commun.* **1998**, *28*, 677-685.
[26] J. S. Kim, J. H. Pang, I. Y. Yu, W. K. Lee, I. H. Suh, J. K. Kim, M. H. Cho, E. T. Kim, D. Y. Ra, *J. Chem. Soc., Perkin Trans. 2* **1999**, 837-846.
[27] C. Bocchi, M. Careri, A. Casnati, G. Mori, *Anal. Chem.* **1995**, *67*, 4234-4238.
[28] A. Arduini, A. Casnati, in *Macrocyclic Synthesis: a Practical Approach*, Oxford University Press (Ed.: D. Parker), Oxford, **1996**, p. 145-173.
[29] J. Guillon, J. M. Leger, P. Sonnet, C. Jarry, M. Robba, *J. Org. Chem.* **2000**, *65*, 8283-8289.
[30] F. Arnaud-Neu, G. Ferguson, S. Fuangswasdi, A. Notti, S. Pappalardo, M. F. Parisi, A. Petringa, *J. Org. Chem.* **1998**, *63*, 7770-7779.
[31] H. Matsumoto, S. Shinkai, *Chem. Lett.* **1994**, 2431-2434.
[32] H. Matsumoto, S. Shinkai, *Tetrahedron Lett.* **1996**, *37*, 77-80.
[33] A. Ikeda, T. Tsudera, S. Shinkai, *J. Org. Chem.* **1997**, *62*, 3568-3574.
[34] R. Ostaszewski, T. W. Stevens, W. Verboom, D. N. Reinhoudt, F. M. Kaspersen, *Recl. Trav. Chim. Pays-Bas* **1991**, *110*, 294-298.
[35] I. Bitter, A. Grun, G. Toth, B. Balazs, L. Toke, *Tetrahedron* **1997**, *53*, 9799-9812.
[36] C. Fischer, G. Sarti, A. Casnati, B. Carrettoni, I. Manet, R. Schuurman, M. Guardigli, N. Sabbatini, R. Ungaro, *Chem. Eur. J.* **2000**, *6*, 1026-1034.
[37] M. Saadioui, Z. Asfari, J. Vicens, N. Reynier, J. F. Dozol, *J. Incl. Phenom.* **1997**, *28*, 223-244.
[38] M. Saadioui, N. Reynier, J. F. Dozol, Z. Asfari, J. Vicens, *J. Incl. Phenom.* **1997**, *29*, 153-165.
[39] B. Pipoosananakaton, M. Sukwattanasinitt, N. Jaiboon, N. Chaichit, T. Tuntulani, *Tetrahedron Lett.* **2000**, *41*, 9095-9100.
[40] P. Thuéry, M. Nierlich, M. Saadioui, Z. Asfari, J. Vicens, *Acta Crystallogr., Sect. C: Cryst. Struct. Commun.* **1999**, *C55*, 443-445.
[41] A. Arduini, A. Casnati, L. Dodi, A. Pochini, R. Ungaro, *J. Chem. Soc., Chem. Commun.* **1990**, 1597-1598.
[42] A. Arduini, A. Casnati, M. Fabbi, P. Minari, A. Pochini, A. R. Sicuri, R. Ungaro, *Supramol. Chem.* **1993**, *1*, 235-246.
[43] H. Yamamoto, T. Sakaki, S. Shinkai, *Chem. Lett.* **1994**, 469-472.
[44] A. Arduini, L. Domiano, A. Pochini, A. Secchi, R. Ungaro, F. Ugozzoli, O. Struck, W. Verboom, D. N. Reinhoudt, *Tetrahedron* **1997**, *53*, 3767-3776.
[45] S. Caccamese, A. Notti, S. Pappalardo, M. F. Parisi, G. Principato, *Tetrahedron* **1999**, *55*, 5505-5514.
[46] F. Arnaud-Neu, S. Caccamese, S. Fuangswasdi, S. Pappalardo, M. F. Parisi, A. Petringa, G. Principato, *J. Org. Chem.* **1997**, *62*, 8041-8048.
[47] A. F. D. de Namor, R. M. Cleverley, M. L. Zapata-Ormachea, *Chem. Rev.* **1998**, *98*, 2495-2525.
[48] L. Prodi, F. Bolletta, M. Montalti, N. Zaccheroni, A. Casnati, F. Sansone, R. Ungaro, *New J. Chem.* **2000**, *24*, 155-158.
[49] F. Allain, H. Virelizier, C. Moulin, C. Jankowski, J.-F. Dozol, J. C. Tabet, *Spectroscopy* **2000**, *14*, 127-139.
[50] V. W. W. Yam, K. L. Cheung, L. H. Yuan, K. M. C. Wong, K. K. Cheung, *Chem. Commun.* **2000**, 1513-1514.
[51] F. Arnaud-Neu, N. Deutsch, S. Fanni, M. J. SchwingWeill, A. Casnati, R. Ungaro, *Gazz. Chim. Ital.* **1997**, *127*, 693-697.
[52] Y. Kubo, S. Maeda, S. Tokita, M. Kubo, *Nature* **1996**, *382*, 522-524.
[53] Y. Kubo, *J. Incl. Phenom.* **1998**, *32*, 235-249.
[54] Y. Kubo, S. Obara, S. Tokita, *Chem. Commun.* **1999**, 2399-2400.
[55] Y. Kubo, *Synlett* **1999**, 161-174.
[56] C. Hill, J. F. Dozol, V. Lamare, H. Rouquette, S. Eymard, B. Tournois, J. Vicens, Z. Asfari, C. Bressot, *J. Incl. Phenom.* **1994**, *19*, 399-408.
[57] J.-F. Dozol, V. Lamare, N. Simon, R. Ungaro, A. Casnati, in *"Calixarene Molecules for Separation"*, (Eds.: G. J. Lumetta, R. D. Rogers, A. S. Gopalan), Washington, **2000**, p. 12-25.
[58] L. A. J. Chrisstoffels, F. de Jong, D. N. Reinhoudt, S. Sivelli, L. Gazzola, A. Casnati, R. Ungaro, *J. Am. Chem. Soc.* **1999**, *121*, 10142-10151.
[59] T. G. Levitskaia, D. M. Macdonald, J. D. Lamb, B. A. Moyer, *Physical Chemistry Chemical Physics* **2000**, *2*, 1481-1491.

[60] G. Arena, A. Casnati, A. Contino, L. Mirone, D. Sciotto, R. Ungaro, *Chem. Commun.* **1996**, 2277-2278.
[61] R. J. W. Lugtenberg, Z. Brzozka, A. Casnati, R. Ungaro, J. F. J. Engbersen, D. N. Reinhoudt, *Anal. Chim. Acta* **1995**, *310*, 263-267.
[62] H. Yamamoto, S. Shinkai, *Chem. Lett.* **1994**, 1115-1118.
[63] E. Bakker, *Anal. Chem.* **1997**, *69*, 1061-1069.
[64] H. F. Ji, R. Dabestani, G. M. Brown, R. A. Sachleben, *Chem. Commun.* **2000**, 833-834.
[65] Z. Asfari, M. Nierlich, P. Thuéry, V. Lamare, J.-F. Dozol, M. Leroy, J. Vicens, *Anal. Quim. Int. Ed.* **1996**, *92*, 260-266.
[66] P. Thuery, M. Nierlich, E. Lamare, J. F. Dozol, Z. Asfari, J. Vicens, *J. Incl. Phenom.* **2000**, *36*, 375-408.
[67] Z. Asfari, J. P. Astier, C. Bressot, J. Estienne, G. Pepe, J. Vicens, *J. Incl. Phenom.* **1994**, *19*, 291-300.
[68] G. Pèpe, J. P. Astier, J. Estienne, C. Bressot, Z. Asfari, J. Vicens, *Acta Crystallogr., Sect. C: Cryst. Struct. Commun.* **1995**, *C51*, 726-729.
[69] A. Arduini, W. M. McGregor, D. Paganuzzi, A. Pochini, A. Secchi, F. Ugozzoli, R. Ungaro, *J. Chem. Soc., Perkin Trans. 2* **1996**, 839-846.
[70] Z. Asfari, J. Harrowfield, A. N. Sobolev, J. Vicens, *Aust. J. Chem.* **1994**, *47*, 757-762.
[71] R. Abidi, Z. Asfari, J. M. Harrowfield, A. N. Sobolev, J. Vicens, *Aust. J. Chem.* **1996**, *49*, 183-188.
[72] Z. Asfari, R. Abidi, F. Arnaud, J. Vicens, *J. Incl. Phenom.* **1992**, *13*, 163-169.
[73] Z. Asfari, C. Bressot, J. Vicens, C. Hill, J. F. Dozol, H. Rouquette, S. Eymard, V. Lamare, B. Tournois, *Anal. Chem.* **1995**, *67*, 3133-3139.
[74] J. S. Kim, I. H. Suh, J. K. Kim, M. H. Cho, *J. Chem. Soc., Perkin Trans. 1* **1998**, 2307-2311.
[75] P. Thuéry, M. Nierlich, E. Lamare, J. F. Dozol, Z. Asfari, J. Vicens, *J. Incl. Phenom.* **2000**, *36*, 375-408.
[76] F. Arnaud-Neu, Z. Asfari, B. Souley, J. Vicens, *New J. Chem.* **1996**, *20*, 453-463.
[77] A. D'Aprano, J. Vicens, Z. Asfari, M. Salomon, M. Iammarino, *J. Sol. Chem.* **1996**, *25*, 955-970.
[78] Z. Asfari, C. Naumann, J. Vicens, M. Nierlich, P. Thuéry, C. Bressot, V. Lamare, J. F. Dozol, *New J. Chem.* **1996**, *20*, 1183-1194.
[79] F. Arnaud-Neu, Z. Asfari, B. Souley, J. Vicens, *An. Quim. Int. Ed.* **1997**, *93*, 404-407.
[80] V. Lamare, J. F. Dozol, S. Fuangswasdi, F. Arnaud-Neu, P. Thuéry, M. Nierlich, Z. Asfari, J. Vicens, *J. Chem. Soc., Perkin Trans. 2* **1999**, 271-284.
[81] T. J. Haverlock, P. V. Bonnesen, R. A. Sachleben, B. A. Moyer, *Radiochim. Acta* **1997**, *76*, 103-108.
[82] T. J. Haverlock, P. V. Bonnesen, R. A. Sachleben, B. A. Moyer, *J. Incl. Phenom.* **2000**, *36*, 21-37.
[83] Z. Asfari, V. Lamare, J. F. Dozol, J. Vicens, *Tetrahedron Lett.* **1999**, *40*, 691-694.
[84] M. Grunder, J. F. Dozol, Z. Asfari, J. Vicens, *J. Radioanal. Nucl. Chem.* **1999**, *241*, 59-67.
[85] J. S. Kim, W. K. Lee, D. Y. Ra, Y. I. Lee, W. K. Choi, K. W. Lee, W. Z. Oh, *Microchem. J.* **1998**, *59*, 464-471.
[86] H. F. Ji, G. M. Brown, R. Dabestani, *Chem. Commun.* **1999**, 609-610.
[87] H. H. Zeng, B. Dureault, *Talanta* **1998**, *46*, 1485-1491.
[88] C. Pérèz-Jiménez, L. Escriche, J. Casabo, *Anal. Chim. Acta* **1998**, *371*, 155-162.
[89] A. Ikeda, S. Shinkai, *J. Am. Chem. Soc.* **1994**, *116*, 3102-3110.
[90] A. Ikeda, S. Shinkai, *Chem. Rev.* **1997**, *97*, 1713-1734.
[91] K. N. Koh, K. Araki, S. Shinkai, Z. Asfari, J. Vicens, *Tetrahedron Lett.* **1995**, *36*, 6095-6098.
[92] Wenger, S., Ph.D. thesis, Université de Strasbourg, **1997**.
[93] J. S. Kim, W. K. Lee, K. No, Z. Asfari, J. Vicens, *Tetrahedron Lett.* **2000**, *40*, 3345-3348.
[94] P. Thuéry, M. Nierlich, F. Arnaud-Neu, B. Souley, Z. Asfari, J. Vicens, *Supramol. Chem.* **1999**, *11*, 143-150.
[95] V. S. Talanov, G. G. Talanova, R. A. Bartsch, *Tetrahedron Lett.* **2000**, *41*, 8221-8224.
[96] Z. Asfari, P. Thuéry, M. Nierlich, J. Vicens, *Tetrahedron Lett.* **1999**, *40*, 499-502.
[97] J. S. Kim, W. K. Lee, W. Sim, J. W. Ko, M. H. Cho, D. Y. Ra, J. W. Kim, *J. Incl. Phenom.* **2000**, *37*, 359-370.
[98] M. Saadioui, Z. Asfari, J. Vicens, *Tetrahedron Lett.* **1997**, *38*, 1187-1190.
[99] D. M. Rudkevich, J. D. Mercer-Chalmers, W. Verboom, R. Ungaro, F. de Jong, D. N. Reinhoudt, *J. Am. Chem. Soc.* **1995**, *117*, 6124-6125.
[100] W. Aeungmaitrepirom, Z. Asfari, J. Vicens, *Tetrahedron Lett.* **1997**, *38*, 1907-1910.
[101] B. Pulpoka, Z. Asfari, J. Vicens, *J. Incl. Phenom.* **1997**, *27*, 21-30.
[102] B. Pulpoka, Z. Asfari, J. Vicens, *Tetrahedron Lett.* **1996**, *37*, 6315-6318.

[103] Symmetrical *1,3-alternate* calix[4]arene-biscryptand has been reported: *see* B. Pulpoka, Z. Asfari, J. Vicens, *Tetrahedron Lett.* **1996**, *37*, 8747-8750.
[104] C. D. Hall, N. Djedovic, Z. Asfari, B. Pulpoka, J. Vicens, *J. Organomet. Chem.* **1998**, *571*, 103-106.
[105] B. Pulpoka, M. Jamkratoke, T. Tuntulani, V. Ruangpornvisuti, *Tetrahedron Lett.* **2000**, *41*, 9167-9171.
[106] L. Baldini, C. Bracchini, R. Cacciapaglia, A. Casnati, L. Mandolini, R. Ungaro, *Chem. Eur. J.* **2000**, *6*, 1322-1330.
[107] A. Fersht, in *Enzymes structure and mechanism*, W.H. Freemon, New York, **1985**.
[108] A. Dougas, *Bioorganic Chemistry*, 3rd ed. (Ed.: Springer-Verlag), New York, **1996**, p. 184-204.
[109] Z. Asfari, J. Vicens, *J. Incl. Phenom.* **2000**, *36*, 103-118.
[110] Z. Asfari, C. Naumann, G. Kaufmann, J. Vicens, *Tetrahedron Lett.* **1996**, *37*, 3325-3328.
[111] Z. Asfari, C. Naumann, G. Kaufmann, J. Vicens, *Tetrahedron Lett.* **1998**, *39*, 9007-9010.
[112] J.-D. van Loon, L. C. Groenen, S. S. Wijmenga, W. Verboom, D. N. Reinhoudt, *J. Am. Chem. Soc.* **1991**, *113*, 2378-2382.
[113] K. Paek, H. Ihm, *Chem. Lett.* **1996**, 311-312.
[114] H. Ihm, H. Kim, K. Paek, *J. Chem. Soc., Perkin Trans. 1* **1997**, 1997-2003.
[115] J. Guenot, R. Lamartine, J. L. Royer, *An. Quim. Int. Ed.* **1998**, *94*, 332-334.
[116] W. Wasikiewicz, G. Rokicki, E. Rozniecka, J. Kielkiewicz, Z. Brzozka, V. Böhmer, *Monatsh. Chem.* **1998**, *129*, 1169-1181.
[117] Y. Y. Chen, Z. L. Zhong, Y. S. Li, B. L. Zhang, X. R. Lu, *Chin. Chem. Lett.* **1997**, *8*, 213-214.
[118] B. Tomapatanaget, B. Pulpoka, T. Tuntulani, *Chem. Lett.* **1998**, 1037-1038.
[119] X. B. Hu, Y. Y. Chen, X. R. Lu, *Chin. Chem. Lett.* **1997**, *8*, 777-778.
[120] I. Higler, H. Boerrigter, W. Verboom, H. Kooijman, A. L. Spek, D. N. Reinhoudt, *Eur. J. Org. Chem.* **1998**, 1597-1607.
[121] Z. T. Li, G. Z. Ji, S. D. Yuan, A. L. Du, H. Ding, M. Wei, *Tetrahedron Lett.* **1998**, *39*, 6517-6520.
[122] Z. T. Li, G. Z. Ji, C. X. Zhao, S. D. Yuan, H. Ding, C. Huang, A. L. Du, M. Wei, *J. Org. Chem.* **1999**, *64*, 3572-3584.
[123] Z. T. Li, X. L. Zhang, X. D. Lian, Y. H. Yu, Y. Xia, C. X. Zhao, Z. Chen, Z. P. Lin, H. Chen, *J. Org. Chem.* **2000**, *65*, 5136-5142.

Chapter 21

METAL-ION COMPLEXATION BY NARROW RIM CARBONYL DERIVATIVES

FRANCOISE ARNAUD-NEU,[a] M. ANTHONY McKERVEY[b], MARIE-JOSÉ SCHWING-WEILL[a]

[a]*Laboratoire de Chimie-Physique, UMR 7512 au CNRS, ECPM-ULP, 25, rue Becquerel, 67087 Strasbourg Cedex 2, France.*
[b]*School of Chemistry, The Queen's University of Belfast, David Keir Building, Belfast BT9 5AG, N-Ireland.*
e-mail: farnaud@chimie.u-strasbg.fr

1. Introduction

Cation binding and transport by unsubstituted calixarenes or their derivatives with ethyleneoxy arms or bridges (calixcrowns - see Chapter 20) were the main focus of studies of calixarene coordination chemistry prior to 1985 [1-3]. One early study [4], however, involved derivatives with carbonyl-group containing substituents on the narrow rim and the realisation that simple ester functions on the narrow rim were effective ligating groups for metal cations led to a comprehensive physicochemical evaluation of calix[4], [6], and [8]arenes with carbonyl-containing substitutents. Calix[4] and [6]arene derivatives proved the more useful [4]. In 1993, the study was extended to include several calix[5]arene derivatives [5]. Initially, the ester derivatives were of the type shown in Figure 1. These narrow rim derivatives were readily prepared by exhaustive alkylation of the parent calixarenes using an alkyl bromoacetate. Narrow rim calix[4]arene dialkylamides for alkali and alkaline earth metal ion complexation were investigated at the same time [6].

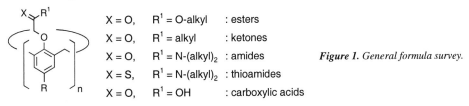

X = O, R¹ = O-alkyl   : esters
X = O, R¹ = alkyl   : ketones
X = O, R¹ = N-(alkyl)₂ : amides
X = S, R¹ = N-(alkyl)₂ : thioamides
X = O, R¹ = OH   : carboxylic acids

Figure 1. General formula survey.

Over the past 15 years an extensive body of facts, predominantly concerning the alkali and alkaline earth metal ions, for calixarene derivatives with narrow rim ester, ketone, amide, thioamide and carboxylic acid groups and various combinations thereof has been developed (Figure 1) [7, 8]. This has concerned :
- stability and association constant measurements in single solvents,
- calorimetry of complexation in single solvents,
- extraction of metal salts, mostly picrates, from water into an organic solvent, usually dichloromethane (CH_2Cl_2) or chloroform ($CHCl_3$),
- transport rates of metal salts through bulk liquid or supported liquid membranes,

- NMR spectroscopic measurements on complex formation,
- X-ray structure determinations of crystalline complexes.

1.1. CALIX[n]ARENE ESTERS AND KETONES

Calixarene esters and ketones (Figure 1) exhibit a broad spectrum of affinity for alkali metal cations which is a size-related phenomenon, a conclusion illustrated graphically in Figure 2 for the ester series in extraction of alkali metal picrates from water into CH_2Cl_2. For n = 4, there is a distinct preference for Na^+; n = 5 produces the highest extraction levels for all cations but discriminates poorly between the larger cations; the larger n = 6 derivative favours Cs^+, Rb^+ and K^+ with a significant selectivity over Na^+; n = 7, 8 give very poor extractants. Changing the nature of the alkoxy residue in the ester on the narrow-rim and/or the *para* substitutent on the wide rim can be an effective way of modulating extraction efficiency and selectivity. The combined effect of these two variables on Na^+/Cs^+ selectivity of calix[6]arene esters is illustrated in Figure 3.

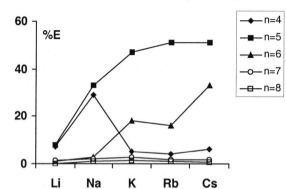

Figure 2. Extraction of alkali metal picrates by p-tert-butylcalix[n]arene ethyl esters from water into CH_2Cl_2.

Cation extraction efficiency in the ester series is a function of calixarene size and conformation. The latter is largely unexplored with the larger systems but in the n = 4 series the ethyl ester was found to be Na^+ selective in the *cone* conformation and K^+ selective in the *partial cone* and *1,2-alternate* conformations, observations important for the design of new ion-selective calixarene receptors.

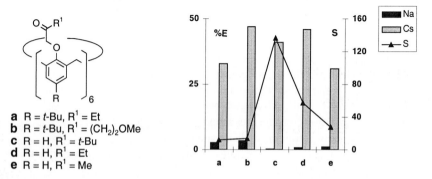

Figure 3. Cs^+ and Na^+ Extraction percentages (from water into CH_2Cl_2) and selectivities $S_{Cs/Na}$ = %E(Cs^+)/%E(Na^+) of calix[6]arene hexaethyl esters [8].

Many of the trends in extraction are also observed in the stability constants, which range for the alkali metal cation complexes of the esters from log K = 2 - 6 in methanol (MeOH) and 2 - 6.5 in acetonitrile (MeCN). These values are comparable to those found for dibenzo-18-crown-6, but much lower than those found for cryptands. There is also a significant substituent effect on K values for Na^+ and K^+ in the n = 4 series in MeOH, especially with substituents containing heteroatoms, where the selectivity $S_{Na/K}$ ranges from ~ 25×10^3 for the phenacyl to 25 for the trifluoroethyl ester.

1.2. CALIX[n]ARENE AMIDES

Extraction studies with the first calix[4]arene tertiary amide **16** showed that it possessed a high affinity for Na^+ and K^+. **16** proved to be of unusual significance inasmuch as it was the first derivative for which details of the organisation and disposition of the binding sites in complexation with KSCN were revealed through X-ray analysis. Numerous calixarene amides have since been studied broadly and it is apparent that tertiary amides have a high affinity for alkali, alkaline earth and several transition metal ions. Substituent effects have also been used to probe the effect on ion complexation of the local environment of the amide carbonyl group. Simple tertiary amides show a selectivity for Na^+ in both extraction and stability constant values in MeOH, mimicking closely the behaviour of tetraesters and tetraketones, though the levels of extraction and log β values are higher for Li^+, Na^+ and K^+. However, amides tend to be less selective for Na^+ over K^+ than the tetraethyl ester.

1.3. CALIX[n]ARENE CARBOXYLIC ACIDS

The -CO_2H group differs significantly from those in the other derivatives discussed above in that it is ionisable. Thus, cation binding can be controlled by pH variation. By 1994 the tetraacid (Figure 1), its *p*-sulfonated analogue and several mixed derivatives with at least one carboxylic acid moiety had been investigated extensively with respect to their acidity and the stability of their alkali, alkaline earth, selected lanthanide metal and uranyl ion complexes. In general, calixarene carboxylic acids bind alkali metal cations more strongly than the non-ionisable esters and ketones. The alkaline earth metal cations are more strongly bound than the alkali metal cations and the n = 4 acids are selective for Ca^{2+}. The Ca^{2+}/Mg^{2+} selectivity displayed by the tetraacid itself is much higher than that of the tetra(diethyl amide). Both tetra- and hexa-acids efficiently extract lanthanide ions.

2. Developments since 1994

2.1. CALIXARENE ESTERS

Since 1994, work has involved extraction, complexation and transport studies of calix[4], [5] and [6]arene ethyl esters, as well as solution thermodynamics of the complexation by calix[4]arene ethyl esters. New issues have concerned the influence of conformational forms of a given derivative and of chemical modification of the methylene bridges between adjacent phenyl moieties, and a new field has been opened concerning the photochemical behaviour of calix[4]arene ester sodium complexes.

2.1.1. Binding by tetra(ethyl ester) 1

^{1}H NMR and ^{23}Na NMR studies of Na$^+$ complexation by **1** in a 50:50 (v:v) mixture of CD$_3$CN and CDCl$_3$ [9] gave evidence for a 1:1 complex, with a formation constant ~10^5, similar to that reported earlier [4]. An upper limit of the dissociation rate constant was estimated as 3 Hz. The existence of a 1:2 sodium:calixarene complex was demonstrated by ^{23}Na NMR. For conversion of the 1:1 to the 1:2 complex, $\Delta H_c = -(16 \pm 5)$ kJ mol^{-1} and $\Delta S_c = -(28 \pm 17)$ J K^{-1} mol^{-1}. The 1:2 complex may be an intermediate in the transfer of Na$^+$ in a 1:1 complex to another ligand, and could consist of an outer sphere association of the polar part of a second calixarene molecule with the complexed Na$^+$ [9], allowing exchange of the Na$^+$ between one complexation cavity and another.

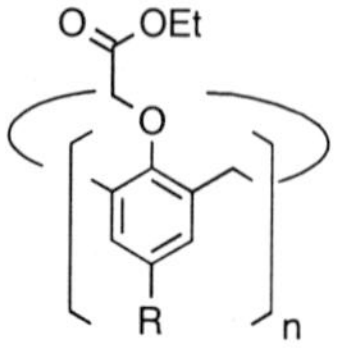

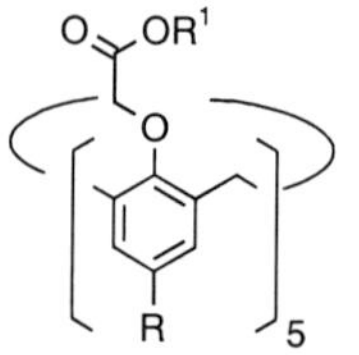

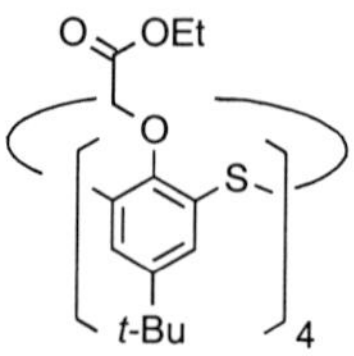

1 R = *t*-Bu, n = 4
2 R = *t*-Bu, n = 5
3 R = *t*-Bu, n = 6
4 R = *t*-Bu, n = 8
5 R = H, n = 6

6 R = Bn, R^1 = Me (*cone*)
7 R = Bn, R^1 = Et (*1,2-alt*)
8 R = *t*-Oct, R^1 = Me (*cone*)

9

Study of the dissolution of sodium halides in CHCl$_3$ has revealed the necessity for not only cation binding substituents, such as ester groups, but also for halide receptors, such as urea units, to be bound to the calixarene ionophore [10].

The Ag$^+$, Na$^+$ and K$^+$ complexes of **1** in the *1,3-alternate* conformation are stronger than those of the *cone* conformer, and spectroscopy indicates that the cation is bound to one of the two ionophoric binding sites composed of two oxygenic ligands and two benzene rings, presumably because of both electrostatic and "cation-π" interactions [11]. Intramolecular exchange between the two binding sites appears to occur for Ag$^+$, Na$^+$ and K$^+$. This has been proposed as a model of metal transport in π-base cavities, particularly in fullerenes, and intercalation of metal cations into graphite. The crystal structure of the sodium complex is known [12] and has been modelled through molecular mechanics (MM), the computation exploring the effect of placing the cation in different starting positions on the energy-minimised geometry of the complex [13].

Limited studies of transition metals have shown that tetraester **1** extracts Fe(III) more efficently than Co(II) or Cu(II) from water into CHCl$_3$ [14]. The extraction shows a complicated pH dependence. **1** does not bind Fe^{3+} in *N,N*-dimethylformamide (DMF).

2.1.2. Binding by pentaesters

Binding properties of the calix[5]arene derivatives **6** and **8** (*cone*) and **7** (*1,2-alternate*) have been assessed by liquid-liquid extraction of the alkali metal picrates from water into CH$_2$Cl$_2$, and by the determination of the stability constants of the alkali metal cation complexes by spectrophotometry or, when possible, by competitive potentiometry with Ag^+ as auxiliary cation [15]. Comparison with previous results for pentaesters with a *p-tert*-butyl group reveals a strong dependence of the complexation properties upon the nature of the substituents on the functional groups and on the *para* position [5]. In extraction, a plateau selectivity for larger cations is displayed by the three ligands, as for the *p-tert*-butyl analogues, though the discrimination remains significant. The *p*-benzyl derivatives are more selective than their *p*-alkyl counterparts, possibly because of an increase in rigidity. The *p-tert*-octyl is a stronger ligand than the *p*-benzyl derivative in MeOH and MeCN, and stability constants mirror the extraction data.

2.1.3. Binding by hexaesters

The *p-tert*-butylcalix[6]arene esters (ethyl ester **3** and *tert*-butyl ester) are selective towards K^+ and Cs^+ in the alkali metal group [16]. Extraction was hardly affected by the nature of the alkyl group in the ester moiety. ^1H NMR studies showed that the ligand conformation changes upon complexation: the free *tert*-butyl ester has the *1,2,3-alternate* conformation but adopts a C_6 symmetrical *cone* conformation upon K^+ or Cs^+ complexation. The situation is more complicated with the ethyl ester. Progressive replacement of the ester groups by methoxy groups show that the percentage extraction decreases with increasing number of methoxy groups, in line with the better cation affinity of esters relative to ether groups.

2.1.4. Transport studies

Preliminary results on the transport of Na^+, K^+ and Cs^+ thiocyanates by the *p-tert*-butylcalix[4], [6] and [8]arene ethyl esters through CH_2Cl_2 [4] were encouraging enough to prompt a more extensive study published in 1995 [17]. This concerned the mediated transport of alkali thiocyanates through a bulk liquid CH_2Cl_2 solution by *p-tert*-butyl tetra-, penta-, hexa-, and octaesters **1 - 4** and by the hexaester **5**. The transport rates showed that all derivatives, except for n = 8, are efficient and selective neutral ionophores for alkali metal cations. The most efficient carrier, by far, was the pentamer. The tetra- and pentaesters were selective for Na^+, and the hexaester for Cs^+. The Cs^+/Na^+ selectivity of the hexaester was increased by dealkylation in the *para* position. The same trend was observed in extraction experiments into CH_2Cl_2. Attempts to correlate the transport rates with the extraction data according to the theory [18] for diffusion-controlled transport processes, were not fully convincing, but the diffusion controlled nature of the process was demonstrated for a few systems. Results similar to those for the liquid membrane [17] have been reported for alkali metal ion transport through phospholipid bilayer membranes [19,20]. Transport is selective for Na^+, with a permeability of 17 with respect to K^+, and higher for Li^+, Rb^+, Cs^+. From ^{23}Na NMR spectra, it was deduced that the transport mechanism is of the carrier type, and that the transport rate is comparable to that with monensin.

2.1.5. Solution thermodynamics

The first report on solution thermodynamics of complexation and the thermodynamic parameters of complexation of alkali cations by **1** in solvents MeOH and MeCN was published in 1991 [21]. A detailed discussion of solution thermodynamics of calixarene binding is contained in Chapter 19.

2.1.6. Photochemical behaviour of the p-tert-butylcalix[4]arene tetraethyl ester Na^+ complex

Decomplexation of sodium salt complexes of the *p* tert-butyl tetraethyl ester **1** in various solvents under irradiation with a low pressure Hg lamp [22] shows an anion and solvent dependence. In optimal cases, the decomplexation leads to free tetraester with no decomposition nor conformational change. Toluene and $CHCl_3$ solvents aid reaction but it is inhibited in MeOH and MeCN. Iodide and thiocyanate are the most effective anions, and perchlorate and tetraphenylborate the least. A detailed study [23] showed photodecomplexation to result from the photodestruction of the anions. The $NaPh_4B$ complex prepared in $CHCl_3$ undergoes photodecomplexation to the free tetraester and (insoluble) sodium chloride. There is evidence that solvent inclusion plays a subtle role in stabilising the tetraester complexes.

2.1.7. Miscellaneous ester derivatives

The *cone, partial-cone* and *1,3-alternate* conformers of *p-tert*-butyltetrathiacalix[4]arene tetraethyl ester **9** are also useful ionophores [24]. Alkali metal picrate extraction from water into CH_2Cl_2 revealed that the *1,3-alternate* conformer preferentially extracted larger cations such as K^+ and Rb^+, whereas the *cone* conformer was Na^+ selective. The *partial-cone* conformer is less effective, but also shows a preference for K^+. It was concluded that the cavity size of the three conformers follows the sequence : *cone* < *partial-cone* < *1,3-alternate*. The extraction ability of the *cone* is lower than that of **1** for both nitrobenzene and CH_2Cl_2 solvents. The stability constants ($\log \beta$) at 25°C of the 1:1 alkali complexes, determined by ^1H NMR in 50% (v:v) $CDCl_3:CD_3OD$ have values of 2.8 and 4.0 for the *cone* conformers with S or CH_2, respectively, as bridging entities between the phenolic moieties. The fact that the former is more selective than the latter suggests that it may be a good candidate for sensor systems.

New polymers with several calix[4]arene tetraethyl esters grafted onto carbon atoms of the polymeric backbone by one of their *para* positions, appear to form Na^+ complexes involving 1:1 association of Na^+ with each calixarene tetraester unit [25, 26].

2.2. CALIXARENE KETONES

The potential of calix[5] and [6]-arenes as platforms for multifunctional receptors has recently been explored through examination of several ketone derivatives [27]. They possess larger frameworks than their calix[4]arene counterparts, yet retaining the capability to adopt true *cone* conformations. Complexation and extraction data reveal dramatic effects of both calix size and substitution. The most striking aspects of alkali metal cation extraction in the n = 5 series (Figure 4) is the strong dependency on the nature of the alkyl group adjacent to the ketonic carbonyl on the narrow rim, and to a smaller extent the nature of the *para* substituent on the wide rim. *tert*-butyl groups on both rims as in **11** produce the highest levels of extraction for all cations but little discrimination after Li^+. The levels of extraction are comparable to those observed with pentaamides and tetraamides. In contrast, the combination of a

10 R = *t*-Bu, R^1 = Me, n = 5
11 R = *t*-Bu, R^1 = *t*-Bu, n = 5
12 R = H, R^1 = Me, n = 5
13 R = H, R^1 = *t*-Bu, n = 5
14 R = *t*-Oct, R^1 = Me, n = 4
15 R = *t*-Oct, R^1 = Me, n = 6

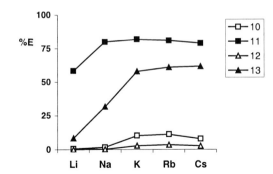

Figure 4. Extraction of alkali picrates by calix[5]arene pentaketones from water into CH_2Cl_2.

para hydrogen atom on the wide rim and a methyl ketone on the narrow rim as in **12** produces the lowest levels of extraction in the entire n = 5 series.

In contrast with the effect of a methyl ketone moiety in the n = 4 series, there is no strong preference for Na^+ in extraction. The stability constants (log β) for **10**, **12** and **13** in MeOH broadly reflect the trends in extraction (Table 1). The beneficial effects of narrow rim *tert*-butyl groups explain why **12** is consistently a weaker binder than **13**.

TABLE 1. Log β values for cation complexation in MeOH (T = 298 K, I = 0.01 Et_4NCl or Et_4NClO_4)

Ketone	Li^+	Na^+	K^+	Rb^+	Cs^+
10	< 1.5	< 1.5	3.34 ± 0.06	3.6 ± 0.2	3.22 ± 0.08
12	< 1.5	< 1.5	3.1 ± 0.2	3.2 ± 0.3	3.1 ± 0.2
13	< 1.5	4.8 ± 0.2	5.67 ± 0.01	5.81 ± 0.08	5.84 ± 0.03

The thermodynamic parameters (Table 2) for complexation of Cs^+ reveal that the increase in stability constant (Δ log β = 2.74) accompanying the replacement of methyl (compound **12**) by *tert*-butyl groups (compound **13**) is the result of a large change in $-\Delta H_c$ from 14 to 35 kJ mol^{-1}. The large enthalpy contribution for $Cs13^+$ may reflect differences in solvation of the ligand for steric reasons, better preorganisation for complexation and differences in the basicity of the carbonyl oxygen atoms in **12** and **13**. The small entropy term, even slightly negative compared to the value for $Cs12^+$, suggests the importance of solvation differences. The log β values for **10** are also completely consistent with its performance in extraction, both data showing a slight preference for Rb^+. X-Ray diffraction measurements have been used to probe the solid state conformations of some of these ketones and their complexes. The analysis of the $Na^+ClO_4^-$ and $Rb^+ClO_4^-$ complexes of **11** revealed that the cations lie deeply embedded within the cavities lined by the ethereal and carbonyl oxygen atoms, the cavities having distorted *cone* conformations, reproducible by MM calculations.

TABLE 2. Thermodynamic parameters in MeOH for Cs^+ complexes with ligands **10**, **12** and **13** (T = 298 K).

	10	12	13
$-\Delta G_c$ (kJ mol^{-1})	18.4 ± 0.5	18 ± 1	33.3 ± 0.2
$-\Delta H_c$ (kJ mol^{-1})	15 ± 2	14 ± 2	35.0 ± 0.1
$T\Delta S_c$ (kJ mol^{-1})	3 ± 2	4 ± 3	-2 ± 1
ΔS_c (JK^{-1} mol^{-1})	10 ± 7	13 ± 10	-7 ± 3

In contrast to the tetra and pentaesters and the tetramethyl ketone, pentaketones extract alkaline earth metal picrates from water into CH_2Cl_2 to some extent, depending on the ketone substituent [28]. Complexation results in MeOH confirmed their affinity for Ba^{2+} (S' = $\beta(Ba^{2+})/\beta(Ca^{2+}) = 10^2$ for ligand **13**).

The tetra- and hexamethyl ketones **14** and **15** with *tert*-octyl substituents on the wide rim can be used to extract silver (I) and palladium (II) from aqueous nitric acid solutions into $CHCl_3$ [29]. An extraction agent with a high selectivity for silver over palladium was sought which would permit the production of highly purified palladium by removing small or trace amounts of silver impurities. **14** was indeed found to be capable of extracting silver over palladium (in 100-fold excess) from ~ 1M nitric acid.

2.3. CALIXARENE AMIDES

The *cone p-tert*-butylcalix[4]arene tetraamides **16** and **17** typify derivatives studied before 1994 [30]. Their ability to discriminate between metal cations has been ascribed to their preorganized structure, illustrated by the X-ray structures of the potassium [6] and strontium complexes of **16** [31]. A much wider range is now known with either full or partial substitution of the phenolic hydrogen atoms.

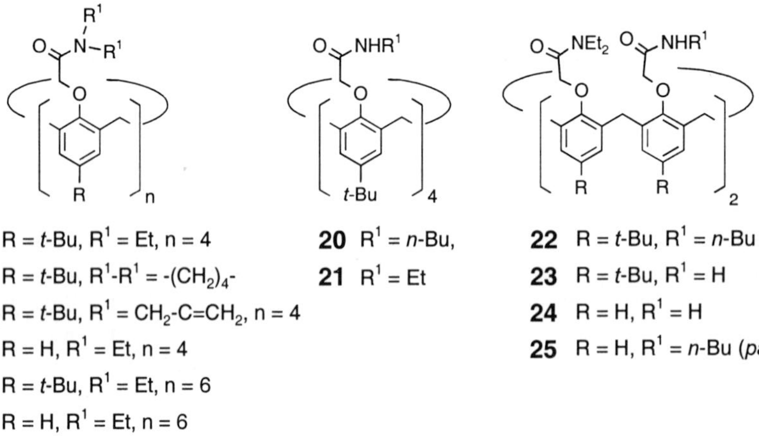

16 R = *t*-Bu, R¹ = Et, n = 4
17 R = *t*-Bu, R¹-R¹ = -(CH₂)₄-
18 R = *t*-Bu, R¹ = CH₂-C≡CH₂, n = 4
19 R = H, R¹ = Et, n = 4
26 R = *t*-Bu, R¹ = Et, n = 6
27 R = H, R¹ = Et, n = 6

20 R¹ = *n*-Bu,
21 R¹ = Et

22 R = *t*-Bu, R¹ = *n*-Bu
23 R = *t*-Bu, R¹ = H
24 R = H, R¹ = H
25 R = H, R¹ = *n*-Bu (*paco*)

2.3.1. Alkali and alkaline earth metal cation complexation

Recent studies of the extracting properties of eight new *p-tert*-butylcalix[4]arene amides (R^1 = propyl, butyl, pentyl, hexyl, octyl, allyl, propargyl, benzyl) [32] have shown that the extraction levels of alkali picrates are dependent on the nature of the amide substituents. Whereas no drastic changes were noticed for alkyl substituents with similar electronic properties, remarkable differences were observed in the extraction profiles of derivatives bearing substituents containing multiple bonds or having cyclic structures. Thus, the selectivities of calix[4]arene amides can be tuned by appropriate substituent choice. The propargyl derivative was the most selective of the present group with S = %E(Na⁺)/%E(K⁺) = 6.7, a performance confirmed in complexation (S' = $\beta(NaL)/\beta(KL)$ = 1258 instead of 126 for **16**). The influence of R^1 is even more important for Sr^{2+}. While the extraction level is still high for the alkyl derivatives, it drops greatly with allyl and benzyl and approaches zero with the propargyl derivative.

Dealkylation as in the calix[4]arene tetradiethyl amide **19** [33] results in a substantial decrease in the extraction of alkali and alkaline earth picrates without change in the selectivity pattern, which still favours Na^+ and Ca^{2+}. This effect was confirmed in alkali metal cation complexation in MeOH, $\Delta\log \beta$ ranging from 0.5 to 1.8. The stabilisation of the Na^+, K^+ and Rb^+ complexes is enthalpy controlled, whereas the stabilisation of the Li^+ complex is entropy driven, as with **16**. The decrease in stability upon dealkylation results from a decrease in $-\Delta H$ which is not compensated for by an increase in entropy. For alkaline earth metal cations, no clear conclusions arose as only lower limits of $\log \beta$ and $T\Delta S_c$ could be obtained in some cases. Unexpected however, were the more favourable enthalpy values found, especially with Ba^{2+}.

Changing the type of the amide functions (primary, secondary, tertiary) [33] was expected to modify the donor properties of the amide carbonyl groups and to allow intra- and intermolecular hydrogen bonding in the primary and secondary derivatives which could affect the cation binding. Since the demonstration that the tetra *n*-butylamide **20** was inefficient in extraction of alkali and alkaline earth picrates [34], no systematic investigation of this effect had been undertaken. The four calix[4]arenes bearing mixed tertiary/secondary (**22** and **25**) and tertiary/primary amide functions (**23** and **24**) [33] and the fully substituted tetra(ethyl amide) **21** [35] are all poor extractants of alkali and alkaline earth metal picrates from water into CH_2Cl_2. Compound **25** in the *partial cone* conformation and the secondary amide **21** are completely ineffective. However, *cone* **22** and **23** display extraction patterns similar to that of **16**, with selectivity for Na^+ and Ca^{2+}. Complexation in MeOH also appears to be strongly dependent on the basicity of the carbonyl oxygen atoms : the replacement of two distal tertiary by two secondary amides leads to a dramatic decrease in stability and the replacement of the secondary by primary amides to a further decrease (Figure 5).

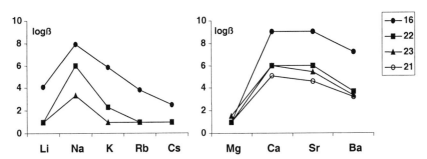

Figure 5. Stability of alkali and alkaline earth metal ion complexes with calix[4]arene amides in MeOH.

For **21**, lower complexation enthalpies than those of **16** can be explained by stronger solvation of the former ligand due to the presence of secondary amides. However, the selectivity still remains in favour of Na^+ within the alkali metal series and for Ca^{2+} and Sr^{2+} within the alkaline earth metal series. All these ligands, with the exception of **25**, form very stable complexes with Ca^{2+} and Sr^{2+}, resulting in noticeable selectivities *vs.* alkali metal cations. Sr^{2+}/Na^+ selectivity, very important in the treatment of radioactive waste, is 125 for **23**. Another important result is the different behaviour of these compounds in extraction and in complexation : compound **23**, in particular, is a poor extractant, but an excellent complexing agent. Different factors may explain this behaviour : an unfavourable conformation of the ligands in the organic phase, and/or less effective ion-ligand interaction in the complexes. ^{1}H NMR studies in CDCl$_3$ [33] suggest the existence of intramolecular hydrogen bonds. However, according to molecular dynamics (MD) simulations in CDCl$_3$ and MeOH [33], the lack of preorganisation should not be important since hydrogen bonding does not lead to a marked energy stabilisation. It is clear from the results obtained in MeOH that there is effective complexation even with secondary and primary amides. However, the complexes which should form during the extraction process are certainly not lipophilic enough to diffuse into the organic phase.

Very few studies highlight the importance of ligand conformation in the tetraamide series. The X-ray structure of the K$^+$ complex of *1,3-alternate* **16** [36] reveals a binuclear complex with a similar environment for the two cations, each being in interaction with four converging oxygen atoms and in close contact with two phenolic rings. In solution, ^{1}H and ^{13}C NMR results were consistent with the formation of both 1:1 and 2:1 complexes, the latter being the more stable.

Calix[4]arene amides: solution thermodynamics. The thermodynamic parameters for complexation of alkali and alkaline earth cations by **16** and **17** in MeOH [32] show that with both ligands and all alkali cations (except Li$^+$), the complexation process is enthalpy-controlled (Figure 6). The entropy terms are favourable for Rb$^+$ with **16** and **17** and Cs$^+$ with **16** and unfavourable for the other cations. Both amides show the same trends along the series, *i.e.* an exothermic maximum for Na$^+$ and a decrease in ΔS_c from Li$^+$ to Na$^+$ followed by an increase for Rb$^+$ and Cs$^+$. A weakening of the interactions between the donor sites and the cations with the increasing size of the latter may explain the enthalpy changes. Opposing entropy effects due to decreasing solvation of the cation along the series as well as the opening of the structure of the complexes with the increasing size of the cations and to decrease in degrees of freedom for the systems as cation size increases may be responsible for the entropy trends observed along the series. The enthalpy terms are always more favourable with **16** but the entropy terms are more favourable with **17**, in agreement with the more important desolvation of the cations imposed by the higher steric hindrance and greater rigidity of **17**. Comparison with **1** shows that (i) the similarly varying but less favourable enthalpy terms with **1** may be explained by the lower basicity of the ester *vs.* the amide carbonyls, and (ii) the lower Na$^+$/K$^+$ selectivity of amides is mainly due to the high enthalpy term with K$^+$.

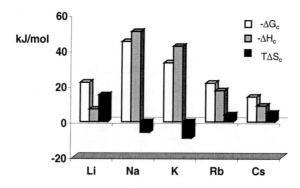

Figure 6. *Thermodynamic parameters for the complexation of alkali cations by **16** in MeOH.*

Only a lower limit of 6 could be attributed to log β for Li⁺, Na⁺ and K⁺ complexes of **16** in MeCN, where the stability order was reduced to the following sequence : K⁺ > Rb⁺ > Cs⁺. The favourable enthalpies of complexation, which could be measured for all cations, reveal an exothermic maximum for Na⁺. Lower limits only could be given for the complexation entropies of the first three cations, with the lowest limit being for Na⁺. As known with **1**, the higher stability of the complexes in MeCN *vs.* MeOH is due to an enthalpic effect, $-\Delta H_c$ and $T\Delta S_c$ being respectively higher and lower in MeCN than in MeOH. Gutmann donor numbers indicate the ligand and the cations should be less solvated in MeCN than in MeOH. Consequently the cation-ligand interactions are favoured, leading to more favourable enthalpies. However, the values of the transfer activity coefficients log $\gamma_{MeOH \rightarrow MeCN}$ indicate that K⁺, Rb⁺ and Cs⁺ are slightly more solvated in MeCN than in MeOH. Consistently, Rb⁺ and Cs⁺ complexes with **1** are not more stable in MeCN than in MeOH. These results suggest that the lower solvation of amide derivatives in MeCN may be responsible for the high stability of these complexes in this solvent. MD simulations [37] showed that carbonyls are less solvated in MeCN than in water, allowing the complexes to adopt a 'closed' type structure with the cation effectively shielded from the solvent. Li⁺ shows unusual behaviour whatever the ligand and the solvent, with entropies of complexation always higher than for the other cations. As expected for these highly solvated cations, the complexation of alkaline earth metal cations in MeOH is entropy controlled, although in some cases, the enthalpy terms are also favourable.

Calix[4]arene amides: transport studies. **16** and **17** are better transporting agents for alkali metal thiocyanates than their ester counterparts, and **17** is more efficient for Li⁺, Na⁺ and K⁺ than is **16** [17]. Both compounds are selective for K⁺. The very low extent of transport of Na⁺ is noteworthy considering the well established selectivity for Na⁺ in both extraction and complexation [30]. Both **16** and **17** appear to be selective receptors rather than selective carriers [18].

Binding by calix[6]arene amides. It has been previously shown that the extraction selectivity of penta- and hexaesters is shifted towards the larger cations. In the amide series, the *p-tert*-butylcalix[6]arene hexadiethyl amide **26** appeared to show little selectivity in either the alkali or alkaline earth metal series (at least from Mg²⁺) (Figure 7) [38]. Similar trends were observed in the complexation of alkali cations although it was difficult to establish the stoichiometry of the species formed (1:1 or 2:1), because of the very low solubility of this compound in MeOH and insignificant spectral changes upon complexation [38]. With Ca²⁺, Sr²⁺ and Ba²⁺, log β could only be estimated as higher than 6. In the solid state, a 1:1 complex with SrPic₂ has been isolated [39]. *p*-Dealkylation of **26** to give **27** led to an important decrease in the extraction level of all cations except Cs⁺. Thus, **27** presents remarkable Cs⁺/Na⁺ and M²⁺/Na⁺ extraction selectivities (M²⁺= Ca²⁺, Sr²⁺, Ba²⁺) which are duplicated in complexation in MeOH [38]. Hexaamides bearing OMe, OOct and OBn substituents in *p*-position [40] show similar extraction behaviour to **26** and **27**, *i.e.* low extraction levels and no selectivity (except for Cs⁺) within the alkali metal series, and increase in the efficiency with increasing cation size for the alkaline earth metals (Figure 7). The most flexible derivatives (*p*-OMe and *p*-H), although the least efficient, are the most selective.

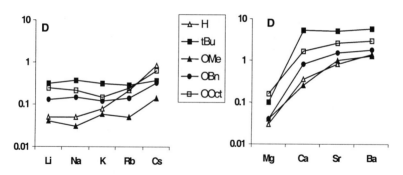

Figure 7. Extraction of metal picrates from water into CH$_2$Cl$_2$ by p-substituted calix[6]arene diethyl amides.

Hexaamides **26** and **27** have been studied in two other solvents of different polarity : DMF and tetrahydrofuran (THF) [28]. In DMF, spectral changes are large enough to establish the stoichiometry of the complexes of Na$^+$, Cs$^+$ and alkaline earth metal cations and evaluate their stability constants (Table 3). In most cases, there is unambiguous formation of 1:1 species. However, for **26** with Na$^+$ and Mg^{2+}, the spectral changes suggest the formation of a binuclear 2:1 species in addition to the 1:1 species.

TABLE 3. Stability constants (log β) of alkali and alkaline earth metal complexes with **26**, **27** and **16** in DMF, T = 298 K, (I = 0.01M Et$_4$NClO$_4$)

Ligands	Species	Na$^+$	Cs$^+$	Mg^{2+}	Ca^{2+}	Sr^{2+}	Ba^{2+}
26	1:1	2.6	2.7	5.7	4.8	4.7	3.6
	2:1	4.9	-	11.9	-	-	-
27	1:1	1.5	3.1	3.35	1.5	1.4	1.5
16	1:1	5.5	< 1	1.3	6.0	4.7	2.3

The results show : (i) a regular decrease in stability along the alkaline earth metal series, contrasting with the peak selectivity for Ca^{2+} displayed by **1**, and (ii) an important decrease in stability upon p-dealkylation for all cations except Cs$^+$; the determination of the thermodynamic parameters of complexation of Sr^{2+} showed that the lower affinity of **27** originates from a strongly unfavourable entropy term unbalanced by a more favourable enthalpy change. This can be explained by the greater flexibility of **27** resulting in a loss of freedom and hence of entropy upon complexation. The stabilities and selectivities are lower in DMF than in MeOH, especially for the alkaline earths; **27** is not selective even in DMF. Upon addition of either ligand, solutions of Na$^+$-picrate in THF undergo spectral changes which are characteristic of full separation of the tight ion pair and of formation of 1:1 complexes. With alkaline earth metal ions, there is evidence for the formation of both 1:1 and 2:1 complexes with **26** and only 1:1 complexes with **27**.

Binding by calix[8]arene amides. Early studies [6,34] of calix[8]arene octaamides showed their low extraction efficiency. More recently synthesised [40] wide rim (H, *t*-Bu, O-Bn, O-Me, O-Pent, O-Oct, 2-, 3- and 4-picolyloxy, 5- and 6-bipyridyloxy) derivatives, excepting the p-unsubstituted derivative which is selective for Cs$^+$ (S = ratio of distribution coefficients = D(Cs$^+$)/D(Na$^+$) = 31), are also poor and unselective extractants of alkali metal picrates.

In contrast, these ligands exhibit an affinity for alkaline earth metal ions which increases with the ionic radius. This difference in behaviour for the two series of cations leads to remarkable selectivities, especially for Sr^{2+}/Na^+ (see chapter 35). Figure 8 allows the comparison of D values for some important metal picrates by representative calix[4]-, [6]- and [8]arene amides.

Complexation of both series of cations has been followed by UV-absorption spectrophotometry in MeOH using either a direct or a competitive method with an auxiliary ligand (1-(2-pyridylazo)-2-naphthol). Formation of 1:1 complexes has been found in all cases, accompanied by 2:1 species except with Ca^{2+} and the p-OMe derivative. 2:1 species are isolated in the solid state. The stability of the complexes with p-OMe and p-OBn derivatives is higher than with their analogues **26** and **27**. X-ray data show the ether oxygen in the *para* position is unbound [41] and the high efficiency of the alkoxy or aryloxy compounds may be due to its electron donor effect.

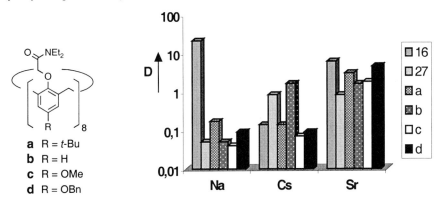

Figure 8. *Extraction of Na^+, Cs^+ and Sr^{2+} picrates from water into CH_2Cl_2 by calix[n]arene amides*

2.3.2. Other cation complexation

Studies of the complexation of europium by ligand **16** [42] and of various lanthanide ions by both **16** and **17** [43] stimulated further work [44,45] on complexation by calixarene amides. The aim was principally to exploit the luminescence properties of some of these cations, a topic discussed in detail in Chapter 31. Gd^{3+} has been shown to be complexed by a tetra primary amide derivative, but the low stability found (log β = 3) precluded the use for MRI enhancement agents, despite the high relaxivity of the complex at high field [46]. In the lanthanum complex of ligand **16** in the *partial cone* conformation, the calixarene is capable of simultaneous first and second sphere coordination with the cation [47].

The X-ray structures of complexes of transition and other metal ions (Fe^{2+}, Ni^{2+}, Cu^{2+}, Zn^{2+} and Pb^{2+}) with the tetraamide **16** are all more or less similar with the metal ion coordinated to the eight oxygens of the ligand [48], except for Ni^{2+} where the ligand undergoes significant conformational rearrangement to accommodate the metal ion in a distorted octahedral environment. Ag(I), Au(III), Pd(II) and Pt(IV) are extracted from highly acidic aqueous solutions into $CHCl_3$ by the *p-tert*-octylcalix[4]arene tetradiethyl amide [49]. Details of the interactions of transition metals with calixarenes are given in Chapter 28.

2.4. CALIXARENE CARBOXYLIC ACIDS

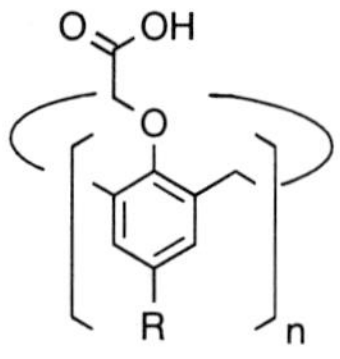

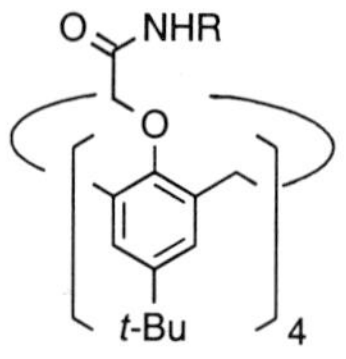

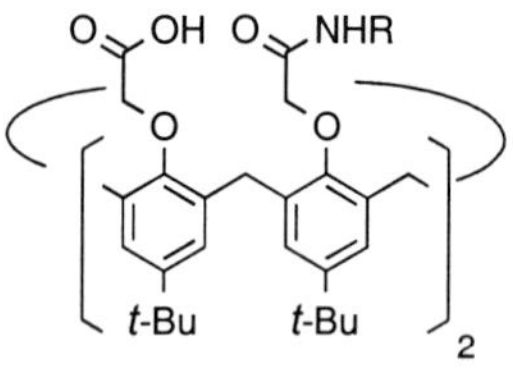

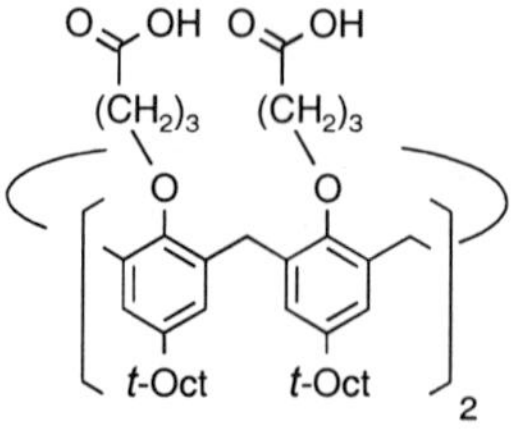

28 R = t-Bu, n = 4   **34** R = CH$_2$CO$_2$H   **35** R = CH$_2$CO$_2$H   **36**
29 R = t-Oct, n = 4
30 R = H, n = 4
31 R = t-Bu, n = 6
32 R = t-Oct, n = 6
33 R = t-Bu, n = 8

2.4.1. Binding of alkali cations

A very detailed investigation [50] of the extraction of Na$^+$ and K$^+$, separately or together, from water into C$_2$H$_4$Cl$_2$ and CHCl$_3$ by *cone* **28** (with chloride, acetate or nitrate as counterion) led to the following conclusions : (i) binuclear complexes LH$_2$M$_2$ form with both cations in CHCl$_3$, while the complexes are insoluble in C$_2$H$_4$Cl$_2$; binuclear species are unusual with ester and amide analogues, which generally form 1:1 complexes; ^{1}H NMR data suggest the location of one cation inside the cavity and the other outside, coordinated to one carboxylate; (ii) interference of non-exchangeable metal ions from the glass wall takes place at low metal concentrations; (iii) **28** exhibits a high Na$^+$/K$^+$ selectivity of 5.5×10^3 (expressed as the ratio of the extraction constants).

2.4.2. Binding of f-elements

In the complexation of three cations representative of the lanthanide series (Pr^{3+}, Eu^{3+} and Yb^{3+}) and of Th^{4+} by **28** and **33** both ligands behave as tetraanionic species [51]. All the lanthanides give mononuclear ML complexes, accompanied by either some of their expected protonated MH$_z$L or methoxy species ML(OCH$_3$)$_z$ according to the pH range. Binuclear complexes M$_2$L are also present with some of their methoxy forms. With thorium, the presence of simple binuclear species was never established. These complexes are generally more stable than their homologues with alkaline earth metal cations, with the exception of [Eu**28**]$^{3+}$ which is less stable than the calcium complex (log β = 21.6 vs 22.4). The strong solvation of Eu^{3+} may explain not only this stability order but also the formation of binuclear complexes with cations possibly located outside the ligand cavity. Thorium complexes are more stable than lanthanide complexes, due to higher charge density of the cation. Europium complexes of both ligands have similar stability, whereas Pr^{3+} and Yb^{3+} complexes of **33** are slightly more stable than those of **28**. This difference, which is never huge, could be attributed to the sterically less hindered structure of **33**. No real intra-series selectivity is observed but the Th^{4+}/Eu^{3+} selectivity is higher than 10^6 with both ligands. The mixed *p-tert*-butylcalix[4]arene mono carboxylic acid **49** forms both dinuclear and dimeric complexes but diacid **50** forms only mononuclear species. **49**, which behaves as a tetra-acid, is by far the best complexing agent for Th^{4+} among all the calixarene acids studied, showing significant Yb^{3+}/Eu^{3+} and Th^{4+}/Eu^{3+} selectivities in appropriate pH ranges.

Similar stoichiometries were found for the complexes of La^{3+} with **28**, its *p*-dealkylated counterpart **30** and the two related derivatives **34** and **35** where two or four glycine units have been introduced in the tetraacid [52]. This study, undertaken to test the potential of these ligands in radioimmunotherapy, La^{3+} being used as model for $^{225}Ac^{3+}$, showed stronger binding by **35** where the four carboxylates and the four amide carbonyls could possibly interact with the cation, as suggested by MD simulations (see Chapters 17, 18).

Extensive studies have been made of lanthanide ion extraction by calix[n]arene carboxylates, and in the extraction, for example, of lanthanide nitrates by both **28** and **31** and their more lipophilic *p-tert*-octyl counterparts **29** and **32**, the pH dependence was consistent with an ion exchange mechanism [53,54]. Comparing the calixarenes with acyclic tri- and mono-acid analogues, the extractibility order was: hexaacid > triacid > tetraacid > monoacid. Owing to ligand dimerisation, a complex of the type $M(LH_2)(LH_3)$ was formed with the tetraacid instead of a 1:1 complex as with the octaacid. Although the calixarenes have a greater separation efficiency than the monomer, the selectivity is not affected by the ring size. Study of the extraction of lanthanide chlorides into $CHCl_3$ by the same ligands showed there was no effect of Li^+ or K^+ on the extraction of lanthanides by the octaacid, and no effect of K^+ on the extraction of the heavier lanthanides by the tetraacid [55]. However, the pH dependence of the extraction was no longer consistent with a simple ion exchange process. In the presence of Na^+, the selectivity of the tetraacid shifted to the heavier lanthanides, with smaller ionic radii.

Studies specifically related to Am(III)/Ln(III) separation by tetra- and hexaacids, and related mixed amide-acid species are described in Chapter 35.

Extraction of UO_2^{2+} and Th^{4+} from water into $CHCl_3$ by ligand **28** [56] takes place via the formation of 2:2 and 1:1 (M:L) complexes, respectively. The presence of Na^+ in the aqueous phase promotes the extraction owing to the formation of heterocomplexes 2:2:2 and 1:1:1 in the case of UO_2^{2+} and 1:1:1 for Th^{4+}. The Th^{4+}/UO_2^{2+} separation factor is ~1000 whether or not alkali metal cations are present.

Direct study of $^{225}Ac^{3+}$ extraction from water into $CHCl_3$ by calixarenes **28** and **31** [57] (see above) has shown that extraction becomes significant after pH 2 and maximal around pH 4 and is consistent with the involvement of 1:1 complexes. After pH 6-7 there is a dramatic decrease in extraction. The ligands are highly selective for $^{225}Ac^{3+}$ *vs.* alkali, alkaline earth and zinc metal ions in weakly acidic to neutral media.

2.4.3. Binding of heavy metal ions
The extraction of Cu^{2+} from water into $CHCl_3$ by the tetraacid **29** and its homologue **36** with longer spacers between the phenolic oxygens and two of the carboxylic functions is of similar efficiency but shows a marked dependence on the presence of Na^+ for the latter, extraction being diminished in its absence [58]. The fact that $pH_{0.5}$ for Cu^{2+} is lower in the presence of Na^+ indicates co-extraction of both cations, an assumption confirmed by experiments with extractants saturated with sodium. The better efficiency of **36** with respect to **29** may be due to the location of Cu^{2+} farther from Na^+ than in **29**, thus reducing the electronic repulsion. This co-extraction mechanism was further supported by extraction experiments of Na^+ by **29**, which showed "self-co-extraction" of this cation by this ligand [59].

Ligand **29** was also shown [60] to quantitatively extract Hg^{2+} and Pb^{2+} at pH < 2.5 and 2.8, respectively, and could be applied for analytical purposes (See Chapter 34.). A resin-immobilised calix[4]arene acid presents a good Pb/Zn selectivity for [61].

2.5. MIXED SUBSTITUTED CALIXARENES

2.5.1. Combinations with ester functions

Di-*tert*-butylester-diphosphine oxide *p-tert*-butylcalix[4]-arenes **37** extract alkali metal picrates from water into CH_2Cl_2 with a peak selectivity for Na^+ [62], as for the tetra-ester analogues, and with efficiency intermediate between those for the homo substituted analogues. The stability constants (log β) in MeOH of the 1:1 Na^+ and K^+ complexes, determined by UV spectrophotometry, are respectively 2.9 and 2.4. The corresponding diethyl ester-diphosphine oxide forms a 1:1 complex with Ag^+ [63].

37 R = Et or *t*-Bu

Of the *cone*, the *partial-cone* and the *1,3-alternate* conformations of dipyridine-diethyl ester calix[4]arenes, a strong cation affinity, comparable to that of the tetraester analogue, is shown in picrate extraction by the *cone* [64]. It binds not only Na^+ but also Li^+, and the extraction sequence is $Na^+ > K^+ > Rb^+ > Li^+$. The *partial-cone* conformer shows a lower degree of extraction, but a higher selectivity for K^+. Of the three conformers, the *1,3-alternate* showed the highest extraction ability for K^+.

38

2.5.2. Combinations with amide functions

Calixarenes with amides and ionisable substituents. Ligand **39** forms a 1:1 complex with K^+ (β = 380) in CD_3CN-CD_2Cl_2 (5:1) mixtures [36]. In this medium the *cone* conformation of the ligand is maintained in the complex in contrast to the situation in the solid state where the complexed ligand adopts the *1,3-alternate* conformation. This ligand also extracts lanthanide picrates from water (pH = 5.8) into CH_2Cl_2 [65] with high efficiency. The lack of extraction below pH 2 suggests deprotonation of two OH units is necessary for efficient extraction. A Ln(L-2H)Pic stoichiometry for the extracted complex was assumed on the basis of the crystal structure [66]. Compound **40** has been shown to form 1:1 complexes with K^+ (log β = 4.61) and Ba^{2+} (log β = 3.5) in MeCN, where the cation is sandwiched between the two crowns [67]. With Na^+ there is evidence for a 2:1 complex, each cation being located in one crown. With this ligand there is exclusive complexation at the crown sites, unlike the fully substituted counterpart which binds a cation (K^+ or Na^+) at the amide site also.

In principle, dianionic ligands should be able to extract alkaline earth ions as neutral complexes and to promote the extraction into the organic phase. Extraction of alkaline earth nitrates from water into $CHCl_3$ containing ligand **41** in the *cone* conformation [68], occurs in the order $Ca^{2+} > Sr^{2+} > Ba^{2+} > Mg^{2+}$. The highest selectivity for Ca^{2+} is reached at pH 5. Competitive extraction of Ca^{2+} from a mixture of four alkaline earths confirms the exceptional performance of this ligand even at higher pH. Potentiometry has provided evidence for the formation of both mononuclear and binuclear complexes of **41** and its *p*-unsubstituted counterpart with alkali metal cations [69]. *Para*-dealkylation does not lead to important changes either in the stoichiometry of the complexes or their stability. The pH-dependent selectivity is in favour of K^+ for **41** and Na^+ for the derivative unsubstituted in *p*-position but, unlike the fully substituted parent derivatives, the discrimination is low, indicating that a size effect is not the predominant factor in the complexation due to the presence of the two symmetrical negative charges which may introduce a structural distortion. Complexes of the alkaline earth metal ions are much stronger, with mononuclear and binuclear as well as protonated complexes formed. With Ba^{2+}, bis-ligand species are also observed. Ca^{2+} and Sr^{2+} are quantitatively complexed from pH 6-7, Ca^{2+} being better complexed. Both ligands present a very high Sr^{2+}/Na^+ selectivity in the pH range 5-7.

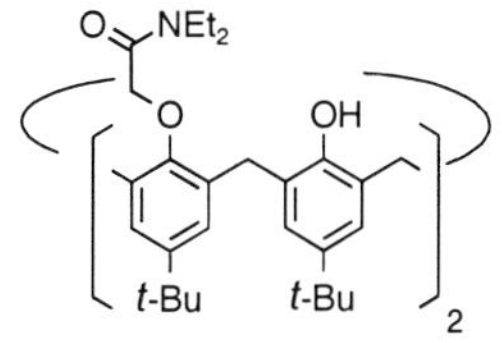

39
40 R = Benzo15C5 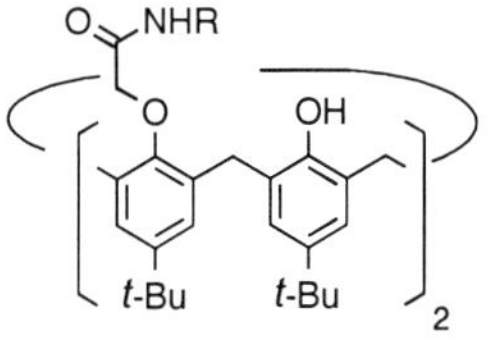
41 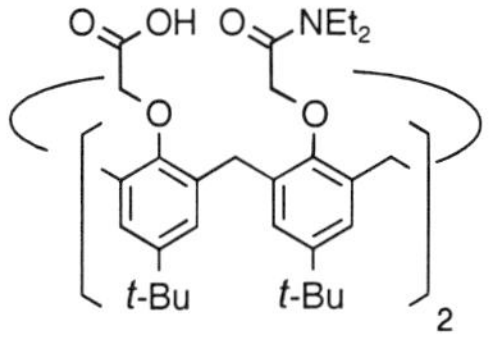
42 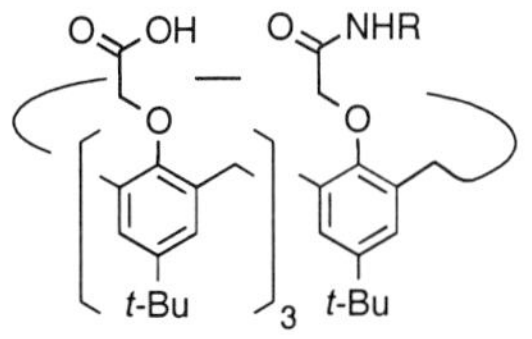
43 LH_3 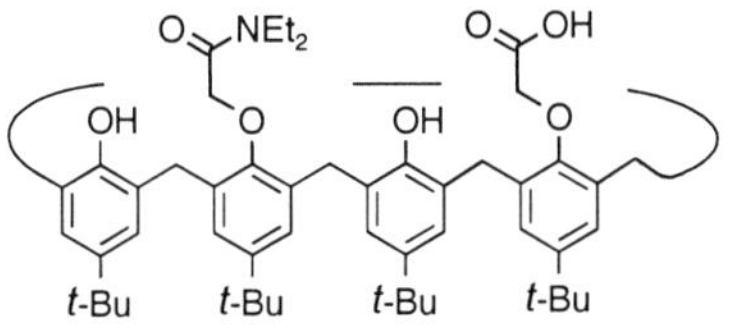
44 *cone* 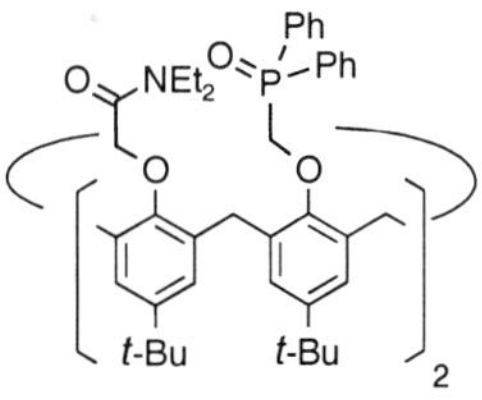
45 *paco*

In the trianionic calixarenes **42**, synthesised for trivalent lanthanides, the amide groups in the fourth position bear chromophoric groups and Eu^{3+} and Tb^{3+} complexes have been studied for their luminescence properties [70, 71]. In complexes of related water soluble calixarenes [72], luminescence lifetimes are enhanced owing to the shielding of the cation by the calixarene predicted by MM and MD simulations [71]. The amide **43**, also bearing two OH and one acid group, extracts La^{3+}, Eu^{3+}, Tm^{3+} and Lu^{3+} from water (pH 5.8) into CH_2Cl_2 in the sequence : $La^{3+} > Eu^{3+} > Tm^{3+} > Lu^{3+}$ [73,74]. This order is consistent with the formation of dimers with the larger cations (Sm^{3+}, Eu^{3+}) and of a monomer with the smaller cations (Lu^{3+}). Uranyl cation, as well as La^{3+}, Lu^{3+} and Hg^{2+} are quantitatively extracted at pH> 6. At physiological pH, Sr^{2+}, Pb^{2+}, Bi^{3+} and to a lesser extent Y^{3+} and Ag^+ are also quantitatively extracted, unlike alkali, alkaline earth and transition metal ions, which are very poorly extracted. This ligand still extracts uranyl ion when grafted on a recyclable polymeric resin [74].

Calixarenes with amides and neutral substituents. The diamide-bis(diphenyl)phosphine oxide) has been extensively studied in both *cone* **44** and *partial cone* **45** conformations. **44** is a good extractant for alkali metal picrates from water into CH_2Cl_2, with a peak selectivity for Na^+ (%E = 65.1), predictable from the behaviour of the parent compound **16** [62]. With alkaline earth metal cations, it is less effective and a plateau-like selectivity is observed for the three larger cations (%E = 12.5-15.8). Complexation shows the same selectivity sequences. For a given metal, **44** always gives less stable complexes than **16**.

For **45**, similar trends are seen in the extraction of alkali metal picrates from water into $C_2H_4Cl_2$, except with Li^+ which is better extracted than Na^+ [75]. The ligand causes full separation of the tight ion pair formed by the alkali picrates in THF, except in the case of Cs^+, thus confirming its strong affinity for these cations. Association constants confirm the trends observed in extraction. Transport experiments of alkali metal thiocyanates through a bulk liquid membrane of $C_2H_4Cl_2$ were analysed by means of a model assuming pure diffusion, allowing the evaluation of mass transfer coefficients. The transport rates depended on the extraction efficiency and the size of the transported complex. The transport selectivity paralleled the extraction selectivity. Compound **44** has also been shown to extract silver nitrate efficiently from water into CH_2Cl_2 in the presence of copper nitrate in great excess [63], via a 1:1 complex. As expected from the presence of both amide and phosphine oxide groups, both ligands **44** and **45** are good extractants of rare earth nitrates (La, Eu, Er, Y) in the presence of aluminium nitrate 0.9 M [76]. The extracted complex appears to be of 1:1 stoichiometry. With **44**, the extraction levels increased with the nitric acid content but in acidic conditions, **44** is less efficient than CMPO (N,N-di-isobutylcarbamoylmethyl)octylphenylphosphine oxide) because of the trapping of H^+ inside the *cone*. Competitive extraction experiments between eleven rare earths in the presence of aluminium nitrate, performed in order to assess the intra-group selectivity, showed that change of *cone* to *partial-cone* reduces extraction, suggesting that the two phosphine oxide groups are involved in the complexation. However, the selectivity is low and similar for the two ligands. The extraction level increased with the solvent polarity without any change in the selectivity. No extraction was observed in the presence of $NaNO_3$, as expected from the efficient extraction of Na^+ by these ligands. Competitive transport through supported liquid membranes of NPHE (o-nitrophenyl)hexyl ether showed that the flux changes parallel the distribution coefficients.

2.5.3. Combinations with carboxylic acid functions
The present section is devoted to the mixed heterocarboxylic acids. These compounds were studied for their binding ability towards alkali metal, rare-earth and uranyl cations, mainly by liquid-liquid extraction and FAB-MS (**46-48**) or, for the water soluble **49**, by potentiometry, calorimetry, UV-Vis, IR, ESR and NMR spectroscopic techniques.

In the first liquid-liquid extraction study of alkali metal cations by **46** (LH_2), from water into $C_2H_4Cl_2$ [77], the extracted species were identified as MLH for Na^+, K^+, Rb^+ and Cs^+, and as ML_2 for Li^+. The fact that the observed extraction selectivity ($Na^+ > K^+ > Rb^+ > Cs^+$) is different from that of the "normal" carboxylic acids and favours Na^+ led to the conclusion that the calixarene moiety participates in cation coordination via the phenolic oxygens. The results parallel those for the formation of the corresponding 1:1 complexes in MeOH [78]. A further investigation [79], using radioanalytical methods, extended the study to the water-$CHCl_3$ system and to higher pH. This study confirmed results for the water-$C_2H_4Cl_2$ system and the higher extraction efficiency with $CHCl_3$ as diluent, implying the simultaneous formation of ML_2 and ML. Ligand **47** also extracts Na^+ from water into $C_2H_4Cl_2$, and with a higher efficiency than **46** : ML_2 complexes are formed exclusively. ^{1}H-NMR confirms the stoichiometries derived from the extraction studies. The extraction of rare earth cations (Y^{3+}, La^{3+}, Er^{3+}) by **46**, from water into

$C_2H_4Cl_2$, in the presence of alkali metal cations, [80] reveals that four protons are exchanged in the extraction process, leading to the heteronuclear complex $LnML_2$, with $M^+ = Na^+, K^+$. The intra-series selectivity is rather poor. In the absence of alkali metal cations, the extraction is slight, from which it was concluded that the interactions between the lanthanide cation and the phenolic oxygens are weak or non-existent. Ligand **46** also extracts U(VI) from water into $CHCl_3$ or $C_2H_4Cl_2$ [81]. In the absence of alkali metal cations, the extent of extraction is low, and UO_2^{2+} is extracted as a 1:2 complex. In the presence of alkali metal cations, the extent of extraction is greatly enhanced due to the formation of heteronuclear complexes of stoichiometry 1:2:2 and 1:1:2 ($UO_2^{2+} : M^+ : LH_2$) when $M^+ = Na^+$, and 1:1:2 exclusively when $M^+ = K^+$, and diluent = $CHCl_3$. With Na^+, MS spectra confirm the existence of both homo and heteronuclear complexes. The corresponding calix[6]arene **48**, (LH_3), has been shown to be also a powerful extractant of U(VI), in the presence of sodium cations : three protons are exchanged in the extraction process from water into CH_2Cl_2, leading to the formation of $LNaUO_2$, whereas two protons only are exchanged in the extraction process into benzene, leading to the formation of $LHUO_2$. Application of **48** to the selective determination of uranium in urine samples in presence of sodium and/or plutonium has also been investigated [82-84].

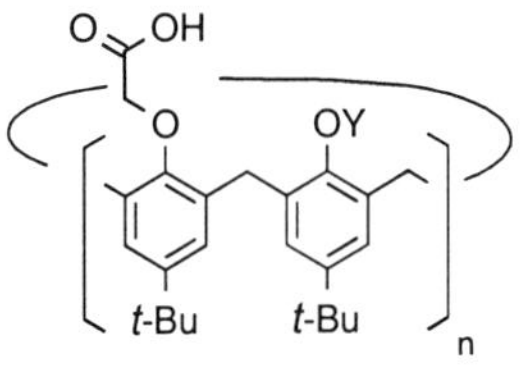

46 n = 2, Y = Me
47 n = 2, Y = $(CH_2)_2OMe$
48 n = 3, Y = Me

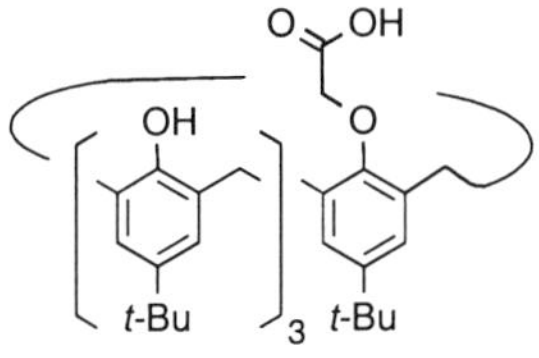

49 LH_4

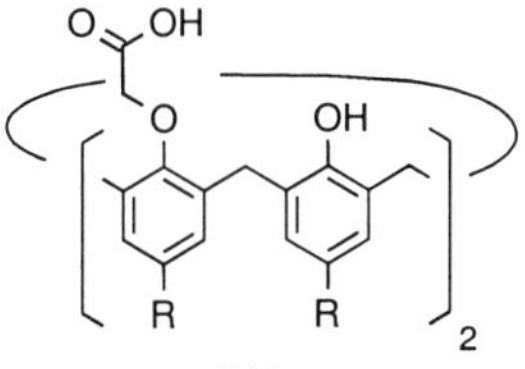

50 R = t-Bu
51 R = SO_3Na

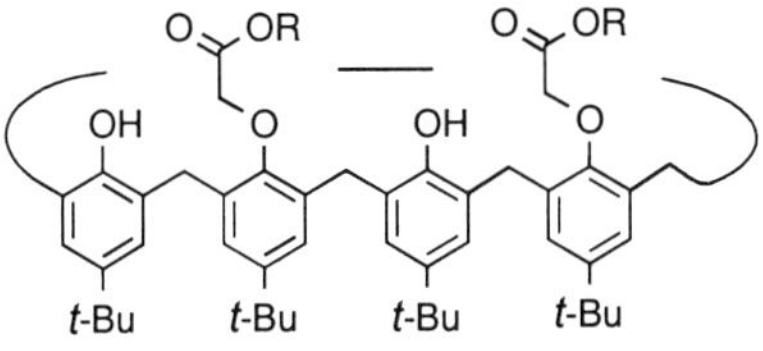

52 R = Et
53 R = t-Bu

Log β values of 6.86 and 6.58 (phenolate groups), and 3.66 and 3.00 (carboxylate groups) have been found for the protonation constants of the water-soluble, *cone* form *p*-sulfonated calix[4]arene **51** (LH_4) [85]. The lowering of the first protonation constant with respect to its value in *p*-hydroxybenzene sulfonic acid appears to be due to a less favourable enthalpic contribution, and may be ascribed to hydrogen bonding by a water molecule bridging the two opposite phenolate oxygens. This conclusion is further corroborated by the thermodynamic changes associated with the second protonation step. This is one of the few examples in the literature of an evaluation of the influence of microsolvation upon the complexing characteristics of a macrocyclic ligand. The favourable entropy value associated with the third protonation step indicates that very likely the third proton entering the anionic system bridges the two carboxylate groups. Entrance of the fourth proton destroys the H bonds, and leads to unfavourable ΔH_c but favourable ΔS_c. With copper(II), two species, CuL^{2-} and $CuLOH^{3-}$, were identified over the investigated pH range, with indications for a third species $CuL(OH)_2^{4-}$. Spectroscopic measurements further characterised the complexes.

The mixed *p-tert*-butylcalix[4]arenes **52** and **53** containing one carboxylic acid, one ester and two phenolic functions on the narrow rim were shown by potentiometric studies to form pH-dependent mononuclear or binuclear assemblies with alkali (excluding lithium) and alkaline earth metal cations [86]. In MeOH, the binuclear species $M_2L_2H_4$ are formed with alkali and alkaline earth cations (charges omitted) but in ethanol only mononuclear species ML, MLH and MLH_2 were found.

3. Conclusions

Although calixarenes are now well established members of the family of macrocycles, their full exploitation in supramolecular chemistry requires a better understanding of the many structural and environmental factors which influence the binding of metal cations by organic functional groups in solution. The new advances highlighted in this chapter confirm that the carbonyl group in its many manifestations continues to play an important role in the development of calixarene receptors. Not only do these studies help clarify the details of the complexed state, but they provide important pointers for future improvements in the design of selective metal ion receptors with predetermined properties and applications.

4. References and Notes

[1] a) R. M. Izatt, J. D. Lamb, R. T. Hawkins, P. R. Brown, S. R. Izatt, J. J. Christensen, *J. Am. Chem. Soc.* **1983**, *105*, 1782-1785; b) S. R. Izatt, R. T. Hawkins, J. J. Christensen, R. M. Izatt, *J. Am. Chem. Soc.* **1985**, *107*, 63-66

[2] a) V. Bocchi, D. Foina, A. Pochini, R. Ungaro, G. D. Andreetti, *Tetrahedron* **1982**, *38*, 373-378; b) R. Ungaro, A. Pochini, G. D. Andreetti, P. Domiano, *J. Incl. Phenom.* **1985**, *3*, 35-42; c) R. Ungaro, A. Pochini, G. D. Andreetti, F. Ugozzoli, *ibid.*, **1985**, *3*, 409-420.

[3] C. Alfieri, E. Dradi, A. Pochini, R. Ungaro, G. D. Andreetti, *J. Chem. Soc., Chem. Commun.* **1983**, 1075-1077.

[4] F. Arnaud-Neu, E. M. Collins, M. Deasy, G. Ferguson, S. J. Harris, B. Kaitner, A. J. Lough, M. A. McKervey, E. Marques, B. L. Ruhl, M. J. Schwing-Weill, E. M. Seward, *J. Am. Chem. Soc.* **1989**, *111*, 8681-8691.

[5] G. Barrett, M. A. McKervey, J. F. Malone, A. Walker, F. Arnaud-Neu, L. Guerra, M. J. Schwing-Weill, C. D. Gutsche, D. R. Stewart, *J. Chem. Soc., Perkin Trans. 2* **1993**, 1475-1479.

[6] A. Arduini, E. Ghidini, A. Pochini, R. Ungaro, G. D. Andreetti, F. Ugozzoli, *J. Incl. Phenom.* **1988**, *6*, 119-134.

[7] M. J. Schwing-Weill, M. A. McKervey, in *Calixarenes, a Versatile Class of Macrocyclic Compounds*, J. Vicens, V. Boehmer, Eds., Topics in Inclusion Science Vol. 3, Kluwer Academic Publishers, Dordrecht, **1991**, pp. 149-172.

[8] M. A. McKervey, M. J. Schwing-Weill, F. Arnaud-Neu, in "*Comprehensive Supramolecular Chemistry*", J. M. Lehn and G. W. Gokel Eds., **1996**, vol.1, pp.537-603.

[9] Y. Israeli, C. Detellier, *J. Phys. Chem. B* **1997**, *101*, 1897-1901.

[10] J. Scheerder, J. P. M. van Duynhoven, J. F. J. Engbersen, D. N. Reinhoudt, *Angew. Chem. Int. Ed. Engl.* **1996**, *35*, 1090-1093.

[11] A. Ikeda, S. Shinkai, *J. Am. Chem. Soc.* **1994**, *116*, 3102-3110.

[12] A. Ikeda, H. Tsuzuki, S. Shinkai, *Tetrahedron Lett.* **1994**, *35*, 8417-8420.
[13] P. Kane, D. Fayne, D. Diamond, S. E. J. Bell, M. A. McKervey, *J. Mol. Model.* **2000**, *6*, 272-281.
[14] M. Yilmaz, H. Deligoz, *J. Macromol. Sci., Pure Appl. Chem.* **1994**, *A31*, (suppls 182), 137-144.
[15] F. Arnaud-Neu, Z. Asfari, B. Souley, J. Vicens, M. Nierlich, P. Thuéry, *J. Chem. Soc., Perkin Trans. 2* **2000**, 495-499.
[16] H. Otsuka, K. Araki, S. Shinkai, *Tetrahedron* **1995**, *51*, 8757-8770.
[17] F. Arnaud-Neu, S. Fanni, L. Guerra, W. M. McGregor, K. Ziat, M. J. Schwing-Weill, G. Barrett, M. A. McKervey, D. Marrs, E. M. Seward, *J. Chem. Soc., Perkin Trans. 2* **1995**, 113-118.
[18] J. P. Behr, M. Kirch, J. M. Lehn, *J. Am. Chem. Soc.* **1985**, *107*, 241-246.
[19] T. Jin, M. Kinjo, T. Koyama, Y. Kobayashi, H. Hirata, *Langmuir* **1996**, *12*, 2684-2689.
[20] T. Jin, M. Kinjo, Y. Kobayashi, H. Hirata, *J. Chem. Soc., Faraday Trans.* **1998**, *94*, 3135-3140.
[21] A. F. Danil de Namor, N. A. de Sueros, M. A. McKervey, G. Barrett, F. Arnaud-Neu, M. J. Schwing-Weill, *J. Chem. Soc., Chem. Commun.* **1991**, 1546-1548.
[22] G. Barrett, D. Corry, B.S. Creaven, B. Johnston, M. A. McKervey, A. Rooney, *J. Chem. Soc., Chem. Commun.* **1995**, 363-364.
[23] N. Pitarch, A. Walker, J. F. Malone, J. J. McGarvey, M. A. McKervey, B. Creaven, D. Tobin, *Gazz. Chim. Ital.* **1997**, *127*, 717-721.
[24] N. Iki, F. Narmi, T. Fujimoto, N. Morohashi, S. Miyano, *J. Chem. Soc., Perkin Trans. 2* **1998**, 2745-2750.
[25] H. Deligoz, M. Yilmaz, *J. Polym. Sci. A. Polym. Chem.* **1995**, *33*, 2851-2853.
[26] H. Deligoz, M. Yilmaz, *React. and Functional Polymers* **1996**, *31*, 81-88.
[27] S. E. J. Bell, J. K. Browne, V. McKee, M. A. McKervey, J. F. Malone, M. O'Leary, A. Walker, F. Arnaud-Neu, O. Boulangeot, O. Mauprivez, M. J. Schwing-Weill, *J. Org. Chem.* **1998**, *63*, 489-501.
[28] F. Arnaud-Neu, S. Bollender, L. Monnier, M. J. Schwing-Weill, 4th International Conference on Calixarenes, Parma (Italy), August 31 – September 4, **1997**.
[29] K. Ohto, E. Murakami, T. Shinohara, K. Shiratsuchi, K. Inoue, M. Iwasaki, *Anal. Chim. Acta* **1997**, *341*, 275-283.
[30] F. Arnaud-Neu, M. J. Schwing-Weill, K. Ziat, S. Cremin, S. J. Harris, M. A. McKervey, *New J. Chem.* **1991**, *15*, 33-37.
[31] N. Muzet, G. Wipff, A. Casnati, L. Domiano, R. Ungaro, F. Ugozzoli, *J. Chem. Soc., Perkin Trans. 2* **1996**, 1065-1075.
[32] F. Arnaud-Neu, G. Barrett, S. Fanni, D. Marrs, W. McGregor, M. A. McKervey, M. J. Schwing-Weill, V. Vetrogon, S. Wechsler, *J. Chem. Soc., Perkin Trans. 2* **1995**, 453-461.
[33] F. Arnaud-Neu, S. Barboso, F. Berny, A. Casnati, N. Muzet, A. Pinalli, R. Ungaro, M. J. Schwing-Weill, G. Wipff, *J. Chem. Soc., Perkin Trans. 2* **1999**, 1727-1738.
[34] S. K. Chang, S. K. Kwon, I. Cho, *Chem. Lett.* **1987**, 947-948.
[35] F. Arnaud-Neu, S. Barboso, S. Fanni, M. J. Schwing-Weill, V. McKee, M. A. McKervey, *Ind. Eng. Chem. Research* **2000**, *10*, 3489-3492.
[36] P. D. Beer, M. G. B. Drew, P. A. Gale, P. B. Leeson, M. I. Ogden, *J. Chem. Soc., Dalton Trans.* **1994**, 3479-3485.
[37] A. Varnek, G. Wipff, *J. Phys. Chem.* **1993**, *97*, 10840-10848.
[38] S. Fanni, F. Arnaud-Neu, M. A. McKervey, M. J. Schwing-Weill, K. Ziat, *Tetrahedron Letters* **1996**, *37*, 7975-7978.
[39] A. Casnati, F. Ugozzoli, R. Ungaro, unpublished results.
[40] F. Arnaud-Neu, S. Barboso, M. J. Schwing-Weill, A. Casnati, F. Ugozzoli, R. Ungaro, J. F. Dozol, *Euradwaste'99*, Luxemburg, November **1999**.
[41] A. Casnati, K. Rissanen, F. Ugozzoli, R. Ungaro, unpublished results.
[42] N. Sabbatini, M. Guardigli, A. Mecati, V. Balzani, R. Ungaro, E. Ghidini, A. Casnati, A. Pochini, *J. Chem. Soc., Chem. Commun.* **1990**, 878-879.
[43] F. Arnaud-Neu, *Chem. Soc. Rev.* **1994**, 235-241.
[44] E. M. Georgiev, J. Clymire, G. L. McPherson, D. M. Roundhill, *Inorg. Chim. Acta* **1994**, *227*, 293-296.
[45] H. Matsumoto, S. Shinkai, *Chem. Lett.* **1994**, 901-904.
[46] E. M. Georgiev, D. M. Roundhill, *Inorg. Chim. Acta* **1997**, *258*, 93-96.
[47] P. D. Beer, M. G. B. Drew, M. I. Ogden, *J. Chem. Soc., Dalton Trans.* **1997**, 1489-1491.
[48] P. D. Beer, M. G. B. Drew, P. B. Leeson, M. I. Ogden, *J. Chem. Soc., Dalton Trans.* **1995**, 1273-1283.
[49] K. Ohto, H. Yamage, H. Murakami, K. Inoue, *Talanta*, **1997**, *44*, 1123-1130.
[50] G. Montavon, G. Duplâtre, Z. Asfari, J. Vicens, *New J. Chem.* **1996**, *20*, 1061-1069.
[51] F. Arnaud-Neu, S. Cremin, S. J. Harris, M. A. McKervey, M. J. Schwing-Weill, P. Schwinté, A. Walker, *J. Chem. Soc., Dalton Trans.* **1997**, 329-334.

[52] M. H. B. Grote Gansey, W. Verboom, F. C. J. M. van Veggel, V. Vetrogon, F. Arnaud-Neu, M. J. Schwing-Weill, D. N. Reinhoudt, *J. Chem. Soc., Perkin Trans. 2* **1998**, 2351-2360.
[53] R. Ludwig, K. Inoue, T. Yamato, *Solvent Extr. Ion Exch.*, **1993**, *11*, 311-330.
[54] K. Ohto, M. Yano, K. Inoue, T. Yamamoto, M. Goto, F. Nakashio, S. Shinkai, T. Nagasaki, *Anal. Sci.*, **1995**, *11*, 893-902.
[55] K. Ohto, M. Yano, K. Inoue, T. Nagasaki, M. Goto, F. Nakashio, S. Shinkai, *Polyhedron* **1997**, *16*, 1655-1661.
[56] G. Montavon, G. Duplâtre, Z. Asfari, J. Vicens, *Solvent Extr. Ion Exch.* **1997**, *15(2)*, 169-188.
[57] X. Chen, M. Ji, D. R. Fisher, C. M. Wai, *Chem. Commun.* **1998**, 377-378.
[58] K.Ohto, K. Shiratsuchi, K. Inoue, M. Goto, F. Nakashio, S. Shinkai, T. Nagasaki, *Solvent Extr. Ion Exch.* **1996**, *14(3)*, 459-478.
[59] K. Ohto, H. Ishibashi, K. Inoue, *Chem. Lett.* **1998**, 631-632.
[60] N. T. K. Dung, R. Ludwig, *New J. Chem.* **1999**, *23*, 603-607.
[61] K. Ohto, Y. Tanaka, K. Inoue, *Chem. Lett.* **1997**, 647-648.
[62] F. Arnaud-Neu, J. K. Browne, D. Byrne, D. J. Marrs, M. A. McKervey, P.O'Hagan, M. J. Schwing-Weill, A. Walker, *Chem. Eur. J.* **1999**, *5*, 175-186.
[63] M. R. Yaftian, M. Burgard, A. Elbachiri, D. Matt, C. Wieser, C.B. Dieleman, *J. Incl. Phenom.* **1997**, *29*, 137-151.
[64] F. Bottino, S. Pappalardo, *J. Incl. Phenom.* **1994**, *19*, 85-100.
[65] P. D. Beer, M. G. B. Drew, P. Kan, P. B. Leeson, M. I. Ogden, G. Williams, *Inorg. Chem.* **1996**, *35*, 2202-2211.
[66] P. D. Beer, Z. Chen, P. A. Gale, J. A. Heath, R. J. Knubley, M. I. Ogden, *J. Incl. Phenom.* **1994**, *19*, 343-359.
[67] P. D. Beer, M. G. B. Drew, R. I. Knubley, M. I. Ogden, *J. Chem. Soc., Dalton Trans.* **1995**, 3117-3123.
[68] M. Ogato, K. Fujimoto, S. Shinkaï, *J. Am. Chem. Soc.* **1994**, *116*, 4505-4506.
[69] F. Arnaud-Neu, S. Barboso, A. Casnati, A. Pinalli, M. J. Schwing-Weill, R. Ungaro, submitted.
[70] D. M. Rudkevich, W. Verboom, E. Van der Tol, C. J. Van Staveren, F. M. Kaspersen, J. W. Verhoeven, D. N. Reinhoudt, *J. Chem. Soc., Perkin Trans. 2* **1995**, 131-134.
[71] F. C. J. M. Van Veggel, D. N. Reinhoudt, *Recl. Trav. Chim., Pays-Bas* **1995**, *114*, 387-394.
[72] F. J. Steemers, H. G. Meuris, W. Verboom, D. N. Reinhoudt, E. B. van der Tol, J. W. Verhoeven, *J. Org. Chem.* **1997**, *62*, 4229-4235.
[73] P. D. Beer, M. G. B. Drew, A. Grieve, M. Kan, P. B. Leeson, G. Nicholson, M. I. Ogden, G. Williams, *Chem. Commun.* **1996**, 1117-1118.
[74] P. D. Beer, M. G. B. Drew, D. Hesek, M. Kan, G. Nicholson, P. Schmitt, P. Sheen, G. Williams, *J. Chem. Soc. Dalton Trans.* **1998**, 2783-2785.
[75] M. R. Yaftian, M. Burgard, D. Matt, C. Wieser, C. Dieleman, *J. Incl. Phenom.* **1997**, *27*, 127-140.
[76] M. R. Yaftian, M. Burgard, C. Wieser, C. B. Dieleman, D. Matt, *Solvent extr. Ion Exch.*, **1998**, *16(5)*, 1131-1149.
[77] J. Soedarsono, M. Burgard, Z. Asfari, J. Vicens, *Solvent Extr. Ion Exch.* **1995**, *13(4)*, 755-769.
[78] F. Arnaud-Neu, G. Barrett, S. J. Harris, M. Owens, M. A. McKervey, M. J. Schwing-Weill, P. Schwinté, *Inorg. Chem.* **1993**, *32*, 2644-2650.
[79] G. Montavon, G. Duplâtre, N. Barakat, M. Burgard, Z. Asfari, J. Vicens, *J. Incl. Phenom.* **1997**, *27*, 155-158.
[80] J. Soedarsono, A. Hagege, M. Burgard, Z. Asfari, J. Vicens, *Ber.* **1996**, *100*, 477-481.
[81] G. Montavon, G. Duplâtre, Z. Asfari, J. Vicens, *J. Radioanal. Nucl. Chem.* **1996**, *210*, 87-103.
[82] N. Baglan, C. Dinse, C. Cossonet, R. Abidi, Z. Asfari, M. Leroy, J. Vicens, *J. Radioanal. Nucl. Chem.* **1997**, *226*, 261-265.
[83] C. Dinse, C. Cossonnet, N. Baglan, Z. Asfari, J. Vicens, *Radioprotection* **1997**, *32*, 659-671.
[84] C. Dinse, N. Baglan, C. Cossonnet, J. F. Le Du, Z. Asfari, J. Vicens, *J. Alloys Compds.* **1998**, 271-273, 778-781.
[85] G. Arena, R. P. Bonomo, A. Contino, F. G. Gulino, A. Magri, D. Sciotto, *J. Incl. Phenom.* **1997**, *29*, 347-363.
[86] Arnaud-Neu, G. Barrett, G. Ferguson, J. F. Gallagher, M. A. McKervey, M. Moran, M. J. Schwing-Weill, P. Schwinté, *Supramolecular Chem.* **1996**, *7*, 215-222.

Chapter 22

PHASE TRANSFER EXTRACTION OF HEAVY METALS

D. MAX ROUNDHILL, JIN YU SHEN

Department of Chemistry
Texas Tech University, Lubbock, Texas 79409-1061, U.S.A.
e-mail: ULDMR@ttacs.ttu.edu

1. Introduction

From both environmental and economic viewpoints, metal ion extraction is an important process. The challenge is to find complexants that selectively extract and release metals from mixtures, where they may be present in both cationic and anionic forms, as well as as neutral (oxides, sulfides) species. Typical problems are those which arise when toxic metal residues from the mining and processing of minerals leak or are leached into soil aquifers [1]. An approach to complexant design for metal cation extraction is to use the calixarene platform as a unit upon which to attach heavy metal specific functionalities. Calixarenes are chosen because of both their preorganization and their availability of multiple sites for incorporating ligating functionalities [2]. The chosen derivatives depend on the metal ion that is targeted. For the lighter transiton metal ions and lanthanides and actinides, derivatives with nitrogen or oxygen donor atoms are often favored. Therefore, carboxylates, amines, amides, and sulfonamides are commonly appended to the calixarene. However, for second and third row transiton metal ions, and for the heavy post-transition metal ions, soft donor atoms such as phosphorus and sulfur are preferable [3-7]. Thus, functionalities such as phosphines, thiolates, thioethers or dithiocarbamoyls are appended to the calixarene rims. These latter two are chosen because of their strong affinity for metals such as Hg(II), Pd(II), Au(III) and Ag(I) and because uncharged groups often provide higher selectivity than charged. The strong complexant ability of the thiolate group, for example, is accompanied by low selectivity, and reaction kinetics (uptake and release) can also often be slow with charged ligands.

For cations, the metal to be extracted can be directly bound to a host ligating group, but for oxyanions and uncharged oxides there is often no direct interaction between metal centre and host. For an oxycation such as the uranyl ion, however, both factors are involved, since the complexant not only directly binds to the metal center but can also hydrogen bond to the oxo groups of the uranyl cation. Despite there being numerous examples of hosts and complexants for cations, there are fewer for anion hosts [8-11]. Recently a number of chemically modified calixarenes have been synthesized that are hosts for simple anions [12]. From an environmental viewpoint a series of anions for which selective hosts would be useful are the oxyanions. One such anion is chromate and a calix[4]arene with ammonium functionalities on its lower rim is an extractant for anionic Cr(VI) [13]. Anionic Cr(VI) is important because of its high toxicity [14-17], and because of its presence in soils and waters [18]. Two other environmentally important tetrahedral oxyanions are pertechnecate and selenate. Pertechnecate is a concern because it is a radioactive fission product. Selenate is also a toxic product of nuclear reactions. Hosts are needed for these anions that show extraction selectivities due to differences in structure, hydrogen bonding and charge. A range of approaches to the design and use of calixarenes as anion complexants is described in Chapter 23.

2. Phase Transfer Extractabilities

A useful extractant must be both a selective complexant and an effective phase transfer reagent. Calixarenes can function in both of these capacities [19,20]. An important property which may be relevant to the kinetics of extraction by calixarenes is their amphiphilicity, which causes them to concentrate at interfaces [21]. They have found application as extractants for most metals, including heavy and transition metal ions and those of the lanthanides and actinides, with both thermodynamic and structural analyses being based on these works [22-26].

2.1. TRANSITION METAL IONS

Although simple (phenolic) calixarenes will form lipid-soluble complexes (see, for example, Chapter 30), strongly basic conditions are usually required and little selectivity is exhibited, so that nearly all calixarenes of practical utility as extractants are those functionalised with groups intended to endow selectivity and to allow use under a wide variety of conditions.

2.1.1. Oxygen donor groups
Among this group are the simple calixarenes themselves. Cu(II) is extracted from an alkaline ammonia solution into chloroform or benzene as a 1:1 complex with *p-tert*-butyl-calix[n]arene (n = 4, 6, 8) [27,28]. It has been proposed that the calixarene binds to the copper in its tetraammine form [29]. Fe(III) is also extracted [30,31].

The narrow rim can be readily functionalized to incorporate ketone, carboxylate, ester and ether groups. (The nature of such molecules and their binding of alkali and alkaline earth metal ions are discussed in detail in Chapter 21) Ketone substituted calix[4]arenes can be used to extract Ag(I) and Pd(II) from acidic nitrate solutions [32,33]. Although a calix[4]arene with both methoxy and ketone functionalities has a strong selectivity for Na(I) extraction, it will also extract Ni(II) among other metals [34]. Both carboxylates and their esters will extract transition metal cations, though the complexes of the carboxylate anions are the more stable. Calix[4]arenes extract Cu(II) in the presence of Na(I), although sodium complexation occurs in more acidic solutions and as the sodium concentration is increased, the extraction of copper is less effective [35,36]. Calixarene carboxylates can also be used to extract Cu(II), Ni(II) and Co(II) from nitric acid solutions [37]. Tetrakis(ethoxycarbonylmethyl)-*p-t*-butyl-calix[4]arene extracts Re(VII) as [ReO$_4$]⁻ (perrhenate) [38].

In a detailed study of the liquid-liquid extraction of transition metal ions with calix[4]arene and calix[6]arene carboxylates, it was found that they consistently show higher extractions than their monomeric aromatic carboxylate counterpart. The extractability of the transition metal ions Cu(II), Ni(II) and Co(II), unlike that of alkali and alkaline earth metal ions, does not correlate with ring size, and increases in the order: monomer < calix[4]arene < calix[6]arene. The extraction equilibrium constants for the calix[4]arene and calix[6]arene are 2 and 4 orders of magnitude larger than are found for the monomeric analog[39]. Palladium(II) can be extracted by calix[6]arene hexaacetate [40]. A combination of the calixarene carboxylate **1** and an amine has been used to extract transition metal ions [41], and a recent crystal structure shows that Cu(II) can associate with this calixarene as a CuN_4 complex [42]. Polymer-supported calixarene carboxylates can be used for the extraction of Fe(III). An extension to these studies is the use of thiacalix[4]arene carboxylates as extractants. Their selectivity reflects the presence of both the sulfur and carboxylate-oxygen donor atoms, and is influenced by the calixarene conformation. The *cone* conformer of the tetracarboxylate extracts Cu(II), Co(II) and Fe(III) [43].

Calixarene ethers with a wide range of functional groups extract transition metals, especially Ag(I). The preferred conformations are the *1,3-alternate* and *partial-cone* with the silver(I) bound to a phenolic oxygen in the inverted phenyl unit and carbons in the proximal phenyls. With the *1,3-alternate* conformer of **2**, the Ag(I) is believed to tunnel across an aromatic cavity between the two binding sites [44]. Capping a calix[6]arene with a 1,3,5-tribenzyl moiety affords a rigid cavity that has a high affinity for silver(I) [45]. This compound **3** has a closed ionophoric cavity which, as indicated by the temperature invariant NMR spectrum, appears to be quite rigid. A calix[6]arene epoxide **4** extracts Cu(II), Ni(II), Co(II) and Fe(III), with the latter being the least extracted [46]. By contrast, a polymeric polyether with an epoxy group at the end of the chain selectively extracts Fe(III) in the presence of Cu(II), Ni(II) and Co(II) [47]. Since calixcrowns are excellent extractants for Group I cations (see Chapter 20), this property can be used to extract [TcO$_4$]$^-$ (pertechnecate) as its Cs(I) salt. Both ^{99}Tc and ^{137}Cs radionuclides can therefore be removed from nuclear waste with a single extractant [48].

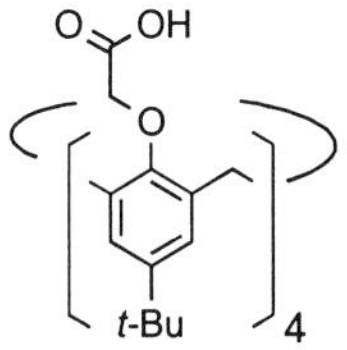

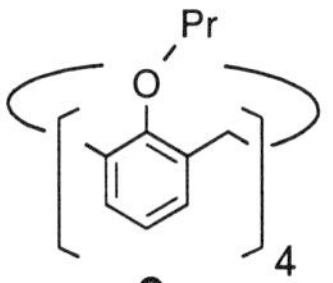

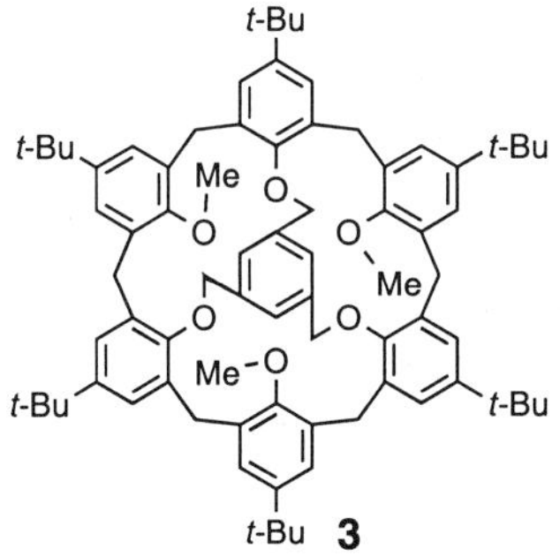

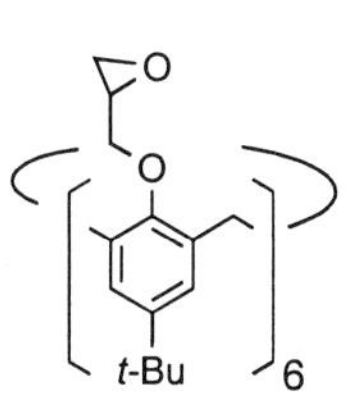

A calix[4]arene with sulfonate groups on the narrow rim is a better extractant for Fe(III) than is the unsubstituted analogue [31]. In the pH range 8-10 and in the presence of trioctylmethylammonium chloride, calixarene sulfonates selectively and quantitatively extract Mn(II) from other metal ions. Cu(II), Ni(II), Co(II) and Fe(III) are not extracted to any measureable extent, nor are alkaline earth, lanthanide and uranium(VI) ions [49]. The manganese is extracted into chloroform as a reddish-purple 1:1 complex [50]. A similar system can be used to extract Ti(IV) and Fe(III) from a zirconium ore [51].

2.1.2. Nitrogen donor groups
The calixarene rims can be readily functionalised to incorporate amide, amine, hydroxamate, azo and pyridyl groups, all of which have been used in the search for selective metal ion extractants. These groups may be the primary metal binding site, or part of a more sophisticated chelate system, possibly including the phenolic oxygen atoms of the parent calixarene.

For the extraction of the heavy metals Pd(II), Pt(IV) and Au(III), calix[4]arenes with both carboxylic acid and amide functionalities on the narrow rim are more effective than those that have only the former [52]. Calixarene amines extract Ag(I) into nitrobenzene [53]. Calixarenes with both amine and carboxylic acid functionalities are extractants for metals of the platinum group [54]. Both cyclic and acyclic calix[4]arene amides are effective for these metals and Ag(I), although there are differences in selectivity between them [55]. A calix[4]arene with amide and phosphine oxide groups in alternate positions of the narrow rim is a good extractant for Ag(I), a property ascribed to the polar amide group [56]. Other examples of Ag(I) extraction by calixarene amides show that the analogue with hydrogen rather than t-butyl groups on the wide rim is a much better extractant [57]. Calixarene amides also extract copper(II) [58]. Calix[n]arenes **5** (n = 4, 6, 8) with N-glycyl-ester acetamide functionalities appended to their narrow rim have been investigated as extractants for metal ions, with the derivative having n = 8 being a good extractant for Ag(I) [59].

Amides and amines have also been used as extractants for transition metal oxides and oxyanions. These are targeted because by proton transfer at nitrogen they can be converted into cations, and because the NH functionality is a potential site for hydrogen bonding with the oxygens of the oxyions. Calix[4]arene amides and amines **6-9** extract chromium(VI), molybdenum(VI), rhenium(VII) and selenium(VI) from acidic aqueous solution into chloroform [13,60,61]. With longer alkyl chain substituents on the amide, extraction into toluene and isooctane is observed, with the conformation of the calixarene being important in determining the oxyanion extraction efficiency [62].

Calix[n]arene hydroxamates have been used as extractants for transition metal ions, and especially for Fe(III). For example, narrow-rim-substituted calix[n]arenes **10** (n = 4, 6) have been used for the liquid-liquid extraction of Cu(II), Fe(III) and Pd(II), with the higher extraction being observed for the derivatives having n = 6. By contrast, the structurally similar amide **11** shows selectivity toward Pd(II) and Pt(IV) [63]. Calixarene hydroxamates will even extract iron(III) from acid solutions [64]. Calixarene **12** with a single dioxime group on the narrow rim also selectively extracts Fe(III) [31]. Calixarene hydroxamates also extract silver, cobalt and chromium in the selectivity sequence Ag(I)>Co(II)>Cr(III) [65]. The chelating properties of calixarenes with hydroxamic acid groups and proline hydroxamic acid groups have been compared to those of siderophores, both being strong binding agents for Fe(III) and V(V) [66]. These calixarenes with hydroxamate groups attached can be immobilized onto silica, and the resulting material used as a solid-phase adsorbent [67]. Calixarene hydroxamates have also been used for the extraction and spectrophotometric determination of Zr(IV) [68]. In some cases Fe(III) is not the preferred metal ion extracted, and a calix[6]arene extractant with a hydroxypropylamino substituent on the narrow rim **13** is found to extract metal ions in the sequence Cu(II)>Co(II) >Ni(II) >Fe(III) [46]. Ion chromatography has been used to monitor the extraction of metal ions from spiked cellulose paper using perfluorooctanohydroxamic acid and a wide-rim-fluorinated calix[4]arene hydroxamate reagent in supercritical carbon dioxide [69].

5

R = H, Et

6

7

8 n = 2
9 n = 4

10

11

12

13

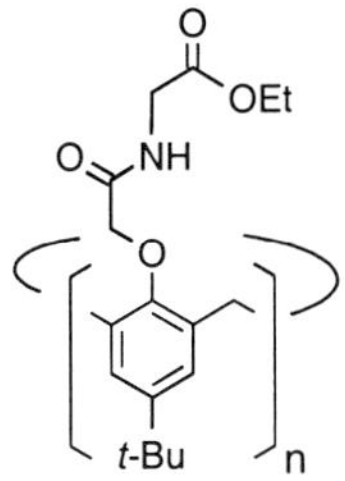

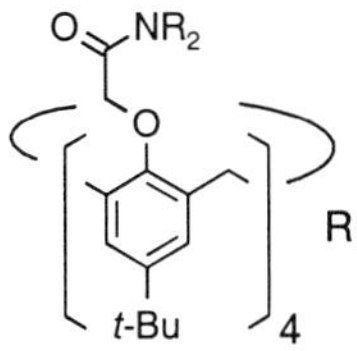

Calix[6]arene **14** with azo groups appended to the wide rim is a selective extractant for Ag(I), in addition to Hg(I) and (II). Interestingly the monomeric analog **15** also extracts Cr(III) and Cu(II) [70]. A wider range of metal ions is extracted with a calix[4]-arene with diazo groups coupled to the calixarene. The (4-phenylazophenylazo)-calix[4]arene with two azo groups is the most effective extractant [71]. Azocalixarenes in the *trans* form show a higher binding ability for metal ions than in the *cis* form [72]. This is a useful feature because it offers the possibility of designing a photoswitch system to sequentially bind and release the metal ion. Calix[4]arenes **16** with Schiff base moieties appended to their narrow rim can be used to extract Cu(II), Co(II), Fe(II), Mn(II) and Ni(II). The compounds with the longer chain lengths (n = 3, 4) are the better extractants [73]. Similar Schiff-base-substituted calixarenes been incorporated into polyethylene glycol chains to prepare extractants for transition metal ions and Hg(II) [74].

14

15

16 m = 2, 3, 4

17

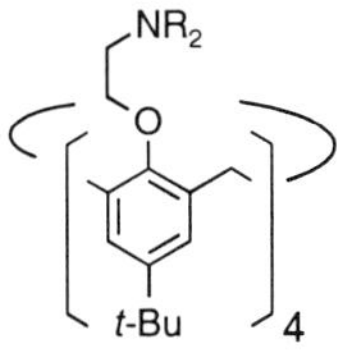

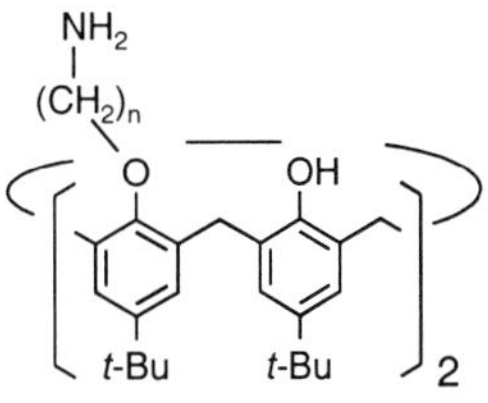

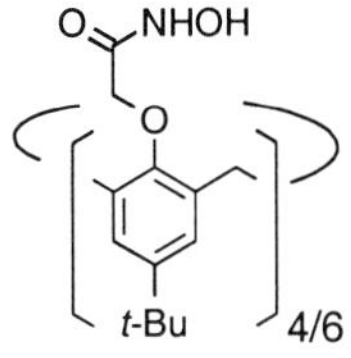

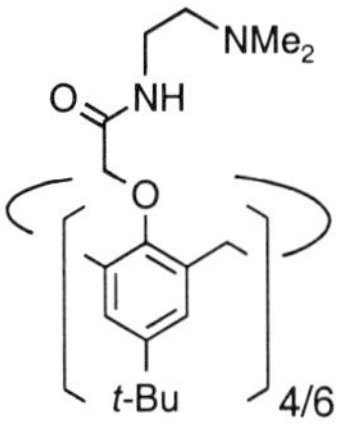

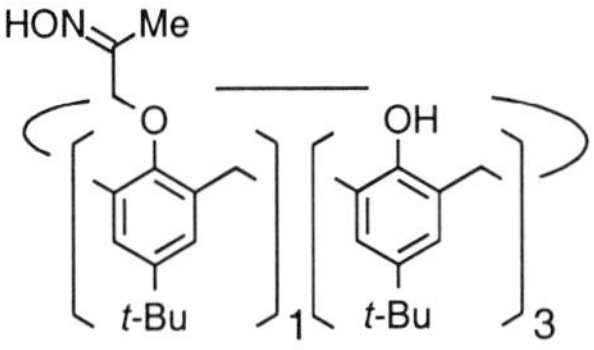

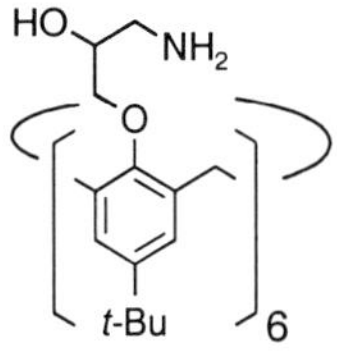

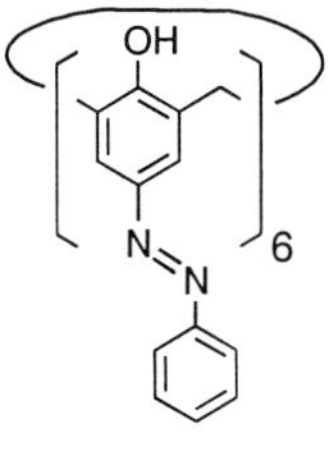

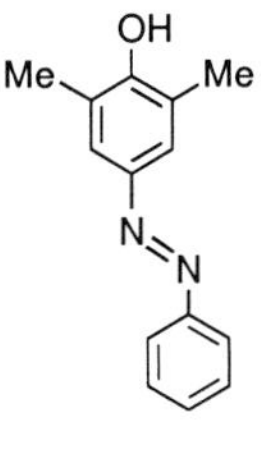

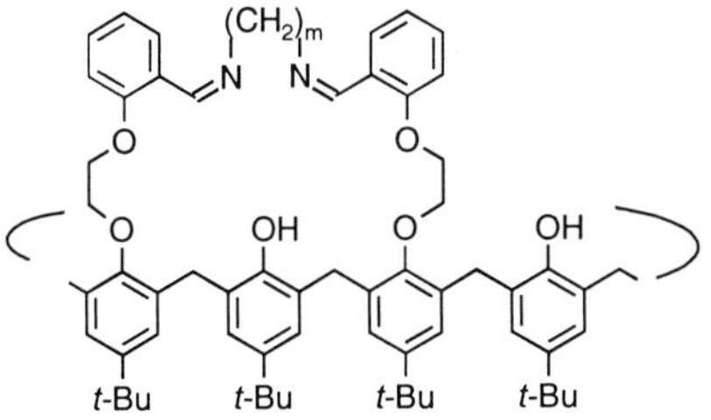

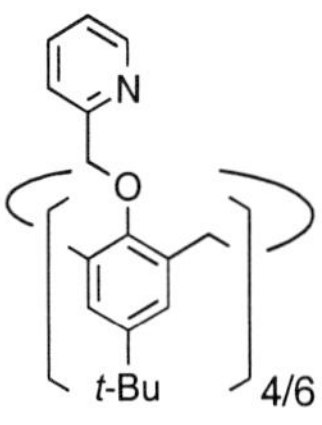

Calixarenes **17** with pyridyl groups bound to the narrow rim can be used to extract Cu(II) and Ag(I), and are stable even at high pH [75]. When the pyridyl moiety is incorporated onto a metacyclophane cage, a high Ag(I) affinity is observed with the formation of a 1:1 complex [76].

Other types of stategies have been used with calixarenes to induce them to be extractants for transition metal ions. One such approach uses a *syn*-dihydroxy[2.n]metacyclophane **18** (see Chapter 11) as a building block to obtain a new calix[4]arene family that extracts Ag(I), Mn(II) and Cr(III) [77]. In another approach, it has been recognized that since silver selectively binds to an alkene group, calix[4]arenes **19** with allyl groups on the narrow rim are extractants for Ag(I) and Tl(I) [78].

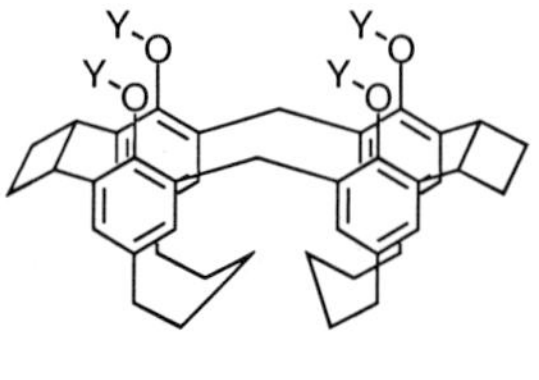

18

Y = CH_2CO_2t-Bu, CH_2COt-Bu, CH_2COPh or [pyridyl-N-CH_2]

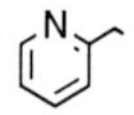

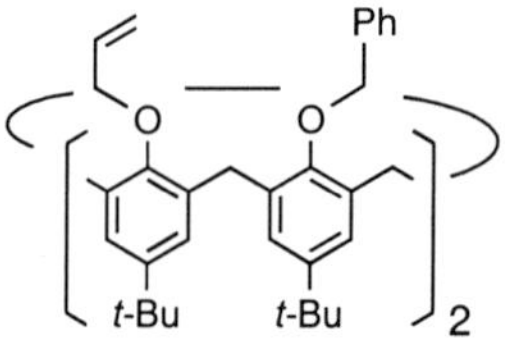

19

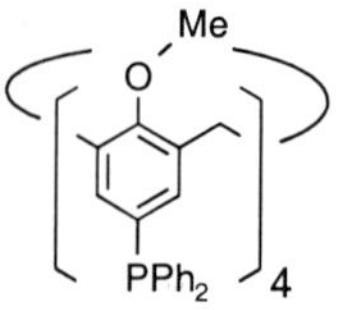

20

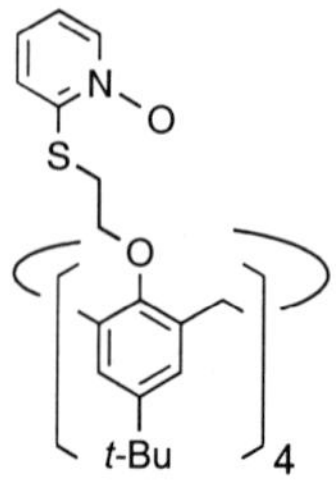

21

For the liquid-liquid extraction of the heavier soft metals, soft ligands appended to the calixarene have been used. In general sulfur derivatives have been used in preference to phosphines or phosphites because of their higher stability to hydrolysis and oxidation, two important aspects if long term stability of the extractant is to be achieved. One example of the use of a tertiary phosphine has, however, led to a wide rim functionalized calix[4]arene **20** that is an extractant for Cu(II) and Ni(II), along with the post-transition metal ions Zn(II), Cd(II) and Hg(II) [79]. A similar calixarene with sulfur derivatives on the wide rim extracts a range of transition metal ions [80].

Sulfur-containing functional groups are often favored when designing selective extractants for the platinum metals group. Many of these derivatives have uncharged sulfur donor centers as although the anionic thiolate group itself binds strongly with these metal ions, it is not a good choice because it is both non-selective in its complexation properties and readily air-oxidized under ambient conditions to the disulfide (noted above as a disadvantage with some P donor ligands). The N,N-dimethyldithiocarbamoyl group appended to the narrow calixarene rim of a calix[4]arene yields a compound that extracts Pd(II). Platinum is not extracted under ambient conditions, and kinetic selectivity has been offered as an explanation. In support of this premise, irradiation of a mixture of Pt(IV) and Pt(II) with a mercury lamp results in extraction of platinum into chloroform [81]. A 2-mercaptopyridine-N-oxide substituted calix[4]arene **21** does, however, extract Pt(II) into chloroform under thermal conditions[82].

Supercritical carbon dioxide is becoming a favoured fluid for liquid-liquid extractions. In order to take advantage of this fluid it is necessary to design extractants whose metal complexes are soluble in it. One method of accomplishing this is to incorporate a fluoroalkyl chain into the complexant structure. Such a strategy has been used with calixarenes to prepare extractants **22** with fluoroalkylthio functionalities of the hydroxamate and thiourea types. These compounds extract Fe(III) and Au(III), respectively [83,84]. Identification of the extractants by HPLC with a detector wavelength of 280 nm allows for the distribution constants to be evaluated in this fluid [85].

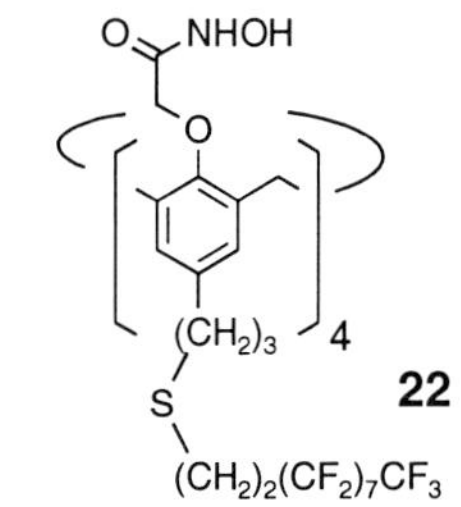

2.2. POST-TRANSITION METAL IONS

Many of the calixarenes with sulfur groups that extract transition metal ions of the platinum group also extract the heavy post-transition metals.

2.2.1. Lead, Cadmium and Mercury
In a series of studies on the use of polyethylene glycol-based aqueous biphasic systems as metal ion extractants, the addition of the simple unsubstituted calixarenes was found to have no significant enhancement of the partitioning of Cd(II), Co(II) and U(VI) [86]. Calixcrowns **23** with both carboxylate and amide functionalities do, however, extract lead in addition to mercury. These macrocyclic, hydrophobic extractants with oxygen donor atoms can selectively recognize Pb(II), Cd(II), Zn(II) and Hg(II) due to size compatibility. The crown-6 derivatives extract the larger, and the crown-5 derivatives the smaller cations. The calix[4]arene carboxylate itself, does, however, quantitatively extract both Pb(II) and Hg(II) above pH 2.8 and 2.6, respectively [87]. The extraction and ^{1}H NMR data show that both 1:1 and 2:1 M:L species may be extracted [88]. Calix[4]arenes **24** with N-(X)-sulfonyl carboxamide groups of "tunable" acidity have excellent selectivity for the extraction of Pb(II) as compared to most alkali, alkaline earth, and transition metal ions [89]. These compounds, however, fail to extract Pb(II) in the presence of Hg(II). Subsequently, it has been found that these N-(X)-sulfonyl carboxamide calix[4]arenes extract Hg(II) from acidic nitrate solutions, and that they have a high selectivity over metals such as Pb(II), Ag(I) and Pd(II) [90]. Selective binding of **25** to Hg(II) is signalled by a shortening of the fluorescence lifetime of the ligand [91]. (See Chapter 32)

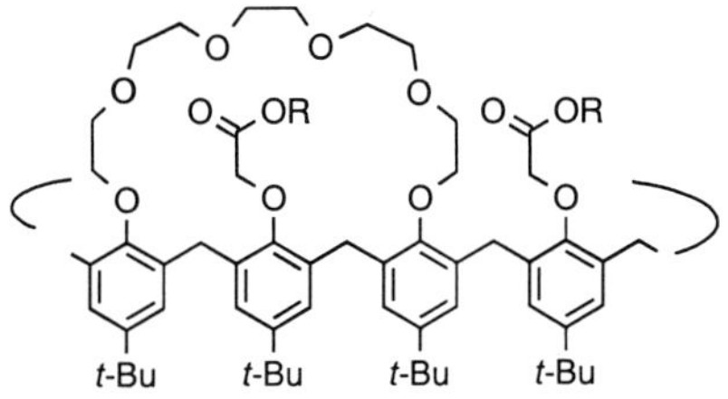

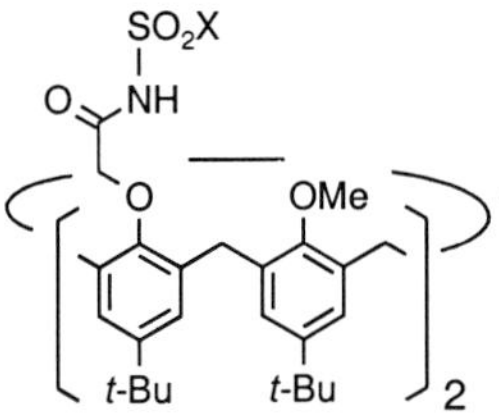

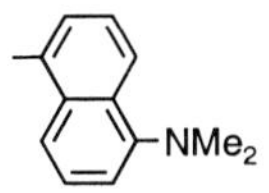

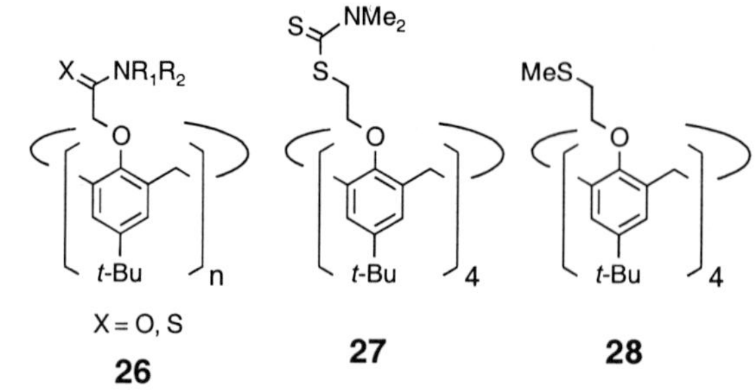

Calixarene phosphine oxide derivatives can be used to obtain lead-selective electrodes [92]. (see Chapter 34) Calix[n]arene thioamides **26** extract both Pb(II) and Cd(II), in addition to Cu(II) and Ag(I). All thioamides extract Pb(II), but only the calix[5]arene extracts Cd(II) [93]. A number of calixarenes with sulfur-containing functional groups are excellent extractants for Hg(II), although in some cases they extract neither Pb(II) nor Cd(II). Among the latter category are calix[4]arenes **27,28** with either N,N-dimethyl dithiocarbamoyl or thioether groups attached to either their narrow or wide rim. Since these complexants are uncharged, the metal ion will not be bound so tightly that it cannot be subsequently displaced and recovered. Both the N,N-dimethyl dithiocarbamoyl and methylthioether calix[4]arene derivatives are good extractants for Hg(II) as well as Pd(II) and Au(III) [94]. This selectivity is observed whether the substituents are on the narrow or the wide rims [95, 96]. By a combination of electronic and ^{1}H NMR spectroscopy it has been concluded that the metal ion binds at one or more of the sulfur sites [97]. The degree of extraction of Hg(II) is dependent on whether the wide rim has hydrogen or tert-butyl groups appended to the para positions [98]. The extractabilties of these N,N-dimethyl dithiocarbamoyl and methylthioether substituted calix[4]arenes do not correlate with metal ion size but a comparison has been made with that of thiacrowns to extract these heavy metal ions into an organic layer [99].

Mercaptothiacalix[4]arenes (see Chapter 6) are a new class of compound with multiple metal binding sites to sulfur, and these are beginning to be used for liquid-liquid phase metal extractions [100].

2.2.2. Gallium and Indium

Calix[n]arenes (n = 4, 6) with either carboxylate or hydroxamate groups attached extract gallium(III) and indium(III) into an organic phase. The higher extraction achieved with the hydroxamate derivative is attributed to the chelate effect of this group [35].

2.3. LANTHANIDE AND ACTINIDE IONS

Two reviews have been published on the use of calixarenes for the extraction of actinides [101,102], but they are not in readily accessible sources. The topic is, however, analysed in detail in Chapter 35 of this volume. The extraction of uranium is of interest both in the context of recovery from nuclear wastes, and in the recovery from natural sources. One possible source of uranium is seawater. A common form of uranium is the linear uranyl ion UO_2^{2+} and a preferred complexant may coordinate to either four or five sites in the equatorial plane. Calix[n]arene (n = 5, 6) sulfonates are reported to be [103,104] excellent "uranophiles" with very high stability constants for uranyl ion binding, and very high selectivities against other metal ions. This selectivity is attributed to the rigid skeleton of these calixarenes which can provide the preorganized coordination geometry for the binding of the uranyl ion, but cannot accommodate a square planar, octahedral or tetrahedral geometry favored by other metal ions. These calixarenes have been used for the selective removal of the uranyl ion from seawater. Calix[n]arene carboxylates (n = 4, 6, 8) are also good extractants for the uranyl ion, and can carry it across both a liquid interface and a polymer/liquid-crystal composite membrane [105,106]. These compounds also extract lanthanide ions [107]. For extraction into toluene the sequence: Eu(III)>Nd(III)>Yb(III)>Er(III)>La(III) is followed. Addition of excess Na(I) results in greater extraction and higher selectivity [108,109]. These calixarene carboxylates are better extractants for lanthanide(III) ions than are their structurally analogous monomeric aromatic carboxylic acids [110]. The equilibrium extraction constants for the calix[6]arene carboxylate with chloroform as the organic phase are some 4 orders of magnitude greater than are found for the monomer. The effect of Na(I) and K(I) on the extraction of U(VI) by calixarene carboxylates has been studied in some detail. For the calix[4]arene dicarboxylate **29** the extraction of U(VI) increases upon addition of Group I ions due to the formation of heteronuclear complexes. The extracted species have stoichiometries 1:2:2 and 1:1:2 for UO_2^{2+}:M+:LH_2 [111-113]. Similar effects are observed with the tetracarboxylate derivative **1**. The tetracarboxylate both in the presence or absence of Na(I) selectively extracts Th(IV). In competitive extractions, separation factors of 1000 have been observed for Th(IV) over U(VI) [114]. The calix[6]arene hexacarboxylate **30** also can be used to selectively extract thorium(IV) from the nuclear waste species U(VI), Cs(I), Pb(II), Sr(II), and Ce(IV) [115]. In the absence of Na(I), calix[4]arene tetracarboxylic acid **1** preferentially extracts the light lanthanides with greater ionic radii, with the reverse being observed in the presence of Na(I). The difference is proposed to be a consequence of the initial sodium binding leading to a complexed calix[4]arene carboxylate of different size than the calix[4]arene carboxylate itself, and that this sodium(I) complex acts as a preorganized ligand for lanthanide ions [116]. Calixarene carboxylates also show selectivity for the extraction of Am(III). Extractions are carried out from acid solutions, and selectivity over La(III), Na(I), K(I), Mg(II), Ca(II) and Zn(II) is observed [117,118]. The extraction of U(VI) by calix[6]arene carboxylates is unaffected by the presence of Zn(II), Cu(II), Mg(II), and Ca(II) [119]. U(VI) is also extracted by a bis-calix[4]arene **31** where the connection is via two (2-carboxy-m-phenylene)dimethylene units [120].

Calixarenes with hydroxamate groups appended have been used as extractants for U(VI). The calix[6]arene hydroxamate **10** is the most effective extractant for U(VI). Extraction is quantitative from a solution at pH 5, is competitive against the carbonate ion, itself a strong complexant for U(VI), and is unaffected by competing metal ions, except Fe(III) [121,122].

Calix[4]arene hydroxamates also extract Th(IV), but the selectivity is decreased in the presence of Fe(III) [123].

Both calixarene amides and phosphoramides are also good extractants for actinides and lanthanides. The 1,3-bis-(diethyl amide) calix[4]arene derivative **32** is a good extractant for transferring lanthanide(III) ions into dichloromethane. The lanthanide cation is encapsulated in an eight-coordinate environment consisting of six oxygen donor atoms from the calixarene, along with oxygens from either water or the added picrate anion [124,125]. Calixarene amides are also good extractants for U(VI) [126]. The structure of the U(VI) complex with the extractant that has the diethylamide and carboxylate substituents in 1,3-positions on the narrow rim shows it to be a dimer. In addition to U(VI), this particular calix[4]arene quantitatively extracts La(III), Lu(III), and Hg(II) from solutions having a pH of 6 or higher [127]. Calixarenes having both amide and carboxylate groups have also been used for the separation of actinide ions such as Am(III) from lanthanides [128,129]. A similar extractant for actinide(IV) ions is calix[4]arene **33** with 3-hydroxy-2-pyridone substituents on the narrow rim [130].

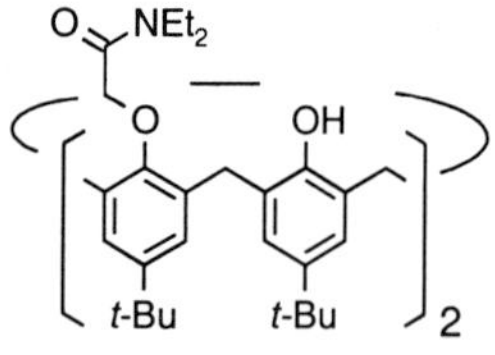

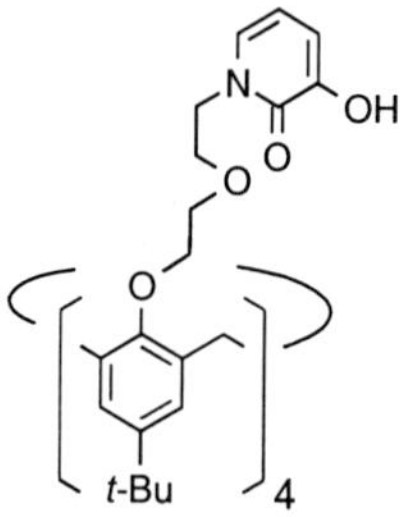

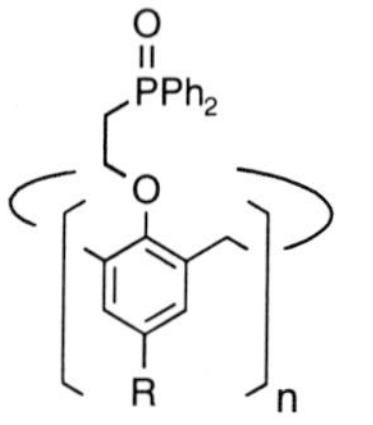

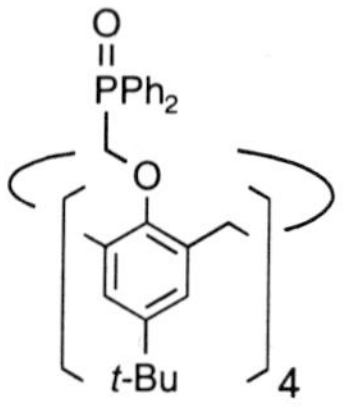

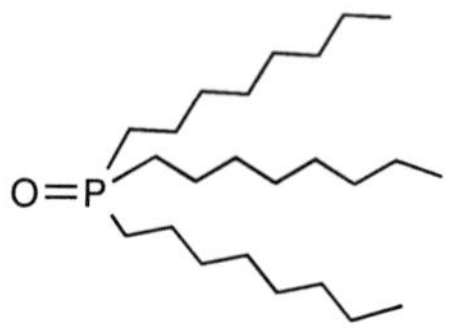

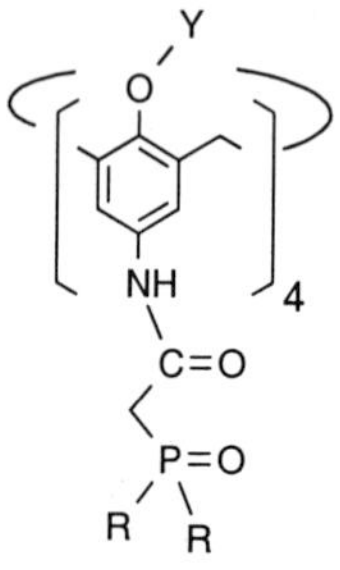

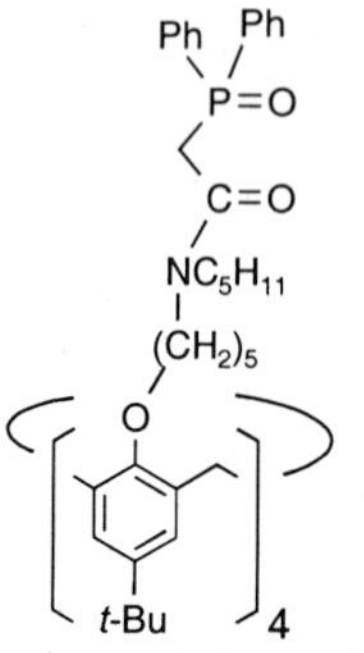

Calix[n]arene phosphine oxides **34**, **35** are good extractants for lanthanides and actinides, including Eu(III), Th(IV), Pu(IV), and Am(IV). The extraction efficiency is dependent on the ring size n, and on the wide rim substituent [131]. The analogous calix[4]arene phosphine oxide **35** with a shorter alkyl chain is a good extractant for lanthanide ions. Such compounds are superior to the acyclic TOPO extractant **36** [132, 133]. The distribution coefficients increase with the polarity of the organic phase [134]. The reader is referred to Chapter 35 for a more detailed coverage of such extractions with calixarenes of the dialkylcarbamoylmethyldiarylphosphine oxide (CMPO) type (*e.g.* **37**, **38**). These derivatives are much better extractants than is the monomeric CMPO itself [135-141]. For Eu(III), the distribution coefficient for the calixarene is up to 5 orders of magnitude greater than CMPO itself.

3. References and Notes

[1] T. E. Clevenger, *Water, Air, and Soil Pollution* **1990**, *50*, 241-254.
[2] D. M. Roundhill, *Prog. Inorg. Chem.* **1995**, *43*, 533-592.
[3] P. L. H. M. Cobben, R. J. M. Egberink, J. G. Bomer, P. Bergveld, W. Verboom, D. N. Reinhoudt, *J. Am. Chem. Soc.* **1992**, *114*, 10573-10582.
[4] G. Drasch, L. V. Meyer, G. Kauert, *Fresenius' Z. Anal. Chem.* **1982**, 311, 571-577.
[5] S. Ichinoki, T. Morita, M. Yamazaki, *J. Liq. Chromatogr.* **1983**, *6*, 2079-2093.
[6] S. Ichinoki, T. Morita, M. Yamazaki, *J. Liq. Chromatogr.* **1984**, *7*, 2467-2482.
[7] G. Bozsai, M. Csanady, *Fresenius' Z. Anal. Chem.* **1979**, *297*, 370-373.
[8] D. H. Busch, *Chem. Rev.* **1993**, *93*, 847-860.
[9] L. F. Lindoy. *The Chemistry of Macrocyclic Ligand Complexes*, Cambridge Univ. Press, Cambridge, U.K., 1989.
[10] G. Gokel. *Crown Ethers and Cryptands*, Monographs in Supramolecular Chemistry, No. 3, J. Fraser Stoddart. ed., Royal Society of Chemistry, **1991**.
[11] F. C. J. M. van Veggel, W. Verboom, D. N. Reinhoudt, *Chem. Rev.* **1994**, *94*, 279-299.
[12] H. C. Visser, D. M. Rudkevich, W. Verboom, F. de Jong, D. N. Reinhoudt, *J. Am. Chem. Soc.* **1994**, *116*, 11554-11555.
[13] E. M. Georgiev, N. Wolf, D. M. Roundhill, *Polyhedron* **1997**, *16*, 1581-1584.
[14] D. Burrows, *Chromium: Metabolism and Toxicity*, CRC Press, Boca Raton, FL., 1983.
[15] V. Bianchi,, A. Zantedeschi, A. Montaldi, F. Majone, *Toxicol. Lett.* **1984**, *23*, 51-59.
[16] S. De Flora, K. E. Wetterhahn. *Life Chem. Rep.* **1989**, *7*, 169-244.
[17] D. M. Stearns, L. J. Kennedy, K. D. Courtney, P. H. Giangrande, L. S. Phieffer, K. E. Wetterhahn, *Biochemistry*, **1995**, *34*, 910-919.
[18] P. R. Wittbrodt, C. D. Palmer, *Environ. Sci. Technol.* **1995**, *29*, 255-263.
[19] S. J. Harris, M. A. McKervey, D. P. Melody, J. Woods, J. M. Rooney, *Eur. Pat. Appl.* **1984**, EP 151,527. *Chem. Abstr. 103*, 216,392x.
[20] A. T. Yordanov, D. M. Roundhill. *Coord. Chem. Rev.* **1998**, *170*, 93-124.
[21] F. Berny, N. Muzet, L. Troxler, G. Wipff, NATO ASI Ser., Ser. C **1999**, 527 (*Supramolecular Science: Where It Is and Where It Is Going*), 95-124.
[22] M-J. Schwing, M. A. McKervey, *Top. Inclusion Sci.* **1991**, *3*, 149-172.
[23] M. Pietraszkiewicz. *Wiad, Chem.* **1998**, Volume Date **1997**, (*Chemia Supramolekularna*), 69-89.
[24] R. Ludwig, *JAERI Rev.* **1995**, 95 022, 55 pp.
[25] A. F. D. de Namor, R. M. Cleverley, M. L. Zapata Ormachea, *Chem. Rev.* **1998**, *98*, 2495-2525.
[26] K. Ohto, H. Tanaka, H. Ishibashi, K. Inoue, *Solvent Extr. Ion Exch.* **1999**, *17*, 1309-1325.
[27] I. Yoshida, S. Fujii, K. Ueno, S. Shinkai, T. Matsuda, *Chem. Lett.* **1989**, 1535-1538.
[28] X. Shi, S. Ding, Y. Yang, D. Xu, Y. Chen, Z. Huang, *Wuji Huaxue Xuebao* **1993**, *9*, 423-426.
[29] S. Ding, X. Shi, Z. Zhu, G. Lu, D. Xu, Y. Chen. *Youkuangye* **1994**, *13*, 180-184.
[30] G. Lu, X. Cao, Z. Zhu, J. Qiao, *Tongji Daxue Xuebao, Ziran Kexueban* **1997**, *25*, 723-727.
[31] M. Yilmaz, H. Deligoz, *Sep. Sci. Technol.* **1996**, *31*, 2395-2402.
[32] K. Ohto, E. Murakami, K. Shiratsuchi, K. Inoue, M. Iwasaki, *Chem. Lett.* **1996**, 173-174.
[33] K. Ohto, E. Murakami, T. Shinohara, K. Shiratsuchi, K. Inoue, M. Iwasaki, *Anal. Chim. Acta* **1997**, *341*, 275-283.
[34] Z. Zhu, R. Xu, G. Lu, Y. Xu, B. Chen, Y. Chen, *He Huaxue Yu Fangshe Huaxue* **1994**, *16*, 173-177.
[35] K. Ohto, K. Maruishi, T. Shinohara, K. Inoue, *Solvent Extr. Res. Dev., Jpn.* **1996**, *3*, 255-260.
[36] K. Ohto, K. Shiratsuchi, K. Inoue, M. Goto, F. Nakashio, S. Shinka, T. Nagasaki, *Solvent Extr. Ion Exch.* **1996**, *14*, 459-478.
[37] T. Toh, T. Kakoi, M. Goto, F. Nakashio, *Proc. Symp. Solvent Extr.* **1995**, 39-40.
[38] Z. Zhou, Y. Xing, Y. Wu, *J. Incl. Phenom.* **1999**, *34*, 219-231.

[39] T. Kakoi, T. Toh, F. Kubota, M. Goto, S. Shinkai, F. Nakashio, *Anal. Sci.* **1998**, *14*, 501-506.
[40] V. J. Mathew, S. M. Khopkar, *Talanta* **1997**, *44*, 1699-1703.
[41] Y. Masuda, Y. Suzuki, *Proc. Symp. Solvent Extr.* **1994**, 69-70.
[42] Z. Asfari, J. M. Harrowfield, P. Thuery, M. Nierlich, J. Vicens, *Aust. J. Chem.* **1999**, *52*, 403-407.
[43] N. Iki, N. Morohashi, F. Narumi, T. Fujimoto, T. Suzuki, S. Miyano, *Tetrahedron. Lett.* **1999**, *40*, 7337-7341.
[44] A. Ikeda, S. Shinkai, *J. Am. Chem. Soc.* **1994**, *116*, 3102-3110.
[45] H. Otsuka, Y. Suzuki, A. Ikeda, K. Araki, S. Shinkai, *Tetrahedron* **1998**, *54*, 423-446.
[46] U. S. Vural, *Sep. Sci. Technol.* **1996**, *31*, 787-798.
[47] H. Deligoz, M. Tavasli, M. Yilmaz, *J. Polym. Sci., Part A: Polym. Chem.* **1994**, *32*, 2961-2964.
[48] M. Grunder, J. F. Dozol, Z. Asfari, J. Vicens. *J. Radioanal. Nucl. Chem.* **1999**, *241*, 59-67.
[49] M. Nishida, M. Sonoda, D. Ishii, I. Yoshida, *Chem. Lett.* **1998**, 289-290.
[50] M. Nishida, M. Sonoda, D. Ishii, I. Yoshida, *Bunseki Kagaku* **1998**, 47, 853-859.
[51] M. Nishida, M. Sonoda, D. Ishii, I. Yoshida, Y. Nakashima, *Nippon Kagaku Kaishi* **1999**, 583-588.
[52] R. Ludwig, S. Tachimori, *Solvent Extr. Res. Dev., Jpn.* **1996**, *3*, 244-254.
[53] A. F. Danil de Namor, M. T. Goitia, A. R. Casal, F. J. Sueros Velarde, M. Isabel Barja Gonzalez, J. A. Villanueva Salas, M. L. Zapata Ormachea, *Phys. Chem. Chem. Phys.* **1999**, *1*, 3633-3638.
[54] R. Ludwig, S. Tachimori. *Proc. Symp. Solvent Extr.* **1995**, 41-42.
[55] K. Ohto, H. Yamaga, E. Murakami, K. Inoue, *Talanta* **1997**, *44*, 1123-1130.
[56] M. R. Yaftian, M. Burgard, A. El Bachiri, D. Matt, C. Wieser, C. B. Dieleman, *J. Incl. Phenom.* **1997**, *29*, 137-151.
[57] F. Arnaud-Neu, S. Barboso, F. Berny, A. Casnati, N. Muzet, A. Pinalli, R. Ungaro, M.-J. Schwing-Weill, G. Wipff, *J. Chem. Soc., Perkin Trans. 2*, **1999**, 1727-1738.
[58] N. J. van der Veen, R. J. M. Egberink, J. F. J. Engbersen, F. J. C. M. van Veggel, D. N. Reinhoudt, *Chem. Commun.* **1999**, 681-682.
[59] B. Konig, T. Fricke, K. Gloe, C. Chartroux. *Eur. J. Inorg. Chem.* **1999**, 1557-1562.
[60] N. J. Wolf, E. M.. Georgiev, A. T. Yordanov, B. R. Whittlesey, H. F. Koch, D. M. Roundhill, *Polyhedron* **1999**, *18*, 885-896.
[61] W. Aeungmaitrepirom, A. Hagege, Z. Asfari, L. Bennouna, J. Vicens, M. Leroy, *Tetrahedron Lett.* **1999**, *40*, 6389-6392.
[62] O. M. Falana, H. F. Koch, D. M. Roundhill, G. J. Lumetta, B. P. Hay, *Chem. Commun.* **1998**, 503-504.
[63] T. Nagasaki, S. Shinkai, *Bull. Chem. Soc. Jpn.* **1992**, *65*, 471-475.
[64] H. Deligoz, A. Pekacar, M. A. Özler, M. Ersöz, *Sep. Sci. Technol.* **1999**, *34*, 3297-3304.
[65] H. Deligoz, M. Yilmaz, *Solvent Extr. Ion Exch.* **1995**, *13*, 19-26.
[66] J. D. Glennon, E. Home, P. O'Sullivan, S. Hutchinson, M. A. McKervey, S. J. Harris, *Met. Based Drugs* **1994**, *1*, 151-160.
[67] S. Hutchinson, G. A. Kearney, E. Horne, B. Lynch, J. D. Glennon, M. A. McKervey, S. J. Harris, *Anal. Chim. Acta* **1994**, *291*, 269-275.
[68] Y. K. Agrawal, M. Sanyal, P. Shrivastav, S. K. Menon. *Talanta* **1998**, *46*, 1041-1049.
[69] M. O' Connell, T. O' Mahony, M. O' Sullivan, S. J. Harris, W. B. Jennings, J. D. Glennon, *Spec. Publ. R. Soc. Chem.* **1999**, *239* (*Advances in Ion Exchange for Industry and Research*), 137-143.
[70] E. Nomura, H. Taniguchi, S. Tamura, *Chem. Lett.* **1989**, 1125-1126.
[71] H. Deligoz, E. Erdem. *Solvent Extr. Ion Exch.* **1997**, *15*, 811-817.
[72] F. Hamada, T. Masuda, Y. Kondo, *Supramol. Chem.* **1995**, *5*, 129-131.
[73] R. Seangprasertkij, Z. Asfari, F. Arnaud, J. Vicens, *J. Org. Chem.* **1994**, *59*, 1741-1744.
[74] A. Yilmaz, S. Memon, M. Yilmaz, *J. Polym. Sci. Part A: Polym. Chem.* **1999**, *37*, 4351-4355.
[75] S. Shinkai, T. Otsuka, K. Araki, T. Matsuda, *Bull. Chem. Soc. Jpn.* **1989**, *62*, 4055-4057.
[76] T. Yamato, M. Haraguchi, J-i. Nishikawa, S. Die, *J. Chem. Soc., Perkin Trans. 1* **1998**, 609-614.
[77] Y. Okada, Y. Kasai, J. Nishimura, *Synlett* **1995**, 85-89.
[78] D. Couton, M. Mocerino, C. Rapley, C. Kitamura, A. Yoneda, M. Ouchi, *Aust. J. Chem.* **1999** *52*, 227-229.
[79] F. Hamada, T. Fukugaki, K. Murai, G. W. Orr, J. L. Atwood, *J. Incl. Phenom.* **1991**, *10*, 57-61.
[80] F. Hamada, Y. Kondo, S. Suzuki, S. Ohnoki, P. K. Unieja, J. L. Atwood, *Int. J. Soc. Mater. Eng. Resour.* **1994**, *2*, 37-46.
[81] A. T. Yordanov, J. T. Mague, D. M. Roundhill, *Inorg. Chim. Acta* **1995**, *240*, 441-446.
[82] A. T. Yordanov, D. M. Roundhill, *Inorg. Chim. Acta* **1997**, *264*, 309-311.

[83]   J. D. Glennon, S. Hutchinson, S. J. Harris, A. Walker, M. A. McKervey, C. C. McSweeney, *Anal. Chem.* **1997**, *69*, 2207-2212.
[84]   J. D. Glennon, S. J. Harris, A. Walker, C. C. McSweeney, M. O' Connell, *Gold Bull.* **1999**, *32*, 52-58.
[85]   G. Czerwenka, L. Scheubeck, *Fresenius' Z. Anal. Chem.* **1975**, *276*, 34-40.
[86]   R. D. Rogers, C. B. Bauer, *J. Radioanal. Nucl. Chem.* **1996**, *208*, 153-161.
[87]   N. T. K. Dung, R. Ludwig, *New J. Chem.* **1999**, *23*, 603-607.
[88]   K. Ohto, Y. Fujimoto, K. Inoue, *Anal. Chim. Acta* **1999**, *387*, 61-69.
[89]   G. G. Talanova, H. S. Hwang, V. S. Talanov, R. A. Bartsch, *Chem. Commun.* **1998**, 419-420.
[90]   G. G. Talanova, V. S. Talanov, R. A. Bartsch, *Chem. Commun.* **1998**, 1329-1330.
[91]   G. G. Talanova, N. S. A. Elkarim, V. S. Talanov, R. A. Bartsch, *Anal. Chem.* **1999**, *71*, 3106-3109.
[92]   F. Cadogan, P. Kane, M. A. McKervey, D. Diamond, *Anal. Chem.* **1999**, *71*, 5544-5550.
[93]   F. Arnaud Neu, G. Barrett, D. Corry, S. Cremin, G. Ferguson, J. F. Gallagher, S. J. Harris, M. A. McKervey, M-J. Schwing Weill, *J. Chem. Soc., Perkin Trans.* 2 **1997**, 575-579.
[94]   A. T. Yordanov, J. T. Mague, D. M. Roundhill, *Inorg Chem.* **1995**, *34*, 5084-5087.
[95]   A. T. Yordanov, B. R. Whittlesey, D. M. Roundhill, *Inorg. Chem.* **1998**, *37*, 3526-3531.
[96]   A. T. Yordanov, O. M. Falana, H. F. Koch, D. M. Roundhill. *Inorg. Chem.* **1997**, *36*, 6468-6471.
[97]   A. T. Yordanov, D. M. Roundhill, *Inorg. Chim. Acta* **1998**, *270*, 216-220.
[98]   A. T. Yordanov, D. M. Roundhill, J. T. Mague, *Inorg. Chim. Acta* **1996**, *250*, 295-302.
[99]   A. T. Yordanov, D. M. Roundhill. *New J. Chem.* **1996**, *20*, 447-451.
[100] P. Rao, M. W. Hosseini, A. de Cian, J. Fischer, *Chem. Commun.* **1999**, 2169-2170.
[101] X. Yu, *Ziran Zazhi* **1990**, *13*, 262-266.
[102] T. Kakoi, M. Goto, *Nippon Kaisui Gakkaishi* **1997**, *51*, 319-324.
[103] S. Shinkai, H. Koreishi, K. Ueda, O. Manabe, *J. Chem. Soc., Chem. Commun.* **1986**, 233-234.
[104] S. Shinkai, H. Koreishi, K. Ueda, T. Arimura, O. Manabe, *J. Am Chem. Soc.* **1987**, *109*, 6371-6376.
[105] S. Shinkai, Y. Shiramama, H. Satoh, O. Manabe, T. Arimura, K. Fujimoto, T. Matsuda, *J. Chem. Soc., Perkin Trans.* 2 **1989**, 1167-1171.
[106] H-F. Du, A-Y. Zhang, Z-X. Yang, Z-M. Zhou, *J. Radioanal. Nucl. Chem.* **1999**, *241*, 241-243.
[107] R. Ludwig, K. Inoue, S. Shinkai, *S. Proc. Symp. Solvent Extr.* **1991**, 53-60.
[108] R. Ludwig, K. Inoue, T. Yamato, *Solvent Extr. Ion Exch.* **1993**, *11*, 311-330.
[109] F. Kubota, T. Kakoi, M. Goto, S. Furusaki, F. Nakashio, T. Hano, *J. Membr. Sci.* **2000**, *165*, 149-158.
[110] K. Ohto, M. Yano, K. Inoue, T. Yamamoto, M. Goto, F. Nakashio, S. Shinkai, T. Nagasaki, *Anal. Sci.* **1995**, *11*, 893-902.
[111] G. Montavon, G. Duplatre, Z. Asfari, J. Vicens, *J. Radioanal. Nucl. Chem.* **1996**, *210*, 87-103.
[112] N. Baglan, C. Dinse, C. Cossonnet, R. Abidi, Z. Asfari, M. Leroy, J. Vicens, *J. Radioanal. Nucl. Chem.* **1997**, *226*, 261-265.
[113] C. Dinse, C. Cossonnet, N. Baglan, Z. Asfari, J. Vicens, *Radioprotection* **1997**, *32*, 659-671.
[114] G. Montavon, G. Duplatre, Z. Asfari, J. Vicens, *Solvent Extr. Ion Exch.* **1997**, *15*, 169-188.
[115] D. D. Malkhede, P. M. Dhadke, S. M. Khopkar,. *J. Radioanal. Nucl. Chem.* **1999**, *241*, 179-182.
[116] K. Ohto, M. Yano, K. Inoue, T. Nagasaki, M. Goto, F. Nakashio, S. Shinkai, *Polyhedron* **1997**, *16*, 1655-1661.
[117] R. Ludwig, K. Kunogi, N. Dung, S. Tachimori, *Chem. Commun.* **1997**, 1985-1986.
[118] X. Chen, M. Ji, D. R. Fisher, C. M. Wai, *Chem. Commun.* **1998**, 377-378.
[119] C. Dinse, N. Baglan, C. Cossonnet, J. F. Le Du, Z. Asfari, J. Vicens, *J. Alloys Compd.* **1998**, 271-273, 778-781.
[120] P. Schmitt, P. D. Beer, M. G. B. Drew, P. D. Sheen, *Tetrahedron Lett.* **1998**, *39*, 6383-6386.
[121] T. Nagasaki, S. Shinkai, T. Matsuda, *J. Chem. Soc., Perkin Trans. 1* **1990**, 2617-2618.
[122] T. Nagasaki, S. Shinkai, *J. Chem. Soc., Perkin Trans.* 2 **1991**, 1063-1066.
[123] L. Dasaradhi, P. C. Stark, V. J. Huber, P. H. Smith, G. D. Jarvinen, A. S. Gopalan, *J. Chem. Soc., Perkin Trans.* 2 **1997**, 1187-1192.
[124] P. D. Beer, M. G. B. Drew, M. Kan, P. B. Leeson, M. I. Ogden, G. Williams, *Inorg. Chem.* **1996**, *35*, 2202-2211.
[125] P. D. Beer, M. G. B. Drew, A. Grieve, M. Kan, P. B. Leeson, G. Nicholson, M. I. Ogden, G. Williams, *Chem. Commun.* **1996**, 1117-1118.
[126] M. J. Kan, G. Nicholson, I. Horn, G. Williams, P. D. Beer, P. Schmitt, D. Hesek, M. G. B. Drew, P. Sheen. *Nucl. Energy* (*Br. Nucl. Energy Soc.*) **1998**, *37*, 325-329.

[127] P. D. Beer, M. G. B. Drew, D. Hesek, M. Kan, G. Nicholson, P. Schmitt, P. D. Sheen, G. Williams, *J. Chem. Soc., Dalton Trans.* **1998**, 2783-2786.
[128] N. T. K. Dung, K. Kunogi, R. Ludwig, *Bull. Chem. Soc. Jpn.* **1999**, *72*, 1005-1011.
[129] R. Ludwig, T. K. D. Nguyen, K. Kunogi, S. Tachimori. JAERI Conf **1999**, 99 004 (Pt. 2, Proceedings of the NUCEF International Symposium, **1998**), 477-485.
[130] T. N. Lambert, L. Dasaradhi, V. J. Huber, A. S. Gopalan, *J. Org. Chem.***1999**, *64*, 6097-6101.
[131] J. F. Malone, D. J. Marrs, M. A. McKervey, P. O'Hagan, N. Thompson, A. Walker, F. Arnaud Neu, O. Mauprivez, M-J. Schwing Weill, J.-F. Dozol, H. Rouquette, N. Simon, *Chem. Commun.* **1995**, 2151-2153.
[132] M. R. Yaftian, M. Burgard, C. B. Dieleman, D. Matt, *J. Membr. Sci.***1998**, *144*, 57-64.
[133] F. Arnaud Neu, J. K. Browne, D. Byrne, D. J. Marrs, M. A. McKervey, P. O'Hagan, M. J. Schwing Weill, A. Walker, *Chem. Eur. J.* **1999**, *5*, 175-186.
[134] M. R. Yaftian, M. Burgard, D. Matt, C. B. Dieleman, F. Rastegar, *Solvent Extr. Ion Exch.* **1997**, *15*, 975-989.
[135] F. Arnaud Neu, V. Böhmer, J-F. Dozol, C. Gruettner, R. A. Jakobi, D. Kraft, O Mauprivez, H. Rouquette, M-J. Schwing Weill, N. Simon, W. Vogt, *J. Chem. Soc., Perkin Trans. 2* **1996**, *6*, 1175-1182.
[136] L. H. Delmau, N. Simon, J-F. Dozol, S. Eymard, B. Tournois, M-J. Schwing Weill, F. Arnaud Neu, V. Böhmer, C. Gruttner, C. Musigmann, A. Tunayar, *Chem. Commun.* **1998**, 1627-1628.
[137] T. N. Lambert, G. D. Jarvinen, A. S. Gopalan, *Tetrahedron Lett.* **1999**, *40*, 1613-1616.
[138] S. Barboso, A. G. Carrera, S. E. Matthews, F. Arnaud Neu, V. Böhmer, J-F Dozol, H. Rouquette, M-J. Schwing Weill, *J. Chem. Soc., Perkin Trans. 2* **1999**, 719-724.
[139] L. H. Delmau, N. Simon, M-J. Schwing Weill, F. Arnaud Neu, J-F. Dozol, S. Eymard, B. Tournois, C. Gruttner, C. Musigmann, A. Tunayar, V. Böhmer. *Sep. Sci. Technol.* **1999**, *34*, 863-876.
[140] S. E. Matthews, M. Saadioui, V. Böhmer, S. Barboso, F. Arnaud Neu, M-J. Schwing Weill, A. G. Carrera, J-F. Dozol, *J. Prakt. Chem.* **1999**, *341*, 264-273.
[141] H. Boerrigter, T. Tomasberger, A. S. Booij, W. Verboom, D. N. Reinhoudt, F. de Jong, *J. Membr. Sci.* **2000**, *165*, 273-291.

Chapter 23

CALIXARENE-BASED ANION RECEPTORS

SUSAN E. MATTHEWS, PAUL D. BEER

Inorganic Chemistry Laboratory, University of Oxford, South Parks Road, Oxford OX1 3QR
e-mail: susan.matthews@chem.ox.ac.uk., paul.beer@chem.ox.ac.uk.

1. Introduction

Anions play a number of fundamental roles in biological and chemical processes and thus the development of selective synthetic anion receptors is an area of current importance. The use of anions as nucleophiles, bases, redox-active agents and phase transfer catalysts has led to the desire for receptors which enable stabilization and separation through co-ordination. The increasing problem of environmental anion pollutants such as phosphate and nitrate, which lead to eutrophication, and radioactive pertechnecate, a product of the nuclear fuel cycle, is also an area of concern. Biochemically, anions are essential to normal metabolic function, both ATP and DNA being anionic, thus the development of anion binding mimetics would enable investigation of basic biological processes. A number of disease states including cystic fibrosis, cancer and Alzheimer's involve misregulation of anion function, offering the long term goal of a medical role for anion receptors [1].

The development of synthetic anion receptors has been slow in comparison to cation receptors. This is due, in the main, to a number of unique properties of anions that need to be addressed in the design of receptors. These include the negative charge, which is often delocalised over a number of atoms, and the size and shape of anions. In contrast to cations, anions are larger and have diverse topology, being spherical, linear, planar, tetrahedral or octahedral. Binding of anions is also affected by their pH dependence and solvation. For a given anion and cation of comparable size, for example fluoride and potassium, the anion is more strongly hydrated and thus more difficult to desolvate - a pre-requisite for binding [1].

The first reported synthetic anion receptor was the polyammonium cryptand of Simmons and Park [2]. This compound bound halide anions within a cavity through a combination of electrostatic interactions and directional hydrogen bonding. Further development of organic based systems was slow until the seminal work of Lehn *et al.* who reported a range of polyammonium macrocycles [3] and cryptates [4] with a variety of binding selectivities. The small operational pH range of these receptors was overcome by the design of receptors based on the guanidinium motif [5] or quaternized nitrogen atoms [6]. In recent years, organic based receptors have been developed which rely solely on hydrogen bond donors such as amides [7].

Inorganic approaches to anion receptors have been based on the properties of metals, which allow interactions either through favourable electrostatics or orbital overlap [1d]. However, such systems also offer the opportunity of sensing the binding event through redox, colorimetric or luminescent responses. Of particular interest are receptors based on Lewis acid metal-anion orbital overlap. The pioneering work of Newcomb on the design of tin based macrocycles [8] has led to the development of Lewis acidic hosts based on boron [9], silicon [10], germanium [11] and mercury [12]. The 1970s saw great interest in the design of cascade complexes in which bound anions are bridged between positively charged metal centres [13]. More recently a number of approaches to anion binding through electrostatic interactions have been investigated using, for example, the cobaltocenium moiety [14] or metal ion cornered macrocyclic receptors [15,16].

This review describes the development of synthetic anion receptors based on calixarenes. It is convenient to divide calixarene based anion receptors into inorganic and organic classes, *i.e.*, where the anion recognition site is primarily of inorganic or organic origin. The recent development of receptors for the binding of ion-pairs is also discussed.

2. Inorganic-Based Systems

2.1. METALLOCENE CALIXARENE RECEPTORS

Acyclic and simple macrocyclic systems using the cobaltocenium/cobaltocene and ferrocene/ferrocenium redox couples, combined with amides, to electrochemically sense the binding of anions have been extensively investigated [14].

2.1.1. Cobaltocenium-based receptors

Calixarenes substituted by cobaltocenium, a positively charged organometallic pH independent redox active unit, show great potential as selective anion receptors. Here binding of the anion takes place via both favourable electrostatic interactions and amide hydrogen bonding. ^{1}H NMR studies show that the 1,3-upper-rim-difunctionalised receptor **1** is an effective receptor of halides, nitrate and hydrogen sulphate as well as a series of dicarboxylates [17]. The dicarboxylates form very strong (K_{ass} 11510 M^{-1} for adipate in (CD$_3$)$_2$CO)) 1:1 host:guest complexes in which the anion is bound between the two cobaltoceniums [18]. In cyclic voltammetric studies significant cathodic perturbations (ΔE = 55 mV) of the reversible redox wave were seen for chloride, nitrate and adipate, exemplifying the use of this receptor for electrochemical anion recognition.

In contrast to the dicarboxylate selective receptor, the 1,3-upper-rim-bridged receptor **2a** shows selectivity for monocarboxylates based on the topology of the anion binding site [19,20]. Binding of a range of anions occurs but the largest association constants were found for monocarboxylates, particularly for acetate (K_{ass} 41520 M^{-1} in (CD$_3$)$_2$SO). The solid state structure of the chloride complex shows the calixarene to be held rigidly in a *pinched cone* with the cobaltocenium moiety pulling together the two phenyl units and arranging the amides for maximum hydrogen bonding interactions with the halide guest. Electrochemical studies with this receptor show large ca-

1

2
a: R = CH$_3$ R' = H
b: R = H R' = Tosyl
c: R = Tosyl R' = H

thodic perturbations (up to 155 mV) of the redox wave for monocarboxylates and dihydrogen phosphate.

The anion binding capabilities of 1,3-bis(cobaltocenium) functionalised receptors can be significantly altered by the distribution of lower rim substituents [20,21]. With the isomeric receptors, **1b** and **1c**, marked differences in both anion binding strength and selectivity are observed. For **1b**, in which the lower rim tosyl substituents are *para* to the amide moieties, large association constants are reported for monocarboxylates and dihydrogen phosphate with only weak binding of halides, nitrate and hydrogen sulphate. Molecular modelling studies (MM2) suggest that the large tosyl substituents at the lower rim force the upper rim into a rigid arrangement in which the cobaltocenium moieties are held close together. This is seen in the X-ray structure of the chloride complex. In contrast, the more flexible *1,3-distal* isomer **1c** is able to accommodate a larger range of anions with selectivity for dihydrogen phosphate. In agreement with the magnitude and selectivity of association constants observed with ^{1}H NMR studies, large cathodic perturbations of the redox wave are observed for receptor **1b** for all anions, with particular selectivity for the monocarboxylates.

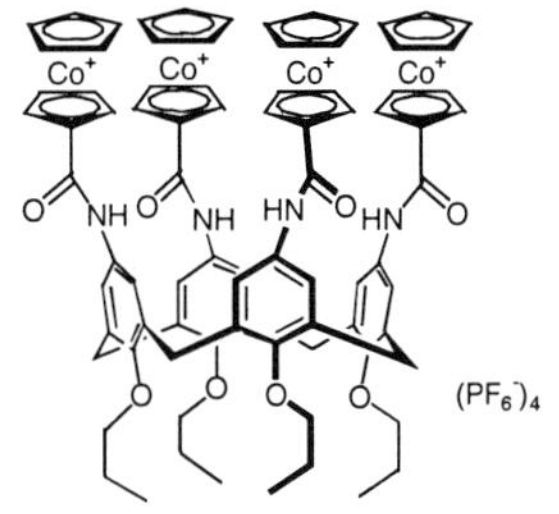

Initial studies have also been performed on a calix[4]arene with four cobaltocenium moieties at the upper rim **3** which exhibits selective dihydrogen phosphate binding in $(CD_3)_2SO$ [14].

2.1.2. Ferrocene-based receptors

Calix[4]arene amides appended with ferrocene moieties provide neutral redox active anion receptors. The lower-rim-bridged receptor **4** shows exceptional selectivity for dihydrogen phosphate in the presence of an excess of other anions [22]. A large cathodic perturbation of the ferrocene/ferrocenium redox wave was observed for dihydrogen phosphate which was retained even in the presence of a ten fold excess of hydrogen sulphate or chloride anions which themselves only exhibit weak cathodic perturbations.

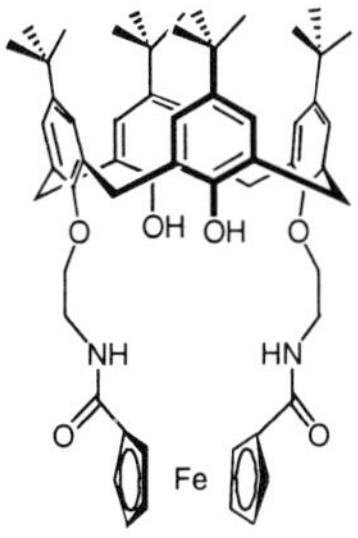

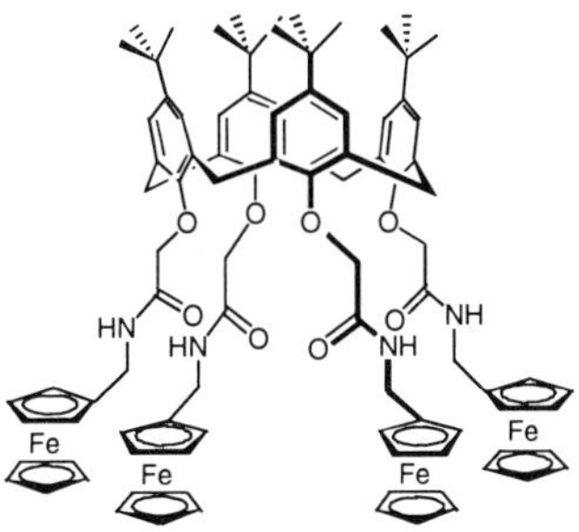

In a later report, lower rim ferrocene substituted calix[4]- and [5]-arenes and a 1,3 disubstituted calix[4]arene were evaluated [23]. The 1,3 disubstituted compound failed to bind a range of anions, which may be due to hydrogen bonding of the amide moieties to the unsubstituted phenols. However, both fully substituted receptors (*e.g.* **5**) bound chloride in a 1:1 ratio. The calix[4]arene receptor bound more strongly due to a better spatial fit between the halide anion guest and host binding site. Although binding of dihydrogen phosphate and hydrogen sulphate occurred the stoichiometry was not simple. Cyclic voltammetric studies showed cathodic perturbations for all three anions, with particularly pronounced shifts for dihydrogen phosphate in CH_2Cl_2 / CH_3CN.

The effect of solvent on anion binding selectivity has been studied with the bis(calix[4]arene)ferrocenoyl receptor **6** [24]. ^{1}H NMR studies with chloride, benzoate and acetate binding in CD_2Cl_2, CD_3CN and $(CD_3)_2CO$ revealed large differences in binding strength and selectivity which could not be related to either the relative permittivity or dipole moment of the solvent. However, if the Gutman acceptor number is taken into consideration, two trends in binding could be observed. As the acceptor number decreases the magnitude of binding increases and the selectivity changes in favour of smaller, harder anions.

Recently, ferrocene has been appended to a cavitand. Preliminary binding studies on receptor **7** show 1:1 complexation of chloride through co-operativity of the four binding sites [25].

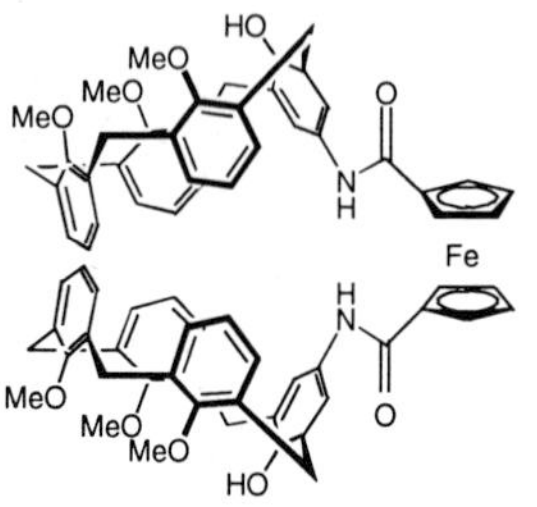

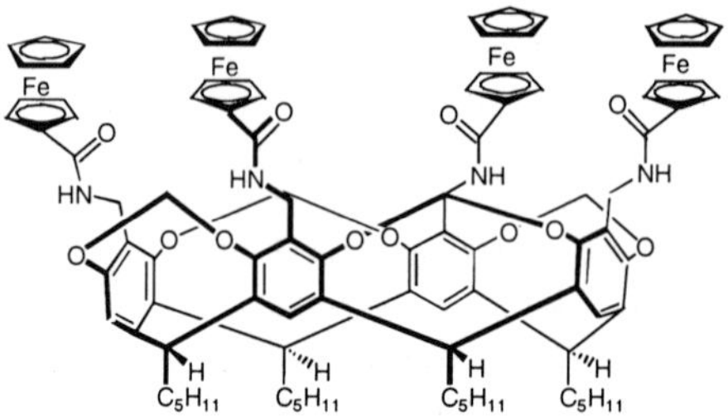

2.1.3. π-Metallated cationic hosts

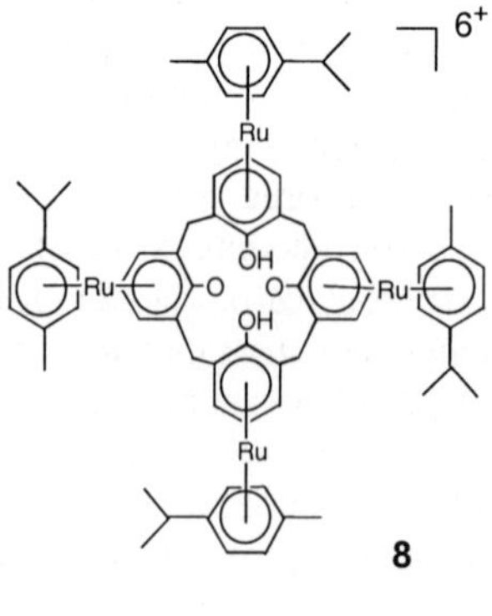

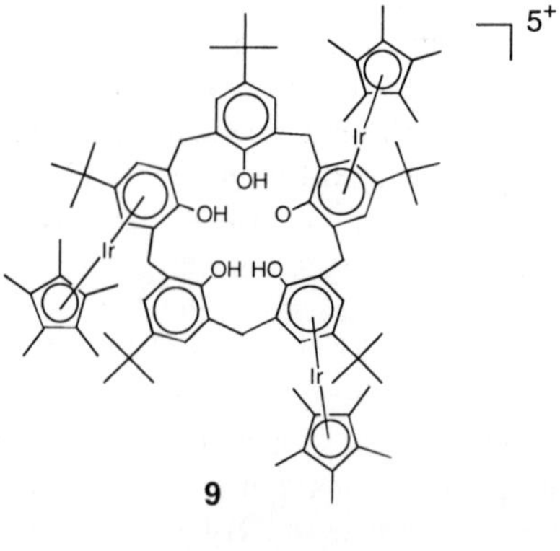

A new class of receptors [26] involves the incorporation of positively charged redox active transition metal centres directly onto the calixarene aromatic rings provides a cationic host suitable for anion inclusion. Initial studies on the treatment of calix[4]arene with [Rh(η-C$_5$Me$_5$)Cl(μ-Cl)$_2$] resulted only in the formation of 1,3 disubstituted compounds which did not include anions [27]. However, modification of reaction conditions led to the development of the ruthenium receptor **8** which showed binding of a BF_4^- anion deep in the calixarene cavity in the solid state. Anion metathesis with iodide, perrhenate and dihydrogen phosphate was easily accomplished [28,29]. ^{1}H NMR investigations were undertaken to assess the degree of anion affinity in water and the selectivity chloride>bromide>iodide observed. Interestingly, similar reactions with calix[5]arene resulted only in the formation of the tri-substituted receptor **9**. However, this also proved suitable for anion binding, leading the authors to suggest that two adjacent functionalised aromatic rings are required for anion binding.

2.2. SYSTEMS BASED ON CALIXARENES AS LIGANDS

2.2.1. Ruthenium(II) and rhenium(I) bipyridyl based systems

Numerous anion receptors have been derived from calix[4]arene amides with appended transition metal complexes. Charged ruthenium (II) or neutral rhenium (I) bipyridine units offer a number of sensing opportunities through electrochemical, redox and luminescence methods [14].

Early studies [32] concerned the binding properties of the Ru(II)-bis(bipyridine) complex of a calix[4]-arene substituted at the lower rim with two pyridyl units **10**. This receptor showed 1:1 binding of chloride and bromide anions. Calix[4]arenes bridged at the lower rim with bipyridine units have also been prepared and converted to the ruthenium (II) complex [33,34]. These receptors showed selectivity for dihydrogen phosphate over the halides. A marked preference was achieved with **11a** which has a smaller anion binding cavity. Electrochemical studies on receptor **11a** in CH$_3$CN confirmed binding of the anion within the amide cavity as only significant perturbations of the substituted bipyridine reduction wave were observed. A marked cathodic shift of 175 mV was seen for dihydrogen phosphate, which was unaltered by addition of a ten-fold excess of chloride or bromide anions offering the possibility of selectively sensing dihydrogen phosphate binding by cyclic voltammetry. Anion binding was also demonstrated through observation of the fluorescence spectrum. On addition of dihydrogen phosphate, the MLCT band undergoes a hypsochromic shift of 16 nm in (CD$_3$)$_2$SO with concomitant increase in emission intensity. The authors proposed that these changes were due to increased rigidity on complexation of the anion, which reduces non-radiative decay.

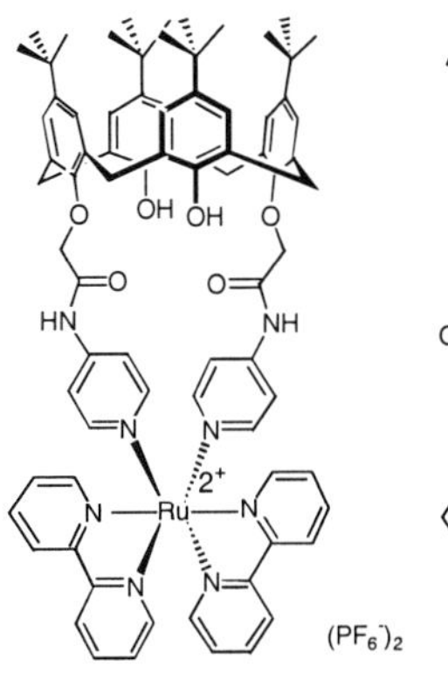

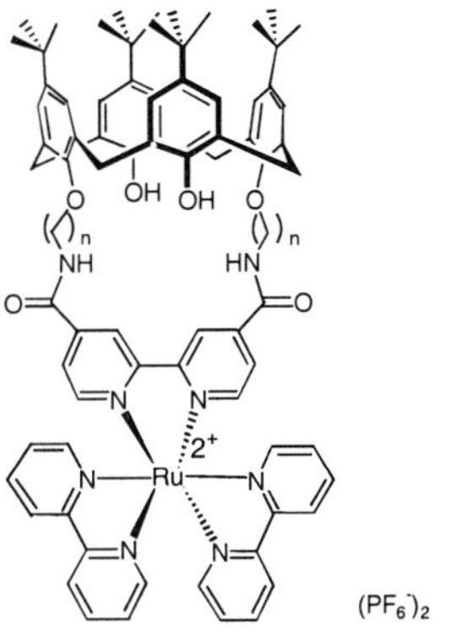

This system has been developed further by incorporation of quinone moieties to prepare switchable and selective luminescent sensitive receptors [35]. The Ru(II) and Re(I) bipyridinecalix[4]diquinones **12a** and **12b** were compared with the similar calix[4]arene derivatives **13**. All showed strong and selective binding of acetate over chloride and dihydrogen phosphate. However, both the topology of the binding site and choice of transition metal centre have a pronounced effect on the association constant (K$_{ass}$ acetate 9990 M^{-1} for **12a**, 1790 M^{-1} for **12b** and 4060 M^{-1} for **13a** in (CD$_3$)$_2$SO) Studies of MLCT luminescence of **12a** and **12b** showed remarkable emission retrieval on anion addition, for example up to 500% with **12a** and acetate in CH$_3$CN. This was proposed to be a consequence of anion binding substantially reducing the electron

12

a: M = Ru(Bpy)$_2^{2+}$ (PF$_6^-$)$_2$
b: M = Re(CO)$_3$Cl

13

14

a: R$_1$ = Tosyl, R$_2$ = H
b: R$_1$ = H, R$_2$ = Tosyl

M = Ru(Bpy)$_2^{2+}$ (PF$_6^-$)$_2$
M = Re(CO)$_3$Cl

transfer rate constant through a decrease in the interaction between the metal bipyridine centre and the quinone moieties.

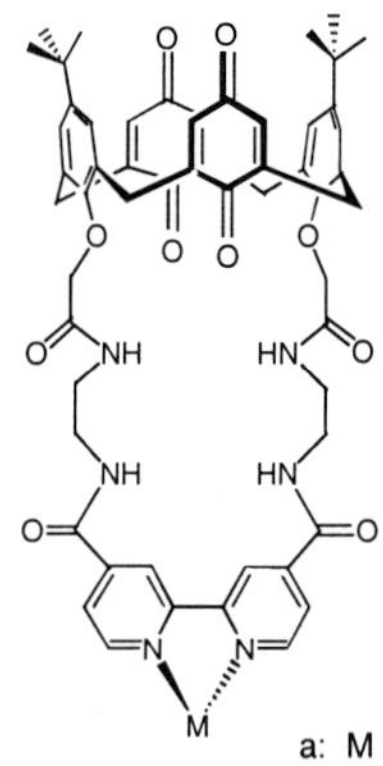

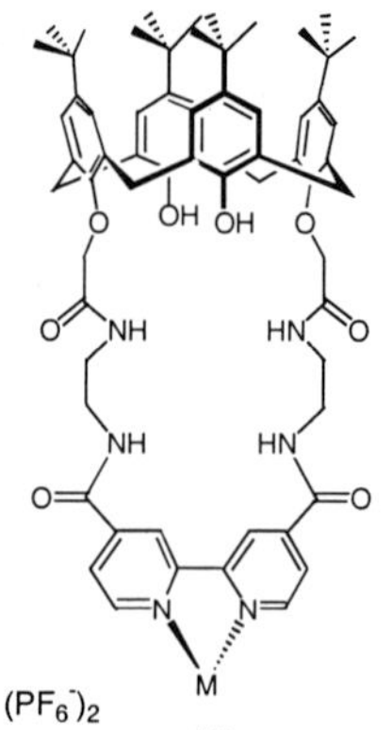

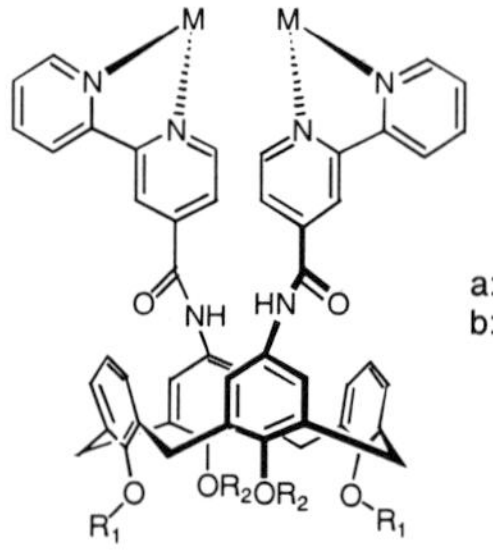

Upper rim substitution of calix[4]arenes with transition metal bipyridine complexes has also provided useful anion receptors [21]. 1,3 substitution leads to receptors that selectively bind dihydrogen phosphate over chloride. Significantly different binding capabilities are displayed by the two isomers **14a** and **14b**, as with **1b** and **1c**, indicating that lower rim substitution patterns can affect conformation of the receptor site at the upper rim.

Unusually carboxylate anions are bound preferentially by a bis(calix[4]arene) receptor with a Ru(II)bipyridine bridge at the upper rim **15** [36]. Here the close proximity of the two calix[4]arene units favours the binding of benzoate and phenylacetate over that of dihydrogen phosphate.

With the cavitand Ru(II)bipyridine receptor **16**, association constants, showing 1:1 binding with a slight preference for carboxylates over chloride, were determined by fluorescence titrations [25].

15

16

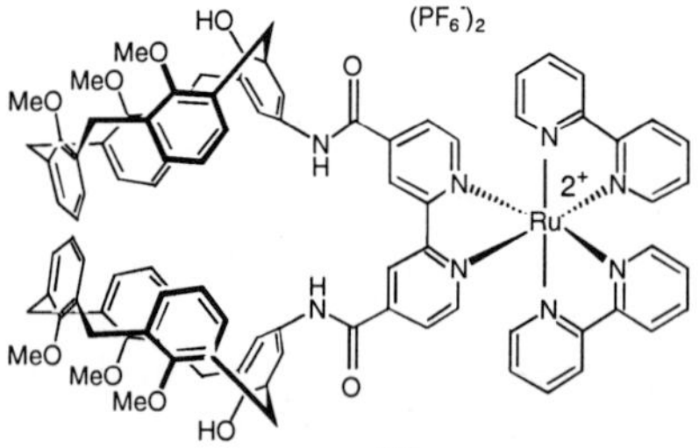

2.2.2. Other transition-metal-based systems

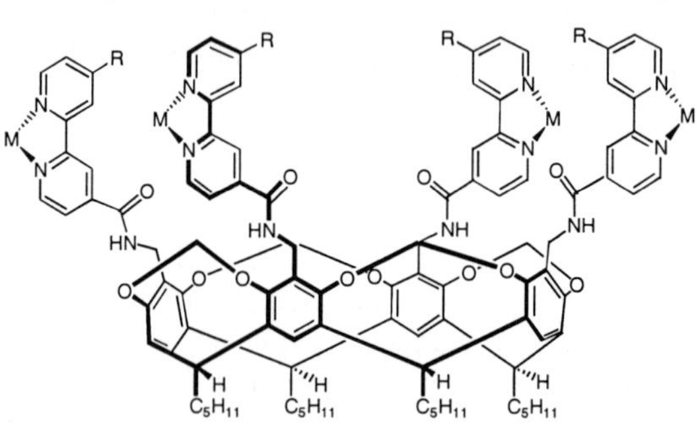

In a host based on a phosphonito-calixresorcin[4]arene cavitand [37-39], complexation of the phosphorus donor atoms with gold, silver or copper alters the nature of the calixarene bowl to enable anion inclusion. The solid state structure of the tetracopper complex shows not only four chloride anions bound at the periphery, but also one to be included by being bound to three of the copper atoms. ^{31}P NMR studies gave only one signal for the complex, indicating rapid migration of chloride between the four copper atoms. Halide exchange enabled the preparation of both iodide and bromide derivatives. The iodide derivative showed μ-4 binding of the fifth halide atom, a result of the larger size of the anion allowing interactions with all four copper atoms. On replacing the gold atoms with silver, μ-4 complexes were obtained with chloride, bromide and iodide. With these latter receptors it proved possible both to exchange the anionic guest and to, on treatment with nitrate, prepare an empty cavity which acted as an anion binder. In an interesting application, receptor **17** was used to alter the mechanism of an aliphatic nucleophilic substitution. Treatment of an alkyl iodide with the chloride complex led to halide exchange with retention of configuration, due to the preferential binding of iodide within the host cavity.

A "cascade" complex of nickel(II) azide with **18** has been reported in the solid state. Two nickel ions are bound within two lower rim triazacyclononane units with three azide 1,1' end-on bridging ligands between them [40]. Recently, a number of metallo-phosphodiesterase models based on calix[4]arene with multidentate-ligand-arm substituents have been developed *e.g.*, **19** [41]. These are fully discussed in Chapter 27.

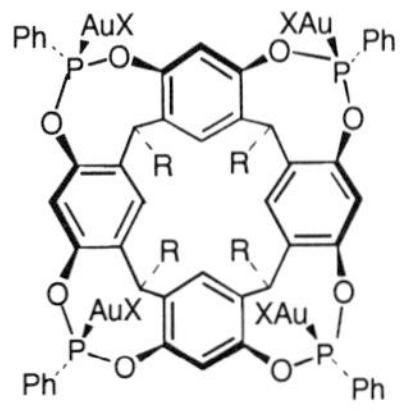

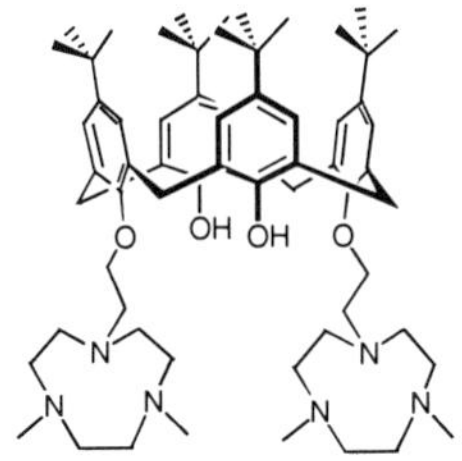

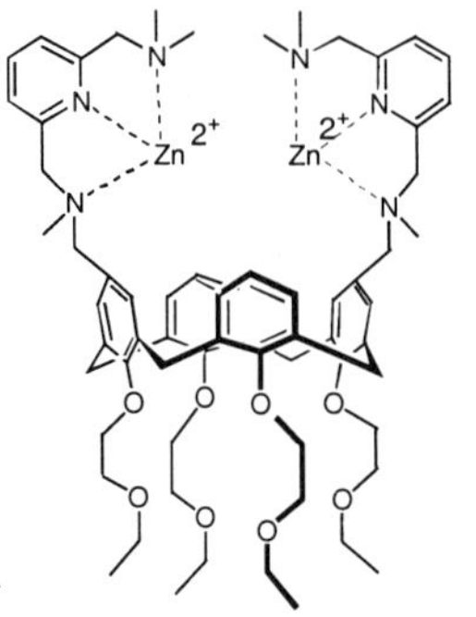

2.2.3. Non-transition metal systems
Functionalised acyclic and macrocyclic uranyl ion complexes of salicylaldimines are effective anion receptors functioning *via* orbital overlap with the Lewis acid centre (U) and also hydrogen bonding interactions [42,43]. Initial ^{1}H NMR studies on the monotopic receptor **20** showed selective binding of dihydrogen phosphate over chloride, hydrogen sulphate and perchlorate in $(CD_3)_2SO$ [44].

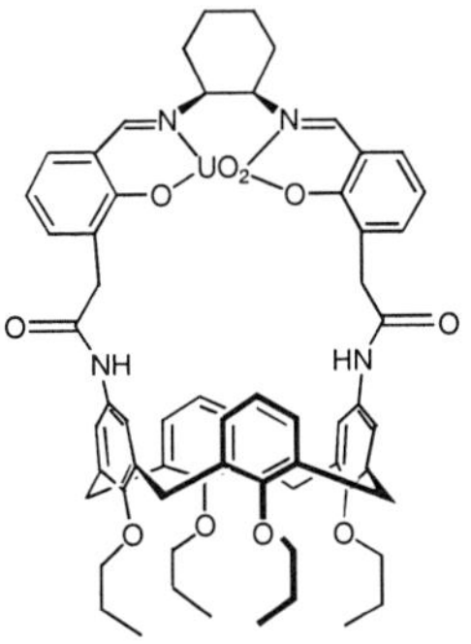

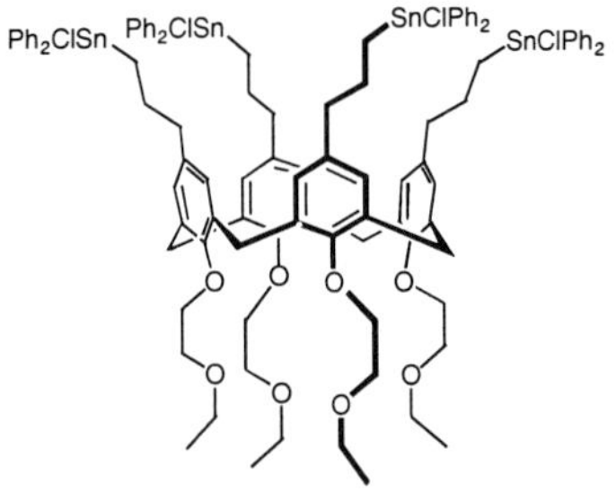

The Lewis acidic calix[4]arene **21**, incorporating four diphenylmonochlorotin(IV) moieties at the upper rim, has also been reported [45]. Not surprisingly, no co-operativity of binding was observed due to the flexibility of the ligands. Instead a ^{119}Sn NMR study of chloride binding showed very weak 1:4 host:guest binding which could be enhanced at low temperature (253 K).

3. Organic-Based Systems

Naturally occurring sulphate and phosphate binding proteins show exceptional selectivity (of over 10^4) for the binding of the respective anion. This is achieved solely through hydrogen bonding; the sulphate binding protein of *Salmonella typhimurium* employing seven bonds both from the peptide backbone and from side chains, the phosphate protein utilising twelve. Secondary stabilisation of the complex is achieved through a macrodipole effect. Thus, a range of synthetic receptors has been prepared using hydrogen bond donor units such as amides, ureas and thioureas to bind anions. Receptors based solely on hydrogen bonds are generally only effective in non-polar organic solvents but the directional nature of H-bonds offers advantages in receptor design [1].

3.1. AMIDE BASED RECEPTORS

One of the earliest examples of a calix[4]arene based anion receptor incorporated a sulfonamide hydrogen bonding unit [46]. In this case, selectivity for the tetrahedral hydrogen sulphate anion over both planar and spherical anions was seen in CDCl$_3$. For compound **22c**, where a second amide unit is included in the side chain, this selectivity is particularly marked (K_{ass} HSO$_4^-$ 103400 M^{-1}, Cl$^-$ 1250 M^{-1}, NO$_3^-$ 513 M^{-1}).

A bis calix[4]arene receptor **23** has been prepared incorporating four amide linkages. ^{1}H NMR titrations in CD$_2$Cl$_2$ showed a pronounced selectivity for binding of fluoride over chloride, hydrogen sulphate and dihydrogen phosphate by an order of magnitude. This selectivity can be considered to be a function of anion size, the cavity being too small for binding of tetrahedral anions [47].

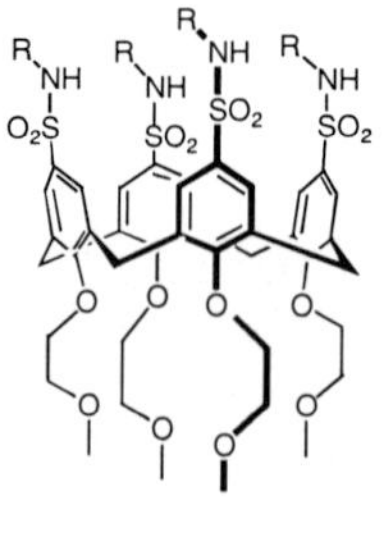

22
a : R = H
b : R = t-butyl
c : R = CH$_2$CH$_2$NHC(O)Me

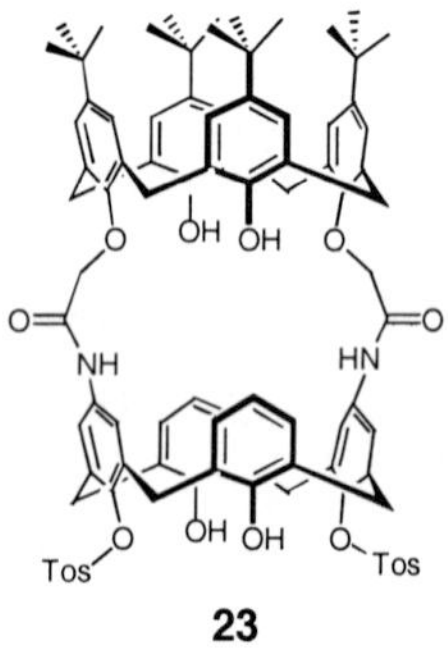

Synthesis of the possibly tunable anion receptor **24** is based upon 1,3 functionalisation of the upper rim of a calix[4]arene [48]. Such functionalisation forces the calix[4]arene into a *pinched cone* conformation providing an ideal receptor arrangement for binding of carboxylates. Introduction of electron withdrawing groups, in the terminal portion of the amide, increases the acidity of the amide proton and thus is effective in increasing anion binding strength *e.g.*, K_{ass} benzoate 5160 M^{-1} for **24a**, 107 M^{-1} for **24b** in CDCl$_3$. Introduction of more electron withdrawing groups as in **24c** results in a loss of anion binding ability, proposed to be due to steric crowding.

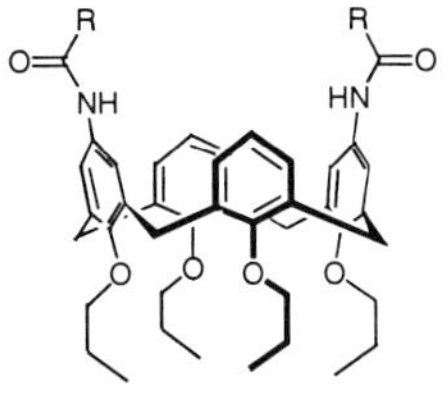

24
a: R = CH₂Cl
b: R = CHCl₂
c: R = CCl₃

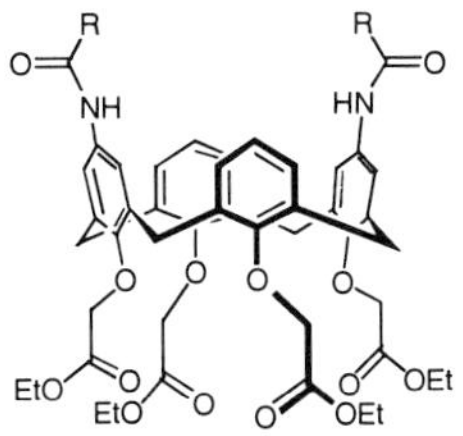

25

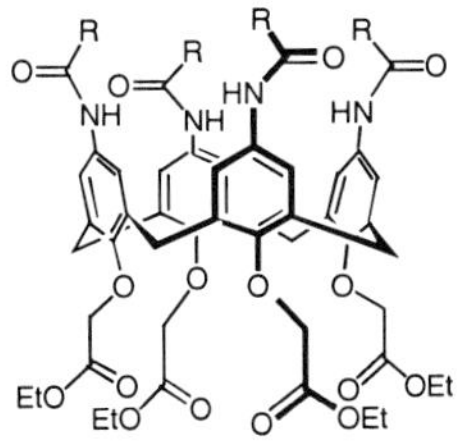

26
a : R = C₆F₅
b : R = C₅H₄N

However, this has been disputed on the basis of extensive study of anion binding by both simple activated ureas and calixarene based systems [49]. They suggested that the lack of anion binding ability of **24c** is due to the absence of an acidic hydrogen on the carbon α to the carbonyl. These authors have prepared a number of calix[4]arene based receptors in which the amide is substituted with either a perfluorinated aryl or a pyridyl unit. The bidentate receptors **25** were evaluated for the binding of alkyl and aryl dicarboxylates and in the case of **25a** showed pronounced selectively based on the length of the dianion, binding adipate and terephthalate over glutarate and isophthalate in $(CD_3)_2CO$. For the quadridentate ligand **26** selective binding of squarate is observed, although this selectivity is solvent dependent, being more pronounced in CD_3CN than $(CD_3)_2CO$. Binding of anions by **26b** was not evaluated due to association in solution.

3.2. UREA BASED RECEPTORS

3.2.1. Calix[4]arene

A number of receptors have structures based on urea units attached to calix[4]arene, for example through four-carbon spacers on the lower rim [50]. Both the bidentate **27** and quadridentate **28** receptors were evaluated through ¹H NMR and mass spectral studies. In general, higher association constants were obtained for the bidentate receptors with a selectivity for chloride > bromide > cyanide in $CDCl_3$ for all receptors. The higher association constants with these anions were considered to be due to their stronger hydrogen bonding acceptor properties and better steric complementarity. Surprisingly,

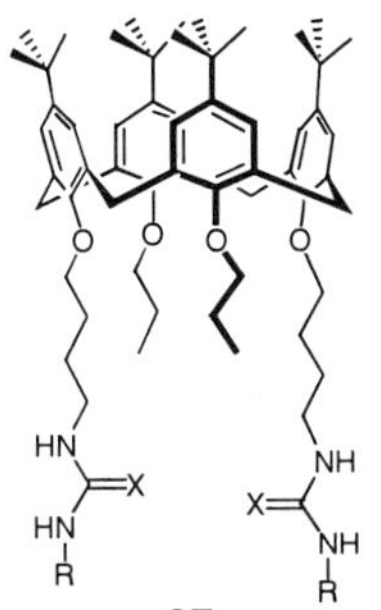

27

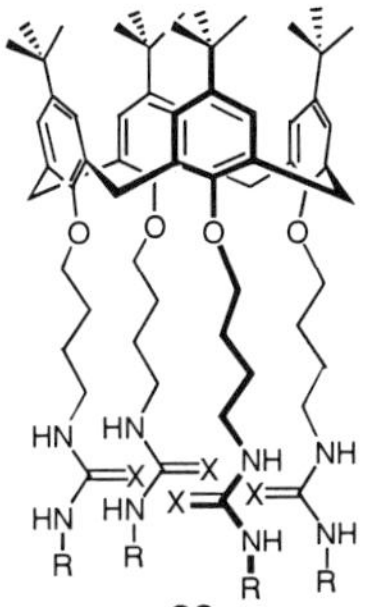

28
a : X = O, R = phenyl
b : X = O, R = propyl
c : X = O, R = t-butyl
d : X = S, R = phenyl

no binding of fluoride was observed. In a separate study [51], the anion binding capabilities of this system were investigated using Monte Carlo simulations. Here, the general selectivity through the halide group was confirmed but fluoride was shown to bind effectively and it was proposed that the lack of experimental binding was due to the presence of adventitious water molecules. Despite the higher acidity of the NH protons in thioureas, ligand **27d** showed reduced binding of all anions. This is possibly due to the formation of very strong host intra- and inter-molecular hydrogen bonds.

In a recent communication [52], anion binding by the lower rim urea groups of a calix[4]diquinone **29** was reported. The combination of the receptor capabilities of the urea moieties with the electron accepting quinones enables analysis of binding capabilities by both electrochemical and ^{1}H NMR techniques. In comparison to the non-quinone receptor **30**, small spherical anions are bound more weakly in $CDCl_3$, but for larger anions binding is higher in the calix[4]diquinone. This is particularly pronounced with hydrogen sulphate and is thought to be due to hydrogen bonding interactions of the anion with the quinone. This is confirmed by electrochemical studies where a large cathodic shift is observed in the presence of hydrogen sulphate.

Both uni- **31** and bidentate **32** upper rim receptors have been prepared and their interactions with monocarboxylates evaluated [53]. The monofunctional receptors showed good binding in $(CD_3)_2SO$ with preference for butyrate and benzoate. The authors proposed that in the case of acetate, which has the weakest binding, the anion is bound in an exocyclic manner solely through hydrogen bonding. For butyrate and benzoate, which have higher association constants, they propose that binding occurs within the cavity of the calixarene enabling further stabilisation by CH_3/π and π/π interactions. The bidentate receptor gives the highest association constant with acetate, possibly through interactions with all four NH groups of the urea moieties.

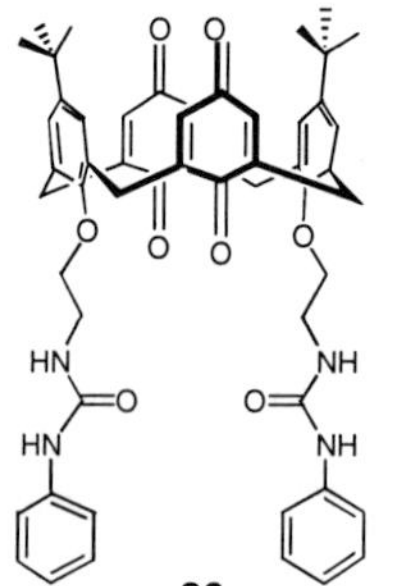

29

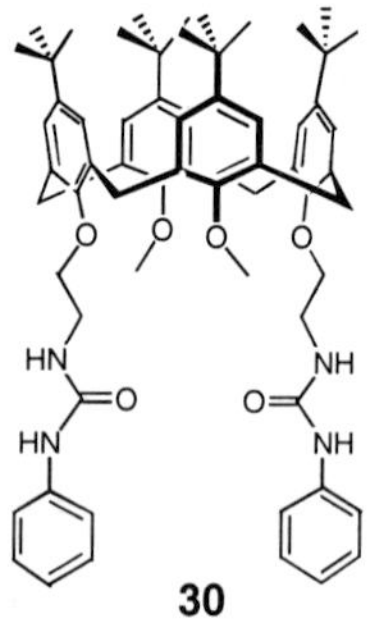

30

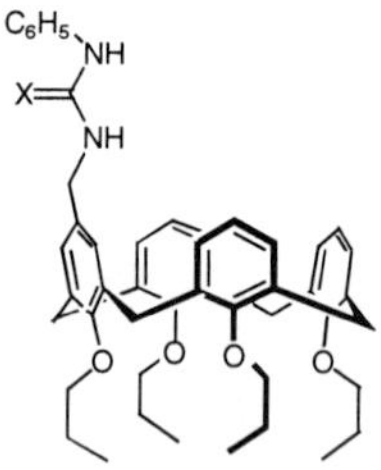

31

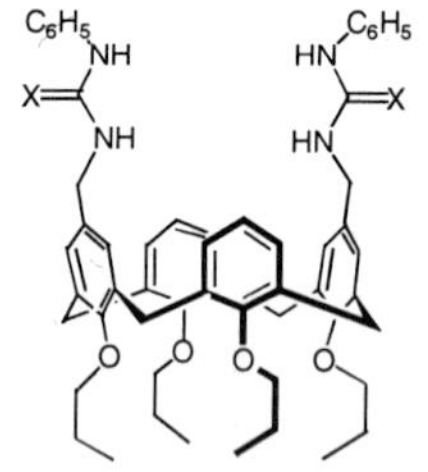

32

a : X = O
b : X = S

3.2.2. Calix[6]arenes

Thiourea and urea derivatives of calix[6]arene **33** have been prepared which show remarkable anion binding selectivity [54]. The incorporation of three urea units at the lower rim of a calix[6]arene combined with three methyl substituents gives a flattened *cone* conformation with effective C_3 symmetry. Binding of small spherical anions was demonstrated by FAB mass spectral and ^{1}H NMR studies in $CDCl_3$. Surprisingly, bromide is bound more effectively than chloride, despite its lower acidity, possibly because the cavity size is more complementary. With tricarboxylate anions it was proposed that larger association constants would be observed through secondary electrostatic interactions between the slightly positively charged NH and the partially negatively charged

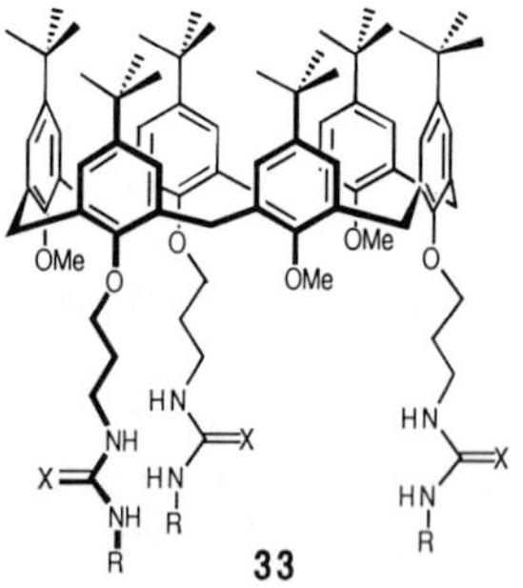

33

a : X = O, R = phenyl
b : X = S, R = phenyl

oxygen of the carboxylate. Indeed greatest binding was seen for the C_3 planar guest 1,3,5-benzenetricarboxylate anion, which can form six hydrogen bonds without distortion. Non planar 1,3,5 cyclohexanetricarboxylate is bound more weakly but significantly more strongly than the spherical monoanionic halides.

3.2.3. Resorcinarene cavitands

Ureas and thioureas have also been appended to aminomethyl cavitands [55]. These cavitands provide a more rigid framework with reduced risk of dimerisation due to hydrogen bonding. All receptors showed 1:1 binding with halides, with a small selectivity for chloride. The thiourea derivatives bound more strongly due to the higher acidity of the NH bonds, this being particularly apparent with **34d** in which an electron withdrawing substituent is incorporated into the urea. Co-operativity of the urea moieties in binding is evidenced by a reduction, of one order of magnitude, in the association constant for a tri-substituted derivative.

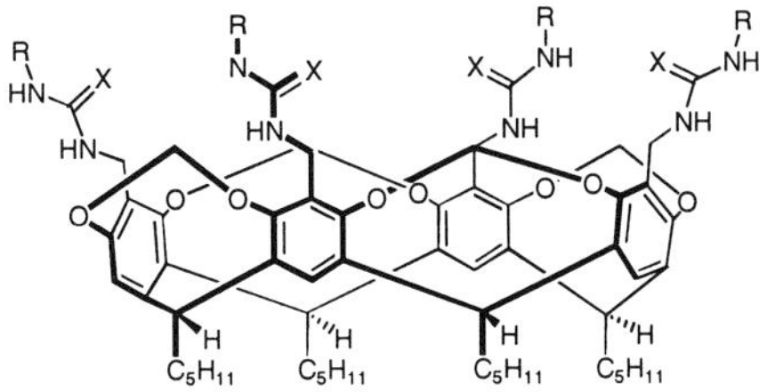

34
a : X = O   R = t-octyl
b : X = O   R = C_6H_4F
c : X = S   R = t-octyl
d : X = S   R = C_6H_4F

3.3. ALCOHOL BASED SYSTEMS

Anion receptors incorporating chiral alcohol moieties have recently been described [56]. 1,3 disubstituted receptors **35** show selectivity for carboxylates over halides due to complementarity of fit, as previously seen for amides. With a chiral guest such as N-lauryl-L-phenylalanine a marked difference in the association constant with the two receptors is observed, as a result of steric preferences. With a similar tetrasubstituted compound **36**, selectivity for halides is observed with inclusion of perfluorinated units proving essential for binding.

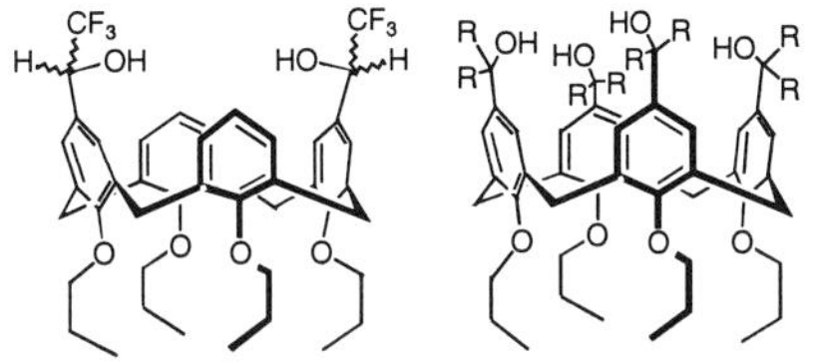

3.4. CHARGED SYSTEMS

A number of receptors have been prepared in which a combination of electrostatic forces and secondary binding elements are responsible for anion binding. Examples are provided by polyaza crown ether derivatives of calix-[4]arene featuring ammonium moieties [57]. Receptor **37a** is effective in binding nitrate *via* a three point interaction with the host, whereas only weak binding is observed for the halides. The important role of hydrogen bonding in anion recognition is evidenced by the absence of binding with the analogous **37b**.

Oxyanions may be extracted from water into $CHCl_3$ by calix[4]arenes **38** substituted with alkylammonium units [58,59]. Selective extraction of chromate and dichromate in the presence of competing monoanions and dianions was observed but the specific nature of the extracted species was not determined.

Recently, pyridinium calix[4]arene derivatives were reported [60]. Both 1,3 functionalised and 1,3 bridged compounds proved effective anion binders in ^{1}H NMR studies. Receptor **39** showed 2:1 binding in $(CD_3)_2SO$ with selectivity for dihydrogen phosphate over chloride and bromide and very weak binding of hydrogen sulphate. This is easily rationalised in terms of anion basicity. Selectivity for chloride over bromide was also seen for the bridged receptor **40**. Cathodic perturbations of the one electron reduction wave for pyridinium were observed with both chloride and dihydrogen phosphate.

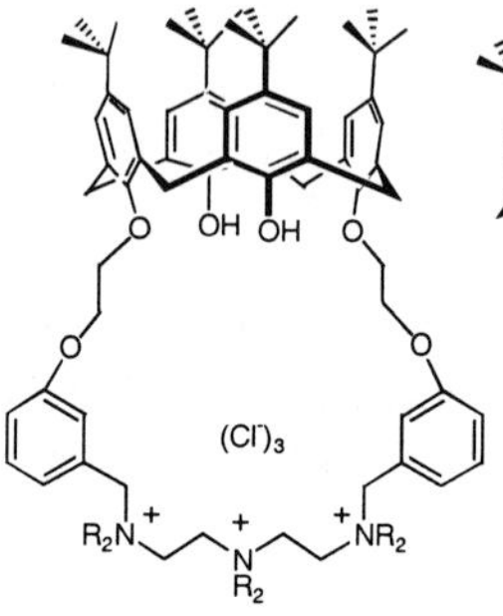

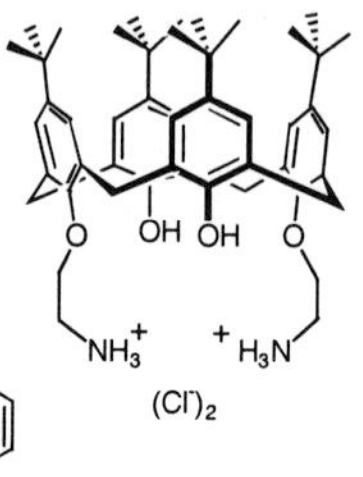

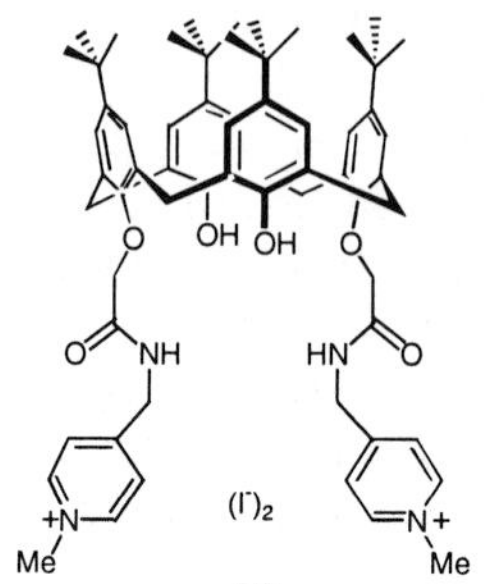

3.5. CALIX[4]PYRROLES

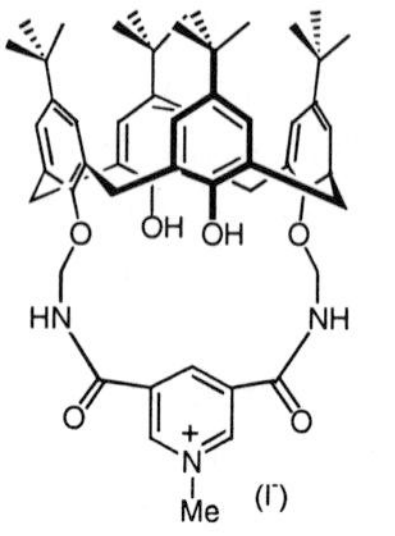

The anion binding properties of calix[4]pyrroles, a type of porphyrinogen known over 100 years, were first reported in 1996 [61]. Unlike calixarenes, which cannot bind anions directly due to poor electrophilicity, calix[4]pyrroles can bind anions selectively through interaction with the pyrrolic hydrogens [61]. Initial solid state studies showed the free receptors **41** and **42** preferentially adopt the *1,3-alternate* conformation. In the presence of anions, a *cone* like conformation is attained to enable all four NH units to be involved in hydrogen bonding to the guest. Analysis of bond lengths showed fluoride to be bound more closely to the plane of the pyrroles than chloride, suggesting stronger binding. This was confirmed by ^{1}H NMR binding studies in CD_2Cl_2 where a marked preference for fluoride over other halides, dihydrogen phosphate and hydrogen sulphate was observed.

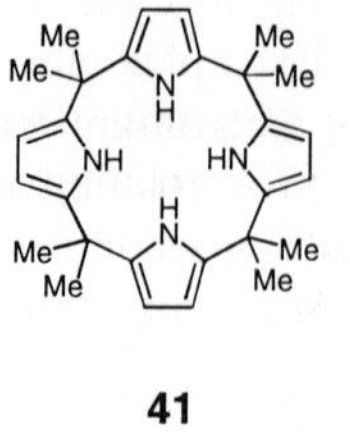

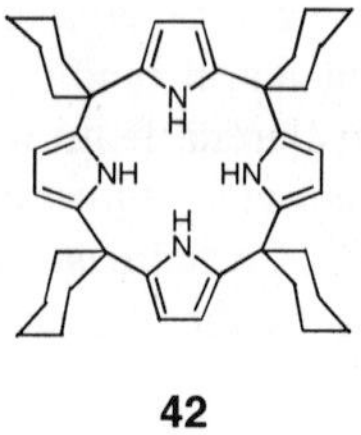

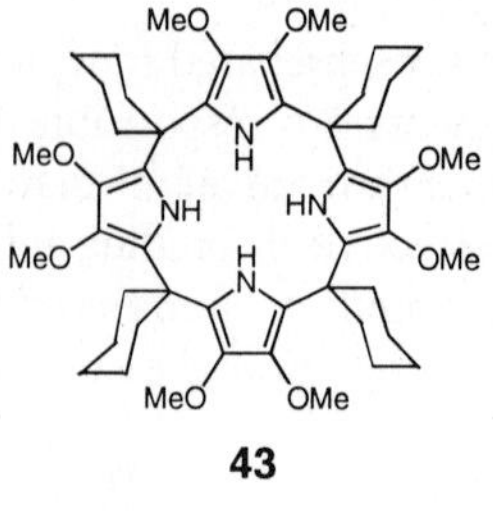

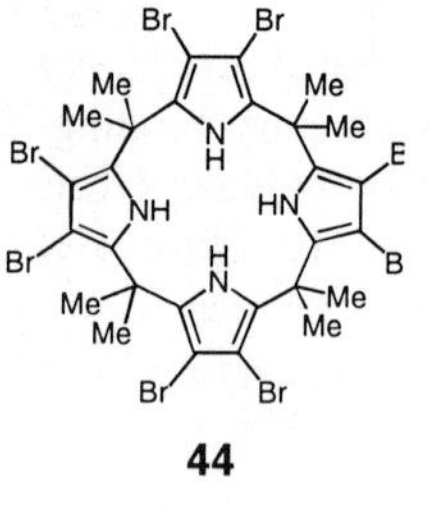

Interestingly, in an independent study [62], Monte Carlo simulations suggested that even larger association constants for fluoride should be observed.

Tuning of the anion binding strength through alteration of the electron density was achieved by substitution at the carbon rim. Incorporation of methoxy groups in **43** leads to decreased binding, due to reduced acidity of the pyrrolic NH whereas bromo substituents in **44** led to enhanced binding through electron withdrawing effects [63]. Deeper cavities resulted in decreased affinity, particularly for smaller anions, but enhanced selectivity [64]. Recently binding of isophthalate through a two point interaction with a rigid calix[4]pyrrole dimer has been reported [65].

The commercial applications of this system have been extensively evaluated in the development of electrochemical [66], fluorescent [67] and colorimetric [68,69] sensors of anion binding. Of particular interest is the use of calix[4]pyrroles as solid supports for the HPLC separation of anions [70]. Derivatistion enables coupling to an amino propyl silica gel and subsequent retention and separation of a range of simple anions and the biologically important phosphates, AMP, ADP and ATP.

Preliminary results on the anion binding properties of calix[6]pyrroles have recently been reported [71]. The larger annulus of this host allows binding of chloride and bromide with a six fold selectivity for chloride in phase transfer experiments. Solid state structures show the originally D_{2d} host attaining, on complexation, a D_{3d} symmetry with the anion lying slightly out of the plane.

3.6. WATER SOLUBLE RECEPTORS

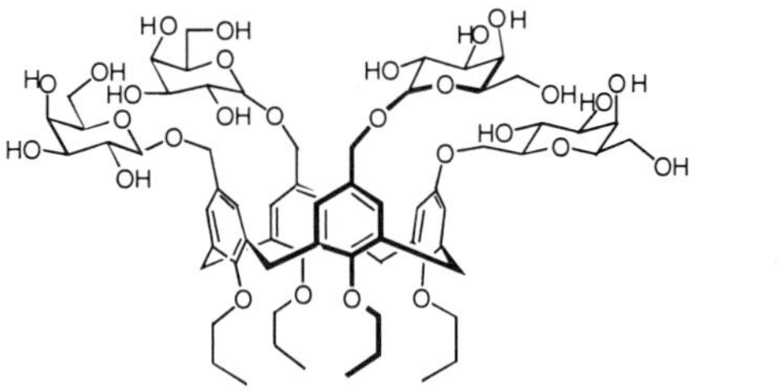

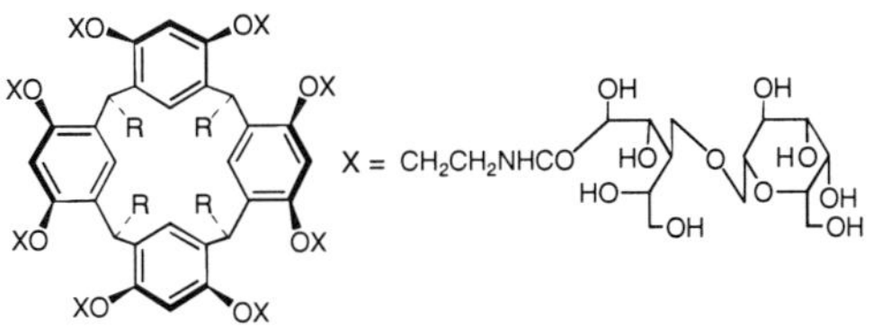

The recognition and sensing of anions in water is an area of particular importance, both for biological species and environmental polluants. Functionalisation of calix[4]arenes and resorcin[4]arenes with carbohydrate moieties results in receptors which show considerable water solubility. A number of calixsugars have been developed [72] and their binding characteristics studied. Neutral guests such as carbohydrates and N-protected amino acids failed to bind. However, 1:1 complexation of dihydrogen phosphate was seen for **45**, offering opportunities for the binding of larger, phosphate containing, biological substrates.

Octa(galactose) derivatives of resorcin[4]arenes **46** have been evaluated as host materials using UV titrations in water [73]. Whilst a range of cationic and anionic guests are bound a marked preference is seen for the anionic species, 8-aniline-1-napthalene sulphonate and methyl orange.

Water soluble cavitands (see Chapters 9 and 24) have also been developed for anion binding and functionalisation with hexamethylenetetramine units gives a cavitand showing particularly good affinity for aromatic dicarboxylates [74]. Similar properties result from the introduction of amidinium substituents to a resorcin[4]arene cavitand [75]. Receptor **47** binds both isophthalates and biologically important phosphates. 1:1 binding of phosphates is seen in a D_2O/tris buffer solution with selectivity based on charge; the triphosphate being bound stronger than the monophosphate and adenine based phosphates being bound strongest.

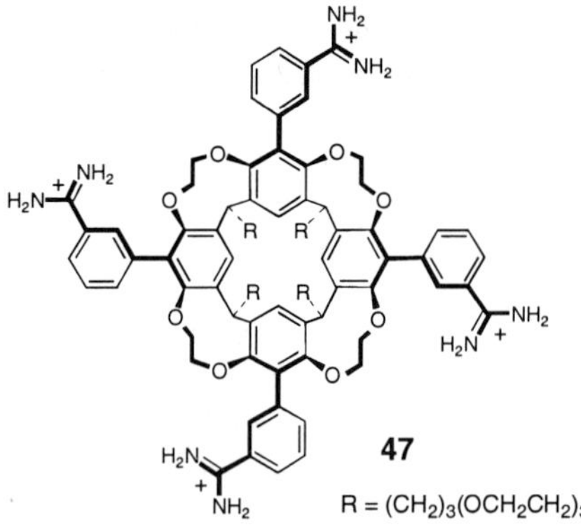

4. Ditopic Calixarene Receptors For Anion Recognition

The simultaneous binding of an anion and its counter cation, ion pair recognition, in a neutral bifunctional receptor utilises the positive charge on the cation to enhance the strength of anion binding and enable solubilisation and extraction of metal salts. This area has recently attracted increasing attention and is of particular interest as it often combines a number of binding strategies in one molecule. Thus, receptors have been prepared that invoke electron deficient centres, hydrogen bonding and the electrostatic attraction of the counter cation to enhance anion binding.

Initial approaches were based on calixarenes, suitably functionalised for cation binding, being further derivatised to allow attachment of a zinc porphyrin for anion binding through either covalent bonding [76] or self assembly [77,78]. More recently, the calixarene framework has been altered to provide two separate binding sites for anions and cations to give true heteroditopic receptors.

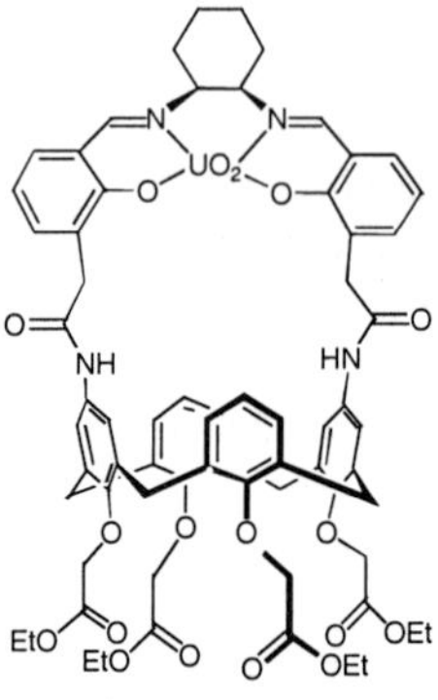

One of the earliest examples of heteroditopic receptors were based on the Lewis acidic salophen system. Two types of receptors **48** and **49** were developed for complexation and transport of NaH_2PO_4 and CsCl respectively [44,79]. In both cases, the anion is bound through a combination of hydrogen bonding and interaction with the uranyl centre, whereas complexation of cations is achieved by ester functionalisation

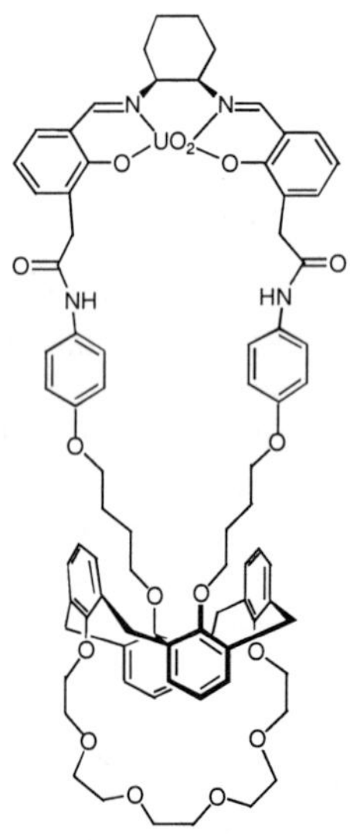

(Na) or incorporation of a crown fragment (Cs) at the lower rim. For **48**, selective dihydrogen phosphate binding over chloride, hydrogen sulphate and perchlorate was demonstrated by ^{1}H NMR studies in (CD$_3$)$_2$SO. The ability to extract ion pairs was confirmed by FAB mass spectral studies showing complexation of both dihydrogen phosphate and sodium. The *1,3-alternate* receptor **49** was found to transport caesium chloride across a supported liquid membrane more efficiently than model compounds containing either the anion or cation binding sites.

In an extension of the bis(calix[4]arene) receptor system, a rhenium (I) bipyridine heteroditopic receptor **50** has been developed [80]. Alkali metal cation binding, by the ester moieties at the lower rim, in a 1:2 host:guest arrangement was demonstrated by ^{1}H NMR studies in CD$_3$CN. Enhanced binding of iodide, in the presence of lithium, sodium and potassium was observed. (K_{ass} iodide 40 M^{-1} **50**, 322 M^{-1} **50** + **Na$^+$**) This was proposed to be due to both rigidification at the lower rim, which alters the topology of the binding site, and increased acidity of the amide protons *via* through bond effects.

The simple organic heteroditopic receptor **51** has also been reported. This incorporates amides for anion binding in combination with benzo-15-crown-5 units for cation complexation [81]. In the presence of potassium or ammonium, effective binding of anions is observed in CD$_3$CN with selectivity for the tetrahedral anions, dihydrogen phosphate and hydrogen sulphate over chloride and nitrate. Thus it was proposed that the formation of 1:1 sandwich complexes by the crown units with potassium and ammonium cations preorganises the anion binding site into a pseudotetrahedral arrangement allowing steric complementarity with tetrahedral anions.

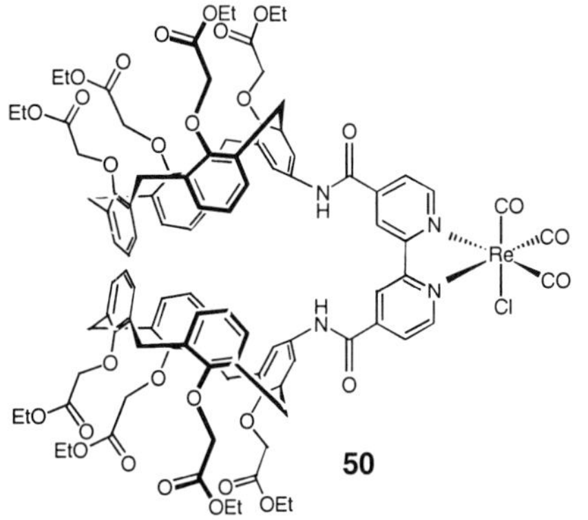

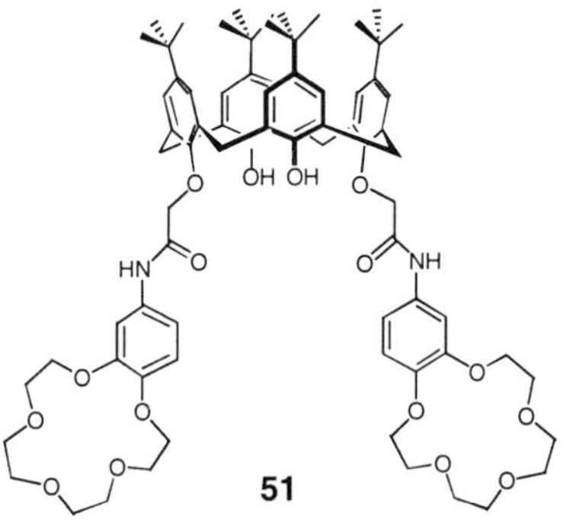

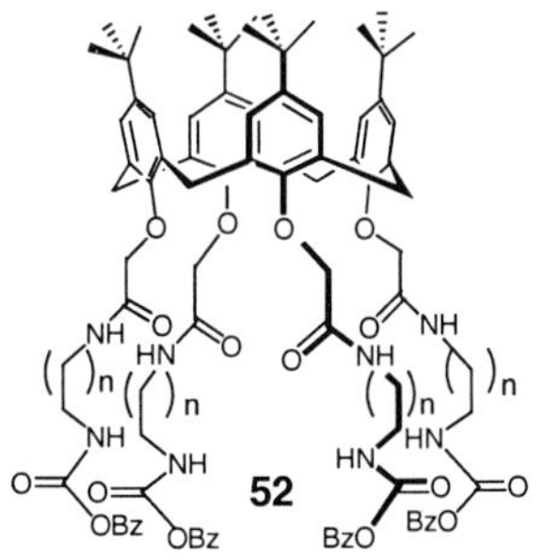

The amide based receptors **25** and **26** (*vide supra*) [49] also incorporate an additional cation binding array at the lower rim. The effect of simultaneous cation binding on the anion recognition properties of these receptors and a novel type of heteroditopic receptor **52** has been investigated. In the presence of sodium, the selectivity of **25** for dianions is abolished and the receptor shows strong binding for spherical halide anions. The complex formed with the halide is 1:1 whereas in the absence of sodium it is of a 1:2 host:guest nature. For the quadridentate ligand **26** anion selectivity is again altered with enhanced bromide binding. This is a consequence of preorganisation of the anion binding site, to favour spherical anions, on binding of the cation. More complex effects are observed for receptor **52** in which both the anion and cation binding site are incorporated on the lower rim. ^{1}H NMR titrations, in the presence and absence of sodium, show anion binding to be enhanced by simultaneous cation complexation. Selectivity for a given anion appears to depend upon the length of spacer between the two sites, chloride and dihydrogen phosphate giving large association constants with **52a,b** and bromide with **52c**.

Urea based ditopic receptors have also been developed [82]. With receptor **53**, anion complexation is not observed on addition of Bu$_4$N$^+$ halide salts in CDCl$_3$, due to strong hydrogen bonding interactions between the ureas forcing the calixarene into a *pinched cone* conformation. Addition of sodium cation and its complexation by the ester functionalities at the lower rim, however, allows the calixarene to adopt a *cone* conformation and disrupts the hydrogen bonding allowing anions to bind. Similarly, the tetraurea derivative **54** is also unsuitable for binding of anions, due to the formation of hydrogen bonded dimers. This dimerisation can be broken by pre-treatment with sodium perchlorate and subsequent significant binding and selectivity for chloride observed.

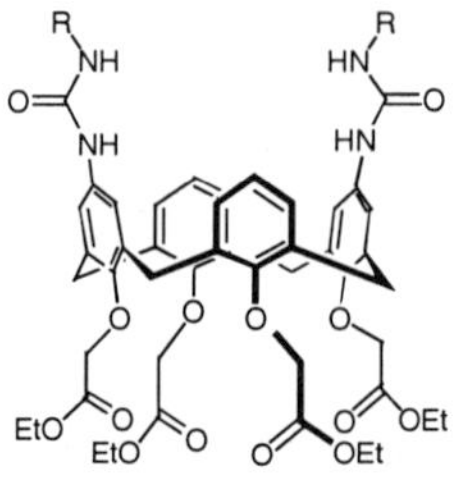

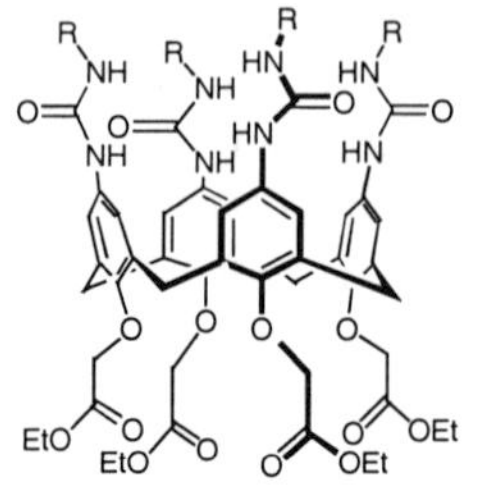

More recently a series of heteroditopic receptors have been prepared in which the upper rim is monofunctionalised with a urea or thiourea whereas the lower rim is modified by amides which are known to complex sodium [83]. Simple complexation studies showed the solubilisation of a range of sodium salts in CHCl$_3$ by both types of receptor. In addition, ^{1}H NMR studies in (CD$_3$)$_2$SO were undertaken on solutions of receptors in the presence or absence of sodium. 1:1 binding was observed for all anions with selectivity for carboxylates. This selectivity was less pronounced than in earlier studies [53] possibly due to a reduced cavity size, through steric repulsion of the lower rim substituents, leading to decreased π/π stabilising interactions. Inclusion of sodium greatly enhanced the binding of anions by receptors where the urea moiety is directly appended to the upper rim due to both rigidification and an increased electron withdrawing effect on the urea NH.

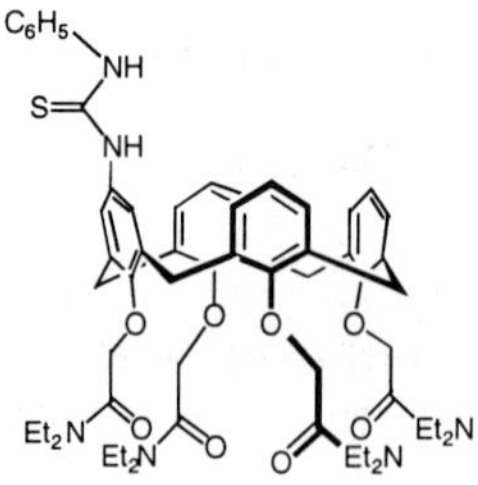

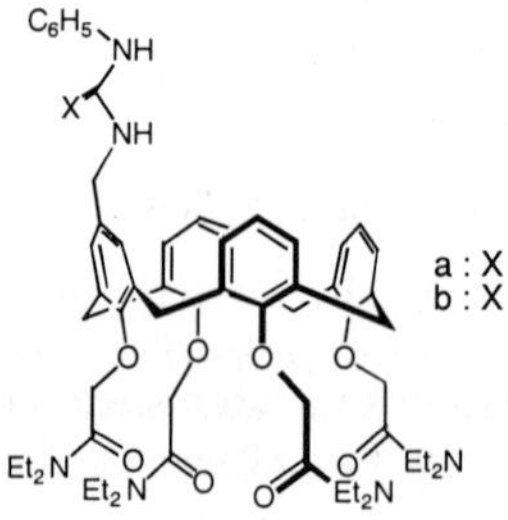

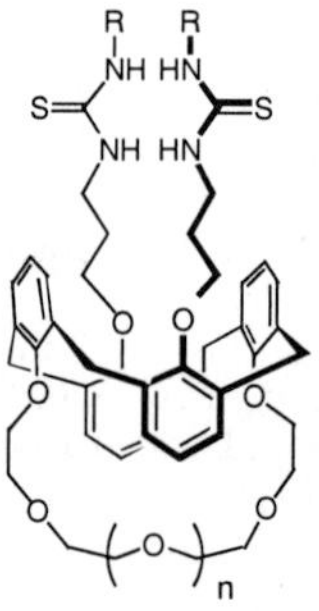

1,3-Alternate calix[4]arenes have also been difunctionalised at the phenolic OH positions with thiourea moieties in the development of ditopic receptors *e.g.* **57** in which a caesium or potassium cation can be complexed by a crown unit [82]. These compounds show increased transport of caesium or potassium chloride in comparison to monotopic anion and cation receptors.

5. Conclusions

The selective complexation of anions by synthetic receptors offers a great challenge to the modern chemist. The last ten years have seen a great expansion in the design of synthetic anion receptors. In particular, this review has shown that calixarenes are excellent platforms for the incorporation of numerous ligands for co-operative binding of anions. Their unique topology offers not only the potential for designed cavities, through selective functionalisation, but also the fine tuning of binding due to the hydrophobic nature of the cavity. The development of calixarene based systems has enabled greater understanding of the complex interactions involved in binding, thus offering the prospect of a number of commercial applications as extractants and sensors.

6. References and Notes

[1] For general reviews on synthetic anion receptors see: a) F. P. Schmidtchen, M Berger, *Chem. Rev.* **1997**, *97*, 1609-1646. b) J Scheerder, J. F. Engbersen, D. N. Reinhoudt, *Recl. Trav. Chim. Pays-Bas.* **1996**, *115*, 307-320. c) M. M. G. Antonisse, D. N. Reinhoudt, *Chem. Commun.* **1998**, 443-448. d) P. D. Beer, D. K. Smith, *Prog. Inorg. Chem.* **1997**, *46*, 1-96.
[2] C. H. Park, H. E. Simmons, *J. Am. Chem. Soc.* **1968**, *90*, 2431-2432.
[3] M. W. Hosseini, J-M. Lehn, *J. Am. Chem. Soc.* **1982**, *104*, 3525-3527.
[4] R. J. Motekaitis, A. E. Matrell, B. Dietrich, J-M. Lehn, *Inorg. Chem.* **1982**, *21*, 4253-4257.
[5] B. Dietrich, D. L. Fyles, T. M. Fyles, J-M. Lehn, *Helv. Chim. Acta* **1979**, *62*, 2763-2787.
[6] F. P. Schmidtchen, *Chem. Ber.* **1980**, *113*, 864-874.
[7] R. A. Pascal, J. Spergel, D.Van Engen, *Tetrahedron Lett.* **1986**, *27*, 4099-4102.
[8] Y. Azuma, M. Newcomb, *Organometallics* **1987**, *6*, 145-150.
[9] H. E. Katz, *J. Am. Chem. Soc.* **1985**, *107*, 1420-1421.
[10] M. E. Jung, H. Xiu, *Tetrahedron Lett.* **1988**, *29*, 297-300.
[11] S. Aoyagi, K. Tanaka, I. Zicmane, Y. Takeuchi, *J. Chem. Soc., Perkin Trans 2* **1992**, 2217-2220.
[12] X. Yang, C. B. Knobler, M. F. Hawthorne, *Angew. Chem., Int. Ed Engl.* **1991**, *30* 1507-1508.
[13] R. Robson, *Aust. J. Chem.* **1970**, 2217-2224.
[14] P. D. Beer, *Chem. Commun.* **1996**, 689-696.
[15] M. Fujita, J. Yazaki, K. Ogura, *J. Am. Chem. Soc.* **1990**, *112*, 5645-5647.
[16] P. J. Stang, D. H. Cao, *J. Am. Chem. Soc.* **1994**, *116*, 4981-4982.
[17] P. D. Beer, M. G. B. Drew, C. Hazlewood, D. Hesek, J. Hodacova, S. E. Stokes, *J. Chem. Soc., Chem. Commun.* **1993**, 229-231.
[18] P. D. Beer, D. Hesek, J. E. Kingston, D. K. Smith, S. E. Stokes, M. G. B. Drew, *Organometallics* **1995**, *14*, 3288-3295.
[19] P. D. Beer, M. G. B. Drew, D. Hesek, K. C. Nam, *Chem. Commun.* **1997**, 107-108.
[20] P. D. Beer, D. Hesek, K. C. Nam, M. G. B. Drew, *Organometallics* **1999**, *18*, 3933-3943.
[21] P. D. Beer, M. G. B. Drew, D. Hesek, M. Shade, F. Szemes, *Chem. Commun.* **1996**, 2161-2162.
[22] P. D. Beer, Z. Cheng, A. J. Goulden, A. Grayon, S. E. Stokes, T. Wear, *J. Chem. Soc., Chem. Commun.* **1993**, 1834-1836.
[23] P. A. Gale, Z. Cheng, M. G. B. Drew, J. A. Heath, P. D. Beer, *Polyhedron* **1998**, *17*, 405-412.
[24] P. D. Beer, M. Shade, *Chem. Commun.* **1997**, 2377-2378.
[25] I. Dumazet, P. D. Beer, *Tetrahedron Lett.* **1999**, *40*, 785-788.
[26] J. L. Atwood, K. T. Holman, J. W. Steed, *Chem. Commun.* **1996**, 1401-1407.

[27]   J. W. Steed, R. K. Juneja, R. S. Burkhalter, J. L. Atwood, *J. Chem. Soc., Chem. Commun.* **1994**, 2205-2206.
[28]   J. W. Steed, R. K. Juneja, J. L. Atwood, *Angew. Chem., Int. Ed. Eng.* **1994**, *33*, 2456-2457.
[29]   M. Staffilani, K. S. B. Hancock, J. W. Steed, K. T. Holman, J. L. Atwood, R. K. Juneja, R. S. Burkhalter, *J. Am. Chem. Soc.* **1997**, *119*, 6324-6335.
[32]   P. D. Beer, C. A. P. Dickson, N. Fletcher, A. J. Goulden, A. Grieve, J. Hodacova, T. Wear, *J. Chem. Soc., Chem. Commun.* **1993**, 828-830.
[33]   P. D. Beer, Z. Cheng, A. J. Goulden, A. Grieve, D. Hesek, F. Szemes, T. Wear, *J. Chem. Soc., Chem. Commun.* **1994**, 1269-1271.
[34]   F. Szemes, D. Hesek, Z. Cheng, S. W. Dent, M. G. B. Drew, A. J. Goulden, A. R. Graydon, A. Grieve, R. J. Mortimer, T. Wear, J. S. Weightman, P. D. Beer, *Inorg. Chem.* **1996**, *35*, 5868-5879.
[35]   P. D. Beer, V. Timeshenko, M. Maestri, P. Passaniti, V. Balzani, *Chem. Commun.* **1999**, 1755-1756.
[36]   P. D. Beer, M. Shade, *Gazz. Chim. Ital.* **1997**, *127*, 651-652.
[37]   W. Xu, J. J. Vittal, R. J. Puddephatt, *J. Am. Chem. Soc.* **1993**, *115*, 6456-6457.
[38]   W. Xu, J. P. Rourke, J. J. Vittal, R. J. Puddephatt, *J. Chem. Soc., Chem. Commun.* **1993**, 145-147.
[39]   W. Xu, J. J. Vittal, R. J. Puddephatt, *J. Am. Chem. Soc.* **1995**, *117*, 8362-8371.
[40]   P. D. Beer, M. G. B. Drew, P.B. Leeson, K. Lyssenko, M. I. Ogden, *J. Chem. Soc., Chem. Commun.* **1995**, 929-930.
[41]   P. Molenveld, J. F. J. Engbersen, D. N. Reinhoudt, *Chem. Soc. Rev.* **2000**, *29*, 75-86.
[42]   D. M. Rudkevich, Z. Brzozka, M. Palys, H. C. Visser, W. Verboom, D. N. Reinhoudt, *Angew. Chem., Int. Ed. Engl.* **1994**, *33*, 467-468.
[43]   D. M. Rudkevich, W. Verboom, Z. Brzozka, M. J. Palys, W. P. R. V. Stauthamer, G. J. van Hummel, S. M. Franken, S. Harkema, J. F. J. Engbersen, D. N. Reinhoudt, *J. Am. Chem. Soc.* **1994**, *116*, 4341-4351.
[44]   D. M. Rudkevich, W. Verboom, D. N. Reinhoudt, *J. Org. Chem.* **1994**, *59*, 3683-3686.
[45]   M. T. Blanda, M. A. Herren, *Chem. Commun.* **2000**, 343-344.
[46]   Y. Morzherin, D. M. Rudkevich, W. Verboom, D. N. Reinhoudt, *J. Org. Chem.* **1993**, *58*, 7602-7605.
[47]   P. D. Beer, P. A. Gale, D. Hesek, *Tetrahedron Lett.* **1995**, *36*, 767-770.
[48]   B. R. Cameron, S. J. Loeb, *Chem. Commun.* **1997**, 573-574.
[49]   I. Stibor, D. S. M. Hafeed, P. Lhotak, J. Hodacova, J. Koca, M. Cajan, *Gazz. Chim. Ital.* **1997**, *127*, 673-685.
[50]   J. Scheerder, M. Fochi, J. F. J. Engbersen, D. N. Reinhoudt, *J. Org. Chem.* **1994**, *59*, 7815-7820.
[51]   N. A. McDonald, E. M. Duffy, W. L. Jorgensen, *J. Am. Chem. Soc.* **1998**, *120*, 5104-5111.
[52]   K. C. Nam, S. O. Kang, H. S. Jeong, S. Jeon, *Tetrahedron Lett.* **1999**, *40*, 7343-7346.
[53]   A. Casnati, M. Fochi, P. Minari, A. Pochini, M. Reggiani, R. Ungaro, D. N. Reinhoudt, *Gazz. Chim. Ital.* **1996**, *126*, 99-106.
[54]   J. Scheerder, J. F. G. Engbersen, A. Casnati, R. Ungaro, D. N. Reinhoudt, *J. Org. Chem.* **1995**, *60*, 6448-6454.
[55]   H. Boerrigter, L. Grave, J. W. M. Nissink, L. A. J. Christoffels, J. H. van de Maas, W. Verboom, F. de Jong, D. N. Reinhoudt, *J. Org. Chem.* **1998**, *63*, 4174-4180.
[56]   N. Pelizzi, A. Casnati, R. Ungaro, *Chem. Commun.* **1998**, 2607-2608.
[57]   T. Rojsajjakul, S. Veravong, G. Tumcharern, R. Seangprasertkij-Magee, T. Tuntulani, *Tetrahedron* **1997**, *53*, 4669-4680.
[58]   E. M. Georgiev, N. Wolf, D. M. Roundhill, *Polyhedron* **1997**, *16*, 1581-1584.
[59]   N. J. Wolf, E. M. Georgiev, A. T. Yordanov, B. R. Whittlesey, H. R. Koch, D. M. Roundhill, *Polyhedron* **1999**, *18*, 885-896.
[60]   P. D. Beer, M. G. B. Drew, K. Gradwell, *J. Chem. Soc., Perkin Trans. 2* **2000**, 511-519.
[61]   P. A. Gale, J. L. Sessler, V. Kral, V. Lynch, *J. Am. Chem. Soc.* **1996**, *118*, 5140-5141.
[62]   W. P. van Hoorn, W. L. Jorgensen, *J. Org. Chem.* **1999**, *64*, 7439-7444.
[63]   P. A. Gale, J. L. Sessler, W. E. Allen, N. A. Tvermoes, V. Lynch, *Chem. Commun.* **1997**, 665-666.
[64]   P. Anzenbacher Jr, K. Juriskova, V. M. Lynch, P. A. Gale, J. L. Sessler, *J. Am. Chem. Soc.* **1999**, *121*, 11020-11021.
[65]   W. Sato, H. Miyaji, J. L. Sessler, *Tetrahedron Lett.* **2000**, 41, 6731-6736.
[66]   J. L. Sessler, A. Gebauer, P. A. Gale, *Gazz. Chim. Ital.* **1997**, *127*, 723-726.
[67]   H. Miyaji, P. Anzenbacher Jr, J. L. Sessler, E. R. Bleasdale, P. A. Gale, *Chem. Commun.* **1999**, 1723-1724.
[68]   P. A. Gale, L. J. Twyman, C. I. Handlin, J. L. Sessler, *Chem. Commun.* **1999**, 1851-1852.
[69]   H. Miyaji, W. Sato, J. L. Sessler, *Angew. Chem., Int. Ed.* **2000**, *39*, 1777-1780.
[70]   J. L. Sessler, P. A. Gale, J. W. Genge, *Chem. Eur. J.* **1998**, *4*, 1095-1099.

[71]   G. Cafeo, F. H. Kohnke, G. L. La Torre, A. J. P. White, D. J. Williams, *Chem. Commun.* **2000**, 1207-1208.
[72]   A. Dondoni, A. Marra, M.-C. Scherrmann, A. Casnati, F. Sansone, R. Ungaro, *Chem. Eur. J.* **1997**, *3*, 1774-1782.
[73]   T. Fujimoto, C. Shimizu, O. Hayashida, Y. Aoyama, *Gazz. Chim. Ital.* **1997**, *127*, 749-752.
[74]   D-R. Ahn, T. W. Kim, J-I. Hong, *Tetrahedon Lett.* **1999**, *40*, 6045-6048.
[75]   L. Sebo, F. Diederich, V. Gramlich, *Helv. Chim. Acta* **2000**, *83*, 93-113.
[76]   T. Nagasaki, H. Fujishima, M. Takeuchi, S. Shinkai, *J. Chem. Soc., Perkin Trans. 1* **1996**, 1883-1888.
[77]   D. M. Rudkevich, A. N. Shivanyuk, Z. Brzozka, W. Verboom, D. N. Reinhoudt, *Angew. Chem., Int. Ed. Engl.* **1995**, *34*, 2124-2126.
[78]   A. N. Shivanyuk, D. M. Rudkevich, D. N. Reinhoudt, *Tetrahedron Lett.* **1996**, *37*, 9341-9344.
[79]   D. M. Rudkevich, J. D. Mercer-Chalmers, W. Verboom, R. Ungaro, F. de Jong, D. N. Reinhoudt, *J. Am. Chem. Soc.* **1995**, *117*, 6124-6125.
[80]   P. D. Beer, J. B. Cooper, *Chem. Commun.* **1998**, 129-130.
[81]   P. D. Beer, M. G. B. Drew, R. J. Knubley, M. I. Ogden, *J. Chem. Soc., Dalton Trans.* **1995** 3117-3123.
[82]   J. Scheerder, J. P. M. van Duynhoven, J. F. G. Engbersen, D. N. Reinhoudt, *Angew. Chem., Int. Ed. Engl.* **1996**, *35*, 1090-1093.
[83]   N. Pelizzi, A. Casnati, A. Friggeri, R. Ungaro, *J. Chem. Soc., Perkin Trans. 2* **1998**, 1307-1311.
[84]   L. A. J. Chrisstoffels, F. de Jong, D. N. Reinhoudt, S. Sivelli, L. Gazzola, A. Casnati, R. Ungaro, *J. Am. Chem. Soc.* **1999**, *121*, 10142-10151.

Chapter 24

WATER-SOLUBLE CALIXARENES

ALESSANDRO CASNATI,[a] DOMENICO SCIOTTO,[b] GIUSEPPE ARENA[b]

[a] *Dipartimento di Chimica Organica e Industriale, Università degli Studi di Parma, Parco Area delle Scienze 17/A, I-43100 Parma, Italy.*
e-mail: casnati@unipr.it
[b] *Dipartimento di Scienze Chimiche, Università degli Studi di Catania, Viale A. Doria 8, I-95125 Catania, Italy. e-mail: dsciotto@dipchi.unict.it, garena@dipchi.unict.it.*

1. Introduction

One of the main reasons for studying water-soluble calixarenes is to exploit hydrophobic effects in order to enhance complexation of apolar guests in their cavity. Extensive studies of cyclodextrins [1] and water-soluble cyclophanes [2] have shown that hydrophobic effects are mainly responsible for the inclusion of apolar molecules in the cavity of these macrocycles. The use of relatively simple synthetic receptors facilitates the study of weak interactions (*e.g.* cation-π [3], π-π [4], CH-π [5]), which are increasingly seen to be important in molecular recognition phenomena in water. Since water is the solvent where most biological processes take place, synthetic water-soluble receptors can be used in mimicry of natural processes such as specific recognition of bioactive molecules (antigens, microbial/viral pathogens, nucleotides) or enzymatic transformation of substrates.

2. Synthesis and Properties

The solubility of calixarenes and resorcarenes in water is very limited, though the latter do dissolve under basic conditions, where four out of eight hydroxy groups are deprotonated [6,7]. To engender water solubility generally, calixarenes and resorcarenes need to be functionalised with groups containing positive or negative charges, or with neutral but highly hydrophilic moieties.

2.1. CHARGED WATER-SOLUBLE RECEPTORS

The alkali metal salts of *cone* **1a** were the first examples of water-soluble calixarenes reported in the literature [8]. However, their solubility is limited (up to 5×10^{-3} M), as is that (up to 10^{-3} M) of salts of the carboxylatocalix[n]arenes (**2**) [9]. Phosphonic acid groups on the upper rim of calix[n]arenes can also lead to compounds soluble under basic conditions [10]. Aminocalixarenes **3** yield water-soluble macrocycles when protonated at sufficiently low pH. Quaternary ammonium substituents, as in conformationally mobile **3b** [11,12], *cone* (**4a**) and *1,3-alternate* (**4b**) [13] provide solubility independent of pH.

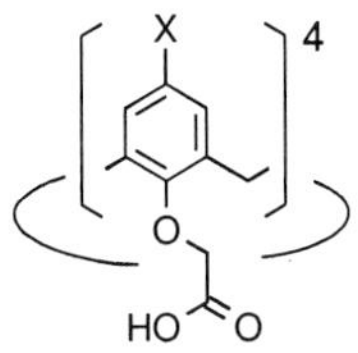

1a: X = *t*-Bu
1b: X = H

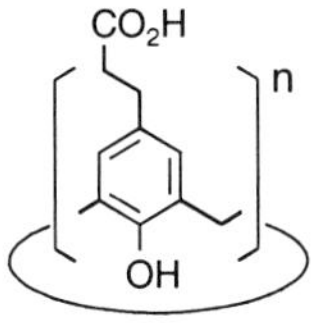

2: n = 4-8

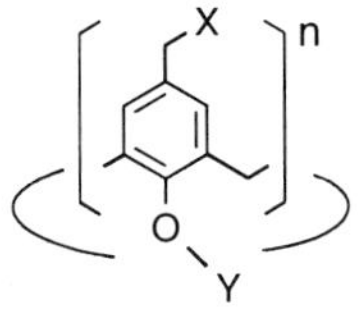

3a: X = N(allyl)$_2$, Y = H, n = 4-8
3b: X = $^+$NMe$_3$, Y = Me, n = 4

The greatest solubility for calixarenes results when upper rim sulfonate groups are introduced following synthetic procedures [14,15] now usefully modified [16]. *p*-Sulfonatocalixarenes **5** are characterised by a remarkable solubility in water that is nearly independent of the pH. Compound **5a** undergoes rapid inversion between *cone* conformers in water, though the inversion rate constant can be greatly diminished by guest inclusion [17].

Tetramethoxy *p*-sulfonatocalix[4]arene **5e** and *p*-trimethylammoniomethyl calix[4]arene **3b** are also conformationally mobile but contrary to calculations which predict an increase of polarity along the series *1,3-alternate* < *partial cone* < *cone*, it was observed that the amount of *1,3-alternate* conformer increases when increasing the percentage of water in a methanol-water solution. The *1,3-alternate* is the conformation where hydrophobic interactions are minimised and where the charged (sulfonate or ammonium) moieties are directed toward the bulk of the solvent, giving rise to a sort of unimolecular micelle [11]. The *1,3-alternate* conformation is found to be the most stable one also in the solid state [18].

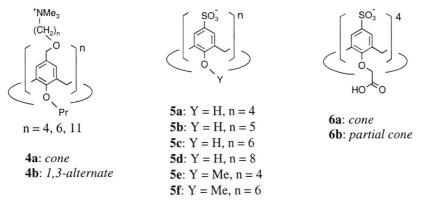

n = 4, 6, 11

4a: *cone*
4b: *1,3-alternate*

5a: Y = H, n = 4
5b: Y = H, n = 5
5c: Y = H, n = 6
5d: Y = H, n = 8
5e: Y = Me, n = 4
5f: Y = Me, n = 6

6a: *cone*
6b: *partial cone*

With larger alkoxy groups at the lower rim, it is possible to stop the conformational interconversion. Thus, isomers fixed in the *cone* **6a** [19] and *partial cone* **6b** [20] conformations by the presence of four carboxymethyl groups were synthesised. These isomers show different acidity of the carboxylic residues and different complexing properties (*vide infra*), indicating a strict stereochemical control of the recogniton properties.

The inclusion properties of water-soluble calixarenes are strongly affected by their aggregation, which depends on the number of aromatic subunits, the type of hydrophilic head groups, the presence of lipophilic chains and the conformation. For example, tetraammonium derivatives fixed in the *cone* or *1,3-alternate* structure form different kind of aggregates. The isomers **4a**, having a *cone*-shaped hydrophobic surface, form globular micelles, while the cylindrical hydrophobic surfaces of *1,3-alternate* compounds **4b** aggregate in a two-dimensional lamella which induces the formation of vesicles [13]. In order to distinguish between inclusion by the host or by its aggregate, it is important to operate below the critical micelle concentration (CMC).

In the solid state, water-soluble calixarenes show quite novel behaviour, due to their amphiphilic character and to their ability to complex ions and neutral molecules. The shape of calix[4]arenes, as well as their amphiphilic character, direct the packing of the molecules in the crystal lattice. The pentasodium salt of **5a** forms a bilayered structure where organic and aqueous layers alternate in a clay-like structure [21]. Solid complexes of pentasodium salt of **5a** with different amounts of lanthanide ion and pyridine N-oxide contain different self-organised aggregates ranging from spherical to helical tubular structures [22] (See Chapter 30). Some X-ray crystal structures of *p*-sulfonatocalixarenes **5** show [23,24] that a water molecule inside the calixarene cavity is hydrogen-bonded to the π cloud of the aromatic nuclei [25], a *motif* which may be commonly encountered in biological systems (*e.g.* the hydrophobic pockets of proteins).

7a: Y = H
7b: Y = OH
7c: Y = Me

8a: R = (CH$_2$)$_3$OPO$_3$H
8b: R = (CH$_2$)$_2$Ph, X = OH
8c: R = CH$_3$, X = OH

9a: R = CH$_3$
9b: R = *n*-C$_5$H$_{11}$
9c: R = *n*-C$_{11}$H$_{23}$

10: R = CH$_3$

Sulfonate groups have also been used to render resorcarenes water-soluble. Sodium 2-formylethane-1-sulfonate condenses with resorcinol or pyrogallol to give resorcarenes **7**, having 2-ethyl sulfonates at the so-called "feet" (carbon bridges between aromatics) of the macrocycle and able to complex amino acids [26], sugars [27], and nucleosides [28]. Methylene-bridged resorcarenes have been successfully solubilised in water either by the functionalisation with 3-propyl phosphate groups at their feet (**8a**) [29] or by introduction of four hydroxy groups at the upper rim (**8b,c**) [30] followed by deprotonation. Recently, cationic cavitands **9** and **10** were obtained by reaction of bromomethyl resorcarenes with pyridine [31] or hexamethylenetetramine [32], respectively (See Chapter 9).

The synthesis of water-soluble macrocycles related to calixarenes and possessing *endo*- [33-36] or *exo*- [37] OH groups has been reported. Unlike *p*-sulfonatocalixarenes **5**, they are synthesised by direct condensation of formaldehyde with salts of the appropriate sulfonated aromatic units, namely chromotropic acid, hydroxynaphthalene sulfonic acid and 5-sulfonatotropolone (See Chapter 1). Proof of their structure by X-ray crystallography is so far lacking, though they appear to be versatile inclusion agents (*vide infra*).

Chiral water-soluble calix[4]arenes were recently synthesised by introducing amino acids at the upper [38,39] or lower [40] rim. Resorcarenes bearing L- or D-proline are soluble at neutral pH [41].

2.2. NEUTRAL WATER-SOLUBLE RECEPTORS

The use of neutral water-soluble receptors is particularly advantageous for *in vivo* applications. Highly charged molecules cross cell membranes with difficulty and charged hosts must repel guests of the same charge. Thus, the use of neutral, hydrophilic substituents has been proposed as an alternative means of rendering calixarenes water-soluble. Alcohols or polyols are by far the most used functional groups.

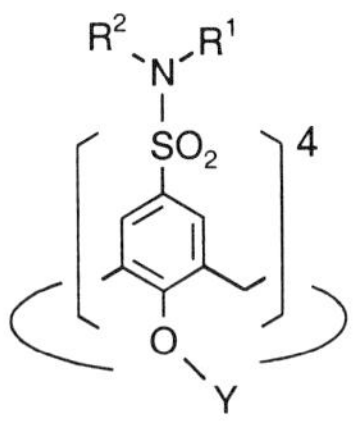

11: $R^1 = R^2 = CH_2CH_2OH$, $Y = H$
12: $Y = CH_2CH_2OCH_2CH_3$
or $Y = CH_2CO_2CH_2CH_3$
a: $R^1 = H$, $R^2 = CH_2CH_2OH$
b: $R^1 = R^2 = CH_2CH_2OH$
c: $R^1 = H$, $R^2 = C(CH_2OH)_3$

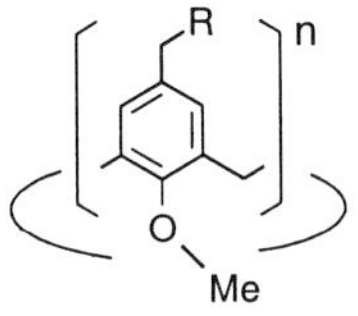

$R = C[CONHC(CH_2OH)_3]$
13a: n = 4
13b: n = 8

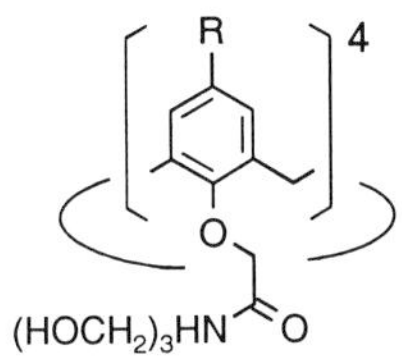

14a: R = H
14b: R = *t*-Bu

Coupling of diethanolamine with *p*-chlorosulfonylcalix[4]arene, for example, gave the octaol **11** sufficiently soluble for the acidity constants of the phenolic units to be determined [42], while the highly water-soluble 36- (**13a**) and 72-silvanol (**13b**) derivatives of calix[4]- and calix[8]arene, respectively, were obtained by introducing three tris(hydroxymethyl)aminomethane (TRIS) moieties onto each aromatic nucleus. In aqueous solution, transmission electron microscopy (TEM) images of **13a** showed the presence of small spheres of a diameter indicating the aggregation of six macrocycles [43].

In a systematic study aimed at determining the number of alcoholic functions per aromatic nucleus necessary to solubilise a calixarene in water a series of tetramers fixed in the *cone* structure and bearing four (**12a**), eight (**12b**) or twelve (**12c**) hydroxy groups were prepared [44,45]. Water solubility increases by nearly two orders of magnitude as every four additional hydroxy groups are introduced, reaching a maximum of 2×10^{-1} M for **12c**. Introduction of TRIS moieties at the lower rim of calix[4]arenes fixed in the *cone* structure also gives high water solubility (0.01 M for **14a**) without hindering the lipophilic cavity with bulky groups and leaves the *para* positions free for the introduction of binding groups which may cooperate with the calixarene cavity in molecular recognition processes. However, the presence of *tert*-butyl groups at the upper rim (**14b**) remarkably reduces the solubility to less than 10^{-5} M [46].

Resorcarene-based cavitands (See Chapter 9) have also been functionalised with TRIS though the compound **15a** with four of these moieties at the feet of the macrocycle is not soluble in water [47]. As already observed for similar systems [48], treatment of the polyol-amide **15a** with HCl provides a rearrangement to the tetrahydrochloride of ammonium resorcarene **15b**, which is highly water-soluble [47]. High water solubility was also obtained by linking carbohydrates to the hydroxy groups or to the feet of resorcarenes [49-51], whilst glycosylated calixarenes show limited solubility [52-54]. The molecular recognition properties of these hosts are reported in detail in Chapter 27.

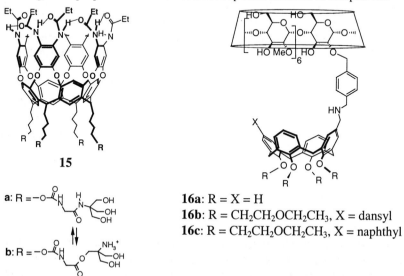

Coupling calix[4]arenes and cyclodextrins (CD) by, for example, linking the primary side of β-cyclodextrin to one of the carboxyl groups of compound **1b** [55], or by connecting the secondary face with the upper rim like in **16** [56-58], also affords water solubility. Compounds **16b** and **16c** give rise to unusual aggregates. The combination of information given by several techniques such as light scattering, TEM and fluorescence indicates that **16b** and **16c** form vesicular bilayers and fibres, respectively. The different aggregation is due to a different behaviour of the dansyl or naphthyl fluorophores. Self-inclusion of the dansyl groups occurs in compound **16b** (Figure 1a), while the naphthyl moieties of **16c** insert into the cavity of a neighbouring cyclodextrin unit (Figure 1b).

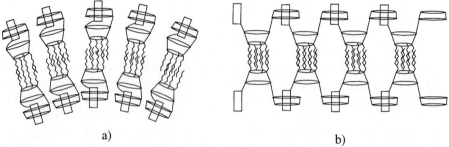

*Figure 1. Schematic representation of the aggregation originating (a) vesicular bilayers and (b) fibers, given by **16b** and **16c**, respectively.*

2.3. ACID-BASE PROPERTIES

Determination of pK_a values of the phenolic OH groups of water-soluble calixarenes presents several difficulties due to the frequent presence of salt impurities in the samples. For the first time, the determination of pK_a values of p-sulfonatocalix[4]arene **5a** [59], revealed the presence of a very acidic OH group and a correlation between the dissociation of the hydroxyl groups and the conformational mobility of the macrocycle. The unusual acidity of this proton is not a peculiar characteristic of p-sulfonatocalix[4]arene since water-soluble p-(4-trimethylammoniophenylazo)calix[4]arene has also low pK_a values [60]. More precise measurements (Table 1) on the salt-free, neutral water-soluble calix[4]arene **11** and its acyclic analogue **17** [42] showed that the first deprotonation of **11** takes place at very low pH in comparison with that of **17**.

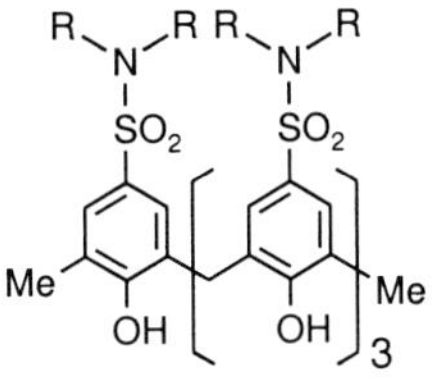

17
R = CH₂CH₂OH

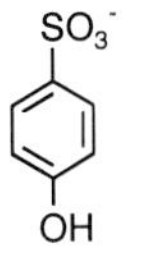

18

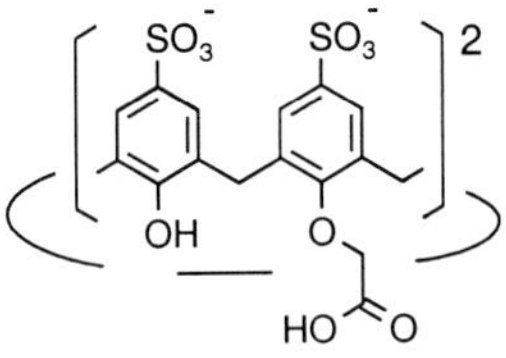

19

Semiempirical calculations indicate that the monoanion is optimally stabilised by hydrogen bond formation with the two adjacent hydroxy groups, which are in turn stabilised by a bifurcated hydrogen bond with the opposite hydroxy groups [61]. The remarkable pK_{a1} difference between calix[4]arenes and their acyclic analogues is ascribed solely to the formation of strong intramolecular hydrogen bonds. After some discussion in the early nineties [16,62,63] questioning the exact pKa values of the OH groups of p-sulfonatocalix[4]arene **5a**, there does now appear to be agreement. The first phenolic OH group of **5a**, although not having the characteristic of a *super-acidic* proton, is still much more acidic (pK_a = 3.34) [62,64,65] than that of the analogous acyclic compound **17** [42] as well as that of the monomeric hydroxybenzenesulfonate **18** (Table 1). No

TABLE 1. $pK_a^§$ values for water-soluble calix[n]arenes.

Compound	pK_{a1}	pK_{a2}	pK_{a3}	pK_{a4}	Ref.
5a	3.34	11.5			[64]
5a	3.26	11.8			[65]
11	1.8	9.7	~ 12.5	>14	[42]
17	4.71	8.27	11.62		[42]
5b	4.31	7.63	10.96		[66]
5c	3.45	5.02	>11		[16]
5c	3.37	4.99			[67]
5c	3.44	4.76			[68]
5d	4.10	4.84			[16]
19	3.00†	3.66†	6.58	6.86	[69]
6a	3.03†	3.27†	3.97†	4.57†	[20]
6b	2.71†	3.35†	4.61†	5.11†	[20]
18	8.62				[70]

§All pK_a values refer to the dissociation of phenolic OH groups, unless otherwise stated. †pK_a values referring to the dissociation of carboxylic acid groups.

other titratable protons were detected in the pH range 2.5-11; the second hydroxy group was found to dissociate in the basic region ($pK_a = 11.5$).

$\Delta H°$ and $\Delta S°$ values for each of the two protonation steps [62,64] revealed the remarkably exothermic value for the equilibrium $H_2L^{6-} + H^+ \rightleftarrows H_3L^{5-}$ ($\Delta H° = -25.9$ kJ mol^{-1}; $pK_a = 11.5$), which supports the postulate of very strong *circular* hydrogen bonding resulting from a favourable charge delocalisation of the type indicated for **B** (Figure 2). The enthalpy value ($\Delta H° = 2.6$ kJ mol^{-1}; $pK_a = 3.34$) for the equilibrium $H_3L^{5-} + H^+ \rightleftarrows H_4L^{4-}$ is an indication of a less favourable proton-oxygen interaction in the species having all hydroxy groups protonated (**A** in Figure 2).

In conclusion, both the species with fully protonated hydroxy groups (**A**) and the mono oxy-anion (**B**) can hydrogen-bond but intramolecular hydrogen bonds are stronger for the mono oxy-anion owing to the presence of a negative charge. The acid-base characteristics of *p*-sulfonatocalix[5]arene **5b** (Table 1) are quite different from those of *p*-sulfonatocalix[4]arene **5a** [66]. The pentamer, even though more acidic than the monomer **18**, is less acidic than the tetramer

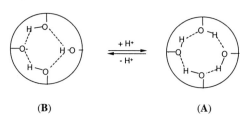

Figure 2. Schematic representation of circular hydrogen bonding in the tetrahydroxycalix[4]arene (A) and in one of the possible resonance structures of its anion (B).

and, in addition, has a proton ($pK_a = 7.63$) having acid-base characteristics that are encountered neither for the tetramer nor for the hexamer. Replicate measurements [16,67,68] have confirmed that the hexamer **5c** possesses only two ionisable OH groups in the pH range 2.5-7 (Table 1) and that the remaining four hydroxy groups are only weakly acidic. Note that *p*-sulfonatocalix[6]arene **5c** possesses two protons that are more acidic than that of the monomeric compound **18**. The $\Delta H°$ and $\Delta S°$ values associated with the two dissociation steps [67] indicate that both the bis- and the mono-oxy anions are stabilised by hydrogen bonding and that again the unusually low detected pK_a values (especially for the first pK_a) are connected with a remarkable hydrogen bonding resulting from a favourable charge delocalisation. pK_a values for *p*-sulfonatocalix[8]arene **5d** (Table 1) [16] seem to indicate that the two tetrameric units originating from the *pinched* conformation [71] are fairly independent from one another.

Very few data can be found for the acid-base properties of chemically modified calixarenes in water. To the best of our knowledge, pK_a values have only been determined for *p*-sulfonatocalix[4]arenes bearing two distal [69] or four [20] carboxymethyl groups at the lower rim. *p*-Sulfonatocalix[4]arene-1,3-dicarboxylic acid **19** was shown to have four ionisable protons (Table 1) in the pH range 2.5-11 [69]. The first two pK_a values refer to the dissociation of the carboxylic groups, whereas the last two refer to the dissociation of the hydroxy groups. As found for the parent *p*-sulfonatocalix[4]arene **5a**, for this ligand the dissociation of the phenolic hydroxyls takes place at significantly lower pH values than that of the monomer **18**. Enthalpy and entropy values for the dissociation process are consistent with the pK_a lowering of the hydroxy groups being the result of microsolvation effects, *i.e.* a water molecule bridging the two phenolate oxygens [69,72]. The acid-base characteristics of **6a** and **6b** were determined together with the $\Delta H°$ and $\Delta S°$ values for the four protonation steps of each ligand [20]. The *cone* and *partial cone* isomers were found to have significantly different pK_a values (Table 1). This is a nice example of the remarkable influence of stereochemistry on the acid-base properties of conformational isomers. Both the ^{1}H NMR and calorimetric results strongly support the presence of intramolecular hydrogen bonding in the *cone* isomer, which has all four carboxylate groups on the same side; on the contrary, for the *partial cone* isomer no evidence for the occurrence of such bonding was obtained.

3. Complexation of Neutral Molecules

Most of the work concerning the complexation of neutral molecules by water-soluble calixarenes was carried out in the eighties and has been already critically reviewed [71,73,74]. The pioneering work [9] on the complexation of aromatic hydrocarbons by hosts **2** and **3a** has however to be mentioned, since it disclosed a rough correlation between binding constants and host-guest complementarity. Calix[4]arenes are too small to host durene or naphthalene, calix[5]- or -[6]arenes prefer naphthalene, anthracene and phenanthrene, while calix[7]- and [8]arenes are selective for pyrene, the association constants [75] being in the range 10^2 - 1.5×10^4 M.

Cyclotetrachromotropylene shows selectivity for chrysene and pyrene, thus indicating that its cavity is larger than that of calix[4]arene [35,76a]. Monosubstituted benzenes C_6H_5X (X = H, OMe, CH_2OH, CHO, NO_2) are complexed with binding constants ranging from 64 to 700 M^{-1}. The reasons for the observed selectivity for nitrobenzene and anisole are not entirely clear, since several factors such as π-π, CH-π or SO_3^--HO interactions seem to play an important role [76b]. Binding constants of cyclotetrachromotropylene with monosubstituted methanes CH_3X (X = OH, COOH, CN, SOMe, COMe) are in the range 2-40 M^{-1}.

Studies of the complexation of nitriles (acetonitrile), ketones (acetone and butanone) and alcohols (ethanol and propanol) using water-soluble calix[4]arenes **5a**, **6a**, **1b** showed that hosts **5a** and **6a** are able to complex all the guests studied (K_{ass} = 16-64 M^{-1}), the former being more selective for ketones and the latter for alcohols. This indicates that there is a charge assistance in the hydrophobic binding of these guests by sulfonated hosts **5a** and **6a**. On the other hand host **1b**, which lacks the sulfonate groups at the upper rim, complexes CH_3CN only, but more efficiently than **5a** and **6a** where negative interactions between the CN dipole and the sulfonate groups may occur. The small methanol molecule is not included by either **5a** nor **6a**, thus suggesting that the binding of ethanol and propanol is due to the synergy of CH-π and SO_3^--HO interactions [77].

The β-cyclodextrin calix[4]arene couples **16b** and **16c** were used to study the binding of neutral analytes by fluorescence spectroscopy. Compound **16c** complexes a series of neutral organic guests such as steroids, terpenes and other natural products, while **16b** is not able to include any of these guests. The sensitivity for steroids, which is approximately 10 times higher than that for terpenes, suggests that the cyclodextrin cavity is mainly involved in the complexation [58]. However, the selectivity for norethindrone over adamantanol observed for **16c** is opposite to that of native β-cyclodextrin and indicates that the calixarene moiety is useful to expand the hydrophobic cavity of the cyclodextrin (CD). CPK models indicate that **16b** self-includes the fluorophore into the CD cavity, likely due to strong intramolecular hydrogen bonds between the sulfonamide group and the secondary hydroxy groups. On the other hand, **16c** gives rise to a less stable, intermolecularly-locked, fibre-like structure. Upon addition of a guest, the fluorophore is displaced and the fibre-like structure evolves into vesicles.

Measurements of the rate constants and the activation parameters for the inclusion of a persistent radical (benzyl *tert*-butyl nitroxide) in a water-soluble calixarene showed that EPR can be a valuable technique for a detailed study of complexation phenomena in these systems. The benzyl residue is selectively bound by **5a**, while the NO group of the radical is exposed to bulk water. Thermodynamic parameters show that the complexation is enthalpy-driven ($\Delta H° = -16.4$ kJ mol^{-1}) and that the low binding constant ($K_{ass} = 12.5$ M^{-1}) is due to a negative entropy term ($\Delta S° = -33.9$ J mol^{-1} K^{-1}) originating from the loss of degrees of freedom of the guest molecule. The difference between $\Delta S^{\neq}$ for association and $\Delta S°$ is small (-13.4 J mol^{-1} K^{-1}), while the corresponding enthalpic difference is large (31.7 kJ mol^{-1}) indicating that, in the transition state, the solvation is closer to that of the complex rather than that of the free host and guest [78].

4. Complexation of Anionic Guests

Very few examples of anion complexation by water-soluble calixarenes have been reported so far. This is probably due to the fact that anion recognition is a rather new field in supramolecular chemistry and, moreover, to the fact that anions are more highly hydrated than cations of comparable size and, therefore, their complexation in water is a remarkably difficult task. In the case of 1-anilino-8-naphthalenesulfonate (ANS) [56,79] and 2-*p*-toluidino-6-naphthalenesulfonate (TNS) [56,80] the lipophilic residue of the guest is included inside the calixarene cavity. Compound **9** complexes *p*-toluenesulfonate and *p*-cresol with K_{ass} of 5.2×10^2 M^{-1} and 1.1×10^2 M^{-1}, respectively. The higher association constant found for *p*-toluene sulfonate indicates that there is a positive effect due to the presence of opposite charges on the host and guest in addition to CH-π interactions.

An even more important synergistic effect between electrostatic and CH-π interactions is displayed by ligand **10**, which shows selectivity for benzene dicarboxylates over monocarboxylates ($K_{ass} = 10^2$-10^4 M^{-1}) [32]. Among 1,3-benzene dicarboxylates, the efficiency significantly increases in the series **20 < 21 < 22**, indicating that CH-π interactions between the methyl or methoxy groups in *meta* to carboxylates and the host cavity may remarkably contribute to binding [32].

A series of π-metalated calix[n]arenes (n = 4, 5) were synthesised, among which compounds **23** are water-soluble due to the presence of six positive charges [81]. The cavities of π-metalated calixarenes are therefore electron-poor and able to complex anions both in the solid state and in water. The X-ray crystal structures of compounds **23a**, **23c** and **23d** show that a BF$_4^-$, SO$_4^{2-}$ and I$^-$ anion is complexed in the calixarene cavity, respectively, tetrafluoroborate being the most deeply included one. Acetate, phosphate and sulfate anions are not bound by host **23b**, due to their high hydrophilicity. An interesting inversion of the selectivity expected on the basis of the order of the free energy of hydration (Hofmeister series) is observed for halide ions (K_{ass} of Cl$^-$ > Br$^-$ > I$^-$), due to size complementarity between the guest and the calixarene cavity [81,82].

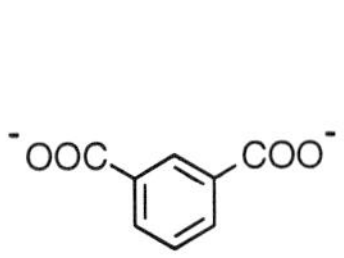

20

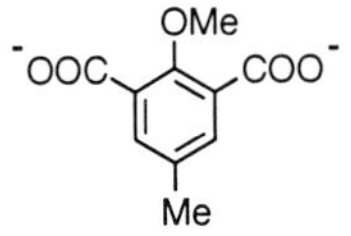

21

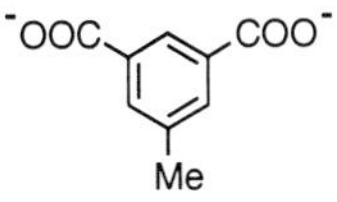

22

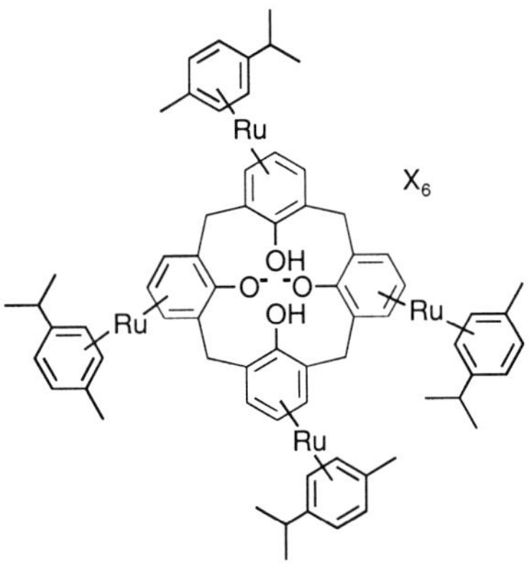

23a: X = BF$_4^-$
23b: X = CF$_3$SO$_3^-$
23c: X$_6$ = 4HSO$_4^-$ + SO$_4^{2-}$
23d: X = I$^-$

5. Complexation of Cationic Guests

5.1. METAL ION COMPLEXATION

Evidence has been obtained for remarkable selectivity in certain instances of metal ion binding by sulfonatocalix[n]arenes [83]. Preferential binding of uranyl ion over transition metal ions has been of major interest and discussion in terms of structural coordination chemistry is contained in Chapter 30. The remarkable solid state structures of various lanthanide ion complexes of sulfonatocalix[4]arene **5a** are also discussed there, while structures of transition metal complexes [84] are discussed in Chapter 28. The coordination chemistry of sulfonatocalixarenes is complicated by their multidentate and ambidentate nature, and a full analysis of their solution behaviour is very difficult. Both *p*-sulfonatocalix[n]arenes (**5a-c**) and their carboxymethoxy derivatives (**24a-c**) form with Cu^{2+}, Zn^{2+}, Ni^{2+} and UO$_2^{2+}$ stable 1:1 complexes in neutral or basic conditions. Log K_{ass} values for UO$_2^{2+}$ with the tetramers are smaller by about 16 units than those for pentamers and hexamers. This was attributed to a coordination-geometry selectivity. *p*-Sulfonatocalix[6]arene **5c**, as well as its carboxymethoxy derivative **24c**, exhibits extremely high selectivity factors ($10^{10.6}$ -10^{17}) for UO$_2^{2+}$ against Cu^{2+}, Zn^{2+} and Ni^{2+}. In solution, complexation occurs *via* the hydroxy or the carboxylate groups only while in the solid state the situation is different [66,84,85].

p-Sulfonatocalix[4]arene **5a** [85] gives a complex with lead(II) having a 2:1 metal:ligand stoichiometry. All lead ions are coordinated both by sulfonate oxygens and water molecules. In the solid state the involvement of the sulfonato groups in cation coordination seems to be a rather general phenomenon, independent of the dimension of the calixarene and the metal ion, while in solution other donor sites, such as phenoxide and carboxylate ions, seem to be preferred.

Further, in the Cu^{2+} complexes of two isomeric derivatives **6a** and **6b** [20] the speciation cannot be simply described in terms of a simple 1:1 stoichiometry. Diprotonated and monoprotonated as well as [CuL]$^{6-}$ species are detected for both ligands; in addition a hydroxo species is also detected for the *partial cone* isomer. For the complexes of the dicarboxylated derivative of *p*-sulfonatocalix[4]arene **19** with Cu^{2+} [86], it is not possible to describe the system in terms of a [CuL]$^{6-}$ species only. Sulfonato or carboxylic groups have been successfully introduced onto the calix[4]arenes-bis-crown-6 backbone [87], and all the investigated derivatives form 1:1 complexes with caesium in neutral and basic media. A few resorcarene based ligands have been reported to selectively complex caesium or strontium, yielding entities large enough to be "cut-off" by nanofiltration membranes [88-90].

Water-soluble neutral lanthanide complexes, such as those of **25** with Tb^{3+} or Eu^{3+}, may be employed for imaging or bioassay purposes [91]. The two calix[4]arenes **25** are functionalised with either a 2-pyridyl disulfide or a chrysene moiety as sensitiser and **25b** forms a highly luminescent Eu^{3+} complex.

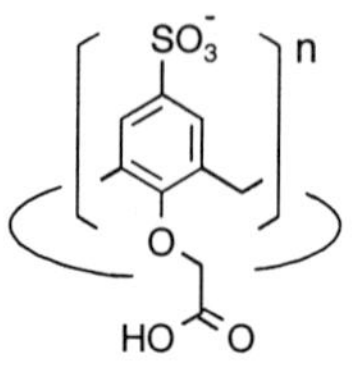

24a: n = 4 (*1,3-alternate*)
24b: n = 5
24c: n = 6

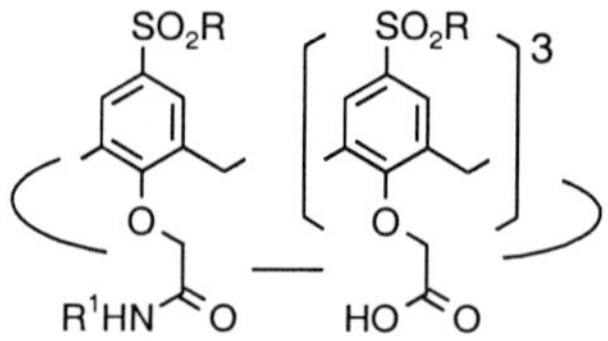

25a: R^1 = CH$_2$CH$_2$-S-S-Pyridine
25b: R^1 = CH$_2$-Chrysene

Sulfonated calixarenes also form interesting supramolecular entities with metal ions complexes. For example, hexamethoxy-*p*-sulfonatocalix[6]arene **5f** gives a strong 1:1 complex in aqueous solution with cobalt(III) sepulchrate [92], its formation being both enthalpically and entropically favoured though the latter is prevailing. The X-ray crystal structure shows that in the solid state the complex has a 2:1 guest:host stoichiometry and that cobalt(III) sepulchrate is not included by the host. Compound **5c** has been shown to form inclusion complexes with a variety of ferrocene derivatives in solution [93,94]. Voltammetric data reveal that, in all cases, the oxidised forms of the ferrocene guests are more tightly bound than the reduced ones; however, as indicated by the inclusion of neutral ferrocene guests, these interactions have to be viewed as an example of non-classical hydrophobic interaction [95]. The redox chemistry of cobaltocenium guests can be used to select the appropriate host between **5c** and β-CD [96]. In addition it was demonstrated that while **5c** binds both ferrocene and cobaltocene guests, its O-methylated analogue of **5f** incorporates neither one of these guests, thus showing the importance of intramolecular hydrogen bonding for the preorganisation and binding of molecular guests [97].

5.2. COMPLEXATION OF QUATERNARY AMMONIUM CATIONS

The complexation of quaternary ammonium cations (Quats) by synthetic receptors has attracted particular attention in the last few years, especially after the discovery that neurotransmitter acetylcholine is bound to acetylcholine esterase thanks to cation-π interactions of the activated methyl groups with some of the 14 aromatic residues present in a narrow pocket of the enzyme with no negative charge in the immediate proximity [98], thus highlighting the crucial role of weak cation-π interactions in the recognition process.

Trimethylanilinium (TMA) is unselectively included in the cavity of the conformationally mobile *p*-sulfonatocalix[4]arene **5a** in aqueous solution [99]. At neutral pH the two complexes of Figure 3 with different orientation of the guest (aromatic nucleus or charged methylammonium group) in the calixarene cavity are present.

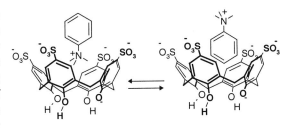

Figure 3. Unselective inclusion of TMA in host 5a.

The ditopic receptor **19**, which is rigidified in the *cone* structure by one water molecule bridging the two opposite phenolate oxygen atoms of the lower rim, is able to incorporate selectively the aromatic portion of TMA (Figure 4a) [72]. Thus host preorganisation seems to play an important role in determining selective guest inclusion. Both **5a** and **5c** encapsulate a series of Quats including choline and acetylcholine and it seems that the binding properties of these receptors result from the synergistic operation of electrostatic interactions and hydrophobic effects [100]. These findings indicate the possibility of inducing selectivity in the recognition of ditopic aromatic methylammonium cations by using conformationally immobilised calix[4]arene receptors.

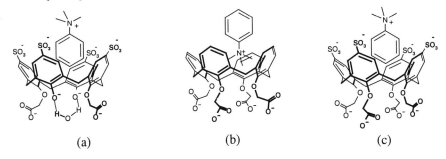

Figure 4. Selective mode of inclusion of TMA by: (a) host 19; (b) host 1b; (c) host 6a.

Hosts **1b** and **6a** are both preorganised in the *cone* structure, but differ by four sulfonate groups. This enables evaluation of the role of the negative charges on the complexation of three ditopic methylammonium cations, namely *N,N,N*-trimethylanilinium (TMA), benzyltrimethyl ammonium (BTMA) and *p*-nitrobenzyltrimethylammonium (BTMAN), at neutral pH [101]. In contrast to the conformationally mobile receptor **5a**, host **1b** specifically binds the trimethylammonium charged group of TMA (Figure 4b), whereas **6a** recognises only the aromatic ring (Figure 4c). The introduction of a spacer between the charged polar group and the aromatic residue (BTMA) or an electron-withdrawing group (BTMAN) does not alter this selectivity for the host **1b**, whereas host **6a** selectively recognise the $N^+(CH_3)_3$ group of BTMAN but unselectively binds BTMA, both by the methylammonium group and the aromatic moiety.

The binding constants show that the inclusion is favoured by the presence of the sulfonate groups, which act as anchoring points for the polar moiety of the guests, regardless the stereochemistry of the inclusion complex. Thermodynamic parameters (Table 2) show that all the inclusion processes are enthalpically favoured and entropi-

TABLE 2. Log K_{ass} values and thermodynamic parameters for the complex formation of TMA, BTMA and BTMAN with hosts **1b** and **6a**; (pH 7.0, at 25°C).

REACTION	log K_{ass}	$\Delta G°$ (kJ mol^{-1})	$\Delta H°$ (kJ mol^{-1})	$\Delta S°$ (J mol^{-1} deg^{-1})
1b + TMA = 1b×TMA	2.2	-12.6	-20.5	-26.7
6a + TMA = 6a ×TMA	3.3	-18.8	-36.4	-59
1b + BTMA = 1b×BTMA	1.7	-10.0	-17.2	-24
6a + BTMA = 6a ×BTMA	3.2	-18.4	-26.8	-28.1
1b + BTMAN = 1b×BTMAN	1.6	-9.2	-20.9	-39
6a + BTMAN = 6a ×BTMAN	3.4	-19.3	-26.8	-25

cally unfavoured, thus indicating that attractive interactions rather than entropically favourable desolvation are the major driving force for the inclusion of these small guests. However, because of this selective yet opposite mode of binding, it was not possible to assess the relative importance of electrostatic or cation-π interactions. The inclusion of the symmetrical tetramethylammonium cation (TEMA) was compared with that of TMA, extending the study to the water-soluble calix[4]arene receptors **19**, **26a-c** and **6b** [102]. Log K_{ass} values for complex formation are reported in Table 3.

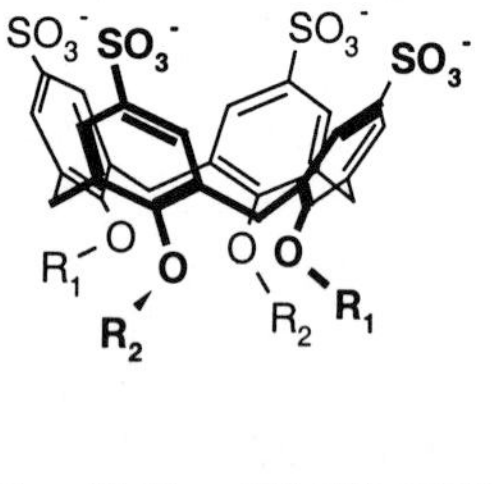

26a: R_1 = H, R_2 = CH$_2$CH$_2$OCH$_2$CH$_3$
26b: R_1 = R_2 = CH$_2$CONMe$_2$
26c: R_1 = R_2 = CH$_2$CH$_2$OCH$_2$CH$_3$

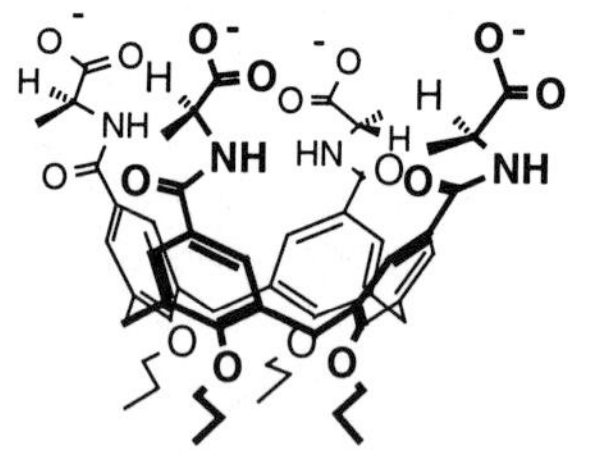

27

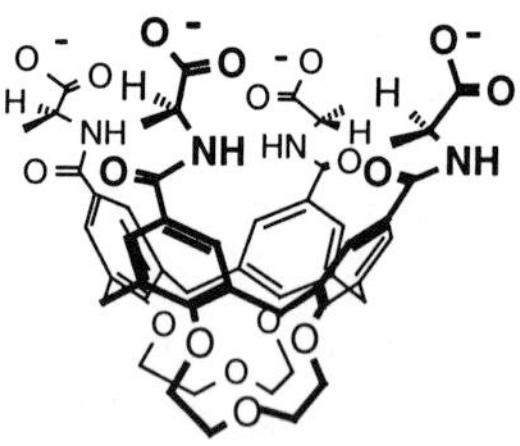

28

These results show how crucial are these subtle conformational and steric effects in determining efficiency and selectivity in the complexation of Quats. In fact the symmetrical and sterically more demanding TEMA is efficiently complexed by all the hosts investigated but not by the somewhat similar host **26c**, which has a narrower cavity. The ditopic guest TMA is selectively complexed *via* the aromatic ring by all the hosts in the *cone* conformation bearing sulfonate groups at the upper rim, whereas the *partial cone* isomer **6b** complexes TMA in an unselective fashion.

Calorimetrically derived $\Delta H°$ and $\Delta S°$ values for **6a**•TEMA and **1b**•TEMA indicate that complexation is enthalpy-driven and thus cation-π interactions are the driving force of the processes. However, these data also highlight that the larger values of K_{ass} found for **6a**•TEMA, in comparison with **1b**•TEMA, are due to a less unfavourable entropic term, which results from a larger desolvation of the system due to the interaction of the TEMA positive charge with the negative charges of the sulfonate groups [102].

TABLE 3. Log K_{ass} values for the complex formation of trimethylanilinium (TMA) and tetramethylammonium (TEMA) with different hosts (pH 7.3 at 25°C).

Host	TMA log K_{ass}	Included moiety	TEMA log K_{ass}
1b	2.2	$N^+(CH_3)_3$	2.4
5a	4.6	Ar or $N^+(CH_3)_3$	4.9
6a	3.4	Ar	3.6
6b	3.1	Ar or $N^+(CH_3)_3$	3.7
19	3.4	Ar	2.6
26a	3.4	Ar	2.6
26b	3.3	Ar	3.6
26c	2.4	Ar	No inclusion

A study of the inclusion of TEMA, TMA, BTMA and BTMAN by the tetrasulfonatoresorcarene **7a** shows the importance of supporting calorimetric data. Specific interactions that are not expressed in ΔG^0 values are nicely revealed by ΔH^0 and ΔS^0 values [103].

The complexation constants obtained with TMA and BTMA in water at neutral pH using the chiral calix[4]arene receptor **27**, bearing four L-alanine units at the upper rim, are significantly lower than those obtained with the tetrasulfonate derivative **6a** [38]. Since this might have been partly due to the residual C_{2v}–C_{2v} conformational mobility [104], the receptor platform was rigidified by synthesising bis-crown-3 **28**. Compound **28** proves to be much more efficient than the *mobile cone* analogue **27**, pointing out the remarkable influence of the degree of the rigidity of the *cone* calixarene platform on the recognition properties. The same receptor **28** shows also interesting inclusion properties towards α-amino acids [38] (See Chapter 27).

N-Methylquinuclidinium and other ammonium ions form complexes with **15b** in slow exchange at 295 K on the 600 MHz ^1H NMR time scale. This rarely encountered kinetic stability is attributed to large conformational changes which take place with host **15b** upon complexation [47].

The inclusion of quaternary ammonium salts in the apolar cavity of calixarenes has also been studied in the solid state. The X-ray crystal structure of the complex **5a**•[NMe$_4$]$_5$•4H$_2$O shows a bound tetramethylammonium ion within the hydrophobic cavity [105]. The overall structure shows that the remaining four tetramethylammonium ions and the water molecules are situated in the region between the calixarene bilayers. In the solid state, choline is included into receptor **5a** via the charged ammonium group whereas the hydroxyl points between the two sulfonates [100]. Multiple cation-π interactions, analogous to those of biological systems, were observed in the solid state between quaternary ammonium ions and the aromatic rings of calixarenes having different size and protonation degree [106-108]. They are also found between the aromatic rings of tetraethylresorc[4]arene and the quaternary trimethylammonium group of acetylcholine [109] or of 3-phenylpropionic acid choline ester [110]. Two head-to-head arranged tetraethylresorcarenes have been shown to be linked by eight water molecules and to incorporate a tetraethylammonium ion within the cavity [111].

Acknowledgements: This work has been partially supported by CNR (M.U.R.S.T.- Chimica Legge 95/95 "Agenti di contrasto, di shift e sonde luminescenti" and Progetto Bilaterale "Nuovi leganti macrociclici per separazioni mediante membrane liquide supportate e cromatografia") and by M.U.R.S.T. ("Supramolecular Devices" Project and Progetto "Sistemi per la rimozione dall'ambiente di specie indesiderate"). We are indebted to Professor R. Ungaro for helpful discussions.

6. References and Notes

[1]   M. V. Rekharsky, Y. Inoue, *Chem. Rev.* **1998**, *98*, 1875-1917.
[2]   F. Diederich, *Cyclophanes*, Royal Society of Chemistry (Ed.: J. F. Stoddart), Cambridge, **1991**.
[3]   J. C. Ma, D. A. Dougherty, *Chem. Rev.* **1997**, *97*, 1303-1324.
[4]   C. A. Hunter, *Chem. Soc. Rev.* **1994**, 101-109.
[5]   M. Nishio, M. Hirota, Y. Umezawa, *The CH-π Interaction*, Wiley-VCH, New York, **1998**.
[6]   H.-J. Schneider, D. Güttes, U. Schneider, *Angew. Chem., Int. Ed. Engl.* **1986**, *25*, 647-649.
[7]   H.-J. Schneider, D. Güttes, U. Schneider, *J. Am. Chem. Soc.* **1988**, *110*, 6449-6454.
[8]   A. Arduini, A. Pochini, S. Reverberi, R. Ungaro, *J. Chem. Soc., Chem. Commun.* **1984**, 981-982.
[9]   C. D. Gutsche, I. Alam, *Tetrahedron* **1988**, 4689-4694.
[10] M. Almi, A. Arduini, A. Casnati, A. Pochini, R. Ungaro, *Tetrahedron* **1989**, *45*, 2177-2182.
[11] T. Nagasaki, K. Sisido, T. Arimura, S. Shinkai, *Tetrahedron* **1992**, *48*, 797-804.
[12] S. Shimizu, K. Kito, Y. Sasaki, C. Hirai, *Chem. Commun.* **1997**, 1629-1630.
[13] S. Arimori, T. Nagasaki, S. Shinkai, *J. Chem. Soc., Perkin Trans. 2* **1995**, 679-683.
[14] S. Shinkai, T. Tsubaki, T. Sone, O. Manabe, *Tetrahedron Lett.* **1984**, *25*, 5315-5318.
[15] S. Shinkai, K. Araki, T. Tsubaki, T. Arimura, O. Manabe, *J. Chem. Soc., Perkin Trans. 1* **1987**, 2297-2299.
[16] J. P. Scharff, M. Mahjoubi, *New J. Chem.* **1991**, *15*, 883-887.
[17] S. Shinkai, K. Araki, M. Kubota, T. Arimura, T. Matsuda, *J. Org. Chem.* **1991**, *56*, 295-300.
[18] J. L. Atwood, S. G. Bott, in *Calixarenes, a Versatile Class of Macrocyclic Compounds*, Kluwer Academic Publishers (Eds.: J. Vicens, V. Böhmer), Dordrecht, **1991**, pp. 199-210.
[19] A. Casnati, Y. Ting, D. Berti, M. Fabbi, A. Pochini, R. Ungaro, D. Sciotto, G. G. Lombardo, *Tetrahedron* **1993**, *49*, 9815-9822.
[20] G. Arena, R. P. Bonomo, R. Cali, F. G. Gulino, G. G. Lombardo, D. Sciotto, R. Ungaro, A. Casnati, *Supramol. Chem.* **1995**, *4*, 287-295.
[21] A. W. Coleman, S. G. Bott, S. D. Morley, C. D. Means, K. D. Robinson, H. Zhang, J. L. Atwood, *Angew. Chem., Int. Ed. Engl.* **1988**, *27*, 1361-1362.
[22] G. W. Orr, L. J. Barbour, J. L. Atwood, *Science* **1999**, *285*, 1049-1052.
[23] P. C. Leverd, P. Berthault, M. Lance, M. Nierlich, *Eur. J. Org.* **2000**, 133-139.
[24] A. Drljaca, M. J. Hardie, C. L. Raston, *J. Chem. Soc., Dalton Trans.* **1999**, 3639-3642.
[25] J. L. Atwood, F. Hamada, K. D. Robinson, G. W. Orr, R. L. Vincent, *Nature* **1991**, *349*, 683-684.
[26] K. Kobayashi, M. Tominaga, Y. Asakawa, Y. Aoyama, *Tetrahedron Lett.* **1993**, *34*, 5121-5124.
[27] K. Kobayashi, Y. Asakawa, Y. Aoyama, *Supramol. Chem.* **1993**, *2*, 133-135.
[28] K. Kobayashi, Y. Asakawa, Y. Kato, Y. Aoyama, *J. Am. Chem. Soc.* **1992**, *114*, 10307-10313.
[29] A. R. Mezo, J. C. Sherman, *J. Org. Chem.* **1998**, *63*, 6824-6829.
[30] J. R. Fraser, B. Borecka, J. Trotter, J. C. Sherman, *J. Org. Chem.* **1995**, *60*, 1207-1213.
[31] M. H. B. Grote Gansey, F. K. G. Bakker, M. C. Feiters, H. P. M. Geurts, W. Verboom, D. N. Reinhoudt, *Tetrahedron Lett.* **1998**, *39*, 5447-5450.
[32] D. R. Ahn, T. W. Kim, J. I. Hong, *Tetrahedron Lett.* **1999**, *40*, 6045-6048.
[33] B. L. Poh, C. S. Yue, *Tetrahedron* **1999**, *55*, 5515-5518.
[34] B. L. Poh, C. S. Lim, K. S. Khoo, *Tetrahedron Lett.* **1989**, *30*, 1005-1008.
[35] B. L. Poh, L.-S. Koay, *Tetrahedron Lett.* **1990**, *31*, 1911-1914.
[36] B. L. Poh, Y. Y. Ng, *Tetrahedron* **1997**, *53*, 8635-8642.
[37] P. E. Georghiou, Z. P. Li, M. Ashram, *J. Org. Chem.* **1998**, *63*, 3748-3752.
[38] F. Sansone, S. Barboso, A. Casnati, D. Sciotto, R. Ungaro, *Tetrahedron Lett.* **1999**, *40*, 4741-4744.

[39] F. Sansone, S. Barboso, A. Casnati, M. Fabbi, A. Pochini, F. Ugozzoli, R. Ungaro, *Eur. J. Org. Chem.* **1998**, 897-905.
[40] M. S. Peña, Y. L. Zhang, S. J. Thibodeaux, M. L. McLaughlin, A. M. de la Peña, I. M. Warner, *Tetrahedron Lett.* **1996**, *37*, 5841-5844.
[41] Y. Matsushita, T. Matsui, *Tetrahedron Lett.* **1993**, *34*, 7433-7436.
[42] S. Shinkai, K. Araki, P. D. J. Grootenhuis, D. N. Reinhoudt, *J. Chem. Soc., Perkin Trans. 2* **1991**, 1883-1886.
[43] G. R. Newkome, Y. Hu, M. J. Saunders, F. R. Fronczek, *Tetrahedron Lett.* **1991**, *32*, 1133-1136.
[44] M. H. B. Grote Gansey, W. Verboom, D. N. Reinhoudt, *Tetrahedron Lett.* **1994**, *35*, 7127-7130.
[45] M. H. B. Grote Gansey, F. J. Steemers, W. Verboom, D. N. Reinhoudt, *Synthesis* **1997**, 643-648.
[46] M. Segura, F. Sansone, A. Casnati, R. Ungaro, *submitted for publication*.
[47] a) T. Haino, D. M. Rudkevich, J. Rebek Jr., *J. Am. Chem. Soc.* **1999**, *121*, 11253-11254; b) T. Haino, D. M. Rudkevich, A. Shivanyuk, K. Rissanen, J. Rebek Jr., *Chem. Eur. J.* **2000**, *6*, 3797-3805.
[48] P. L. Anelli, M. Brocchetta, S. Canipari, P. Losi, G. Manfredi, C. Tomba, G. Zecchi, *Gazz. Chim. Ital.* **1997**, *127*, 135-142.
[49] T. Fujimoto, C. Shimizu, O. Hayashida, Y. Aoyama, *Gazz. Chim. Ital.* **1997**, *127*, 749-752.
[50] O. Hayashida, C. Shimizu, T. Fujimoto, Y. Aoyama, *Chem. Lett.* **1998**, 13-14.
[51] T. Fujimoto, C. Shimizu, O. Hayashida, Y. Aoyama, *J. Am. Chem. Soc.* **1997**, *119*, 6676-6677.
[52] A. Marra, A. Dondoni, F. Sansone, *J. Org. Chem.* **1996**, *61*, 5155-5158.
[53] A. Marra, M. C. Scherrmann, A. Dondoni, A. Casnati, P. Minari, R. Ungaro, *Angew. Chem., Int. Ed. Engl.* **1994**, *33*, 2479-2481.
[54] A. Dondoni, A. Marra, M. C. Scherrmann, A. Casnati, F. Sansone, R. Ungaro, *Chem. Eur. J.* **1997**, *3*, 1774-1782.
[55] F. D'Alessandro, F. G. Gulino, G. Impellizzeri, G. Pappalardo, E. Rizzarelli, D. Sciotto, G. Vecchio, *Tetrahedron Lett.* **1994**, *35*, 629-632.
[56] E. van Dienst, B. H. M. Snellink, I. von Piekartz, J. F. J. Engbersen, D. N. Reinhoudt, *J. Chem. Soc., Chem. Commun.* **1995**, 1151-1152.
[57] J. Bügler, N. A. J. M. Sommerdijk, A. J. W. G. Visser, A. van Hoek, R. J. M. Nolte, J. F. J. Engbersen, D. N. Reinhoudt, *J. Am. Chem. Soc.* **1999**, *121*, 28-33.
[58] J. Bügler, J. F. J. Engbersen, D. N. Reinhoudt, *J. Org. Chem.* **1998**, *63*, 5339-5344.
[59] S. Shinkai, K. Araki, H. Koreishi, T. Tsubaki, O. Manabe, *Chem. Lett.* **1986**, 1351-1354.
[60] S. Shinkai, K. Araki, J. Shibata, D. Tsugawa, O. Manabe, *Chem. Lett.* **1989**, 931-934.
[61] P. D. J. Grootenhuis, P. A. Kollman, L. C. Groenen, D. N. Reinhoudt, G. J. van Hummel, F. Ugozzoli, G. D. Andreetti, *J. Am. Chem. Soc.* **1990**, *112*, 4165-4176.
[62] G. Arena, R. Cali, G. G. Lombardo, E. Rizzarelli, D. Sciotto, R. Ungaro, A. Casnati, *Proceedings of the XII National Meeting on Calorimetry and Thermal Analysis* **1990**, Bari (Italy), p. 147.
[63] J. P. Scharff, M. Mahjoubi, R. Perrin, *C. R. Acad. Sci.* **1990**, *t. 311, Series II*, 73-77.
[64] G. Arena, R. Cali, G. G. Lombardo, E. Rizzarelli, D. Sciotto, R. Ungaro, A. Casnati, *Supramol. Chem.* **1992**, *1*, 19-24.
[65] I. Yoshida, N. Yamamoto, F. Sagara, D. Ishii, K. Ueno, S. Shinkai, *Bull. Chem. Soc. Jpn.* **1992**, *65*, 1012-1015.
[66] J. W. Steed, C. P. Johnson, C. L. Barnes, R. K. Juneja, J. L. Atwood, S. Reilly, R. L. Hollis, P. H. Smith, D. L. Clark, *J. Am. Chem. Soc.* **1995**, *117*, 11426-11433.
[67] G. Arena, A. Contino, G. G. Lombardo, D. Sciotto, *Thermochim. Acta* **1995**, *264*, 1-11.
[68] J. L. Atwood, D. L. Clark, R. K. Juneja, G. W. Orr, K. D. Robinson, R. L. Vicent, *J. Am. Chem. Soc.* **1992**, *114*, 7558-7559.
[69] G. Arena, R. P. Bonomo, A. Contino, F. G. Gulino, A. Magri, D. Sciotto, *J. Incl. Phenom.* **1997**, *29*, 347-363.
[70] A. E. Martell, R. M. Smith, *Critical Stability Constants*, Plenum Press, New York, **1977**.
[71] C. D. Gutsche, *Calixarenes*, The Royal Society of Chemistry (Ed.: J. F. Stoddart), Cambridge, **1989**.
[72] G. Arena, A. Casnati, L. Mirone, D. Sciotto, R. Ungaro, *Tetrahedron Lett.* **1997**, *38*, 1999-2002.
[73] S. Shinkai, in *Calixarenes, a Versatile Class of Macrocyclic Compounds*, Kluwer Academic Publishers (Eds.: J. Vicens, V. Böhmer), Dordrecht, **1991**, pp. 173-198.
[74] A. Pochini, R. Ungaro, in *Comprehensive Supramolecular Chemistry, Vol. 2*, Pergamon Press (Ed.: F. Vögtle), Oxford, **1996**, pp. 103-142.
[75] F. Diederich, K. Dick, *J. Am. Chem. Soc.* **1984**, *106*, 8024-8036.

[76]   a) B. L. Poh, C. S. Lim, L.-S. Koay, *Tetrahedron* **1990**, *46,* 6155-6160; b). B. L. Poh, C. M. Tan, *J. Incl. Phenom.* **1994**, *18,* 93-99.
[77]   G. Arena, A. Contino, F. G. Gulino, A. Magri, D. Sciotto, R. Ungaro, *Tetrahedron Lett.* **2000**, *41,* 9327-9330..
[78]   P. Franchi, M. Lucarini, G. F. Pedulli, D. Sciotto, *Angew. Chem., Int. Ed.* **2000**, *39,* 263-266.
[79]   Y. Shi, Z. Zhang, *J. Chem. Soc., Chem. Commun.* **1994**, 375-376.
[80]   Y. Shi, Z. Zhang, *J. Incl. Phenom.* **1994**, *18,* 137-147.
[81]   M. Staffilani, K. S. B. Hancock, J. W. Steed, K. T. Holman, J. L. Atwood, R. K. Juneja, R. S. Burkhalter, *J. Am. Chem. Soc.* **1997**, *119,* 6324-6335.
[82]   J. W. Steed, R. K. Juneja, J. L. Atwood, *Angew. Chem., Int. Ed. Engl.* **1994**, *33,* 2456-2457.
[83]   S. Shinkai, H. Koreishi, K. Ueda, T. Arimura, O. Manabe, *J. Am. Chem. Soc.* **1987**, *109,* 6371-6376.
[84]   C. P. Johnson, J. L. Atwood, J. W. Steed, C. B. Bauer, R. D. Rogers, *Inorg. Chem.* **1996**, *35,* 2602-2610.
[85]   A. T. Yordanov, O. A. Gansow, M. W. Brechbiel, L. M. Rogers, R. D. Rogers, *Polyhedron* **1999**, *18,* 1055-1059.
[86]   F. Hajek, E. Graf, M. W. Hosseini, X. Delaigue, A. De Cain, J. Fischer, *Tetrahedron Lett.* **1996**, *37,* 1401-1404.
[87]   L. Nicod, S. Pellet-Rostaing, F. Chitry, M. Lemaire, *Tetrahedron Lett.* **1998**, *39,* 9443-9446.
[88]   E. Gaubert, H. Barnier, L. Nicod, R. A. Favre, J. Foos, A. Guy, C. Bardot, M. Lemaire, *Sep. Sci. Technol.* **1997**, *32,* 585-597.
[89]   L. Nicod, F. Chitry, E. Gaubert, M. Lemaire, H. Barnier, *J. Incl. Phenom.* **1999**, *34,* 141-151.
[90]   S. Pellet-Rostaing, L. Nicod, F. Chitry, M. Lemaire, *Tetrahedron Lett.* **1999**, *40,* 8793-8796.
[91]   F. J. Steemers, H. G. Meuris, W. Verboom, D. N. Reinhoudt, E. B. vanderTol, J. W. Verhoeven, *J. Org. Chem.* **1997**, *62,* 4229-4235.
[92]   R. Castro, L. A. Godinez, C. M. Criss, S. G. Bott, A. E. Kaifer, *Chem. Commun.* **1997**, 935-936.
[93]   L. Zhang, A. Macias, T. Lu, J. I. Gordon, G. W. Gokel, A. E. Kaifer, *J. Chem. Soc., Chem. Commun.* **1993**, 1017-1019.
[94]   L. Zhang, A. Macias, R. Isnin, T. Lu, G. W. Gokel, A. E. Kaifer, *J. Incl. Phenom.* **1994**, *19,* 361-370.
[95]   D. B. Smithrud, T. B. Wyman, F. Diederich, *J. Am. Chem. Soc.* **1991**, *113,* 5420-5426.
[96]   Y. Wang, J. Alvarez, A. E. Kaifer, *Chem. Commun.* **1998**, 1457-1458.
[97]   J. Alvarez, Y. Wang, M. GomezKaifer, A. E. Kaifer, *Chem. Commun.* **1998**, 1455-1456.
[98]   J. L. Sussmann, M. Harel, F. Frolow, C. Gefner, A. Goldman, L. Toker, I. Silman, *Science* **1991**, *253,* 872-879.
[99]   S. Shinkai, K. Araki, T. Matsuda, N. Nishiyama, H. Ikeda, L. Takasu, M. Iwamoto, *J. Am. Chem. Soc.* **1990**, *112,* 9053-9058.
[100] J. M. Lehn, R. Meric, J. P. Vigneron, M. Cesario, J. Guilhem, C. Pascard, Z. Asfari, J. Vicens, *Supramol. Chem.* **1995**, *5,* 97-103.
[101] G. Arena, A. Casnati, A. Contino, G. G. Lombardo, D. Sciotto, R. Ungaro, *Chem. Eur. J.* **1999**, *5,* 738-744.
[102] G. Arena, A. Casnati, A. Contino, F. G. Gulino, D. Sciotto, R. Ungaro, *J. Chem. Soc., Perkin Trans.* 2 **2000**, 419-423.
[103] G. Arena, A. Contino, T. Fujimoto, D. Sciotto, Y. Aoyama, *Supramol. Chem.* **2000**, *11,* 279-288.
[104] A. Arduini, M. Fabbi, M. Mantovani, L. Mirone, A. Pochini, A. Secchi, R. Ungaro, *J. Org. Chem.* **1995**, *60,* 1454-1457.
[105] J. L. Atwood, L. J. Barbour, P. C. Junk, G. W. Orr, *Supramol. Chem.* **1995**, *5,* 105-108.
[106] J. M. Harrowfield, M. I. Ogden, W. R. Richmond, B. W. Skelton, A. H. White, *J. Chem. Soc., Perkin Trans.* 2 **1993**, 2183-2190.
[107] J. M. Harrowfield, W. R. Richmond, A. N. Sobolev, A. H. White, *J. Chem. Soc., Perkin Trans.* 2 **1994**, 5-9.
[108] J. M. Harrowfield, W. R. Richmond, A. N. Sobolev, *J. Incl. Phenom.* **1994**, *19,* 257-276.
[109] K. Murayama, K. Aoki, *Chem. Commun.* **1997**, 119-120.
[110] K. Murayama, K. Aoki, *Chem. Lett.* **1998**, 301-302.
[111] K. Murayama, K. Aoki, *Chem. Commun.* **1998**, 607-608.

Chapter 25

RECOGNITION OF NEUTRAL MOLECULES

ARTURO ARDUINI,[a] ANDREA POCHINI,[a] ANDREA SECCHI,[a]
FRANCO UGOZZOLI[b]

[a]*Dipartimento di Chimica Organica e Industriale, Università di Parma,
Parco Area delle Scienze 17/A, I-43100 Parma, Italy.
e-mail: pochini@unipr.it*
[b]*Dipartimento di Chimica Generale ed Inorganica Chimica Analitica
Chimica Fisica, Università di Parma and
Centro di Studio per la Strutturistica Diffrattometrica del CNR
Parco Area delle Scienze 17/A, I-43100 Parma, Italy.*

1. Introduction

Molecules having convergent concave surfaces are frequently potent receptors and provide useful models for natural systems such as the binding sites of enzymes [1,2]. Using its cavity as binding site, a receptor (host) can perform the recognition of a substrate (guest) on the base of the structural complementarity of the two molecular species. The strength of binding is controlled by the kinds and number of simultaneous interactions, by the preorganisation of the host and by solvation effects. The complexation of organic guests in lipophilic solvents generally requires solvents which, because of their molecular size, cannot enter the cavity [3,4]. The kinetics of the complexation-decomplexation processes determines the field of application of these systems. Fast exchange in host-guest recognition, for example, can give receptors useful for sensor devices as well as in molecular separations (see Chapters 34 and 36).

It is well known that calix[4]arenes, particularly when in their *cone* conformation, possess an intramolecular cavity which can host neutral guest molecules of complementary size [5; see also Chapter 16]. Larger calix[n]arenes are also known to form inclusion complexes but because of the usually greater conformational flexibility of these macrocycles and the relatively limited range of studies so far made of them (see, however, Chapters 3 - 5), the factors controlling their inclusion selectivity are not well understood.

The first class of synthetic receptors having concave surfaces, which was studied as receptors for neutral molecules both in apolar solution and in the solid state, was that derived from calixarene[4]resorcinols [3]. In contrast, the class of calix[4]arenes showed good inclusion properties, in particular with aromatic guests, in the solid state but not in solution [5,6]. Only recently was it verified that in these media the calixarene cavity can host neutral organic molecules of complementary size. It was also established that the efficiency of the recognition process is strongly determined by the rigidity of the hosts and by the nature of the guest, which should bear acid CH groups (see ahead).

On this basis it has been hypothesised that specific CH-π interactions stabilise the complexes formed [7]. This chapter is mainly focussed on recent developments in the understanding of recognition of these neutral organic molecules by calix[4]arene receptors both in apolar media and in the solid state. In relation to solid state structures, a broad overview is presented of X-ray crystallographic determinations of inclusion compounds with guests having acid CH groups. A combined analysis using crystallographic and nuclear magnetic resonance measurements relating in particular to solid inclusion complexes of *p-t*-butyl-calix[4]arene is given in Chapter 16.

2. Recognition Processes in Apolar Media

Unlike cavitands derived from calix[4]resorcinols, tetrahydroxycalix[4]arenes show very weak interactions with neutral organic guests in apolar media [5,6]. However, since, in solution, the complexation energy gain derives from the balance of the enthalpy of the weak specific host-guest interactions (which take place also in the solid state) and desolvation processes, and also from the entropy balance of the whole recognition process, it was considered that the complexes might be stabilised by greater pre-organisation of the hosts. The simple procedure of locking the *cone* conformation by introduction of large alkyl substituents on the narrow rim, however, is insufficient for this purpose, seemingly because a limited degree of conformational flexibility is retained. Most tetra-O-alkylated calix[4]arene *cone* isomers adopt a *flattened cone* (*pinched cone*) [8] conformation of at most C_{2v} symmetry in the solid state [9], and this conformation has also been shown to be the most stable by molecular modeling [10], but in solution, the ^{1}H NMR spectrum usually reflects an effective C_{4v} symmetry as a result of a fast interchange between two C_{2v} structures (Figure 1).

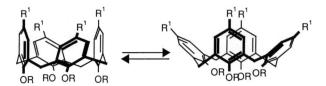

Figure 1. Dynamic stereochemistry of tetraalkoxycalix[4]arenes in the cone conformation.

Direct experimental evidence of this residual conformational flexibility was obtained [11] by studying the temperature dependence of the ^{1}H NMR spectrum of *cone* tetrakis(n-octyloxy)calix[4]arene which, in CD_2Cl_2, indicates C_{4v} symmetry at room temperature. On decreasing the temperature to 213 K, its spectrum changes, indicating a C_{2v} structure. These changes clearly show the freezing of the molecular motion into a preferred *flattened cone* conformation. Thus, it is possible that the lack of complexing efficiency could be partially ascribed to this residual mobility of these hosts.

Molecular modelling suggests that a possible means of rigidifying and further functionalising calix[4]arene is by the linkage, at the upper rim of the calix, of the distal aromatic nuclei with rigid bridges containing π-donor groups. Thus, new cavitands (*e.g.* **1**) were designed to improve the complexation ability by introducing more efficient additional binding sites using pyridine containing bridges. By comparing the data obtained with **1** with those of its *iso*-phthaloyl analogue, it was possible to investigate the effect of the pyridine basic group located in close proximity to the calixarene cavity on complexation [12]. It was found that while **1** binds acetonitrile, nitromethane and malononitrile guests in CCl$_4$ with K$_{ass}$ = 36(10), 57(5), and 79(20) M^{-1} respectively, in parallel to the acidity of the guest, its *iso*-phthalic analogue, where the nitrogen atom is not present, shows no complexation [12].

Other approaches to binding enhancement have involved selective 1,2 (proximal) narrow rim functionalisation [13] and, recently, a one-step synthesis of a series of calix[4]arene biscrowns (**2-6**) (Figure 2) where the conformational flexibility can be tuned by varying the length of the bridging unit [14].

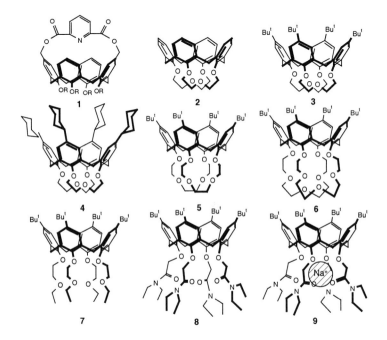

Figure 2. Calix[4]arene hosts having different rigidity.

In fact, while *p-t*-butylcalix[4]arene-biscrown-5 **6** is flexible and adopts a *flattened cone* conformation in the solid state [15], calix[4]arene-biscrown-3 derivatives (*e.g.* **4**) possess a rigid *cone* structure, though of course it cannot be of higher than C$_{2v}$ symmetry. The recognition properties of rigidified *p-t*-butylcalix[4]arene-*bi*scrown-3 **3** in solution were evaluated in apolar organic solvents (CDCl$_3$, CCl$_4$) using CH$_3$NO$_2$ as guest [16]. By adding variable amounts of guest to a solution of **3** a significant upfield shift of the CH protons of the guest and fast exchange conditions were observed. These shifts clearly show an interaction of the acidic protons of the guest with the π electrons of the calixarene cavity, while excluding a possible interaction of these protons with the crown-ether region, which should result in a downfield shift.

The analysis of the data showed the formation of 1:1 complexes and provided quantitative information on the host-guest interactions (see Table 1). To gain further insight into the role of rigidity on the molecular recognition properties of calix[4]arene receptors, comparison was made with the more mobile tetrakis-(2-ethoxyethoxy)-*p-t*-butylcalix[4]arene **7** and *p-t*-butylcalix[4]arene-biscrown-5 **6**. Interestingly, with both hosts no variation of the chemical shift of the guest was observed, while with **5** complexation did occur but only in CCl$_4$ and with a lower binding constant than with the more rigid **3** (see Table 1).

These results demonstrate the importance of rigidity in determining the complexation properties of the π donor cavity of calix[4]arenes. The groups present at the upper rim of calix[4]arene-biscrown-3 strongly affect the K$_{ass}$ values. In fact, increasing the extension of the cavity increases the complexation efficiency toward nitromethane, as verified by comparing the K$_{ass}$ values for **2**, **3** and **4**. In contrast, malononitrile is bound with comparable efficiency by **2**, **3**, and **4**, probably for steric reasons. As expected, K$_{ass}$ values are greater in the less polar CCl$_4$, where even **5** is able to include nitromethane.

TABLE 1. Association constants (K$_{ass}$, M^{-1}) of 1:1 complexes of nitromethane and malononitrile with rigidified calix[4]arenes at 300 K.

Calix[4]arene	CH$_3$NO$_2$		CH$_2$(CN)$_2$
	CDCl$_3$	CCl$_4$	CDCl$_3$
2	5 ± 2	28 ± 7	17 ± 2
3	27 ± 4	230 ± 60	6 ± 2
4	36 ± 8	123 ± 25	23 ± 5
5	a	50 ± 10	a
9	34 ± 7	a	a

^{a} not determined.

Another interesting observation, which confirms the importance of rigidity and, consequently, of the preorganization of the *cone* conformer of calix[4]arenes, was obtained by the comparison between *p-t*-butylcalix[4]arene tetraamide **8** and its sodium picrate complex **9** [17] in the complexation of nitromethane in CDCl$_3$. In fact, while the conformationally mobile **8** does not show any significant complexation, its rigid sodium complex **9**, through an allosteric effect, binds CH$_3$NO$_2$ (see Table 1).

Further confirmation came from research which showed the possibility of utilising as efficient hosts, for guests bearing acid CH, calix[4]arenes partially alkylated at the lower rim, *e.g.* 1,3-dialkoxy derivatives (Figure 3) [18]. Using carbon tetrachloride

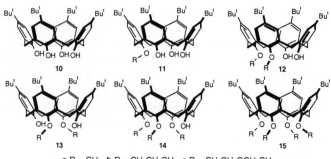

Figure 3. *Alkoxy derivatives of p-t-butylcalix[4]arene.*

as solvent and, *e.g.*, acetonitrile as guest, no complexation was observed with the tetra-alkoxy derivative of *p-t*-butylcalix[4]arene **15**, whereas association constant values at 298 K of 39 M^{-1} for tetrahydroxy derivative **10**, rising to 157 M^{-1} with the 1,3-diethoxy-ethoxy **13c** and to 232 M^{-1} with the monoethoxyethoxy **11c** derivative, were observed (Table 2) [18]. These data indicate that complexation efficiency increases when the calixarene host bears both alkoxy and hydroxy groups, thus suggesting a key role of hydrogen bonding in the host rigidification.

TABLE 2. Association constants (K_{ass}, M^{-1}) of the 1:1 complexes of calix[4]arenes **10**, **11c**, and **13a-c** with neutral guests in CCl_4 at 298 K.

Calix[4]arenes	CH_3CN	$ClCH_2CN$	C_2H_5CN	CH_3NO_2	$C_2H_5NO_2$
10	39	27	18	52	21
11c	232	109	61	284	7.5
13c	157	-	19	129	19
13a	80	65	14	76	-
13b	152	94	20	132	20

Experimental measures of the rigidity of these structure are not known so far, but an indirect explanation could be found in the studies on the conformational distribution and interconversion of lower rim partially alkylated calix[4]arenes [19]. In particular, the partially methylated derivatives **11a-13a** are far less mobile than either the tetrahydroxy **10** or the tetramethoxycalix[4]arene **15a**. The 1,3-dimethoxy ether **13a**, for example, adopts a *cone* conformation both in solution and in the solid state and does not show any sign of coalescence in the NMR spectrum at temperatures up to 125 °C. The order of barriers calculated for the *cone-inverted cone* interconversion is monomethoxy **11a** > 1,2-dimethoxy **12a** > 1,3-dimethoxy **13a** > trimethoxy derivative **14a** (Figure 3) [19].

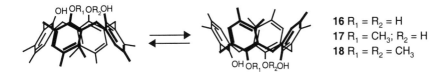

Figure 4. Cone-to-cone ring inversion of dimethylcalix[4]arene derivatives.

Further experimental data on the energy barriers come from studies on the racemization of partially methylated chiral calix[4]arenes (Figure 4). Calix[4]arenes (*e.g.* **16**) bearing on the macrocycle *meta*-substituted phenols such as 3,4-dimethylphenol are chiral in their *cone* conformation and their *cone-to-cone* inversion converts one enantiomer into its mirror image. Thus, investigations of this thermal racemization by measurement of rate constants as a function of temperature enabled the experimental determination of the energy barriers for the chiral derivatives, which were verified to be **17** > **18** > **16** [20], in agreement with the theoretical calculations [19]. These data suggest that the factors which prevent the *cone*-to-*cone* ring inversion probably also operate in the rigidification of the *cone* structure.

Further analysis of the data summarised in Table 2 involved study of the correlation between the guest acidity and the association constants in the recognition of neutral CH_3X and CH_2XY species by rigidified calix[4]arenes. Comparison of the K_{ass} of CH_3CN (pK_a-DMSO = 31.3 [21]) and CH_3NO_2 (pK_a-DMSO = 17.2 [21]), measured in CCl_4, shows that these two guests are bound with very similar efficiency. This indicates, in agreement with the results obtained in the solid state, that the complexation of the CH_3X guests is not determined by their acidity. Conversely, as verified in the solid state (see ahead), data on the values of K_{ass} measured in CCl_4 for hosts **13b** with guests of the type CH_2XY strongly suggest their dependence on the acidity of the CH_2 group (see Table 3) [22].

TABLE 3. $\Delta G°$ (KJ/mol, 300 K) of 1:1 complexes of host **13b** with CH_2XY guests (standard deviations are in brackets) in CCl_4.

Guests	pK_a	MR_x	$\Delta G°$(KJ/mol)
FCH_2CN	25	8.67	-8.9(0.5)
$ClCH_2CN$	26	13.38	-11.34(0.05)
$BrCH_2CN$	29	16.45	-10.9(0.4)
CH_3CH_2CN	31	14.83	-8.2(0.3)
CH_2Cl_2	35	12.61	-5.2(0.3)
$BrCH_2Cl$	38	15.59	-5.0(0.3)
CH_2Br_2	41	18.82	-6.4(0.3)

In fact, by plotting $\Delta G°$ vs pK_a of the guest (Figure 5), a good linear correlation, $\Delta G° = m \cdot pK_a + C$ was observed (m = 0.58 ± 0.06, C = -26 ± 2, r^2 = 0.973) considering all guests except CH_2Br_2 and FCH_2CN.
The big deviation from linearity shown for these guests suggests that the guest acidity cannot be considered as the sole factor responsible for the formation of the host-guest adducts. In fact, careful examination of the plotted data (Figure 5) shows that guests

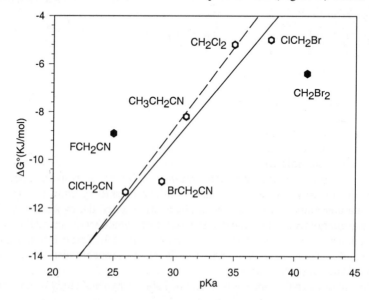

Figure 5. $\Delta G°$ for complex formation by host **13b** in CCl_4 vs. guest pK_a.

having very similar polarizability (ClCH$_2$CN, CH$_3$CH$_2$CN and CH$_2$Cl$_2$), give a better linear correlation ($m = 0.67 \pm 0.02$, $C = -28.8 \pm 0.6$, $r^2 = 0.999$), which may mean that dispersive interactions can substantially affect the recognition phenomena. For guests having the more polarizable Br atom (BrCH$_2$CN, BrCH$_2$Cl and CH$_2$Br$_2$), a higher than anticipated association constant results, while the lower polarizability of the fluorine atom of FCH$_2$CN results in a lower value (Table 5) [23]. Therefore, a two-parameter empirical equation including an index of the guest polarizability was devised, and a satisfactory correlation $\Delta G° = a \cdot pK_a + b \cdot MR_x + C$ ($a = 0.62 \pm 0.05$, $b = -0.7 \pm 0.1$, and $C = -18 \pm 1$, $r^2 = 0.982$), where the MR$_x$ is the *solute molar refraction*, was obtained [24]. This can be rationalised by assuming that CH$_2$XY guests interact with hosts having a C_{2v} symmetry, such as **13b**, *via* CH-π interactions with the two opposite "pinched" rings, and *via* dispersive interactions (of X and Y) with the two opposite "winged" aromatic rings. A significant observation is that the CH$_3$CN guest is bound more efficiently than CH$_2$ClCN, which is the best bound among the CH$_2$XY guests. A larger contribution of the entropic term in determining the $\Delta G°$ could explain these results.

Another means to improve the recognition efficiency of neutral organic species is to anchor suitable binding sites at one of the two rims of the calixarene platform. In most cases, the additional binding sites have consisted of hydrogen-bond donor or acceptor groups which thus interact with the potential guest through multipoint hydrogen bonding. Following early work [25], many examples of calix[4]arene-based hosts able to recognise neutral species have been reported.

One of the common features of these hosts, *e.g.*, **19**, is that the recognition process is driven by the interactions of the introduced binding groups, while the cavity plays only a minor role [26]. A calix[4]-arene-based receptor bearing two adamantylamide groups and forming a large cavity (see, *e.g.*, **20** in Figure 6b) is able to include, *e.g.*, a water molecule *via* the co-operative action of the calixarene cavity and the amide groups [27]. Clear evidence that the hydrogen bonding groups linked at the upper rim and the cavity are able to

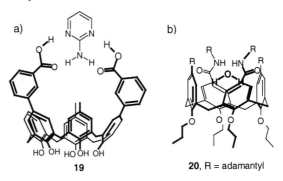

*Figure 6. Complexes of a) upper rim functionalised calix[5]-arene **19** with 2-aminopyrimidine and of b) upper rim functionalised calix[4]arene **20** with water.*

co-operate in binding has been obtained recently [28]. In particular, a rigidified calix[4]arene bearing an methylenephenylureido additional binding site at the upper rim was exploited to recognise low molecular weight amides (Figure 7).

The complexation properties of receptor **21** were evaluated in CDCl$_3$ by ^{1}H NMR measurements and the association constants for the 1:1 complex formation calculated (Table 4). Calix[4]arene **21** is able to efficiently recognise amides bearing the NH$_2$ or the NHR group (**22a-b**, **23a-b**, and **24a-b**), whereas a substantial decrease of the K$_{ass}$ was observed with guests bearing *N,N*-dimethylamino groups (**22c**, **23c**, and **24c**).

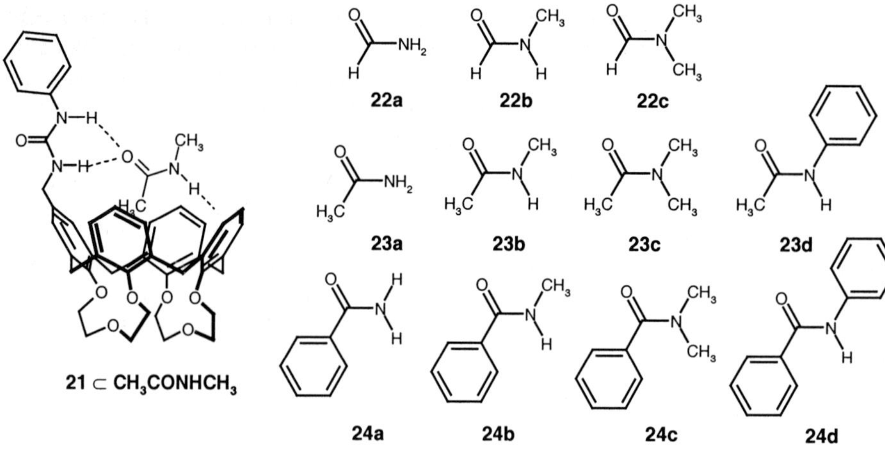

Figure 7. Proposed structure of the complex **21** ⊂ CH₃CONHCH₃ and the set of amides considered as guests.

TABLE 4. Association constants (K_{ass}, M⁻¹, T = 300 K) of 1:1 complexes of host **21** with amide guests **22-24** (R¹CONR²R³) in CDCl₃ (standard deviations are in brackets).

R¹	R²=R³=H	R²=CH₃, R³=H	R²=R³=CH₃	R²=C₆H₅, R³=H
H	**22a** 746(161)	**22b** 204(8)	**22c** 38(7)	-
CH₃	**23a** 342(25)	**23b** 261(24)	**23c** 56(30)	**23d** 198(10)
C₆H₅	**24a** 96(10)	**24b** 245(17)	**24c** < 10	**24d** < 10

Complexation is strongly affected by the presence of the bulky phenyl group on the guest nitrogen (**23d** and **24d**). The results obtained imply the co-operative action of the two host binding sites (the cavity and the ureido group) in the recognition process. The substantial upfield shift experienced by all the protons of the guest and the downfield shift of the ureido NH protons seen in NMR measurements are explicable in terms of hydrogen bond formation between the ureido NH and the carbonyl group of the guest. In addition, since neither the model components, calix[4]arene*bis*crown-3 nor benzylphenylurea C₆H₅NHCONHCH₂C₆H₅, showed binding ability toward the same series of guests, the presence of NH-π interactions between the guests and the aromatic cavity of calixarene was surmised. These data represent one of the few pieces of evidence for a role of such interactions in molecular recognition processes.

New possibilities for the recognition of organic species can be envisaged for complexes of calix[4]arene where transition metals like W, Mo, Ta, *etc.* are bound at the lower rim of the calix [29]. In this context, it is interesting to point out that the metal not only rigidifies the macrocycle, but it can also act as an acid co-ordination centre. Several solid state complexes between these metallo-calixarenes and small guests like acetonitrile, water and DMF have been reported. In these complexes, the guest molecule is co-ordinated to the transition metal and is included into the cavity (see Chapters 28, 29 and 30). However, the exploitation of this for the recognition of neutral organic molecules in solution is unexplored [30].

The chemistry of cavitands derived from calix[4]resorcinols has also been extensively explored and the recognition properties of the hosts synthesised studied in the solid state, the gas phase and solution. The "parent" calix[4]resorcinols possess a rigid cone structure having C_{4v} symmetry imposed by hydrogen bond formation between the eight hydroxy groups. When the OH groups are substituted, the flexibility of the skeleton increases. However, the introduction of bridging groups at the OH builds up the walls of the macrocycle and creates cavitands (see Chapter 9) [3].

A new approach to the recognition of neutral organic molecules involves a class of calix[4]resorcinol receptors which are able to fold upon themselves by means of intramolecular hydrogen bonds and to bind guests of complementary size bearing the amide functional group, with slow exchange kinetics in the NMR time-scale (see Chapter 8) [31].

3. Inclusion Compounds in the Solid State

Numerous crystallographic studies have been made of complexes of neutral organic molecules having acidic CH groups with calix[4]arenes, calix[4]resorcinols and thiacalix[4]arenes in the *cone* conformation. Data retrieved from the Cambridge Structural Database (CSD) concern inclusion resulting fortuitously from the choice of recrystallisation solvent as well as studies deliberately focussed on inclusion complex formation [32]. The influences of molecular symmetry and guest acidity were important issues in the latter group. As discussed in Chapter 16, solid state structural determinations are frequently complicated by "disorder" problems which have often not been fully resolved, so that the following discussion must be qualified by recognition of the fact that an averaged symmetry, higher than that of one particular component of a disordered array, may have been necessarily assumed.

To be able to compare in a consistent manner the structural features of known complexes, suitable geometrical descriptors which define the host geometry, and the position and the orientation of the guest inside the cavity are required.

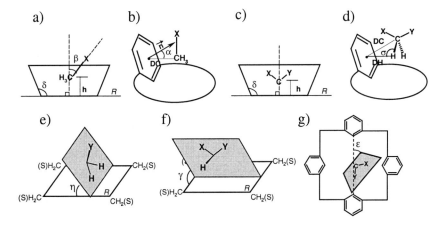

Figure 8. Geometrical descriptors of host-guest calix[4]arene complexes.

The shape of the cavity of the host and consequently its symmetry are defined through the angle δ between each benzene ring and the **R** plane defined by the bridging methylene carbons of the calixarene or of the sulfur atoms of thiacalixarene (Figure 8a).

The orientation of each guest having a CH_3 group inside the host cavity is defined by the parameters **DC**, α and **h** (Figure 8a,b). Beside these, other geometrical descriptors are necessary to define the orientation in the cavity of the guest: for acetonitrile and nitromethane, these are the angle β between the normal to the **R** plane and a vector defined by the CH_3-X moiety of the guest (Figure 8a).

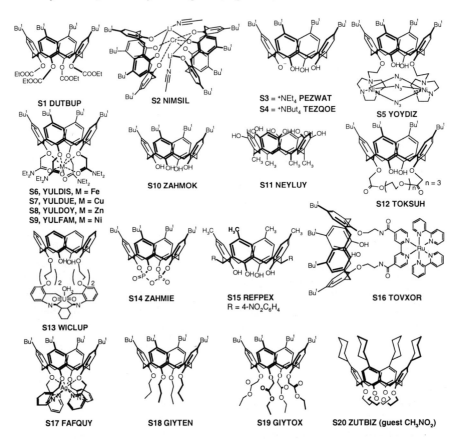

Figure 9. Host structures of acetonitrile ⊂ calix[4]arene complexes (unless otherwise specified). The acronyms denote reference codes for the Cambridge Structural Database.

For CH_2XY guests, in addition to the geometrical parameters **h** and **DC** (Figure 8c,d), parameters defining hydrogen atom positions: **DH**, (distance between each hydrogen atom and the centroid of the nearest aromatic ring), σ (angle between CH and the line connecting the hydrogen atoms to the centroid of the closest aromatic ring; Figure 8d), the angles γ and η, respectively formed by the planes containing XCY and HCH moieties with the **R** plane (Figure 8e,f), and the angle ε between the plane XCY and a line joining two distal C(1) carbons of calixarene (Figure 8g) were analysed. The parameter ε was also used to define the orientation in the aromatic cavity of the NO_2 group in nitromethane complexes.

Structures of the hosts able to bind acetonitrile are shown in Figure 9 and in Table 5 the most significant geometrical parameters of their complexes, together with the data for the complexes of *p*-cyclohexylcalix[4]arene-biscrown-3 and *p-t*-butylthiacalix[4]arene with nitromethane. The final crystallographic R factor for each complex is also given to provide a test for the reliability of the reported geometrical parameters. The structures of the hosts able to include dichloromethane, chloroacetonitrile and malononitrile are presented in Figure 10 and the geometrical parameters calculated for their complexes are summarised in Table 6.

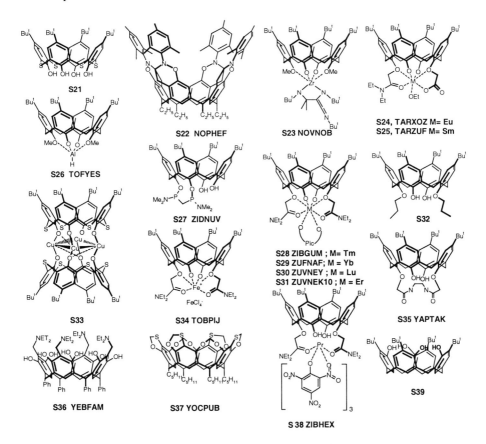

Figure 10. Hosts structures of dichloromethane ⊂ calix[4]arenes complexes (unless otherwise specified).

3.1. COMPLEXES WITH ACETONITRILE AND NITROMETHANE

The **DC** values, which must be in some way be indicative of the strength of the host-guest interactions, show that only few of these distances are over the sum of the van der Waals radii (*ca.* 3.8 Å) [7]. If the preferred host symmetry is C_4 and the host-guest distances are strictly related to the intermolecular forces, the cooperative effect exerted by the four aromatic nuclei present in the receptor site can be estimated as the sum of DC_1

+ DC_2 + DC_3 + DC_4 (ΣDC_{1-4}). The lowest values of ΣDC_{1-4} were observed with **S7** and **S1** (14.198 and 14.280 Å), having C_4 and nearly C_4 symmetry, respectively, whereas **S11** and **S14**, both with C_2 symmetry, show the highest values (15.186 and 15.423 Å).

TABLE 5. Geometrical parameters of the acetonitrile⊂ calix[4]arenes complexes (unless otherwise specified)[16,32-48].

Hosts	CSD	Ref.	R	$\Sigma\delta_{1-4}$ (Å)	β (°)	h (Å)	ΣDC_{1-4} (Å)	$\Sigma\alpha_{1-4}$ (°)
S1	DUTBUP	33	0.0560	458.324	0.000	2.603	14.280	8.588
S2	NIMSIL	34	0.0873	478.847	27.882	2.585	14.458	26.8051
S3	PEZWAT	35	0.0640	485.350	0.374	2.531	14.268	36.344
S3	PEZWAT	35	0.0640	477.418	0.000	2.689	14.428	22.073
S4	TEZQOE	36	0.1200	493.518	8.560	2.604	14.708	43.23
S4	TEZQOE	36	0.1200	491.806	6.712	2.746	14.960	31.494
S5	YOYDIZ	37	0.0870	477.197	6.970	2.699	14.429	18.743
S6	YULDIS	38	0.0810	462.876	5.995	2.756	14.348	4.18
S7	YULDUE	38	0.0840	459.396	8.176	2.701	14.198	4.514
S8	YULDOY	38	0.0796	461.998	6.275	2.730	14.358	4.099
S9	YULFAM	38	0.0893	463.452	13.330	2.786	14.448	12.481
S10	ZAHMOK	39	0.0520	492.184	0.000	2.714	14.868	34.008
S11	NEYLUY	40	0.0362	501.444	25.900	2.837	15.186	44.497
S12	TOKSUH	41	0.0880	475.200	0.000	2.608	14.430	26.266
S13	WICLUP	42	0.0550	488.259	22.648	2.598	14.591	59.745
S14	ZAHMIE	43	0.0810	515.498	49.099	2.840	15.423	67.799
S15	REFPEX	44	0.0540	480.077	37.696	2.741	14.660	28.445
S16	TOVXOR	45	0.1032	476.872	33.546	2.740	14.650	30.804
S16	TOVXOR	45	0.1032	474.922	31.672	2.590	14.347	25.035
S17	FAFQUY	46	0.0570	466.220	0.000	2.760	14.668	5.980
S18	GIYTEN	47	0.0530	459.625	5.237	2.701	14.419	8.276
S19	GIYTOX	48	0.1059	462.367	37.038	2.647	14.458	22.513
S20		32	0.0594	464.267	14.371	2.747	14.517	9.067
S20[a]	ZUTBIZ	16	0.0950	469.212	2.087	2.742	14.679	14.167
S21		32	0.1030	472.224	0.000	2.556	14.596	38.300
S21[a]		32	0.0961	476.084	0.000	2.518	14.592	44.59

[a]Nitromethane as guest.

The presence among these hosts of calixarenes bearing, on the aromatic rings, groups with different electron donating ability like the monoanion **S3** and **S4** or the tetraethylcarbonate **S1**, can afford useful information on the importance of the basicity of these rings on the recognition process in the solid state. The monoanions **S3** and **S4** show a lower ΣDC_{1-4} (for the presence of two different complexes in the two crystal structures 14.268 or 14.428 Å and 14.708 or 14.960 Å respectively) compared with that of the neutral **S10** (14.868 Å), suggesting a dependence of these interactions on the aromatic ring basicity. However, host **S1** which bears four electron withdrawing ethylcarbonate groups, shows the lowest ΣDC_{1-4} (14.280 Å).

The effect of the acidity of the guest on the intermolecular interactions was deduced from comparison of the ΣDC_{1-4} of two guests having very different acidity, acetonitrile and nitromethane [21], in their complexes with two different hosts *i.e.* p-cyclohexylcalix[4]arene-biscrown-3 **S20** and *p-t*-butyltetrathiacalix[4]arene **S21**. In both cases, similar parameters for the two guests were observed. In particular, acetonitrile in **S20** gives the lowest ΣDC_{1-4}, suggesting that acidity of the guest is not very important in determining the intermolecular distances in comparison with, *e.g.*, the steric requirement of the guest (Figure 11a) [32].

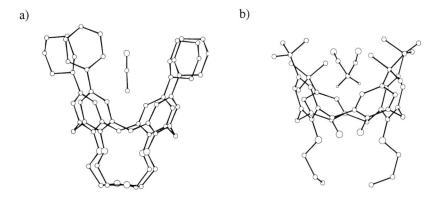

Figure 11. *X-ray crystal structures of a)* $S20 \supset CH_3CN$ *and b)* $S32 \supset CH_2(CN)_2$.

The correlation of δ_{1-4} with ΣDC_{1-4} is plotted in Figure 12. Complexes of hosts **S1** and **S7**, where ΣDC_{1-4} is the lowest, show also the lowest $\Sigma\delta_{1-4}$, whereas **S11** and **S14** where ΣDC_{1-4} is the highest, show the highest $\Sigma\delta_{1-4}$. As a general trend, in the majority of the complexes a roughly linear correlation between ΣDC_{1-4} and $\Sigma\delta_{1-4}$ was observed, suggesting that the shape of the aromatic cavity may determine the host-guest interactions, even if some complexes (all the 1,3-dialkoxy calixarenes derivatives **S5**, **S12**, **S13** and **S16** and all the formally anionic calixarene derivatives **S2**, **S3** and **S4**) show shorter distances in comparison with the other calixarene hosts.

Since directionality, monitored by the α angles, more than distances is indicative of these intermolecular interactions [7], we have also analysed this parameter. Hosts having C_4 symmetry show α angles in the range 0-10°, confirming a good directionality of these interactions, with the methyl carbon of the guest placed over the centroids of the aromatic rings of the cavity. It thus appears that the guest methyl group finds its energy minimum when its carbon atom is localised near the level of the *para* aromatic carbon plane of the host, whereas with the thiacalix[4]arenes, which have larger cavity, the carbon atom lies at an intermediate position between their *meta* and *para* planes (Figure 8a-b).

In addition with calix[4]arenes, the lowest distances **DC** and lowest α angle values are found when the δ angles of the aromatic rings range between 114-115°. This suggests that the strongest interactions occur when the α angles approaches 0°, that is the CH bond of the guest is aligned along the perpendicular to the centroid of the aromatic ring of the host. When the angle β is 0°, that is the guest is perpendicular to the methylene (or sulfur) **R** plane of the calix, a CH bond of the guest can be perpendicular to the aromatic ring when δ angle of the host equals the sp^3 CCH angle, that is, about 109°.

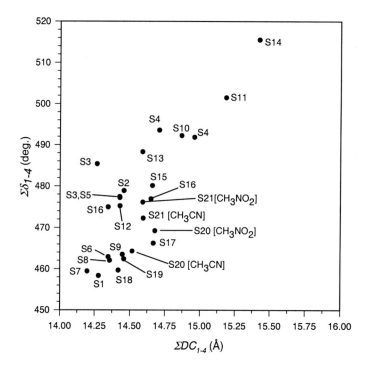

Figure 12. Correlation between ΣDC_{1-4} and $\Sigma\delta_{1-4}$ in acetonitrile $\subset$ calix[4]arene complexes (unless otherwise specified).

Whilst there no obvious correlation between simple characteristics of the host or guest and the strength of inclusion, it is clear, nonetheless, that what is important is the specific position of the guest methyl group inside the cavity, a position which must facilitate the strongest possible intermolecular interactions with all the aromatic walls of the host.

3.2. COMPLEXES WITH DICHLOROMETHANE, CHLOROACETONITRILE AND MALONONITRILE

Table 6 summarises the geometrical parameters calculated for the complexes of hosts **S21-39** with dichloromethane, chloroacetonitrile and malononitrile [32,49-62]. The analysis of the data reveals that only hosts **S39** and **S21** have an ε angle which approaches 45°, whereas for all the other hosts this angle is almost 0°, indicating that these guests preferably orient their XCY plane parallel to two distal aromatic rings of the host and point their two hydrogen atoms toward these aromatic nuclei.

On the basis of the $\boldsymbol{\Sigma DC_{1-4}}$ values, these complexes can be divided into two series: the first one with values ranging from 14.884 up to 15.512 Å (hosts **S22-31**, except **S25**), and the second 15.812 to 16.625 Å (hosts **S25, S33-38**).

TABLE 6. Geometrical parameters of the dichloromethane ⊂ calix[4]arenes complexes (unless otherwise specified).[32,49-62].

Hosts	CSD	Ref.	R	$\delta_1+\delta_3$ (°)	DC_1+DC_3 (Å)	ΣDC_{1-4} (Å)	DH_1+DH_3 (Å)	ε (°)	η (°)	γ (°)	h (Å)
S21[c]	NOPHEF	49	0.1078	238.814	-	16.625	6.038	44.380	90	90	3.582
S22	NOVNOB	50	0.1033	231.493	7.088	15.356	5.253	14.170	88.611	84.685	2.896
S23	TARXOZ	51	0.0700	219.294	6.919	15.129	5.019	9.734	87.937	87.615	2.858
S24	TARXUF	52	0.0826	221.086	6.968	14.884	5.073	2.110	89.229	87.689	2.838
S25	TOFYES	52	0.0764	221.048	-	15.812	5.434	13.406	48.472	76.867	3.199
S26	ZIDNUV	53	0.0770	224.543	7.133	15.512	5.095	5.325	87.456	87.334	2.968
S27	ZIBGUM	54	0.0800	226.683	7.232	15.230	5.376	7.247	84.227	85.122	2.936
S28	ZUVNAF	55	0.0657	217.608	6.911	15.122	5.015	2.227	88.273	89.115	2.825
S29	ZUVNEJ	56	0.0849	218.310	6.797	14.877	4.905	3.944	88.583	86.842	2.778
S30	ZUVNEK10	56	0.0932	219.093	6.949	15.149	5.052	2.413	87.415	89.183	2.832
S31		57	0.0572	218.219	6.891	15.097	4.996	2.891	87.913	87.944	2.817
S32[a]		32	0.1317	222.774	6.846	14.919	4.924	2.507	84.691	83.389	2.716
S32[b]		32	0.0471	219.945	7.045	15.078	5.176	5.036	84.524	88.983	2.857
S33		58	0.0460	218.720	7.280	15.440	5.400	3.962	89.64	89.18	2.960
S33		58	0.0460	222.180	7.390	15.480	5.500	0.732	89.58	89.66	2.980
S34	TOBPIJ	57	0.0879	225.918	-	16.121	[d]	2.236	[d]	88.981	3.274
S35	YAPTAK	59	0.0770	225.691	-	15.833	5.969	7.936	66.609	84.28	3.165
S36	YEBFAM	60	0.0850	257.020	-	15.674	6.150	16.647	75.802	75.351	3.007
S37	YOCPUB	61	0.0720	255.293	-	15.894	6.155	29.413	82.128	61.418	3.020
S38	ZIBHEX	52	0.1235	230.631	-	16.263	5.861	0.249	46.856	77.622	3.295
S39		62	0.0761	253.068	-	15.470	[d]	44.380	[d]	90	3.003

[a] malononitrile as guest; [b] chloroacetonitrile as guest; [c] the data were not retrievable from CSD, therefore they were calculated on the structural data obtained in our labs; [d] hydrogen atoms not defined.

For complexes of dichloromethane with hosts belonging to the first series, the γ and η angle values are almost 90°, *i.e.*, the CH_2 "side" of the guest is inserted furthest into the cavity, and no significant tilting of the guest relative to the **R** plane is observed. In contrast, with hosts **S25, S33-38** the guests are usually tilted. In all these complexes, as expected, the δ angle values of the hosts indicate a preferred C_2 symmetry and a *pinched cone* conformation of the calix, which thus has an elliptical cavity, complementary to the symmetry of the guest. More interesting is the observation that the complexes found in the CSD which show the shortest distances with dichloromethane are generally *p-t*-butylcalix[4]arenes having two anionic distal phenolic rings (**S23-24, S26, S28-29**, and **S30-31**), and only two such hosts are undeprotonated (**S27** and **S32**). Also interesting is that it emerges that the aromatic rings which interact with the acid CH_2 group of the guest are not the anionic but the dialkoxy substituted ones. As with acetonitrile, geometric factors seem to overcome electronic, so that the less basic alkoxy substituted aromatic rings, which are almost perpendicular to the methylene plane of the calix, are those which interact with the guest CH groups. The other two rings interact with the chlorine atoms *via* van der Waals attractions, thus further stabilising the host-guest complex. The greater polarizability of the anionic aromatic ring probably maximises the London dispersion forces.

An interesting modification of the binding mode was observed with *p-t*-butylthiacalix[4]arene (**S21**), which has C_4 symmetry and includes dichloromethane through one chlorine atom which enters into the cavity. Conversely, its copper (II) complex (**S33**), probably as consequence of its C_2 symmetry and of its anionic structure, binds the guest through CH_2-π aromatic interactions [58].

In these complexes of dichloromethane and related species, only the two "pinched" distal rings interact with the hydrogen atoms of the guests. In this regard, the correlation between $δ_1 + δ_3$ and $DC_1 + DC_3$ was studied, as were the distances **DH** between the guest hydrogen atoms and the nearest aromatic ring, most structures being of sufficient quality for these H-atom positions to have been defined.

With the exception of the copper derivative of thiacalix[4]arene (**S33**) and calix[4]resorcinol (**S22**), an almost linear correlation between $δ_1 + δ_3$ and $DC_1 + DC_3$ for all the calix[4]arene hosts was found (Figure 13). Similar results were obtained for the correlation of $δ_1 + δ_3$ with $DH_1 + DH_3$. Very interesting is the comparison of the structural parameters obtained using 1,3-dipropoxy-*p-t*-butylcalix[4]arene (**S32**) as host and chloroacetonitrile and malononitrile, which have similar size and shape but different acidity of the methylene groups [21], as guests (see Figure 11b). The more acid malononitrile shows shorter $DC_1 + DC_3$ (respectively 6.846 vs. 7.045 Å) and $DH_1 + DH_3$ (respectively 4.924 vs. 5.176 Å) distances in comparison with chloroacetonitrile, indicating a higher contribution of hydrogen bonding in these CH_2 aromatic interactions [32].

The values of σ (Figure 8d), indicative of the directionality of the host-guest interactions, range from 160 to 170°, in good agreement with those obtained for CH-π cyclopentadienyl anion interactions, where σ is about 165° [63].

These data show that the intermolecular interactions between the aromatic cavity of the calix[4]arene hosts and the molecules having acid CH groups are different for CH_3X and CH_2XY guests. There is no evidence for hydrogen-bond-like characteristics of CH_3X interactions. These results are in agreement with the INS (Inelastic Neutron

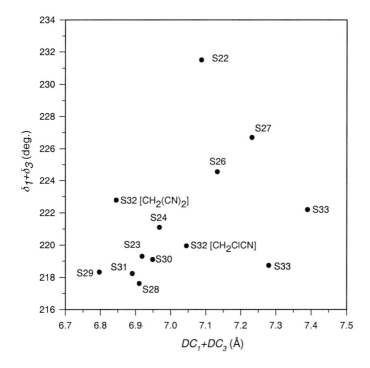

Figure 13. Correlation between $\Sigma DC1+3$ and $\Sigma\delta1+3$ in dichloromethane $\subset$ calix[4]arene complexes (unless otherwise specified).

Scattering) studies performed in the solid state of calix[4]arene complexes with these guests, which show that the guest methyl group behaves as an almost free quantum rotor [64]. This affords an entropic advantage to these interactions. Conversely, with CH_2XY guests, free rotation is partially inhibited by complex formation. This entropic cost is probably partially compensated by a higher value of the enthalpic contribution as shown by the dependence of the CH-aromatic distances on the acidity of the guest.

Further studies on the CH, NH and OH-π aromatic interactions are required to better understand these intermolecular forces and to succeed in the preparation of more efficient and selective receptors for neutral organic molecules.

4. References and Notes

[1] J.-M. Lehn, *Supramolecular Chemistry: Concepts and Perspectives*, VCH, Weinheim, **1995**.
[2] J.H. Hartley, T.D. James, C.J. Ward, *J. Chem. Soc., Perkin Trans 1*, **2000**, 3155-3184.
[3] D. J. Cram, J. M. Cram, *Container Molecules and Their Guests - Monographs in Supramolecular Chemistry*, Vol. 4 (Ed. J.F. Stoddart), The Royal Society of Chemistry, Cambridge **1994**;
[4] A. Collet in *Comprehensive Supramolecular Chemistry*, Vol.2 (Eds.: J. L. Atwood, J. E. D. Davies, D. D. McNicol, F. Vögtle), Pergamon, Oxford, **1996**, 325-365.
[5] C. D. Gutsche, *Calixarenes Revisited - Monographs in Supramolecular Chemistry*, Vol. 6 (Ed. J.F. Stoddart), The Royal Society of Chemistry, Cambridge **1998**.

[6] A. Pochini, R. Ungaro in F. Vögtle (Vol. Ed.), *Comprehensive Supramolecular Chemistry*, Elsevier, **1996**, Vol. 2, Ch. 4, pp. 103-142.
[7] a) M. Nishio, M. Hirota,Y. Umezawa in A.P Marchand (Ed.), *The CH/π Interaction*, Methods in Stereochemical Analysis, No. 11, Wiley-VCH, **1998** and references therein; b) Y. Umezawa, S. Tsuboyama, K. Honda, J. Uzawa, M. Nishio, *Bull. Chem. Soc. Jpn.* **1998**, *71*, 1207-1213; c) D. Braga, F. Grepioni, E. Tedesco, *Organometallics* **1998**, *17*, 2669-2672; d) Further information about CH-π interactions is available in the following web site: http://www.tim.hi-ho.ne.jp/dionisio/
[8] For indirect evidence in solution see: A. Yamada, T. Murase, K. Kikukawa, T. Arimura, S. Shinkai, *J. Chem. Soc. Perkin Trans. 2* **1991**, 793-797.
[9] G. D. Andreetti, F. Ugozzoli, A. Pochini, R. Ungaro in *Inclusion Compounds, Vol. 4* (Eds. J. L. Atwood, J. E. Davies, D. D. McNicol), Oxford University Press, Oxford, **1991**.
[10] P. D. J. Grootenhuis, P. A. Kollman, L. C. Groenen, D. N. Reinhoudt, G. J. van Hummel, F. Ugozzoli, G. D. Andreetti, *J. Am. Chem. Soc.* **1990**, *112*, 4165-4176.
[11] A. Arduini, M. Fabbi, M. Mantovani, L. Mirone, A. Pochini, A. Secchi, R. Ungaro, *J. Org. Chem.* **1995**, *60*, 1454-1457.
[12] A. Arduini, W. M. McGregor, A. Pochini, A. Secchi, F. Ugozzoli, R. Ungaro, *J. Org. Chem.* **1996**, *61*, 6881-6887.
[13] A. Arduini, A. Casnati, L. Dodi, A. Pochini, R. Ungaro, *J. Chem. Soc., Chem. Commun.* **1990**, 1597-1598.
[14] A. Arduini, L. Domiano, A. Pochini, A. Secchi, R. Ungaro, F. Ugozzoli, O. Struck, W. Verboom, D. N. Reinhoudt, *Tetrahedron* **1997**, *53*, 3767-3776.
[15] G. Pèpe, J.-P. Astier, J. Estienne, C. Bressot, Z. Asfari, J. Vicens, *Acta Crystallogr., Sect. C* **1995**, *51*, 726-729.
[16] A. Arduini, W. M. McGregor, D. Paganuzzi, A. Pochini, A. Secchi, F. Ugozzoli, R. Ungaro, *J. Chem. Soc. Perkin Trans. 2*, **1996**, 839-846.
[17] A. Arduini, E. Ghidini, A. Pochini, R. Ungaro, G. D. Andreetti, L. Calestani, F. Ugozzoli, *J. Incl. Phenom.* **1988**, *6*, 119-134.
[18] S. Smirnov, V. Sidorov, E. Pinkhassik, J. Havlicek, I. Stibor, *Supramol. Chem.* **1997**, *8*, 187-196.
[19] W. P. Van Hoorn, M. G. H. Morshuis, F. C. J. M. van Veggel, D. N. Reinhoudt, *J. Phys. Chem. A* **1998**, *102*, 1130-1138.
[20] T. Kusano, M. Tabatabai, Y. Okamoto, V. Böhmer, *J. Am. Chem. Soc.* **1999**, *121*, 3789-3790.
[21] a) F. G. Bordwell, *Acc. Chem. Res.* **1988**, *21*, 456-463; b) K. Yoshimura, Y. Fukazawa, *Tetrahedron Lett.* **1996**, *37*, 1435-1438.
[22] A. Arduini, G. Giorgi, A. Pochini, A. Secchi, F. Ugozzoli, *Tetrahedron* in press.
[23] Probably also the conclusions of Fugazawa et al. who studied the binding abilities of monomethoxy-monodeoxy-*p*-*t*-butylcalix[4]arene towards guests having acidic CH_2 groups can be interpreted in this context although no polarisability effects were observed (see ref. 21b).
[24] (a) M. H. Abraham, G. S. Whiting, R. M. Doherty, W. J. Shuely, *J. Chem. Soc., Perkin Trans 2* **1990**, 1451-1460; (b) M. H. Abraham, J. C. McGowan, *Chromatographia* **1987**, *23*, 243-246.
[25] J. D. Van Loon, R. G. Janssen, W. Verboom, D. N. Reinhoudt, *Tetrahedron Lett.* **1992**, *33*, 5125.
[26] See *e.g.* a) T. Haino, K. Matsumura, T. Harano K. Yamada, Y. Saijyo, Y. Fukazawa, *Tetrahedron* **1998**, *54*, 12185-12196; b) T. Haino, Y. Katsutani, H. Akii, Y. Fukazawa, *Tetrahedron Lett.* **1998**, *39*, 8133-8136.
[27] E. Pinkhassik, V. Sidorov, I. Stibor, *J. Org. Chem.* **1998**, *63*, 9644-9651.
[28] A. Arduini, A. Pochini, A. Secchi, *J. Org. Chem.* **2000**, in press.
[29] C. Wieser, C. B. Dieleman, D. Matt, *Coordin. Chem. Rev.* **1997**, *165*, 93-161.
[30] (a) J. A. Acho, L. H. Doerrer, S. J. Lippard, *Inorg. Chem.* **1995**, *34*, 2542-2556; (b) O. Sénèque, M.-N. Rager, M. Giorgi, O. Reinaud, *J. Am. Chem. Soc.* **2000**, *122*, 6183-6189.
[31] (a) D. M. Rudkevich, G. Hilmersson, J. Rebek Jr, *J. Am. Chem. Soc.* **1998**, *120*, 12216-12225; (b) D. M. Rudkevich, J. Rebek, Jr., *Eur. J. Org. Chem.* **1999**, 1991-2005; (c) A. Shivanyuk, K. Rissanen, S. K. Körner, D. M. Rudkevich, J. Rebek, Jr., *Helv. Chim. Acta* **2000**, *83*, 1778-1790.
[32] A. Arduini, F. F. Nachtigall, A. Secchi, A. Pochini, F. Ugozzoli, *Supramol. Chem.* **2000**, in press.
[33] M. A. McKervey, E. M. Seward, G. Ferguson, B. L. Ruhl, *J. Org. Chem.* **1986**, *51*, 3581-3584.
[34] V. C. Gibson, C. Redshaw, W. Clegg, M. R. J. Elsegood, *J. Chem. Soc., Chem. Commun.* **1997**, 1605-1606.
[35] J. M. Harrowfield, M. I. Ogden, W. R. Richmond, B. W. Skelton, A. H. White, *J. Chem. Soc., Perkin Trans.2* **1993**, 2183-2190.

[36] R. Abidi, M. V. Baker, J. M. Harrowfield, D. S.-C. Ho, W. R. Richmond, B. W. Skelton, A. H. White, A. Varnek, G. Wipff, *Inorg. Chim. Acta* **1996**, *246*, 275-286.
[37] P. D. Beer, M. G. B. Drew, P. B. Leeson, K. Lyssenko, M. I. Ogden, *J. Chem. Soc., Chem. Commun.* **1995**, 929-930.
[38] P.D.Beer, M. G. B. Drew, P. B. Leeson, M. I. Ogden, *J. Chem. Soc., Dalton Trans.* **1995**, 1273-1283.
[39] W. Xu, R. J. Puddephatt, L. Manojlovic-Muir, K. W. Muir, C. S. Frampton, *J. Incl. Phenom.* **1994**, *19*, 277-290.
[40] L. R. MacGillivray, J. L. Atwood, *J. Am. Chem. Soc.* **1997**, *119*, 6931-6932.
[41] Z. Zhen-Lin, C. Yuan-Yin, L. Xue-Ran, L. Bao-Sheng, C. Liao-Rong, *Jiegou Huaxue (J. Struct. Chem.)* **1996**, *15*, 358.
[42] A. M. Reichwein, W. Verboom, S. Harkema, A. L. Spek, D. N. Reinhoudt, *J. Chem. Soc., Perkin Trans. 2* **1994**, 1167-1172.
[43] O. Aleksiuk, F. Grynszpan, S. E. Biali, *J. Incl. Phenom.* **1994**, *19*, 237-256.
[44] S. E. Biali, V. Böhmer, S. Cohen, G. Ferguson, C. Grüttner, F. Grynszpan, E. F. Paulus, I. Thondorf, W.Vögt, *J. Am. Chem. Soc.* **1996**, *118*, 12938-12949.
[45] F. Szemes, D. Hesek, Zheng Chen, S. W. Dent, M. G. B. Drew, A. J. Goulden, A. R. Graydon, A. Grieve, R. J. Mortimer, T. Wear, J. S. Weightman, P. D. Beer, *Inorg.Chem.* **1996**, *35*, 5868-5879.
[46] A. F. Danil de Namor, O. E. Piro, L. E. Pulcha Salazar, A. F. Aguilar-Cornejo, N. Al-Rawi, E. E. Castellano, F. J. Sueros Velarde, *J. Chem. Soc., Faraday Trans.* **1998**, *94*, 3097-3104.
[47] W. Verboom, O. Struck, D. N. Reinhoudt, J. P. M. van Duynhoven, G. J. van Hummel, S. Harkema, K. A. Udachin, J. A. Ripmeester, *Gazz. Chim. Ital.* **1997**, *127*, 727-739.
[48] M. Pitarch, A. Walker, J. F. Malone, J .J. McGarvey, M. A. McKervey, B. Creaven, D. Tobin, *Gazz. Chim. Ital.* **1997**, *127*, 717-721.
[49] H. Akdas, L. Bringel, E. Graf, M. H. Hosseini, G. Mislin, J. Pansanel, A. De Cian, J. Fischer, *Tetrahedron Lett.* **1998**, *39*, 2311.
[50] K. Airola, V.Böhmer, E. F. Paulus, K. Rissanen, C. Schmidt, I. Thondorf, W. Vogt, *Tetrahedron* **1997**, *53*, 10709-10724.
[51] L. Giannini, A. Caselli, E. Solari, C. Floriani, A. Chiesi-Villa, C. Rizzoli, N. Re, A. Sgamellotti, *J. Am. Chem. Soc.* **1997**, *119*, 9709-9719.
[52] P. D. Beer, M. G. B. Drew, A. Grieve, M. Kan, P. B. Leeson, G. Nicholson, M. I. Ogden, G. Williams, *Chem. Commun.* **1996**, 1117-1118.
[53] M. G. Gardiner, G. A. Koutsantonis, S. M. Lawrence, P. J. Nichols, C. L. Raston, *Chem. Commun.* **1996**, 2035-2036.
[54] I. Shevchenko, H. Zhang, M. Lattman, *Inorg. Chem.* **1995**, *34*, 5405-5409.
[55] P.D. Beer, M. G. B. Drew, A. Grieve, M. I. Ogden, *J. Chem. Soc., Dalton Trans.* **1995**, 3455-3466.
[56] P.D.Beer, M. G. B. Drew, M. Kan, P. B. Leeson, M. I. Ogden, G. Williams, *Inorg. Chem.* **1996**, *35*, 2202-2211.
[57] P. D. Beer, M. G. B. Drew, P. B. Leeson, M. I. Ogden, *Inorg. Chim. Acta* **1996**, *246*, 133-141.
[58] G. Mislin, E. Graf, M. H. Hosseini, A. Bilyk, A. K. Hall, J. M. Harrowfield, B. W. Skelton, H. White *Chem. Commun.* **1999**, 373-374.
[59] V. Böhmer, G. Ferguson, J. F. Gallagher, A. J. Lough, M. A. McKervey, E. Madigan, M. B. Moran, J. Phillips, G. Williams, *J. Chem. Soc., Perkin Trans.1* **1993**, 1521-1527.
[60] D. A. Leigh, P. Linnane, R. G. Pritchard, G. Jackson, *Chem. Commun.* **1994**, 389-390.
[61] R. C. Helgeson, C .B. Knobler, D. J. Cram, *Chem. Commun.* **1995**, 307-308.
[62] G. Sartori, C. Porta, F. Bigi, R. Maggi, F. Peri, E. Marzi *Tetrahedron* **1997**, *53*, 3287-3300.
[63] S. Harder, *Chem. Eur. J.* **1999**, *5*, 1852-1861.
[64] R. Caciuffo, G. Amoretti, C. J. Carlile, C. Ferrero, S. Geremia, B. Paci, M. Prager, F. Ugozzoli, *Physica B* **1997**, *234-236,*, 115-120.

Chapter 26

COMPLEXATION OF FULLERENES

ZHEN-LIN ZHONG, ATSUSHI IKEDA, SEIJI SHINKAI
Department of Chemistry and Biochemistry
Graduate School of Engineering, Kyushu University
Fukuoka 812-8581, Japan. e-mail: seijitcm@mbox.nc.kyushu-u.ac.jp

1. Introduction

Fullerenes [1], which were discovered in 1985 [1b] and isolated in macroscopic quantities in 1990 [1c], can be regarded as the third carbon allotrope after graphite and diamond. Because of their unique physical and chemical properties, such as efficient singlet oxygen sensitising ability, strong electron accepting character, photoinduced charge separation ability, and superconductivity upon binding with metals, fullerenes have attracted intense research interest in various fields. In parallel, calixarenes [2] may be regarded as the third generation of potent host molecules after cyclodextrins and crown ethers. Since fullerenes have globular structures with external π-electron surfaces, and calixarenes have cavity structures with internal π -electron surfaces, a "marriage" of the two third generations by guest-host complexation, e.g., through π– π interactions is an obvious possibility. In 1992, a water-soluble calix[8]arene derivative was shown to extract [60]fullerene (C_{60}) from an organic into an aqueous phase [3]. In 1994, it was discovered that *p-tert*-butylcalix[8]arene selectively includes C_{60} from carbon soot and forms a precipitate with 1:1 stoichiometry, and provides a novel and very useful method to obtain C_{60} in large quantity and in high purity [4,5]. This finding greatly stimulated research into the host-guest chemistry of fullerenes. It was found that a variety of macrocyclic host molecules, such as calixarenes or calixarene analogues [6], cyclodextrins [7], cyclotriveratrylenes [8], and azacrown ethers [9], form complexes with fullerenes. Other host systems include a supramolecular box composed by hydrogen-bonded hydroquinones [10], a saddle-shaped Ni(II) macrocycle [11], and metalloporphyrins [12]. In the recent years, several excellent review articles concerning both general supramolecular fullerene chemistry [13] and specifically calixarene–fullerene interactions [6] have been published. This chapter provides an overview of the various aspects of host-guest complexation of fullerenes with calixarene-based receptors, including those in the solid state, in films, and in organic and aqueous solutions.

2. Complexation in the Solid State

2.1. ISOLATION OF FULLERENES BY COMPLEXATION WITH CALIXARENES

tert-Butylcalix[8]arene **1** reacts with C_{60} in toluene to form a precipitate having 1:1 **1**/C_{60} stoichiometry (Fig. 1) [4,5]. Since this complex is sparingly soluble in most organic solvents, it is very easy and efficient to purify C_{60} in this way. With carbon-arc soot, calixarene **1** selectively "includes" C_{60} in a mixture of fullerene homologues and thus can precipitate it from toluene. In particular, the complexation is highly selective for C_{60} over [70]fullerene (C_{70}) (and other higher fullerenes). HPLC analysis showed that C_{60} can be obtained in 96% purity from a carbon soot containing 72 wt% of C_{60}, 13 wt% of C_{70}, and 15 wt% of others in a single pass. The purity of the precipitate can be improved to 99 wt% by recrystallisation from toluene, and to 99.8 wt % by one more recrystallisation [5]. High purity C_{60} can be recovered as a precipitate by dispersing this 1:1 complex into chloroform, since C_{60} is only sparingly soluble in chloroform and is displaced from the calixarene as a result of the relatively strong ClC-H⋯aromatic π-ring interactions [14] between the solvent and the calixarene. The recovery of high purity C_{60} is essentially quantitative if mother liquor solutions are recycled back to the crude fullerene mixture along with calixarene **1** recovered from chloroform degradation of the complex [4]. C_{70} can also be enriched up to 87% purity by forming a fullerene-rich 2:1 complex with *p-tert*-butylcalix[6]arene (**2h**) from C_{60}-depleted fullerene mixture [4].

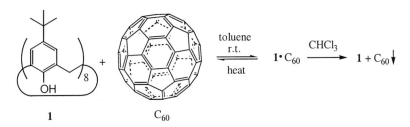

Figure 1. C_{60} can be easily separated from carbon-arc soot by specific inclusion complex formation with *p-tert*-butylcalix[8]arene in toluene [4,5].

In 1999, it was shown that *p*-benzylcalix[5]arene (**2d**) and *p*-benzylhomooxacalix[3]arene (**3b**) form 2:1 solid complexes with C_{60} from a fullerite solution in toluene [15]. Addition of dichloromethane to the complex gives C_{60} with >99.5% purity. This represents an effective one step method for purification of C_{60}, although the benzylcalixarenes (**2d**) and (**3b**) are not readily available in large quantity while *p-tert*-butylcalix[8]arene is.

	n	R	X
2a	4	I	Bn
2b	4	Br	*n*-Pr
2c	4	*t*-Bu	CH_2COOEt
2d	5	Bn	H
2e	5	H	H
2f	5	*t*-Bu	H
2g	5	Allyl	H
2h	6	*t*-Bu	H
2i	6	H	H
2j	6	*t*-Bu	CH_2COOEt
2k	8	H	H
2l	8	*t*-Bu	Me
2m	8	SO_3H	H

	R	X
3a	H	H
3b	Bn	H
3c	*t*-Bu	H
3d	Br	H
3e	OMe	H
3f	*t*-Bu	CH_2COOEt
3g	Me	CH_2COOEt
3h	Br	CH_2COOEt

2.2. STRUCTURES OF CALIXARENE–FULLERENE COMPLEXES

Much attention has been paid and significant advances have been made in understanding the structural and supramolecular nature of the $\mathbf{1}\cdot C_{60}$ complex, although with some conflicting results. In the IR region, calixarene **1** gives a v_{OH} band at 3240 cm^{-1} in the free form and at 3304 cm^{-1} in the $\mathbf{1}\cdot C_{60}$ complex, indicating that the intramolecular hydrogen-bonding network is only partially disrupted or weakened. In the visible region the broad absorption band at 450–550 nm of free C_{60} disappears in the complex, suggesting that the interaction among C_{60} molecules is suppressed by the isolation effect of the host **1**. In CP-MAS ^{13}C NMR spectra, significant upfield shifts (up to 2.5 ppm) of the *tert*-butyl carbons and a downfield shift of 2.7 ppm of the ArCH$_2$Ar carbons were observed upon complexation with C_{60} [5]. Other CP-MAS ^{13}C NMR spectroscopic investigations [16] on the greenish microcrystalline $\mathbf{1}\cdot C_{60}$ complex obtained from CS_2 showed a significant upfield shift (1.4 ppm) of the C_{60} ^{13}C NMR resonance and sharpening of all the calixarene signals upon complexation. These results not only signal supramolecular interactions between **1** and C_{60}, but also indicate complexation-induced conformational changes of the calixarene. Other information comes from the result that *p-H*-calix[8]arene (**2k**) and *O*-methylcalix[8]arene (**2l**) do not form a complex with C_{60} as a precipitate from toluene, indicating the importance of the *tert*-butyl groups for CH–π interactions [17] with C_{60} and the importance of the OH groups in constraining the calixarene to a relatively rigid structure through intramolecular hydrogen-bonding. A "ball and socket" nanostructure (Fig. 2A) in which C_{60} is bound to the upper (*tert*-butyl) edge of the calixarene [4,5], was initially proposed and appeared to be consistent with the results of studies on Langmuir films (see Section 3) of the complex. Later, on the basis of combined spectroscopic and X-ray powder diffraction studies supported by molecular mechanics calculations, a micelle-like structure (Fig. 3) with a trimeric aggregate of fullerenes surrounded by three host molecules each in the double *cone* conformation, was proposed [18,19]. This model gives an explanation of the fact that there are two different types of *tert*-butylphenyl groups in the ratio *ca.* 2:6 shown by the CP-MAS ^{13}C NMR spectra [16]. Changes in optical absorption of $\mathbf{1}$–C_{60} mixed solutions over a period of time have been interpreted [20] in terms of two consecutive processes taking place between **1** and C_{60} in toluene: (1) molecular inclusion, and (2) growth and sedimentation of the insoluble $C_{60}/\mathbf{1}$ clusters. Despite multiple efforts to obtain single crystals of $\mathbf{1}\cdot C_{60}$, the structural details of this complex are still unknown.

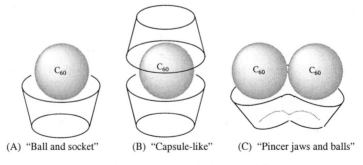

(A) "Ball and socket"   (B) "Capsule-like"   (C) "Pincer jaws and balls"

Figure 2. Schematic representation of the structure types of calixarene–fullerene complexes.

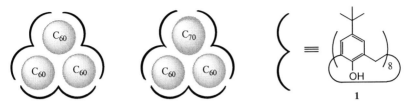

Figure 3. Schematic representation of the molecular mechanics-generated structures of $(1)_3 \cdot (C_{60})_3$ and $(1)_3 \cdot (C_{60})_2 \cdot C_{70}$ [18].

The first structurally authenticated calixarene–fullerene complex was reported in 1997 (Fig. 4) [21a]. X-ray analysis showed that the purple prisms obtained by slow concentration of a solution of calix[5]arene **4** and an equimolar amount of C_{60} in CS_2 are a $(4)_2 \cdot C_{60}$ complex in which the guest fullerene molecule is encapsulated within a cavity composed of the two host molecules (Fig. 2B). Due to the positional disorder of the two iodine atoms over the *para* positions of the five phenolic units, the complex has *pseudo* D_{5d} symmetry in the solid state. There are 144 short (< 4.0 Å) interactions between the aromatic carbons in the two hosts and the carbons in the guest, suggesting strong binding in this complex. While the diiodocalix[5]arene **4** forms a 2:1 complex with C_{60}, calix[5]arenes **5** and **6** without iodine atoms on the *para* positions form 1:1 complexes (Fig. 2A) in the solid state [22]. In all these complexes, all the calixarene hosts cover a portion of the globular surface of C_{60} with their bowl-like interior surfaces, and inhibit the free rotation of C_{60}. The calixarenes have a circular *cone* conformation with almost the same five inclination angles of the phenyl rings (average of the five angles, **4**: 133.7°, **5**: 134.8°, **6**: 135.7°) to the globular guest, indicating a guest-fitting effect of C_{60} on the conformation of the calixarene hosts, since calix[5]arenes usually exist in an oval *cone* conformation in other crystalline systems [23].

4: X = I
5: X = Me
6: X = H

7: R = CH_2CH_2Ph

p-Benzylcalix[5]arene (**2d**) was also found to form capsule-like 2:1 complex (with 8 toluene molecules in it) with C_{60} from toluene (Fig. 5) [15]. The complex has D_{5d} symmetry if the orientation of the benzyl groups is ignored, which means that two calixarenes are bound to a single fullerene in a *trans*-arrangement, and that the fullerene guest is located just at the centre of the capsular cavity, with two opposite five-membered rings on the top and the bottom of the cavity composed of the two calixarenes. The closest interactions between the fullerene and the calixarene are at the van der Waals limit (most of the C···C distances are about 3.5 Å). Ignoring the different substituent groups, the structural characteristics of this and the above-mentioned diiodocalix[5]arene complex are very similar.

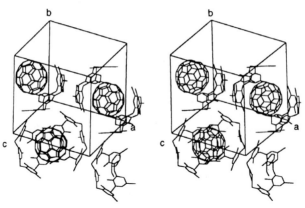

Figure 4. A stereoscopic view of the crystal packing of $(4)_2 \cdot C_{60}$ (reproduced by permission of WILEY-VCH Verlag GmbH from Angew. Chem., Int. Ed. Engl. 1997, 36, 259–260) [21a].

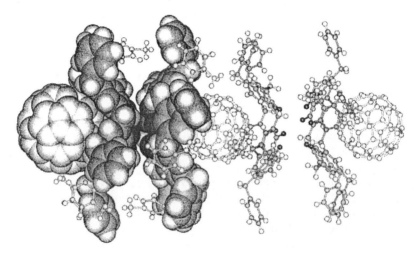

Figure 5. Columnar structure for successive $[(2d)_2 \cdot C_{60}] \cdot 8$ toluene entities (reproduced by permission of WILEY-VCH Verlag GmbH from Chem. Eur. J. 1999, 5, 990–996) [15].

Both *p-tert*-butylcalix[6]arene (**2h**) and calix[6]arene (**2i**) form fullerene-rich 1:2 complexes with C_{60} or C_{70} [4,24]. The structures of (calix[6]arene)·$(C_{60})_2$ and (calix[6]arene)·$(C_{70})_2$ were determined by X-ray diffraction studies [24]. In marked contrast to the nearly perfect *cone* conformation of calix[5]arenes in their complexes with C_{60}, calix[6]arene in the above two complexes is in double-*cone* conformation and associates with two fullerenes, like the jaws of a pincer acting on two adjacent spheres (Fig. 2C). The two shallow cavities of calix[6]arene interact weakly with a small portion of the π-cloud surface of their fullerene guests, and allow the fullerene to be in close proximity to six other fullerene molecules, with a fullerene-fullerene distance of ~10-11 Å.

Homooxacalix[3]arenes **3c** and **3d** form 1:1 complexes with C_{60} in the solid state, having the "ball and socket" structure shown in Fig. 2A, in which the host molecules are in a perfect *cone* conformation with C_{3v} symmetry [25]. In the structure of **3d**·C_{60}, a six-membered ring of C_{60} is disposed parallel to the mean plane composed of the three phenolic oxygens of the calixarene, and three six-membered rings around the above-mentioned six-membered ring at the bottom position of C_{60} are approximately parallel to the three phenolic rings of the host, where the closest distance is 3.615 (6) Å. In addition to the π–π interactions, the very short distance (3.29 Å) of the dibenzyl ether oxygens to the six-membered ring at the bottom of C_{60} indicates σ–π or charge transfer interactions between the oxygens and C_{60}. There are also close intermolecular interactions between C_{60} molecules with closest C–C distance of 3.51 (3) Å.

p-Benzylhomooxacalix[3]arene (**3b**), however, forms 2:1 complex with C_{60} from toluene [15]. This complex has D_{3d} symmetry, and the conformational characters of the oxacalixarene and the relationship between the host and guest are very similar to those in the above **3d**·C_{60} complex.

O-benzyl-*p*-iodocalix[4]arene (**2a**) co-crystallises with C_{60} from an *o*-dichlorobenzene/propan-2-ol mixture forming a well-ordered solid-state structure with intercalation of C_{60} into a calixarene bilayer formed by many van der Waals contacts [26], but there is no specific interaction between the calixarene and C_{60} in this structure. Structures of clathrates of C_{60} with *O*-propyl-*p*-bromocalix[4]arene (**2b**) [27] and resorcin[4]arene **7** [28] were also reported. In all these cases, C_{60} is outside the aromatic cavities of calix[4]arenes or resorcin[4]arene, apparently because the size of the cavities is not large enough for binding C_{60}.

In summary, the main members of calixarene family (homooxacalix[3]arene and calix[5,6,8]arene) except calix[4]arene were found to form inclusion complexes with C_{60} (also C_{70} in some cases). Without consideration of the packing patterns in the solid state, there are three basic complex types as shown in Fig. 2. Small calixarenes (homooxacalix[3]arene and calix[5]arene) prefer the *cone* conformation and thus form either 1:1 complexes with fullerene like a ball in a socket or 2:1 complexes like a ball in a capsule; large calixarenes (calix[6]arene, and maybe calix[8]arene) prefer double-*cone* conformations and thus form 1:2 complex with two fullerene molecules in each of the shallow cavities. The structures clearly imply that the attractive force between the host and the guest mainly comes from the π-π interactions. Even subtle change in the calixarene structure may change the equilibrium between the competitive calixarene–fullerene, fullerene–fullerene, and calixarene–calixarene interactions, leading to a different structure of the complex with the fullerene in the solid state.

3. Films of Calixarene-Fullerene Complexes

Not very long after the report of the *p*-*tert*-butylcalix[8]arene–fullerene complex (**1**·C_{60}) as a precipitate [4,5], Langmuir films of calixarene–fullerene complexes at the water-air interface were described [29,30,31]. By repeated deposition of very dilute (*ca.* 4 × 10^{-5} M) chloroform solutions of calixarene–fullerene complexes (**1**·C_{60} and **1**·C_{70}) onto an unbuffered water subphase and evaporation of the solvent, Langmuir films of the complexes at the water–air interface were successfully prepared [29]. Based on the π–A isotherms and Brewster angle microscope (BAM) images, the authors observed that both free calixarene **1** and the complex **1**·C_{60} form monolayer films at not very high surface pressure, and start to form double layers with a pressure increase. The effective molecular diameters for **1** and **1**·C_{60} were estimated to be ~ 15.6 ± 1.4 Å and 17.3 ± 1.9 Å, respectively. Based on the effective diameter values and the very similar characteristics of the films of the free calixarene and the complex, it was concluded that the C_{60} molecule was indeed inside the calix[8]arene molecule. However, the [70]fullerene complex **1**·C_{70} was found to have a high tendency to form clusters and multilayers ever at very low surface pressure, *i.e.*, at the gas–solid coexistence stage, suggesting that the complex dissociates so that the C_{70} molecule is not inside the basket formed by the calixarene molecule in the film. This result is consistent with the facts that the calixarene–C_{70} bond is weaker than the calixarene–C_{60} bond and that C_{70} forms 1:1 or 1:2 complexes with calixarene **1** depending on the conditions [32].

For Langmuir films formed by spreading CCl_4 solutions of the free calixarene **1**, the complex **1**·C_{60}, or the equimolar mixture of the calixarene and the fullerene onto a water subphase, spreading surface pressure–molecular area isotherms showed that the limiting molecular areas of the **1**·C_{60} complex (58 ± 2 Å^2) and the equimolar mixture (59 ± 2 Å^2) are almost identical, and are similar to that of the free calixarene **1** (75 ± 2 Å^2) in the Langmuir films [30,31]. Based on the very small molecular area values, it was suggested that, at the air–water interface, *p-t*-butylcalix[8]arene adopts not a *cone* conformation but a "*pleated-loop*" conformation [19] perpendicular to the water–air interface, somewhat like an upright, flexible wheel, in all above films. The interaction between the hydrophobic C_{60} and the calixarene aromatic backbone cavity does not change the overall conformation, but partially hinders the fluctuations of the phenolic rings, leading to the decrease in both the molecular flexibility and the interfacial fluidity of the calixarene. The presence of the complex in the films prepared from either **1**·C_{60} complex or equimolar mixture is supported by UV–VIS absorption spectra of Langmuir–Blodgett films and collapsed material transferred onto quartz plates, which show two bands located at λ_{max} = 264 and 338 nm and a shoulder at 288 nm. A promotion effect of the water–air interface on the formation of the calixarene–fullerene complex was also observed. However, it is not apparent how the "pleated-loop" calixarene complexes with the fullerene.

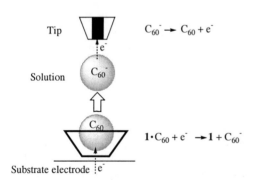

Figure 6. Reduction of the fullerene breaks the calixarene–fullerene complex in the particle film showed by SECM–DCM experiment [33].

The electrochemical reduction of 3–5 μm particle films of the complex was studied using the scanning electrochemical microscope (SECM) combined with a quartz crystal microbalance (QCM) for several electrolytes in acetonitrile [33]. With the generation/collection mode of the SECM monitoring the flux of electrochemically active species in and out of the particle film, and the QCM measuring small changes in the mass of the substrate resulting from ion incorporation or dissolution processes, the combination of both techniques [34] provides a powerful tool for determining the effects of reduction upon the particulate film of **1**·C_{60} complex. The experiments showed that the complex breaks apart upon reduction of the fullerene, with the fullerene anion escaping from the calixarene basket into the solution, leaving the calixarene as an insoluble film on the electrode surface (Fig. 6). This result indicates that the calixarene–C_{60} interaction is weakened by the injection of an electron into the fullerene, supporting the view that the attractive force between calixarene and fullerene mainly comes from the electron-sharing of the electron-poor fullerene with the electron-rich calixarene. Complexation of the fullerene by calixarene **1** within the film results in a negative shift of the peak potential of the first cathodic wave by about 400 mV compared to the reduction of a pure C_{60} film [33]. The electrochemical behaviour of calixarene-fullerene complexes, **2i**·$(C_{60})_2$ [35], **1**·$(C_{70})_2$ [36] and as well as **1**·C_{60} [37], in films on glassy carbon electrode were also reported.

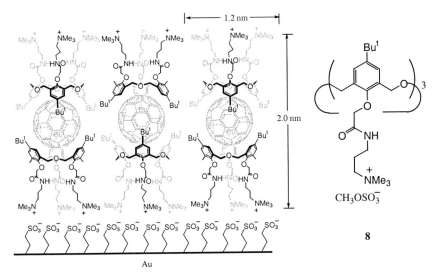

Figure 7. Cationic calixarene–fullerene complex monolayer film on the anion-covered gold surface.

A 2:1 complex of a water-soluble cationic homooxacalix[3]arene **8** and C_{60} [52] forms a monolayer (or at least, monolayer-like, nanometre-sized) film on an anion-covered gold surface (Fig. 7) [53]. Knowledge that the cationic calixarene **8** is capable of solubilizing C_{60} into water by forming a capsule-like $(8)_2 \cdot C_{60}$ complex [52] suggested that this cationic complex would be adsorbed onto an anionic surface, forming a monolayer or nanometre-sized thin film. A quartz crystal microbalance (QCM) [54] with a gold electrode coated on each side of the QCM resonator was used to monitor the adsorption process. The resonator was first exposed to an aqueous solution containing 2-mercaptoethanesulphonate to form the anionic surface on the gold electrode, and then immersed in an aqueous solution of the cationic calixarene **8** or the $(8)_2 \cdot C_{60}$ complex. In the case of **8**, the QCM frequency decreases linearly with the number of adsorption cycles, indicating that the layers grow indefinitely. In the case of $(8)_2 \cdot C_{60}$ complex, in contrast, the frequency change is saturated after the second cycle, indicating that the surface, once covered by the complex, grows no further. Atomic force microscopy and scanning electron microscopy analyses showed that the surface was smooth and very similar to that of the bare gold surface without any cluster-like domains of the $(8)_2 \cdot C_{60}$ complex. The thickness of the film was estimated to be 2.1 nm from the QCM frequency change, and the molecular area of the complex in the film was estimated to be 1.0 nm^2 from the first reduction peak current of its cyclic voltammogram. These values are in good agreement with the molecular size (2.0 nm × 1.2 nm) obtained by a computational method (Discover 98), implying that the adsorbed complex forms a smooth monolayer film. Very interestingly, when the $(\mathbf{8})_2 \cdot C_{60}$ complex-coated electrode (1.8 cm^2) was photoirradiated (wavelength 300~510 nm), a reversible photoinduced electron current wave (*ca.* 640 nA) was observed [53].

4. Complexation in Organic Solutions

While the X-ray crystal structures of calixarene–fullerene complexes in the solid state give rich information on how calixarene hosts interact with fullerene guests, studies in solution can give more information on the strength of the host–guest interaction by determination of association constants. While the interaction is inherently weak since fullerenes have apolar and smooth globular structures, calixarenes (or other host molecules) with preorganised complementary shapes can bind fullerenes through the accumulation of weak van der Waals interactions and weak charge-transfer interactions to form stable complexes even in organic solutions, overcoming the competitive solvent–calixarene and solvent–fullerene interactions.

At the beginning of our studies on the complexation in solutions, a survey of the literature showed that the charge-transfer interaction with *N,N*-dialkylaniline derivatives was the sole non-covalent interaction useful to capture [60]fullerene in organic solvents [38]. Therefore, we decided to introduce *N,N*-dialkylaniline units or *m*-phenylenediamine units into calix[n]arenes. When they are introduced into the upper rim of a calix[6]arene, the size of the extended cavity becomes comparable with that of a fullerene molecule. Furthermore, the treatment of a calix[6]arene is much easier than that of a calix[8]arene. Taking these advantages into consideration, we designed and synthesised **9** and **10** and compared their K_{ass} values with those of reference compounds **11** and **12** [39]. The K_{ass} values determined in toluene from the spectroscopic absorption change were 7.9 M^{-1} for **9**, 110 M^{-1} for **10**, 0.20 M^{-1} for **11**, and 0.43 M^{-1} for **12**.

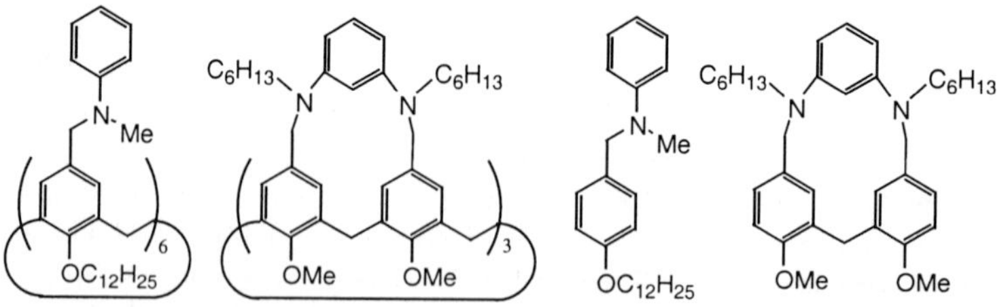

Comparison of the data for **9 ~ 12** establishes that K_{ass} of reference compound **12** is only 2-fold larger than that of **11**, whereas the K_{ass} of the host compound **10** is 14-fold larger than that of **9**. The large K_{ass} enhancement of **10** relative to **9** is accounted for by (i) the greater donor ability of the *m*-phenylenediamine unit relative to the *N,N*-dialkylaniline one and (ii) the higher preorganisation brought about by linking two calix[6]arene phenyl units. The proposed structure including the co-operative action of donor groups is shown in Fig. 8. These results successfully demonstrate that stable inclusion complexes of C_{60} in solution can be formed with host molecules in which donor groups such as N,N-dialkylaniline or m-phenylenediamine are preorganised on an appropriate platform. Undoubtedly, calix[6]arene is one of the best platforms for this purpose.

Later, it was found that C_{60} and excess cyclotriveratrylene (CTV: **13**) in toluene form micelle-like aggregates which are spectrophotometrically detectable [8b]. This implies that the CTV–C_{60} interaction is not so strong as to disperse C_{60} particles discretely into toluene solution but does exist even in toluene solution.

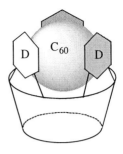

Figure 8. Proposed structure **10**·C_{60} *including the co-operative action of donor groups.*

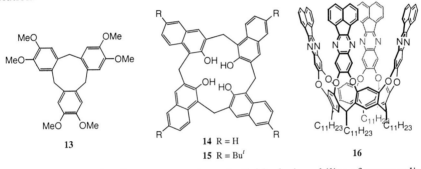

This finding stimulated us to screen the potential inclusion ability of many calixarene derivatives we have synthesised so far. After examination of 28 calixarene derivatives we discovered that OH-unsubstituted calix[5]arenes, calix[6]arenes, and homooxacalix[3]arenes do interact with C_{60} in toluene [41]. These findings support the view that C_{60} can be included by a few selected calixarenes. From the extensive screening, we concluded that the primary prerequisite for C_{60} inclusion is that the OH groups on the lower rim are not substituted so that they can adopt a *cone* conformation. The secondary prerequisite is a suitable ring size for the interaction is observed only for unmodified calix[5]arene, calix[6]arene, and homooxacalix[3]arene but not for calix[4]arene. Previously, we theoretically estimated the most stable conformations of unmodified calix[n]arenes [42], which showed good agreement with those determined by X-ray crystallographic analyses [43,44]. From comparison of the structures of the most stable conformers with the C_{60} globular structure we noticed that the cavities of these calixarenes are too small to deeply "include" C_{60} but have an inclination of the benzene rings appropriate for a multi-point contact with the C_{60} surface: that is, they can interact as a partial "cap" for C_{60} (Fig. 9). This result caused us to consider that the major driving-force for C_{60} inclusion is a π–π interaction (including the charge-transfer-type interaction) and/or a solvophobic effect. To fully exploit this effect, calixarenes are required to be preorganised in a *cone* conformation through intramolecular hydrogen-bonds among OH groups.

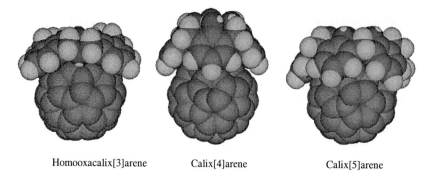

Homooxacalix[3]arene   Calix[4]arene   Calix[5]arene

Figure 9. Energy-minimised [with MM3 (92)] structures of homooxacalix[3]arene, calix[4]arene, and calix[5]arene show that calix[5]arene and homooxacalix[3]arene can provide a benzene ring inclination suitable to C_{60}-binding whereas the upper-rim edge of calix[4]arene is two narrow to accept C_{60}.

TABLE 1. Association constants (K_{ass}) for calixarene–C_{60} complexation in toluene determined by UV-VIS titration method at room temperature.

Compound	Calix[n]arene	Upper rim substituent	Lower rim substituent	K_{ass} / dm^3 mol^{-1}	Reference
2d	5	Bn	OH	2800 ± 200 [a]	15
2e	5	H	OH	30 ± 2	45
2f	5	t-Bu	OH	9 ± 1	45
2g	5	Allyl	OH	292 ± 15	45
4	5	see the formula	OH	2120 ± 110	21
5	5	see the formula	OH	1673 ± 70	21
6	5	see the formula	OH	588 ± 70	21
3a	3	H	OH	9.1 ± 1	25
3c	3	t-Bu	OH	35.6 ± 0.3	25
3c	3	t-Bu	OH	35 ± 5	42
3e	3	OMe	OH	20.7 ± 0.9	25
3d	3	Br	OH	14.9 ± 2.0	25
3b	3	Bn	OH	100 [b]	15
9	6	see the formula	OC$_{12}$H$_{25}$	7.9	39
10	6	see the formula	OMe	110	39
14	4	see the formula	OH	661 ± 51	46
15	4	see the formula	OH	708 ± 91	46

[a] K_2 for formation of 2:1 complex is also reported to be 230 ± 50 M^{-1}.
[b] K_2 for formation of 2:1 complex.

The association constants summarised in Table 1 show that calix[5]arenes bind C_{60} more tightly than other calixarenes. This is consistent with the fact that X-ray crystallography (see Section 2) shows that calix[5]arene can provide more points for short-range interactions with C_{60}.

The conformational freedom remaining in calixarene ester derivatives can be lost in a *cone* conformation enforced by complexation with appropriate metal cations [43,44]. If the above proposals are correct, the conformationally-mobile ester derivatives should not include C_{60} whereas their metal-frozen *cone* conformers should. To test this hypothesis, we thoroughly investigated the spectroscopic properties of C_{60} in the presence of **2c**, **2j**, **3f**, **3g** and **3h**. We confirmed that the calix[n]aryl esters become able to interact with C_{60} only when they are appropriately preorganised into a *cone* conformation by complexed metal cations [42]. Typical metal complexes which accept C_{60} are (**2j** + Cs^+) and (**3f** + Li^+). **2j** tends to adopt a *1,2,3-alternate* conformation while it changes into a C_{3v}-symmetrical *cone* in the presence of Cs^+ (Fig. 10) [43]. **3f** is immobilised in a *cone* but the ether-containing ring is flexible. Li^+ bound to the lower-rim ester-groups restricts the molecular motion and rigidifies the *cone* skeleton [44]. Furthermore, ^1H NMR spectroscopic studies showed that the inclination angles of (**2j** + Cs^+) and (**3f** + Li^+) are very close to those of **2i** and **3c**, respectively, both of which can accept C_{60} [25,41].

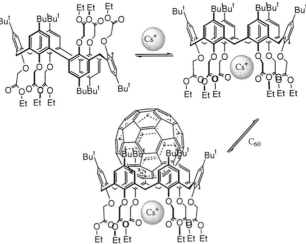

Figure 10. Complexation of $2_6 \cdot Bu^t \cdot Es$ with Cs^+ creates a cone cavity for C_{60} inclusion.

While calix[4]arene and resorcin[4]arene cannot form fullerene inclusion complexes because their upper rims are too narrow, their analogues or derivatives with extended cavities can. For example, calix[4]naphthalenes (**14, 15**) can effectively bind C_{60} in toluene, benzene, and CS_2 solutions [46]. Compared to calix[4]arene, calix[4]naphthalene has a deeper and larger cavity, and thus can include the fullerene by multiple π-π interactions. The same is true for resorcin[4]arenes, the resorcin[4]arene-based deeper cavitand **16** binding C_{60} in toluene with a K_{ass} of 900 ± 250 M^{-1} [47].

The association constants of calixarene–fullerene complexes are strongly solvent dependent [21], the K_{ass} values of **4**·C_{60}, for example, increasing from 308 M^{-1} in *o*-dichlorobenzene to 660 M^{-1} in carbon disulfide, 750 M^{-1} in tetralin, 1840 M^{-1} in benzene, and 2120 M^{-1} in toluene. The association constant increases with decrease of the solubility of C_{60}, suggesting that the more weakly solvated guest is more strongly bound by the host. The thermodynamic parameters for complex formation by C_{60} with calix[5]arene receptors show that the entropy change in chloroform is negative, but in toluene is positive, suggesting tight solvation around the guest in toluene [21b]. However, interpretation of high-precision density measurements indicated that in complexation of C_{60} with *p*-benzylcalix[5]arene (**2d**) in toluene, two solvent molecules are displaced from the cavity of the host [40]. These results proved that complex formation competes with solvation of both the host and guest.

Direct calorimetric measurements at 25 °C of the enthalpy changes ($\Delta H°$) for inclusion of C_{60} and C_{70} by p-t-butylhomooxacalix[3]arene (**3c**) in toluene gave values of -10.8 ± 0.3 kJ mol^{-1} and -9.1 ± 0.2 kJ mol^{-1}, respectively [48]. These values are comparable with the values (-11.9 ± 0.5 kJ mol^{-1} and -12.5 ± 0.5 kJ mol^{-1}, respectively) estimated from the temperature dependence of spectrophotometrically determined stability constants, and thus support the reliability of the absorption spectral method which is generally used.

Complexation by calixarenes in solutions usually causes an absorbance increase at around 430–440 nm in the visible spectrum of C_{60}, and, when the binding is strong enough, leads to a colour change from magenta to pale yellow [21,22,25,41,47]. The complexation also causes negative shifts in the reduction and oxidation potentials of C_{60} [49]. In some cases, complexation-induced chemical shift changes in the ^{1}H NMR spectra of calixarenes [41,47] or the ^{13}C NMR spectra of C_{60} [21,50] can also be observed.

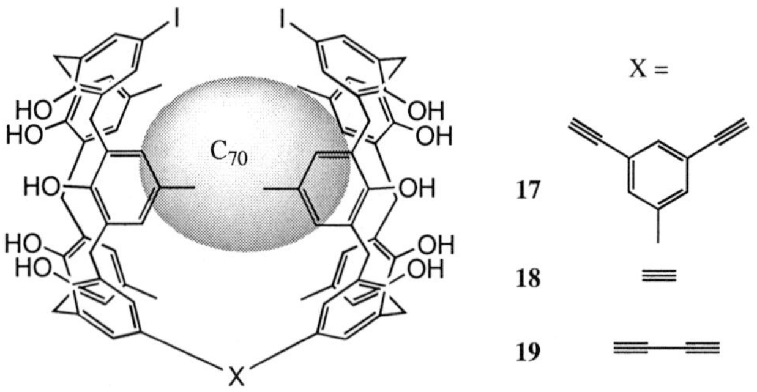

5. Complexation by Calixarene Dimers: Towards Complete Inclusion of Fullerenes

Characterisation of a calixarene–fullerene complex with 2:1 stoichiometry (($\mathbf{4}$)$_2$·C_{60}) in the solid state [21a] indicated that covalent bridging of two calix[5]arene units as in **17**–**19**, with well-defined cavities, might create more effective fullerene receptors [51]. This provided a dramatic increase in the association constants for 1:1 complexes of fullerenes in solution. For example, the K_{ass} value of the complex of host **17** with C_{60} in toluene is as high as 76000 ± 5000 M^{-1}, which is more than 36 times larger than the corresponding value (2100 ± 100 M^{-1}) of the monomeric host **4**. All three receptors bind C_{70} preferentially over C_{60}. Amongst them, **18** has the highest C_{70}/C_{60} selectivity (10.2 in toluene), while **17** has the highest binding ability for C_{70} ($K_{ass} = 163000$ M^{-1} in toluene). Upon complexation by **17**, the largest complexation-induced up-field shift (1.77 ppm) in the ^{13}C NMR spectrum was observed for the carbon atoms at the poles of C_{70}, indicating that the elliptical guest adopts an orientation with its poles residing deepest within the host cavity [51]. Butenyl-bridged biscalix[n]arenes (n = 5, 6) also show higher binding abilities to C_{60} and C_{70}, compared with the monocalixarenes [45].

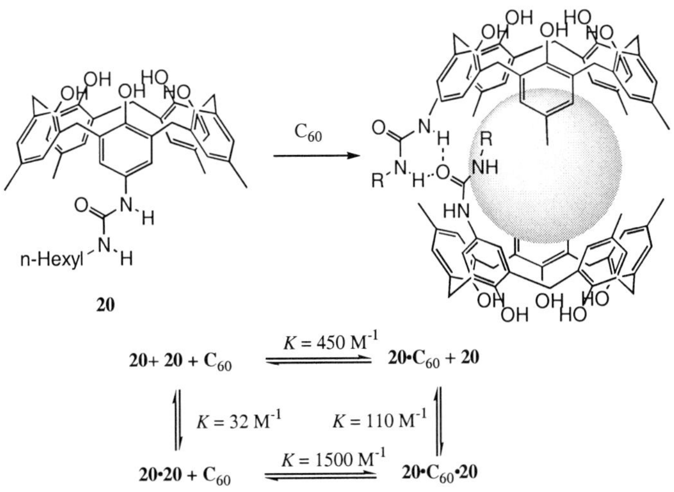

Figure 11. Schematic presentation of the association process in the hydrogen-binding dimeric system [55].

A molecular container for fullerenes self-assembling through hydrogen-bonding (Fig. 11) has been recently developed [55]. Based on the chemical shift change of the NH proton in the urea-functionalised calix[5]arene **20** accompanying the hydrogen-bonding, the formation constant of **20·20** dimer was determined to be 32 M^{-1}. In the presence of a small amount of C_{60}, the constant increased significantly to 104 M^{-1}, indicating the formation of the ternary complex **20·C_{60}·20**. ^{13}C NMR data provided further evidence for the formation of the ternary complex. In the presence of excess host **20**, the complexation-induced up-field shift value of the C_{60} signal was found to be 0.82 ppm ([C_{60}] = 0.92 mM, [**20**] = 3.7 mM, in $CDCl_3$: CS_2 = 2:1 v/v). When trifluoroacetic acid was added, however, the shift value decreased to 0.21 ppm, presumably because the 2:1 complex was decomposed to a 1:1 complex by the cleavage of the hydrogen bonds. Using this self-assembling host, the binding and release of the fullerenes can be easily controlled.

The pyridylhomooxacalix[3]arene **21** with a fixed *cone* conformation self-assembles into a dimeric capsule molecule **22** in the presence of PdL^{2+} (Fig. 12) [56]. There is unequivocal evidence that C_{60} is encapsulated in this cavity. In $CDCl_2CDCl_2$ at 25°C, a mixture of C_{60} (5.8 mM: ^{13}C-enriched) and **22** (5.0 mM) gave two separate peaks at 142.87 and 140.97 ppm in the ^{13}C spectrum. The peak at higher magnetic field is ascribable to C_{60} included in the cavity of **22** (Fig. 12). The ^1H NMR spectrum also showed separate peaks for free **22** and complex **22·C_{60}**. From the ratio of the peak intensities, K_{ass} was estimated to be 54 M^{-1}. Upon complexation, the phenyl rings of the host became a little more flattened to create a suitable cavity for the guest. Comparison of the NMR spectra of **22·C_{60}** and the lithium complex of **21** indicates that they have very similar phenyl ring inclination angles. We thus examined the binding ability of the Li^+ complex of the molecular capsule to C_{60}, and found that the K_{ass} is increased by as much as 38-fold to 2100 M^{-1} [56c]. The Na^+ complex, however, shows very poor binding ability to the fullerene, apparently because the larger size of Na^+ impedes the tilting of the phenyl rings. To the best of our knowledge, this is the first example for total encapsulation of C_{60} in a cage compound.

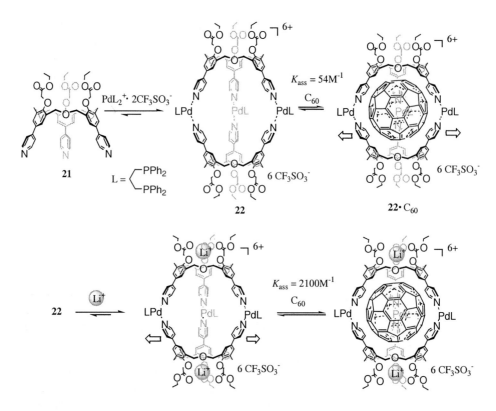

Figure 12. Proposed structures of self-assembled molecular capsule 22 and its complex with C_{60} [56].

6. Water-Soluble Calixarene–Fullerene Complexes

One of the most interesting properties of fullerenes is their biological activity [57]. Their spherical shape, hydrophobicity, and electronic effects make fullerenes very appealing in medicinal chemistry. Owing to their peculiar geometrical shape, fullerenes can act as HIV protease inhibitors by occupying the quasispherical hydrophobic cavity of the active site [58]. As singlet oxygen ($^1O_2$) photosensitiser, they can cleave DNA selectively at guanine residues, showing potentials for use in diagnostics and photodynamic therapy [59,60]. As free radical scavengers, they show neuroprotective properties [61] and anti-apoptotic activities [61b,62]. Endohedral metallofullerenes have potentials for use as new diagnostic or therapeutic radiopharmaceuticals [63]. In spite of these great potentials, medicinal applications have been greatly limited by the very low solubility of fullerenes in water. Although their water solubility can be improved by chemical introduction of hydrophilic appendages, solubilisation by supramolecular interactions is especially attractive because the fullerene can remain its original structure. Undoubtedly, water-soluble calixarenes are very suitable for this purpose.

As early as 1992, it was known that the water-soluble calix[8]arene **23** (14-fold excess to C_{60}) can extract C_{60} from organic (*e.g.*, toluene) phase into aqueous solution [3]. An effective mechanochemical method for aqueous solubilisation of crystalline fullerenes by supramolecular complexation with sulfonatocalix[8]arene **2m** (and also γ-cyclodextrin) was reported in 1999 [64]. Reaction of an equimolar mixture of C_{60} and the sulfonatocalix[8]arene in a high-speed vibration mill for just 10 minutes resulted in a complex which could be dissolved in water to a concentration of 0.13 mM. Fullerene dimer (C_{120}) can also be made water-soluble by this method.

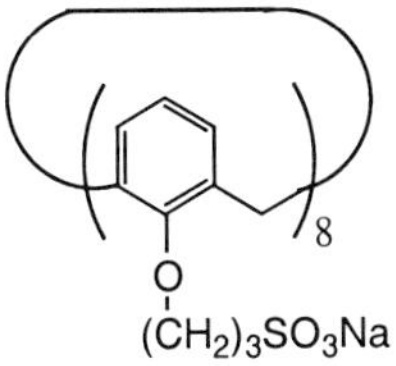

23

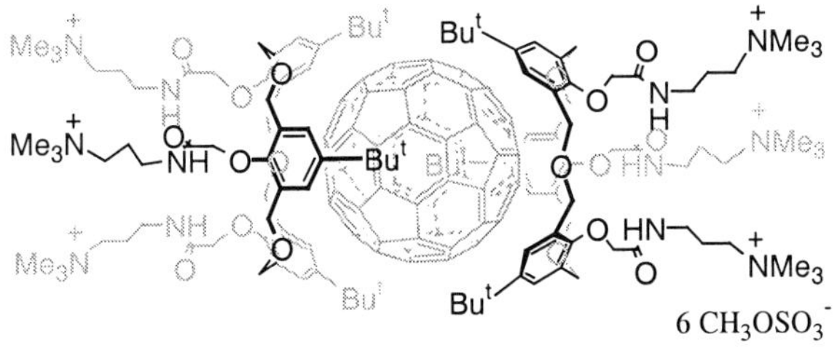

$(8)_2 \cdot C_{60}$

Extraction of solid C_{60} with an aqueous solution of the cationic homooxacalix[3]arene **8** (0.5 mM) yields a water-soluble 2:1 complex (($8)_2 \cdot C_{60}$) [52]. From dynamic light-scattering (DLS) measurements, the particle size of the complex was estimated to be 1.4 nm, which is comparable with the size of the 2:1 complex as simulated by a computational method, suggesting that the complex does not form large aggregates under the experimental conditions. Very interestingly, this calixarene-fullerene solution shows efficient activity in the photo-cleavage of DNA. Under visible light irradiation (for 6 h), about 65% of ColE1 supercoiled plasmid was converted to nicked DNA in the presence of the complex, whereas no conversion was observed in the dark or in the absence of the complex. Presumably, C_{60} included in "cationic" **8** is transported onto "anionic" DNA with the aid of electrostatic interactions and cleaves it with the aid of photoirradiation. This is one of a few successful examples [60] in which an "unmodified" fullerene is directly supplied to DNA.

Very recently, we investigated the properties of the excited states of C_{60} in this water-soluble complex, and found that the triplet-triplet (T-T) absorption band shifts to shorter wavelengths and the decay rate increases very much, compared with that of bare C_{60} in benzene [65].

7. Miscellaneous

To enhance the binding between calixarenes and fullerenes, one approach is to create a better-preorganised cavity in the calixarenes (*e. g.*, by extending the cavity or linking two calixarenes to form a capsule-like structure) as discussed in Sections 3 and 4; another possible approach is to connect a fullerene moiety onto a calixarene host through a flexible linkage. In 1994, we reported the calixfullerene **24** which can form an intramolecular inclusion complex in the solid state, but shows no clear evidence for the inclusion of the C_{60} moiety in the calixarene moiety [66]. Reexamining the structure of **24** in the light of later results concerning calixarene–fullerene interactions, we surmised that the possible reasons are that the calix[8]arene moiety (especially because one of the OH groups is substituted) has only very weak binding ability, and that the C_{60} moiety was connected onto the lower rim, opposite to the cavity. We thus designed and synthesised a novel calixfullerene (**25**) in which a C_{60} moiety was connected onto the upper rim of a *cone*-conformation homooxacalix[3]arene moiety through a triethylene glycol chain [67]. In CDCl$_3$, **25** gives a complicated ^{1}H NMR spectrum with two separate sets of peaks which can be assigned to the self-complexed conformer and the uncomplexed conformer (Fig. 13), respectively. Upon addition of CD$_3$CN to the solution, the intensity of the signals for the self-complexed conformer increases significantly. In CDCl$_3$, **25** exists mainly as the uncomplexed conformer; in CDCl$_3$: CD$_3$CN = 1:1 (v/v), however, it exists only as the self-complexation conformer. This result is consistent to the relatively high solubility of C_{60} in chloroform and very poor solubility in acetonitrile. In chloroform, the strong solvation interaction suppresses the calixarene–fullerene interactions leading to the lack of complexation; in acetonitrile, the weak solvation interaction facilitates the self-complexation. Molecular weight measurements, coupled to the concentration independence of the solution distribution, show that there is no intermolecular complexation and support this notion of a self-complexation equilibrium.

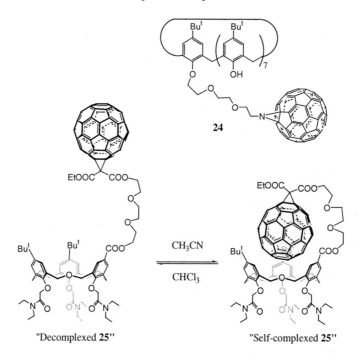

Figure 13. Schematic representation of the complexation–decomplexation exchange in calixfullerene **25**.

Other covalently-linked calixarene–fullerene conjugates [68,69] show attractive potential for applications in modification and control of the fullerene properties. Since these reports do not directly involve supramolecular calixarene–fullerene binding, discussion is not given here in detail.

Two of the ultimate goals of supramolecular chemistry are the catalysis and control of chemistry reactions. Studies of calixarene–fullerene interactions for these purposes are still very limited, but are worthy of notice. The complex of *p-t*-butylcalix[8]-arene (**1**) with C_{60} can be used for the functionalisation of the fullerene [70]. Gaseous reagents (Cl_2, Br_2, and amines) react with C_{60} in the solid $1 \cdot C_{60}$ complex at dramatically higher reaction rates and give simpler products, in comparison with the reactions with bare C_{60}. This represents a novel and effective method for functionalisation of C_{60}. $(C_{70})_2 \cdot 2h$ complex is capable of mediating the electron transfer for electrocatalytic reduction of nitrite on a glassy carbon electrode surface [71].

8. Conclusions

The foregoing studies have clearly demonstrated that a major driving force for formation of calixarene–fullerene complexation is the π–π interaction between the large complementary surfaces of the host and guest. Calixarenes with a preorganised *cone* (or double-*cone*) conformation and a proper inclination of the benzene rings can form inclusion complexes with fullerenes. The binding ability can be greatly enhanced by extension of the cavity through introduction of electron-rich moieties onto the upper rim, or by creating a capsule-like cavity through dimerisation of the calixarene with an appropriate linkage. Covalently connecting the host and guest is another approach to facilitate their interactions.

The supramolecular interaction between calixarenes and fullerenes has already been shown to have important potential applications, such as in purification of fullerenes, medicinal uses, molecular devices, and catalysis. With continuing progress in the research of fullerenes and supramolecular chemistry, we believe that the calixarene–fullerene interaction will find more and more applications.

9. References and Notes

[1] a) F. W. McLafferty (Ed.), *Acc. Chem. Res.* **1992**, *25*, 97–176 (Special issue on buckminsterfullerenes); b) H. W. Kroto, J. R. Health, S. C. O'Brien, R. F. Smally, *Nature* **1985**, *318*, 162–163; c) W. Krätschmer, I. D. Lamb, K. Fostiropoulus, D. R. Huffman, *Nature* **1990**, *347*, 354–358.

[2] For comprehensive reviews see: a) C. D. Gutsche, *Calixarenes*, Royal Society of Chemistry, Cambridge, **1989**; b) J. Vicens, V. Böhmer (Eds.), *Calixarenes*, Kluwer Academic Press, Dordrecht, **1991**; c) V. Böhmer, *Angew. Chem. Int. Ed. Engl.* **1995**, *34*, 713–745; d) A. Ikeda, S. Shinkai, *Chem. Rev.* **1997**, *97*, 1713–34; d) C. D. Gutsche, *Calixarenes Revisited*, Royal Society of Chemistry, Cambridge, **1998**.

[3] R. M. Williams, J. W. Verhoeven, *Recl. Trav. Chim. Pays–Bas* **1992**, *111*, 531–532.

[4] J. L. Atwood, G. A. Koutsantonis, C. L. Raston, *Nature* **1994**, *368*, 229–231.

[5] T. Suzuki, K. Nakashima, S. Shinkai, *Chem. Lett.* **1994**, 699–702.

[6] a) S. Shinkai, A. Ikeda, *Pure & Appl. Chem.* **1999**, *71*, 275–280; b) S. Shinkai, A. Ikeda, *Gazz. Chim. Ital.* **1997**, *127*, 657–662.

[7] a) T. Andersson, K. Nilsson, M. Sundahl, G. Westman, O. Wennerström, *J. Chem. Soc., Chem. Commun.* **1992**, 604–606; b) T. Andersson, G. Westman, G. Stenhagen, M. Sundahl, O. Wennerström, *Tetrahedron Lett.* **1995**, *36*, 597–600; c) Z. Yoshida, H. Takekuma, S. Takekuma, Y. Matsubara, *Angew. Chem. Int. Ed. Engl.* **1994**, *33*, 1597–1599; d) S. Takekuma, H. Takekuma, T. Matsumoto, Z. Yoshid, *Tetrahedron Lett.* **2000**, *41*, 1043–1046.

[8] a) J. W. Steed, P. C. Junk, J. L. Atwood, M. J. Barnes, C. L. Raston, R. S. Burkhalter, *J. Am. Chem. Soc.* **1994**, *116*, 10346–10347; b) J. L. Atwood, M. J. Barnes, M. G. Gardiner, C. L. Raston, *J. Chem. Soc., Chem. Commun.* **1996**, 1449–1450; c) H. Matsubara, T. Shimura, A. Hasegawa, M. Semba, K. Asano, K. Yamamoto, *Chem. Lett.* **1998**, 1099–1100; d) H. Matsubara, A. Hasegawa, K. Shiwaku, K. Asano, M. Uno, S. Takahashi, K. Yamamoto, *Chem. Lett.* **1998**, 923–924; e) J.-F. Nierengarten, L. Oswald, J.-F. Eckert, J.-F. Nicould, N. Armaroli, *Tetrahedron Lett.* **1999**, *40*, 5681–5684.

[9] F. Diederich, J. Effing, U. Jonas, L. Jullien, T. Plesnivy, H. Ringsdorf, C. Tilgen, D. Weistein, *Angew. Chem. Int. Ed. Engl.* **1992**, *31*, 1599–1602.

[10] a) O. Ermer, C. Röbke, *J. Am. Chem. Soc.* **1993**, *115*, 10077–10082; b) O. Ermer, *Helv. Chim. Acta* **1991**, *74*, 1339–1351.

[11] P. C. Andrews, J. L. Atwood, L. J. Barbour, P. J. Nichols, C. L. Raston, *Chem. Eur. J.* **1998**, *4*, 1384–1387.
[12] a) M. M. Olmstead, D. A.Costa, K. Maitra, B. C. Noll, S. L. Phillips, P. M. Van Calcar, A. L. Balch, *J. Am. Chem. Soc.* **1999**, *121*, 7090–7097 and references therein; b) K. Tashiro, T. Aida, J. Zheng, K. Kinbara, S. Saigo, S. Sakamoto, K. Yamaguchi, *J. Am . Chem. Soc.* **1999**, *121*, 9477–9478.
[13] a) M. J. Hardie, C. L. Raston, *Chem. Commun.* **1999**, 1153–1163; b) A. L. Balch, M. M. Olmstead, *Coord. Chem. Rev.* **1999**, *185–186*, 601–617; c) F. Diederich, M. Gómez-López, *Chem. Soc. Rev.* **1999**, *28*, 263–277.
[14] J. L. Atwood, S. Bott, C. Jones, C. L. Raston, *J. Chem. Soc., Chem. Commun.* **1992**, 1349–1351.
[15] J. L. Atwood, L. J. Barbour, P. J. Nichols, C. L. Raston, C. A. Sandoval, *Chem. Eur. J.* **1999**, *5*, 990–996.
[16] R. M. Willams, J. M. Zwier, J. W. Verhoeven, *J. Am. Chem. Soc.* **1994**, *116*, 6965–6966.
[17] a) G. D. Andreetti, O. Ori, F. Ugozzoli, A. Alfieri, A. Pochini, R. Ungaro, *J. Incl. Phenom.* **1988**, *6*, 523–536; b) R. Ungaro, A. Pochini, G. D. Andreetti, P. Domian, *J. Chem. Soc., Perkin Trans. 2* **1985**, 197–201; c) G. D. Andreetti, A. Pochini, R. Ungaro *J. Chem. Soc., Perkin Trans. 2* **1983**, 1773–1779.
[18] C. L. Raston, J. L. Atwood, P. J. Nichols, I. B. N. Sudria, *Chem. Commun.* **1996**, 2615–2616.
[19] C. D. Gutsche, A. E. Gutsche, A. I. Karaulov, *J. Incl. Phenom.* **1985**, *3*, 447–451.
[20] O. P. Dimitriev, Z. I. Kazantseva, N. V. Lavrik, *J. Incl. Phenom.* **1999**, *35*, 85–91.
[21] a) T. Haino, M. Yanase, Y. Fukazawa, *Angew. Chem. Int. Ed. Engl.* **1997**, *36*, 259–260; b) M. Yanase, M. Matsuoka, Y. Tatsumi, M. Suzuki, H. Iwamoto, T. Haino, Y. Fukazawa, *Tetrahedron Lett.* **2000**, *41*, 493–497.
[22] T. Haino, M. Yanase, Y. Fukazawa, *Tetrahedron Lett.* **1997**, *38*, 3739–3742.
[23] a) J. F. Gallagher, G. Ferguson, V. Böhmer, D. Kraft, *Acta Cryst., C (Cr. Str. Comm.)* **1994**, *50*, 73–77; b) L. J. Atwood, R. K. Juneja, P. C. Junk, K. D. Robinson, *J. Chem. Cryst.* **1994**, *24*, 573–576; c) M. Coruzzi, G. D. Andreetti, V. Bocchi, A. Pochini, R. Ungaro, *J. Chem. Soc., Perkin Trans. 2* **1982**, 1133–1138.
[24] J. L. Atwood, L. J. Barbour, C. L. Raston, I. B. N. Sudria, *Angew. Chem. Int. Ed. Engl.* **1998**, *37*, 981–983.
[25] K. Tsubaki, K. Tanaka, T. Kinoshita, K. Fuji, *Chem. Commun.* **1998**, 895–896.
[26] L. J. Barbour, G. William, J. L. Atwood, *Chem. Commun.* **1997**, 1439–1440.
[27] L. J. Barbour, G. W. Orr, J. L. Atwood, *Chem. Commun.* **1998**, 1901–1902.
[28] K. N. Rose, L. J. Barbour, G. W. Orr, J. L. Atwood, *Chem. Commun.* **1998**, 407–408.
[29] R. Castillo, S. Ramos, R. Cruz, M. Martinez, F. Lara, J. Ruiz-Garcia, *J. Phys. Chem.* **1996**, *100*, 709–713.
[30] P. Lo Nostro, A. Casnati, L. Bossoletti, L. Dei, P. Baglioni, *Colloids Surf. A* **1996**, *116*, 203–209.
[31] L. Dei, P. Lo Nostro, G. Capuzzi, P. Baglioni, *Langmuir* **1998**, *14*, 4143–4147.
[32] T. Suzuki, K. Nakashima, S. Shinkai, *Tetrahedron Lett.* **1995**, *36*, 249–252.
[33] D. E. Cliffel, A. J. Bard, S. Shinkai, *Anal. Chem.* **1998**, *70*, 4146–4151.
[34] D. E. Cliffel, A. J. Bard, *Anal. Chem.* **1998**, *70*, 1993–1998 and references therein.
[35] T. Liu, M. Li, N. Li, Z. Shi, Z. Gu, X. Zhou, *Electroanalysis* **1999**, *11*, 1227–1232.
[36] H. Lou, N. Li, W. He, Z. Shi, Z. Gu, X. Zhou, *Electroanalysis* **1998**, *10*, 576–578.
[37] H. Luo, N. Li, W. He, Z. Shi, Z. Gu, X. Zhou, *J. Solid State Electrochem.* **1998**, *2*, 253–256.
[38] Y.-P. Sun, C. E. Bunker, B. Ma, *J. Am. Chem. Soc.* **1994**, *116*, 9692–9699 and references therein.
[39] K. Araki, K. Akao, A. Ikeda, T. Suzuki, S. Shinkai, *Tetrahedron Lett.* **1996**, *37*, 73–76.
[40] N. S. Isaacs, P. J. Nichols, C. L. Raston, C. A. Sandova, D. J. Young, *Chem. Commun.* **1997**, 1839–1840.
[41] A. Ikeda, M. Yoshimura, S. Shinkai, *Tetrahedron Lett.* **1997**, *38*, 2107–2110.
[42] A. Ikeda, Y. Suzuki, M. Yoshimura, S. Shinkai, *Tetrahedron* **1998**, *54*, 2497–2508.
[43] H. Otsuka, K. Araki, S. Shinkai, *Tetrahedron* **1995**, *51*, 8757–8770.
[44] H. Matsumoto, S. Nishio, M. Takeshita, S. Shinkai, *Tetrahedron* **1995**, *51*, 4647–4654.
[45] J. Wang, C. D. Gutsche, *J. Am. Chem. Soc.* **1998**, *120*, 12226–12231.
[46] a) S. Mizyed, P. E. Georghiou, M. Ashram, *J. Chem. Soc., Perkin Trans. 2* **2000**, 277–280; b) P. E. Georghiou, S. Mizyed, S. Chowdbury, *Tetrahedron Lett.* **1999**, *40*, 611–614.
[47] F. C. Tucci, D. M. Rudkevich, Jr. J. Rebek, *J. Org. Chem.* **1999**, *64*, 4555–4559.
[48] A. Ikeda, M. Kawaguchi, Y. Suzuki, T. Hatano, M. Numata, S. Shinkai, A. Ohta, M. Aratono, *J. Incl. Phenom.* **2000**, *38*, 163-170.
[49] M. Kawaguchi, A. Ikeda, S. Shinkai, *J. Incl. Phenom.* **2000**, *37*, 253-258.
[50] Y. Fukazawa, S. Usui, *J. Synth. Org. Chem. Jpn.* **1997**, *55*, 1124–1133.
[51] T. Haino, M. Yanase, Y. Fukazawa, *Angew. Chem. Int. Ed. Engl.* **1998**, *37*, 997–998.
[52] A. Ikeda, T. Hatano, M. Kawaguchi, H. Suenaga, S. Shinkai, *Chem. Commun.* **1999**, 1403–1404.

[53] T. Hatano, A. Ikeda, T. Akiyama, S. Yamada, M. Sano, Y. Kanekiyo, S. Shinkai, *J. Chem. Soc. Perkin Trans. 2* **2000**, 909–912.
[54] Y. Kanekiyo, K. Inoue, Y. Ono, M. Sano, S. Shinkai, D. N. Reinhoudt, *J. Chem. Soc. Perkin Trans. 2* **1999**, 2719–2722.
[55] M. Yanase, T. Haino, Y. Fukazawa, *Tetrahedron Lett.* **1999**, *40*, 2781–2784.
[56] a) A. Ikeda, M. Yoshimura, S. Shinkai, *Chem. Lett.* **1998**, 587–588; b) A. Ikeda, M. Yoshimura, H. Udzu, C. Fukuhara, S. Shinkai, *J. Am. Chem. Soc.* **1999**, *121*, 4296–4297; c) A. Ikeda, H. Udzu, M. Yoshimura, S. Shinkai, *Tetrahedron* **2000**, *56*, 1825–1832.
[57] For reviews, see: a) T. D. Ros, M. Prato, *Chem. Commun.* **1999**, 663–669; b) A. W. Jensen, D. I. Schuster, *Bioorg. Med. Chem.*, **1996**, *4*, 767–779; c) E. Nakamura, H. Tokuyama, S. Yamago, T. Shiraki, Y. Sugiura, *Bull. Chem. Soc. Jpn.* **1996**, *69*, 2143–2151.
[58] a) S. H. Friedman, D. L. DeCamp, R. P. Sijbesma, G. Srdanov, F. Wudl, G. L. Kenyon, *J. Am. Chem. Soc.* **1993**, *115*, 6506–6509; b) R. Sijbesma, G. Srdanov, F. Wudl, J. A. Castoro, C. Wilkins, S. H. Friedman, D. L. DeCamp, G. L. Kenyon, *J. Am. Chem. Soc.* **1993**, *115*, 6510–6512; c) S. H. Friedman, P. S. Ganapathi, Y. Rubin, G. L. Kenyon, *J. Med. Chem.* **1998**, *41*, 2424–2429.
[59] a) Y.-Z. An, C.-H. B. Chen, J. L. Anderson, D. S. Sigman, C. S. Foote, Y. Rubin, *Tetrahedron*, **1996**, *52*, 5179–5180; b) A. S. Boutorin, H. Tokuyama, M. Takasugi, H. Isobe, E. Nakamura, C. Helene, *Angew. Chem., Int. Ed. Engl.* **1994**, *33*, 2462–2465; c) E. Nakamura, H. Tokuyama, S. Yamago, T. Shiraki, Y. Sugiura, *Bull. Chem. Soc. Jpn.* **1996**, *69*, 2143–2151; d) H. Tokuyama, S. Yamago, T. Nakamura, T. Shiraki, Y. Sugiura, *J. Am. Chem. Soc.* **1993**, *115*, 7918–7919; e) N. Higashi, T. Inoue, M. Niwa, *Chem. Commun.* **1997**, 1507–1508.
[60] Y. N. Yamakoshi, T. Yagami, S. Sueyoshi, M. Miyata, *J. Org. Chem.* **1996**, *61*, 7236–7237.
[61] a) A. Hirsch, I. Lamparth, H. R. Karfunkel, *Angew. Chem., Int. Ed. Engl.* **1994**, *33*, 437–438; b) L. L. Dugan, D. M. Turetsky, C. Du, D. Lobner, M. Wheeler, C. R. Almli, C. K.-F. Shen, T.-Y. Luh, D. W. Choi, T.-S. Lin, *Proc. Natl. Acad. Sci. U.S.A.* **1997**, *94*, 9434–9439.
[62] Y. L. Huang, C. K. Shen, T. Y. Luh, H. C. Yang, K. C. Hwang, C. K. Chou, *Eur. J. Biochem.* **1998**, *254*, 38–43.
[63] a) L. J. Wilson, D. W. Cagle, T. P. Trash, S. J. Kennel, S. Mirzadeh, J. M. Alford, G. L. Ehrhardt, *Coord. Chem. Rev.* **1999**, *192*, 199–207; b) D. W. Cagle, T. P. Thrash, M. Alford, L. P. F. Chibante, G. J. Ehrhardt, L. J. Wilson, *J. Am. Chem. Soc.* **1996**, *118*, 8043–8047.
[64] K. Komatsu, K. Fujiwara, Y. Murata, T. Braun, *J. Chem. Soc., Perkin Trans. 1* **1999**, 2963–2966.
[65] S. D. M. Islam, M. Fujitsuka, O. Ito, A. Ikeda, T. Hatano, S. Shinkai, *Chem. Lett.* **2000**, 78–79.
[66] M. Takeshita, T. Suzuki, S. Shinkai, *J. Chem. Soc., Chem. Commun.* **1994**, 2587–2588.
[67] A. Ikeda, S. Nobukuni, H. Udzu, Z. Zhong, S. Shinkai, *Eur. J. Org. Chem.* **2000**, *19*, 3287–3293.
[68] a) M. Kawaguchi, A. Ikeda, I. Hamachi, S. Shinkai, *Tetrahedron Lett.* **1999**, *40*, 8245–8249; b) M. Kawaguchi, A. Ikeda, S. Shinkai, *J. Chem. Soc., Perkin Trans. 1* **1998**, 179–184; c) A. Ikeda, S. Shinkai, *Chem. Lett.* **1996**, 803–804.
[69] A. Soi, A. Hirsch, *New J. Chem.* **1999**, 1337–1339.
[70] P. -F. Fang, Y.-Y. Chen, X.-R. Lu, L. Zhu, *Synth. Commun.* **1999**, *29*, 3547–3554.
[71] T. Liu, M. X. Li, N. Q. Li, Z. J. Shi, Z. N. Gu, X. H. Zhou, *Talanta* **2000**, *50*, 1299–1305.

Chapter 27

CALIXARENES IN BIOORGANIC AND BIOMIMETIC CHEMISTRY

FRANCESCO SANSONE, MARGARITA SEGURA, ROCCO UNGARO
*Dipartimento di Chimica Organica e Industriale dell'Università,
Parco Area delle Scienze 17/A, I-43100 Parma, Italy.
e-mail: ungaro@ipruniv.cce.unipr.it*

1. Introduction

Bioorganic chemistry [1,2] is a broad field, including enzyme, medicinal, and natural products chemistry, and is closely related to bioinorganic chemistry [3], where the emphasis is more on the metal ions which are coordinated to proteins and play a very important role in enzyme catalysis. The term "biomimetic chemistry", introduced by R. Breslow in 1972 [4], refers to a particular aspect of those two fields, namely to chemical processes which imitate a biochemical reaction. Bioorganic chemistry has been termed both a "natural" and an "unnatural" science [1] and its natural side is well established as natural products chemistry, enzymology, biological recognition and chemical ecology [5], whereas the unnatural part is wider and more subject to changes since, from time to time, it enriches itself with new fields and concepts such as "molecular appreciation" [6] or supramolecular science [7].

Cyclodextrins have long been used in medicinal [8] and biomimetic chemistry [9], and other cavity containing synthetic macrocycles, such as cyclophanes, are popular in host-guest chemistry and enzyme mimics [10]. Calixarenes are emerging as a new class of synthetic hosts enjoying interest in several areas of bioorganic and biomimetic chemistry.

2. Recognition of Biorelevant Species

2.1. PEPTIDOCALIXARENES

In biological systems, the cooperative action of peptide hydrogen bonds plays an important role in organisation, assembly and molecular recognition processes [11]. The use of macrocycles as platforms for anchoring amino acids and small peptides has been part of many efforts to model such behaviour [12]. Although calixarenes possess a suitable cavity-shaped architecture, only recently have they been functionalised with these chiral units.

Full substitution of calix[4]- and -[6]arenes at the wide rim with L-cysteine subunits was early shown to give compounds soluble in both aqueous acid and base and able to bind fluorescent molecules such as sodium 8-anilino-1-naphthalenesulfonate (ANS) or [2-[5-(dimethylamino)-1-naphthalenesulfonamido]ethyl]trimethylammonium perchlorate (DASP) [13]. Selective introduction of two (*e.g.* **1a-c**) and four (*e.g.* **1d**) L-alanine and L-alanyl-L-alanine groups into calix[4]arenes locked in the *cone* conformation has since been described as part of a project aimed at obtaining *hybrid* calix[4]arene receptors [14].

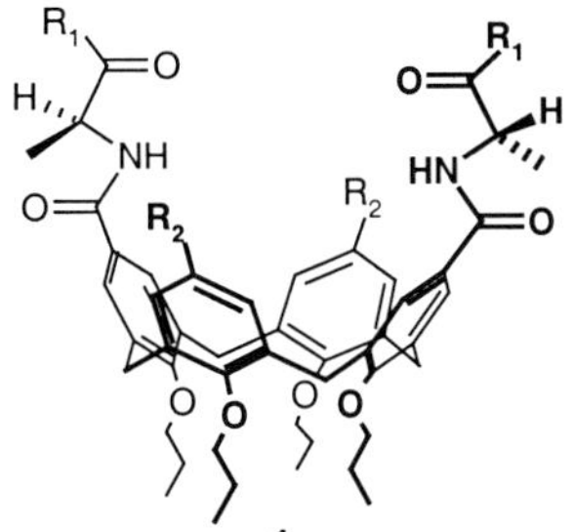

a: R_1 = OCH_3; R_2 = H
b: R_1 = $NHNH_2$; R_2 = H
c: R_1 = HN–CH(CH$_3$)–C(O)–NHNH$_2$; R_2 = H
d: R_1 = HN–CH(CH$_3$)–C(O)–OCH$_3$; R_2 = –NH–C(O)–CH(CH$_3$)–NH–C(O)–CH(CH$_3$)–OCH$_3$

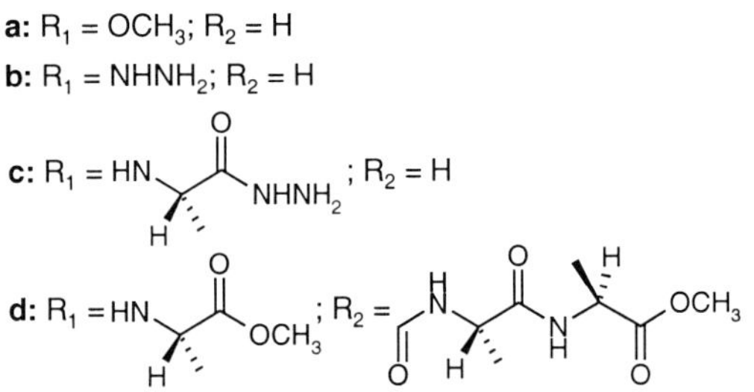

The X-ray crystal structure of **1a** shows three different conformations in the solid state but none of which has *intrachain* hydrogen bonding which might cause collapse of the binding cavity [14]. Preliminary complexation studies indicate that **1b** and **1c** are able to extract *N*-acetylated D-alanine and D-alanyl-D-alanine into organic media.

Since hosts **1a-d** retain some conformational mobility, they were rigidified by inserting two short di(ethyleneglycol) units in proximal (1,2) positions at the narrow rim [15]. The water soluble peptidocalix[4]arene **2** is able to complex α-amino acids and their methyl esters at neutral pH ($10 < K_{ass} < 7 \times 10^2$ M^{-1}). The apolar side of the guest is included into the host cavity and the strongest binding is observed with aromatic amino acids.

The chiral homocalix[4]arenes with L-phenylalanine (**3a**) and L-phenylglycine (**3b**) pendent groups at the narrow rim efficiently extract lipophilic amino acid esters as well as their Z-carboxylates into a dichloromethane phase and transport them through liquid membranes [16].

A water-soluble chiral *p-t*-butylcalix[4]arene podand having four L-alanine units was used as mobile phase additive in capillary electrophoresis for the resolution of racemic (±)-1,1'-binaphthyl-2,2'-diyl hydrogen phosphate (BNHP) [17]. Similar examples employing different amino acids have been reported [18].

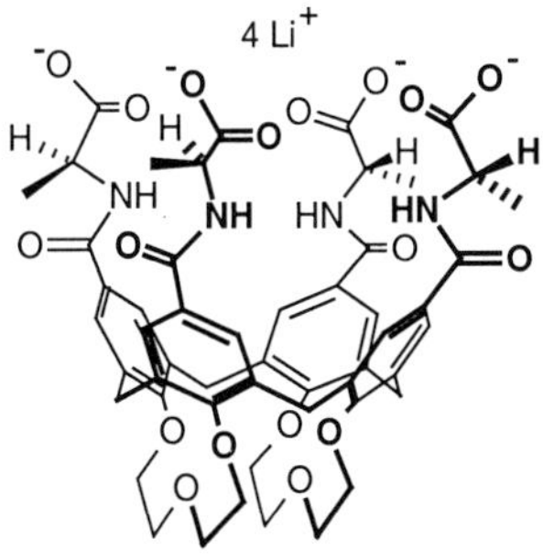

a: R_1 = –C(O)–CH(R)–NH–C(O)–OMe; R = CH_2Ph
b: R_1 = –C(O)–CH(R)–NH–C(O)–OMe; R = Ph

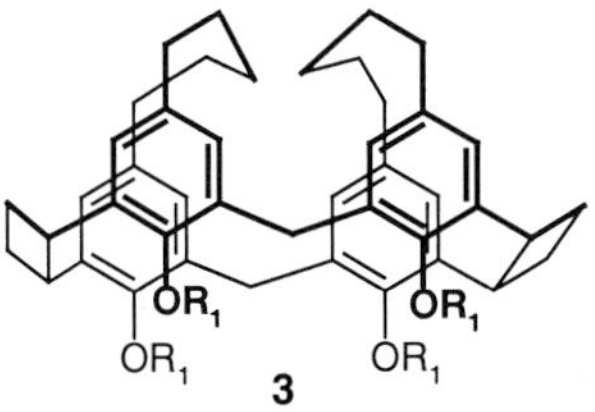

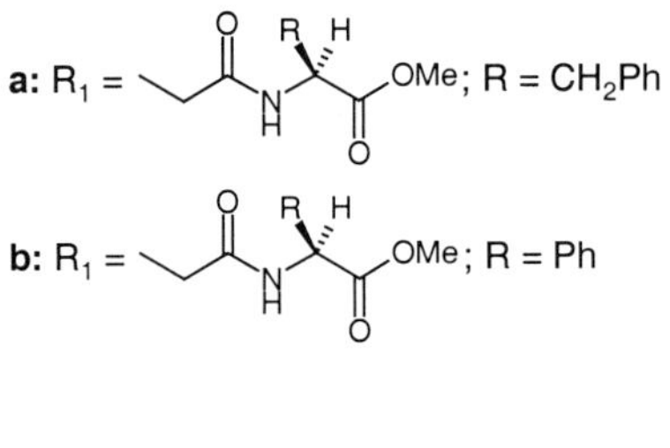

In molecular capsules formed by self-assembly of wide rim urea-calix[4]arenes, built from aromatic and aliphatic amino acids [19], the chirality of the amino acids has a dramatic effect on the dimerization behaviour of the calixarenes, giving rise to chiral capsules capable of discriminating between enantiomeric guests.

Cyclic peptides have been anchored on the narrow rim of calix[4]arenes to reduce their conformational flexibility, thus improving their molecular recognition properties [20]. The cysteine-bearing calix[4]arene **4** forms a 1:1 complex with 4-nitrophenyl phosphate with a binding constant of 3.9×10^3 M^{-1} in DMSO [20a].

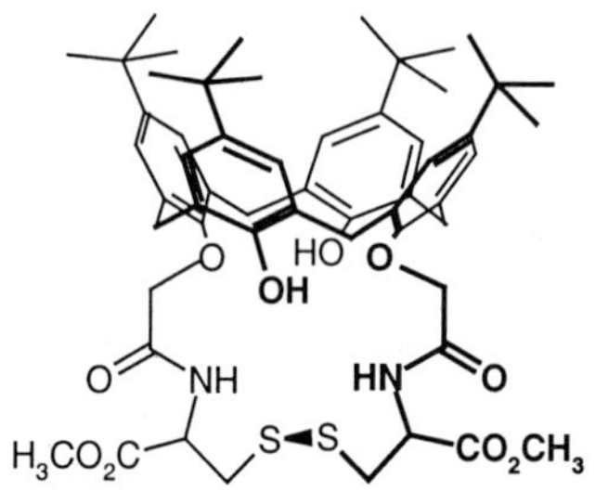

A peptide library of ca. 50,000 members based on a calix[4]arene was recently reported [21]. To create this library (e.g. **5**), an appropriate calixarene derivative was attached to a polystyrene resin *via* a polymethylene chain inserted at the wide rim, and 15 amino acids (AA_1-AA_4: L- and D-Ala, L- and D-Leu, L- and D-Pro, L- and D-Phe, L-Ser, L-Thr, L-Lys, L-Asn, L- and D-Gln, β-Ala) were used as building blocks in the combinatorial synthesis. Screening of the library, using four types of dye-labelled oligopeptides as guests, indicated a selectivity in the binding of specific amino acid sequences.

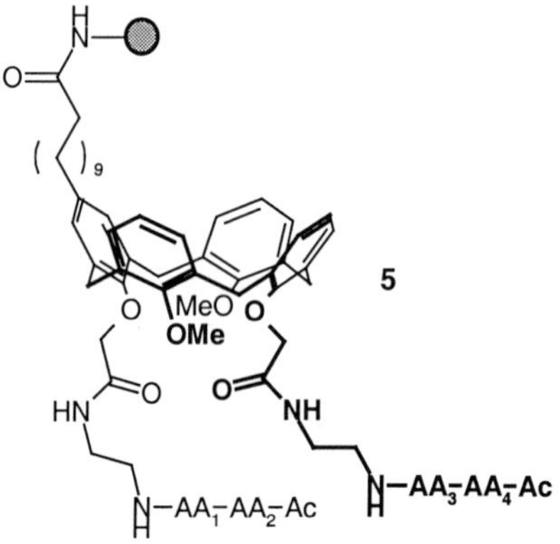

Peptidoresorc[4]arenes with helical chirality are also known. Coupling amino acid derivatives to the rigid bowl-shaped macrocycles provides a new family of *de novo* proteins, the *caviteins* **6** [22] (see also Chapter 10). The X-ray crystal structure of compound **7** shows that the four L-proline ethyl esters are positioned around the outer edge of the host cavity in a clockwise helical conformation [23].

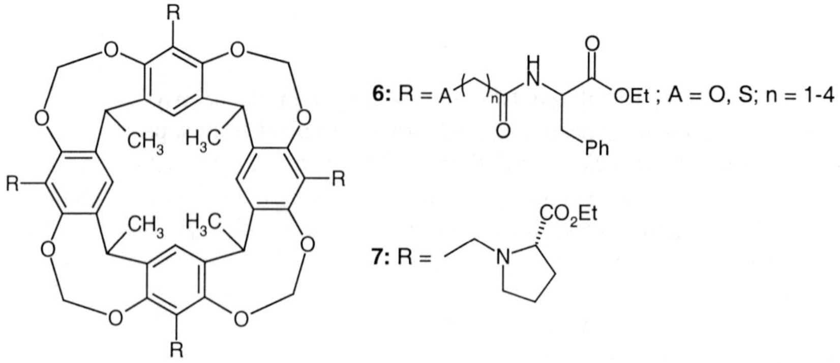

2.2. GLYCOCALIXARENES

Because of its great biological relevance [24], the subject of carbohydrate recognition by synthetic receptors has recently attracted much attention [25]. The affinity of the carbohydrate-binding proteins for single units of saccharides is usually weak but, in these processes, multiple identical sugar residues of the cell-surface glycoproteins interact with proteins bearing several sugar binding sites [24]. This phenomenon, termed the *cluster effect* [26], considerably increases the binding affinity and can be the basis for the development of new drugs, vaccines and antibody mimics. For this reason, polyglycosylated molecules have been recently synthesised and investigated [27].

Thus, the conditions necessary to introduce furanose and pyranose moieties at the narrow and wide rims of calix[4]arenes have been explored [28]. Under standard Mitsunobu conditions [29], calix[4]arene reacts with diisopropylidene-α-furanose and with tetra-*O*-acetylated-α,β-glucopyranose to afford mono- and bisglycosylated compounds. Starting from di- and tetrahydroxymethyl-tetra-*n*-propoxycalix[4]arene, galactose and lactose units were linked at the wide rim with the formation of totally β-selective anomeric bonds (*e.g.* **8** and **9**) [28b]. Glycocalix[4]arene **8** shows poor binding properties towards peptides and carbohydrates, however [28b].

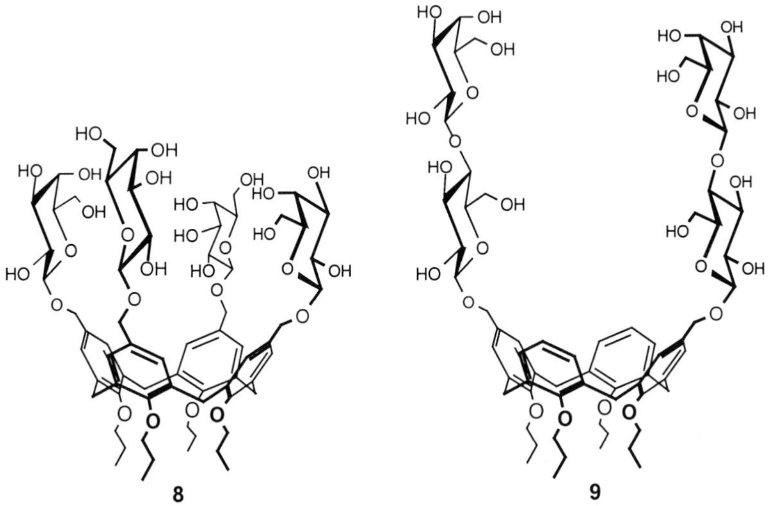

The calixarenes **10a** [30] and **10b** [31], bearing four thiosialosyl and sixteen *N*-acetylgalactosamine units, respectively, bind strongly to plant lectins, particularly wheat germ agglutinin and *Vicia villosa* agglutinin (VVA), as evidenced by turbidimetric experiments in water. The corresponding monomeric carbohydrates inhibit the interaction between the glycocalixarenes and lectins, thus demonstrating the sugar specificity of the binding. Compound **10b** and some of its analogues were also successfully tested as inhibitors of the binding to VVA by asialoglycophorin, a natural glycoprotein of human erythrocytes, used as an antigen. Other glycocalixarenes have been synthesised [32] but no binding or other properties have been reported.

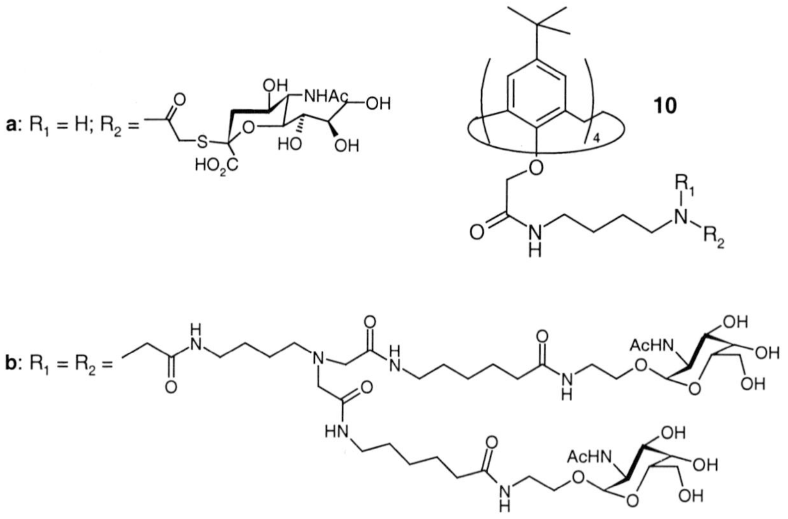

Sugar clusters based on the resorc[4]arene **11a** [33] have been obtained from the octaamine **11b** which reacts with several glycoside lactones to give a series of highly water soluble (>0.1 M) macrocyclic sugar clusters (**11c-f**). These compounds are almost irreversibly adsorbed on the surface of a quartz plate, which acts as a simplified model of a multivalent receptor site [33a]. The corresponding carbohydrate monomers do not have the same properties towards the quartz plate and a selectivity factor glycoocta-mer/monomer larger than 10^4 was observed, thus disclosing a remarkable cluster effect. Further experiments were performed with lectins having four sugar binding sites, both in water solution [33b] and at the interface between the bulk water and a monolayer assembly of the sugar clusters [33c]. Concanavalin A (ConA) and peanut lectin (PNA), known as glucoside- and galactoside-binding proteins, respectively, and a sialic acid-specific lectin [33d], a model of the sugar-binding proteins of influenza viruses, were tested. ConA shows a specific binding ($K_{ass} = 1 \times 10^6$ M^{-1}) with the octaglucose derivative **11c**, which is inhibited by an excess of glucose, while PNA selectively recognises the galactose terminal residues of cluster **11d**. Evidence of a strong binding between sialic acid cluster **11e** and sialic acid-specific lectin was obtained from Surface Plasmon Resonance experiments. Thus, compound **11e** can be considered, potentially, as an excellent virus-binder. Bovine Serum Albumin, a protein with no carbohydrate binding sites, is inert towards sugar cluster resorcarenes, thus showing the absence of nonspecific interactions.

Compounds **11c-e** are also able to form stable 1:1 complexes in water with several guests which are included in the apolar cavity ($10^3 < K_{ass} < 10^5$ M^{-1}) [33a,c]. Interestingly, in the presence of lectins, they give rise to receptor-guest-[sugar cluster] ternary systems with substrates that otherwise would not show any affinity for such proteins. This characteristic aspect suggests a potential use of sugar cluster resorcarenes as molecular delivery systems for transporting drugs or probes directly to a particular receptor site, *e.g.* at the surface of viruses [33f]. Resorc[4]arenes **11e** and **11f** complex phosphate salts in water, acting as a kind of *macrosolvent* for these species [33e]. NMR studies reveal the inclusion of phosphate anions through hydrogen bonding which causes the agglutination of the hosts after the rapid formation of submicrometer-sized particles. The ability of these hosts to interact with the phosphate group has been exploited in the binding of nucleotides in water (K_{ass} = 10^3-10^4 M^{-1}). Remarkably, the corresponding nucleosides do not show any detectable affinity for the sugar clusters.

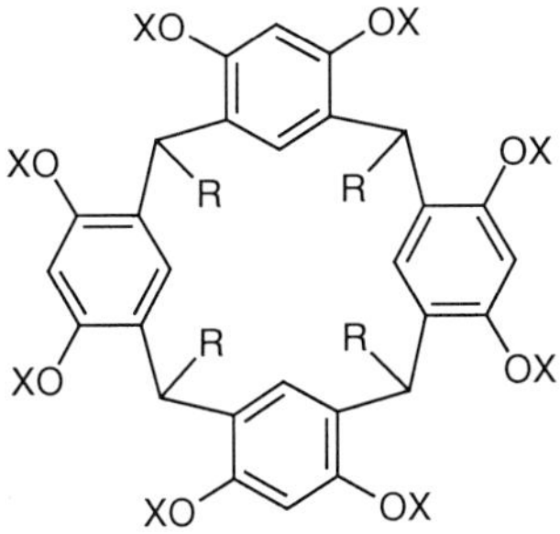

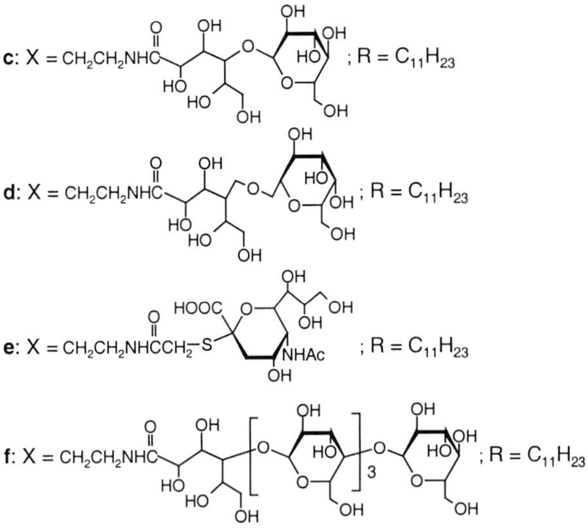

2.3. OTHER HOSTS

Amino acids, small peptides, carbohydrates and other biorelevant species have also been complexed by achiral calixarene hosts. Both arginine and lysine are bound by water-soluble p-sulfonato-calix[n]arenes [34a,b], which could be good candidates as heparin mimics. Only weak electrostatic interactions between the amino acids and the host systems are observed with large calixarenes (n = 6, 8) due to their conformational flexibility, whereas the calix[4]arene derivative shows a pH-selective recognition for both amino acids. A weaker binding is observed at pH 8 since repulsion between the carboxylate group of the guest and the sulfonate groups of the host occurs at this pH. In solution the data fit a 1:1 stoichiometry whereas in the solid state a 2:1 L-lysine-calixarene complex is observed, one guest molecule being inside and one outside the calixarene cavity [34c].

In D$_2$O at pD 7.3, ^{1}H NMR experiments show that binding of native L-α-amino acids by conformationally mobile *cone* sulfonatocalix[4]arenes occurs by insertion of the aromatic (L-Phe, L-Tyr, L-His and L-Trp) or the aliphatic (L-Ala, L-Val and L-Leu) apolar side chains of the guests into the calixarene cavity [35].

Compounds **12** and **13** are good carriers for the membrane transport of aromatic amino acids (for example, Phe *vs* Trp), and the efficiency and selectivity are larger for the wide rim derivatives **13** due to the participation of the apolar cavity in the recognition process [36].

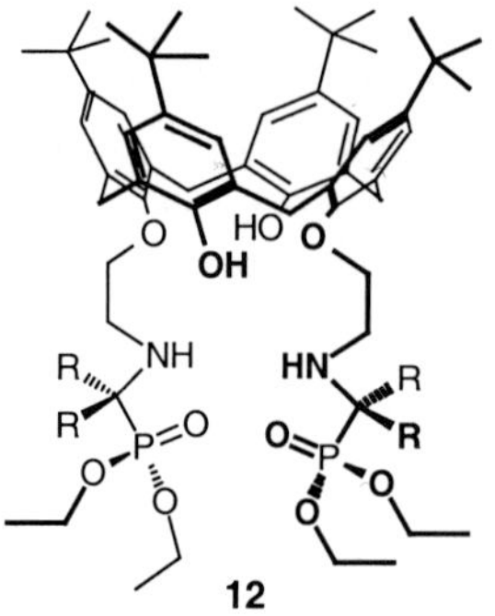

12 R = CH₃ **13**

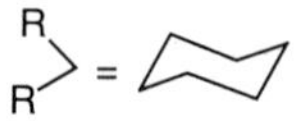

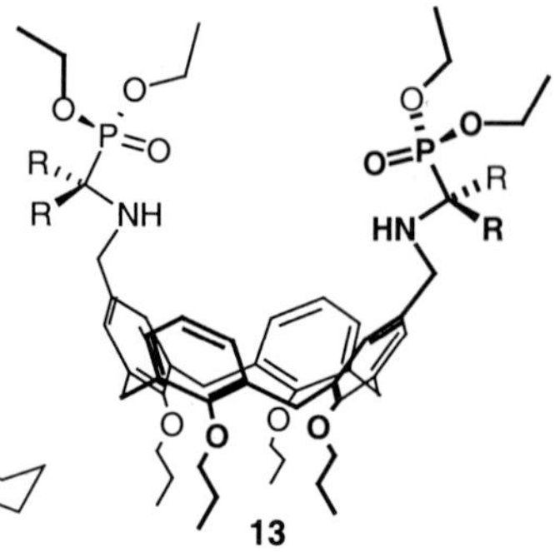

A pseudo-C_2-symmetrical homooxacalix[3]arene is able to bind α-amino acid ethyl esters (74% ee for L-phenylalanine) through hydrogen bonding interactions between the oxygen atoms of the host and the ammonium ions of the guests [37]. Several dipeptides (Ala-Ala, Ala-Glu, Ala-Ser and Ser-Leu) are complexed by an achiral water soluble resorc[4]arene which incorporates four aminoformyl residues [38].

Extensive studies of the complexation of carbohydrates by resorc[4]arenes **11a** and **11g**, which occurs through a combination of hydrogen bonding and apolar forces, have been recently reviewed [39]. Complementary studies on sugar binding by calixarene boronic acid derivatives, which form reversible covalent adducts with carbohydrates, have been performed [40].

Calixarene-based receptors have also been used for the recognition of other bio-relevant species. Thus, pyrimidine (uracil, thymine and cytosine) and xanthine (theobromine, theophylline and dyphylline) bases are separated on a HPLC stationary RP-18 phase, coated with resorc[4]arene **11a** [41a,b], or simply extracted into different organic solvents by similar compounds [41c]. Dopamine is extracted in preference to other catecholamine neurotransmitters (adrenaline and noradrenaline) by calix[6]arene hexaesters and a homooxacalix[3]arene triether incorporated in PVC liquid membranes [42], while substrates like vitamins B_2 (riboflavin) and B_{12} (cyanocobalamin) [43a,b] and several nucleosides such as cytidine, uridine, thymidine [43c] are complexed by resorc[4]arene **11a** and related receptors [44].

Particularly interesting is the recognition of nucleotides, and even DNA, by calix[n]arenes containing four, six or eight permethylammonium groups at the wide rim, with association constants up to 7×10^4 M^{-1} [45]. The main host-guest interaction is ion-pairing, since the nucleobase and the sugar moiety are only partially included into the calixarene cavity. In the case of DNA, a stabilisation of the double-strands is observed, in contrast to some other macrocyclic polyamines, indicating the additional contribution of van der Waals-type interactions.

3. Bioactive Calixarenes

3.1. VANCOMYCIN MIMICS AND ENZYME INHIBITORS

Vancomycin is a potent natural glycopeptide antibiotic active towards gram-positive bacteria, for more than thirty years in clinical protocols. The mode of action of this family of antibiotics has been clarified during the years and it is based on the selective recognition and strong complexation of the D-alanyl-D-alanine terminal part of cell wall mucopeptide precursors [46]. Vancomycin consists of a heptapeptide covalently linked to aromatic groups which form an hydrophobic pocket able to accommodate the substrate. A peripheral ammonium group properly orients the guest, which then interacts with the antibiotic mainly through hydrogen bonding (Figure 1).

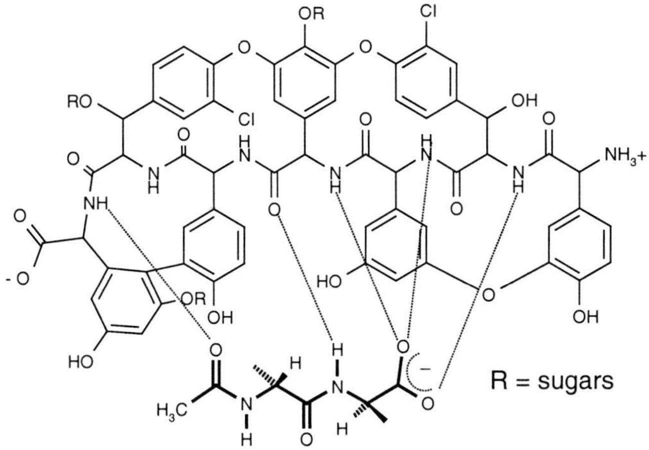

Figure 1. Schematic representation of the complex between Vancomycin and N-acetyl-D-alanyl-D-alanine.

Macrobicyclic calix[4]arenes (*e.g.* **14**) bearing peptide bridges containing a basic nitrogen group protonated at physiological pH, can be considered Vancomycin mimics [47].

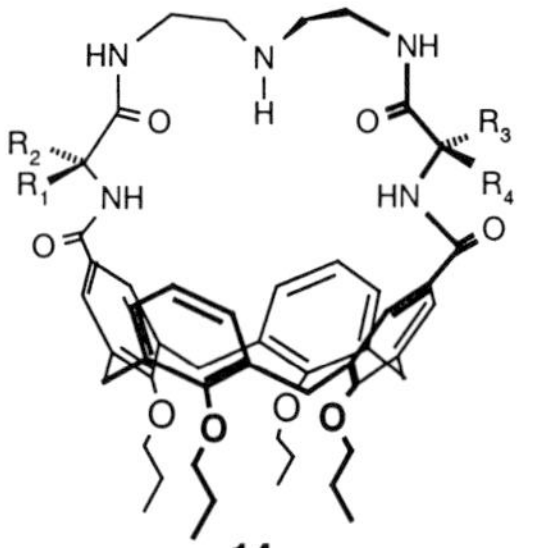

14

a: $R_1 = R_3 = H; R_2 = R_4 = CH_3$
b: $R_1 = R_4 = H; R_2 = R_3 = CH_3$
c: $R_1 = R_4 = H; R_2 = CH_3; R_3 = CH_2Ph$
d: $R_1 = R_3 = H; R_2 = R_4 = CH_2Ph$

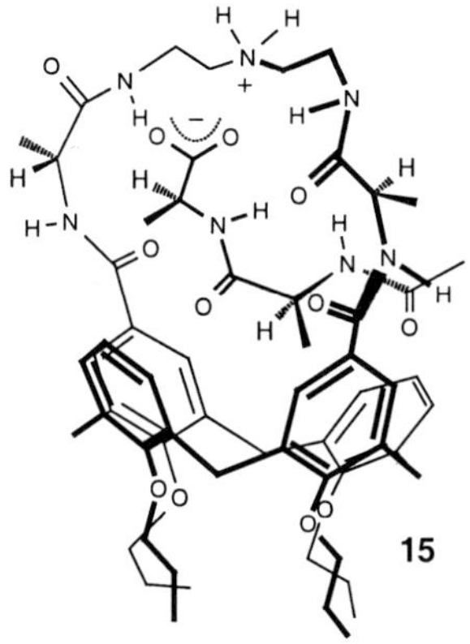

15

Microbiological studies *in vitro* were performed to evaluate the antibacterial properties of these receptors. The activity, expressed as minimum inhibitory concentration (MIC), is from moderate to good against gram-positive bacteria, including methicillin-resistant strains, and in some cases it is very close to that of Vancomycin (Table 1). Remarkably, hosts **14** show the same selectivity pattern of the natural antibiotic, as no biological activity (MIC>64) against gram-negative and cell-wall lacking bacteria, yeasts or fungi is observed. NMR [47c] and ESI-MS [47b] data indicate the formation of a 1:1 complex between **14a** and *N*-acetyl-D-alanyl-D-alanine and justify a binding model (*e.g.* **15**) where electrostatic, hydrogen-bonding and CH_3/π interactions could be involved. The trend of the observed MIC values is also in agreement with this model since no activity is observed with derivatives bearing phenylalanine units (*e.g.* **14c,d**), which probably hinder the penetration of the guest inside the peptide loop of the host. Since no antimicrobial activity is observed with cleft-like peptidocalix[4]arenes **1**, the results obtained with macrobicyclic receptors **14** show the importance of rigidity and preorganization of the hosts in determining their biological and binding properties.

TABLE 1. Minimum Inhibitory Concentration (MIC, µg/µl) of macrobicyclic peptidocalix[4]arenes **14** in comparison with Vancomycin; nt: not tested.

Organism	Vancomycin	14a	14b	14c	14d
S. aureus 853	1	8	8	>64	>64
S. aureus 4543	2	8	4	>64	>64
S. aureus 1131	2	4	nt	nt	>64
S. pneumoniae 4636	0.5	16	8	>64	32
E. faecalis 3807	2	16	16	>64	32
E.coli	>64	>64	>64	>64	>64
A. laidlawii	>64	>64	>64	>64	>64
S. cerevisiae	>64	>64	>64	>64	>64

The design of synthetic agents, which recognise and bind to specific regions of a protein surface and/or disrupt biologically key protein-protein interactions, constitutes an interesting approach to enzyme inhibitors. The immune system is able to generate a great number of antibodies with high sequence and structural selectivity in binding to a wide range of large protein surfaces (>600 Å^2). These features were the basis of the design of new protein surface receptors (*e.g.* **16**) in which four peptide loops are attached to a central calix[4]arene platform locked into the semirigid *cone* conformation [48].

In this way, all cyclopeptide binding units are projected in the same region and can strongly interact with the protein surface due to cooperative effects. Thus, the receptor **16a** (with GlyAspGlyAsp), having a surface >400 Å^2, shows a great affinity for cytochrome C ($K_{ass} = 3 \times 10^6$ M^{-1}), whose surface is positively charged due to the presence of lysine and arginine residues, while the single cyclopeptide loop shows no binding affinity for the protein [48a]. In phosphate buffer, Fe(III)-cytochrome C is rapidly reduced by excess ascorbate, but inhibition of this process is observed in the presence of **16a**. Probably the macrocycle binds to a region of cytochrome C close to the heme edge, thus blocking the approach of the reducing agents and causing a 10-fold decrease in the rate of Fe(III)-cytochrome C reduction, in a similar manner to that observed with cythocrome C/cytochrome C peroxidase complex [48b]. Variation of the peptide sequence confirms the importance of host-guest complementarity both in terms of charge and position of the binding groups, as shown by a small improvement in enzyme inhibition using Gly-Asp-Asp-Gly sequence (**16b**). Evidence that hydrophobic surfaces may also provide a favourable interaction with the protein has been achieved [48b]. Receptor **16a** is also able to induce a slow binding inhibition of bovine pancreatic α-chymotrypsin [48d].

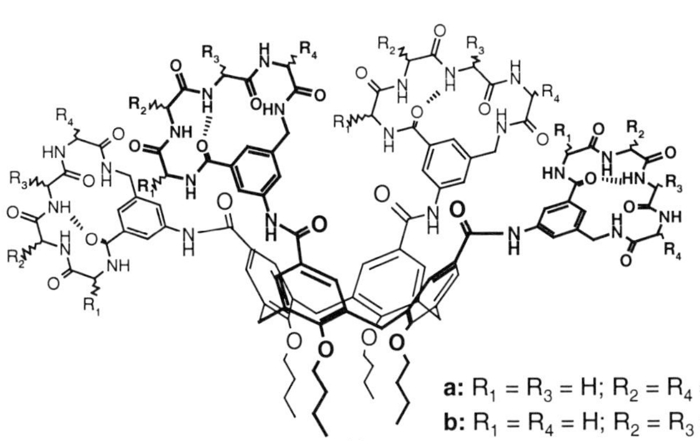

a: $R_1 = R_3 = H$; $R_2 = R_4 = (S)\text{-}CH_2CO_2H$
b: $R_1 = R_4 = H$; $R_2 = R_3 = (S)\text{-}CH_2CO_2H$

16

3.2. MEMBRANE ACTIVE CALIXARENES

The transport of ions across biological membranes mainly occurs *via* channel or carrier mechanisms [49]. Resorcarenes and calixarenes that exhibit characteristics of ion channels have been reported. A resorc[4]arene amphiphile with long alkyl chains penetrates into a soybean lecithin bilayer and provides the channel pore for the passage of K$^+$ and Na$^+$ with a permeability ratio P_{K+}/P_{Na+} ~3 [50]. In *1,3-alternate* calix[4]arenes bearing dodecyl chains and azacrowns (*e.g.* **17**), which are inserted in the bilayer membrane, the function of the calixarene platform is, however, to organise the lipophilic tails at both sides rather than acting as a tube for the ions [51].

17: $R_1 = H$, *t*-Bu

$R_2 = $

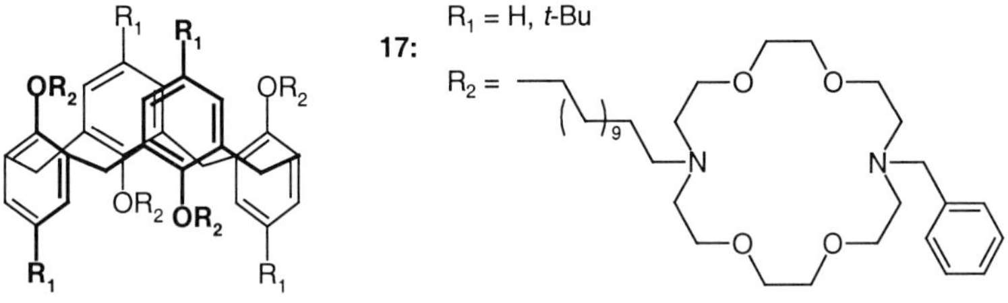

Ionophoric calixarene ethers [52a] and esters [52b] are also able to perform ion transport through phospholipid bilayer membranes according to the carrier mechanism. The selectivity observed in the case of esters [52b] follows the order found in binding studies, the tetramer being selective for Na$^+$, the pentamer for K$^+$ and both the hexamer and heptamer for Cs$^+$.

In sharp contrast to cation channels, nature has provided few high affinity ligands that modulate chloride channels. These are the key transport proteins that allow the rapid entry and exit of chloride across the plasma membrane [53], contributing to the normal function of skeletal and smooth muscle cells and to the absorption and secretion phenomena in the physiology of epithelia. In this way, chloride channel modulators may serve as effective pharmaceuticals for treating respiratory, cardiovascular and gastrointestinal disorders. Tetramethoxy-tetrasulfonato-calix[4]arene acts as an inhibitor of the plasma membrane outwardly rectifying chloride channels (ORCC), resulting in a potent blocker (block periods >>100 seconds) at a concentration as low as 3 nM [54]. This compound also inhibits p64, a chloride channel from kidney microsomes [55] and also induces a fast and reversible block of the volume-regulated anion channels (VRAC) in endothelial cells, especially at acidic pH [56]. In fact, a sharp drop in the inhibition efficiency is observed in the pH range 6-9. This behaviour has been explained assuming a strong interaction between the calixarene and a positively charged site inside the channel pore involving one or more histidine residues which, instead, are not protonated at basic pH.

3.3. OTHER PHARMACOLOGICAL APPLICATIONS

In 1955, *p-t*-octylcalix[8]arene (at that time considered to be a tetramer) functionalised with polyoxyethylene units at the narrow rim was reported to be active against experimental tuberculosis after parenteral administration [57]. Two compounds were studied, the Macrocyclon, having an average length of 12.5 ethylene oxide units in the hydrophilic chain, and HOC-60, functionalised with *ca*. 60 of such units [58]. The growth of *Mycobacterium tuberculosis* inside infected cultivated macrophages was inhibited by the short chain Macrocyclon, that behaved as bacteriostatic, while it was stimulated by the long chain HOC-60 [59]. Recently, it was shown that the effects of both compounds, that enter macrophages by endocytosis and are lysosomotropic [59], are host cell mediated [60]. Lipid metabolism seems indeed to be affected by administering the calix[8]arene-based compounds to experimental animals infected with *M. tuberculosis*. This mechanism of action, involving directly the host cell and specifically its macrophages, is completely different from that of other drugs currently used against tuberculosis, and these compounds are promising since the resistance to conventional chemotherapeutic agents is increasing.

In addition to these examples, many other calixarene derivatives have been proposed and tested as potential bioactive compounds. Their properties are described in several patents [61]. For example, calixarenes functionalised at the wide rim with sulfonic acid, sulfonate and sulfonamide groups were tested as anticoagulant and antithrombotic agents *in vitro* and *in vivo* [61a]. In some cases they exhibit properties close to those of heparin and coumarin, currently used as anticoagulants in antithrombotic therapy. Furthermore, treatment with these compounds only slightly increases the bleeding, checked as an indicator of the hemorrhagic potential, and this is an important result considering that heparin and coumarin anticoagulants have the tendency to produce hemorrhage as side effect.

Calixarene derivatives are also active as antiviral, antimicrobial and antifungal agents [61b-d]. A remarkable activity is found in the treatment of infections by enveloped viruses like HIV, herpes simplex and influenza viruses. The calixarenes seem to bind to some components of the virus envelope during its adhesion to and entry into the infectable cells [61b].

The potential use of radioactive $^{225}Ac^{3+}$ complexes of calixarene-based ligands in the treatment of cancer cells by radiotherapy has been recently explored [62].

4. Biomimetic Catalysis

Biomimetic catalysis has long been the focus of bioorganic and bioinorganic chemistry [1,2,63]. The object of enzyme mimicry is to design artificial systems which have efficiency and selectivity comparable with those of the natural ones. A common feature, which biomimetic catalysts and enzymes share, is that the catalytic transformation occurs on a bound substrate following Michaelis-Menten saturation kinetics [64].

Many macrocyclic compounds, and especially water-soluble cyclodextrins [9] and cyclophanes [10], have been used as binding units for artificial enzymes. Calixarenes have shown a variety of host-guest properties and, therefore, are attractive as building blocks for the construction of supramolecular catalytic systems [65]. Supramolecular catalysis by calix[4]crowns, in particular, is discussed in Chapter 20 of this book.

Both calixarenes [66,67] and resorcarenes [68,69] have been the basis of synthetic molecular catalysts. A recent example [70] is the calix[6]arene-based acetylcholine esterase mimic **18**.

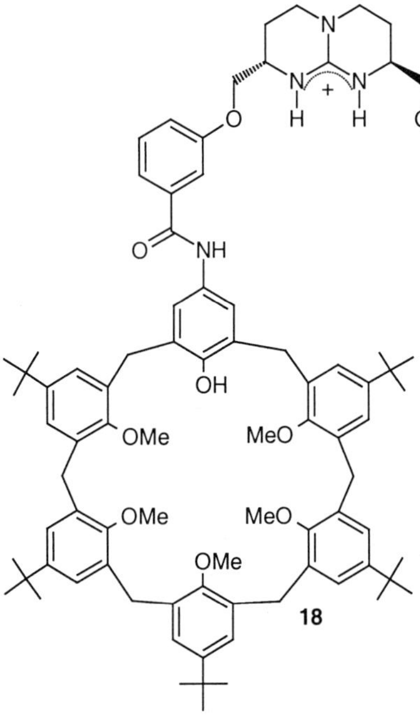

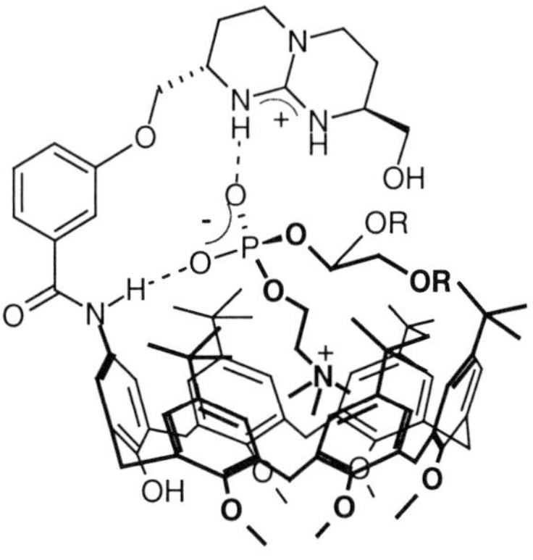

This receptor is able to bind strongly in chloroform (K_{ass} = 7.3×10^4 M^{-1} at 298 K) dioctanoyl-L-α-phosphatidylcholine (DOPC; R = COC$_7$H$_{15}$), a transition state analogue for the hydrolysis of esters and carbonates, thanks to the coordination of the phosphate monoanion to the guanidinium group and to the inclusion of the trimethylammonium ion into the calixarene cavity (**18-DOPC**). Receptor **18** is also able to catalyse the methanolysis of choline p-nitrophenylcarbonate in CHCl$_3$:MeOH 99:1 (v/v), in the presence of (i-Pr)$_2$EtN/(i-Pr)$_2$EtNHClO$_4$ buffer, showing turnover, inhibition by DOPC, and a 76-fold rate enhancement, in comparison with the background reaction (no catalyst added) [71].

In analogy to natural systems, where many enzymes catalyse phosphate ester hydrolysis by the intervention of two or more metal ions [72], several artificial dinuclear and trinuclear metallo-enzymes have been proposed [73]. The concept of multi-centre catalysis has also important consequences in asymmetric synthesis [74]. Calix[4]arenes offer the unique opportunity of linking up to four metal complexes in close proximity (see Chapter 29) and tetraalkoxycalix[4]arenes in the *cone* conformation have been used as platforms for the synthesis of a number of efficient phosphoesterase mimics.

The dinuclear Zn(II)complex **19a**-[Zn]$_2$ is able to catalyse with high efficiency the intramolecular transesterification of the RNA model substrate 2-(hydroxypropyl)-*p*-nitrophenyl phosphate (HPNP). A 23,000-fold rate enhancement is observed at pH 7 and 25 °C in the presence of 0.48 mM of **19a**-[Zn]$_2$ in comparison to the background reaction [75]. The mononuclear zinc complex **19b**-[Zn] is 50 times less active, thus enphasizing the role of cooperativity of the two metal centres. Moreover, the catalyst **19a**-[Zn]$_2$ shows a saturation kinetics which indicates a strong affinity for the substrate whereas no saturation is observed for the mononuclear calix[4]arene complex **19b**-[Zn], up to 6 equiv. of HPNP (pH 7) [75].

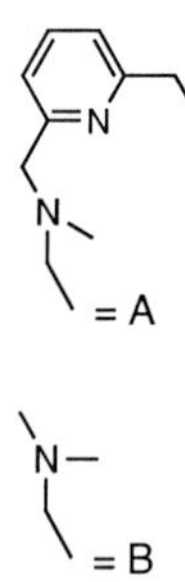

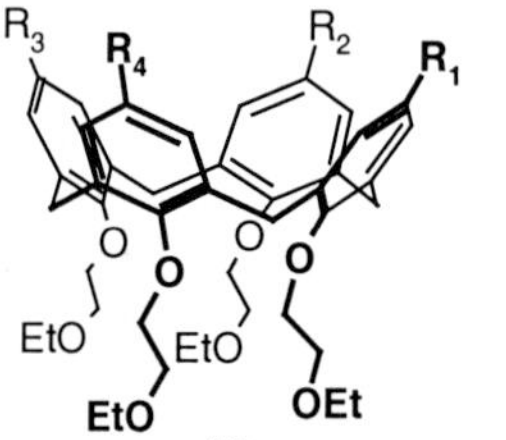

a: R$_1$ = R$_3$ = A; R$_2$ = R$_4$ = H
b: R$_1$ = A; R$_2$ = R$_3$ = R$_4$ = H
c: R$_1$ = R$_3$ = A; R$_2$ = R$_4$ = B
d: R$_1$ = R$_2$ = R$_3$ = A; R$_4$ = H

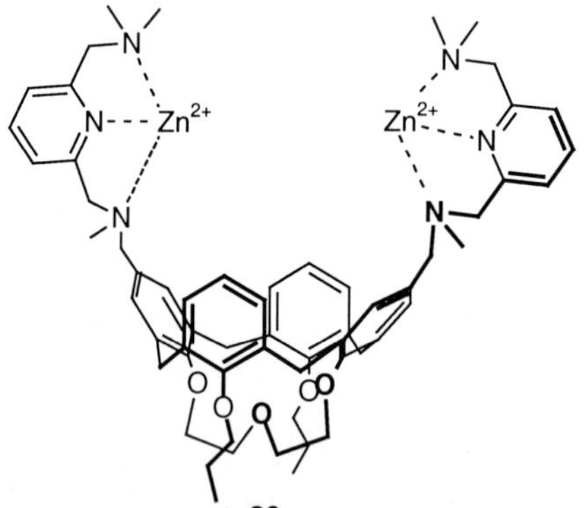

The addition of two dimethylamino groups on the *para* position of the remaining distal aromatic nuclei (**19c**) does not improve the activity of the catalyst but changes the optimum pH of the reaction from 7.5 (in the case of **19a**-[Zn]$_2$) to 6.8 (in the case of **19c**-[Zn]$_2$). This is an evidence that one dimethylamino group may act as intramolecular general base catalyst in the deprotonation of the hydroxyl group of the substrate, according to the representation shown in Figure 2a. Interestingly, the conformationally more rigid Zn(II) catalyst **20** is a factor of 8 less efficient than **19a**-[Zn]$_2$ and also binds the substrate less strongly [76]. This indicates that the rapid, low energy conformational changes of the calix[4]arene backbone play an important role in determining the catalytic activity of the artificial metallo-enzyme, probably allowing it to reach the best geometrical arrangement for catalysis and facilitating the dynamic binding of the substrate and transition state. Both complexes **19a**-[Zn]$_2$ and **19c**-[Zn]$_2$ are not active towards the DNA model substrate ethyl-*p*-nitro-phenylphosphate (EPNP), which is more difficult to cleave. Nevertheless, a dinuclear copper complex, containing chelating bisimidazolyl groups that mimic the natural histidine residues, is able to catalyse the hydrolysis of EPNP (rate enhancement 2.7×10^4, 35% EtOH, pH 6.4) [77].

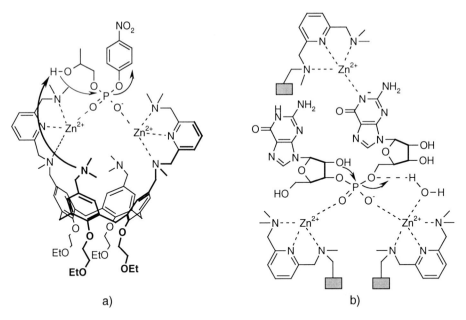

Figure 2. a) Possible mechanism for HPNP transesterification catalysed by 19c-[Zn]₂. b) Schematic representation of possible mechanism for RNA dinucleotide GpG cleavage by receptor 19d-[Zn]₃.

Even more spectacular results were obtained with the trinuclear zinc synzyme **19d-[Zn]₃** [76,78]. This complex, 0.48 mM at pH 7, is able to catalyse the cleavage of HPNP substrate showing a 32,000-fold rate acceleration. This corresponds to a reduction of the half-life from approximately 300 days for the uncatalysed reaction to 13 minutes. Compared with the dinuclear complex **19a-[Zn]₂**, the third Zn(II) centre in **19d-[Zn]₃** effects a 40% higher rate enhancement in HPNP cleavage [76]. More interestingly, the trinuclear Zn(II) complex is capable of RNA dinucleotide (3',5'-NpN) cleavage, with high rate enhancement and significant nucleobase specificity. Rate accelerations on the order 10^4-10^5 are observed with large difference between the pseudo first order rate constants ($k_{obs}/10^5$ s^{-1}) of the RNA dinucleotide cleavage: GpG (72) > UpU (8.5) > ApA (0.44) [78]. The heterotrinuclear (Zn₂Cu) complex similar to **19d-[Zn]₃** is more active and both systems show turnover catalysis. The higher reactivity in the reactions of GpG and UpU dinucleotides compared with ApA is due to enhanced binding to **19d-[Zn]₃**, whereas the higher reactivity of GpG compared with UpU is due to a higher rate of conversion. A possible mechanism for catalysis is shown in Figure 2b. One Zn(II) centre orients the RNA dinucleotide within the catalytic site by strongly interacting with the deprotonated nitrogen atom of the guanine (or uridine) nucleobase, while the other two metal centres activate the phosphoryl group by double Lewis acid coordination. Elimination of the leaving group may be assisted by a Zn(II) bound water molecule. The bidentate adenyl group imposes a less favourable substrate orientation within the catalyst **19d-[Zn]₃**, compared to unidentate binding of a deprotonated uracil or guanine group, thus explaining the lower binding affinity of ApA dinucleotide. These results are quite promising in regard to developing synthetic catalysts able to perform sequence-selective cleavage of RNA [78].

Acknowledgements: Financial support from EU (TMR grant FMRX-CT98-0231), M.U.R.S.T. (Ministero dell'Università e della Ricerca Scientifica e Tecnologica) (Supramolecular Devices Project) and University of Parma is gratefully acknowledged.

5. References and Notes

[1] R. Breslow, *J. Chem. Ed.* **1998**, *75*, 705–717.
[2] H. Dugas, *Bioorganic Chemistry* 3rd Ed., Springer-Verlag, New York, **1996**.
[3] S. J. Lippard, J. M. Berg, *Principles of Bioinorganic Chemistry*, University Science Books, Mill Valley CA, **1994**.
[4] R. Breslow, *Chem. Soc. Rev.* **1972**, *1*, 553-580.
[5] K. Mori, *Acc. Chem. Res.* **2000**, *33*, 102–110.
[6] Ref.1, p. 709.
[7] *Supramolecular Science: Where It Is and Where It Is Going*, Eds.: R. Ungaro, E. Dalcanale, NATO ASI Series, Kluwer Academic Publisher, Dordrecht, **1999**.
[8] a) K. Uekama, T. Irie, in *Comprehensive Supramolecular Chemistry*, Eds.: J. Atwood, J. E. D. Davies, D. D. MacNicol, F. Vögtle, Pergamon Press, Vol. 3, **1996**, pp. 401–481; b) K. Uekama, F. Hirayama, T. Irie, *Chem. Rev.* **1998**, *98*, 2045–2076.
[9] R. Breslow, *Chem. Rev.* **1998**, *98*, 1997–2011.
[10] a) Y. Murakami, *Top. Curr. Chem.* **1983**, *115*, 108–155; b) F. Diederich, *Cyclophanes*, Monographs in Supramolecular Chemistry, Ed.: F. J: Stoddart, Royal Society of Chemistry, **1991**, pp. 264–297.
[11] R. E. Bobine, S. L. Bender, *Chem. Rev.* **1997**, *97*, 1359–1472.
[12] a) H.-J. Schneider, F. Eblinger, M. Sirish, *Adv. Supramol. Chem.* **2000**, *6*, 185–216 and references therein; b) W. C. Still, *Acc. Chem. Res.* **1996**, *29*, 155–163 and references therein.
[13] T. Nagasaki, T. Tajiri, S. Shinkai, *Recl. Trav. Chim. Pays-Bas* **1993**, *112*, 407–411.
[14] F. Sansone, S. Barboso, A. Casnati, M. Fabbi, A. Pochini, F. Ugozzoli, R. Ungaro, *Eur. J. Org. Chem.* **1998**, 897–905.
[15] F. Sansone, S. Barboso, A. Casnati, D. Sciotto, R. Ungaro, *Tetrahedron Lett.* **1999**, *40*, 4741–4744.
[16] Y. Okada, Y. Kasai, J. Nishimura, *Tetrahedron Lett.* **1995**, *36*, 555–558.
[17] M. Sánchez Peña, Y. Zhang, S. Thibodeaux, M. L. McLaughlin, A. Muñoz de la Peña, I. M. Warner, *Tetrahedron Lett.* **1996**, *37*, 5841–5844.
[18] a) With Gly, L-Ala, L-Val and L-Phe: E. Nomura, M. Takagaki, C. Nakaoka, M. Uchida, H. Taniguchi, *J. Org. Chem.* **1999**, *64*, 3151–3156; b) With Gly and His: Y. Molard, C. Bureau, H. Parrot-Lopez, R. Lamartine, J.-B. Regnouf-de-Vains, *Tetrahedron Lett.* **1999**, *40*, 6383–6387; c) With R-Phg and S-Leu: L. Frkanec, A. Vinjevac, B. Kojic-Prodic, M. Zinic, *Chem. Eur. J.* **2000**, *6*, 442–453.
[19] R. K. Castellano, C. Nuckolls, J. Rebek, Jr., *J. Am. Chem. Soc.* **1999**, *121*, 11156–11163.
[20] a) X. Hu, J. He, A. S. C. Chan, X. Han, J.-P. Cheng, *Tetrahedron: Asymmetry* **1999**, *10*, 2685–2689; b) X. Hu, A. S. C. Chan, X. Han, J. He, J.-P. Cheng, *Tetrahedron Lett.* **1999**, *40*, 7115–7118.
[21] H. Hioki, T. Yamada, C. Fujioka, M. Kodama, *Tetrahedron Lett.* **1999**, *40*, 6821–6825.
[22] B. C. Gibb, A. R. Mezo, A. S. Causton, J. R. Fraser, F. C. S. Tsai, J. C. Sherman, *Tetrahedron* **1995**, *51*, 8719–8732.
[23] O. D. Fox, N. K. Dalley, R. G. Harrison, *J. Incl. Phenom.* **1999**, *33*, 403–414.
[24] M. Mammen, S.-K. Choi, G. M. Whitesides, *Angew. Chem., Int. Ed. Engl.* **1998**, *37*, 2754–2794 and references therein.
[25] A. P. Davis, R. C. Wareham, *Angew. Chem., Int. Ed. Engl.* **1999**, *38*, 2978–2996.
[26] Y. C. Lee, R. T. Lee, *Acc. Chem. Res.* **1995**, *28*, 323–327.
[27] a) J. E. Kingery-Wood, K. W. Williams, G. B. Sigal, G. M. Whitesides, *J. Am. Chem. Soc.* **1992**, *114*, 7303–7305; b) S. Sakai, T. Sasaki, *J. Am. Chem. Soc.* **1994**, *116*, 1587–1588; c) N. V. Bovin, H.-J. Gabius, *Chem. Soc. Rev.* **1995**, 413–421; d) P. R. Ashton, S. E. Boyd, C. L. Brown, N. Jayaraman, S. A. Nepogodiev, J. F. Stoddart, *Chem. Eur. J.* **1996**, *2*, 1115–1128; e) C. Ortiz-Mellet, J. M. Benito, J. M. García Fernández, H. Law, K. Chmurski, J. Defaye, M. L. O'Sullivan, H. N. Caro, *Chem. Eur. J.* **1998**, *4*, 2523–2531.

[28] a) A. Marra, M.-C. Scherrmann, A. Dondoni, A. Casnati, P. Minari, R. Ungaro, *Angew. Chem., Int. Ed. Engl.* **1994**, *33*, 2479–2481; b) A. Dondoni, A. Marra, M.-C. Scherrmann, A. Casnati, F. Sansone, R. Ungaro, *Chem. Eur. J.* **1997**, *3*, 1774–1782.
[29] O. Mitsunobu, *Synthesis* **1981**, 1–28.
[30] S. J. Meunier, R. Roy, *Tetrahedron Lett.* **1996**, *37*, 5469–5472.
[31] R. Roy, J. M. Kim, *Angew. Chem., Int. Ed. Engl.* **1999**, *38*, 369–372.
[32] a) A. Marra, A. Dondoni, F. Sansone, *J. Org. Chem.* **1996**, *61*, 5155–5158; b) A. Dondoni, M. Kleban, A. Marra, *Tetrahedron Lett.* **1997**, *38*, 7801–7804; c) C. Felix, H. Parrot-Lopez, V. Kalchenko, A. W. Coleman, *Tetrahedron Lett.* **1998**, *39*, 9171–9174; d) C. Saitz-Barria, A. Torres-Pinedo, F. Santoyo-González, *Synlett* **1999**, *12*, 1891–1894.
[33] a) T. Fujimoto, C. Shimizu, O. Hayashida, Y. Aoyama, *J. Am. Chem. Soc.* **1997**, *119*, 6676–6677; b) T. Fujimoto, C. Shimizu, O. Hayashida, Y. Aoyama, *J. Am. Chem. Soc.* **1998**, *120*, 601–602; c) O. Hayashida, C. Shimizu, T. Fujimoto, Y. Aoyama, *Chem. Lett.* **1998**, 13–14; d) Y. Aoyama, Y. Matsuda, J. Chuleeraruk, K. Nishiyama, K. Fujimoto, T. Fujimoto, T. Shimizu, O. Hayashida, *Pure Appl. Chem.* **1998**, *70*, 2379–2384; e) O. Hayashida, M. Kato, K. Akagi, Y. Aoyama, *J. Am. Chem. Soc.* **1999**, *121*, 11597–11598; f) K. Fujimoto, T. Miyata, Y. Aoyama, *J. Am. Chem. Soc.* **2000**, *122*, 3558–3559.
[34] a) N. Douteau-Guével, A. W. Coleman, J.-P. Morel, N. Morel-Desrosiers, *J. Phys. Org. Chem.* **1998**, *11*, 693–696; b) N. Douteau-Guével, A. W. Coleman, J.-P. Morel, N. Morel-Desrosiers, *J. Chem. Soc., Perkin Trans. 2* **1999**, 629–633; c) M. Selkti, A. W. Coleman, I. Nicolis, N. Douteau-Guével, F. Villain, A. Tomas, C. de Rango, *Chem. Commun.* **2000**, 161–162.
[35] G. Arena, A. Contino, F. G. Gulino, A. Magrì, F. Sansone, D. Sciotto, R. Ungaro, *Tetrahedron Lett.* **1999**, *40*, 1597–1600.
[36] I. S. Antipin, I. I. Stoikov, E. M. Pinkhassik, N. A. Fitseva, I. Stibor, A. I. Konovalov, *Tetrahedron Lett.* **1997**, *38*, 5865–5868.
[37] K. Araki, K. Inada, S. Shinkai, *Angew. Chem., Int. Ed. Engl.* **1996**, *35*, 72–74.
[38] W. Zielenkiewicz, O. Pietraszkiewicz, M. Wszelaka-Rylik, M. Pietraszkiewicz, G. Roux-Desgranges, A. H. Roux, J.-P. E. Grolier, *J. Solution Chem.* **1998**, *27*, 121–134.
[39] Y. Aoyama, in *Comprehensive Supramolecular Chemistry*, Vol. 2, Eds: J. L. Atwood, J. E. D. Davies, D. D. MacNicol, F. Vögtle, Pergamon Press, **1996**, pp. 279–307.
[40] T. D. James, K. R. A. S. Sandanayake, S. Shinkai, *Angew. Chem., Int. Ed. Engl.* **1996**, *36*, 1910–1922.
[41] a) O. Pietraszkiewicz, M. Pietraszkiewicz, *Polish J. Chem.* **1998**, *72*, 2418–2422; b) O. Pietraszkiewicz, M. Pietraszkiewicz, *J. Incl. Phenom.* **1999**, *35*, 261–270; c) O. Pietraszkiewicz, Z. Brzózka, M. Pietraszkiewicz, *Polish J. Chem.* **1999**, *73*, 2043–2052.
[42] K. Odashima, K. Yagi, K. Tohda, Y. Umezawa, *Bioorg. Med. Chem. Lett.* **1999**, *9*, 2375–2378.
[43] a) Y. Aoyama, Y. Tanaka, H. Toi, H. Ogoshi, *J. Am. Chem. Soc.* **1988**, *110*, 634–635; b) K. Kurihara, K. Ohto, Y. Tanaka, Y. Aoyama, T. Kunitake, *J. Am. Chem. Soc.* **1991**, *113*, 444–450; c) K. Kobayashi, Y. Asakawa, Y. Kato, Y. Aoyama *J. Am. Chem. Soc.* **1992**, *114*, 10307–10313.
[44] The formation of a 1:2 complex between methylresorc[4]arene and L-(-)-norephedrine was recently reported: W. Iwanek, R. Fröhlich, M. Urbaniak, C. Näther, J. Mattay, *Tetrahedron* **1998**, *54*, 14031–14040.
[45] Y. Shi, H.-J. Schneider, *J. Chem. Soc., Perkin Trans. 2* **1999**, 1797–1803.
[46] D. H. Williams, B. Bardsley, *Angew. Chem., Int. Ed. Engl.* **1999**, *38*, 1172–1193 and references therein.
[47] a) A. Casnati, E. Di Modugno, M. Fabbi, N. Pelizzi, A. Pochini, F. Sansone, G. Tarzia, R. Ungaro, *Bioorg. Med. Chem. Lett.* **1996**, *6*, 2699–2704; b) F. Sansone, M. Arduini, A. Casnati, E. Di Modugno, M. Fabbi, N. Pelizzi, R. Ungaro, unpublished results; c) L. Frish, F. Sansone, A. Casnati, R. Ungaro, Y. Cohen, *J. Org. Chem.* **2000**, *65*, 5026–5030.
[48] a) Y. Hamuro, M. C. Calama, H. S. Park, A. D. Hamilton, *Angew. Chem., Int. Ed. Engl.* **1997**, *36*, 2680–2683; b) Q. Lin, H. S. Park, Y. Hamuro, A. D. Hamilton, in Ref. 5, pp. 197–204; c) A. D. Hamilton, Y. Hamuro, Patent US5770387, **1998** (*CA.* 129: 95721e) ; d) H. S. Park, Q. Lin, A. D. Hamilton, *J. Am. Chem. Soc.* **1999**, *121*, 8–13.
[49] W. D. Stein, *Channels, Carriers and Pumps: An Introduction to Membrane Transport*, Academic Press, San Diego, **1990**.
[50] Y. Tanaka, Y. Kobuke, M. Sokabe, *Angew. Chem., Int. Ed. Engl.* **1995**, *34*, 693–694.

[51] J. de Mendoza, F. Cuevas, P. Prados, E. S. Meadows, G. W. Gokel, *Angew. Chem., Int. Ed. Engl.* **1998**, *37*, 1534–1537.
[52] a) T. Jin, M. Kinjo, T. Koyama, Y. Kobayashi, H. Hirata, *Langmuir* **1996**, *12*, 2684–2689; b) T. Jin, M. Kinjo, Y. Kobayashi, H. Hirata, *J. Chem. Soc., Faraday Trans.* **1998**, *94*, 3135–3140.
[53] a) R. A. Frizzell, D. R. Halm, in *Channels and Noise in Epithelial Tissues*, Eds.: F. Bronner, S. I. Helman, W. van Driessche, Academic Press, New York, **1990**, *37*, 215–242; b) Q. Al-Awqati, *Curr. Op. Cell. Biol.* **1995**, *7*, 504–508.
[54] a) A. K. Singh, C. J. Venglarik, R. J. Bridges, *Kidney Int.* **1995**, *48*, 985–993; b) J. L. Atwood, R. J. Bridges, R. K. Juneja, A. K. Singh, Patent US5489612, **1996** (*CA. 124*: 250965e).
[55] J. C. Edwards, B. Tulk, P. H. Schlesinger, *J. Membrane Biol.* **1998**, *163*, 119–127.
[56] a) G. Droogmans, J. Prenen, J. Eggermont, T. Voets, B. Nilius, *Am. J. Physiol.* **1998**, *275*, C646–C652; b) G. Droogmans, C. Maertens, J. Prenen, B. Nilius, *Brit. J. Pharmacol.* **1999**, *128*, 35–40.
[57] J. W. Cornforth, P. D. Hart, G. A. Nicholls, R. J. W. Rees, J. A. Stock, *Br. J. Pharmacol. Chemother.* **1955**, *10*, 73–86.
[58] J. W. Cornforth, E. D. Morgan, K. T. Potts, R. J. W. Rees, *Tetrahedron* **1973**, *29*, 1659–1667.
[59] a) P. D. Hart, *Science* **1968**, *162*, 686–689; b) J. A. Armstrong, P. D. Hart, *J. Exp. Med.* **1971**, *134*, 713–740.
[60] P. D. Hart, J. A. Armstrong, E. Brodaty, *Infect. Immun.* **1996**, *64*, 1491–1493.
[61] a) K. M. Hwang, Y. M. Qi, S.-Y. Liu, T. C. Lee, W. Choy, J. Chen, Patent US5409959, **1995** (*CA. 123*: 959c); b) K. M. Hwang, Y. M. Qi, S.-Y. Liu, W. Choy, J. Chen, Patent US5441983, **1995** (*CA. 123*: 275992d); c) S. J. Harris, Patent WO95/19974, **1995** (*CA.* **1996**, *124*: 55584c); d) M. Tanaka, A. Kikuchi, Patent JP7187930, **1995** (*CA. 123*: 220827y); e) J. Schlessinger, I. Lax, J. E. Ladbury, P. C. Tang, Patent US5783568, **1998** (*CA.* **1996**, *124*: 185562e); f) T. Yo, K. Fujiwara, M. Otsuka, Patent JP10203906, **1998** (*CA. 129*: 132541u); g) T. Ghosh, Patent EP954965, **1999** (*CA. 131*: 318951z).
[62] M. H. B. Grote Gansey, A. S. de Haan, E. S. Bos, W. Verboom, D. N. Reinhoudt, *Bioconjugate Chem.* **1999**, *10*, 613–623.
[63] a) *Molecular Design and Bioorganic Catalysis*, Eds.: C. S. Wilcox, A. D. Hamilton, NATO ASI Series, Kluwer Academic Publisher, Dordrecht, **1996**, *478* and references therein; b) J. K. M. Sanders, *Chem. Eur. J.* **1998**, *4*, 1378–1383; c) J. K. M. Sanders, in Ref. 5, pp. 273–286.
[64] A. Fersht, *Enzyme Structure and Mechanism*, W. H. Freeman & Co., New York, **1985**.
[65] R. Cacciapaglia, L. Mandolini, in *Calixarenes in Action*, Eds.: L. Mandolini, R. Ungaro, Imperial College Press, Singapore, **2000**, pp. 242–265.
[66] S. Shinkai, S. Mori, H. Koreishi, T. Tsubaki, O. Manabe, *J. Am. Chem. Soc.* **1986**, *108*, 2409–2416.
[67] C. D. Gutsche, I. Alam, *Tetrahedron* **1988**, *44*, 4689–4694.
[68] N. Pirrincioglu, F. Zaman, A. Williams, *J. Chem. Soc., Perkin Trans. 2* **1996**, 2561–2562.
[69] a) H.-J. Schneider, U. Schneider, *J. Org. Chem.* **1987**, *52*, 1613–1615; b) U. Schneider, H.-J. Schneider, *Chem. Ber.* **1994**, *127*, 2455–2469.
[70] J. O. Magrans, A. R. Ortiz, M. A. Molins, P. H. P. Lebouille, J. Sánchez-Quesada, P. Prados, M. Pons, F. Gago, J. de Mendoza, *Angew. Chem., Int. Ed. Engl.* **1996**, *35*, 1712–1715.
[71] F. Cuevas, S. Di Stefano, J. O. Magrans, P. Prados, L. Mandolini, J. de Mendoza, *Chem. Eur. J.* **2000**, *6*, 3228–3234.
[72] a) N. Sträter, W. N. Lipscomb, T. Klabunde, B. Krebs, *Angew. Chem., Int. Ed. Engl.* **1996**, *35*, 2024–2055; b) D. E. Wilcox, *Chem. Rev.* **1996**, *96*, 2435–2458.
[73] a) J. Chin, *Curr. Op. Chem. Biol.* **1997**, *1*, 514–521; b) R. Krëmer, *Coord. Chem. Rev.* **1999**, *182*, 243–261.
[74] H. Steinhagen, G. Helmchen, *Angew. Chem., Int. Ed. Engl.* **1996**, *35*, 2339–2342.
[75] P. Molenveld, S. Kapsabelis, J. F. J. Engbersen, D. N. Reinhoudt, *J. Am. Chem. Soc.* **1997**, *119*, 2948–2949.
[76] P. Molenveld, W. M. G. Stikvoort, H. Kooijman, A. L. Spek, J. F. J. Engbersen, D. N. Reinhoudt, *J. Org. Chem.* **1999**, *64*, 3896–3906.
[77] P. Molenveld, J. F. J. Engbersen, H. Kooijman, A. L. Spek, D. N. Reinhoudt, *J. Am. Chem. Soc.* **1998**, *120*, 6726–6737.
[78] P. Molenveld, J. F. J. Engbersen, D. N. Reinhoudt, *Angew. Chem., Int. Ed. Engl.* **1999**, *38*, 3189–3191.

Chapter 28

COORDINATION CHEMISTRY AND CATALYSIS

STÉPHANE STEYER[a], CATHERINE JEUNESSE[a], DOMINIQUE
ARMSPACH[a], DOMINIQUE MATT[a], JACK HARROWFIELD[b]

[a]*Laboratoire de Chimie Inorganique Moléculaire, UMR 7513 CNRS,
Université Louis Pasteur, 1 rue Blaise Pascal, F-67008 Strasbourg Cedex,
France. e-mail: dmatt@chimie.u-strasbg.fr*
[b]*Research Centre for Advanced Mineral and Materials Processing
University of Western Australia, Nedlands, 35 Stirling Highway, Crawley
WA 6009 Australia. e-mail: jmh@chem.uwa.edu.au*

1. Introduction

Calixarenes [1] provide unique platforms for the assembly not only of selective, multidentate monotopic ligands for transition metal ions but also of polytopic ligands which can maintain several metal centres in close proximity. Many opportunities thereby arise for the control and understanding of metal ion reactivity and much recent work has focused on the use of calixarenes as unique ligands in homogeneous catalyst systems, including those designed for asymmetric synthesis [2]. Certain aspects of this catalysis chemistry are discussed in some detail in Chapters 9, 27 and 29, and some non-catalytic applications of transition metal complexes of calixarenes are described in Chapters 23, 32 and 34. An earlier review provides a background to the present chapter [3].

2. Oxygen and Sulfur Donors Derived from *p*-Substituted Calix[n]arenes

2.1. COMPLEXES HAVING THE METAL CENTRE BOUND TO PHENOLIC OXYGEN ATOMS

Remarkable chemistry, such as the catalysis of the reduction of dinitrogen to nitride by a Nb complex of calix[4]arene [4,5] (see Chapter 29), falls within a broader context of studies of calix[4]arene anions as terminal ligating groups. This function is seen in the complex **1** [6], which is dimeric as a result of bridging by two solvent-derived oxygen atoms and K$^+$ ions involved in π-bonding to calixarene phenyl groups. Similarly, alkali metal ion bridging is seen in the complex **2**, where activation of a methylene CH bond has led to oxidative coupling of two calixarene units. Reacting the diimido complex **3** with *p-t*-butylcalix[4]arene resulted in ring-opening of the metallacycle and deprotonation of the four phenol units to give the imido complex **4** [7]. Use of a bridging diimide ligand in a similar reaction led to the binuclear complex **6**.

Reaction of *p-t*-butylcalix[4]arene with the mixed imido/amino complexes [M(NBu^t)$_2$(NHBu^t)$_2$] (M = Mo, W) gave respectively complexes **7** and **8**, while performing the reaction with [M(NMes)$_2$Cl$_2$(dme)] (Mes = 2,4,6-Me$_3$-C$_6$H$_2$) afforded the related calixarenes **9** and **10**, respectively [8]. Calixarenes **7-10** are able to include molecules such as CH$_3$CN, CNBu^t or H$_2$O.

Anions from partially methylated calix[4]arenes, such as the dianion **11** (not drawn) derived from *p-t*-butylcalix[4]arene 1,3-dimethylether, give complexes such as **12**, which reacts with LiNH^{t}Bu or LiNHMes (Mes = 2,4,6-trimethylphenyl) to give imido complexes **13** and **14** containing Ti in a distorted trigonal bipyramidal environment [9].

1

2

3

p-Bu^t-calix[4]-(OH)$_4$

4

5

2 p-Bu^t-calix[4]-(OH)$_4$

6

7 M = Mo
8 M = W

9 M = Mo
10 M = W

12

13 R = Bu^t
14 R = Mes

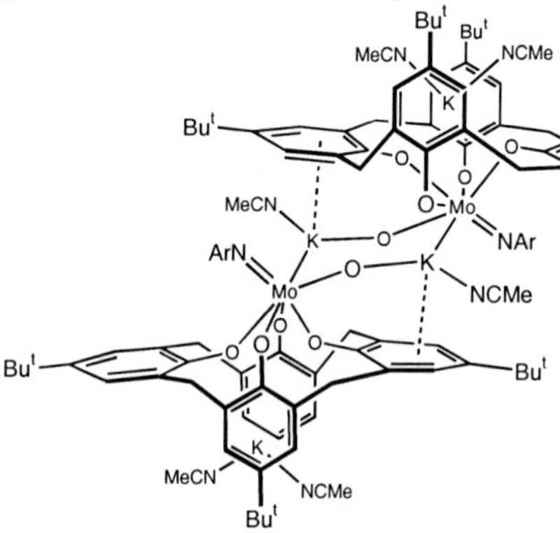

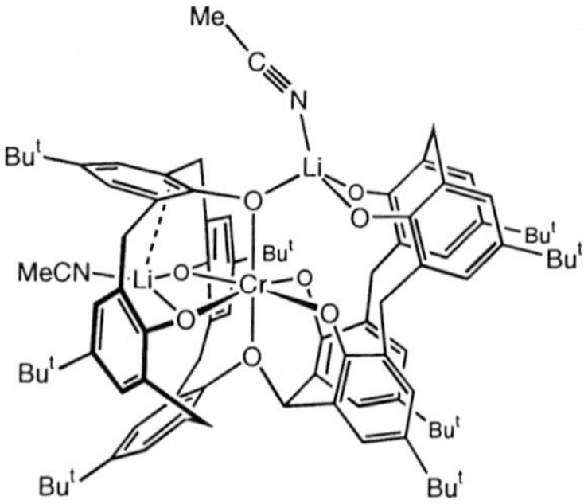

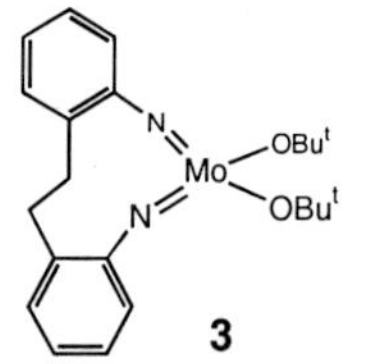

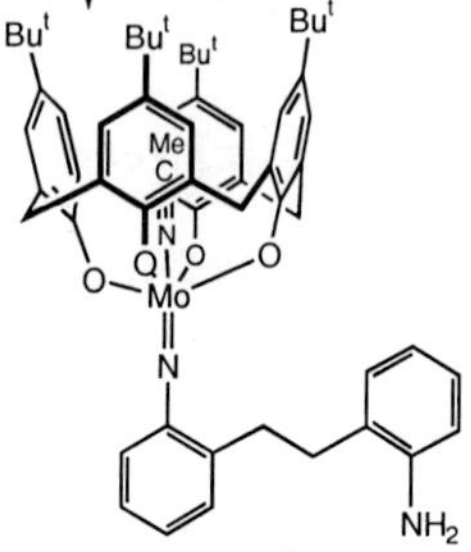

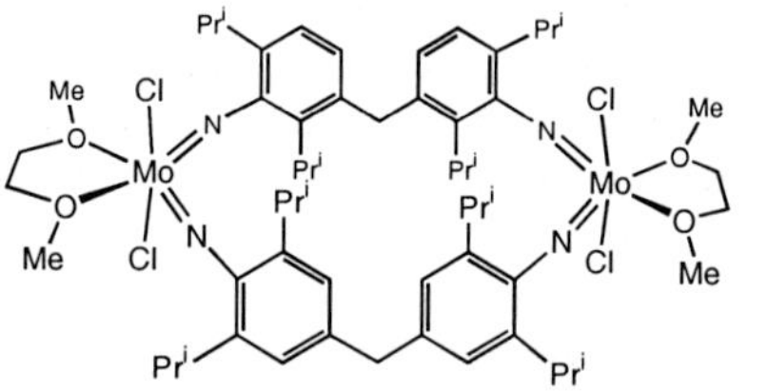

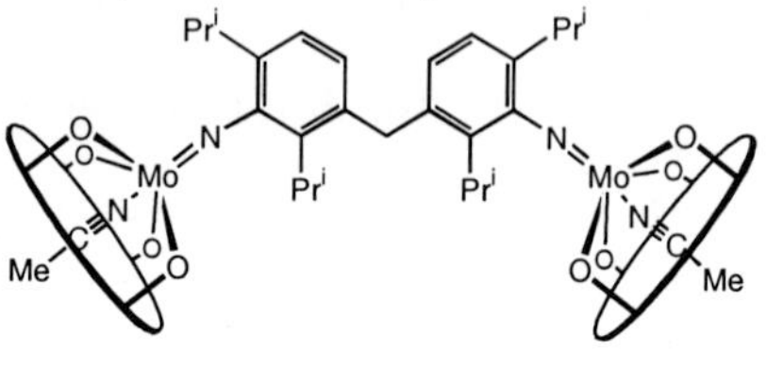

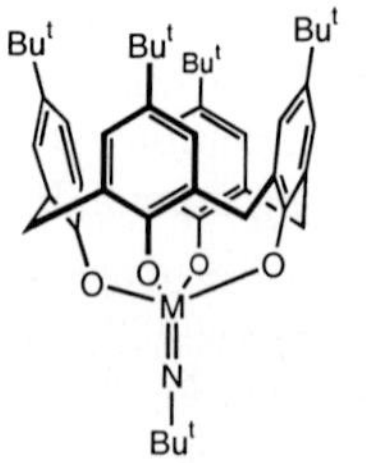

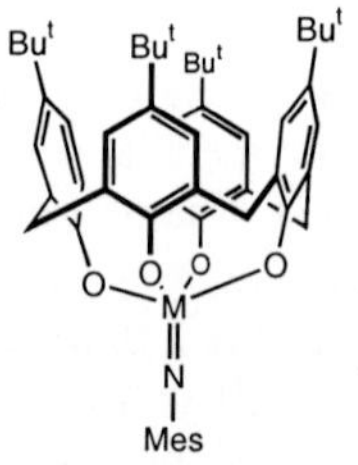

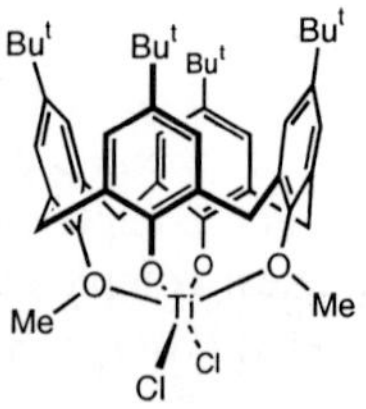

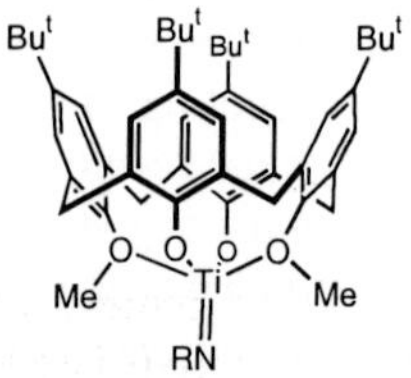

1,2-Alternate calixarene **15** simply chelates Ti(IV) in **16**. Reduction with Na gives a species which catalyses cyclotrimerisation of terminal alkynes to 1,2,4-trisubstituted benzenes with regioselectivities > 95 % [10].

15

16 R = Cl
17 R = Me
18 R = CH_2Ph

19 R = Me
20 R = CH_2Ph

21 R = Me
22 R = CH_2Ph

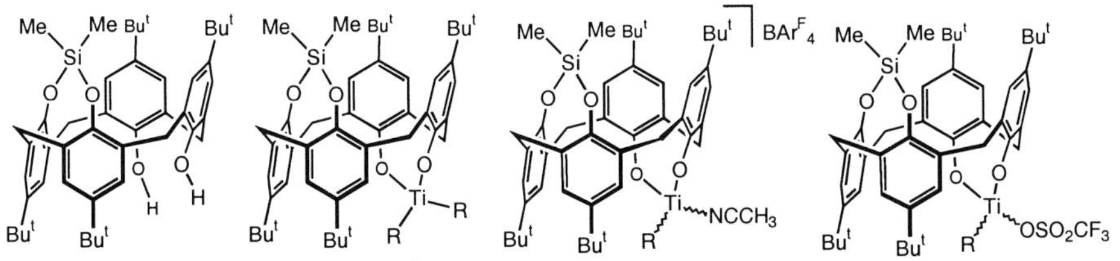

Compounds **17** and **18** undergo alkyl abstraction with $[Ph_3C][B\{3,5\text{-}(CF_3)_2C_6H_3\}_4]$ in CH_3CN to give cationic solvento complexes **19** and **20**, respectively [11]. The triflato species **21** and **22** result when $[Ph_3C][CF_3SO_3]$ is used. Other Ti complexes **26-28** involving chelation by 1,2-phenoxy units result from reaction of $TiCl_4$ with **23-25** but conformational changes are associated with formation of **27** and **28**, the latter existing in solution as a mixture of *cone* and *1,2-alternate* species [11]. With methylalumoxane as activator, complexes **26-28** show modest ethene polymerisation activities.

26 **27** **28**

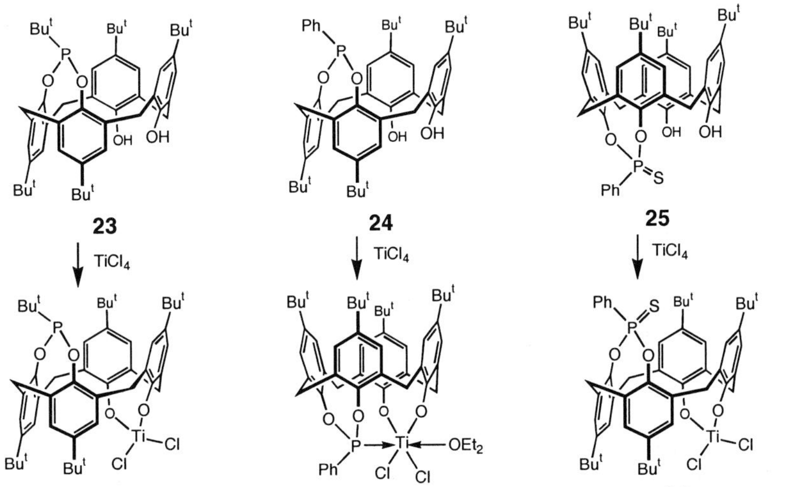

The centrosymmetric, tetranuclear complex **29** [12] has a core structure similar to that found in a calix[6]arene/Ti complex [13] and is a sandwich species like those (**30**, **31**) involving triply-bonded, bimetallic units obtained by reacting *p-t*-butylcalix[4]arene with $[M_2(NMe_2)_6]$ (M = Mo, W) [14,15]. Reaction of **30** with pyridine to give **32** results in conversion of the calixarene units into terminal ligands, and thermal treatment of **32** leads to reversible loss of $NHMe_2$ [16].

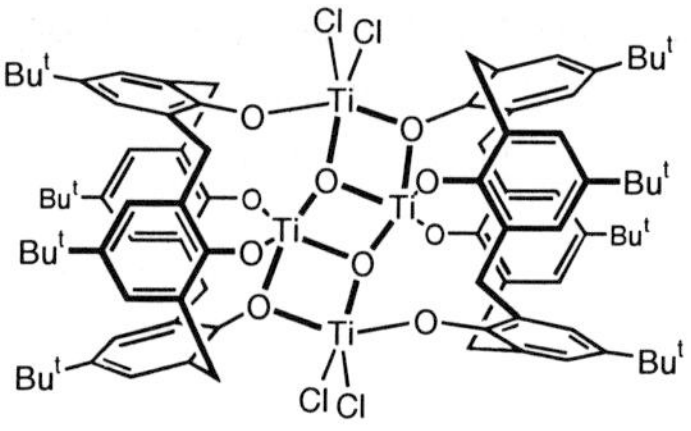

29

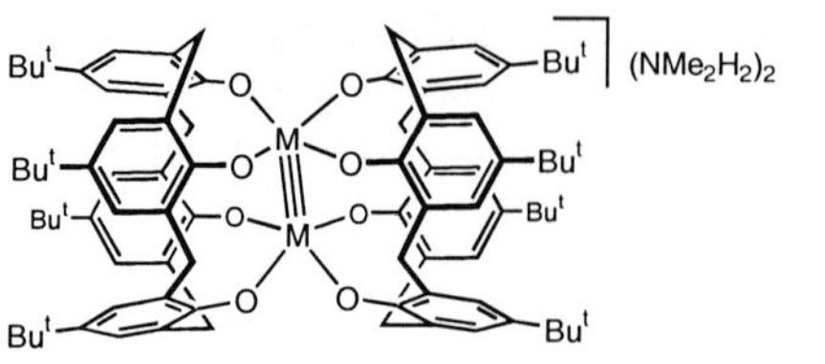

30 M = Mo
31 M = W

The product **33** can also be obtained by reaction of the calixarene with [Mo(O^{t}Bu)$_6$] and the closely related W complex **34** can be obtained simply by heating the W analogues of **32** and **33** in benzene. Sandwich structures like those of **29-34** also occur in transition metal complexes of *p*-*t*-butyltetrathiacalix[4]arene [17] (See Chapter 6).

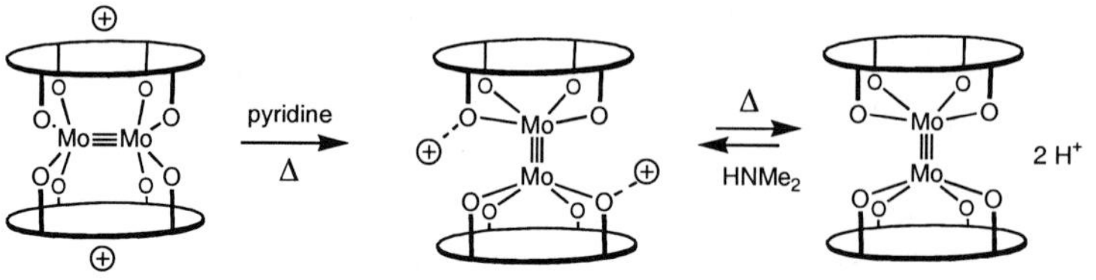

30 ⊕ ≡ NMe$_2$H$_2$$^+$ **32**   **33**

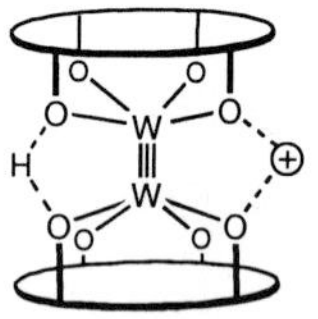

34   **35** R = *n*-C$_6$H$_{13}$ **36**   Tp = $\left(\begin{array}{c} N\diagdown N \end{array} \right)_3 BH$

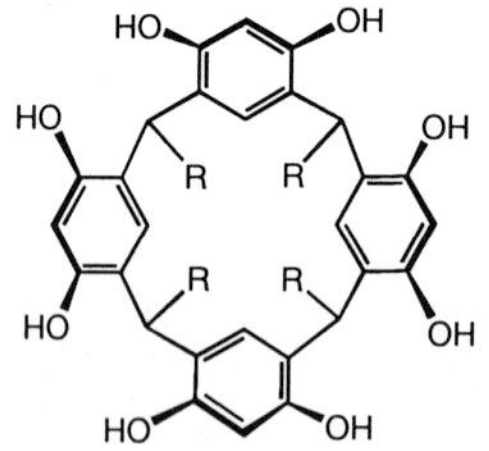

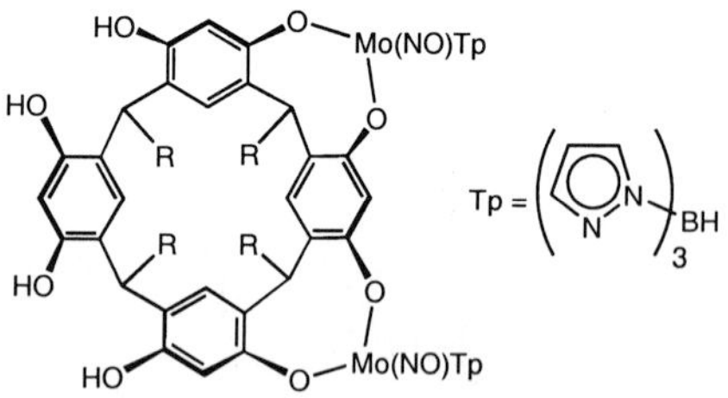

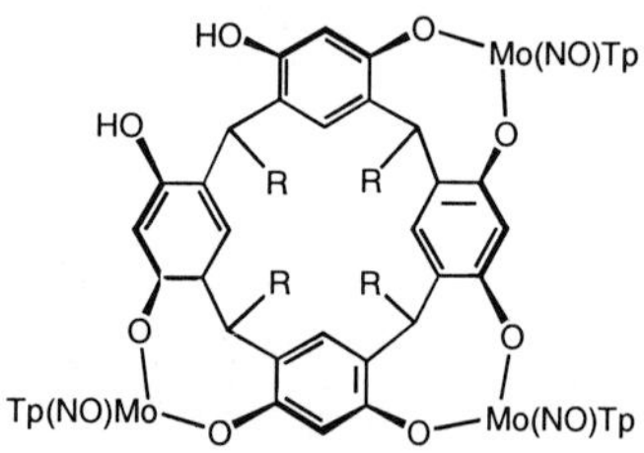

37   **38**

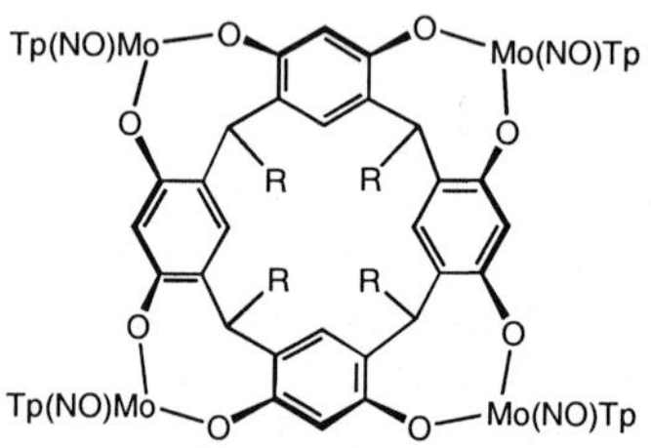

Multimetal coordination of the resorcinarene **35** results from reaction with [MoTp(NO)I$_2$] (Tp = hydridotris(pyrazol-1-yl)borate), **36-38** being characterised in the product mixture [18].

2.2. COMPLEXES HAVING A METAL CENTRE BOUND TO OXYGEN OR SULFUR ATOMS BELONGING TO PENDENT FUNCTIONAL GROUPS

The palladium complex **40** of ligand **39** has the bound metal held above the calix cavity and NMR evidence indicates that ligands L such as acetonitrile and 4-phenylpyridine (in

41) are inserted into the cavity [19,20]. There is a marked preference for 4-phenylpyridine over its 3- isomer. The more constrained ligand **42** forms complexes **43** and **44**. NMR studies show a fluxional process for **44** where the pyridine moves "in and out of" or through the cavity.

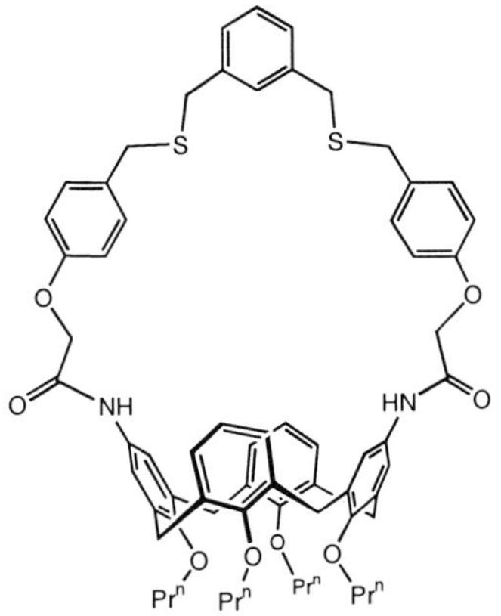

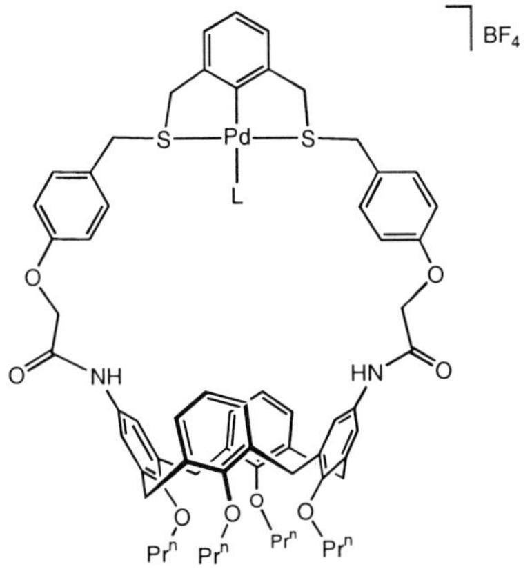

40 L = CH$_3$CN
41 L = 4-phenylpyridine

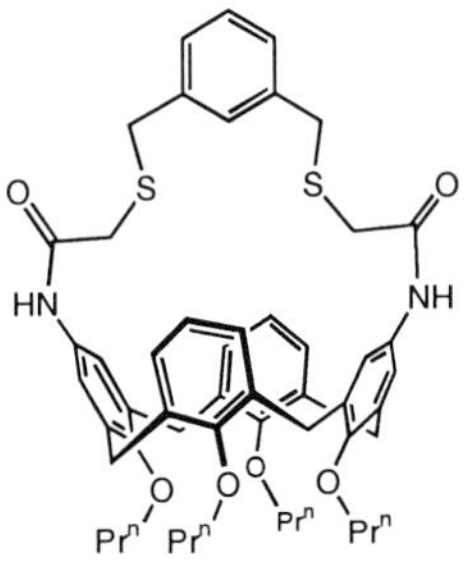

39

42

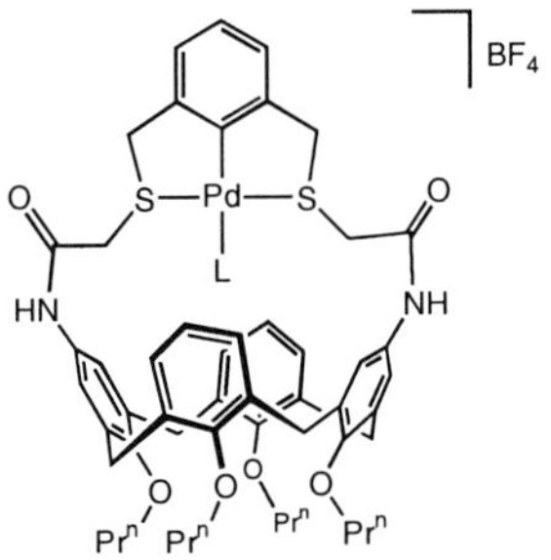

43 L = CH$_3$CN
44 L = pyridine

Podand calixarenes with oxygen donor groups have been widely used in solvent extraction (see Chapters 21, 22) of transition metals and the X-ray crystal structure of the tetramide complex **45** shows eight-coordinate Cu(II) interacting more strongly with carbonyl oxygen atoms (Cu-O 1.926(6) Å) than with etheral (Cu-O 2.963(6) Å) [21]. In **46**, the six-coordinate Fe is bound to 2 phenolic oxygen (Fe-O 1.802(8), 1.830(7) Å), 2 carbonyl (Fe-O 2.058(7), 2.031(7) Å) and two etheral oxygen atoms (Fe-O 2.230(7), 2.312(7) Å) [22].

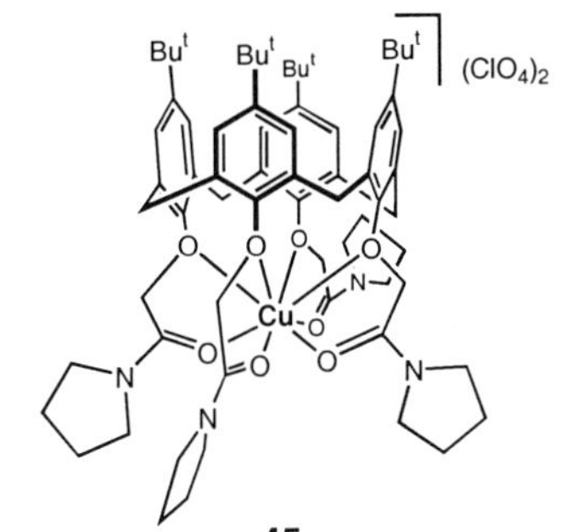

45

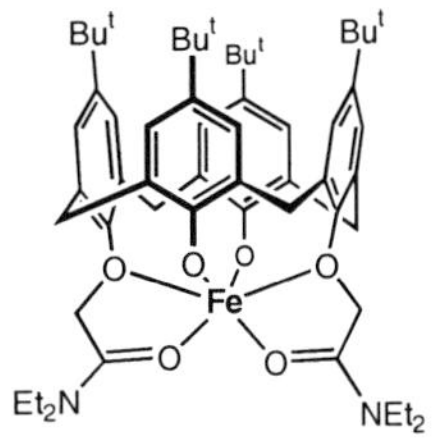

46

3. Calixarene Complexes with Metals Bound to Nitrogen Donor Atoms

3.1. ATTEMPTS TO PREPARE METALLO-ENZYME MODELS

Upper-rim functionalisation of calix[4]arene with up to three tridentate metal-binding units has provided Cu(II) and Zn(II) complexes which function as catalysts for phosphate ester hydrolysis and provide useful models of phosphoesterase enzymes [23-25]. Their chemistry is discussed in detail in Chapter 27. A monometallic enzyme pocket model has been explored in the Cu(I) complexes **48** and **49** of ligand **47** [26], the metal ion binding to 3 pyridine units resulting in the induction of a chiral, though rapidly inverting (at room temperature) structure for the complex [27].

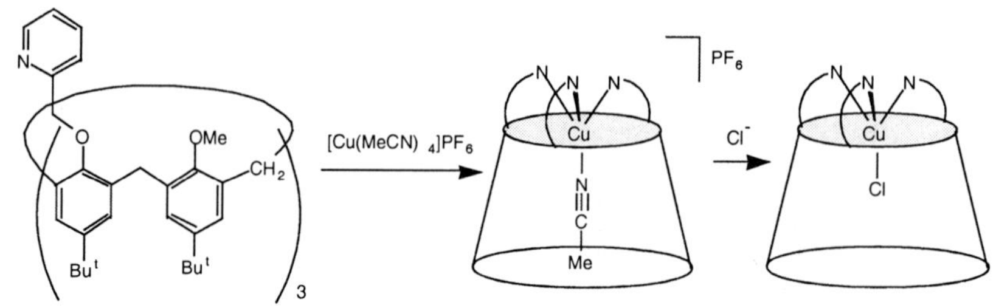

3.2. COMPLEXES FORMED FROM RESORCINARENES CONTAINING PENDENT NITROGEN LIGANDS

Binding of transition metal ions can be used to assemble capsules (a topic covered more generally in Chapters 7 and 8) from appropriately substituted calixarenes. Square planar cations with adjacent coordination positions blocked by a chelate, for example, can be used to enforce orthogonal linking of calixarenes [28], and reactions of resorcinarenes **50** and **51** with [Pd(dppp)(CF$_3$SO$_3$)$_2$] (dppp = Ph$_2$P(CH$_2$)$_3$PPh$_2$) gave **52** and **54**, respectively, while reaction of **50** with [Pt(dppp)(CF$_3$SO$_3$)$_2$] gave **53**. The presence of one triflate anion within these D_{4h}-symmetry capsules was inferred by ^{19}F NMR and confirmed by a crystal structure determination.

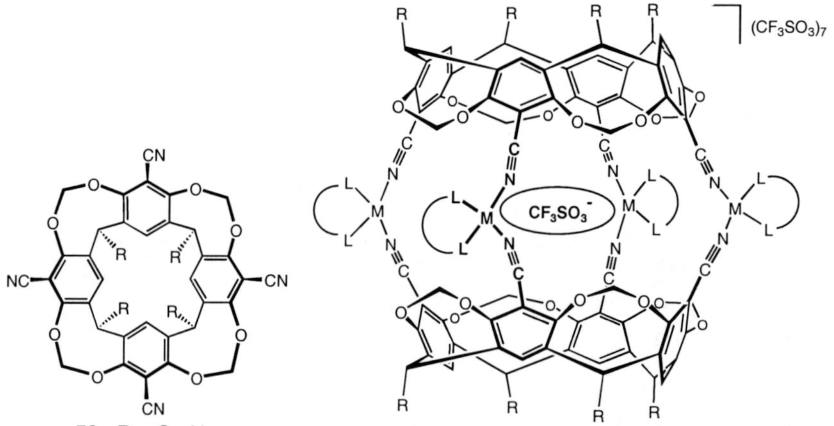

50 R = C$_{11}$H$_{23}$
51 R = C$_6$H$_{13}$

52 R = C$_{11}$H$_{23}$, M = Pd, L$_2$ = Ph$_2$P(CH$_2$)$_3$PPh$_2$
53 R = C$_{11}$H$_{23}$, M = Pt, L$_2$ = Ph$_2$P(CH$_2$)$_3$PPh$_2$
54 R = C$_6$H$_{13}$, M = Pd, L$_2$ = Ph$_2$P(CH$_2$)$_3$PPh$_2$

The crystal structure of the paramagnetic capsule **55** shows *N,N*-cis-fac coordination of iminodiacetate units to Fe(II), the alternative *N,N*-trans-fac coordination being incompatible with formation of a capsule [29].

55

56

57

59 Δ,Δ,Δ,Δ-isomer

58 ≡

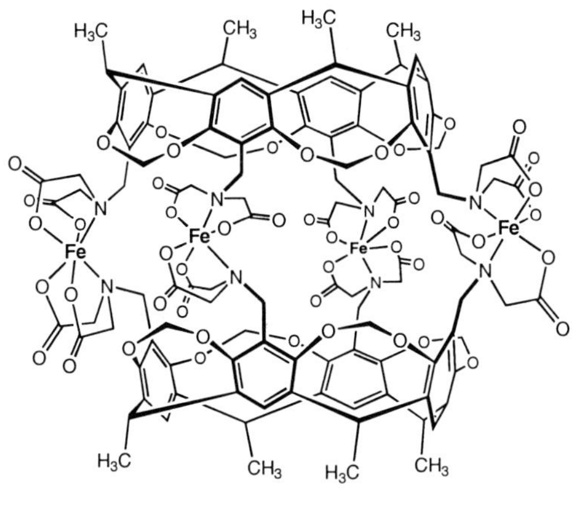

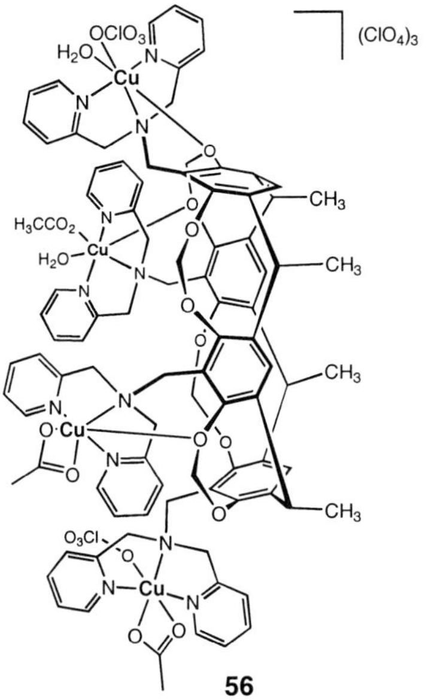

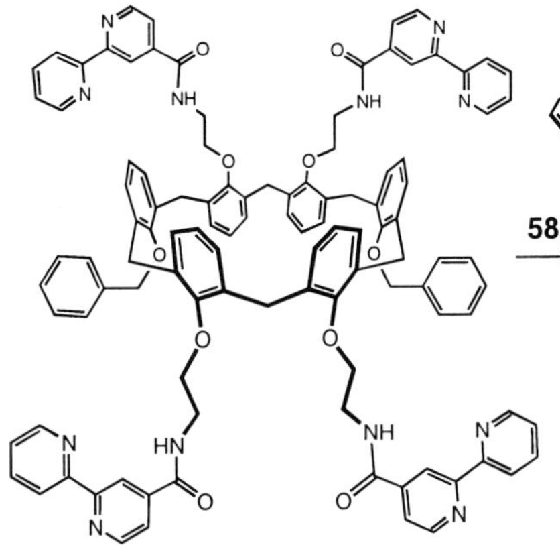

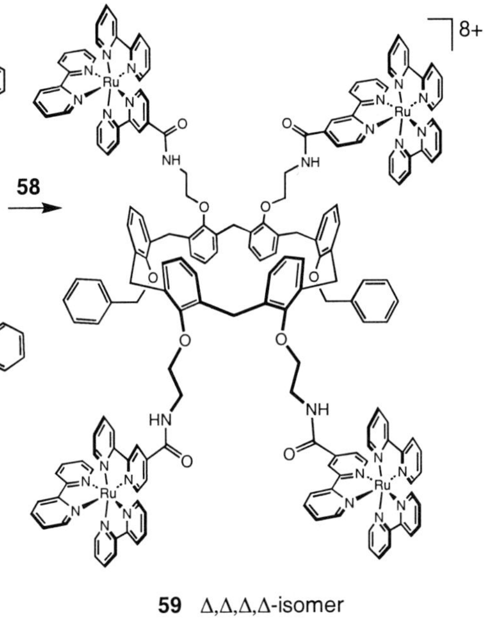

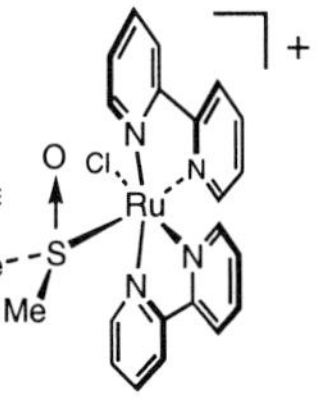

Depending on the synthesis conditions, the capsule may be obtained with six water molecules or organic species such as pentane, cyclohexane, benzene, fluorobenzene and bromobenzene included. In the tetranuclear cavitand **56**, all four Cu(II) units are different, though the tridentate arms of the ligand are all bound meridionally [30].

3.3. OTHER COMPLEXES OBTAINED FROM CALIX[n]ARENES SUBSTITUTED BY NITROGEN-CONTAINING LIGANDS

Attachment of photo- and/or electro-active transition metal units to calixarenes using *N,N*-chelating entities such as 2,2'-bipyridine has been widely investigated, and this chemistry and its applications are discussed in Chapters 23, 32 and 34. Isomerism associated with the metal units can be a complication but can be avoided by the use of resolved precursors, as in the synthesis of **59** from **57** and the Δ-isomer of **58** [31].

Under the appropriate stoichiometry, *cone* **60** reacts efficiently with [Pd(dppp)(CF$_3$SO$_3$)$_2$] to give a species characterised as **61** on the basis of spectroscopic and osmometric measurements [32].

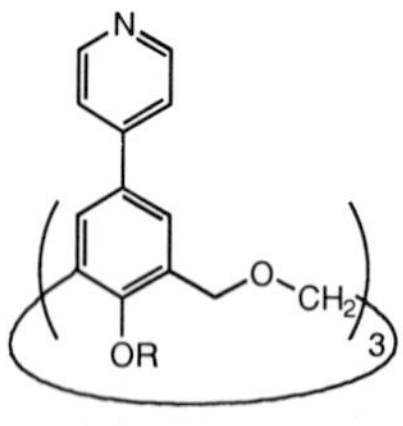

60 R = CH$_2$CO$_2$Et

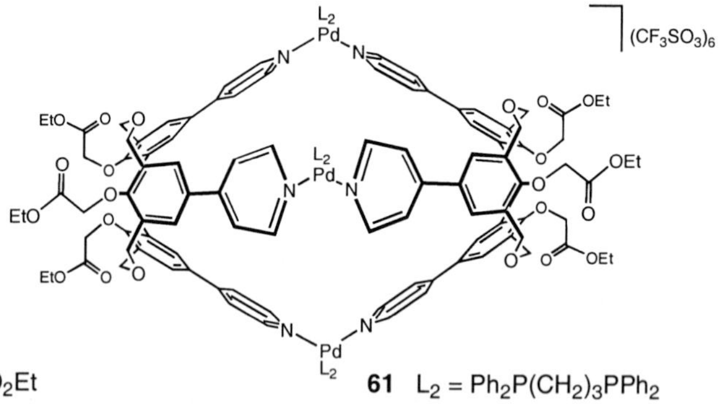

61 L$_2$ = Ph$_2$P(CH$_2$)$_3$PPh$_2$

X-ray crystallography has shown that divergent nitrile arrays on a *1,3-alternate* calixarene **62** can be used to form a linear coordination polymer **63** with AgAsF$_6$ [33].

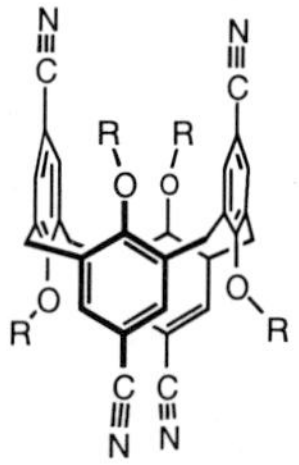

62 R = CH$_2$CH$_2$OCH$_3$

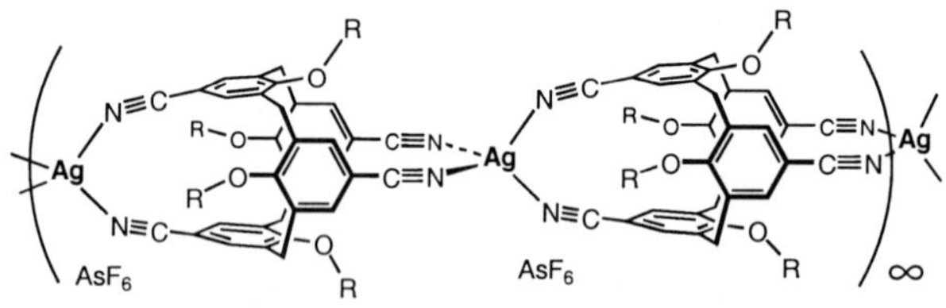

polymeric **63** R = CH$_2$CH$_2$OCH$_3$

Convergent chelate arm arrays enforced in a *cone* conformation make Ag(I) eight-coordinate in a square-antiprismatic geometry (Ag-O 2.923(3), Ag-N 2.483(5)Å) in **64** [34], though for Cu(I) in the similar complex **65**, spectroscopic measurements indicate that the metal ion is bound to four N atoms only [35]. The histidyl podands **66** and **67** both bind a single Co(II) but the exact form of the complexes is unknown [36].

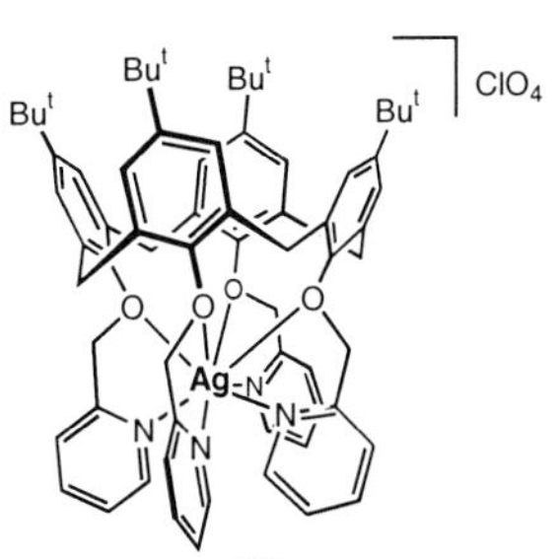

COORDINATION CHEMISTRY AND CATALYSIS 521

64

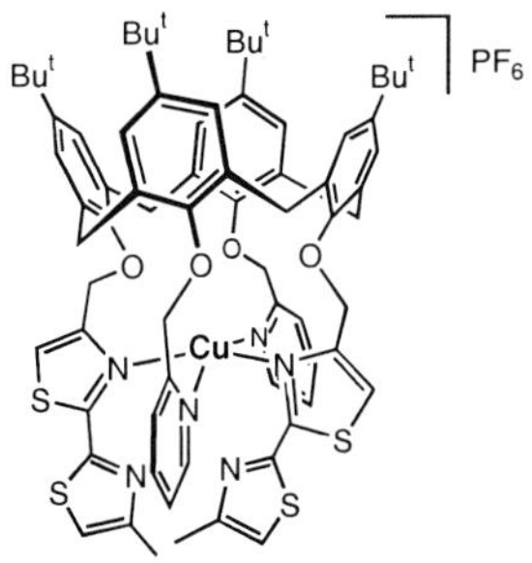

65

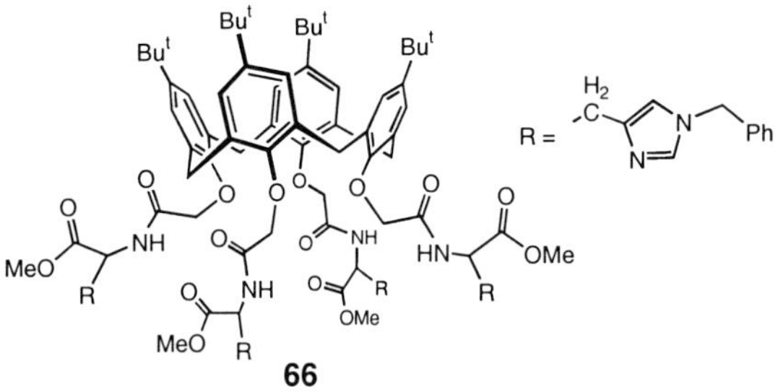

66

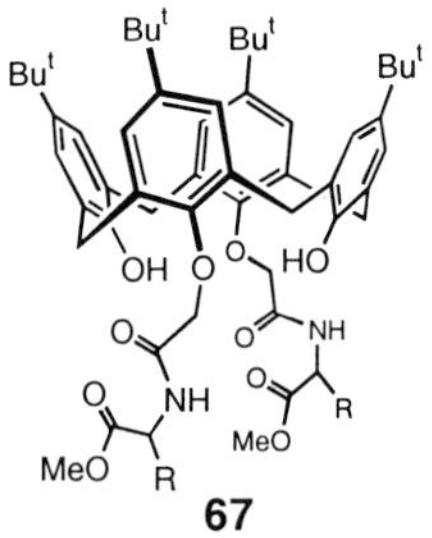

R = -CH$_2$-imidazole-N-CH$_2$-Ph

67

4. Calixarene Complexes Obtained from Phosphorus Ligands

4.1. COMPLEXES FROM CALIXARENES WITH PHOSPHORUS ATOMS TETHERED TO THE LOWER RIM

Monophosphinite **68** reacts with [MCl$_2$(PhCN)$_2$] (M = Pd, Pt) to form the isomorphous MCl$_2$L$_2$ complexes **70** and **71** [37]. The selective formation of *trans* compounds can be assigned to the bulkiness of the ligand. In the solid state the phosphino groups adopt the position that causes minimal steric perturbation to the calixarene units and the two calixarenes moieties adopt a divergent spatial array. Each phosphinated aryl ring is pushed towards the centre of the cavity. The C···C separation between the *p*-carbon atoms of the phenol ring and the phosphinated one is only 4.64 Å. Reaction of diphosphinite **69** with [PtCl$_2$(PhCN)$_2$] afforded sparingly soluble oligomeric material. The formation of tetrameric structures has previously been established with soluble forms of **69** [38].

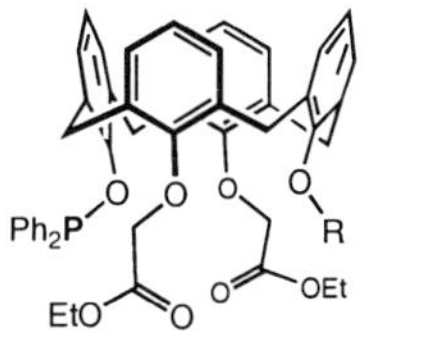

4.64 Å

68 R = H
69 R = PPh$_2$

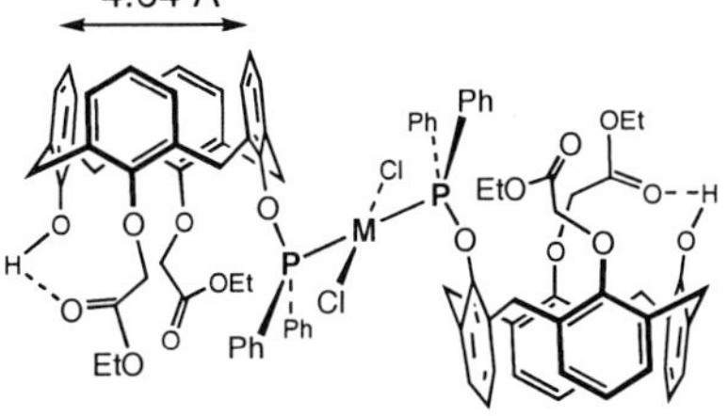

70 M = Pd
71 M = Pt

The tetraphosphinite ligands **72** and **73** readily form chelate complexes involving proximal phosphinites [39]. Thus, the homodimetallic complexes **74** and **75** have been obtained in high yield from the corresponding [MCl$_2$(COD)] precursor (M = Pd, Pt, COD = cycloocta-1,5-diene) and **72**. The dirhodium complex **76** was obtained from [Rh(CO)$_2$Cl]$_2$, but the exact relative stereochemistry about the two metal centres could not be determined. A remarkable downfield shift was observed for the *endo*-ArC*H* proton of the metallomacrocyclic units of complexes **74-76**. Similar observations were made within related but somewhat larger metallo-macrocycles (vide infra) [40].

Reaction of **72** with NiCl$_2$ gave the monometallic chelate complex **77**, whatever the stoichiometry used [39]. This complex is useful for the preparation of heterobimetallic systems, such as the Ni-Mo complex **78** (Scheme 1a). The monometallic palladium complex **79** was obtained from [PdCl$_2$(COD)] using a 1Pd:1L stoichiometry, but this complex slowly disproportionates into **72** and the dipalladium complex **80** (Scheme 1b).

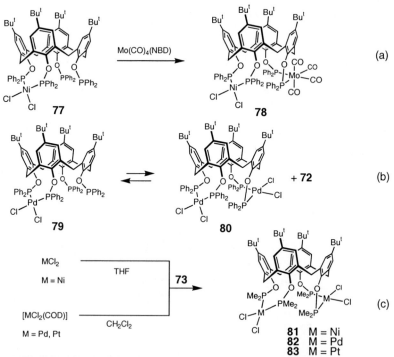

Scheme 1. Formation of bimetallic calixarene complexes.

Bis-chelate complexes (for M = Ni, Pd, Pt) were easily obtained from the less sterically demanding and more basic tetraphosphinite **73** (Scheme 1c) [39]. The nickel centres in **81** undergo planar-tetrahedral interconversion. As shown by X-ray diffraction studies of **82** and **83**, the P-M bonds point away from the cavity.

Despite the long separation between two adjacent phosphorus centres, the reaction of **84** with [PtCl$_2$(PhCN)$_2$] afforded the bis-chelate complex **85** in high yield. When [PdCl$_2$(PhCN)$_2$] was used as a starting compound, the related di-palladium complex **87** was formed in 91 % yield along with some oligomeric material [41]. In the presence of SnCl$_2$, complex **85** catalyses the hydroformylation of styrene, but the catalytic activity of this system is much lower than that of conventional PtCl(SnCl$_3$)(diphos) catalysts. However, a slightly higher regioselectivity (+ 10 %) towards branched/aldehyde was observed. The catalytic activity of **87** in the hydroalkoxycarbonylation of styrene (PhCH=CH$_2$ + CO/ROH, 130 °C, 140 bar) is weak owing to alcoholysis of the PO bonds [41].

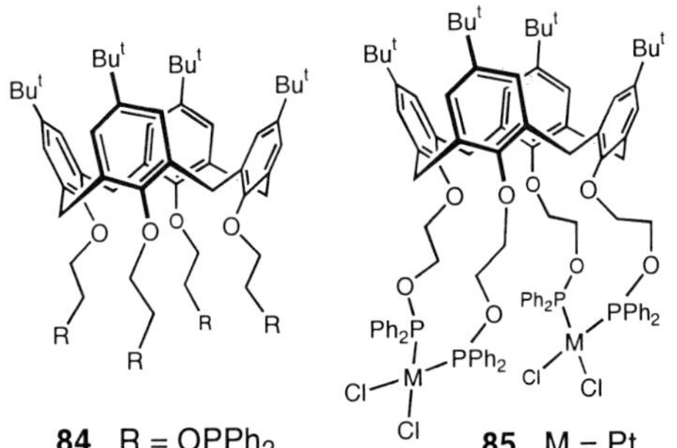

84 R = OPPh$_2$
86 R = PPh$_2$

85 M = Pt
87 M = Pd

The activities of [RhCl$_2$(NBD)]$_2$/**84** and [RhCl$_2$(NBD)]$_2$/**86** mixtures towards the hydroformylation of styrene were found to be slightly lower than the usual Rh/phosphine systems. The aldehyde selectivities are comparable to the best rhodium catalysts [41].

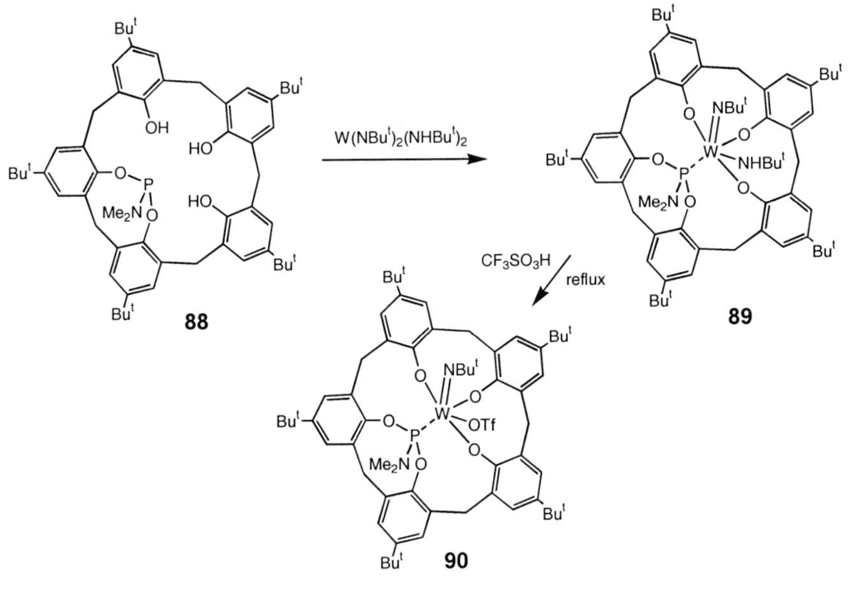

Positioning of a tungsten unit at the entrance of a calix[5]arene was achieved by reacting the aminophosphite **88** with [W(NBu^t)$_2$(NHBu^t)$_2$], giving **89** [42]. The tungsten atom adopts a square-pyramidal geometry, while the calixarene backbone is rather flattened. The phosphorus lone pair points towards the sixth coordination site with a P···W distance of 3.15 Å that is far outside the usual range found for P-W bonds. Nevertheless, both atoms are close enough to exhibit a small through-space coupling of 43 Hz in the NMR spectrum. Substitution of the butylamido ligand by OTf gave complex **90**. The change of electron density at the metal centre induces a significant shortening of the P-W bond (2.74 Å in **90**) and the imido ligand is now located inside the calix cavity.

The chelate complexes [(syn-**91**)MCl$_2$] (M = Pd, **92**; M = Pt = **93**), [(syn-**91**)M(CH$_3$)Cl] (M = Pd, **94**; M = Pt = **95**), [(syn-**91**)Pd(CH$_3$)(CH$_3$CN)]CF$_3$SO$_3$ (**96**) and [(syn-**91**)$_2$Pd(0)] (**97**), all display a *cis* arrangement of the phosphorus atoms [43], thus leaving two adjacent sites on Pd available for catalytic action.

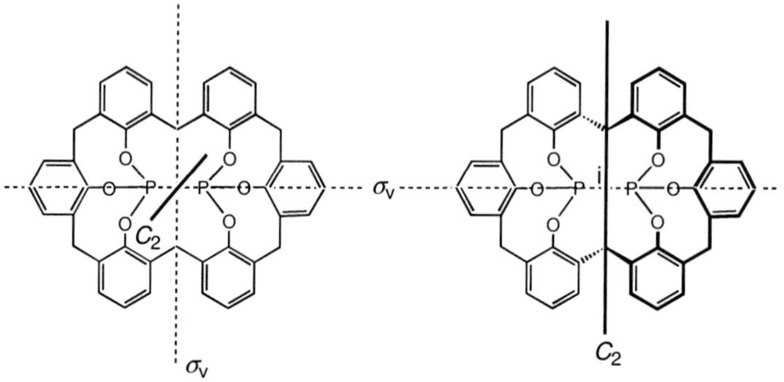

91 syn

91 anti

The structure of **92** (Figure 1) showed that the ligand bite angle in this chiral complex is 94°. Both complexes **94** and **96** react with CO to afford the insertion products **98** and **99**, respectively. Only the cationic complex **99** reacts further with ethylene to yield the five-membered ring species **100**. This behaviour made complex **99** a potential candidate for the catalytic copolymerization of CO/ethylene. Remarkably, complex **99** is the first reported complex based on a diphosphite that displays activity in the co-polymerization of ethylene and carbon monoxide. Its activity is close to that of other Pd/phosphine complexes [43].

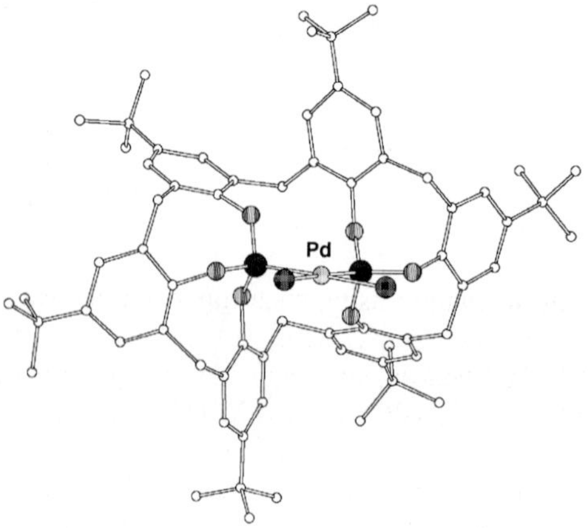

*Figure 1. Molecular structure of **92**.*

Very high regioselectivities in favour of *n*-nonanal were found in the rhodium-catalysed hydroformylation of 1-octene using **101** [44]. A reaction time of eight hours (1-octene: Rh = 4000, P(H$_2$-CO) = 20 bar, 100 °C) resulted in 63 % conversion and an aldehyde selectivity of 61 % (12 % *n*-octenes and 27 % octane). The *n/iso* ratio was 200:1, representing the highest regioselectivity observed to date. Molecular modelling shows that the P_2-chelated rhodium atom is blocked between the two sterically demanding 2,6-di-Bu^t-phenoxy groups in such a manner that the olefin approach is difficult. This explains the relatively low reaction rate. Furthermore, the steric hindrance strongly favours the formation of the 1-octyl-rhodium intermediate over the 2-octyl complex, hence the high proportion of *linear* nonanal. It was verified that removal of the two Bu^t substituents increases the activity, but this was also accompanied by a selectivity drop.

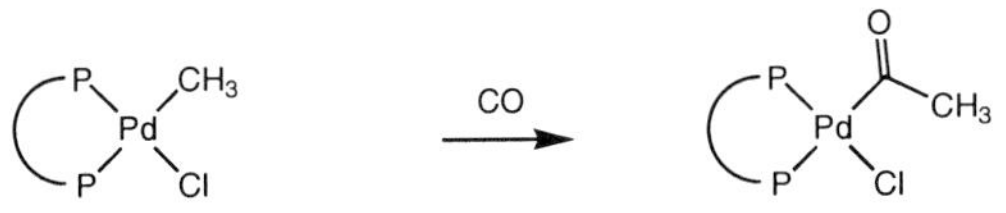

94 (PP = *syn*-**98**)   **98**

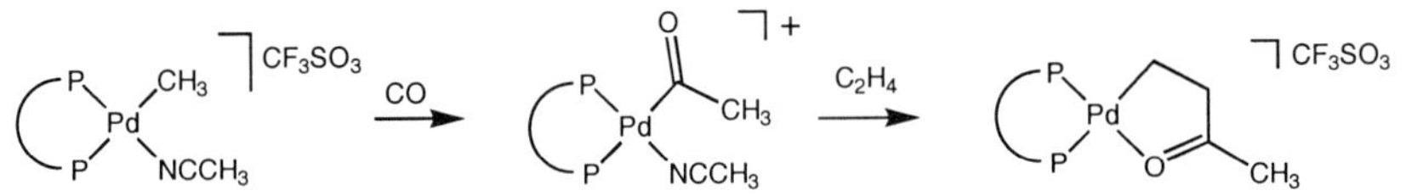

96   **99**   **100**

The coordinative properties of calixarenes bearing two distally-positioned-CH$_2$PPh$_2$ arms have been investigated towards a series of precious metals [45]. Diphosphines of this family are able to behave as *trans*-spanning ligands, as *e.g.* in the cationic complexes **102** and **103**. While the metal centre in **102** appears to be in a linear PMP arrangement, the silver atom of **103** is probably in a T-shaped P_2O_{amide} coordination environment.

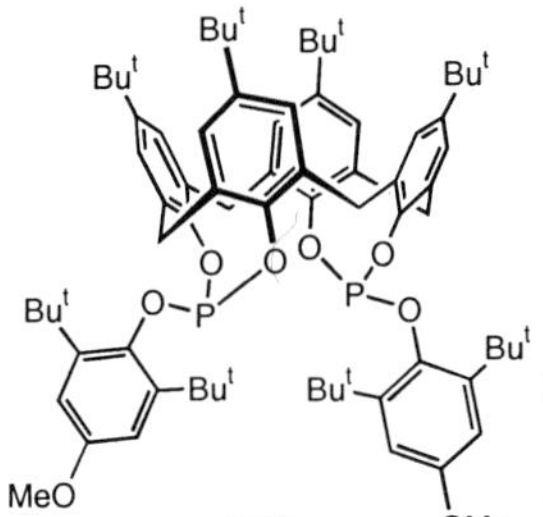

101 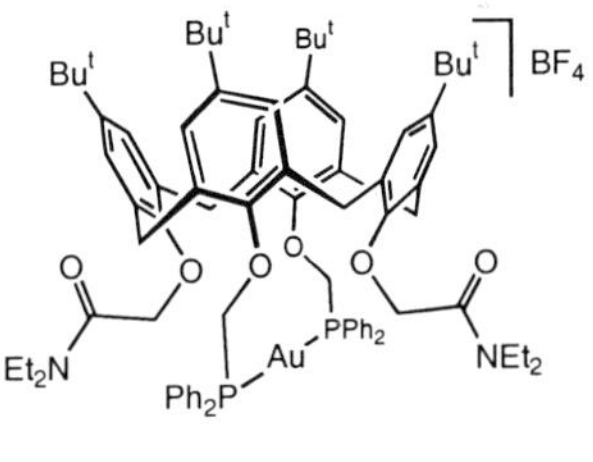   **102** 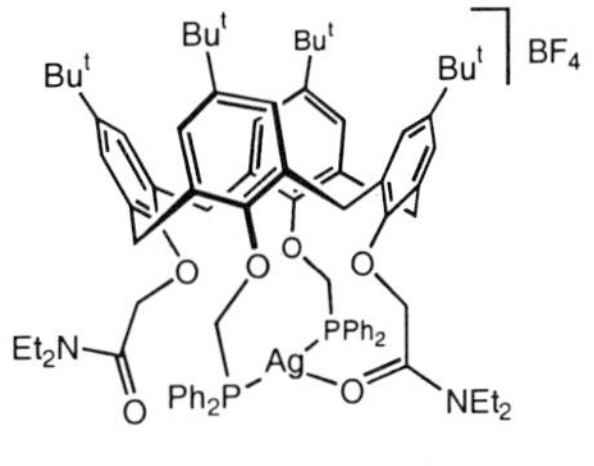   **103**

The rate of chelate formation using such diphosphines was found to strongly depend on the nature of the functional groups linked to the other two phenolic oxygen atoms. Diphosphines **104-107** bearing side arms with oxygen donor functionalities, react with the hydrido complex [PtHCl(PPh$_3$)$_2$]BF$_4$, to afford a mixture of *trans* complexes (Scheme 2). With diphosphines bearing weakly coordinating side groups, compounds of type A with a remaining PPh$_3$ ligand are formed slowly but preferentially, whereas with strong donors such as -C(O)NEt$_2$ both PPh$_3$ are rapidly substituted (type B complexes are the major compounds). In these reactions, A-type complexes are first formed and then are slowly converted into complexes of type B. With **104**, which has weak ether donors, no reaction at all takes place at room temperature. It is plausible that the phosphine moieties on their own cannot initiate the substitution reaction, probably for steric reasons. The substitution must therefore be initiated by the sterically less demanding side-functions. The ease of formation of B-type complexes increases with the donor properties of the auxiliary function. With diamide **107**, only formation of a B-type complex (**107B = 108**) was observed. In all the complexes formed, the platinum-hydrogen bond points inside the calixarene cavity. Because of the steric protection of the hydride, the Pt-H bond displays a rather high stability towards potential reagents, such as e. g. MeO$_2$CC≡CCO$_2$Me, which normally insert into the Pt-H bond.

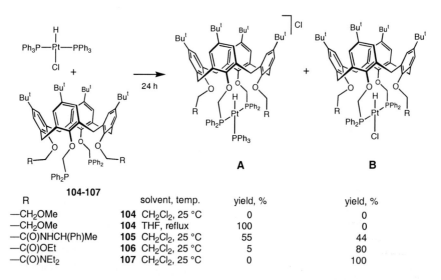

R		solvent, temp.	yield, %	yield, %
—CH$_2$OMe	104	CH$_2$Cl$_2$, 25 °C	0	0
—CH$_2$OMe	104	THF, reflux	100	0
—C(O)NHCH(Ph)Me	105	CH$_2$Cl$_2$, 25 °C	55	44
—C(O)OEt	106	CH$_2$Cl$_2$, 25 °C	5	80
—C(O)NEt$_2$	107	CH$_2$Cl$_2$, 25 °C	0	100

Scheme 2. Anchimeric assistance of functional side groups during the formation of calixarene chelate complexes.

Under forcing conditions and with the help of the auxiliary group, the direction of the Pt-H vector can be modified. The reaction of **108** with AgBF$_4$ results in the formation of complex **109** in which the Pt-H bond points towards the exterior of the cavity. In this structure, the metal plane closes the cavity [45]. In this arrangement, the hydride becomes also much more reactive. With MeO$_2$CC≡CCO$_2$Me, the platinum alkenyl insertion product **110** is formed. The high *trans* influence of the Pt-C(alkenyl) bond labilises the platinum amide bond and favours the substitution at the free co-ordination site by the opposite amide. This results in a fast pendulum motion, the exchange possibly occurring via a transient penta-coordinated platinum species. Substitution of the co-ordinated amide of **109** by strong donors allows the repositioning of the hydride inside the cavity, as illustrated by the reaction with 4,4'-bipyridine, leading to **11**. In the related complex **112** no pendulum motion was observed, indicating a strong metal-amide bond [45]. Complex **112** was tested as hydroformylation catalyst of styrene, but its properties were unexceptional.

Reaction of the hexahomotrioxocalix[3]arene-derived C_{3v}-symmetrical triphosphine **113** with [Rh(acac)(CO)$_2$] under 20 bar CO/H$_2$ results in formation of a trigonal bipyramidal hydrido carbonyl complex with the Rh-H vector directed inside the cavity (**114**) [46].

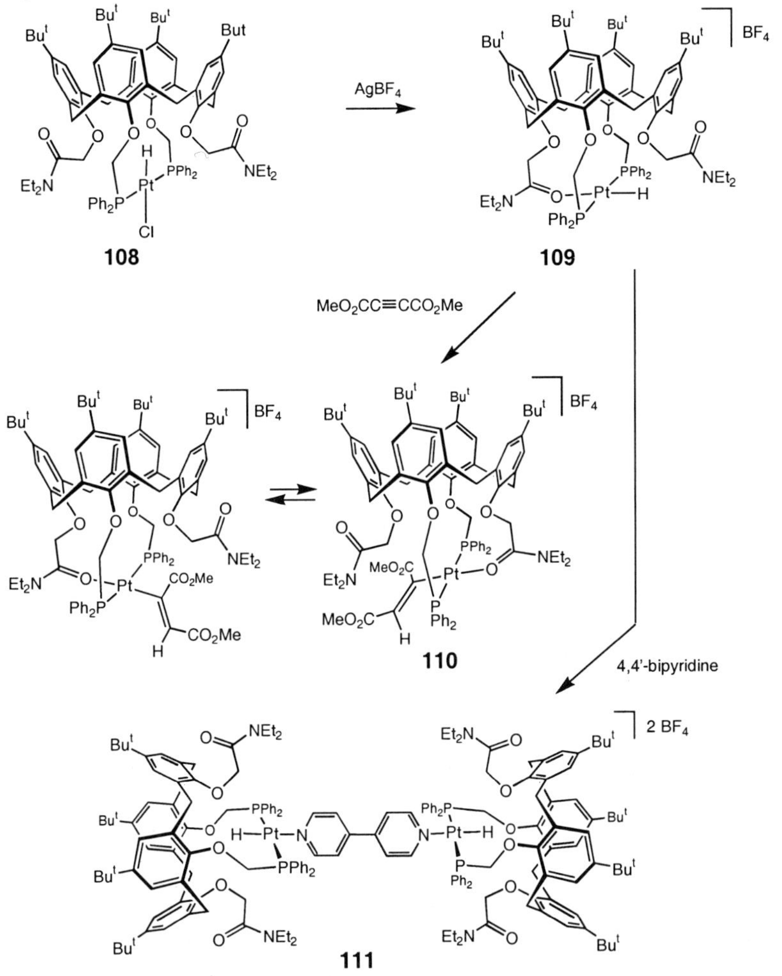

C_{3v}-Symmetrical complexes derived from **113** were also obtained for gold (complex **115**) and molybdenum (**116**) [46].

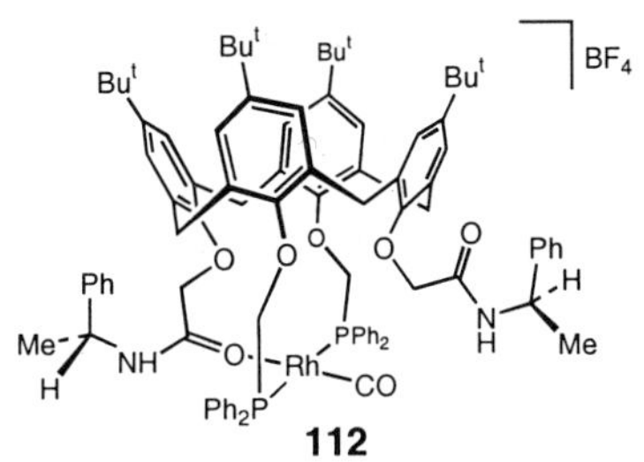

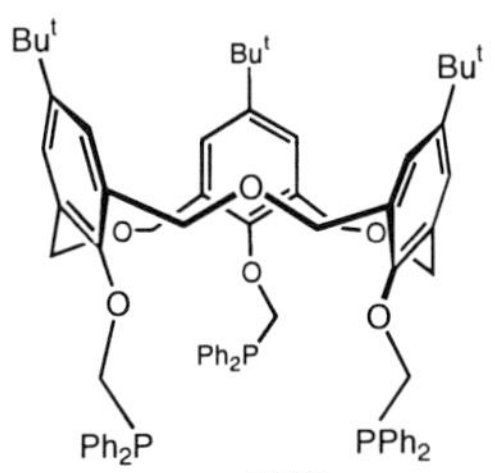

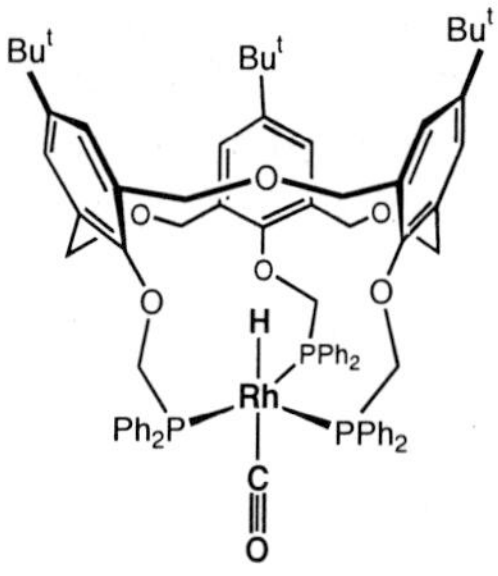

114

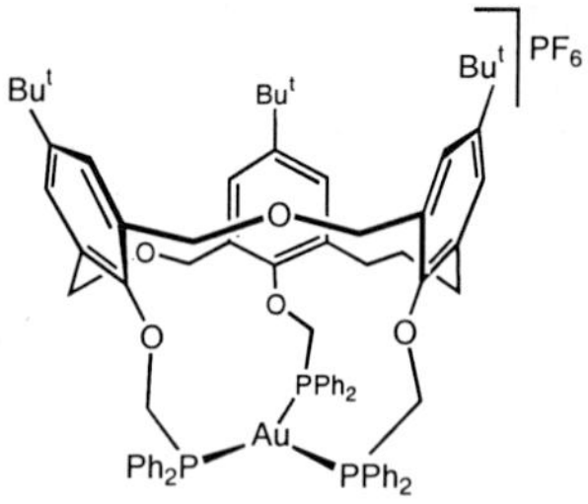

115

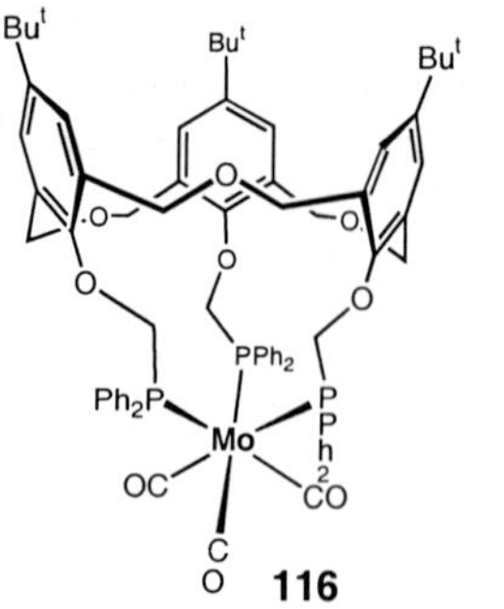

116

Synthesis of multinuclear species can be achieved using tetrapodes **117** and **118** [47]. The addition of four equiv. of [AuCl(SC_4H_8)] to **117** gave the tetragold complex **119**. Di-auration of **117** was observed when the latter was reacted with 2 equiv. of [AuCl(SC_4H_8)].

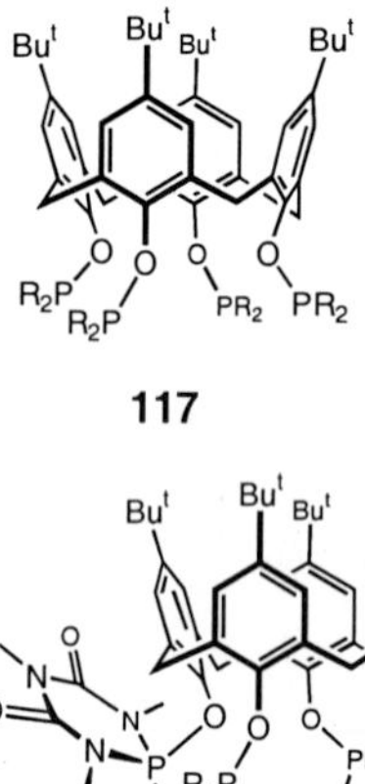

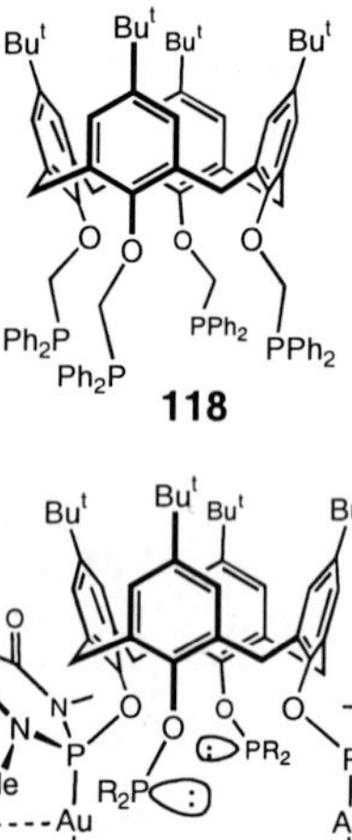

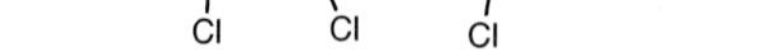

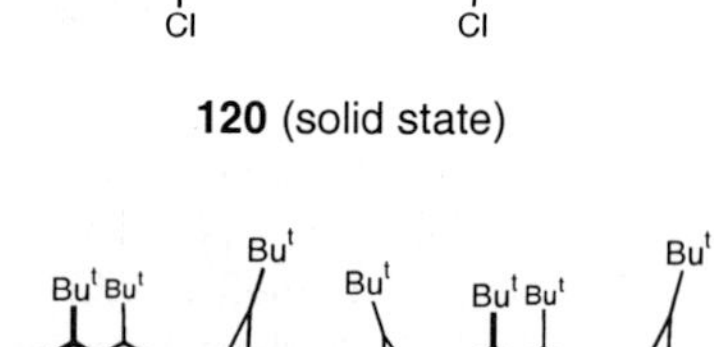

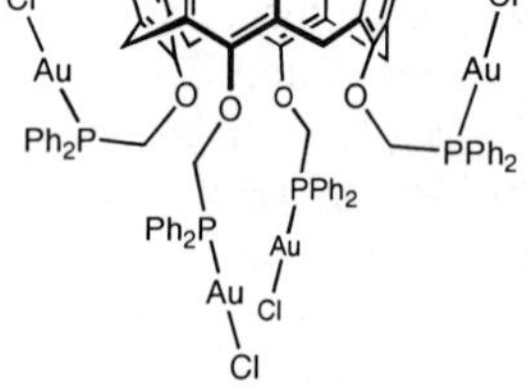

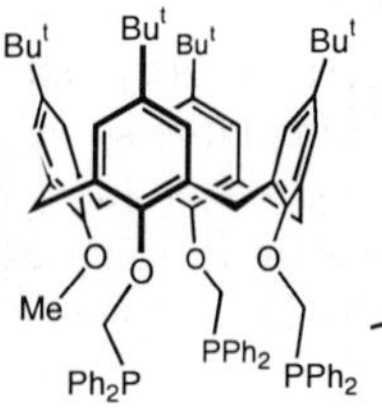

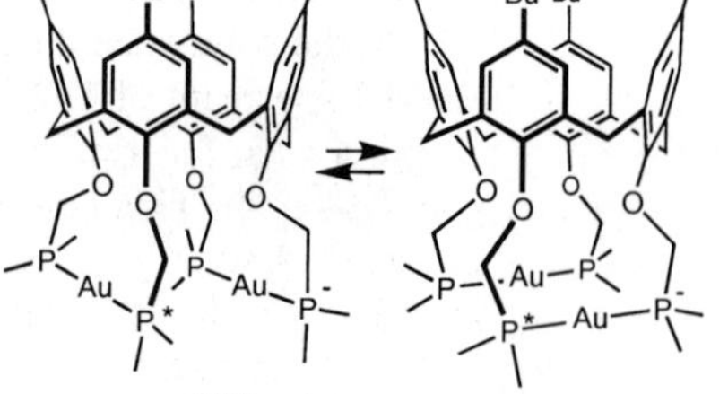

The gold moieties of the complex formed (**120**) are bonded to two distal P atoms. Interestingly, the P lone pairs of the uncomplexed phosphorinane moieties point roughly

117

118

119

120 (solid state)

121

122

124

in the same direction so that the two methyl groups of these heterocycles are formally inequivalent. As a result of short intermolecular Au···Au' contacts (3.288(1) Å), a loose polymeric structure emerges in the solid state. Reaction of **118** with [AuCl(SC$_4$H$_8$)] yields the complex **121** having an apparent C_{4v}-symmetry in solution [47]. In the solid state, two opposite Au-Cl units are oriented parallel to their appended phenoxy rings, whilst the other two lie approximately orthogonal to the calixarene axis. Reacting the tridentate ligand **122** with excess [AuCl(SC$_4$H$_8$)] led to the trimetallic complex **122**•(AuCl)$_3$ (**123**, not drawn). Treatment of **118** with excess [Au(solvent)$_2$]BF$_4$ gave the cationic digold complex **124** [47]. This complex displays dynamic behaviour in solution, the two gold atoms rotating simultaneously on the P$_4$ surface. This motion constitutes a rare example in which two metal ions move in a concerted way on an organic backbone.

Another potential application of ligand **118** is the preparation of heterometallic complexes [40]. Reaction of **118** with [PtCl$_2$(COD)] afforded the chelate complex **125** where the metal is selectively bonded to two proximal phosphines. This complex was further reacted with two equiv. of [AuCl(SC$_4$H$_8$)] to afford the PtAu$_2$ complex **126**. Complex **125** could not be isolated since polymerization occurred during concentration. A simple way to prevent this phenomenon consists in either complexing the P donors or oxidising them, this latter reaction leading then to complex **127**. Facial P$_3$ coordination of **118** and **122** is found in complexes **128-130** [40]. The ruthenium complex **130** binds reversibly two molecules of acetonitrile.

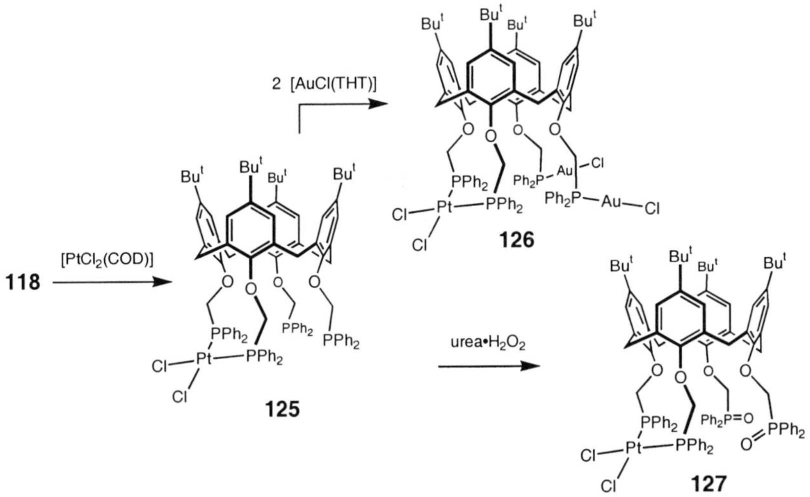

The chelate complex **132** was obtained by reacting diphosphine **131** with [PtCl$_2$(COD)] [40]. In the solid state the PPt vectors point away from the calixarene axis. A feature of the NMR spectrum of **132** is the large splitting of the A and B parts of the ArCH$_2$ group lying between the two phosphines ($\Delta\delta$ = 3.6 ppm !), with the axial-CH signal appearing at 7.34 ppm. This could arise from a ring current generated by the metallo-macrocycle that opposes the external field. An interaction between the axial CH atom and *both* neighbouring oxygen atoms may also be involved. Interestingly, in the solid state the axial ArCH is exactly apically sited above the platinum atom, with a H···Pt separation of only 2.5 Å, so that the deshielding could possibly also be caused by an interaction with the metal d$_{z^2}$ orbital.

Complex **132** displays dynamic behaviour in solution, which can be rationalised as follows: (*i*) a fast flip-flop motion of both hydroxyl groups at low temperature, alternately forming hydrogen bonds with each of the neighbouring ether oxygen atoms; (*ii*) a reversible inversion of the phenol rings through the lower-rim annulus, triggered by cleavage of the hydrogen bonds at higher temperature (Scheme 3).

Calixarenes have been used in the construction of "molecular squares" [48]. The reaction of **133** with 4,4'-bipyridine gave **134** and reaction of **133** with the bisheterodiaryliodonium salt **135** gave another square, **136**, which contains two platinum atoms in the central macrocyclic structure.

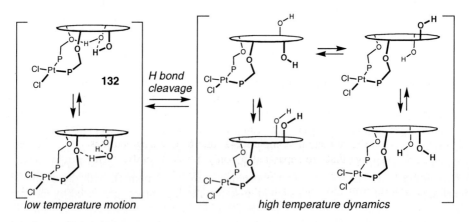

Scheme 3. Dynamics of **132**.

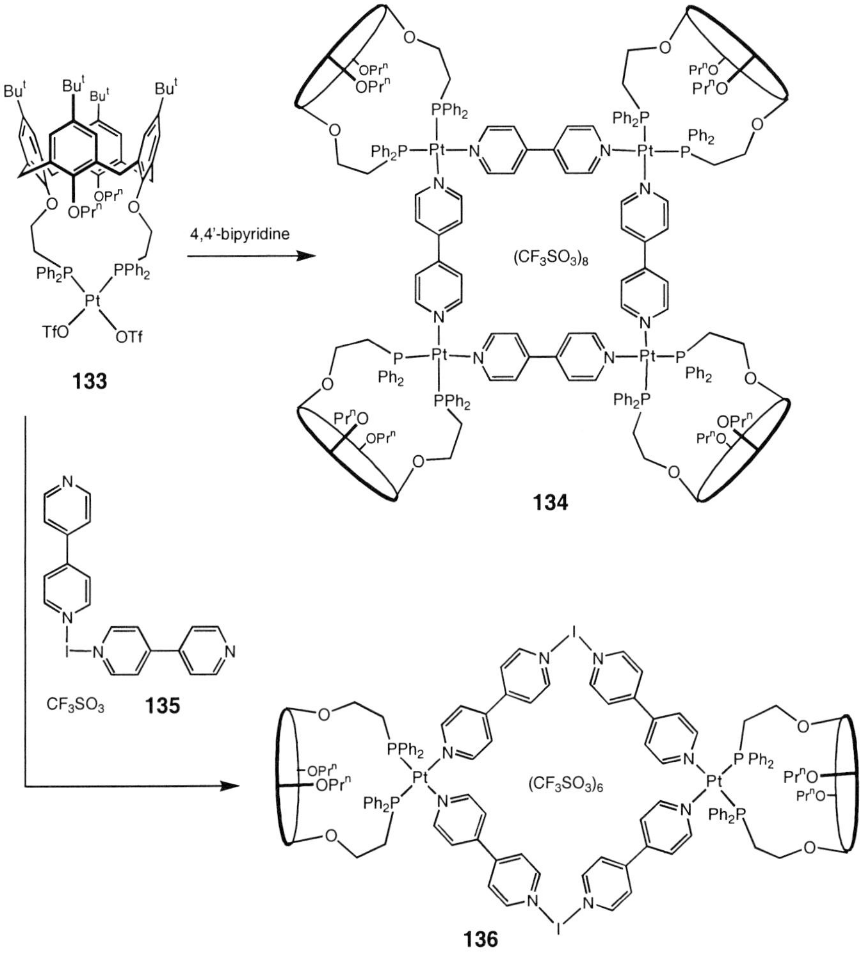

4.2. COMPLEXES OF CALIXARENES WITH UPPER RIM p-SUBSTITUENTS

Entrapment of organometallic fragments inside a calix[4] cavity was achieved with the cage compound **137** [49]. In the complexes **138-141**, the diphosphine acts as a *trans*-spanning ligand which positions the metal centre above the wider entrance. The Pd-Me moiety of complex **139** and the Pt-H bond of **140** are located inside the calix cavity. Upon formation of the ruthenium complex **141**, one M-CO fragment nests itself between two parallel aryl rings that are separated by only 5.5 Å. Clearly, such complexes open new perspectives for intra-cavity catalysis. With ligand **142** which contains somewhat longer phosphorus arms, the binuclear complexes **143-145** were obtained [50]. The reaction of diphosphite **146** with [Pd(η^3-Me-allyl)(THF)$_2$]BF$_4$ afforded the mononuclear chelate complex **147** [50]. The activity of **145** in the hydroformylation of olefins is comparable to that of conventional Rh-PPh$_3$ systems.

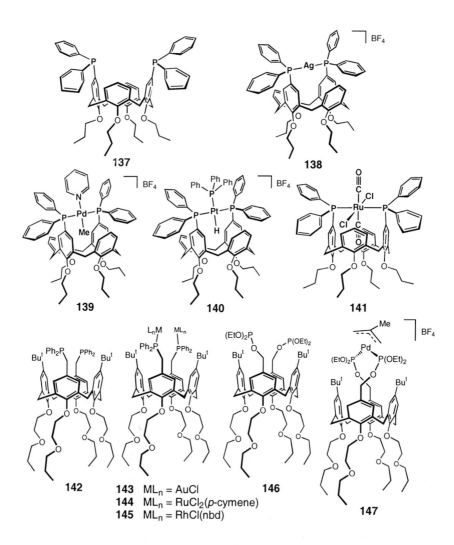

5. Metallocalixarenes with Metal Centres Bonded to π-Donor Units

5.1. Complexes Having a Metal π-Bound to a Phenoxy Ring

Reactions of neutral calixarene precursors and cationic metal arene or metal cyclopentadienyl complexes have provided a series of cationic π-metalated calixarenes [51]. The properties of such complexes notably for anion binding have been detailed in Chapter 23 and will not be further discussed here.

The reaction of AgClO$_4$ with C-methylcalix[4]resorcinarene **148** results in the formation of complex **149**·2C$_6$H$_6$ [52]. The dinuclear cation is twofold symmetric, each silver ion being bonded to two OH groups, one carbon atom of an aromatic resorcinarene ring (π interaction) and one carbon atom of a benzene molecule (π interaction). Reacting calix[6]arene with AgClO$_4$ afforded the dimeric complex **150**.

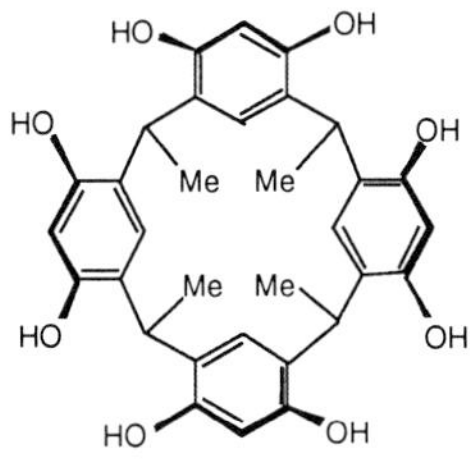

148

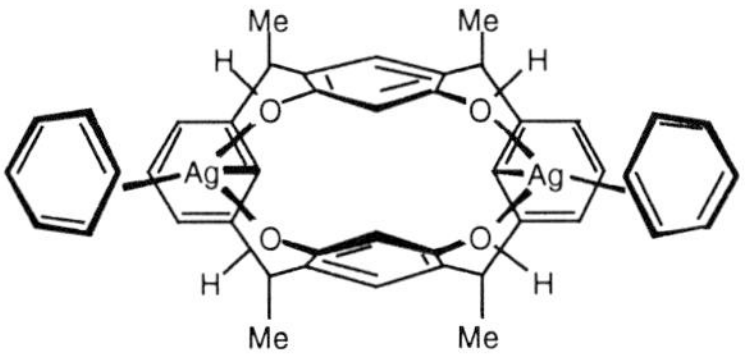

149

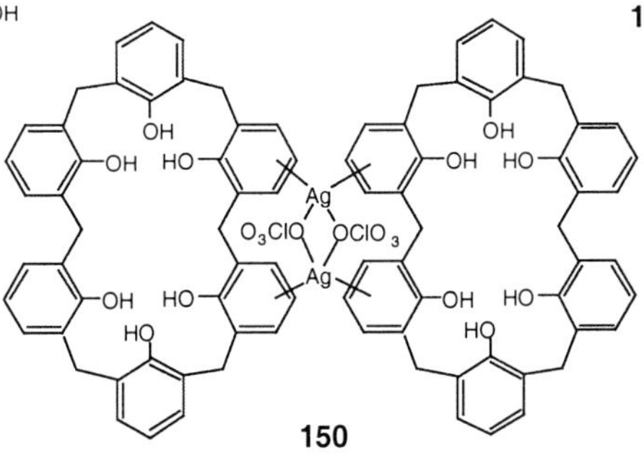

150

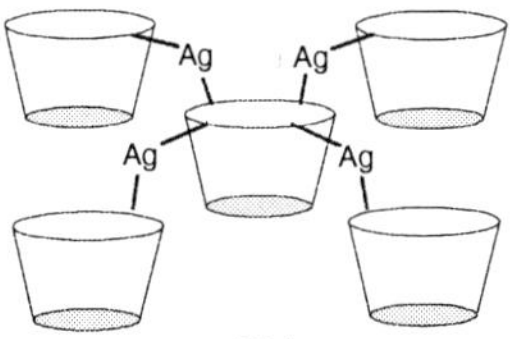

151

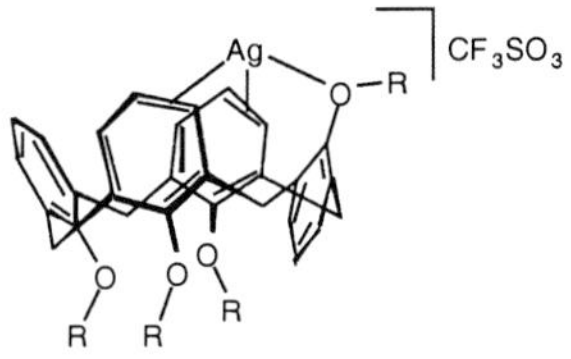

152 R = CH₂C≡CH

The reaction of AgClO₄ with calix[4]arene was also examined. This leads to [Ag₂(calixarene)(ClO₄)] **151**, a complex that possesses a polymeric structure based on cation π interactions. In the solid state each calixarene is π-bonded to four silver atoms, two of which lie above the cavity and the other two outside. Silver ions lying above a calixarene unit are connected to the outer face of a neighbouring cavity (and *vice versa*) so as to generate a two-dimensional polymeric framework. All calixarene units have the same orientation. In the partial *cone* calixarene complex **152**, Ag(I) is bound to both "hard" and "soft" donor centres as in **149** [53].

6. Special Uses of Calixarenes in Catalysis

Calixarenes have recently been used as additives in the zirconium-BINOL catalysed enantioselective allylation of aldehydes. The addition of *p-t*-butylcalix[4]arene to a mixture of BINOL, ZrCl₄(THF)₂, and allyltributyltin (Scheme 4) strongly activates the catalytic system [54]. Much higher ee's (up to 96 % in the case of n-C₇H₁₅CHO) were obtained compared to the system without additives. Characterization of the species responsible for these outstanding results has not been achieved.

Modified calix[n]arenes (n = 4, 5) may also be used as external donors in late-generation Ziegler-Natta catalysis [55]. A general result is that the addition of calixarenes to Ti/Al catalysts significantly increases the amount of isotactic polypropylene in the bulk polymerization of propylene. These promising results will certainly trigger new research in the ever evolving field of olefin polymerisation.

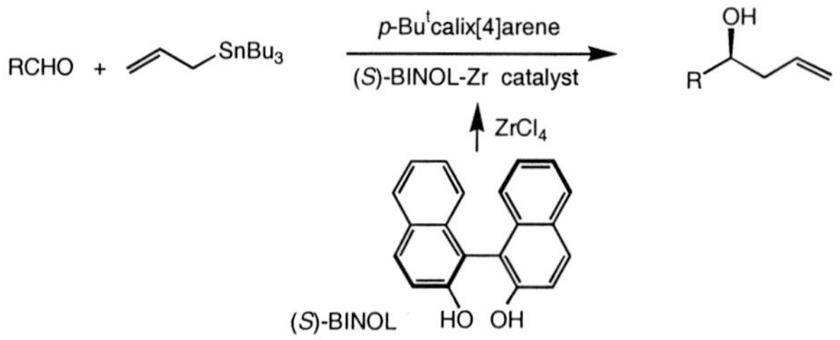

Scheme 4. Calixarene-assisted allylation of aldehydes.

7. Conclusion

The particular research reviewed therein illustrates some new possibilities arising from the combination of calixarenes and transition metal centres. Most of these topics rely on two major properties associated with calixarene units, namely the possibility to generate sophisticated coordination spheres by multiple functionalisation and the ability to entrap organic or organometallic fragments inside their cavities or pockets. The fascinating discoveries made recently in the fields of homogeneous catalysis, anion complexation, and supramolecular sensors as well as large cage compound synthesis fully justify further research in the rapidly expanding field of metallo-calixarenes.

8. References and Notes

[1] C. D. Gutsche, *Calixarenes Revisited* : C. D. Gutsche in Monographs in Supramolecular Chemistry (Ed. J. F. Stoddart), Royal Society of Chemistry, Cambridge, 1998.
[2] V. Böhmer, *Angew. Chem. Int. Ed.* **1995**, *34*, 713-745.
[3] C. Wieser, C. B. Dieleman, D. Matt, *Coord. Chem. Rev.* **1997**, *165*, 93-161.
[4] C. Floriani, *Chem. Eur. J.* **1999**, *5*, 19-23.
[5] A. Zanotti-Gerosa, E. Solari, L. Giannini, C. Floriani, A. Chiesi-Villa, C. Rizzoli, *J. Am. Chem. Soc.* **1998**, *120*, 437-438.
[6] V. C. Gibson, C. Redshaw, W. Clegg, M. R. J. Elsegood, *Chem. Commun.* **1997**, 1605-1606.
[7] V. C. Gibson, C. Redshaw, W. Clegg, M. R. J. Elsegood, *Chem. Commun.* **1998**, 1969-1970.
[8] U. Radius, J. Attner, *Eur. J. Inorg. Chem.* **1999**, 2221-2231.
[9] U. Radius, A. Friedrich, *Z. Anorg. Allg. Chem.* **1999**, *625*, 2154-2159.
[10] O. V. Ozerov, F. T. Ladipo, B. O. Patrick, *J. Am. Chem. Soc.* **1999**, *121*, 7941-7942.
[11] O. V. Ozerov, N. P. Rath, F. T. Ladipo, *J. Organomet. Chem.* **1999**, *586*, 223-233.
[12] W. Clegg, M. R. J. Elsegood, S. J. Teat, C. Redshaw, V. C. Gibson, *J. Chem. Soc., Dalton Trans.* **1998**, 3037-3039.
[13] G. D. Andreetti, G. Calestani, F. Ugozzoli, A. Arduini, E. Gidini, A. Pochini, R. Ungaro, *J. Incl. Phenom.* **1987**, *5*, 123-126.
[14] U. Radius, J. Attner, *Eur. J. Inorg. Chem.* **1998**, 299-303.
[15] M. H. Chisholm, K. Folting, W. E. Streib, D.-D. Wu, *Chem. Commun.* **1998**, 379-380.
[16] M. H. Chisholm, K. Folting, W. E. Streib, D.-D. Wu, *Inorg. Chem.* **1999**, *38*, 5219-5229.
[17] G. Mislin, E. Graf, M. W. Hosseini, A. Bilyk, A. K. Hall, J. M. Harrowfield, B. W. Skelton, A. H. White, *Chem. Commun.* **1999**, 373-374.
[18] F. S. McQuillan, T. E. Berridge, H. Chen, T. A. Hamor, C. J. Jones, *Inorg. Chem.* **1998**, *37*, 4959-4970.
[19] B. R. Cameron, S. J. Loeb, *Chem. Commun.* **1996**, 2003-2004.
[20] B. R. Cameron, S. J. Loeb, G. P. A. Yap, *Inorg. Chem.* **1997**, *36*, 5498-5504.

[21] F. Arnaud-Neu, G. Barrett, D. Corry, S. Cremin, G. Ferguson, J. F. Gallagher, S. J. Harris, M. A. McKervey, M.-J. Schwing-Weill, *J. Chem. Soc., Perkin Trans. 2* **1997**, 575-579.
[22] P. D. Beer, M. G. B. Drew, P. B. Leeson, M. I. Ogden, *Inorg. Chim. Acta* **1996**, *246*, 133-141.
[23] P. Molenveld, S. Kapsabelis, J. F. J. Engbersen, D. N. Reinhoudt, *J. Am. Chem. Soc.* **1997**, *119*, 2948-2949.
[24] P. Molenveld, J. F. J. Engbersen, H. Kooijman, A. L. Spek, D. N. Reinhoudt, *J. Am. Chem. Soc.* **1998**, *120*, 6726-6737.
[25] P. Molenveld, J. F. J. Engbersen, D. N. Reinhoudt, *J. Org. Chem.* **1999**, *64*, 6337-6341.
[26] S. Blanchard, L. Le Clainche, M.-N. Rager, B. Chansou, J.-P. Tuchagues, A. F. Duprat, Y. Le Mest, O. Reinaud, *Angew. Chem. Int. Ed.* **1998**, *37*, 2732-2735.
[27] S. Blanchard, M.-N. Rager, A. F. Duprat, O. Reinaud, *New J. Chem.* **1998**, 1143-1146.
[28] P. Jacopozzi, E. Dalcanale, *Angew. Chem. Int. Ed.* **1997**, *36*, 613-615.
[29] O. D. Fox, N. K. Dalley, R. G. Harrison, *Inorg. Chem.* **1999**, *38*, 5860-5863.
[30] O. D. Fox, N. K. Dalley, R. G. Harrison, *Inorg. Chem.* **2000**, *39*, 620-622.
[31] D. Hesek, Y. Inoue, S. R. L. Everitt, M. Kunieda, H. Ishida, M. G. B. Drew, *Tetrahedron Asymmetry* **1998**, *9*, 4089-4097.
[32] A. Ikeda, M. Yoshimura, F. Tani, Y. Naruta, S. Shinkai, *Chem. Lett.* **1998**, 587-588.
[33] G. Mislin, E. Graf, M. W. Hosseini, A. De Cian, N. Kyritsakas, J. Fischer, *Chem. Commun.* **1998**, 2545-2546.
[34] A. F. Danil de Namor, O. E. Piro, L. E. Pulcha Salazar, A. F. Aguilar-Cornejo, N. Al-Rawi, E. E. Castellano, F. J. Sueros Velarde, *J. Chem. Soc., Faraday Trans.* **1998**, *94*, 3097-3104.
[35] S. Pellet-Rostaing, J.-B. Regnouf-de-Vains, R. Lamartine, B. Fenet, *Inorg. Chem. Commun.* **1999**, *2*, 44-47.
[36] Y. Molard, C. Bureau, H. Parrot-Lopez, R. Lamartine, J.-B. Regnouf-de-Vains, *Tetrahedron Lett.* **1999**, *40*, 6383-6387.
[37] P. Faidherbe, C. Wieser, D. Matt, A. Harriman, A. De Cian, J. Fischer, *Eur. J. Inorg. Chem.* **1998**, 451-457.
[38] C. Loeber, D. Matt, P. Briard, D. Grandjean, *J. Chem. Soc., Dalton Trans.* **1996**, 513-524.
[39] M. Stolmàr, C. Floriani, A. Chiesi-Villa, C. Rizzoli, *Inorg. Chem.* **1997**, *36*, 1694-1701.
[40] C. B. Dieleman, C. Marsol, D. Matt, N. Kyritsakas, A. Harriman, J.-P. Kintzinger, *J. Chem. Soc., Dalton Trans.* **1999**, 4139-4148.
[41] Z. Csòk, G. Szalontai, G. Czira, L. Kollár, *J. Organomet. Chem.* **1998**, *570*, 23-29.
[42] M. Fan, H. Zhang, M. Lattman, *Chem. Commun.* **1998**, 99-100.
[43] F. J. Parlevliet, M. A. Zuideveld, C. Kiener, H. Kooijman, A. L. Spek, P. C. J. Kamer, P. W. N. M. van Leeuwen, *Organometallics* **1999**, *18*, 3394-3405.
[44] R. Paciello, L. Siggel, M. Röper, *Angew. Chem. Int. Ed.* **1999**, *38*, 1920-1923.
[45] C. Wieser, D. Matt, J. Fischer, A. Harriman, *J. Chem. Soc., Dalton Trans.* **1997**, 2391-2402.
[46] C. B. Dieleman, D. Matt, I. Neda, R. Schmutzler, A. Harriman, R. Yaftian, *Chem. Commun.* **1999**, 1911-1912.
[47] C. B. Dieleman, D. Matt, I. Neda, R. Schmutzler, H. Thönnessen, P. G. Jones, A. Harriman, *J. Chem. Soc., Dalton Trans.* **1998**, 2115-2121.
[48] P. J. Stang, D. H. Cao, K. Chen, G. M. Gray, D. C. Muddiman, R. D. Smith, *J. Am. Chem. Soc.* **1997**, *119*, 5163-5168.
[49] C. Wieser-Jeunesse, D. Matt, A. De Cian, *Angew. Chem. Int. Ed.* **1998**, *37*, 2861-2864.
[50] I. A. Bagatin, D. Matt, H. Thönnessen, P. G. Jones, *Inorg. Chem.* **1999**, *38*,
[51] M. Staffilani, K. S. B. Hancock, J. W. Steed, K. T. Holman, J. L. Atwood, R. K. Juneja, R. S. Burkhalter, *J. Am. Chem. Soc.* **1997**, *119*, 6324-6335.
[52] M. Munakata, L. P. Wu, T. Kuroda-Sowa, M. Maekawa, Y. Suenaga, K. Sugimoto, I. Ino, *J. Chem. Soc., Dalton Trans.* **1999**, 373-378.
[53] W. Xu, J. J. Vittal, R. J. Puddephatt, *Can. J. Chem.* **1996**, *74*, 766-774.
[54] S. Casolari, P. G. Cozzi, P. Orioli, E. Tagliavini, A. Umani-Ronchi, *Chem. Commun.* **1997**, 2123-2124.
[55] R. A. Kemp, D. S. Brown, M. Lattman, J. Li, *J. Molec. Cat. A: Chem.* **1999**, *149*, 125-133.

Chapter 29

METAL REACTIVITY ON OXO SURFACES MODELED BY CALIX[4]ARENES

CARLO FLORIANI, RITA FLORIANI-MORO

Université de Lausanne
Institut de Chimie Minérale et Analytique, BCH 3307
CH-1015 Lausanne, Switzerland.
e-mail: carlo.floriani@icma.unil.ch

1. Introductory Remarks: Scope and Limits of the Chapter

The general purpose of this chapter is to show the potentiality of calix[4]arenes as ancillary ligands in coordination and organometallic chemistry of transition metals, in the domains of both metal reactivity and catalysis. Therefore, the main focus is on the metalation of calix[4]arenes and their O-alkylated derivatives where it produces compounds potentially useful for exploring the chemical reactivity of the metal centres. Reports concern only compounds which have a well established synthesis and are already fully characterized spectroscopically and structurally. A particular emphasis is placed on modeling studies of heterogeneous metal-oxo surfaces using metalla-calix[4]arenes.

These specific foci mean that the chapter will not cover: *i*) transition metal complexation by protonated or functionalized forms of calix[4]arenes; *ii*) the metalation of calix[4]arenes using non-transition metals; *iii*) chemical curiosities derived from the metalation of calix[4]arenes, some recent reviews covering these areas very well [1].

2. Introduction

Modeling studies of oxo-surfaces [2] that bind metals appropriate to drive catalytic or stoichiometric reactions date back many years [3]. An interesting approach involves the use of a pre-organized set of oxygen donor atoms in a quasiplanar arrangement for binding metal ions. In such a context, the calix[4]arene skeleton [1], with its unique geometric and electronic properties, becomes prominent [4]. Calix[4]arenes have only recently been exploited for binding metals, [1,4,5] and even more recently as ancillary ligands for supporting reactive metals in organometallic chemistry [1]. Let us first have a look at their geometric properties. In the metalla-calix[4]arenes, the four oxygen donor atoms pre-organized in a quasiplanar geometry have a major effect in determining the set and the relative energy of the frontier orbitals available at the metal for substrate activation (see Figure 1). The four lowest-lying orbitals, which can accommodate up to eight electrons, lie on two orthogonal planes, thus favoring a *facial* over a *meridional* structure when such a fragment is bound to three additional ligands. The *facial* compared to the *meridional* arrangement of the three low-lying frontier orbitals favors, among other things, the formation of a metal substrate multiple bond [4], *i.e.* stabilization of alkylidenes and alkylidynes, and affects migration-insertion reactions [4]. The metal-oxo molecular models outlined above have a quite remarkable potential for studying the metal activity in an unusual environment. Some of the possibilities are: *ii*) the generation and the chemistry of M-C, M=C, M≡C functionalities; *ii*) the interaction with alkenes, alkynes, hydrocarbons and hydrogen; *iii*) the activation of small molecules like N_2, CO, O_2, and CO_2; *iv*) the support of metal-metal bonded functionalities; and *v*) the generation of highly reactive low-valent metals.

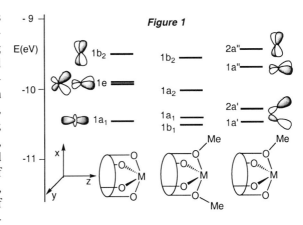

Figure 1

3. Metalation of Calix[4]arenes and O-Alkylated Calix[4]arenes using Transition Metals

A preliminary question is how to make the starting materials for entering organometallic and coordination chemistry in the area of calix[4]arenes. The main synthetic approaches include direct metalation using early transition metal halides; the metathesis reaction between metal halides and the sodium or lithium derivatives; and the direct metalation using homoleptic alkyl or aryl derivatives.

Calix[4]arene complexes of transition metals are well known [1,5]. Until recently [6-17], however, it was very difficult to discern general methods for the synthesis of compounds suitable for studying metal reactivity. The synthetic methods mentioned in this section concern exactly this. The metalation of calix[4]arene using a preformed metal functionality, namely metal-oxo, metal-imido, metal-cyclopentadienyl, etc., is reported in a separate section (see section 6). Note that this chapter does not provide an exhaustive list of metallated calix[4]arenes, since this is available in previous reviews [1].

4. Low-Valent Metalla-Calix[4]arenes

4.1. METAL-METAL BONDED DIMERS

A set of oxygen donor atoms, providing both σ and π donation to a metal center, is not expected to stabilize a low oxidation state of a metal [18]. This is, however, a synthetic advantage since very reactive, unstable, low-valent metalla-calix[4]arenes can be generated *in situ* and intercepted by an appropriate substrate. In the absence of a suitable substrate, however, the reactive fragment can collapse to form metal-metal bonded dimers, though the formation of metal-metal bonds has been observed so far in the case of V and VI group metals only. The most complete sequences so far reported have been for tungsten, molybdenum and niobium.

In the case of tungsten [10,19], complex **1** has been used as starting material. In the absence of any intercepting substrates like olefins, acetylenes, etc. (see the next section), it undergoes controlled reduction (Scheme 1) leading to a number of structurally interesting compounds, **2, 4-6**.

Scheme 1

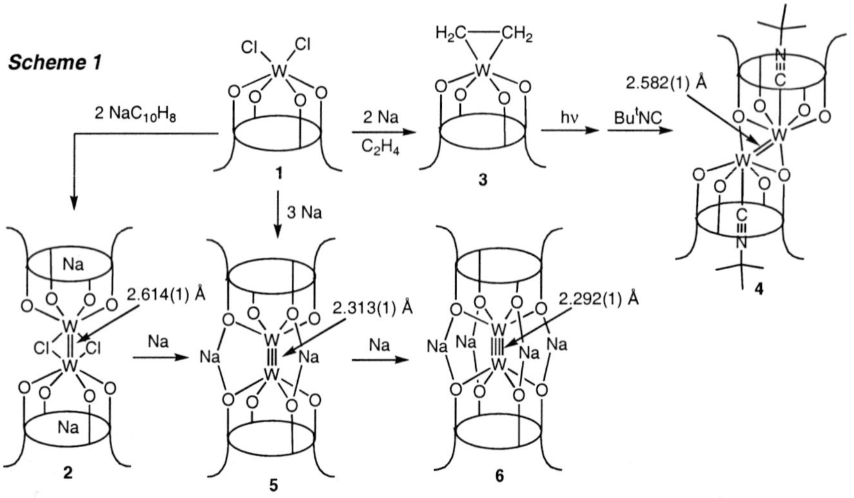

The different genesis of **2** and **4**, both containing a W=W double bond in an edge-sharing bioctahedron [20], clearly indicates that the W=W functionality requires the support of two bridging ligands, either chlorides (see **5**) or oxygens of adjacent metalla-calix[4]arene moieties (see **4**). Among the properties of the electron-rich metalla-calix[4]arene dimers mentioned above, it should be emphasized that they occur in an ion-pair form, with alkali metal cations tightly bound to the calix[4]arene fragments. The alkali metal cations have very different binding sites, either displaying a bridging bonding mode across the lower rims of two adjacent calix[4]arenes, as in **5** and **6**, or being complexed inside the cavity by η^6 or η^3 interactions with the arene rings, as in **4**. Both dimers **5** and **6** contain a tungsten-tungsten multiple bond [2.313(1) and 2.292(1) Å in **5** and **6**, respectively], consistent with a triple bond in **5** and in fact a quadruple bond in **6**.

A different synthetic approach to M-M bonded metalla-calix[4]arenes does not make use of any redox chemistry [21]. Such an approach reveals the possibility of other bonding modes of the M_2 unit to the calix[4]arene [21a,b]. The synthesis has been performed reacting $[M_2(NMe_2)_6]$ (M = Mo, W) or $[Mo_2(OBu^t)_6]$ with calix[4]arene. In neither series of compounds reported above has any particularly interesting reactivity of the M-M bonded functionality been disclosed so far [21].

A stark contrast is provided in the case of the stepwise reduction of $[p\text{-}Bu^t\text{-calix}[4]\text{-}(O_4)]_2Nb_2(Cl)_2]$, **7**, reported in Scheme 2 [7,22,23]. This reduction has been carried out under argon with a controlled amount of sodium metal at each step. Analogous results have been obtained using other alkali metals, though the chemistry has been mainly investigated in the case of the sodium derivatives. Complexes **8-10** are all diamagnetic. In the case of **8**, the ^{1}H NMR spectrum shows a two-fold symmetry in toluene at low temperature, while in coordinating solvents, like pyridine, a four-fold symmetry is observed according to the transformation of a tight to a separated ion-pair form. The structural features of **8-10** are very much similar to those reported for the tungsten analogues, though the reactivity is very different. In particular, complex **9** reacts with a variety of substrates, like CO, N_2, alkenes, and H_2 (see below).

Scheme 2

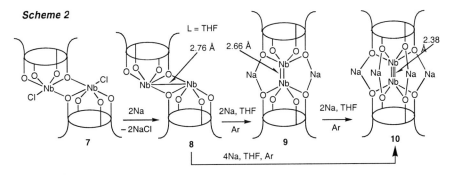

4.2. ALKENE, DIENE, AND ALKYNE COMPLEXES

The reduction of a high-valent metalla-calix[4]arene in the presence of olefins, dienes, acetylenes, is probably the best, even though not the only, synthetic method for the corresponding complexes.

Although the reduction of **1** with alkali metals in THF leads to metal-metal bonded species [24], suitable substrates can intercept the d^2 [{p-Bu^t-calix[4]-(O)$_4$}W] fragment, preventing the formation of such dimers, which represent a "thermodynamic sink" [24].

All η^2-olefin species **3**, **11-13** (Scheme 3) exhibit an effective C_{4v} symmetry in solution (NMR), in agreement with a free rotation of the olefin. The C_2H_4 ligand in **3** gives rise to a single signal both in ^{1}H NMR and in the ^{13}C NMR (at 70 ppm). The W-C [2.124(9) Å] and C-C [1.40 Å] bond distances in complex **11** are in good agreement with the metallacyclopropane formulation [24b].

The interaction between the [W(calix)] metal fragment and the ethylene moiety has been analyzed using extended Hückel calculations [24b]. The presence of two orthogonal d_π orbitals equally available for the interaction with the π system of C_2H_4 suggests a small activation barrier (only 3 kcal mol^{-1}) for the rotation about the z axis. This finding is in good agreement with the ^{1}H NMR of **3** and **11-13**, indicating an apparent C_{4v} symmetry of the calix[4]arene moiety even at low temperatures [24b]. The metal-assisted olefin chemistry of W-calix[4]arene has another significant peculiarity that justifies the comparison of metalla-calix[4]arenes with a metal-oxo surface. This is the electron-transfer-catalyzed dimerization of ethylene and propylene to the corresponding metallacyclopentane [24b]. Unlike many organometallic systems [25], the reductive coupling of an olefin (see Scheme 4) is not a reversible reaction and only occurs in the presence of small amounts of a reducing agent, in our case sodium metal. This enables regiochemically controlled isomers synthesis of metallacyclopentanes.

Scheme 4

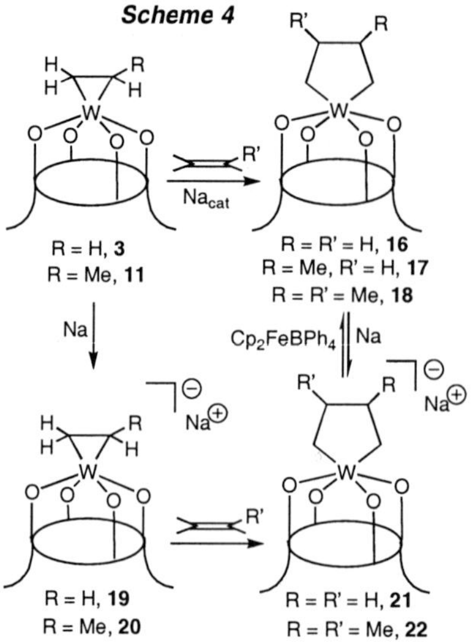

Related results in this area have been reported studying the synthesis and the reactivity of a zirconium-butadiene functionality bonded to the p-Bu^t-calix[4](O)$_2$(OMe)$_2$ dianion [26]. The synthesis of the parent compound used in this study, **24**, is displayed, along with the analogous diphenyl derivative, in Scheme 5. The reaction can be carried out either via the preliminary isolation of **23** or it can be performed in situ with the corresponding Mg-butadiene derivative. In the solid state, the bonding mode of the butadiene ligands in both complexes **24** and **25** can be described as π^2, η^4. A variety of bonding modes of butadiene to zirconium ranging from π^2, η^4 to σ^2, π, η^4 can be found in complexes having sub-units other than [Cp$_2$Zr] [27]. The structural and spectroscopic characteristics of the Zr-diene fragment in **24** and **25** are more closely related to the [(tmtaa)Zr] [28] derivatives [tmtaa = dibenzotetramethyltetraaza[14]annulene dianion] than the corresponding [Cp$_2$Zr] species. In particular, the s-cis form is the only butadiene conformer detected in the solid state; heating of **24** or **25** did not result in any cis-trans isomerization [29].

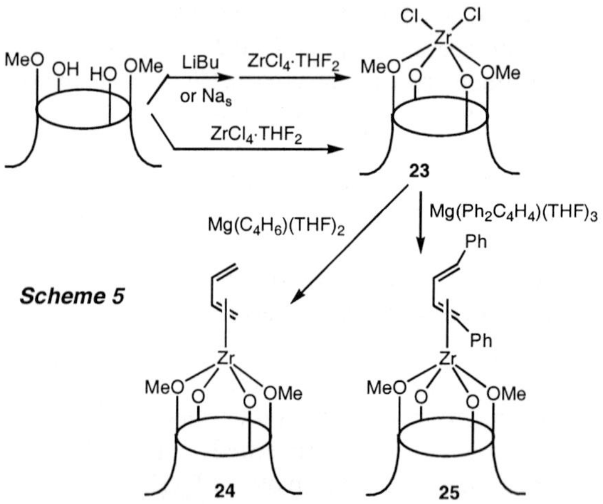

Scheme 5

Scheme 6 displays reactions where **24** behaves as a source of a Zr(II) derivative [26]. They can be formally viewed as oxidative additions to the [p-Bu^t-calix[4](OMe)$_2$(O)$_2$Zr] fragment.

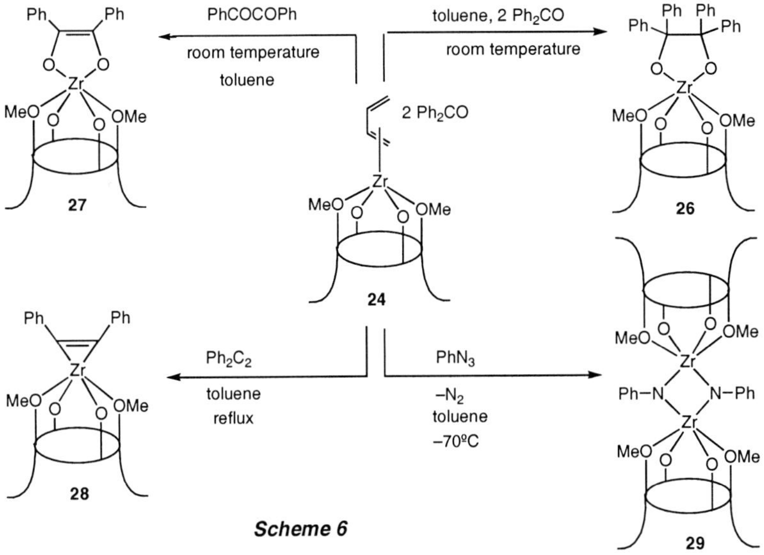

Scheme 6

Although the bonding mode of butadiene to zirconium is π^2, η^4, as shown by the X-ray structure, under certain conditions it behaves like a σ^2, π, η^4 butadiene, thus the zirconium-butadiene functionality behaves as if it contains two Zr-C σ bonds. This is seen in the reactions of Scheme 7.

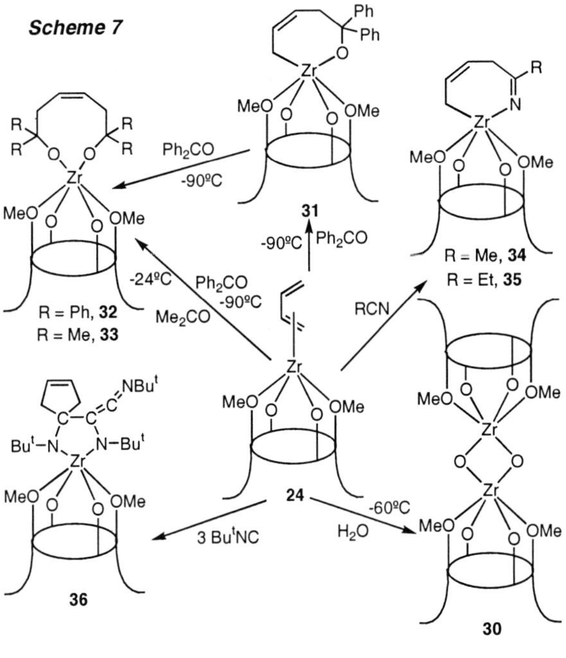

The reaction of **24** with ketones has a particular dependence on the temperature [26]. When it was performed at room temperature (see Scheme 6, compounds **26** and **27**), the displacement of butadiene by the incoming substrate was observed. In contrast, when the reaction was carried out at low temperature and with a 1:1 molar ratio (Scheme 7), the insertion of the ketone unit was observed in one of the two Zr-C σ bonds, leading to the seven-membered metallacyle in **31**. The reaction then proceeded further with a second mole of Ph_2CO to give **32**. The chemistry and structural features of tantalum-butadiene derivatives [8] parallel those of the analogous zirconium derivative **24** [26].

Scheme 7

542

5. The Organometallic Functionalization of the Metalla-Calix[4]arenes

5.1. THE SYNTHESIS AND THE CHEMISTRY OF M-C σ BONDS

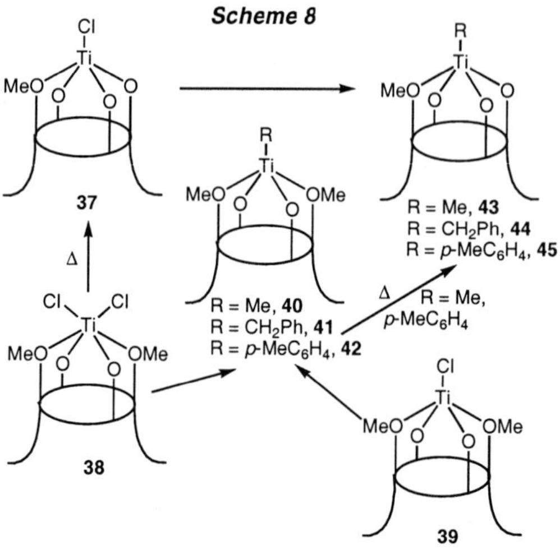

The existence, properties, and chemical behavior of the M-C σ bond functionality has been extensively explored in the case of group IV and V metals. The starting material used in the case of titanium(III) and titanium(IV) are those reported in Scheme 8, namely **37**, **38**, and **39** [11]. They undergo alkylation or arylation using conventional organometallic methodology (Scheme 8) [11]. The organometallic derivatives in Scheme 8 have a basic structure which is exemplified by that of **42**. Complex **42** has, however, an additional structural peculiarity, since the presence of the *p*-tolyl at the metal gives rise to a very interesting solid state assembly [11]. In all cases, the calix[4]arene fragment displays C_s symmetry in the solution ^{1}H NMR spectrum.

The trianion in Scheme 8 is a suitable ancillary ligand for assuring stability to the titanium(IV)-carbon functionality, and similarly the dianion stabilises the titanium(III)-carbon functionality. Attempts to synthesize the [TiR$_2$] fragment bonded to the O$_4$ oxo-matrix *via* the alkylation of **38** failed, due to the preliminary reduction of titanium(IV) to titanium(III). Thus, only a single Ti-C functionality over the O$_4$ set environment has been identified. All organometallic derivatives **40-45** are particularly appropriate for studying the chemistry of the Ti-C bond in titanium(III) and titanium(IV) derivatives. Significantly, the [(DMSC)TiCl$_2$] complex [DMSC = *1,2-alternate* dimethylsilyl-bridged *p*-Bu^t-calix[4]-arene] under reducing conditions is an excellent highly regioselective catalyst for the cyclotrimerization of terminal alkynes (Scheme 9) [30,31].

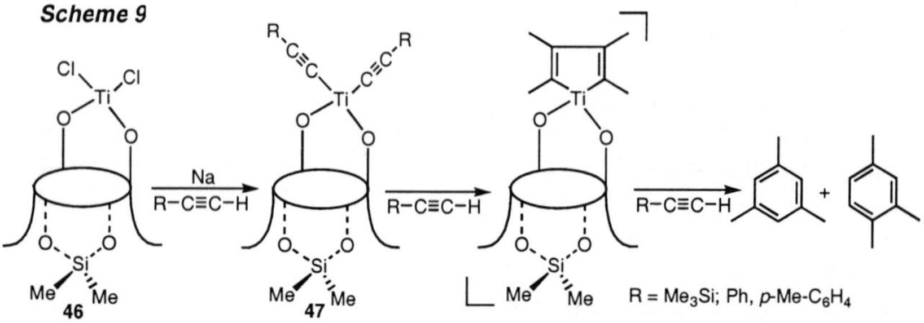

A detailed study has been published on the synthesis, rearrangement and migration insertion reactions of Zr-C functionalities anchored to a calix[4]arene moiety [32]. This approach has been developed through the synthesis and the organometallic functionalization of **23**, which give rise to two classes of derivatives related by the base-induced demethylation of one of the methoxy groups.

The alkylation of **23** (path *b*, Scheme 10) [32] has been performed conventionally using lithium or Grignard reagents but the preliminary isolation of **23** is not compulsory. The synthesis of the zirconium alkyl or aryl derivatives can be carried out *in situ* directly from *p*-Bu^t-calix[4](OMe)$_2$(OH)$_2$ (path *a*, Scheme 10). The aryl derivative **51** is more thermally sensitive than the corresponding alkyl compounds **48-50**, and refluxing in benzene for 36 h gave the corresponding benzyne derivative, **52**, via toluene elimination [33]. The analogous α-elimination, leading to the corresponding alkylidene [34], has never been observed in the case of the attempted thermal decomposition of **48-50**. Note that **48-51** undergo base-induced demethylation of one of the methoxy groups when they are treated with a relatively strong base.

Scheme 10

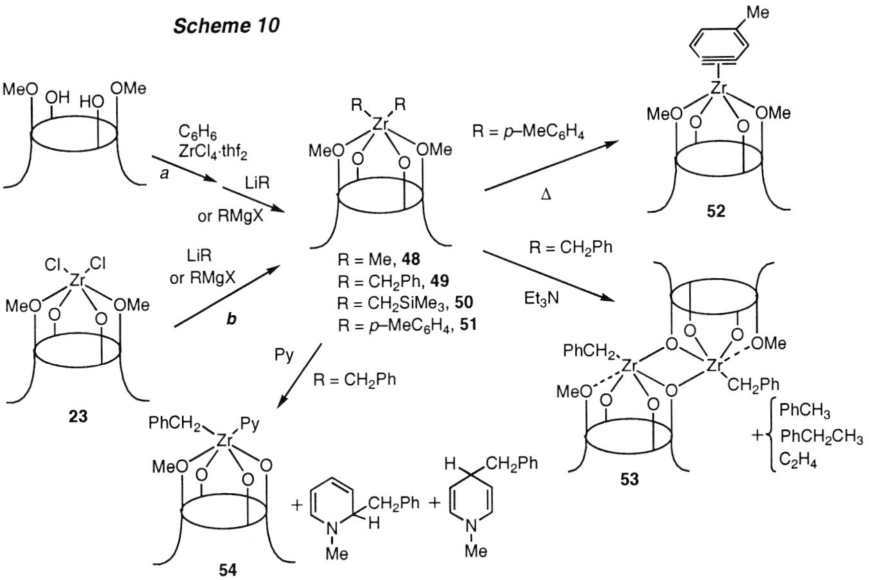

Compounds **48-51** are particularly appropriate precursors for the corresponding mono-alkyl cationic species, which, especially in the case of [Cp$_2$Zr] chemistry [35], are still receiving much attention as polymerisation catalysts. The generation of cationic species has been achieved via two major routes: *i*) one-electron oxidation using [Cp$_2$Fe]$^+$BPh$_4^-$ to generate **55** in toluene, or **56** in a coordinating solvent like THF [35]; *ii*) reaction with a Lewis acid, like [B(C$_6$F$_5$)$_3$], to give **57** (Scheme 11) [35,36].

The reaction of **55** with pyridine emphasizes the action of a strong base in the demethylation of one of the methoxy groups. The pyridine functions as the acceptor of the methyl cation and thus forms **54** and PyMe$^+$BPh$_4^-$. The zirconium assists in this demethylation of one of the methoxy groups. A better synthetic approach to the organometallic derivatives of the monomethoxycalix[4]arene-zirconium fragment has been achieved starting from the corresponding monochloro derivative [{[*p*-Bu^t-calix[4]-(O)$_3$(OMe)}Zr(Cl)]$_2$ [32]. The results mentioned above show the variety of the Zr-C bonds we can build up over a calix[4]arene oxo-surface, *i.e.* Zr-C σ functionalities in

Scheme 11

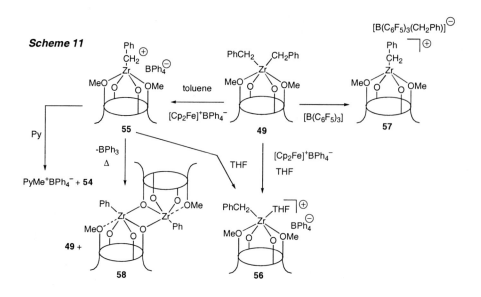

[ZrR$_2$], [ZrR] and [ZrR]$^+$, and π functionalities in [Zr-aryne]. The O$_4$-macrocycle is a unique chemical environment for such Zr-C fragments, because of a number of peculiarities: *i*) the topology of the O$_4$-Zr moiety is approximately square-pyramidal, with the metal out of the "O$_4$" plane; *ii*) the metal orbitals available for binding functionalities or assisting their transformations are located on the free face of the [ZrO$_4$] hemi-sandwich, unlike in the case of [Cp$_2$Zr] derivatives [37] (Figure 2); *iii*) the partial methylation of the phenoxo groups not only provides a variable range of charges for an O$_4$ set, but allows one to tune the ratio between strong σ, π donor (the phenoxo) and weak σ donor oxygens (the methoxy).

To characterise the special role of calix[4]arene as a supporting ligand compared to cyclopentadienyl and phenoxo anions, a detailed investigation was carried out on the migratory insertion of carbon monoxide and isocyanides in the Zr-C σ bonds [38]. The starting materials of this study were **48**, **49**, and **51** (Scheme 12). They undergo a sequential series of insertion reactions occurring with the formation of C-C bonds and giving organic fragments which remain bonded to the calix[4]arene-zirconium moiety.

Migratory insertion of CO gives Zr-C moieties susceptible to further migratory insertions of ketones and isocyanides. The formation of the η^2-ketones **59-61** probably occurs via the intermediacy of an η^2-acyl, followed by the migration of the second alkyl group to the η^2-acyl carbon [37,39,40]. The η^2-metal bonded ketones **59-61** maintain some of the migratory insertion properties typical of the metal-carbon σ-bond, as shown in Scheme 12 [41].

At this stage two results, which have few precedents in the literature, should be emphasized: *i*) the reactivity of the metal-carbon bond in the η^2-metal bonded ketones **59-61** towards inserting groups [41], a reaction which can be of synthetic utility; *ii*) the potential use of **59-61** as a source of zirconium(II) in displacement reactions with appropriate substrates.

Scheme 12

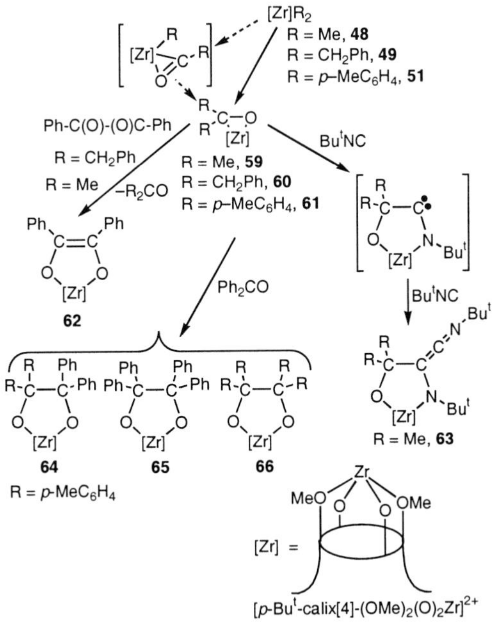

Scheme 13

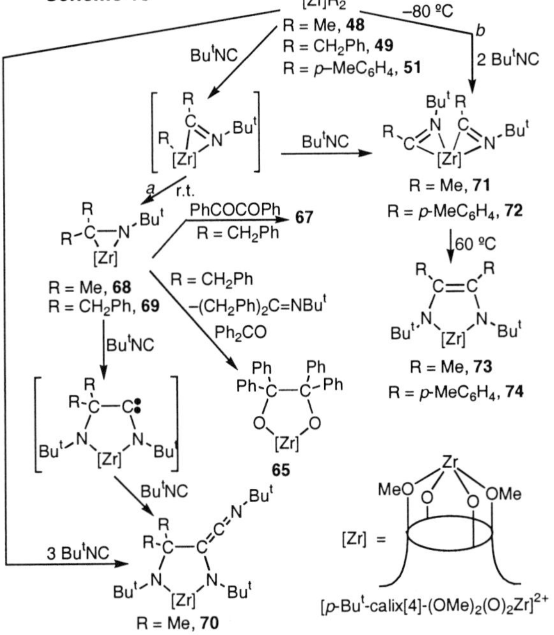

In the absence of a strong driving force derived, as in the case of CO, from the oxophilicity of the metal, the reaction with Bu^{t}NC [39,42] allows almost all of the steps of the migratory insertion of the isocyanide into the Zr-C bonds to be singled out (Scheme 13) [38].

The reaction of **48**, **49**, and **51** with Bu^{t}NC exemplifies the migratory aptitude of the alkyl and aryl groups bonded to zirconium. A complete range of migratory insertion pathways have been observed as a function of the temperature and the nature of R [38].

The competitive pathway dominant at low temperature for **48** and **51** is the preferential migration of the second alkyl group to an incoming Bu^{t}NC [42], to give bis-η^2-iminoacyl

71 and **72**. Regardless of the Zr:Bu^{t}NC stoichiometric ratio, the reaction proceeds via pathway *b* at low temperature for **48** and **51**, while in the case of **49**, which is unreactive at low temperature, the only compound formed at room temperature, via the pathway *a*, is always **69**. In the case of the *p*-tolyl substituent, pathway *b* is preferred, since only very limited formation of the η^2-imine occurs, even under forcing conditions. Gentle heating (60 °C) induces the intramolecular coupling [43] of the two η^2-iminoacyl derivatives **71** and **72** to the corresponding ene-diamido complexes, **73** and **74**. The intramolecular coupling followed first order kinetics in both cases. The activation parameters [$\Delta H^{\ddagger}$ (kJmol^{-1}), E$_a$ (kJmol^{-1})] obtained from the kinetic measurements are 106±2, 109±2 for complex **71** and 113±8, 116±8 for complex **72** [38].

All the reactions displayed in Schemes 12 and 13 belong to four general classes, namely: *i*) the migratory insertion of carbon monoxide or isocyanides into a metal-alkyl bond with the consequent formation of η^2-acyl or η^2-iminoacyl functionalities; *ii*) the alkyl- aryl- migration to an η^2-acyl or η^2-iminoacyl metal-bonded moiety, leading to the corresponding η^2-ketone or η^2-imine; *iii*) the intramolecular coupling of two η^2-iminoacyls to an ene-diamido ligand; *iv*) the migratory insertion of ketones and isocyanides into the M-C bond of η^2-metal bonded ketones and imines.

For V(III), the genesis and reactivity of the V–C functionality anchored to the dimethoxy-*p*-Bu^t-calix[4]arene dianion) have been studied. Oxidative demethylation of the bound ligand provides an entry into the analogous vanadium(IV) organometallic chemistry [14].

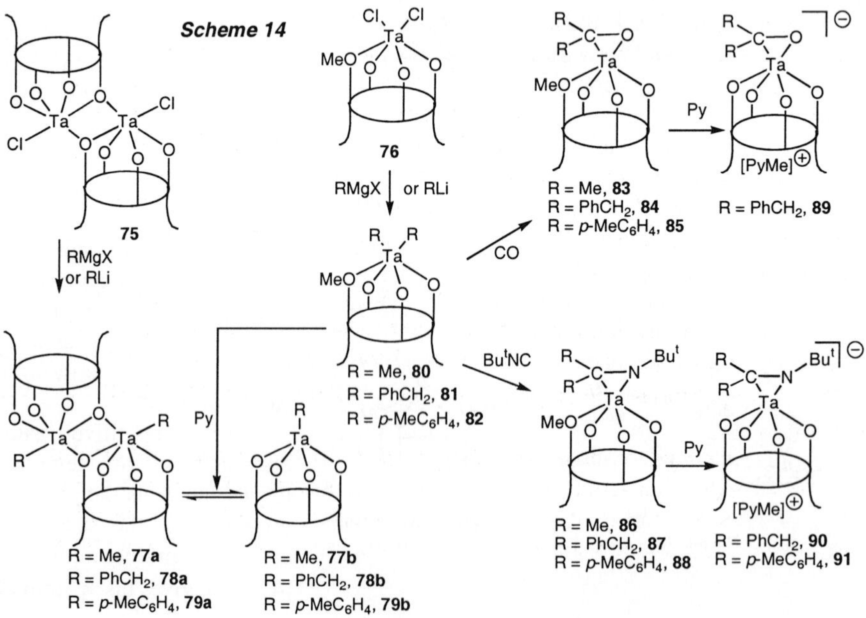

A rich chemistry of the metal carbon σ bond has been developed in the case of tantalum(V) anchored to the tetraoxo matrix, either as the calix[4]arene tetraanion or the monomethoxy calix[4]arene trianion [8]. In particular, the synthesis and the chemical reactivity of tantalum σ- and π-bonded to organic fragments are known. The base-induced demethylation of the methoxy-calix[4]arene allows the transfer from one model ligand to the other one without affecting the organic functionality bonded to tantalum. Formation of Ta alkyls and their migratory insertion reactions with CO and RNC are shown in Scheme 14. Although **78** and **79** dimeric in the solid state, solutions of **77-79** involve monomer-dimer equilibria. In the reaction of **80-82** with both CO or Bu^{t}NC, the migration, under mild conditions (*i.e.* room temperature), of both alkyl or aryl groups to form η^2-ketones **83-85** and η^2-imino derivatives, **86-88**, has been observed. Unlike [Cp$_2$M] or polyphenoxo derivatives of Ta, migration of the second alkyl or aryl group to the intermediate η^2-acyl or η^2-iminoacyl derivatives is very fast, precluding the interception of the precursor [39, 43,44]. As with the [(η^8-C$_8$H$_8$)ZrR$_2$] derivatives [45], such a pathway is assisted by the presence on the metal of the three facial frontier orbitals. Complex **76** can also be used a starting material to synthesize complexes which contain unsaturated carbon functionalities, namely alkenyl and allyl derivatives [8].

5.2. THE SYNTHESIS AND THE CHEMISTRY OF METAL-ALKYLIDENE AND –ALKYLIDYNE FUNCTIONALITIES

The approach to the metal-alkylidene [46] and metal-alkylidyne [47] chemistry has essentially been focused on the attempt to make a bridge between homogeneous and heterogeneous systems [48] using a pre-organized quasi-planar tetraanionic O$_4$ set of oxygen donor atoms derived from the deprotonated form of calix[4]arene. The choice of tungsten, which has played a major role in both heterogeneous and homogeneous systems since the discovery of the metathesis reaction, and whose ligands of preference contain oxygen donor atoms [46c,49], has been considered almost compulsory [50]. One might wonder what might be the consequence of using the calix[4]arene as ancillary ligand in metal-alkylidene and metal-alkylidyne chemistry. The metal bonded to the nearly planar calix[4]arene skeleton in its cone conformation displays three frontier orbitals, one σ and two π, particularly appropriate for stabilizing the M-C multiple bond functionality (see Section 2). Thus, alkylidenes and alkylidynes may form spontaneously from conventional alkylation reactions. The other unique role of calix[4]arene, which makes comparisons with the heterogeneous metal-oxide systems [2] valuable, is the surrounding of the metal where the oxygen donor atoms can assist the protonation-deprotonation of alkylidynes and, in general, their reaction with electrophiles. The so-called macrocyclic stabilization [51] helps significantly in keeping the metal-calix[4]arene fragment resistant to protic acids and strong electrophiles. The use of a tetraanionic macrocycle leads to anionic alkylidynes, thus providing the best entry to functionalized alkylidenes. Anionic alkylidynes allow one to set up a novel redox chemistry in this field of research.

This section covers: *i*) the genesis of W-anionic alkylidynes; *ii*) their reversible protonation and deprotonation reactions; *iii*) their metalation with carbophilic metals leading to dimetallic alkylidenes; *iv*) their unusual transformation into functionalized alkylidenes using appropriate electrophiles; *v*) their oxidative coupling to μ^2-η^2:η^2-acetylene derivatives [50].

The first attempts to obtain alkyl derivatives from **1** in its reaction with 1 or 2 equiv of Li or Mg alkylating agents were completely unsuccessful. On the other hand, when a 3:1 molar ratio was used, alkylidyne derivatives **92-94** were readily obtained (Scheme 15). The reaction solvents played an important role both in the selection of the reaction path (α elimination *vs* reduction) and in the separation of Mg and Li halides. The best results were obtained at low temperature in toluene [50].

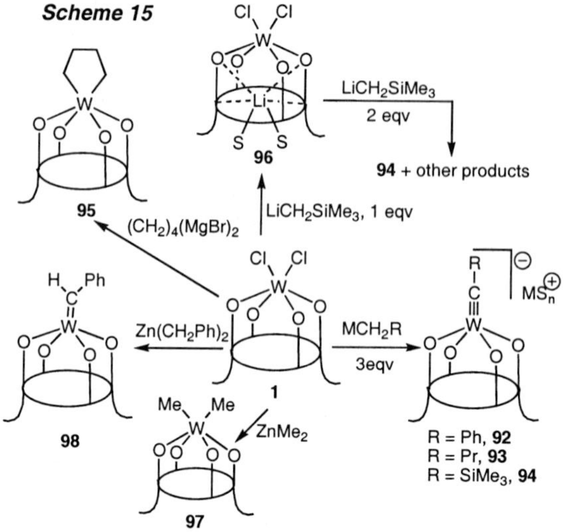

M≡C functionalities may be obtained by loss of H_2 from W(IV) primary alkyls [52]. This raises the question of the actual mechanism of the direct generation of alkylidynes described above, especially considering the fact that using 1 or 2 equiv of MCH_2R (M = Li, Mg; R = Ph, $SiMe_3$, Bu^n), reduced, rather than alkylated, species were obtained. Detailed study of the reaction of **1** with $LiCH_2SiMe_3$ led to the conclusion that alkylation is an alternative to the reduction path [53]. Using less reducing alkylating agents in a 2:1 ratio, dialkyls/alkylidenes could

Scheme 15

Scheme 16

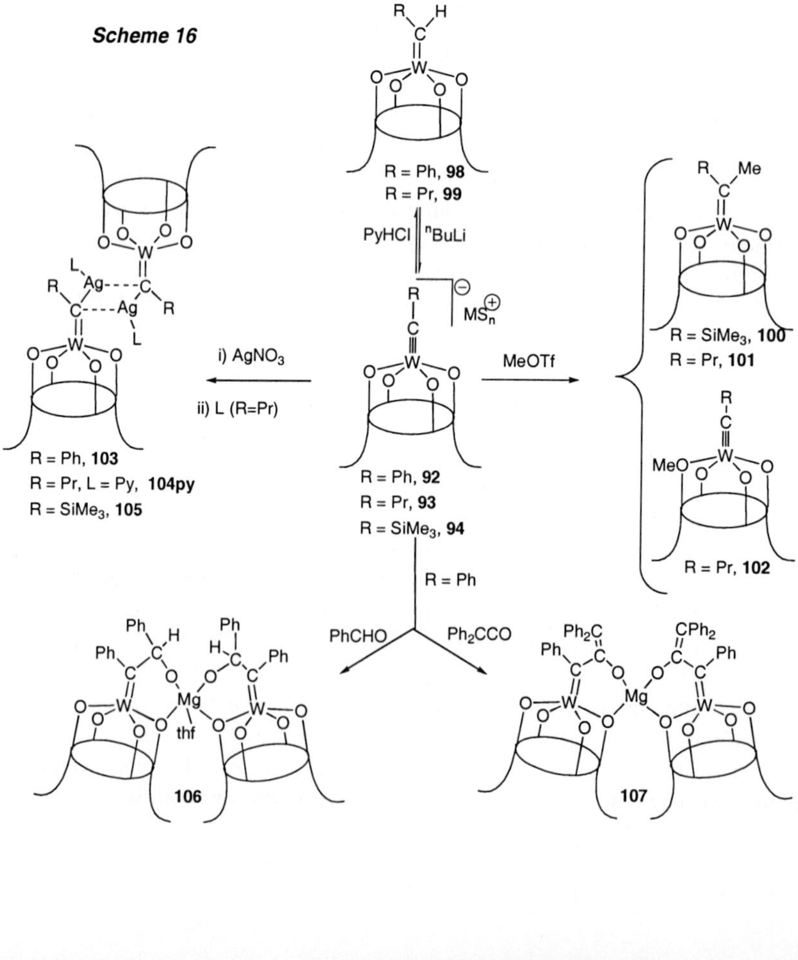

be prepared (see complexes **95**, **97**, and **98** in Scheme 15) [50,54]. The reaction of **1** with Zn(CH$_2$Ph)$_2$ led to the phenyl alkylidene **98**, and with ZnMe$_2$ to the clean formation of the dimethyl derivative **97**. Complexes **92-94** were found to be thermally and photochemically very stable. Their NMR spectra in coordinating solvents showed a C_{4V} symmetric calix[4]arene moiety. The alkylidyne carbon gave a signal at 260-300 ppm in ^{13}C NMR spectra.

The electron-rich, anionic alkylidyne species react readily with electrophiles (Scheme 16) [50]. The (reversible) protonation of **92-94** led cleanly to the corresponding alkylidenes without any other major change in the coordination sphere of the metal. Although the protonation of alkylidynes is known, it is not reversible and occurs with important modifications in the coordination geometry and changes in the nature of the donor atoms around the metal [55]. In the ^{1}H NMR spectra the calix[4]arene moiety appeared to be of C_{4V} symmetry.

A synthetically useful derivatization of the anionic alkylidynes **92-94** is their metalation, which was performed using the carbophilic Ag$^+$. Alkylidynes can also react with neutral electrophiles, such as a carbonyl functionality, provided it is sterically accessible. The reaction between **93** and benzaldehyde and diphenylketene (Scheme 16) illustrates the possibility of using anionic alkylidynes as organometallic Grignard reagents.

The heteroatom at the alkylidene carbon gives such complexes Fischer carbene [56] character and increases the possible use of the metal-alkylidene [46] synthon both in organic and organometallic synthesis. Anionic tungsten-alkylidene derivatives, such as **92** [50], are the appropriate starting materials for entering the area of functionalized metal-alkylidenes [43,57,58]. Two major complementary synthetic routes [59] have been devised to this purpose, the first exemplified in the reaction of **92** with a variety of electrophiles, such as PH$_3$SnCl, COCl$_2$ and ClPR$_2$ (Schemes 17 and 18), while another interesting approach [59] to functionalized alkylidenes is shown in Scheme 18. It resembles the methodology leading to functionalized carbynes [60]. Oxidation of **92** with I$_2$ led to **116** [50, 59], where iodine can be potentially replaced by a number of organic or organometallic nucleophiles.

Scheme 17

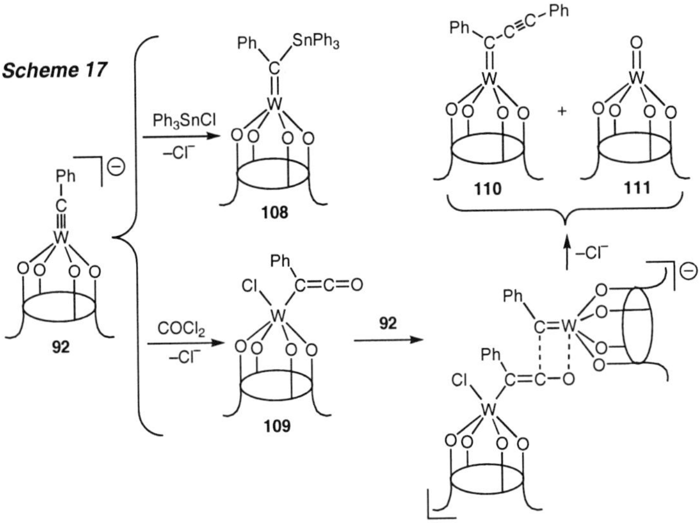

Another relevant synthetic methodology leading to metal-alkylidenes and alkylidynes has been discovered in studying the olefin rearrangements assisted by the metal-oxo surface [24]. Once more advantage is taken of the metal having the appropriate frontier orbitals for establishing a σ and two π bonds along the axial direction. This study addressed: i) the deprotonation of η^2-olefins to give 1-metallacyclopropenes rearranging, in the case of terminal olefins, to alkylidynes; ii) the photochemical rearrangements of the metallacyclopentane and the derived metallacylopentene to the corresponding alkylidene and alkylidyne; and iii) the assistance of the basic O-donor atoms in reactions with electrophiles, in close analogy with a metal-oxo surface [1].

Scheme 18

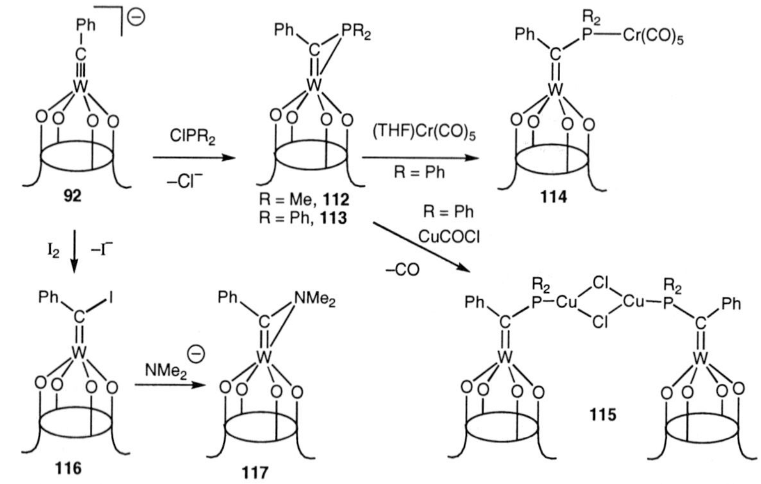

Scheme 19

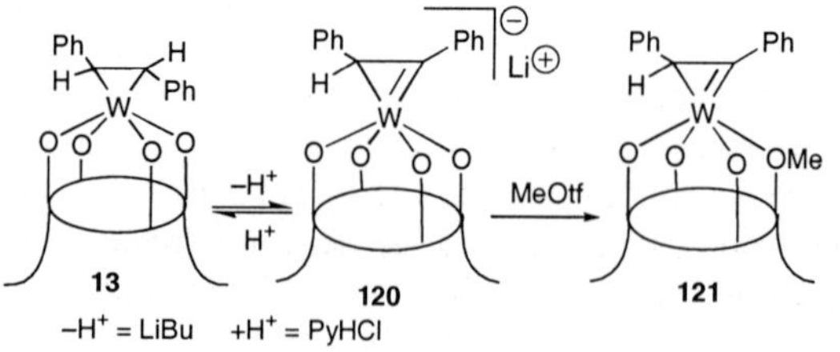

The electron-deficient nature of calixarene-supported tungsten(VI) makes α-carbon atoms susceptible to proton abstraction, the resulting carbanion being stabilised by donation to the metal and charge delocalisation onto the oxygen atoms [18]. The reaction of complex **13** with BuLi in toluene at low temperature (-80 °C) led to the clean deprotonation of the 1,2-disubstituted η^2-olefin to give an anionic 1-metallacyclopropene, **120** (Scheme 19).

The attempted deprotonation of the terminal olefin in **3** led to the alkylidyne **118**, most likely via an anionic 1-metallacyclopropene intermediate, analogous to **120**, undergoing an irreversible 1,2 proton shift. Protonation (PyHCl) of **118** gave the corresponding alkylidene **119**. Complexes **118** and **119** were identified by their characteristic spectroscopic features, *i.e.* a signal at 283.7 ppm, with a J_{CW} = 283.7 Hz, in the ^{13}C NMR spectrum of **118**, and a quartet at 9.93 ppm (J = 7.8 Hz, 1H) in the ^{1}H NMR of **119**. The outcome of this deprotonation-protonation sequence is the isomerization of an η^2-olefin to an alkylidene. Such a rearrangement, proposed [61] to occur in heterogeneous systems, has seldom been observed in solution [62].

A fundamental point both in molecular and in surface chemistry related to the olefin rearrangements mentioned above concerns the involvement of donor atoms (from the ancillary ligand or the surface) in acid/base reactions [63].

A further significant example of acid-base assisted rearrangements of W-C bonds has been reported for the metallacycle **16** (see Scheme 4). It undergoes, under mild conditions, some remarkable transformations like the deprotonation (LiBu) to the metallacyclopentene **122**, which can be reversibly protonated (PyHCl) back to the starting material (Scheme 20). Both acid-base interrelated metallacycles **16** and **122**, which are thermally stable, rearrange when irradiated (Xe lamp) to the alkylidene **123** and alkylidyne **124**, respectively. Although the photochemical generation of alkylidenes from dialkyls is well established [64], it has never been observed on a metallacycle, where such a reaction consists in an isomerization. More interesting still is the rearrangement of **122** to **124**, which represents the first example of the photochemical generation of an alkylidyne.

Although the generation of M-C, M=C, and M≡C functionalities directly from hydrocarbons has been recognized for a long time as a superior feature of heterogeneous over homogeneous catalysts, the investigation of the chemistry of the d^2-[{*p*-Bu^t-calix[4]-(O)$_4$}W] fragment, and in particular of η^2-olefin species, led to the discovery of a variety of olefin rearrangements which are very closely related to those often supposed to occur on metal oxides or other active surfaces. Such rearrangements are driven by light, acids, bases or occur under reducing conditions. This means that they can be controlled and, in perspective, used to generate *in situ* active species from inert precursors. These rearrangements lead to metallacycles, alkylidenes and alkylidynes, where the organometallic fragment derives from one of the cheapest building blocks of chemistry, *i.e.* ethylene. The protonation or deprotonation reactions of alkylidene and alkylidynes, respectively, beyond their synthetic value as a means to modify the ligand and access to new complexes, shed new light on the underestimated role of coordinated donor atoms in acid/base reactions, both in homogeneous and surface chemistry. Although some of these transformations are known for different metal fragments, the occurrence both on a single fragment and on a metal-oxo-surface is unique and unprecedented.

Although the alkylidene- and alkylidyne-metal functionalities have a very relevant impact on preparative chemistry and catalysis [46,47,57], few syntheses are known.

Metalla-calix[4]arenes allow a direct synthesis of metal-alkylidenes and -alkylidynes from among the most common organic functionalities, namely from olefin, acetylenes (see above), ketones and aldehydes. This latter entry in the field opens much wider possibilities in the use of such functionalities as synthons in organic synthesis [57]. In addition, this novel synthetic methodology has been applied to niobium [65], having a limited number of alkylidene [66] and alkylidyne [67] derivatives, by the use, as ancillary ligand, of p-Bu^t-calix[4]arene tetraanion [7].

The active compound able to manage the chemistry outlined above is a Nb^{III}-calix[4]arene dimer **9**, which contains an Nb=Nb double bond [7,22]. This section covers in particular: *i)* the genesis of Nb-alkylidenes from ketones and aldehydes; *ii)* their reversible deprotonation-protonation to alkyl and bridging alkylidenes; *iii)* the use of **9** in the McMurry [68] synthesis of olefins (see Scheme 21).

The reactivity of **9** is particularly pronounced with a variety of organic substrates, causing their overall reduction by four electrons. Among them, particularly interesting is the metathesis of the Nb=Nb double bond with the >C=O ketonic functionality, which corresponds to a four electron reduction of the carbonyl with complete C=O cleavage [7]. The reaction proceeds with the formation of the corresponding alkylidenes **125-131** and the oxo-niobium(V) derivative, **132**. The synthesis of alkylidenes has some novel features. It is not very sensitive to the substituent at the carbonyl functionality, occurring equally well with aromatic, mixed or aliphatic ketones. The synthetic value of the reaction lies in the easy separation of the alkylidene from the corresponding oxo-compound, **132**, thus allowing the introduction of the metal-alkylidene functionality, via the presence of a ketonic group in a variety of organic substrates. Aldehydes usually behave differently and in a more complex manner in their reaction with low-valent metals but with the [Nb=Nb] bond, the presence of the hydrogen at the carbonyl functionality does not induce a different pathway [69].

Scheme 21

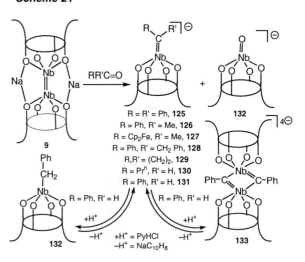

The synthetic method leading to Nb-alkylidenes and Nb-alkylidynes is particularly successful, due to a quite remarkable difference in the reaction rate of **9** with ketones or aldehydes, *vs* the subsequent reaction of the alkylidene with ketones and aldehydes [70]. The former reaction takes a few minutes at –40 °C, while the latter one occurs in hours at room temperature [71]. Such a kinetic difference allows the use of **9** in the stepwise synthesis of non-symmetric olefins, in a McMurry type reaction [68]. Taking advantage of the reactivity of the [Nb=Nb] bond in complex **9**, it is possible to achieve the formation of metal-carbido functionalities, related to metalla-alkylidenes, directly from carbon monoxide and under very mild conditions [72]. The synthetic sequence is shown in Scheme 22. The stepwise reduction of CO has been followed with the isolation and the structural characterization of all the intermediates shown. The reductive cleavage of carbon monoxide by using organometallic fragments represents a discrete homogeneous analogue of the CO dissociation to carbide and oxide on many metal surfaces [73,74].

Other synthetic approaches have been explored for binding an alkylidene functionality to a metalla-calix[4]arene. Among them, particular mention is deserved by the reaction of diazoalkanes with coordinatively unsaturated metalla-calix[4]arenes. The synthesis and the reactivity of an unusual high-spin (5.2 BM at 292 K) iron(II)-carbene has been recently reported [12].

Scheme 22

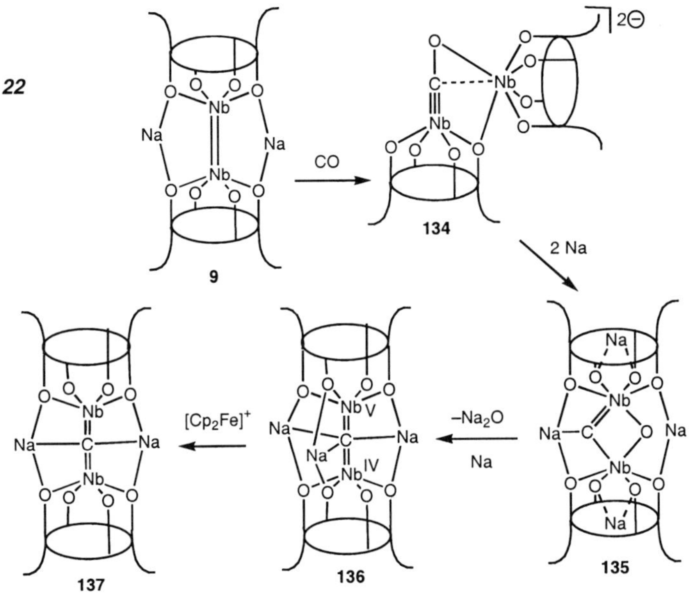

6. Metal-Ligand Multiple Bonds other than Metal-Carbon: Metal-Oxo, Metal-Imido, Metal-Nitrido Functionalities

6.1. METAL-OXO DERIVATIVES

The synthesis of M=O functionalities bonded to calix[4]arene anions is usually made starting from the corresponding oxochlorides, as was the case for Mo=O [75] and W=O [9]. In the latter case, an interesting phenomenon of self-assembly into a columnar structure was observed [76]. A similar self-assembling into polymeric linear structures has been observed for the related phenoxo derivatives [{p-Bu^t-calix[4]-(O)$_4$}W(OAr)$_2$] [77]. In some cases the M=O functionality derives from a two-electron oxidation of the corresponding metal-reduced species, as for oxo-niobium(V) derivatives such as **132**. A detailed study has been reported on the oxidation of vanadium(III)-monomethoxy-calix[4]arene, with dioxygen, quinones, and epoxides, leading in the latter case to an oxovanadium(V) derivative [13].

6.2. METAL-IMIDO DERIVATIVES

The synthesis of metal-imido-calix[4]arenes is easily made from the reaction of calix[4]arene in its protic or deprotonated form with the corresponding metal-imido halides, or metal-amido-imido derivatives [5c,78]. The M=NR functionality can be made using a more interesting methodology *via* a redox reaction at the metal centers. Two examples are shown in Scheme 23, where the starting material contains a metal having in both cases a d^2-configuration [13,23].

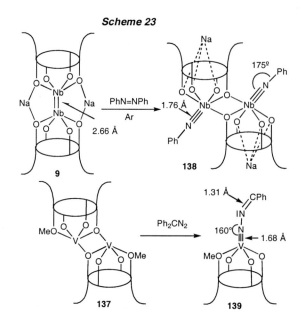

Scheme 23

6.3. METAL-NITRIDO DERIVATIVES: DINITROGEN ACTIVATION AND REDUCTION [23]

A complete and very detailed report on stepwise reduction of dinitrogen by the use of Nb(III)-calix[4]arene dimer **9** has been made [23]. The Nb(III)-calix[4]arene **9** is a powerful reducing agent of the N-N multiple bonds. The reaction of **9** with dinitrogen leading to the formation of the dinuclear metalla-hydrazone, implies a four-electron reduction of N_2 (Scheme 24). This reaction is, however, strongly dependent on the nature of the solvent used, which should be either THF or DME. The reduction in hydrocarbons (see below) follows a different pathway. The complete cleavage of the residual

Scheme 24

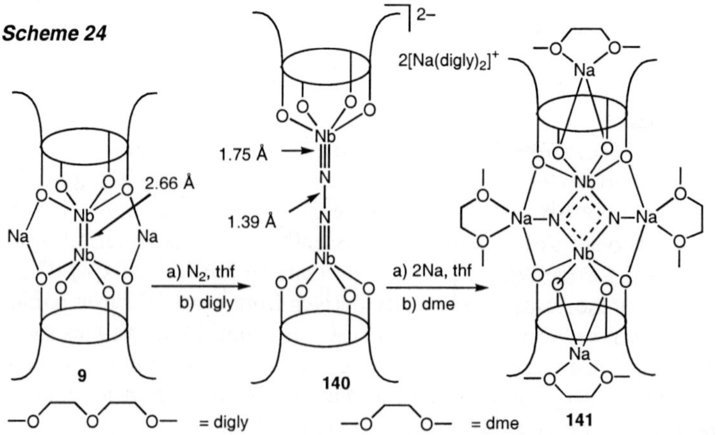

N-N single bond in **140** is achieved when two additional electrons are provided to the system in the reaction with sodium metal (Scheme 24). Such a reaction leads to the formation of a dinuclear μ-nitrido species, which in solution is in equilibrium with the corresponding monomeric form (see Figure 3).

It has been shown that this first step in the cleavage of the N-N single bond follows two different pathways, according to the reaction solvents, which are summarized in Figure 3. Regardless of the solvent used, THF or DME, the first two electrons reduce the two Nb(V) to Nb(IV), with the loss of Nb≡N triple bond. At this stage, the Nb-N-N-Nb skeleton undergoes a *transoid* rearrangement (THF) followed by the cleavage to the monomeric nitrido species in equilibrium in solution with the corresponding dimer. In the case of DME, the Nb-N-N-Nb skeleton undergoes a *cisoid* arrangement followed by the formation of an Nb-Nb bond. Thus, the two electrons introduced in the system are temporarily stored in the metal-metal bond. At the same time, the N_2 molecule moves from an end-on to a side-on bonding mode, thus being preorganized to the cleavage to the dimeric nitrido species **141**, the latter occurring upon heating. The experimental sequence related to the pathway *b* in Figure 3 is shown in Scheme 25. The bonding rearrangement of N_2 does not affect its reduction degree, the N-N bond remaining very long (1.42 Å). The isolation of **142** reveals not only the reorientation of N_2 over a metal-oxo surface, but also the possible use of the two electrons stored in the Nb-Nb bond for introducing a further functionality at the metals near the reduced N_2 moiety. The isolation of **143** with O_2 or PyO exemplifies such a possibility (Scheme 25).

Figure 3

Calix[4]arene tetraanion around Nb omitted

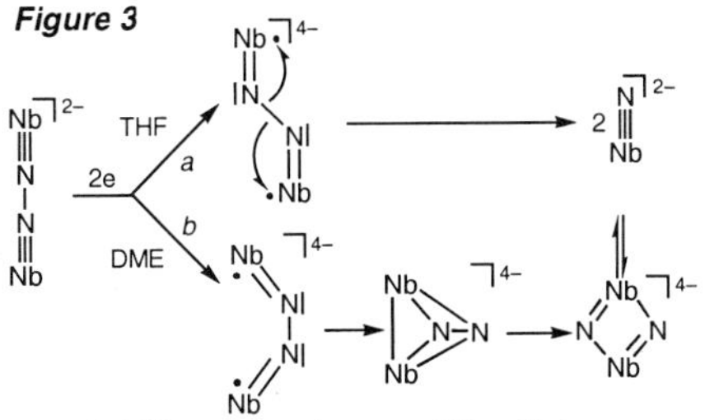

The active species [Nb=Nb], **9**, can perform the six-electron reduction of N_2 without the addition of any further reducing agent, when the reaction is carried out in toluene. The reaction leads to the trinuclear μ_2-bis-nitrido species, which is rather labile in the presence of solvents binding alkali cations, thus leading to the compounds **145** and **146** (Scheme 26). How the solvent may so greatly affect the reduction pathway of N_2 using metalla-calix[4]arenes has been studied in detail [23]. The bifunctionality of the complexes used is such that the solvation of the alkali cation can be an important

Scheme 25

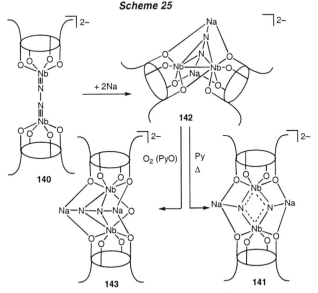

driving force and at the same time the presence of tight-ion pair or separated-ion forms can affect the kinetic pathways.

Some major advances in metal-dinitrogen chemistry [79,80] have been described herein. For the first time, it has been shown that the active species performing the reduction of N_2 is a very reactive M=M functionality, and that an ancillary ligand containing exclusively oxygen donor atoms can be successfully employed. The presence of alkali metal ions in the active bifunctional species magnifies the role of the solvent in dinitrogen reduction assisted by transition metal complexes. In particular, the solvent allows one to select either the four- (THF or DME) or the six-electron (toluene) reduction of dinitrogen to hydrazine or ammonia, respectively. In the former case, fine tuning by the solvent (DME vs THF) drives the ultimate two-electron reduction of N_2 to nitride through two different pathways, with the interception of key intermediates.

Scheme 26

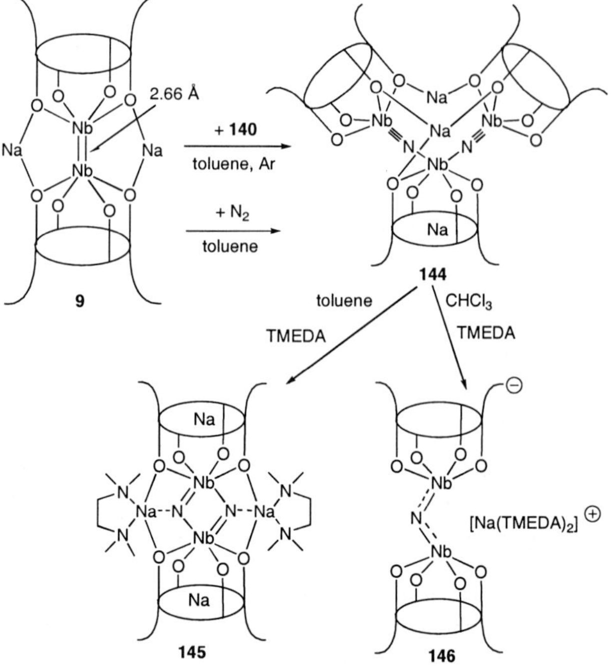

Particularly relevant in this context is the rearrangement of the end-on to the μ-η^2-η^2 bonding mode of dinitrogen over a metal-oxo surface modeled by Nb-calix[4]arene fragment. A comprehensive summary of the different stepwise pathways leading to the reduction of dinitrogen to nitrides is given in Figure 4 [23]. The results obtained in dinitrogen activation [79,80] show the potentiality of using metallacalix[4]arenes in driving reactions both in coordination and organometallic chemistry.

Figure 4

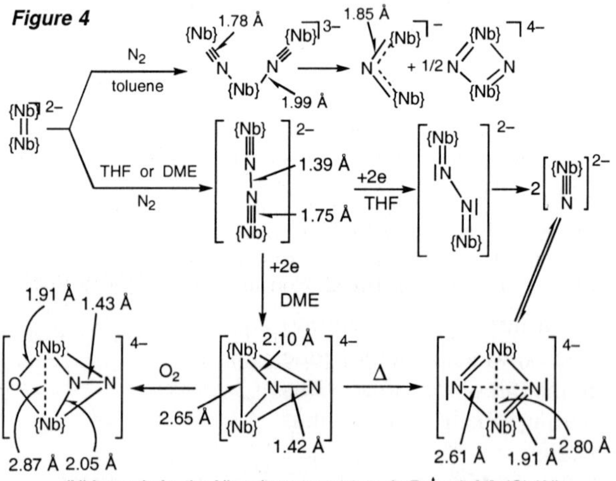

{Nb} stands for the Nb-calixarene moiety, {p-Bu^t-calix[4]–(O)$_4$}Nb

7. Outlook

A novel generation of catalysts are expected to come out of the chemistry of metallacalix[4]arenes. The pre-organised oxo-surface of calix[4]arene anions offers a unique opportunity for making molecular model compounds competitive with the well known heterogeneous oxo-catalysts. They can challenge heterogeneous systems as catalysts for hydrocarbon rearrangements, dinitrogen reduction and cleavage, oxo-transfer processes.

8. References and Notes

[1] a) C. D. Gutsche, *Calixarenes*, The Royal Society of Chemistry, Cambridge, U.K., **1989**; b) C. D. Gutsche, *Calixarene Revisited*, The Royal Society of Chemistry, Cambridge, U.K., **1998**; c) *Calixarenes, A Versatile Class of Macrocyclic Compounds* (Eds.: J. Vicens, V. Böhmer), Kluwer, Dordrecht, the Netherlands, **1991**; d) V. Bohmer, *Angew. Chem.* **1995**, *107*, 785-817; *Angew. Chem. Int. Ed. Engl.* **1995**, *34*, 713-745; e) C. Wieser, C. B. Dieleman, D. Matt, *Coord. Chem. Rev.* **1997**, *165*, 93-161; f) D. M: Roundhill, "*Metal Complexes of Calixarenes*" In *Progr. Inorg. Chem.* **1995**, 533-592; g) A. Ikeda, S. Shinkai, *Chem. Rev.* **1887**, *97*, 1713-1734.

[2] a) J. M. Thomas, W. J. Thomas, *Principles and Practice of Heterogeneous Catalysis*, VCH, Weinheim, Germany, **1997**; b) *Mechanisms of Reactions of Organometallic Compounds with Surfaces* (Eds. D. J. Cole-Hamilton, J. O. Williams), Plenum, New York, **1989**; c) H. H. Kung, *Transition Metal Oxides: Surface Chemistry and Catalysis*, Elsevier, Amsterdam, the Netherlands, **1989**; d) R. Hoffmann, *Solid and Surfaces, A Chemist's View of Bonding in Extended Strucures*, VCH, Weinheim, Germany, **1988**; e) *Catalyst Design, Progress and Perspectives* (Ed. L. Hegedus), Wiley, New York, **1987**; f) G. C. Bond, *Heterogeneous Catalysis, Principles and Applications, IInd Ed.*, Oxford University Press, New York, **1987**.

[3] For related molecular approaches to oxo-surfaces binding organometallic functionalities see: a) M. H. Chisholm, *Chemtracts-Inorganic Chemistry* **1992**, *4*, 273-301; b) W. Kläui, *Angew. Chem.* **1990**, *102*, 661-671; *Angew. Chem. Int. Ed. Engl.* **1990**, *29*, 627-637; c) F. J. Feher, T. A. Budzichowski, *Polyhedron* **1995**, *14*, 3239-3253; d) T. Nagata, M. Pohl, H. Weiner, R. G. Finke, *Inorg. Chem.* **1997**, *36*, 1366-1377; e) M. Pohl, D. K. Lyon, N. Mizuno, K. Nomiya, R. G. Finke, *Inorg. Chem.* **1995**, *34*, 1413-1429.

[4] C. Floriani, *Chem. Eur. J.* **1999**, *5*, 19-23.

[5] a) M. G. Gardiner, S. M. Lawrence, C. L. Raston, B. W. Skelton, A. H. White, *Chem. Commun.* **1996**, 2491-2492; b) M. G. Gardiner, G. A. Koutsantonis, S. M. Lawrence, P. J. Nichols, C. L. Raston, *Chem. Commun.* **1996**, 2035-2036; c) V. C. Gibson, C. Redshaw, W. Clegg, M. R. J. Elsegood, *J. Chem. Soc., Chem. Commun.* **1995**, 2371-2372; d) J. A. Acho, L. H. Doerrer, S. J. Lippard, *Inorg. Chem.* **1995**, *34*, 2542-2556; e) M. M. Olmstead, G. Sigel, H. Hope, X. Xu, P. P. Power, *J. Am. Chem. Soc.* **1985**, *107*, 8087-8091; f) X. Delaigue, M. W. Hosseini, E. Leize, S. Kieffer, A. Van Dorsselaer, *Tetrahedron Lett.* **1993**, *34*, 7561; g) F. Hajek, E. Graf, M. W. Hosseini, A. De Cian, J. Fischer, *ibid.* **1997**, *38*, 4555; h) J. L. Atwood, S. G. Bott, C. Jones, C. L. Raston, *J. Chem. Soc., Chem. Commun.* **1992**, 1349; i) J. L. Atwood, P. C. Junk, S. M. Lawrence, C. L. Raston, *Supramol. Chem.* **1996**, *7*, 15.

[6] F. Corazza, C. Floriani, A. Chiesi-Villa, C. Guastini, *J. Chem. Soc., Chem. Commun.* **1990**, 1083-1084.

[7] A. Caselli, E. Solari, R. Scopelliti, C. Floriani, *J. Am. Chem. Soc.* **1999**, *121*, 8296-8305.

[8] B. Castellano, E. Solari, C. Floriani, N. Re, A. Chiesi-Villa, C. Rizzoli, *Chem. Eur. J.* **1999**, *5*, 722-737.

[9] F. Corazza, C. Floriani, A. Chiesi-Villa, C. Rizzoli, *Inorg. Chem.* **1991**, *30*, 4465-4468.

[10] L. Giannini, E. Solari, C. Floriani, N. Re, A. Chiesi-Villa, C. Rizzoli, *Inorg. Chem.* **1999**, *38*, 1438-1445.

[11] A. Zanotti-Gerosa, E. Solari, L. Giannini, C. Floriani, N. Re, A. Chiesi-Villa, C. Rizzoli, *Inorg. Chim. Acta* **1998**, *270/1-2*, 298-311.

[12] M. Giusti, E. Solari, L. Giannini, C. Floriani, A. Chiesi-Villa, C. Rizzoli, *Organometallics* **1997**, *16*, 5610-5612.

[13] B. Castellano, E. Solari, C. Floriani, R. Scopelliti, N. Re, *Inorg. Chem.* **1999**, *38*, 3406-3413.

[14] B. Castellano, E. Solari, C. Floriani, N. Re, A. Chiesi-Villa, C. Rizzoli, *Organometallics* **1998**, *17*, 2328-2336.

[15] J. Hesschenbrouck, E. Solari, C. Floriani, N. Re, C. Rizzoli, A. Chiesi-Villa, *J. Chem. Soc., Dalton Trans.* **2000**, 191-198.

[16] L. Giannini, E. Solari, A. Zanotti-Gerosa, C. Floriani, A. Chiesi-Villa, C. Rizzoli, *Angew. Chem., Int. Ed. Engl.* **1996**, *35*, 85-87.
[17] Unpublished results.
[18] a) *Early Transition Metal Clusters with π-Donor Ligands*, (Ed. M. H. Chisholm), VCH, New York; **1995**; b) W. A. Nugent, J. M. Mayer, *Metal-Ligand Multiple Bonds*, Wiley, New York; **1988**.
[19] L. Giannini, E. Solari, A. Zanotti-Gerosa, C. Floriani, A. Chiesi-Villa, C. Rizzoli, *Angew. Chem., Int. Ed. Engl.* **1997**, *36*, 753-754.
[20] F. A. Cotton, R. A. Walton, *Multiple Bonds Between Metal Atoms*, 2nd *Ed.*, Oxford University Press, New York, **1993**; pp 597 and 603.
[21] a) M. H. Chisholm, K. Folting, W. E. Streib, D.-D. Wu, *Chem. Commun.* **1998**, 379-380; b) M. H. Chisholm, K. Folting, W. E. Streib, D.-D. Wu, *Inorg. Chem.* **1999**, *38*, 5219-5229; c) J. A. Acho, T. Ren, J. W. Yun, S. J. Lippard, *Inorg. Chem.* **1995**, *34*, 5226-5233; d) J. A. Acho, S. J. Lippard, *Inorg. Chim. Acta* **1995**, 229, 5-8.
[22] A. Zanotti-Gerosa, E. Solari, L. Giannini, C. Floriani, A. Chiesi-Villa, C. Rizzoli, *J. Am. Chem. Soc.*, **1998**, *120*, 437-438.
[23] A. Caselli, E. Solari, R. Scopelliti, C. Floriani, N. Re, C. Rizzoli, A. Chiesi-Villa, *J. Am. Chem. Soc.* **2000**, *122*, 3652-3670.
[24] a) L. Giannini, E. Solari, C. Floriani, A. Chiesi-Villa, C. Rizzoli, *J. Am. Chem. Soc.* **1998**, *120*, 823-824; b) L. Giannini, G. Guillemot, E. Solari, C. Floriani, N. Re, A. Chiesi-Villa, C. Rizzoli, *J. Am. Chem. Soc.* **1999**, *121*, 2797-2807.
[25] a) J. E. Hill, P. E.; Fanwick, I. P. Rothwell, *Organometallics* **1992**, *11*, 1771-1773; b) J. E. Hill, G. J. Balaich, P. E. Fanwick, I. P. Rothwell, *Organometallics* **1991**, *10*, 3428-3430; c) T. Takahashi, M. Tamura, M. Saburi, Y. Uchida, E.-I. Negishi, *J. Chem. Soc., Chem. Commun.* **1989**, 852; d) G. Erker, P. Czisch, C. Krüger, J. M. Wallis, *Organometallics* **1985**, *4*, 2059-2060; e) R. R. Schrock, S. McLain, J. Sancho, *Pure Appl. Chem.* **1980**, *52*, 729-732; f) S. McLain, C. D. Wood, R. R. Schrock, *J. Am. Chem. Soc.* **1979**, *101*, 4558; g) G. Ingrosso, In *Reactions of Coordinated Ligands*, (Ed. P. S. Braterman), Plenum, New York, **1986**; Vol. 1; Chapter 10.
[26] A. Caselli, L. Giannini, E. Solari, C. Floriani, N. Re, A. Chiesi-Villa, C. Rizzoli, *Organometallics* **1997**, *16*, 5457-5469.
[27] a) M. D. Fryzuk, T. S. Haddad, S. J. Rettig, *Organometallics* **1989**, *8*, 1723-1732; b) G. M. Diamond, M. L. H. Green, N. M. Walker, J. A. K. Howard, *J. Chem. Soc., Dalton Trans.* **1992**, 2641; c) J. Blenkers, B. Hessen, F. van Bolhuis, A. J. Wagner, J. H. Teuben, *Organometallics* **1987**, *6*, 459-469; d) H. Yasuda, A. Nakamura, *Angew. Chem., Int. Ed. Engl.* **1987**, *26*, 723-742.
[28] L. Giannini, E. Solari, C. Floriani, A. Chiesi-Villa, C. Rizzoli, *Angew. Chem., Int. Ed. Engl.* **1994**, *33*, 2204-2206.
[29] a) G. Erker, K. Engel, U. Korek, P. Czisch, H. Berke, P. Caubère, R. Vanderesse, *Organometallics* **1985**, *4*, 1531-1536; b) G. Erker, J. Wicker, K. Engel, C. Krüger, *Chem. Ber.* **1982**, *115*, 3300-3310.
[30] a) M. T. Reetz, in M. Schlosser (ed.), *Organometallics in Synthesis*, Wiley, New York, **1994**, Chapter 3; b) N. A. Petasis, D. K. Fu, *J. Am. Chem. Soc.* **1993**, *115*, 7208-7214; c) S. L. Buchwald, R. B. Nielsen, *Chem. Rev.* **1988**, *88*, 1047; d) C. H. Zambrano, P. E. Fanwick, I. P. Rothwell, *Organometallics* **1994**, *13*, 1174.
[31] O. V. Ozerov, F. T. Ladipo, B. O. Patrick, *J. Am. Chem. Soc.* **1999**, *121*, 7941-7942.
[32] L. Giannini, A. Caselli, E. Solari, C. Floriani, A. Chiesi-Villa, C. Rizzoli, N. Re, A. Sgamellotti, *J. Am. Chem. Soc.* **1997**, *119*, 9198-9210.
[33] a) S. L. Buchwald, R. D. Broene, In *Comprehensive Organometallic Chemistry II*, (Eds. E. W. Abel, F. G. A. Stone, G. Wilkinson), Pergamon, Oxford, **1995**, Vol. 12; Chapter 7.4 and references therein; b) G. Erker, M. Albrecht, C. Krüger, S. Werner, *J. Am. Chem. Soc.* **1992**, *114*, 8531; c) S. L. Buchwald, B. T. Watson, J. C. Huffman, *J. Am. Chem. Soc.* **1986**, *108*, 7411-7413; d) S. L. Buchwald, R. B. Nielsen, *Chem. Rev.* **1988**, *88*, 1047-1058.
[34] M. D. Fryzuk, S. S. H. Mao, *J. Am. Chem. Soc.* **1993**, *115*, 5336.
[35] a) A. S. Guram, R. F. Jordan, In *Comprehensive Organometallic Chemistry II*, (Eds. E. W. Abel, F. G. A. Stone, G. Wilkinson), Pergamon, Oxford, **1995**, Vol. 4; Chapter 12; b) R. F. Jordan, *Adv. Organom. Chem.* **1991**, *32*, 325-387.
[36] X. Yang, C. L. Stern, T. J. Marks, *J. Am. Chem. Soc.* **1991**, *113*, 3623.
[37] K. Tatsumi, A. Nakamura, P. Hofmann, P. Stauffert, R. Hoffmann, *J. Am. Chem. Soc.* **1985**, *107*, 4440.
[38] L. Giannini, A. Caselli, E. Solari, C. Floriani, A. Chiesi-Villa, C. Rizzoli, N. Re, A. Sgamellotti *J. Am. Chem. Soc.* **1997**, *119*, 9709-9719.
[39] L. D. Durfee, I. P. Rothwell, *Chem. Rev.* **1988**, *88*, 1059-1079 and references therein.

[40] a) P. E. Fanwick, L. M. Kobriger, A. K. McMullen, I. P. Rothwell, *J. Am. Chem. Soc.* **1986**, *108*, 8095; b) J. Arnold, T. D. Tilley, A. L. Rheingold, *J. Am. Chem. Soc.* **1986**, *108*, 5355; c) B. D. Martin, S. A. Matchett, J. R. Norton, O. P. Anderson, *J. Am. Chem. Soc.* **1985**, *107*, 7952.
[41] a) S. Smuck, G. Erker, S. Kotila, *J. Organometal. Chem.* **1995**, *502*, 75-86; b) M. Bendix, M. Grehl, K. Fröhlich, G. Erker, *Organometallics* **1994**, *13*, 3366-3369; c) G. Erker, M. Mena, C. Krüger, R. Noe, *Organometallics* **1991**, *10*, 1201.
[42] L. R. Chamberlain, L. D. Durfee, P. E. Fanwick, L. Kobriger, S. L. Latesky, A. K. McMullen, I. P. Rothwell, K. Folting, J. C. Huffman, W. E. Streib, R. Wang, *J. Am. Chem. Soc.* **1987**, *109*, 390-402.
[43] L. R. Chamberlain, L. D. Durfee, P. E. Fanwick, L. M. Kobriger, S. L. Latesky, A. K. McMullen, B. D. Steffey, I. P. Rothwell, K. Folting, J. C. Huffman, *J. Am. Chem. Soc.* **1987**, *109*, 6068-6076.
[44] a) L. R. Chamberlain, B. D. Steffey, I. P. Rothwell, J. C. Huffman, *Polyhedron* **1989**, *8*, 341-349.
[45] Unpublished results.
[46] a) J. Feldman, R. R. Schrock, *Progr. Inorg. Chem.* **1991**, *39*, 1; b) R. R. Schrock, *Alkylidene Complexes of the Earlier Transition Metals in Reactions of Coordinated Ligands*, (Ed.: P. S. Braterman), Plenum Press, New York, **1986**, Chapter 3; c) R. R. Schrock, *Acc. Chem. Res.* **1990**, *23*, 158-165; d) R. H. Grubbs, S. J. Miller, G. C. Fu, *Acc. Chem. Res.* **1995**, *28*, 446.
[47] H. Fischer, P. Hofmann, F. R. Kreissl, R. R. Schrock, U. Schubert, K. Weiss, *Carbyne Complexes*, VCH, Weinheim, Germany, **1988**.
[48] a) J. Corker, F. Lefebvre, C. Lecuyer, V. Dufaud, F. Quignard, A. Choplin, J. Evans, J.-M. Basset, *Science* **1996**, *271*, 966-969; b) G. P. Niccolai, J.-M. Basset, *Appl. Catal., A*, **1996**, *146*, 145-156; c) V. Vidal, A. Theolier, J. Thivolle-Cazat, J.-M. Basset, J. Corker, *J. Am. Chem. Soc.* **1996**, *118*, 4595-4662.
[49] a) R. R. Schrock, in *Reactions of Coordinated Ligands*, (Ed. P. S. Braterman), Plenum Press, New York, **1986**; Vol. 1; b) W. E. Buhro, M. H. Chisholm, *Adv. Organomet. Chem.* **1987**, *27*, 311.
[50] L. Giannini, E. Solari, S. Dovesi, C. Floriani, N. Re, A. Chiesi-Villa, C. Rizzoli, *J. Am. Chem. Soc.* **1999**, *121*, 2784-2796.
[51] a) *Stereochemistry of organometallic and inorganic compounds. Volume 2: Stereochemical and Stereophysical Behaviour of Macrocycles* (Ed. I. Bernal), Elsevier, Amsterdam, **1987**; b) B. Dietrich, P. Viout, J.-M. Lehn, *Macrocyclic Chemistry*, VCH, Weinheim, **1993**.
[52] a) K.-Y. Shih, K. Totland, S. W. Seidel, R. R. Schrock, *J. Am. Chem. Soc.* **1994**, *116*, 12103-12104; b) R. R. Schrock, S. W. Seidel, N. Mösch-Zanetti, D. A. Dobbs, K.-Y. Shih, W. M. Davis, *Organometallics* **1997**, *16*, 5195-5208; c) R. R. Schrock, *Acc. Chem. Res.* **1997**, *30*, 9-16.
[53] R. R. Schrock, D. N. Clark, J. H. Wengrovius, S. M. Rocklage, S. F. Pedersen, *Organometallics* **1982**, *1*, 1645-1651 and references therein.
[54] L. Giannini, S. Dovesi, E. Solari, C. Floriani, A. Chiesi-Villa, C. Rizzoli *Angew. Chem., Int. Ed.* **1999**, *38*, 807-810.
[55] a) Ref. 55, Chapter 4; b) Ref. 55, Chapter 5; c) G. R. Clark, K. Marsden, W. R. Roper, L. J. Wright, *J. Am. Chem. Soc.* **1980**, *102*, 6570-6571; d) L. Weber, G. Dembeck, H.-G. Stammler, B. Neumann, M. Schmidtmann, A. Müller, *Organometallics* **1998**, *17*, 5254-5259; e) H. P. Kim, S. Kim, R. A. Jacobson, R. J. Angelici, *Organometallics* **1984**, *3*, 1124-1126; f) R. A. Doyle, R. J. Angelici, *Organometallics* **1989**, *8*, 2207-2214; g) M. Green, A. G. Orpen, I. D. Williams, *J. Chem. Soc., Chem. Commun.* **1982**, 493.
[56] K. H. Dötz, H. Fischer, P. Hofmann, F. R. Kreissl, U. Schubert, K. Weiss, *Transition Metal Carbene Complexes*, VCH, Weinheim, **1983**.
[57] *Comprehensive Organometallic Chemistry, Vol. 12* (Eds.: E. W. Abel, F. G. A. Stone, G. Wilkinson), Pergamon, Oxford, **1995**, a) Chapter 5.3; b) Chapter 5.4; c) Chapter 5.5.
[58] a) F. R. Kreissl, J. Ostermeier, C. Ogric, *Chem. Ber.* **1995**, *128*, 289-292; b) Th. Lehotkay, K. Wurst, P. Jaitner, F. R. Kreissl, *J. Organometal. Chem.* **1996**, *523*, 105-110; c) S. Schmidt, J. Sundermeyer, F. Möller, *J. Organometal. Chem.* **1994**, *475*, 157-166; d) L. Weber, G. Dembeck, H.-G. Stammler, B. Neumann, M. Schmidtmann, A. Müller, *Organometallics* **1998**, *17*, 5254-5269 and references therein; e) H. P. Kim, S. Kim, R. A. Jacobson, R. J. Angelici, *Organometallics* **1984**, *3*, 1124-1126; f) R. A. Doyle, R. J. Angelici, *Organometallics* **1989**, *8*, 2207-2214; g) G. M. Jamison, P. S. White, J. L. Templeton, *Organometallics* **1991**, *10*, 1954-1959.
[59] S. Dovesi, E. Solari, R.Scopelliti, C. Floriani, *Angew. Chem., Int. Ed.* **1999**, *38*, 2388-2391.
[60] T. Desmond, F. J. Lalor, G. Ferguson, M. Parvez, *J. Chem. Soc., Chem. Commun.* **1983**, 457-459.
[61] Y. Iwasawa, H. Hamamura, *J. Chem. Soc., Chem. Commun.* **1983**, 130.
[62] a) G. A. Miller, N. J. Cooper, *J. Am. Chem. Soc.* **1985**, *107*, 709-711. b) J. S. Freundlich, R. R. Schrock, C. C. Cummins, W. M. Davis, *J. Am. Chem. Soc.* **1994**, *116*, 6476; c) J. S. Freundlich, R. R. Schrock, W. M. Davis, *J. Am. Chem. Soc.* **1996**, *118*, 3643.

[63] R. R. Schrock, S. W. Seidel, N. C. Mösch-Zanetti, K.-Y. Shih, M. B. O'Donoghue, W. M. Davis, W. M. Reiff, W. M. *J. Am. Chem. Soc.* **1997**, *119*, 11876.
[64] a) P. A. Van der Schaaf, A. Hafner, A. Mühlebach, *Angew. Chem., Int. Ed. Engl.* **1996**, *35*, 1845; b) L. R. Chamberlain, I. P. Rothwell, K. Folting, J. C. Huffman, *J. Chem. Soc., Dalton Trans.* **1987**, 155; c) L. R. Chamberlain, I. P. Rothwell, *J. Chem. Soc., Dalton Trans.* **1987**, 163; d) L. R. Chamberlain, A. P. Rothwell, I. P. Rothwell, *J. Am. Chem. Soc.* **1984**, *106*, 1847-1848.
[65] D. E. Wigley, S. D. Gray, In *Comprehensive Organometallic Chemistry II*, (Eds.: E. W. Abel, F. G. A. Stone, G. Wilkinson), Pergamon, Oxford, **1995**; Vol. 2; Chapter 2.
[66] a) R. R. Schrock, J. D. Fellmann, *J. Am. Chem. Soc.* **1978**, *100*, 3359-3370; b) R. R. Schrock, L. W. Messerle, C. D. Wood, L. J. Guggenberger, *J. Am. Chem. Soc.* **1978**, *100*, 3793-3800; c) G. A. Rupprecht, L. W. Messerle, J. D. Fellmann, R. R. Schrock, *J. Am. Chem. Soc.* **1980**, *102*, 6236-6244; d) J. K. Cockcroft, V. C. Gibson, J. A. K. Howard, A. D. Poole, U. Siemeling, C. Wilson, *J. Chem. Soc., Chem. Commun.* **1992**, 1668-1670; e) I. De Castro, J. De La Mata, M. Gomez, P. Gomez-Sal, P. Royo, J. M. Selas, *Polyhedron* **1992**, *11*, 1023-1027; f) F. Biasotto, M. Etienne, F. Dahan, *Organometallics* **1995**, *14*, 1870-1874; g) A. Antiñolo, A. Otero, M. Fajardo, C. García-Yebra, R. Gil-Sanz, C. López-Mardomingo, A. Martín, P. Gomez-Sal, *Organometallics* **1994**, *13*, 4679-4682; h) T. S. Kleckley, J. L. Bennett, P. T. Wolczanski, E. B. Lobkovsky, *J. Am. Chem. Soc.* **1997**, *119*, 247-248.
[67] a) R. N. Vrtis, C. P. Rao, S. Warner, S. J. Lippard, *J. Am. Chem. Soc.* **1988**, *110*, 2669-2670; b) R. N. Vrtis, L. Shuncheng, C. P. Rao, S. G. Bott, S. J. Lippard, *Organometallics* **1991**, *10*, 275-285; c) E. M. Carnahan, S. J. Lippard, *J. Am. Chem. Soc.* **1992**, *114*, 4166-4167; d) P. N. Riley, R. D. Profilet, P. E. Fanwick, I. P. Rothwell, *Organometallics* **1996**, *15*, 5502-5506 and references therein; e) F. Huq, W. Mowat, A. C. Skapski, G. Wilkinson, *J. Chem. Soc., Chem. Commun.* **1971**, 1477; f) W. Mowat, G. Wilkinson, *J. Chem. Soc., Dalton Trans.* **1973**, 1120; g) A. E. Ogilvy, P. E. Fanwick, I. P. Rothwell, *Organometallics* **1987**, *6*, 73.
[68] J. E. McMurry, *Acc. Chem. Res.* **1983**, *16*, 405.
[69] M. H. Chisholm, J. C. Huffman, E. A. Lucas, A. Sousa, W. E. Streib, *J. Am. Chem. Soc.* **1992**, *114*, 2710-2712.
[70] For the reductive cleavage of C-O of ketones see: M. H. Chisholm, C. Folting, J. A. Klang, *Organometallics* **1990**, *9*, 602-606 and 609-613, and references therein.
[71] The reaction between an early transition metal-alkylidene and a ketone to yield a metal-oxo group and an olefin was first noted by Schrock: R. R. Schrock, *J. Am. Chem. Soc.* **1976**, *98*, 5399.
[72] A. Caselli, E. Solari, R. Scopelliti, C. Floriani, *J. Am. Chem. Soc.* **2000**, *122*, 538-539.
[73] a) M. W. Roberts, *Chem. Soc. Rev.* **1977**, *6*, 373-391; b) E. L. Muetterties, J. Stein, *Chem. Rev.* **1979**, *79*, 479-490; c) D. F. Shriver, M. J. Sailor, *Acc. Chem. Res.* **1988**, *21*, 374-379; d) B. C. Gates, *Angew. Chem., Int. Ed. Engl.* **1993**, *32*, 228-229; e) M. L. Colaianni, J. G. Chen, W. H. Weinberg, J. T. Jr. Yates, *J. Am. Chem. Soc.* **1992**, *114*, 3735-3743.
[74] a) R. L. Miller, P. T. Wolczanski, A. L. Rheingold, *J. Am. Chem. Soc.* **1993**, *115*, 10422-10423; b) D. R. Neithamer, R. E. LaPointe, R. A. Wheeler, D. S. Richeson, G. D. Van Duyne, P. T. Wolczanski, *J. Am. Chem. Soc.* **1989**, *111*, 9056-9072; c) M. H. Chisholm, C. E. Hammond, V. J. Johnston, W. E. Streib, J. C. Huffman, *J. Am. Chem. Soc.* **1992**, *114*, 7056-7065 and references therein; d) J. C. Peters, A. L. Odom, C. C. Cummins, *Chem. Commun.* **1997**, 1995-1996 and references therein; e) C. C. Cummins, *Chem. Commun.* **1998**, 1777-1786.
[75] F. Corazza, C. Floriani, A. Chiesi-Villa, C. Guastini, *J. Chem. Soc., Chem. Commun.* **1990**, 640-641.
[76] B. Xu, T. M. Swager, *J. Am. Chem. Soc.* **1993**, *115*, 1159-1160.
[77] A. Zanotti-Gerosa, E. Solari, L. Giannini, C. Floriani, A. Chiesi-Villa, C. Rizzoli, *Chem. Commun.* **1996**, 119-120.
[78] a) U. Radius, J. Attner, *Eur. J. Inorg. Chem.* **1999**, 2221-2231; b) V. C. Gibson, C. Redshaw, W. Clegg, M. R. J. Elsegood, *Chem. Commun.* **1997**, 1605-1606.
[79] For general review of transition metal dinitrogen complexes see: a) A. D. Allen, R. O. Harris, B. R. Loescher, J. R. Stevens, R. N. Whiteley, *Chem. Rev.* **1973**, *73*, 11; b) J. Chatt, J. R. Dilworth, R. L. Richards, *Chem. Rev.* **1978**, *78*, 589; c) R. A. Henderson, G. J. Leigh, C. J. Pickett, *Adv. Inorg. Chem. Radiochem.* **1983**, *27*, 197; d) G. J. Leigh, *Acc. Chem. Res.* **1992**, *25*, 177; e) G. J. Leigh, *New J. Chem.* **1994**, *18*, 157; f) M. Hidai, Y. Mizobe, *Chem. Rev.* **1995**, *95*, 1115; g) A. E. Shilov, *Metal Complexes in Biomimetic Chemical Reactions;* CRC: Boca Raton, FL, **1997**; h) F. Tuczek, *Angew. Chem., Int. Ed. Engl.* **1998**, *37*, 2636.
[80] a) C. E. Laplaza, M. J. A. Johnson, J. C. Peters, A. L. Odom, E. Kim, C. C. Cummins, G. N. George, I. J. Pickering, *J. Am. Chem. Soc.* **1996**, *118*, 8623-8638 and references therein; b) G. K. B. Clentsmith, V. M. E. Bates, P. B. Hitchcock, F. G. N. Cloke, *J. Am. Chem. Soc.* **1999**, *121*, 10444-10445.

Chapter 30

PHENOXIDE COMPLEXES OF f-ELEMENTS

PIERRE THUÉRY[a], MARTINE NIERLICH[a], JACK HARROWFIELD[b], MARK OGDEN[c]

[a] *CEA / Saclay, SCM (CNRS URA 331), Bât. 125*
91191 Gif-sur-Yvette, France. e-mail: thuery@drecam.cea.fr
[b] *Research Centre for Advanced Mineral and Materials Processing*
University of Western Australia, Nedlands, 35 Stirling Highway, Crawley WA 6009. Australia. e-mail: jmh@chem.uwa.edu.au
[c] *A. J. Parker Collaborative Research Centre for Hydrometallurgy*
School of Applied Chemistry, Curtin University of Technology
Bentley WA 6845, Australia. e-mail: mark@power.curtin.edu.au

1. Introduction

The donor properties of the calixarenes, now a family of macrocycles for which a remarkably extensive range of ring sizes is available [1], make them effective ligands for all groups of metal ions, including the lanthanide (rare earth) [2] and actinide ions [3], often grouped together as "f-element ions" [4]. This chapter concerns the crystal structures and basic aspects of the coordination chemistry of lanthanide and actinide ion complexes with calixarenes unsubstituted on their lower rim, the "phenolic calixarenes". Relatively recent but broader reviews dealing with the complexing properties of calixarenes and their many derivatives are available [5] and other chapters in the present text describe important aspects of the solution coordination chemistry of both lanthanide and actinide cations with sophisticated calixarene derivatives.

Even for phenolic calixarenes, the site of metal ligation is not necessarily obvious and the *p*-sulfonatocalixarenes, for example, form complexes in which binding with sulfonate oxygens appears to be preferred to that with phenolic oxygens [6–11]. Polyhapto binding to the phenyl rings has been structurally proven for Cr(0) [12], Ru(II), Rh(I), Ir(I) [13], Cs(I) [14] and Ag(I) [15], having been surmised in some instances prior to these structural studies [16]. There is also spectroscopic evidence for its occurrence with a variety of other metals [17]. For phenolic calixarene complexes of the lanthanides and actinides, however, there is as yet no conclusive evidence for binding other than with oxygen donor atoms [2]. Very largely, studies of actinide binding to calixarenes concern uranium(VI) as the uranyl ion, UO_2^{2+} [3, 18–31], with but a single example of a structurally characterised complex of another actinide [32]. Thus, it is obviously premature to generalise about calixarene/actinide interactions, though it is of course expected that there should be some parallel between the heavier, synthetic actinides and the rare earths [4, 33]. The uranyl ion seemingly differs from the tripositive rare earths in the degree to which it enhances the acidity of bound phenol units, so that a common mode of uranyl ion binding is through at least four deprotonated phenol units, giving an anionic complex, whereas the tripositive rare earth ions are best known to cause triple deprotonation, thus giving neutral complexes.

The phenolic calixarenes are considered members of the class of aryloxide ligands, [34,35] as the coordination chemistry of neutral phenols is very limited compared to that of their conjugate bases. However, it is certainly not unusual to find complexes in which both phenol and phenoxide species are present and neutral phenolic donors are well known in complexes of multidentate and macrocyclic ligands with mixed donor atoms [36]. The difference in donor capacity between phenol and phenoxide is clearly apparent in the case of a calixarene in which one phenolic group has been functionalised with a pendent tridentate pyridine-2,6-dicarboxylate derivative. Under neutral conditions, Eu(III) binds at the tridentate pyridine site, as indicated by the relatively strong luminescence of the complex [2], but addition of base results in binding to phenoxide donors and release of the pyridine unit, as reflected in the strong quenching of the metal luminescence and confirmed by a crystal structure determination [37]. Nonetheless, while it may be that some degree of deprotonation is essential for strong coordination of a polyphenol to a metal ion, proton chelation (hydrogen bonding) is an important determinant of many calixarene properties. In the dipotassium complex of the dianion of *p-t*-butylcalix[8]arene, for example, the metal is bound only to residual phenolic groups and not to the formally deprotonated donors [38]. Otherwise, possibly the most important characteristics of calixarenes as aryloxide ligands are the limitations on both the number and stereochemical array of aryloxide oxygen donor atoms available to a given metal ion, that result from their macrocyclic structure.

The calixarenes are macrocycles but differ from most ligands of this type [39] in that the phenolic oxygen donor atoms are pendent from rather than constituents of the macrocyclic ring. There is a close resemblance of calix[4]arene systems to various biologically important macrocycles such as beauvericin, in the sense that the tilting of small rings (in the calixarenes, the phenyl rings) can provide a mechanism for adjusting the size of the donor atom array [40]. This may be a mechanism which serves to reduce the selectivity of the calixarenes, as has indeed been suggested in the case of hexahomotrioxacalix[3]arene ligands [41]. However, the only truly quantitative data on the selectivity of phenoxide-form calixarenes concern kinetic and not equilibrium measurements [42]. Presumably, the "chelate effect" [43] may play some role in determining the stability of calixarene complexes but again there are essentially no quantitative data available that allow a proper analysis of this factor.

Study of lanthanide and actinide binding by calixarenes has generally been based on exploration of the potential of the macrocyclic ligands in selective extraction of the metals or upon anticipation of useful spectroscopic properties of the complexes [2, 5, 44]. An earlier review [2] of the chemistry of lanthanide ion complexes with a more limited selection of phenolic calixarenes includes some discussion of their spectroscopic properties. The particular characteristics of calixarenes can give useful control of the electronic properties of lanthanide ions. Calixarene anions are reducing agents and coordination to the weakly oxidising Eu(III) ion results in the intrusion of a charge transfer state between levels associated with the ligand and emitting states of the metal so that luminescence from Eu(III) complexes of phenolic calixarenes is very weak, whereas that from complexes of the much more weakly oxidising Tb(III) ion is strong [2, 45–49]. A point that needs firm emphasis in any discussion of lanthanide chemistry, even one limited to the common Ln(III) species, is that each member of the series is different from the others, so that generalisation on the basis of a few particular studies is notoriously unreliable. It is common in structural studies to commence with investigations of La, Gd and Lu as the "beginning", "middle" and "end" of the series, but even this can be unduly superficial [50]. In fact, very few investigations of calixarene coordination have covered the complete lanthanide series [51] (and, of course, none has covered the complete series of actinides).

There is an extensive literature on lanthanide and actinide complexes of functionalised calixarenes [28,52] but we have deliberately chosen to focus upon the chemistry of the unfunctionalised systems in the belief that their relative simplicity reveals important aspects of the metal/ligand interactions with greater clarity. We do include discussion of some systems which are functionalised in the sense that potential donor atoms may be present within the (sometimes expanded) macrocyclic ring or within a *p*-substituent of the phenyl rings (*p*-sulfonatocalixarenes). Of course it does not necessarily follow that functionalisation of a calixarene excludes retention of some interactions with phenolic groups [28]. The work on uranyl systems has been largely stimulated by early reports [18] that the efficacy of several (mostly water-soluble) calixarene derivatives as uranyl ion binding agents justified their description not just as "uranophiles" [53] but even as "superuranophiles". Unfortunately, it has been a universal experience that the uranyl ion complexes of these particular functionalised calixarenes are especially recalcitrant towards crystallisation in a form suited to diffraction studies. The reasons for their failure to act as effective agents for *in vivo* uranium removal [54] therefore remain obscure.

2. An Overview of the Coordination Chemistry of the Complexes

Whilst it is often contended that the coordination chemistry of U(VI) is well understood, this is because the species referred to is the uranyl ion, UO_2^{2+} [4, 55]. The coordination chemistry of polyuranates [55, 56] is certainly less familiar and although it would seem that the "equatorial" donor sites about a linear uranyl unit can be only 4, 5 or 6 in number, it is not trivial to predict the form adopted with a given ligand [57]. With unidentate alkoxide ligands, four units bind to give the metal octahedral coordination [35], so that perhaps this should be the form expected for a calixarene complex. In relation to selective binding, it may well be significant that U(VI) in the uranyl ion can accept 5 or 6 ligands to give pentagonal and hexagonal bipyramidal geometries rarely observed for other metals [58] but it is not clear whether a ligand bearing aryloxide donors should favour these high coordination numbers. For the lanthanides, expectations as to the coordination mode of a flexible multidentate ligand are more difficult to define than for the uranyl ion, since the lanthanides are characterised by their high and variable coordination numbers, often associated with rather irregular coordination geometry [4, 45]. In their alkoxide/aryloxide chemistry, however, they share a characteristic with Th(IV) in that complexes seemingly resulting from partial hydrolysis of initially formed species are commonly obtained [34]. Coupled with the tendency of aryloxide ligands to act as bridges, this results in oligomeric species and an interesting question with regard to the coordination chemistry of lanthanides and calixarenes is that of the extent to which oligomerisation reactions may be controlled by the size of the macrocycle. It has long been known, of course, that bidentate aryloxides (catecholates and their derivatives) give novel rare earth complexes, one of their special features being the stabilisation of higher oxidation states such as that of Ce(IV) [59].

The fact that the water soluble p-sulfonatocalixarenes are extensively ionised under acidic conditions [6–11,18] may warrant neglect of hydrolytic oligomerisation as a factor in their coordination chemistry but this is not justified in general. In the investigations of calixarene coordination with all groups of metals, only rarely have stringent precautions been applied to avoid the possibility of hydrolysis occurring. Since the objective of such work has often been simply to obtain single crystals suitable for structural characterisation by X-ray diffraction, syntheses have frequently involved long periods of crystal growth during which at least adventitious contamination of the reaction mixture with water has been possible. This undoubtedly means as well that some structurally characterised calixarene complexes are kinetically, rather than thermodynamically, determined products of the reactions. Nonetheless, there are also instances, such as in the synthesis of the binuclear lanthanide ion complexes of p-t-butylcalix[8]arene and of various uranyl ion complexes of different calixarenes [2, 3], where the use of hydrated reagents or crystallisation in the presence of water does not change the nature of the product (and does not involve hydroxo- or oxo-bridge formation).

At present, there is a poor understanding of the solution chemistry involving phenolic calixarenes and f-element ions, especially in so far as it relates to synthesis. Even neglecting systems in which slow reactions may continue over extended periods (perhaps months), there is a lack of the most basic equilibrium data, such as the acidity constants for calixarenes in various solvents [49] and the stability constants defining the speciation of reactants such as $UO_2(NO_3)_2 \cdot 6H_2O$ when dissolved in a solvent such as methanol, which might enable a detailed characterisation of the complex species present in solution. Crude estimates of the "effective" equilibrium constants for the lanthanide(III)/p-t-butylcalix[8]arene system in dimethylformamide (dmf) [2] indicated that both ML and M_2L species were present and indeed both have been isolated and structurally characterised [51, 60, 61], though the ML species incorporates a nitrate ligand not considered in the analysis. ^{1}H NMR spectroscopy has been used to show that binuclear lanthanide ion complexes of p-t-butylhexahomotrioxacalix[3]arene dissociate in dimethylsulfoxide (dmso) [41], though in general it has been found that in poorly coordinating solvents the spectra of dissolved crystalline complexes are consistent with their solid state structures [2, 41]. Only in this sense, however, can the solution chemistry of these complexes be said to be well-defined. Even for water soluble complexes obtained with p-sulfonatocalixarenes, where at least some of the complications of reactant speciation might be expected to be minimised, the system is still apparently resistant to any detailed analysis of the solution equilibria. For this reason, this review is largely focused on the coordination chemistry of the calixarene complexes as revealed in their solid state structures.

Although many phenolic calixarenes are characterised by their rather low solubility in all solvents, reactions with Brønsted bases such as tetra-alkylammonium hydroxides, triethylamine, diazabicyclo-octane (DABCO), pyridine, imidazole and alkali metal hydrides, hydroxides and carbonates usually leads to major increases in solubility in solvents of moderate to high polarity suitable for the dissolution of metal ion reactants. The extent of calixarene deprotonation prior to complexation under conditions of synthesis is often unclear, though it is possible, for example, to isolate particular tetraalkylammonium derivatives of many calixarenes as solids [62], which, depending on the alkyl group of the ammonium moiety, can be dissolved readily in a variety of solvents. For the *p*-sulfonatocalixarenes, which are highly water-soluble, reaction with lanthanides and actinides is readily achieved by addition of almost any simple compound (chloride, nitrate, perchlorate, even a freshly-precipitated hydroxide), though the products are not necessarily easily crystallised, and co-ligands such as pyridine-*N*-oxide have sometimes been introduced to this end [6–11]. It has often been convenient to use stable, crystalline solvates such as $[Ln(dmso)_8](ClO_4)_3$ or $[Ln(NO_3)_3(dmso)_n]$ ($n = 3, 4$) to conduct addition of a precisely known amount of the metal ion [2]. Such complexes of dipolar, aprotic solvents are also useful in minimising the water content of reaction mixtures and $[UO_2(dmso)_5](ClO_4)_2$ is another compound which has been used with this aim in mind [19]. Anhydrous triflates, $Ln(CF_3SO_3)_3$, have also been exploited to the same end [41].

The actinide and lanthanide metal ions are generally very labile in their simple solvated forms [63] and it is usually clear during synthesis that mixture of suitable alkaline solutions of calixarene and metal ion results in rapid reaction, though at room temperature it is commonly observed that at least several minutes can be required to reach effective completion. The metal ions Eu(III) and U(VI) are particularly convenient to use, as phenoxide binding is usually signalled by the development of a bright yellow colour with the former and a deep red or deep orange one with the latter. The subsequent crystallisation of reaction products (in a form suitable for structure determination by single-crystal X-ray diffraction) may be immediate or can take several months, and considerable effort has been invested in the development of crystallisation procedures.

As the chemistry of f-element complexes of calixarenes has been developed, the variety of product stoichiometry and structure has increased, so that it is difficult to generalise about forms. The *p*-sulfonatocalixarenes appear remarkably ineffective as ligands in the conventional sense, rarely binding directly other than through sulfonate oxygens [6–11], usually considered rather poor donors [64], though this may depend to some extent on the lanthanide ion [65]. While unidentate coordination of *p*-alkylcalixarenes is known, in most instances these calixarenes function as multidentate chelates towards lanthanides and actinides. Where calixarene donor atoms predominate in the primary coordination sphere, the lanthanide ions show either 6- (rarely), 7- or 8-coordination, while uranyl ion accepts 3, 4 or 5 but (so far) never 6 donors in its equatorial garland. In very broad terms, as the ring size of the calixarene increases, the nuclearity of the metal unit bound to it increases but there is no regularity to this which is presently understood.

Considered as supramolecular synthons, metal complexes of the calixarenes offer prospects which have only just begun to be realised [11]. In some cases, it is clear that the "free" ligand conformation is puckered by metal coordination in a way which should enhance inclusion properties, and the effects here may be quite subtle. This is illustrated by the facts that the Eu(III) complex of partially-reduced calix[4]arene shows a "koilate" structure in the solid state which is not observed in the analogous complex of *p*-*t*-butylcalix[4]arene [66] and that the Eu(III) complex of *p*-*t*-butyldihomooxacalix[4]arene includes two solvent molecules within the calixarene *cone* [67], rather than one as usual. The simultaneous complexation of a metal of another type, leading to heterometallic species, is another issue remaining to be fully explored, as is the prospect of using metal centres not encircled or enclosed within a calixarene unit (and so "externally" rather than "internally" coordinated) as bridges to build up oligomeric or polymeric structures [26]. An infrequently noted feature of some calixarene complexes is their chirality [2]. The binuclear lanthanide complexes of both calix[8]arene and *p-t*-butylcalix[8]arene, for example (along with some of their transition metal analogues [68]), provide two cavities oriented so that the molecule has (at most) C_2 symmetry and thus raise the prospect that a bridging, difunctional inclusion guest could generate helical polymers.

3. Crystal Structures and Solid State Coordination Chemistry of Lanthanide(III) Ion Complexes

3.1. COMPLEXES WITHOUT DIRECT COORDINATION OF THE PHENOLIC UNITS

This chemistry concerns the complexes of the water-soluble *p*-sulfonatocalixarenes, and the crystallisation of these complexes has thus generally been from aqueous media. Though this is no guarantee that the species present in the solid state are representative of those in the original solutions, analysis of the first structural results has nonetheless provided a remarkable range of plausible models for various modes of solvation/coordination of lanthanide ions in aqueous solution [6–11]. Some of the striking features first evidenced with these ligands were the clay-like, bilayer structure of the alkali metal complexes [6] and the inclusion of water apparently through $H_2O\cdots\pi$ aromatic hydrogen bonds [69]. Subsequent investigations of the lanthanide ion complexes [7–11,70] showed that the layered structure, with an "up-down" ordering of calixarenes (in distorted *cone* conformations) in the hydrophobic bilayers, was a robust feature, retained in most of these complexes. This arrangement influences the bonding mode and stoichiometry, and is replaced in only some cases by complex spherical or tubular architectures [11]. Three main types of complexes have been exemplified, the differences observed in stoichiometry and bonding mode for different lanthanide ions with the same ligand being attributed to the relative weakness and low specificity of coordinate bonds to lanthanide ions [10].

The first type corresponds to a simple intercalation of the metal-containing species in the aqueous interlayer without bonding of the calixarene within the primary coordination sphere of the cation. The second is a 1:1 "external" complexation by a sulfonato group, while the third involves a metal atom bridging between calixarenes. The resulting structures, which may exhibit more than one of these interaction modes, typically involve an extended hydrogen bond network. The addition of pyridine-*N*-oxide (py-*N*-O) as a ligand extends the structural features with inclusion of the aromatic group in the calixarene cavity, so that second and third sphere coordination of the metal cation can be exploited to produce extensive supramolecular architectures.

The "supercomplex" $Na_8[Tb_4(py\text{-}N\text{-}O)_4(H_2O)_{18}(p\text{-sulfonatocalix[5]arene} - 5H)_4]$ **1** (Figure 1) is particularly noteworthy [10] since it arises from the centrosymmetric assemblage of four calixarenes and four eight-coordinate cations, two of them triply bridging. Second sphere coordination of two cations is provided in this case by encapsulation of first sphere py-*N*-O molecules in the calixarene cavities.

The bridging mode of complexation of the co-ligand py-N-O has been used to design architectures in which the arrangement of adjacent p-sulfonatocalix[4]arenes is "up-up" and not "up-down" as in the bilayers [11]. Depending on the stoichiometry of the three components, it has been possible to crystallise self-organized, nanometre-scale spheres and tubules. A 2:2:1 molar ratio for Na_5(p-sulfonatocalix[5]arene – 5H), py-N-O and $La(NO_3)_3$, respectively, produces a spherical assembly **2**, built from 12 calixarene molecules, with an overall diameter of ca. 28 Å. The outer surface of this assembly is coated with the 48 sulfonato groups, whereas the corresponding phenolic groups (12 of which are deprotonated) define the surface of the inner cavity, ca. 15 Å in diameter, which contains 30 water molecules and two sodium ions, bound together and to the surface by an extended hydrogen bonding network. The bridging metal ions are located outside, in aqueous interstitial spaces, with their coordinated py-N-O ligands encapsulated in the calixarene cavities. A different 2:8:1 molar ratio for the three components results in a tubular assembly, with a diameter of ca. 28 Å. The calixarene molecules are located, in a helical arrangement, on the surface of a cylinder, which is lined on the outer surface by sulfonato groups and on the inner one by phenolic groups. The La(III) ions and their py-N-O ligands, held by the calixarene as in **2**, join adjacent cylinders. Other La(III) ions and hydrated sodium ions occupy the internal channel.

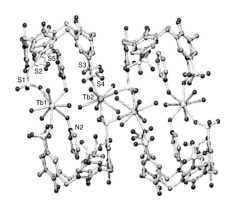

Figure 1. $[Tb_4(py\text{-}N\text{-}O)_4(H_2O)_{18}(p\text{-}sulfonatocalix[5]\text{-}arene – 5H)_4]$ **1** [10].

3.2. COMPLEXES INVOLVING SIMPLE PHENOXIDE COORDINATION

As both phenol and phenoxide oxygen donor atoms have more than one electron pair available for coordinate bonding, cases other than the "simple" donation of one pair can and do arise, giving bridges between two or more metal ions. In the lanthanide ion complexes of the calixarenes, it is uncommon to find the "simple" coordination mode only. However, some examples are known, one being the sole instance of phenoxide coordination of a lanthanide ion to a sulfonated calixarene [9]. Here, it may be that the very limited observed extent of interaction of the phenol/phenoxide groups with lanthanide ions is simply a consequence of the rather acidic conditions of synthesis. In the case of the 1:1 complex of Eu(III) with the dianion of p-t-butylcalix[8]arene, in which the macrocycle behaves as a simple chelate, it may be that oligomerisation via phenoxide bridging is inhibited by the bulk of the calixarene molecule, since bridging is observed when a second lanthanide ion is included within the macrocyclic ring [2, 48, 51, 60, 61]. In the case of the unidentate coordination of the dianion of p-t-butylcalix[6]-arene to Eu(III) [71], the coordinate interaction appears too weak to sustain the approach of another cation to the phenoxide entity, presumably because steric factors would become considerable, due to the particular conformation adopted by the calixarene.

A peculiarity of homoazacalixarenes is the possibility of internal deprotonation, enabling the formation of phenoxide-bound metal complexes without addition of an extra basic agent. This is exemplified by the formation of the 1:1 complex [Nd(p-chloro-N-benzylhexahomotriazacalix[3]arene)-(NO$_3$)$_3$] 3 (Figure 2) [72]. This complex is monomeric and the neodymium ion is bound to the three phenoxide oxygen atoms (via "simple" coordination) and six oxygen atoms from bidentate nitrate ions. The cation is located completely outside the calixarene cavity, at 1.59 Å from the mean plane of the phenoxide oxygen atoms. In the uncomplexed ligand, the three protons are located on the phenolic groups and involved in hydrogen bonds with the amine groups whereas, in the complex, they are located on the nitrogen atoms (i.e. the complexing moiety is a zwitterion) and involved in hydrogen bonds with the phenoxide oxygen atoms. In this sense, the phenoxide groups are already bridging and this may explain why they each interact with but a single metal ion. In addition, as in the case of the uranyl complexes of azacalixarenes (Section 4.2), the presence of positive charges in the calixarene framework may generally inhibit a close approach of cations to the ligand.

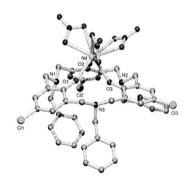

Figure 2. [Nd(p-chloro-N-Benzylhexahomotriaza calix[3]arene)(NO$_3$)$_3$] 3 [72].

3.3. OLIGONUCLEAR COMPLEXES INVOLVING PHENOXIDE BRIDGING

One of the significant contrasts between the uranyl and lanthanide ions complexes of phenolic calixarenes presently known [2, 3] is the relative importance of oligomerisation via phenoxide bridging in the solid state. Most lanthanide ion complexes show either this form of bridging or this form in conjunction with oxyanion (oxide/hydroxide) bridging (see Section 3.4). A common structure has been found for the complete series of lanthanide(III) ions [excluding Ce, where complications arise due to the formation of Ce(IV)] complexes of (p-t-butylcalix[8]arene – 6H) as both dmf and dmso solvates [51, 60]. The two metal cations enveloped by the macrocycle are each bound in a "simple" manner to two phenoxide and one phenol units while bridging to one another through two phenoxide units. One neutral ligand (dmf or dmso) bridges the two metal ions, while two more coordinate to each cation in a "simple", unidentate manner (one of each pair being drawn into the partial cone structure of the ligand) making the lanthanide ions overall eight-coordinate. The apparent strength of the binding of two lanthanide ions within the chiral, "two-bladed propeller" conformation of a calix[8]arene is confirmed by the later observations of the same structure for the Eu(III) complexes of p-nitrocalix[8]arene [48] and calix[8]arene itself [73].

Phenoxide bridging also appears to be important in a number of complexes better described as dimeric rather than as binuclear, though the metal-metal separations are similar. Thus, structural studies of the Eu(III) complexes have established an Eu$_2$L$_2$ formulation in the solid state when L is the trianion derived from p-t-butylcalix[n]arene where n = 4, 5 and 9 [2, 49, 73, 74], from p-t-butyldihomooxacalix[4]arene [67] and from one of the diastereomers of a hexahydrocalix[4]arene [66], as well as from a monoether derivative of *p-t*-butylcalix[4]arene [37]. The Eu(III) ion is either 6-, 7- or 8-coordinate in these species, with solvent molecules being coordinated by being drawn through the calixarene cavity in the case of *p-t*-butyldihomooxacalix[4]arene and *p-t*-butylcalix[5, 9]arenes. In all cases, the ligand conformation can be described as conical, though pinched in the case of the *p-t*-butylcalix[9]arene complex and rather strongly distorted (flattened) in the case of the *p-t*-butylcalix[5]arene complex. Spectroscopic evidence supports the adoption of the same structure by the complexes of Gd(III) and Tb(III) with *p-t*-butylcalix[5]arene [49].

Structure determinations have been performed for both the La(III) and Lu(III) complexes of the trianion of *p-t*-butylhexahomotrioxacalix[3]arene as their dmso solvates [41]. These centrosymmetric complexes are again doubly-phenoxide-bridged dimers, with the calixarene in a *cone* conformation and a solvent molecule both included and coordinated to the metal ion. Isostructural with the Y(III) [41] and Sc(III) [75] complexes is [Lu(*p-t*-butylhexahomotrioxacalix[3]arene – 3H)(dmso)]$_2$ **4** (Figure 3).

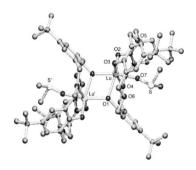

Figure 3. *[Lu(p-t-butylhexahomotrioxacalix[3]arene – 3H)(dmso)]$_2$, 4 [41].*

The metal ion is surrounded by six oxygen atoms (two phenoxides, two bridging phenoxides, one ether and one from dmso). In the corresponding La(III) complex, however, the metal is eight-coordinate, the three ether oxygen atoms being involved. In all homooxacalixarene complexes, the cation is close to the phenoxide mean plane and slightly displaced towards the calixarene cavity [particularly so with La(III)]. ^{1}H NMR spectra provided evidence of dynamic behaviour involving cleavage of one μ-phenoxo linkage in solution and showed that, whereas the dimeric structure is retained in non-coordinating solvents, monomeric species M(L – 3H)(dmso)$_n$ (n = 2, 3) are formed in dmso [41]. Those *p-t*-butylhexahomotrioxacalix[3]arene ligands have been shown by ^{1}H NMR measurements to bind selectively to tripositive metal ions (La^{3+}, Lu^{3+}, Y^{3+}, Sc^{3+}) over alkali metal ions (Li$^+$, Na$^+$, K$^+$).

A significant feature of the coordination of homooxacalixarenes to the lanthanide ions is the involvement of the ether oxygens, as this is not seen in uranyl ion complexes (Section 4.1), which may be a consequence of the more flexible coordination geometry of the tripositive lanthanides. The family of dimeric, phenoxide-bridged lanthanide(III) ion complexes of calixarenes provides a potentially useful group of divergent receptor units for construction of new solid lattices, both single and double occupancy of the calixarene cavities having been found. Retention of the same overall structure even after partial reduction of the calixarene [66] suggests that some level of cavity shaping is possible, the preferred retention of a proton at the reduced ring site indicating that different hydrogen bonding interactions might arise in the different species. Further, the two unidentate dmf ligands on each 7-coordinate Eu(III) ion in the complex of the reduced calix[4]arene are responsible for a polymeric structure in the solid resulting from the inclusion of the N(CH$_3$)$_2$ terminus of a bound dmf molecule of one dimer unit within the cavity of another. This provides an adventitious example of the supramolecular chemistry which may be open to exploitation in such systems.

3.4. OLIGONUCLEAR COMPLEXES INVOLVING BOTH PHENOXIDE AND OXYANION BRIDGING

The formal increase in the ring size of *p-t*-butylcalix[*n*]arenes with *n* passing from 4 to 9 does not involve a recognisably simple progression in the nature of the bound lanthanide species, though this is perhaps unsurprising given the different conformations adopted by the ligands. It is also likely that the full range of lanthanide complex structures with the various calixarenes is not yet fully explored. Some indication of this can be seen in the Eu(III) complexes of *p-t*-butylcalix[9]arene, where the change in reaction solvent from acetone to acetonitrile results in the isolation of [Eu$_7$(OH)$_8$(dmso)$_4$(*p-t*-butylcalix[9]arene – 6H)(*p-t*-butylcalix[9]arene – 7H)] **5**, rather than the relatively simple dimeric complex mentioned in Section 3.3. Thus, under appropriate conditions, *p-t*-butylcalix[9]arene appears to exhibit a characteristic which may be termed "cluster keeping" [3] in that its complexes with Eu(III) involve small oxyanion-bridged metal ion clusters sandwiched between two ligand molecules. While the forms adopted with other lanthanides are unknown, this characteristic of enveloping a small group of metal ions to form a lipophilic capsule is obviously of some potential utility in, for example, the production of catalysts based on the particular clusters as active sites immobilised in an oxide matrix [76]. Similarly, *p-t*-butylcalix[7]arene gives [Eu$_4$(OH)$_2$(CO$_3$)(*p-t*-butylcalix[7]arene – 4H)$_2$(dmso)$_6$] **6** [73] (Figure 4).

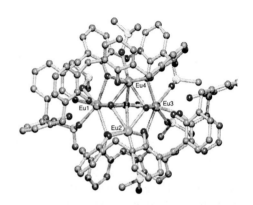

Figure 4. [Eu$_4$(OH)$_2$(CO$_3$)(p-t-butylcalix[7]arene – 4H)$_2$(dmso)$_6$](t-Bu groups omitted) 6 [73].

Another "cluster keeper" is *p-t*-butyltetrathiacalix[4]arene (see also Chapter 6) and its unique ligating properties are clearly illustrated by the nature of its lanthanide ion complexes and the differences between their structures and those of their analogues with *p-t*-butylcalix[4]arene described above. The binding of the sulfur atoms of the thiacalixarene is an obvious source of some differences but it is associated with extensive phenoxide bridging and, in all but one case involving Sm(III) where a bridging water molecule is found, with further bridging involving hydroxide ions. The first of these complexes to be structurally characterised was the species [Nd$_4$(OH)(*p-t*-butyltetrathiacalix[4]arene – 4H)$_2$(dmf)$_6$(dmso)$_2$](NO$_3$)$_3$ [77]. The core of the molecule is a square of four neodymium ions surrounding an hydroxide anion and coplanar with it, a unit which is unusual but nevertheless well known [78]. The calixarene entities are fully deprotonated and each phenoxide bridges two metal ions, every metal ion also being bound to two sulfur atoms from different calixarenes and to two oxygen atoms from the neutral ligands dmf and/or dmso. The metal ions are thus 9-coordinate and an identical structure has been found when Nd is replaced by Pr and Gd [79]. For La and Ce, [Ln$_3$(OH)(p-t-butyltetrathia-calix[4]arene − 2.5H)$_2$(dmf)$_3$(H$_2$O)$_3$](NO$_3$)$_3$ species [79], which may again be regarded as resulting from partial hydrolysis of the metal cations, are obtained. Their structure closely resembles that of the Nd species except that only three of the four metal sites are now occupied. For Sm(III), the species isolated can be described as a complex in which only half the metal sites in the Nd complex are occupied and the bridging hydroxide ion has been replaced by a bridging water molecule, the metal retaining 9-coordination [79]. For the Eu(III) complex, there is a return towards full occupancy of the Nd structure (as found with the following element, Gd). In this case, three 9-coordinate metal ions are present, bound to a triply-bridging hydroxide anion, though with only four phenoxide oxygen atoms bridging, making one Eu different to the other two, an inequivalence also found in the La and Ce complexes. The composition of the crystalline solid shows a 1:1 metal:ligand ratio as a result of the presence of a molecule of uncoordinated calixarene within the lattice (a situation which has been observed before, as in the Eu(III) complex of p-t-butylcalix[6]arene [71] and is rather common with crown ethers, for example [80]).

The 3:2 metal:ligand stoichiometry is found again at the end of the lanthanide series but is associated with the presence of three hydroxide ligands, perhaps reflecting the greater acid enhancing strength of the smaller lanthanide ions on bound ligands such as water. The complex ion entity found in both the Yb and Lu compounds has the composition [Ln$_3$(p-t-butyltetrathiacalix[4]-arene − 3H)$_2$(OH)$_3$(H$_2$O)$_2$(dmf)$_2$] **7** (Figure 5) and has a structure unlike that of any of the species described above in that the two *cone*-form calixarene entities are not diametrically opposed but instead are at an angle of ~120°, so that the complex can no longer be termed a "sandwich" [79].

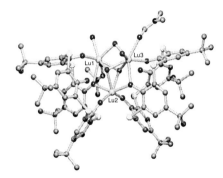

Figure 5. [Lu$_3$(p-t-butyltetrathiacalix[4]arene − 3H)$_2$(OH)$_3$(H$_2$O)$_2$(dmf)$_2$] 7 [79].

4. Crystal Structures and Solid State Coordination Chemistry of Actinide Ion Complexes

4.1. URANYL ION COMPLEXES INVOLVING SIMPLE PHENOXIDE COORDINATION

Uranyl ion complexes have been obtained and characterised with a wide range of calix[n]arenes, with n continuous from 4 to 9, and also n = 12. Including homooxa- and homoazacalixarenes, it can be said that the series begins with n = 3, though this is misleading in the sense that these macrocycles are significantly larger in ring size than a calix[4]arene. The complexes can be described as either "internal", when three or more phenoxide oxygen atoms are bound to one uranyl ion, or "external", when only one or two phenoxide units are bound to uranium and the metal is not effectively encircled by the calixarene. In all cases, the charge transfer excitation between reducing phenoxide and oxidising U(VI) is responsible for the characteristic colours of the complexes, either dark red or dark orange. Unlike the lanthanides, the uranyl ion forms many complexes in which phenoxide/phenol oxygen atoms are the only donors in the primary coordination sphere (together with neutral solvent molecules in some cases), and it is this group which we consider first.

A pseudo-trigonal coordination geometry is observed for the uranyl ion in the "internal" complex [UO$_2$(p-t-butylhexahomotrioxacalix[3]arene – 3H)](HNEt$_3$) **8** (Figure 6a) and the analogous complex in which HDABCO$^+$ replaces HNEt$_3^+$ [27]. In both, the uranyl ion is located at the centre of the calixarene, along its pseudo-trigonal axis, and it is bound to the three phenoxide groups with relatively short bond lengths [mean value 2.20(3) Å]. In contrast with the lanthanide ion complexes, none of the ether groups is bonding. This unusual coordination geometry has never been taken into account in the design of "uranophiles". It may be noted, however, that six-coordination in the equatorial plane of uranyl ion is commonly achieved with small "bite" ligands such as nitrate and carbonate, and very many complexes of nitrate ion in particular are known to involve a bidentate nitrate ion in sites occupied in analogous species by unidentate, monoatomic ligands (examples include [CoCl$_4$]$^{2-}$ and [Co(NO$_3$)$_4$]$^{2-}$, [PbCl$_6$]$^{4-}$ and [Pb(NO$_3$)$_6$]$^{4-}$, [CeCl$_6$]$^{2-}$ and [Ce(NO$_3$)$_6$]$^{2-}$ [81]). Thus, the design of a tris-(dithiocarbamate) "uranophile" by Tabushi [82] might be considered a fortuitous attempt to exploit pseudo-trigonal coordination. With respect to the uncomplexed ligand [83], the *cone* shape in **8** is deeper due to an increase in O⋯O distances upon coordination, and the counter cation is included in the cavity.

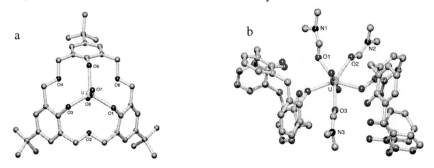

Figure 6. a) [UO$_2$(p-t-butylhexahomotrioxacalix[3]arene – 3H)], **8** [27]; b) [UO$_2$(calix[4]arene – H)$_2$(dmf)$_3$], **9**[31].

Despite evidence that any complex of uranyl ion with p-t-butylcalix[4]arene is very readily hydrolysed [3], the complex (Figure 6b) with calix[4]arene, [UO$_2$(calix[4]arene – H)$_2$(dmf/dmso)$_3$].calix[4]arene, **9**, [31] is sufficiently stable to be crystallised from solvents which have not been rigorously dried. The observed structure reinforces the belief that a simple calix[4]arene should be too small to encapsulate uranyl ion in that it is clearly of an "external" nature, with the solid having a 1:3 metal:ligand stoichiometry, though with one calixarene not directly coordinated to uranium. The calix[4]arene behaves simply as a unidentate ligand and the binding of just two such species appears to be insufficient for uranyl ion to retain the coordination number seen in its simple alkoxides, so that three neutral ligands accompany the two phenoxides, an observation interpretable in terms of the need to achieve a critical degree of "charge return" to any cation by ligand electron pair donation and the assumption that two negatively charged donors would be the equivalent of more than two neutral species involving the same donor atom. The calixarene in **9** is not extensively deprotonated and the formally phenoxide oxygen donor atom is involved in intramolecular hydrogen bonding, which results in relatively long U-O bonds (mean value 2.377(3) Å). Interestingly, this complex provides an unusual example of inclusion within the cavity of one of the bound calix[4]arene ligands of a dmf molecule oriented as in many *p-t*-butylcalixarene adducts, rather than within the lattice as in most solvates of calix[4]arene [62a].

Expansion of the dimensions of calix[4]-arene by the replacement of methylene bridges with sulfur, a change which corresponds to increasing the edge length of the square defined by the bridging atoms by approximately 0.5 Å, appears to result in a ligand which very closely matches the function of four unidentate aryloxide ligands. The coordination sphere of the uranyl ion in [UO_2(*p-t*-butyltetrathiacalix[4]-arene – 4H)] ($HNEt_3$)$_2$ **10** (Figure 7) [31] is very close to a regular octahedral geometry. This thiacalixarene may thus be regarded as the "perfect" polyphenoxide ligand for the uranyl ion. The U-O(phenoxide) distances are 2.27 Å, only slightly longer than the approximate bond length estimated for an unrestrained aryloxide donor [3].

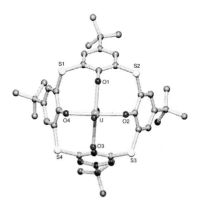

Figure 7. [UO_2(*p-t*-butyltetrathiacalix[4]-arene – 4H)] **10** [31].

An alternative means of expanding the size of a calix[4]arene is to introduce a homooxa bridge, though this necessarily reduces the symmetry. Thus, the uranyl ion complex of the tetra-anion of *p-t*-butyldihomooxacalix[4]arene [19] also shows binding to four phenoxide oxygen atoms, with U-O(phenoxide) distances very similar to those in **10** but with significant angular distortions. The widest O-U-O angle is that facing the extended bridge but there is only a weak contact, if any, with the ether function [U...O distance 3.534(8) Å, to be compared with a mean value of 2.26(1) Å for the phenolic oxygen atoms]. The calixarene is in the *cone* conformation, as in its uncomplexed form, and includes one triethylammonium cation in its cavity. With respect to calix[4]arene, which does not give such an "internal" uranyl ion complex, the role of the ether group is considered more to enlarge the macrocycle than to provide an extra bonding site. If the ideal aryloxide ligand for uranyl ion is indeed one with four donor centres capable of being arranged at a common distance of ~2.2–2.3 Å from uranium, then this of course is as expected.

Further expansion of the calix[4]arene skeleton as in *p-t*-butyltetrahomodioxa-calix[4]arene provides a tetra-anion which binds uranyl ion in a more regular square planar environment. In the complex [UO_2(*p-t*-butyltetrahomodioxacalix[4]arene – 4H)]($HNEt_3$)$_2$ [30], the ether-oxygen atoms are clearly not involved in coordination and the mean U-O(phenoxide) bond length of 2.28 Å is very similar to that in **10**. The two triethylammonium counter-ions are hydrogen bonded, one to a phenolic oxygen atom, and the other to an uranyl oxo atom. The latter is further included in the calix-arene cavity. The propensity of uranyl ion to act as a hydrogen-bond acceptor has been proposed as a basis for the design of "stereognostic" selective ligands [84]; in the present and other similar cases, the inclusion of the counter ion, which behaves as a second sphere ligand, may act in a comparable way.

In this context, it is unsurprising to find an irregular coordination geometry in the complex [UO$_2$(*p-t*-butylcalix[5]arene – 4H)](HNEt$_3$)$_2$ [23]. The cation is located at the centre of the lower rim coordination site, perpendicular to the mean oxygen atoms plane and bonded to the five oxygen atoms, four of which are deprotonated. Binding of the fifth, phenolic donor is enforced by the ligand conformation and this is associated with a very significant lengthening of one of the U-O(phenoxide) bonds (2.836(8) Å).

A return to the "ideal" case of four phenoxide donors (see Section 2) becomes possible with *p-t*-butylcalix[6]arene through the formation of the "external" centrosymmetric dimeric complex [(UO$_2$)(H$_3$O)(*p-t*-butylcalix[6]arene – 4H)]$_2$(HNEt$_3$)$_2$ in which two separate uranyl units are sandwiched between two face-to-face calixarenes in a distorted *pinched cone* conformation [22], a result in discord with the predictions of molecular modelling [57]. The cations are bound to two phenoxide groups from each calixarene and a hydroxonium cation is included in each calixarene cavity. In the related heterotrimetallic complex [(UO$_2$Cl$_2$)$_2$(Cs)(*p-t*-butylcalix[6]arene – 4H)](HNEt$_3$)$_3$, the two 4-coordinate uranyl ions are bound to two phenoxide oxygens and two chlorine atoms, and the Cs$^+$ ion is bound to two phenolic oxygen atoms and two aromatic rings [24]. This Cs$^+$ ion has been shown by NMR spectroscopy to play a role in organising the ligand for uranyl ion complexation. Note, however, that a different complex between uranyl ion and *p-t*-butylcalix[6]arene can be obtained in a different solvent system (see Section 4.3).

Regardless of the form of the complex, *p-t*-butylcalix[6]arene appears to be unable to complex UO$_2^{2+}$ in an "internal" fashion. This result is of interest when related to the "uranophile" properties of *p*-sulfonatocalix[6]arene [18]. Those properties have been attributed to the presence of a nearly planar array of six phenolic oxygen atoms suitable for this ion. However, the present results indicate that not only is it unnecessary to have six phenoxide donors to effectively bind uranyl ion but that calix[6]arenes are too large and too conformationally maladjusted to bind six (or even five) pendent donors to one cationic centre (six-coordination in particular would require smaller "bite" ligands). In fact, the octa-anion of *p*-sulfonatocalix[6]arene in its sodium complex has been found to crystallise in a double partial *cone* conformation [85], making it improbable that *p*-sulfonatocalix[6]arene could be considered as preorganised for uranyl complexation through six phenoxide or even six seemingly more strongly ligating (see Section 3.1) sulfonate oxygen atoms. It is likely that either a metal-ion induced reorganisation occurs or that the complexation involves at least partly the sulfonato groups (since the complexes have the colour typical of U(VI)/O(phenoxide) interactions). However, even this latter hypothesis seems improbable, since the photophysical properties of the Eu(III) complexes of *p-t*-butyl, *p*-nitro- and *p*-sulfonatocalix[8]arene in a basic medium (NEt$_3$) show that the structures are similar in the three cases [46] and spectroscopic evidence indicates that Fe(III) binds to the phenolic donors [86], indicating that mixed bonding modes are not frequent. In the absence of any crystal structure of an uranyl complex in this family of "uranophiles", the origin of this property is not presently well understood.

In relation specifically to the action of calix[6]arene as a ligand for U(VI), the recent characterisation [87] of a true hexa-aryloxide species derived from *p-t*-butylcalix[6]arene is particularly significant. Indirect reaction *via* UCl$_4$ as a uranium source under oxidising conditions enables the isolation of a complex of composition [U(*p-t*-butylcalix[6]arene − 3H)$_2$]. This is an "internal" complex but one in which each calixarene provides only three phenoxide donors and adopts the common 1,2,3-*alternate* conformation, with the uranium atom being octahedrally six-coordinate. This is further evidence that a hexagonal-planar array of phenoxide donors is favoured neither by the presence of U(VI) nor by the conformational flexibility of the calixarene.

A single uranyl ion binds to the tetra-anion of *p-t*-butylcalix[7]arene to give [UO$_2$(*p-t*-butylcalix[7]arene − 4H)](HNEt$_3$)$_2$, with a coordination environment for the uranyl ion closely similar to that found in the complex of *p-t*-butyldihomooxacalix[4]arene, once again demonstrating that four phenoxide donors held at a mean separation of ~2.28 Å provide a relatively stable complex species [25]. The conformation of *p-t*-butylcalix[7]arene in this complex is closely similar to that in its pyridine solvate, suggesting some preorganisation for uranyl ion complexation. However, the consequences of changing the *t*-butyl substituents to benzyl on the nature of the uranyl complex (see Section 4.3) suggest great caution should be applied in the interpretation of such limited data.

4.2. URANYL ION PHENOXIDE SPECIES INVOLVING BRIDGING AND CHELATING OXYANIONS

The zwitterionic homoazacalixarenes known to date appear unable to form "internal" complexes by providing more than two phenoxide oxygen donor atoms [88]. In both [UO$_2$(*p*-chloro-*N*-benzylhexahomotriazacalix[3]arene)(NO$_3$)$_2$] **11** (Figure 8) as well as in [UO$_2$(*p*-methyl-*N*-benzyltetrahomodiazacalix[4]arene)(NO$_3$)$_2$] [30], the cation is bound to two phenoxide groups only. The two nitrate ions are bidentate in **11**, giving a six-coordinate geometry, but only one of them is bidentate in the second compound, resulting in a five-coordinate environment. As in the neodymium(III) complex **3**, the phenolic protons have been transferred to the amine functions, giving the zwitterionic form of the macrocycle. The resulting complexes are neutral, which is a rare occurrence in this family of uranyl compounds. The difference in coordination mode between oxa- and azacalixarenes is likely due to the presence of positive charges within the calixarene framework and near the lower rim in the latter preventing a close approach of the uranyl ion, the repulsive interactions being minimised by the "external" coordination mode.

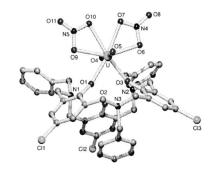

Figure 8. *[UO$_2$(p-chloro-N-benzylhexahomotriazacalix[3]arene)(NO$_3$)$_2$] **11** [30].*

Both nitrate anion and the isoelectronic carbonate anion are distinguished by the variety of their coordination modes when bound to heavy metals [89] and in distinction to the chelating and unidentate modes seen in the homoazacalixarene complexes just described, bridging is observed in the complexes of uranyl ion with the "large" calixarenes, p-t-butylcalix[9]arene and p-t-butylcalix[12]arene. p-t-butylcalix[9]arene gives [(UO$_2$)$_2$(p-t-butylcalix[9]arene – 5H)(CO$_3$)](HNEt$_3$)$_3$ **12** (Figure 9a) [3,73], in which two uranyl entities are bound to three phenoxide oxygen donors each and are bridged by a doubly-chelating carbonate ion (a coordination mode also giving oligomerisation in the species [(UO$_2$)$_3$(CO$_3$)$_6$]$^{6-}$ [90]). As in the complex of p-t-butylcalix[7]arene, a trimeric subunit is left uncomplexed. **12** is structurally similar to the binuclear uranyl complex of an acyclic analogue of p-t-butylcalix[6]arene, where a nitrate ion replaces carbonate [91], indicating the high stability of the present configuration.

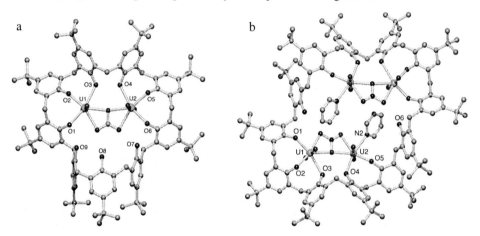

Figure 9. a) [(UO$_2$)$_2$(p-t-butylcalix[9]arene – 5H)(CO$_3$)] **12** [73]; b) [(UO$_2$)$_4$(p-t-butylcalix[12]arene – 8H)-(NO$_3$)$_2$(C$_5$H$_5$N)$_2$] **13** [29].

The largest calixarene for which a uranyl ion complex has yet been structurally characterised is p-t-butylcalix[12]arene. The tetranuclear complex of the calixarene octa-anion, [(UO$_2$)$_4$(p-t-butylcalix[12]arene – 8H)(NO$_3$)$_2$(C$_5$H$_5$N)$_2$](HNEt$_3$)$_2$ **13** (Figure 9b) [29], can be viewed as resulting from the fusion of two complex cores as observed in **12** for example, in which five phenolic oxygen atoms only (four of which deprotonated) are linked to each uranyl dimer pair, the coordination spheres being completed by a doubly bridging nitrate ion and a pyridine molecule. The two uranium atoms within each dimer are inequivalent in that a pyridine ligand on one replaces a phenolic oxygen atom on the other, but both are 5-coordinate in their equatorial planes.

Bridging of metallocomplex species by a hydroxide ion is found in complexes formed by uranyl ion with the tetra-anion of p-t-butylcalix[8]arene. Two forms of this binuclear, singly hydroxy-bridged species, [(UO$_2$)$_2$(p-t-butylcalix[8]arene – 4H)(OH)](HNEt$_3$)$_2$·(OH) [20] and [(UO$_2$)$_2$(p-t-butylcalix[8]arene – 4H)(OH)]$_2$(HNEt$_3$)$_5$·(OH)$_3$ [21], have been structurally characterised. In each, the same cation environment is found, with uranyl bound to four phenolic oxygen atoms (two of which are deprotonated) and a bridging hydroxyl ion. The same 2:1 stoichiometry has been reported in the case of the water-soluble p-sulfonatocalix[8]arene [18f]. These complexes could be described as those of the diuranyl cation [(UO$_2$)$_2$OH]$^{3+}$ and in this sense they are related to the species described in Section 4.3.

4.3. OLIGOURANYL SPECIES INVOLVING SIMPLE AND BRIDGING PHENOXIDE DONORS

The solid state chemistry of "uranates", formally the products of base addition to uranyl ion, has long been known but is relatively unfamiliar in solution [56]. A simple change of solvent (to dmso) in the preparation of a complex with *p-t*-butylcalix[6]arene (see Section 4.1), however, results in the crystallisation of a trinuclear product in which three phenoxide-bound uranyl units are linked by a triply-bridging oxide ion [3, 73]. This is also the first system presently encountered in which phenoxide oxygen atoms display the bridging ability so commonly seen in lanthanide ion complexes of calixarenes. Each uranyl entity achieves 5-coordination by being bound to the triply-bridging oxygen atom, two bridging and one unidentate phenoxide oxygen atoms and one oxygen atom from a dmso ligand, the overall composition of the complex being $[(UO_2)_3(O)(p\text{-}t\text{-butylcalix[6]arene} - H)_2(dmso)_2(p\text{-}t\text{-butylcalix[6]arene} - 2H)(dmso)_3]$ and the structure showing threefold symmetry, although the exact distribution of residual protons on the calixarene moieties is unknown.

A similar seemingly minor change in preparative conditions, here the replacement of NEt$_3$ as base by [N(CH$_3$)$_4$]OH, results in the isolation of a different oligouranyl species from the reaction involving *p-t*-butylcalix[8]arene [3, 73]. Formally, $[(UO_2)_3(OH)(p\text{-}t\text{-butylcalix[8]arene} - 6H)(dmso)_2](NMe_4)$ **14** (Figure 10) can be seen as resulting from the replacement of hydrogen in the hydroxide bridge of the diuranyl derivative with a UO$_2$(dmso)$_2$ moiety. The resultant cluster differs from that found in the *p-t*-butylcalix[6]arene derivative in that one uranyl oxygen on each of two uranium atoms is involved in a bridge to the third uranium atom. In this case, the phenoxide oxygen atoms are not involved in any bridging interactions.

Figure 10. $[(UO_2)_3(OH)(p\text{-}t\text{-butylcalix[8]-}$ $arene - 6H)(dmso)_2]$**14** [73].

Yet another variation on the combination of three uranyl units with an oxide anion, in reality the combination of six uranyl units with two oxide ions, is seen as the consequence of yet more seemingly minor variations in procedure, here the replacement of *t*-butyl groups by benzyl on a calix[7]arene and the use of DABCO as a base. At the expense of a significant conformational change relative to that of *p-t*-butylcalix[7]arene, perhaps associated with the fact that the ligand is now fully deprotonated, *p*-benzylcalix[7]arene reacts with uranyl ions in the presence of DABCO to give a centrosymmetric hexanuclear dimer, $[(UO_2)_3(p\text{-benzylcalix[7]arene} - 7H)(O)(CH_3CN)]_2$ (HDABCO)$_6$ **15** (Figure 11a) [26]. The three independent uranyl ions are located in different environments. U1 and U3 form with their symmetry-related counterparts U1' and U3' a tetranuclear core held by two μ$_3$-oxo bridges [mean U—O bond length 2.21(6) Å], an arrangement which has previously been described [92]. U1 is bound to the four phenoxide groups belonging to the tetrameric subunit in which it is included, whereas U3 is bound to one phenoxide group (bridging to U1) from each calixarene unit, in an "external" fashion, and to the nitrogen atom of an acetonitrile molecule. The environment of U2 is more unusual, since its four-coordinate equatorial environment comprises the three phenoxide groups of the trimeric subunit and an oxo atom from the uranyl ion U1O$_2$. The latter bonding mode corresponds to a dissymetric μ-oxo bridge and it results in a U1—O bond length of 1.88(3) Å, longer than the usual one (*ca.* 1.77 Å), the U2—O bond length being 2.32(2) Å. Such a bonding mode has rarely been observed in molecular chemistry [93] but has now been twice observed in calixarene complexes. This result shows that calix[7]arene is able to complex one uranyl ion in each of its two subunits but the two cations are so close to one another in the resulting structure that direct bonding between them occurs.

4.4. "Non-Uranyl" Complexes of Uranium Ions with Calixarenes

Reactions of UCl$_4$ with *p-t*-butylcalix[*n*]arenes, *n* = 4, 5, 6 [87] have provided the only fully characterised "non-uranyl" complexes of uranium with calixarenes. Oxidation during the syntheses or subsequently results in the isolation of mixed valence (U$^{VI}_2$, U^V) [{UCl(*p-t*-butylcalix[4]arene − 4H)}$_3$O], **16**, U(V)-containing [pyH]$_2$[{U(*p-t*-butyl-calix[5]arene − 5H)}$_2$O] (py = pyridine), **17**, and its U(VI) analogue, and the U(VI) species [U(*p-t*-butylcalix[6]arene − 3H)$_2$], **18** (Figure 11b), (see Section 4.1). None is markedly sensitive to hydrolysis, though their solution chemistry is incompletely established. While the structure of **18** is unique, and provides the first example of characterisation of a U(VI) hexa-aryloxide by X-ray crystallography, the structures of **16** and **17** show oligomerisation to occur in ways familiar from the structures of species such as the uranyl ion complexes of *p-t*-butylcalix[6]arene and *p-t*-butylcalix[8]arene.

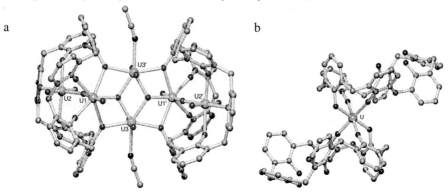

Figure 11. a) [(UO$_2$)$_3$(*p-benzylcalix[7]arene* − 7H)(O)(CH$_3$CN)]$_2$ **15** (*p*-benzyl groups omitted) [26]; b) [U(*p-t*-butylcalix[6]arene − 3H)$_2$] **18** [87].

4.5. Complexes of Other Actinides

The only structurally characterised complex of an actinide other than uranium is the complex formed by Th(IV) with *p-t*-butylcalix[8]arene [32]. The structure shows considerable similarities to those of the lanthanides with larger calixarenes and it could, in hindsight, be regarded as the first illustration of the capacity of calixarenes to act as "cluster keepers". The structure can also be considered typical of a highly charged metal aryloxide in that a small hydrolytic Th$_4$(OH)$_3$ cluster is sandwiched by lipophilic aryloxide donor ligands, some of the phenoxide donor atoms being involved in bridging interactions. The complex is interesting in that the calixarene ligands adopt two different conformations, perhaps a consequence of differing degrees of deprotonation, though presumably also dependent to some extent on the shape of the cluster being enveloped.

5. References and Notes

[1]   D. R. Stewart, C. D. Gutsche, *J. Am. Chem. Soc.* **1999**, *121*, 4136–4146.
[2]   J.-C. G. Bünzli and J. M. Harrowfield in *Calixarenes: a Versatile Class of Macrocyclic Compounds* (Eds.: J. Vicens, V. Böhmer), Kluwer, Dordrecht, **1991**, pp. 211–231 and references therein.
[3]   J. M. Harrowfield, *Gazz. Chim. Ital.* **1997**, *127*, 663–671, and references therein.
[4]   S. Cotton, *Lanthanides and Actinides*, Macmillan Education, London, **1991**.
[5]   a) D. M. Roundhill, *Progr. Inorg. Chem.* **1995**, *43*, 533–592; b) M. A. McKervey, F. Arnaud-Neu, M. J. Schwing-Weill in *Comprehensive Supramolecular Chemistry*, Vol. *1* (Vol. Ed.: G. Gokel), Pergamon, Oxford, **1996**, pp. 537–603; c) C. Wieser, C. B. Dieleman, D. Matt, *Coord. Chem. Rev.* **1997**, *165*, 93–161; d) S. Shinkai, *Chem. Rev.* **1997**, *97*, 1713–1734.
[6]   J. L. Atwood, S. G. Bott in *Calixarenes: a Versatile Class of Macrocyclic Compounds* (Eds.: J. Vicens, V. Böhmer), Kluwer, Dordrecht, **1991**, pp. 199–210, and references therein.
[7]   J. L. Atwood, G. W. Orr, N. C. Means, F. Hamada, H. Zhang, S. G. Bott, K. D. Robinson, *Inorg. Chem.* **1992**, *31*, 603–606.
[8]   J. L. Atwood, G. W. Orr, K. D. Robinson, F. Hamada, *Supramol. Chem.* **1993**, *2*, 309–317.
[9]   J. L. Atwood, G. W. Orr, K. D. Robinson, *Supramol. Chem.* **1994**, *3*, 89–91.
[10] J. W. Steed, C. P. Johnson, C. L. Barnes, R. K. Juneja, J. L. Atwood, S. Reilly, R. L. Hollis, P. H. Smith, D. L. Clark, *J. Am. Chem. Soc.* **1995**, *117*, 11426–11433.
[11] (a) G. W. Orr, L. J. Barbour, J. L. Atwood, *Science* **1999**, *285*, 1049–1052 ; (b) M.J. Hardie, C.L. Raston, *J. Chem. Soc., Dalton Trans*, **2000**, 2483 – 2492..
[12] a) H. Iki, T. Kikuchi, H. Tsuzuki, S. Shinkai, *Chem. Lett.* **1993**, 1735–1738; b) H. Iki, T. Kikuchi, H. Tsuzuki, S. Shinkai, *J. Incl. Phenom.* **1994**, *19*, 227–236.
[13] M. Staffilani, K. S. B. Hancock, J. W. Steed, K. T. Holman, J. L. Atwood, R. K. Juneja, R. S. Burkhalter, *J. Am. Chem. Soc.* **1997**, *119*, 6324–6335.
[14] a) J. M. Harrowfield, M. I. Ogden, W. R. Richmond, A. H. White, *J. Chem. Soc., Chem. Commun.* **1991**, 1159–1160; b) R. Asmuss, V. Böhmer, J. M. Harrowfield, M. I. Ogden, W. R. Richmond, B. W. Skelton, A. H. White, *J. Chem. Soc., Dalton Trans.* **1993**, 2427–2433; c) U. C. Meier, C. Detellier, *Can. J. Phys. Chem. A* **1999**, *103*, 3825–3829.
[15] a) A. Ikeda, S. Shinkai, *J. Am. Chem. Soc.* **1994**, *116*, 3102–3110; b) A. Ikeda, H. Tsuzuki, S. Shinkai, *J. Chem. Soc., Perkin Trans. 2* **1994**, 2073–2080; c) W. Xu, R. J. Puddephatt, K. W. Muir, A. A. Torabi, *Organomet.* **1994**, *13*, 3054–3062; d) P. Thuéry, M. Nierlich, B. Souley, Z. Asfari, J. Vicens, *Supramol. Chem.* **1999**, *11*, 143–150; e) M. Munakata, L. P. Wu, T. Kuroda-Sowa, M. Maekawa, Y. Suenaga, K. Sugimoto, I. Ino, *J. Chem. Soc., Dalton Trans.* **1999**, 373–378.
[16] H. Goldmann, W. Vogt, E. Paulus, V. Böhmer, *J. Am. Chem. Soc.* **1988**, *110*, 6811–6817.
[17] a) K. Kimura, K. Tatsumi, M. Yokohama, M. Ouchi, M. Mocerino, *Anal. Commun.* **1999**, 36, 229–230; b) D. Couton, M. Mocerino, C. Rapley, C. Kitamura, A. Yoneda, M. Ouchi, *Aust. J. Chem.* **1999**, 52, 227–229; c) N. J. van der Veen, R. J. M. Egberink, J. F. J. Engbersen, F. J. C. M. van Veggel, D. N. Reinhoudt, *Chem. Commun.* **1999**, 681–682.
[18] See, for example: a) S. Shinkai in *Calixarenes: a Versatile Class of Macrocyclic Compounds* (Eds.: J. Vicens, V. Böhmer), Kluwer, Dordrecht, **1991**, pp. 173–198, and references therein; b) R. Perrin, S. Harris, *ibid.*, pp. 235–259; c) S. Shinkai, H. Koreishi, K. Ueda, T. Arimura, O. Manabe, *J. Am. Chem. Soc.* **1987**, *109*, 6371–6376; d) T. Nagasaki, S. Shinkai, T. Matsuda, *J. Chem. Soc., Perkin Trans. 1* **1990**, 2617–2618; e) T. Nagasaki, S. Shinkai, *J. Chem. Soc., Perkin Trans. 2* **1991**, 1063–1066; f) T. Nagasaki, K. Kawano, K. Araki, S. Shinkai, *J. Chem. Soc., Perkin Trans. 2* **1991**, 1325–1327; g) K. Araki, N. Hashimoto, H. Otsuka, T. Nagasaki, S. Shinkai, *Chem. Lett.* **1993**, 829–832.
[19] J. M. Harrowfield, M. I. Ogden, A. H. White, *J. Chem. Soc., Dalton Trans.* **1991**, 979–985.
[20] P. Thuéry, N. Keller, M. Lance, J. D. Vigner, M. Nierlich, *Acta Crystallogr., Sect. C* **1995**, *51*, 1570–1574.
[21] P. Thuéry, N. Keller, M. Lance, J. D. Vigner, M. Nierlich, *New J. Chem.* **1995**, *19*, 619–625.
[22] P. Thuéry, M. Lance, M. Nierlich, *Supramol. Chem.* **1996**, *7*, 183–185.
[23] P. Thuéry, M. Nierlich, *J. Incl. Phenom.* **1997**, *27*, 13–20.

[24] P. C. Leverd, P. Berthault, M. Lance, M. Nierlich, *Eur. J. Inorg. Chem.* **1998**, 1859–1862.
[25] P. Thuéry, M. Nierlich, M. I. Ogden, J. M. Harrowfield, *Supramol. Chem.* **1998**, *9*, 297–303.
[26] P. Thuéry, M. Nierlich, B. Souley, Z. Asfari, J. Vicens, *J. Chem. Soc., Dalton Trans.* **1999**, 2589–2594.
[27] P. Thuéry, M. Nierlich, B. Masci, Z. Asfari, J. Vicens, *J. Chem. Soc., Dalton Trans.* **1999**, 3151–3152.
[28] P. D. Beer, M. G. B. Drew, D. Hesek, M. Kan, G. Nicholson, P. Schmitt, P. D. Sheen, G. Williams, *J. Chem. Soc., Dalton Trans.* **1998**, 2783–2785.
[29] P. C. Leverd, I. Dumazet-Bonnamour, R. Lamartine, M. Nierlich, *Chem. Commun.* **2000**, 493–494.
[30] P. Thuéry, M. Nierlich, J. Vicens, B. Masci, H. Takemura, *Eur. J. Inorg. Chem.* **2001**, 637-643.
[31] Z. Asfari, A. Bilyk, J. Dunlop, A. K. Hall, J. M. Harrowfield, M. W. Hosseini, B. W. Skelton, A. H. White, *Angew. Chem., Int. Ed. Engl.*, in press.
[32] J. M. Harrowfield, M. I. Ogden, A. H. White, *J. Chem. Soc., Dalton Trans.* **1991**, 2625–2632.
[33] a) M. H. B. Grote-Gansey, W. Verboom, F. C. J. M. van Veggel, V. Vetrogon, F. Arnaud-Neu, M.-J. Schwing-Weill, D. N. Reinhoudt, *J. Chem. Soc., Perkin Trans 2* **1998**, 2351–2360; b) N. T. K. Dung, K. Kunogi, R. Ludwig, *Bull. Chem. Soc. Jpn.* **1999**, *72*, 1005–1011.
[34] a) R. C. Mehrotra, A. Singh, *Progr. Inorg. Chem.* **1997**, *46*, 239–454 and references therein; b) M. H. Chisholm, I. P. Rothwell in *Comprehensive Coordination Chemistry, Vol. 2* (Eds.: G. Wilkinson, R. D. Gillard, J. A. McCleverty), Pergamon, Oxford, **1987**, Ch. 15.3.
[35] a) W. G. van der Sluys, A. P. Sattelberger, *Chem. Rev.* **1990**, *90*, 1027–1040; b) K. G. Caulton, L. G. Hubert-Pfalzgraf, *Chem. Rev.* **1990**, *90*, 969–995, specifically concern uranium alkoxides and aryloxides.
[36] a) D. E. Fenton, P. A. Vigato, *Chem. Soc. Rev.* **1988**, *17*, 69–90; b) V. Alexander, *Chem. Rev.* **1995**, *95*, 273–342.
[37] P. Froidevaux, J. M. Harrowfield, A. N. Sobolev, *Inorg. Chem.*, **2000**, *39*, 4678–4687.
[38] N. P. Clague, W. Clegg, S. J. Coles, J. D. Crane, D. J. Moreton, E. Sinn, S. J. Teat, N. A. Young, *Chem. Commun.* **1999**, 379–380.
[39] B. Dietrich, J.-M. Lehn and P. Viout, *Macrocycle Chemistry*, VCH, Weinheim, **1993**.
[40] M. Dobler, *The Ionophores and Their Structures*, Wiley, New York, **1980**, Ch. 6.
[41] C. E. Daitch, P. D. Hampton, E. N. Duesler, T. M. Alam, *J. Am. Chem. Soc.* **1996**, *118*, 7769–7773.
[42] a) R. M. Izatt, J. D. Lamb, R. T. Hawkins, P. R. Brown, S. R. Izatt, J. J. Christensen, *J. Am. Chem. Soc.* **1983**, *105*, 1782–1785; b) S. R. Izatt, R. T. Hawkins, J. J. Christensen, R. M. Izatt, *J. Am. Chem. Soc.* **1985**, *107*, 63–66.
[43] R. D. Hancock, A. E. Martell in *Coordination Chemistry - A Century of Progress* (Ed.: G. B. Kauffman), ACS, Washington, **1994**, Ch. 20.
[44] See, for example: F. J. Steemers, H. G. Meuris, W. Verboom, D. N. Reinhoudt, E. B. van der Tol, J. W. Verhoeven, *J. Org. Chem.* **1997**, *62*, 4229–4235.
[45] C. Piguet, J.-C. G. Bünzli, *Chem. Soc. Rev.* **1999**, *28*, 347–358.
[46] J.-C. G. Bünzli, F. Besançon, F. Ihringer in *Calixarene Molecules for Separation* (Eds.: G. Lumetta, A. Gopalan, R. D. Rogers), ACS Symposium Series, ACS, Washington (D. C.), in press.
[47] J.-C. G. Bünzli, P. Froidevaux, J. M. Harrowfield, *Inorg. Chem.* **1993**, *32*, 3306–3311.
[48] J.-C. G. Bünzli, F. Ihringer, P. Dumy, C. Sager, R. D. Rogers, *J. Chem. Soc., Dalton Trans.* **1998**, 497–503 and references therein.
[49] L. J. Charbonnière, C. Balsiger, K. J. Schenk, J.-C. G. Bünzli, *J. Chem. Soc., Dalton Trans.* **1998**, 505–510.
[50] P. C. Junk, C. J. Kepert, W.-M. Lu, B. W. Skelton, A. H. White, *Aust. J. Chem.* **1999**, *52*, 437–457 and references therein.
[51] a) J. M. Harrowfield, M. I. Ogden, A. H. White, *Aust. J. Chem.* **1991**, *44*, 1237–1247; b) J. M. Harrowfield, M. I. Ogden, A. H. White, *Aust. J. Chem.* **1991**, *44*, 1249–1262.
[52] See, for example: a) P. D. Beer, M. G. B. Drew, A. Grieve, M. I. Ogden, *J. Chem. Soc., Dalton Trans.* **1995**, 3455–3466; b) P. D. Beer, M. G. B. Drew, P. B. Leeson, M. I. Ogden, *Inorg. Chim. Acta* **1996**, *246*, 133–141; c) P. D. Beer, M. G. B. Drew, M. I. Ogden, *J. Chem. Soc., Dalton Trans.* **1997**, 1489–1491; d) R. Ludwig, K. Kunogi, N. Dung, S. Tachimori, *Chem. Commun.* **1997**, 1985–1986; e) G. Montavon, G. Duplâtre, Z. Asfari, J. Vicens, *Solv. Extr. Ion Exch.* **1997**, *15*, 169–188; f) P. Schmitt, P. D. Beer, M. G. B. Drew, P. D. Sheen, *Tetrahedron Lett.* **1998**, *39*, 6383–6386; g) M. J. Kan, G. Nicholson, I. Horn, G. Williams, P. D. Beer, P. Schmitt, D. Hesek, M. G. B. Drew, P. Sheen, *Nuclear Energy* **1998**, *37*, 325–329; h) X. Chen, M. Ji, D. R. Fisher, C. M. Wai, *Chem. Commun.* **1998**, 377–378; i) C. Dinse, N. Baglan, C. Cossonnet, J. F. Le Du, Z. Asfari, J. Vicens, *J. Alloys Comp.* **1998**, *271–273*, 778–781; j) T. N. Lambert, G. D. Jarvinen, A. S. Gopalan, *Tetrahedron Lett.* **1999**, *40*, 1613–1616; k) H. Du, A. Zhang, Z. Yang, Z. Zhou, *J. Radioanal. Nucl. Chem.* **1999**, *241*, 241–243; l) F. Arnaud-Neu, J. K. Browne, D. Byrne, D. J. Marrs, M. A. McKervey, P. O'Hagan, M.-J. Schwing-

Weill, A. Walker, *Chem. Eur. J.* **1999**, *5*, 175–186 and references therein. See also ref. 5 and Chapters 21 and 31 in the present text.

[53] a) I. Tabushi, Y. Kobuke, T. Nishiya, *Nature (London)* **1979**, *280*, 665–666; b) *Tetrahedron Lett.* **1979**, *37*, 3515–3518. See also c) A. H. Alberts, D. J. Cram, *J. Am. Chem. Soc.* **1977**, *99*, 3880–3882.

[54] M. Archimbaud, M. H. Henge-Napoli, D. Lilienbaum, M. Desloges, C. Montagne, *Rad. Prot. Dosim.* **1994**, *53*, 327–330.

[55] K. W. Bagnall in *Comprehensive Coordination Chemistry, Vol. 3* (Eds.: G. Wilkinson, R. D. Gillard, J. A. McCleverty), Pergamon, Oxford, **1987**, Ch. 40.

[56] F. A. Cotton, G. Wilkinson, *Advanced Inorganic Chemistry*, 4th Ed., Wiley, New York, **1980**, p. 1029.

[57] P. Guilbaud, G. Wipff, *J. Incl. Phenom.* **1993**, *16*, 169–188.

[58] A. F. Waters, A. H. White, *Aust. J. Chem.* **1996**, *49*, 27–34, 35–46, 87–98, 117–135, 147–154.

[59] a) S. R. Sofen, K. Abu-Dari, D. P. Freyberg, K. N. Raymond, *J. Am. Chem. Soc.* **1978**, *100*, 7882–7887; b) S. R. Sofen, S. R. Cooper, K. N. Raymond, *Inorg. Chem.* **1979**, *18*, 1611–1616; c) K. N. Raymond, W. L. Smith, F. L. Weitl, P. W. Durbin, E. S. Jones, K. Abu-Dari, S. R. Sofen, S. R. Cooper in *Lanthanide and Actinide Chemistry and Spectroscopy* (Ed.: N. M. Edelstein), ACS Symposium Series No. 131, ACS, Washington, **1980**, Ch. 7, pp. 143–172.

[60] B. M. Furphy, J. M. Harrowfield, D. L. Kepert, B. W. Skelton, A. H. White, F. R. Wilner, *Inorg. Chem.* **1987**, *26*, 4231–4236.

[61] J. M. Harrowfield, M. I. Ogden, W. R. Richmond, A. H. White, *J. Chem. Soc., Dalton Trans.* **1991**, 2153–2160.

[62] See, for example: a) J. M. Harrowfield, M. I. Ogden, W. R. Richmond, B. W. Skelton, A. H. White, *J. Chem. Soc., Perkin Trans 2* **1993**, 2183–2190; b) J. M. Harrowfield, W. R. Richmond, A. N. Sobolev, A. H. White, *J. Chem. Soc., Perkin Trans 2* **1994**, 5–9; c) J. M. Harrowfield, W. R. Richmond, A. N. Sobolev, *J. Incl. Phenom.* **1994**, *19*, 257–276.

[63] S. F. Lincoln, A. E. Merbach, *Adv. Inorg. Chem.* **1995**, *42*, 1–88.

[64] G. A. Lawrance, *Adv. Inorg. Chem.* **1987**, *34*, 145–194.

[65] D. L. Faithfull, J. M. Harrowfield, M. I. Ogden, B. W. Skelton, K. Third, A. H. White, *Aust. J. Chem.* **1992**, *45*, 583–594 and references therein.

[66] A. Bilyk, J. M. Harrowfield, B. W. Skelton, A. H. White, *J. Chem. Soc., Dalton Trans.* **1997**, 4251–4256.

[67] Z. Asfari, J. M. Harrowfield, M. I. Ogden, J. Vicens, A. H. White, *Angew. Chem. Int. Ed. Engl.* **1991**, *30*, 854–856.

[68] a) G. E. Hofmeister, F. E. Hahn, S. F. Pedersen, *J. Am. Chem. Soc.* **1989**, *111*, 2318–2319; b) G. E. Hofmeister, E. Alvarado, J. A. Leary, D. I. Yoon, S. F. Pedersen, *J. Am. Chem. Soc.* **1990**, *112*, 8843–8851.

[69] J. L. Atwood, F. Hamada, K. D. Robinson, G. W. Orr, R. L. Vincent, *Nature (London)* **1991**, *349*, 683–684.

[70] A. Drljaca, M. J. Hardie, J. A. Johnson, C. L. Raston, H. R. Webb, *Chem. Commun.* **1999**, 1135–1136.

[71] L. M. Engelhardt, B. M. Furphy, J. M. Harrowfield, D. L. Kepert, A. H. White, F. R. Wilner, *Aust. J. Chem.* **1988**, *41*, 1465–1476.

[72] P. Thuéry, M. Nierlich, J. Vicens, H. Takemura, *J. Chem. Soc., Dalton Trans.* **2000**, 279–283.

[73] C. D. Gutsche, J. M. Harrowfield, X. Delaigue, M. I. Ogden, B. W. Skelton, D. R. Stewart, A. H. White, manuscript in preparation.

[74] B. M. Furphy, J. M. Harrowfield, M. I. Ogden, B. W. Skelton, A. H. White, F. R. Wilner, *J. Chem. Soc., Dalton Trans.* **1989**, 2217–2221.

[75] C. E. Daitch, P. D. Hampton, E. N. Duesler, *Inorg. Chem.* **1995**, *34*, 5641–5645.

[76] A broad overview of the potential applications of materials containing small metal clusters is given in *Science* **1996**, *271*, 920–941.

[77] A. Bilyk, A. K. Hall, J. M. Harrowfield, M. W. Hosseini, B. W. Skelton, A. H. White, *Aust. J. Chem.* **2001**, in press.

[78] a) V. A. Igonin, S. V. Lindeman, Yu. T. Struchkov, O. I. Shchegolikhina, Yu. A. Moldotsova, Yu. A. Pozdniakova, A. A. Zhdanov, *Russ. Chem. Bull.* **1993**, *42*, 168–173; b) M. G. Davidson, J. A. K. Howard, S. Lamb, C. W. Lehmann, *Chem. Commun.* **1997**, 1607–1608.

[79] A. Bilyk, J. Dunlop, A. K. Hall, J. M. Harrowfield, M. W. Hosseini, B. W. Skelton, A. H. White, manuscript in preparation.

[80] R. D. Rogers, C. D. Bauer in *Comprehensive Supramolecular Chemistry, Vol. 1* (Vol. Ed.: G. Gokel), Pergamon, Oxford, **1996**, Ch. 8, pp. 315–355.

[81] For an overview of nitrate ion coordination chemistry, see: B. J. Hathaway in *Comprehensive Coordination Chemistry* (Eds.: G. Wilkinson, R. D. Gillard, J. A. McCleverty), Pergamon, Oxford,

1987, Vol. 2, Ch. 15.5. Particular discussions of the issue may be found in: a) F. A. Cotton, J. G. Bergman, *J. Am. Chem. Soc.* **1964**, *86*, 2941–2942; b) T. J. King, N. Logan, A. Morris, S. C. Wallwork, *J. Chem. Soc., Chem. Commun.* **1971**, 554 and references therein.

[82] a) I. Tabushi, Y. Kobuke, A. Yoshizawa, *J. Am. Chem. Soc.* **1984**, *106*, 2481–2482; b) I. Tabushi, Y. Kobuke, *Israel J. Chem.* **1985**, *25*, 217–227.

[83] K. Suzuki, H. Minami, Y. Yamagata, S. Fujii, K. Tomita, Z. Asfari, J. Vicens, *Acta Crystallogr., Sect. C* **1992**, *48*, 350–352.

[84] a) T. S. Franczyk, K. R. Czerwinski, K. N. Raymond, *J. Am. Chem. Soc.* **1992**, *114*, 8138–8146; b) P. H. Walton, K. N. Raymond, *Inorg. Chim. Acta* **1995**, *240*, 593–601.

[85] J. L. Atwood, D. L. Clark, R. K. Juneja, G. W. Orr, K. D. Robinson, R. L. Vincent, *J. Am. Chem. Soc.* **1992**, *114*, 7558–7559.

[86] J.-P. Scharff, M. Mahjoubi, R. Perrin, *New J. Chem.* **1993**, *17*, 793–796.

[87] P. C. Leverd, M. Nierlich, *Eur. J. Inorg. Chem.*, **2000**, 1733–1738.

[88] Solution complexation of uranyl ion by homoazacalixarenes has been described by: H. Takemura, K. Yoshimura, I. U. Khan, T. Shinmyozu, T. Inazu, *Tetrahedron Lett.* **1992**, *33*, 5775–5778.

[89] See the review by B. J. Hathaway quoted in reference 81.

[90] P. G. Allen, J. J. Bucher, D. L. Clark, N. M. Edelstein, S. A. Ekberg, J. W. Gohdes, E. A. Hudson, N. Kaltsoyannis, W. W. Lukens, M. P. Neu, P. D. Palmer, T. Reich, D. K. Shuh, C. D. Tait and B. D. Zwick, *Inorg. Chem.* **1995**, *34*, 4797–4807.

[91] P. Thuéry, M. Nierlich, *J. Chem. Soc., Dalton Trans.* **1997**, 1481–1482.

[92] a) M. Åberg, *Acta Chem. Scand., Sect. A* **1976**, *30*, 507–514; b) A. Perrin, J. Y. Le Marouille, *Acta Crystallogr., Sect. B* **1977**, *33*, 2477–2481; c) A. Perrin, *J. Inorg. Nucl. Chem.* **1977**, *39*, 1169–1172; d) A. J. Zozulin, D. C. Moody, R. R. Ryan, *Inorg. Chem.* **1982**, *21*, 3083–3086; e) A. J. Zozulin, D. C. Moody, R. R. Ryan, *Inorg. Chem.*, **1982**, *21*, 3083–3086; f) G. van den Bossche, M. R. Spirlet, J. Rebizant, J. Goffart, *Acta Crystallogr., Sect. C* **1987**, *43*, 837–839; g) U. Turpeinen, R. Hämäläinen, I. Mutikainen, O. Orama, *Acta Crystallogr., Sect. C* **1996**, *52*, 1169–1171.

[93] a) J. C. Taylor, A. Ekstrom, C. H. Randall, *Inorg. Chem.* **1978**, *17*, 3285–3289; b) D. Rose, Y. D. Chang, Q. Chen, J. Zubieta, *Inorg. Chem.* **1994**, *33*, 5167–5168.

Chapter 31

LUMINESCENT PROBES

NANDA SABBATINI[a], MASSIMO GUARDIGLI[a], ILSE MANET[b], RAYMOND ZIESSEL[c]

[a]*Dipartimento di Scienze, Farmaceutiche dell'Università, Via Belmeloro, 6, 40126 Bologna, Italy.*
[b]*Dipartimento di Chimica Organica "A. Mangini", Via S. Donato, 15, 40127 Bologna, Italy.*
[c]*Laboratoire de Chimie, d'Electronique et de Photonique Moléculaires, ECPM, 25 rue Becquerel, 67087 Strasbourg Cedex 02, France.*
e-mail: ziessel@chimie.u-strasbg.fr

1. Introduction

Molecular devices are defined as structurally organized and functionally integrated chemical systems. The design of such devices requires the selection of molecular components with given properties and suitable for incorporation into an organized system capable of performing the desired function. The components may be photo-, electro-, iono-, magneto-, thermo-, mechano-, or chemoactive depending on whether they respond to photonic, electronic or ionic input, to magnetic field or heat, or undergo a change in mechanical properties or a chemical reaction. If the design of the molecular device involves the incorporation of photoactive components, it is possible that both ground- and excited-state properties of the components change upon formation of the organized system giving rise to novel properties. In particular, if the system exhibits luminescence properties different from those of the separate components, a luminescent molecular device has been developed [1].

In this chapter, two types of luminescent molecular devices will be discussed. The first derives from encapsulation complexes of luminescent lanthanide ions in which a light-conversion process gives rise to intense metal luminescence, and the second from molecular receptors acting as sensors via luminescence properties that change upon substrate recognition, due to the varying efficiency of a photoinduced electron transfer process in competition with the radiative decay process.

Recently reviewed [2,3] developments in the field of supramolecular chemistry relating to devices of the first type concern the design of new molecules capable of including a lanthanide ion, due to a cavity which is complementary in size, shape and binding sites with the ion. If the encapsulating ligand incorporates chromophores it may act as *antenna* by absorbing light and transferring the excitation energy to the emitting metal ion [4,5]. An efficient antenna is expected to lead to lanthanide luminescence much more intense than that obtained upon metal excitation since lanthanide ions are characterized by very low molar absorption coefficients. Among the possible exploitations of lanthanide complexes with encapsulating ligands as luminescent devices, we focus on the use of Eu^{3+} and Tb^{3+} complexes as luminescent labels in bioaffinity assays.

These assays rely on the selective-recognition capability of biomolecules such as antibodies or nucleic acids [6]. Eu^{3+} and Tb^{3+} ions are particularly suitable for this application, because they show intense luminescence in the visible and possess long-lived emitting states. These characteristics render the complexes of these ions particularly important for time-resolved luminescence bioaffinity assays, since they allow a sensitivity enhancement of the assay via minimizing interference of the short-lived, background luminescence in the UV of the biological species and of scattered excitation light [4,5].

For devices of the second type, much research has focused on the design of new molecules incorporating a redoxactive luminophore together with other iono- and redoxactive components. Photoinduced electron transfer involving the luminophore is often observed and competes with the radiative decay process. It is worthwhile pointing out that many factors control the rate of electron transfer and thus the luminescence intensity. Numerous studies have shown that the rate of electron transfer can depend on the molecular conformation since it increases with decreasing separation of the redoxactive components, reaching a maximum when these are in orbital contact [7]. This can be exploited to design a reversible photoionic switch provided the molecular conformation can be modulated by external stimulation. We will describe some examples of such switches in which cation or anion binding modifies the luminescence intensity due to conformational changes.

2. Luminescence of Lanthanide Ions Included in Calixarene-Based Receptors

Complexes of lanthanide ions with neutral, unidentate ligands are usually very labile in aqueous solution. Different behavior is shown by complexes of ligands encapsulating the metal ion, which can show remarkable kinetic inertness in water [2,3]. This property is crucial for the application of lanthanide complexes as luminescent labels in bioaffinity assays, because the label should not exchange the lanthanide ion with the Ca^{2+} and Mg^{2+} ions present in biological materials and should not be affected by the hydroxide ions. Moreover, complexes of encapsulating ligands offer an efficient shielding of the lanthanide ion from interaction with water molecules quenching the metal luminescence [4,5]. The possibility of using calixarenes as encapsulating ligands for lanthanide complexation has been explored considering their capability to complex alkali and alkaline-earth metal ions and the generally similar complexation properties of such ions and those of lanthanide ions. In particular, many calixarene-based ligands have been designed and synthesized in order to obtain luminescent lanthanide complexes. These ligands can be roughly divided into four classes: (i) unmodified calixarenes, (ii) neutral, podand-type functionalized calixarenes, (iii) calixarenes with additional ionisable groups, and (iv) calixcrowns and calixcryptands.

2.1. PHOTOPHYSICAL PROPERTIES

In the following, we present and discuss some photophysical properties of complexes obtained upon encapsulation of luminescent lanthanide ions (mainly Eu^{3+} and Tb^{3+}) by different types of functionalized calixarenes. Metal luminescence properties have been studied mostly for complexes of calix[4]arenes, although a few results are available for complexes of higher calix[n]arenes (n = 5-8). Some properties of the luminescent states of these compounds are gathered in Table 1. Figures 1-3 show schematically the calixarene ligands of the lanthanide complexes examined.

TABLE 1. Lifetimes and quantum yields of the luminescent metal excited states of some calixarene-based Eu^{3+} and Tb^{3+} complexes[a].

Ligand	Eu^{3+} complex Lifetime (ms)	Eu^{3+} complex Quantum yield	Tb^{3+} complex Lifetime (ms)	Tb^{3+} complex Quantum yield	Solvent
1	0.65	0.0002	1.5	0.20	H_2O
4	0.65	0.04	1.9	0.12	CH_3CN
6	1.20	0.03	0.70	0.001	CH_3CN
7	1.60	0.15	0.80	0.002	CH_3CN
8	1.60	0.16	0.80	0.007	CH_3CN
9	0.69	0.19	-[b]	-[b]	CH_3OH
17	0.50	0.0005	1.0	0.01	CH_3CN
18	-	0.003	-	0.002	H_2O
19	-	0.0017	-	0.16	CH_3CN
20	-	0.006	-	0.27	CH_3CN
21	-	0.015	-	0.017	CH_3CN
22	0.65	<0.01	1.4	0.02	CH_3CN
23	-	-	-	0.13	H_2O
24	1.17	0.017	1.11	0.16	CH_3CN
25	-	<0.001	0.76	0.07	CH_3CN
28	-	-	-	0.02	H_2O
29	-	-	-	0.20	H_2O
31	-[b]	-[b]	0.21	0.05	THF
32	0.68	0.11	-[b]	-[b]	CH_3OH
33	-[b]	-[b]	3.30	0.15	CH_3CN
34	0.65	0.0002	1.85	0.01	CH_3CN
35	0.95	0.18	1.85	0.32	CH_3CN
36	1.38	0.32	1.88	0.35	CH_3CN
37	1.24	0.23	1.86	0.39	CH_3CN
38	1.29	0.28	1.83	0.37	CH_3CN

(a) The Table refers only to complexes for which quantum yield values are available.
(b) No metal luminescence was observed.

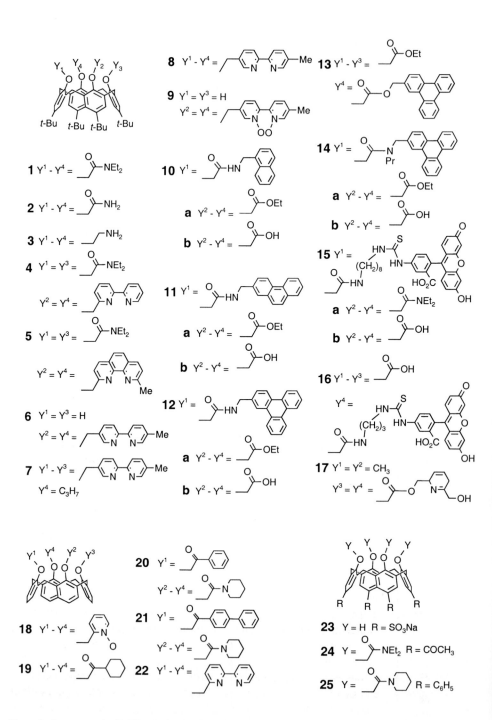

Figure 1. Structures of calix[4]arene ligands used for lanthanide ion complexation.

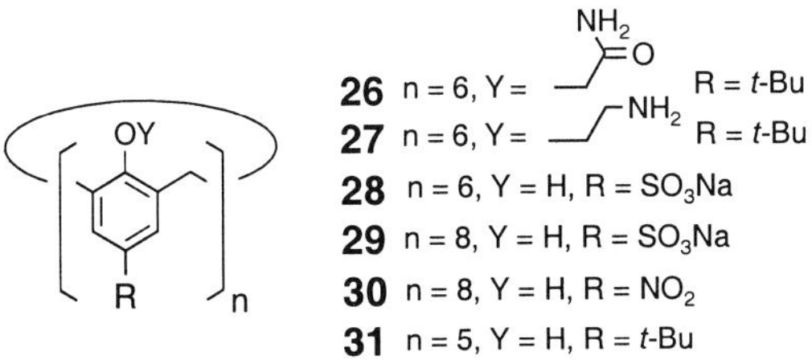

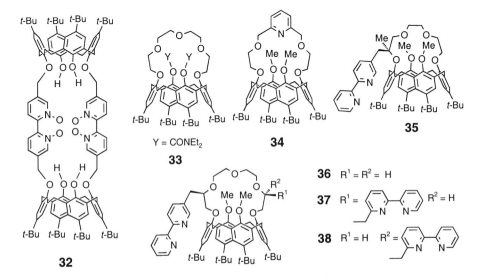

Figure 2. Structures of calix[n]arene ligands (n > 4) used for lanthanide ion complexation.

Figure 3. Structures of calixcrowns and calixcryptands used for lanthanide ion complexation.

The first Eu^{3+} and Tb^{3+} complexes of a calixarene-derived ligand to be studied were the complexes of the *p-tert*-butylcalix[4]arene tetraacetamide ligand **1**. Eu**1** and Tb**1** were water soluble, unlike the free ligand and the K^+ complex, and their photophysics was studied in water [8]. The absorption spectra of the complexes show two bands due to ligand-centred transitions (λ_{max} = 273 and 282 nm, λ_{max} ~ 1100 $M^{-1}cm^{-1}$). Comparison of the absorption and metal luminescence excitation spectra indicated that ligand-to-metal energy transfer from the calix[4]arene moiety to the metal ion takes place. About one water molecule was found to coordinate the metal ion, showing that this ligand efficiently shields the metal ion from solvent molecules (as is known, the Eu^{3+} and Tb^{3+} aquo ions coordinate about 9 to 10 water molecules). The lifetime of Tb**1** in D_2O showed no temperature dependence. This observation has been explained considering that the energy of the lowest ligand triplet excited state, which is about 4000 cm^{-1} higher than that of the metal emitting state, renders metal-to-ligand back energy transfer inefficient. The quantum yields of Tb**1** are very high (0.20 in H_2O), as a result of the low efficiency of nonradiative deactivation processes of the metal emitting state and the efficient ligand-to-metal energy transfer. In fact, the efficiency of the energy transfer from the ligand singlet excited state to the metal emitting state is 0.35. For Eu1, the lifetimes in H_2O and D_2O showed that vibronic coupling plays a role in the deactivation of the metal emitting state. The lifetime values in the deuterated solvent show that thermally activated nonradiative decay processes of the Eu^{3+} emitting state are not efficient. The rate constant of the radiative decay process is 500 s^{-1}, a rather low value compared to other encapsulation complexes of the Eu^{3+} ion that are characterised by values ranging up to 1000 s^{-1}. The quantum yield has a very low value, attributed to inefficient ligand-to-metal energy transfer due to deactivation of the ligand excited states via ligand-to-metal charge-transfer LMCT excited states.

The Tb^{3+} complexes of ligands 2, 3, 26, 27 were strongly luminescent in methanol, while the Eu^{3+} complexes showed only weak emission [9], though no metal luminescence quantum yields were reported. For all the Tb^{3+} complexes a biexponential decay of the metal luminescence has been observed and has been attributed to the presence of short-lived ligand luminescence and long-lived metal luminescence. In all cases addition of water caused quenching of the luminescence, this process being less efficient for complexes of ligands 26 and 27 based on calix[6]arenes. This effect has been explained considering that the larger, more flexible calix[6]arenes encapsulate the metal ion more efficiently.

With Tb^{3+} in water, ligands 28 and 29 and ligand 23 gave 1:1 and 2:1 (ligand:metal) species, respectively [10]. Comparison of the metal luminescence excitation spectra with the absorption spectra indicated that metal emission followed energy transfer from the phenol moiety of the ligand to the metal ion. The luminescence intensities were pH-dependent and reached the highest values at pH>11. Interestingly, quite high quantum yields, 0.13 and 0.20, were obtained upon excitation in the ligand-centered bands for Tb23 and Tb29, respectively. The rather low quantum yield value, 0.02, for Tb28 has been attributed to the worse shielding ability of this ligand. For Tb23 and Tb28, note that the quantum yields were determined in the presence of a ligand excess which affects the efficiencies of light absorption by the complex and therefore also the quantum yield.

Considering that calixarenes have rather low molar absorption coefficients, more efficient chromophores were attached to calixarene moieties in order to obtain strongly absorbing ligands for Eu^{3+} and Tb^{3+} complexation. Complexes of ligand 18 containing four pyridine-N-oxide units as chromophores showed metal luminescence upon ligand excitation in methanol [11]. The absorption spectra of the complexes exhibited one broad band with maximum at 263 nm. Compared to the ligand absorption spectrum this band is slightly blue shifted (ca. 400 cm^{-1}). The quantum yields in methanol were of the order of 10^{-3} for both Eu18 and Tb18. The metal luminescence was completely quenched upon addition of water, suggesting the lability of the complexes in this solvent.

The Eu^{3+} and Tb^{3+} complexes of ligands 20, 21 containing phenyl and diphenyl groups as chromophores have 1:1 stoichiometries in both acetonitrile and methanol [12]. The absorption and metal luminescence excitation spectra were similar, except for Tb21, suggesting that both the phenol groups and the phenyl and biphenyl chromophores are involved in the energy transfer process. This may mean that in the case of Tb21 energy transfer occurs mainly from the phenol units rather than from the biphenyl chromophore. The metal luminescence quantum yields upon ligand excitation were high for Tb19 and Tb20, and much lower for Tb22, where the ligand contains the biphenyl chromophore.

For Eu^{3+}, the highest quantum yields, *ca.* 0.06, were obtained for the complexes of ligands **20** and **21** in CH_3CN.

A 1:1 stoichiometry was found for the Eu^{3+} and Tb^{3+} complexes of the ligands **24** and **25** [13]. The absorption and metal luminescence excitation spectra were similar and confirmed the energy transfer from the ligand to the metal ion. The quantum yields upon ligand excitation in acetonitrile were low for the Eu^{3+} complexes, while quite high values, 0.16 and 0.07, were found for Tb**24** and Tb**25**, respectively. It should be noticed that the quantum yields were measured using as standards the Tb^{3+} and Eu^{3+} aquo ions excited in the metal-centred bands at 376 nm and 394 nm, respectively. Therefore, considering that the quantum yield of the standard should be obtained at the excitation wavelength used for the sample, one has to be cautious with the values obtained for the complexes of ligands **19-25**.

Several calix[4]arenes carrying the 2,2'-bipyridine (bpy) chromophore have been synthesised. Ligands **4** and **22** containing two bpy and two amide or four bpy units, respectively, with the bpy units ligated to the calixarene moiety through C^6, were obtained. These ligands gave 1:1 complexes with Eu^{3+} and Tb^{3+} in acetonitrile with stability constants of the order of 10^5 M^{-1}. The photophysical properties of these complexes were studied in solutions containing an excess of the Eu^{3+} or Tb^{3+} salts [14, 15]. The absorption spectra of the ligands were characterized by intense bands in the UV with maxima at 290 nm due to bpy absorption. Upon complexation of these ligands, the absorption bands were red shifted with maxima at 305 nm and the molar absorption coefficients decreased. Comparison of the absorption and the metal luminescence excitation spectra indicated that energy transfer from the bpy moiety to the metal ion takes place. Good quantum yields were obtained for the complexes of ligand **4**. The lower quantum yields of the complexes of ligand **22** were attributed to a less efficient ligand-to-metal energy transfer due to steric hindrance the four bpy units may undergo when approaching the metal ion.

No metal luminescence upon ligand excitation was found in acetonitrile solutions containing the Eu^{3+} or Tb^{3+} ions and ligand **5** [15]. Ligands **6-8** in which the bpy units are linked to the calixarene moiety through C^5 were synthesized and their Eu^{3+} and Tb^{3+} complexes were isolated [16]. The absorption spectra of ligands **6-8** in anhydrous acetonitrile [14] show the typical absorption bands of the bpy chromophore with maxima at 288 nm and the molar absorption coefficients are proportional to the number of bpy units in the ligand. Complexation of these ligands caused a red shift of the absorption bands and a decrease of the molar absorption coefficients. For all complexes metal luminescence upon bpy excitation was observed. The Eu^{3+} complexes showed longer lifetimes and higher quantum yields than the Tb^{3+} complexes. Interestingly, Eu**7** and Eu**8** gave particularly high quantum yields. The very low quantum yields of the Tb^{3+} complexes have been attributed to nonradiative deactivation of the metal emitting state via ligand excited states.

Recently, the podand **9** and the barrel-shaped cryptand **32** based on calix[4]arenes bearing $bpyO_2$ chromophores were shown to form complexes with Eu^{3+} and Tb^{3+} ions stable in methanol and water [17]. The better complexation properties of ligand **9** compared to ligand **6** were explained by the better complexation abilities of $bpyO_2$ with respect to bpy. The photophysical properties were studied in methanol since the complexes were barely soluble in water. The absorption spectra of the complexes of these two ligands showed a band with maximum at 260 nm and a broad shoulder at ca. 290 nm. The metal excitation spectra of Eu**9** and Eu**32** were similar to the absorption spectra indicating ligand-to-metal energy transfer from the bpyO$_2$ units. Comparison of the lifetime in hydrogenated and deuterated solvents indicated that no vibronic coupling due to solvent molecules occurs. A small contribution to nonradiative deactivation of the metal excited state to the ground state was attributed to vibronic coupling involving the OH oscillators of the ligand. Nonradiative deactivation via thermally activated back-energy transfer is also negligible. The authors report high quantum yield values for Eu**9** and Eu**32**, and estimate that the efficiencies of the ligand-to-metal energy transfer are 0.19 and 0.11 for Eu**9** and Eu**32**, respectively, suggesting that these values may represent lower limits for this quantity. The conformational rigidity of ligand **32** with respect to ligand **9** may account for the different efficiencies of energy transfer. The Tb^{3+} complexes did not show metal luminescence at room temperature probably because of an efficient back-energy transfer to low-lying ligand excited states, as observed also for other complexes containing the bpyO$_2$ chromophore.

The calix[4]arene ligands **10-14** contain one naphthalene, phenanthrene or triphenylene chromophore and three carboxylate and carboxylic ester functions. The Eu^{3+} and Tb^{3+} complexes of this series of ligands are not soluble in water and their photophysical properties were studied in methanol [18]. Metal luminescence was observed upon excitation of the chromophore, and compared with the metal luminescence obtained by direct excitation in the metal-centred bands taking in consideration corrections necessary for the differences in intensity of the excitation light as a function of wavelength. Metal luminescence intensities clearly increase upon excitation in the ligand compared to excitation in metal-centered states. For both Eu^{3+} and Tb^{3+} complexes the strongest increase is observed for the ligands incorporating the triphenylene chromophore. The Eu^{3+} and Tb^{3+} complexes of ligands **13**, **14a** and **12b**, **14b** were characterized by short lifetimes in methanol. Deaeration of the solutions did not change the lifetime values of the metal emitting states but led to an increase of the intensity of the metal luminescence. This behaviour was taken to indicate [18] that ligand-to-metal energy transfer is so slow that the ligand excited states can be quenched by oxygen.

Recently, calixarenes bearing three diethylamide functions or three carboxylate groups and a fluorescein chromophore were used for the complexation of Nd^{3+} and Er^{3+} [19]. The metal luminescence excitation spectra for the Nd^{3+} complexes of ligands **15b** and **16** in DMSO and [D$_4$]methanol and ligand **15a** in acetonitrile are similar to the absorption spectrum of the ligand indicating ligand-to-metal energy transfer. The metal luminescence intensity upon excitation in the fluorescein band at 500 nm is higher than the metal luminescence intensity upon excitation in the calixarene band at 287 nm. Due to the poor solubility of the Er^{3+} complexes in [D$_4$]methanol, the excitation spectra could be recorded only in DMSO. The metal excitation spectra of Er**15b** and Er**16** have evidenced the absorption bands of the fluorescein chromophore. Excitation in the calixarene absorption bands did not lead to metal luminescence. According to the authors this proves that sensitized emission is due only to the fluorescein. All the complexes show emission originating both in fluorescein-centred excited states and metal excited states. The authors claim that the lack of ^{1}O$_2$ infrared emission at 1300 nm upon ligand excitation in both Nd^{3+} and Er^{3+} complexes indicates that the intramolecular ligand-to-metal energy transfer is fast. The lifetimes of Nd**15b** and Nd**16**, analogously to those of Er**15b** and Er**16**, were equal within the experimental error and in both cases longer in DMSO than in [D$_4$]methanol.

Eu^{3+} and Tb^{3+} complexes of calixcrowns **33** and **34** and of ligand **17** having a 1:1 stoichiometry were obtained in acetonitrile [20]. Similarity of the metal luminescence excitation spectra and the absorption spectra showed that ligand-to-metal energy transfer occurs. Tb**33** showed a long lifetime and quite a high quantum yield. Considering the energy of the lowest ligand triplet excited state, 26700 cm^{-1}, it has been suggested that thermally activated nonradiative decay is inefficient. Eu**33** was luminescent neither at 300 K nor at 77 K. This behavior has been attributed to efficient nonradiative decay via LMCT excited states probably involving the ether oxygens. The quantum yields of Tb**34** and Tb**17** were one order of magnitude lower than that of Tb**33**. Considering that the lowest ligand triplet excited state lies at 24 000 cm^{-1}, it has been suggested that thermally activated nonradiative decay is most likely inefficient and the lower quantum yields may be due to little efficient ligand-to-metal-energy transfer. Eu**34** and Eu**17** gave low lifetimes and very low quantum yields. This has been attributed to nonradiative deactivations of the ligand and metal excited states via low-lying LMCT excited states involving the ether oxygens.

In order to increase the molar absorption coefficients of the calixcrown **35**, the bpy chromophore was incorporated in ligands **36-38**. In acetonitrile, 1:1 complexes form with Eu^{3+} and Tb^{3+} [21]. The absorption spectra of the ligands and the complexes were characterised by intense bands in the UV due to bpy absorption. For all complexes, metal luminescence upon bpy excitation was observed. The Tb^{3+} complexes of the ligands **36-38** were characterised by rather long lifetimes and high quantum yields (see Table 1). These values suggested that thermally activated nonradiative decay via ligand triplet excited states is inefficient, consistent with the energy of the lowest ligand triplet excited states ranging from 22500 cm^{-1} to 24000 cm^{-1}. For Eu^{3+} complexes of the ligands **36-38**, quite long lifetimes and high quantum yields were measured. The lower quantum yield of Eu**38** has been ascribed to the smaller crown rendering the ion more exposed to water molecules present in the Eu^{3+} salt used for titration or in the solvent.

Ligand **30**, *p*-nitrocalix[8]arene, was synthesised in order to study the influence of the substitution of the *tert*-butyl groups of the *p-tert*-butylcalix[8]arene by the stronger electron-acceptor nitro groups on the photophysical properties of the homo- and heteronuclear lanthanide ion complexes of this ligand [22]. This ligand formed binuclear complexes with the lanthanide ions upon sixfold deprotonation of the ligand with N(Et)$_3$. The fully protonated ligand **30** exhibited a broad absorption band with maximum at 315 nm and molar absorption coefficient at this wavelength equal to 80000 M^{-1} cm^{-1} in DMF. Complexation resulted in a red shift of this band and a twofold increase of the molar absorption coefficients. Of the homodinuclear complexes with Lu^{3+}, Tb^{3+}, Gd^{3+} and Eu^{3+}, only the Eu^{3+} complex exhibited metal luminescence upon ligand excitation. Rather short lifetimes were obtained for this complex. The absence of Tb^{3+} luminescence has been attributed to efficient deactivation of the metal emitting state via isoenergetic ligand-centered excited states. For the heterodinuclear complexes (Eu,Gd)⊂**30** and (Eu,Tb)⊂**30**, only Eu^{3+} luminescence was recorded upon ligand excitation. The heterodinuclear complexes (Eu,Nd)⊂**30** and (Eu,Ho)⊂**30** were characterised by lower lifetime values of the Eu^{3+} emitting state because of quenching by energy transfer to the other lanthanide ion.

Recently, *p-tert*-butylcalix[5]arene, **31**, was studied for complexation of lanthanide ions [23]. Dimeric complexes of the threefold deprotonated ligand with Eu^{3+}, Tb^{3+} and Gd^{3+} could be isolated from THF in the presence of the base NaH. Excitation of the ligand-centred excited states resulted in ligand-centred emission with maximum at 320 nm and metal luminescence was observed only in the case of Tb**31**, for which a quantum yield of 0.05 was obtained in THF. The quenching of the Eu^{3+} luminescence has been attributed to LMCT states lying at low energy.

The intensity of the metal luminescence obtained upon ligand excitation is the photophysical property of main interest in this research dealing with the antenna effect in lanthanide complexes. This quantity is evaluated on basis of the product of the metal luminescence quantum yield upon excitation at a certain wavelength and the molar absorption coefficient of the ligand in the complex at the same wavelength. For the complexes treated above, introduction of chromophores in the calixarene moiety leads to intense absorption, the molar extinction coefficients ranging from 15000 to 50000 $M^{-1}cm^{-1}$. Anyway, the factor affecting mainly the luminescence intensity is the metal luminescence quantum yield, which is of the order of 10^{-3}-10^{-4} for complexes showing very weak luminescence and 0.5 for the most luminescent ones.

2.2. APPLICATIONS

Finally, we illustrate the most important areas where the luminescence of lanthanide complexes in solution has been successfully used. One of these concerns the analysis of biological substances by immunological and nucleic acid hybridization assays involving the reaction between specific immunoreagents and between complementary nucleic acids, respectively. These methods characterized by high sensitivity and specificity have been applied, first, in clinical studies of substances occurring in very low concentrations, whose detection could not be performed with specific chemical methods.

In order to determine the amount of analyte using these methods, a suitable label must be introduced in one of the substrates of the biospecific reaction. Radioisotopes, enzymes and luminescent compounds are the most used labels. Compared to other analytical techniques, those based on luminescent labels are safer and relatively fast and require simple methods of measurements [24,25]. However, an important drawback is the presence of a background luminescence of biological species. For example, background luminescence is a major problem with serum because it exhibits an intense luminescence extending over a wide range of wavelengths (300-600 nm), which overlaps the luminescence spectra of most organic fluorophores. The use of luminescent lanthanide labels instead of luminescent organic compounds allows elimination of this drawback thanks to the large Stokes shift and the long emitting state lifetimes of lanthanides. In fact, measurements can be performed using a time-resolved method, which introduces a delay time of 200-400 μs between the excitation and the luminescence measurements. In this way, only the lanthanide luminescence is measured because the emitting state lifetimes of the lanthanides are much longer than those of organic molecules. This method is characterized by very high sensitivity, even higher than that obtained with radioisotopes.

Immunoassays are classified according to several criteria, one of which is based on the presence of a separation step of the immunocomplex formed from excess reagents, thus defining heterogeneous or homogeneous immunoassays, respectively [26]. Up to now, heterogeneous immunoassays have been used more because of their high sensitivity, which is achieved via the physical separation of the fraction carrying the labeled immunocomplex from other interfering biological species present in the sample. Interestingly, a related system based on a luminescent lanthanide cryptate suitable for homogeneous immunoassays has been recently commercialized. For example, this system has been applied in the quantitative determination of the antigen prolactin [27]. As mentioned before, luminescent lanthanide complexes have been also used for the examination and determination of nucleic acids in nucleic acid hybridization assays [24, 28-30]. Whereas in immunological assays the detection reagent is usually an antibody, in the nucleic acid hybridization assays the reagent is a specific sequence of DNA or RNA nucleotides labelled with a suitable lanthanide complex. This technique can be used for the detection of the presence of genetic diseases and pathogenic organisms such as viruses and bacteria. A luminescent Eu^{3+} cryptate has been used as label in DNA hybridization assays based on indirect labelling techniques [31,32]. The DNA probe contains a hapten which is detected by the labelled binding protein specific for this hapten. The property which is exploited is the permanent luminescence of the Eu^{3+} cryptate which allows performance of the assay without an extra step for the formation of the luminescent complex.

Nowadays, much research aims at the development of homogeneous assays because the hybridization kinetics in solution is more favorable and simplified analytical procedures are attained. Homogeneous assays require a special labeling strategy of the DNA probes, in which the signal is generated or altered only when the labelled probe is hybridized to the target. Different approaches to homogeneous assays for the detection of mutations in DNA sequences via time-resolved luminescence measurements of Eu^{3+} and Tb^{3+} complexes are reported in Refs. [33-35]. Luminescent lanthanide labels have also been applied in luminescence imaging techniques for the localization of proteins and nucleic acid sequences in tissue sections, cells and chromosomes [36]. In fluorescence imaging using conventional organic fluorophores, the sensitivity is sometimes limited by the background signal emitted by the organic compounds of the biological sample itself. The use of time-resolved imaging techniques in conjunction with suitable luminescent lanthanide labels can thus allow a gain in sensitivity.

3. Photoinduced Electron Transfer in Cation and Anion Receptors

The methodology for oligopyridine and calix[4]arene derivatization allows the incorporation within a new molecular device of redox- and photoactive transition metal fragments [16,37]. Compounds **39** and **40** (Figure 4) were built as potential sensors for cations and consist of three discrete components; namely, (i) an electron donor luminophore, tris(2,2'-bipyridine)ruthenium(II) or fac-chlorotricarbonyl-(2,2'-bipyridine) rhenium(I) moieties, respectively, (ii) an electron acceptor, two quinone moieties, and (iii) a chelator, a pendant 2,2'-bipyridine (bpy) unit, joined together by a calix[4]-diquinone platform. NMR studies made with **39** and **40** indicate that the quinoid walls of the macrocyclic receptor rotate through the calixarene annulus, a common situation for calix[4]diquinones [38]. The phenolic compounds **41** and **42** in a *cone* conformation, act as reference compounds because they contain the same luminophore but lack the electron acceptor quinones. The rhenium derivatives, being neutral complexes, are interesting in analysis of the importance of electrostatic interaction upon cation binding.

Excitation of **41** and **42** in deoxygenated acetonitrile gave rise to luminescence from triplet MLCT excited states with maximum around 600 nm characterized by lifetimes of 1100 and 45 ns and luminescence quantum yields of 0.085 and 0.003, respectively. As expected, very efficient luminescence quenching in **39** and **40** is observed which has been attributed to intramolecular photoinduced electron transfer from the d-metal centers to the nearby quinone. Molecular dynamics simulations made for **39** in acetonitrile solution indicate that the coordinated bpy ligands are in orbital contact with the quinoid walls of the macrocyclic receptor, the latter existing in a *partial cone* or a preferential *1,3-alternate* conformation [39]. The luminescence lifetimes are 6 and 5 ns for compounds **39** and **40** respectively, and a similar luminescence quantum yield of 4×10^{-4} has been determined.

Addition of an inorganic salt, such as $KClO_4$ or $Ba(ClO_4)_2$, to a deoxygenated acetonitrile solution of the phenolic compounds **41** and **42** had no observable effect on the photophysical properties of the pendent metal complexes. However, for the calix[4]-diquinone-derived rhenium(I) complex **40**, addition of salt caused almost complete luminescence quenching and in the presence of excess $Ba(ClO_4)_2$ the triplet lifetime was reduced from 5 ns to 85 ps. This has been attributed to photoinduced electron transfer from the triplet excited state of the rhenium(I) complex to a quinone, favoured by the barium cations because of a substantial increase in the thermodynamic driving force. The rate of intramolecular electron transfer ($k_{ET} = 1.2 \times 10^{10}$ s^{-1}) is increased 70-fold upon binding the cation.

Addition of different types of cations to a solution of **39** in deoxygenated acetonitrile caused a progressive increase in luminescence quantum yield to a plateau value, which approached but never quite reached that measured for the phenolic derivative.

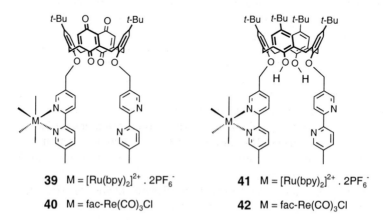

Figure 4. Structures of some luminescent calix[4]arene cation receptors.

Cation complexation causing luminescence intensity enhancement from the appended tris(2,2'-bipyridine)ruthenium(II) subunit is in marked contrast with the situation found for the rhenium derivative **40**. The inorganic monocations are characterised by binding constants of the order of 100 M^{-1} and by triplet lifetimes of *ca*. 160 ns, the dications form less stable complexes with stability constants of the order of 10 M^{-1} possessing triplet lifetimes around 700 ns, while the trications form weak complexes but with triplet lifetimes around 1000 ns. The stability constant is controlled, to a large extent, by electrostatic repulsion associated with the appended tris(2,2'-bipyridine)ruthenium(II) subunit bearing two positive charges. This is illustrated by the following stability constants for complexation of barium perchlorate to **39** (K = 16 ± 3 M^{-1}) and to **40** (K = 2810 ± 20 M^{-1}). Detailed NMR studies and molecular dynamics simulations made with the barium perchlorate complex indicate that the cation binds to the lower rim of the calix[4]diquinone receptor [39] where it is held by coordination to the four oxygen atoms and two nitrogen atoms provided by the free bpy. A schematic representation of the process is given in Figure 5.

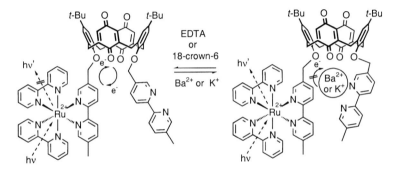

Figure 5. Scheme of the processes determining the luminescent response to cations in ligand 39.

The bound cation forces the macrocycle to adopt a *cone* conformation and repels the appended tris(2,2'-bipyridine)ruthenium(II) moiety. This has the effect of minimizing orbital contact between chromophore and quinone to such an extent that electron transfer is markedly inhibited and the luminescence of the chromophore is restored. Consequently, the luminescence intensity enhancement that accompanies cation binding to **39** compared to **40** can be attributed to an electrostatically-driven conformational change that serves to separate the redox-active components. This effect occurs in addition to restriction of fluctuational motions of the quinoid walls known to take place upon cation complexation [40-42]. Additionally, with barium perchlorate it has been demonstrated that addition of excess 18-crown-6, this being a more avid cation complexer, extinguished luminescence from (**39**)⊂Ba^{2+} due to extraction of the cation, exhibiting the reversibility of all the processes.

In summary, the most interesting aspect of these studies concerns the remarkable effect that binding a cation to an electron acceptor has on the luminescence quantum yield due to changes in photoinduced electron transfer rate. The same cation can either amplify or attenuate electron transfer, depending on whether or not the electron donor is charged. With a neutral rhenium(I) complex as donor, coordination of a cation increases the thermodynamic driving force for light-induced electron transfer to the quinone walls of the receptor. Replacing the rhenium(I) donor with a dicationic ruthenium(II) subunit completely changes the behavior since the rate of electron transfer now decreases due to a conformational change that accompanies coordination of a cation. In this latter system the adventitious cation forces the partners apart by about 5Å which is sufficient to close-down the electron-transfer pathway. Furthermore, the magnitude of the effect depends only on the charge of the cation.

Recognition of halide, dihydrogen phosphate and hydrogen sulphate anions by tris(2,2'-bipyridine)ruthenium(II)-functionalized calix[4]arenes has also been studied via fluorescence measurements [43]. The binding of the anion causes a blue shift of the ruthenium MLCT emission band and a significant increase of the emission intensity, which may be a consequence of the bound anion rigidifying the receptor, and inhibiting the non-radiative decay processes. Ruthenium(II) and rhenium(I) bipyridine calix[4]diquinone receptors have shown to selectively bind acetate anions, and the recognition process has been detected via effects on luminescence intensity originating in photoinduced electron transfer [44]; such electron transfer processes are very similar to those described in the complexes of ligand **39** and in Ref. [39].

Calix[4]arene-linked tris(2,2'-bipyridine)ruthenium(II) complexes can act as luminescent pH sensors [45], whose sensing action relies on the equilibrium between phenolic and phenolate groups in the calixarene moiety. Formation of phenolate anion(s) causes photoinduced intramolecular electron transfer to take place from the phenoxide ion to the tris(2,2'-bipyridine)ruthenium(II) moiety, thus quenching the luminescence. Emission is restored by protonation of the phenolate groups. Photoinduced electron transfer has been observed in a non-covalent assembly between a calix[4]arene-substituted Zn(II) metallo-porphyrin and an adventitious benzoquinone substrate [46]. Recognition of benzoquinone by the calixarene moiety determines a decrease in the fluorescence intensity of the metallo-porphyrin, attributed to a singlet-singlet electron transfer to benzoquinone, even if it is not clear whether electron transfer occurs through bonds, through solvent or through both pathways.

4. References and Notes

[1]   J.-M. Lehn, *Supramolecular Chemistry*, VCH, Weinheim, 1995.
[2]   J.-M. Lehn, *Science*, **1985**, *227*, 849-856.
[3]   J.-M. Lehn, *Angew. Chem., Int. Ed. Engl.* **1988**, *27*, 89-112.
[4]   N. Sabbatini, M. Guardigli, I. Manet, in K. A. Gschneider, L. Eyring, Eds., *Handbook on the Physics and Chemistry of Rare Earths*, Elsevier Science Publishers B.V., Amsterdam, 1996, *Vol. 23*, pp. 69-119.
[5]   N. Sabbatini, M. Guardigli, I. Manet, in D.C. Neckers, D.H. Volman, G. von Bunau, Eds., *Advances in Photochemistry*, Wiley, New York, 1997, *Vol. 23*, p. 213-278.
[6]   P. G. Sammes, G. Yahioglu, (1996) *Natural Products Report*, 1-28.
[7]   G. L. Closs, J. R. Miller, *Science* **1988**, *240*, 440-447.
[8]   N. Sabbatini, M. Guardigli, A. Mecati, V. Balzani, R. Ungaro, E. Ghidini, A. Casnati, A. Pochini, *J. Chem. Soc., Chem. Commun.* **1990**, 878-879.
[9]   E. M. Georgiev, J. Clymire, G. L. McPherson, D. M. Roundhill, *Inorg. Chim. Acta* **1994**, *227*, 293-296.
[10] N. Sato, M. Goto, S. Matsumoto, S. Shinkai, *Tetrahedron Lett.* **1993**, *34*, 4847-4850.

[11] S. Pappalardo, F. Bottino, L. Giunta, M. Pietraszkiewicz, J. Karpiuk, *J. Incl. Phenom.* **1991**, *10*, 387-392.
[12] N. Sato, S. Shinkai, *J. Chem. Soc., Perkin Trans. 2* **1993**, 621-624.
[13] N. Matsumoto, S. Shinkai, *Chem. Lett.* **1994**, 901-904.
[14] N. Sabbatini, M. Guardigli, I. Manet, R. Ungaro, A. Casnati, C. Fischer, R. Ziessel, G. Ulrich, *New J. Chem.* **1995**, *19*, 137-140.
[15] A. Casnati, C. Fischer, M. Guardigli, A Isernia,. I. Manet, N. Sabbatini, R. Ungaro, *J. Chem. Soc., Perkin Trans. 2* **1996**, 395-399.
[16] G. Ulrich, R. Ziessel, *Tetrahedron Lett.* **1994**, *35*, 6299-6302.
[17] L. Prodi, S. Pivari, F. Bolletta, M. Hissler, R. Ziessel, *Eur. J. Inorg. Chem.* **1998**, 1959-1965.
[18] F. J. Steemers, W. Verboom, D. N. Reinhoudt, E.B. Van der Tol, J. Verhoeven, *J. Am. Chem. Soc.* **1995**, *117*, 9408-9414.
[19] M. P. Oude Wolbers, F. C. J. M. Van Veggel, F. G. A. Peters, E. S. E. Van Beelen, J. W. Hofstraat, F. A. J. Geurts, D. N. Reinhoudt, *Chem. Eur. J.* **1998**, *4*, 772-780.
[20] N. Sabbatini, A. Casnati, C. Fischer, R. Girardini, M. Guardigli, I. Manet, G. Sarti, R. Ungaro *Inorg. Chim. Acta* **1996**, *252*, 19-24.
[21] C. Fisher, G. Sarti, A. Casnati, B. Carrettoni, I. Manet, R. Schuurman, M. Guardigli, N. Sabbatini, R. Ungaro, *Chem. Eur. J.* **2000**, *6*, 1026-1034.
[22] J.C.G. Bünzli, F. Ihringer, *Inorg. Chim. Acta* **1996**, *246*, 195-205.
[23] L.J. Charbonniere, C. Balsiger, K.J. Schenk, J.C.G. Bünzli, *J. Chem. Soc., Dalton Trans.* **1998**, 505-510.
[24] J-C. G. Bünzli in *Lanthanide Probes in Life, Chemical and Earth Sciences: Theory and Practice*, J-C. G. Bünzli, G. R. Choppin (Eds.), Elsevier, Amsterdam, 1989, Ch. 7, pp. 219-290.
[25] A. Mayer, S. Neuenhofer, *Angew. Chem., Int. Ed. Engl.* **1994**, *33*, 1044-1072.
[26] a) I. A. Hemmilä, *J. Alloys Compounds* **1995**, *225*, 480-485; b) I. A. Hemmilä, in *Chemical Analysis*, J.D. Winefordner, I.M. Kolthoff (Eds.), Wiley, G New York, **1991**, Vol. 117.
[27] G. Mathis, *Clin. Chem.* **1993**, *39*, 1953-1959.
[28] E. Soini, *Trends Anal. Chem.* **1990**, *9*, 90-108.
[29] E. P. Diamandis, T. K. Christopoulos, *Anal. Chem.* **1990**, *62*, 1149A-1157A.
[30] P. Hurskainen, *J. Alloys Compounds* **1995**, *225*, 489-491.
[31] O. Prat, E. Lopez, G. Mathis, *Anal. Biochem.* **1991**, *195*, 283-289.
[32] E. Lopez, C. Chypre, B. Alpha, G. Mathis, *Clin. Chem.* **1993**, *39*, 196-201.
[33] A. Oser, G. Valet, *Angew. Chem., Int. Ed. Engl.* **1990**, *29*, 1167-1169.
[34] J. Coates, P. G. Sammes, G. Yahioglu, R. M. West A. J. Garman, *J. Chem. Soc., Chem. Commun.* **1994**, 2311-2312.
[35] J. Coates, P. G. Sammes, R. M. West, A. J. Garman, *J. Chem. Soc., Chem. Commun.* **1995**, 1107-1108.
[36] L. Seveus, M. Väisälä, S. Syrjänen, M. Sandberg, A. Kuusisto, R. Harju, J. Salo, I. Hemmilä, H. Kojola, E. Soini, *Cytometry* **1992**, *13*, 329-338.
[37] G. Ulrich, R. Ziessel, I. Manet, M. Guardigli, N. Sabbbatini, F. Fraternali, G. Wipff, *Chem. Eur. J.* **1997**, *3*, 1815-1822.
[38] P. D. Beer, P. A. Gale, Z. Chen, M. G. B. Drew, J. A. Heath, M. I. Ogden, H. R. Powell, *Inorg. Chem.* **1997**, *36*, 5880-5893.
[39] A. Harriman, M. Hissler, P. Jost, G. Wipff, R. Ziessel, *J. Am. Chem. Soc.* **1999**, *121*, 14-27.
[40] S. Shinkai, S. Araki, M. Kubota, T. Arimura, T. Matsuda, *J. Org. Chem.* **1991**, *56*, 295-300.
[41] J. Blixit, C. Detellier, *J. Am. Chem. Soc.* **1995**, *117*, 8536-8540.
[42] M. Gomez-Kaifer, P. A Reddy,. C. D. Gutsche, L. Eschegoyen, *J. Am. Chem. Soc.* **1997**, *119*, 5222-5229.
[43] P. D. Beer, *Chem. Commun* **1996**, 689-696.
[44] P. D. Beer, V. Timoshenko, M. Maestri, P. Passaniti, V. Balzani, *Chem. Commun.* **1999**, 1755-1756.
[45] R. Grigg, J. M. Holmes, S. K. Jones, W. D. J. Amilaprasadh Norbert, *J. Chem. Soc., Chem. Commun.* **1994**, 185-187.
[46] T. Arimura, C. T. Brown, S. L. Springs, J. L. Sessler, *Chem. Commun.* **1996**, 2293-2294.

Chapter 32

TURNING IONOPHORES INTO CHROMO- AND FLUORO-IONOPHORES

RAINER LUDWIG

Freie Universität Berlin,
Institut für Chemie/Anorg. u. Analyt. Chemie
Fabeckstr. 34 -36, 14195 Berlin, Germany.
e-mail: rludwig@mail.chemie.fu-berlin.de

1. Introduction

This chapter concerns the use of selective, calixarene-based ligands for optical signalling in chemical analysis [1]. Although chromophoric ligands of many types are well-known, calixarenes provide a number of novel and useful examples. Exploitation *via* optical sensing of the recognition of both ions and neutral molecules by macrocycles in general is based on their ease of application in, for example, flow-cells or optical fibre tips [2-7]. Macrocyclic chromophores or fluorophores may be employed in devices of various levels of sophistication, nonetheless [7]. A simple system for cation detection, discussed in Section 2, uses an auxiliary H^+-sensitive chromophore or fluorophore mixed in the sensing membrane with a neutral ligand. A newer approach is to have the chromophoric group as part of the ligand, with guest binding causing either a bathochromic or hypsochromic shift of the absorption (Section 3). The high sensitivity of fluorescence detection is one of the reasons for the design of calixarenes bearing fluorophoric groups (Section 4), the binding response being an increase or decrease in the fluorescence emission, a change in the monomer/excimer emission ratio, or a change in the emission wavelength.

2. Optical Sensing with Auxiliary Chromophores and Fluorophores

2.1. COLORIMETRY

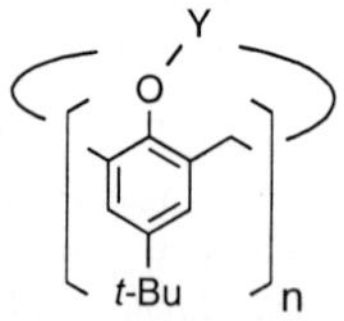

An advantage of the use of auxiliary chromophores and fluorophores is that the ligand selectivity remains as established by independent direct measurements. The sensor function depends on the electroneutrality of the membrane phase, where deprotonation of the auxiliary dye (*e.g.*, a Nile Blue derivative) accompanies uptake of a cation. Consequently, the method is restricted to ionic analytes. Examples are the uptake of Na^+ and K^+ from aqueous solutions by the ester derivatives of *p-t*-butyl-calix[4]- and -[6]arene (1_4, 1_6, Y = $CH_2CO_2C_2H_5$), respectively, with the response seen as a decrease in absorption of the dye ETH 5294 (9-(diethylamino)-5-octadecanoylimino-5H-benzophenox-

1_n

azine) at 660 nm (linear range 10^{-6} to 10^{-2} M Na^+) [8]. Using dyes with different spectral properties in two membranes, both cations can be analysed simultaneously with < 2% error [9]. The Na^+-selectivity is greater with methoxyethylester substituents in the molecule (**1₄**, Y = $CH_2CO_2CH_2CH_2OCH_3$; $logK_{Na,K}$ = -3.06, Li^+, NH_4^+, group II: no response, NaTFPB/DBS, TFPB = tetrakis[3,5-bis(trifluoromethyl)phenyl]borate, DBS = dibutylsebacate) [10,11].

The selectivity of a phosphine oxide derivative (**1₄**, Y = $CH_2CH_2P(O)Ph_2$) as a ligand for Mg^{2+} discriminating against alkali metal ions is higher in an optode using ETH 5294 ($logK_{Ca,M}$ = -4.77 (Na), -3.84 (K), -3.65 (NH_4), -2.3 (Li), -1.4 (Mg), -1.1 (Ba), NaTFM/DBS, TFM = trifluoromethansulfonate), than in an ion-selective electrode (ISE). The different spacer length between ether and carbonyl oxygens contributes to the selectivity [10,11].

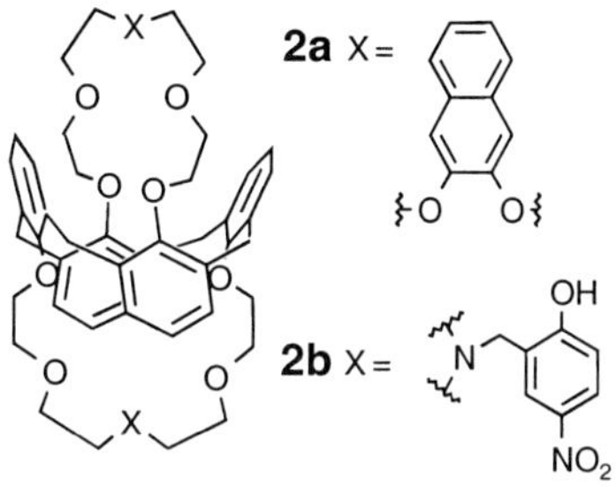

In the same way, Cs^+ can be analysed with an optode composed of 1,3;2,4-calix[4]bisnaphthyl-crown-6 **2a** and ETH 5294 ($logK_{Cs,M}$ = -2.1 (Na), -2.3 (K), -2.5 (Ca), -3.2 (Li), -2.8 (Mg), linear range 10^{-2}-10^{-6} M, response time 5 min, NaTFPB/NPOE, separate solution method; NPOE = o-nitrophenyl ether) [12]. This ligand belongs to the group of highly Cs^+-selective cage-like structures engaging not only in O-M^+ but also π-M^+ interactions (*vide infra* and Chapters 20, 35).

Anion analysis by competitive binding has been demonstrated for the displacement of 4-nitrophenoxide from calix[4]pyrrole. A relative binding order of $F^- > Cl^- > H_2PO_4^- > Br^-$ was observed, the absorption at 432 nm being a measure of the "released" phenoxide [13]. Calixarenes as anion receptors are discussed in detail in Chapter 23.

Where *neutral* molecules are readily converted to charged derivatives, analysis based on auxiliary dye systems can again be applied. Aldehydes, for example, may be analysed after in-situ conversion to pyridinium acetohydrazone ions by an optode composed of **1₆** (Y = $CH_2CO_2C_2H_5$), dynamic range 4×10^{-5} to 0.2 M butyraldehyde, ETH 5294, KTFPB/DOS; DOS = dioctylsebacate = bis(2-ethylhexyl)sebacate) [14].

2.2. FLUORIMETRY

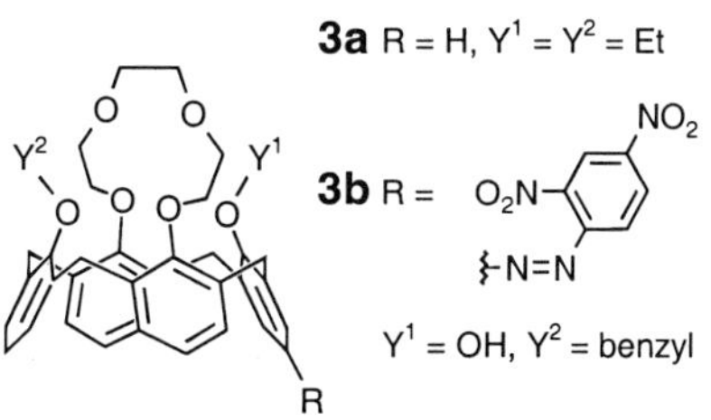

H^+-sensitive fluorophores in combination with Na^+-selective calixarenes such as crown **3a** [15,16] allow the sensitive and miniaturized analysis required for intracellular measurements ($logK_{Na,M}$ <-3.06 (K), <-2.4 (Ca), <-3.9 (Mg), detection limit (d.l.) 10^{-7} M Na^+, response time 10 s, ETH 5294, KTFPB/DOS) [17]. The fluorescence intensity ratio method is advantageous over the single-wavelength emission method. Slightly less sensitive are membranes based on the inner filter effect, where the H^+-sensitive chromophore absorbs the emission of the inert fluorophore in the absence of Na^+ [17,18].

Due to the protonation equilibria involved in these systems, the pH has to be exactly determined or controlled. Instead of the protonation degree, that of intramolecular hydrogen bonding can be used to signal Na^+-complexation in organic or mixed solvents: Na^+ interrupts the intra- and allows the inter-molecular hydrogen bonding of diamide derivative **4** with a flavin guest, the emission of which decreases in consequence (d. l. 3×10^{-4} M Na^+ in $CHCl_3$:MeCN 30/1) [19]. In mixed aqueous/organic solution, the recognition by *p*-sulfonatocalix[6]arene of *acetylcholine* in preference to primary and secondary amines can be detected as an increase in fluorescence intensity due to displacement of a fluorescent cation for which the fluorescence is quenched in its bound form (d.l. 10^{-6} M acetylcholine, H_2O/MeOH 1:1) [20]. The neutral medium is advantageous over the alkaline one required for calix[4]resorcinol as alternative host.

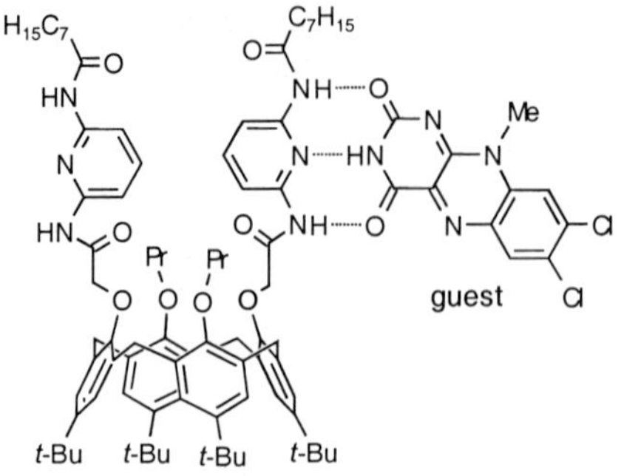

3. Calixarenes Incorporating Chromophoric Groups

3.1. Indoaniline as a Chromophoric Group

The presence of one indoaniline group in **5a** does not reduce the selectivity towards Na^+ typical for calix[4]arenes with carbonyl and ether oxygen atoms, and Na^+ binding produces a bathochromic shift of 42 nm along with increased absorption [21]. The chromophore **5b** with two indoaniline groups instead of one binds Ca^{2+} selectively, as the presence of two quinone carbonyl groups favours divalent cation chelation [22]. The binding is accompanied by a bathochromic shift (110 nm, EtOH 99%), because the cation stabilizes the excited state of the chromophore group. The compatible size of cavity and cation contributes to the high selectivity over Mg^{2+} ($\log(K_{ass,Ca}/K_{ass,Mg}) = 4.9$, EtOH). An interference by K^+ in binding ($\log(K_{ass,Ca}/K_{ass,K}) = 1.4$), does not affect the optical response significantly, due to a smaller shift (28 nm) and a smaller increase in molar absorbance. The formation of 1:1-complexes was confirmed in all cases.

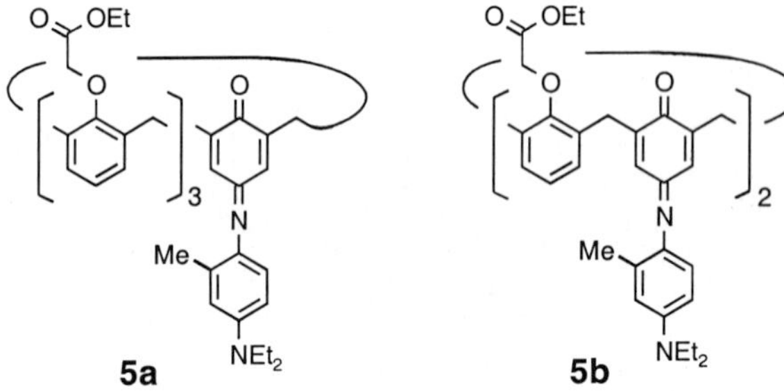

Is it possible, then, to design a logical device on the molecular level based on Ca^{2+} and K^+-binding? It has been achieved with the crown-5 derivative **6a**, which strongly binds Ca^{2+} ($\log K_{ass} = 6.26$, EtOH) and, at the crown unit, also K^+ ($\log K_{ass} = 6.86$), but only Ca^{2+}, which interacts with the carbonyl groups induces a colour change

[23]. Reversible switching of the 734 nm absorbance can thus be induced by competitive cation binding at the micromolar level.

With one phenolic unit of hexahydroxycalix[6]arene converted to an indoaniline group, UO_2^{2+} can be detected with by a bathochromic shift (59 nm) induced in the presence of a base [24]. The shift induced by group I + II ions amounts to 10 nm (Sr^{2+}) or less.

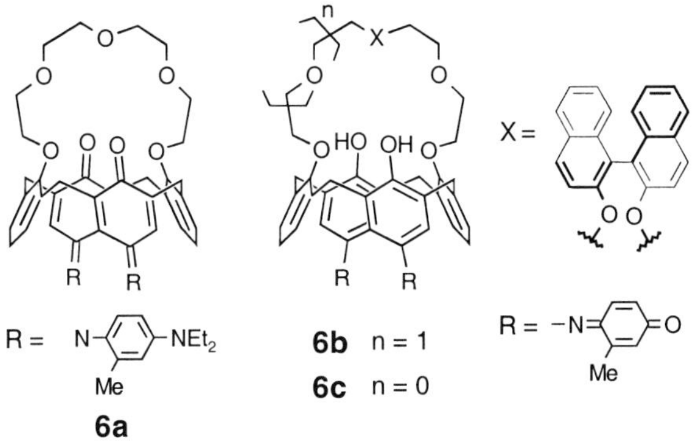

Butylamines are sensed in the order *t*- > *s*-> *n*- by calix[4]arene **6b**, which incorporates two indophenol groups ($logK_{ass}$ = 3.0 (*t*-), 2.8 (*s*-), 2.5 (*i*-), 2.0 (*n*-butylamine), EtOH) [25]. The discrimination is ascribed to steric and electrostatic effects induced by the 1,1'-binaphthyl-crown-6 bridge at the lower rim.

When the crown bridge in **6b** is changed to an unsymmetrical one in **6c**, chiral recognition of (R)-isomers of amines and amino acids is visualized as a colour change from red to blue. A 23 nm bathochromic shift to 538 nm and a new absorption at 652 nm result from proton transfer and hydrophobic effects, while (S)-isomers are not bound [6,26].

3.2. NITROPHENOL AS A CHROMOPHORIC GROUP

The introduction of four 3-hydroxy-4-nitrobenzyl groups onto the lower rim *via* ester linkages renders 1_4 a Li^+-selective ligand, with a bathochromic shift from 350 to 425 nm (THF) resulting from Li binding. Higher sensitivity is achieved with **7a**, with one instead of four nitrophenol and three ethyl ester groups (sensitivity 10^{-5} M Li^+, $logK_{Li,Na}$ = -1.5) [27,28]. The colouring due to deprotonation of the nitrophenol moiety is assisted by an auxiliary base and occurs in homogeneous phase as well as upon solvent extraction into butanol. In the latter case, the leaching of the complex diminishes with increasing hydrophobicity, *e.g.* n-octadecyl instead of *t*-Bu groups.

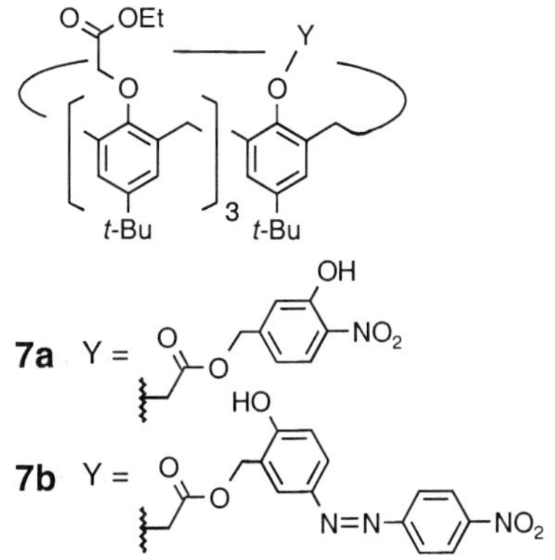

Appended to mono- or bis-azacrown-5 moieties in *1,3-alternate* calix[4]arenes such as **2b**, the nitrophenol group translates the K^+-selectivity of the macrocycle observed in transport experiments and ISEs into an optical signal, because the OH-group engages in cation exchange [29,30].

3.3. THE NITROPHENOL GROUP AS PART OF THE RING

De-*t*-butylated 1_4 with two nitro groups at the upper and two ethyl ester groups at the lower rim, is selective for Ca^{2+} over Na^+, K^+ and Mg^{2+} as detected by a new absorption at 405-441 nm in polar protic solvents [31]. Li^+ gives a weaker response. With a bridging dicarboxamide instead of the ester groups, the rigidified chromophore forms more stable complexes with amines. Since the mechanism is host-guest proton transfer accompanied by electrostatic and H-bond interactions, aliphatic primary amines are recognized most sensitively.

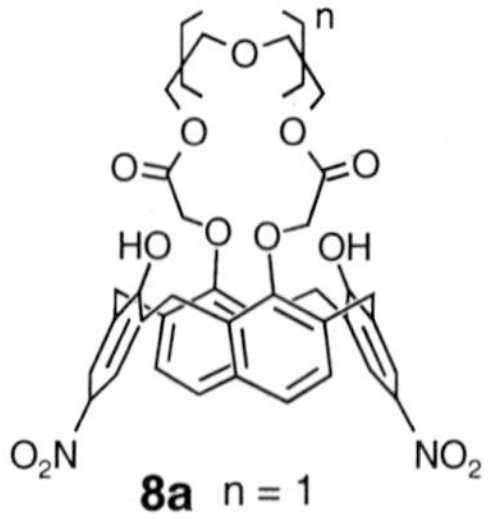

An order of butylamine binding stability (n- >> s- > t-) opposite to the one with **6b** (*vide supra*) is observed with bridged ligand **8a** [32]. The crown moiety improves the complex stability, while a 93 nm bathochromic shift results from proton transfer from the nitrophenol units to the primary amine. The discrimination between amines by crown-bridged calixarene derivatives depends upon various factors in addition to basicity and steric bulk of the amine, and is complicated by the possibility of binding at different, *exo* and *endo*, sites [33].

8a n = 1
8b n = 2

3.4. THE NITROPHENYLAZOPHENOL GROUP AS PART OF THE RING

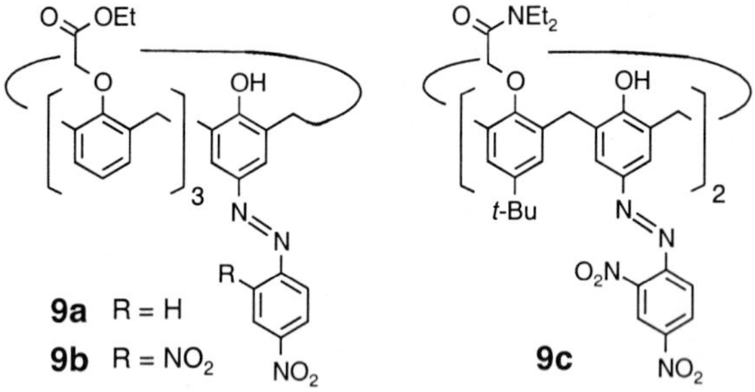

In **9a**, the binding of Li^+ to the azophenol oxygen as well as the ether oxygen atoms facilitates deprotonation by an auxiliary base, resulting in the appearance of a new absorption band at 584 nm [34]. Complexation of Li^+ is strongest among alkali metal ions in homogeneous solution and in solvent extraction. The Li^+/Na^+ selectivity was interpreted in terms of cavity size effects and differences in coordination modes. In plasticized PVC-optodes however, the selectivity of the related **9b** changes towards Na^+, which was interpreted as a result of differences in cation solvation (optode: linear range 10^{-4} to 10^{-2} M (*cf.* ISE d.l. $10^{-5.3}$), 630 nm, $\log K_{Na,M}$ = -2.8 (Li), -1.3 (K), -3.1 (NH$_4$), -2.5 (Ca), KTFPB/NPOE) [35].

9a R = H
9b R = NO$_2$
9c

It is known that amide groups with their higher carbonyl dipole moments compared to ester groups render calixarenes selective for alkaline earth ions [36]. Together with two azophenol groups, this turns **9c** into a Ca^{2+}-selective chromophore, a 168 nm bathochromic shift to 605 nm resulting from deprotonation (d.l. 10^{-4} M, $\log K_{Ca,M}$ = -2.3 (Sr), -2.9 (Mg), -2.8 (Na), extraction into CHCl$_3$) [37]. For maximum colour changes due to metal binding, the optimal pH range for measurement is 6 - 7.

Incorporation of an azophenol group into calixcrowns (Chapter 20) transfers the high selectivities of ISEs into that of chromoionophores. For example, ligand **3b** shows Na^+/K^+ selectivity in solvent extraction of $10^{3.1}$ [38]. With the larger crown-5 moiety, the calix[5]arene bearing two azophenol groups responds selectively to Cs^+ with a new absorption at 475 nm due to deprotonation ($\log K_{Cs,M}$ = -2.2 (K), -0.8 (Rb), Li^+, Na^+ no effect, d.l. 10^{-5} M, methoxyethanol:water 9/1) [39].

3.5. NITROPHENYLAZOPHENOL APPENDED TO CALIX[4]ARENES

Interestingly, the chromophoric ester derivative **7b** shows high Li$^+$-selectivity, attributable to ion pairing upon release of the acidic proton [40]. The selectivity determined from the 520 nm absorbance (log$K$ $_{Li,Na}$ = -1.9, THF/H$_2$O) is improved compared with the nitrophenol derivatives investigated earlier (3.2) and retained in PVC-based optodes [41]. The auxiliary weak base used to support deprotonation was demonstrated to be an advantage for NH$_3$- and volatile amine-sensing [41-43]. The addition of a Li$^+$-salt increases the apparent acidity of the labile proton and improves the sensitivity towards the base as well as the response time (d.l.: 20 ppb N(Me)$_3$, 2 ppm NH$_3$, resp. time < 2 min, 1:1-complex Li$^+$/ligand). The sensitivity decreases when four instead of one nitrophenylazophenol groups and thus an excess of ionisable groups are present (log$K_{Li,Na}$ = -1.5).

3.6. OTHER APPROACHES TO CHROMOIONOPHORES

Bisazobiphenyl groups bridging the upper rim of calixarenes allow detection of the binding of amines at a 0.2 mM level by color changes: yellow to orange (38-40 nm, monoamines), yellow to red (98-100 nm, di- and triamines) [44]. In various ligands derived from calix[8]arene by fourfold bridging of the *p*-positions, the bathochromic shifts are caused by host-guest transfer of up to two protons, provided the phenolic OH-groups are not substituted. Hence, mono- and diamines are distinguished from each other, but not higher ones. Aromatic amines with lower basicity are not recognized. The rigid and bulky bridges cause the formation of endo-complexes. With the same strategy applied to calix[4]arene [45], monoamines cause a 80 nm shift (red colour) [46].

Recently, a crowned calix[4]arene was utilized as the Na$^+$-selective moiety and combined with 2 *p*-terphenoquinone derivatives bearing thienyl groups (similar to **6a** but with crown-4, R= 2-(3,5-di-*t*-butyl-4-oxo-2,5-cyclohexadiene-1-ylidene)-2,5-dihydrothiophen-5-ylidene) [47]. The resulting ligand (selectivity Na$^+$/K$^+$ > 255, MeCN) combines chromophoric and redoxactive properties which can be applied for Na$^+$-sensing. The purple to blue colour change upon complexation results from a conformational inversion: the through-space exciton coupling in the uncomplexed *1,3-alternate* conformer (angle between the chromophores 43°) and in the Na$^+$-complex (*cone* conformer, angle 106°) differs by their energy states and leads to a reversal in the extinction coefficients of the 540 and 579 nm absorptions.

Calix[4]arenes with one or two 2,4,6-triphenylpyridino groups at the upper rim respond with bathochromic shifts most strongly to Li$^+$ ($\Delta\lambda$ = 265 nm in acetone) and Na$^+$ (326 nm), respectively [48]. The recognition of Li$^+$ is selective but the positive halochromism is the reverse of the expected negative one.

Chromophoric groups such as azophenol, nitrophenol or anthraquinone in **10a-e** cause pronounced bathochromic shifts upon anion complexation to the calix[4]pyrrole ring [49]. In **10b**, the largest change from yellow to red colour is caused by F$^-$ ($\Delta\lambda$ = 57 nm) with charge transfer from the bound anion to the e$^-$-deficient nitrobenzene groups. A similar mechanism operates in **10d,e** bearing anthraquinone groups [50].

The bathochromic shifts of the 467 nm absorption for **10d** (0.05 mM in CH$_2$Cl$_2$) amount to 51 nm (F$^-$, most sensitive), 34 nm (Cl$^-$) and 30 nm (H$_2$PO$_4^-$). Even more dramatic are the effects in **10e** with a colour change from red to blue (87 nm) for F$^-$. The selectivity follows the order F$^-$ > Cl$^-$ > H$_2$PO$_4^-$ >> Br$^-$, I$^-$, HSO$_4^-$ Combining quinoxaline and pyrrole groups in a calix[4]-like array, the optical response of these quinoxpyrroles (**11**) to F⁻-binding can be monitored sensitively [50].

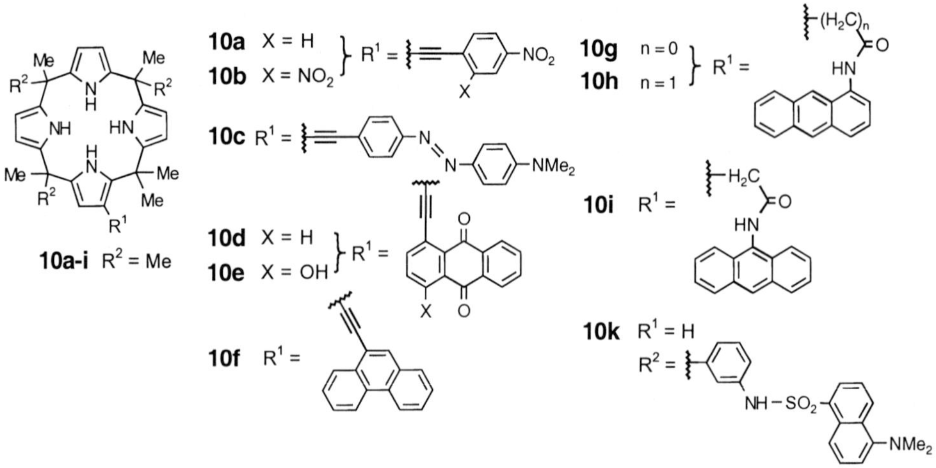

Further improvements in anion binding stability could provide the basics for improved chromo-/fluoroionophores as well. E.g., replacing β-pyrrolic hydrogens by the e⁻-withdrawing F-atoms increases the anion affinity and sensitivity of dipyrrolylquinoxaline, a building block of **11**, towards F⁻ and $H_2PO_4^-$ (contg. H: $logK$ = 4.26 (F⁻), 1.70(Cl⁻), 1.78 ($H_2PO_4^-$); contg. F: 4.79 (F⁻), 2.25(Cl⁻), 4.24 ($H_2PO_4^-$), CH_2Cl_2, fluorescence titr.) [51]. Both the color change (yellow to orange) and the fluorescence quenching upon complexation can be utilized. Other approaches such as extended-cavity calix[4]pyrrole [52] and use of calix[6]pyrrole with the ability of light halide ion inclusion [53] may also provide molecular backbones for new chromophores with high selectivity.

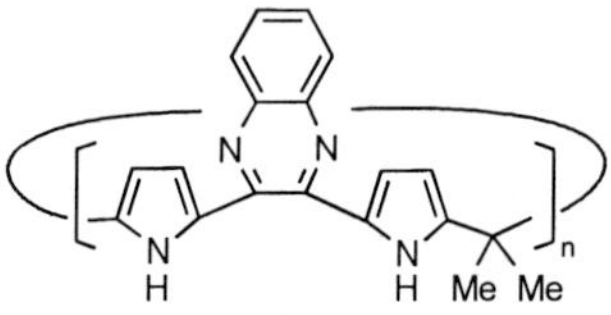

4. Calixarenes Incorporating Fluorophoric Groups

4.1. Cation Sensing by Increased Fluorescence Emission

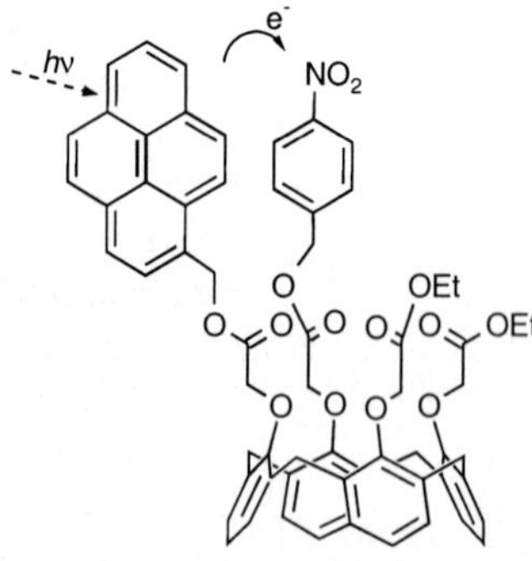

By attaching a pyrene and a nitrophenol group adjacent to the metal binding sites in **12**, intramolecular quenching occurs (as shown) in the absence of metal ions, while the fluorescence intensity increases up to 8 times when complexation of Na⁺ separates the groups [54]. The Na⁺-selectivity is retained due to the ester groups ($logK_{ass}$ for MSCN = 4.3 (Na), 2.9 (K), 2.2 (Cs), 1.9 (NH_4), 1.2 (Li), Et_2O:MeCN 97/3). This photoinduced electron transfer (PET) as the quenching mechanism plays a crucial role in photosynthesis, but it allows also the sensitive detection of host-guest interactions in supramolecular systems [2]. A recent example employs one or two Ru(III)-tris(2,2'-bibyridine) complexes tethered to the phenolic oxygens, the fluorescence emission of which is quenched by PET from the triplet state to adjacent quinone groups which are part of the calix[4]dibenzoquinone [55]. The quenching occurs due to co-facial orientation and close contact of the redox partners in the free ligand. Upon complexation of metals such as alkaline earth or rare earth ions, which stabilize the *cone* conformation, they become separated (as proved by a sharp decrease in the rate of PET and by structural investigations) and the fluorescence emission strongly increases (L contg. one Ru(bpy)$_3$ and one bpy group each: Φ = 0.0004 (uncomplexed), 0.052 (Ca^{2+}), 0.057 (Ba^{2+}), 0.054 (Sr^{2+}), 0.077 (La^{3+}), 0.074 (Gd^{3+}), 0.07 ([Zn^{2+}]$_2$-complex), 0.012 (Na$^+$), MeCN).

Binding of Na$^+$ to the oxygen donor atoms 'rigidifies' calix[4]arene as demonstrated for the crown-4 derivative **13a** [56]. The reduced molecular motion enhances the fluorescence intensity [57] up to 5 times. The number and position of the *p*-phenyl groups contribute to the sensitivity, with **13a** being the most effective structure for Na$^+$.

A recent approach to Na$^+$-sensing employs intramolecular fluorescence energy transfer from a pyrene donor to an anthroyloxy acceptor pair based on the calixarene **1$_4$** backbone (selectivity Na$^+$/K$^+$ = 59, MeOH:THF 15/1) [58]. A good overlap of donor emission and acceptor absorption spectra and the presence of carbonyl groups are crucial for efficient signalling and Na$^+$-selectivity, respectively.

Acylmethoxynaphthalene units in combination with ester groups were used on the calix[4]arene backbone (**14**) to monitor changes in absorption and fluorescence spectra for selectively sensing Na$^+$ [59]. In the absence of metal ions, intersystem crossing from the lowest excited singlet to the triplet state quenches the emission. Metal binding to the acyl group increases the energy of the nπ^* excited state. The most stable complexes are formed with Na$^+$, while the largest optical responses (absorption wavelength shift and emission intensity) are observed for Ca^{2+} and Li$^+$.

Two recent examples (**15,16**) utilizing PET employ calix[4]arenes as a molecular platform with anthracene groups attached to the crown-6 group for that mechanism. The Cs$^+$-selectivity in these ligands results from the crown-6 moiety and the *1,3-alternate* conformation with restricted rotational freedom, similar to **2a**. In the absence of complexed ions, PET from benzocrown oxygens or aza-crown nitrogens to the excited singlet state of anthracene groups occurs (*e.g.* **15**: Φ = 0.03 (Cs$^+$), Φ = 0.003 (no M$^+$), logK= 3.76 (Cs$^+$), 2.41 (K$^+$), <d.l. (Na$^+$, Li$^+$); d.l. 10^{-5} M Cs$^+$, CH$_2$Cl$_2$:MeOH 1/1) [60]. Both the quantum yield and the sensitivity/selectivity dramatically increase in **16b** : Φ = 0.625 (Cs$^+$), 0.49 (Rb$^+$), 0.405 (K$^+$), 0.07 (Na$^+$), 0.012 (no M$^+$), log$K$ = 6.6 (Cs$^+$), 5.48 (Rb$^+$), 4.06 (K$^+$), 1 μM in CH$_2$Cl$_2$/MeOH 1/1) [61]. The enhancement in emission is the

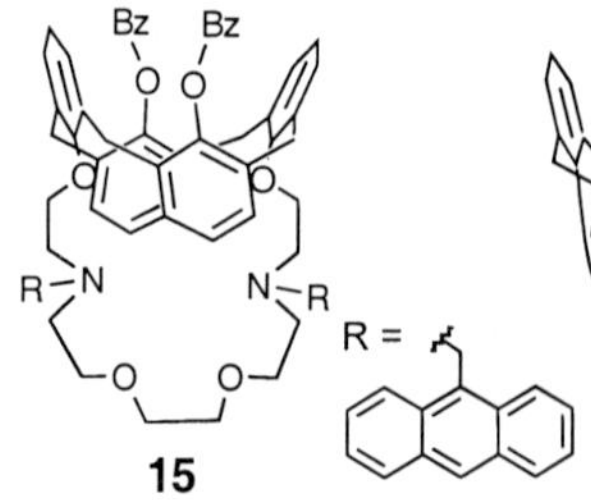

largest response on Cs$^+$-binding of all ligands known today. The fact that **16b** is more sensitive than **16c** is interpreted by a stronger interaction of Cs$^+$ with the benzo oxygens in the former case in agreement with shorter Cs$^+$-O distances found in a related series of crystal structures.

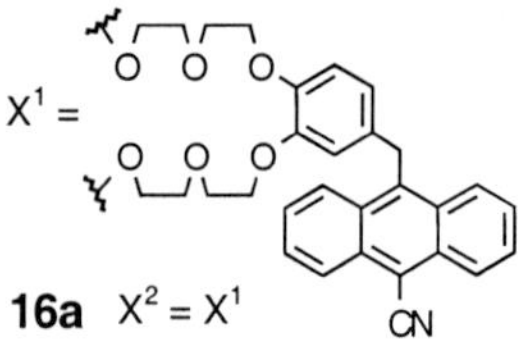

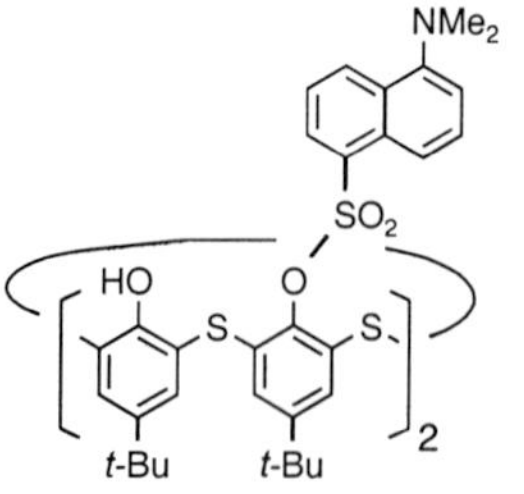

The dansyl-modified thiacalixarene **17** responds with a fluorescence increase sensitively to Cd^{2+} ($\Delta I/I_0 = 0.42$) over Al^{3+} (0.23), Cr^{3+} (0.2), Zn^{2+} and Cu^{2+} (0.17) (<0 for alkali metal, Co^{2+} and Ni^{2+} ions; H$_2$O, 489 nm, 3x10^{-4} M metal, pH 7) [62]. Here, the environment of the dansyl group determines the emission intensity, which increases inside the hydrophobic calixarene cavity. Selectivity towards Cd^{2+} is explained as due to its nature as a 'soft' cation.

Based on an different principle however, the highly UO$_2^{2+}$-selective p-sulphonatocalix[6]arene bearing one lissamine rhodamine B group was applied for capillary electrophoretic analysis [63]. The high thermodynamic stability of the UO$_2^{2+}$-complex and its slow dissociation kinetics allow the complex to be monitored during electrophoresis (emission 593 nm, excitation 532 nm) in ligand-free buffer (pH 8.3) without dissociation. The complex at this pH is [L·UO$_2$]$^{9-}$, thus migrating with a different velocity than the parent ligand.

4.2. DETECTION OF OTHER ANALYTES BY INCREASED FLUORESCENCE

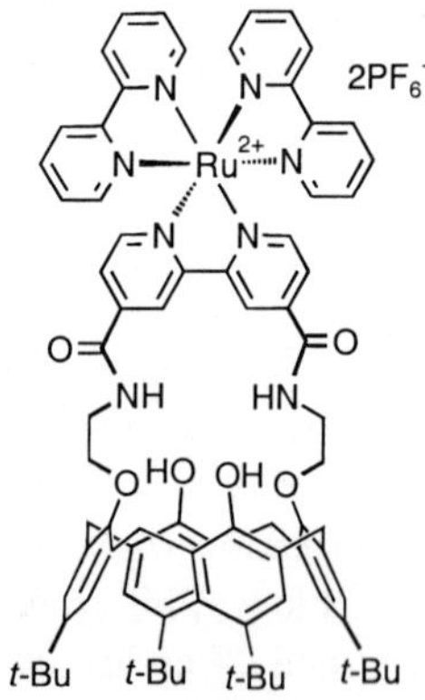

Ru(II)-bipyridine groups at the upper or lower rim of calix[4]arenes have been used to monitor anion binding [64, 65]. Anion binding occurs due to hydrogen bonding, the ligands being selective for H$_2$PO$_4^-$ over Cl$^-$ (e.g. **18**: logK_{ass} = 4.45 (H$_2$PO$_4^-$), 3.2 (Cl$^-$), DMSO). The increase in fluorescence emission, caused by molecular rigidification, parallels the order of stability and is accompanied by a 16 nm blue shift. Alternatively, electrochemical recognition of the binding is possible (see Chapter 23).

Appending one trisbipyridineruthenium(II) complex via a methylene spacer to **1$_4$**, leaving the remaining Y unsubstituted (Y = H), results in a fluorescent pH-sensor for pH 7 to 9, above which PET from a phenoxide anion quenches the emission (pK_a = 8.1) [66].

The selective recognition of sugars by phenylboronic acids [e.g. 67] can also be monitored with fluorophoric groups, e.g. due to reduced excimer formation between two of them and reduced PET from a neighbouring nitrogen lone pair [68]. Combined with the calix[4]arene platform, new fluorescent hosts with tunable selectivity are possible [68].

The first chiral fluorescent host based on calixarenes (the enantiomer pair of **19a**) showed increased excimer emission upon amino acid binding. The chiral discrimination may improve with a higher degree of molecular preorganization [69].

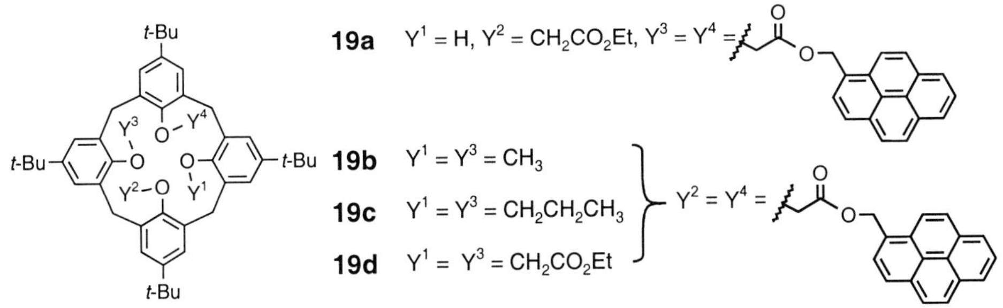

19a $Y^1 = H$, $Y^2 = CH_2CO_2Et$, $Y^3 = Y^4 = $ [pyrenyl ester]

19b $Y^1 = Y^3 = CH_3$
19c $Y^1 = Y^3 = CH_2CH_2CH_3$ $Y^2 = Y^4 = $ [pyrenyl ester]
19d $Y^1 = Y^3 = CH_2CO_2Et$

4.3. DETECTION OF ANALYTES BY DECREASED FLUORESCENCE

The strong decrease of $^1\pi\pi^*$ fluorescence emission of **20** upon complexation with Cs^+ and Ag^+ has been attributed to the strength of the cation/aromatic π interaction [70]. The ligand has no fluorophoric groups other than the calixarene skeleton, so the intrinsic photophysical properties are observed. In contrast, Na^+ and K^+ increase the emission intensity and the luminescence life time more than 2-fold. The related crown-6 derivative behaves similarly, with the exception that Na^+ is not bound.

When four anthracene groups are tethered to the calix[4]arene backbone via ester linkages, the resulting fluorophore **21** retains selectivity for Na^+ over K^+, as proved by NMR spectral changes [71,72]. The overall fluorescence emission at 391, 418 and 443 nm decreases when Na^+ is complexed. This behaviour changes when two pyrene groups are contained in a rigidified molecule (Section 4.4, *vide infra*).

The high selectivity for Hg^{2+} in solvent extraction even in the presence of a 100-fold excess of other heavy metal ions converts an N-(X)sulfonyl carboxamide derivatised calix[4]arene [73] into the fluorescence sensor **22** [74]. Electron transfer from dan-

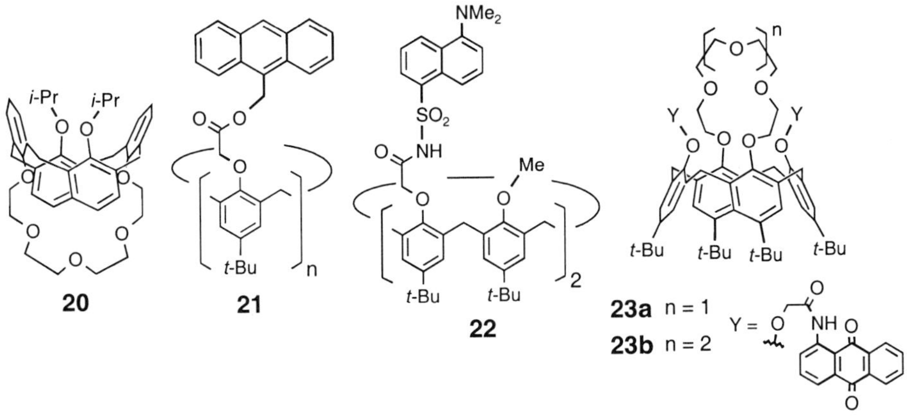

23a n = 1
23b n = 2   Y = [anthraquinone amide]

syl to bound Hg^{2+} quenches the emission. The decrease in 520 nm emission intensity is proportional to the extractability into $CHCl_3$ (pH_{50}=1.9, 0.25 mM ligand).

Bridging calix[4]arene by a dioxotetraaza group with an anthracene group appended allows transition metal detection [75]. Quenching cations [76] such as Ni^{2+} decrease the 276 nm absorption as well as the emission. On the other hand, chelation-induced enhancement [76] in fluorescence emission is observed for non-quenching ions such as Zn^{2+}.

The fluorophoric groups in calix[4]pyrroles **10f** [49] and **10g** – **10k** [50,77,78] respond to anion binding due to hydrogen bond interactions. The complexation with preference for F^- (*e.g.* **10g**: $logK_{1.1}$ = 4.94 (F^-), 3.69 (Cl^-), 3.01 (Br^-), 4.2 ($H_2PO_4^-$), CH_2Cl_2, d.l. < $5x10^{-5}$ M F^-) characteristic for this cyclic structure reduces the 446 nm emission (**10g**) due to quenching, which is most effective in conjugated bond systems, compared with **10h** and **10i**.

4.4. DETECTION OF ANALYTES BY CHANGES IN THE MONOMER /INTRAMOLECULAR EXCIMER EMISSION RATIO

The solvent polarity influences the conformation of fluorophore **19b**. The amount of *cone* conformer, which is more polar than the *partial cone*, increases in polar solvents [79]. Simultaneously, the ratio of monomer:excimer emission at 380 and 480 nm respectively increases, because the pyrene groups interact stronger in the *partial cone* conformer. Added alkali metal ions also influence the ratio in the order of complex stability ($logK_{ass}$ = 5.34 (Na^+), 4.73 (Li^+), 4.06 (K^+), MSCN, Et_2O).

Compounds like **19b** and **19c** are also sensitive towards hydrogen bond donors and thus form complexes with carboxylic acids such as CF_3COOH [80]. The monomer:excimer emission ratio increases with the acidity of the COOH group and a plot of $(I_0-I)/I_0$ *vs*. pK_a is linear for **19c** but not for **19b**.

With its four carbonyl and ether oxygens, **19d** combines Na^+-selectivity with a 30-times increase of monomer:excimer emission ratio upon complexation (Na/K selectivity = 154) [81]. This is the result of the larger distance between the pyrene groups caused by conformational changes upon complexation.

More preorganized ligands such as crown-4 derivatives of calix[4]arene can be successfully converted to Na^+-selective fluorophores (Section 4.1). With two pyrene groups, **13b** ($logK_{ass}$ = 5.48 (Na^+), 4.84 (Li^+), <d.l. (K^+), $MClO_4$, MeCN) is characterized by a large increase in molar absorption, accompanied by an increase in the monomer/excimer fluorescence intensity ratio [82]. The conformational changes influencing the ratio are similar to the preceding example.

With both the crown bridge and the fluorophoric groups at the lower rim, ligands **23a,b** respond with a monomer:excimer ratio *decrease* to both Na^+ and K^+ [83]. This is interpreted in terms of cooperative binding by crown and carbonyl oxygens.

The fluorimetric detection of guanidinium with high selectivity over primary ammonium ions was achieved with **24**, bearing the C_3-symmetrical binding sites adjacent to pyrene groups, the 478 nm excimer emission of which is quenched and the monomer emission increased upon binding [84]. At high concentration (*e.g.* 50:1 over **24**), *t*-butylammonium ions interfere but can be masked by adding 18C6.

If the homotrioxacalix[3]- instead of the calix[6]-skeleton is used (**25**), primary ammonium ions are recognized and detected by decreased 480 nm excimer emission [85]. The *cone* conformer forms stronger complexes with the ammonium ions than the *partial cone* but the monomer:excimer emission ratio of the latter is much more sensitive to complexation due to its pseudo-C_3 symmetry. Alkali metal ions are sufficiently discriminated one from the other by the *partial cone* conformer.

The calix[4]resorcinarene backbone was employed recently to anchor one to four 2-(pyren-1-yl)ethyl residues onto the methylene bridges [86]. The resulting solvated excimer emission is expected to be sensitive not only to solvent effects as demonstrated, but also to guest binding, which can occur due to the presence of eight acetyl groups.

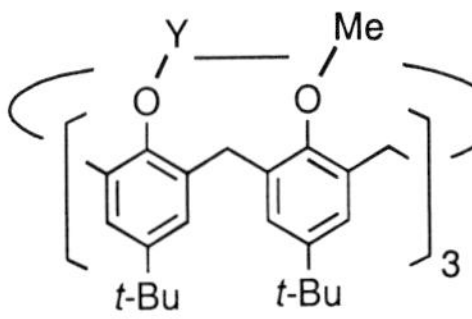

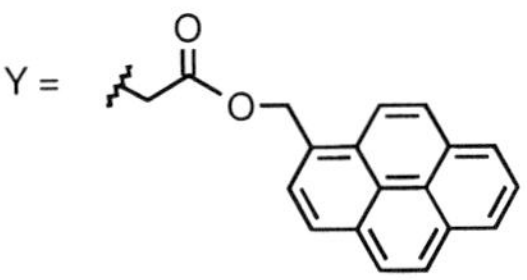

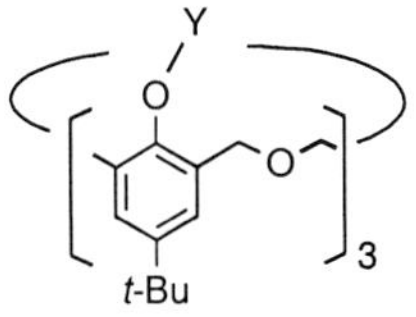

4.5. DETECTION OF ANALYTES BY AN EMISSION WAVELENGTH SHIFT

The binding of Li^+ to the lower rim of benzothiazole-functionalised **26**, which is selective amongst the alkali metal ions is accompanied by deprotonation and seen as new absorption (359 nm) and emission bands (422 nm) [87]. The methoxy group distal to phenoxide does not contribute to binding but stabilises the complexed ligand as the *partial cone* conformer.

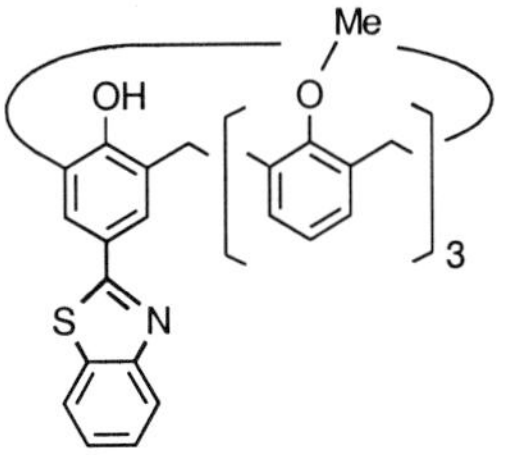

A conjugated poly(phenylene bithiophene) **27** with a calix[4]arene-crown-4 moiety and an appropriately long chain length (M_n = 9600, degree of polymerization ≈ 7) was shown to selectively respond to Na^+ and switch the fluorescence emission wavelength from 548 to 498 nm above a critical concentration (*ca.* 1 mM) due to energy migration within the complexed polymer [88].

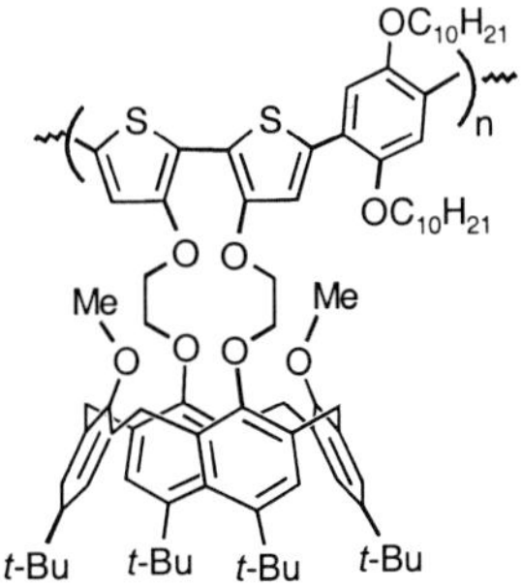

6. References and Notes

[1] R. Ludwig, *Fresenius' J. Analyt. Chem.* **2000**, *367*, 103-128.
[2] A. P. de Silva, H. Q. N. Gunaratne, T. Gunnlaugsson, A. J. M. Huxley, C. P. McCoy, J. T. Rademacher, T.E. Rice, *Chem. Rev.* **1997**, *97*, 1515-1566.
[3] T. Hayashita, M. Takagi in *Comprehensive Supramolecular Chemistry, vol. 1, Molecular Recognition: Receptors for Cationic Guests* (Vol. Ed. G. W. Gokel), Pergamon Press, New York, Oxford, **1996**, 635-669.
[4] T. Hayashita, N. Teramae, T. Kuboyama, S. Nakamura, H. Yamamoto, H. Nakamura, *J. Incl. Phenom.* **1998**, *32*, 251-265.
[5] M. Pietraszkiewicz, in *Comprehensive Supramolecular Chemistry, vol. 10: Supramolecular Technology* (Vol. Ed. D. N. Reinhoudt), Pergamon Press, New York, Oxford, **1996**, 225-266.

[6] Y. Kubo, *Syn. Lett.* **1999**, 161-174.
[7] P. Bühlmann, E. Pretsch, E. Bakker, *Chem. Rev.* **1998**, *98*, 1593-1687.
[8] W. H. Chan, A. W. M. Lee, C. M. Lee, K. W. Yau, K. Wang, *Analyst* **1995**, *120*, 1963-1967.
[9] W. H. Chan, A. W. M. Lee, D. W. J. Kwong, Y.-Z. Liang, K.-M. Wang, *Analyst* **1997**, *122*, 657-661.
[10] S. O'Neill, P. Kane, M. A. McKervey, D. Diamond, *Anal. Commun.* **1998**, *35*, 127-131.
[11] S. O'Neill, S. Conway, J. Twellmeyer, O. Egan, K. Nolan, D. Diamond, *Anal. Chim. Acta* **1999**, *398*, 1-11.
[12] H. Zeng, B. Dureault, *Talanta* **1998**, *46*, 1485-1491.
[13] P. A. Gale, L. J. Twyman, C. I. Handlin, J. L. Sessler, *Chem. Commun.* **1999**, 1851-1852.
[14] W. H. Chan, X. J. Wu, *Analyst* **1998**, 2851-2856.
[15] H. Yamamoto, T. Sakaki, S. Shinkai, *Chem. Lett.* **1994**, 469-472.
[16] H. Yamamoto, S. Shinkai, *Chem. Lett.* **1994**, 1115-1118.
[17] M. Shortreed, E. Bakker, R. Kopelman, *Anal. Chem.* **1996**, *68*, 2656-2662.
[18] X. Yang, K. Wang, C. Guo, *Anal. Chim. Acta* **2000**, *407*, 45-52.
[19] H. Murakami, S. Shinkai, *J. Chem. Soc. Chem., Commun.* **1993**, 1533-1535.
[20] K. N. Koh, K. Araki, A. Ikeda, H. Otsuka, S. Shinkai, *J. Am. Chem. Soc.* **1996**, *118*, 755-758.
[21] Y. Kubo, S.-i. Hamaguchi, K. Kotani, K. Yoshida, *Tetrahedron Lett.* **1991**, *32*, 7419-7420.
[22] Y. Kubo, S. Hamaguchi, A. Niimi, K. Yoshida, S. Tokita, *J. Chem. Soc., Chem. Commun.* **1993**, 305-307.
[23] Y. Kubo, S. Obara, S. Tokita, *Chem. Commun.* **1999**, 2399-2400.
[24] Y. Kubo, S. Maeda, M. Nakamura, S. Tokita, *J. Chem. Soc., Chem. Commun.* **1994**, 1725-1726.
[25] Y. Kubo, S. Maruyama, N. Ohhara, M. Nakamura, S. Tokita, *J. Chem. Soc., Chem. Commun.* **1995**, 1727-1728.
[26] Y. Kubo, S. Maeda, S. Tokita, M. Kubo, *Nature* **1996**, *382*, 522-524; *Synlett* **1999**, 161-174.
[27] M. McCarrick, B. Wu, S. J. Harris, D. Diamond, G. Barrett, M. A. McKervey, *J. Chem. Soc., Chem. Commun.* **1992**, 1287-1289.
[28] M. McCarrick, B. Wu, S. J. Harris, D. Diamond, G. Barrett, M. A. McKervey, *J. Chem., Soc. Perkin Trans. 2* **1993**, 1963-1968.
[29] J. S. Kim, I. Y. Yu, A. H. Suh, D. Y. Ra, J. W. Kim, *Synth. Commun.* **1998**, *28*, 2937-2944.
[30] J. S. Kim, Abstr. 5th Int. Conf. Calixarene Chem. (Perth 19-23.9.), **1999**, L-24.
[31] I. Mohammed-Ziegler, M. Kubinyi, A. Grofesik, A. Grün, I. Bitter, *J. Mol. Struct.* **1999**, *480-481*, 289-292.
[32] Q.-Y. Zheng, C.-F. Chen, Z.-T. Huang, *Tetrahedron* **1997**, *53*, 10345-10356.
[33] M. Gattuso, A. Notti, S. Pappalardo, M. F. Parisi, *Tetrahedron Lett.* **1998**, *39*, 1969-1972.
[34] H. Shimizu, K. Iwamoto, K. Fujimoto, S. Shinkai, *Chem. Lett.* **1991**, 2147-2150.
[35] K. Toth, B. T. T. Lan, J. Jeney, M. Horvath, I. Bitter, A. Grün, B. Agai, L. Töke, *Talanta* **1994**, *41*, 1041-1049.
[36] M. A. McKervey, M. J. Schwing-Weill, F. Arnaud-Neu, in *Comprehensive Supramolecular Chemistry, Vol.1: Receptors for Cationic Guests.* (Vol. Ed. G. W. Gokel), Pergamon Press, New York, Oxford, **1996**, 537-603.
[37] N. Y. Kim, S.-K. Chang, *J. Org. Chem.* **1998**, *63*, 2362-2364.
[38] H. Yamamoto, K. Ueda, K. Sandanayaka, S. Shinkai, *Chem. Lett.* **1995**, 497-498.
[39] J. L. M. Gordon, V. Böhmer, W. Vogt, *Tetrahedron Lett.* **1995**, *36*, 2445-2448.
[40] M. McCarrick, S. J. Harris, D. Diamond, G. Barret, M. A. McKervey, *Analyst* **1993**, *118*, 1127-1130.
[41] R. Grady, T. Butler, B. D. MacCraith, D. Diamond, M. A. McKervey, *Analyst* **1997**, *122*, 803-806.
[42] M. McCarrick, S. J. Harris, D. Diamond, *J. Mater. Chem.* **1994**, *4*, 217-221.
[43] M. Loughran, D. Diamond, *Food Chem.* **2000**, *69*, 97-103.
[44] H. M. Chawla, K. Srinivas, *J. Chem. Soc., Chem. Commun.* **1994**, 2593-2594; idem, *Tetrahedron Lett.* **1994**, *35*, 2925-2928
[45] H. M. Chawla, K. Srinivas, *J. Org. Chem.* **1996**, *61*, 8464-8467.
[46] Some caution should be expressed, however, with respect to the assumed molecular forms. As indicated by references [44], there is uncertainty as to the structures of the calix[8]arene derivatives, and simple molecular models indicate that intramolecular bridging of calix[4]arene by a bis-azodiphenyl linker [45] would require major bond distortions.
[47] K. Takahashi, A. Gunji, D. Guillaumont, F. Pichierri, S. Nakamura, *Angew. Chem.* **2000**, *112*, 3047-3050; *Int. Ed.*, *39*, 2925-2928.
[48] I. Bitter, A. Gruen, G. Toth, A. Szoellosy, G. Horvath, B. Agai, L. Toke, *Tetrahedron* **1996**, *52*, 639-646.
[49] H. Miyaji, W. Sato, J. L. Sessler, V. M. Lynch, *Tetrahedron Lett.* **2000**, *41*, 1369-1373.

[50]   H. Miyaji, W. Sato, J. L. Sessler, V. M. Lynch, *Angew. Chem.* **2000**, *112*, 1847-1850, *Int. Ed. 39*, 1827-1780; J. L. Sessler, B. Andrioletti, P. Anzenbacher, C. T. Black, P. A. Gale, K. Juriskova, H. Miyaji, A. Try, Abstr. 5th Int. Conf. Calixarene Chem. (Perth 19-23.9.), **1999**, L-11.
[51]   P. Anzenbacher Jr., A. C. Try, H. Miyaji, K. Jursikova, V. M. Lynch, M. Marquez, J. Sessler, *J. Am. Chem. Soc.* **2000**, *122*, 10268-10272.
[52]   S. Camiolo, P. A. Gale, *Chem. Commun.* **2000**, 1129-1130.
[53]   G. Cafeo, F. H. Kohnke, G. L. La Torre, A. J. P. White, D. J. Williams, *Chem. Commun.* **2000**, 1207-1208.
[54]   I. Aoki, T. Sakaki, S. Shinkai, *J. Chem. Soc., Chem. Commun.* **1992**, 730-732.
[55]   A. Harriman, M. Hissler, P. Jost, G. Wipff, R. Ziessel, *J. Am. Chem. Soc.* **1999**, *121*, 14-27.
[56]   H. Matsumoto, S. Shinkai, *Chem. Lett.* **1994**, 2431-2434.
[57]   H. Shizuka, K. Takada, T. Morita, *J. Phys. Chem.* **1980**, *84*, 994-999.
[58]   T. Jin, *Chem. Commun.* **1999**, 2491-2492.
[59]   I. Leray, F. O'Reilly, J. L. H. Jiwan, J. P. Soumillion, B. Valeur, *Chem. Commun.* **1999**,795-796.
[60]   H.-F. Ji, G. M. Brown, R. Dabestani, *Chem. Commun.* **1999**, 609-610; H.-F. Ji, R. Dabestani, G. M. Brown, *Photochem. Photobiol.* **1999**, *70*, 882-886.
[61]   H.-F. Ji, R. Dabestani, G. M. Brown, R. A. Sachleben, *Chem. Commun.* **2000**, 833-834.
[62]   Narita, M., Y. Higuchi, F. Hamada, H. Kumagai, *Tetrahedron Lett.* **1998**, *39*, 8687-8690.
[63]   Q. Lu, J. H. Callahan, G. E. Collins, *Chem. Commun.* **2000**, 1913-1914.
[64]   P. D. Beer, *Chem. Commun.* **1996**, 689-696; P. D. Beer, Z. Chen, A. J. Goulden, A. Grieve, D. Hesek, F. Szemes, T. Wear, *J. Chem. Soc., Chem. Commun.* **1994**, 1269-1271.
[65]   P. D. Beer, R. J. Mortimer, N. R. Stradiotto, F. Szemes, J. S. Weightman, *Anal. Commun.* **1995**, *32*, 419-421.
[66]   R. Grigg, J. M. Holmes, S. K. Jones, W. D. J. A. Norbert, *J. Chem. Soc., Chem. Commun.* **1994**, 185-187.
[67]   R. Ludwig, T. Harada, K. Ueda, T. D. James, S. Shinkai, *J. Chem. Soc., Perkin Trans. 2* **1994**, 697-702.
[68]   T. D. James, P. Linnane, S. Shinkai, *Chem. Commun.* **1996**, 281-287.
[69]   T. Jin, K. Monde, *Chem. Commun.* **1998**, 1357-1358.
[70]   L. Prodi, F. Bolletta, M. Montalti, A. Casnati, F. Sansone, R. Ungaro, *New J. Chem.* **2000**, *24*, 155-158.
[71]   C. Perez-Jimenez, S. J. Harris, D. Diamond, *J. Chem. Soc., Chem. Commun.* **1993**, 480-483.
[72]   C. Perez-Jiminez, S. J. Harris, D. Diamond, *J. Mater. Chem.* **1994**, *4*, 145-151.
[73]   G. G. Talantova, H.-S. Hwang, V. S. Talantov, R. A. Bartsch, *Chem. Commun.* **1998**, 1329-1330.
[74]   G. G. Talantova, N. S. A. Elkarim, V. S. Talantov, R. A. Bartsch, *Anal. Chem.* **1999**, *71*, 3106-3109.
[75]   F. Unob, Z. Asfari, J. Vicens, *Tetrahedron Lett.* **1998**, *39*, 2951-2954.
[76]   E. U. Akkaya, M. E. Huston, A. W. Czarnik, *J. Am. Chem. Soc.* **1990**, *112*, 3590-3593.
[77]   H. Miyaji, P. Anzenbacher, J. L. Sessler, E. R. Bleasdale, P. A. Gale, *Chem. Commun.* **1999**, 1723-1724.
[78]   P. A. Gale, E. R. Bleasdale, G. Z. Chen, L. J. Twyman, P. Anzenbacher, H. Miyaji, J. L. Sessler, Abstr. 5th Int. Conf. Calixarene Chem. (Perth 19-23.9), **1999**, L-10.
[79]   I. Aoki, H. Kawaba, K. Nakashima, S. Shinkai, *J. Chem. Soc., Chem. Commun.* **1991**, 1821-1823.
[80]   I. Aoki, T. Sakaki, S. Tsutsui, S. Shinkai, *Tetrahedron Lett.* **1992**, *33*, 89-92.
[81]   T. Jin, K. Ichikawa, T. Koyama, *J. Chem. Soc., Chem. Commun.* **1992**, 499-501.
[82]   H. Matsumoto, S. Shinkai, *Tetrahedron Lett.* **1996**, *37*, 77-80.
[83]   Y.-B. He, H. Huang, L.-Z. Meng, C.-T. Wu, T.-X. Yu, *Chem. Lett.* **1999**, 1329-1330.
[84]   M. Takeshita, S. Shinkai, *Chem. Lett.* **1994**, 1349-1352.
[85]   M. Takeshita, S. Shinkai, *Chem. Lett.* **1994**, 125-128.
[86]   Y. Hayashi, T. Maruyama, T. Yachi, K. Kudo, K. Ichimura, *J. Chem. Soc., Perkin Trans. 2* **1998**, 981-988.
[87]   K. Iwamoto, K. Araki, H. Fujishima, S. Shinkai, *J. Chem. Soc., Perkin Trans. 1* **1992**, 1885-1887.
[88]   K. B. Crawford, M. B. Goldfinger, T. M. Swager, *J. Am. Chem. Soc.* **1998**, *120*, 5187-5192.

Chapter 33

MONO- AND MULTI-LAYERS

ANDREW LUCKE[a], CHARLES J. M. STIRLING[a], VOLKER BÖHMER[b]

[a] *Department of Chemistry, The University, Sheffield S3 7HF United Kingdom. e-mail: c.stirling@sheffield.ac.uk*
[b] *Johannes Gutenberg-Universität, Fachbereich Chemie und Pharmazie, Abteilung Lehramt Chemie, Duesbergweg 10-14, D-55099 Mainz, Germany, e-mail: vboehmer@mail.uni-mainz.de*

1. Introduction

Thin films incorporating or consisting of calixarenes have been widely investigated. The capacity of such films to adsorb both ions and neutral species has engendered particular interest in their use in sensor devices. Semiconductors, non-linear optical and pyroelectric materials or switchable systems for data storage are further examples of their potential use [1].

Both mono- and multi-layer films can be formed from calixarenes and resorcarenes, using a variety of techniques. This chapter deals with the fundamental basis of calixarene thin film applications, including film formation and characterisation as well as studies of selectivity in small molecule interactions. Real and potential applications are considered in more detail in Chapters 26, 32, 34 and 36.

The formation of monolayers by adsorption or (self) assembly at interfaces usually requires long chain molecules with different functional groups at each end (*e.g.* stearic acid for Langmuir-Blodgett-films, alkane thiols for assembly on gold.) Especially when fixed in the *cone* conformation calix[4]arenes possess two ends, the phenolic hydroxy groups (and residues attached to them) and the p-positions (and their substituents) which can be independently functionalized (see Chapter 2). The same is true for the all-cis

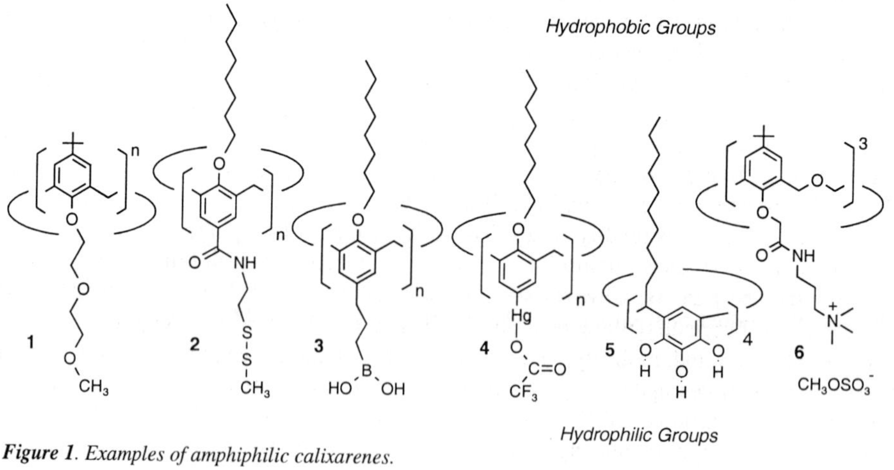

Figure 1. Examples of amphiphilic calixarenes.

Z. Asfari et al. (eds.), Calixarenes 2001, 612–626.

(*rccc*) isomers of resorcarenes, where the residues R introduced by the aldehyde and the OH-groups point in different directions and can undergo independent chemical reactions. Among the known synthetic macrocyclic molecules, calixarenes and resorcarenes may be considered especially suitable for the controlled construction of mono- and multilayers.

An indication of the structural variations in calixarenes used for thin film production is given in Figure 1, from which it is apparent that essentially all sites on the calixarene framework have been the subject of substitution. The homo-oxa calixarene **6** is of particular interest as a material whose use in the formation of ultrathin films of C_{60} is an indicator of the high level of sophistication now involved in the use of calixarenes for film formation [2].

2. Thin Film Formation and Characterisation

The focus here is on use of Langmuir-Blodgett and self-assembly (see also Chapter 8) procedures [3,4] as those which are best understood at the molecular level.

2.1. LANGMUIR-BLODGETT (LB) FILMS

Amphiphilic calixarenes, *viz.*, those bearing both hydrophobic and hydrophilic substituents are generally suitable for the formation of Langmuir-Blodgett (LB) films. This involves the spreading of a solution of the substrate in a volatile solvent onto the surface of an aqueous subphase. After evaporation of the solvent, a disordered monolayer is usually formed on the water surface which can be compressed by a barrier to form stable ordered monolayers.

Surface pressure / area diagrams (π/A-diagrams) thus obtained (*e.g.* by measuring the area of the film as a function of the increasing pressure exerted by the barrier) give direct information about the structure of the monolayer. The limiting area per molecule reflects the conformation of the molecules and/or their orientation at the surface, while the collapse pressure characterises the stability of the layers.

These Langmuir films can then be transferred to an appropriate solid support, *e.g.*, a glass slide or a silicon wafer, by dipping under constant pressure (Blodgett technique) to form an LB monolayer. Repetition of the dipping procedure leads to multilayers, since one layer is usually transferred per passage of the carrier through the surface (this means a double-layer is deposited by dipping in and pulling out the carrier). The transfer ratio (decrease of the area of the Langmuir film on the water surface per area of the solid support) gives information of the density of the coverage, while the number of layers in a multilayer is readily controlled by the number of dippings.

Parent calixarenes themselves may be considered amphiphilic molecules [5]. Some controversy exists, however, as to whether truly stable LB monolayers are formed by simple *p-t*-butylcalix[n]arenes alone [6,7]. Larger calixarenes especially may retain some conformational mobility, though this is not a factor which has yet been very systematically investigated [8]. 4-Alkylphenylazo substituted calix[4/6]arenes, with more acidic OH-groups and thus greater amphilicity than alkylcalixarenes, form stable monolayers showing increasing limiting areas and collapse pressures for increasing pH [9].

The amphiphilic properties can be strongly enhanced by substituents (see Figure 1 and further examples below). Surfactant properties are readily detected even in such relatively simple derivatives as the *cone* tetra(ethylester) **7a** [10]. Surface area measurements show that such a symmetrically substituted derivative orients at the interface with the *cone* axis perpendicular to the interface, but with less symmetrical derivatives **8** involving substitution of calix[4]arene with 1 or 4 ethoxycarbonylmethyl groups at the narrow rim in association with 1 or 4 dodecoyl groups at the wide rim, both this orientation and the monolayer stability appear to depend on the substitution pattern [11].

7a n = 4 R = *t*-Bu	**9** R = -CHO
7b n = 6 R = *t*-Bu	**10** R = -C(=S)-N(Me)(NMe$_2$)
7c n = 8 R = *t*-Bu	
8 n = 4 R = C(O)-C$_{11}$H$_{23}$	**11** R = -C(=NOH)-NH$_2$
a n = 4	
b n = 5	
c n = 6	

Limiting areas of 128, 151 and 190 Å^2 per molecule for **9a-c** and 119, 144 and 181 Å^2 for **10a-c** [12] roughly correspond to the area expected from models for a "planar" calixarene and suggest that the base of each calixarene is parallel and the aliphatic chains are perpendicular to the surface.

The stability of monolayers of **4** can be greatly increased by malonic acid or 1,2-dithiols in the subphase (through formation of Hg-O-COCH$_2$CO-O-Hg or Hg-S-(CH$_2$)$_2$-S-Hg bridges) [13], while monolayers of **2** become more stable after UV-irradiation (crosslinking via disulfide bridges) [14]. Monolayers of the boronic acid based calix[6]arene **3** directly show an extraordinary cohesiveness [15] which can be explained by the formation of oligoborate bridges between the calix[6]arene molecules.

12

	R
a	C$_{11}$H$_{23}$
b	C$_{17}$H$_{35}$
c	C$_{10}$H$_{20}$-OH
d	C$_{10}$H$_{20}$-SH
e	C$_{10}$H$_{20}$-S-C$_{10}$H$_{21}$
f	CH$_2$-C$_8$F$_{17}$

13

	R	Y
a	C$_{11}$H$_{23}$	CH$_2$COOH
b	C$_{11}$H$_{23}$	CH$_2$CH$_2$-OH
c	C$_{11}$H$_{23}$	CH$_2$-C(O)-OEt
d	C$_{11}$H$_{23}$	CH$_2$-C(O)-NEt$_2$
e	C$_{11}$H$_{23}$	Si(CH$_3$)$_3$
f	C$_{11}$H$_{23}$	C(O)-CH$_3$

Resorcarenes (**12**) and analogous pyrogallol derivatives (*e.g.* **5**) may be considered molecules with a greater inherent amphiphilicity than simple calixarenes. In the *bowl* (*cone*) form of the *rccc* isomer, they present a hydroxylated face with 8 or 12 OH groups at the wide rim opposite to a lipophilic side involving the alkyl chains of the aldehyde used in the initial synthesis. They, and their cavitand derivatives (see Chapter 9), have been widely investigated for their surfactant properties by LB techniques [16-19]. Their amphiphilicity - inverse of that displayed by simple calixarenes can be altered again by extended substitution, *e.g.* **13a-d** [18,20]. Modification of the hydroxyl groups by trimethylsilylation or acetylation, for example, led to much lower collapse pressures for monolayers formed from **13e** and **13f**, and increased the observed area per molecule [17]. Bilayer formation was observed for **13a** at higher pressure, while mixed monolayers with *p*-hexylazobenzene show the incorporation of this nematic liquid crystalline dye between the octadecyl chains [21].

Remarkable behaviour was shown by **12c** when compressed on the water subphase [19]. Contact angle, grazing angle FTIR, and X-ray reflectometry experiments suggested that at low pressures, the hydroxylated legs curl over to embed themselves in the water surface while high pressures enforce highly ordered structures [22] in which either the hydroxyl groups surrounding the bowl-rim or those at the ends of the legs are forced out of the aqueous phase (Figure 2). Area per molecule measurements best fit the model in which the bowl rims and the leg ends alternate in the surface [19].

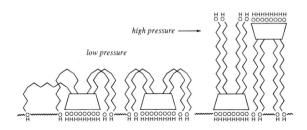

Figure 2. Langmuir film behaviour of 12c.

2.2. SELF ASSEMBLED FILMS

Labile interactions are increasingly exploited for the rational construction of supramolecular systems [23]. Important examples are associated with chemisorptive formation of films and layers on surfaces involving hydrogen bonding between hydroxyl substituents and metal oxides, and weak coordination of donor groups to pure metal surfaces, the best known cases of the latter being thiol(ate) [24] and thioether [25] interactions with gold. As for LB films, self-assembled films can be studied by the use of a wide range of techniques, including X-ray photoelectron spectroscopy (XPS), X-ray and neutron reflectometry, electron microscopy, atomic force microscopy (AFM) [26], surface plasmon resonance (SPR) [27] and ellipsometry, as well as deceptively simple methods such as contact angle measurement [3].

In the case of the resorcarene **12e**, XPS showed that its adsorption on gold involved the sulfur atoms of the thioether arms being held on the surface, though as initially deposited, the film appeared not to be sufficiently compact to inhibit heterogeneous electron transfer between the gold surface and $[Fe(CN)_6]^{3-}$ in solution. Annealing at 60 °C, however, resulted in inhibition of this process, interpreted as indicative of a transition from a disordered to an ordered array as shown in Figure 3 [25]. The final

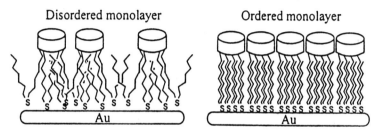

Figure 3. Schematic representation of a disordered monolayer of 12e obtained at room temperature and an ordered monolayer deposited at 60°C.

monolayers are thus oriented with the hydrophilic bowl towards the surrounding solvent. Ordered monolayers of **12e** have a hexagonal packing [28,29] and this has also been observed for self-assembled monolayers (SAMs) derived from the calixarene thioether derivatives **14a** and **14b** [30].

Close packing of the aliphatic chains may assist in these ordering processes, since eight alkyl chains assume a similar space than the resorcarene or calixarene "head group". However, self-assembled monolayers on gold are also formed with the resorcarene thiol **12d**, and contact angle measurements indicate that these, too, have the hydroxylated rims remote from the gold surface [31].The formation of SAMs from mercapto-*t*-butylcalix[4]arene on gold coated QCMs demonstrates that long aliphatic spacers are not necessarily required for monolayer formation [32].

Instances where the resorcarene hydroxyl groups or their derivatives provide a mechanism for surface attachment occur with resorcarenes **12a,b** and **15**. The *E* isomer of **15** can be adsorbed on colloidal silica (Figure 4) and inhibits particle aggregation. Irradiation of the coated particles with ultra-violet light, however, leads to isomerisation of the azo unit to its Z form and subsequent formation of relatively large particle aggregates. Since the isomerisation is photo-reversible, the coated particles can be subjected to light-induced dispersion-sedimentation cycles [33]. The adsorption strength can be modified by *O*-substituents as in **13a,b** [20], especially for carboxymethylated derivatives on aminated silica substrates [34]. The large difference thus obtained in the cross-sectional areas between the resorcarene skeleton (1.7 nm^2) and the azobenzene chains (1.0 nm^2) also ensures an efficient photoisomerisability of the azogroups [35].

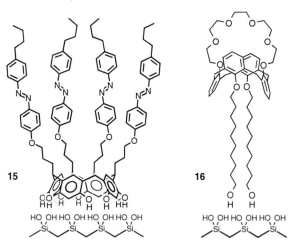

Figure 4. Binding of resorcarene **15** and calix[4]arene **16** via self-assembly on colloidal silica.

In a similar way the amphiphilic, Cs$^+$ selective *1,3-alternate* calix[4]arene crown-6 **16** self-assembles on silica surfaces to provide colloidal particles with an affinity for Cs$^+$ [36]. In another calixarene application based on this principle, variation in the crown link and the length of hydrophobic tails on calix[4]arenes has also been used to selectively control micelle formation from potassium and caesium dodecylsulfates [37].

The "sugar cluster" substitution found in **17** enables the formation of stable monolayers of this resorcarene on quartz surfaces [38]. The octa-galactose or -glucose face of **17** binds to quartz at least 10^4 times more strongly than do the simple monosaccharides themselves. This binding to the surface creates a host cavity in which guests, such as 8-

anilinonaphthalene-1-sulfonate, otherwise not adsorbed by quartz, can be effectively strongly bound to this surface [39].

Calixarene derived sugar dendrimers **18,** on the other hand, bind to polystyrene surfaces via hydrophobic forces with the carbohydrate residues directed towards the aqueous solution [40]. The coatings thus obtained may be used in biological tests/assays.

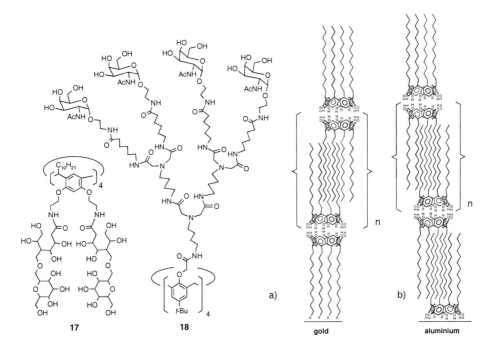

*Figure 5. Putative structures of multilayering of resorcarene **12a** (a) on a self-assembled monolayer of resorcarene **12d** on gold; (b) on aluminium.*

Multilayers can be also formed by self-assembly, although their thickness is not as easily controlled as with the LB-technique. Monolayers, both of the "bowl up" kind as formed by **12d** on gold and of the "bowl down" type as formed by **12a** on aluminium (presumably with a surface oxide layer), can initiate multilayer formation. Thus, when a gold-thiol monolayer of **12d** was exposed to a solution of resorcarene **12c** in hexane, spontaneous formation of multilayers (up to 40) of the resorcarene on top of the original monolayer was observed [31,41]. Contact angle, grazing angle FTIR, X-ray and neutron reflectometry experiments all strongly suggest bowl to bowl hydrogen bonding and deep interdigitation of the alkyl chains maximizing van der Waals contacts. Such interdigitation was also seen in the X-ray crystal structures of **12a** and related compounds [38,25b,42]; the reason may again be that four aliphatic chains occupy a smaller space than the resorcarene head groups. When a multilayer was immersed in ethanol, it was dispersed by disruption of the bowl to bowl hydrogen bonds and the monolayer was reformed.

2.3. FURTHER TECHNIQUES

Various other techniques for thin film formation from calixarenes, including deposition from the vapour, have been exploited. Calixarene films have been formed by spin coating, a completely general technique of practical importance for the formation of all kinds of films. Here a solution of the material is applied to a rapidly rotating plate and spread over its surface by centrifugal forces. This is a pragmatic technique for practical applications more readily assessed in terms of function rather than structure of the film [43]. In fact little is known about the precise structure of such films, but an orientation can be achieved sometimes after film formation, *e.g.* for molecules with an dipole moment, by application of a strong electric field.

Complex combinations of self-assembly and molecular recognition processes can be used to construct thin films of sophisticated composition [2,44] (see Chapter 26).

3. Properties of Calixarene- and Resorcarene-based Thin Films

3.1. IMPREGNATION OF SURFACES

Mono-or multilayers attached to the surface of a solid material may drastically change its properties. Some particularly interesting systems are provided by the macrocycles derived from resorcinol or pyrogallol and perfluoroaldehydes [45]. Thus, a four-fold LB-layer of **11f** shows a very high water contact angle, while a multilayer self-assembled from a hydrophilic solvent such as acetone shows a low contact angle. This suggests, that in the former case the hydrophobic tails are in the surface, while the opposite orientation - "bowls-in-the-surface" – is present in the latter case.

Perfluoroalkyl chains are evidently able to interdigitate, albeit to a lesser extent than perhydroalkyl ones and avoid presenting the hydrophobic perfluoroalkyl surface to the hydrophilic solvent. Such resorcarenes adhere readily to the hydrophilic surfaces of paper cotton, cloth and wood, conferring a hydrophobic surface and in the case of the perfluoro compounds one which is oleophobic as well. These effects are manifested by tiny coverages of the order of 1 to 5 layers and are clearly of commercial potential [45].

In combination with the *cis-trans* isomerisation of azobenzene functions, the surface properties of solids can be also switched by irradiation, which has been demonstrated by the light-driven motion of liquid droplets (*e.g.* olive oil, tetrachloroethane) on a silica surface coated with a resorcarene similar to **15** [46].

3.2. ADSORPTION BY FILMS

Adsorption of guest ions or molecules by calixarenes in thin films raises possibilities of applications in sensors and separative media (see Chapters 26, 32, 34, 36). Layers of calixarenes are two and three dimensional supramolecular arrays and may offer adsorption properties differing from those anticipated from those exhibited by the isolated molecules in solution.

3.2.1. Adsorption of ions
The per(alkoxycarbonylmethyl) derivatives of t-butyl calix[n]arenes **7a,b,c** (n = 4,6,8 respectively) form strong complexes with the alkali metal cations in homogeneous solutions (see Chapter 22). LB monolayers of these calixarenes show properties dependent

upon the alkali metal salt present in the aqueous subphase [10]. Typically, for the conformationally fixed calix[4]arene **7a**, the π-A curves were much expanded when the underlying aqueous solution contained NaCl but not when it contained KCl or LiCl. Coordination of a Na^+ ion apparently causes either conformational change, increasing the effective cross-sectional area of the calixarene, or produces electrostatic repulsion between bound ions, or both. For chloride as counterion, the effects fall in the order K^+ > Rb^+ > Na^+ > Li^+ for **7b**, and Rb^+ > K^+ > Na^+ > Li^+ for **7c**, in good agreement with data from experiments in solution for the selectivity in ion binding.

Similar studies of monolayers of *t*-butylcalix[6]arene [47] and its hexa-amide derivative **21b** [48] revealed somewhat more complicated behaviour, interpreted for **21b** in terms of rather dramatic conformational changes resulting from the binding of different alkali metal cations (plus chloride anion), so that limiting molecular area estimates could not be used as a measure of relative cation affinities. Nonetheless, it can be said that monolayers of *t*-butylcalix[6]arene show a preference for Cs^+ binding, while **21b** strongly prefers guanidinium cation (with thiocyanate anion) over the alkali metal cations (with chloride). Cs^+-selectivity was also found for monolayers formed by 1,3-dioctyl-calix[4]-crown-6 (compare structure **16**) [49].

19 X = OH	**22a** Y = H	
20 X = NHOH	**22b** Y = CH_2-C(O)-OEt	
21 X = NEt_2	**22c** Y = CH_2-C(O)-CH_3	
a n = 4 b n = 6	**22d** Y = CH_2-CO_2H	
	22e Y = CH_2-CH_2-CH_2-NH_2	

In addition to thorough investigations of the alkali metal salts, monolayer binding of lanthanide (rare earth) cations by **19** and **20** has also been demonstrated [50]. Instability of the monolayers may explain some of the seemingly complex results (noted above) found for the effects of alkali metal salts in the aqueous subphase. With a calix[4]arene, *p*-substituted by two octadecanoyl residues, stable monolayers are formed but surface light scattering measurements show no specific effects of alkali metal salts on either static or dynamic properties of the monolayers. The phenolic hydroxyl groups are appreciably acidic (pK_1 ~6.4, pK_2 ~9.2) in the monolayer and this may explain structural expansion when the subphase contains polycations derived from species such as spermine or polyethyleneimine [51].

XPS studies of LB multilayers (16 - 17 layers) of calix[8]arenes **22a-c** deposited on *n*-type silicon have revealed the incorporation of both anions and cations, with an apparent excess of the former, on dipping the multilayer into aqueous salt solutions. While ion inclusion may not be uniform throughout the layers, there is a preference for larger species such as Cs^+ and I^- [52]. The reversibility of the ion inclusion suggests that applications of such films in ion sensors may be possible. In contrast, irreversible changes in the UV-absorbance are observed when multilayers from the *p*-nitrophenyl-azo-substituted resorcarene **12a** are soaked in metal salt solutions [53].

A novel sensor system was based on a SAM-coated microcantilever. A thiol derivative of 1,3-alternate calix[4]benzocrown-6 similar to **16** assembled on gold enabled the detection of caesium ions in the concentration range of 10^{-11}-10^{-7} M in the presence of high concentrations of potassium ions [54].

3.2.2. Adsorption of neutral species
Very extensive investigations have been made of the use of calixarenes and their derivatives in films responsive to the presence of neutral species in both liquid and gaseous phases. Applications of such films as sensors have been largely based on their incorporation into the quartz crystal microbalance (QCM) and field-effect transistor (FET) devices [1a,c,55,56]. The following discussion gives some idea of the diversity of systems investigated.

Resorcarene **12a** forms a monolayer on SnO_2 and the binding of simple sugars as guests within the monolayer can then be monitored electrochemically [16]. Interestingly the observed affinity is different from that observed for extraction experiments from the aqueous phase into CCl_4. SPR measurements have been used to investigate selective binding to a bilayer formed by adsorption of **17** to a self-assembled monolayer of octane thiol on gold (deposited on silicon) [39]. The bilayer binds lectins, polysaccharides and poly(vinyl alcohol) in water and in some instances marked discrimination (*e.g.*, in the binding of concanavalin A) results from the change of the terminal saccharide from galactose to glucose. Recently, it has been shown that these two resorcarenes are highly discriminatory in their interactions with cell walls, an effect which may allow selective delivery of bioactive guests to cell membrane surfaces [57].

Selective adsorption of a series of poly-oxygenated compounds from dilute aqueous solutions by SAMs of the resorcarene tetrathiol **12d** on gold [31] has been observed. Binding of glucuronic and glutaric acids, butyrolactone and poly(vinylpyrrolidone) was weak, but binding of glucuronolactone, vitamin C or other related lactones was strong and was attributed to an intra-complex reaction resulting in acylation of phenolic hydroxyl groups subsequent to initial inclusion. For self-assembled multilayers, 7-40 layers thick, of resorcarene **12a**, steric effects may explain the reversible binding of glutaric acid when neither 2,2-dimethyl- nor 1,3-dimethyl-glutaric acid (nor adipic acid) show appreciable interactions [31].

The selective binding of low molecular weight amines from aqueous solution by thin films of a *1,3-alternate* calix[4]arene-crown-6 and calix[6]arene hexaester **7b** was established using a QCM sensor [58]. The selectivity of **7b** followed an order of decreasing steric bulk, *i.e.,* propylamine > butylamine > 1-phenylethylamine > heptylamine ~ hexylamine > sec-butylamine >> triethylamine > dicyclohexylamine ~ diisobutylamine.

Changes in surface area resulting from the binding of the D- and L-amino acids alanine, valine, leucine and tryptophan present in an aqueous subphase to Langmuir monolayers of chiral resorc[4]arenes with $(S)(+)$-2-phenylethylamine or $(1R,2S)(-)$-norephedrine substituents (introduced via Mannich reactions) were sufficient to allow discrimination between the enantiomers of all the acids [59].

Langmuir monolayers formed from a mixture of equimolar amounts of *p-t*-butylcalix[8]arene and C_{60} on a water subphase exhibited the same π-A isotherms as a Langmuir monolayer of the isolated C_{60} complex [8b] suggesting formation of 1:1 complexes stabilised at the air-water interface. Spreading isotherm measurements were interpreted as indicating that the calix[8]arene has a "pleated loop" conformation which is retained in its complex with C_{60}. Related chemistry is discussed in greater detail in Chapter 26.

Self-assembly of a thiol (and carboxylate) functionalised diacetylene on a gold surface followed by photochemical cross-linking and activation of carboxyl units enabled immobilisation of *p*-alkyl-calix[n]arenes (n = 4,6) and their use to adsorb volatile organic compounds such as benzene, n-heptane and trichloroethene [60]. The results were consistent with specific interactions between the calixarene cavities and the volatile molecules. The studies have been extended to bilayers formed with calix[4]arenes substituted at the wide rim by phenylazo or dibenzylamino groups [61]. QCM sensors have been used to monitor similar adsorptive behaviour by thin films of the calixarenes only [62,63]. For tetrachloroethene, the mass change for calixarenes increased in the order p-t-butyl-calix[8]- > p-i-propyl-calix[4]- > p-t-butyl-calix[4]arene. The selectivity of p-t-butyl-calix[4]arene increased in the order benzene < chloroform < tetrachloroethene < toluene. For tetrachloroethene alone, QCM studies [8a] also showed that the formation of the trimethylsilyl ether of p-t-butyl-calix[4]arene dramatically reduced adsorption, presumably as a result of repulsions between the narrow rim substituents closing down the cavity volume.

XPS studies [64,65] of the adsorption of volatile organic compounds by monolayers formed from the rigid cavitand **23** showed binding energies to increase in the order $CCl_4 \sim CHCl_3 < C_2HCl_3 \sim$ toluene $\ll C_2Cl_4$ (tetrachloroethene). The largest response and highest selectivity was for C_2Cl_4. By comparison, self-assembled monolayers of di-decyl sulfide and octadecanethiol gave very much smaller responses to C_2Cl_4, interpreted by the lack of binding cavities at the surface. SPR measurements [28,66] to compare the response of the sulfide cavitand **23** and octadecanethiol monolayers to the same organic vapours, indicated a response which was at least a factor of two greater for the cavitand monolayer for each of the compounds but twelve times greater for C_2Cl_4 [28]. Comparison of extended, quinoxaline-derived cavitands **24** and **25** which exist in the so-called *vase* and *kite* conformations, respectively, showed the *kite* form to have reduced sensitivity towards organic pollutant gases both as gases and in solution [67]. Related cavitands **26** where the phosphoryl group provides an additional hydrogen bond acceptor have been used as selective alcohol sensors (QCM) [68]; the diastereomers with the O=P-group pointing towards the cavity show distinctly stronger responses, due to synergistic interactions with the included alcohol.

	A	R	Y
23	H	$C_{10}H_{20}$-S-$C_{10}H_{21}$	CH_2
24	H	C_6H_{13}	(quinoxaline)
25	CH_3	C_6H_{13}	(quinoxaline)
26	H, Br	$C_{11}H_{23}$	CH_2

23 - 25   **26**   Ar = Ph, O-Ph

SAMs of various cavitands with thioether chains (analogous to **12e**) were studied for the binding of neutral molecules from water [69] and the recognition of steroids by similar monolayers with receptors composed of calix[4]arene/resorcarene (2:1) was reported [70] (see also Chapter 9).

3.3. FILM PERMEABILITY

The formation of monolayers of amphiphilic calix[n]arenes on aqueous subphases with areas corresponding to a hexagonal packing of the calixarenes suggested their deposition as Langmuir-Blodgett films on macroporous supports. Assuming the orientation of the calixarene molecules is retained they can be regarded as providing a pore for the selective permeation of small molecules across the film. For this reason, LB films derived from the mercurated calix[6]arene **4** have been described as "perforated monolayers" [14], and development of these materials led to selective gas membranes (He or N_2 versus SF_6) functioning with the deposition of as few as two monolayers of the calixarene on a particular substrate [71].

Although it may be difficult to establish whether passage occurs through or between the calixarene pores [72], composite films based on calix[6]arenes in particular show permeation properties which are distinctly different from membranes prepared analogously using a "conventional" surfactant [73] and seem to be now at the point of true practical applications [74]. Even a single layer of calixarene **11** (similar in structure to **4** with $Hg(O_2CCF_3)$ replaced by $C(=NOH)NH_2$) deposited on a polymer/cyclodextrin membrane support can alter the relative permeance for He over N_2 by nearly two orders of magnitude[75]. Very recently the decrease of the permeability for He of a membrane composed of six layers of **11** in the presence of moisture (the lower permeability for N_2 was not changed) was attributed to the involvement of two distinct permeation pathways, through molecular pores and through transient gaps [76].

3.4. NONLINEAR OPTICAL PROPERTIES

Compounds showing nonlinear optical (NLO) properties are of interest for applications such as optical switching, communications and the frequency doubling of low cost laser sources. Appropriate molecules must be noncentrosymmetric or form noncentrosymmetric crystals. Calix[n]arenes as donor-acceptor aromatic molecules can be fixed in the noncentrosymmetric *cone* conformation and several calix[4]arenes bearing *p*-nitro substituents have been found to display significant optical nonlinearity [77-79]. Their macroscopic properties were examined in thin films spun cast with polymethylmethacrylate onto appropriate substrates and then subjected to poling by a strong electric field. The *p*-nitrocalix[4]arene-tetrapropylether content could be increased even to 100% (!) and the quality of the films increased by crystallisation at 130-140°C in a dc electric field. Highly stable films with a noncentrosymmetric structure were obtained, while similar nitrocalix[4]arene derivatives with longer or branched ether chains gave no crystalline phase which explains their limited NLO stability [80].

Incorporation of the NLO-active calix[4]arene into the polyimide **27** [81] gave smooth, physically stable, transparent thin films, features important for practical applications (*e.g.* frequency doubling at 820 nm).

Monolayers of calix[4]arene based chromophores with NLO properties have been obtained also by covalent attachement on silica surfaces as illustrated by structure **28** [82].

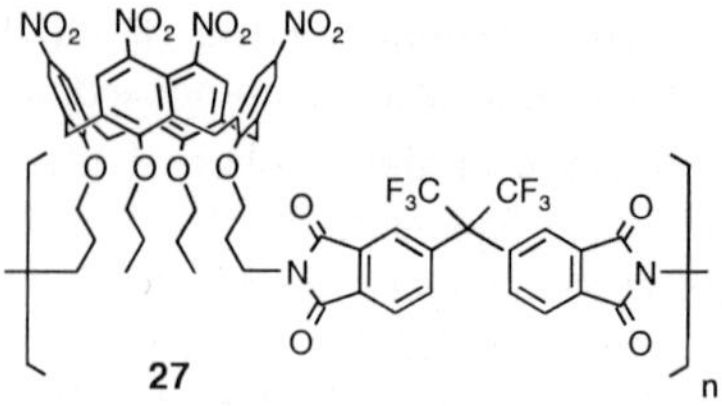

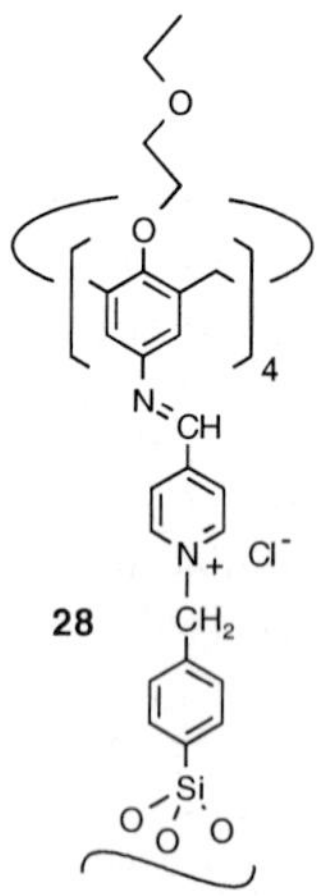

3.5. PYROELECTRIC PROPERTIES

A non-centrosymmetric structure is also necessary to observe the so-called pyroelectric effect, a temperature-dependent electric polarisation of compounds or materials. The alternate layer Langmuir-Blodgett deposition technique can be used to arrange molecules in the required way. An LB multilayer constructed from carboxyl and aminofunctionalised calix[8]arenes **22d** and **22e** in an alternating **22d/22e/22d/22e** pattern on an aluminium surface was prepared by dipping through the monolayer of **22e** and removing the carrier through the monolayer of **22d**, both kept at a surface pressure of 25 mN m^{-1}. They showed very high pyroelectric coefficients as compared to other fatty acid/fatty amine LB films [83]. The pyroelectric response remained constant over ~5000 heating / cooling cycles while the temperature range could be dramatically increased by introducing metal ions into the LB matrix [84]. The effect of the pendent chain structure on the pyroelectric effect has been studied [85] and the effect of admixture with polysiloxanes [86].

3.6. LUMINESCENCE AND FLUORESCENCE OF MONOLAYERS

Some functionalised calixarenes and their metal complexes are known to display luminescence in solution (see Chapter 31) and for LB monolayers of calix[4]arenes **19** and **20** at the airwater interface, complexation with lanthanide ions such as Sm(III), Eu(III), Tb(III) or Dy(III) from the water subphase provided luminescent films [50]. The fluorescence emission spectra could be tuned by changing the surface pressure of the monolayers. Terbium luminescence was also observed for LB-mono- and multilayers of **13a,b** [87].

Monolayers of the Na$^+$-selective fluoroionophore **29**, covalently fixed to a glass surface as indicated showed a fluorescence response (decrease of the monomer and increase of the excimer band of pyrene) to sodium acetate in methanol [88].

3.7. OTHER APPLICATIONS OF CALIXARENE THIN FILMS

Thin films of calixarenes have been examined for use in semiconductor type materials for electronic and sensor devices. Electrical measurements on thin layers of *p-i-*propylcalix[6]arene on Al, Cr and Au electrodes [1b,89] showed the films to be highly insulating but, after iodine doping, to become conducting. Another potential application of calixarenes in semiconductor materials is the formation of cadmium sulfide (CdS) nanoparticles. LB films of cadmium salts of **17** and **19** on reaction with hydrogen sulfide [90] provided CdS particles of mean diameter 1.5 ± 0.3 nm, a size which did not depend on the type of calixarene or the number of LB layers. These CdS nanoparticles have been shown to behave as "quantum dots" [91,92].

Self-assembled monolayers bound through sulfur onto gold (see Figure 3) have been prepared also from the DMF complex of a desymmetrized carcerand composed of a calix[4]arene and a cavitand (similar to **23**) [29]. They may function as molecular switching devices [28]. Within the carceplex, the encapsulated DMF molecule can switch via rotation between two orientations (see Chapters 9 and 10). If such devices could be used in digital data storage materials, they could greatly increase storage capacity.

The *rccc* resorcarene **30** obtained with 3,7,12-trimethylcholanal was recently incorporated into lipid bilayers. It forms pores with ion-channel properties, with long lasting stable open states (2.5-4.5 s) as compared to similar channels formed by **12a** [93]. This example shows again the inexhaustible possibilities offered by calixarenes and resorcarenes.

4. References and Notes

[1] See, for example, a) M. M. G. Antonisse, R. J. W. Lugtenberg, R. J. M. Egberink, J. F. J. Engbersen, D. N. Reinhoudt (Ch. 3) and D. Diamond, T. Grady, S. O'Neill, P. Kane (Ch. 7) in J. P. Desvergne, A. W. Czarnik (Eds) *Chemosensors of Ion and Molecule Recognition* NATO ASI Series C, Vol. 492, Kluwer, Dordrecht, **1997**; b) R. B. Chaabane, M. Gamoudi, G. Guillard, C. Jouve, F. Gaillard, R. Lamartine, *Synth. Metals* **1994**, *66*, 49-54; c) R. Mlika, I. Dumazet, M. Gamoudi, R. Lamartine, B. H. Ouada, N. Jaffrezic-Renault, G. Guillard, *Proc. Electrochem. Soc.* **1997**, *97*, 33-49; d) P. J. Kenis, O. F. J. Noordman, S. Houbrechts, G. J. van Hummel, S. Harkema, F. C. J. M. van Veggel, K. Clays, J. F. J. Engbersen, A. Persoons, N. F. van Hulst, D. N. Reinhoudt, *J. Am. Chem. Soc.* **1998**, *120*, 7875-7883; e) D. Lacey, T. Richardson, F. Davis, R. Capan, *Mater. Sci. Eng. C* **1998**, *C8*, 377-384, and references therein.
[2] T. Hatano, A. Ikeda, T. Akiyama, S. Yamada, M. Sano, Y. Kanekiyo, S. Shinkai, *J. Chem. Soc., Perkin Trans 2* **2000**, 909-912.
[3] A. Ulman *An Introduction to Ultrathin Organic Films: From Langmuir-Blodgett to Self-Assembly*, Academic Press, New York, **1991**.
[4] a) G. Picard, I. Nevernov, D. Alliata, L. Pazdernik, *Langmuir* **1997**, *13*, 264-276; b) V. Böhmer, A. Shivanyuk in L. Mandolini, R. Ungaro (Eds) *Calixarenes in Action*, Imperial College Press, London, **2000**, pp. 203-240.
[5] a) Early studies indicate the formation of Langmuir monolayers from t-butyl calix[6]arene while the unsubstituted calix[6]arene forms vesicles: M. A. Markowitz, R. Bielski, S. L. Regen, *Langmuir* **1989**, *5*, 276-278. b) Stable monolayers are formed from p-stearyl calix[4]arene only on (slightly) basic subphases, indicating, that deprotonation may be necessary to make the narrow rim sufficiently hydrophilic: Y. Nakamoto, G. Kallinowski, V. Böhmer, W. Vogt, *Langmuir* **1989**, *5*, 1116-1117.
[6] A. R. Esker, L-H. Zhang, C. E. Olsen, K. No, H. Yu, *Langmuir* **1999**, *15*, 1716-1724 and references therein.
[7] For Langmuir films of *t*-butylcalix[8]arene and its mixtures with C_{60} and C_{70} see: R. Castillo, S. Ramos, R. Cruz, M. Martinez, F. Lara, J. Ruiz-Garcia, *J. Phys. Chem.* **1996**, *100*, 709-713.
[8] See, however, a) F. L. Dickert, O. Schuster, *Adv. Mater.* **1993**, *5*, 826-829; (b) L. Dei, P. LoNostro, G. Capuzzi, P. Baglioni, *Langmuir* **1998**, *14*, 4143-4147.
[9] J. C. Tyson, J. L. Moore, K. D. Hughes, D. M. Collard, *Langmuir* **1997**, *13*, 2068-2073.
[10] Y. Ishikawa, T. Kunitake, T. Matsuda, T. Otsuka, S. Shinkai, *J. Chem. Soc., Chem. Commun.* **1989**, 736-738.
[11] G. Merhi, M. Munoz, A. W. Coleman, *Supramol. Chem.* **1995**, *5*, 173-177.
[12] P. Dedek, V. Janout, S. Regen, *J. Org. Chem.* **1993**, *58*, 6553-6555.
[13] M. A. Markowitz, V. Janout, D. G. Castner, S. L. Regen, *J. Am. Chem. Soc.* **1989**, *111*, 8192-8200.
[14] M. D. Connor, V. Janout, I. Kudelka, P. Dedek, J. Zhu, S. L. Regen, *Langmuir* **1993**, *9*, 2389-2397.
[15] R. A. Hendel, V. Janout, W. Lee, S. L. Regen, *Langmuir* **1996**, *12*, 5745-5746.
[16] K. Kurihara, K. Ohto, Y. Tanaka, Y. Aoyama, T. Kunitake, *J. Am. Chem. Soc.* **1991**, *113*, 444-450.
[17] K. Ichimura, N. Fukushima, M. Fujimaki, S. Kawahara, Y. Matsuzawa, Y. Hayashi, *Langmuir* **1997**, *13*, 6780-6786.
[18] W. Moreira, P. L. Dutton, R. Aroca, *Langmuir* **1994**, *11*, 4148-4152; *Langmuir* **1995**, *11*, 3137-3144.
[19] F. Davis, A. L. Lucke, K. A. Smith, C. J. M. Stirling, *Langmuir* **1998**, *14*, 4180-4185.
[20] E. Kurita, N. Fukushima, M. Fujimaki, Y. Matsuzawa, K. Kudo, K. Ichimura, *J. Mater. Chem.* **1998**, *8*, 397-403.

[21]   Y. Matsuzawa, T. Seki, K. Ichimura, *Chem. Lett.* **1998**, 411-412.
[22]   Compare also: M. Conner, I. Kudelka, S. L. Regen, *Langmuir* **1991**, *7*, 982-987.
[23]   L. F. Lindoy, I. Atkinson *Self Assembly in Supramolecular Systems*, Monographs in Supramolecular Chemistry No. 4, J. F. Stoddart (Ed.), Royal Society of Chemistry, Cambridge, **2000**.
[24]   See, for example, H. Adams, F. Davis, C. J. M. Stirling, *J. Chem. Soc., Chem. Commun.* **1994**, 2527-2529.
[25]   See, for example, a) K-D. Schierbaum, T. Weiss, E. U. Thoden van Velzen, J. F. J. Engbersen, D. N. Reinhoudt, W. Göpel, *Science* **1994**, *265*, 1413-1415; b) E. U. Thoden van Velzen, J. F. J. Engbersen, D. N. Reinhoudt, *J. Am. Chem. Soc.* **1994**, *116*, 3597-3598.
[26]   See for instance: a) M. Namba, M. Sugawara, P. Bühlmann, Y. Umezawa, *Langmuir* **1995**, *11*, 635-638; b) H. Schönherr, G. J. Vansco, B.-H. Huisman, F. C. J. M. van Veggel, D. N. Reinhoudt, *Langmuir* **1997**, *13*, 1567-1570 and *Langmuir* **1999**, *15*, 5541-5546.
[27]   See for example: a) A. V. Nabok, A. K. Hassan, A. K. Ray, O. Omar, V. I. Kalchenko, *Sens. Actuators B* **1997**, *45*, 116-121; b) A. K. Hassan, A. K. Ray, A. V. Nabok, S. Panigrahi, *IEE Proc. Sci. Measurem. Techn.* **2000**, *147*, 137-140.
[28]   E. U. Thoden van Velzen, J. F. J. Engbersen, P. J. de Lange, J. W. G. Mahy, D. N. Reinhoudt, *J. Am. Chem. Soc.* **1995**, *117*, 6853-6862.
[29]   B-H. Huisman, F. C. J. M. van Veggel, D. N. Reinhoudt, *Pure Appl. Chem.* **1998**, *70*, 1985-1992.
[30]   B-H. Huisman, E. U. Thoden van Velzen, F. C. J. M. van Veggel, J. F. J. Engbersen, D. N. Reinhoudt, *Tetrahedron Lett.*, **1995**, *36*, 3273-3276.
[31]   a) F. Davis, C. J. M. Stirling, *Langmuir* **1996**, *12*, 5365-5374; b) F. Davis, M. Gerber, N. Cowlam, C. J. M. Stirling, *Thin Solid Films* **1996**, *284-285*, 678-682.
[32]   M. T. Cygan, G. E. Collins, T. D. Dunbar, D. L. Allara, C. G. Gibbs, C. D. Gutsche, *Anal. Chem.* **1999**, *71*, 142-148.
[33]   M. Ueda, N. Fukushima, K. Kudo, K. J. Ichimura, *J. Mater. Chem.* **1997**, *7*, 641-645.
[34]   S.-K. Oh, M. Nakagawa, K. Ichimura, *Chem. Lett.* **1999**, 349-350.
[35]   a) M. Fujimaki, Y. Matsuzawa, Y. Hayashi, K. Ichimura, *Chem. Lett.* **1998**, 165-166; b) M. Fujimaki, S. Kawahara, Y. Matsuzawa, E. Kurita, Y. Hayashi, K. Ichimura, *Langmuir* **1998**, *14*, 4495-4502.
[36]   A. M. Nechifor, P. A. Philipse, F. de Jong, J. P. M. van Duynhoven, R. J. M. Egberink, D. N. Reinhoudt, *Langmuir* **1996**, *12*, 3844-3854.
[37]   G. Capuzzi, E. Fratini, F. Pini, P. Baglioni, A. Casnati, J. Texeira, *Langmuir* **2000**, *16*, 186-194.
[38]   T. Fujimoto, C. Shimizu, O. Hayashida, Y. Aoyama, *J. Am. Chem. Soc.* **1997**, *119*, 6676-6677.
[39]   O. Hayashida, C. Shimizu, T. Fujimoto, Y. Aoyama, *Chem. Lett.* **1998**, 13-14.
[40]   R. Roy, J. M. Kim, *Angew. Chem.* **1999**, *111*, 380-384.
[41]   F. Davis, C. J. M. Stirling, *J. Am. Chem. Soc.* **1995**, *117*, 10385-10386.
[42]   D. E. Hibbs, M. B. Hursthouse, K. M. A. Malik, H. Adams, C. J. M. Stirling, F. Davis, *Acta Cryst. C* **1998**, *54*, 987-992.
[43]   A. Lucke, C. J. M. Stirling in L. Mandolini, R. Ungaro (Eds) *Calixarenes in Action* Imperial College Press, London, **2000**, pp. 172-202.
[44]   See also: L. Zhang, L. A. Godinez, T. Lu, G. W. Gokel, A. E. Kaifer, *Angew. Chem.* **1995**, *107*, 236-239.
[45]   a) F. Davis, C. Capaccioli, E. Ashkenazi, N. H. Williams, C. J. M. Stirling, unpublished work; (b) F. Davis, C. J. M. Stirling, UK Patent application No. 97916571.9-2115.
[46]   a) K. Ichimura, S.-K. Oh, M. Nakagawa, *Science* **2000**, *288*, 1624-1626; b) S.-K. Oh, K. Ichimura, M. Nakagawa, *Mol. Cryst. Liqu. Cryst.* **2000**, *345*, 635-640.
[47]   L. Dei, A. Casnati, P. LoNostro, P. Baglioni, *Langmuir* **1995**, *11*, 1268-1272.
[48]   L. Dei, A. Casnati, P. LoNostro, A. Pochini, R. Ungaro, P. Baglioni, *Langmuir* **1996**, *12*, 1589-1593.
[49]   G. Capuzzi, E. Fratini, L. Dei, A. Casnati, R. Gilles, P. Baglioni, *Colloids Surfaces A* **2000**, *167*, 105-113.
[50]   R. Ludwig, H. Matsumoto, M. Takeshita, K. Ueda, S. Shinkai, *Supramol. Chem.* **1995**, *4*, 319-327.
[51]   L-H. Zhang, A. R. Esker, K. No, H. Yu, *Langmuir* **1999**, *15*, 1725-1730.
[52]   F. Davis, L. O'Toole, R. Short, C. J. M. Stirling, *Langmuir* **1996**, *12*, 1892-1894.
[53]   A. K. Hassan, A. V. Nabok, A. K. Ray, F. Davis, C. J. M. Stirling, *Thin Solid Films* **1998**, *327-329*, 686-690.
[54]   H.-F. Ji, E. Finot, R. Dabestani, T. Thundat, G. M. Brown, P. F. Britt, *Chem. Commun.* **2000**, 457-458.
[55]   W. Göpel, *Sens. Actuators B*, **1995**, *B24*, 17-32.
[56]   For recent considerations of general issues, see a) R. M. Crooks, A. J. Ricco, *Acc. Chem. Res.* **1998**, *31*, 219-227; b) A. J. Ricco, R. M. Crooks, G. M. Osbourn, *Acc. Chem. Res.* **1998**, *31*, 289-296 and associated articles.

[57] K. Fujimoto, T. Miyata, Y. Aoyama, *J. Am. Chem. Soc.* **2000**, *122*, 3558-3559.
[58] X. C. Zhou, S. C. Ng, H. S. O. Chan, S. F. Y. Li, *Sens. Actuators B* **1997**, *42*, 137-144.
[59] M. Pietraszkiewicz, P. Prus, W. Fabianowski, *Polish J. Chem.* **1998**, *14*, 1068-1075.
[60] D. L. Dermody, R. M. Crooks, T. Kim, *J. Am. Chem. Soc.* **1996**, *118*, 11912-11917. See also ref. [44].
[61] D. L. Dermody, Y. Lee, T. Kim, R. M. Crooks, *Langmuir* **1999**, *15*, 8435-8440.
[62] A. Dominik, H. L. Roth, K-D. Schierbaum, W. Göpel, *Supramol. Sci.* **1994**, *1*, 11-19.
[63] K-D. Schierbaum, A. Gerlach, M. Haug, W. Göpel, *Sens. Actuators A* **1992**, *31*, 130-137.
[64] T. Weiss, H-D. Schierbaum, E. U. Thoden van Velzen, D. N. Reinhoudt, W. Göpel, *Sens. Actuators B* **1995**, *26*, 203-207.
[65] J. Rickert, T. Weiss, W. Göpel, *Sens. Actuators B* **1996**, *31*, 31-50.
[66] B-H. Huisman, R. P. H. Kooyman, F. C. J. M. van Veggel, D. N. Reinhoudt, *Adv. Mater.* **1996**, *8*, 561-564.
[67] J. Hartmann, P. Hauptmann, S. Levi, E. Dalcanale, *Sens. Actuators B* **1996**, *35*, 154-157.
[68] R. Pinalli, F. F. Nachtigard, F. Ugozzoli, E. Dalcanale, *Angew. Chem., Int. Ed.* **1999**, *38*, 2377-2380.
[69] A. Friggeri, F. C. J. M. van Veggel, D. N. Reinhoudt, R. P. H. Kooyman, *Langmuir* **1998**, *14*, 5457-5463.
[70] A. Friggeri, F. C. J. M. van Veggel, D. N. Reinhoudt, *Chem. Eur. J.* **1999**, *5*, 3595-3602.
[71] M. D. Conner, S. L. Regen, *Adv. Mater.* **1994**, *6*, 873-874.
[72] P. Dedek, A. S. Webber, V. Janout, R. A. Hendel, S. L. Regen, *Langmuir* **1994**, *10*, 3943-3954.
[73] W. Lee, R. A. Hendel, P. Dedek, V. Janout, S. L. Regen, *J. Am. Chem. Soc.* **1995**, *117*, 6793-6794, 10599-10600.
[74] (a) R. A. Hendel, E. Nomura, V. Janout, S. L. Regen, *J. Am. Chem. Soc.* **1997**, *119*, 6909-6918; (b) R. A. Hendel, L-H. Zhang, V. Janout, M. D. Conner, T. Hsu, S. L. Regen, *Langmuir* **1998**, *4*, 6545-6549.
[75] L-H. Zhang, R. A. Hendel, P. G. Cozzi, S. L. Regen, *J. Am. Chem. Soc.* **1999**, *121*, 1621-1622.
[76] X. Yan, J. T. Hsu, S. L. Regen, *J. Am.Chem. Soc.* **2000**, *122*, 11944-11947.
[77] E. Kelderman, L. Derhaeg, G. J. T. Heesink, W. Verboom, J. F. J. Egbersen, N. F. van Hulst, A. Persoons, D. N. Reinhoudt, *Angew. Chem., Int. Ed.* **1992**, *31*, 1075-1077.
[78] E. Kelderman, G. J. T. Heesink, L. Derhaeg, T. Verbiest, P. T. A. Klaase, W. Verboom, J. F. J. Egbersen, N. F. van Hulst, K. Clays, A. Persoons, D. N. Reinhoudt, *Adv. Mater.* **1993**, *5*, 925-930.
[79] For further calixarene derivatives with NLO properties see: E. Kelderman, L. Derhaeg, W. Verboom, F. J. Engbersen, S. Harkema, A. Persoons, D. N. Reinhoudt, *Supramol. Chem.* **1993**, *2*, 183-190.
[80] P. J. A. Kenis, O. F. J. Noordman, H. Schonherr, E. G. Kerver, B. H. M. Snellink-Ruël, G. J. van Hummel, S. Harkema, C. P. J. M. van der Vorst, J. Hare, S. J. Picken, F. J. Engbersen, N. F. van Hulst, G. J. Vansco, D. N. Reinhoudt, *Chem. Eur. J.* **1998**, *4*, 1225-1234.
[81] P. J. A. Kenis, O. F. J. Noordman, N. F. van Hulst, J. F. J. Engbersen, D. N. Reinhoudt, B. H. M. Hams, C. P. J. M. van der Vorst, *Chem. Mater.* **1997**, *9*, 596-601.
[82] X. Yang, D. McBranch, B. Swanson, D. Li, *Angew. Chem.* **1996**, *108*, 572-575; D. Li, X. Yang, D. McBranch, *Synth. Metals* **1997**, *86*, 1849-1850.
[83] T. Richardson, M. B. Greenwood, F. Davis, C. J. M. Stirling, *Langmuir* **1995**, *11*, 4623-4625
[84] C. M. McCartney, T. Richardson, M. A. Pavier, F. Davis, C. J. M. Stirling, *Thin Solid Films* **1998**, *327-329*, 431-434.
[85] C. M. McCartney, T. Richardson, M. B. Greenwood, N. Cowlam, F. Davis, C. J. M. Stirling, *Supramol. Sci.* **1997**, *4*, 385-390.
[86] D. Lacey, T. Richardson, F. Davis, R. Capan, *Mat. Sci. Engin. C* **1999**, *S1*, 377-384.
[87] P. J. Dutton, L. Conte, *Langmuir* **1999**, *15*, 613-617.
[88] N. J. van der Veen, S. Flink, M. A. Deij, R. J. M. Egberink, F. C. J. M. van Veggel, D. N. Reinhoudt, *J. Am. Chem. Soc.* **2000**, *122*, 6112-6113.
[89] R. B. Chaabane, M. Gamoudi, G. Guillard, *Synth. Metals* **1994**, *66*, 231-233.
[90] A. V. Nabok, T. Richardson, F. Davis, C. J. M. Stirling, *Langmuir* **1997**, *13*, 3198-3201.
[91] A. V. Nabok, A. K. Ray, A. K. Hassan, M. Titchmarsh, F. Davis, T. Richardson, A. Starovoitov, S. Bayliss, *Mat. Sci. Eng. C, Biomimetic Supramol. Syst.* **1999**, *8*, 171-177.
[92] A. V. Nabok, T. Richardson, C. McCartney, N. Cowlam, A. K. Ray, F. Davis, V. Gacem, A. Gibaud, *Thin Solid Films* **1998**, *327-329*, 510-514.
[93] N. Yoshino, A. Satake, Y. Kokube, *Angew. Chem.* **2001**, *113*, 471-473, *Angew. Chem., Int. Ed.* **2001**, *40*, 457-459 and references cited there.

Chapter 34

SENSOR APPLICATIONS

FRANCIS CADOGAN, KIERAN NOLAN, DERMOT DIAMOND
The Biomedical and Environmental Sensor Technology (BEST) Centre, School of Chemical Sciences, Dublin City University, Glasnevin, Dublin 9, Ireland.
e-mail: Dermot.Diamond@dcu.ie

1. Introduction

This chapter focuses on the exploitation of host-guest interactions in calixarene systems for the development of chemical sensors. To fulfil this aim, it is necessary to define what we mean by the term *sensor*. A topic of ongoing debate between chemists is whether the sensing molecule (host) or the whole device is the sensor. The IUPAC Commission on General Aspects of Analytical Chemistry has defined chemical sensors as follows [1]:

"A chemical sensor is a device that transforms chemical information, ranging from the concentration of a specific sample component to total composition analysis, into a useful analytical signal."

On the basis of this definition the whole device is the sensor, for without immobilisation of the host and transduction of the interaction, it would not be possible to convert the host-guest interaction into a useful analytical signal. The physical immobilisation of the selective agent, and the transduction process have distinct effects on the performance of a sensing device, especially in relation to selectivity, sensitivity and response time. We will therefore discuss different modes of measurement and immobilisation using calixarene derivatives.

Cation recognition is undoubtedly the most studied and best developed application of calixarenes in sensing devices, and for this reason is the focus of most of the present discussion. Molecules such as **1-10** (Figure 1), predominantly derivatives of calix[4]-arene in its *cone* conformation with, for example, ester [2], ether [3], ketone [4], carboxylic acid [5], amide [6], crown ether [7] and hemispherand [8] substituents, have been extensively studied over the years. Of particular note in this area is the work pioneered in 1986 [9-11] on the use largely of fixed *cone* calix[4]arene "tetraester" derivatives **1** as ligands with convergent functional groups suitable for encapsulating guest cations. Many of these molecules showed a significant affinity for the alkali metal cations [12] and, upon screening as potential selective agents for use in PVC (polyvinylchloride)-membrane ion-selective electrodes (ISEs), excellent selectivities in favour of Na^+ were obtained [11]. Subsequently, an array of related derivatives was studied [13,14] and the resulting ISEs were used successfully for the analysis of Na$^+$ in blood using both steady-state and flow-injection analysis (FIA) techniques [15-17]. Other work of note in this area includes a comprehensive study of some twenty calix[4]arenes of types **1, 2, 6** and **7** [18] and the use of conformational locking of **9** (R = *t*-Bu, Y = *i*-Pr, n = 1) in the *1,3-alternate* form to engender selectivity for K$^+$ over Na$^+$ [19,20].

In general, particularly important recent developments in calixarene-based sensor systems have concerned ISEs, ion-selective field effect transistors (ISFETs), electrolyte-insulator-semiconductor sensors (EIS), quartz-crystal microbalance (QCM) piezoelectric sensors and optodes. In complement to the discussion here, some of the material in Chapters 9, 21, 23, 26, 32 and 33 provides useful background and amplification.

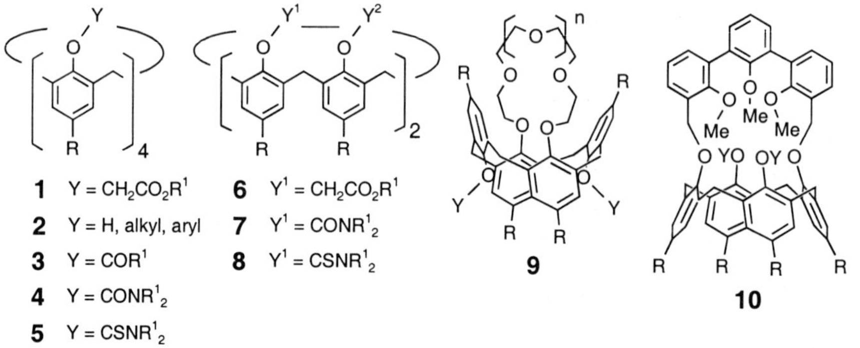

1 Y = CH$_2$CO$_2$R^1 **6** Y^1 = CH$_2$CO$_2$R^1
2 Y = H, alkyl, aryl   **7** Y^1 = CONR$^1_2$
3 Y = COR^1 **8** Y^1 = CSNR$^1_2$
4 Y = CONR$^1_2$
5 Y = CSNR$^1_2$

Figure 1. Molecular survey of typical calixarene ionophores (R very commonly but not invariably, is t-Bu).

2. Electrochemical Sensors

2.1. POTENTIOMETRY

Potentiometry [21] is one of the oldest electroanalytical methods and enables estimations of physicochemical quantities such as activity coefficients, pH values, dissociation constants and solubility products, by either direct or indirect methods [21,22]. The relationship between the concentration of the analyte and an observed potential is expected to be given by the Nernst Equation but, due to the failure of this to describe accurately the processes of many modern ISEs, particularly in the case of ions of different charge, other mathematical treatments have been suggested [23].

While ISEs can show extremely high selectivity, the detection limits are typically around the micromolar range, limiting their use and preventing the measurement of true selectivity values. Recently, however, modified methods have been developed which appear to dramatically improve the detection limits of these devices [24-27], and open up new areas of application (*e.g.* trace level detection of metal ions). Conventionally, an ISE consists of an ion-selective membrane, an internal filling or electrolytic contact solution (or a solid contact, *e.g.*, a hydrogel, in the case of solid state ISEs) and an internal reference electrode. The idea of this new work is to introduce a concentration gradient in the membrane whereby there is a constant flux of primary ions towards the internal filling solution. This gradient is set up by using an appropriate internal filling electrolyte, *i.e.*, one with low enough primary ion activity and high enough interfering ion activity. A study of such a system for Pb analysis has been carried out using a calix[4]-arene derivative, tetrakis(dimethylthiocarbamoylmethyl)-*p-t*-butylcalix[4]arene, (**5**, R = *t*-Bu, R^1 = CH_3) [26],], with dramatic improvement in the limit of detection.

The responses of five electrodes based on this ionophore are shown graphically in Figure 2. All five responses were recorded in a background of 10^{-4} M acetate buffer, pH 4.7. Electrodes A, B, B' and C all had identical membrane compositions, electrode D a different plasticiser and different active component concentrations. Electrode A was a conventional ISE with 10^{-3} M $PbCl_2$ internal filling solution and gave a detection limit ~10^{-7} M Pb^{2+}. Electrode B showed an improved, although non-linear response with a detection limit of 10^{-11} M Pb^{2+}. The internal filling solution in this case was higher in interferent ion activity than Pb^{2+} activity (10^{-2} M Na_2EDTA, 10^{-3} M $PbCl_2$ and 10^{-2} M NaCl) thus inducing an inward flux of Pb^{2+} ions. Electrode B' used the same internal filling solution as electrode B but the membrane thickness in this case was reduced from the conventional 200 µm to 50 µm. The thinner membrane displayed an even more elevated response, which has been

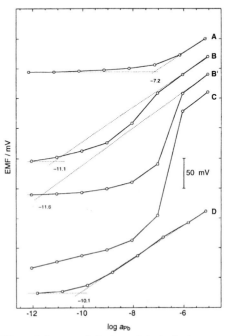

*Figure 2. Response behaviour of electrodes incorporating **5**, R = t-Bu, R^1 = Me.*

explained as a depletion of analyte ions at the surface of the electrode due to a strong inward flux at intermediate activity levels [26]. The thicker membrane is less influenced by this effect. Electrode C, once again had a lower Pb^{2+} activity and a higher Na^+ activity (10^{-2} M Na_2EDTA, 10^{-3} M $PbCl_2$ and 2 M NaCl) and thus the internal filling electrolyte produced a larger gradient towards the inner membrane interface. The response seen was again more extreme than the previous examples, *i.e.*, the depletion at intermediate activities was more pronounced. This problem may be overcome by decreasing the concentration of the active components (the calixarene and the anionic site) by an order of magnitude (as ion fluxes within the membrane are directly dependent on these concentrations [25]. Thus electrode D was found to give reliable measurements with a detection limit of 10^{-10} M. This result is of immense importance to the field of potentiometry and the use of calixarenes in such devices as it shows that selectivity values and electrode performances need to be reassessed in order to obtain the maximum from these devices. Considering the body of knowledge already present through the literature, the low cost of these devices and their ability to perform rapid analysis, this area should further expand resulting in even more demand for calixarene derivatives and other ionophoric compounds.

2.2. RECENT DEVELOPMENTS IN POTENTIOMETRIC SENSOR FABRICATION

As with all sensor technologies, the aim in potentiometry is to develop small rugged solid state devices with excellent selectivity and low cost. To do this it is necessary to eliminate the internal filling solution during fabrication. This goal to date has been achieved through the use of materials such as hydrogels doped with an appropriate salt in order to decouple the charge transfer between the selective membrane and the Ag/AgCl internal reference. These *pseudo* solid-state sensors can be easily mass-produced via techniques currently employed in the electronics/semiconductors industry, such as thin/thick film deposition techniques, drop-on-demand liquid handling and the use of photopolymerisable polymers [28]. For the first time, potentiometric sensors are being mass produced in a miniaturised format [29]. Figure 3 shows a commercially produced flow cell with an array of different sensing devices in one package, four of which are potentiometric, including one based on a calixarene derivative of type **1** [30]. The next step for these devices is the transition from traditional potentiometric methods towards the new, improved limit of detection electrodes discussed earlier.

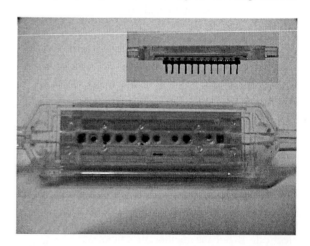

Figure 3. Sensor-array flow cell manufactured by the SenDx Corporation, Carlsbad, CA. The array contains solid state PVC membrane electrodes for Na^+, K^+, Ca^{2+} and H^+. Dimensions, ca. 350 x 15 x 5 mm.

2.3. THE ION SELECTIVE AGENT

The selective membrane of an ISE generally consists of a cocktail of different components which have been studied and optimised for a specific target or sample matrix. Generally, plasticised PVC is used as the membrane matrix, as it can be readily manipulated and used to support the selective components (commonly a charged site, and an ionophore, either charged or neutral). While many groups of compounds have been investigated and employed as the sensing agents in ISEs, calixarenes are particularly well suited to the job. This is due to their relative ease of chemical modification, both in terms of the nature and 3-D arrangement of binding sites. (See Chapters 1-6.)

Criteria for identifying compounds useful as ionophores in ISEs, particularly for cation analysis, have been established on the basis of more than 20 years' research in the area [31,32]. Those identified include acyclic and macrocyclic compounds, the most notable of which being valinomycin for the analysis of potassium ions in blood, while crown ethers and acyclic amides also feature strongly. To function successfully within an ISE, the ideal ionophore must:

(i) be capable of selectively binding with the primary or target ion.
(ii) be preferentially retained within the membrane phase.
(iii) be able to diffuse freely in the direction of the potential gradient across the membrane.
(iv) have only a moderately large stability constant to ensure reversibility.
(v) display fast ion transfer kinetics between the aqueous and membrane phases.

These criteria outline the features of an ideal ionophore or sensing agent and as such define the structural features needed and therefore influence ligand design. For instance, criterion 1 can be satisfied by the optimum spatial arrangement of suitable polar ligating groups that will facilitate strong interaction with the target ion, replacing solvating water molecules and hence enabling extraction of the ions into the membrane. By providing structure and sufficient rigidity to the binding cavity, this helps control or limit the species which can interact with the polar groups. Oxygen-donor ligands such as amides, esters, ketones, carboxylic acids or phosphine oxides will bind many cations. When arranged in a rigid cavity, they bind both strongly and selectively. Criterion 2 is satisfied by the inclusion of sufficient non-polar lipophilic groups to ensure retention of the ionophore within the organic membrane while not restricting the movement of the complex.

Crown ethers and cryptands [33] were the first macrocycles to be recognised as usefully meeting these criteria and, as noted previously, it was not until 1986 that the first publication concerning calixarenes (types **1**, **3**) appeared [11]. The wide structural range (see Figure 1) and suitability of calixarenes for such applications, however, has since led to the development of an extensive literature, particularly that concerning potentiometric ISEs [13,34-37].

2.3.1. Calixarenes as cation ionophores in ion selective electrodes (ISEs)

The bulk of this work has targetted group 1 and 2 metal ions due to their clinical significance. Polymer membrane pH-ISEs incorporating either calix[4]arene or *p-t-*butyl-calix[4]arene as ionophore and tetradecylammonium nitrate as the ionic additive have recently been reported [38]. The resultant electrodes work over a wider pH range than traditional tridecylamine-based electrodes, with the added advantage that the plasticised membrane has a much lower impedance than glass. In extension of early work referred to in Section 1, covalent attachment of calix[4]arene esters **1** (R = *t*-Bu, R^1 = allyl) to siloxysilane oligomers via hydrosilylation reactions before incorporation into a silicone rubber matrix resulted in ISEs sucessfully used to determine sodium activites in urine [39]. Silicone rubber may also be used as a replacement for PVC in solid-state ISEs and enhanced electrochemical performance in blood analysis has been reported for solid state devices based on the tetraester **1** (R = *t*-Bu, R^1 = Et) [40]. A sodium-selective solid state ISE based on room temperature vulcanising-type silicone rubber and this calixarene has been reported to exhibit electrochemical performance similar to conventional PVC based systems [41].

More recently, the emphasis in the production of ionophores has shifted to heavy metal ions. Lead-selective PVC electrodes can be based on di- and tetrathioamide functionalised calix[4]arenes **5** and **8** [42]. The electrode containing the tetrathioamide functionalised calix[4]arene **5** (R = *t*-Bu, R^1 = Me) is claimed to have good selectivity (-log K^{pot}_{Pbj} > 3) against the alkali metals, Cu^{2+}, Zn^{2+} and Cd^{2+} ions and is now commercially available as a selective agent for Pb^{2+}. (See Section 2.1.). The selectivity against Cu^{2+} is particularly surprising, given the well-known affinity of thio groups for Cu^{2+} ions. While this ligand performs well, it is susceptible to hydrolytic or oxidative cleavage.

Calix[n]arene phosphine oxides **11**, first reported in 1995 as a new series of cation receptors for the extraction of certain lanthanides and actinides from simulated nuclear waste [43],were intially studied as possible analytically useful agents for the detection of europium [44]. Following this investigation, it was shown [45] that **11a** (R = *t*-Bu) has the ability to discriminate in favour of calcium ions, against magnesium and alkali metal ions, when incorporated into a PVC-membrane electrodes. Further investigations of these compounds led to the finding that the calix[6]arene homologue **11c** (R = *t*-Bu) showed excellent response characteristics towards Pb^{2+} [46]. This was an important finding as the stability of these compounds to adverse pH conditions meant that they would be suitable for use in the harsh industrial environments. Figure 4 illustrates the effect of changing the calixarene cavity size on its ionophoric behaviour, and it is clear that the ligand selectivity varies dramatically with the ring size. A subsequent comparative study of **11c** and a selection of commercially available Pb^{2+} ligands found this ligand provided improved selectivity against Cu^{2+}, Cd^{2+} and Ag^{2+} ions.

Calixarenes such as **9** (R = *t*-Bu, Y = *i*-Pr, n = 2, *1,3-alternate* conformation), **9** (R = *t*-Bu, Y = Me, n = 2), **12** and **13a,b** have been used in potentiometric devices for K^+ [47], Cs^+ [48], Cd^{2+}, Cu^{2+} [49], and Ag^+ [50].

11a n = 4
b n = 5
c n = 6

12a X = SMe
b X = SC(S)NEt$_2$
c X = O(CH$_2$)$_2$OCH$_2$C(S)NMe$_2$

13a X = SMe, Y = H, Me
b X = SC(S)NEt$_2$, Y = H, Me
c X = O(CH$_2$)$_2$OCH$_2$C(S)NMe$_2$, Y = Pr

2.3.2. Anion discrimination
To date there has been no report in the literature concerning functioning anion-sensing devices. Much work is under way on the development of new anion-selective calixarenes (See Chapter 23), and this may well give rise to sensing devices analogous to those described above.

2.3.3. Organic ion discrimination by ISEs based on calixarenes
While few reports of potentiometric discrimination of organic ions by synthetic hosts have been made [51], discrimination of non-polar structures of organic ions using a PVC-based ISE incorporating a hexakis(decyloxycarbonylmethyl)calix[6]arene derivative **14a** has been shown [52]. Potentiometric selectivities for non-branched primary amines, (*e.g.* 2-phenylethylamine) in their protonated forms were observed and were explained on the basis of bonding between the -NH$_3^+$ group of the guest and the ester carbonyl oxygen atoms of the host. The ligand selectivity in favour of dopamine against

adrenaline and noradrenaline is very interesting from a bioanalytical viewpoint. The fact that K^+ ions proved to be a major interferent forced further research into the use of calix[6]arene compounds into this area. Recently, high selectivity for dopamine has been found using the homooxacalix[3]arene triether host **15**, and not the calix[6]arene compounds. The results showed excellent selectivity for dopamine over other neurotransmitters as well as over K^+ and Na^+ [53].

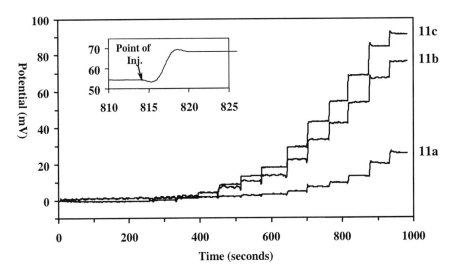

*Figure 4. Response of electrodes based on ligands **11a**, **11b** and **11c** to Pb^{2+} ions ($8 \times 10^{-7} \rightarrow 6 \times 10^{-2}$ M) in a constant background of 1×10^{-2} M $Ca(NO_3)_2$. The inset shows an expanded view of the response of L3 and demonstrates the very rapid dynamic response ($t_{95\%} < 10s$).*

2.4. VOLTAMMETRY AND AMPEROMETRY

Classical electroanalysis uses voltammetry and amperometry in contrast to the potentiometric methods already described. These techniques force redox reactions to occur at the electrode surface under the influence of an externally poised potential. To date, the number of papers on the use of calixarenes in sensing devices based on these techniques is small in comparison to potentiometric or optical procedures. A certain amount of work has been carried out on chemically modified electrodes using calixarenes, while the use of polymeric membrane ion sensors is relatively unexplored. The former involves the immobilisation of the calixarene on the surface of the working electrode, in an attempt to improve the performance of the sensor. The presence of the calixarene on the electrode surface is expected to selectively increase the concentration of the target analyte at the electrode surface, much in the same way as is achieved in classical stripping voltammetry. The use of polymeric calixarene esters e.g., **16**, has been studied for the analysis of Pb^{2+}, Cu^{2+} and Hg^{2+} ions [54,55]. The polymeric calixarene was incorporated into a carbon paste and packed into an electrode body. The electrode was left on open circuit in order to allow accumulation of the ions at the surface of the electrode before being stripped using anodic differential pulse voltammetry. While the typical peak currents increased with increasing accumulation time, the interference of Na^+ and K^+ ions had a dramatic effect on the Pb^{2+} signal in particular, due to competition for the calixarene binding sites. Another approach using amperometric detection and calixarenes as electrode surface coatings was found to provide enhanced selectivity towards neurotransmitters such as dopamine and adrenaline while excluding common electroactive interferents such as ascorbic acid, uric acid and amphetamines [56]. The use of calixarenes in the electrochemical analysis of anions and cations, using electroactive substituents such as ferrocenyl and metal-bipyridine units [57,58] has also been investigated (see Chapter 23) but as yet no working sensing devices have been produced. Little work has been carried out using calixarenes in polymer membranes with voltammetric techniques. However, the ease at which these materials can be engineered into sensing arrays and the resurgence of interest in the area of voltammetry at the interface of two immiscible electrolyte solutions (ITIES) [59], suggests that calixarenes could be investigated for use in this area.

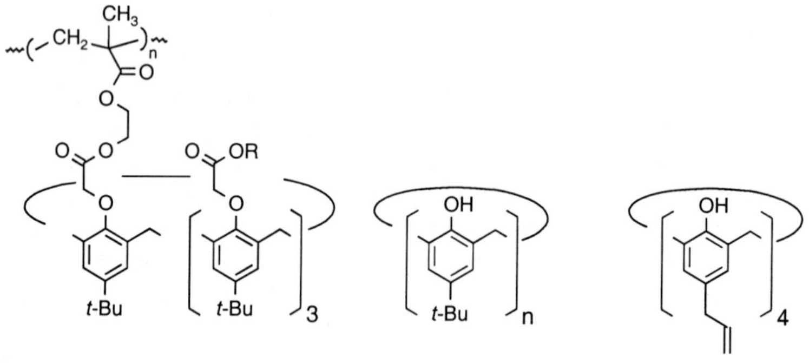

16   **17** n = 8-12   **18**

2.5. Ion Selective Field Effect Transistors (ISFETs) and Electrolyte-Insulator-Semiconductor Sensors (EIS)

ISFET sensors differ from traditional potentiomeric devices in their mode of transduction. A Field-Effect Transistor (FET) is essentially a capacitor where one plate is a conducting channel in which the density of charge carriers depends on the potential applied to the other plate [60]. If the conducting plate has a surface coating able to interact with small molecules, the surface potential depends upon these interactions and hence they may be detected by a change in current through the conducting plate. Where the interacting small "molecules" are ions and the interactions are selective, an ISFET is obtained [60]. They offer some advantages over ISEs, namely their small dimensions (typically 2x2 mm) and their ability to be incorporated into integrated circuits (ICs) for mass production. Over the last decade, a large number of papers has been published using these devices incorporating calixarene derivatives such as **1-8, 13** and **14** as well as large "parent" calixarenes **17** as the sensing agents [61-66]. While most ISFET research using calixarene sensing layers has focused on cation selectivity, some work on anions has also been carried out [60]. Electrolyte-Insulator-Semiconductor (EIS) sensors have also appeared in the literature [67,68]. These capacitive devices, like ISFETs, are well suited to miniaturisation into ICs but offer certain advantages. Namely, their planar geometry allows them to be coated more effectively resulting in increased long term stability [69,70]. Calixarenes such as **17** have been deposited on the EIS surface, with the resulting devices showing selective linear responses to certain cation targets [67,68]. Problems have been experienced (*e.g.* in the encapsulation process for ISFETs) at the mass fabrication stage for all these devices and as yet no commercial applications exist.

3. Calixarene Immobilisation into Polymer Films

In many cases, for calixarenes to be used in sensor devices, they must be immobilised into organic membranes (films/coatings) which are hydrophobic in nature. Furthermore, if aqueous analysis is desired, the calixarenes must be water-insoluble, otherwise immobilisation would be undermined by dissolution of the ionophore from the device.

For ISEs, most commonly used polymer for membrane construction is PVC. The membranes are normally prepared by dissolution of ionophore, PVC, plasticiser and a lipophilic anion salt (*e.g.*, for cation analysis, tetraphenylborates) into an organic solvent such as THF which is then poured onto a glass/quartz surface and the solvent allowed to evaporate. The resulting film prepared from the above procedure has excellent mechanical strength and flexibility [13]. Typical mass ratio compositions of these films are; plasticiser 250; PVC 125 : ionophore 7 : anion 4-5 [55]. It is quite important to note that the plasticiser is in such high concentration that it essentially acts as a solvent. The ionophore and lipophilic anion salt are usually present in small concentrations with their respective ratios being 2:1. Studies have been carried out as to the necessity of the presence of a lipophilic anion salt within the membrane. It has been discovered that the absence of such a material within the membrane results in loss of ion permselectivity. Change in the plasticiser has minimal effect on membrane selectivity and sensitivity. Typical plasticisers used to prepare these membranes are nitrophenyl-octyl ether (NPOE) and diethylsebacate [71]. One of the major drawbacks with the above-described immobilisation technique is the problem of leaching. It has been found that both the ionophore and lipophilic anion leach from the membranes over time, adversely affecting the response of the membrane and leading to sensor failure. Furthermore, the plasticiser can also leach from the membrane, causing it to become fragile and again shortening the life span of the sensor.

An alternative approach to immobilisation involves the coating of electrode surfaces used in voltammetric sensors for ions. This technique requires a coating at the surface of the electrode, which is sensitive/selective for analytes in solution. One of the first applications involving calixarenes used a calix[4]arene tetraester **1** polymerised via reaction of wide-rim acryloyl substituents and suspended in a carbon paste. These systems are simply prepared by the suspension of the calixarene polymer into a carbon powder which is mixed with paraffin oil to give a carbon paste that is used to coat the electrode surface [13,55]. Calixarenes bound via ester links to polymethacrylate derivatives have also found application in this way.

Other matrices also used in chemical sensing devices are sol-gels. It is essential that the calixarenes to be used then are acid- and base-stable, since the preparation of the gels may involve strongly acidic or basic media and the final gel itself can retain acidic or basic properties. In these situations, the calixarenes can be present as an additive, or they can be covalently linked into the sol-gel. Calixarenes , *e.g.*, **18**, have also been successfully immobilised onto capillary walls used in capillary electrophoresis [72].

4. Optical Sensors; Bulk Optodes

Sensors based on optical transduction have been extensively investigated since the late 1980s. Optical transduction is an attractive alternative to electrochemical methods since there is less noise pickup in signal transmission over long distances, and it is possible to analyse the full spectrum with one probe instead of one channel of electrochemical information as is the case with electrochemical devices. The theoretical principles of optical sensors are very similar to those used to define ISE and membranes of similar (quite often identical) composition to that of ISE membranes are used in optical sensors. Optical sensors must convert a host-guest molecular recognition event into an optical response (colour change). One way to achieve this is to co-immobilise a lipophilic acidochromic dye in the membrane along with the ligand. The complexation of the metal ion by the ligand results in the expulsion of a proton from the dye to the sample phase in order to maintain overall charge neutrality. As a result, optical sensors operate as a function of both analyte ion and proton equilibrium between sample solution and the membrane. Thus, if the sample pH is known (this value can be controlled using a buffer) then the activity of the analyte ion can simply be determined by absorbance change of the acidochromic dye (*e.g.* ETH5294, Nile Blue) in the sensor membrane. This class of optical sensor is quite often referred to as a bulk optode, since the analyte must be transferred from the sample/membrane interface into the membrane. Figure 5 displays a schematic diagram of two classes of bulk optodes, one a flow cell and the other a coated optical fibre.

It is essential that the chromoionophores used in bulk optodes possess high molar absorptivity values (greater than 5×10^4 L cm^{-1} M^{-1}), good chemical and photostability and not bind with any ion present in the sample as, their spectral changes must be due solely to proton transfer [73]. Calixarenes have received much interest as potential ionophores in bulk optodes since the early 1990s and a detailed discussion of this and related work is given in Chapter 32.

ETH 5294   Nile Blue   Acridine

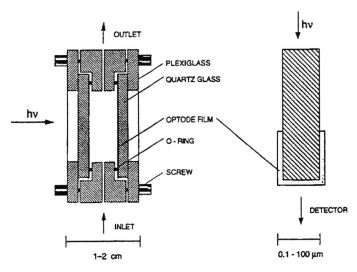

Figure 5. Schematic diagram of a flow cell bulk optode (left), schematic diagram of a membrane coated optical fibre bulk optode (right).

5. New Approaches to Sensors using Calixarenes

Though most of the discussion above has concerned the sensing of charged particles (cations and anions) and this is still an issue of interest, a vast range of analytical problems concerns the sensing of neutral molecules and great promise lies in recent efforts to exploit calixarenes in this way. As illustrated in Chapters 16 and 25 in particular, an understanding of the factors which determine selectivity in the interactions of calixarenes with neutral molecules is slowly deepening. Efforts to use calixarenes in neutral molecule sensors, summarised in Section 5.2 below, are yet to result in practical devices but they are perhaps the basis for the most significant future developments of calixarene-based sensors.

5.1. CHIRAL RECOGNITION (See also Chapters 27 and 36.)

One of the great problems faced by the pharmaceutical industry involves the quantitation of undesired enantiomers in drug raw material. Quite often only one enantiomer of a chiral compound is actually a bioactive therapeutic, the other enantiomer in some cases is non-bioactive or if bioactive it may cause undesirable side-effects. It is therefore essential that final product be properly analysed for enantiomeric purity. Sensing devices which can differentiate between different enantiomers are gaining greater importance in the pharmaceutical industry since these can offer alternative cheap, real time monitoring of drug raw materials from synthesis to final tablet product. Calixarenes, such as the (S)-dinaphthylprolinol ester **19** [73], have been reported which are capable of selective host guest interactions with many amine/alcohol based small molecule bioactives. Prediction as to what enantiomers are preferably bound by specific chiral calixarenes has not been well elucidated and such prediction is extremely difficult, not only in relation to host guest interactions, but also concerning enantiomeric excess in synthesis where either a chiral catalyst or auxiliary are used.

Initial studies carried out with **19** have shown excellent chiral discrimination between the enantiomers of phenylglycinol, phenylethylamine and norephedrine. The mechanism of transduction in this example is based on the fluorescence quenching of the naphthylprolinol "handles" of the calixarene upon complexation with the *R*-enantiomer of phenylglycinol [74]. This calixarene has been successfully immobilised on a capillary wall used in capillary electrophoresis showing very good resolution in the separation of the enantiomers of phenylglycinol. A novel example of chiral discrimination is the interaction of the chromogenic binaphthylcrown calixarene **20** with the enantiomers of 1-phenylethylamine. Interaction of the calixarene with the *R*-enantiomer gives a shift in absorption of the calixarene to 538 and 650 nm, whereas the *S*-enantiomer causes a shift to 515 nm [75]. This type of system should find application in the future in optical sensing devices (as discussed in Chapter 32).

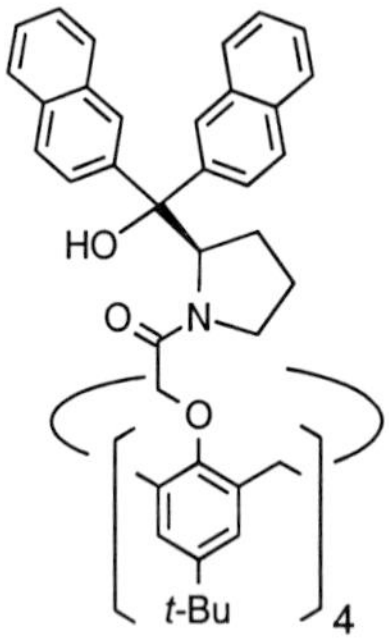

19

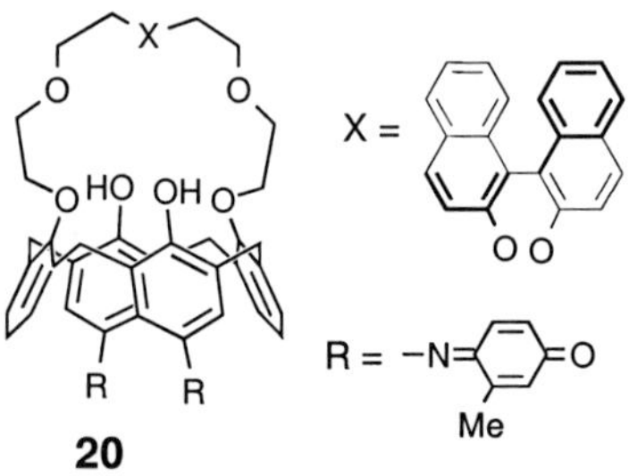

20

5.2. THE QUARTZ CRYSTAL MICROBALANCE (QCM) IN NEUTRAL-MOLECULE SENSING

Minute mass changes due to adsorption on a polymer-coated piezoelectric quartz crystal can be readily detected as changes in the phonon resonance frequency of the crystal [76]. Selective adsorption can therefore lead to detection of particular species present in both gas and solution phases. If an array of piezoelectric quartz crystals is used and each crystal in the array is coated with a different polymer type, then theoretically it should be possible for each crystal to interact with specific components within the sample.

Unfortunately, available polymers give only limited selectivity. An alternative approach is the incorporation of macrocyclic species which bind selectively to specific classes of molecules. Calixarenes as different as **21** and **22** [13] have been used for the detection of low molecular weight amines or their protonated forms and a recent study [77] indicates that both basicity and shape of the amine are key factors in determining selectivity. While amines are neutral, the responses in these systems are associated with proton transfer reactions and a better example of true detection of and discrimination between neutral species in seen in the responses of a QCM coated with trialkylsilyl and tosyloxyethyl ethers of calix[8]arene **17** (n = 8; R = *t*-Bu) to n-heptane and tetrachloroethylene [78]. QCM electrodes coated with tetramercapto-*p*-*t*-butylcalix[4]arene **23** show superior uptake of alkylbenzenes as compared to monolayers of *p*-*t*-butylcalix[4]-

arene [79], though the limits of detection obtained were too high for any practical use. A motivating factor in this work was the prospect of monitoring environmental organic volatiles and this was also true for investigations of the effect of different modes of film preparation from calix[6]arene on the sensitivity to various substituted benzenes [79] and of organic vapour diffusion into nanporous thin films containing the functionalised resorcarenes **24** and **25** [80]. A different environmental issue applies in the investigation of composite resorcarene/polyphenylenesulfide films in QCM sensors for ozone [81]. Much more developmental work is required in this field, and a wider range of analytes and a more diverse number of calixarenes need to be tested before conclusions on their potential applications can be made.

21 R = SO_3H

22 R = $-N=N-\langle\rangle-NO_2$

23

24 A = H, R = C_5H_{11}, Y = $-P(=O)(O\text{-}i\text{-}Pr)_2$

25 R = $C_{11}H_{23}$, Y = H, A = $-N=N-\langle\rangle-NO_2$

6. References and Notes

[1] (a) A. Hulanicki, S. Glab, F. Ingman, *IUPAC Discussion Paper*, Commission V. 1., July, **1989**. (b) A Hulanicki, S. Glab, F. Ingman, *Pure & Appl. Chem.* **1991**, *63*, 1247-1250. (c) U. E. Spichiger-Keller, *"Chemical Sensors for Medical and Biological Application"*, Wiley-VCH, **1998**.

[2] F. Arnaud-Neu, G. Barrett, S. Cremin, M. Deasy, G. Ferguson, S. J. Harris, A. J. Lough, L. Guerra, M. A. McKervey, M. J. Schwing-Weill, P. Schwinte, *J. Chem. Soc., Perkin Trans. 2* **1992**, 1119-1125.

[3] V. Bocchi, D. Foina, A. Pochini, R. Ungaro, C. D. Andreetti, *Tetrahedron*, **1982**, *38*, 373-378.

[4] G. Ferguson, B. Kaiter, M. A. McKervey, E. M. Seward, *J. Chem. Soc., Chem. Commun.* **1987**, 584-585.

[5] F. Arnaud-Neu, G. Barrett, S. J. Harris, M. A. Mc Kervey, M. Owens, M. J. Schwing-Weill, P. Schwinte, *Inorg. Chem.* **1993**, *32*, 2644-2650.

[6] N. Muzet, G. Wipff, A. Casnati, L. Domiano, R. Ungaro, F. Ugozzoli, *J. Chem. Soc., Perkin Trans. 2* **1996**, 1065-1075.

[7] E. Ghidini, F. Ugozzoli, R. Ungaro, S. Harkema, A. Abu El-Fadl, D. N. Reinhoudt, *J. Am. Chem. Soc.* **1990**, *112*, 6979-6985.

[8] D. N. Reinhoudt, P. J. Dijkstra, P. J. A. in't Veld, K. E. Bugge, S. Harkema, R. Ungaro, E. Ghidini, *J. Am. Chem. Soc.* **1987**, *109*, 4761-4762.

[9] D. Diamond, G. Svehla, *Trends Anal. Chem.* **1987**, *6*, 46-49.

[10] D. Diamond, G. Svehla, E. M. Seward, M. A. McKervey, *Anal. Chim. Acta* **1988**, *204*, 223-231.

[11] D. Diamond, *Anal. Chem. Symp. Ser.* M. R. Smyth and J. G. Vos Eds., Elsevier, **1986**, *25*, 155-161.

[12] F. Arnaud-Neu, E. M. Collins, M. Deasy, G. Ferguson, S. J. Harris, B. Kaitner, A. J. Lough, M. A. McKervey, E. Marques, B. L. Ruhl, M. J. Schwing-Weill, E. M. Seward, *J. Am. Chem. Soc.* **1989**, *111*, 8681-8691.

[13] D. Diamond, M. A. McKervey, *Chem. Soc. Rev.* **1996**, 15-24.

[14] T. Grady, A. Cadogan, T. McKittrick, S. J. Harris, D. Diamond, M. A. McKervey, *Anal. Chim. Acta* **1996**, *336*, 1-12.
[15] M. Telting Diaz, F. Regan, D. Diamond, M. R. Smyth, *J. Pharm. Biomed. Anal.* **1990**, *8*, 695-700.
[16] M. Telting Diaz, F. Regan, D. Diamond, M. R. Smyth, *Anal. Chim. Acta* **1991**, *251*, 149-155.
[17] D. Diamond, R. J. Forster, *Anal. Chim. Acta* **1993**, *276*, 75-86.
[18] T. Sakaki, T. Harada, G. Deng, H. Kawabata, Y. Kawahara, S. Shinkai, *J. Incl. Phenom.* **1992**, *14*, 285-302.
[19] H. Yamamoto, S. Shinkai, *Chem. Lett.* **1994**, 1115-1118.
[20] E. Bakker, *Anal. Chem.* **1997**, *69*, 1061-1069.
[21] E. P. Serjeant, *"Potentiometry and Potentiometric Titrations"*, John Wiley and Sons, **1984**.
[22] J. Koryta, J. Dvorák, *"Principles of Electrochemistry"*, Wiley, **1987**.
[23] M. Nägele, E. Bakker, E. Pretsch, *Anal. Chem.* **1999**, *71*, 1041-1048.
[24] T. Sokalski, A. Ceresa, T. Zwickl, E. Pretsch, *J. Am. Chem. Soc.* **1997**, *119*, 11347-11348.
[25] T. Sokalski, T. Zwickl, E. Bakker, E. Pretsch, *Anal. Chem.* **1999**, *71*, 1204-1209.
[26] T. Sokalski, A. Ceresa, M. Fibbioli, T. Zwickl, E. Bakker, E. Pretsch, *Anal. Chem.* **1999**, *71*, 1210-1214.
[27] E. Bakker, D. Diamond, A. Lewenstam, E. Pretsch, *Anal. Chim. Acta* **1999**, *393*, 11-18.
[28] E. Lindner, V. V. Cosofret, R. P. Buck, T. A. Johnson, R. B. Ash, M. R. Neuman, W. J. Kao, J. M. Anderson, *Electroanalysis* **1995**, *7*, 846-851.
[29] E. Lindner, R. P. Buck, *Anal. Chem.* **2000**, *72*, 336A-345A.
[30] A. Lynch, D. Diamond, P. Lemoine, J. McLaughlin, M. Leader, *Electroanalysis*, **1998**, 10, 1096-1100.
[31] H. M. Widner, *'Analytical Methods and Instrumentation'* **1993**, *1*, 3-16.
[32] D. Ammann, W. E. Morf, P. Anker, P. C. Meier, E. Pretsch, W. Simon, *ISE Rev.*, **1983**, *5*, 3-92.
[33] E. Linder, K. Toth, M. Horrath, E. Pungor, B. Agai, I. Bitter, L. Toke, Z. Hell, *Fresenius' Z. Anal. Chem.* **1985**, *322*, 157-163.
[34] R. J. Forster, A. Cadogan, M. Telting Diaz, D. Diamond, *Sens. Actuators B* **1991**, *4*, 325-331.
[35] K. M. O'Conner, D. W. M. Arrigan, G. Svehla, *Electroanalysis* **1995**, *7*, 205-215.
[36] D. Diamond, *J. Incl. Phenom.* **1994**, *19*, 149-166.
[37] P. Bühlmann, E. Pretsch, E. Bakker, *Chem. Rev.* **1998**, *98*, 1593-1687.
[38] V. V. Egorov, Y. V. Sin'kevich, *Talanta* **1999**, *48*, 23-38.
[39] Y. Tsujimura, T. Sunagawa, M. Yokoyama, K. Kimura, *Analyst* **1996**, *121*, 11, 1705-1709.
[40] E. Malinowska, V. Oklejas, R. W. Hower, R. B. Brown, M. E. Meyerhoff, *Sens. Actuators, B* **1996**, *B33*, 161-167.
[41] H. J. Lee, H. J. Oh, G. Ciu, G. S. Cha, H. Nam, *Anal. Sciences* **1997**, *13* (Supplement), 289-294.
[42] E. Malinowska, Z. Brzózka, K. Kasiura, R. J. M. Egberink, D. N. Reinhoudt, *Anal. Chim. Acta* **1994**, *298*, 253-258.
[43] J. F. Malone, D. J. Mars, M. A. McKervey, P. O'Hagan, N. Thompson, A. Walker, F. Arnaud-Neu, O. Mauprivez, M. J. Schwing-Weill, J. F. Dozol, H. Rouquette, N. Simon, *J. Chem. Soc., Chem. Commun.* **1995**, 2151-2153.
[44] T. Grady, S. Maskula, D. Diamond, D. J. Marrs, M. A. McKervey, P. O'Hagan, *Anal Proc.* **1995**, *32*, 471-473.
[45] T. McKittrick, D. Diamond, D. J. Marrs, P. O'Hagan, M. A. McKervey, *Talanta* **1996**, *43*, 1145-1148.
[46] F. Cadogan, P. Kane, M. A. McKervey, D. Diamond, *Anal. Chem.* **1999**, *71*, 5544-5550.
[47] A. Casnati, A. Pochini, R. Ungaro, D. Bocchi, F. Ugozzoli, R. J. M. Egberink, H. Struijk, R. Lugtenberg, F. de Jong, D. Reinholdt, *Chem. Eur. J.* **1996**, *2*, 436-445.
[48] C. Bocchi, M. Careri, A. Casnati, G. Mori, *Anal. Chem.* **1995**, *67*, 4234-4238.
[49] P. L. H. M. Cobben, R. J. M. Egberink, J. B. Bomer, P. Bergveld, W. Verboom, D. N. Reinholdt, *J. Am. Chem. Soc.* **1992**, *114*, 10573-10582.
[50] E. Bakker, *Sens. Actuators, B* **1996**, *35*, 20-25.
[51] K. Odashima, *J. Incl. Phenom.* **1998**, *32*, 165-178.
[52] K. Odashima, K. Yagi, K. Tohda, Y. Umezawa, *Anal. Chem.* **1993**, *65*, 1074-1083.
[53] K. Odashima, K. Yagi, K. Tohda, Y. Umezawa, *Bioorg. Med. Chem. Lett.* **1999**, 9, 2375-2378.
[54] D. W. M. Arrigan, G. Svehla, S. J. Harris, M. A. McKervey, *Electroanalysis* **1994**, *6*, 97-106.
[55] K. M. O'Connor, D. W. M. Arrigan, G. Svehla, *Electroanalysis* **1995**, *7*, 205-215.
[56] J. Wang, J. Liu, *Anal. Chim. Acta* **1994**, *249*, 201-206.
[57] P. A. Gale, Z. Chen, M. G. B. Drew, J. A. Heath, P. D. Beer, *Polyhedron* **1998**, *17*, 4, 405-412.
[58] H. Cano-Yelo Bettaga, J. C. Moutet, G. Ulrich, R. Ziessel, *J. Electroanalytical Chemistry* **1996**, *406*, 247-250.
[59] S. Jadhav, E. Bakker, *Anal. Chem.* **1999**, *71*, 3657-3664.

[60] (a) M. M. G. Antonisse, R. J. W. Lugtenberg, R. J. M. Egberink, J. F. J. Engbersen, D. N. Reinhoudt, *"Chemosensors of Ion and Molecular Recognition"*, Eds. J. P. Desvergne, A. W. Czarnik. Kluwer, **1997**, 23-35; (b) G. Horowitz, *Adv. Mater.* **1998**, *10*, 365-377.
[61] R. J. W. Lugtenberg, R. J. M. Egberink, J. F. J. Egberson, D. N. Reinhoudt, *Anal. Chim. Acta* **1997**, *357, 3*, 225-229.
[62] R. Mlika, I. Dumazet, M. Gamoudi, R. Lamartine, H. Ben Ouada, N. Jaffrezic-Renault, G. Guillaud, *Anal. Chim. Acta* **1997**, *354*, 283-289.
[63] R. Mlika, H. Ben Ouada, N. Jaffrezic-Renault, I. Dumazet, R. Lamartine, M. Gamoudi, G. Guillaud, *Sens. Actuators, B* **1998**, *B47*, 43-47.
[64] R. J. W. Lugtenberg, R. J. M. Egberink, A. van den Berg, J. F. J. Egberson, D. N. Reinhoudt, *J. Electroanal. Chem.* **1998**, *452*, 69-86.
[65] R. Ben Chaabane, M. Gamoudi, G. Guillaud, C. Jouve, R. Lamartine, A. Bouazizi, H. Maaref, *Sens. Actuators, B* **1996**, *31*, 41-44.
[66] R. Mlika, I. Dumazet, H. Ben Ouada, N. Jaffrezic-Renault, R. Lamartine, M. Gamoudi, G. Guillaud, *Sens. Actuators, B* **2000**, *62*, 8-12.
[67] R. Mlika, H. Ben Ouada, M. A. Hamza, M. Gamoudi, G. Guillaud, N. Jaffrezic-Renault, *Synth. Met.* **1997**, *90*, 173-179.
[68] R. Mlika, H. Ben Ouada, R. Ben Chaabane, M. Gamoudi, G. Guillaud, N. Jaffrezic-Renault, R. Lamartine, *Electrochim. Acta* **1998**, 43, 841-847.
[69] M. J. Schöning, M. Thust, M. Müller-Veggian, P. Kordos, H. Lüth, *Sens. Actuators, B* **1998**, *47*, 225-230.
[70] M. J. Schöning, Ü. Malkoc, M. Thust, A. Steffen, P. Kordo, H. Lüth, *Sens. Actuators, B* **2000**, *65*, 288-290.
[71] E. Bakker, P. Bühlmann, E. Pretsch, *Chem. Rev.* **1997**, *97*, 3083-3132.
[72] Z. Wang, Y. Chen, H. Yuan, Z. Huang, G. Liu, *Electrophoresis* **2000**, *21*, 1620-1624.
[73] S. O'Neill, S. Conway, J. Twellmeyer, O. Egan, K. Nolan, D. Diamond, *Anal. Chim. Acta* **1999**, *398*, 1-11.
[74] T. Grady, T. Joyce, M. R. Smyth, S. J. Harris, D. Diamond, *Anal. Comm.* **1998**, *35*, 123-125.
[75] Y. Kubo, S. Maeda, S. Tokita, M. Kubo, *Nature* **1996**, *382*, 522-524.
[76] G. Sauerbery, *Z. Phys.* **1959**, *155*, 208-222.
[77] X. C. Zhou, S.C. Ng, H. S. O. Chan, S. F. Y. Li, *Sens. Actuators, B* **1997**, *31*, 137-144.
[78] F. L. Dickert, R. Sikorski, *Materials Sci. Eng. C* **1999**, *10*, 39-46.
[79] M. T. Cygan, G. E. Collins, T. D. Dunbar, D. L. Allara, C. G. Gibbs, C. D. Gutsche, *Anal. Chem.* **1999**, *71*, 142-148.
[80] S. Munoz, T. Nakamoto, T, Morizumi, *Sens. Mater.* **1999**, *11*, 427-435.
[81] A.V. Nabok, A.K. Hassan, A.K. Ray, *J. Mater. Chem.* **2000**, *10*, 189-194.
[82] A.V. Nabok, A.K. Hassan, A.K. Ray, J. Travis, M. Hofton, A. Dalley, *IEEE Proc.: Sci. Meas. Technol.* **2000**, *147*, 153-157.

Chapter 35

CALIXARENES FOR NUCLEAR WASTE TREATMENT

FRANCOISE ARNAUD-NEU,[a] MARIE-JOSÉ SCHWING-WEILL[a], JEAN-FRANCOIS DOZOL[b]

[a] *Laboratoire de Chimie-Physique, UMR 7512 au CNRS, ECPM-ULP, 25, rue Becquerel, 67087 Strasbourg Cedex 2, France.*
[b] *CEA Cadarache, DCC/DESD/SEP, F-13108 St Paul Lez Durance Cedex, France. e-mail: jean-francois.dozol@cea.fr*

1. Introduction

The fission of uranium or more generally of transuranium elements in reactors is an important energy source, arguably not very polluting for the environment. In use, the nuclear fuel undergoes increasingly important modifications until its replacement is considered to be necessary for the correct operation of the reactor, which occurs well before its total exhaustion in fissile matter.

The spent fuels contain both fissile and fertile residues of significant energy value and radioactive products, alpha, beta, gamma and neutron emitters, which make them very irradiant and which release an important amount of heat (mainly due to ^{90}Sr and ^{137}Cs, after five years of cooling). The radiotoxicity of these fuels can last for millions of years. Reprocessing is of great interest to recover uranium and plutonium, which can be recycled in the form of new nuclear fuels, and to optimize the conditioning of waste.

The fuel is stored in pools for at least three years before its reprocessing. From the pool, fuel assemblies are routed to a shearing and dissolution unit before extraction of uranium and plutonium by tri-*n*-butylphosphate (TBP) (PUREX process). During the back extraction operations, uranium and plutonium are separated and then are purified by one or two supplementary solvent extraction cycles in order to obtain very pure products that can be used again. Acidic solutions arising from PUREX process contain minor actinides (neptunium, americium, curium) and 99 % of the non-gaseous fission products. Some of these radionuclides have long half-lives : ^{93}Zr (1.5×10^6 y), ^{107}Pd (7×10^6 y), ^{129}I (1.7×10^7 y), ^{135}Cs (3×10^6 y), ^{237}Np (2.14×10^6 y). The present strategy is to vitrify these high level activity wastes for disposal or long term storage. An alternative strategy, studied in France, Japan and Russia, is to separate the minor actinides and the long lived fission products from these wastes and destroy them by transmutation. This would also involve the recovery of other fission products, ^{90}Sr (27.7 y) and ^{137}Cs (30 y), which would remove about 90 % of the heat released by high activity wastes, allowing their volume to be reduced and facilitating their disposal or their intermediate storage.

Reprocessing operations and particularly dismantling facilities generate medium level activity liquid wastes. A great part of these high salinity solutions has to be disposed of in stable geological formations or in long term storage after embedding due to the presence of these long lived nuclides, mainly actinides, ^{90}Sr and ^{137}Cs. It would be desirable to remove them from the contaminated liquid wastes before embedding in order to allow a large part of these wastes to be directed to a surface repository, and a very small part containing most of the long lived nuclides to be disposed of or stored after conditioning.

In order to minimise the volume of medium level activity waste or to separate long lived nuclides from high level activity waste, it is of paramount importance to find compounds able to remove mainly strontium, cesium and actinides with high efficiency and selectivity, from acidic or high salinity media, and to release these cations, if possible, in deionized water. In the framework of research and developments on nuclear waste management, the Commission of the European Community supported two successive projects on the complexation and extraction properties of new classes of calixarenes [1,2].

Basic extraction and complexation studies as well as extraction tests and transport through a supported liquid membrane (SLM) [3] on solutions (NaNO$_3$ 4M, HNO$_3$ 1M) simulating evaporator concentrate (radioactive liquid waste arising from reprocessing operations concentrated by evaporation, spiked with different nuclides depending on the studied compounds, ^{22}Na, ^{85}Sr, ^{137}Cs, ^{237}Np, ^{239}Pu and ^{241}Am) were performed.

2. Calixarene Derivatives for Cesium Separation

Initially two classes of extractants were tested for the removal of strontium and cesium : dicarbollides and crown ethers. Dicarbollides, boron cluster species with a pi-bonded trivalent cobalt are lipophilic anions forming neutral compounds with cations [4,5]. Unfortunately the efficiency of this cation exchanger strongly decreases as the acidity or the sodium concentration of the solutions to be treated increases. The most suitable crown ethers for the extraction of cesium are benzo-crown-7 derivatives (e.g. B21C7 = benzo-21-crown-7) [6]. However, their extraction ability is insufficient to allow the extraction of cesium from high salinity or acidic solutions without adding a synergistic agent [7].

p-t-butylcalix[n]arenes and *p-t*-pentylcalix[n]arenes (n = 4, 6, 8) were the first calixarenes shown to transport alkali metal hydroxides through thick membranes and were found to be efficient carriers of cesium [8]. Calixarenes bridged by polyether loops, the "calixcrowns" (see chapter 20) were early found to be good complexants for the alkali metal ions [9-11], as were later the bis-crowns derived from *1,3-alternate* calix[4]arene [11,12].

2.1. CALIX[4]ARENE-MONO- AND BIS-CROWNS

2.1.1. Extraction and complexation results
The efficiency and selectivity of calixarene-mono-crowns depend on many factors such as the size of the crown unit and of the calixarene, and the conformation. The selectivity is shifted towards the larger alkali metal ions as the number of oxygen atoms increases. Calix[4]arene-crown-4 ligands are selective for Na$^+$ [13] and calix[4]arene-crown-5 for K$^+$ [14,15], whereas calix[4]arene-crown-6 [16] and calix[5]arene-crown-x (x = 4, 5, 6) [17] are selective for Cs$^+$.

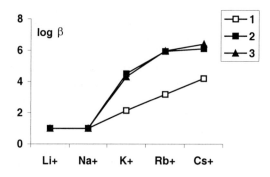

1 x = 6
7 x = 7

2 Y = n-Pr
3 Y = i-Pr
4 Y = n-Oct
5 Y = -(CH$_2$)$_2$-OH
6 Y = allyl

8
a Y = n-Oct, R = H
b Y = n-Oct, R = 4-t-Oct
c Y = n-Oct, R = 4-t-Oct, 5-NO$_2$
d Y = n-Pr, R = H
e Y = allyl, R = 4-t-Oct

9

Initial evaluation of the ionophoric properties of calix[4]arene-crown-6 ligands by the picrate extraction method and determination of association constants in wet CHCl$_3$ revealed the high affinity of these compounds for Cs$^+$ cation [16]. Complexation data (log β) in methanol confirmed the higher efficiency and the remarkable Cs$^+$/Na$^+$ selectivity (>10^5) of the *1,3-alternate* ligands as compared to the conformationally mobile **1** (Figure 1) [16].

Figure 1. Stability constants (log β) for complexes of alkali metal cations and calix[4]arene-crowns in MeOH.

Calorimetry showed (Figure 2) that the high efficiency of **3** toward cesium is controlled by the high enthalpy term (-ΔH) (figure 2a). The entropy term, less negative than with 18C6, is consistent with the preorganization of the calixarene in the *1,3-alternate* conformation, where only a small part of the crown moiety remains flexible. Only this "reduced" flexibility is lost upon complexation of the large Cs$^+$, which fits the cavity very well. On going towards Rb$^+$ and K$^+$, -ΔH decreases and TΔS even becomes positive for the smaller K$^+$, suggesting a less efficient interaction of these cations with the crown. The high Cs$^+$/Na$^+$ selectivity of **3** finds its origin in the good size match between cesium and the crown and the impossibility of the crown part wrapping around the smaller cations. With both ligands **1** and **3**, formation of the Cs$^+$ complexes is enthalpy controlled (-ΔH > 0, TΔS < 0) (Figure 2b). However the enthalpy term is less favourable for **1**, whereas the entropy is more. This can be explained in terms of solvation, the *1,3-alternate* ligand **3** being more preorganized and less polar, and hence less solvated than the mobile **1** for which desolvation during complexation enhances TΔS and decreases -ΔH.

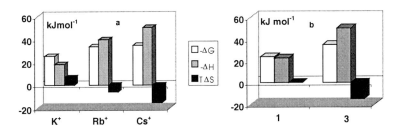

Figure 2. Thermodynamic parameters for the complexation of alkali metal ions by **3** (a) and of cesium complexes with ligands **1** and **3** in methanol.

For dialkyl calix[4]arene-crown-6 derivatives, there is no significant influence of the nature of the alkyl groups. However, if the substituents are functional groups like esters, amides or phosphonates, the effect of the functional group is predominant over the crown effect, leading in most cases to selectivities in favour of Na^+ [1].

Although potential ditopic receptors, calix[4]arene-bis-crowns **10** and **11** form exclusively 1:1 complexes in methanol and acetonitrile [18]. Extraction and complexation results have shown their high efficiency for cesium. However, the presence of a second crown unit on the calixarene does not improve significantly the selectivity as compared to the related mono-crown derivatives.

The binding of the larger cations is less affected by the substitution in the crown linkages than that of the smaller ones. This results for compounds **13-16** in an increase in the Cs^+/Na^+ selectivity which can be interpreted in terms of loss of flexibility of the crown chain, the substituents rigidifying the crown and preventing optimal interaction between the binding sites and the smaller cations. In particular, the introduction of two 1,2 phenylene groups as in **15** leads to important modifications of the nature and the stability of the complexes. Firstly, there is evidence for binuclear complexes in addition to the mononuclear species with K^+, Rb^+ and Cs^+ [19]. A binuclear complex has been isolated in the solid state with Cs^+ [19]. Secondly, the stability constant decreases for Na^+ and increases for K^+, Rb^+ and Cs^+. The increase in stability is entirely due to more favourable entropy terms, explicable by the greater rigidity of **15** and its smaller loss of freedom upon complexation. This results in an enhancement of the selectivity, which becomes higher than 2×10^5 in acetonitrile. On the basis of computational studies, this feature has been interpreted as largely due to the greater cost of desolvating Na^+ cation [19].

The efficiency and the selectivity of dialkyl calix[4]arene dibenzo-crown-x (x = 7, 8) ligands and calixarene-dibenzo-crown-x with two butyl groups attached on the crown ether linkage were established by following the transport of alkali metal ions through a bulk liquid membrane. With the former ligands, the flux values increased from lithium to cesium; they were higher for the bis-crown-7 than for the bis-crown-8. The increase in complexation ability with respect to the unsubstituted derivatives was attributed to the flatness and lipophilicity induced by the presence of benzo units on the crown [20-24]. It was also confirmed that the presence of *p-t*-butyl groups prevented the extraction of alkali metal cations.

2.1.2. Extraction of radioactive solutions
Extraction studies into NPHE (1,2-nitrophenyl hexyl ether) of fifteen calix[4]arene crowns (Table 1) led to the following main conclusions [10,12,25-30] :
- (i) For alkoxy-crown-6 ligands the extraction is mainly dependent on the conformation. The conformationally mobile methoxy derivative displays a much lower efficiency and selectivity than the calixarenes locked in the *1,3-alternate* conformation. The exceptional selectivity of these compounds for cesium can be explained by stabilisation of the complexes by interactions between cesium and the four benzene units [31-34].
- (ii) The extraction is to lesser extent linked to the crown size. The best results are obtained for two classes of calixarenes with polyethylene glycol bridges containing six oxygen atoms. Distribution coefficients obtained with calixarene-bis-crown-5 and 7 are much lower than with bis-crown-6 derivatives. Modification of the crown size obtained by replacing *o*-benzo (**14a**) by *p*-benzo (**13**) generates a strong decrease in the distribution coefficients. These observations highlight the importance of the complementarity between the crown and the cesium cation.

For all tested calixarenes, cesium distribution coefficients increase with the acidity of the aqueous phase, until a maximum is reached for an acidity of about 2M, and then decrease. However the behaviour of all the calixarenes is not identical. Calixarene-bis-crowns display the lowest distribution coefficients whatever the acidity of the aqueous phase. The distribution coefficients of the two alkyl-calixarene-mono crowns, rela-

TABLE 1. Distribution coefficients (D) of cesium and sodium and Cs/Na selectivity S: Aqueous feed solution : MNO_3 5×10^{-4} M - HNO_3 1 M. Organic solution : 10^{-2} M extractant in NPHE (organic to aqueous phase ratio o/a = 1, T=25°C).

Extractants	D_{Na}	D_{Cs}	$S_{Cs/Na}$	Compounds	D_{Na}	D_{Cs}	$S_{Cs/Na}$
1	3×10^{-3}	4×10^{-2}	13	10	2×10^{-3}	0.4	200
5	$< 10^{-3}$	4.2	> 4200	11	1.3×10^{-2}	19.5	1500
2	2×10^{-3}	19.5	9750	12	$< 10^{-3}$	0.3	300
3	$< 10^{-3}$	28.5	> 28500	13	$< 10^{-3}$	2×10^{-2}	20
4	$< 10^{-3}$	33	> 33000	14a	1.7×10^{-3}	32.5	19000
9	$< 10^{-3}$	31	> 31000	15	$< 10^{-3}$	23	> 23000
7	4×10^{-3}	7×10^{-3}	1.75	16	$< 10^{-3}$	29.5	> 29000
n-decyl-B21C7	1.2×10^{-3}	0.3	250	17	$< 10^{-3}$	7×10^{-2}	70
t-Bu-B21C7	1.2×10^{-3}	0.3	250				

tively high in low acidity medium, increase to a maximum. The presence of benzo or naphthyl groups on the crown enhances the extraction of cesium : around the maximum, distribution coefficients higher than 100 are reached with **14a** and **16**. The presence of a second benzene group on the crown (**15**) further increases this efficiency. Distribution coefficients between 200 and 300 are obtained for an acidity ranging from 2 to 4 M [35]. The same trend is observed for the extraction of cesium from acidic medium containing large amounts of sodium, *i.e.* a strong increase in the distribution coefficients, especially for **15,** and to a lesser extent for **13** and **16**.

Recent studies on the extraction of alkali metals by 12 calix[4]arene-crown-6 ligands confirm the previous results and highlight the role of benzene substituents of the crown on the Cs/Na selectivity (Table 2) [36]. In 1,2-dichloroethane a selectivity higher than 6.4×10^5 was achieved with **9**.

Several transport experiments through SLMs were performed at different calixarene-crown-6 concentrations. At 0.05 M in NPOE (2-nitrophenyl octyl ether), quantitative transport of cesium was performed within 24 h with very low transport of sodium [16]. The lipophilic character of the calixarenes and the use of nitrophenyl-alkyl ethers enable stable membranes to be prepared, to a lesser extent with calixarene-bis-crowns-6, and to be reused several times [37].

Extraction and transport studies through SLM were also carried out on a real waste arising from dissolution of a MOX fuel (Burn up 34650 MWJ/tU) where uranium and plutonium were previously extracted by TBP (PUREX process). The efficiency and the selectivity of calixarene-mono or bis-crown-6 derivatives were confirmed since high distribution ratios were achieved for cesium and at a lesser extent for rubidium, the other cations not being extracted [38].

16 was tested for the removal of cesium from a solution simulating a Hanford tank supernatant. At a concentration 10^{-2} M in extractant, the selectivity Cs^+/Na^+ exceeded 10^4 in four different diluents. In the same conditions, this selectivity was at best 200 with the di(*t*-butylbenzo)-21-crown-7. In spite of a relatively high Cs^+/K^+ selectivity in the range 122-935, the high concentration of potassium limited the extraction of cesium [39]. In the USA, the decontamination of radioactive wastes requires not only a high Cs^+/Na^+ selectivity, since these wastes contain 5-7 M Na^+, but also an important Cs^+/K^+ selectivity, as they can also contain 1 M K^+. The use of bis-deoxygenated calix[4]arene-crown-6 derivatives was proposed. In comparison to classical

TABLE 2. Distribution coefficients (D) and cesium selectivities (S) for the extraction of alkali metal nitrates by calix[4]arene-crown ethers in 1,2-dichloroethane (o/a = 1, T=25°C).

Extractants	$D_{Li}(\times 10^5)$	$D_{Na}(\times 10^5)$	D_K	D_{Rb}	D_{Cs}	$S_{Cs/Na}(\times 10^4)$	$S_{Cs/K}$	$S_{Cs/Rb}$
4	2.24	28.8	0.0182	0.294	4.91	1.7	270	17
8a	0.986	3.92	0.0142	0.216	4.04	10	290	18
9	0.976	<0.25	0.0124	0.178	1.61	>64	130	9.0
8b	0.657	4.6	0.0147	0.322	4.96	11	340	15
8c	0.668	0.43	0.00822	0.119	2.29	55	280	19
6	2.39	30.9	0.0160	0.302	5.48	1.8	340	18
8d	3.71	14.2	0.00642	0.0694	1.19	0.84	190	17
8e	0.61	5.85	0.0150	0.356	6.40	11	430	18
11	3.85	44.1	0.0187	0.302	4.54	1.0	240	15
15	1.44	7.96	0.0193	0.262	4.54	5.7	240	17
16	0.38	5.79	0.0194	0.308	5.47	9.4	280	18
14b	1.56	10.2	0.0188	0.254	4.22	4.1	220	17

crown ethers, cesium distribution coefficients decrease by roughly an order of magnitude. However, owing to a larger decrease in the extraction of the other alkali metals, the Cs^+/K^+ selectivity exceeds 4000 for some calixarenes [40].

2.2. CALIX[n]ARENE-CROWNS

The 1,3;4,6 calix[6]arene-bis-crown-4 (**18**) in *1,2,3-alternate* conformation forms 1:1 complexes with alkali metal cations. This is likely a consequence of cooperative binding of one cation by the two crowns [41]. The Cs^+/Na^+ selectivity, *ca* 1500, is lower than that observed for the different calix[4]arene-crown-6 ligands but higher than that measured with the related calix[8]arene-bis-crown-3 (**19**) [42].

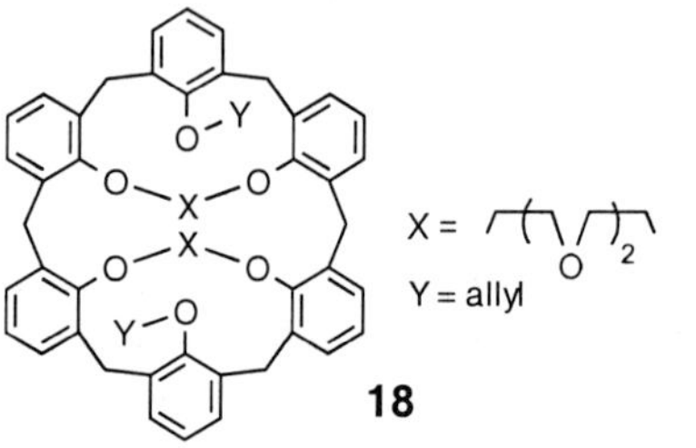

3. Calixarenes for Strontium Removal

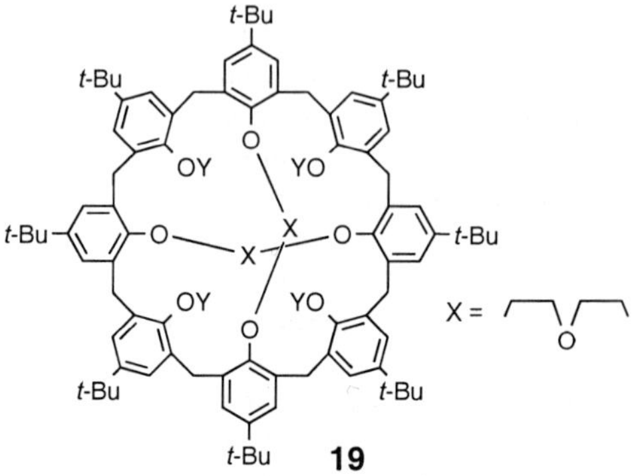

Calix[4]arene amides are efficient ligands for alkaline earth metal ions, but they also strongly bind Na^+ [43]. It was observed that Sr^{2+}/Na^+ selectivity in calixarene amides increased on going from calix[4]- to calix[6]arenes and was strongly dependent on the substituent on the upper rim, the *p*-dealkylated hexaamide **20b** being much more selective than its *t*-butyl counterpart **20a** [44].

3.1. COMPLEXATION RESULTS

In contrast to alkali metals, alkaline earth metal picrates are efficiently extracted by the hexamer and octamer derivatives with a selectivity profile in favour of the larger cations (D > 5 for Ba^{2+} and **23a-d**) [45]. The Sr^{2+}/Na^+ selectivity displayed by all these calixarenes, especially the octamers, increases according to the sequence: **22b** ≈ **23a** < **24a** < **24c**. Compound **24c** (S = 51.3) is much better than DC18C6 (= dicyclohexyl-18-crown-6) (S = 12) and **20b** (S = 16.4).

Compounds **24a**, **24c**, **22a** and **22b** form very weak 1:1 complexes with alkali metal ions (log β < 2) and very stable complexes with alkaline earth metal ions in methanol. Additional 2:1 complexes were also found with Ca^{2+} and Sr^{2+}, except with Ca^{2+} and **24a**. The formation of these complexes is in agreement with crystallographic data showing that these ligands are able to accommodate two metal ions in their cavity [46].

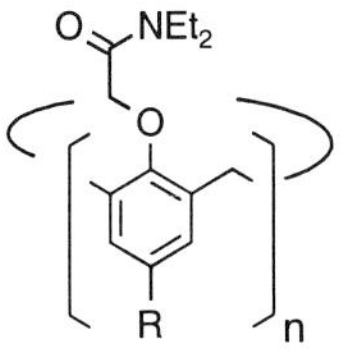

```
     20a  n = 6, R = t-Bu           a  R = Me
       b  n = 6, R = H              b  R = Oct
     21   n = 7, R = Oct            c  R = Bn
     22a  n = 8, R = t-Bu           d  R = C₅H₁₁
       b  n = 8, R = H

                              23  n = 6
                              24  n = 8
```

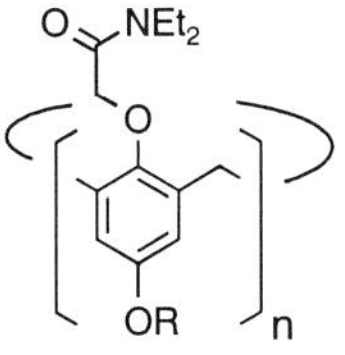

(Structural formulas showing calixarene derivatives with O-NEt₂ groups and OR substituents)

3.2. EXTRACTION RESULTS

For all the calixarenes, except **23c**, strontium distribution coefficients are by far higher than those obtained with DC18C6 or D*t*BuC18C6 (= di(*t*-butylcyclohexyl)-18-crown-6) (Table 3) [45]. Calix[8]arene derivatives present a high Sr^{2+}/Na^+ selectivity due to strong strontium extraction (D > 6) and negligible sodium extraction (D < 10^{-3}). The relatively low distribution coefficients of **24a** and **22b** can be explained by a less lipophilic character. The presence of seven or eight *t*-butyl substituents (**21** and **22a**) seems to prevent the extraction of any cation. While **23a** has a great affinity for strontium, **23b** and **23c** display low strontium distribution coefficients, which can be explained by the presence of bulky groups on the upper rim, reducing the calixarene mobility. The strontium extraction as a function of acidity confirms the higher efficiency of **24b** as compared to DC18C6. With the latter used at a concentration tenfold higher than **24b**, the distribution coefficients do not exceed 1, whereas they reach 30 with **24b**, for nitric acid concentrations ranging from 2 to 4 M.

TABLE 3. Distribution coefficients (D) of ^{85}Sr and ^{22}Na and strontium over sodium selectivity, S. Aqueous solution : $M(NO_3)_n$ 5x10^{-4}M ; HNO_3 1 M. Organic solution : 10^{-2} M extractant in NPHE (o/a = 1, T =25°C).

Extractants	D_{Na}	D_{Sr}	$S_{Sr/Na}$
23a	<10^{-3}	2.9	>2900
23b	<10^{-3}	0.80	>800
23c	<10^{-3}	0.13	130
21	4×10^{-2}	1.1	28
22a	precipitation		
22b	<10^{-3}	8.3	1400
24a	<10^{-3}	6.5	>6500
24b	<10^{-3}	30	>30000
24c	<10^{-3}	24	>24000
24d	<10^{-3}	20	>20000
DC18C6	6x10^{-3}	0.28	47
Dt-Bu-C18C6	7x10^{-3}	0.42	60

4. Calixarenes for Removal of Lanthanides and Actinides

Charge, form, size and coordination number of the lanthanide (Ln) and actinide (An) cations influence their binding of ligands and hence their extractability. In aqueous solution, the predominant species for both the lanthanide and transplutonium elements are derivatives of Ln^{3+} and An^{3+}, while for uranium, neptunium and plutonium, various derivatives of An^{3+}, An^{4+}, AnO_2^+ and AnO_2^{2+} may coexist. In general, An(IV) and An(VI) species are more readily extracted than Ln(III), An(III) or An(V) [47].

Phosphoryl group oxygen is a good donor atom towards both actinides and lanthanides, and the PUREX process is based on the binding of TBP to uranyl ion, UO_2^{2+}. Trialkylphosphine oxides such as trioctylphosphine oxide (TOPO) are better donors than their phosphate analogues and extractants such as TBP and related chelating ligands such as dihexyl-N,N-diethylcarbamoylmethylphosphonate (DHDECMP) and octyl-phenyl-N,N-diisobutylcarbamoylmethylphosphine oxide (CMPO) are particularly useful for the extraction of Ln^{3+} and An^{3+} from acidic solutions [48-50]. Extraction efficiency depends upon the lipophilicity of the ligand and its complexes (see Chapter 22) as well as upon the binding efficiency of the ligand, so that unidentate ligands such as TOPO are less efficient in media where there is significant competition from simple anionic ligands or where protonation of the extractant is favoured. The effectiveness of CMPO in strongly acidic media is believed to derive from both its chelate action and the difficulty of protonating both sites within the one molecule.

Extractants of both the DHDECMP and CMPO types extract all actinides, regardless of their form, from highly saline radioactive wastes. Subtle control of the extractant efficiency results from variation in the P(R) groups of CMPO, and for the TRUEX process, used in the USA for the removal of actinides from radioactive waste, octyl-phenyl-CMPO, which gives high extraction coefficients over a wide acidity range, was chosen [51]. Replacement of the phenyl group by alkyl reduces the extraction selectivity between the lanthanides [52].

The extraction equilibria from nitric acid medium involve three molecules of CMPO with Am^{3+} and only two molecules with UO_2^{2+} and Pu^{4+} [53]. The cation extraction is in competition with that of nitric acid. The synthesis of calixarenes bearing monodentate or bidentate ligands including P=O group was based on the fact that several molecules of these ligands are co-ordinated to the cations during its extraction. Thus, the attachment of several groups such as phosphine oxide, carbamoyl phosphonate or carbamoyl phosphine oxide to a basic scaffold should lead to their synergistic action.

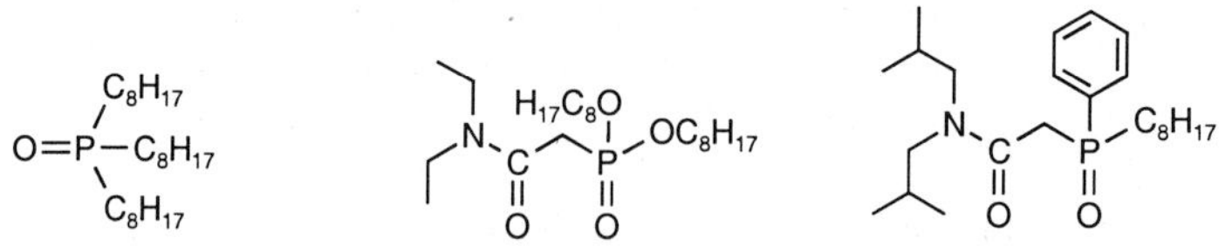

TOPO   DHDECMP   CMPO

4.1. "TOPO-LIKE" CALIXARENES

4.1.1. Lower rim substitution
The extraction of europium and thorium nitrates from 1M HNO_3 into CH_2Cl_2 has been studied for calixarenes **25a-j**, differing by their condensation degree n (n = 4,5,6,8), the type of the *p*-substituent (*t*-butyl or H) and the length of the $(CH_2)_m$ spacer between the phenolic oxygen and the phosphorus atom (m = 1,2,4) [54-59]. The results show that :

- (i) all calixarenes are more efficient than TOPO and CMPO since the ligand concentrations necessary to reach a given extraction percentage are 10 to 100 times lower with the calixarenes than with the classical extractants;

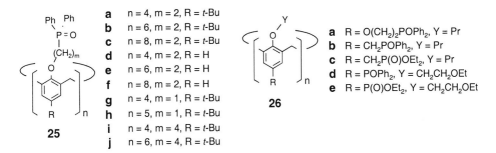

- (ii) in both series (**25a-c** and **25d-f**), the calix[6]arene is the best extractant;
- (iii) thorium(IV) is better extracted than europium(III), as calixarenes concentrations are at least 10 times higher to extract europium to a certain extent than to extract thorium;
- (iv) *p*-dealkylation leads to an increase in the extraction efficiency;
- (v) replacement of the phenyl groups on the phosphorus atom by *n*-butyl groups considerably decreases the extraction efficiency;
- (vi) the extraction of thorium increases with m for n=4 (compounds **25g**, **25a** and **25i**).

The form of plots of log D vs. log $[L]_{org}$, $[L]_{org}$ being the total ligand concentration in the organic phase, depends on the composition of the extracted species and, under appropriate experimental conditions, can be linear with a slope equal to the metal:ligand ratio. With TOPO and CMPO the slope is 3, implying a stoichiometry of 3 calixarenes per metal in the extracted species, as expected from previous studies. As an exception with CMPO and europium, the slope is only 2.5, as previously observed with americium [51]. For the extraction of thorium by calixarenes **25a,b,d-f**, the slopes are close to 2. For the extraction of europium, they are also close to 2 (compounds **25a** and **25g**) or even higher, up to 2.5 with derivative **25f**. These rather unexpected values have been confirmed by saturation experiments through equilibration of the organic phase with thorium. No experimental explanation of these values has been given : this may be due either to the coextraction of HNO_3 or to the simultaneous formation of 1:1 and 1:2 complexes [55].

In contrast with the preceding study, the slopes of log D vs. log $[L]_{org}$ for the extraction of La^{3+}, Eu^{3+}, Er^{3+} and Y^{3+} between a nitrate aqueous solution and $C_2H_4Cl_2$ solutions of **25g** were found close to 1 [59]. Furthermore, the 1:1 stoichiometry has been confirmed by FAB^+ mass spectrometry of the picrate complex. It was also shown that the distribution coefficients increase with the polarity of the diluents : $CHCl_3$ $CH_2Cl_2 < C_2H_4Cl_2 <$ nitrobenzene. The competitive extraction of eleven rare earth metal ions M^{3+} (M= La, Pr, Nd, Sm, Eu, Gd, Dy, Ho, Er, Yb and Y) showed that compound **25g** exhibits a higher extraction efficiency and a better intra-series separation ability than TOPO. Extraction results into NPHE concerning some actinides [54] are summarized in Table 4. The highest distribution coefficients were obtained with the dealkylated calix[8]arene, which is the only extractant able to remove trivalent americium. Most of these products were more effective for the removal of tetravalent

TABLE 4 : Distribution coefficients (D) of neptunium, plutonium and americium. Aqueous feed solution : NaNO$_3$ 4 M - HNO$_3$ 1M. Organic solution : Extractant in NPHE (o/a = 1, T =25°C).

Extractants	Concentration in NPHE (M)	^{237}Np	^{239}Pu	^{241}Am
25g	10^{-3}	1.05	9.5	0.87
25a	10^{-3}	1.5	22.0	<0.01
25b	10^{-3}	9.2	63.5	4.0
25c	10^{-3}	6.0	31	0.09
25d	10^{-3}	1.3	24	0.2
25f	10^{-3}	18	>100	76.0
25i	10^{-3}	0.85	13	0.09
CMPO	10^{-2}	0.85	22	1.2

plutonium. As observed with lanthanides, the length of spacer plays a role, the extraction of plutonium and americium being maximum for m = 2. When m = 1, the cavity formed by the four phosphine oxide groups is too small to host the cation. On the contrary, the lesser preorganisation of the larger calixarenes (m = 4), due to the flexibility of four arms, can explain the decrease of their extracting abilities.

The logarithms of the stability constants of the 1:1 complexes of **25a** with Pr^{3+}, Eu^{3+} and Er^{3+} in methanol are respectively 4.8, 4.9 and 5.1, and show nearly no selectivity within the lanthanide series [58]. The dealkylated compound **25d** also forms a 1:1 complex with europium, more stable than that of **25a** (5.6), in line with extraction results. The interpretation of the data gives evidence for an additional 1:2 complex.

4.1.2. Upper rim substitution
Calix[4]arene phosphine oxides **26a-b, 26d** and calix[4]arene phosphonates **26c** and **26e** are less efficient in the same experimental conditions than the lower rim homologues **25a-j** [1,60].

4.2. "CMPO-Like" Calixarenes

4.2.1. Upper rim substitution
a) Extraction into dichloromethane or chloroform and stability constant determinations with lanthanides (III) and thorium (IV)
The calix[4]- and calix[5]arenes **27** and **28** substituted on the upper rim by carbamoylmethylphosphine oxide functions are all, except the methoxy derivative **27a**, fixed in a rigid *cone* conformation.
The main conclusions of the extraction studies of Eu(III) and Th(IV) nitrates from aqueous 1M HNO$_3$ solutions into CH$_2$Cl$_2$ are the following [56-58,61] :
- (i) The number of carbon atoms of the alkyl chain at the lower rim has very little influence on the extraction efficiency of both cations, with only one exception, related to the methoxy compound **3a** which is less efficient with europium than the other tetramers (35% extraction against an average of 65%).
- (ii) All calixarenes are considerably more efficient than CMPO, and even more efficient than the lower rim "TOPO-like" calixarenes : the concentration now necessary to reach a 50% cation extraction is lowered by a factor 10 with respect to the best "TOPO-like" calixarenes, and ranges from 10^{-3} to 10^{-4} for thorium and from 10^{-2} to 10^{-3} M for europium.

- (iii) The increase in the cavity dimensions from a calix[4]- to a calix[5]arene does not affect significantly the extent of extraction for any of both cations.
- (iv) The stoichiometry of the extracted complex, according to the values of the slopes of the linear log D vs. log [L] graphs, is 1:1 for the extraction of thorium(IV); but in the case of the extraction of europium (III), the slopes are near to 2, as found with the lower rim "TOPO-like" calixarenes in the same experimental conditions.

This value of 2 was not confirmed by the saturation experiments through equilibration of the organic phase which led to the value 1. This apparent contradiction was lifted in a further study employing radioactive europium tracer at concentration 10^{-6} M in aqueous solutions 0.01 M in HNO_3 and 4 M in $NaNO_3$, and solutions of calixarene **27e** in $CHCl_3$ at concentrations from 10^{-7} to 10^{-1} M : a slope value of 1 was obtained for the lower calixarene concentrations, and of 2 when the ligand to cation concentration ratio exceeded 100 [57,62].

The spectrophotometric determination (competition method with 1-(2-pyridylazo)-2-naphthol, PAN) of the stoichiometries and stabilities of the complexes of **27c** formed in MeOH solutions (ionic strength 0.05 M in $NaNO_3$), confirmed the formation of only 1:1 complexes of Th(IV) (log β_{11} = 6.4) and showed the presence of 1:1 and 1:2 complexes of europium(III) (log β_{11} = 6.2 and log β_{12} = 11.1), in agreement with the extraction results. These stability constants are higher than those obtained in the same experimental conditions for CMPO, viz 5.1 (log β_{11}) and 9.4 (log β_{12}) for thorium and 3.6 (log β_{11}) and 5.5 (log β_{12}) for europium [58]. Confimation of the formation of both types of complexes was given by the ES/MS spectrum of a solution of calixarene **27c** and europium (analytical concentration ratio 2) which presents several peaks attributable to *mono*- and *bis*-ligand complexes, without and with bound nitrate anions [58]. When followed by direct spectrophotometric titration in methanol, a method which allows higher cation:ligand concentration ratios to be reached, the complexation of europium by **27b** and **27c** shows the formation of 1:1 and binuclear 2:1 complexes, with the same value of log β_{11} around 6. The formation of binuclear complexes could imply that the upper cavity of the calixarene is large enough to enable the coordination of two cations and indeed, the X-ray structure of the lanthanum complex with **27b,** isolated in the solid state in presence of an excess cation, shows that each cation is coordinated to the PO and CO groups of two adjacent units of the calixarene [63].

The monomeric subunits **29a-c** and the acyclic compounds **30a-j** (dimers to pentamers) were studied in order to detect a possible macrocyclic effect in the calixarenes [58]. The results reveal that :
- (i) going from the monomer to the tetramer produces an increase in the extraction of thorium;
- (ii) the pentamers extract thorium no better than the tetramers and europium less than the tetramers;
- (iii) the acyclic compounds are less efficient than the corresponding calix[4]arenes, especially for thorium (15% with **30g** and 60% with **27c**);
- (iv) all compounds form 1:1 and 1:2 species with europium in MeOH, as do the corresponding calixarenes; (v) the monomers form more stable complexes than CMPO (for instance, log β_{11} = 4.5 and 3.6 with **29b** and CMPO respectively);
- (vi) the introduction in an acyclic compound of a supplementary phenolic unit induces an increase in the stability of the 1:1 complexes, from the di- to the pentamer;

- (iv) the stabilities of the europium complexes of the calixarene **27c** and of the acyclic analogue **30g** (log β_{11} = 6.2 and 5.6 respectively) show a little effect of the macrocyclic structure.

b) Extraction into NPHE and transport experiments with lanthanides and actinides
All the "CMPO-like" calixarenes were soluble at a very low concentration in NPHE (10^{-3} M). None of the classical extractants is able to remove actinides from acidic media at this concentration, whereas the dimer extracts plutonium and the trimer extracts americium, in agreement with the stoichiometries of the corresponding CMPO complexes [62].

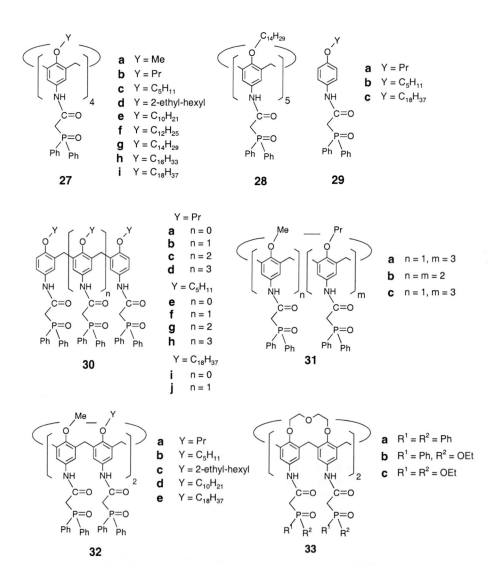

Problems of solubility and precipitation were encountered with "CMPO-like" calix[4]arenes bearing hexa- and octadecyloxy chains. With all the others, the distribution coefficients for plutonium(IV), europium(III) and americium(III) exceeded 100, even when used at a concentration as low as 10^{-3}M (Table 5). These values are much higher that those obtained with any other extractant particularly with CMPO (10^{-2} M). Increasing the size of calixarenes (from **27g** to **28a**) leads to a slight decrease of americium distribution coefficients but to an important enhancement of neptunium extraction, which could be related to the linear shape of the NpO_2^+ with an interatomic distance Np-O of 0.198 nm.

For transport experiments, water soluble diphosphonic acids were used as stripping agents. The permeability of cations through a SLM is calculated from both the decrease of radioactive cation in feed solution (P_f) and the increase of radioactive cation in the stripping one (P_s). In ideal case the two values are similar. A value of P_s lower than that of P_f indicates a slow kinetics of decomplexation. In spite of a low concentration of calixarenes in the membrane (10^{-3}M), high permeability values were achieved, ranging from 3 to 7 cm h^{-1}. The highest permeabilities were obtained with calixarenes bearing long alkyl chains (**27e-g**), in agreement with their high lipophilicity. With **27f**, more than 90 % of americium was transported and 92% and 99.7% of plutonium were transferred through the membrane, after 2 and 6 hours, respectively, by **32e**. The very close values of P_f and P_s indicate a good stripping.

The strong decrease of distribution coefficients along the lanthanide series suggests a size recognition (Figure 3) [61,57]. With classical extractants like di-(2-ethyl hexyl) phosphoric acid (HDEHP), industrially used for the separation of different lanthanides, distribution coefficients increase with the atomic number of lanthanides, in agreement with the increasing charge density resulting from the lanthanide contraction. In contrast, for "CMPO-like" calix[4]arenes, the reverse trend is observed, due to a good adjustment between the size of the lightest cations and that of the cavity formed by the CMPO moieties. Owing to their important discrimination ability, these calixarenes, initially designed for the extraction of both lanthanides and actinides, could be used for a selective separation of actinides from almost all lanthanides, in one step, from high

TABLE 5. Distribution coefficients (D) of europium, neptunium, plutonium, americium. Aqueous feed solution : $NaNO_3$ 4 M - HNO_3 1 M. Organic solution : extractant (10^{-3} M) except CMPO (10^{-2}M) in NPHE (o/a = 1, T = 25°C).

Extractants	^{152}Eu	^{237}Np	^{239}Pu	^{241}Am
27c	> 100	2	100	> 100
27d	> 100	2	20	4.5
27e	> 100	2	90	> 100
27f		4	> 100	> 100
27g	> 100	2	80	> 100
28a	> 100	12	> 100	61
32c	> 100	2	> 100	> 100
32d	> 100	3	> 100	> 100
32e	> 100	3	> 100	> 100
29c	< 0.001	0.5	0.3	< 0.001
30i	1.2	0.9	23	1.8
30j	13	2	20	17
CMPO		0.85	22	1.2

activity liquid waste arising from PUREX process. Comparison of extraction behaviour of calixarenes and their acyclic counterparts demonstrates the role of the calixarene structure in the size recognition of lanthanides and actinides.

The trend in extraction is the same for the different lanthanides whatever the acidity, *i.e.* a strong decrease of the distribution coefficients with increasing atomic number. For all the lanthanides, with the exception of cerium, an extraction maximum is reached for a 2 M nitric acid concentration, close to that of fission product solutions. Cerium, present in solution at the +IV oxidation state, has a smaller ionic radius as compared to trivalent cations, and thus is less adapted to the calixarene cavity.

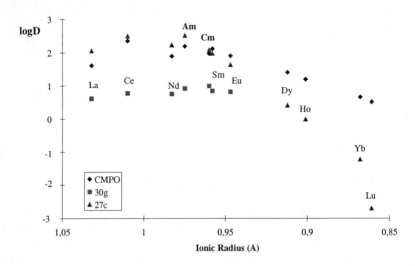

Figure 2. *Extraction of lanthanides americium and curium by CMPO (0.25 M), the acyclic tetramer* **30g** *(10^{-3} M) and* **27c** *(10^{-3} M) in NPHE from nitric acid 1.5 M (o/a = 1, T =25°C).*

c) Structural modifications of the calixarenes
In order to better understand which structural features of "CMPO-like" calix[4]arenes are responsible for their exceptional behaviour, structural modifications have been made in compounds **27**. The following conclusions can be drawn from the extraction and stability studies :
- (i) Removal of the -CH_2-C=O moieties between the NH groups and the P atoms, or addition of a methylene group between the upper rim annulus and the amide functions decreases the extent of europium(III) extraction [62].
- (ii) Replacement of the phenyl by hexyl groups on the phosphorus atom does not decrease the extraction efficiency, but leads to a complete loss of selectivity [65,66].
- (iii) Progressive replacement of the CMPO groups on the upper rim by *t*-Bu leads to poorer extractants than **27b**, and even than the acyclic analogues **30b** and **30c** [62].
- (iv) Replacement of the propoxy groups of **27b** by one to four methoxy, leading to a progressive increase in the structural flexibility of the calixarene, has no great influence on the efficacy of the thorium (IV) extraction, but reveals a regular decrease of the europium (III) extraction, with the exception of compound **31b** with two methoxy substituents in adjacent positions, which extracts europium better than **27b** [64]. Complexes

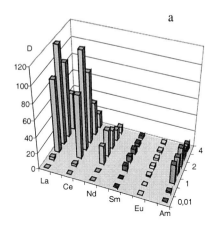 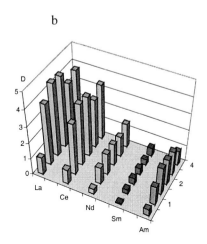

Figure 4. Extraction of La, Ce, Nd, Sm, Eu and Am by **33a** (10^{-4}M) in NPHE (o/a = 1,T =25°C) as a function of the HNO_3 concentration (0.01, 0.10, 1.0, 1.5, 2.0, 3.0 and 4.0 M) Concentration of lanthanides: a : 10^{-6}M – b : 10^{-5}M.

of **27a-b**, **31b** and **32a**, formed in methanol at 25°C and ionic strength 0.05 M in $NaNO_3$, have 1:1 and 2:1 (cation:ligand) stoichiometries with both thorium and europium. The formation of a supplementary 1:2 complex of thorium with **32a** was shown [58].

- (v) The rigidified compound **33a** locked in a nearly ideal, C_{4v}-symmetrical conformation by the presence at the lower rim of a bis-crown ether displays [66] distribution coefficients for light lanthanides and actinides close to 1000 at a concentration 10^{-3}M; tests performed at 10^{-4} M in calixarene and with lanthanide 10^{-6} M confirmed the exceptional affinity of this rigidified calixarene for some trivalent lanthanides or actinides (Figure 4): lanthanum and cerium distribution coefficients are higher than 100; the extraction of neodymium and americium is comparable, whereas it becomes negligible from europium; keeping the same extractant concentration and increasing the lanthanide concentration (10^{-5} M) enhances the competition between the cations and leads to lower distribution coefficients without increasing the intra lanthanide selectivity.

- (vi) Replacement in the calixarene structure of phosphine oxide moieties by phosphinate or phosphonate groups as in DHDECMP leads to a drastic decrease of the distribution coefficients, especially for the latter (D < 0.1) [2]. Again rigidification yields a strong increase in the extraction ability of the calixarenes, since **33b** displays distribution coefficient s comparable to those of "CMPO-like" calixarenes. Moreover, this compound is particularly interesting, since the highest extraction is achieved for americium. A strong decrease of extracting ability is observed when phosphinates are replaced by phosphonates (D < 6 with **33c**) (Figure 5).

- (vii) NMR spectroscopy, particularly its use to measure the significant reduction of the relaxation times of the solvent brought about by Gd^{3+}, has proved valuable for unravelling the solution structures of paramagnetic lanthanide chelates with a variety of ligands [67]. The encapsulation of Gd^{3+} is accompanied by an increase of the relaxation time of the solvent protons because solvent molecules are removed from the first coordination sphere of the paramagnetic metal ion. The formation of calixarene Gd^{3+} complexes was followed by recording the relaxation times of the acetonitrile ^{1}H peak when increasing amounts of ligands were added to Gd(ClO$_4$)$_3$ solutions [67]. The curve recorded for **27b** does not exhibit the expected break at a 1:1 ligand:Gd^{3+} ratio but reaches a plateau only for a 1:3 ratio. This behaviour has been ascribed to the formation of oligomers. The titration curve of **33a** significantly differs from that of **27b** by a much higher relaxivity. It seems that the rigid skeleton of **33a** favours the formation of oligomeric species even in presence of a large excess of ligand.

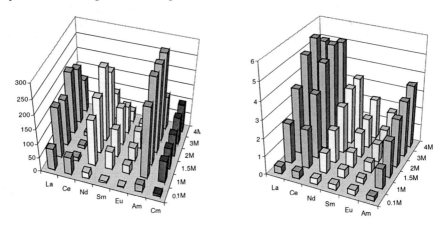

Figure 5. Distribution coefficients of lanthanides (10^{-5}M) and americium (trace level) for **33b** (left) and **33c** (right) (10^{-3}M in NPHE) as a function of the concentration of HNO$_3$ (o/a = 1, T =25°C).

4.2.2. Lower rim substitution

The grafting of CMPO residues at the lower rim of the *p-t*-butyl, *p*-H *and p-t*-octyl-calix[4]arenes led to compounds **34a-i**, with varying number of CH$_2$ spacers between the phenolic oxygens and the CMPO functions [68].

Extraction studies from 1 M HNO$_3$ aqueous solutions into dichloromethane show that **34a-d** are highly efficient for thorium extraction, even more than their upper rim counterpart **27a**. Lanthanides, however, are extracted to a much lesser extent. With all cations, the optimum number of CH$_2$ spacers is 3 or 4. The extraction level is close for La^{3+} and Eu^{3+} but decreases for Yb^{3+}. The slopes of the linear plots log D *vs.* log [L]$_{org}$ are close to 1, indicating that the cation is extracted as a 1:1 species, in contrast with the upper rim counterparts. The extraction level is very similar for compounds **34a-d** and **34h,i**, but *p*-dealkylation (compounds **34e-g**) leads to a decrease in extraction efficiency. The determination of the nature and the stability constants of the complexes formed in MeOH gives evidence for the formation of 1:1 complexes only, of similar stability within the lanthanides series and whatever the ligand (log β= 6.5-7) and, curiously, if one compares with the extraction results, with a much lower stability with thorium (log β = 5.3-5.5) [58].

The distribution coefficients of both trivalent lanthanides and actinides, and of plutonium (IV), from 1 M HNO$_3$ into NPHE, are low except for **34c** (m = 4); but even for this compound, they are lower than with upper rim "CMPO-like" calix[4]arenes [54]. Derivatives **34a**, **34b** and **34d** (m = 2,3, and 5, respectively) display a low intra lanthanide selectivity and a low affinity for actinides in comparison to lanthanides. The distribution coefficients of lanthanides decrease with increasing atomic number for **34a**. On the contrary for **34b** and **34d** a slight increase of the distribution coefficients is observed along the lanthanide series. Compound **34c** presents an extraction maximum for neodymium, samarium and europium, followed by a sharp decrease for the heaviest lanthanides.

Comparison of the calixarenes with the same spacer length (m = 4) shows that the *p*-substitution of H by butyl or octyl appreciably increases the distribution coefficients (increase of lipophilicity) and the selectivity. The affinity for europium over the other lanthanides is pronounced for **34i**, whereas high americium distribution coefficients are obtained for **34c**.

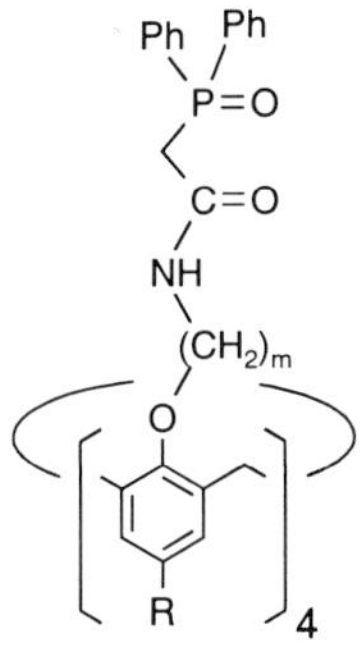

	R = *t*-Bu	R = H	R = *t*-Oct
a	m = 2	**e** m = 2	**h** m = 3
b	m = 3	**f** m = 3	**i** m = 4
c	m = 4	**g** m = 4	
d	m = 5		

34

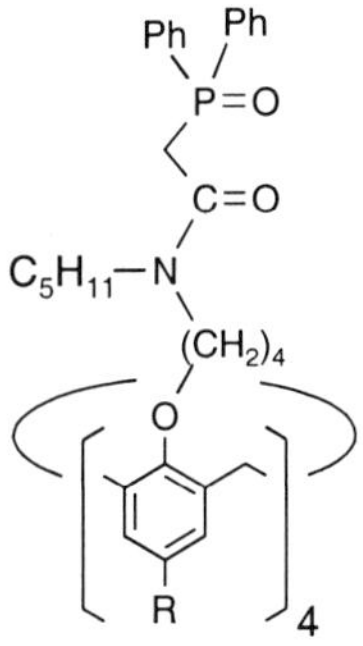

35

4.3. MISCELLANEOUS CALIXARENES

4.3.1. Other phosphorylated calixarenes
An analogue of **34c**, **35** also displays a high selectivity for thorium(IV) *vs.* europium(III) in extraction from an 4 M $NaNO_3$/1% HNO_3 aqueous solution into chloroform [69].

4.3.2. Calixarenes bearing acid functional groups
Acid derivatives **36a,b** and **37** are able to extract 98% and 94% of thorium(IV) and uranyl ions, respectively, into chloroform at pH = 4. The stoichiometry of the extracted thorium complexes with **36a-b** is 1:1. However, the primary hydroxamate **36a** is the most efficient extractant and the best candidate for the extraction of actinides from from acidic waste solutions [70].

The extraction of f-elements from acidic aqueous solutions into chloroform by calix[n]arene acid derivatives increases from the calix[4]arenes **38a-b** to the calix[6]arenes **39a-c**, corresponding to the balance between the cavity size, the molecular flexibility and the number of donor atoms [71-75]. Introduction of mixed functionalities into calix[6]arenes, *e.g.* carboxylic acid and amide groups, in a symmetrical manner like in **40** or asymmetrical manner as in **41**, led to higher extractibilities of both series of cations and to higher intra-group selectivities, owing to the cooperative binding of the different ionophilic groups. For instance, with **40**, the separation factor $D_{Am(III)}/D_{Nd(III)}$ is 66 and $D_{Am(III)}/D_{Nd(III)}$ is 118, at pH = 3 [72]. With both ligands, the extracted americium(III) and lanthanide(III) complexes have the stoichiometry 1:2 (metal:ligand) different from the 1:1 complexes formed between americium(III) and the hexamers **39a-c** and between Ln(III) and **39b** [75].

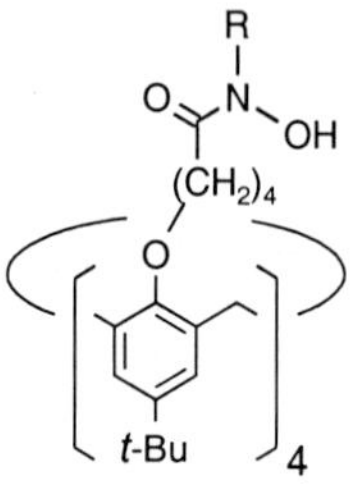

36
a R = H
b R = Me

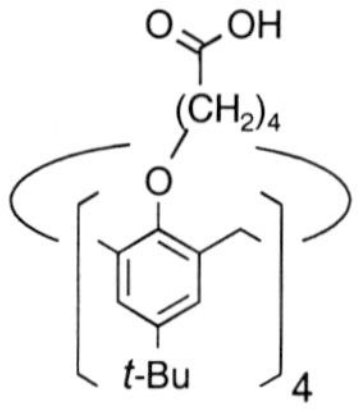

37

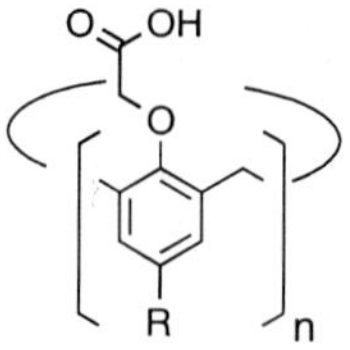

38 n = 4
39 n = 6

a R = t-Bu
b R = $C_{18}H_{37}$
c R = t-Oct

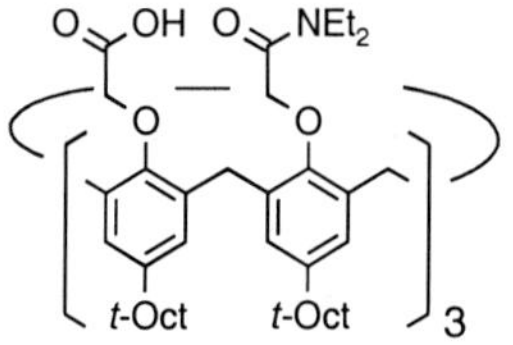

40

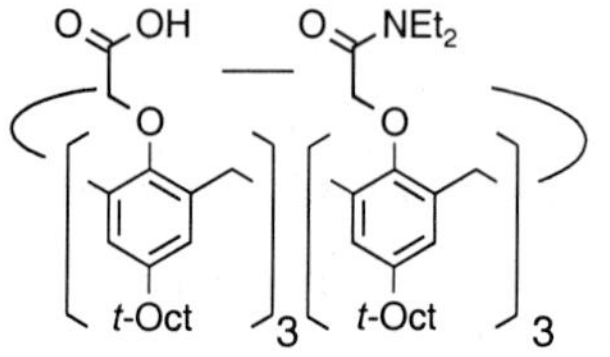

41

Acknowledgements: The authors thank the European Commission for financial support of a large part of the studies described in this chapter.

5. References and Notes

[1] J. F. Dozol, V. Böhmer, M. A. McKervey, F. Lopez Calahorra, D. N. Reinhoudt, M. J. Schwing, R. Ungaro, G. Wipff, *New macrocyclic extractants for radioactive waste treatment : ionizable crown ethers and functionalized calixarenes*, Contract F12W-CT-0062, EUR 17615 EN (**1997**).

[2] J. F. Dozol, F. Arnaud, V. Böhmer, A. Costero, J. De Mendoza, J. F. Desreux, M. J. Schwing, R. Ungaro, F. C. J. M. Van Veggel, G. Wipff, *Extraction and selective separation of long lived nuclides by functionalized macrocycles*, Contract F14W-CT-960022, EUR 19605 EN (2000).

[3] T. B. Stolwijk, E. J. R. Sudhölter, D. N. Reinhoudt, *J. Am. Chem. Soc.* **1987**, *109*, 7042-7047.

[4] M. Hawthorne, T. Andrews, *J. Chem. Soc., Chem. Commun.* **1965**, 443-444.

[5] J. Rais, M. Kyrs, M. Pivokonva, *J. Inorg. Nucl. Chem.* **1968**, *34*, 611-619.

[6] C. J. Pedersen, *J. Am. Chem. Soc.* **1967**, *89*, 7017-7036.

[7] W. J. Mc Dowell, B. A. Moyer, G. N. Case, F. I. Case, *Solv. Extr. Ion Exch.* **1986**, *4*, 217-236.

[8] S. R. Izatt, R. T. Hawkins, J. J Christensen, R. M. Izatt, *J. Am. Chem. Soc.* **1985**, *107*, 63-66.

[9] C. Alfieri, E. Dradi, A. Pochini, R. Ungaro, G. D. Andreetti, *J. Chem. Soc., Chem. Commun.* **1983**, 1075-1077.

[10] R. Ungaro, A. Casnati, F. Ugozzoli, A. Pochini, J. F. Dozol, C. Hill, H. Rouquette, *Angew. Chem., Int. Ed. Engl.* **1994**, *33*, 14, 1506-1509.

[11] E. Ghidini, F. Ugozzoli, R. Ungaro, S. Harkema, A. Abu El Fadl, D. N. Reinhoudt, *J. Am. Chem. Soc.* **1990**, *112*, 6979-6985.

[12] Z. Asfari, C. Bressot, J. Vicens, C. Hill, J. F. Dozol, H. Rouquette, S. Eymard, V. Lamare, B. Tournois, *Anal. Chem.* **1995**, *67*, 3133-3139.

[13] H. Yamamoto, S. Shinkai, *Chem. Lett.* **1994**, 1115-1118.

[14] A. Casnati, A. Pochini, R. Ungaro, C. Bocchi, F. Ugozzoli, R. J. M. Egberink, H. Struijk, R. Lugtenberg, F. de Jong, D. N. Reinhoudt, *Chem. Eur. J.* **1996**, *2,* 436-445.
[15] F. Arnaud-Neu, N. Deutsch, R. Ungaro, A. Casnati, M. J. Schwing-Weill, *Gazz. Chim. Ital.* **1997**, *127*, 693-697.
[16] A. Casnati, A. Pochini, R. Ungaro, F. Ugozzoli, F. Arnaud, S. Fanni, M. J. Schwing, R. J. M. Egberink, F. De Jong, D. N. Reinhoudt, *J. Am. Chem. Soc.* **1995**, *117*, 2667.
[17] F. Arnaud-Neu, R. Arnecke, V. Boehmer, S. Fanni, J. L. M. Gordon, M. J. Schwing-Weill, W. Vogt, *J. Chem. Soc., Perkin Trans. 2* **1996**, 1855, 1860.
[18] F. Arnaud-Neu, Z. Asfari, B. Souley, J. Vicens, *New J. Chem.* **1996**, *20*, 453-463.
[19] V. Lamare, J. F. Dozol, S. Fuangswasdi, F. Arnaud-Neu, P. Thuéry, M. Nierlich, Z. Asfari, J. Vicens, *J. Chem. Soc., Perkin Trans. 2* **1999**, 271-284.
[20] J. S. Kim, M. H. Cho, I. Y. Yu, J. H. Pang, E. T. Kim, I. H. Suh, M. R. Oh, D. Y. Ra, N. S. Cho, *Bull. Korean Chem. Soc.* **1997**, *18*, 677-680.
[21] J. S. Kim, A. Ohki, M. H. Cho, J. K. Kim, D. Y. Ra, N. M. Cho, R. A. Bartsch, K. W. Lee, W. Z. Oh, *Bull. Korean Chem. Soc.***1997**, *18*, 1014-1017.
[22] J. S. Kim, J. H. Pang, I. H. Suh, D. W. Kim, D. W. Kim, *Synth. Commun.* **1998**, *28,* 677-685.
[23] J. S. Kim, I. H. Suh, J. K. Kim, M. H. Cho, *J. Chem. Soc., Perkin Trans. 1* **1998**, *15*, 2307-2311.
[24] J. S. Kim, I. Y. Yu, J. H. Pang, J. K. Kim, Y. I. Lee, K. W. Lee, W. Z. Oh, *Microchem.* **1998**, J. 58, 225-235.
[25] C. Hill, J. F. Dozol, V. Lamare, H. Rouquette, S. Eymard, B. Tournois, J. Vicens, Z. Asfari, C. Bressot, R. Ungaro, A. Casnati, *J. Incl. Phenom.* **1994**, *19*, 399-408.
[26] R. Ungaro, J. F. Dozol, F. Arnaud, M. J. Schwing, G. Wipff, D. N. Reinhoudt, Fourth Conference of the European Commission on the Management and Disposal of Radioactive Waste, Luxembourg (**1996**) pp. 119-133.
[27] J. F. Dozol, Z. Asfari, C. Hill, J. Vicens, French patent number 92 14245, Nov. 26 1992, international number WO 94 12502.
[28] J. F. Dozol, H. Rouquette, R. Ungaro, A. Casnati, French patent number 93 04566, Apr. 19 **1993**, international number WO 94 24138.
[29] J. F. Dozol, V. Lamare, C. Bressot, R. Ungaro, A. Casnati, J. Vicens, Z. Asfari, French patent number 97 02490, Mar. 3 **1997**, international number WO98 39321.
[30] T. J. Haverlock, R. A. Sachleben, P. V. Bonnesen, B.A. Moyer, *J. Incl. Phenom.* **2000**, *36,* 21-37.
[31] F. Ugozzoli, O. Ori, A. Casnati, A. Pochini, R. Ungaro, D. N. Reinhoudt, *Supramol. Chem.* **1995**, *5*, 179-184.
[32] T. Fujimoto, R. Yanagihara, K. Kobayashi, Y. Aoyama, *Bull. Chem. Soc. Jpn.* **1995**, *68*, 2113-2124.
[33] A. Ikeda, S. Shinkai, *Tetrahedron Lett.* **1992**, 7385-7388.
[34] A. Ikeda, H. Tsuzuki, S. Shinkai, *Tetrahedron Lett.* **1994**, 8417-8420.
[35] J. F. Dozol, N. Simon, V. Lamare, H. Rouquette, S. Eymard, B. Tournois, D. De Marc. *10th Symposium on Separation Science and Technology for Energy Applications*, Gatlinburg, Tennessee, October **1997** *Sep. Sci. Technol* **1999**, *34*, 877-909.
[36] R. A. Sachleben, P. V. Bonnesen, T. Descazeaud, T. J.Haverlock, A. Urvoas, B. Moyer, *Solv. Extr. Ion Exch.* **1999**, *17*, 1445-1459.
[37] C. Hill, J. F. Dozol, H. Rouquette, S. Eymard, B. Tournois. *J. Membrane Sci.* **1996**, *114*, 73-80.
[38] J. F. Dozol, N. Simon, H. Rouquette, S. Eymard, B.Tournois, V. Lamare, M. Lecomte, M. Masson, C. Viallesoubranne, *Global 95*, **1995**, Versailles, France.
[39] T. J. Haverlock, P. V. Bonnesen, R. A. Sachleben, B. A. Moyer, *Radiochimica Acta* **1997**, *76,* 103-108.
[40] R. A. Sachleben, A. Urvoas, J. C. Bryan, T. J. Haverlock, B. J. Hay, B. A. Moyer, *Chem. Commun.* **1999**, 1751-1752.
[41] M. T. Blanda, D. B. Farmer, J. D. Brodbelt, B. J. Goolsby, *J. Am. Chem. Soc.* **2000**, *122*, 1486-1491.
[42] C. Geraci, G. Chessari, M. Piatelli, P. Neri, *Chem. Commun.* **1997**, 921-922.
[43] a) A. Arduini, A. Pochini, S. Reverberi, R. Ungaro, G. D. Andreetti, F. Ugozzoli, *Tetrahedron* **1986**, *42*, 2089-2100; b) F. Arnaud-Neu, M. J. Schwing-Weill, K. Ziat, S. Cremin, S. J. Harris, M. A. McKervey, *New J. Chem.* **1991**, *15*, 33-37; c) N. Muzet, G. Wipff, A. Casnati, L. Domiano, R. Ungaro, F. Ugozzoli, *J. Chem. Soc., Perkin Trans. 2* **1996**, 1065-1075.
[44] S. Fanni, F. Arnaud-Neu, M. A. McKervey, M. J. Schwing-Weill, K. Ziat, *Tetrahedron Lett.* **1996**, *37*, 7975-7978.
[45] F. Arnaud-Neu, S. Barboso, M. J. Schwing-Weill, A. Casnati, R. Ungaro, J. F. Dozol, Euradwaste 1999 *Radioactive Waste Management Strategies and Issues,* Fifth European Commission Conference on Radioactive Waste Management and Disposal and Decommissioning, Luxembourg, C. Davies Ed., **2000**.

[46] F. Ugozzoli, unpublished results.
[47] K. N. Raymond, W. L. Smith, *Struct. Bonding* (Berlin) **1981**, *43*, 159-186.
[48] W. W. Schulz, J. D. Navratil, *Sep. Sci. Technol.* **1984-1985**, *19*, 927-941.
[49] E. P. Horwitz, D. G. Kalina, A. C. Muscatello, *Sep. Sci. Technol.* **1981**, *16*, 403-416.
[50] D. G. Kalina, E. P. Horwitz, L. Kaplan, A. C. Muscatello, *Sep. Sci. Technol.* **1981**, *16*, 1127-1145.
[51] E. P. Horwitz, K. A. Martin, H. Diamond, L. Kaplan, *Solv. Extr. Ion Exch.* **1986**, ?, 449-494.
[52] M. N. Litvina, M. K. Chmutova, B. F. Myasoedov, M. I. Kabachnik, *Radiochemistry* **1996**, *38*, 6, 494-499.
[53] E. P. Horwitz, H. Diamond, K. A. Martin, *Solvent Extr. Ion Exch.* **1987**, *5*, 447-470.
[54] J. F. Malone, D. J. Marrs, M. A. McKervey, P. O'Hagan, N. Thompson, A. Walker, F. Arnaud-Neu, O. Mauprivez, M. J. Schwing-Weill, J. F. Dozol, H. Rouquette, N. Simon, *J. Chem. Soc., Chem. Commun.* **1995**, 2151-2153.
[55] F. Arnaud-Neu, J. K. Browne, D. Byrne, D. J. Marrs, M. A. McKervey, P. O'Hagan, M. J. Schwing-Weill, A. Walker, *Chem. Eur. J.* **1999**, *5*, 175-185.
[56] J. F. Dozol, F. Lopez-Calahorra, M. A. McKervey, V. Böhmer, R. Ungaro, M. J. Schwing, F. Arnaud, D. N. Reinhoudt, G. Wipff, *Fourth European Conference of the European Commission on Management and Disposal of Radioactive Waste*, EUR 17543 EN, **1997**, pp. 104-118.
[57] M. J. Schwing-Weill, F. Arnaud-Neu, *Gazz. Chim. Ital.* **1997**, *127*, 687-692.
[58] F. Arnaud-Neu, S. Barboso, D. Byrne, L. J. Charbonnière, M. J. Schwing-Weill, G. Ulrich, Symposium series No.757 *Calixarenes for Separations*, The American Chemical Society, G. J. Lumettta, R. D. Rogers, A. S. Gopalan, Eds., Chapter 12, **2000**, pp.150-164.
[59] M. R. Yaftian, M. Burgard, D. Matt, C. B. Dieleman, F. Rastegar, *Solv. Extr. Ion Exch.* **1997**, *15*, 975-989.
[60] S. Barboso, Thèse de Doctorat de l'Université Louis Pasteur, Strasbourg (France), **1999**.
[61] F. Arnaud-Neu, V. Böhmer, J. F. Dozol, C. Grüttner, R. A. Jakobi, D. Kraft, O. Mauprivez, H. Rouquette, M. J. Schwing-Weill, N. Simon.,W. Vogt, *J. Chem. Soc. Perkin Trans. 2* **1996**, 1175-1182.
[62] L. Delmau, Thèse de Doctorat de l'Université Louis Pasteur, Strasbourg (France), **1997**.
[63] S. Cherfa, Thèse de Doctorat de l'Université de Paris-Sud (France), **1998**.
[64] S. E. Matthews, M. Saadioui, V. Böhmer, S. Barboso, F. Arnaud-Neu, M. J. Schwing-Weill, A. Garcia Carrera, J. F. Dozol, *J. Prakt. Chem.* **1999**, *341*, 264-273.
[65] L. Delmau, N. Simon, M. J. Schwing-Weill, F. Arnaud-Neu, J. F. Dozol, S. Eymard; B. Tournois, V. Böhmer, C. Grüttner,C. Musigmann, A. Tunayar, *Chem. Commun.* **1998**, 1627-1628.
[66] L. Delmau, N. Simon, M. J. Schwing-Weill, F. Arnaud-Neu, J. F. Dozol, S. Eymard, B. Tournois, C. Grüttner,C. Musigmann, A. Tunayar, V. Böhmer, *Sep. Sci. Technol.* **1999**, *34*, 863-876.
[67] A. Arduini, V. Böhmer, L. Delmau, J. F. Desreux, J. F. Dozol, A. Garcia Carrera, B. Lambert, C. Musigmann, A. Pochini, A. Shivanyuk, F. Ugozzoli, *Chem. Eur. J.* **2000**, *6*, 2135-2144.
[68] S. Barboso, A. Garcia Carrera, S. E. Matthews, F. Arnaud-Neu, V. Böhmer, J. F. Dozol, H. Rouquette, M. J. Schwing-Weill, *J. Chem. Soc., Perkin Trans. 2* **1999**, 719-723.
[69] T. N. Lambert, G. D. Jarvinen, A. S. Gopalan, *Tetrahedron Lett.* **1999**, 1613-1616.
[70] L. Dasaradhi, C. Stark, V. J. Huber, P. H. Smith, G. D. Jarvinen, A. S. Gopalan, *J. Chem. Soc., Perkin. Trans. 2* **1997**, 1187-1192.
[71] R. Ludwig, K. Inoue, T. Yamato, *Solv. Extr. Ion Exch.* **1993**, 11(2), 331-348.
[72] R. Ludwig, K. Kunogi, N. Dung, S. Tachimori, *Chem. Commun.* **1997**, 1985-1986.
[73] R. Ludwig, S. Tachimori, T. Yamato, *Nukleonika* **1998**, *43*, 161-174.
[74] R. Ludwig, T. K. D. Nguyen, K. Kunogi, S. Tachimori, JAERI Conf., *Proceedings of the 2nd NUCEF International Symposium, Safety Research and Development of Base Technology on Nuclear Fuel Cycle*, November 16-17 1998, Hitachinaka, Ibaraki, Japan.
[75] N. T. K. Dung, K. Kunogi, R. Ludwig, *Bull. Chem. Soc. Jpn.* **1999**, *72*, 1005-1011.

Chapter 36

CALIXARENES AS STATIONARY PHASES

ROBERT MILBRADT, VOLKER BÖHMER

Johannes Gutenberg-Universität, Fachbereich Chemie und Pharmazie, Abteilung Lehramt Chemie, Duesbergweg 10-14, D-55099 Mainz, Germany.
e-mail: vboehmer@mail.uni-mainz.de

1. Introduction

Since the early work of Gutsche, calixarenes have been considered as potential host molecules which, due to their conical shape, should be able to form inclusion complexes with various guest molecules [1]. With the exception of a single publication in 1983, [2] however, applications of this inclusion chemistry in chromatography have only been described in relatively recent literature dating from 1993. Calixarenes and resorcarenes, in particular, are now the subjects of increasing attention.

In contrast to crown ethers and cyclodextrins, other macrocycles with established applications in chromatography, the host-guest interactions of calixarenes with analytes are not determined solely by their macrocyclic structure and their hydrophobic cavities. Additional substituents and functional groups easily attached at their rims provide the potential to modify these interactions or to act as entirely different interaction sites. To date, calixarenes and resorcarenes have been used in gas chromatography (GC), liquid chromatography (LC, HPLC, RP-HPLC, IC), electrokinetic chromatography (EKC) and capillary electrochromatography (CEC). We start this short review [3] by a general survey of the structures of calixarene derivatives used in the different techniques which is given in Figure 1.

2. Gas Chromatography

The first investigations dealing with calixarenes in chromatography [4] used *p-tert*-butylcalix[8]arene **1d** and its octamethoxyethylether **1h** deposited from THF solutions as stationary phases on silanised Chromosorb W in gas-solid-chromatography (GSC) for the separation of alcohols, chlorinated hydrocarbons and aromatics [2]. The best separations were obtained with the unsubstituted calixarene **1d** and were attributed to the possibility of interactions with the free OH-groups. No indication was found of a role for inclusion.

GSC-investigations with *p-tert*-butylcalix[4]arene **1a** deposited from dichloromethane solution on the same support material (silanised Chromosorb W) showed a strong influence of the phenolic hydroxyls on the overall interaction mechanism for homologous series of sorbates (alkanes, alkenes, halogenated hydrocarbons, aromatics, ethers and alcohols).

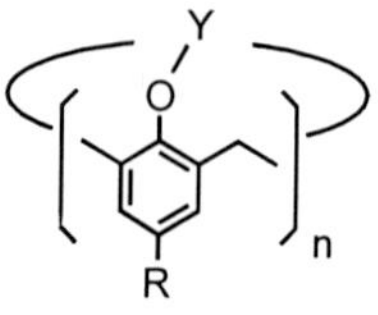

		n	Y	R
1	a-d	4,5,6,8	H	t-Bu
	e	4	SiMe$_3$	t-Bu
	f	4	Spacer †—⌇	t-Bu
	g	6	(CH$_2$)$_3$—⌇	t-Bu
	h	8	(CH$_2$)$_2$OCH$_3$	t-Bu
	i	8	PO(OEt)$_2$	t-Bu
	j,k,m	4,6,8	H	SO$_3^-$
	l	6	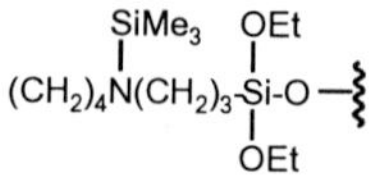 (CH$_2$)$_4$N(CH$_2$)$_3$-Si(SiMe$_3$)(OEt)$_2$-O—⌇	SO$_3^-$
	n-q	4,5,7,8	H	(CH$_2$)$_2$COOH
	r	8	H	C(CH$_3$)$_2$C$_{11}$H$_{23}$
	s	4	H	H

† short hydrophilic spacer, not specified by the authors[21]

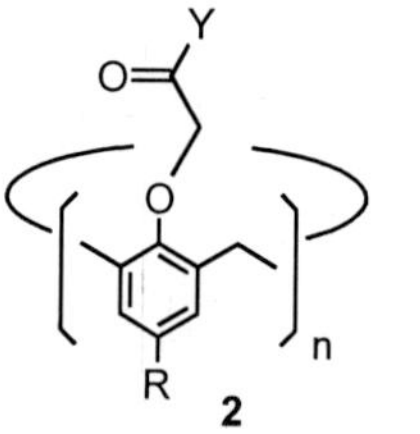

		n	Y	R
2	a	4	OEt	t-Bu
	b,c	4,6	OEt	(CH$_2$)$_3$—⌇
	d,e	4	NEt$_2$; NHOH	(CH$_2$)$_3$-S-(CH$_2$)$_3$—⌇
	f	4	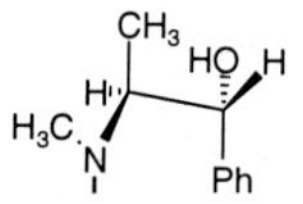 H$_3$C-N(H)-CH(CH$_3$)-CH(OH)-Ph	(CH$_2$)$_3$—⌇
	g,h	4	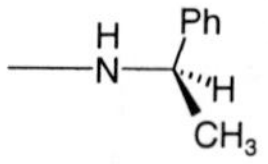 -N(H)-C*(Ph)(H)(CH$_3$)	t-Bu
	i-l	4,5,6,8	O-Spacer—⌇	t-Bu
	m	4	OH	CH$_2$NHCH$_2$-Polymer
	n	4	(S)-NH-C*(CH$_3$)COOH	t-Bu
	o	4	(S)-NH-C*(i-Pr)COOH	t-Bu
	p	4	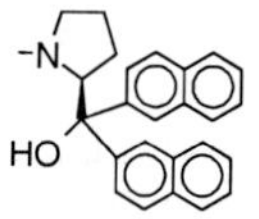 N-pyrrolidinyl-C(OH)(binaphthyl)	t-Bu

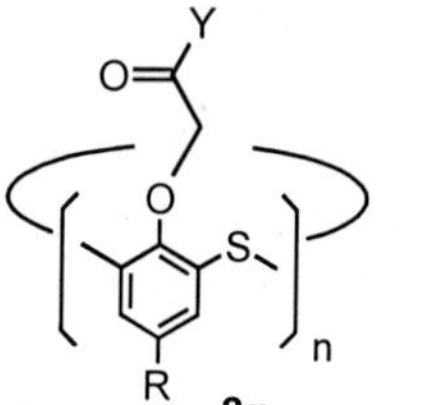

3
a Y = Bz
b Y = CH$_2$COOEt

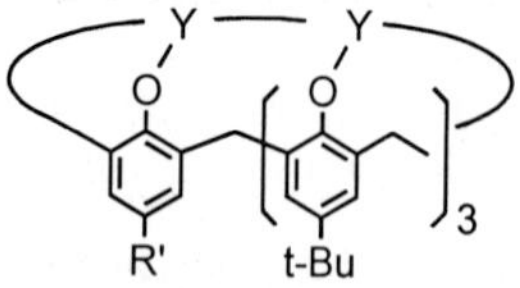

R' = CH$_2$-C$_6$H$_4$-C(CH$_3$)$_2$-(CH$_2$)$_{10}$—⌇

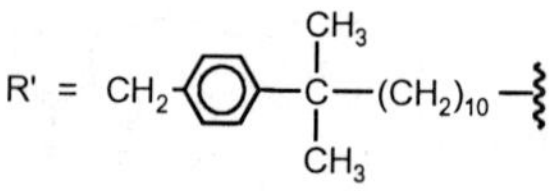

Figure 1. Survey of stationary phases.

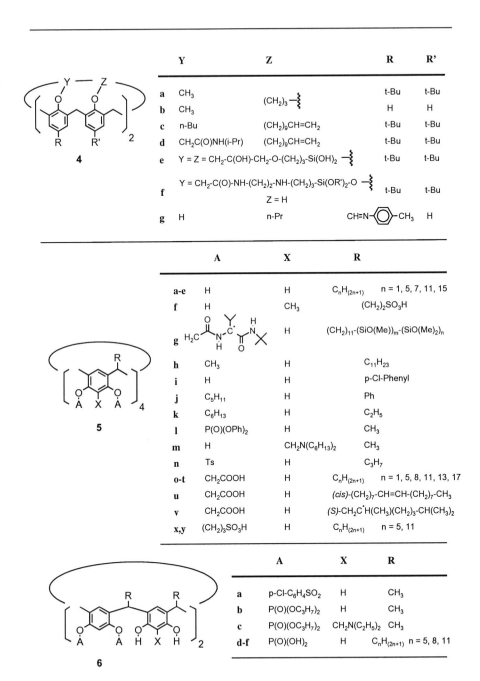

Figure 1. Continued.

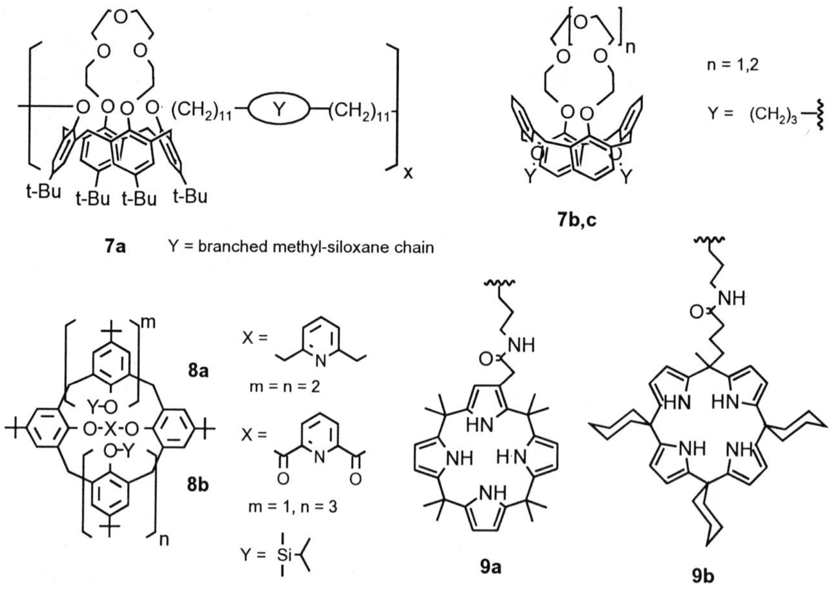

Figure 1. Continued.

Here, evidence was obtained for the formation of inclusion complexes of the calixarene with benzene, its lower *n*-alkyl derivatives (methyl to *n*-butyl), *p*-xylene, *p*-ethyltoluene, m-xylene, dichloromethane, trichloromethane, methanol and ethanol [5].

p-tert-Butylcalix[n]arenes (n = 4,5,6,8) **1a,b,c,d**, as well as tetrasilylated *p-tert*-butylcalix[4]arene **1e**, were coated via deposition from a dichloromethane solution onto a chemically bonded methyl silicone phase. With these stationary phases, the separation of both cyclic and acyclic alkanes and alkenes (including isomeric forms), alkyl substituted benzenes, chloromethanes, alcohols and ethers was obtained [6]. Due to the low solubility of the unsubstituted calixarenes in the liquid stationary phase the efficiency of these highly heterogeneous systems was rather poor, though they did show enhanced selectivity for analytes like aromatics, alkyl substituted aromatics and compounds analogous to di- and trichloromethane. Complete silylation of the phenolic OH-groups (**1e**) led to a reduction in selectivity.

Importantly, in some cases the retention behaviour was observed to depend on the geometry of the analytes (*e.g. o-, m-* or *p*-dialkyl substituted benzenes), presumably due to different inclusions into the calixarene cavity. For instance *p*-xylene showed a prolonged and *o*-xylene a reduced retention time compared with those on an untreated stationary phase.

Calix[4]arene crown-5 telomers **7a** in which the calixarene units are connected via siloxane bridges were used as stationary phases for the separation of the regioisomeric chlorophenols, dihydroxybenzenes and xylenes [7]. A fused-silica capillary column, which was coated with a 0.5 % solution of **7a** in dichloromethane by evaporation of the solvent (static method), showed a high efficiency (plate number $N \approx 4500$ m^{-1}), a low glass transition temperature and good chemical and thermal stability, making separations possible within a wide temperature range (ca. 80 to 300 °C).

The separation of these analytes (regioisomeric chlorophenols, dihydroxybenzenes and xylenes) was also performed on stationary calixarene polysiloxane phases which were prepared starting from **4c,d** via static coating with a 0.5 % solution in dichloromethane [8]. With the calixarene siloxane telomers, mixed phases were examined which contained lipophilic calixarenes **4c,d** dissolved in a polysiloxane-matrix. Mixed calixarene-crown ether and calixarene-cyclodextrin phases gave positive as well as negative synergistic effects which resulted in a partially different elution order.

The polymeric siloxane phases **4a,b** with grafted calix[4]arenes were compared in capillary gas chromatography with substituted aromatics and alicyclic cis- and trans-isomers as analytes [9]. The *tert*-butyl groups in **4a** lead to increased retention times but do not significantly affect the selectivity.

A.D-Bridged and A.C-bridged isopropyldimethylsilylcalix[6]arene **8a,b** were dissolved in OV-1701 and used as stationary phases in isothermal capillary gas chromatography for the separation of positional isomers of monosubstituted phenols and other aromatic compounds. The retention of all the solutes investigated was longer on the the A.C-bridged calix[6]arene **7e** phase probably due to interactions with the carbonyl functions [10].

To obtain the stationary phases **3a,b** the respective calix[4]arenes were synthesized by 3+1-fragment condensation, transformed into the tetrabenzyl- and tetra(ethoxycarbonylmethoxy)-derivatives and immobilized together with a polysiloxane matrix on a fused silica capillary. **3a,b** were used for the separation of isomeric hydrocarbons, polycyclic aromatic hydrocarbons (PAH) and regioisomeric aromatic compounds ($N \approx 4000$ m^{-1}) [11].

p-(2'-Methyltridecyl-2')calix[8]arene **1r** was dissolved in a polysiloxane matrix and employed for the separation of *n*-alkanes, regioisomeric substituted benzenes, quinolines, methyl- and chloro-naphthalenes, and indole ($N \approx 3000$ m^{-1}). In contrast to the results with **1a-e** [6] a preference of the host molecule for π-electron-donors was not observed [12].

Alkylated resorcarenes were also tested as stationary phases in gas chromatography. *O*-alkyl- and -phenyl-resorcarenes **5j,k** were used for the coating of capillaries [13]. With these phases, successful separations of regioisomeric substituted benzenes, long chain aliphatics (C10-14) as well as aliphatic ketones and alcohols (C8) and aldehydes (C9) were possible ($N \approx 2000$-3600 m^{-1}). With mixed phases of resorcarenes and cyclodextrins or resorcarenes and liquid crystalline aromatic esters, different relative retention times and elution orders resulted.

Gas chromatographic separations of chiral compounds with calixarene derivatives were first performed with stationary phases based on thiacalix[4]arenes **2g** containing chiral groups derived from (S)-1-phenylethylamine [14]. It was possible to separate various racemic mixtures of protected amino acids, alcohols, and amines. Interestingly the phases based on the corresponding *classical* calix[4]arene **2h** with methylene bridges did not show this enantioselectivity, probably due to its higher melting point (**2g**: 118 °C, **2h**: 323 °C).

Separations of enantiomers of protected amino acids were obtained with a stationary phase based on an undecenyl-resorcarene with 8 L-valine-*tert*-butylamide-residues (**5g**) which was covalently bound on a polysiloxane matrix [15]. A direct influence of the resorcarene cavity on the enantiomer separation was not observed. However, decreased selectivities were found for amino acids with aromatic residues compared with a resorcarene free system. If the chiral resorcarene **5g** was simply dissolved in the polysiloxane matrix, no enantioselectivity was observed.

3. Liquid Chromatography

Calixarenes and resorcarenes have been used for the separation of both ionic and neutral analytes by liquid chromatography. For this purpose, they were either covalently linked to a support material or simply adsorbed on a stationary phase. They have also been applied as water soluble additives to the mobile phase.

3.1. STATIONARY PHASES WITH COVALENTLY LINKED CALIXARENES

Ethylesters of *p*-allylcalix[n]arenes (n = 4,6) were immobilized by addition of triethoxysilane under catalysis with hexachloroplatinic acid and reaction of the resulting (triethoxysilyl)propyl groups with activated silica. These phases (**2b,c**) were employed for the separation of alkali and alkaline earth metal chlorides [16]. With suitably conditioned columns a significant selectivity of **2b** for sodium resulted. Calix[6]arene hexaester **2c** showed only weak selectivity for cesium, rubidium and potassium. In both cases the efficiency was rather poor. Preliminary tests with a *p-tert*-butylcalix[4]arene-tetraester **2a** which was simply adsorbed on a RP-18-phase did not show any retention of the metal ions.

2d and **2e** were prepared by the treatment of the *p*-allyl calixarene with mercaptopropyltriethoxysilane and cumene hydroperoxide and subsequent reaction with activated silica. The calix[4]arene-tetraacetamide **2d** was used for the separation of alkali and alkaline earth metal ions and of amino acid esters. For the latter a RP-analogous retention behaviour resulted [17]. The fixed calix[4]arene-tetrahydroxamate **2e** enabled the selective pre-column concentration of traces of lead in water samples collected from a river polluted by an industrial effluent discharge [18].

The selective separation of lead ions from a large excess of zinc ions was possible also with the fixed calix[4]arene tetracarboxylic acid **2m**, prepared by the reaction of the corresponding *p*-chloromethyl derivative with polyallylamine [19]. The lead concentration could be enriched 200-fold.

1,3-Crown ether derivatives (crown-5, crown-6) of calix[4]arenes in the *1,3-alternate* conformation which were fixed on silica via their *p*-allyl residues (**7b,c**), could be employed for the selective separation of potassium or cesium ions respectively from other alkaline metal ions (**7b**: α_{K^+/Na^+} = 3.29; α_{K^+/Cs^+} = 1.76; **7c**: α_{Cs^+/Na^+} = 2.66; α_{Cs^+/K^+} = 1.82) [20].

p-tert-Butylcalix[4]arene **1f** fixed on silica via the narrow rim with an unspecified hydrophilic spacer was applied for the HPLC-separation of regioisomeric nitroanilines (N ≈ up to 18000 m^{-1}), nucleic bases, nucleosides and of cis/trans isomers of proline-containing dipeptides. Similar fixed carboxylic acid derivatives of *p-tert*-butylcalix[n]arenes (n = 4,5,6,8) **2i-l** were also used for studies with isomeric methyluracils and estradiols [21]. For these new HPLC-phases, a RP-analogous retention behaviour and a dependence of the selectivity from the cavity size (n) were detected.

A patented column material that contains, for instance, a hexapropylether of *p-tert*-butylcalix[6]arene covalently linked to silica (**1g**, derived from the hexaallylether), was used for the separation of PAH and fullerenes and showed higher selectivity and lower consumption of solvent than conventional RP-18-phases [22, 23].

p-tert-Butylcalix[n]arenes (n = 4,6) which were linked to silica via the narrow rim by longer spacer groups (3-glycidoxypropyltriethoxysilane, γ-(ethylendiamino)-propyltriethoxysilane) **4e,f** served for the separation of regioisomeric, disubstituted benzenes, PAH, purine- and pyrimidine bases and nucleosides (N ≈ 19000 m^{-1}) [24].

Calix[4]pyrroles (compare chapter 13), condensation products from ketones (*e.g.* acetone, cyclohexanone) and pyrroles, were fixed on silica (**9a,b**) and used as stationary phases in HPLC with selectivity for anions and neutral analytes (**9a**: N ≈ 39000 m^{-1}, **9b**: N ≈ 14600 m^{-1}). Several anions were tested as their tetrabutylammonium salts. Further studies included nucleotides, oligonucleotides, *N*-protected amino acids and fluorinated biphenyls [25].

Few examples are known of chiral HPLC stationary phases based on calixarenes or resorcarenes. A fixed calix[4]arene bearing L(-)-ephedrine residues (**2f**) was used for the separation of *R*(-)- and *S*(+)-1-phenyl-2,2,2-trifluorethanol [26].

3.2. MODIFIED STATIONARY PHASES WITH NON-COVALENTLY LINKED CALIXARENES

Undecylresorcarene **5d** and its octamethylether **5h** were used as dynamic coatings to modify stationary RP-18-phases in HPLC [27] The column material obtained (**5d**, N ≈ 6000 m^{-1}; **5h**, N ≈ 2000 m^{-1}) showed good stability over several months and gave reproducible measurements. Regioisomeric substituted phenols were separated. Their retention times decreased due to the coating compared to the untreated RP-phase in the order: free RP-phase > octamethylether > resorcarene. The resorcarene-coated RP-phases were also applied for the separation of pyrimidine bases. Here an increase of the retention times resulted due to increased interactions between the analytes and the OH-groups of the adsorbed resorcarene. Furthermore, a strong dependence of the elution behaviour from the pH value and the solvent was observed in the case of cytosine.

A RP-18-phase modified with undecylresorcarene **5d** was employed also for the separation of linear and cyclic alcohols, diols and sugars in an aqueous medium [28]. Due to the increased polarity of the modified phase most of the alcohols showed decreased retention times as mentioned above [27]. However, the retention times for 1,2-, 1,3- and cis-1,4-cyclohexandiol increased remarkably compared to those measured with untreated RP-phases. This behaviour was explained by host-guest interaction through hydrogen bonds.

Ion exchange resins coated with calix[n]arenesulfonates (n = 4,6,8) **1j,k,m** served as stationary phases for the separation of fullerenes (C_{60}, C_{70}) [29]. It was shown that the modified column material functions according to a RP-mechanism which is almost independent of the ring size of the calixarenes.

3.3. ADDITIVES TO MOBILE PHASES

Several functionalised calix[n]arenes (n = 4,8) [30] and resorcarenes [31] were tested as mobile phase additives in RP-HPLC. An octaphosphorylated *p-tert*-butylcalix[8]arene **1i**, a 5,17-bis(*N*-tolyliminomethyl)-25,27-dipropoxy-calix[4]arene **4g** and various resorcarenes with different alkyl residues (heptyl, pentadecyl) **5c,e** and different substituents at the phenolic OH-groups (octatosylate **5n**, tetra-*p*-chlorobenzenesulfonate **6a**, tetra- and octaphosphate **6b, 5l**) were used. As well, an aminomethylated resorcarene **5m** and a mixed bis-(aminomethyl)-tetraphosphate **6c** were tested. As analytes, various substituted benzenes (alkylbenzenes, aldehydes, substituted phenols, carboxylic acids) were studied. In all cases, the presence of the macrocyclic host molecules led to lower retention times. An increased selectivity was observed in many measurements but not generally.

Water soluble calix[6]arene hexasulfonates **1k** have been used also as mobile phase additives in a RPLC-system [32]. For three test mixtures of regioisomeric disubstituted benzenes a decrease of the retention times resulted. This was explained by an enhanced solubility of the analytes in the mobile phase due to host-guest interactions. (In the case of the isomeric nitrophenols examined here, this acceleration was later shown to be due to a pH-effect [33]) In most cases an increased selectivity compared with a calixarene-free system was observed. Nonetheless, one drawback was the high UV-absorption of the calixarenes that disturbed the detection. To solve this problem, the macrocycles were fixed via their phenolic OH-groups on normal silica gel so that they now themselves formed a reverse phase. A silylated dialkylamino bridge was used as a linker (**1l**) [34]. Among the regioisomeric substituted aromatics examined here only those with polar and hydrogen bond forming groups could be well separated.

4. Electrophoretic Separations with Calixarenes

4.1. DIFFERENT SEPARATION TECHNIQUES

Capillary electrophoresis (CE) is now established as a highly sensitive method for the separation of ionic, water soluble compounds. It permits the analysis of very small liquid volumes (nl), and its detection limits are in the fmol-range. The separation takes place within a fused silica capillary (50-100 µm i.d.) due to the electrophoretic mobility of the charged analytes in the electric field (20-100 kV m^{-1}).

For a complete understanding, one has to take into account the so-called *electroosmotic flow* (EOF) that occurs in capillary systems. This EOF results from the movement of the solvent within the capillary due to the electric field. The capillary surface (silanol groups) is more or less negatively charged depending on the pH of the electrolyte used. Next to this surface a layer of mobile (solvated) cations must be present which migrate under the influence of the electric field towards the cathode. Due to the small cross-section of the capillary, this leads to a plug-like flow of the whole solution by which also uncharged analytes are transported.

So-called *electrokinetic chromatography* (EKC) represents an extension of CE by which also uncharged analytes normally without electrophoretic mobility can be separated. For this purpose a *pseudostationary phase* is added to the electrolyte (aqueous phase). Originally, these additives were micelle forming agents such as anionic (or less frequently) cationic surfactants, explaining the name *micellar EKC* (MEKC). More recently, appropriately functionalised, charged, water-soluble host compounds such as cyclodextrins or calixarenes have been used as pseudostationary phases.

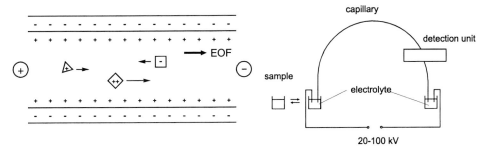

Figure 2. Schematic representation of a CE-apparatus and capillary section.

Usually, the pseudostationary phases are negatively charged under the operating conditions and migrate under the influence of the electric field in the opposite direction to the EOF with a relative velocity which may be lower or higher than that of the EOF. The separation of the uncharged analytes is caused then by their distribution between the free electrolyte phase, which is moving towards the cathode, and the pseudostationary phase, for which this movement is at least slower. Very high plate numbers (see below) are reached thereby.

In contrast to EKC a fixed stationary phase which is located on the capillary surface or on inert support particles (packed capillaries) is used in *capillary electrochromatography* (CEC). While the transport of the uncharged analytes along the capillary is accomplished only by the EOF, their separation is based upon the distribution between the fixed stationary phase and the electrolyte. *Figure 3* illustrates the different separation methods schematically.

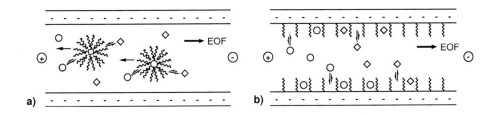

Figure 3. Schematic representation of the retention mechanism for a) EKC (e.g. MEKC) and b) CEC.

4.2. ELECTROPHORETIC SEPARATIONS WITH CALIXARENES

Calix[6]arene hexasulfonate **1k** was the first calixarene used as additive to modify the selectivity in electrokinetic chromatography [35]. The retention behaviour of regioisomeric chlorophenols, dihydroxybenzenes and toluidines was studied at different pH. Separations occured in the range of pH = 7-8, where the analytes were nearly uncharged. The interaction with the pseudostationary phase enabled in all cases the separation of the *o*-, *m*- and *p*-isomers.

With the calix[4]arene tetrasulfonate **1j** as a pseudostationary phase, a baseline separation of the three isomeric nitrophenols, dihydroxybenzenes and aminophenols was achieved at pH-values between 4 and 5 ($N \approx 3 \cdot 10^5 - 8.8 \cdot 10^5 \text{ m}^{-1}$) [36]. A test mixture consisting of 8 phenols was separated within *ca.* 8 minutes. Under these acidic conditions, the aminophenols existed in their protonated forms and possessed their own electrophoretic mobility in the direction of the detector which exceeded the EOF. Thus, negative capacity factors were observed.

Calix[6]arene hexasulfonate **1k** was also studied as pseudostationary phase for the separation of neutral analytes (*e.g.* lipophilic vitamins) in electrokinetic chromatography with mass spectrometric (electrospray-ionisation) detection (EKC-MS) [37]. A substantial advantage of the ionic calixarene compared with sodium dodecylsulfate (SDS) was its high electrophoretic mobility under acidic conditions. (A similar mobility was observed for β-cyclodextrin-sulfobutylether) The pseudostationary phase migrated out of the detection unit towards the injection end of the capillary and thus the background disturbance disappeared. The selectivity strongly depended on the concentration of the pseudostationary phase.

The UV-absorption of the calixarenes was used as the basis of an indirect detection method [38]. For this purpose, calix[n]arene sulfonates (n = 4,6) **1j,k** were applied as pseudostationary phases for the separation and detection of *UV-transparent* compounds like amino acids, biogenic amines and inorganic anions and cations. The signals were detected as negative peaks due to a diminished absorbance from a high background.

p-(Carboxyethyl)calix[n]arenes (n = 4,5,7,8) **1n-q** were studied as pseudostationary phases for the separation of simple and polar-substituted PAH [39]. The best results in terms of efficiency and selectivity were achieved with the acid derivatives of the *medium sized* calix[5]- and -[7]arenes **1o,p** ($N \approx 200000 \text{ m}^{-1}$). Unfortunately, the respective *p*-(carboxyethyl)calix[6]arene was not investigated here. If calix[6]arene sulfonate **1k** was employed, no retention of the PAH was observed.

Various resorcarenes (methyl-, pentyl-, undecyl-, *p*-chlorphenyl-) **5a,b,d,i** and an unsubstituted calix[4]arene **1s** were chosen as pseudostationary phases for the separation of PAHs ($N \approx 1.4 \cdot 10^5 - 3.2 \cdot 10^5 \text{ m}^{-1}$) [40]. With undecyl resorcarene **5d** the separation of a test mixture of 12 PAHs succeeded, and a RP-HPLC analogous retention behaviour was observed. With calix[4]arene **1s** and methyl resorcarene **5a**, no separation occured. These results suggested that the residues R at the methylene bridges of the resorcarenes were responsible for the interaction with the hydrophobic analytes and that host-guest phenomena within the aromatic cavity could be neglected.

Resorcarenes **5o-y** and **6d,e,f** show a better water solubility at lower pH. They served as pseudostationary phases for the separation of functionalised homologous and isomeric amines, which were detected by fluorescence [41]. Extremely high efficiencies and selectivities were achieved with the undecyl substituted compound **5r** ($N \approx$ up to $3 \cdot 10^6 \text{ m}^{-1}$), while longer alkyl chains at the bridging methine carbons are not advantageous.

With the sulfopropylresorcarenes **5x,y** [42], which are highly water soluble even at low pH, and the tetraphosphates **6d,e,f** a discontinuous system could be developed in which the pseudostationary phase migrates out of the detection window due to its very high electrophoretic mobility. Thus, the detection of the analytes is possible by their

UV-absorption without the background signals of the resorcarenes. In all these cases the separation behaviour was analogous to RP-HPLC.

For the separation of the nitrophenol isomers, a 2-methylresorcarene with ethylsulfonate-residues **5f** was used [43]. In comparison with to a β-cyclodextrin-sulfobutyl-ether-phase an inverse order of elution resulted. For the separation of a mixture of different *p*-substituted phenols, both phases showed a similar selectivity. In addition, the authors observed in the case of the three isomeric nitrophenols an increasing resolution by the use of a mixed phase system consisting of the charged sulfoethylresorcarene **5f** and a neutral α-cyclodextrin.

Water soluble *p-tert*-butylcalix[4]arenes with 4 amino acid residues (L-alanine, L-valine) **2n,o** served as chiral selectors for the separation of racemic binaphthyl derivatives in CE [44]. The best baseline separation of the enantiomeric pairs was obtained at pH 11 with the valine-derivative and with a combination of sodium dodecylsulfate and the alanine-derivative. Here, RP-analogous retention behaviour was found again.

Finally, a silica capillary coated with tetra-(*S*)-di-2-naphthylprolinol-calix[4]arene **2p** as chiral stationary phase was tested for capillary electrochromatography (CEC) [45]. It was possible to separate the racemic mixture of *R*(-)- and *S*(+)-2-phenylglycinol.

5. Summary and Outlook

Various calixarenes and resorcarenes have been shown to function as chromatographic selectors in gas and liquid chromatography as well as in electrophoretic separations. The choice of derivative of the macrocycles depends upon the separation method. Both dissolution in a matrix (*e.g.*, siloxane, RP-phase, aqueous electrolyte) and permanent attachment to a support material (*e.g,* polymer, silica, capillary surface, particles) allow their effective utilisation.

In gas chromatography, it seems to be crucial that the calixarenes form a homogeneous phase with the stationary support. Best results were obtained with low melting calixarenes and resorcarenes with alkyl or siloxane chains which were well embedded into, *e.g.*, a polysiloxane stationary phase.

Although there have been indications of guest inclusion within the cavity of the macrocycle or hydrogen bonding with the phenolic OH-groups it is not proven beyond doubt in all cases that the cyclic structure of the calixarenes is the determining factor for the separation of the analytes. For this purpose the comparison with, *e.g.*, linear oligomers of substituted phenols or alkyl substituted resorcinols has to be examined.

The large variety of calixarene applications in liquid chromatography makes it difficult to discern general rules for a separation mechanism. Functionalised calixarenes fixed at the narrow rim have served for the separation of ions. Separations of uncharged analytes have been performed with calixarenes bound via either the wide or the narrow rim and often reverse-phase behaviour has been observed. Adsorption of calixarenes or resorcarenes on a RP-stationary phase leads to an increased polarity of the column, and thus decreased retention times in most cases. Decreased retention times were also observed when water soluble calixarenes were introduced as mobile phase additives.

Water soluble calixarenes and resorcarenes have also been used in electrokinetic chromatography for the separation of neutral analytes, again showing reverse-phase behaviour. In some cases, the cavity size determined the selectivity but in the case of long chain substituted resorcarenes the analytes seemed to interact with the hydrophobic chains and the cavity served simply as a support. In any case, the uniform molecular character of these pseudostationary phases leads to sharper separations, in comparison to micelles, which differ in size and charge and consequently in their migration rate.

For all separation techniques (GC, LC, EKC, CEC) examples are known where calixarenes substituted with chiral residues can be used as selectors for the separation of enantiomers. Again, it remains unproved whether the cyclic calixarene structure is decisive for the enantioselectivity or if it simply acts as a support.

The use of calixarenes as stationary phases in chromatography has developed only recently. In the future, one might expect a more detailed understanding of the retention mechanisms which are responsible for the separation of analytes. Interesting results will be possible wherever the shape of the calixarene cavity or the geometrically restricted arrangement of the residues at the rims present an advantage in selectivity or efficiency compared with conventional stationary phases. The introduction of tailor-made residues to the calixarene macrocycle may lead to specific stationary phases for difficult separation problems.

6. References and Notes

[1] Consider the inclusion of toluene in the cavity found in the first crystal structure reported for a calix[4]arene: G. D. Andreetti, R. Ungaro, A. Pochini, *J. Chem. Soc., Chem. Commun.* **1979**, 1005-1007.

[2] A. Mangia, A. Pochini, R. Ungaro, G.D. Andreetti, *Anal. Lett.* **1983**, *16*, 1027-1036.

[3] Two short reviews (in Chinese) have appeared previously: a) L. Lin, C. Wu, *Fenxi Huaxue (Chinese Journal of Analytical Chemistry)* **1997**, *25*, 850-856; b) X. Xiao, Y. Feng, S. Da, *Huaxue Tongbao* **1998**, 32-35.

[4] The use of *p-tert*.-butylcalix[4]arene as stationary phase for GC was actually first described in 1982: E. Smolková-Keulemansová, L. Feltl, Abstracts of the 2nd International Symposium on Clathrate Compounds and Molecular Inclusion Phenomena, Parma, **1982**, 45.

[5] P. Mnuk. L. Feltl, *J. Chromatogr. A.* **1995**, *696*, 101-112.

[6] P. Mnuk. L. Feltl, V. Schurig, *J. Chromatogr. A.* **1996**, *732*, 63-74.

[7] Z.-L. Zhong, C.-P. Tang, C.-Y. Wu, Y.-Y. Chen, *J. Chem. Soc., Chem. Commun.* **1995**, 1737-1738.

[8] a) W. Y. Zhang, S. W. Zhang, C. Zhang, C. Y. Wu, Z. L. Zhong, *Chem. J. Chin. Univ.* **1997**, *18*, 1296-1299; b) W. Zhang, C. Wu, J. Wang, S. Zhang, *Sepu* **1997**, *15*, 204-205; c) L. Lin, Z. Yan, X. Su, C. Wu, *Sepu* **1998**, *16*, 208-210; d) L. Lin, C. Y. Wu, Z. Q. Yan, X. Q. Yan, X. L. Su, H. M. Han, *Chromatographia* **1998**, *47*, 689-694; e) X. H. Lai, L. Lin, C. Y. Wu, *Chromatographia* **1999**, *50*, 82-88; f) H. Ye, L. Lin, C. Wu, *Fenxi Huaxue* **1999**, *27*, 1087-1090; g) J. Xing, C. Y. Wu, T. Li, Z. L. Zhong, Y. Y. Chen, *Anal. Sci.* **1999**, *15*, 785-789; h) L. F. Zhang, L. Chen, X. R. Lu, C. Y. Wu, Y. P. Chen, *J. Chromatogr. A.* **1999**, *840*, 225-233; i) X. D. Yu, L. Lin, C. Y. Wu, *Chromatographia* **1999**, *49*, 567-571.

[9] H. J. Lim, H. S. Lee, I. W. Kim, S. H. Chang, S. C. Moon, B. E. Kim, J. H. Park, *Chromatographia* **1998**, *48*, 422-426.

[10] J. H. Park, H. J. Lim, Y. K. Lee, J. K. Park, B. E. Kim, J. J. Ryoo, K.-P Lee, *J. High Resolut. Chromatogr.* **1999**, *22*, 679-682.

[11] a) Z. Zeng, J. Wang, X. Tang, S. Tang, X. Lu, *Fenxi Huaxue* **1998**, *26*, 1060-1064; b) Y. Y. Chen, X. H. Tang, X. R. Lu, Z. R. Zeng, J. L. Wang, *Gaodeng Xuexiao Huaxue Xuebao* **1998**, *19*, 717-719.

[12] B. Gross, J. Jauch, V. Schurig, *J. Microcolumn Sep.* **1999**, *11*, 313-318.

[13] a) H. B. Zhang, Y. Ling, R. J. Dai, Y. X. Wen, R. N. Fu, J. L. Gu, *Chem. Lett.* **1997**, 225-226; b) D. Q. Xiao, Y. Ling, Y. X. Wen, R. N. Fu, J. L. Gu, R. J. Dai, A. Q. Luo, *Chromatographia* **1997**, *46*, 177-

182; c) H. B. Zhang, R. J. Dai, Y. Ling, Y. X. Wen, S. Zhang, R. N. Fu, J. L. Gu, *J. Chromatogr. A.* **1997**, *787*, 161-169; d) J. Zhang, T. Zhang, G. Lu, R. Fu, Z. Zhao, *Fenxi Huaxue* **1999**, *27*, 85-88.
[14] N. Iki, F. Narumi, T. Suzuki, A. Sugawara, S. Miyano, *Chem. Lett.* **1998**, 1065-1066.
[15] J. Pfeiffer, V. Schurig, *J. Chromatogr. A.* **1999**, *840*, 145-150.
[16] J. D. Glennon, K. O'Connor, S. Srijaranai, K. Manley, S. J. Harris, M. A. McKervey, *Anal. Lett.* **1993**, *26*, 153-162.
[17] a) J. D. Glennon, E. Horne, K. O'Connor, G. Kearney, S. J. Harris, M. A. McKervey, *Anal. Proc.* **1994**, *31*, 33-35; b) R. Brindle, K. Albert, S. J. Harris, C. Tröltzsch, E. Horne, J. D. Glennon, *J. Chromatogr. A.* **1996**, *731 (1-2)*, 41-46; c) J.D. Glennon, E. Horne, K. Hall, D. Cocker, A. Kuhn, S. J. Harris, M. A. McKervey, *J. Chromatogr. A.* **1996**, *731 (1-2)*, 47-55; d) J. D. Glennon, B. Lynch, K. Hall, S. J. Harris, P. O'Sullivan, *Spec. Publ. - R. Soc. Chem.* **1997**, *196*, Nr. Progress in Ion Exchange, 153-159.
[18] M. P. O'-Connell, J. Treacy, C. Merly, C. M. M. Smith, J. D. Glennon, *Anal. Lett.* **1999**, *32*, 185-192.
[19] K. Ohto, Y. Tanaka, K. Inoue, *Chem. Lett.* **1997**, *7*, 647-648.
[20] G. Arena, A. Casnati, A. Contino, L. Mirone, D. Sciotto, R. Ungaro, *Chem. Commun.* **1996**, *19*, 2277-2278.
[21] a) S. Friebe, S. Gebauer, G.J. Krauss, G. Goermar, J. Krueger, *J. Chromatogr. Sci.* **1995**, *33*, 281-284; b) S. Gebauer, S. Friebe, G. Gubitz, G.J. Krauss, *J. Chromatogr. Sci.* **1998**, *36*, 383-387; c) S. Gebauer, S. Friebe, C. Scherer, G. Gubitz, G.J. Krauss, *J. Chromatogr. Sci.* **1998**, *36*, 388-394.
[22] a) U. Menyes, U. Roth, C. Troeltzsch, *Eur. Pat. Appl.* 786661 A2, **1997**; b) U. Menyes, U. Roth, *Eur. Pat. Appl. EP 952110*, **1999**; c) U. Meynes, U. Roth, Thomas Jira, *Eur. Pat. Appl. EP 952134*, **1999**; d) U. Menyes, A. Haak, T. Sokoliess, T. Jira, U. Roth, C. Troltzsch, *GIT Spez. Sep.* **1999**, *19*, 17-19.
[23] For further patents decribing the use of stationary calixarene phases in adsorption chromatography for the separation of regioisomers and optical active substances see: a) K. Iwata, S. Moriguchi, (Showa Denko Kk, Japan) *Jpn. Kokai Tokkyo Koho 05264531 A2* **1993**; b) K. Iwata, H. Kimizuka, H. Suzuki, (Showa Denko Kk) *Jpn. Kokai Tokkyo Koho 06058920 A2* **1994**.
[24] a) W. Xu, J. S. Li, Y. Q. Feng, S. L. Da, Y. Y. Chen, X. Z. Xiao, *Chromatographia* **1998**, *48*, 245-250; b) X. Z. Xiao, Y. Q. Feng, S. L. Da, Y. Zhang, *Chromatographia* **1999**, *49*, 643-648.
[25] J. L. Sessler, P. A. Gale, J. W. Genge, *Chem. Eur. J.* **1998**, *4*, 1095-1099.
[26] L. O. Healy, M. M. McEnery, D. G. McCarthy, S. J. Harris, J. D. Glennon, *Anal. Lett.* **1998**, *31*, 1543-1551.
[27] a) O. Pietraszkiewicz, M. Pietraszkiewicz, *Pol. J. Chem.* **1998**, *72*, 2418-2422; b) M. Pietraszkiewicz, O. Pietraszkiewicz, M. Kozbial, *Pol. J. Chem.* **1998**, *72*, 1963-1970; c) O. Pietraszkiewicz, M. Pietraszkiewicz, *J. Incl. Phenom.* **1999**, *35*, 261-270.
[28] Y. Kikuchi, Y. Aoyama, *Nippon Kagaku Kaishi* **1995**, 966-970.
[29] a) T. Takeuchi, J. Chu, T. Miwa, *Analusis* **1996**, *24*, 271-274; b) T. Takeuchi, J. Chu, T. Miwa, *Kuromatografi*, **1996**, *17*, 142-145.
[30] a) O. I. Kalchenko, J. Lipkowski, V. I. Kalchenko, M. A. Vysotsky, L. N. Markovsky, *J. Chromatogr. Sci.* **1998**, *36*, 269-273; b) O. I. Kalchenko, J. Lipkowski, R. Nowakowski, V. I. Kalchenko, M. A. Vysotsky, L. N. Markovsky, *Mol. Recognit. Inclusion, Proc. Int. Symp. 9th*, Editor(s): A. W. Coleman. Publisher: Kluwer, Dordrecht, Neth., **1998**, 377-380; c) O. I. Kalchenko, A. V. Solovyov, J. Lipkowski, V. I. Kalchenko, *J. Chem. Res. S* **1999**, 60-61.
[31] a) J. Lipkowski, O. I. Kalchenko, J. Slowikowska, V. I. Kalchenko, O. V. Lukin, L. N. Markovsky, R. Nowakowski, *J. Phys. Org. Chem.* **1998**, *11*, 426-435; b) O.I. Kalchenko, A. V. Solovyov, V. I. Kalchenko, J. Lipkowski, *J. Incl. Phenom.* **1999**, *34*, 259-266.
[32] J. H. Park, Y. K. Lee, N. Y. Cheong, M. D. Jang, *Chromatographia* **1993**, *37*, 221-223.
[33] J. S. Millership, M. A. McKervey, J. A. Russell, *Chromatographia* **1998**, *48*, 402-406.
[34] Y. K. Lee, Y. K. Ryu, J. W. Ryu, B. E. Kim, J. H. Park, *Chromtographia* **1997**, *46*, 507-510.
[35] D. Shohat, E. Grushka, *Anal. Chem.* **1994**, *66*, 747-750.
[36] a) X.-B. Hu, X.-R. Ran, T. Zhao, J.-K. Cheng, *Chem. J. Chin. Univ.* **1997**, *18*, 1616-1617; b) T. Zhao, X. Hu, J. Cheng, X. Lu, *Anal. Chim. Acta* **1998**, *358*, 263-268; T. Zhao, X.B. Hu, J.K. Cheng, X.R. Lu, *J. Liq. Chromatogr.* **1998**, *21*, 3111-3124.
[37] Y. Tanaka, Y. Kishimoto, K. Otsuka, S. Terabe, *J. Chromatogr. A* **1998**, *817*, 49-57.

[38] L. Arce, A. S. Carretero, A. Rios, C. Cruces, A. Fernandez, M. Valcárcel, *J. Chromatogr. A* **1998**, *816*, 243-249.
[39] S. X. Sun, M. J. Sepaniak, J. S. Wang, C. D. Gutsche, *Anal. Chem.* **1997**, *69*, 344-348.
[40] a) K. Bächmann, A. Bazzanella, I. Haag, K. Y. Han, R. Arnecke, V. Böhmer, W. Vogt, *Anal. Chem.* **1995**, *67*, 1722-1726; b) K. Bächmann, A. Bazzanella, B. Göttlicher, I. Haag, K.-Y. Han, R. Arnecke, V. Böhmer, *GIT Spezial* **1995**, *15*, 96-103.
[41] a) A. Bazzanella, H. Mörbel, K. Bächmann, R. Milbradt, V. Böhmer, W. Vogt, *J. Chromatogr. A* **1997**, *792*, 143-149; b) A. Bazzanella, K. Bächmann, R. Milbradt, V. Böhmer, W. Vogt, *Electrophoresis*, **1999**, *20*, 92-99.
[42] R. Milbradt, V. Böhmer, unpublished results.
[43] P. Britz-Mckibbin, D. D. Y. Chen, *Anal. Chem.* **1998**, *70*, 907-912.
[44] a) M. S. Peña, Y. Zhang, S. Thibodeaux, M. L. McLaughlin, A. Muñoz de la Peña, I. M. Warner, *Tetrahedron Lett.* **1996**, *37*, 5841-5844; b) M. S. Peña, Y. Zhang, I. M. Warner, *Anal. Chem.* **1997**, *69*, 3239-3242.
[45] T. Grady, T. Joyce, M. R. Smyth, S. J. Harris, D. Diamond, *Anal. Commun.* **1998**, *35*, 123-125.

SUBJECT INDEX

Acetylcholine esterase mimic 508, 600
Actinides (f-elements) (see also "complexation", "nuclear waste treatment")
- binding 571ff
- removal 649ff
Aggregation see "self assembly"
Allosteric systems 376, 460, 487
Amphiphilicity of calixarenes 612ff
Analysis using calixarenes (also "sensors")
- amperometry 633
- of anions 599, 632
- bioaffinity assays 583, 592f
- of blood 630
- capillary electrophoresis 670
- of cations 631f, 634
- chromatography 662, 666ff
- of chiral species 637
- colorimetry 598, 636
- flow injection analysis (FIA) 628
- fluorimetry 371, 599
- of ionic analytes 668
- of neutral species 632, 638
- potentiometry 628
- spectrophotometry 598, 636
- voltammetry 378, 633, 635
Anion receptors see "receptors"
Annelated calixarenes 149
Assemblies of calixarenes 155ff see "self-assembly"
- at interfaces 327ff, 612ff
Atomic Force Microscopy (AFM) 491, 615

Benzyne hemicarceplex 214
Bicyclo-calixarenes 149
Bioactive calixarenes 503
- antibacterial 503
- anticoagulant, antithrombotic 506
- bioaffinity assays 583, 592f
- biomimetic catalysis 507
- *in vivo* applications 443
- ion-channel mimics 505
- pharmacological applications 490, 506
- radiopharmaceuticals 399, 506
- vancomycin mimics 502ff
Biomimetic/-organic chemistry 496ff
Brewster Angle Microscopy (BAM) 482

Calixarenes see mainly under particular "calix[n]arenes"
- definition 1f
Calix[3]arenes see mainly under "homoaza-", "homo-" and "homooxacalixarenes"
- -benzofurans 257f
- -furan[3]pyrrole 254
- -indoles 256f
- silaphosphinines 261
- -triazine, S-linked 259
Calix[4]arenes 26ff
- 1,2-difunctionalisation 28f
- 1,3-difunctionalisation 27f

- chiral (dissymmetric) 15f, 17f
- Claisen rearrangement 35
- conformations 2, 31, 281ff
- cone-to-cone inversion 282
- crown-ether derivatives 319f, 334ff, 365ff, 643ff, 666 (see also direct entry)
- electrophilic substitution 33ff
- hydrogenation of aromatic rings 273ff
- ipso-substitution 32f
- MM calculations 281ff
- mercapto derivatives 120ff (see direct entry)
- modification of methylene bridges 36
- monoalkylation 27
- narrow (lower) rim substitution 27ff
- partially hydroxylated 283
- podands 322f, 385ff, 627
- rearrangments 35
- simulation (MD, FEP) of cation binding 316f, 321f, 325, 330
- tetrafunctionalisation 30f
- tetraalkoxy conformers 286
- thia-derivatives 20, 110 (see also "thiacalix[4]arenes")
- trifunctionalisation 29f
- wide (upper) rim substitution 32ff
Calix[5]arenes 54ff
- conformational properties 54, 60ff, 290f
- crown-ether derivatives 57f, 61, 65ff
- fragment condensation 55
- fullerene complexes 63
- host-guest complexes 62ff
- lower (narrow) rim manipulation 56f
- MM-calculations 290f
- one-step synthesis 54
- selective functionalisation 56ff
- simulation (MD, FEP) of cation binding 321f
- upper (wide) rim manipulation 58
Calix[6]arenes 71ff
- bridged 73ff, 84ff
- capped 81ff, 86
- concave reagents from 80f
- conformational properties 75, 77f, 291f
- exhaustive O-alkylation 71
- MM-calculations 291f
- protection of OH functions 72
- ring designation 71
- regioselective modification 71ff
- simulation (MD, FEP) of cation binding 321f
Calix[7]arenes 89f
- exhaustive functionalisation 90f
- synthesis 2f
Calix[8]arenes
- alternate O-alkylation 92ff
- bridged 99ff
- capped 103
- chiral 104
- complexing properties 105f, 568, 576ff
- conformations of partially substituted 95
- crown-ether derivatives 100ff

- crystal structures 99
- double 98, 104
- isomeric derivatives 92f, 95f
- *meta*-bromo substituted 92
- molecular mitosis 8
- mono-spirodienone 98
- selective functionalisation 92ff, 97ff
- structure assignment of *O*-alkyl derivatives 96

Calix[n]arenes n>8
- f-element complexes 576
- synthesis 4, 8f

Calix[n]arene crowns see "crown ethers of " individual calix[n]arenes
Calixbarreland 135
Calixbenzofurans 257
Calixcatenan 151
Calixcrowns see "crown ethers of " individual calix[n]arenes
Calixcryptands 376, 380
Calixcyclohexanole, -hexanone 276f
Calixfullerene 492
Calixfurans 252f
Calixhydroquinones 269
Calixindoles 256f
Calixnaphthols 5, 271
Calixphyrins 251
Calixpyridines 255
Calixpyrroles
- receptors 432f, 608, 669
- synthesis 250ff, 432f

Calixquinones 266ff
Calixresorc(in)arenes see "resorc(in)arenes"
Calixspherands 321, 628
- simulation (MD, FEP) of cation binding 321

Calixthiophenes 253
Calixtube 135
Calixureas 258f
Capsules 158ff, 199ff (see also "(hemi)carcerands", "self assembly")
- chiral 160
- crystal structures 159f, 166, 174
- fullerenes included 174, 482, 489
- enlarged 161, 166
- homo- and heterodimeric 160
- metal-ion bridged 173f, 519f
- paramagnetic 519
- polycaps 167
- reactions within 164
- reversible formation 201
- strapped 161
- urea-bridged 159

Carbamoylmethyl phosphine oxide (CMPO) 650
- CMPO-like calixarenes 652ff
- CMPO-like cavitands 191f

Carceplexes, Carcerands 194f, 199ff (see also "hemicarceplexes, -carcerands")
- acetal-bridged 200ff
- benzylthia-bridged 204ff
- biscapsules 203f
- carceroisomerism 194, 206

- cavity reactions 213
- charged hydrogen-bonds 201f, 208f
- chiral 211
- formation mechanism 200f
- gating 212
- guest exchange in 201
- lower symmetry 205ff,
- multiple cavities 202ff
- portals 207f, 208
- template formation 200f, 206
- template ratio 201
- twistomers 206f

Catalysis
- biomimetic 378ff, 507ff
- by metallacalixarenes 515, 523ff
- by metallocalixarenes 81f, 518

Catenanes 151
Cation-π interactions 65, 317, 335, 365, 374, 450, 452, 457, 539, 561
Cavitands 181ff, 201, 465
- in analysis 190f
- amide derivatives 191
- as anion receptors 193f
- bridge variation 182ff
- as building blocks 194ff
- as cation receptors 183, 191ff
- caviteins 185, 189, 498
- CMP(O) derivatives 191f
- crown-ether derivatives 185
- deepened cavity 176ff, 186ff, 215f
- fenced 185
- functionalisation 184ff
- larger assemblies from 194ff
- as neutral molecule receptors 183, 190
- in nuclear waste treatment 191f
- phosphorus bridges 182ff
- self-complementary 177
- self-folding 187f
- synthesis 181f
- thioamide derivatives 191
- water-soluble 189

Cavity
- deep, in capsules 176f
- enclosed 158
- enforced 181, 199
- π-donor 460
- nanosize 194f

Cesium effect 373
CH-π-interactions 175, 190, 305, 430, 463, 475, 478f, 505
CH acidity of guests 460, 462
Chiral recognition 160, 497, 637f
Chiral calixarenes
- inherently chiral 67, 104, 241
- derivatives 245, 285, 287, 620, 637f

Chromatography
- calixarenes as stationary phases 663ff
- electrokinetic 670
- electrophoretic 497, 671ff
- gas 662ff

- liquid (HPLC) 668ff
- separation of calixarenes 9

Chromoionophores 598ff
- azophenol based 602f
- indoaniline based 600f
- nitrophenol based 601ff

Claisen rearrangement
- calix[4]arenes 35
- double calixarenes 138ff,

Cluster keepers 570f

Complexation of
- actinides 571ff
- alkali metal ions 312ff, 334ff, 355ff, 367, 373ff, 385ff, 538f,
- alkaline-earth metal ions 313ff, 354, 392ff, 561f
- amines 308
- biomolecules 442, 502
- (alkyl)ammonium ions 65, 67, 140, 158, 183, 243, 317, 320, 373, 450ff
- anions 193f, 323, 421ff, 448f, 501
- f-element cations 414ff, 561ff
- fullerenes 476ff
- lanthanides 566ff
- neutral molecules 187, 306ff, 373, 447ff, 457ff, 465
- sugars 499f, 606
- thermodynamics, see direct entry
- transition metal cations 449f

Coordination chemistry 513ff, 536ff (see also "complexation", "metalla calixarenes")

Constrictive binding 212

Crown ethers of calix[4]arenes 365ff
- alkali metal cation complexes 372
- azacrowns 369
- azocrowns 370, 376
- biscrowns (1,2;3,4) 372ff, 459
- biscrowns (1,3;2,4) 365ff
    - unsymmetrical 376f
    - ditopic receptors 377
- cesium separation 643ff
- chiral 371,
- complexation selectivity 368, 371f, 375f
- conformations 366ff, 459
- crystal structures 373f
- mixed metal complexes 377
- molecular machines 380
- monocrowns 365ff, 587
    - fluorescent derivatives 369
    - in ISEs 628
    - as stationary phases 666
- quinones 367
- quantum mechanical calculations 340ff
- simulation (MD) of cation binding 319ff, 325ff
- as transacylase mimics 378ff

Crown ethers of calix[5]arenes 57ff, 61f
- complexation properties 65f

Crown ethers of calix[6]arenes 74ff, 648
- monocrowns 74ff
- three legged crowns 81ff
- biscrowns 84ff

Crown ethers of calix[8]arenes 99ff, 648
- monocrowns 99f
- biscrowns 101ff
    - cesium complexation 105f, 648

Crownopaddlanes 231

Crystal structures
- actinide complexes 571ff
- calixarenes 99, 102, 157, 160, 466ff
- capsules 159f, 166, 174
- f-element complexes 561ff
- lanthanide complexes 567ff
- inclusion complexes 113ff, 298, 302, 306ff, 465ff
- mercapto calixarenes 120, 123, 125ff
- resorc(in)arenes 159, 166, 174f
- thiacalixarenes 113ff, 120,
- transition metal complexes 127, 524

CYANEX 192f

Dendrimers 147, 616
Density Functional Theory (DFT) 336
Double calixarenes 131ff (see also "(hemi)carcerands")
- 1,3-alternate 136, 142f
- amide-bridged 131
- doubly-bridged 132ff, 139ff
- 'head-to-tail'-linkage 143f
- multiple bridges 135, 141f
- metal-metal bonded dimers 538f
- narrow rim linkage 131ff
- porphyrin-bridged 138, 141
- rearrangement synthesis 139f
- singly-bridged 131f, 136ff
- special structures 149ff
- spiro-bridged 134
- template synthesis 135
- tubular 135, 143
- wide rim linkage 136ff

Electrolyte-Insulator-Semiconductor (EIS) sensors 627
endo-Calixarenes, see individual "calix[n]arenes"
Enzyme models 378f, 502ff, 507ff, 518
exo-Calixarenes 6, 8, 11, 90, 271, 288, 316 (see also "resorc(in)arenes")

Extraction of (see also "simulation")
- actinides 414ff, 649ff
- alkali metal ions 386ff
- alkaline earth metal ions 395ff, 400ff, 649
- lanthanides 399, 414ff, 649ff
- post-transition metal ions 413f
- transition metal ions 388, 408ff
- oxy-anions 410, 432

Free Energy Perturbation (FEP) Simulations 314ff, 320, 323, 325
- alchemical route 315

- binding selectivities 315
Field Effect Transistors (FET),(ion-selective ISFET, chemically modified CHEMFET) 141, 191, 620, 628, 634f
Films (thin) 613ff
- adsorption by 618ff
  - of ions 618f
  - of neutral species 620f
- of calixarenes 612f
- fluorescence 623
- of fullerene complexes 481ff
- Langmuir-Blodgett 481, 613ff
- luminescence 623
- nanofiltration membranes 450
- NLO properties 622
- polymeric 622, 635
- permeability 621f
- pyroelectric properties 623
- of resorc(in)arenes 612ff
- self assembled 615
- in sensors 623
- by spin coating 618
- surface modification 618
- vapour deposition 618
Fluoroionophores 598, 604ff
Force fields (see also "MM", "MD")
- Amber 282ff, 314
- BOSS 314
- CHARMm 281ff, 314
- MM2, MM3 281ff
- TRIPOS 287f
Fullerene-Calixarene-Complexes
- AFM-studies 491
- capsules 174, 489f
- calix[5]arenes 479, 486, 488f
- calix[6]arenes 487
- calix[8]arenes 478ff
- conformations of bound calixarenes 480, 483
- films/monolayers of calixarene adducts 482ff
- homooxacalix[3]arenes 489f, 491
- isolation and purification of C_{60} / C_{70} 477f
- mechanochemical synthesis 492
- modelling 479, 487
- solution studies 485ff, 493
- crystal structures 479, 481
- water-soluble 491

Glycocalixarenes 499ff
Guest-determining step 200
Guest exchange 201

Hemicarcerands 199
Hemicarceplexes 199, 207ff
- decomplexation 212f
- lower symmetry 210ff
- template formation 209
- template ratio 210
Heterocalixarenes 250ff - see also individual entries under "calix(heterocycle)"
- bridges other than C, S 259ff

- calix[2]uracil[2]arene 258
- cationic 256
- conformational properties 260
- metallacalixarenes 262
- mixed macrocycles 254f
Homoazacalixarenes 7, 235, 245ff
- complexes 247, 568, 575
- definition 235
- formation 245
- properties 247
- zwitterionic form 247, 568
Homocalix[n]pyridines 220
Homocalixarenes 219ff
- alkylation 227, 229
- conformational properties 225ff, 227f, 230
- ionophoric properties 231ff
- synthesis 5f, 219ff
  - convergent methods 221ff
  - one-pot methods 219ff
Homo-oxacalixarenes 3, 235ff
- complexation in solution 243f
- conformational behaviour 241f
- crown-ether derivatives 239f
- fullerene complexes 483, 486, 490f
- narrow (lower) rim substitution 239
- solid state properties 242
- wide (upper) rim substitution 238
Hydrogen bonding
- assemblies by see "self assembly"
- charged H-bonds 201f, 208
- circular 446
- guest binding by 64, 463f
- homodromic pattern 284, 446
- host rigidification by 461
- intramolecular 187f, 228ff, 237, 241f, 247, 288f, 446, 450
- self-folding cavitands 187f
- in water-soluble calixarenes 446
Hydrogenation of calix[4]arenes 273ff
- complete 273, 275
- face selectivity 276
- partial 274f, 276
- stereochemisty 273f, 276

Inclusion complexes of calix[4]arenes (crystal structures) with
- acetonitrile 467
- alkanes 306f
- amines 307f
- haloalkanes 120, 306, 470
- nitriles 476, 470
- nitromethane 467
Ionophores
- from calixarenes 385ff, 598ff
- from homocalixarenes 231
Ion-Selective Electrodes (ISEs) 627ff
- for butylammonium ions 68f
- lead-selective 629, 633
- polymer membranes 631, 635

Ion-Selective Field-Effect Transistors (ISFETs) 628, 634f
Ion transport 372, 387, 389, 395

Koilands, koilates 128, 135

Langmuir-Blodgett (LB) Films 613ff, see "films"
Lanthanides
- binding 561f, 584ff
- removal 649ff

Liquid-Liquid Extraction (LLE) 312, 323, 325, 329 - see "extraction"
Luminescence
- biological applications 592
- in films 623
- fluorophores 598ff
- of lanthanides 584ff
- molecular probes 583ff
- of monolayers 623
- photoswitches 593ff

Membrane transport 92, 372, 501, 505, 642ff
Mercaptocalixarenes 121ff
- conformations, 122
- Newman-Kwart rearrangement 119, 123
- mercury complexes 125f
- crystal structures 125
- synthesis 123ff
Metacyclophanes 2
- conformation 280
- MM calculations 281
Metalla/metallo-calixarenes 262, 464, 514ff, 532f, 536ff, 552ff
Molecular Dynamics (MD) simulations 314ff, 334
- assemblies at liquid-liquid interface 324ff
- binding modes 316f
- cation complexes 316ff, 322, 328
    - alkali cations 316ff, 321
    - ammonium ions 320
    - lanthanide ions 321f
    - uranyl ions 321f
- force fields 314
- interface crossing 325f
- liquid-liquid extraction 323ff
- solvent demixing 326f
- selectivities
    - extraction 323ff
    - binding 319
- topomerisation pathways 282
Molecular Mechanics (MM) calculations 280ff, 334
- alkoxycalix[4]arenes
- aminocalix[4]arenes
- calix[4]arenes 281ff
- calix[5]arenes 290f
- calix[6]arenes 77, 291f,
- force fields 281
- larger calixarenes 291f
- [1₄]metacyclophanes 280f
- mercaptocalix[4]arenes 285f

Molecular machines 227, 379
Molecular Recognition (see also "complexation", "inclusion")
- of neutral molecules in solution and in solids
- 458ff, 502ff
- by water-soluble calixarenes 440ff
Monolayers 612ff see "films"
Monte Carlo simulations 314
Multicalixarenes 144ff
- branched oligomers 147ff
- calixarene/cyclodextrin coupling 444
- conformations 131, 136, 139, 143
- dendrimeric 147
- inclusion by 140, 142
- linear oligomers 144ff
- macrocyclic oligomers 146ff
- metal complexation 131, 133ff, 135, 143
- non-covalently linked 669
- oligomeric 144
- tubular 146
Multilayers 612ff see "films"

Nanostructures 155, 166, 170f, 175, 478
- nanofiltration membranes 450
- nanoparticles 623
- nano-scale hosts 215
- nanotubes 142
Neutron reflectometry 615
Newman-Kwart rearrangement 119
NH-π interactions 464
Non-Linear Optical materials 622
Nuclear waste treatment 191f, 642ff
- CMPO derivatives in 652ff
- crown-6 derivatives 643ff
- hydroxamate derivatives 659f
- PUREX process 642, 647, 650, 656
- removal of actinides 414f, 649ff
- removal of cesium 371f, 375, 409, 643ff
- removal of lanthanides 414f, 649ff
- removal of strontium 393, 648f
- removal of technetium 375, 409
- transport through SLMs 643, 647
- TRUEX process 650

Optodes 602, 636f (see also "sensors")
Organometallic calixarenes (see also "transition metals")
- alkyls 543
- alkene, alkyne complexes 540
- alkylidenes, alkylidynes 548
- catalysis 513, 544
- diene complexes 540
- dinitrogen reduction by 553, 555ff
- hydrocarbon rearrangements 541
- hydroformylation 525
- η^2 ketone complexes 547
- migration-insertion reactions 537
- polyhapto-metal species 335ff, 422ff, 448, 515, 533f, 561

Oxidation of calixarenes 266ff (see also "spirodienones", "quinones")
Oxo-surfaces from calixarenes 536ff
- metallation by transition metals 537
- dimers with metal-metal bonds 538f
- metal-imido/-nitrido/-oxo derivatives 553ff
- molecular orbitals in 537, 544
- quasi-planar array 537

Peptidocalixarenes 496ff, 502ff
Perforated monolayers 622
Phase transfer extraction see "extraction"
Photoreactive calixarenes
- cleavage of DNA 491
- induced electron transfer 593ff
- luminescent response to cations 595f
- photophysical properties 584f
π-basic/donor cavity 143, 191, 460
π-basic tube 369, 375
π-π interactions 63, 334, 430, 436, 476, 481, 486, 493
Podand calixarenes 385ff
- carbonyl functionalities 385ff
- complexation/extraction of
    - alkaline earth metal ions 387, 392f, 397, 400, 402
    - alkali metal ions 386ff, 400ff
    - f-element metal ions 397ff, 401ff, 414ff
    - oxyanions 407f, 410
    - post-transition metal ions 397, 400f, 413f
    - transition metal ions 388, 397, 408ff
- metal ion transport 387, 389, 395
- N-donor groups 410ff
- O-donor groups 408ff
- P-donor groups 521ff
- S-donor groups 414, 516
- thermodynamics of complexation/extraction 346ff, 385, 391ff
Pseudostationary phase 670
PUREX process 642, 647, 650, 656
Pyroelectricity 623

Quantum Mechanical (QM) calculations 334ff
- alkali cation binding by
    - tetramethoxycalix[4]arene 337ff
    - dimethoxycalix[4]arenecrown-6 340ff
    - calix[4]arene-bis-crown-6 342f
- computational methods 335ff
- GEOMOS, Gaussian 98 programs. 337
Quantum dots 623
Quartz-Crystal Microbalance (QCM) 190, 483f, 620f, 628, 638f

Reactions
- inside a cavity 213f
- of substituents of calix[4]arenes 36ff
Receptors (see also "complexation", "extraction")
- for anions 421ff
    - alcohol-based 431

- amide-based 428
- charged 424, 431f
- ditopic 434ff
- metallocene-based 422ff
- switchable 425
- urea-based 429ff, 436
- for alkylammonium ions 65ff, 450ff
- for ion pairs 434 ff
- for organic molecules 458ff
    - rigidified 459, 464
- water soluble 440ff
    - charged 440ff
    - neutral 443ff
Resorc(in)arenes
- bridge substituents 43f
- conformations 10ff, 288
- electrophilic substitution 41ff
- hydroxyl group reactions 39ff
- in membaranes 624
- MM calculations 288f
- as pseudo-stationary phase 672
- stereoisomerism 10f, 182, 288f
- synthesis 9ff,
- in thin films 612, 614ff, 618, 620f
- water-soluble 442
Rotaxanes 138, 242

Saturated calixarenes 273f
Scanning Electrochemical Microscope (SECM) 483
Selectivity
- of amide binding 620
- in anion recognition/sensing 421ff, 603ff
- for cadmium(II) 606
- for calcium ion 602
- in calixarene oxidation 267
- in carbohydrate recognition 499
- in carboxylate systems 422, 424ff, 428ff, 434
- in cation recognition/sensing 598ff, 626
- chiral discrimination 607
- in complexation 346, 385ff
- computations 315, 323
- Cs^+/Na^+ 106, 319f, 338f, 375, 602, 643ff
- in extraction of
    - alkali metal ions 386f, 390ff, 402
    - alkaline earth metal ions 392, 400, 402
    - if-element ions 399, 401ff
    - post-transition metal ions 400f
    - transition metal ions 388, 401
- in halide binding 423ff, 427ff, 435f
- for guanidinium ion 608
- for $H_2PO_4^-$ 423, 425f, 432f, 435
- for HSO_4^- 270, 428, 430
- for mercury(II) 607
- for lead(II) 414
- for lithium ion 601
- Li^+/K^+ 232
- Li^+/Na^+ 232, 602
- Na^+/K^+ 602
- for organic guests 457, 631

SUBJECT INDEX

- for sodium ion 599, 608
Self Assembly 155ff, 327
- capsule/carcerand formation 158ff, 201f
- combinatorial libraries 169f
- of films 615ff
- on gold 473, 615f
- by hydrogen-bonding 155ff
- at interfaces 327ff, 615ff
- of larger aggregates 166f, 169ff
- by metal ion coordination 171ff, 489f, 518ff
- of molecular squares 173, 530f
- monolayers 615
- rosette formation 167
- solvophobic forces in 175ff
- template effects in 159, 169, 201f
- on silica surfaces 616
Sensors 626ff
- for anions 632
- arrays 630
- for alkylammonium ions 68
- for chiral species 637f
- for neutral molecules 638f
- optical 598ff, 635
- optodes 635f
- for ozone 639
- for Pb(II) 629
- potentiometric 627ff
- for uranyl ion 601
- voltammetric 632
Separations
- of calixarenes 9
- by gas chromatography 663, 666ff
- by liquid chromatography 668ff
- by electrophoretic techniques 670ff
Simulations see "MD" and "FEP"
Solid State Studies 296ff
- f-element coordination 561ff
- guest dynamics 300ff
- lattice distortion by guests 298
- long-chain guests 306
- NMR spectroscopy 299ff, 303ff
- order/disorder problems 296f
- space group assignment 297
- spatial averaging of inclusion 298
- thermogravimetry 305
Solvent Extraction see "extraction"
Spirodienones 98, 270ff,
Super-acidic phenolic protons 445f
Superuranophiles 563
Supported Liquid Membranes (SLMs) 192, 372, 642, 646
Surface area measurement 613f
Surface Plasmon Resonance (SPR) 615
Synthesis of calixarenes
- acid-catalysed 8
- base induced 2ff
- convergent 14ff
- formation mechanism 7
- fragment-condensation 14ff, 55, 90, 254
- multi-step 14ff, 254

Template effects 201, 206
Thermodynamics
- association/complexation constants
  - alkali metal cations 356, 372, 391, 393, 396
  - alkaline earth metal cations 396
  - alkylammonium cations 244, 453
  - amides 464
  - neutral guests 460ff
- binding (free) enthalpies
  - calculation 339
  - for univalent cations 356, 391, 394
- extraction-complexation cycles 355ff, 361
- ion-pair formation 351, 354
- solvent-solute interactions 347, 350
- transfer functions 346ff, 350ff
Thiacalix[4]arenes 20, 110ff
- coordination chemistry 126ff
- crystal structures 113ff
- functionalisation
  - narrow (lower) rim 111f
  - wide (upper) rim 113
- mercapto derivatives 119ff
- Newman-Kwart rearrangement 119
- sulfinyl/sulfonyl derivatives 110f, 114f
Thiacalix[5,6]arenes 20
Thiacalix[4]thiophenes 262
Thin films - see "films"
Topomerisation pathways 282
Transition metals (see also "catalysis", "metalla calixarenes", "organometallic calixarenes", "podand calixarenes")
- complexes with
  - O- and S-donor calixarenes 514ff
  - N-donor calixarenes 521ff
  - phosphorus containing calixarenes 522ff
- metalation by 537ff
- multinuclear species 127, 515, 517, 519ff, 529ff
- sandwich complexes 127, 516f
Trioctylphosphine oxide (TOPO) 650
- TOPO-like calixarenes 650ff
Transmission Electron Microscopy (TEM) 189
TRUEX process 650
Twistomers 206f

Uranophiles 321, 414, 561, 563, 572

Water-soluble calixarenes 440ff
- acid-base properties 445f
- fullerene complexes 490f
Water-soluble cavitands 189
Water-soluble receptors
- charged 433f
- neutral 443f
- in mobile phases 670

X-ray Photoelectron Spectroscopy (XPS) 615ff

Zinke synthesis 2